Lecture Notes on Data Engineering and Communications Technologies

Volume 43

Series Editor

Fatos Xhafa, Technical University of Catalonia, Barcelona, Spain

The aim of the book series is to present cutting edge engineering approaches to data technologies and communications. It will publish latest advances on the engineering task of building and deploying distributed, scalable and reliable data infrastructures and communication systems.

The series will have a prominent applied focus on data technologies and communications with aim to promote the bridging from fundamental research on data science and networking to data engineering and communications that lead to industry products, business knowledge and standardisation.

**** Indexing: The books of this series are submitted to ISI Proceedings, MetaPress, Springerlink and DBLP ****

More information about this series at http://www.springer.com/series/15362

D. Jude Hemanth · Utku Kose
Editors

Artificial Intelligence and Applied Mathematics in Engineering Problems

Proceedings of the International Conference on Artificial Intelligence and Applied Mathematics in Engineering (ICAIAME 2019)

Springer

ICAIAME'19

Editors
D. Jude Hemanth
Department of ECE
Karunya University
Coimbatore, Tamil Nadu, India

Utku Kose
Department of Computer Engineering,
Faculty of Engineering
Suleyman Demirel University
Isparta, Isparta, Turkey

ISSN 2367-4512 ISSN 2367-4520 (electronic)
Lecture Notes on Data Engineering and Communications Technologies
ISBN 978-3-030-36177-8 ISBN 978-3-030-36178-5 (eBook)
https://doi.org/10.1007/978-3-030-36178-5

This Springer imprint is published by the registered company Springer Nature Switzerland AG
The registered company address is: Gewerbestrasse 11, 6330 Cham, Switzerland

Preface

On behalf of the proceedings editors and the organization committee, it is with deep honor that I write this Preface to the Proceedings of the International Conference on Artificial Intelligence and Applied Mathematics in Engineering (ICAIAME 2019) held in Antalya, Manavgat (Turkey). The objective of the conference was to promote in academia, industry, organizations and governments, progress and expansion of knowledge concerning artificial intelligence and mathematical modeling techniques for advancing day-to-day affairs to make better and smart living. The conference provided the opportunity to exchange ideas on machine learning, deep learning, robotics, algorithm design for intelligent solutions, image processing, prediction and diagnosis applications, operations research, discrete mathematics and general engineering applications, to experience the state-of-the-art technologies, identify solutions and build collaborations for real-time implementations. In this context, the event provided a three-day, enjoyable scientific environment for all authors and participants to share–discuss their research results and experiences with an international audience.

Based on reviews from the scientific committee and the external reviewers, a total of 197 papers have been accepted to be presented within around 40 parallel sessions. The proceedings are published by Springer under the Springer Series: Lecture Notes on Data Engineering and Communications Technologies, and the extended versions of papers with post-processing review will be published under some reputable journals. In terms of international scope, ICAIAME 2019 included contributions by 18 different countries such as Algeria, China, Cyprus, Denmark, England, France, India, Iraq, Jordan, Kuwait, Lebanon, Mexico, Pakistan, Palestine, Switzerland, Trinidad Tobago, Turkey and USA. It is great to see the outcomes of the research by the authors have found their way to the literature, thanks to valuable efforts shown in that remarkable event.

In addition to the contributed papers, a total of six invited keynote presentations were delivered by top experts in artificial intelligence and applied mathematics. Dr. Çetin Elmas highlighted the importance of 'Artificial Intelligence in Project Management,' Dr. Jude Hemanth covered technical aspects of 'Innovative Artificial

Intelligence Approaches for Medical Image Analysis,' Dr. Paniel Reyes Cárdenas discussed 'Diagrammatic Reasoning, Topological Mathematics and Artificial Intelligence,' Dr. Ali Allahverdi enlightened the audience with regard to 'How to Publish Your Paper in a Reputable Journal,' Dr. Ender Özcan discussed 'Recent Progress in Selection Hyper-Heuristics for Intelligent Optimisation' and finally Dr. Ekrem Savaş elaborated the topic titled 'Some Sequence Spaces defined by Invariant Mean.'

The success of ICAIAME 2019 depends completely on the effort, talent and energy of researchers in the field of computer-based systems who have written and submitted papers on a variety of topics. Praise is also deserved for the organizing and scientific committee members, and external reviewers, who have invested significant time in analyzing and assessing multiple papers, as holding and maintaining a high standard of quality for this conference. The ICAIAME will act as strong base for researchers and scientists in the form of that excellent reference book.

I would like to thank all authors and participants for their contributions.

Anand Nayyar

Preface

Artificial intelligence (AI) is an exciting field of knowledge that had an explosion of sophistication and technical nuance in the last few years. Let us consider only how the state of AI was purely hypothetical in many ways only 50 years ago and now we have not only developments that were envisaged in the wildest imaginations, but developments that were not even expected. No doubt that AI is a field that has incited us to question about the nature of what we define as intelligence and the limits of our concepts about it.

However, though the discipline of AI in itself is essentially transdisciplinary, there is an important connection with philosophy that has not always been underlined properly: on the one hand because we need to every now and again stop and think the meaning of the achievements we have gotten thus far. On the other hand, philosophy becomes important to even question what we want to achieve. We need philosophy of AI to relate the achievements and plans that we engineer with the highest purposes of humankind. Indeed, no discipline of knowledge is alien to human ethical issues and AI is not the exception.

One of the important lessons we have learned in the last few decades is, in my opinion, the ability to acknowledge that AI does not need to be necessarily modeled in human intelligence, and that human minds have aspects that cannot be translated into modeling due to its own very nature of being self-conscious in ways that artificial systems are not. But the illustration also works for us: There are advantages that AI has given us that make humans recognize that we can flourish by integrating to our life developments that are exclusive of AI systems and we could not do by ourselves. For example, the world of mass communications has indeed made people be easily connected and promoted an encounter of cultures that otherwise can have little or no dialogue at all. In this way, AI has made us more human and we can so give flesh to people who were not visible to us before.

A prominent aspect of the discussions between mathematicians, engineers, designers and philosophers is acknowledging that AI has grown in such a way that illustrates us for having new ideas that are informing ethics, aesthetics, art, experimental sciences such as chemistry and metaphysics, medicine and even philosophy.

Philosophy of AI is then an important activity within the disciplines of AI: Engineers need the motivation to strive for better and deeper understanding of the capacities of managing information. A philosophical dose of thought helps the engineer to understand that her or his contribution is absolutely valuable and crucial to the growth of humanity, and that technical advances are always a step forward in developing our humanity. However, the philosophical dose of the engineer also helps her or him to acknowledge that there are ethical responsibilities to humanity, to truth and to the advancement of AI. The drive that has led us to where we are has been an unrestricted desire for knowledge much more than economical rewards, for example.

This Springer edited collection at hand that came from the International Conference on Artificial Intelligence and Applied Mathematics in Engineering (ICAIAME 2019) is a great example of a sincere desire to have a dialogue guided by truth and openness, and the human exchange of ideas has been a model to follow for other disciplines, since the mathematicians and engineers are less prone to be affected by other egoistic interests but by a thirst of knowledge and inquiry. All the contributions connect in fascinating and innovative ways.

Paniel Reyes Cárdenas

Organization

International Conference on Artificial Intelligence and Applied Mathematics in Engineering 2019

Web: http://www.icaiame.com
E-Mail: `info@icaiame.com`

Briefly About

International Conference on Artificial Intelligence and Applied Mathematics in Engineering (ICAIAME 2019) held within **April 20-21-22, 2019,** at the **Antalya, Manavgat (Turkey)**, which is the pearl of the Mediterranean, heaven corner of Turkey and the fourth most visited city in the world.

The main theme of the conference, which was held at Bella Resort & Spa with international participations along a three-day period, is solutions of artificial intelligence and applied mathematics in engineering applications. The languages of the ICAIAME 2019 are English and Turkish.

Scope/Topics

Conference Scope/Topics (as not limited to):
In Engineering Problems:

- Machine Learning Applications
- Deep Learning Applications
- Intelligent Optimization Solutions
- Robotics/Soft Robotics and Control Applications
- Hybrid System-Based Solutions
- Algorithm Design for Intelligent Solutions
- Image/Signal Processing Supported Intelligent Solutions
- Data Processing-Oriented Intelligent Solutions

- Prediction and Diagnosis Applications
- Linear Algebra and Applications
- Numerical Analysis
- Differential Equations and Applications
- Probability and Statistics
- Operations Research and Optimization
- Discrete Mathematics and Control
- Nonlinear Dynamical Systems and Chaos
- General Engineering Applications

Honorary Chairs

İlker Hüseyin Çarikçi	Rector of Süleyman Demirel University, Turkey
İbrahim Diler	Rector of Applied Sciences University of Isparta, Turkey
Ekrem Savaş	Rector of Uşak University, Turkey

General Chair

Tuncay Yiğit	Süleyman Demirel University, Turkey

Conference Chairs

İsmail Serkan Üncü	Applied Sciences University of Isparta, Turkey
Utku Köse	Süleyman Demirel University, Turkey

Organizing Committee

Mehmet Gürdal	Süleyman Demirel University, Turkey
Anar Adiloğlu	Süleyman Demirel University, Turkey
Şemsettin Kilinçarslan	Süleyman Demirel University, Turkey
Akram M. Zeki	International Islamic University Malaysia, Malaysia
Bogdan Patrut	Alexandru Ioan Cuza University of Iasi, Romania
Hasan Hüseyin Sayan	Gazi University, Turkey
Kemal Polat	Bolu Abant İzzet Baysal University, Turkey
Halil İbrahim Koruca	Süleyman Demirel University, Turkey
Uğur Güvenç	Düzce University, Turkey
Okan Bingöl	Applied Sciences University of Isparta, Turkey
Yusuf Sönmez	Gazi University, Turkey
Ali Hakan Işik	Mehmet Akif Ersoy University, Turkey
Hamdi Tolga Kahraman	Karadeniz Technical University, Turkey
Ercan Nurcan Yilmaz	Gazi University, Turkey
Cemal Yilmaz	Gazi University, Turkey
Asım Sinan Yüksel	Süleyman Demirel University, Turkey

Muhammed Maruf Öztürk	Süleyman Demirel University, Turkey
Mehmet Kayakuş	Akdeniz University, Turkey
Mevlüt Ersoy	Süleyman Demirel University, Turkey
Gürcan Çetin	Muğla Sıtkı Koçman University, Turkey
Osman Özkaraca	Muğla Sıtkı Koçman University, Turkey
Ferdi Saraç	Süleyman Demirel University, Turkey
Hamit Armağan	Süleyman Demirel University, Turkey
Murat Ince	Süleyman Demirel University, Turkey
Gül Fatma Türker	Süleyman Demirel University, Turkey

Secretary

Nilgün Şengöz	Mehmet Akif Ersoy University, Turkey

Accommodation/Venue Desk

Doctorant Merve Aydemir	Süleyman Demirel University, Turkey

Travel/Transportation

Şadi Fuat Çankaya	Süleyman Demirel University, Turkey

Web/Design/Conference Sessions

Ali Topal	Süleyman Demirel University, Turkey

Scientific Committee

Tuncay Yiğit (General Chair)	Süleyman Demirel University, Turkey
Utku Köse (Committee Chair)	Süleyman Demirel University, Turkey
Ahmet Bedri Sözer	Fırat University, Turkey
Ali Öztürk	Düzce University, Turkey
Anar Adiloğlu	Süleyman Demirel University, Turkey
Aslanbek Naziev	Ryazan State University, Russia
Ayhan Erdem	Gazi University, Turkey
Çetin Elmas	Gazi University, Turkey
Daniela Elena Popescu	University of Oradea, Romania
Eduardo Vasconcelos	Goias State University, Brazil
Eşref Adali	İstanbul Technical University, Turkey
Hüseyin Demir	Samsun 19 May University, Turkey
Hüseyin Merdan	TOBB University Economy and Technology, Turkey
Igbal Babayev	Azerbaijan Technical University, Azerbaijan

Igor Litvinchev	Nuevo Leon State University, Mexico
İbrahim Üçgül	Süleyman Demirel University, Turkey
İbrahim Yücedağ	Düzce University, Turkey
Jose Antonio Marmolejo	Panamerican University, Mexico
Junzo Watada	Universiti Teknologi PETRONAS, Malaysia
Marwan Bikdash	North Carolina Agricultural and Tech State University, USA
Mehmet Ali Akçayol	Gazi University, Turkey
Mehmet Gürdal	Süleyman Demirel University, Turkey
Melih Günay	Akdeniz University, Turkey
Mustafa Alkan	Gazi University, Turkey
Nuri Özalp	Ankara University, Turkey
Oktay Duman	TOBB University Economy and Technology, Turkey
Ömer Akin	TOBB University Economy and Technology, Turkey
Recep Demirci	Gazi University, Turkey
Resul Kara	Düzce University, Turkey
Reşat Selbaş	Applied Sciences University of Isparta, Turkey
Sabri Koçer	Necmettin Erbakan University, Turkey
Sadık Ülker	European University of Lefke, Cyprus
Sergey Bushuyev	Kyiv National University, Ukraine
Yusuf Öner	Pamukkale University, Turkey
Ahmet Cüneyd Tantuğ	İstanbul Technical University, Turkey
Akram M. Zeki	International Islamic University Malaysia, Malaysia
Alexandrina Mirela Pater	University of Oradea, Romania
Ali Hakan Işik	Mehmet Akif Ersoy University, Turkey
Arif Özkan	Kocaeli University, Turkey
Aydın Çetin	Gazi University, Turkey
Bogdan Patrut	Alexandru Ioan Cuza University of Iasi, Romania
Cemal Yilmaz	Gazi University, Turkey
Devrim Akgün	Sakarya University, Turkey
Ender Ozcan	University of Nottingham, England
Ercan Nurcan Yilmaz	Gazi University, Turkey
Erdal Kiliç	Samsun 19 May University, Turkey
Erman Erkan	Atılım University, Turkey
Ezgi Ülker	European University of Lefke, Cyprus
Gültekin Özdemir	Süleyman Demirel University, Turkey
Hasan Hüseyin Sayan	Gazi University, Turkey
Hüseyin Şeker	Northumbria University, England
İlhan Koşalay	Ankara University, Turkey
İsmail Serkan Üncü	Applied Sciences University of Isparta, Turkey

J. Anitha	Karunya University, India
Jude Hemanth	Karunya University, India
Kemal Polat	Bolu Abant İzzet Baysal University, Turkey
M. Kenan Döşoğlu	Düzce University, Turkey
Mehmet Karaköse	Fırat University, Turkey
Mehmet Sıraç Özerdem	Dicle University, Turkey
Muharrem Tolga Sakalli	Trakya University, Turkey
Murat Kale	Düzce University, Turkey
Okan Bingöl	Applied Sciences University of Isparta, Turkey
Özgür Aktunç	St. Mary's University, USA
Sedat Akleylek	Samsun 19 May University, Turkey
Selami Kesler	Pamukkale University, Turkey
Selim Köroğlu	Pamukkale University, Turkey
Tiberiu Socaciu	Stefan cel Mare University of Suceava, Romania
Tolga Ovatman	İstanbul Technical University, Turkey
Ümit Deniz Uluşar	Akdeniz University, Turkey
Abdulkadir Karaci	Kastamonu University, Turkey
Ali Şentürk	Applied Sciences University of Isparta, Turkey
Arif Koyun	Süleyman Demirel University, Turkey
Barış Akgün	Koç University, Turkey
Deepak Gupta	Maharaja Agrasen Institute of Technology, India
Dmytro Zubov	University of Information Science and Technology "St. Paul The Apostle", Macedonia
Erdal Aydemir	Süleyman Demirel University, Turkey
Fatih Gökçe	Süleyman Demirel University, Turkey
Gür Emre Güraksin	Afyon Kocatepe University, Turkey
Iulian Furdu	Vasile Alecsandri University of Bacau, Romania
Mehmet Kayakuş	Akdeniz University, Turkey
Mehmet Onur Olgun	Süleyman Demirel University, Turkey
Muhammed Hanefi Calp	Karadeniz Technical University, Turkey
Mustafa Nuri Ural	Gümüşhane University, Turkey
Okan Oral	Akdeniz University, Turkey
Osman Palanci	Süleyman Demirel University, Turkey
Paniel Reyes Cardenas	Popular Autonomous University of the State of Puebla, Mexico
S. T. Veena	Kamaraj Engineering and Technology University, India
Serdar Biroğul	Düzce University, Turkey
Serdar Çiftçi	Harran University, Turkey
Ufuk Özkaya	Süleyman Demirel University, Turkey
Veli Çapali	Uşak University, Turkey

Vishal Kumar	Bipin Chandra Tripathi Kumaon Institute of Technology, India
Anand Nayyar	Duy Tan University, Vietnam
Simona Elena Varlan	Vasile Alecsandri University of Bacau, Romania
Ashok Prajapati	FANUC America Corp., USA
Katarzyna Rutczyńska-Wdowiak	Kielce University of Technology, Poland
Nabi Ibadov	Warsaw University of Technology, Poland
Özkan Ünsal	Süleyman Demirel University, Turkey

Keynote Speaks

Çetin Elmas
Gazi University, Turkey
"Artificial Intelligence in Project Management"

Ekrem Savaş
Usak University, Turkey
"Some Sequence Spaces Defined By Invariant Mean"

Ali Allahverdi
Kuwait University, Kuwait
"How to Publish Your Paper in a Reputable Journal"

Jude Hemanth
Karunya University, India
"Innovative Artificial Intelligence Approaches for Medical Image Analysis"

Ender Ozcan
University of Nottingham, England
"Recent Progress in Selection Hyper-heuristics for Intelligent Optimisation"

Paniel Reyes Cárdenas
Popular Autonomous University of the State of Puebla, Mexico
"Diagrammatic Reasoning, Topological Mathematics and Artificial Intelligence"

Acknowledgement

As the editors, we would like to thank Dr. Gül Fatma TÜRKER (Süleyman Demirel University, Turkey) for her valuable efforts on pre-organization of the book content and the Springer team for their great support to publish the book.

Contents

State and Trends of Machine Learning Approaches in Business: An Empirical Review

Samia Chehbi-Gamoura[1(✉)], Ridha Derrouiche[1], Halil-Ibrahim Koruca[2], and Umran Kaya[3]

[1] EM Strasbourg Business School, University of Strasbourg, HuManiS (EA 7308), Strasbourg, France
{samia.gamoura, ridha.derrouiche}@em-strasbourg.eu

[2] Department of Industrial Engineering, Süleyman Demirel University, Isparta, Turkey
halilkoruca@sdu.edu.tr

[3] Department of Industrial Engineering, Antalya Bilim University, Antalya, Turkey
umran.kaya@antalya.edu.tr

Abstract. Strong competition is imposing to enterprises an incessant need for extracting more business values from collected data. The business value of contemporary volatile data derives from the meanings mainly for market tendencies, and overall customer behaviors. With such continuous urge to mine valuable patterns from data, analytics have skipped to the top of research topics. One main solution for the analysis in such context is 'Machine Learning' (ML). However, Machine Learning approaches and heuristics are plenty, and most of them require outward knowledge and deep thoughtful of the context to learn the tools fittingly. Furthermore, application of prediction in business has certain considerations that strongly affects the effectiveness of ML techniques such as noisy, criticality, and inaccuracy of business data due to human involvement in an extensive number of business tasks. The objective of this paper is to inform about the trends and research trajectory of Machine Learning approaches in business field. Understanding the vantages and advantages of these methods can aid in selecting the suitable technique for a specific application in advance. The paper presents a comprehensively review of the most relevant academic publications in the topic carrying out a review methodology based on imbricated nomenclatures. The findings can orient and guide academics and industrials in their applications within business applications.

Keywords: Machine learning · Analytics · Artificial intelligence · Business information systems

1 Introduction

Since the early 2000's, Business Information Systems (BIS) offer data collections and analytical approaches to enterprises [1]. As such, they merge fundamental theories of management, processes and Information Systems (IS) theory with engineering technologies to manage the flows organization of data.

D. J. Hemanth and U. Kose (Eds.): ICAIAME 2019, LNDECT 43, pp. 1–16, 2020.
https://doi.org/10.1007/978-3-030-36178-5_1

In BIS, a number of approaches with adaptive and learning abilities are being integrated to mine the business operations and improve interacting with stakeholders [2]. These approaches, commonly named Machine Learning (ML) approaches, constitute a new gold source for identifying new business values for organizations [3]. In the academic side, research scientists believe that ML methods present today an avoidable opportunity for enterprises to process their data-business intelligently instead of the traditional recorded and unexploited data sets [4]. This is especially true with the rapid growth of Big Data phenomenon [5]. Furthermore, unlike traditional information systems, contemporary and future BIS will require the integration of mature and scalable ML techniques in all levels of business processes [6] such as opinion mining [7], risk management [8], recommendation systems [9], Business Process Management (BPM) [10], and so forth.

Technically, many applied software programs have been developed to enable different types of learning algorithms [11]. These algorithms have successfully proven to be of a great business value in containing valuable implicit regularities that can be revealed autonomously [12].

To ease the use of ML techniques in BIS concerns, it is indispensable to recapitulate the empirical observations acquired on these approaches from the existing works [13]. A methodical literature review can be a powerful tool to understand thoroughly the evolution trends and answer numerous research questions about the applicability of these methods [14]. However, academic research is lacking in extensive reviews that classify all the business activities in one panoramic view. The existing reviews have almost of them focused on a specific concern, such as decision systems [1], knowledge management [15], accounting [11], product design and engineering [16], and so forth.

The methodology we provide in this paper is an extensive analysis of the academic literature regarding the use of ML approaches. In order to perform, the proposed approach identifies five nomenclatures based on five areas of items as categories. Then, a set of three structured findings are extracted and justified empirically based on ML applications on enterprises scopes. Furthermore, the methodology highlights a number of research gaps that require serious consideration and more efforts from academic community. The main purpose is to situate the current status and the potential trends in the application of ML approaches in BIS.

The remainder of the paper is arranged as follows: Sect. 2 provides a background overview of BIS and ML applications. Section 3 details the methodology followed in the analysis of the literature. Findings and empirical results are outlined in Sect. 4. Section 5 concludes the paper with the relevant outcomes and opened views.

2 Background and Research Gaps

Machine Learning approaches are the class of computational methods that automate the acquisition of knowledge from experience [17]. One object in applying such methods in BIS is to develop tools, heuristics, and techniques that can supplement domain expertise in engineering and modeling tasks [11] such as stock prediction [18], supplier management [19], and bankruptcy prediction [20]. Such methods have the ability to relieve the workload of human workers [14], provide unseen patterns [6], and diminish irregularities in human errors [21].

ML applications in BIS have certain considerations that influence the performance of the applied learning technique [22]. Business data is characterized by noise as inaccuracies due to human involvement in an extensive number of business tasks [16] in addition to the accumulated unstructured data in Big Data environment [23]. In fact, this impreciseness strongly influences the effectiveness of ML methods [12]. To clarify the motives of using such advanced methods in business, we provide the main features and challenging points related to the business and management field.

2.1 Business Information Systems: Overview and Challenges

For decades, BIS have provided management with powerful and improved computational abilities with the major concern of studying the business information and its sourcing, movements and usage within organizations. BIS incorporates both manual tasks and automated computing tasks. The main objective of such systems is to advance the services and capabilities of programs and humans to enable them to extract business valuable aspects.

The landscape of management and applications in BIS goes from simply operations of printing press to more advanced word-wide, mobile and cloudified operations [19]. Some specific integrated systems as BPM [24], Product Lifecycle Management (PLM) [25], Customer Relationship Management (CRM) [26], Supplier Relationship Management (SCM) [27] illustrate clearly the need to use ML as a required support in contemporary Big Data Analytic (BDA) environments. To perform, BIS are required to operate data analytics to investigate three main landmarks as classified in [11]: (1) Past: What has occurred, also called 'descriptive analytics', (2) Present: is the improved way to do through 'prescriptive analytics', and (3) Future: What will happen, namely 'predictive analytics'.

The above challenges are likely to be encouraged by the advent of Big Data to constitute reasons of integrating ML approaches in different BIS platforms and solutions such as predictive maintenance [28], predictive scheduling [29], predictive marketing [30], etc. On the other hand, one of the motivations of integrating ML methods is the outperforming of those methods in many other common challenges in the other applicative fields such as medicine [31], chemical physics [32], neurosciences [33], and communication technologies [34], and many others. The other significant motivation is the availability of data in the advent of Big Data environment while most of academics think that the coming back of ML algorithm is almost encouraged by Big Data paradigm. Definitely, because of the massive records of business data and the need of enterprises to seek for business value.

2.2 Machine Learning in Business: Overview and Role

The examination of the academic bibliography in ML business applications includes an abundant methods, sub-methods, heuristics, and techniques. Commonly, these approaches are divided into three subdomains: supervised learning, unsupervised learning, and reinforcement learning.

- Supervised Machine Learning (SML): The main objective of those methods focuses on classification and prediction concerns [9] where models are made for expecting a variable that supports one of a pre-arranged set of values [30]. Classification defines the assignment of data into predefined groups (classes) and learns the relationship between the other variables and the target class [6]. Classification and prediction purposes can use the same approaches but are differentiated in the handled data [28]. If the approach is applied to existing records, it has a classification purpose [2]. But when applied to a new data for which the class is unknown, it becomes a prediction [35]. The main advantage of these methods is their robustness in processing large data sets [19]. However, one disadvantage of these methods is that when a problem is easy to classify and its boundary function is more complex than it needs to be, this function is expected to over-fit [35]. Likewise, when a problem is too complex and the function is not prevailing sufficiently, the boundary under-fits [36]. Figure 1 illustrates a simple model of classification and prediction method with 2 variables and 2 classes.

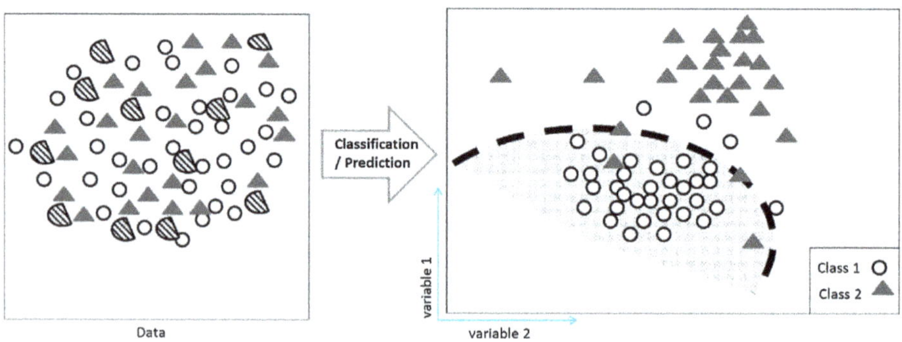

Fig. 1. Illustration of supervised machine learning mechanism (2 variables and 2 classes).

- Unsupervised Machine Learning (UML): In unsupervised learning, data are grouped and classified without labialization [37]. Two important branches exist:
  - Clustering: Is a common class of approaches that are used in several fields, including ML [38]. The method classifies a set of objects into different clusters (groups), so that the data in each group are characterized by one or more similarities of traits [39]. The power of this approach is in the ability of identifying thick and thin regions by finding correlations among data attributes, and then, discovering complete distribution patterns within a reasonable amount of time [40]. However, as clustering is basically a statistical algorithm, thus its major drawback is in its slackness in bulky databases due to memory restrictions and the extensive running times [41].
  - Association-rules: This class of approaches defines rules that administer the relationships among sets of entities [42] and aim to reveal patterns between the variable factors [43]. The powerful ability of association rules is their general applicability and flexibility to be integrated in all business concerns [44].

However, their main drawback is in a huge number of parameters, non-understandable for non-experts, and the acquired rules are often too many with low clarity [45].

- Reinforcement Machine Learning (RML): The reinforcement learning approaches enable learning from feedback received through interactions with an external environment [46]. In these approaches, input/output combinations are not presented, and selected actions are implicitly adjusted [47].

The examination of literature reveals plenty cases of ML applications in business. Table 1 summarizes the most relevant among them with respect to some relevant application fields.

Table 1. Nomenclature 1: business application.

Business field	Reference
Opinion mining	[7]
Risk management	[8]
Recommendation systems	[9]
Business process management (BPM)	[10]
Business administration	[14]
Business decision	[1]
Business knowledge management (BKM)	[15]
Enhanced accounting	[11]
Product design and engineering	[16]
Stock prediction	[18]
Supplier management	[19]
Bankruptcy prediction	[20]
Business process management (BPM)	[24]
Product lifecycle management (PLM)	[25]
Customer relationship management (CRM)	[26]
Supplier relationship management (SCM)	[27]
Smart manufacturing	[48]
Predictive maintenance	[28]
Predictive scheduling	[29]
Predictive marketing	[30]
Logistics management	[47]

3 Proposed Methodology

3.1 Proposed Classification Framework

Before elaborating the review strategy in our bibliography analysis, we pro-pose a classification model through which we construct five matrices: (1) Business Application (Table 2), (2) business area (Table 3), (3) data environment concept (Table 4), (4) technical platform (Table 5), and (5) machine learning approaches (Table 6). Each table embeds multiple hierarchical levels in nomenclatures of levels, depending on the content of the publications we examined.

Table 2. Nomenclature 1: business application.

Level 1	Information technology and information systems		
	Level 2	Enterprise information systems (EIS)	
		Level 3	Advanced information systems research (AISR)
		Level 3	ERP
	Level 2	Management information systems (MIS)	
		Level 3	BPM
		Level 3	Workflow management
		Level 3	Data management research
	Level 2	Transaction processing systems (TPS)	
		Level 3	Data privacy/Security management
		Level 3	Data accuracy management
	Level 2	Knowledge management	
		Level 3	Assessment
		Level 3	Ontology management (OM)
		Level 3	Natural language processing (NLP)
Level 1	Design/engineering/manufacturing		
	Level 2	Product design/engineering/manufacturing	
		Level 3	Product design/engineering
		Level 3	Product manufacturing
		Level 3	Product information management
	Level 2	Manufacturing systems	
		Level 3	Factory/machines design/engineering
		Level 3	Factory/machines manufacturing
		Level 3	Factory/machines information management
		Level 3	Job shop scheduling
		Level 3	Factory/machines design/engineering
		Level 3	Fault diagnostics
Level 1	Hybrid business applications?		

3.2 Proposed Review Method

The conducting of the literature review analysis follows the procedure of data preparation in the sub-sequent section as depicted in Fig. 2.

Step 1: Data Collection: By means of Harzing's Publish or Purish V.5® tool [49], we queried out the academic publications in the topic of ML and BIS (from 2010 to 2016). Parameters of Harzing's search request are enumerated in Table 7.

Table 3. Nomenclature 2: business area.

Level 1	Accounting
Level 1	Healthcare
Level 1	Industry
Level 1	Economy
Level 1	Marketing
Level 1	Commerce
Level 1	Management

Table 4. Nomenclature 3: data environment concept.

Level 1	Big data analytics (BDA)
Level 1	Competitive intelligence (CI)
Level 1	Business intelligence (BI)
Level 1	Data mining (DM)
Level 1	Traditional data analytics (DBA)
Level 1	Hybrid data environment?

Table 5. Nomenclature 4: technical platform.

Level 1	On-promise
Level 1	Web-based
Level 1	Big data
Level 1	Cloud computing
Level 1	Grid computing
Level 1	Hybrid technical context?

Step 2: Data Filtering: After sorting the records (publications), we cleansed and filtered the database by removing the undesirable columns (ranks, ISSN, types) and rows (citations, books, reports, white papers, patents), publications with topics that are not linked to the enterprises systems (for example [50]). We kept only interesting columns (cites, authors, title, year, source, publisher, URL) and rows (journals and conferences).
Step 3: Data Aggregation: Each for was filled a form for each row (publication) of the nomenclatures in the matching cells. The purpose of using this matching process is to determine which research (publication) is used and in which item in the nomenclatures.

At the end of this procedure, five filled nomenclatures were the base of a cross-analysis. The objective is having a reading of these date following three main axes in ML-BIS research: (1) Chronology and trends of research, (2) Scope and purpose of research, and (3) Industrial and academic impact of research. The empirical results are conducted to answer those three main axes in the next section.

Table 6. Nomenclature 5: machine learning approaches.

Level 1	Supervised learning		
	Level 2	Classification/prediction	
		Level 3	Supervised ANN
		Level 3	Fisher's linear discriminant
		Level 3	Regression
		Level 3	Polynomial regression
		Level 3	Linear regression
		Level 3	Maximum entropy
		Level 3	k-nearest neighbor (k-NN)
		Level 3	Decision trees (DT)
		Level 3	Conditional random fields (CRF)
		Level 3	Naive Bayes classifier
		Level 3	Bayesian networks
		Level 3	SVM
		Level 3	Case-bases reasoning (CBR)
		Level 3	Hidden Markov models
Level 1	Unsupervised learning		
	Level 2	Clustering	
		Level 3	k-means
		Level 3	Mixture models
		Level 3	Hierarchical cluster analysis
	Level 2	Unsupervised ANN	
		Level 3	Hebbian learning
		Level 3	Generative adversarial networks (GAN)
	Level 2	Association-rule	
		Level 3	FP-growth
		Level 3	Apriori algorithm
Level 1	Reinforcement learning		
	Level 2	Sarsa	
	Level 2	Q-learning	
Level 1	Hybrid approach?		
Level 1	All approaches (review papers)		

4 Empirical Analysis and Findings

This section presents the empirical results based on the aforementioned methodology. Through three key findings, we summarized publishing evolution by frequency, types and categories, distribution of data by scopes and purposes, and finally the distribution of techniques by business fields.

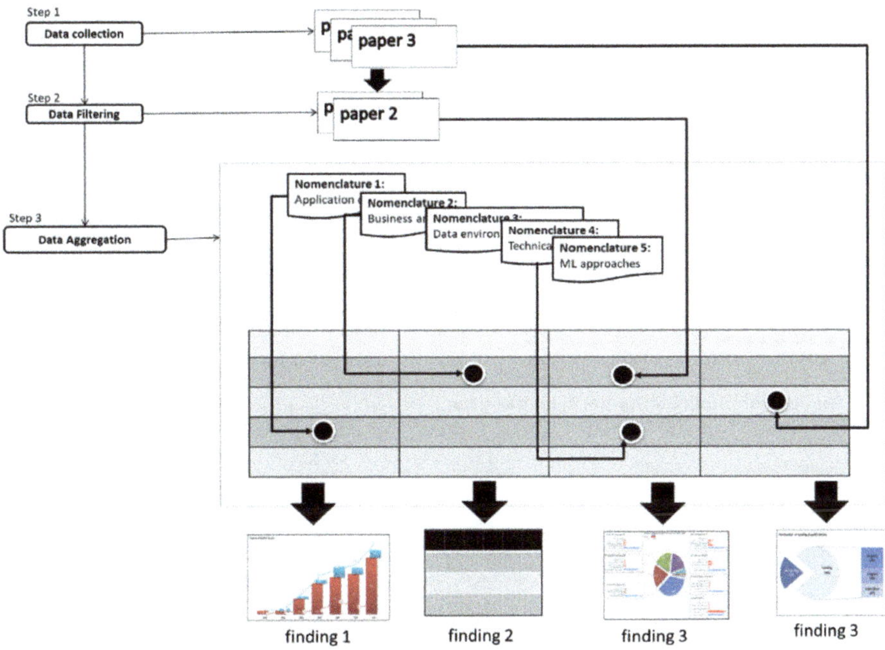

Fig. 2. Proposed method of review.

Table 7. Search request in Harzing's Publish or Purish V.5® [49].

Query date	2019-01-22
Maximum papers number	1000
Any of the words	{Machine learning, Business, Information, System}
Publication type	Journal
Years	2010–2016
Publisher	All
Search engine	Google scholar engine
Location	Title, abstract, key words

4.1 Finding 1: Chronological Evolution of ML-BIS Research

In Fig. 3, we illustrate the tendencies of evolution (numbers and rates) of publications chronologically per types, categories, and frequencies of cites for the periods 2010–2018.

As clearly illustrated in Fig. 3, the frequency of papers have taken more attention from year to year: It was jumped from 1%–0% to 1%–14% between 2010 and 2013, and then 3%–26% from 2014 to 2016, to finally increase more than the double (84%) in 2018. Following the trajectory of tendency line, the growth is going faster since 2016. Therefore, this lead us to think that ML use in BIS will continue taking more consideration in the next years.

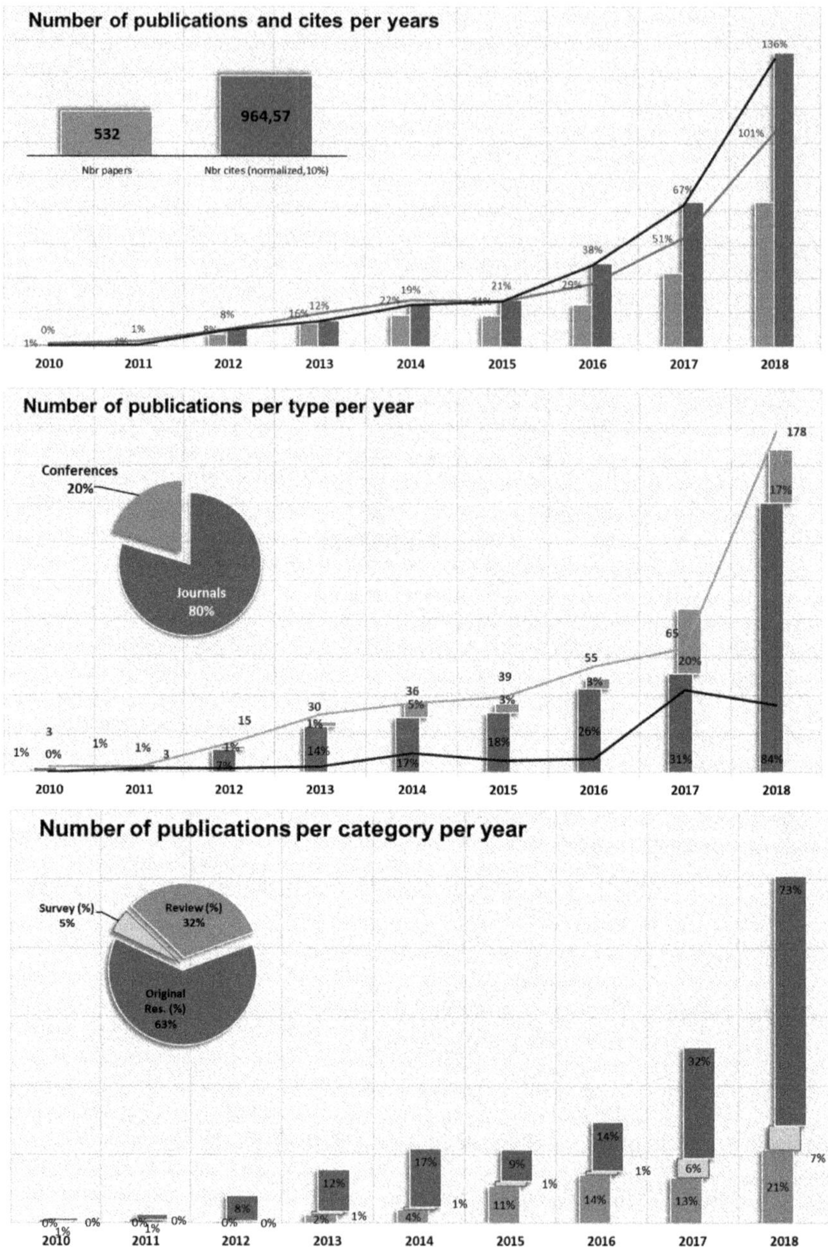

Fig. 3. Year–wise growth of publications (cites, types, categories) (finding 1).

About the bars graph about frequency of papers by categories (review papers, surveys or original researches), we notice the proportion of use of original researches is twice higher than literature studies (63%–32%), as also illustrated in the pie graph. However, when examining in details by years, this proportion kept true except for the year 2015 (11%–9%). This seems rational, as for any new paradigm in a research field; researches must go through a profound analysis of literature before applications in the beginning stage. Although ML methods are not new, their use in the business environment remains an innovative topic. However, the surveys do not exceed 5% overall and 1% per year, due to the scarcity of practical cases.

4.2 Finding 2: Scope and Purpose of ML-BIS Researches

Table 8 illustrates the rate (number) of publications by journals with the business applications (scopes).

Table 8. Distribution of publications by business scopes (finding 2).

Journal *	Rates of Publications (%)	Accuracy of assessment	Analysis/Study of Current/future trends	detection of faults	Detection of Markets opportunities	Detection of performance's factors	Detection of privacy/security Violation	Improvement and Facilitating management	improvement of management	Improvement of Risk Management	Improvement of information flows	Prediction of Bankruptcy	Prediction of Business Failures	Prediction of Customer behavior	Recognition of Language
IJML	3	✓	✓												
ASE	7		✓				✓	✓				✓			
COT	3									✓			✓		
ATM	25	✓	✓			✓	✓	✓	✓	✓					
CAI	19	✓	✓				✓	✓		✓					
DKE	1				✓										
DSS	11	✓	✓	✓				✓						✓	✓
EIS	18		✓				✓				✓				
EJIS	5		✓				✓								
ITJ	8		✓											✓	

** List of abbreviations of journals are provided in Table 9 (Appendix)*

4.3 Finding 3: Distribution of ML Techniques in BIS Research

In this finding, we are interested to study the influence of the business application area and construct an influence graph among these criteria. The synthesis graph in Fig. 4 aims to reveal interdependencies among these criteria and their assets. To do this, we superposed the nomenclatures 1 (Business Application, Level 2) and 5 (Machine Learning Approaches, Level 2) for only original researches (132 publications).

As shown in Fig. 4, the 'Enterprise Information Systems (EIS)' is the most popular area with 36% while the areas of Management Information systems (MIS)', 'Transaction Processing Systems (TPS)', and 'Decision Support Systems (DSS)' are the following fields with a rate halved to 19%. The other fields of 'Knowledge Management (KM)', 'Product Design/Engineering/Manufacturing', and 'Manufacturing Systems (MS)' are ranked lastly with insignificant rates of respectively 4%, 2%%, and 1%. Firstly, in EIS we find mainly 'Advanced Information Systems Research (AISR)' and 'Enterprise Resource Planning (ERP)'. Secondly, in Management Information Systems we find primarily Business Process Management (BPM), Workflow Management (WM), and Data Management Research (DMR). Thirdly, in Transaction Processing Systems (TPS), we note predominantly the sub-fields 'Data Privacy/Security Management' (DPSM), Data Accuracy Management (DAM), and Data Integrity/Compliance Management (DTCM)'. And lastly, in 'Decision support systems (DSS)', we figure out mostly 'Advanced Aid-Decision Support Systems (AADS)' ('Predictive Aid-Decision (PAD)', 'Opinion Mining & Sentiment Analysis OMSA)', 'Risk Management RM)', 'Business Intelligence (BI)', 'Information Filtering System (IFS)' ('Recommender System (RS)'), and 'Content Discovery Platform (CDP)').

To summarize, we see that classification and prediction techniques are the most used ML approaches in BIS, including mostly supervised learning with a majority rates of 98% in EIS, 96% in MIS, 87% in TPS, and 69% in DSS.

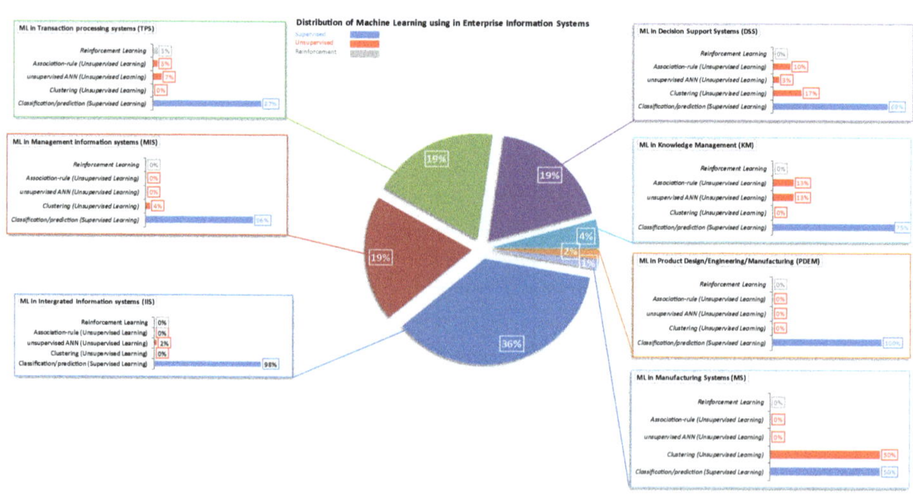

Fig. 4. Distribution of using machine learning techniques in business information systems (finding 3).

5 Conclusion and Discussion

In this paper, we have examined the evolution of Machine Learning approaches in the relevant research works of Business Information Systems through a comprehensive literature review. By this paper, we contribute in drawing a guideline for academic and practitioners regarding the application of the ML techniques in resolving business concerns.

Throw the deep analysis of the examined literature in this paper; we submit that the application of machine learning to enterprises systems is just at its beginning and will certainly spread in the near and far futures. In addition, results figured out that classification and prediction techniques are the most predominant ML approaches that have conducted research in BIS during the last ten years.

The main outcome in this paper led our thinking to understand that business information systems are moving to mutate into data-driven models in the middle and long terms future.

Appendix

Table 9. Abbreviations table of journals (with editors) included in the literature review.

Journal	Editor	Abbreviation
International Journal of Machine Learning and Computing	IEEE	IJML
ACM sIGKDD Explorations Newsletter	Taylor & Francis	ASE
Communications of the ACM	Elsevier	COT
ACM Transactions on Management Information Systems	World Scientific	ATM
Computing and informatics	Springer	CAI
Data & Knowledge Engineering	Elsevier	DKE
Decision Support Systems	Elsevier	DSS
Enterprise Information Systems	Emerald Insight	EIS
European Journal of Information Systems	Elsevier	EJIS
Information and Technology Journal	IEEE	ITJ

References

1. Aluri, A., Price, B.S., McIntyre, N.H.: Using machine learning to cocreate value through dynamic customer engagement in a brand loyalty program. J. Hospitality Tourism Res. **43**(1), 78–100 (2019)
2. Magomedov, S., Pavelyev, S., Ivanova, I., Dobrotvorsky, A., Khrestina, M., Yusubaliev, T.: Anomaly detection with machine learning and graph databases in fraud management. Int. J. Adv. Comput. Sci. Appl. **9**(11), 33 (2018)
3. Walsh, T.: How machine learning can help solve the big data problem of video asset management. J. Digital Media Manag. **6**(4), 370–379 (2018)
4. Akhtar, P., Frynas, J.G., Mellahi, K., Ullah, S.: Big data-savvy teams' skills, big data-driven actions and business performance. Br. J. Manag. **30**(2), 252–271 (2019)

5. Raguseo, E.: Big data technologies: an empirical investigation on their adoption, benefits and risks for companies. Int. J. Inf. Manag. **38**(1), 187–195 (2018)
6. Yogeshwar, J., Quartararo, R.: How content intelligence and machine learning are transforming media workflows. J. Digital Media Manag. **7**(1), 24–32 (2018)
7. Li, Z., Tian, Z.G., Wang, J.W., Wang, W.M.: Extraction of affective responses from customer reviews: an opinion mining and machine learning approach. Int. J. Comput. Integr. Manuf. **16**, 1–13 (2019)
8. De Paula, D.A., Artes, R., Ayres, F., Minardi, A.M.A.F.: Estimating credit and profit scoring of a Brazilian credit union with logistic regression and machine-learning techniques. RAUSP Manag. J. (2019)
9. Nilashi, M., Ibrahim, O., Ahmadi, H., Shahmoradi, L., Samad, S., Bagherifard, K.: A recommendation agent for health products recommendation using dimensionality reduction and prediction machine learning techniques. J. Soft Comput. Decis. Support Syst. **5**(3), 7–15 (2018)
10. Mendling, J., Decker, G., Richard, H., Hajo, A., Ingo, W.: How do machine learning, robotic process automation, and blockchains affect the human factor in business process management? Commun. Assoc. Inf. Syst. **43**, 297–320 (2018)
11. Appelbaum, D., Kogan, A., Vasarhelyi, M., Yan, Z.: Impact of business analytics and enterprise systems on managerial accounting. Int. J. Account. Inf. Syst. **25**, 29–44 (2017)
12. Deanne, L., Chang, V.: A review and future direction of agile, business intelligence, analytics and data science. Int. J. Inf. Manag. **36**(5), 700–710 (2016)
13. Eitle, V., Buxmann, P.: Business analytics for sales pipeline management in the software industry: a machine learning perspective. In: Proceedings of the 52nd Hawaii International Conference on System Sciences (2019)
14. Li, Y., Jiang, W., Yang, L., Wu, T.: On neural networks and learning systems for business computing. Neurocomputing **275**, 1150–1159 (2018)
15. Sumbal, M.S., Tsui, E., See-to, E.W.: Interrelationship between big data and knowledge management: an exploratory study in the oil and gas sector. J. Knowl. Manag. **21**(1), 180–196 (2017)
16. Ireland, R., Liu, A.: Application of data analytics for product design: sentiment analysis of online product reviews. CIRP J. Manufact. Sci. Technol. **23**, 128–144 (2018)
17. Ehret, M., Wirtz, J.: Unlocking value from machines: business models and the industrial internet of things. J. Mark. Manag. **33**(1–2), 111–130 (2017)
18. Pahwa, N., Khalfay, N., Soni, V., Vora, D.: Stock prediction using machine learning a review paper. Int. J. Comput. Appl. **5**, 163 (2017)
19. Hong, J.S., Yeo, H., Cho, N.W., Ahn, T.: Identification of core suppliers based on e-invoice data using supervised machine learning. J. Risk Financ. Manag. **11**(4), 70 (2018)
20. Mai, F., Tian, S., Lee, C., Ma, L.: Deep learning models for bankruptcy prediction using textual disclosures. Eur. J. Oper. Res. **274**(2), 743–758 (2019)
21. Mihalis, G., et al.: A multi-agent based system with big data processing for enhanced supply chain agility. J. Enterprise Inf. Manag. **29**(5), 706–727 (2016)
22. Jennifer, L., et al.: Expediting expertise: supporting informal social learning in the enterprise. In: Proceedings of the 19th International Conference on Intelligent User Interfaces. ACM (2014)
23. Sun, Z., Sun, L., Strang, K.: Big data analytics services for enhancing business intelligence. J. Comput. Inf. Syst. **58**(2), 162–169 (2018)
24. Fosso Wamba, P.S.: Big data analytics and business process innovation. Bus. Process Manag. J. **23**(3), 470–476 (2017)
25. Nagorny, K., Lima-Monteiro, P., Barata, J., Colombo, A.W.: Big data analysis in smart manufacturing: a review. Int. J. Commun. Netw. Syst. Sci. **10**(3), 31 (2017)

26. Ahmad, A.K., Jafar, A., Aljoumaa, K.: Customer churn prediction in telecom using machine learning in big data platform. J. Big Data **6**(1), 28 (2019)
27. Cavalcante, I.M., Frazzon, E.M., Forcellini, F.A., Ivanov, D.: A supervised machine learning approach to data-driven simulation of resilient supplier selection in digital manufacturing. Int. J. Inf. Manag. **49**, 86–97 (2019)
28. Yamato, Y., Fukumoto, Y., Kumazaki, H.: Predictive maintenance platform with sound stream analysis in edges. J. Inf. Process. **25**, 317–320 (2017)
29. Cardin, O., Trentesaux, D., Thomas, A., Castagna, P., Berger, T., El-Haouzi, H.B.: Coupling predictive scheduling and reactive control in manufacturing hybrid control architectures: state of the art and future challenges. J. Intell. Manuf. **28**(7), 1503–1517 (2017)
30. Pauwels, K., Joshi, A.: Selecting predictive metrics for marketing dashboards-an analytical approach. J. Mark. Behav. **2**(2–3), 195–224 (2016)
31. Obermeyer, Z., Emanuel, E.J.: Predicting the future—big data, machine learning, and clinical medicine. N. Engl. J. Med. **375**(13), 1216 (2016)
32. Behler, J.: Perspective: Machine learning potentials for atomistic simulations. J. Chem. Phys. **145**(17), 170901 (2016)
33. Segler, M.H., Waller, M.P.: Neural-symbolic machine learning for retrosynthesis and reaction prediction. Chem. Eur. J. **23**(25), 5966–5971 (2017)
34. Zibar, D., Piels, M., Jones, R., Schäeffer, C.G.: Machine learning techniques in optical communication. J. Lightwave Technol. **34**(6), 1442–1452 (2016)
35. Alexander, D.K., Liebrock, L.M., Neil, J.C.: Authentication graphs: analyzing user behavior within an enterprise network. Comput. Secur. **48**, 150–166 (2015)
36. Malik, G., Rathore, A., Vij, S., Malik, G., Rathore, A., Vij, S.: Utilizing various machine learning techniques to classify data in the business domain. Int. J. **4**, 118–122 (2017)
37. Sabharwal, S., Nagpal, S., Aggarwal, G.: Empirical analysis of metrics for object oriented multidimensional model of data warehouse using unsupervised machine learning techniques. Int. J. Syst. Assur. Eng. Manag. **8**(2), 703–715 (2017)
38. Täuscher, K., Laudien, S.M.: Understanding platform business models: a mixed methods study of marketplaces. Eur. Manag. J. **36**(3), 319–329 (2018)
39. Kim, M.S., Choi, E.S., Lee, J.Y., Kang, M.S.: A study on the analysis of stability indicators in financial statements using fuzzy c-means clustering. Int. J. Appl. Eng. Res. **12**(20), 9863–9865 (2017)
40. Hong, Y., Lee, J.C., Ding, G.: Volatility clustering, new heavy-tailed distribution and the stock market returns in South Korea. Int. J. Inf. Bus. Manag. **11**(2), 317–325 (2019)
41. Tan, K.H., et al.: Harvesting big data to enhance supply chain innovation capabilities: an analytic infrastructure based on deduction graph. Int. J. Prod. Econ. **165**, 223–233 (2015)
42. Sinaga, F., Sarno, R.: Business process anomali detection using multi-level class association rule learning. IPTEK J. Proc. Ser. **2**(1) (2016)
43. Frédéric, S., St-Pierre, J., Biskri, I.: Mining and visualizing robust maximal association rules on highly variable textual data in entrepreneurship. In: Proceedings of the 8th International Conference on Management of Digital EcoSystems. ACM (2016)
44. Amatriain, X., Pujol, J.M.: Data mining methods for recommender systems. In: Ricci, F., Rokach, L., Shapira, B. (eds.) Recommender Systems Handbook, pp. 227–262. Springer, Boston, MA (2015). https://doi.org/10.1007/978-1-4899-7637-6_7
45. Kamsu-Foguem, B., Rigal, R., Mauget, F.: Mining association rules for the quality improvement of the production process. Expert Syst. Appl. **40**(4), 1034–1045 (2013)
46. Okdinawati, L., Simatupang, T.M., Sunitiyoso, Y.: Multi-agent reinforcement learning for value co-creation of collaborative transportation management (CTM). Int. J. Inf. Syst. Supply Chain Manag. **10**(3), 84–95 (2017)

47. Li, X., Zhang, J., Bian, J., Tong, Y., Liu, T.Y.: A Cooperative Multi-Agent Reinforcement Learning Framework for Resource Balancing in Complex Logistics Network, arXiv:1903.00714 (2019)
48. Wang, J., Ma, Y., Zhang, L., Gao, R.X., Wu, D.: Deep learning for smart manufacturing: Methods and applications. J. Manuf. Syst. **48**, 144–156 (2018)
49. Harzing.com. Harzing's Publish or Purish. Harzing.com, 01 January 2019. https://harzing.com/resources/publish-or-perish. Accessed 22 Jan 2019
50. Yuan, R., Li, Z., Guan, X., Xu, L.: An SVM-based machine learning method for accurate internet traffic classification. Inf. Syst. Front. **12**(2), 149–156 (2010)

Piecewise Demodulation Based on Combined Artificial Neural Network for Quadrate Frequency Shift Keying Communication Signals

Nihat Daldal and Kemal Polat(✉)

Department of Electrical-Electronic Engineering,
Bolu Abant Izzet Baysal University, Bolu, Turkey
{nihatdaldal, kpolat}@ibu.edu.tr

Abstract. The Quadrate Frequency-shift Keying (QFSK) modulation is one of the most widely used modulation methods for transmitting base band signal in the transition band in digital communication. It is a very common modulation type for fast and easy transmission of data, especially in wireless communication. In QFSK modulation, 4 separate carriers are used, and since each carrier represents 2 bits, the sending speed doubles according to classical digital modulation. In this study, firstly, with QFSK modulation, 8-bit information-bearing module signals have been obtained.

Moreover, the theory has been developed on the demodulation of the QFSK module signal. The aim is to get the base band signal again. For this purpose, all data from 0–255 to 8-bit decimal length has been obtained as QFSK modulated. SNR = 5 dB–10 db–15 dB–20 dB noise has been added to the QFSK signal in order to examine the system performance during the transmission of the modulated signal. The generated signals are given to the developed demodulation system using different foldings. The matrix consisting of QFSK modulation data to be trained in the network of ANN in demodulation is divided into 4 parts. Each piece represents two bits of data. In the case of a column matrix, 4 parts have been applied as an introduction to the neural network model. In the output of ANN, the result matrix to be estimated has been formed. Trained with 5 dB noisy QFSK data of 10-layer network, application of other QFSK signal data to the network for testing, base band data of the signal has been obtained and 100% performance has been achieved. Each piece of modulation signal applied to the ANN network input is classified between 0–3 at the output. The modulation data that carries 8 bits are applied to the network in 4 steps and classified. 4 separate classification data from ANN output are converted to 2 bits of logic. 8-bit demodulation data is obtained from 4 steps. After the formation of the ANN network, the base band digital signal estimation is performed quickly in 4 steps through each byte QFSK modulation signal under different noise coming to the network.

Keywords: Demodulation · Digital modulation · Digital communication · Piecewise ANN · QFSK modulation

D. J. Hemanth and U. Kose (Eds.): ICAIAME 2019, LNDECT 43, pp. 17–24, 2020.
https://doi.org/10.1007/978-3-030-36178-5_2

1 Introduction

With the use of digital technologies, developments have been made in digital communication systems today. In particular, the data is sent in modulated signals, and the noise effect is reduced to a minimum, and the sending speed has been increased with different techniques [1].

The logic signals used in digital communication systems are called base band signals. It is used more in short-range communications. In remote wireless or infra communication, these base band signals must be modulated and converted to high frequency [2]. In this case, the base band is called the transition band signal by the expression of the high-frequency carrier signal bits, and the modulation that provides this state is called the digital transition band modulation [3].

When studying the studies of digital modulation, Xuming and his colleagues [4] identified the convolution network demodulator structure, analyzed the 1024 byte modulation signal entering the system by dividing it into 32 symbols, and obtained the symbol that it carries in the signal by classifying it. Here, under different noises, bask and BPSK signals have been studied. Leader and his colleagues work in multi-channel for distortion analysis in a neural network model developed for BPSK and QPSK signals entering the system via wireless communication in different ways and different artificial neural network modeling of data with is included in the estimation. In this case, the signal is sent to the ANN system with a noise range of 5–45 dB, and the Rayleigh' model is very close to the receiver performance [5].

Landing and Lenin in their studies, they used a deep learning detection method to obtain data in multi-channel short-range wireless news-carriages and performed it as a simulation. In this study, they first obtained the modulated signal from the communication channel and obtained the demodulated data with the feature extraction, the deep learning algorithm with the trained data set [6].

Mohammad and in the study of his colleagues, rayleigh studied the demodulation of the 10 dB–20 dB damped FSK data through the damped communications channel by the deep convolutional neural network structure. Moreover, they compared the other machine learning methods with SVM, LDA, MLP, and QDA. According to these methods, they obtained at least 10 dB earnings [7].

Amini and Balarastaghai they studied Ann-based bfsk demodulation, designed two neural networks based on Elman and TDNN and trained ANN-based neural networks with noisy data. They then applied the noisy bfsk data to Ann network and obtained the base band signal from the output. Ber reported that the rate was quite low [8].

This study, a theory was developed on the classification of modulation signals. The amplitude and period averages of the signal in each region were first segmented on digital and an analog modulation signals. When the average values are given as inputs to the Ann network have been designed at the end, the network has been decided the modulation signal type [9].

As shown in the studies, it is important to develop modulator and demodulator structures in communication systems, especially in order to obtain more flexible and low-cost demodulator structures.

2 Quadrate Frequency Shift Keying (QFSK) Modulation

One of the most widely used modulation types of transition band modulation is the FSK modulation. Moreover, two separate carriers are used for the logic bits in the FSK modulation [10]. For each bit, the carrier is sent with the modulation of the FSK. Moreover, 1-byte data is sent in Step 8.

In transition band modulation, 8 steps for 1-byte data is reduced by grouping bits instead of single bits for increasing data sending speed. For this, quadrate type modulations are used very often.

In these modulation modulations, 4 carriers are used. The QFSK modulation also uses 4 separate carrier signals with the same amplitude but different frequencies. In this case, each carrier represents 2 bits. Thus, 1-byte data is sent in 4 steps with QFSK modulation instead of 8 steps and doubles the speed. In Table 1, QFSK equations are given for 2-bit binary values.

Table 1. QFSK signal equations according to binary values

Lojik value	QFSK signal
00	$5.sin(2.\pi.f1.t)$
01	$5.sin(2.\pi.f2.t)$
10	$5.sin(2.\pi.f3.t)$
11	$5.sin(2.\pi.f4.t)$

When the QFSK signal is obtained, the base band signal is divided into 4 groups in 2 bits. Figure 1 shows the QFSK signals that correspond to an example of 8-bit digital data. Here, the signal amplitude is 5 V, and the frequencies are 5, 10, 15, 20 kHz.

Then the bits are checked, and the carriers in Table 1 are sent according to the bit condition. Figure 1 shows the QFSK signal and its noisy states, which correspond to the 8-bit base band signal.

The modulation of the QFSK signal is also the most common structures of the coherent type demodulator structures. Figure 2 shows the coherence type of demodulator structure. As shown in the demodulator, when the QFSK module signal arrives from the channel, the 2-bit period length is taken as part and multiplied by a carrier and receiver. The Integral of the product is taken. If the integral output is high, 2 bits are produced which are represented by the carrier. At the end of Step 4, 8-bit base band sign is obtained from the output with the collection of bits.

The modulation of the QFSK signal can be seen as a complex structure, and for a stable reception, the coherent-type receiver structure is preferred. The other disadvantage of this structure is that 4 separate carriers used must be produced in the demodulation circuit [8–11]. Using ANN with a different approach to the demodulation system has become more popular lately.

Thus, studies are being carried out for the demodulation of various modulation signals. The ANN structure shows that even high-level noisy signals can be performed without error. In this study, an integrated ANN structure represented by each carrier of the Coherent type demodulator circuit was resolved with an ANN neural network model developed.

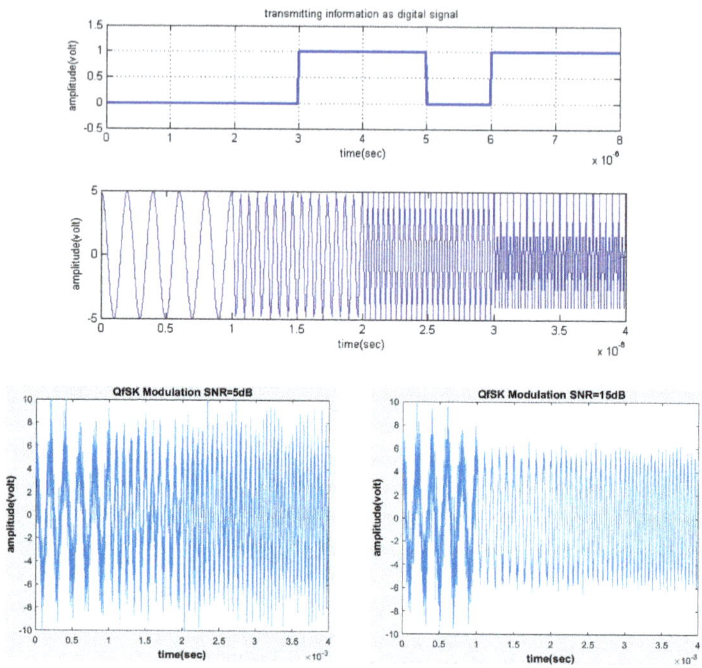

Fig. 1. QFSK signals generated at different SNR values for 8 bit binary numbers

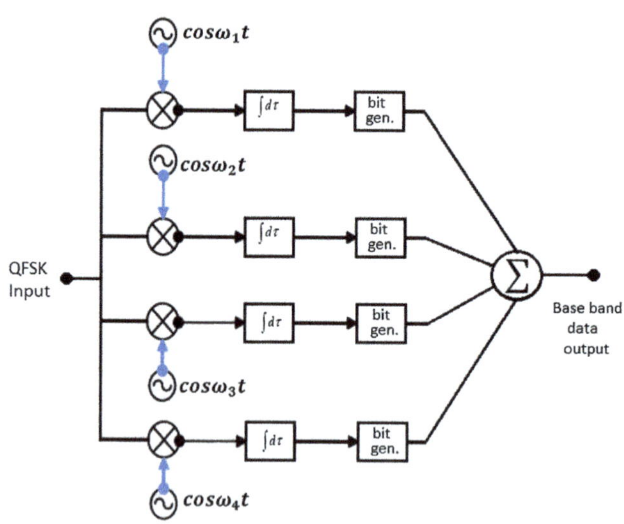

Fig. 2. QFSK demodulator structure

3 Artificial Neural Network Structure

Neural networks are synthetic networks that mimic biological neural networks. The ability to learn from a source of information is one of the most important characteristics of ANN. The information in artificial neural networks is stored in the weights of the connections of the neurons in the network [12].

Therefore, how to determine the weights is important. Because the information is stored in the entire network, the weight value of a node does not mean anything by itself. The weights in the entire network should be optimal. Processing to achieve these weights is called training of the network [13]. Accordingly, for a network to be educable, the weight values must be dynamically changeable within a given rule. Figure 3 shows a basic ANN structure.

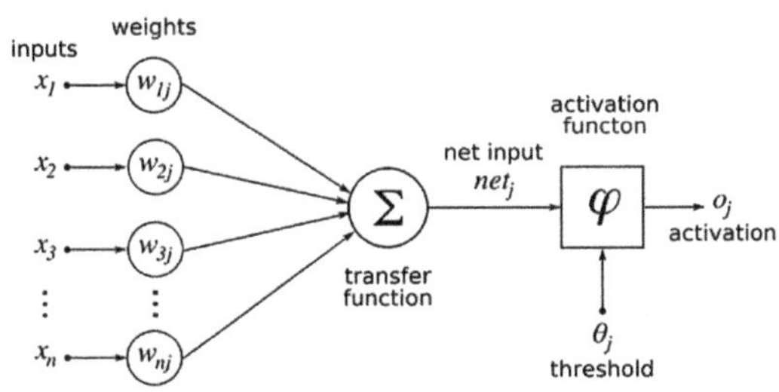

Fig. 3. Basic ANN structure

In this study, the network structure was created and trained for the 5 dB noisy QFSK data matrix. In the study, the signal matrices consisting of 0–255 data to be used for education were divided into 4 parts and named as P1, P2, P3, P4.

Because these matrices represent 2-bit data, they can take 00, 01, 10 or 11. Since 2-bit values cannot be obtained at the output of ANN, these 2-bit data are classified by the name of 0, 1, 2, 3. Then, all matrices to be used for training were classified and the corresponding output matrix 0, 1, 2, 3 was determined by placing each piece at the bottom. Finally, all data of the 5 dB noisy QFSK matrix, which will be entered into the training, is divided into 4 parts, and the ANN network is created and trained by sorting each p part of the Matrix and sorting the corresponding output matrices in the lower part.

In Fig. 4, the data of the 5 dB matrix are divided into 4 parts and transformed into a column matrix. Now the system has learned the output classification corresponding to 4 parts of each 1 byte of data.

With the creation of the ANN network structure, the data coming to the system is classified in 4 steps in the ANN network, the classification values are converted into binary bits, and 8 bits are obtained.

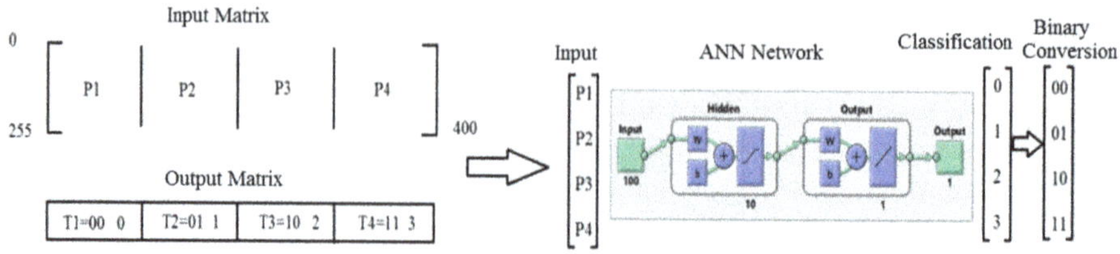

Fig. 4. ANN system modeling

3.1 Performance of the Generated ANN Network

An 8-bit quadrate type modulation signal consists of a total of 4 bits, and each track is 2 bits. For example, the QFSK signal is divided into 4 parts, as shown in Fig. 1, with a 5 dB AWGN noise matrix consisting of 0–255.

Here it is preferable to create and train the neural network that makes it with the loudest signal. When the network is trained for the worst condition, it will be easier to predict the signals at other noise levels. For this reason, 5 dB signal matrix is preferred for training. 5 dB QFSK data matrix consists of 4000 signal samples for 1-byte information.

Since quadrate signals have 4 carriers, each piece represents 2 bits of data. Since 1 byte consists of 4 parts, each part consists of 1000 signal samples. For each value of the matrix containing all values from 0 to 255, the 4-part matrix data is converted to the matrix of the column by adding the parts to the bottom. The data obtained as a result of the training are given in Fig. 5.

Results

	Samples	MSE	R
Training:	714	9.32189e-23	9.99999e-1
Validation:	153	1.00579e-22	9.99999e-1
Testing:	153	9.80619e-23	9.99999e-1

Fig. 5. Creating an ANN network with 0 dB FSK data

As shown, 714 matrix data from the 1020 data given to the network entrance were used in the network Education, 153 data were used for the test, and the following results were obtained in the network. In Fig. 6, the error distribution graph and error differences and performance graph are observed as a result of the creation of the network. The MSE (mean squared error) value is quite low, indicating that the resulting network structure will perform very well.

In Fig. 7, the regression graphs obtained as a result of the creation of the ANN network are given. The value of R^2 is 1, which confirms that network performance is very good.

After this phase, when the random QFSK data which contains different noises are given as input, it is necessary to estimate the base band signal, that is, the digital

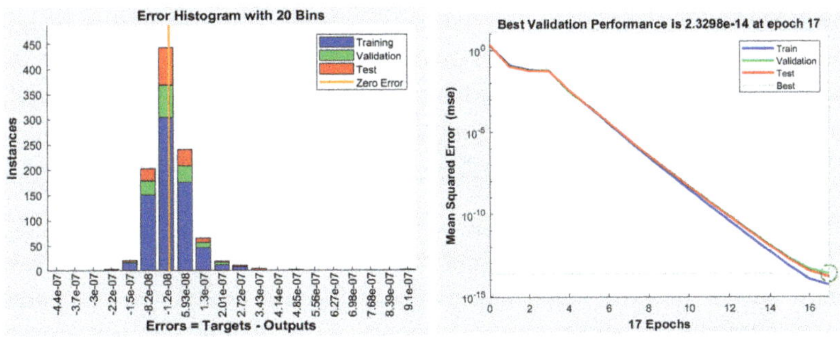

Fig. 6. ANN error and performance graph

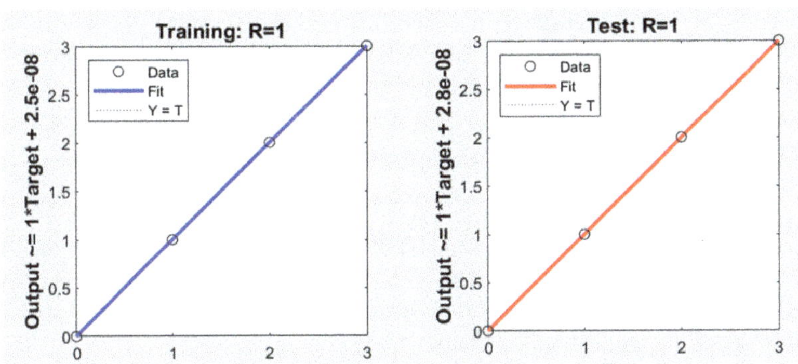

Fig. 7. ANN training and test regression results

information it carries. For this purpose, as an example, the matrix containing 8 bits of data with 15 dB noise signals to the input is primarily applied as an introduction to the ANN system in the case of the column matrix by dividing the pieces into 4 pieces consisting of logic 2 bits.

In this case, the error distribution and regression graph are shown in Fig. 8.

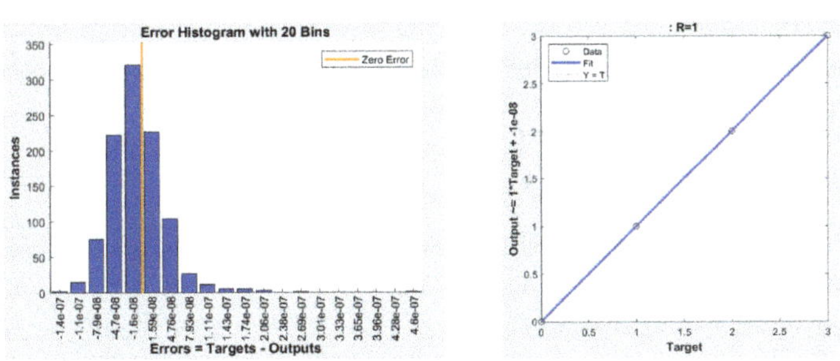

Fig. 8. Estimation results of 15 dB QFSK data

4 Conclusion

In digital communication, the transmission of logic values using a carrier is called transition band modulation. In this study, QFSK modulation from transition band modulation was introduced. In QFSK modulation, modulation signals are generated for each 8-Bit value using 4 carrier signals.

When the modulation signal is sent to the remote point, the noise effects on the signal affect the demodulator performance. The study has been conducted to reveal the base band data that the qfsk signal carries under different noises. For this purpose, an ANN based demodulator structure was proposed, network structure has been created, and 5 dB noisy QFSK signals has been trained.

A 10 dB–15 dB–20 dB noisy 1-byte long QFSK modulation signal was applied as input to the generated network structure. It was observed that the network obtained the base band data at 100% accuracy by 4-step processing. By using 4 network structure parallel instead of a single network, data can be obtained with one operation.

References

1. Ertürk, S.: Sayısal Haberleşme, Birsen Yayınevi, İstanbul, Türkiye (2016)
2. Altun, H., Öztürk, Y., Proakis, J.G., Masoud, S.: Fundamentals of Communication Systems, Nobel Yayın 2010, vol. 4, pp. 545–567
3. Louis, E., Frenzel, J.: Principles of Electronic Communications Systems, 3rd edn, pp. 385–400. McGrawHill companies, New York (2008)
4. Xuming, L. et al.: A deep convolutional network demodulator for mixed signals with different modulation types. In: IEEE 15th International Conference on Dependable, Autonomic and Secure Computing Conference (2017)
5. Onder, M., Akan, A., Doğan, H.: Advanced neural network receiver design to combat multiple channel impairments. Turk. J. Electr. Eng. Comput. Sci. **24**, 3066–3077 (2016)
6. Lanting, F., Lenan, W.: Deep learning detection method for signal demodulation in short range multipath channel. In: 2nd International Conference on Opto-Electronic Information Processing, p. 978 (2017)
7. Mohammad, A., Raddy, N., James, F., Beard, C.: Demodulation of faded wireless signals using deep convolutional neural networks. In: IEEE 8th Annual Computing and Communication Workshop and Conference (2018)
8. Amini, M., Balarastaghi, E.: Improving ANN bfsk demodulator performance with training data sequence sent by transmitter. In: Second International Conference on Machine Learning and Computing (2010)
9. Hossen, A., Wadahi, F., Jervase, J.: Classification of modulation signals using statistical signal characterization and artificial neural networks. Eng. Appl. Artif. Intell. **20**, 463–472 (2007)
10. Gallager, R.: Principles of Digital Communication. Cambridge University Press, Cambridge (2008)
11. Link Budget Analysis: Digital Modulation, Part 2. www.AtlantaRF.com
12. Egrioglu, E., Aladag, Ç., et al.: A new approach based on artificial neural networks for high order multivariate fuzzy time series. Expert Syst. Appl. **36**(7), 10589–10594 (2009)
13. Elmas, Ç.: Yapay Sinir Ağları Kuram Mimari Eğitim Uygulama. Seçkin Yayıncılık, pp. 65–90 (2003)

A New Variable Ordering Method for the K2 Algorithm

Betül Uzbaş[1(✉)] and Ahmet Arslan[2]

[1] Department of Computer Engineering, Faculty of Engineering and Natural Sciences, Konya Technical University, Konya, Turkey
buzbas@ktun.edu.tr

[2] Enelsis Industrial Electronic Systems Research and Development Co. Ltd., Konya, Turkey
ahmetarslan@enelsis.com

Abstract. K2 is an algorithm used for learning the structure of a Bayesian networks (BN). The performance of the K2 algorithm depends on the order of the variables. If the given ordering is not sufficient, the score of the network structure is found to be low. We proposed a new variable ordering method in order to find the hierarchy of the variables. The proposed method was compared with other methods by using synthetic and real-world data sets. Experimental results show that the proposed method is efficient in terms of both time and score.

1 Introduction

Bayesian networks (BN), which are also known as belief networks, are used for providing information about uncertainty. Learning BN structure from data is a difficult process. Chickering pointed out that learning BN structure is NP-hard even when each node is restricted to have at most 2 parents [1].

K2 is an algorithm used for learning the structure of a BN. K2 algorithm assumes that given dataset is ordered and asserts that only variables that come before the given variable X in the dataset may be parent of X variable [2]. The performance of the algorithm depends on the order of the variables. The order of variables must be correct to construct suitable network structure.

The ordering of the variables can be performed by experts. However, it is not possible to find a human expert when performing data mining and machine learning tasks. Thus, BN learning from data algorithms are applied by random ordering of the variables or the given order of a data set and this might cause bad results [2]. In the literature, various methods have been proposed to find variable ordering [2–7].

In this study, a new variable ordering method (ScrRnkK2), that has good performance in sparse data (dataset consists mostly zeros), was proposed to determine the order of the variables in sparse dataset for K2 algorithm. The proposed method was compared with other methods by using synthetic and real-world data sets. The results of the comparisons showed that the ordering

D. J. Hemanth and U. Kose (Eds.): ICAIAME 2019, LNDECT 43, pp. 25–34, 2020.
https://doi.org/10.1007/978-3-030-36178-5_3

of the variables using the proposed method is efficient in terms of the score and operating time of the network structure [8]. The aim of ScrRnkK2 is to find appropriate variable ordering before K2 is applied. It is necessary to have the correct variable ordering to construct a BN that is a correct representation of the domain [9]. Our aim is to find a quick sorting method that has good performance in sparse data. The importance of sorting speed is increasing, especially in social networks that have huge size data set. The BN that are constructed through dataset by sorting ScrRnkK2, has both a high network structure score and a strong performance in terms of sorting speed.

Information on BNs is presented in section two. The method proposed is presented in section three. Experimental results presented in section four. The obtained results are presented as summary in the final section.

2 Bayesian Network

BNs were first introduced by Pearl. BNs are directed acyclic graphs. Each node in the graph represents proposition (or variable), the arcs between the nodes represent direct causal influences and strengths of these influences are quantified by conditional probabilities [10].

There are two fundamental approaches to learning the BN structure:

Constraint-based algorithms: This approach was proposed by Verma and Pearl [11] and the learning of the BN structure is performed by identifying the conditional independence relationships among the variables.

Search and scoring based algorithms: In this process, each candidate structure is scored by evaluating its fitness to the data. The most appropriate BN structure for the data is searched.

The K2 algorithm was developed by Cooper, Heskovits [12] to calculate the score of a network structure. When D represents the dataset and B represents the network structure for BN, the score of the D dataset is found through the B structure that is shown Eq. (1). The K2 algorithm attempts to maximize the $f(i, \pi_i)$ score given in Eq. (2) for each node to increase the score of the network.

$$P(B|D) = P(B) \prod_{i=1}^{n} f(i, \pi_i) \tag{1}$$

$$f(i, \pi_i) = \prod_{j=1}^{q_i} \frac{(r_i - 1)!}{(N_{ij} + r_i - 1)!} \prod_{k=1}^{r_i} \alpha_{ijk}! \tag{2}$$

n is number of node, π_i is set of parents of node i, φ_i is list of all possible instantiations of π_i, q_i is number of φ_i, r_i is number of all possible values of node i. α_{ijk} is number of cases in D in which the node i instantiated with its k_{th} value and the parents of node i in π_i are instantiated with the j_{th} instantiation in φ_{ijk}. $N_{ij} = \sum \alpha_{ijk}$. In order to accelerate the K2 algorithm, $log(f(i, \pi_i))$ score is calculated instead of $f(i, \pi_i)$ score.

The K2 algorithm operates assuming that the variables are ordered in a dataset. If the ordering of the variables is not sufficient, the performance of the network remains low. The ordering can be obtained by experts. However, it may not be possible to find an expert for each problem and in such a case using a random ordering result in a bad performance for the network.

Larranaga et al. used a Genetic Algorithm (GA) that searched for an ordering that was passed on to the K2 algorithm [3]. The chromosomes were coded by the ordering of the nodes. The GA used the fitness function given in Eq. (3).

$$fitness = log(\prod_{i=1}^{n} f(i, \pi_i)) \tag{3}$$

3 Proposed Method

We propose a new method that named ScrRnkK2 (Score Ranking K2). ScrRnkK2 find the priority order of the variables according to their interactions. This method has good performance for sparse data. In ScrRnkK2, Eq. (2) which is used in the K2 algorithm is used for calculating the score of the variables. In the K2 algorithm, this equation is applied to the predecessor nodes to find the parent of a given node, whereas ScrRnkK2 is applied to all the nodes within the network to obtain the score of a given node. The pseudo-code of the K2 algorithm [12] and the proposed ScrRnkK2 method is presented in Fig. 1.

```
procedure K2:
{Input: A set of n nodes, an ordering on the nodes, an
upper bound u on the number of parents a node may
have, and a database D containing m cases.}
{Output: For each node, a printout of the parents of the
node.}
for i := 1 to n do
        πi : = Ø ;
        Pold := f(i, πi); {This function is computed using
        Eq. (2).}
        OKToProceed := true;
        while OKToProceed and | πi | < u do
                let z be the node in Pred(xi) - πi that
                maximizes f(i, πi U {z} );
                Pnew := f(i, πi U {z} );
                if Pnew > Pold then
                        Pold := Pnew;
                        πi := πi U {z}
                else OKToProceed := false;
        end {while} ;
        write ('Node:', xi ,'Parent of this node:' , πi )
end {for};
end {K2};
```

```
procedure ScrRnkK2:
{Input: : A set of n nodes, an upper bound u on the
number of parents a node may have, and a database D
containing m cases.
{Output: ranked dataset}
for i := 1 to n do
        πi : = Ø ;
        Pold := f(i, πi); {This function is computed using
        Eq.(2).}
        OKToProceed := true;
        while OKToProceed and | πi | < u do
                let z be the node in D - πi that
                maximizes f(i, πi U {z} );
                Pnew := f(i, πi U {z} );
                if Pnew > Pold then
                        Pold := Pnew;
                        πi := πi U {z}
                else OKToProceed := false;
        end {while} ;
        Pscore(i) := Pold;
end {for};
variables are ranked according to value of Pscore
end {ScrRnkK2};
```

Fig. 1. The pseudo-code of the K2 algorithm and the ScrRnkK2 method

When we look at the pseudo-code given in Fig. 1, while node z in the K2 algorithm is defined as the node that maximizes the score within $Pred(x_i) - \pi_i$, node z in ScrRnkK2 is defined as the node that maximizes the score within $D - \pi_i$. $Pred(x_i) - \pi_i$ is list of predecessor nodes for a given node other than the parent list and $D - \pi_i$ is list of all the nodes in the network other than the parent list. In other words, Eq. (2) is applied to the predecessor nodes for a given node and it is determined whether there is an interaction between this node and the predecessor nodes. If there is interaction between the nodes, the predecessor node is set as the parent of the related node. In ScrRnkK2, however, this equation is applied to all the nodes in the network for each node. Thus, the interaction score of each node is obtained for all the other nodes in the network. Our idea here is to bring the individual with the highest interaction score to the front in the ordering. The ordering is done by sorting the scores of the nodes in descending order. Furthermore, the maximum number of parents in the K2 algorithm should also be determined in the same way here.

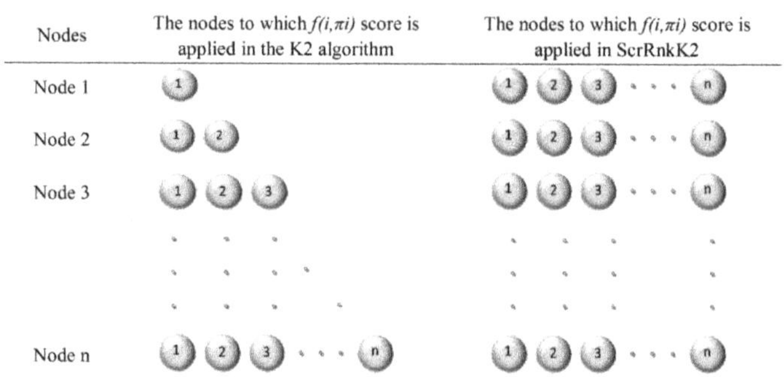

Fig. 2. The nodes to which $f(i, \pi_i)$ score is applied for each node in the network according to the K2 algorithm and ScrRnkK2 method

Figure 2 shows to which nodes $f(i, \pi_i)$ score is applied for each node in ScrRnkK2 and K2 algorithm. As is seen, the formula is applied only to the predecessor nodes for a given node in the K2 algorithm, whereas this formula is applied all nodes in the network in the ScrRnkK2 suggested for finding the ordering and a score is obtained for each node. Afterwards, the variables that represent the nodes are sorted in descending order of their scores in a way that the node with the highest score among these scores comes first in the ordering. Then, the data set is arranged according to this ordering. The K2 algorithm is applied to the ordered data and the bayesian network structure is constructed.

4 Experimental Results

The proposed method was compared with other methods by using synthetic and real-world data sets.

4.1 Synthetic Data Sample

First, a synthetic sparse data set generated by Tetrad IV software package was used to evaluate the performance of ScrRnkK2. Tetrad IV is a Java-based package for Causal Modeling and Statistical Data developed at Carnegie Mellon University.

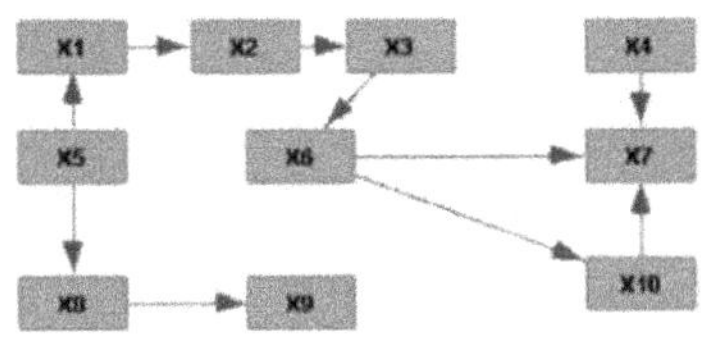

Fig. 3. A small network structure consisting of 10 nodes

A small network structure shown in Fig. 3, which consists of 10 nodes, was constructed using Tetrad IV software. Synthetic sparse data sets that are connected to this network structure and contain 100, 500, 1000, 2000, 3000 records were created using Tetrad IV software. The software generates the data randomly and according to the constructed network. First, the variables in these generated data sets are ordered by using ScrRnkK2, random and GA. It is aimed to find the initially constructed network structure by training the ordered data by using the K2 algorithm. Furthermore, Greedy Equivalence Search (GES) [13] and PC [14] algorithms, that are included in TETRAD, are used to compare proposed method.

In order to find the graph structure of the ordered random variables, 10 different random orderings were generated for each record and of these orderings; the network structure with the best score is presented in Fig. 4.

In order to find the network structure consisting of the variables ordered by using GA, crossover probability was set as 0.5, mutation probability was set as 0.1, population size was set as 10, and the number of iterations was set as 100. Order crossover and swap mutation were used. Ten random initial populations were generated. The network structure that was constructed according to the ordering that yielded the best fitness value obtained as the result of the GA application is added to Fig. 4.

As it can be seen in Fig. 4, each algorithm gets closer to the correct network structure with the increasing number of data that samples the 10-node network in the data set. When the network was trained with 3000 samples, it was seen that the variables ordered by using ScrRnkK2 and GA found the correct network structure. However, random-ordered variables found wrong connections in all 3000 samples. GES and PC algorithms found the right connections but could not find the directions accurately.

The K2 algorithms were run 10 times by using random sorted data and sorted data by GA. As a result of this 10 times running, the network structure that

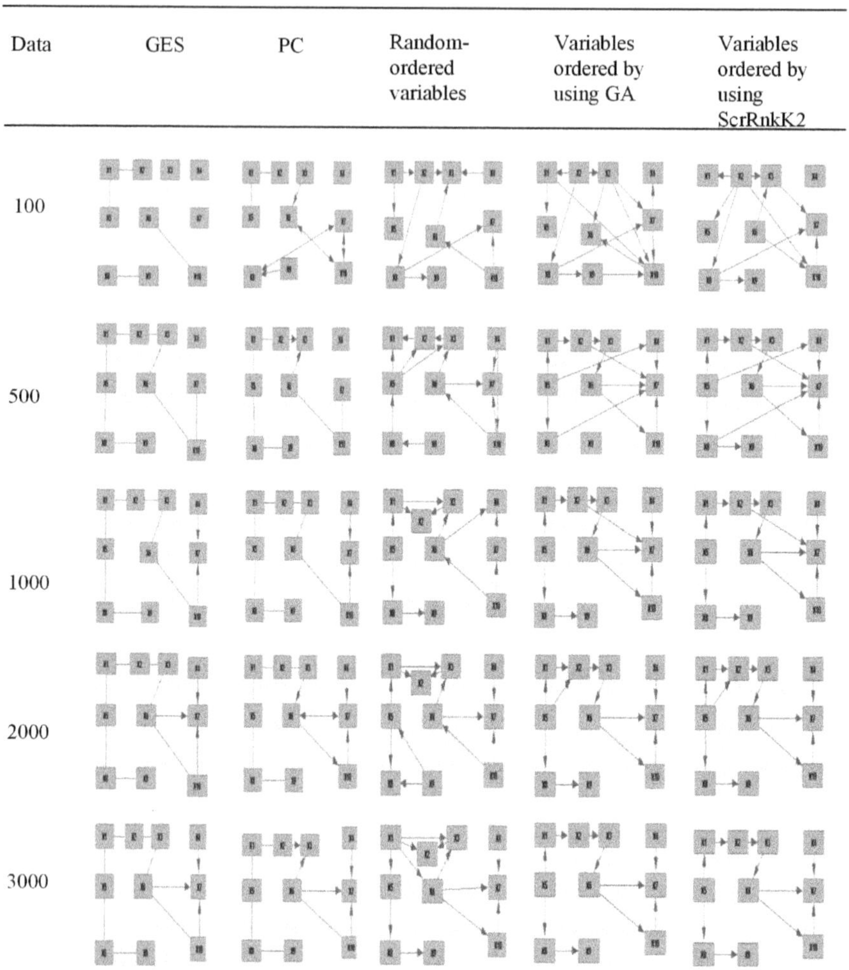

Fig. 4. Comparison of network structures that were obtained from the synthetic dataset

has the best scores is presented in Fig. 4. Even the best score that was obtained through random sorted data appears to be less successful than obtained score through sorted data by ScrRnkK2. The best score that was obtained through sorted data by GA reaches the same success with ScrRnkK2. But network structures that have worse score were present in the 10 times running for the GA.

Also, ScrRnkk2 has much better performance as time. The obtained network structures by using sorted data by ScrRnkK2 and GA were reached the appropriate network in the 3000 samples. ScrRnkK2 was found appropriate order in 9 s, while GA was found in an average of 11 min 19 s.

4.2 Real-World Data Set

A real-world data set (LabData) was compiled by Kubica [15]. The Lab Data consists of co-publication links for members of the Auton Laboratory at Carnegie Mellon University. It includes 115 individuals and 94 links.

First of all, the data set was converted into a format that is suitable for use in our application. For this, a table was created in which individuals were used as column variables and links for events as row variables. If an individual participated in an event, value "1" was assigned, otherwise, value "0" was assigned. The maximum number of parents was set as 10 for the K2 algorithm.

The variables were ordered using GA. For GA, population size was set as 10 and the number of iterations was set as 100. Orderings were found by using GA on 30 different randomly generated initial populations. Network structures were constructed through the K2 algorithm by using these orderings and the scores of these networks were calculated.

The proposed method ScrRnkK2 that is exact solution method produces the same results in each run. GA that is heuristic method can produce a different solution for each run. The K2 score was found as −466.7387 through the sorted data by ScrRnkK2. The K2 scores that were obtained through the sorted data by GA were presented in Table 1. All obtained scores were presented in Fig. 5.

Table 1. The K2 scores of sorted data by GA

No	Score	No	Score
1	−472.68	16	−469.28
2	−472.82	17	−478.36
3	−474.23	18	−470.73
4	−468.73	19	−472.71
5	−470.4	20	−474
6	−471.39	21	−468.98
7	−473.95	22	−470.25
8	−468.77	23	−469.84
9	−469.62	24	−471.63
10	−471.6	25	−470.63
11	−476.05	26	−469.37
12	−468.64	27	−469.81
13	−469.4	28	−471.53
14	−480.29	29	−470.58
15	−466.03	30	−469.7

The average elapsed time using GA was calculated and the obtained result was compared with the elapsed time for sorting the data through the proposed method.

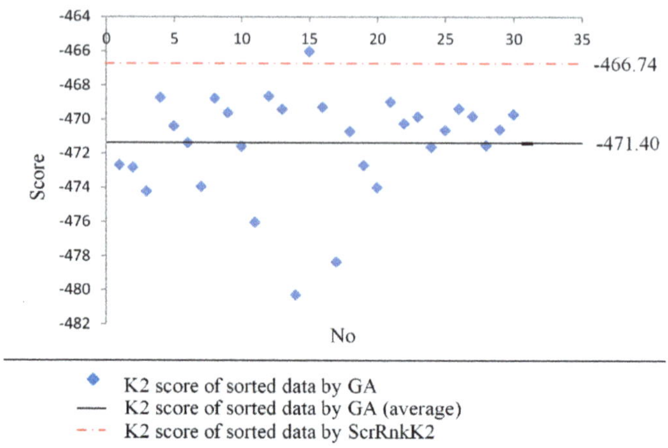

Fig. 5. K2 scores of the variables ordered by using ScrRnkK2 and GA

Sorting time of GA (average): 04:23:38
Sorting time of ScrRnkK2: 00:00:27

As is seen, the score of the network structure constructed through the ordering of the variables by using ScrRnkK2 is higher than the score of the network structure constructed through the ordering of the variables by using GA. While 100 iterations and a population size of 10 were adequate for finding the appropriate network structure for a synthetic dataset with 10 variables in GA, these values were not adequate when the number of variables was 115. Besides, ScrRnkK2 showed 500 times better performance in terms of time. The number of iterations and population size can be increased in order to increase the performance of GA. However, in such a case the application of GA for datasets with a high number of variables becomes difficult in terms of time.

5 Conclusions

BN training is a difficult process. K2 is an algorithm used for learning the structure of BN. The performance of the algorithm depends on priority order of the variables in the dataset. If the ordering is not sufficient, the performance of the constructed network is low.

In this study, a new method, that has good performance in sparse data, was proposed in order to find the appropriate ordering of the variables. The proposed method was compared with other methods by using synthetic and real-world data sets. First, a synthetic sparse data set generated by Tetrad IV. The variables in these generated data sets are ordered by using proposed method, random and GA. It is aimed to find the initially constructed network structure by training the ordered data by using the K2 algorithm. Furthermore, GES and PC algorithms, that are included in TETRAD, are used to compare proposed method. When the network was trained with 3000 samples, it was seen that the variables ordered by using proposed method and GA found the correct network structure. GES and PC algorithms found the right connections but could not find the directions accurately.

Then a real-world data set was used. The variables in these data sets are ordered by using proposed method and GA. When the number of variables increases, it is necessary to increase the iteration number and population size of GA in order to find the appropriate ordering. Increasing the iteration number and population size further decreases the speed of GA. The results of the comparisons showed that the proposed method is better than GA in terms of both time and score.

The BN that was constructed through the data set by sorting ScrRnkK2, has a high network structure score. In addition, it provides good performance in terms of the sorting time. It has good performance in the huge size data set in which speed is important as social networks.

In this study, we assume the data set that is complete. But the data may incomplete, erroneous and inconsistent in real-world. In the future, it is planned to work on missing and incomplete data set.

References

1. Chickering, D.M.: Learning Bayesian networks is NP-complete. In: Learning from Data: Artificial Intelligence and Statistics, Springer, New York (1996)
2. Hruschka, E.R., Ebecken, N.F.F.: Towards efficient variables ordering for Bayesian networks classifier. Data Knowl. Eng. **63**(2), 258–269 (2007). https://doi.org/10.1016/j.datak.2007.02.003
3. Larranaga, P., Kuijpers, C.M.H., Murga, R.H., Yurramendi, Y.: Learning Bayesian network structures by searching for the best ordering with genetic algorithms. IEEE Trans. Syst. Man Cybern. Part A Syst. Hum. **26**, 487–493 (1996)
4. Cowie, J., Oteniya, L., Coles, R.: Particle swarm optimization for learning Bayesian network. In: World Congress on Engineering, London, UK, pp. 71–76 (2007)
5. Hsu, W.H., Guo, H., Perry, B.B., Stilson, J.A.A.: Permutation genetic algorithm for variable ordering in learning Bayesian networks from data. In: Genetic and Evolutionary Computation Conference, New York, USA, pp. 383–390 (2002)
6. Acid, S., de Campos, L.M., Huete, J.F.: The search of causal orderings: a short cut for learning belief networks. In: Lecture Notes in Computer Science, vol. 2143, pp. 216–227 (2001)
7. Lee, J., Chung, W., Kim, E.: Structure learning of Bayesian networks using dual genetic algorithm. IEICE Trans. Inf. Syst. **E91–D**(1), 32–43 (2008). https://doi.org/10.1093/ietisy/e91-d.1.32
8. Akkoç, B.: The use of Bayesian network for social network analysis. Selçuk University (2012)
9. Russel, S.J., Norvig, P.: Articial Intelligence: A Modern Approach. Prentice-Hall, New Jersey (1995)
10. Pearl, J.: Bayesian networks: a model of self-activated memory for evidential Reasoning. In: 7th Cognitive Science Society, Irvine, ABD 1985, pp. 329–334 (1985)
11. Verma, T.S., Pearl, J.: Equivalence and synthesis of causal models. In: Uncertainty in Artificial Intelligence 6, USA, pp. 255–268 (1991)

12. Cooper, G.F., Heskovits, E.: A Bayesian method for the induction of probabilistic networks from data. Mach. Learn. **9**, 309–347 (1992)
13. Chickering, D.M., Meek, C.: Finding optimal Bayesian networks. In: Eighteenth Conference on Uncertainty in Artificial Intelligence, Canada, pp. 94–102 (2002)
14. Spirtes, P., Glymour, C.: An algorithm for fast recovery of sparse causal graphs. Soc. Sci. Comput. Rev. **9**, 62–72 (1991)
15. Kubica, J., Moore, A., Cohn, D., Schneider, J.: Finding underlying connections: a fast graph-based method for link analysis and collaboration queries. In: The Twentieth International Conference on Machine Learning, Washington DC, USA, pp. 392–399 (2003)

A Benefit Optimization Approach to the Evaluation of Classification Algorithms

Shellyann Sooklal(✉) and Patrick Hosein(✉)

Department of Computer Science, The University of the West Indies, St. Augustine, Trinidad
shellyann.sooklal@gmail.com, patrick.hosein@sta.uwi.edu

Abstract. We address the problem of binary classification when applied to non-communicable diseases. In such problems the data are typically skewed towards samples of healthy subjects. Because of this, traditional performance metrics (such as accuracy) are not suitable. Furthermore, classifiers are typically trained with the assumption that the benefit or cost associated with decision outcomes are the same. In the case of non-communicable diseases this is not necessarily the case since it is more important to err on the side of treatment of the disease rather on the side of over-diagnosis. In this paper we consider the use of benefits/costs for evaluation of classifiers and we also propose how the Logistic Regression cost function can be modified to account for these benefits and costs for better training to achieve the desired goal. We then illustrate the advantage of the approach for the case of identifying diabetes and breast cancer.

Keywords: Supervised binary classification · Health analytics · Performance measures · Benefit-based analytics

1 Introduction

Advances in biotechnology have lead to the production and access to a vast amount of data in health care. Medical records are now being digitized and digital machines and sensors contribute to the production of large amounts of data [14]. One of the most popular applications of this type of data is disease diagnosis. Some common machine learning classifiers which have been used for this task are Logistic Regression, Neural Networks, Support Vector Machines, Naive Bayes and also decision trees [8,18,21]. However, data on diseases are normally highly skewed towards the negative class; that is, the class associated with being healthy or disease free. Thus, this results in classifiers being biased towards the majority (negative) class and hence most, or in some cases all, of the minority (positive) class instances are incorrectly classified [8,15,16,18,21,22]. Also, the classifiers assume equal costs for different types of misclassification errors and thus reports high accuracy since the majority class would be correctly classified. However, in the real world, this is not desirable since the cost of misdiagnosing a positive

D. J. Hemanth and U. Kose (Eds.): ICAIAME 2019, LNDECT 43, pp. 35–46, 2020.
https://doi.org/10.1007/978-3-030-36178-5_4

instance is high. For example, misdiagnosing someone as healthy when they have a disease can lead to earlier death or diminished quality of life if untreated. On the other hand, misdiagnosing a healthy person as having a disease can cause the person unnecessary stress and medical expenses. The consequences of the latter case are far fewer than the former therefore misclassification of these different types of errors should be associated with different costs. Similarly, the benefit of correctly classifying someone as positive for a disease is more valuable than correctly classifying a healthy person as healthy.

In this paper we provide a cost-sensitive approach to the disease classification task. However, we consider the case where the algorithm is trying to maximize benefit or reduce cost. We first associate the different types of classifications (true negative, false negative, false positive and true positive) with benefit values and then adjust the cost function of a Logistic Regression (LR) classifier to incorporate these values. We tested our approach using data on two common chronic non-communicable diseases, diabetes and breast cancer. We also formulated and calculated a benefit score for each test case and compared the results to that of the traditional case of maximizing accuracy. We also bench-marked our results against that of traditional LR and LR with SMOTE [3]. Hence, our contributions are a benefit-based LR algorithm as well as a benefit score for evaluating the performance of a classifier. In addition, we illustrate how one can derive such benefit scores.

The rest of the paper is organized as follows. Section 2 gives a brief description of previous approaches to classification with imbalanced data and traditional performance metrics. In Sect. 3 we formulate a metric for benefit-based classification and then in Sect. 4 we describe how it was incorporated into the LR classifier. Section 5 explains how the benefit-based metrics and classifier were applied to diabetes while Sect. 6 considers the breast cancer case.

2 Related Work

Many researchers have already applied different techniques to deal with the issues posed by imbalanced data. Both [7] and [9] presented detailed summaries of the work performed by these researchers and described the main approaches to deal with imbalanced data and also common metrics to evaluate the performance of the classifier. Although, [9]'s summary is almost ten years old, the main points from the review are still relevant today and also matches up to that of [7]. Therefore, instead of looking at individual works, we highlight the main points from both [7] and [9] and focus on one previous work that is similar to ours.

Two popular approaches presented by [7] and [9] to handle imbalanced data are sampling methods and cost-sensitive methods. Sampling methods involve either removing samples in the majority class (under-sampling) or adding copies of instances of the minority class (oversampling) in order to reduce the imbalance of the data. Under-sampling poses the issue of losing important aspects of the majority class whilst oversampling introduces the issue of over-fitting due to replicated data. Some algorithms, such as EasyEnsemble and SMOTE, try to address these issues but they require multiple classifiers and additional processing of the data which adds to the complexity and computation time of the solution. Also, there is still generalization and loss of information in these techniques.

On the other hand, cost-sensitive methods addresses the imbalanced data problem by assigning misclassification costs to incorrect classification. Costs can be applied to the training stage of the classification process where the objective is to minimize the overall cost. [7] found more papers implementing sampling methods than cost-sensitive methods; however, they explained that a reason for this could be due to sampling techniques being easier to implement since it does not required knowledge of the classification algorithms. On the other hand, [9] pointed out that studies have shown cost-sensitive learning to be more effective than sampling techniques and can therefore be used as a viable alternative. Thus, it was the chosen method for our purposes. A cost matrix is typically used to represent the penalties associated with misclassifying a sample. An example of the cost matrix for a binary classification is shown in Table 1.

Table 1. Cost-matrix for cost-sensitive classification

	Predicted no	Predicted yes
Actual no	True negatives, TN (Cost $= 0$)	False positives, FP (Cost > 0)
Actual yes	False negatives, FN (Cost > 0)	True positives, TP (Cost $= 0$)

A cost of zero is normally assigned to correct classification whereas non-zero values are assigned to misclassification. However, our proposed method goes a step further and assigns non-zero values for all four categories, unless it is required for the specific context, and hence, aims at maximizing benefit while reducing cost. Many of the proposed methods use complex classifiers or ensembles of classifiers in order to achieve reasonable results. However, our approach makes a simple change to a LR classifier in order to include benefit values into the learning process. Thus proving that satisfactory results can be achieved without adding complexity and computational costs.

Bahnsen et al. [2] implemented a cost-sensitive LR algorithm for credit scoring. Their solution applied individual cost matrices to each example in the dataset. Their original formulation to include cost into the LR classifier is similar to ours. However, we go a step further in our formulation in order to account for both benefits and cost as well as represent these values as a single metric which influences the LR classifier. Also, manually specifying individual cost matrices for each sample in the dataset is very time consuming and impractical as the size of the dataset grows. This step also becomes impossible if the data was collected independently of the classification step. Our approach uses only one cost matrix that is generalized for all samples in the dataset and we show that this cost matrix can be easily formulated using readily available cost and benefit values.

Some of the most widely used performance metrics for classification are accuracy, precision, sensitivity, specificity, F1-score, receiver operating characteristic (ROC) curve and area under the ROC curve (AUC) [2,5,7,20]. Accuracy, precision and F1-score are all dependent on data distribution and hence are not suitable for imbalanced data [9]. Recall and specificity are not data sensitive but they do not provide any indication of how many of the minority and majority samples were correctly classified. [9] explained that the ROC curve presents an optimistic view of the classifier's performance and do not provide any confidence intervals or statistical significance on the performance. [7] also stated that AUC/ROC can be incoherent.

The main issue with these traditional metrics is that they consider equal benefit for correct classification of each class and similarly, equal cost for misclassification. Therefore, when the data is skewed towards the majority class, these metrics illustrate high percentages for bias classifiers which in turn gives the impression that the classifier is performing well. Instead of using these standard metrics, [2] created a formula for calculating a savings score for credit scoring. Our work includes a benefit performance metric which is more suited for medical diagnosis and is also more generalized for other classification problems. Our benefit metric gives a better representation, than traditional metrics, of how the classifier is performing with both the majority and minority classes.

3 Benefit-Based Performance Metric

We focus on binary classification in which N labeled samples are provided. Each outcome belongs to either class 0 or class 1. If we denote the feature vector of a given sample by $\boldsymbol{x}$ then a classifier will produce a continuous score $s(\boldsymbol{x})$ which is used to determine the class in which it belongs. We assume that the instances for class 0 produce scores that are typically less than those of class 1. One must then determine some threshold t such that if $s(\boldsymbol{x}) \leq t$ the instance is classified as 0 while if $s(\boldsymbol{x}) > t$ then the instance is classified as a 1. For a given classifier, we denote the probability density function of the scores for class 0 instances by $f_0(s)$ and for class 1 scores by $f_1(s)$. We denote the corresponding cumulative distribution functions by $F_0(s)$ and $F_1(s)$ respectively.

Next we associate the corresponding costs and benefits. Let b_{ij} denote the benefit of classifying an instance of class i as class j. We assume that if $i = j$ (in which case the classification was correct) then the benefit is positive (i.e., $b > 0$) while if $i \neq j$ then the benefit is non-positive (or is rather a cost) (i.e., $b \leq 0$). The prior probability of class $j \in \{0, 1\}$ is denoted by π_j.

For a given threshold t, the expected number of instances from class 0 that are correctly classified can be written as $\pi_0 F_0(t) N$ which is the product of the probability that the instance is in class 0 times the probability that the instance is correctly classified (i.e., the score is less than the threshold t) times the total number of instances N. For a given threshold the expected benefit can then be written as

$$\begin{aligned} B(t) = \; & \pi_0 F_0(t) b_{00} + \pi_0 (1 - F_0(t)) b_{01} + \\ & \pi_1 F_1(t) b_{10} + \pi_1 (1 - F_1(t)) b_{11} \end{aligned} \tag{1}$$

Note that since only b_{00} and b_{11} are positive then the expected benefit is maximized when $F_0(t) = 1$ and $F_1(t) = 0$ which occurs when the two distributions do not overlap. Therefore the benefit is upper bounded by $\pi_0 b_{00} + \pi_1 b_{11}$. If a classifier γ obtains an expected benefit of B_γ we define the following performance metric

$$\mu_\gamma \equiv \frac{B_\gamma}{\pi_0 b_{00} + \pi_1 b_{11}} \tag{2}$$

Note that if the classifier's performance is close to the optimal then $\mu_\gamma \approx 1$.

In general, for a given classifier and benefits we would like to maximize the benefit $B(t)$. A necessary condition for optimality can be obtained by taking the derivative of $B(t)$ with respect to t and setting the result to zero. If we do this then we obtain:

$$f_0(t^*)\pi_0(b_{00} - b_{01}) = f_1(t^*)\pi_1(b_{11} - b_{10}) \tag{3}$$

4 Benefit Objective with Logistic Regression

In the previous section we considered how, given the classifier, one could optimize the expected benefit by varying the threshold used by the classifier. However, the classifier itself was not trained with the objective of optimizing the benefit function. In this section we consider how the cost function used for LR can be modified to account for benefits and costs.

In LR, the posterior probability of the positive class is estimated as the logistic sigmoid of a linear function of the feature vector. For a given feature vector $\boldsymbol{x}_i$ this probability is given by

$$p_i = P(y = 1|\boldsymbol{x}_i) = h_\theta(\boldsymbol{x}_i) = g(\boldsymbol{\theta}^T \boldsymbol{x}_i) \tag{4}$$

where $h_\theta(\boldsymbol{x}_i)$ refers to the classification for input $\boldsymbol{x}_i$ given the parameter vector $\boldsymbol{\theta}$. The function $g(\cdot)$ is the logistic sigmoid function given by

$$g(z) = \frac{1}{1 + e^{-z}}. \tag{5}$$

The parameters $\boldsymbol{\theta}$ are determined by minimizing the LR cost function

$$J(\boldsymbol{\theta}) \equiv \frac{1}{N} \sum_{i=1}^{N} J_i(\boldsymbol{\theta}) \tag{6}$$

where

$$J_i(\boldsymbol{\theta}) = -y_i \log(h_\theta(\boldsymbol{x}_i)) - (1 - y_i) \log(1 - h_\theta(\boldsymbol{x}_i)) \tag{7}$$

However this cost function assumes the same cost is associated with different errors (false positives and false negatives). Consider, instead, the following cost function based on maximizing benefits b_{ij}.

$$J^B(\boldsymbol{\theta}) \equiv \frac{1}{N} \sum_{i=1}^{N} J_i^B(\boldsymbol{\theta}) \tag{8}$$

where

$$\begin{aligned} J_i^B(\boldsymbol{\theta}) = & \ y_i[h_\theta(\boldsymbol{x}_i)b_{11} + (1 - h_\theta(\boldsymbol{x}_i))b_{10}] + \\ & (1 - y_i)[h_\theta(\boldsymbol{x}_i)b_{01} + (1 - h_\theta(\boldsymbol{x}_i))b_{00}]. \end{aligned} \tag{9}$$

We can rewrite this as

$$\begin{aligned} J_i^B(\boldsymbol{\theta}) = & \ y_i h_\theta(\boldsymbol{x}_i)(b_{11} - b_{10}) + b_{10} \\ & (1 - y_i)(1 - h_\theta(\boldsymbol{x}_i))(b_{00} - b_{01}) + b_{01} \end{aligned} \tag{10}$$

Since we will be maximizing this function with respect to $\boldsymbol{\theta}$ and b_{10} and b_{01} are constants then they can be removed from the function without affecting the optimal $\boldsymbol{\theta}$. Furthermore we can change this into a minimization problem by multiplying by -1. Finally we can divide the resulting function by $(b_{11} - b_{10})$ and not change the optimal solution. If we do this we obtain the new function as

$$J_i^B = -y_i h_\theta(\boldsymbol{x}_i) - (1 - y_i)(1 - h_\theta(\boldsymbol{x}_i))\eta \tag{11}$$

where we have defined

$$\eta \equiv \frac{b_{00} - b_{01}}{b_{11} - b_{10}}. \tag{12}$$

Note that this has a similar form to the LR cost function but here the error for class 0 instances are scaled by the factor η. We therefore propose using the Logistic Cost function but include this scaling so that we take into account benefits. Hence we use the following:

$$J_i(\boldsymbol{\theta}) = -y_i \log(h_\theta(\boldsymbol{x}_i)) - \eta(1 - y_i) \log(1 - h_\theta(\boldsymbol{x}_i)) \tag{13}$$

when training the LR algorithm.

5 Application to Diabetes Classification

In this section we consider the problem of diabetes classification based on a number of patient features. We first determined appropriate benefit values and used LR with the computed value for η. We compared the results with traditional LR and also LR with SMOTE since SMOTE is a popular method for handling unbalanced data. We also included a wide range of other η values to illustrate the robustness of the approach. We then compared the results based on the benefits based optimization with those obtained with accuracy based optimization.

5.1 Dataset Description

The Pima Indian Diabetes dataset [13,19] was chosen for our experiments. The data contains 9 attributes based on measurements that are normally used to diagnose a patient (for example, BMI, age). It also contains 768 instances (500 for healthy persons, 268 for diabetic persons). 75% of the dataset was randomly selected for the training set and the remaining 25% was used for testing.

5.2 Cost-Based Benefit Model

We determine the benefits b_{ij} which are then used to compute η. We do this by determining benefits with respect to a baseline. The baseline case is the one in which nothing is done for the patient (i.e., as if the person was never a patient). The benefit b_{00} corresponds to the case when the patient has no diabetes and this was predicted correctly. In this case nothing will be done for the patient and so we set $b_{00} = 0$ since no benefit (except for the fact that the patient now knows they do not have diabetes) is achieved. The cost b_{10} corresponds to the case in which the patient has diabetes but it was not predicted. Now note that the baseline is the case where nothing would have been done for the patient anyway so the benefit (when compared to doing nothing) is still zero and so $b_{10} = 0$. Next consider the benefit b_{11} which is the case that the patient has diabetes and this was predicted and so treatment is provided to avoid health complications due to diabetes (which would have occurred in the baseline case). From [12], the annual cost of treatment for a patient with diabetes complications is approximately \$10,000 more than someone who did not have diabetes (or was treated early). Therefore the benefit of this correct detection of diabetes is a savings of \$10,000 so we set $b_{11} = 10$. Finally in the case b_{01} the patient was incorrectly diagnosed as having diabetes when they in fact did not have it. In this case drugs would be prescribed for the patient when they did not need it. Using information from [10], the annual cost of drugs for diabetes treatment for a new patient would be approximately \$1000 and so we use this as the associated cost (since the baseline would not have incurred this cost). Therefore we set $b_{01} = -1$. If we substitute these numbers we obtain $\eta = 0.1$.

What this means is that, when training, errors for the common instance (the patient does not have diabetes) are treated more lightly than errors for those with diabetes. According to [1] the prevalence of diabetes (in the United States) is typically on the order of 10% and hence we can approximate $\pi_0 = 0.9$ and $\pi_1 = 0.1$. We can now use these values to determine our proposed metric 2 as: $\mu = \frac{B}{1}$ where B is the expected benefit obtained and the upper bound is 1.

We can note the following. If a classifier always chooses 0 (no diabetes) then the accuracy will be 90% which is quite high. However the benefit cost for this classifier is 0 while the maximum possible benefit is 1. If a classifier always chooses 1 (has diabetes) then the accuracy now becomes 10% but the expected benefit becomes 0.1. This clearly shows the advantage of considering expected benefit rather than other non-cost dependent approaches.

5.3 Life Expectancy Based Benefit Model

In addition to medical expenses, we can determine benefits based on life expectancy. In the baseline case where nothing is done for a patient without diabetes, their life expectancy does not change, therefore we can set $b_{00} = 0$. However, if we do nothing for a person who has diabetes, that is, they are not diagnosed as being diabetic, then according to [11] and [4] their life expectancy can be reduced by 10 years. Therefore, we set $b_{10} = -10$. [4] also states that diabetics are normally diagnosed years after developing the condition. Based on this fact, we set $b_{11} = 5$ to represent extending a person's life by at least half the amount of not detecting it at all. For the case where the person does not have diabetes but we predicted them as having it, diabetes drugs can be administered. [6] explained that overdose of the insulin taken by type 1 diabetics and also insulin-requiring type 2 diabetics can result in coma or even death. Therefore, we set $b_{01} = -1$ to represent a possibility of negative side effects (including affected lifespan) through the use diabetic medication. Hence using this approach we obtain $\eta = 0.07$ which is close to the value obtained with the cost based mode.

5.4 Numerical Results

Table 2 shows the results obtained for accuracy and benefit score for our models compared to regular LR and LR with SMOTE. For both models, our adjusted algorithm produced results which achieved the highest benefits; however, accuracy was sacrificed in order to get better performance with the minority class.

Table 2. Accuracy and benefits achieved for LR, LR with SMOTE, benefit-based LR using cost-based model and benefit-based LR using life-expectancy model

Algorithm	Accuracy	Cost-based benefit	Life-based benefit
LR	0.81	2.02	-0.27
LR with SMOTE	0.74	2.52	0.54
BB-LR (Cost-based model)	0.34	2.67	NA
BB-LR (Life-based model)	0.34	NA	1.01

In Fig. 1 we plot the performance metric 2 as a function of η. Note that at $\eta = 1$ we obtain the value that we would have obtained using the traditional approach of maximizing accuracy. Note that optimizing with respect to benefit always provides a better expected benefit than optimizing with respect to accuracy.

In Fig. 2 we plot the accuracy as a function of η. Here we find that to achieve improved benefit we must sacrifice accuracy. In other words we should err more on the side of large benefits. Here we find that the accuracy is either the same as or worse than that obtained with the traditional approach of maximizing accuracy.

5.5 Discussion

As shown in Table 2, we can achieve higher benefits by applying η to the LR classifier to account for benefits and misclassification costs. This supports the fact that a simple adjustment can be made to the LR classifier to reduce the

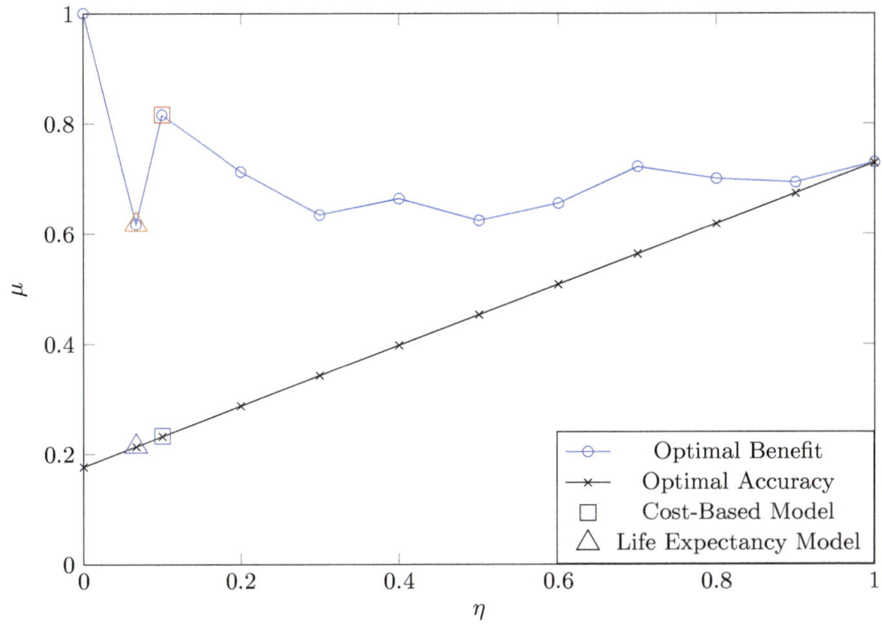

Fig. 1. Benefit scores for pima Indians diabetes dataset

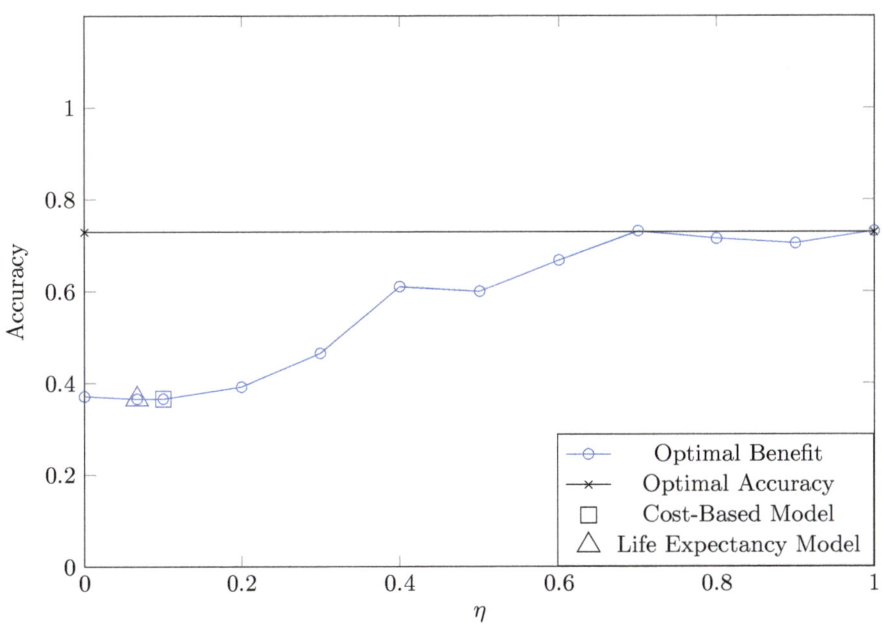

Fig. 2. Accuracy scores for pima Indians diabetes dataset

misclassification of the higher cost class. Thus, reducing complexity and computational costs which are introduced by other methods which handle imbalanced data such as SMOTE.

As explained in Sect. 5.4 above, the benefit-based LR classifier was tested against multiple benefit values and hence multiple values of η. When $\eta = 1.0$, this is identical to setting the classifier to perform accuracy optimization. Hence, as illustrated in Figs. 1 and 2, the benefit and accuracy scores were the same for this case. Furthermore, for the derived value of $\eta = 0.1$, we obtained $B = 3.02$, $\mu = 0.82$, benefit-wrt-accuracy $= 0.23$ and accuracy $= 0.36$. From these values we see that optimizing with respect to benefit did indeed provide a better performance score than with the accuracy optimization approach. As illustrated in Fig. 1, there was a difference of 0.59 and hence an increase of 256.52% in the performance value with respect to benefit than with the accuracy approach. Thus, the benefit approach can save a person on average $2,165.00 more per year than the accuracy approach which only saves a person approximately $850 per year. The accuracy score is very low (36%) and does not truly represent the performance of the classifier with respect to its classification on positive instances. For this scenario, the value of η was very small. This in turn represented a much higher benefit score for the positive class as opposed to the negative class. Therefore, the classifier would have highly skewed the classification to favor the positive class in order to maximize benefit. The performance value was high (82%) since the classification results obtained comprised of the majority of the positive class instances being classified as positive (true positives) but it also comprises of a high number of false positives. These results illustrates how sensitive the η value is in affecting the performance of the classifier.

Similarly, for the second derived value of $\eta = 0.07$ we obtained $B = 1.14$, $\mu = 0.62$, benefit-wrt-accuracy $= 0.21$ and accuracy $= 0.36$. The performance metric is slightly lower than that of the previous case. This can be due to the benefit matrix associated with this value of η. However, when compared to the benefit based on accuracy, the benefit increases by 0.41 or 195.24%, as shown in Fig. 1. This is an improvement in life expectancy by approximately 9 months per person over the accuracy approach which gives a life expectancy of approximately 4.6 months per person. The traditional accuracy value is very low for this scenario (36%), therefore, it does not provide a true reflection of the savings generated by the classifier. From these observations, the benefit approach is more robust than the accuracy approach since it gives a better classification based on benefits and it also gives a more accurate depiction of the performance of the classifier with respect to benefit.

6 Application to Breast Cancer Classification

Similar tests and derivation of η values were performed on a breast cancer dataset (the Breast Cancer Wisconsin (Diagnostic) dataset [17]), as with the diabetes dataset. Also, similar trends were observed in the results. However, due to space limitations we only provide the benefit values used for the experiments (Table 3).

Table 3. Benefit values for cost-based and life-expectancy models for breast cancer

Model	b_{00}	b_{01}	b_{10}	b_{11}	η
Cost-based	0	–30.5	0	74	0.41
Life-expectancy	0	–1	–5	5	0.1

7 Conclusions

Medical datasets are typically skewed towards the negative class (healthy persons). This imbalance in the dataset influences classifiers to be biased towards the majority class. However, misclassification of a disease-affected person is more costly than the misclassification of a healthy person. In addition, the benefits associated with the correct classification of a disease-affected person is higher than that of a healthy person. In this paper, we presented a robust benefit-based LR approach which was sensitive to varying benefits and costs associated with different datasets/diseases. We also presented a benefit score which was used to evaluate the performance of the classifier. This score was successful in illustrating how the classifier performed with regards to overall gain in benefit. We are currently working on extending this work by incorporating benefit-based approaches to other classifiers.

References

1. Association, American Diabetes: Statistics about diabetes, July 2017. http://www.diabetes.org/diabetes-basics/statistics/
2. Bahnsen, A.C., Aouada, D., Ottersten, B.: Example-dependent cost-sensitive logistic regression for credit scoring. In: Proceedings of the 2014 13th International Conference on Machine Learning and Applications ICMLA 2014, pp. 263–269. IEEE Computer Society, Washington, DC, USA (2014). https://doi.org/10.1109/ICMLA.2014.48. http://dx.doi.org/10.1109/ICMLA.2014.48
3. Chawla, N.V., Bowyer, K.W., Hall, L.O., Kegelmeyer, W.P.: Smote: synthetic minority over-sampling technique. J. Artif. Int. Res. **16**(1), 321–357 (2002). http://dl.acm.org/citation.cfm?id=1622407.1622416
4. Diabetes.co.uk: Diabetes life expectancy. https://www.diabetes.co.uk/diabetes-life-expectancy.html
5. Garrido, F., Verbeke, W., Bravo, C.: A robust profit measure for binary classification model evaluation. Expert Syst. Appl. **92**, 154–160 (2018). https://doi.org/10.1016/j.eswa.2017.09.045, http://www.sciencedirect.com/science/article/pii/S0957417417306498
6. Gundgurthi, A., Kharb, S., Dutta, M.K., Pakhetra, R., Garg, M.K.: Insulin poisoning with suicidal intent. Indian J. Endocrinol. Metabol. **16**(Suppl1), S120 – S122 (2012). https://doi.org/10.4103/2230-8210.94254, https://www.ncbi.nlm.nih.gov/pmc/articles/PMC3354941/
7. Haixiang, G., Yijing, L., Shang, J., Mingyun, G., Yuanyue, H., Bing, G.: Learning from class-imbalanced data. Expert Syst. Appl. **73**(C), 220–239 (2017). https://doi.org/10.1016/j.eswa.2016.12.035

8. He, F., Yang, H., Miao, Y., Louis, R.: A cost sensitive and class-imbalance classification method based on neural network for disease diagnosis. In: 2016 8th International Conference on Information Technology in Medicine and Education (ITME), pp. 7–10, December 2016. https://doi.org/10.1109/ITME.2016.0012
9. He, H., Garcia, E.A.: Learning from imbalanced data. IEEE Trans. Knowl. Data Eng. **21**(9), 1263–1284 (2009). https://doi.org/10.1109/TKDE.2008.239
10. Helper, Health Cost: How much does diabetes medication cost? October 2013. http://health.costhelper.com/diabetes-medication.html
11. Huizen, J.: Type 2 diabetes and life expectancy, May 2017. https://www.medicalnewstoday.com/articles/317477.php
12. Institute, Health Care Cost: Issue brief 10: Per capita health care spending on diabetes: 2009–2013, May 2015. http://www.healthcostinstitute.org/files/HCCIDiabetesIssueBrief205-7-15.pdf
13. Kaggle.com: Pima Indians diabetes database. https://www.kaggle.com/uciml/pima-indians-diabetes-database/data
14. Kavakiotis, I., Tsave, O., Salifoglou, A., Maglaveras, N., Vlahavas, I., Chouvarda, I.: Machine learning and data mining methods in diabetes research. Comput. Struct. Biotechnol. J. **15**, 104–116 (2017). https://doi.org/10.1016/j.csbj.2016.12.005, http://www.sciencedirect.com/science/article/pii/S2001037016300733
15. Krawczyk, B., Schaefer, G., Woźniak, M.: A cost-sensitive ensemble classifier for breast cancer classification. In: 2013 IEEE 8th International Symposium on Applied Computational Intelligence and Informatics (SACI), pp. 427–430, May 2013. https://doi.org/10.1109/SACI.2013.6609012
16. Li, L., Chen, M., Wang, H., Li, H.: Cosfuc: a cost sensitive fuzzy clustering approach for medical prediction. In: 2008 Fifth International Conference on Fuzzy Systems and Knowledge Discovery, vol. 2, pp. 127–131, October 2008. https://doi.org/10.1109/FSKD.2008.378
17. Repository U.M.L: Breast cancer wisconsin (diagnostic) data set, November 1995. https://archive.ics.uci.edu/ml/datasets/Breast+Cancer+Wisconsin+(Diagnostic)
18. Santos-Rodríguez, R., García-García, D., Cid-Sueiro, J.: Cost-sensitive classification based on bregman divergences for medical diagnosis. In: 2009 International Conference on Machine Learning and Applications, pp. 551–556, December 2009. https://doi.org/10.1109/ICMLA.2009.82
19. Smith, J.W., Everhart, J., Dickson, W., Knowler, W., Johannes, R.: Using the ADAP learning algorithm to forecast the onset of diabetes mellitus. In: Proceedings of the Annual Symposium on Computer Application in Medical Care, pp. 261–265, November 1988
20. Verbraken, T., Verbeke, W., Baesens, B.: A novel profit maximizing metric for measuring classification performance of customer churn prediction models. IEEE Trans. Knowl. Data Eng. **25**, 961–973 (2013)
21. Zhang, D., Shen, D.: Multicost: multi-stage cost-sensitive classification of alzheimer's disease. In: Suzuki, K., Wang, F., Shen, D., Yan, P. (eds.) Machine Learning in Medical Imaging, pp. 344–351. Springer, Heidelberg (2011)
22. Zhou, Z.H., Liu, X.Y.: Training cost-sensitive neural networks with methods addressing the class imbalance problem. IEEE Trans. Knowl. Data Eng. **18**(1), 63–77 (2006). https://doi.org/10.1109/TKDE.2006.17

Feature Extraction of Hidden Oscillation in ECG Data via Multiple-FOD Method

Ekin Can Erkuş[1] and Vilda Purutçuoğlu[1,2](✉)

[1] Biomedical Engineering, Middle East Technical University, 06800 Ankara, Turkey
eerkus@metu.edu.tr, vpurutcu@metu.edu.tr

[2] Department of Statistics, Middle East Technical University, 06800 Ankara, Turkey

Abstract. Fourier transform (FT) is a non-parametric method which can be used to convert the time domain data into the frequency domain and can be used to find the periodicity of oscillations in time series datasets. In order to detect periodic-like outliers in time series data, a novel and promising method, named as the *outlier detection via Fourier transform* (FOD), has been developed. From our previous studies, it has been shown that FOD outperforms most of the commonly used approaches for the detection of outliers when the outliers have periodicity with low fold changes or high sample sizes. Recently, the multiple oscillation and hidden periodic-like pattern cases for time series data have been investigated and found that the multiple application of FOD, shortly multiple-FOD, can also be a successful method in the detection of such patterns. These empirical results are based on real electrocardiogram (ECG) data where the discrimination of disorders can be helpful for the diagnosis of certain heart diseases in advance. Hereby, in this study, we evaluate the performance of multiple-FOD in different types of simulated datasets which have distinct sample sizes, percentage of outliers and distinct hidden patterns.

Keywords: Outlier detection · Fourier transform · Time series data · Electrocardiogram

1 Introduction

The electrocardiography (ECG) is an imaging modality which is used to observe the electrical actions of the heart [28] and an electrocardiogram is an ECG recording device. Here, the recording operation is performed by the electrodes which have specific locations on the chest in such a way that they can collect the signals generated by the beating of the heart [22]. On the other hand, a regular and non-patient ECG dataset has common units which are called the PQRST structure. This special structure creates a periodic pattern [25]. However, several factors such as heart disorders or malfunctions in the signalling pathway in hearts may change the pattern of the ECG data by suppressing/enhancing or timing of the parts in the PQRST patterns, resulting in altering the regular beat

D. J. Hemanth and U. Kose (Eds.): ICAIAME 2019, LNDECT 43, pp. 47–56, 2020.
https://doi.org/10.1007/978-3-030-36178-5_5

frequency [6,12]. We give some examples of these shapes which are generated from simulated ECG data in Fig. 2.

Accordingly, we propose that the detection of the features in the data which may differ among the types of datasets can be useful if they are applied in the classification of data into categories based on the types of disorders [7,23]. But it is a challenging problem since most of the ECG data have similar behaviours [20] and finding the features in this type of data depends on the detection of their QRS complexes [8]. There are many studies which present a new feature detection algorithm depending on the QRS structures [5,13,15,18,21,29] and these detection methods have a variety of performance conditions depending on the data length, sampling rate, pattern and the data distribution [14,16]. However, there has not been a single best method which could identify patterns in every ECG dataset [24] yet. Hereby, in the detection of disorders in the ECG datasets, we consider that the identification of the periodicity can be a good way to describe the classes of the heart rates and the QRS complexes [3,9] since these data generally form periodic-like behaviours. While searching the periodicity in the data, we use the Fourier transform (FT) which is widely applied as a frequency domain estimator [27,32]. Indeed, there are several studies in which FT is implemented in order to find the heart rate [1,26,30,31]. Among these studies, [1] uses FT to find the frequency of the average heart rate, also known as the average pulse rate, in such a way that the average pulse rate can be computed by locating the global maxima value in the frequency domain. In this study, the symmetric side peaks around the global maximal are revealed, whereas, these side peaks are discarded and treated as noise signals in the calculation. On the other hand, the study of [26] performs FT in order to filter the data in the frequency domain. In their analyses, the detection of the R peaks is found by the window filtering in which the maximum values of windows are detected. However, such an R-peak detection method can only be valid for unvarying heart rates or by using dynamic length windows. Here, although the peaks can be detected by a variety of methods, the average heart rate may have been found by using FT. On the other side, the study [31] implements FT for both filtering in the frequency domain and for detecting the average heart rate. But, there is no feature extraction based on the frequency domain behaviour even though there might have been valuable features in the representation of the frequency domain of the data which they apply. The study of [30] uses FT to estimate the power spectrum of the cardiac pulse data to extract features. However, this study lacks the details of the algorithmic steps and they crop out the valuable features of their data by applying a very narrow band pass filtering to 1–4 Hz as eliminating the detectable heart rates below 60 beats per minute which can be considered as the slow-normal regular heart rate frequency.

Hereby, from our previous studies [11], we have proposed an application of FT to identify the main oscillations of time series data based on the calculation of the periodicity, particularly, for the ECG datasets and we have called it as the *outlier detection via Fourier transform* (FOD). We have further extended that study into the multiple and sequential application of FT's and have found the relationships between the sequentially applied number of FT and the frequency of the main oscillations [9]. We call this special FT as the N-FOD where N refers to the number of sequential applications of FOD. Later, we have focused on finding the features out of the ECG datasets and have come up with the detection of multiple oscillations by using the 2-FOD applications and the detection of the certain hidden patterns in these ECG datasets [10]. From these preliminaries analyses, we have observed that the features of distinct hear rates under single and multiple oscillations of periodic components, namely the PQRST complexes, can be used to discriminate the ECG datasets. But, as these outcomes are based on the limited number of real data, we could not evaluate the performance of our proposal approach under large numbers of subjects and distinct conditions. Hence, in this study, we present the assessments of the reliability of the N-FOD method by using the synthetically generated ECG datasets. For this purpose, we use 6 different synthetically generated ECG datasets by implementing 1000 Monte-Carlo runs for each group. We generate the 3 groups which have regular patterns of the ECG signals, but with different amplitudes of the R peaks. Then, we simulate the ECG data based on the common ventricular tachycardia patterns as can be seen in Fig. 2.

Therefore, we organize the paper as the following plan: In Sect. 2, we describe some background methods and the algorithm behind our proposed features. In Sect. 3, we represent the steps of the data generation and the results of the Monte Carlo applications. Finally, in Sect. 4, we summarize our findings with a direction of our future works.

2 Background Methods and Proposed Feature

The discrete-time Fourier Transform (DTFT) can be used to estimate the representation of the frequency domain in a time series dataset by using the following well-known formula [19]:

$$X(k) = \frac{1}{N} \sum_{n=0}^{N-1} x_{(n)} e^{(-j2\pi k/N)} \tag{1}$$

where $x_{(n)}$ denotes the data in the time domain with the sample size n in the frequency domain. Furthermore, the number of discrete frequencies is shown by N while k stands for the number of cycles in each N-sample. However, according to the Nyquist sampling theorem [17], the sample size of the frequency domain becomes half of the original sample size due to the sampling operation. As the number of sequentially applied FT increases, the sample size decreases by the power of 2 for each step.

From our previous analyses [9], we have observed that the 2-time sequentially applied FT on the periodic-like data often generates periodic responses in the 2nd order frequency domain due to the harmonic components of the periodicity in the time data. Moreover, we have also seen that there appears the neighborhood peaks which are symmetric to the periodic peaks in the 2nd order frequency domain for the time series data with more than one noticeable oscillation frequency as represented in Fig. 1. We name these periodic components in the 2nd order frequency domain as the *main peaks* and call the symmetric and smaller components near to those main peaks as the *side peaks.* In Fig. 1, we indicate a visual application of both types of peaks as an example. Here, we take 2nd, 5th, 8th and 11th marked samples from left to right as the main peaks and classify the rest of the marked samples as the side peaks. In our analyses, we investigate these peaks whether they can be used as a distinguishing factor which generates bands for the behaviour of the periodic data. Thereby, we create two variables γ and $\delta = \gamma^{-1}$ which are based on the mean distance between main and related side peaks. The mathematical expression for the proposed variable γ can be seen as follows.

$$\gamma = \frac{f_{side}}{\frac{M}{2^p}} \tag{2}$$

in which f_{side} is the mean distance between main and associated side peaks of the data in the 2nd frequency domain, M stands for the total number of samples in the 2nd frequency domain and p indicates the pth frequency domain which is set to 2 in this study. Therefore, for our analyses, the expression for γ simply gives the ratio of f_{side} and M which can be considered as a normalized expression of the mean distance between main and side peaks according to the total number of samples. Here, the normalization allows the comparability of the variable among different data types. In order to be able to find more distinctive band ranges for the values of the variables, δ is proposed as the multiplicative inverse of γ such that

$$\delta = \gamma^{-1}. \tag{3}$$

From our outputs in Fig. 1, it is seen that the ranges of both δ and γ values change depending on the single, multiple or no oscillation conditions of the original time domain signal. Therefore, we perform the following procedure in order to compute both values.

1 Estimate the second frequency domain representation of the time domain data using the Fourier transform.
2 Apply the moving average filtering to smooth the data in order to eliminate one-sample spikes.
3 Detect periodic peaks in the second frequency domain, namely, main peaks.
4 Crop N/M samples before and after the main peaks and find symmetric side peaks from the cropped samples where N is the number of samples in the second frequency domain. Here, the divisive constant M is found from the empirical analyses. In this study, it is found as 60.
5 Calculate the mean distance between all main and associated side peaks which are used to compute f_{side}.
6 Calculate γ and δ by using Eqs. 2 and 3, respectively.

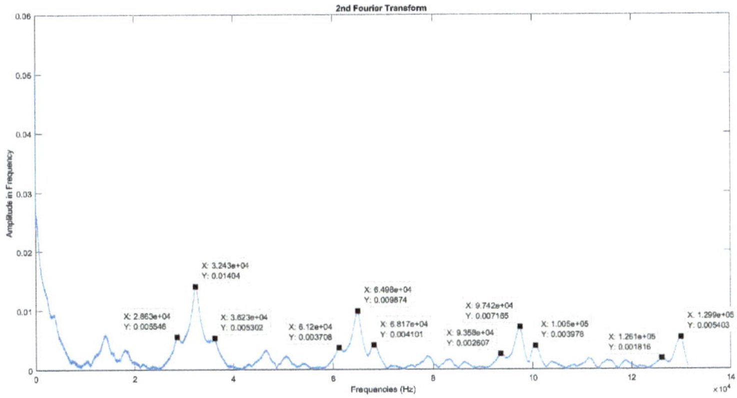

Fig. 1. An example of the second order frequency domain representation of a real ECG dataset with multiple oscillations.

3 Application

3.1 Description of Data Generation

In our assessments, the EGC datasets are generated from the ECG simulator ECGSYN (a realistic ECG waveform generator algorithm) [2,4] which is publicly available and whose latest version can be retrieved from https://physionet.org/physiotools/ecgsyn. These simulated ECG datasets are used for the Monte-Carlo runs with 1000 iterations. That is, 1000 different variations of the data for each group of data types are generated in our calculations.

On the other hand, in order to compare the values of γ and δ, 6 different data types are constructed in such a way that the first two of them (Groups 1 and 2) represent the regular ECG signals, but with a variation in the amplitudes of the R peaks which are selected as 30 and 15 of the mean of the whole data. Additionally, other two datasets' group (Groups 3 and 4) are generated based on the common patterns of the ventricular tachycardia and multiple oscillation cases and finally, the remaining two datasets' groups (Groups 5 and 6) are simulated based on the patterns for the Q Waves. To eliminate the main periodic behaviours of all datasets, we conduct two common oscillations in all data by taking the same heart rates and low frequency sinusoidal oscillations with a period of around 1000 samples. Hence, the amplitudes and the locations of the PQRST components in a stable heart rate will only differ between the datasets' groups. Note that since each generated dataset has a variation and added noises, there are in-group variances between the datasets from the same group too.

On the other side, the ECGSYN MATLAB algorithm used in our analyses has several parameters in order to generate the ECG data. Although each dataset has variations for the reproductivity of the conditions, the parameters of datasets can be found below as in the ECGSYN algorithm supported by the MATLAB programming language. The parameters which are the same among all datasets can be listed as the ECG sampling frequency (256 Hz), approximate number of heart beats per data (10), additive uniformly distributed measurement noise (0.1 mV), mean heart rate (60), standard deviation of the heart rate (1 bpm), low-high filter ratio (0.5), internal sampling frequency (256 Hz), order of the extrema (PQRST pattern) and the angles of the extrema (−70 – 15015100). Whereas, we change certain parameters in order to obtain the variability among datasets' groups which can be seen in Table 1. We plot an example from each group in Fig. 2 for illustration. Here, the figures from 2(a) to (e) indicate as an example of the tabulated values in Table 1 from upwards to downwards.

Table 1. The generated ECG datasets' groups composed of 1000 Monte Carlo runs and generated by using the ECGSYN algorithm and their changing parameters. The values in parentheses are given in the order of the PQRST pattern.

Group of dataset	z position of extrema	Gaussian width of peaks
1. Regular ECG with R amplitude = 15	(1.20 −5.0 15.0 −7.50 0.75)	(0.25 0.10 0.10 0.10 0.40)
2. Regular ECG with R amplitude = 30	(1.20 −5.0 30.0 −7.50 0.75)	(0.25 0.10 0.10 0.10 0.40)
3. Ventricular tachycardia example 1	(0.0 0.0 10.0 0.0 0.0)	(0.25 0.10 0.10 0.10 0.40)
4. Ventricular tachycardia example 2	(1.20 −5.0 1.0 −7.50 0.75)	(0.25 0.10 0.60 0.10 0.40)
5. Q wave example 1	(1.20 −1.0 10.0 −1.0 0.75)	(0.10 0.50 0.20 0.20 0.20)
6. Q wave example 2	(1.20 −2.0 10.0 −0.50 0.75)	(0.10 0.60 0.30 0.10 0.10)

3.2 Results

The results obtained by applying algorithms in order to find the numeric values for proposed features with the generated ECG datasets can be found in Table 2. In this table, we list the mean and variance values of δ and γ variables with respect to each dataset from 1 to 6.

Table 2. Results of the means μ and variances V(.) of the generated features for different datasets.

Dataset	μ_δ	$V(\delta)$	μ_γ	$V(\gamma)$
1	17.1161	0.1393	0.0584	<0.0001
2	17.4298	0.3186	0.0574	<0.0001
3	25.4654	3.9782	0.0402	0.0061
4	24.7889	0.4217	0.0404	<0.0007
5	37.7849	1.5669	0.0265	0.0012
6	32.9132	0.1932	0.0304	<0.0002

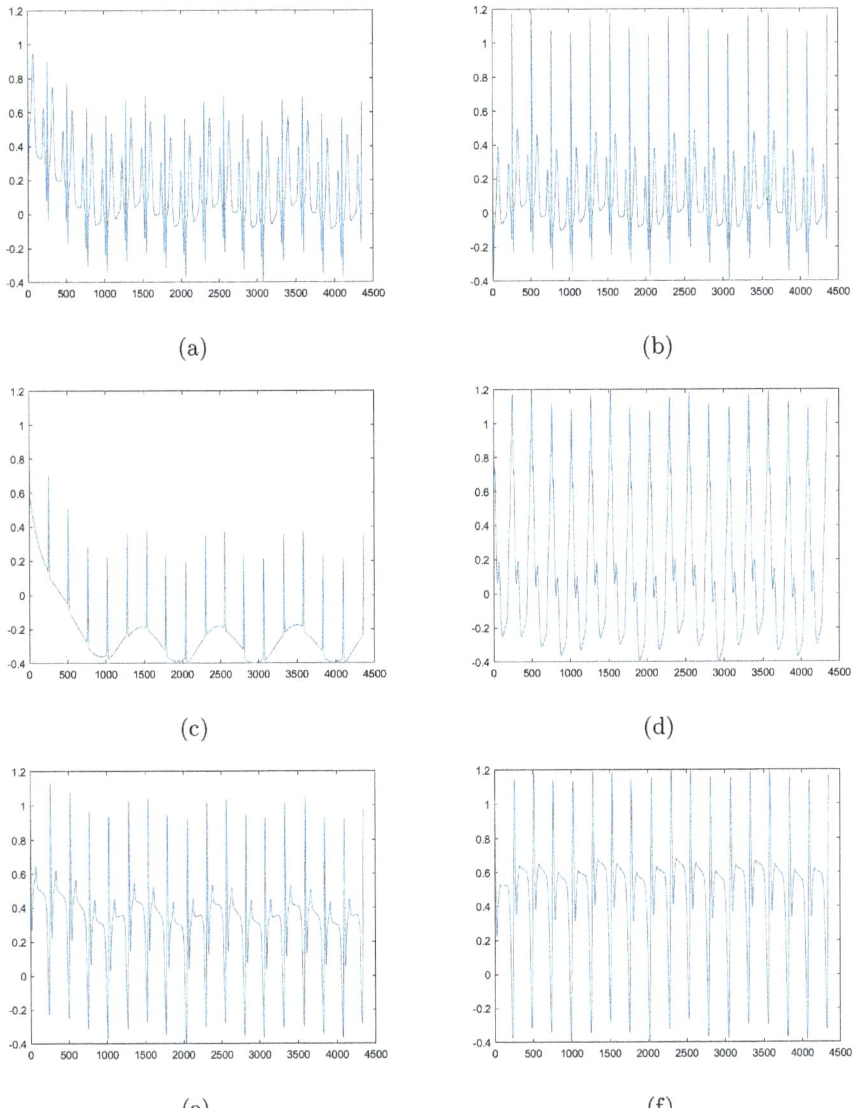

Fig. 2. One example from each ECG group of datasets which is used in this study. (a) Regular ECG whose R peak amplitude is 15 times of the data mean. (b) Regular ECG whose R peak amplitude is 30 times of the data mean. (c) Ventricular tachycardia for the example 1. (d) Ventricular tachycardia for the example 2. (e) Q Wave for the example 1. (f) Q Wave for the example 2.

According to Table 2, it is seen that there are differences between the ranges of band for δ and γ for distinct datasets. The intervals of these bands may enable the researchers to infer the type of the data for the patients whose conditions are not known. Because from the comparisons of statistics in Datasets 1 and 2 with other sets, it is found that the ranges of mean and variances can be used for the diagnosis of the ventricular tachycardia and Q waves. Moreover, since the variances of each dataset's group are comparably low, implying stable outcomes, we can conclude that both δ and γ bands can be taken as a discrimination factor between regular and problematic ECG datasets. Furthermore, by comparing the regular ECG datasets, we observe that as the R peak amplitude decreases, δ increases and thereby, γ decreases too.

4 Conclusion

In this study, we have focused on the reliability of our recently proposed features δ and γ on the synthetically generated ECG datasets in a Monte-Carlo study where each type of datasets is composed of 1000 Monte Carlo runs. In these calculations, three groups of datasets have been generated according to the types of the ECG diagnostic patterns and each group has had two different datasets with distinct parameters. From our analyses, we have found that the proposed features δ and γ have different bands of ranges for each group of dataset, i.e., distinct heart problems. We have considered that the underlying differences between the band of ranges can be used as a pre-diagnostic feature for the ECG data, specifically, in machine learning studies.

Hereby, as a future work, we think to extent this study by detecting other heart diseases whose ECG measurements can be used to diagnose the illnesses. Additionally, we consider to develop a new statistical test in order to capture significant patterns in the ECG disorders under certain confidence intervals.

References

1. Adhikari, A.: Video-analysis inference automated ECG (VID-ECG): improving video-based heart rate detection and exposing security risks of ECG-based biometric authentication. Ph.D. thesis, University of Illinois (2015)
2. Andreotti, F., Behar, J., Zaunseder, S., Oster, J., Clifford, G.D.: An open-source framework for stress-testing non-invasive foetal ECG extraction algorithms. Physiol. Meas. **37**(5), 627–648 (2016). https://doi.org/10.1088/0967-3334/37/5/627
3. Arzeno, N.M.: Discriminating noise from chaos in heart rate variability: application to prognosis in heart failure. Ph.D. thesis, Massachusetts Institute of Technology (2007)
4. Behar, J., Andreotti, F., Zaunseder, S., Li, Q., Oster, J., Clifford, G.D.: An ECG simulator for generating maternal-foetal activity mixtures on abdominal ECG recordings. Physiol. Meas. **35**(8), 1537–1550 (2014). https://doi.org/10.1088/0967-3334/35/8/1537
5. Benitez, D.S., Gaydecki, P., Zaidi, A., Fitzpatrick, A.: A new QRS detection algorithm based on the Hilbert transform. In: Computers in Cardiology 2000 (Cat. 00CH37163), vol. 27, pp. 379–382. IEEE (2000)

6. Bennett, T.D., Combs, W.J., Kallok, M.J., Lee, B.B., Mehra, R., Klein, G.J.: Subcutaneous multi-electrode sensing system, method and pacer, US Patent 5,331,966, 26 July 1994
7. Brown, T.A., Barlow, D.H.: Comorbidity among anxiety disorders: implications for treatment and DSM-IV. J. Consult. Clin. Psychol. **60**(6), 835 (1992)
8. Elhaj, F.A., Salim, N., Harris, A.R., Swee, T.T., Ahmed, T.: Arrhythmia recognition and classification using combined linear and nonlinear features of ECG signals. Comput. Methods Programs Biomed. **127**, 52–63 (2016)
9. Erkuş, E., Purutçuoğlu, V., Purutçuoğlu, E.: Detection of abnormalities in heart rate using multiple Fourier transforms. Int. J. Environ. Sci. Technol. **16**, 5237–5242 (2019)
10. Erkuş, E.C., Purutçuoğlu, V.: Detection of abnormalities in heart rate using multiple Fourier transforms. In: 5th International Conference on Computational and Experimental Science and Engineerins (ICCESEN2018) (2018)
11. Erkuş, E.C., Purutçuoğlu, V., Agraz, M.: Detection of outliers using Fourier transform. In: 10th International Statistics Congress (ISC2017) (2017)
12. Formaggia, L., Quarteroni, A., Veneziani, A.: Cardiovascular Mathematics: Modeling and Simulation of the Circulatory System, vol. 1. Springer, Heidelberg (2010)
13. Friesen, G.M., Jannett, T.C., Jadallah, M.A., Yates, S.L., Quint, S.R., Nagle, H.T.: A comparison of the noise sensitivity of nine QRS detection algorithms. IEEE Trans. Biomed. Eng. **37**(1), 85–98 (1990)
14. Jalaleddine, S.M., Hutchens, C.G., Strattan, R.D., Coberly, W.A.: ECG data compression techniques-a unified approach. IEEE Trans. Biomed. Eng. **37**(4), 329–343 (1990)
15. Kadambe, S., Murray, R., Boudreaux-Bartels, G.F.: Wavelet transform-based QRS complex detector. IEEE Trans. Biomed. Eng. **46**(7), 838–848 (1999)
16. Liao, T.W.: Clustering of time series dataa survey. Pattern Recogn. **38**(11), 1857–1874 (2005)
17. Marks, R.J.I.: Introduction to Shannon Sampling and Interpolation Theory. Springer, Heidelberg (2012)
18. Moraes, J., Freitas, M., Vilani, F., Costa, E.: A QRS complex detection algorithm using electrocardiogram leads. In: Computers in Cardiology, pp. 205–208. IEEE (2002)
19. Oppenheim, A.V.: Discrete-Time Signal Processing. Pearson Education India, Chennai (1999)
20. Özbay, Y., Ceylan, R., Karlik, B.: A fuzzy clustering neural network architecture for classification of ECG arrhythmias. Comput. Biol. Med. **36**(4), 376–388 (2006)
21. Pan, J., Tompkins, W.J.: A real-time QRS detection algorithm. IEEE Trans. Biomed. Eng. **32**(3), 230–236 (1985)
22. Pantelopoulos, A., Bourbakis, N.G.: A survey on wearable sensor-based systems for health monitoring and prognosis. IEEE Trans. Syst. Man Cybern. Part C (Appl. Rev.) **40**(1), 1–12 (2010)
23. Ritchie, J., Spencer, L.: Qualitative data analysis for applied policy research. Qual. Res. Companion **573**(2002), 305–329 (2002)
24. Ruttimann, U.E., Pipberger, H.V.: Compression of the ECG by prediction or interpolation and entropy encoding. IEEE Trans. Biomed. Eng. **11**, 613–623 (1979)
25. Sansone, M., Fusco, R., Pepino, A., Sansone, C.: Electrocardiogram pattern recognition and analysis based on artificial neural networks and support vector machines: a review. J. Healthc. Eng. **4**(4), 465–504 (2013)
26. Savalia, S., Acosta, E., Emamian, V.: Classification of cardiovascular disease using feature extraction and artificial neural networks. J Biosci. Med. **5**(11), 64 (2017)

27. Scargle, J.D.: Studies in astronomical time series analysis. II-statistical aspects of spectral analysis of unevenly spaced data. Astrophys. J. **263**, 835–853 (1982)
28. Sermesant, M., Delingette, H., Ayache, N.: An electromechanical model of the heart for image analysis and simulation. IEEE Trans. Med. Imaging **25**(5), 612–625 (2006)
29. Slimane, Z.E.H., Naït-Ali, A.: QRS complex detection using empirical mode decomposition. Digit. Signal Process. **20**(4), 1221–1228 (2010)
30. Thiagarajan, S., Balleda, K.: HiBeat: a novel highly accurate implementation of cardiac pulse measurement on a multicore architecture. In: 2014 International Conference on High Performance Computing and Applications (ICHPCA), pp. 1–6. IEEE (2014)
31. Vehkaoja, A., Salpavaara, T., Verho, J., Lekkala, J.: EMFI material as wearable heart rate sensor for night time recordings. In: SENSORS, 2010 IEEE, pp. 145–149. IEEE (2010)
32. Worden, K., Tomlinson, G.R.: Nonlinearity in Structural Dynamics: Detection, Identification Modelling. CRC Press, Boca Raton (2019)

Financial Fraud Detection Through Artificial Intelligence

Roman Rodriguez-Aguilar[1], Jose A. Marmolejo-Saucedo[2(✉)], Pandian Vasant[3], and Igor Litvinchev[4]

[1] Escuela de Ciencias Económicas y Empresariales, Universidad Panamericana, Augusto Rodin 498, 03920 Ciudad de México, Mexico
rrodrigueza@up.edu.mx

[2] Facultad de Ingeniería, Universidad Panamericana, Augusto Rodin 498, 03920 Ciudad de México, Mexico
jmarmolejo@up.edu.mx

[3] Universiti Teknologi Petronas, Seri Iskandar, Malaysia
pandian.vasant@tdtu.edu.vn, vasantglobal@gmail.com

[4] Nuevo Leon State University, San Nicolás de Los Garza, Mexico
igorlitvinchev@gmail.com

Abstract. The present work shows the analysis and modeling of a database with information about the various credit card transactions. The objective is to detect transactions that are fraudulent. In the modeling process, the "Ridge and Lasso", "Boosting" and "Random Forest" techniques were applied in the modeling and variables selection. The results show that the accuracy of the models was very high, so the metric "Recall" was chosen as a second criterion for selecting the best model. This metric measures the percentage of positive values of the variable "fraud". It is concluded that the best model is that of "Boosting" with 1,500 trees and a K-Folds of 10 that presented the best results in both training and validation.

Keywords: Fraud · Boosting · Random forest · Ridge · Lasso

1 Introduction

In the world of financial services, there are multiple products that are offered to customers, including credit cards that are the most used credit instruments in Mexico. For a company it is vital to be able to recognize the type of transactions that customers are making and in this way identify fraudulent credit card transactions to prevent customers from paying for items they did not buy.

The database analyzed in the study contains transactions made with credit cards by their holders. The database was chosen from a set available on the "Kaggle" website. It is a set of data that presents the transactions that occurred in two days. There were 492 frauds out of a total of 284,807 transactions where the frauds represent 0.172% of all transactions. It is observed that the data set is unbalanced. This information is contained in a total of 285 thousand records and 31 columns (all continuous). The objective of the study is to determine the best model to be able to identify fraudulent transactions on

D. J. Hemanth and U. Kose (Eds.): ICAIAME 2019, LNDECT 43, pp. 57–72, 2020.
https://doi.org/10.1007/978-3-030-36178-5_6

credit cards. For reasons of confidentiality, the input variables are the result of a transformation carried out through main components (PCA) from which there is no access to how it was built. The only variables that have not been transformed with PCA are 'Time' and 'Quantity'. At the end of the model, a model capable of identifying these fraudulent transactions was obtained.

2 Theoretical Background

2.1 Lasso Regression

Regression Shrinkage and selection with Lasso methodology was developed by Tibshirani (1996) is a method for estimation in linear models that objective is minimize the square errors subject to the sum of the absolute value of coefficients being less than a constant. The main result is that after this optimization process generate some parameters equal cero, hence give interpretable models, and as consequence allows selecting relevant variables in the model [1, 2].

Consider data $(x_i, y_i), i = 1, 2, .., N$, were x_i are the predictor variables and y_i are the dependent variable. Assume that the observations x_{ij} are independent and standardized, then Lasso estimate $(\hat{\alpha}, \hat{\beta})$ is defined by,

$$\left(\hat{\alpha}, \hat{\beta}\right) = min\left\{\sum\nolimits_{i=1}^{N} (y_i - \alpha - \sum\nolimits_j \beta_j x_{ij})^2\right\} \text{ subject to } \sum\nolimits_j \left|\beta_j\right| \leq t. \quad (1)$$

So simplifying, the estimators would be as follows:

$$\left(\hat{\beta}\right) = min\left\{\sum\nolimits_{i=1}^{N} (y_i - \alpha - \sum\nolimits_j \beta_j x_{ij})^2\right\} + \lambda \sum\nolimits_j \left|\beta_j\right| \quad (2)$$

Where $t \geq 0$ is a tuning parameter, Eq. (1) is a quadratic programming problem with linear inequality constraints. The parameter (t) controls the amount of shrinkage that is applied to the estimates, it is clear that Lasso regression will often generate coefficients that are exactly zero. Lasso also shrinks coefficients to reduce variance, but it also shares similarities with better subset regression, where for each number of variables we seek to choose the variables that give the best model. However, the Lasso approach is more scalable and can be calculated more simply [5].

2.2 Bagging (Boostrap Aggregating)

The bootstrap is a way of assessing the accuracy of a parameter estimate or a prediction. Here we use the bootstrap to improve the estimate or prediction itself. The bootstrap mean is approximately a posterior average. Bagging further exploits this connection. Consider first the regression problem we fit a model to our training data $Z = \{(x1, y1), (x2, y2), \ldots, (xn, yn)\}$, obtaining the prediction of $\hat{f}(x)$ at input x [3–5].

Bootstrap aggregation or bagging averages this prediction over a collection of bootstrap samples, thereby reducing its variance. In this case we want to classify in place to predict, therefore the estimation is defined by the greater frequency of all the models estimated with respect to the perception of quality and the selection of the relevant variables is done based on an importance function using a Gini Index [14]. For each bootstrap sample $Z^{*b}, b = 1, 2, ..., B$, we fit our model, giving prediction $\hat{f}^{*b}(x)$. The bagging estimate is defined as:

$$\hat{f}_b(x) = \frac{1}{B}\sum\nolimits_{b=1}^{B} \hat{f}^{*b}(x) \tag{3}$$

The proposed model integrates three approaches to statistical learning, variable selection, classification and resampling for validation of estimates taking into account the feedback of the model when changing data or temporality of the same (Fig. 1).

3 Materials and Methods

Due to the amount of records that are required to be processed in the R software, a virtual server was created in Azure to have a better performance, the server types used were D12 v2 (4 vcpus, 28 GB of memory) and D13 v2 (8 vcpus, 56 GB of memory) [6].

To achieve a better use of the resources of the server, the "h2o" package of R was used for the execution of most of the models. The "h20" package is a scalable Open Source Machine Learning platform that offers implementations with parallel processing of several algorithms, such as Generalized Linear Models, Gradient Boosting Machines (including XGBoost), Random Forests, Deep Neural Networks (Deep Learning), Stacked Ensembles, Naive Bayes, Cox Proportional Hazards, K-Means, PCA, Word2Vec, as well as a fully automatic Machine Learning algorithm (AutoML).

a. Exploratory Analysis

As mentioned above, the database contains 31 columns of which 1 is the flag that identifies the fraud (Class); 28 variables (V1-V28) are numerical variables resulting from the application of principal components in the original variables to maintain the confidentiality of the information. A variable that contains the amount of the transaction (Amount) and finally a time variable that refers the seconds elapsed between each transaction and the first transaction in the data set. The Table 1 contains the type of data contained in each variable.

Table 1. Variables of the database

Variable	Tipo de Dato	Descripción	V15	Numérico	Componente principal obtenido de PCA
Time	Entero	Segundos transcurridos desde la primera transacción	V16	Numérico	Componente principal obtenido de PCA
V1	Numérico	Componente principal obtenido de PCA	V17	Numérico	Componente principal obtenido de PCA
V2	Numérico	Componente principal obtenido de PCA	V18	Numérico	Componente principal obtenido de PCA
V3	Numérico	Componente principal obtenido de PCA	V19	Numérico	Componente principal obtenido de PCA
V4	Numérico	Componente principal obtenido de PCA	V20	Numérico	Componente principal obtenido de PCA
V5	Numérico	Componente principal obtenido de PCA	V21	Numérico	Componente principal obtenido de PCA
V6	Numérico	Componente principal obtenido de PCA	V22	Numérico	Componente principal obtenido de PCA
V7	Numérico	Componente principal obtenido de PCA	V23	Numérico	Componente principal obtenido de PCA
V8	Numérico	Componente principal obtenido de PCA	V24	Numérico	Componente principal obtenido de PCA
V9	Numérico	Componente principal obtenido de PCA	V25	Numérico	Componente principal obtenido de PCA
V10	Numérico	Componente principal obtenido de PCA	V26	Numérico	Componente principal obtenido de PCA
V11	Numérico	Componente principal obtenido de PCA	V27	Numérico	Componente principal obtenido de PCA
V12	Numérico	Componente principal obtenido de PCA	V28	Numérico	Componente principal obtenido de PCA
V13	Numérico	Componente principal obtenido de PCA	Amount	Dinero	Importe de la transacción
V14	Numérico	Componente principal obtenido de PCA	Class	Binario	Target 1 indica fraude, 0 indica transacción normal

Analyzing the time variable, we have two days of information in the database, so we created a new variable "Period" to divide the day into 6 periods of 4 h of each and in this way to explain a profile of client in a better way. Applying a basic descriptive analysis, it can be observed that there is no missing data in any variable, so it will not be necessary to carry out any type of information imputation treatment. Likewise, it is observed that the variable "Class" has 492 observations with value "1" out of a total of 284,807 observations, this is equivalent to 0.17%. So the sample is biased to the population without fraud. However, it is important to perform the analysis in 100% of the population since it is likely that fraudulent transactions have very well identified characteristics that allow them to be detected. If the previously mentioned does not result, then a more balanced sample will be made to obtain said characteristics. On the other hand, it is observed that the ranges of the variables vary between 20 and 150. The variable with more amplitude in the range is the "V5", however, its standard deviation tells us that it is not of the variables with greater dispersion. In addition, in the variable "amount" there are values equal to zero that are rare, however, it is decided to leave them to see if this is an indicative of fraud (Table 2).

Table 2. Statistics

Variable	Mean	Std Dev	Minimum	Maximum	Range	N	N Miss
Periodo	4.1338591	1.4489174	1	6	5	284807	0
V1	3.94E-15	1.9586958	-56.4075096	2.45493	58.86244	284807	0
V2	5.69E-16	1.6513086	-72.7157276	22.057729	94.773457	284807	0
V3	-8.76E-15	1.516255	-48.3255894	9.3825584	57.708148	284807	0
V4	2.81E-15	1.4158686	-5.6831712	16.875344	22.558515	284807	0
V5	-1.55E-15	1.3802467	-113.743307	34.8016659	148.54497	284807	0
V6	2.04E-15	1.3322711	-26.1605059	73.3016255	99.462132	284807	0
V7	-1.70E-15	1.2370936	-43.5572416	120.5894939	164.14674	284807	0
V8	-1.88E-16	1.1943529	-73.2167185	20.0072084	93.223927	284807	0
V9	-3.15E-15	1.0986321	-13.4340663	15.5949946	29.029061	284807	0
V10	1.77E-15	1.0888498	-24.5882624	23.7451361	48.333399	284807	0
V11	9.29E-16	1.020713	-4.7974735	12.0189132	16.816387	284807	0
V12	-1.80E-15	0.9992014	-18.6837146	7.8483921	26.532107	284807	0
V13	1.67E-15	0.9952742	-5.7918812	7.126883	12.918764	284807	0
V14	1.48E-15	0.9585956	-19.2143255	10.5267661	29.741092	284807	0
V15	3.50E-15	0.915316	-4.4989447	8.8777416	13.376686	284807	0
V16	1.39E-15	0.8762529	-14.1298545	17.3151115	31.444966	284807	0
V17	-7.47E-16	0.8493371	-25.1627994	9.2535263	34.416326	284807	0
V18	4.26E-16	0.8381762	-9.4987459	5.0410692	14.539815	284807	0
V19	9.02E-16	0.8140405	-7.2135274	5.5919714	12.805499	284807	0
V20	5.13E-16	0.770925	-54.4977205	39.4209042	93.918625	284807	0
V21	1.47E-16	0.734524	-34.8303821	27.2028392	62.033221	284807	0
V22	8.04E-16	0.7257016	-10.9331437	10.5030901	21.436234	284807	0
V23	5.33E-16	0.6244603	-44.8077352	22.5284117	67.336147	284807	0
V24	4.46E-15	0.6056471	-2.8366269	4.5845491	7.4211761	284807	0
V25	1.43E-15	0.5212781	-10.2953971	7.5195887	17.814986	284807	0
V26	1.70E-15	0.482227	-2.6045506	3.5173456	6.1218962	284807	0
V27	-3.67E-16	0.4036325	-22.5656793	31.6121981	54.177877	284807	0
V28	-1.22E-16	0.3300833	-15.4300839	33.8478078	49.277892	284807	0
Amount	88.3496193	250.1201092	0	25691.16	25691.16	284807	0
Class	0.0017275	0.0415272	0	1	1	284807	0

An analysis of the distribution of the variables was carried out, however, for illustrative purposes, only the variables where inconsistencies were detected, the analysis behind them and the treatment required in each one were exposed. See Figs. 1, 2 3, 4 and 5.

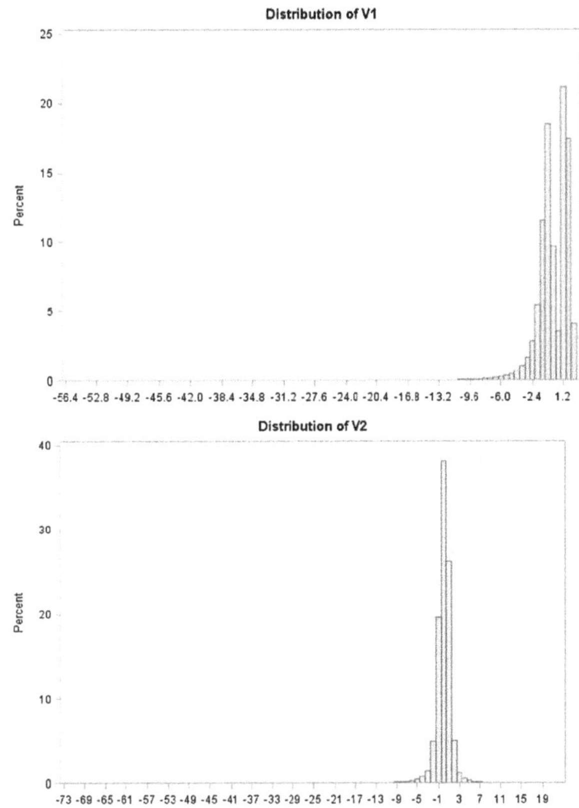

Fig. 1. Distribution of the variable V1 and V2

It is observed that the variables V8, V21, V23, V27 and V29 have a very grouped distribution to a single value which are distributed in the center of the distribution, so they are expected to be variables that help differentiate customers without fraud. On the other hand, a summary analysis was carried out dividing by the objective variable. The results are presented in the following table. The variable "Amount" has a right bias where there are observations that we can consider outliers and that do not affect the target, so the amounts greater than $2,500 will not be considered. There are many extremes that could be considered "outliers". However, after an analysis of the data it was found that around 400 observations of 6,000 observations that are at the ends have

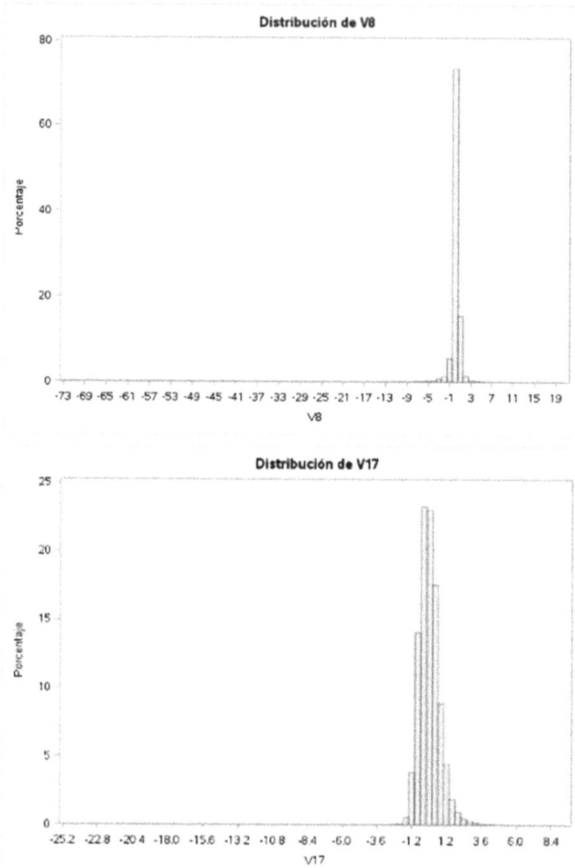

Fig. 2. Distribution of the variable V8 and V17

the target variable "Class" with a value of "1" so they can not be debugged. It is decided to keep the variables as they are, since the records with frauds have characteristics of the tails of the distributions. Making an analysis of the distribution of the variables divided by objective variable, it is observed that the variables V2, V4 and V11 present a different distribution in each category so that by themselves, they can explain well the behavior of the frauds which means that for Yes alone they have an important potential to discriminate the accounts (Figs. 6, 7 and 8).

Although graphically a distinction is observed between the variables mentioned above, in the table below it is observed that the means of the variables are very different between the population with fraud and the population without fraud, especially the variables framed in red (Tables 3a and 3b).

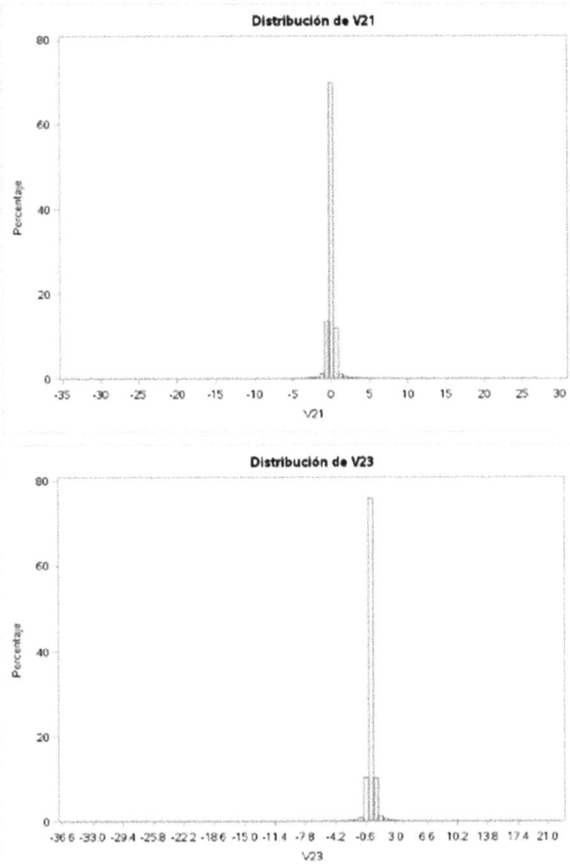

Fig. 3. Distribution of the variable V21 and V23

Finally, as part of the descriptive analysis, a correlation analysis was performed in which no significant correlation was found between the analysis variables. This behavior is expected since the variables are already the result of applying main components.

After performing the exploratory analysis of the variables, no reason was detected to exclude the variables. However, as mentioned, the period variable was created from the time variable, then the period variable will be used and not the time variable.

b. Modeling

To predict the behavior of transactions (fraud or non-fraud), two regression methodologies, Ridge Regression and Lasso Regression, will be used, as well as two methodologies that use Random Forest and Boosting random forests. A comparison of the results will be made to obtain the best classification model.

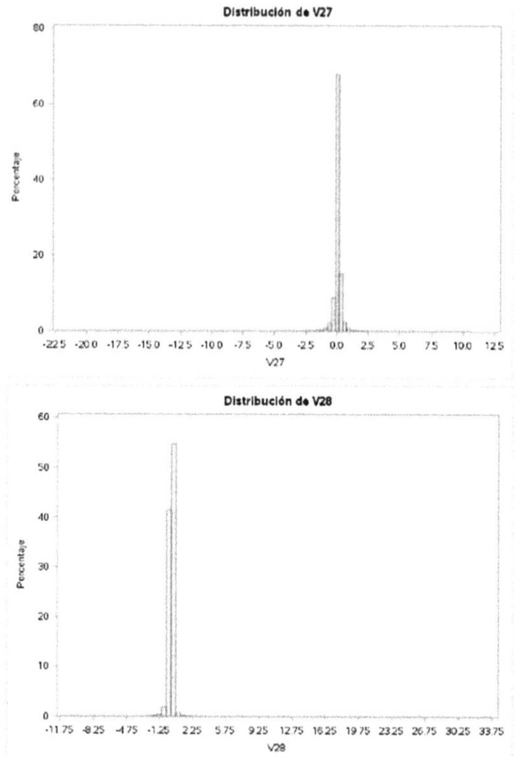

Fig. 4. Distribution of the variable V27 and V29

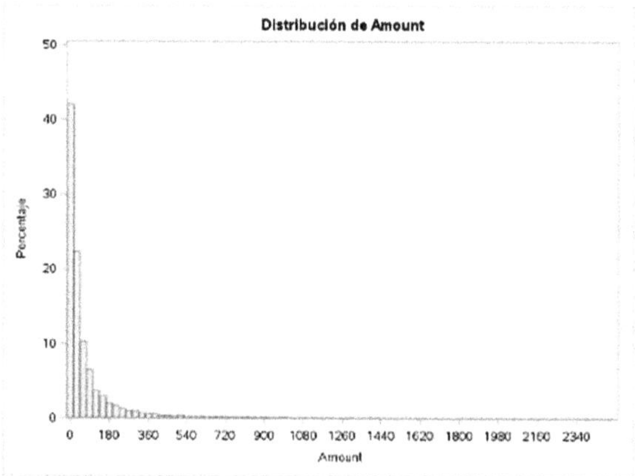

Fig. 5. Distribution of the amount variable

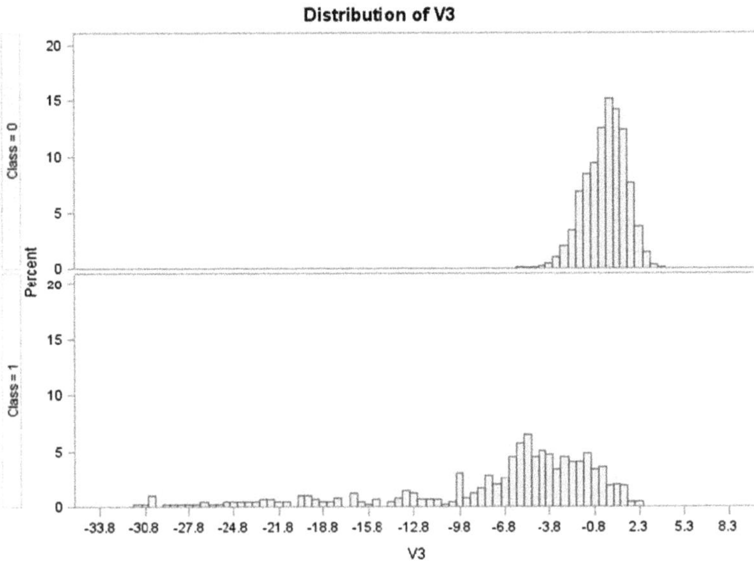

Fig. 6. Distribution of variables with different distribution in the target variable

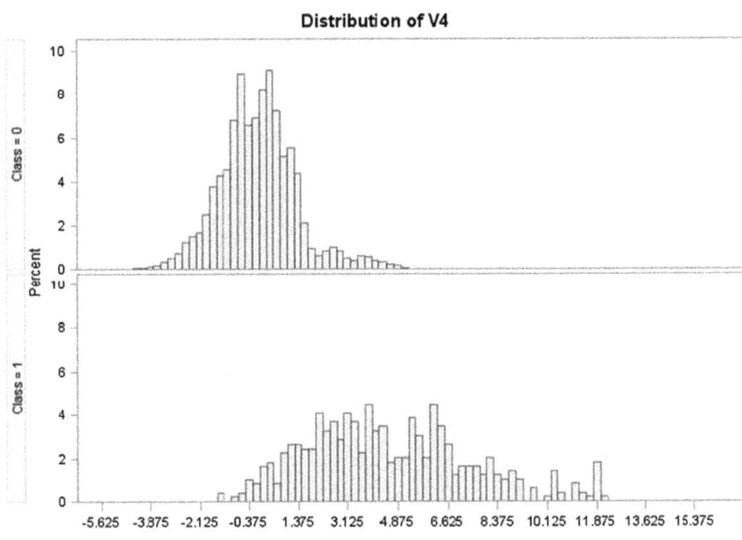

Fig. 7. Distribution of variables with different distribution in the target variable.

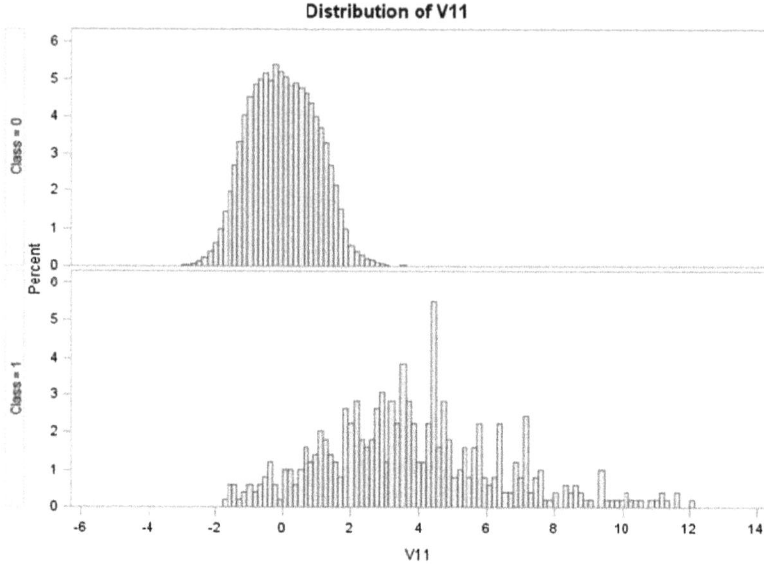

Fig. 8. Distribution of variables with different distribution in the target variable.

In the following diagram you can see the process followed during the exploitation of the data, the modeling and the selection of the optimal model. See Fig. 9.

Optimization of Parameters
Prior to the application of the algorithms in the validation sample, an optimization of its parameters was performed using the Cross Validation technique in the training data set, for which 10 K-folds were used in each configuration of the models. It should be noted that for each type of algorithm the configuration of its parameters was different, although the metric to choose the best performance was the same, this metric is the recall of the fraud variable and is defined as follows.

Recall = #True Positives/#Positives = TP/(TP + FN)[2]

Tree-Based Models
Three different configurations were made for the Random Forest algorithm, where the number of trees was changed, as well as the number of variables. For the boosting model, 3 different configurations were also made, in which the change was the number of trees used, in the following table we can see the results obtained for each of the configurations (Table 4).

Although the Accuracy is higher than 99.9% and it could be thought that the model classifies practically all transactions, we must pay attention to the volume of Frauds (target 1) that can be classified correctly. To get a clearer idea of this, the Recall metric is used, reaching 80.86% accuracy, using both Boosting and Random Forest algorithms.

Table 3a. Sums for each objective variable

Class	N Obs	Variable	Mean	Std Dev	Minimum	Maximum
0	283866	Periodo	4.1352046	1.4485959	1	6
		V1	0.0193739	1.8967954	-56.4075096	2.45493
		V2	0.0114948	1.5425515	-72.7157276	18.9024528
		V3	0.0202443	1.4386448	-33.680984	9.3825584
		V4	-0.0121019	1.3925174	-5.6831712	16.4912172
		V5	0.0171834	1.279539	-23.6697257	34.8016659
		V6	-0.0040665	1.2993353	-26.1605059	16.6140541
		V7	-0.0033669	1.0765038	-31.7649465	21.0145869
		V8	0.0021226	1.1533761	-73.2167185	18.7092545
		V9	0.0053584	1.088448	-6.2907304	15.5949946
		V10	0.0127431	1.0402602	-14.7410964	23.7451361
		V11	-0.006951	1.0027784	-4.7974735	10.0021902
		V12	0.0112252	0.9454566	-15.1449882	7.8483921
		V13	-0.000292174	0.9948257	-5.7918812	4.569009
		V14	0.0112032	0.8961639	-18.3920915	10.5267661
		V15	-0.000464015	0.9144138	-4.3913071	5.8256543
		V16	0.0059986	0.8423942	-10.11556	6.4427979
		V17	0.0115727	0.7492707	-17.0984442	9.2535263
		V18	0.0034123	0.823706	-5.36666	4.2956482
		V19	0.000153048	0.8099884	-4.9327331	5.5919714
		V20	-0.0073838	0.6843166	-23.6468903	16.7564479
		V21	-0.0031941	0.7071388	-34.8303821	22.6148894
		V22	0.0019022	0.7197746	-8.5936416	10.5030901
		V23	0.0022807	0.5549568	-36.6660001	22.0835449
		V24	-0.000038328	0.6053105	-2.8366269	4.0228659
		V25	0.000894303	0.5164873	-7.4957411	7.5195887
		V26	0.000189321	0.4818634	-2.0685609	3.5173456
		V27	-0.000804404	0.387206	-22.5656793	10.5078844
		V28	-0.000134776	0.3218766	-11.7108956	33.8478078
		Amount	82.3490876	183.37419	0	2499.9

Regression Models
In the case of the regression models, the H2o package has a function in which it first chooses an optimal value of lambda and subsequently carries out the Cross Validation process with that lambda value, this function can be used in both Ridge and Lasso. The results obtained can be seen in the Table 5.

Importance of Variables of the Best Models
Then the most important variables are explored for each of the chosen models (Figs. 10, 11, 12 and 13).

In the random forest models the variables V17, V14, V12 and V10 are those that appear among the 5 most important.

Table 3b. Sums for each objective variable

Class	N Obs	Variable	Mean	Std Dev	Minimum	Maximum
1	492	Periodo	3.4776423	1.6793669	1	6
		V1	-4.7719484	6.7836868	-30.55238	2.132386
		V2	3.6237781	4.2912156	-8.4021537	22.057729
		V3	-7.033281	7.1109373	-31.1036848	2.2502096
		V4	4.5420291	2.8733177	-1.3132748	12.1146718
		V5	-3.1512249	5.372468	-22.1055315	11.0950886
		V6	-1.3977367	1.8581236	-6.4062666	6.4741146
		V7	-5.5687311	7.2067728	-43.5572416	5.8025374
		V8	0.5706359	6.7978306	-41.0442609	20.0072084
		V9	-2.5811229	2.5008956	-13.4340663	3.3535249
		V10	-5.6768829	4.897341	-24.5882624	4.0314351
		V11	3.8001729	2.6786045	-1.7022284	12.0189132
		V12	-6.259393	4.6544585	-18.6837146	1.3759414
		V13	-0.1093338	1.1045183	-3.127795	2.8154398
		V14	-6.9717229	4.2789399	-19.2143255	3.442422
		V15	-0.0929287	1.0499153	-4.4989447	2.4713579
		V16	-4.1399457	3.8650347	-14.1298545	3.1396557
		V17	-6.6658364	6.9706177	-25.1627994	6.7393844
		V18	-2.2463081	2.8993659	-9.4987459	3.7903162
		V19	0.6806593	1.5398533	-3.6819036	5.2283418
		V20	0.3723194	1.3466352	-4.1281858	11.0590043
		V21	0.7135884	3.8693044	-22.7976039	27.2028392
		V22	0.0140489	1.4946016	-8.8870171	8.3619852
		V23	-0.040308	1.5796423	-19.2543276	5.46623
		V24	-0.1051303	0.5155765	-2.0280242	1.0914347
		V25	0.0414489	0.7972047	-4.7816055	2.2082092
		V26	0.0516481	0.4716789	-1.1526711	2.7452607
		V27	0.1705748	1.3767658	-7.2634821	3.0523577
		V28	0.0756673	0.5472914	-1.8692905	1.7793639
		Amount	122.2113211	256.68329	0	2125.87

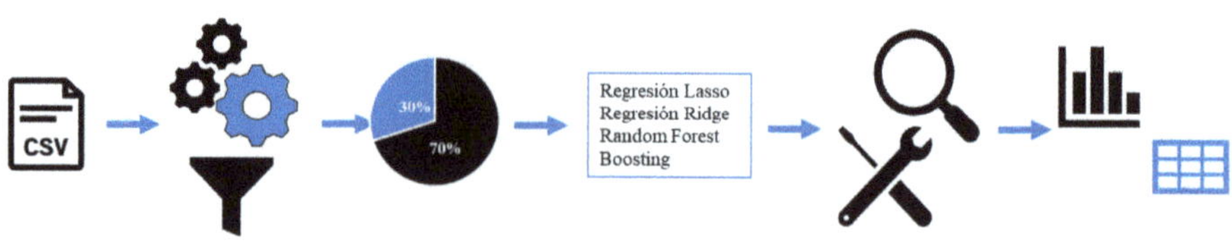

Fig. 9. Modeling construction diagram

Table 4. Summary tree-based models

Minutos	Modelo	K folds	Árboles	Predictores aleatorios	Recall	Accuracy
14.6	Boosting	10	500	-	79.71%	99.95%
44.8	Boosting	10	1500	-	80.86%	99.96%
85.3	Boosting	10	3000	-	80.86%	99.96%
8.1	RandomForest	10	500	6	80.86%	99.96%
22.8	RandomForest	10	1500	6	80.29%	99.95%
44.9	RandomForest	10	3000	6	80.86%	99.96%

Table 5. Summary of Regression models

Minutos	Modelo	K folds	Lambda	Recall	Accuracy
0.91	Ridge	10	0.00002447	77.7%	99.92%
0.55	Lasso	10	0.00003841	78.0%	99.92%

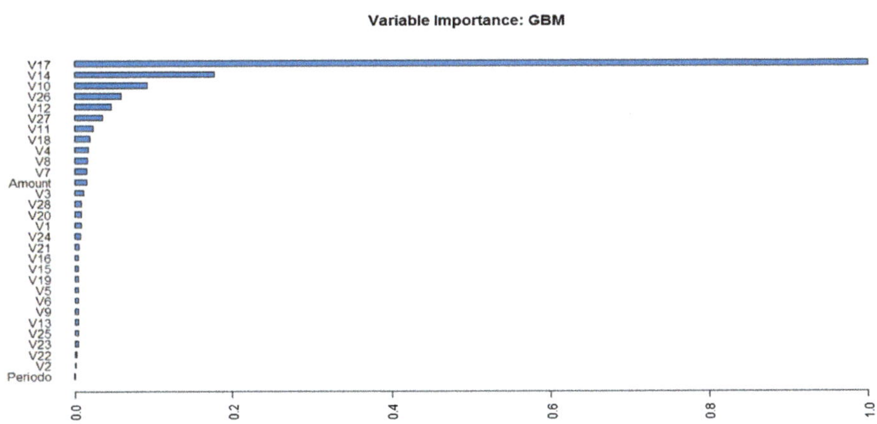

Fig. 10. Importance of variables in Boosting model with 1500 trees

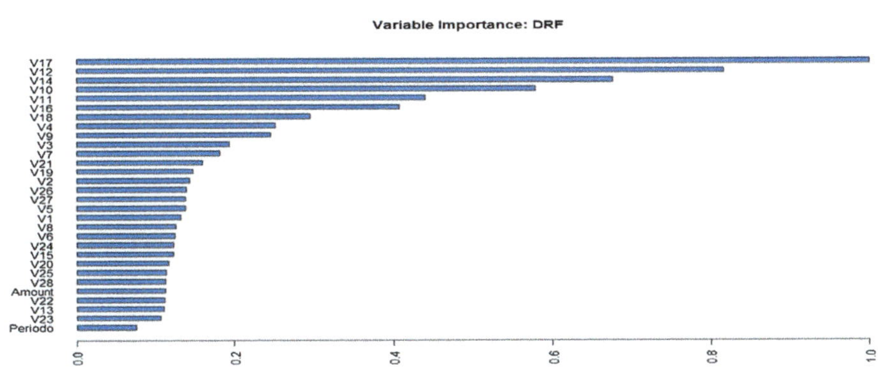

Fig. 11. Importance of variables in Random Forest model with 500 trees

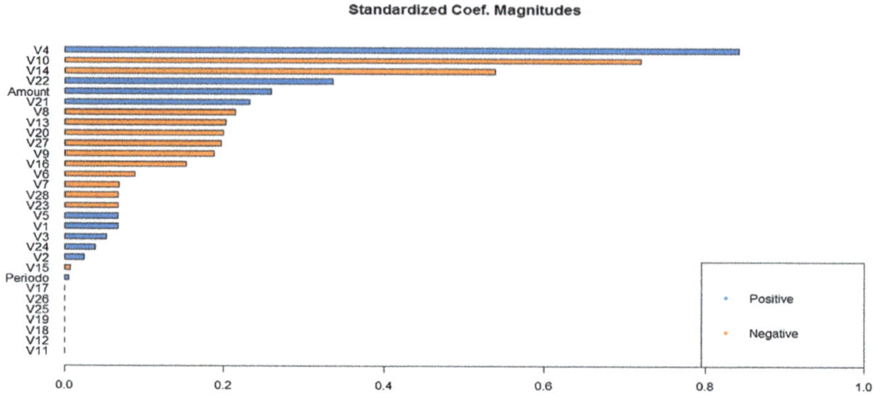

Fig. 12. Importance of variables in Ridge model

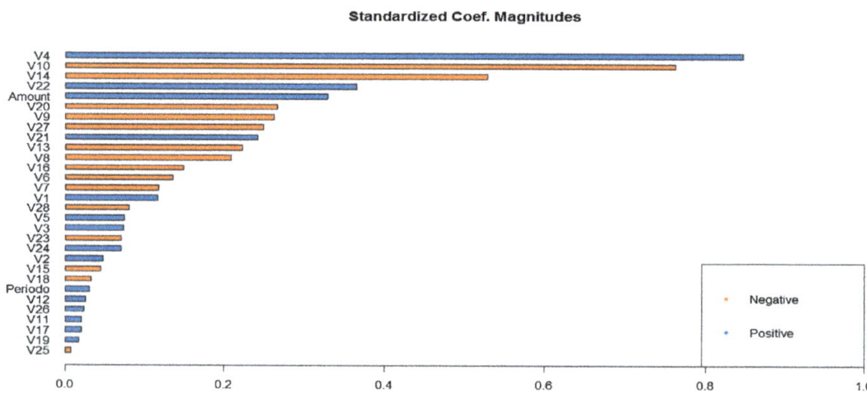

Fig. 13. Importance of variables in Lasso model

In the regression models the variables V10 and V14 are among the 3 most important, both in Ridge and Lasso among higher values take, the probability that there is a fraud is lower. The variables that appear in both random forest models and regression models are: V10 and V14, to try to obtain an interpretation we could use the findings found in the regression models.

4 Results

The models were evaluated within the testing data set (which was left out of the sample with which the model was made) (Table 6).

Table 6. Comparison of best models

Modelo	Recall	Accuracy
Boosting	83.10%	99.96%
RandomForest	79.58%	99.96%
Ridge	79.58%	99.94%
Lasso	79.58%	99.94%

Although in the construction of models, both Boosting and Random Forest have a similar recall of 81.86%, when the validation is done in a set of information outside the sample, the one that showed the best performance was Boosting, which has an improvement of 3 percentage points with respect to the other methodologies (which all presented a recall of 79.58%).

Additionally, to corroborate the results of these models, tree models, regression and traditional neural networks were executed in SAS with the following results (Table 7):

Table 7. Traditional models in SAS

Modelo	Recall	Accuracy
Neural	79.87%	99.92%
Tree	70.47%	99.94%
Regression	61.07%	99.92%

Comparing the traditional models against those executed in the project we can observe that only the results of the neural network model were close to the tested models.

5 Conclusions

After carrying out the study, technical limitations were faced regarding the application of the algorithms through the libraries. The previous thing was solved with the use of a server mounted in Microsoft Azure and the H2o package, we could compare 4 different algorithms Random Forest, Boosting, Regression Ridge and Lasso Regression, having a completely unbalanced sample the accuracy metrics for classification error not a good measure of accuracy since the high number of zeros may be biased. For this reason, it was decided to use the Recall of the fraudulent cases to ensure that the model correctly identifies those transactions that are frauds. In the training, the tree-based models had a better Recall, and for the validation process with data outside the sample, the Boosting model had better results. Since most of the variables were constructed from PCA, it is not possible to perform a variable interpretation by variable and to give a business interpretation or detect defined behavior patterns between transactions. However, it was possible to identify which were the most important variables within the models and verify which are key and appear in most models. For the particular case of the regressions it was possible to know what was the impact (positive or negative) of each of the variables on the target. Despite having obtained a satisfactory model that manages to identify fraudulent accounts, there is a proportion of accounts that none of the models managed to identify. So an improvement that is proposed for the model is to analyze all the frauds that none of our models could identify and make a deeper profiling analysis on those observations. In particular, decision tree techniques could be used in conjunction with some sampling technique to generate a balanced sample, another way would be to use association rules on these clients to try to find relationships between the variables. At the end, there would be two models that would form a segmentation of the clients and in this way the greatest number of fraud cases could be identified. In case no pattern was found using the variables with main components, it would be proposed to decompose the variables to see if any of the raw variables manages to capture the characteristics of these clients. One more scope that could be covered in subsequent studies is related to the cost of misclassifying individuals, for this the organization should provide the corresponding information to make the estimates of resources used.

References

1. Bansal, A., Mehta, K., Arora, S.: Face recognition using PCA and LDA algorithm. In: 2012 Second International Conference on Advanced Computing & Communication Technologies, pp. 251–254. IEEE (2012)
2. Plumbley, M.D., Oja, E.: A nonnegative PCA algorithm for independent component analysis. IEEE Trans. Neural Netw. **15**(1), 66–76 (2004)
3. Li, W., Yue, H.H., Valle-Cervantes, S., Qin, S.J.: Recursive PCA for adaptive process monitoring. J. Process Control **10**(5), 471–486 (2000)
4. Lepoivre, M.R., Avanzini, C.O., Bignon, G., Legendre, L., Piwele, A.K.: Credit card fraud detection with unsupervised algorithms. J. Adv. Inf. Technol. **7**(1) (2016)
5. Chen, M.Y.: Predicting corporate financial distress based on integration of decision tree classification and logistic regression. Expert Syst. Appl. **38**(9), 11261–11272 (2011)
6. Machine Learning Group - ULB. Credit Card Fraud Detection (2008). https://www.kaggle.com/mlg-ulb/creditcardfraud
7. LeDell, E., et al.: R Interface for 'H2O' de CRAN, 25 September 2018. https://Cran.r-Project.org/web/Packages/H2O/H2O.pdf
8. Parra, D.: Métricas de Evaluación, de Profesor Asistente, DCC, PUC CHile (2018). http://dparra.sitios.ing.uc.cl/classes/recsys-2017-2/clase5_metricas.pdf
9. SAS Institute. The UNIVARIATE Procedure (2018). http://support.sas.com/documentation/cdl/en/procstat/66703/HTML/default/viewer.htm#procstat_univariate_syntax01.htm

Deep Learning-Based Software Energy Consumption Profiling

Muhammed Maruf Öztürk(✉)

Suleyman Demirel University, Isparta, Turkey
muhammedozturk@sdu.edu.tr

Abstract. Energy efficient software development is a requirement to meet software quality standards. A great number of works have been done to enhance the level of information related to software energy consumption (SEC). They are generally focused on raw code data. These data can be profiled to predict SEC trends of future versions of a software. However, SEC works lack energy profiling with powerful predictive models. In this work, a deep learning-based SEC model is proposed. The model is than evaluated with 14 open-source projects. The experiment shows that deep learning performs better in SEC profiling than the alternatives such as random forest. Further, contrary to expectations, the success of the profiler is sensitive for the number of hidden layers of deep neural network.

Keywords: Software engineering · Energy consumption model · Green software · Deep learning

1 Introduction

In recent years, practitioners doing experiment on software development have become aware of the importance of green software [23]. In accordance with this awareness, they began to consider green software requirements in software development [3]. Inherently, advance in mobile application increased the number of works related to SEC [16–18, 25]. However, evolution of software systems have led to an expansion in the scope of SEC studies. For instance, embedded and object-oriented software systems are the examples of this expansion [14].

Software has a critical role in energy consumption in terms of hardware level [13]. To reduce energy consumption of hardware resources, component-oriented methods are developed. They can be focused on CPU, network, and data storage. However, the decline in SEC is not sufficient without changing the behaviour of a software. Thus, maximum decline has been detected three percent so far. In order to increase this rate, alternative methods are required.

A software consists of various versions of them. To obtain a better insight into the software quality, SEC should be included in the quality evaluation criteria.

Supported by Intel.

D. J. Hemanth and U. Kose (Eds.): ICAIAME 2019, LNDECT 43, pp. 73–83, 2020.
https://doi.org/10.1007/978-3-030-36178-5_7

Observing SEC in evolving software versions could give useful tips for software developers with respect to green software. To the best of our knowledge, the works related to SEC are generally focused on refactoring of software development models [22]. However there is a need for profiling SEC with reliable models to predict future trends in energy consumption rates.

Predictive models are constructed with common learners such as random forest, naive bayes, and support vector machine [2]. However, with the great computation ability of new generation computers, deep neural network (DNN) has become widespread among researchers developing predictive models [4,8, 19,24]. DNN are used in various fields such as geometric problems, big data, classification, and image processing. In particular, the competitive advantage of DNN has seen in forecasting accuracy [19].

In this work, a deep learning-based SEC profiling method is proposed. The method is then tested on 14 open-source object-oriented projects. According to the obtained results; (1) Proposed method shows competitive performance in terms of energy consumption profiling, (2) The number of hidden layers of DNN is not directly proportional to the accuracy of profiling, (3) Large-scale software systems are much more suitable for establishing DNN-based models.

The rest of the paper is organized as follows. Section 2 mentions related works. The method proposed in this work is in Sect. 3. The tools utilized to establish the experiment, preliminaries of the method, and data sets are also detailed in this section. Experimental findings are given in Sect. 4. Section 5 concludes the results.

2 Related Work

The main objective of this section is to give a historical summary about DNN along with selected works.

Primitive computer systems had not coped with multiple-layer structure in DNN due to the insufficient computational capability. However, in the end of 2009, learner based on DNN was introduced. One of the first designed DNN methods was for auido classification [15]. In subsequent years, some works pointed out that DNN would not be limited with academic research [1]. They have contributed greatly to understanding the underlying mechanism of DNN methods [9].

DNN studies are twofold. First, some of them aim at developing new techniques to improve DNN in terms of theoretical. They generally propose modification on DNN structure. Second, DNN is applied on a specific field such as speech recognition and image classification. In the beginning, deep belief networks (DBN) were widespread. DNN has a feedforward network structure along with directed connections in the layers. In contrast, DBN has undirected connection between layers and they are generally trained with unsupervised methods that accelerates the learning mechanism.

Lee et al. [15] used DBN to classify audio data. Their method achieved higy accuracy in TIMIT data set with a convolutional DBN. Model uncertainty problem is well known among researchers handling with Bayesian approximation.

Gal and Ghahramani investigated Bayesian and deep learning models together to cast dropout training in DNN [10]. They concluded that using DNN in Bayesian models helps reduce RMSE value in training. Chen et al. discussed DNN in terms of big data along with brief description of current issues [5]. Their work was cited by a great number of works including a big data experiment. DNN was also used in the classification of hyperspectral data. In this context, a DNN classifier namely SAE-LR [6] was proposed. Inspite of disadvantageous training time, SAE-LR has completed testing faster than other competitive methods such as SVM and KNN. Medical image processing is another application area of DNN [11]. In such works, having noise-free data is of great importance as much as developing novel methods producing high sensitivity and specificity [20].

With respect to the software engineering, deep learning found various subfields for itself such as software repository, code clone detection, fixing code errors, and vulnerability detection. White et al. [27] showed that deep learning algorithms enhance the quality of software language modeling. They also stated that using DNN to optimize hyperparameters of machine learning methods used for software engineering tasks could mitigate well-known search-based problems. DNN is useful for detecting similar patterns. For instance, a DNN method [26] was developed for code clone detection. It is capable of pairing file and method level components. Code errors can be fixed with the help of DNN. A DNN method namely DeepFix [12] was proposed by Gupta et al. to fix c errors. DeepFix was able to fix 27% of codes completely and 19% codes partially. It has an iterative repair design. Deep language models could help predict sequences of codes. Such a work was performed by Dam et al. [7] to predict long-term dependencies of software code. They achieved 4.7% performance improvement in training. One of the most profiling-focused studies was performed by Romansky et al. [21]. In their work, DNN was utilized to create time series of SEC. However, they did not consider the correlation between source codes.

3 Method

SEC is simply measured by $P = ExT$. However, this measurement brings some requirements. First, the machine in which SEC is recorded should be evaluated in terms of idle case. Considering such a case helps differentiate real consumption from all SEC including idle and individual rates. CPU cycle counts give tips for SEC of a software, and therefore, the tools developed for measuring SEC generally takes useful information from system records including CPU cycle. In the method, Intel Power Gadget version 3.0 has been used for recording SEC. Intel Power Gadget is able to record valuable information related to SEC such as CPU power consumption (watt), temperature of CPU, and frequency values. SEC of processor cores and graphics are called IA and GT energy, respectively. SEC can be represented with Joules and mWh in Intel Power Gadget. Figure 1 is an example of Intel Power Gadget logging screen. It was produced by executing two YouTube social media videos simultaneously for 254 s. In this figure, SEC details is given with processor energy consumption in terms of milliwatt.

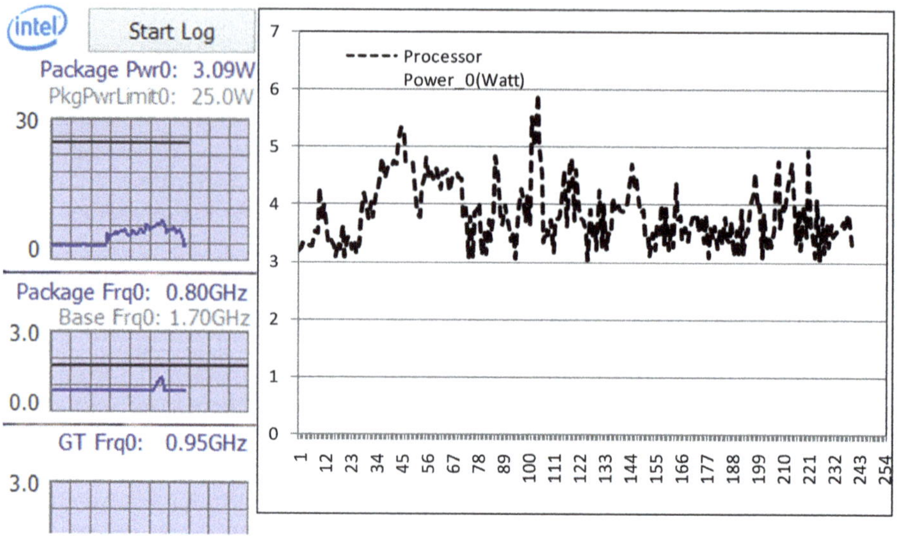

Fig. 1. Intel Power gadget logging screen.

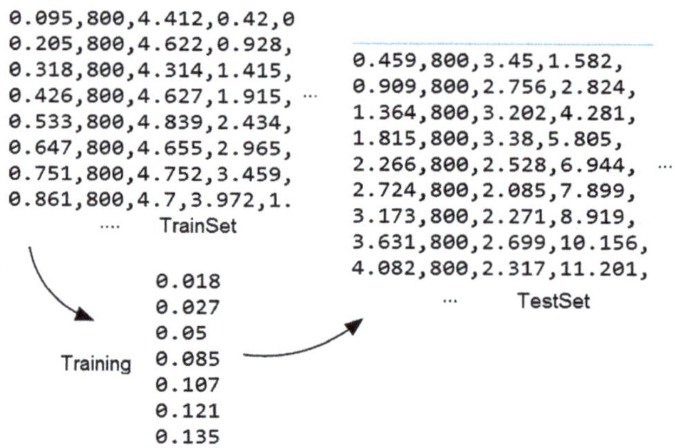

Fig. 2. Training and testing process of a data set produced by Intel Power Gadget. Training data is trained with recorded SEC rates. Subsequently, SEC values of testing data are predicted.

Deep learning codes have been designed with *deepnet* which is a DNN package of R. *deepnet* provides user to implement DNN architecture. 17 features of the data sets are employed for profiling. In the beginning, all the data is exposed to normalization. To predict SEC trends, 5 applications are used to create training data sets. Trained DNN is then tested on 1 application. In these processes, equal number of the features in training and testing should be used. Prediction accuracy is also investigated in terms of the number of hidden layers.

In train function, default settings of *dbn.dnn.train* function is used (learningrate = 0.8, activefunction = sigmoid, numepoch = 3, batchsize = 100, momentum = 0.5). The format of the data used for profiling is given in Fig. 2.

3.1 Deep Learning

SEC data recorded during the experiment is matrix format. If a matrix has n features, a vector x in matrix is as $x = 1...n$. In any matrix, $x_1, ..., x_m$ represents all the instances. This matrix is used as input layer in a DNN. Any hidden layer is denoted with h and $h_1, .., h_m$ includes all hidden layers. If we have t hidden layers, weighted hidden layers are denoted with $h^1, .., h^t$. For each hidden output, Eq. 1 is employed. In this equation, m represents the number of features and w represents related weight. A multi-layer neural network is presented in Fig. 3.

$$W.X = w_1.x_1 + w_2.x_2 + ... + w_m.x_m \tag{1}$$

In a multi-layer neural network, it is decided whether a neuron should be activated or not. This operation called activation function suppresses noise points to help network get useful information. Activation function selected in a DNN could be linear, sigmoid, or scaled version of sigmoid called tanh. In Fig. 4, the active function used in DNN model is shown. The formula of Fig. 4 is given in Eq. 2. In this function, the output becomes 1 if related sigmoid neuron produces 0.5 or higher output value. Otherwise, it outputs 0. If x is 0, any set up rule can not be established. x in Eq. 2 can be calculated via Eq. 3. Here p denotes input data and w is the related weight.

$$f(x) = 1/1 + e^{-x} \tag{2}$$

$$x = \sum_{i=1}^{m} w_i.p_i \tag{3}$$

3.2 Data Sets

To perform the experiment, 14 open source C# projects were retrieved from cyotek and GitHub that they include a great number of open-source projects for developers and researchers. Details of the data sets are given in Table 2. The metrics extracted during the experiment is seen in Table 1 Having a wide variety of projects may have extended the generality of the experiment.

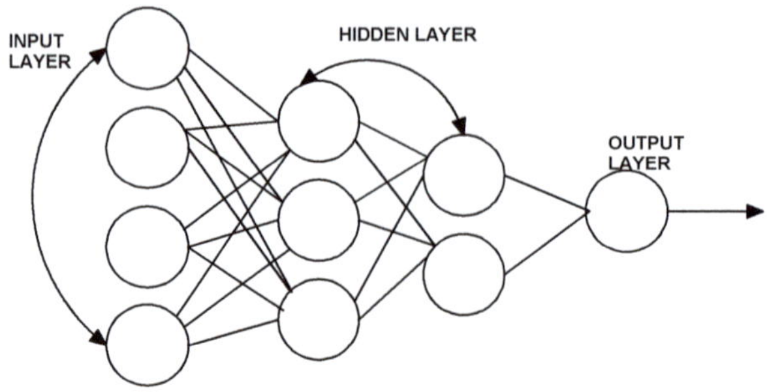

Fig. 3. A multi-layer neural network.

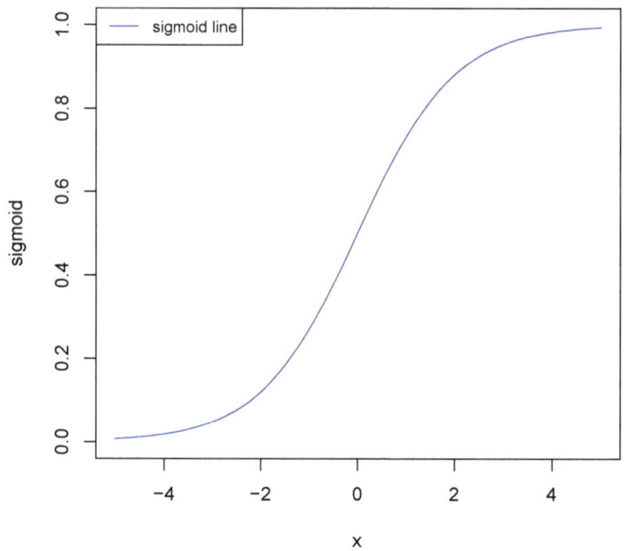

Fig. 4. Active function of hidden unit. In the experiment, sigmoid type is selected.

Table 1. Metric details.

Name	Description
LOC	Lines of Code
BLOC	Blank Lines of Code
SLOC-P	Physical Executable Lines of Code
SLOC-L	Logical Executable Lines of Code
MVG	McCabe VG Complexity
C&SLOC	Code and Comment Lines of Code
CLOC	Comment Only Lines of Code
CWORD	Commentary Words
HCLOC	Header Comment Lines of Code
HCWORD	Header Commentary Words

Table 2. Details of the projects used in the experiment.

Project	LOC	BLOC	SLOC-P	SLOC-L	MVG	C&SLOC	CLOC	CWORD	HCLOC	HCWORD
cmyk	1728	241	1246	886	25	1	241	612	0	8
awesome-dotnet-core	1131	216	859	615	37	3	56	281	0	0
colorDistance	888	148	627	420	20	0	113	249	0	0
corelDraw	2839	364	2231	973	98	1	244	833	0	11
defectFramework	185186	11788	109126	71171	13610	19	64272	274843	235	1430
drawAnimated	331	54	231	127	9	3	46	226	0	15
getText	660	126	463	341	13	0	71	256	0	0
imageBoxTiff	849	86	614	440	8	0	149	289	0	0
listBoxEmpty	450	80	303	167	6	4	67	275	0	24
langTonSim	3932	479	2924	1984	115	7	529	1096	0	1
quadEquation	956	100	626	435	20	4	230	884	10	40
riffPalette	772	133	514	323	22	2	125	470	0	8
sourceSafe	4154	640	3048	1932	216	5	466	1539	0	0
wavePlayResult	370	33	273	193	9	0	64	197	0	0

Table 3. Fomulas of performance parameters.

Name	Formula
True positive rate Sensitivity Recall	TP/(TP + FN)
False positive rate	FP/(FP + TN)
Precision	TP/(TP + FP)
True negative rate Specificity	TN/(TN + FP)
F-measure	$\frac{(2*Recall*Precision)}{(Recall+Precision)}$
Accuracy	$\frac{(TN+TP)}{(TN+FN+FP+TP)}$
G-mean	$\sqrt{sensitivity.specificity}$

4 Results

Performance measures used in the experiment are given in Table 3. Energy consumption rates of colorDist and getText are seen in Figs. 8 and 7, respectively. In a program like colorDist, SEC increases at a stable rate in the beginning of the application. For idle state, SEC dos not increase rapidly. Re-executing the application with some commands lead to an increase as in the beginning. getText is a program which enters an infinitive loop from the beginning of the application. Such an execution produces SEC graph as in Fig. 7. getText has a lower complexity than colorDistance. It can be concluded that SEC lines of software systems having high complexity become unstable. Figure 5 illustrates profiling performed with 200 random instances. The prediction completed with Random

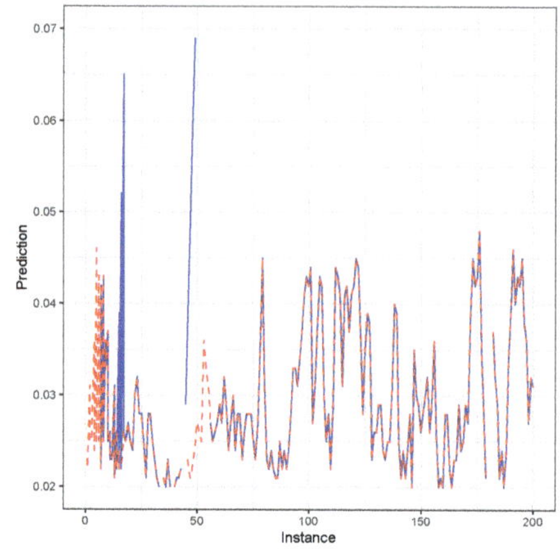

Fig. 5. SEC profiling with random forest predictor. Random instances retrieved from all projects. Real and predicted SEC are represented with red and blue lines, respectively.

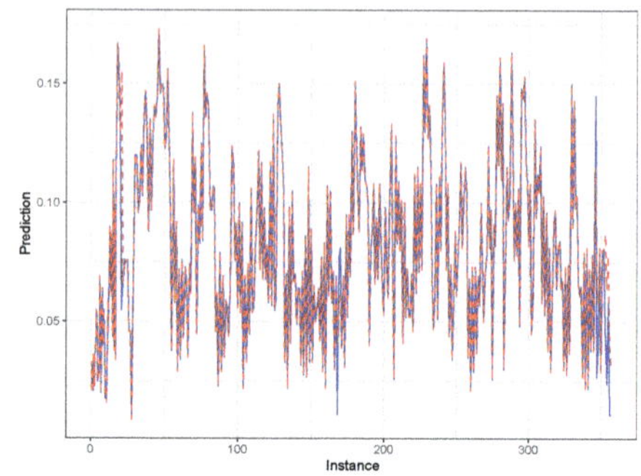

Fig. 6. SEC profiling with DNN. Random instances retrieved from all projects. Real and predicted SEC are represented with red and blue lines, respectively.

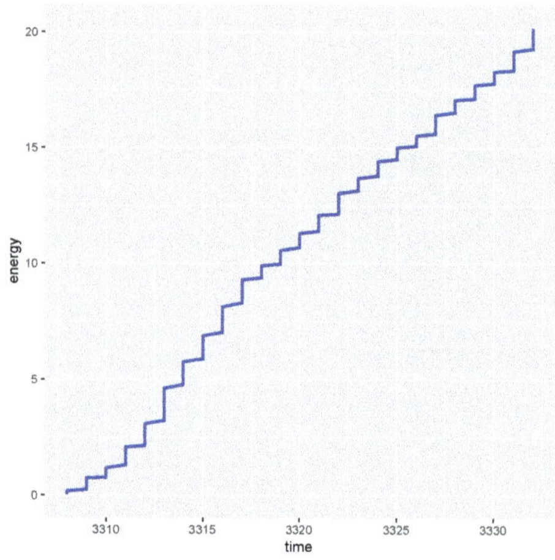

Fig. 7. Cumulative energy consumption rate (joule) of getText. Y axis represents related seconds.

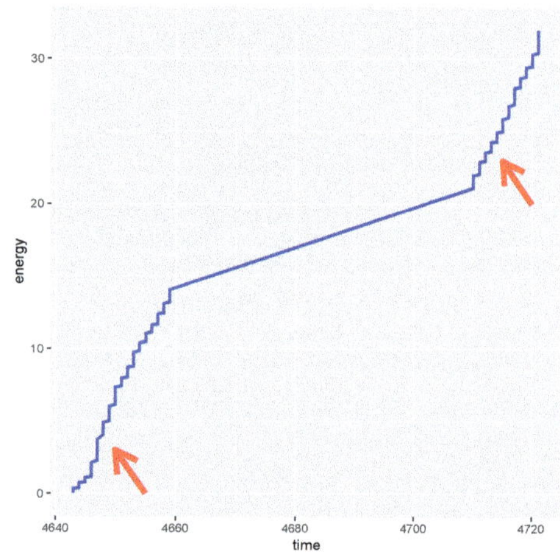

Fig. 8. Cumulative energy consumption rate (joule) of colorDistance. Y axis represents related seconds.

Forest does not perform well especially for the instances 40–60. On the other hand, DNN has an higher accuracy than Random Forest as seen from Fig. 6.

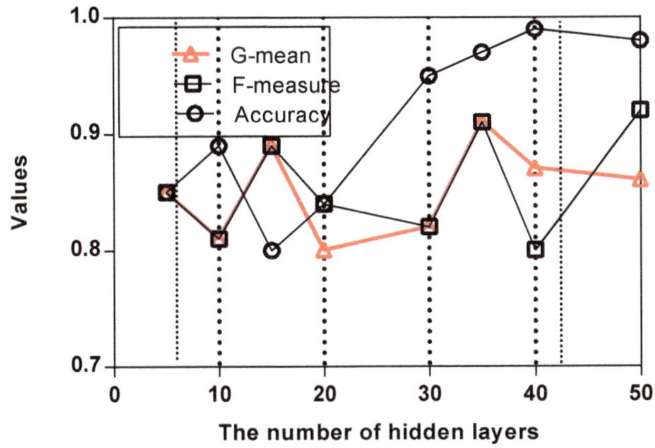

Fig. 9. Profiling performance with respect to three performance parameters.

5 Conclusion

This paper proposes a SEC profiling method based on DNN. The method is then tested on 14 open-source projects along with Random Forest. According to the obtained results, DNN gives higher accuracy than Random Forest. Further, the number of the hidden layers of DNN is crucial for creating reliable models. The accuracy of the profiling is not directly proportional with the number of hidden layers (Fig. 9) that can be determined depending on the type and the scale of the data sets. In the experiment, default configuration of *deepnet* package was used. For instance, sigmoid was used for activation function. Tuning methods may be investigated to go further to increase the success of the profiling. Last, the method presented in this paper should be examined in terms of mobile applications.

References

1. Arel, I., Rose, D.C., Karnowski, T.P., et al.: Deep machine learning-a new frontier in artificial intelligence research. IEEE Comput. Intell. Mag. **5**(4), 13–18 (2010)
2. Beghoura, M.A., Boubetra, A., Boukerram, A.: Green software requirements and measurement: random decision forests-based software energy consumption profiling. Requirements Eng. **22**(1), 27–40 (2017)
3. Chauhan, N.S., Saxena, A.: A green software development life cycle for cloud computing. IT Prof. **15**(1), 28–34 (2013)
4. Chen, K., Chen, K., Wang, Q., He, Z., Hu, J., He, J.: Short term load forecasting with deep residual networks. IEEE Trans. Smart Grid (2018)
5. Chen, X.W., Lin, X.: Big data deep learning: challenges and perspectives. IEEE Access **2**, 514–525 (2014)
6. Chen, Y., Lin, Z., Zhao, X., Wang, G., Gu, Y.: Deep learning-based classification of hyperspectral data. IEEE J. Sel. Top. Appl. Earth Obs. Remote Sens. **7**(6), 2094–2107 (2014)

7. Dam, H.K., Tran, T., Pham, T.: A deep language model for software code. arXiv preprint arXiv:1608.02715 (2016)
8. Ding, X., Zhang, Y., Liu, T., Duan, J.: Deep learning for event-driven stock prediction. In: IJCAI, pp. 2327–2333 (2015)
9. Erhan, D., Bengio, Y., Courville, A., Manzagol, P.A., Vincent, P., Bengio, S.: Why does unsupervised pre-training help deep learning? J. Mach. Learn. Res. **11**, 625–660 (2010)
10. Gal, Y., Ghahramani, Z.: Dropout as a Bayesian approximation: representing model uncertainty in deep learning. In: International Conference on Machine Learning, pp. 1050–1059 (2016)
11. Gulshan, V., Peng, L., Coram, M., Stumpe, M.C., Wu, D., Narayanaswamy, A., Venugopalan, S., Widner, K., Madams, T., Cuadros, J., et al.: Development and validation of a deep learning algorithm for detection of diabetic retinopathy in retinal fundus photographs. JAMA **316**(22), 2402–2410 (2016)
12. Gupta, R., Pal, S., Kanade, A., Shevade, S.: DeepFix: fixing common C language errors by deep learning. In: AAAI, pp. 1345–1351 (2017)
13. Hao, S., Li, D., Halfond, W.G., Govindan, R.: Estimating mobile application energy consumption using program analysis. In: Proceedings of the 2013 International Conference on Software Engineering, pp. 92–101. IEEE Press (2013)
14. Kim, D., Hong, J.E., Yoon, I., Lee, S.H.: Code refactoring techniques for reducing energy consumption in embedded computing environment. Cluster Comput. 1–17 (2016)
15. Lee, H., Pham, P., Largman, Y., Ng, A.Y.: Unsupervised feature learning for audio classification using convolutional deep belief networks. In: Advances in Neural Information Processing Systems, pp. 1096–1104 (2009)
16. Li, D., Halfond, W.G.: An investigation into energy-saving programming practices for android smartphone app development. In: Proceedings of the 3rd International Workshop on Green and Sustainable Software, pp. 46–53. ACM (2014)
17. Li, D., Hao, S., Gui, J., Halfond, W.G.: An empirical study of the energy consumption of android applications. In: 2014 IEEE International Conference on Software Maintenance and Evolution (ICSME), pp. 121–130. IEEE (2014)
18. Li, D., Lyu, Y., Gui, J., Halfond, W.G.: Automated energy optimization of http requests for mobile applications. In: Proceedings of the 38th International Conference on Software Engineering, pp. 249–260. ACM (2016)
19. Lv, Y., Duan, Y., Kang, W., Li, Z., Wang, F.Y., et al.: Traffic flow prediction with big data: a deep learning approach. IEEE Trans. Intell. Transp. Syst. **16**(2), 865–873 (2015)
20. Ng, H.W., Nguyen, V.D., Vonikakis, V., Winkler, S.: Deep learning for emotion recognition on small datasets using transfer learning. In: Proceedings of the 2015 ACM on International Conference on Multimodal Interaction, pp. 443–449. ACM (2015)
21. Romansky, S., Borle, N.C., Chowdhury, S., Hindle, A., Greiner, R.: Deep green: modelling time-series of software energy consumption. In: 2017 IEEE International Conference on Software Maintenance and Evolution (ICSME), pp. 273–283. IEEE (2017)
22. Sahin, C., Pollock, L., Clause, J.: How do code refactorings affect energy usage? In: Proceedings of the 8th ACM/IEEE International Symposium on Empirical Software Engineering and Measurement, p. 36. ACM (2014)
23. Shenoy, S.S., Eeratta, R.: Green software development model: an approach towards sustainable software development. In: 2011 Annual IEEE India Conference (INDICON), pp. 1–6. IEEE (2011)

24. Spencer, M., Eickholt, J., Cheng, J.: A deep learning network approach to ab initio protein secondary structure prediction. IEEE/ACM Trans. Comput. Biol. Bioinform. (TCBB) **12**(1), 103–112 (2015)
25. Wan, M., Jin, Y., Li, D., Gui, J., Mahajan, S., Halfond, W.G.: Detecting display energy hotspots in android apps. Softw. Testing Verification Reliab. **27**(6), e1635 (2017)
26. White, M., Tufano, M., Vendome, C., Poshyvanyk, D.: Deep learning code fragments for code clone detection. In: Proceedings of the 31st IEEE/ACM International Conference on Automated Software Engineering, pp. 87–98. ACM (2016)
27. White, M., Vendome, C., Linares-Vásquez, M., Poshyvanyk, D.: Toward deep learning software repositories. In: Proceedings of the 12th Working Conference on Mining Software Repositories, pp. 334–345. IEEE Press (2015)

Implementation of GIS for the Sanitation System in El-Oued City

Brarhim Lejdel(✉)

University of El-Oued, El Oued, Algeria
brahim-lejdel@univ-eloued.dz

Abstract. The last years, the management of the sanitation system have many problems. Thus, engineers, technicians, and managers have to resolve the daily problem for improving the control and management of the sanitation system. Actually, The use of computers has become essential for most studies for the development and making-decision in sanitation system. The big quantity of data and their diversities in sanitation systems and the multiple interventions on this network have accelerated the using of geographical information systems (GIS). The GIS has a great advantage to implement an efficient system which permits us to improve the management and the control of Sanitation network. In this paper, we will propose a conceptual model for sanitation system. These systems based on a geographical database to respond to the different questions of users, in terms of spatial and descriptive queries. Firstly, we will create a Geographic Information System (GIS) and implement an application that permits to manage and control the sanitation network in the EL-Oued city. This application composed of different user interfaces. We use GIS ArcGIS and its language VBA (Visual Basic for Application) to implement this system.

Keywords: Sanitation system · Geographical Information System · Geographic Database

1 Introduction

The management of sanitation systems is a complex process. Thus, it requires a strong information base which well enriched by geographic and descriptive data and the best coordination between different actors who manage the sanitation network. In developing countries, the management of this system is becoming more difficult because of the lack of financial tools and technical skills. Also, the classical management of the network of the sanitation system (classical topographic plan, manual management ...etc.) is far from citizens' needs. In the last decade, the system that is related to cartographic domain is revolutionized by many software tools, such as GIS (Geographic Information System) and maps language. Several concepts were used to create the different parts of the GIS, like themes, layers, vector, raster...etc.

D. J. Hemanth and U. Kose (Eds.): ICAIAME 2019, LNDECT 43, pp. 84–95, 2020.
https://doi.org/10.1007/978-3-030-36178-5_8

In my country, the geographic information is increasingly used in different areas of the network and especially in the sanitation system. A geographic information system is a fundamental tool of decision-making that can help a decision-maker to make the best decisions. In the past, geographic information is represented in maps that are often old and static. Today, the geographic information is represented in the screen which is updated and dynamic which permits to help decision-makers.

As all public infrastructures, the Sanitation system is an important investment in the country, it requires sustainable management to ensure the continuity of service to users. For this reason, our country (Algeria) creates a National Office of Sanitation (NOS) to manage and control this system, where the sanitation network should be put perfectly under its responsibility. In order to control and manage the Sanitation network, engineers should have many tools (hardware and software) which they help them in their works such as collecting the current data on sanitation network, making necessary interventions of maintenance, seepage detection, and risk treatment, cleaning of different types of equipment (piping, manholes ... etc.). Today, all engineers want to use plans for sanitation network that have to be numeric, accurate and update, which permit to help them in their works.

The main objective of this work is the design and the implementation of GIS dedicated to the management and the control of sanitation network especially for creating a new network, risk treatment and the maintenance of the existing ones.... as the work of [1–7]. Therefore, we should have.

- A central Geographic DataBase: we have to create a geographic database that has enriched by a big spatial data (descriptive and spatial data).
- A system of decision-making: in the form of user interface that can summarize the different activities which can carry out on the sanitation network in the form of reports, graphs, illustrations that can help decision-makers to carry out or to build program of intervention(installation of new network, extended the existing ones, and others....) suitable to the study area.

This paper will be organized as follows. Firstly, we present the Geographical Information System (GIS) (Sect. 2). Then, we describe the conceptual model of our sanitation system in Sect. 3, where the description of the development environment will be presented. Also, we present some experimentation and results (Sect. 4). Finally, we present a conclusion and some future works (Sect. 5).

2 Geographic Information System

2.1 Definition

A Geographic Information System (GIS) is a computer system for capturing, storing, checking, and displaying data related to positions on Earth's surface [8]. GIS can show many different kinds of data on one map. This enables people to more easily see, analyze, and understand patterns and relationships between the different geographic data.

With GIS technology, people can compare the locations of different objects in order to discover how they relate to each other. Thus, A map would help people determine which buildings are most at risk. GIS can use any information that includes location. The location can be expressed in many different ways, such as latitude and longitude, address, or ZIP code. Many different types of information can be superposed and compared using GIS. The system can also include data about people, such as population, income or education level ... [8].

2.2 GIS and Data

Different kinds of data can be stored in the GIS. We have principally two kinds of data; spatial and descriptive data. An example of this kind of information is data collected by satellites that show land use—the location of farms, towns, or forests. The GIS can also include data in table form, such as population information. The GIS technology allows all these different types of data no matter their source or original format to be superposed in a single map.

Putting information into GIS is called data capture. Data that are already in digital forms, such as images gathered by satellites or tables can be stored in the GIS for storing an image in GIS, it should be scanned, or converted into digital form.

The GIS can also store different forms of data which are gathered from various maps and various sources as the topographic or cadastral plan. These different maps have different scales. Scale refers to the relationship (or ratio) between distance on a map and the corresponding distance on the ground [9]. The GIS combines information from different sources in such a way that have the same scale.

2.3 GIS and Maps

Once all of the needed data have been stored in a GIS system, they can be combined to produce a wide variety of layers, depending on which data layers are included. For instance, using GIS technologies, many kinds of data be shown about the same city. Maps can be produced about such information as the average income of citizen, book sales, and voting system ..., etc. Any GIS data layer can be added or omitted into the same map.

The maps in the GIS can be used to show data about the number and density of information. For example, GIS can be used to show how many doctors there are in different areas compared with the population. They can also show what is near what, such as which homes and companies are prone to flooding in urban areas.

With GIS technologies, researchers can also Follow changes in the ground over time. They can use satellite data to study different topics for example how many regions are covered by the sanitation network. A police department can also use GIS technologies to study the ratio of crime over time, to help the decision-maker to determine for example the place of installing a new police office.

In GIS, the screen often contains a large variety of data that they cannot appear same printed map. The GIS technologies sometimes allow users to access information at any time and place. A user can define a location in the digital map to find other information stored in the GIS about this location. For example, a user might click on a school to find how many students there are, how many students per teacher, or what sports exercise in this school.

Using GIS technologies, we can make updates in maps most easy then a traditional map. A new map can be printed or displayed on a screen. All defects of the traditional process during the drawing of the map, which can be the very time-consuming and expensive cost of production will disappear.

GIS technology can be used in different domains. In the trade, users can use a GIS to locate their stores. Biologists can use GIS to follow animal migration patterns. The urban policies can use the GIS to make plans that help them, in the case of a natural disaster such as an earthquake or storm, for example, the maps can show; what are the buildings that are in danger? The authorities can use the GIS to compare population growth in such town or to determine what are the regions that need in the future public services like parking, roads, and electricity. Thus, there are many kinds of data that can be analyzed and treated using GIS technologies.

2.4 GIS and Queries

The scope of spatial analysis varies from a simple query about the spatial phenomenon to complicated combinations of attribute queries and spatial queries.

- Attributes queries: An attribute query is the process of searching and retrieving records of features in a database based on desired attribute values. Such queries are a fundamental part of managing and analyzing GIS data. Typically, this is performed using a criteria-based query language [10].
- Spatial queries: A spatial query is a query that returns features based on their spatial relationship with a query geometry. This includes searching for features given an extent (finding all streets within an envelope) or searching for features based on their relation to other features (finding all farmland intersected by a particular stream) [11].
- Generalization in GIS: Spatial analysis can require the generation of new data from the original data set. For example, assume that you wish to find a relationship between noise level and proximity of motorway within one mile, for all urban parcels. To answer the question, we should combine and superpose the parcel layer and the motorway layer. Then, we compute the urban areas within one mile of the motorway.

3 Modeling the System

3.1 The Applied Rules

During our work, we can apply the following rules of management which should be used in all section of the network:

- The study area contains buildings, roads, and motorways.
- Each building is connected to the sanitation network via connection box and collector.
- Each Manhole may connect four sanitation canals.
- Each canal has a manhole in the extremity.
- The sanitation canal may have one or more seepage points.
- The pumping station has coordinates (X, Y).

3.2 Conceptual Model

Best implementation requires making a good design. To do the best design, we have to choose a suitable modeling language for our project. For this reason, we chose the UML.

Use-Case Diagram. To develop the Use-case diagram, we have to identify the different actors of the system. There are two main actors: the administrator and user. This diagram composed of: (See Fig. 1).

- Authentication: allow to login.
- Consult-data: allow the user to make queries on the geographic database (spatial and descriptive).
- Action Plan: allow the administrator to calculate the number and the length of objects in the selected zone.
- Cleaning: permit to return the number and the date of the cleaning operations, if they are applicable on the sanitation canal or manhole.
- Connection: allow to return the status of building a connection (it return, connected or no).
- Maintenance: Notify the Administrator, what are the canals that need maintenance? The maintenance operation can be applied every six months.
- Edit reports: allow the administrator to print the final report that contains statistics, graph..., etc.
- Risk treatment: permit to calculate the number of buildings near the seep. It based on the seepage diameter.

UML Class Diagram. The class diagram is considered as the most important diagram that is used in the object-oriented approach. It shows the internal structure of the system, where it can provide an abstract representation of system objects that will interact together to realize the use-cases. We can define the following classes: (See Fig. 2).

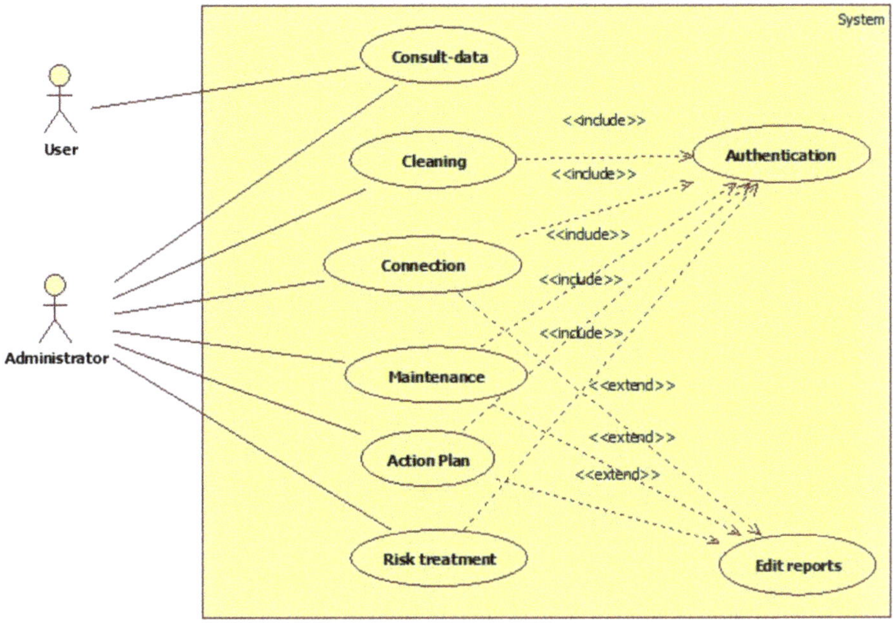

Fig. 1. User case diagram.

Passage to Relational Model the creation of a relational database consists to translate the class diagram and associations into relations, using the rules of passage from Oriented-Object model to the Relational model.

Firstly, we made a long discussion with the engineers of the NOS, to know the different problem which carries out during the managing and the control process of the sanitation network. Then, we organize multiple visits to the sanitation networks in a reality for collecting the diverse data about this network and its components. Thus, our work is the result of these discussions and many analyses carry out on sanitation network. Therefore, in this section, we detailed the rules of management and control process, use-case diagram, and class diagram. Thus, we can obtain the following relations:

#Pumping-station (Id-station, Name-station, Flow-in, Flow-out, Motor-power).
#Sanitation Canal (Section, Diameter, Slope, Type, direction, side project).
#Point (ID_point, terrain, *fe*, Profondeur).
#Manhole (Id-Manh, Name,$fe(n), fe(o), fe(s), fe(e)$, depth, Fab-company, type, Inst-date).
#Client (customer number, name, date of application, address, phone number, craft, street).
#Roads (Id_road, Name_road).
#Connection box (Id-box, Name-Box).
#Building (Id-Build, Name-Build, Surface).

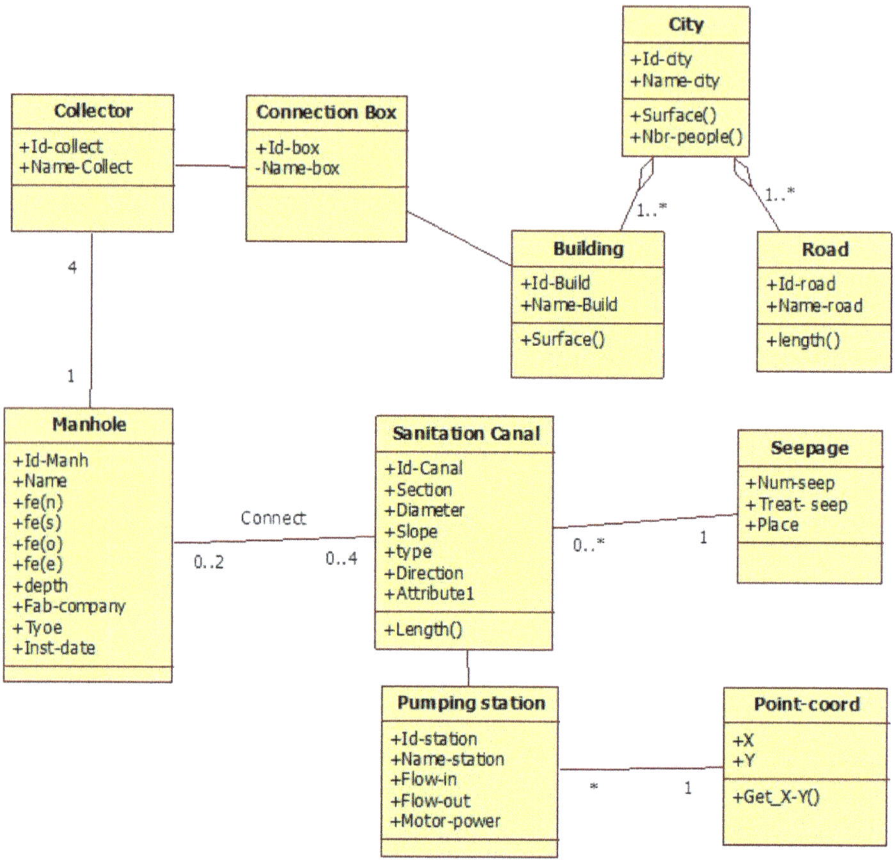

Fig. 2. UML class diagram

#Collector (Id-collect, Name_collect).
#Seepage (Num-seep, Treat-seep, Place).
#City (Id-city, Name-city, People).
#Point-Coord (X, Y).

4 Experimentation and Results

In this section, we will present the principal functions of our application and we introduce some results of applications of certain queries on Geographic DataBase of study zone, which is El-Oued city. As the treatment of large areas in GIS consumes much time perhaps equal to many years. Thus, we have to choose a small site to realize our project. The study area is Kouinine region. It is a town near El-Oued city where the surface is $2,817,835\,\text{m}^2$. The geographical database includes maps of Kouinine Town. We use this GeoDatabase to create the different thematic layers of our application. These layers are sanitation canals, manholes, buildings, collectors, box connectors, and roads ..., etc. Then, we enriched this DataBase with descriptive data.

Our Objective is to implement a user interface under ArcGIS 9.3 using VBA language (Visual Basic for Application).

4.1 ArcGIS 9.3

ArcGIS (ArcMap) is a real tool for decision support that can represent any information on maps with a spatial data [12]. It is used both in the private sector and the public sector to organize geographical territories, optimize the flow in networks, managing the patrimony, managing infrastructure, and prevent the natural risks ...etc. In this work, we used version 9.3. Figure 3 shows the general interface of this system.

4.2 Visual Basic Application (VBA)

VBA (Visual Basic for Application) is a language inspired from Visual Basic that can provide a development environment includes a text editor to write the program, a compiler to create an executable application, and a link editor to create a complex application by breaking it down into different "objects" modules [13]. Figure 4 shows the general interface of used VBA.

4.3 Interfaces of Application

Our application interface composed of many functions which automate the different manual tasks which have been realized by engineers. Before we can use the application, we have to login into the application via the authentication interface (See Fig. 5):

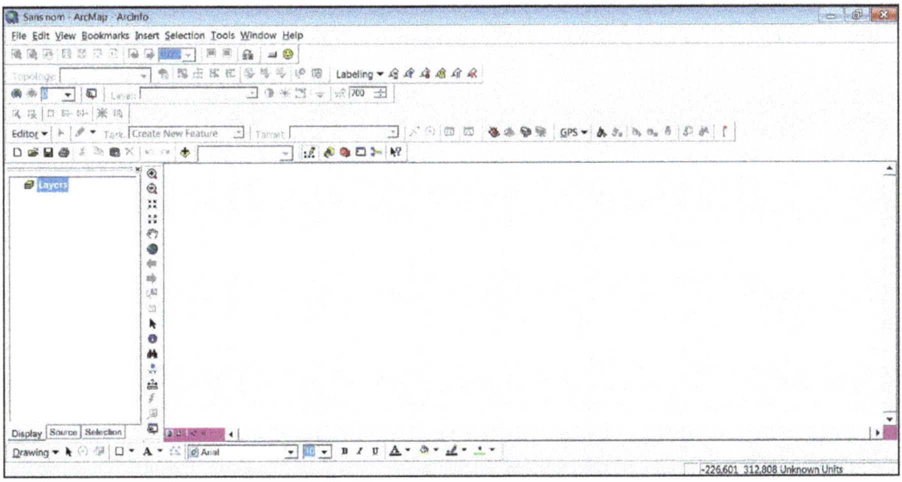

Fig. 3. Interface of ArcGis 9.3.

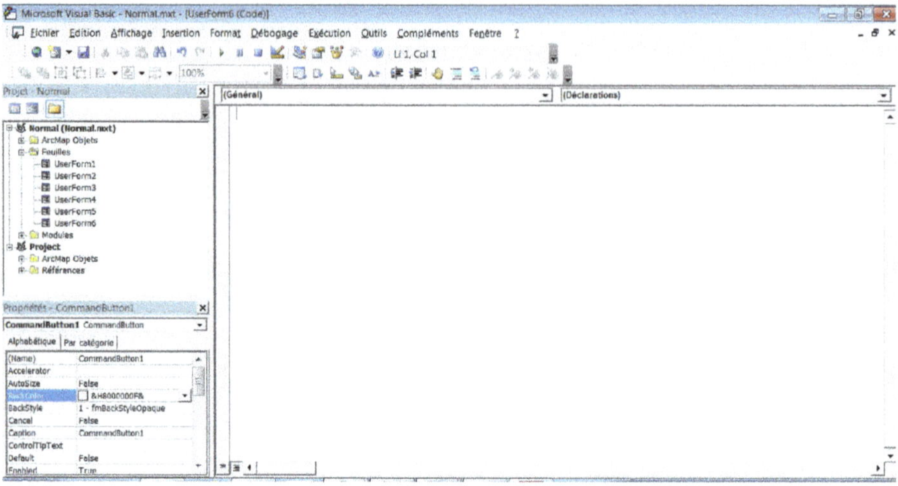

Fig. 4. Interface of VBA.

Fig. 5. The authentication.

4.4 Main Interface of Application

NOS engineers need to execute many functions on sanitation networks to manage and control them. In this work, we implement all the functions requested by engineers. We can group these functions in Main Interface. See Fig. 6.

4.5 Plan of Action

This operation is very important for engineers of NOS. It permits to calculate the number and lengths of objects in a selected area to prepare the operation of maintenance or cleaning. According to the engineers, they manually finish this operation in fifteen (15) days, when they work on the average areas. Now, when they use our application, they need less than a minute to execute the operation and obtain the results. See Fig. 7.

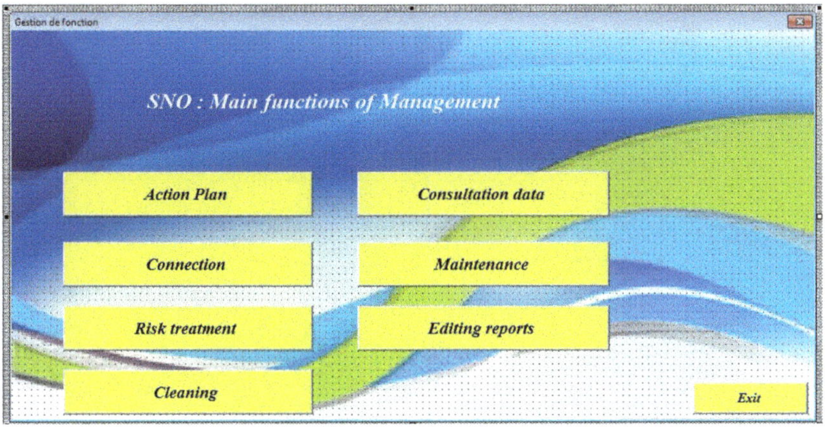

Fig. 6. Main interface of application.

Fig. 7. The results of plan of action.

4.6 Treatment of Risk

This function calculates the number of buildings that can be touched by seepage. The user puts only the diameter of seep into our application. This operation is very important because it helps, for example, healthy agents to intervene quickly in the touched urban areas for protecting the people of against diseases by vaccination . . . , etc. See Fig. 8.

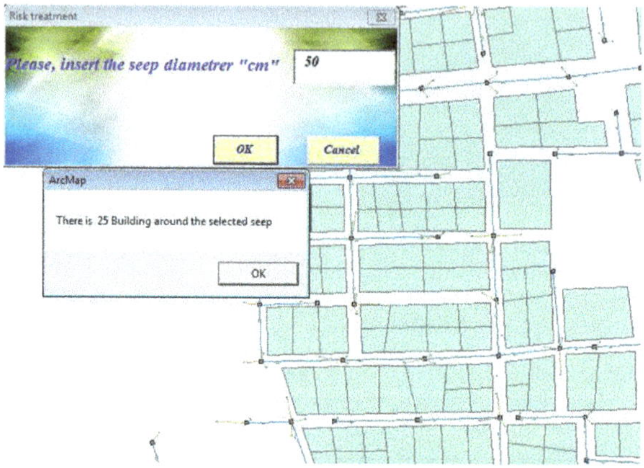

Fig. 8. Operation of treatment risk.

5 Conclusion and Future Works

This work is registered in a policy of better management and control of the sanitation network, using computer tools to help decision-maker to take the best decision. For this, we use the Geographic Information System (GIS). Thus, the developed application can manage and control the sanitation network in flexibility, conviviality by a variety of functionality which responds to different user needs.

In this paper, we implement GIS for sanitation network. First, we create a Geographic Database from a topographic plan of the Kouinine Town and we develop GIS application using ArcGIS with Visual Basic for Application (VBA). The main objective of this application consists to permit the NOS engineers to facilitate the managing and controlling of the sanitation network. However, during the realization of the project, we faced the following challenges.

- Familiarize with the Geographic Information Systems in theory and practice.
- Enrich our knowledge on the ArcGIS and its programming language VBA (Visual Basic for Application).
- Familiarize with the terms used by the NOS to understand their notions and to give them the necessary functions to manage and control the sanitation system.
- Development of many functions of the application,
- Take into account, the ergonomics and the attractiveness in the different user interfaces.

As future work, we will develop the other functions to construct a complete GIS which contains all the necessary functionality. We will also want to develop a GIS of the entire El-OUED city.

References

1. Geographic Information Systems as an Integrating Technology: Context, Concepts, and Definitions. https://www.esri.com/fr-fr/home. Accessed 4 Apr 2019
2. Kholladi, M.K.: Représentation, Modélisation et Manipulation des connaissances spatiales en géomatique. Institut National des Sciences Appliquées de Lyon, France (1991)
3. Kholladi, M.K.: Geographical information system for urban spatial analysis. In: ACIT 2004: The International Arab Conference on Information Technology, pp. 188–193. Mentouri University of Constantine, Algeria (2004)
4. Kholladi, M.K.: GIS for the management and follow-up of the landslides of the Constantine City. In: ACIT 2005: The International Arab Conference on Information Technology, pp. 364–373. Al Isra Private University, Amman (2005)
5. Kholladi, M.K.: GIS for the desertification detection and follow-up in the Biskra region. In: CSIT 2006: The 4th International Multiconference on Computer Science and Information Technology, pp. 401–411. Applied Science Private University, Amman (2006)
6. Laurini, R., Thompson, D.: Fundamentals of Spatial Information Systems. Series Academic Press, London (1992)
7. Laurini, R., Milleret-Raffort, F.: Les bases de données en géomatique. Editions Hermès, Paris, France (1993)
8. GIS (geographic information system). https://www.nationalgeographic.org/encyclopedia/geographic-information-system-gis/. Accessed 4 Apr 2019
9. What is Map Scale? https://www.ga.gov.au/scientific-topics/national-location-information/topographic-maps-data/basics/what-is-map-scale. Accessed 4 Apr 2019
10. Muehrcke, S.: Map Use, Reading, Analysis and Interpretation, 2nd edn. J.P. Publication, USA (1986)
11. How to execute spatial queries. http://resources.esri.com/help/9.3/arcgisengine/dotnet/7b4b8987-a3f0-4954-980f-720e61965449.htm. Accessed 4 Apr 2019
12. ArcGIS Desktop Help 9.3. http://webhelp.esri.com/arcgisdesktop/9.3/index.cfm?TopicName=welcome. Accessed 4 Apr 2019
13. Introduction á la programmation en VBA sur ArcGIS, Ecole national des sciences geographique (2003)

Prediction of Potential Bank Customers: Application on Data Mining

Muhammet Sinan Başarslan[1(✉)] and İrem Düzdar Argun[2(✉)]

[1] Doğuş University, İstanbul, Turkey
mbasarslan@dogus.edu.tr
[2] Duzce University, Düzce, Turkey
iremduzdar@duzce.edu.tr

Abstract. Banking is an important industry, where financial transactions are performed to meet our needs in our everyday lives. Today, banks are frequently used to meet all kinds of financial transactions. In line with the increasing competition, the banks are aiming at acquiring new customers through customer satisfaction. At this point, studies on acquiring new customers by analyzing the customer data have gained importance recently. As a result, data analysis units have been established in the banks. In addition to the banks, these units have also been established for data analysis in customer focused industries such as insurance and telecommunication. In this study, models are established by using classification algorithms to estimate potential bank customers on the bank dataset obtained by telemarketing method in UCI Machine Learning Repository, and the results are compared. Using this comparison result, it is aimed to perform a more detailed and effective data analysis. Various models have been established with various classification algorithms for the estimation of customer acquisition. The classification algorithms used in this study include the C4.5 Decision Tree, Navie Bayes (NB) algorithm, K nearest neighbors algorithm (k-nn), Logistic Regression algorithm (LogReg), Random Forest algorithm (RanFor), and Adaptive Boosting algorithm (AdaBoostM1-Ada). While establishing the classification models, it is aimed to achieve consistency in the performance of the classification models by dividing the test and training data set by two different methods. K-fold Cross Validation and Holdout methods are used for this purpose. In the K-fold cross validation, training and test da-ta sets are separated with 5- and 10-fold cross validation. In the holdout method, the dataset was divided into training and test datasets with the 60–40%, 75–25% and 80–20% training and test separation ratios, respectively. These separations are evaluated for Accuracy (ACC), Precision (PPV), Sensitivity (TPR), and F-measure (F) performance. The performance results are similar in both separation results. According to the Accuracy and F-measure criteria, the classification model established by Random Forest algorithm highest results the other models, whereas the Naive Bayes algorithm gave highest results according to the precision criterion, and the AdaBoostM1 classification algorithm yielded better according to the sensitivity criterion.

Keywords: Data mining · Banking · Customer acquisition · Customer forecasting

D. J. Hemanth and U. Kose (Eds.): ICAIAME 2019, LNDECT 43, pp. 96–106, 2020.
https://doi.org/10.1007/978-3-030-36178-5_9

1 Introduction

Customer acquisition is crucial for businesses. In addition to new customer acquisition, they also carry out a variety of activities on customer retention. These studies are referred to as customer churn analysis. The purpose of this study is to conduct a study on new customer acquisition. Marketing is one of the common uses of data mining. The dataset used in this study has been collected by tele-marketing method [1]. Basically, data mining is used for following marketing activities [2]:

- Determining the customer profile
- Forecasting potential customers
- Basket analysis
- CRM (Customer Relationship Management)
- Sales forecast

Many classifying applications have been carried out on the bank data set. Bach and colleagues used classifier algorithms using artificial neural networks [3]. Sumathi and Sivanandam used a variety of data mining methods to explore the relationships between data [4]. Keramati et al. One of the classification algorithms on the data of a telecommunications company in Iran; they used decision trees, artificial neural networks, nearest neighbors and support vector machines. [5].The result of Nachev's artificial neural networks in the accuracy performance criterion is 90% [6]. Kumari and colleagues studied with decision trees and found that the performance value in the classifier model was 88.67% [7]. Elsalamony Artificial Neural Networks, C4.5 Decision Tree, Navie Bayes [8]. These workshops focus on a single performance criterion.

Classification algorithms are preferred in studies conducted to estimate potential customers. In this study, a model was established with classification algorithms for estimation of customer acquisition. Data mining is used for obtaining usable data from large data sets. A certain standard procedure should be followed against the problems that may arise because of the data size. For this purpose, processes such as KDD (Knowledge Discovery in Databases), SEMMA (Sample, Explore, Modify, Model, Assess) and CRISP-DM (Cross-Industry Standard Process for Data Mining) are used [9].

The SEMMA process has been developed by a proprietary data analysis vendor [10]. CRIPS-DM, however, has been introduced by a consortium in contrast to SEMMA methodology, and is preferred to KDD and SEMMA methodologies [11]. Cross-Industry Standard Process (CRISP) model, which is widely used in data mining studies, was used. The CRISP model consists of six steps:

- Problem Definition
- Data Understanding
- Data Preparation
- Model Building
- Model Evaluation and Selection
- Application of the Model

These six steps start with the problem definition step. This step must be well-defined in order for the analysis to be successful. In the next step, data understanding step, the relationship between descriptor values and variables such as summary information, averages, frequency, etc. of variables on the data set is examined graphically and using various methods. In the data preparation step, data such as missing data, outlier data, and normalization are performed on the data set. After these steps the classification models are carried out in the model building step. The generated model is evaluated in the model performance evaluation and selection step. Figure 1 shows the visuals of these six stages [12].

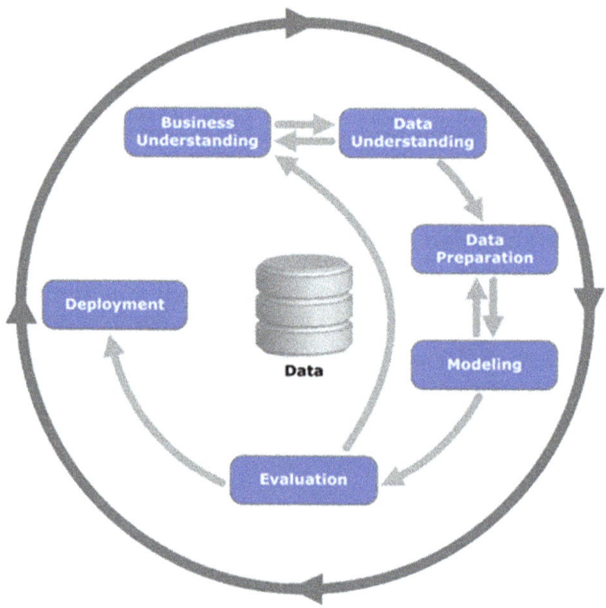

Fig. 1. CRISP Model.

Section 2 data mining and classification algorithms used in the study and performance evaluation criteria of these classification algorithms are given. In Sect. 3, the application was carried out using the CRISP-DM process. Finally, the performances of the classification models set forth in the study are included in the results and discussion headings in Sect. 4.

2 Methodology

In this section, data mining, classification algorithms and performance evaluation criteria of these algorithms will be given.

2.1 Data Mining

As the most general definition, data mining is obtaining usable information from complex data [13]. Data mining is the process of extracting useful information from large amounts of data. The knowledge discovery on the data by combining areas such as statistics, machine learning, database management systems, together is called data mining [14–16].

2.2 Used Classification Algorithms

The classification algorithms used in this study are Adaptive Boosting (AdaBoostM1), Logistic Regression, k nearest neighbors (k-nn), Naive Bayes, Random Forest, and C4.5 Decision Tree. These algorithms will be mentioned in this section.

Adaptive Boosting (AdaBoostM1) Classifier Algorithm. The AdaBoost.M1 algorithm is used in binary classification [17]. In this algorithm, the aim is to create a more efficient algorithm by combining weak learning algorithms. Pre-misclassified samples are assigned a weight to classify more accurately. The weight assigned here provides poor classification. This algorithm is a widely used upgrade algorithm [18].

Logistic Regression Classifier Algorithm. Logistic regression is an algorithm used to analyze data sets with one or more independent variables that determine a result developed by David Cox [19].

k-Nearest Neighbor Classifier Algorithm. The nearest neighbor algorithm is one of the sampling-based learning algorithms. In this algorithm, learning is carried out through training data. In order to find out which class the new data belongs to, there is a k value closer to the model field. While this k value is found, distance calculation methods are used. Newly unknown data have been assigned to the class most similar to the nearest neighbor [20].

Navie Bayes Classifier Algorithm. The Naive Bayes algorithm is based on Bayes' theorem. It is one of the statistical classification methods. This algorithm is an algorithm to find the relationship between the target variable and the argument [21].

Random Forest Classifier Algorithm. Leo Breiman has developed. In this algorithm, instead of creating a single Decision Tree, it has emerged to combine the decisions of trees with a variable of different training sets. In this algorithm, different training clusters are looked at and different. After the calculation of the attributes at all levels, the properties in all formed trees are combined. in this combination, the most used attribute is selected and added to the tree. This is done at all levels of the tree [14].

C4.5 Decision Tree Classifier Algorithm. Developed by Ross Quinlan. C4.5 Decision Tree algorithm can be used in categorical or numerical data set. C4.5 The gain ratio is used in the decision tree algorithm [22].

2.3 Performance Criteria of Classification Models

In the study, the evaluation of the models with classification algorithms was done with the confusion matrix [24]. The confusion matrix is shown in Table 1. The performance criteria obtained from the mess matrix are shown below [23–25].

Table 1. Confusion matrix.

		Actual result		
		Yes	No	Total
Prediction result	Yes	True Positive (TP)	False Positive (FP)	TPos
	No	False Negative (FN)	True Negative (TN)	TNeg
	Total	Pos	Neg	M

Accuracy (ACC) and Error values of classifier models Eqs. (1) and (2), are shown respectively [23].

$$\text{Accuracy (ACC)} = \frac{T_P + T_N}{M} \tag{1}$$

$$\text{Error} = 1 - \text{Accuracy} \tag{2}$$

Precision (Positive Prediction Rate-PPV) and Sensitivity (Real Positive Rate-TPR) values of classifying models Eqs. (3) and (4), are shown respectively [23].

$$\text{Precision (PPV)} = \frac{TT}{TT + FT} \tag{3}$$

$$\text{Accuracy (ACC)} = \frac{T_P + T_N}{M} \tag{4}$$

Specificity and F-measure values of classifying models are given by Eqs. (5) and (6), are shown respectively [23].

$$\text{Specificity} = \frac{FF}{FF + FT} \tag{5}$$

$$F - \text{measure} = \frac{2 \times TPR \times PPV}{TPR + PPV} \tag{6}$$

The classifier models created in the study were created to make estimates on the data set. The performances of these classifier models are divided into data set training and test data. Holdout and k-fold cross validation methods are used in this literature. In Fig. 2, there are 60%–40% training-tests, 75%–25% training-test, 80%–20% training test distinctions [26]. Figure 3 shows 5 times cross validation differences [27].

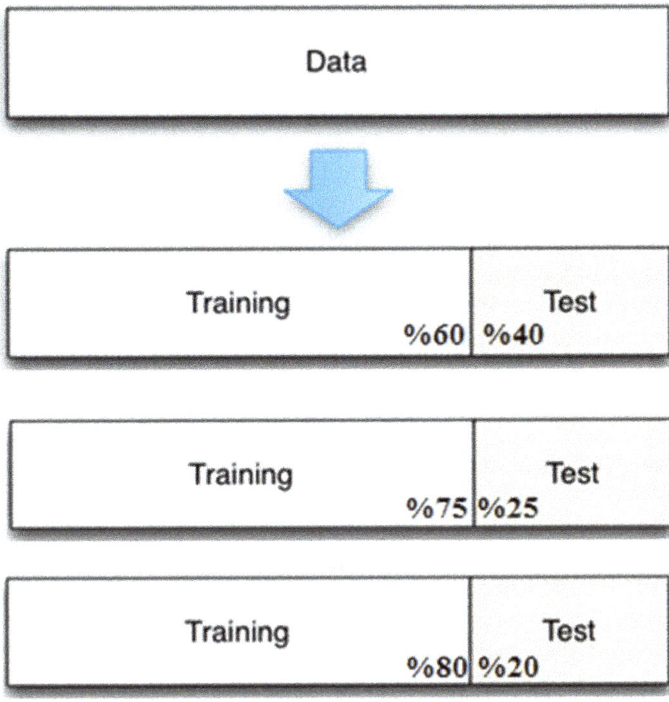

Fig. 2. Holdout with test and training set partition.

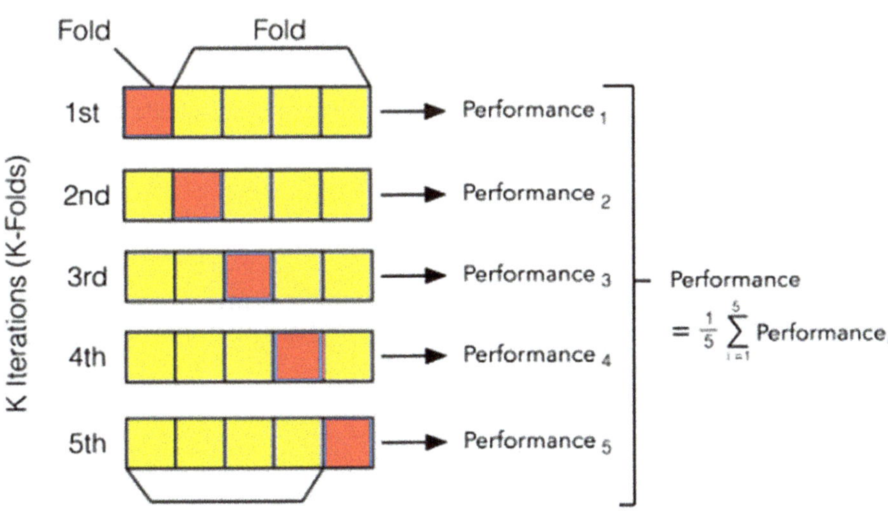

Fig. 3. Cross validation with test and training set partition

3 Application

In this study, the models are established with the classification algorithms on the dataset obtained according to whether the customers called during telemarketing are bank customers in order to determine the best classifier algorithm that gives the best estimation for bank customer acquisition. Within the scope of this study, CRISP model is used to follow a systematic approach for establishing models using classification algorithms. The steps used for preparing the classification models in the CRISP model process include problem definition, data interpretation, data preparation and model setup. The performances of the classification models established after these steps are presented in the conclusion section.

3.1 Problem Definition

Among the data mining methods, classification algorithms are used in studies conducted for customer acquisition to predict the potential customers of the company in question in the related industry. This study was carried out to determine the best estimator for bank customer acquisition by establishing models with classification algorithms on the bank customer dataset obtained by telemarketing method in UCI Machine Learning Repository [28].

3.2 Data Analysis

There are 17 properties and 45211 customer records in the bank data set. Table 2 shows the data types and characteristics of the data set from the UCI.

The statistical analysis results of the dataset obtained with the R program is shown in Fig. 4. This Figure shows the minimum, maximum, median, mean, first and third quartile values of the numerical values. In addition, the frequencies of the values of the categorical features are also shown. Figure 4 shows that categorical attributes such as education, contact, and outcome get unknown values in the data set. It is believed that the inclusion of data sets in this way affects data analysis.

3.3 Data Pre-processing

According to the statistical analysis, there was no missing data in the bank dataset (Fig. 4). For this reason, no pre-processing was performed to complete the missing data.

3.4 Application of Classification Models

In this study, models are established with various classification algorithms using the bank dataset to estimate potential bank customers. The classification algorithms used in the models is Adaptive Boosting (AdaBoostM1- Ada), k nearest neighbors (k-nn), Logistic Regression (LogReg), Naive Bayes (NB), Random Forest (RanFor), and C4.5 Decision Tree. To evaluate the performances of these models, four performance measures are used, namely the Accuracy, Precision, Sensitivity, and F measures.

Table 2. All variables, formats and types related to telecommunication data set

Nu	Attribute	Description	Data type
1	Age	Age	Numerical
2	Job	Job	Categorical
3	Marital	Marital status	Categorical
4	Education	Education	Categorical
5	Default	Credit debt status?	Categorical
6	Balance	Average annual balance	Numerical
7	Housing	Real estate loan status?	Categorical
8	Loan	Personal loan status?	Categorical
9	Contact	Type of communication	Categorical
10	Day	Last day of call	Numerical
11	Month	Last month of call	Numerical
12	Duration	Last call duration	Numerical
13	Campaign	Number of customers called the customer during the campaign	Numerical
14	Pdays	The number of times the customer has been called since the previous campaign	Numerical
15	Previous	How many times the customer has been called before the campaign	Numerical
16	Poutcome	The outcome of the previous marketing campaign	Categorical
17	Customer	Was the customer became a bank customer?	Categorical

```
      age                  job              marital            education        default            balance
Min.   :18.00   blue-collar:9732   divorced: 5207   primary  : 6851   no :44396   Min.   :  -8019
1st Qu.:33.00   management :9458   married :27214   secondary:23202   yes:  815   1st Qu.:     72
Median :39.00   technician :7597   single  :12790   tertiary :13301               Median :    448
Mean   :40.94   admin.     :5171                    unknown  : 1857               Mean   :   1362
3rd Qu.:48.00   services   :4154                                                  3rd Qu.:   1428
Max.   :95.00   retired    :2264                                                  Max.   :102127
                (Other)    :6835
 housing          loan              contact              day               month             duration
no :20081     no :37967     cellular :29285   Min.   : 1.00   may    :13766   Min.   :   0.0
yes:25130     yes: 7244     telephone: 2906   1st Qu.: 8.00   jul    : 6895   1st Qu.: 103.0
                            unknown  :13020   Median :16.00   aug    : 6247   Median : 180.0
                                              Mean   :15.81   jun    : 5341   Mean   : 258.2
                                              3rd Qu.:21.00   nov    : 3970   3rd Qu.: 319.0
                                              Max.   :31.00   apr    : 2932   Max.   :4918.0
                                                              (Other): 6060
    campaign           pdays            previous            poutcome         customer
Min.   : 1.000   Min.   : -1.0   Min.   :  0.0000   failure: 4901     no :39922
1st Qu.: 1.000   1st Qu.: -1.0   1st Qu.:  0.0000   other  : 1840     yes: 5289
Median : 2.000   Median : -1.0   Median :  0.0000   success: 1511
Mean   : 2.764   Mean   : 40.2   Mean   :  0.5803   unknown:36959
3rd Qu.: 3.000   3rd Qu.: -1.0   3rd Qu.:  0.0000
Max.   :63.000   Max.   :871.0   Max.   :275.0000
```

Fig. 4. Statistical results of data set

4 Conclusion and Discussion

Estimation models are established using the Adaptive Boosting (AdaBoostM1), k nearest neighbors, Logistic Regression, Naive Bayes, Random Forest, Part and C4.5 Decision Tree classification algorithms with the bank marketing dataset. The performance of all models implemented within the scope of the study is tested using a training and test dataset separated with holdout and k-fold cross validation methods. It is aimed to check the similarity of the results of two different model performance evaluations The dataset was divided into training and test datasets using the holdout method with the 60–40%, 75–25% and 80–20% separation ratios as shown in Table 3.

Table 3. Accuracy, Precision, F measure values for holdout separation of the bank marketing dataset

Algorithms	ACC			PPV			TPR			F		
	%60	%75	%80	%60	%75	%80	%60	%75	%80	%60	%75	%80
C4.5	0.90	0.91	0.89	0.93	0.93	**0.93**	0.95	0.96	0.95	0.94	0.94	0.94
NB	0.88	0.88	0.88	**0.94**	**0.95**	**0.93**	0.93	0.93	0.92	0.93	0.93	0.93
K-nn	0.86	0.86	0.86	0.91	0.91	0.91	0.93	0.93	0.93	0.92	0.92	0.92
Ada	0.89	0.89	0.89	0.90	0.90	0.90	**0.98**	**0.98**	**0.98**	0.92	0.94	0.94
LogReg	0.90	0.90	0.90	0.91	0.93	0.91	0.97	0.96	0.97	0.94	0.93	0.93
RanFor	**0.91**	**0.92**	**0.93**	0.92	0.92	0.92	0.96	0.96	0.96	**0.95**	**0.95**	**0.96**

In Table 3, the dataset is divided into training and test datasets using the holdout separation method with the 60–40%, 75–25% and 80–20% separation ratios to evaluate the performances of classification algorithms. In these separations, the highest performance is achieved with the Random Forest algorithm according to the accuracy criterion. According to the precision criterion, however, the Bayes algorithm yielded best results, whereas the AdaBoostM1 was highest in sensitivity, and Random Forest algorithm is highest in the F-measure criterion. The reason for not having a single superior classification model in these four different performance criteria is the confusion matrix.

In the K-fold cross validation, training and test data sets are separated with 5- and 10-fold cross validation as shown in Table 4. The results obtained in 5-fold cross validation and 10-fold cross validation is similar to all holdout separation results. In the 5-fold and 10-fold cross validation, the Random Forest algorithm had the best performance in accordance with the accuracy criteria, similar to the holdout method. According to the precision criterion, the Bayes algorithm gave prime results. AdaBoostM1 is best in sensitivity analysis, and Random Forest algorithm was best in F-measure analysis. It is clearly observed that the accuracy analysis, which is the most used performance criterion, is the most distinctive among the algorithms used.

As a result, it can be deduced that the classifier models established are consistent since similar results are obtained in two different separations. Since the performances

Table 4. Accuracy, Precision, F measure values for 5- and 10-fold cross validation separation of the bank marketing dataset

Algorithms	10-fold				5-fold			
	ACC	PPV	TPR	F	ACC	PPV	TPR	F
C4.5	0.89	0.92	0.95	0.94	0.89	0.93	0.96	0.94
NB	0.87	**0.93**	0.92	0.93	0.88	**0.94**	0.92	0.93
K-nn	0.86	0.91	0.93	0.92	0.86	0.91	0.93	0.92
Ada	0.89	0.90	**0.97**	0.93	0.89	0.90	**0.98**	0.94
LogReg	0.90	0.91	0.96	0.93	0.89	0.91	0.90	0.94
RanFor	**0.92**	0.92	0.96	**0.94**	**0.90**	0.92	0.96	**0.95**

are obtained by the confusion matrix, the classifier models established within the scope of the study give different results in each performance criterion. As a result of the similar results obtained in the test and training dataset separation in this study, as a future study to see whether the results are unique to this dataset by testing various datasets under the same conditions.

In the literature study conducted within the scope of the study, it was observed that the models produced are generally evaluated according to a single performance criterion (accuracy). In this study, modellers are evaluated with four different criteria: Accuracy (ACC), Precision (PPV), Sensitivity (TPR), and F-measure (F) performance criteria. In addition, the models generated by C4.5 Decision Tree, Navie Bayes algorithm, K nearest neighbors algorithm, Logistic regression algorithm, Random forest algorithm, and Adaptive Boosting algorithm As a training-test data for evaluating their performance, the data set is divided into hold-out with random discrimination as 60%–40%, 75%–25%, 80%–20% training-test respectively. In this study, k-fold cross validation was performed in addition to the random discrimination on the training and test set, and the results on the performance criteria are compared and compared with each other.

References

1. Baynal, K., Çaliş, A.: Veri Madenciliğinde Kümeleme Analizi ile Bankacılık Sektöründe Bir Uygulama. Beykent Üniversitesi Fen ve Mühendislik Bilimleri Dergisi, vol. 9(1) (2016)
2. Çankiri, S., Kartal, E., Yildirim, K., Gülseçen, S.: Organizasyonlarda Bilgi Yönetimi Sürecinde Veri Madenciliği Yaklaşımı. Fırsatlar ve Tehditler Sempozyumu, İstanbul, Bilgi Çağında Varoluş (2009)
3. Bach, M.P., Juković, S., Dumiči, K., Šarlija, N.: Business client segmentation in banking using self-organizing maps. South East Eur. J. Econ. Bus. **8**(2), 32–41 (2013)
4. Sumathi, S., Sivanandam, S.N.: Introduction to Data Mining Principles. Springer, Heidelberg (2006)
5. Keramati, A., Jafari-Marandi, R., Aliannejadi, M., Ahmadian, I., Mozaffari, M., Abbasi, U.: Improved churn prediction in telecommunication industry using data techniques. Appl. Soft Comput., 994–1012 (2014)
6. Nachev, A.: Application of Data Mining Techniques for Direct Marketing, Sofia (2014)

7. Elsalamony, H.A.: Bank direct marketing analysis of data mining techniques. Int. J. Comput. Appl., 12–22 (2014)
8. Kumari, B., Shrivastava, V.: Evaluation and comparison of performance of different classifiers. Int. J. Emerg. Trend Eng. Basic Sci. (IJEEBS), 604–611 (2015)
9. Erol, Ç.: Sağlık Bilimlerinde R ile Veri Madenciliği, R ile Veri Madenciliği Uygulamaları, Çağlayan Kitabevi, İstanbul, pp. 25–46 (2016)
10. Azevedo, A.I.R.L., Santos, M.F.: KDD, SEMMA and CRISP-DM: a parallel overview, IADS-DM (2008)
11. Dolgun, M.Ö., Ersel, D.: Doğrudan Pazarlama Stratejilerinin Belirlenmesinde Veri Madenciliği Yöntemlerinin Kullanımı. İstatistikçiler Dergisi: İstatistik & Aktüerya **7**, 1–13 (2014)
12. Bakioğlu, F.Ö.K., Kartal, E., Özen, Z., Erol, Ç., Gülseçen, S.: Aspects of students about ınformation technology courses in social science. Procedia - Soc. Behav. Sci., **176**, 148–154 (2015)
13. Özkan, Y.: Veri madenciliği yöntemleri, 2. basım, İstanbul, Türkiye: Papatya Yayıncılık (2013)
14. Han, J., Kamber, M., Pei, J.: Data mining: concepts and techniques. In: Data Management Systems, pp. 230–240. The Morgan Kaufmann Series (2006)
15. Gordon, L.S., Berry, M.J.A.: Mastering data mining: for marketing, sales, and customer relationship management, 2nd ed. Willey Publishing, New York (2004)
16. Ayre, L.B.: Data Mining for Information Professionals, San Diego. California, USA (2006)
17. Freund, Y., Schapire, R.E.: Experiments with a new boosting algorithm. In: Thirteenth International Conference on Machine Learning, pp. 148–156 (1996)
18. Özgür, A., Erdem, H.: Saldırı Tespit Sistemlerinde Kullanılan Kolay Erişilen Makine Öğrenme Algoritmalarının Karşılaştırılması. Bilişim Teknolojileri Dergisi **5**(2), 41–48 (2012)
19. Logistic Regression. https://wiki2.org/en/Logistic_regression+Newton. Accessed 20 Aug 2018
20. Anonymous. http://www.matlabyar.com/wp-content/uploads/edd/2016/03/knnng.png. Accessed 16 Apr 2018
21. Harrington, P.: Machine Learning in Action. Manning, New York (2012)
22. Witten, I.H., Frank, E., Mark, A.H.: Veri Madenciliği: Pratik makine öğrenme araçları ve teknikleri, 3. Baskı, p. 191. Morgan Kaufmann, San Francisco (2011)
23. Japkowicz, N.: Performance evaluation for learning algorithms. International Conference on Machine Learning, Scotland (2012)
24. Clark, M.: An introduction to machine learning: with applications in R
25. Flach, P.: The many faces of ROC analysis in machine learning. ICML Tutorial
26. Avrim, B., Adam, K., John, L.: Beating the hold-out: Bounds for k-fold and progressive cross-validation. In: Proceedings of the Twelfth Annual Conference on Computational Learning Theory. ACM (1999)
27. Kohavi, R.: A study of cross-validation and bootstrap for accuracy estimation and model selection. Ijcai **14**, 1137–1145 (1995)
28. Moro, S., Cortez, P., Rita, P.: A data-driven approach to predict the success of bank telemarketing. Decis. Support Syst. **62**, 22–31 (2014)

The Model Selection Methods for Sparse Biological Networks

Mehmet Ali Kaygusuz[1] and Vilda Purutçuoğlu[1,2(✉)]

[1] Biomedical Engineering, Middle East Technical University, 06800 Ankara, Turkey
makaygusuz1988@gmail.com, vpurutcu@metu.edu.tr
[2] Department of Statistics, Middle East Technical University, 06800 Ankara, Turkey

Abstract. It is still crucial problem to estimate high dimensional graphical models and to choose the regularization parameter in dependent data. There are several classical methods such as Akaike's information criterion and Bayesian Information criterion to solve this problem, but also more recent methods have been proposed such as stability selection and stability approach to regularization selection method (StARS) and some extensions of AIC and BIC which are more appropriate for high dimensional datasets. In this review, we give some overview about these methods and also give their consistency properties for graphical lasso. Then, we evaluate the performance of these approaches in real datasets. Finally, we propose the theoretical background of our proposal model selection criterion that is based on the KL-divergence and the bootstrapping computation, and is particularly suggested for the sparse biological networks.

Keywords: Graphical lasso · Model selection criteria · Sparse networks

1 Introduction

In the construction of biological networks, the main challenge in the calculation is the high dimensional number of random variables for low number of observations. Moreover, the interactions between genes are very sparse which are typically described by the scale-free networks. Therefore, there is no explicit form of estimators of the model parameters in different modeling approaches. Hence, the selection of the best estimates in inference of the fitted mathematical models is one of the interesting questions in systems biology. There are a number of model selection criteria in order to choose the optimal model in these structures. Some of them are already used in ordinary nonlinear models such as the Akaike's information criterion and the Bayesian Information criterion. Whereas, some of them are particularly designed for sparse structural models like the stability selection [27] and stability approach to regularization selection method (StARS) [25]. In all these criteria, the major properties of the fitted model can be named as the *graphical models*. These models basically depend on either the

D. J. Hemanth and U. Kose (Eds.): ICAIAME 2019, LNDECT 43, pp. 107–126, 2020.
https://doi.org/10.1007/978-3-030-36178-5_10

estimate of the covariance matrix or the inverse of the covariance matrix, also called the precision matrix, of the data. These matrices present the physical or functional relationships between the components of the biological networks, which are genes or proteins and also called *the nodes*, via a non-zero entry in the matrix. On the other hand, if two nodes have no interaction, we denote it by a zero entry in the matrix. The theoretical background of these interpretations comes from the conditional independence of the genes that is typically described under the multivariate normal distribution within the family of parametric models. The multivariate normality is one of the common distributional assumptions for the construction of biological networks due to its computational simplicity with respect to other multivariate distributions and the description of the high dimensionality via the normal distribution. Whereas, there are other alternative models which are distribution-free such as ordinary differential equations, multivariate adaptive regression splines or the random forest method [22]. Although the graphical interpretation of all these models is the same when they are implemented for the construction of biological networks, their initial mathematical descriptions change. But, it is still possible to find certain common features in all these models. One common feature is their biological meaning. The majority of the graphical models are based on the description of the systems under the steady-state condition. This condition aims to capture the natural behaviour of the system without stochastic reactions such as transcriptions or translations in the cell. Therefore, they cannot explain the molecular changes in the biological reactions and systems, whereas, they can successfully present the change in concentrations or expressions of the nodes (or systems' elements) and accordingly, systems. These changes are called the *states* in this type of models. As the majority of the available data, such as microarray datasets, are suitable for this sort of models, it is one of the most common modeling types for biological networks.

From the comparative studies within the family of steady-state models, it is seen that the models which belong to the *generalized additive models* [22] are very promising in order to represent the actual behaviour of the systems. Some strong alternates within this group can be listed as the Gaussian graphical model (GGM) [36], copula Gaussian graphical model (CGGM) [18] and conic multivariate adaptive regression splines (CMARS) [5]. In all these models, due to the underlying challenges caused by the high dimensions of the systems, limited sample sizes of measurements and sparse topological structures of the systems, we cannot obtain the estimators of the model parameters explicitly. Moreover because of the natural description of biological networks, all the predictors in a regression model, which are nodes, are highly dependent on each other. This high dependency leads to an additional problem in inference of the model parameters. This special type of regression models, which is also a member of the generalized additive models, is called the *lasso regression* [22]. Briefly, this regression model puts some penalizing constraints on the regression coefficients via different levels of *norms*. Among alternates, L_1 and L_2 norms are the most common ones. Thereby, by calibrating the penalizing constant in the lasso regression, we aim to obtain the optimal penalizing constant and thereby, its associated construction of the biological network. In systems biology, the selection of the optimal penalizing constant is known as the *model selection* and the rule which identifies this constant is called the *model selection criterion*.

Hence, in this study, first of all, we comprehensively describe the major model selection criteria used for the construction of the sparse biological networks in Sect. 3 under the structure of the lasso regression Sect. 2. Then, we assess their performances in different dimensional real datasets in Sect. 4. Finally, we propose the theoretical background of our proposal criterion in Sect. 5. We summarize our works and briefly discuss the outputs in Sect. 6.

2 Penalized Regression

As stated previously, there are different constructions of the regression models for the description of the sparse biological networks. Dempster [17] suggests the setting elements of the inverse covariance matrix to zero. But the study of Meinhausen and Bühlmann [26] makes a new contribution to this problem in the sense that it can combine the estimation of covariance in the lasso regression with the graphical methods. Then, Yuan and Lin [34], Banerjee et al. [4] and Friedman et al. [21] apply the penalized maximum likelihood method with a lasso penalty to estimate the inverse covariance matrix. This special type of model is named as the *graphical lasso* in the literature and it is one of the successful lasso regression models for the representation of biological networks. In this model, let $X = (X_{(1)}, \ldots, X_{(n)})$ be the p-dimensional random vector which is distributed under the multivariate normal distribution with a mean vector μ and a covariance matrix Σ, i.e., $X \sim N_d(\mu, \Sigma)$. In order to generate the networks, we aim to estimate the covariance matrix since there is a direct proportional relationship between the inverse of the covariance matrix and the regression coefficient of the (graphical) lasso regression. Because of this relation, this calculation is also called as the covariance selection problem [17] and its application can be seen in different fields too such as biomedical data science, fMRI studies and climate studies besides its application in systems biology.

Hereby, in the construction of the graphical models, the selection of the regularization parameter in a dependent data that is also known as the penalizing constant is a crucial problem and besides classical approaches, such as Akaike's information criterion (AIC) and Schwartz's Bayesian information criterion (BIC), there are some recent specific selection methods for the graphical models.

In this review, we give definitions and some properties of the graphical lasso in the second part and then, we represent the classical model selection methods, i.e., AIC and BIC. Later, we describe some extended AIC and BIC methods suggested by Chen and Chen [14], Foygel and Drton [20] and Bozdogan [10,11]. Next, we explain the model selection consistency of the lasso model [35] and some specific model selection methods for the graphical lasso which are proposed by Meinhausen and Bühlmann [27]. Finally, we present the stability approach to the regularization selection method which uses subsampling methods [25,29].

2.1 Graphical Lasso

In this section, we give the definition and some properties of the graphical lasso in order to estimate a sparse inverse covariance matrix $\Theta = \Sigma^{-1}$ based on a random sample for a p-dimensional observation. A natural way to estimate this matrix, also called the precision matrix as stated previously, we need to define the following the log-likelihood expression.

$$\log|\Sigma^{-1}| - \mathrm{tr}(S\Sigma^{-1}) \tag{1}$$

where log(.) and tr(.) denote the logarithm and the trace of the given expressions, respectively. S is the sample covariance matrix obtained from the data. Thus, Θ leads to the maximum likelihood estimate via $\hat{\Theta} = S^{-1}$, which will contain no element exactly to zero. Particularly, when the number of features (p) exceeds the number of observations (n), i.e., $p < n$, the ordinary maximum likelihood estimator has some limitations and when S is singular, the maximum likelihood equation cannot be computed explicitly. In order to solve this computational problem and estimate the covariance matrix, Yuan and Lin [34] suggest the idea of the penalized likelihood and Friedman et al. [21] derive the following expression as the correspondence of Eq. (1).

$$\log|\Theta| - \mathrm{tr}(S\Theta) - \lambda\,\|\Theta\|_1 \tag{2}$$

in which S denotes the empirical covariance matrix and $\Theta = \Sigma^{-1}$ is a nonnegative precision matrix whose elements are θ_{ij} $(i, j = 1, \ldots, p)$. This implies that the corresponding variables X_j and X_k are conditionally independent. Furthermore, λ refers to a nonnegative tuning parameter which controls the amount of l_1-shrinkage that implies a semidefinite programming in the variable Θ [9].

On the other hand, Banerjee et al. [4] show that the problem in Eq. (1) is convex and presents an estimation in Eq. (2) that can be solved by optimizing over each row and corresponding W, stated in Eq. (5) in a block coordinate descent fashion. Hence, the gradient to be maximized can be written as

$$\Theta^{-1} - S - \lambda\Gamma(\Theta) = 0. \tag{3}$$

Accordingly the method of the graphical lasso, shortly glasso, uses the block coordinate method for solving S, Θ and Γ in Eq. (3) which have the following expressions.

$$\Theta = \begin{pmatrix} \Theta_{11} & \Theta_{12} \\ \Theta_{21} & \Theta_{22} \end{pmatrix} \qquad \Gamma = \begin{pmatrix} \Gamma_{11} & \Gamma_{12} \\ \Gamma_{21} & \Gamma_{22} \end{pmatrix}. \tag{4}$$

In the above expressions, the matrices are partitioned into two parts. The first part belongs to the first $(p-1)$ rows and columns and the second part is the pth row and column. Then, by using them, we can write

$$\begin{pmatrix} W_{11} & W_{12} \\ W_{21} & W_{22} \end{pmatrix} \cdot \begin{pmatrix} \Theta_{11} & \Theta_{12} \\ \Theta_{21} & \Theta_{22} \end{pmatrix} = \begin{pmatrix} I & 0 \\ 0^T & 1 \end{pmatrix}. \tag{5}$$

where $W\Theta = I$ and $(.)^T$ indicates the transpose of the given expression. So, by considering the pth column, we can get

$$W_{12} - S_{12} - \lambda \Gamma_{12} = 0 \tag{6}$$

with $W_{12} = W_{11}\beta$. This is

$$W_{11}\beta - S_{12} - \lambda \Gamma_{12} = 0. \tag{7}$$

Thus, the stationary equation for the following L_1-regularized quadratic program is found by minimizing the following expression when $\beta \in \Re^{p-1}$ via

$$0.5\beta W_{11}\beta + \beta S_{12} + \lambda \left\|\beta\right\|_1. \tag{8}$$

In the above expression, W_{11} is assumed to be fixed which is analogous to a lasso regression problem of the last variable on the rest, except the cross-product matrix S_{11} that is replaced by the current estimate W_{11}. By reducing the dimension via the sparsity in β, we can perform the element-wise coordinate method to overcome the problem.

$$1/\hat{\theta}_{22} = W_{22} - \hat{\beta}^T \hat{W}_{12}. \tag{9}$$

Finally, $\hat{\theta}_{12}$ might be updated from $\hat{\beta}$ and $\hat{\theta}_{22}$.

3 The Model Selection Methods for Sparse Biological Networks

3.1 Akaike's Information Criterion and Bayesian Information Criterion

Assumed that the data are generated from the true density function f_{Θ_0} with a parameter vector $\Theta_0 = (\Theta_0, \ldots, \Theta_d)^T$. The problems of how to estimate the dimension of the model and to compare models of different dimensions can become very crucial in statistical applications. This problem is called the *model selection* in statistical literature. In this section, we describe two classical model selection methods which are proposed by Akaike [2] and Scwartz [30]. The Akaike's information criterion (AIC), proposed by Akaike [2], is mainly a penalized version of the attained maximum log-likelihood for each model. On the other side, the Schwartz's bayesian information criterion (BIC) [30] has a form of the penalized log-likelihood function with more penalty for additional parameters. From different analyses, it has been seen that BIC can be more consistent than AIC.

Hereby, Akaike [2,3] proposes to choose a model which minimizes the Kullback-Leiber (KL) divergence of the fitted model from the true model. AIC considers the maximum likelihood estimator (MLE) of the parameter vector M and shows that with an additive constant, the estimated KL divergence can be asymptotically expended as

$$\mathrm{AIC}(M) = 2\hat{l}_n(M) - 2\mathrm{dim}(M) \tag{10}$$

where $\hat{l}_n(M)$ is the log-likelihood function based on the sample size n and the parameters M and dim(.) denotes the number of parameters, i.e., dimension, of the model. This expression also emerges to the equation of AIC. Thus, AIC behaves like a penalized likelihood criterion. Therefore, the minimun AIC shows the optimal model.

On the other side, Schwartz [30] suggests a Bayesian approach for the same problem as presented below. Herein, each proposal/candidate model and its model parameters are assigned a prior distribution.

$$\mathrm{BIC}(M) = 2\hat{l}_n(M) - (\log n)\mathrm{dim}(M) \tag{11}$$

Thus, the model which has the highest BIC value is selected as the optimal model. In application, BIC has a stronger penalty for the complexity, particularly, when n >8. For more details about this model selection, we suggest the monograph of Claeskens and Hjort [16].

Accordingly, when M is greater than n, these two information criteria might be inconsistent. To solve this challenge, there are some extensions of the classical AIC and BIC selection criteria such as EBIC which is specifically applied in the genome-wide association studies. In the following parts, we present these extensions.

3.2 Some Extended BIC-based Model Selection Methods

In last fifteen years, the BIC methods became a popular choice for the selection of the sparse models for high dimensional data. Bogdan et al. [7] and Chen and Chen [14] propose an extended BIC for the variable selection under high dimensional regressions. These extensions give more flexibility, particularly, for the selection of large models. Because the asymptotic analysis of the new criterion allows a sub-exponential growth in the number of covariates p as a function of the sample size n. The analyses of Chen and Chen [14] show the consistency in the variable selection under this asymptotic condition. Later, they extend the results to the generalized linear models (GLM) [15] and Foygel and Drton [20] propose that the extended BIC (EBIC) for the graphical lasso and its asymptotic is more consistent.

Let the models be denoted by $M_i, \ldots, M_k$. By using Y as the notation for data $Y_1, \ldots, Y_n$ and $l(Y|M_i, \theta_i)$ is the log-likelihood function by a vector parameter space and the model M_i, the posterior distribution of M is represented as

$$P(Y|M_i) = \int l(Y|M_i, \theta_i) f(\theta_i|M_i) \mathrm{d}\theta_i \tag{12}$$

$$P(M_i|Y) = \frac{\pi(i)P(Y|M_i)}{\sum_{i=1}^{k} \pi(i)P(Y|M_i)} \tag{13}$$

where k is the number of possible models. So, BIC suggests a model which has the highest posterior probability $P(M_i|Y)$. Later, Chen and Chen [14] suggest an alternative EBIC with a high consistency via

$$\mathrm{BIC}_\gamma(J) = -2l_n(\hat{\Theta}_j 1) + |J| \log(n) + 2_\gamma |J| \log(p) \tag{14}$$

where $\hat{\Theta}_j$ is MLE of Θ_j and the first two terms in $\mathrm{BIC}_\gamma(J)$ are the Laplace approximation. Finally, the last term $-2\log(p)$ presents a common constant. Then, Foygel and Drton [20] propose another EBIC for the Gaussian graphical model as follows.

$$\mathrm{BIC}_\gamma(E) = -2l_n(\hat{\Theta}(E)) + |E|\log(n) + 4_\gamma|E|\log(p) \tag{15}$$

in which E shows the set of edges for the candidate graph and $-2l_n\hat{\Theta}(E)$ is the log-likelihood function of the related model as usual. Here, the criterion is controlled by a parameter $\gamma[0,1]$ such that $\gamma = 0$ denotes the ordinal expression of BIC for a given p and n, whereas, when $\gamma > 0$, the criterion works a penalization constant which can be suitable for large graphs.

3.3 Consistency Properties of Extended Bayesian Information Criterion for Graphical Lasso

In this part, we present the consistency properties of EBIC for the graphical lasso which is given by Foygel and Drton [20]. But firstly, we describe some notations and definitions to study the consistency properties of EBIC for the graphical lasso. In a graph, E is the set of edges for p nodes. Therefore, it can be also denoted by a $(p \times p)$-dimensional matrix whose entries (j,k) and (k,j) are the same. Herein, the index pairs (j,j) shows the diagonal entries of the matrix. In Eq. 2, the non-zero entries and off-diagonal entries are denoted as Θ_0 and E_0, respectively. Furthermore, the maximum variance of the individual nodes is stated as below.

$$\sigma^2_{max} = max(\Theta_0^{-1})_{jj}. \tag{16}$$

So, let us define $\Theta = min|(\Theta_0)_e|$ the smallest signal among edges e where $e = (j,k)$ and $(\Theta_0)_e = (\Theta_0)_{jk} = (\Theta_0)_{kj}$. On the other hand, λ_{max} is the maximum eigenvalue of Θ_0. In this expression, $\sigma^2_{max}\lambda_{max}$ is not bigger than the condition number of Θ_0 because of $1/\lambda_{max}(\Theta_0) = \lambda_{max}(\Theta_0^{-1}) \geq \sigma^2_{max}$. Thereby, assuming that n goes to infinity and we have the following asymptotic assumption on the data and the model, we can write the theorem below by using the following expressions.

a. E_0 is decomposable with $|E_0| \leq q$,
b. $\sigma^2_{max}\lambda_{max} \leq C$,
c. $p = O(n^k)$ when $p \to \infty$,
d. $\gamma_0 = \gamma - (1 - 1/4K) > 0$,
f. $(p + 2q)\log p(\lambda^2_{max}/\Theta^2_0) = o(n)$,

where $C, K > 0$ and γ are constant real numbers, p and q denote the integers, and lastly, the edge set E_0, Θ_0, σ^2_{max} and λ_{max} are changed with n.

Theorem 1: We assume that the conditions in Eq. 16 are hold. Let E_0 be the set of all decomposable models E with $|E| \leq q$. While n goes to infinity, the probability tends to 1 via

$$E_0 = \arg\min \mathrm{BIC}_\gamma(E). \tag{17}$$

Therefore, we aim to select the smallest extended BIC for the true model. The proof of this theorem with connections of the following Theorems 2 and 3 can be found in the study of Foygel and Drton [20].

Theorem 2: We assume that the conditions in Eq. 16 are hold and let ε_1 be the set of models E and $|E| < q$. Thus, when n goes to infinity, the probability tends to 1 via

$$l_n(\hat{\Theta}(E)) - l_n(\hat{\Theta}(E_0)) > 2q(\log p)(1+\gamma_0) \qquad \forall E \in \varepsilon_1.$$

Theorem 3: We assume that the conditions in Eq. 16 are hold and let ε_0 be the set of decomposable models E and $|E| \leq q$. Hence, when n goes to infinity, the probability tends to 1 by

$$l_n(\hat{\Theta}(E)) - l_n(\hat{\Theta}(E_0)) > 2q(1+\gamma_0)(|E| - |E_0|)(\log p) \qquad \forall E \in \varepsilon_0(E_0).$$

3.4 Some Extensions of Akaike Information Criterion: CAIC, CAICF and ICOMP

In the literature, similar to the extension of BIC as stated previously, some extensions of AIC are also used to select the right model. Three of them are proposed in the studies of Bozdogan [10,11]. These selection methods are called the extended AIC and aim to propose more consistent model selection. For this purpose, Bozdogan [10] uses the Kullback information and Fisher information matrix in his calculations.

He proposes that if we define a measure which can minimize the distance between the model and the true distribution, its minimum value indicates the optimal estimators fr the related model. The Kullback-Leibler information quantity can satisfy this feature. Thereby, Bozdogan (1987) used this quantity to extend AIC and calls the consistent AIC (CAIC) whose expression as follows.

$$\text{CAIC}(k) = l - 2\log L(\hat{\Theta}_k) + k[(\log n) + 1] \tag{18}$$

where $\log L(\hat{\Theta}_k)$ is log-likelihood function for Θ and k denotes the degrees of freedom of the distribution. By this way, CAIC(k) suggests another and bigger penalizing value for the model with respect to BIC if we compare the final constant term for both criteria.

Another Bozdogan's selection method [10] is named as the consistent AIC with a Fisher information (CAICF(k)). This method penalizes the over-parametrization more strongly via the following selection procedure.

$$\text{CAIC}(k) = l - 2\log L(\hat{\Theta}_k) + k[(\log n) + 2] + \log|I(\hat{\Theta}_k)| \tag{19}$$

in which $\log L(\hat{\Theta}_k)$ describes the likelihood estimation of Θ, as used beforehand, k shows the degrees of freedom of the distribution and $I(\hat{\Theta}_k)\hat{F}^{-1}$ indicates the Fisher information matrix.

Finally, as the extension of CAICF, another new model selection criterion, called the information and complexity (ICOMP) [11], is developed by

Bozdoğan [11]. This novel criterion can penalize the free parameters and the covariance matrix directly as follows.

$$\mathrm{ICOMP} = l - 2\log L(\hat{\Theta}_k) + 2C(\hat{\Sigma}) \tag{20}$$

where $\log L(\hat{\Theta}_k)$ is the maximum likelihood estimation, $\hat{\Theta}_k$ stands for the maximum likelihood estimate of the parameter vector of Θ_k, C represents a real-valued complexity measure and finally, $(\hat{\Sigma}) = (c\hat{o}v)(\hat{\Theta}_k)$ shows the estimated covariance matrix of the parameter vector of the model. This covariance matrix may be estimated in different ways. One of these choices is the computation of the inverse of the Cramer-Rao lower bound matrix. From this way, the estimated inverse Fisher information matrix (IFIM) of the model can be obtained as below.

$$\hat{F}^{-1} = (-E(\partial^2)/(\partial\Theta\partial(\Theta^T))\log L(\Theta))^{-1} \tag{21}$$

$\hat{F}^{-1}$ is the $(s \times s)$-dimensional second-order partial derivatives of the log-likelihood function of the estimated model. Accordingly, a more general form of ICOMP can be written as follows.

$$\mathrm{ICOMP} = l - 2\log L(\hat{\Theta})_k + 2C(\hat{F}^{-1}) \tag{22}$$

when

$$C(\hat{F}^{-1}) = s/2\log[\mathrm{tr}(\hat{F})^{-1}/s] - 1/2\log|\hat{F}^{-1}| \tag{23}$$

is the information complexity of the estimated inverse Fisher information matrix of the model and $s = \dim(\hat{F}^{-1}) = \mathrm{rank}(\hat{F}^{-1})$ while dim(.) refers to the dimension of the given matrix.

3.5 Stability Selection Method and Consistency of Lasso Models

In this section, we present a stability variable selection method for the regression with high dimensional data which is proposed by Meinhausen and Bühlmann [27]. Previously, Zhao and Yu [35] investigate the consistency of the model selection in this type of lasso models. Therefore, below, we initially present the strong and weak *irrepresentable* conditions of the lasso model in order to choose the right model. Because in the study of Meinhausen and Bühlmann [27], it is proposed that the stability selection methods are less sensitive methods for the different choices of the regularization parameter in the variable selection.

We consider a linear regression model as in Eq. 24 and assume that the noise vector is independent identically distributed (iid) via the normal distribution as below.

$$Y = X\beta + \varepsilon \tag{24}$$

in which $Y = (Y_1, \ldots, Y_n)$, X is the $(n \times p)$-dimensional design matrix and $\varepsilon = (\varepsilon_1, \ldots, \varepsilon_n)$. Here, it is accepted that the predictor variables are normalized with $X_{k2} = 1$ for all k ($k = 1, \ldots, p$ for totally p numbers of nodes) while $p \gg n$. Without loss of generality, we might assume that $\beta^n = (\beta_1^n, \ldots, \beta_q^n, \beta_{q+1}^n, \ldots, \beta_p^n)$, $\beta_j^n \neq 0$ for $j = 1, \ldots, q$ and $\beta_j^n = 0$ for $j = q+1, \ldots, p$. Then, let

$\beta^n_{(1)} = (\beta^n_1, \ldots, \beta^n_q)$ and $\beta^n_{(2)} = (\beta^n_q, \beta^n_{q+1}, \ldots, \beta^n_p)$. Now, we write $X_n(1)$ and $X_n(2)$ are the first q and the last $(p-q)$ columns of X_n, respectively, and let $C^n = (1/n)X_n^T X_n$. Finally, C^n can be expressed like

$$X = \begin{pmatrix} C^n_{11} & C^n_{12} \\ C^n_{21} & C^n_{22} \end{pmatrix}. \tag{25}$$

Assuming that C^n_{11} is invertible, we can define the following irrepresentable condition,

Definition (irrepresentable condition): There exists a positive constant vector β

$$|C^n_{21}(C^n_{11})^{-1}\text{sign}(\beta^n_{(1)})| \leq I - \beta \tag{26}$$

where I shows a vector of one with length $(p-q)$.

In the computation, the stability selection [27] has two major advantageous. Firstly, it can decrease the sensitivity of the regularization parameter under noisy data. Secondly, it can produce consistent model if the selection methods are not successful [26]. The control of consistency works in majority of the systems. Whereas, if the model is a lasso, regression, a stronger control of consistency is also suggested as used in the study of Meinhausen and Bülhmann [26] via the neighbourhood stability condition which is equivalent to the irrepresentable condition of Zhao and Yu [35] and Yuan and Lin [34].

In an ordinary model construction, the variable selection implies the selection of the suitable elements in a model by

$$\hat{S}^{\lambda}; \lambda \in \Lambda \tag{27}$$

where Λ describes the set of the continuous/discrete regularization parameters. Here, we have two challenges. Firstly, the correct model S should be one of the alternates of Λ and secondly, the optimal regularization can find the true S from the data. Thus, Meinhausen and Bühlmann [27] suggest that the data can be repeated, e.g., subsampled, many times so that we can choose all structures of variables from this target set.

Definition (stable variables): From a set of regularization parameters, the set of variables is described by

$$\hat{S}^{\text{stable}} = k : \max(\hat{\Pi})^{\lambda} \geq \pi_{\text{thr}} \tag{28}$$

in which $1 > \pi_{\text{thr}} > 0$ with a cut-off probability and these results do not depend on the choice of the regularization λ and the regularization region Λ.

By this way, Meinhausen and Bühlmann [27] can make a huge impact to select a true variable for the graphical lasso since the main idea of this selection method is based on the application of the algorithm to the whole dataset and on the determination of the selected set of variables. In practice, we can do this calculation several times to get subsamples until we arrive the number of the data size ($n/2$) and then we can choose variables from most frequently selected subsamples. Later, some extensions of the stability selection such as the studies of Liu et al. [25], Lim and Yu [24] and Shah and Samworh [31] are developed. Liu et al. [25] suggest a selection method which combines the stability selection and the cross-validation. Lim and Yu [24] propose a new class of the stability selection with an error control which is given by an upper bound on the expected number of falsely selected variables. Finally, Shah and Samworh [31] suggest a stability selection method which is based on the re-sampling methods. Below, we present the mathematical background of Liu et al.'s method [25] with more details.

3.6 Extension of the Stability Selection Method

In the construction of graphical models, since the selection of the true model is challenging due to the high dimension and the sparse structure of the systems, some specific solutions are suggested. One of these alternates is the extension of the stability selection method which is the stability approach to the regularization selection (StARS). It is known that the traditional selection methods such as AIC and BIC, have a parametric assumption and even these assumptions are satisfied, they may not work well in the finite sample setting. For this reason, the bootstrap [19] and the subsampling citePolitisRomanoandWolf1999 of the non-parametric model selection methods, which are computationally feasible for large amount data, are considered, specifically, for the graphical models.

Hereby, StARS suggested by Liu et al. [25] is different from the method of Meinhausen and Bühlmann [27] because of the fact that they use the subsampling methods for the model selection. Accordingly, they draw many random subsamples and unlike of the cross-validation, these subsamples are overlapped. Then, they choose the regularization parameter whose graph is sparse and does not too much variability across sub-samples. Here, the aim of Liu et al. [25] is to control the dissonance between graphs. Finally, the StARS approach selects $\Lambda = 1/\lambda$ based on the stability. When Λ is 0, the graph is empty and when we increase Λ, the variability of the graph increases and thus, the stability decreases. Therefore, this method increases the variability until the point where the graphs become variable as measured by stable. Below, we describe the mathematical details of StARS.

Let $b = b(n)$ be such that $1 < b(n) < n$, we draw N random subsamples $S_1, \ldots, S_n$ from $X_1, \ldots, X_n$ where each of them has the size m. By this way, we can have C_b^n subsamples where C denotes the number of combinations of n by b. To estimate Θ_{st}^{Λ}, Liu et al. [25] use a U-statistics of the order m such that $\hat{\Theta}_{st}^{\Lambda} = 1/n \sum_j^n \psi_{st}^{\Lambda}(S_j)$ where $\Theta_{st}^{b} = p(\psi_{st}^{\Lambda}(X_1), \ldots, X_m) = 1$ and $\psi^{\Lambda}(.)$ is the glasso under Λ.

So, now, let us define the parameter $\xi_{st}^{b}(\Lambda) = \Theta_{st}^{b}(\Lambda)1 - \Theta_{st}^{b}(\Lambda)$ and $\hat{\xi}_{st}^{b}(\Lambda) = \hat{\Theta}_{st}^{b}(\Lambda)1 - \hat{\Theta}_{st}^{b}(\Lambda)$ to be estimated. The estimator for the edge (s,t) is two times of the variance obtained via the Bernoulli indicator and $\xi_{st}^{b}(\Lambda)$ is taken as a measure of the instability of the edge across subsamples with $0 < \xi_{st}^{b}(\Lambda) < 1/2$.

Lastly, let us define the total instability by averaging over all edges via $\hat{D}_b(\Lambda) = \sum_{s<t} \hat{\xi}^b_{st}(\Lambda)/C^p_2$. From this expression, it is seen that on the boundary, $\hat{D}_b(0)$ and $\hat{D}_b(\Lambda)$ will increase when Λ increases in general. But when Λ gets very large, almost all graphs will become dense and $\hat{D}_b(\Lambda)$ will begin to decrease. Therefore, Liu et al. [25] choose Λ by defining $\hat{\Lambda}_s = \sup \Lambda : \bar{D}_b(\Lambda) \leq \beta$ where β is a cut-off point. In this calculation, $\hat{E}, \hat{D}, \hat{\xi}$ and $\hat{\theta}$ depend on the block size b. Moreover, the effective sample size to estimate the selected graph is taken as b in place of n. Later, Müller et al. [28] suggest the generalization of StARS by using the Poisson and the binomial distribution and under the convex non-asymptotic distributional modeling.

3.7 Kullback-Leibler Distance for Model Selection

In 1951, Kullback and Leibler published a paper which is about a scientific propose of the information and its close relationship with the Fisher's sufficient statistics. Now, we call this relation as the Kullback-Leibler (KL) information which is very frequently used in electrical engineering and signal progressing. The underlying relation is one of the basic theorems of the modern science which is based on the Boltzmann's entropy [8]. This idea is associated with the second law of thermodynamics.

Basically, the information of the KL distance is one of the ways for measuring the parametric approximation $f(.,\Theta)$ to the true density g as presented in Eq. 30. Here, let $x = (x_1, \ldots, x_n)^T$ denote a random variable and $\Theta = (\Theta_1, \ldots, \Theta_n)^T$ be a vector of parameters. Furthermore, we assume that the joint distribution of $y = (y_1, \ldots, y_n)^T$ can be obtained from an unknown G which has a density $g(y) = \prod_i g(y_i)$ with respect to μ. Then, let us denote $\hat{\Theta} = \hat{\Theta}(y)$ as MLE of Θ and let us define the information of the KL distance for the model M as

$$KL(g(x), f(.,\Theta(y)) = \int g(x) \log(g(x)/f(x,\Theta)(y)\mathrm{d}_\mu x \tag{29}$$

$$= \int g(x) \log g(x)\mathrm{d}_\mu x - \int g(x) \log f(x|\hat{\Theta}(y))\mathrm{d}_\mu x \tag{30}$$

for multiple $\Theta_1, \Theta_2, \ldots$ parameters under different models as listed by $M_1, M_2, \ldots$ based on the values of the KL information. When we define the second part of Eq. 30 as a target variable, the minimum KL distance which implies the maximum target variable should be selected. Then, by equating the last term on the right hand side of Eq. 30, we can obtain the following expression.

$$T = T(y) = \int g(x) \log f(x|\hat{\Theta}(y))\mathrm{d}_\mu x. \tag{31}$$

In the above equation, when we use the Taylor-series expansion for $\bar{\Theta}$, we can have an approximate of T via

$$T = T(y) = \int g(x) \log f(x|\hat{\Theta}(y))\mathrm{d}_\mu x - 0.5Q + o_{p(1)}. \tag{32}$$

Here, the exact definition of T is the value which minimizes the KL distance or maximizes the target variable $T(y)$. Moreover, in Eq. 32, we may define Q as

$$Q = \hat{\Theta}(y) - \bar{\Theta})^T \hat{J}(y, \hat{\Theta})\hat{\Theta}(y) - \bar{\Theta} \tag{33}$$

where

$$\hat{J}(y, \hat{\Theta}) = -(\partial^2)/(\partial\Theta\partial(\Theta^T)) \log f(y, \Theta). \tag{34}$$

In the application of these expressions, we need to estimate T since T depends on an unknown $g(.)$ and the method of MLE is a natural way to infer T. Thus, its estimate can be approximately obtained by

$$\log f(y|\hat{\Theta}(y)) = \log f(x|\bar{\Theta}) + 0.5(Q_1 - Q_2) + o_p(1). \tag{35}$$

If we write the maximum likelihood estimate of Θ in terms of Θ_1 and Θ_2 for models M_1 and M_2, respectively, we can present the difference of the maximum likelihoods in the logarithmic scale via

$$\log f(y|\hat{\Theta}_1) - \log f(y|\hat{\Theta}_2) = 0.5(Q_1 - Q_2) + o_p(1). \tag{36}$$

Hence, we can obtain the difference of the values of the target variable T under two alternate models as

$$T_1 - T_2 = -0.5(Q_1 - Q_2) + o_p(1) \tag{37}$$

when $\bar{\Theta}_1 = \bar{\Theta}_2$ and $-E(Q_1 - Q_2)$ is asymptotically equal to $-(p_1) - (p_2)$ when p_1 are p_2 the numbers of parameters of the models M_1 and M_2, in order. It is also known that if the corrected logarithm of the maximum likelihood is multiplied by -2 for the convenience, we can find the AIC equation again which can also be shown as follows.

$$\text{AIC} = -2 \log f(y, \hat{\Theta}(y)) + 2p. \tag{38}$$

On the other hand, Sugiura [33] and Hurvich and Tsai [23] suggest that the following expression in Eq. 39, which is called as the corrected AIC (AICC), has a better bias than BIC can find.

$$\text{AICC} = -\text{AIC} + (p + (p+1)(p+2))/(n - p - 2). \tag{39}$$

4 Application via Existing Model Selection Criteria

4.1 Data Description

In this part, we perform a quasi network for the description of gynecological cancers. Under these diseases, we obtain data from cervix, ovarian and endometrial cancers whose core genes can be listed as CTNNB1, TFAM, CEBPM, MAP2K1, MAPK1, TP53, PDIA3, IMP3, ERRB2, CHD4 and MBD3. From the literature, it is evaluated that these genes which have high impact in the selected cancers have connections with each other by generating a complete graph. It means that their adjacency matrices, which present the existence of edges between each pair of genes, are made up of only 1 s. Furthermore, their gene expression values are gathered from the ArrayExpress database from the following datasets. More details about these data can be found in the study of Bahçivanci et al. [6]

- *E-GEOD-9891 Transcription Profiling 285 Human Ovarian Tumour Data (Data 1):* The data are collected from 285 patients who have epithelial ovarian, primary peritoneal or fallopian tube cancers. The patients are diagnosed between 1992 and 2006 and identified in different hospital in Austria, UK and Netherland.
- *E-GEOD-6378 Expression Data from Vulvar, Cervical, Endometrial, Carcinoma Tissue (Data 2):* The dataset has 35 samples which are collected to find candidate biomakers and signatures from different gynecological cancer types. Hereby, 18 cancer samples include 5 cervical, 7 endometrial and 6 vulvar cancers which are hybridized on the Affymetrix platform in order to identify the common features among cancer types.
- *E-GEOD-81248 Expression Data from HEY Cells (Data 3):* The dataset has two samples of distal cells that are exposed. In this dataset, the mRNA expression is captured on the Affymetrix U133 Plus 2 chips and each gene has 12 observations.
- *E-GEOD-14764 A Prognostic Gene Expression Data in Ovarian Cancer (Data 4):* The data are composed of 80 ovarian carcinomas (TOC cohort) for the development of a predictive model. In this dataset, RNAs from 80 frozen ovarian cancer samples are hybridized on the Affymetrix Human Genome U133A Array.

In our analyses, we apply these datasets in two different ways. In the first application, we infer the networks by using merely the listed core genes, resulting in 11-dimensional networks. On the other hand, in the second application, we apply 1000 genes from each dataset in such a way that 11 genes are inside of these 1000 genes. By this way, we aim to evaluate the effect of the dimension in all model selection criteria.

4.2 Accuracy Measures

In the real data analyses, when we compare the performance of the model selection methods, we use the accuracy value and F-measure. These two values are calculated in terms of true positive (TP), true negative (TN), false positive (FP)and false negative (FN) values. The short representations of these terms via actual and predicted classes in a confusion matrix are shown in Table 1. For more detailed description, the study of Bahçivanci et al. [6] and Bülbül et al. [12] can be read.

Table 1. The confusion matrix.

		Actual class	
		Positive	Negative
Predicted class	Positive	TP	FP
	Negative	FN	TN

Accordingly, the accuracy value is defined by the ratio of the correctly classified objects in both labels to the total number of classified objects which is shown by the following expression.

$$\text{Accuracy} = \frac{\text{TP} + \text{TN}}{(\text{TP} + \text{FP} + \text{TN} + \text{FN})}. \tag{40}$$

On the other hand, we can define the F-measure, which is the harmonic mean of the precision, as below.

$$\text{F-measure} = \frac{2\text{TP}}{(2\text{TP} + \text{FP} + \text{FN})}. \tag{41}$$

In the computation, the perfect values of both accuracy measures are taken as one.

4.3 Results

In this section, we present the accuracy value and F-measure of AIC, BIC, extended BIC, CAIC, CAICF, ICOMP and StARS model selection methods for the given datasets. In Table 2, we present the performance of all criteria via small networks which are composed of the selected 11 core genes. From the results, it is seen that CAIC and CAICF have better accuracies than others. In the assessment, we take the mean of F-measures and accuracies of 3 datasets.

Table 2. F-measure (F) and accuracy (ACC) values for four networks based on 11 genes.

Criterion	Data 1		Data 2		Data 3		Data 4	
	F	ACC	F	ACC	F	ACC	F	ACC
AIC	**0.605**	**0.754**	**0.785**	**0.880**	0.091	0.167	0.422	0.593
BIC	0.605	0.754	0.107	0.194	**0.587**	**0.740**	0.422	0.593
EBIC	0.111	0.200	0.091	0.167	0.091	0.167	0.422	0.593
CAIC	0.111	0.200	**0.785**	**0.880**	0.455	0.625	**0.719**	**0.393**
CAICF	**0.630**	**0.773**	**0.785**	**0.880**	**0.587**	**0.740**	0.091	0.167
ICOMP	0.111	0.200	0.174	0.296	0.422	0.593	0.091	0.167
StARS	0.111	0.200	0.174	0.296	0.091	0.167	0.107	0.194

On the other hand, when we compare the findings in Table 3 computed from large networks which have 1000 genes, it is observed that there is no difference among model selection criteria. Moreover, although the accuracies of all criteria are high, their F-measures are lower with respect to their performances in small dimensions. Therefore, we consider an alternative criterion to improve both measures. Our proposal criterion is based on the KL divergence via the Bootstrap estimation. In the following section, we present the theoretical details of the method which has not been yet implemented for sparse networks.

Table 3. F-measure (F) and accuracy (ACC) values for four networks based on 1000 genes.

Criterion	Data 1		Data 2		Data 3		Data 4	
	F	ACC	F	ACC	F	ACC	F	ACC
AIC	0.020	0.999	0.020	0.999	0.020	0.999	0.020	0.999
BIC	0.020	0.999	0.020	0.985	0.020	0.999	0.020	0.999
EBIC	0.020	0.999	0.020	0.999	0.020	0.999	0.020	0.999
CAIC	0.020	0.999	0.020	0.999	0.020	0.999	0.020	0.999
CAICF	0.016	0.993	0.020	0.999	0.016	0.993	0.016	0.993
ICOMP	0.020	0.999	0.020	0.999	0.020	0.999	0.020	0.999
StARS	0.020	0.999	0.020	0.999	0.020	0.999	0.020	0.999

5 Proposed Model Selection Method for Sparse Biological Networks

In this part, we initially introduce the KL divergence for the sparse networks and then the bootstrap estimation of this divergence. Finally, we introduce our bootstrap-based model selection criterion proposed for the sparse networks.

In Eq. 1, we present the log-likelihood function for the covariance estimation which is obtained from the data. As stated previously, we denote $\Theta = \Sigma^{-1}$. Furthermore, as used beforehand, from Eq. 2, we can use the penalized likelihood expression of Friedman et al. [21] for the graphical lasso model.

Thereby, we suggest that the KL divergence may be applied in order to measure the difference between the estimated model and the true data. Here, we assume that the model is taken from the normal distribution via $N(0, \hat{\Theta}_{\lambda}^{-1})$ and the true data is taken from the normal distribution by $N(0, \Theta_0^{-1})$. Accordingly, if we give the KL divergence for the graphical lasso, we can find the following expression.

$$\mathrm{KL}(\Theta_0|\hat{\Theta}_{\lambda}) = 0.5\mathrm{tr}(\Theta_0^{-1}(\hat{\Theta}_{\lambda})) - \log|\mathrm{tr}(\Theta_0^{-1}(\hat{\Theta}_{\lambda})| - m. \tag{42}$$

So, the above expression can be stated by

$$\mathrm{KL}(\Theta_0|\hat{\Theta}) = -2l(\hat{\Theta}_{\lambda}) + n\mathrm{tr}(\hat{\Theta}(\Sigma_0 - S)). \tag{43}$$

In Eq. 43, $-2l(\hat{\Theta}_\lambda)$ is a biased estimation of the KL divergence while the bias can be found by $ntr\hat{\Theta}(\Sigma_0 - S)$ which includes unknown terms Σ_0. For more details of the KL divergence of the Gaussian graphical model, we suggest the study of Abbruzzo et al. [1].

5.1 Bootstrap Estimation of Kullback-Leibler Information Criterion

In this section, we represent the bootstrap estimation of Eq. 30. For this purpose, firstly, we replace x in $\log f(x|(\hat{\Theta})(y))$ with a bootstrap sample $y^* = (y_1^*, \ldots, y_n^*)^T$ and then, we compute the bootstrap expectation E_*. Here, in the computation of $E_* \log f(y^*, \bar{\Theta}(y))$, we use the sampling with replacement with an upper boundary constraint such that it cannot exceed the log-likelihood $\log f(y, \bar{\Theta}(y))$. Moreover, y^* is obtained from the non-parametrically re-sampled distribution of y. But this value is not a good estimator for T. For this reason, Shibata [32] suggest the Efron's bootstrap estimator in such as way that he assumes a re-sampling size m for the number of observation n. He also assumes that the random variable x is independent from y, but, it has the same distribution of y. Therefore, the bias of the log- likelihood of $\log f(y, \bar{\theta}(y))$ with a target variable T can be found as

$$E^y T(y) - \log f(y, \bar{\theta}(y)) = E^y E^x \log f(x, \bar{\theta}(y)) / \log f(y, \bar{\theta}(y)) \tag{44}$$
$$= logf(y, \bar{\theta}(x))/logf(x, \bar{\theta}(x)). \tag{45}$$

Hence, the expected value of a bootstrap estimator can be defined as

$$B_1 = E_* \log f(y, \bar{\Theta}(y^*)) / \log f(y^*, \bar{\Theta}(y^*)) \tag{46}$$
$$T_1 = \log f(y, \bar{\Theta}(y)) + B_1 \tag{47}$$

which is as an estimator of T. In application, the bootstrap estimated value of E_* is obtained by means of the results of the Monte Carlo simulation. Additionally, the expected value of E_* does not have the same bias in Eq. 38 because of the fact that the bootstrap estimated value is dependent on y. On the other side, the study of Cavamaugh and Shumway [13] proposes a different way for the calculation of the bias correction for the model selection of the Gaussian state space model. In order to estimate Q in Eq. 33 by the bootstrap method, the following expression is obtained as the expected value.

$$B_2 = 2E_* \log f(y, \bar{\Theta}(y^*)) / \log f(y, \bar{\Theta}(y)) \tag{48}$$

that is asymptotically equal to $-Q$ and $-B_2$ is used same as p in AIC. But, the study of Shibata [32] proves that both bootstrap bias estimates B_1 and B_2 are asymptotically equivalent under certain assumptions.

Below we list the steps of our bootstrap procedure. In this calculation, we aim to make a bias adjustment for the KL divergence. Here, we suppose that $\hat{\Theta}_n^*$ $(i = 1, \ldots, N)$ shows the number of bootstrapped replicates and E_* is the expectation of the bootstrap distribution $\hat{\Theta}_n^*$. Hence, the bootstrap sample $\hat{y}_n^*$ is applied to compute a bootstrap vector $\hat{\Theta}_n^*$. Thereby, given N individuals measured on p standardized predictors y, our bootstrap procedure is as follows.

- **First step:**
  1. Fit a lasso model to select predictors.
  2. Identify candidate values by using sample covariance.
  3. Calculate $\log f(y, \bar{\Theta}(x))/\log f(x, \bar{\Theta}(x))$.
  4. Finalized selected predictors y.
- **Second step:**
  1. Fit the KL divergence for sparse networks as in step 1.
  2. Estimate marginal sampling distributions $\hat{F}^*(\hat{\beta}_k)$ for $k \in s$.
  3. Draw B bootstrap samples ($B > 1000$).
  4. Use $\hat{F}^*(\hat{\beta}_k)$ to calculate the mean, SE (standard error), CI (confidence interval) for each $\hat{\beta}_k$.
  5. Exclude y:
     If CI for $(\hat{\beta}_k)$ includes 0.
     If $C^*_{var} = (t^*_k)^{-1} = (\text{SE}^*(\beta_k))/|(\text{mean}^*(\beta_k))|$.

6 Conclusion

In our study, we have presented different model selection methods, listed as AIC, BIC, EBIC, CAIC, CAICF, ICOMP and StARS for sparse small and high dimensional biological networks. In these analyses, we have used four real datasets taken freel from the ArrayExpress database. From the results, we have seen that the accuracies of CAIC and CAICF are higher than others for small systems. Whereas, it has been observed that there is no distinction among selected measures when we deal with high dimensional networks. On the other hand, as we observe that the accuracy values become lower for high dimensional systems, we have proposed an alternative model selection criterion for sparse systems. Hereby, we have represented the mathematical details of our proposal method. This proposal approach is based on the KL distance and its correction via the bootstrapping estimation. As the extension of this study, we consider to apply this approach within the Gaussian graphical models and compare its accuracies with other methods.

Acknowledgement. The authors thank to Ms Gül Bahar Bülbül for her help while preparing the tables.

References

1. Abbruzzo, A., Vujacic, I., Wit, E., Mineo, A.M.: Generalized information criterion for model selection in penalized graphical models. Arxiv (2014)
2. Akaike, H.: Information theory and an extension of the maximum likelihood priciple. In: Petrov, B.N., Csaki, F. (eds.) Second International Symposium on Information Theory, pp. 267–281. Akademiai Kiad, Budepest (1973)
3. Akaike, H.: A new look at the statistical model identification. IEEE Trans. Autocontrol **19**, 716–723 (1974)

4. Banerjee, O., El Ghaoui, L., d'Aspremont, L.: Model selection through sparse maximum likelihood estimation for multivariate Gaussian or binary data. J. Mach. Learn. Res. **9**, 485–516 (2008)
5. Ayyildiz, E., Ağraz, M., Purutçuoğlu, V.: MARS as an alternative approach of Gaussian graphical model for biochemical networks. J. Appl. Stat. **44c**(16), 2858–2876 (2017)
6. Bahçivancı, B., Purutçuooğlu, V., Purutçuoğlu, E., Ürün, Y.: Estimation of gynecological cancer networks via target proteins. J. Multidiscip. Eng. Sci. **5**(12), 9296–9302 (2018)
7. Bogdan, M., Ghosh, J.K., Doerge, R.W.: Modifying the Schwarz Bayesian information criterion to locate multiple interacting quantitative trait loci. Genetics **167**, 989–999 (2004)
8. Boltzmann, L.: Uber die Beziehung zwischen dem zweiten Hauptsatze dewr mechanischen Warmetheorie und der Wahrscheinlichkeitsrechnung, respective den Satzenuber das Warmegleichgewicht. Weiner Bericte **76**, 373–435 (1877)
9. Boyd, S., Vanderberghe, L.: Convex Optimization. Cambridge University Press, Cambridge (2004)
10. Bozdogan, H.: Model selection and AIC: the general theory and its analytical extensions. Pscychometrica **52**(3), 345–370 (1987)
11. Bozdogan, H.: A new class of information complexity (ICOMP) criteria with an application to costumer profiling and segmentation. Istanbul Univ. J. Sch. Bus. Adm. **39**(2), 370–398 (2010)
12. Bülbül, G.B., Purutçuoğlu, V., Purutçuoğlu, E.: Novel model selection criteria on sparse biological networks. Int. J. Environ. Sci. Technol. **16**, 1–12 (2019)
13. Cavanaugh, J.E., Shumway, R.H.: A bootstrap variant of AIC for state-space model selection. Stat. Sin. **7**, 473–496 (1997)
14. Chen, J., Chen, Z.: Extended Bayesian information criterian for model selection with large model space. Biometrika **95**, 759–771 (2008)
15. Chen, J., Chen, Z.: Extended BIC for small-n-large-p sparse GLM. Stat. Sin. **22**, 555–574 (2011)
16. Claeskans, G., Hjort, N.L.: Model Selection and Model Everaging. Cambridge Series in Statistical and Probabilistic Mathematics. Cambridge University Press, Cambridge (2008)
17. Dempster, A.: Covariance selection. Biometrics **28**, 157–175 (1972)
18. Dobra, A., Lenkoski, A.: Copula Gaussian graphical models and their application to modeling functional disability data. Ann. Appl. Stat. **5**(2A), 969–993 (2011)
19. Efron, B.: The Jackknife, The Bootstrap and Other Resampling Plans. SIAM [Society for Industrial and Applied Mathematics], Philadelphia (1982)
20. Foygel, R., Drton, M.: Extended Bayesian information criteria for Gaussian graphical models. In: Advances in Neural Information Processing Systems, vol. 23, pp. 2020–2028 (2010)
21. Friedman, J., Hastie, T., Tibshirani, R.: Sparse inverse covariance estimation with the graphical lasso. Biostatistics **9**, 432–441 (2007)
22. Hastie, T., Tibshirani, R., Friedman, J.: The Elements of Statistical Learning: Data Mining, Inference and Prediction. Springer Verlag, New York (2009)
23. Hurvich, C.M., Tsai, C.L.: A corrected Akaike information criterion for vector autoregressive model selection. J. Time Ser. Anal. **14**, 271–279 (1993)
24. Lim, C., Yu, B.: Estimation stability with cross-validation. J. Comput. Graph. Stat. **25**(2), 464–492 (2016)

25. Liu, H., Roeder, K., Wasserman, L.: Stability approach to regulazation selection (STARS) for high dimensional graphical models. In: Proceeding of the Twenty-Third Annual Conference on Neural Information Processing System (NIPS), pp. 1–14 (2010)
26. Meinhausen, N., Buhlmann, P.: High dimensional graphs and variable selection with lasso. Ann. Stat. **34**, 1436–1462 (2006)
27. Meinhausen, N., Bühlmann, P.: Stability selection. J. Roy. Stat. Soc. Ser. A **72**, 417–473 (2010)
28. Müller, C.L., Bonneau, R., Kurtz, Z.D.: Generalized stability approach for regularized graphical models. Arxiv (2016)
29. Politis, D.N., Romano, J.P., Wolf, M.: Subsampling. Springer, Heidelberg (1999)
30. Schwartz, G.: Estimating the dimension of a model. Ann. Stat. **6**, 461–464 (1978)
31. Shah, R.D., Samworth, R.J.: Variable selection with error control: another look at stability selection. J. Roy. Stat. Soc. B **1**(1), 55–80 (2013)
32. Shibata, R.: Bootstrap estimate of Kullback-Leibler information for model selection. Stat. Sin. **7**(2), 375–394 (1997)
33. Sugiura, N.: Further analysis of the data by Akaike's information criterion and the finite correction. Commun. Stat. Theory Methods **A7**, 13–26 (1978)
34. Yuan, M., Lin, Y.: Model selection and estimation in Gaussian graphical model. Biometrika **94**, 19–35 (2007)
35. Zhao, P., Yu, B.: On model selection consistency of lasso. J. Mach. Learn. Res. **7**, 2541–2563 (2006)
36. Whittaker, J.: Graphical Models in Applied Multivariate Statistics. Wiley, New York (1990)

ICS Cyber Attack Analysis and a New Diagnosis Approach

Ercan Nurcan Yılmaz[1], Hasan Hüseyin Sayan[1], Furkan Üstünsoy[1], Serkan Gönen[1], Erhan Sindiren[2](✉), and Gökçe Karacayılmaz[2]

[1] Faculty of Technology, Gazi University, 06500 Ankara, Turkey
erhan.sindiren@gazi.edu.tr
[2] Institute of Informatics, Gazi University, 06680 Ankara, Turkey

Abstract. Artificial Intelligence and Machine Learning technologies have a widespread use in many disciplines thanks to raw data processing and computational power. The capabilities of these technologies will enable the identification of legal and illegal traffic/behavior by classifying these data rapidly and without damaging the continuity of the system through the high amount of network traffic/behavior, which is one of the biggest problems in the field of cyber security. In this respect, artificial intelligence and machine learning technologies will provide valuable contributions in protecting Industrial Control Systems (ICS), which play an important role in the control of critical infrastructures such as electrical power generation-transmission-distribution systems, nuclear power plants, gas and water, against cyber-attack. In this study, it is aimed to reveal the anatomy of the attacks by executing denial of service, Start/Stop, and the man in the middle attacks to PLCs, an important component of ICS. In the test environment created in the study, the attacks on the real two types of PLCs were analyzed. The analyzes focused to obtain the rule sequences, which can be used by the artificial intelligence and machine learning technologies, by benefitting from data sets obtained in the test environment. In this way, a new security approach has been created for ICS. The results also revealed the importance of PLCs' vulnerability to attacks and the continuous monitoring of the network in order to detect and identify the attacks as soon as possible and to protect the ICS and to maintain its continuous functioning.

Keywords: Artificial intelligence · Machine learning · Cyber-attack · PLC attack · Industrial control systems · Start/Stop · Man in the middle · Denial of service · Diagnosing · Detection

1 Introduction

Industrial Control Systems (ICS) serve as the basic infrastructure for controlling or operating any industrial system, including those used in critical infrastructures. The use of ICS has increased as a result of increased industrial process automation, low operating cost and growth in the global economy. Therefore, they have acquired widespread use beyond traditional practices for highly critical infrastructure systems such as nuclear facilities, power generation, water, transportation. After being used in

D. J. Hemanth and U. Kose (Eds.): ICAIAME 2019, LNDECT 43, pp. 127–141, 2020.
https://doi.org/10.1007/978-3-030-36178-5_11

the 1960s, ICS have experienced a great improvement and change with the alterations in technology. ICS have been transformed into client/server systems that use centralized communication protocols to send data from the client, peripherals, to the master. With the Internet of Things (IoT), the ICS sensor and devices connected to the Internet, ICS are transformed from a stand-alone system into a remote accessible system connected to the Internet. With this process, ICS have become an attractive target for hackers [1–3].

There is a number of benefits for Information Systems, including Internet access, scalability, better communication protocols, productivity, cost effectiveness, interoperability among components [4], and remote access, but ICS have never been designed with network connectivity and security aspect in mind [2, 5]. Because in ICS, continuity is emphasized rather than security and it has been tried to provide protection by using special standards [6]. Since the 1990s, however, control systems have been integrated into computer networks, and at the same time more Commercial off the shelves (COTS) products have been used in the ICS [7]. As a result, the ICS server and user interfaces are now accessible via the Internet and cellular networks, which provide many entry points for an attacker. Internet and cellular network connectivity makes it easy for an attacker to have an in-depth knowledge of ICS networks vulnerabilities. Today, most ICS communication protocols do not include message authentication and only use plain text. As a result, they have become more vulnerable to information disclosure attacks such as Man in the Middle (MitM) attack [8–10]. Using hybrid protocols in control networks, security vulnerabilities specific to TCP/IP protocols have also led to new threat environments for ICS.

Cyber-attacks on ICS continue with a significant increase [11]. General techno-logy awareness, the widespread availability of free information and the presence of malicious elements make such attacks easier and more likely. In this context, it is an urgent need to take adequate measures to prevent cyber-attack and to strengthen the defense in depth. At the time of an attack, the organization should be able to deactivate all components affected by the attack. Hence, when the disaster recovery plans are prepared with the reaction to be shown at the time of the attack, the organization should not forget that cyber security is not an instant response but a process. If we accept cyber security as a process, continuous monitoring is becoming the most important component of this process. Common cyber security measures such as limiting physical access, cryptography, patch management, separation of enterprise and production systems (network segmentation), Firewall and Access Control List (ACL) are some of the security methods applied to IT. However, these security practices do not show the same success and impact in in ICS [12].

ICS projects are expected to operate continuously for many years after commissioning without the need for patching and updating [13]. This raises a dilemma between managers providing continuous system operation and adequate security. The application of software updates and patches is usually postponed on the execution environment [14]. Because security measures, such as patches and antivirus applications, can lead to unintended consequences, particularly in the slowing down of the transmission or the prevention of normal operation of the system. As a result, the operational nature of these systems hampers cyber security testing during commissioning and the security conditions of the systems remain unclear. In practice, 100% safety or protection of systems (such as computer systems, PLC, SCADA) is not possible. For this reason, the key component in the field of security is to create the best security model for critical infrastructure systems. Continuous monitoring is one of the most important parts of this security model.

One of the areas investigated in the field of cyber security attacks is machine learning and artificial intelligence [15]. Although machine learning and artificial intelligence technologies provide enormous benefits for simplifying simple tasks and analyzing complex equations, it does not seem possible to realize the intuitive capabilities that a person can have in the near future. However, the raw data computational power is absolutely undeniable, and from this point of view the performance of machine learning and artificial intelligence technologies combined with human intuition seems to be undeniable in the field of cyber security. Today, one of the biggest challenges in the field of cyber security is the detection of abnormalities in the information systems such as malicious behaviors and malicious assets, etc. without additional burden and interruption to the main system. At this phase, machine learning and artificial intelligence technologies emerge to process data. These technologies allow the identification of anomalies in big-data sets within a very short time in accordance with the predefined rules. In our study, we have focused on creating a series of rules that can be used by Artificial Intelligence and machine learning technologies by using the data sets obtained in the test environment and revealing a new understanding of security for ICS.

In the study, a test environment (Testbed) was created for detecting security vulnerabilities of the protocols used in the PLC device which is one of the important components of ICS. In this context, Siemens (s7-1200) and Schneider (M241) PLC devices have been tested. In the test environment, Denial of Service-DoS, Start/Stop and Man in the Middle-MitM attacks were performed and thereafter packets of attacks were captured and analyzed and the patterns of the attacks were found out. Thus, it is aimed that the network will be saved from the attacks with the least damage and to use these patterns effectively to prevent similar attacks.

In diagnostic analysis, in order to preserve the continuity, which is the most important requirement of ICS, the samples were transferred to the diagnostic system by using the mirroring technique without adding additional load to the system. During the diagnostic phase, it has been observed that, continuous monitoring of network traffic can provide significant contributions to diagnose possible attacks and take subsequent measures.

2 Related Works

Some of the studies on ICS security focused on analysis based on simulation models. In this context, Giani et al. have suggested two different test environments for examining ICS. In the first test environment Simulink Mathworks and in the second test environment, Omnet ++ DEVS was used to transfer the network components to the simulation environment for testing the attacks [16]. In their study, Oman and Phillips examined TELNET leakage inputs via a simulation model-based test environment [17].

On the other hand, Byres, Hoffman and Kube examined the identification of vulnerabilities in command lines of protocols used in ICS. As a result of their study, high severity vulnerabilities were identified and urgent solutions were proposed [18]. In their study on the security of ICS, Genge, Graur and Haller were analyzed in depth defence strategies network segmentation, network firewall configuration, IDS/IPS systems and anomaly detection system on the simulation system. They emphasized the importance of the use of defense in depth according to the results of the attack scenarios [19]. Sayegh, Chehab and Elhajj tested the openings of the Omron PLC and HMI equipment and the FINS protocol used by these systems [20]. Simulation and modeling techniques are useful for modeling and testing complex systems. The development of realistic models can help to create scenarios that do not yet exist or are too costly to build. However, one the most important lack of studies based on simulation systems is the difficulty of fully reflecting the actual system, and the fact that the analyzes made do not give the same results in the real system at all times.

It is observed that most of the studies carried out for ICS are directed to simulation and emulation systems, and applications with real systems are quite limited. It is very important to analyze the actual system components in order to determine the damages that may occur in the systems and the precautions to be taken as a result of the attacks against ICS used in the management of critical infrastructure. For this reason, in the study, diagnostic analyzes of attacks were carried out on the testbed including real ICS components.

3 PLC Attack Diagnostic Analysis and Inferences

In this study, diagnostic analysis tests were carried out in a testbed where a real control systems were involved. The main focus was to determine the vulnerabilities of the PLC device and the solutions for dealing with these vulnerabilities.

As depicted in Fig. 1, Denial of Service (DoS), Start/Stop and Man in the Middle (MitM) attacks were carried out on PLCs. The analysis of each attack method consists of three basic stages. In the first stage, attacks on PLC were carried out and their effects on the system were measured. The second stage is the observation phase based on the analysis of the packets of the attacks. The last stage is the diagnosis and detection phase in which the attacks are identified by creating attack-patterns for protecting the system from similar attacks.

3.1 Testbed Environment

During the first phase of the analysis, the DoS, MitM and Start/Stop attacks for industrial control systems were carried out with Kali Metasploit, Ettercap and Armitage attack vehicles and Wireshark network traffic analysis tools. The testing process was carried out on unsegmented network topology as shown in Fig. 2 and segmented network topology as shown Fig. 3.

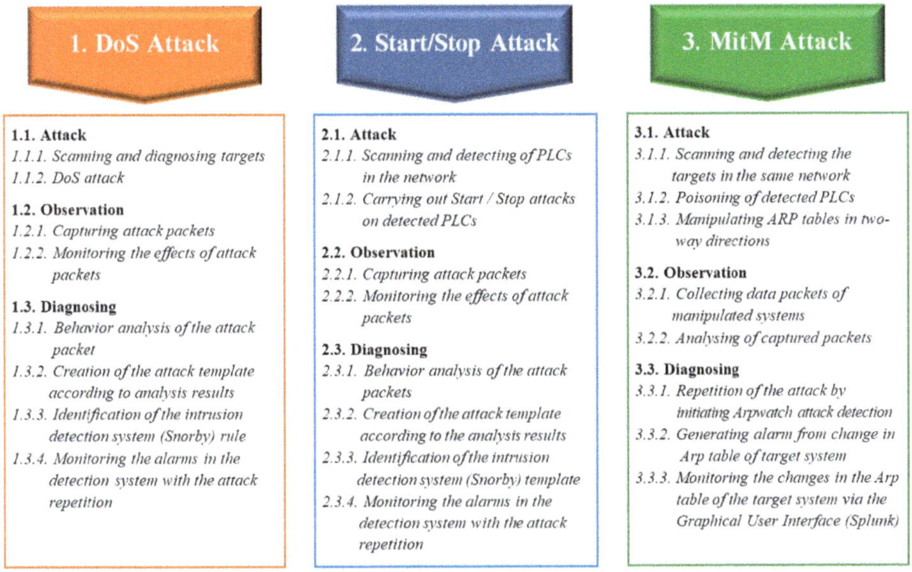

Fig. 1. Phases of reconnaissance, attack and diagnosing for PLC devices.

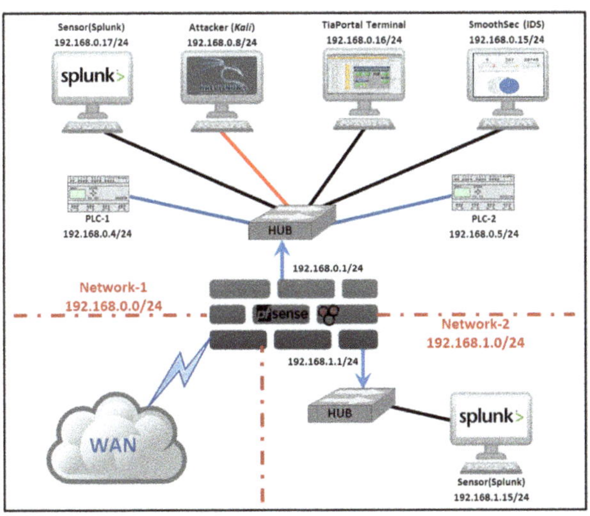

Fig. 2. Unsegmented network topology (diagnose).

Network segmentation is a security measure that is often used to better protect hardware or applications that are vital in information technologies and networks and to isolate them from other parts of the network. Therefore, in a segmented network topology, the PLCs are configured in a separate network to isolate PLCs from other components as shown in Fig. 3.

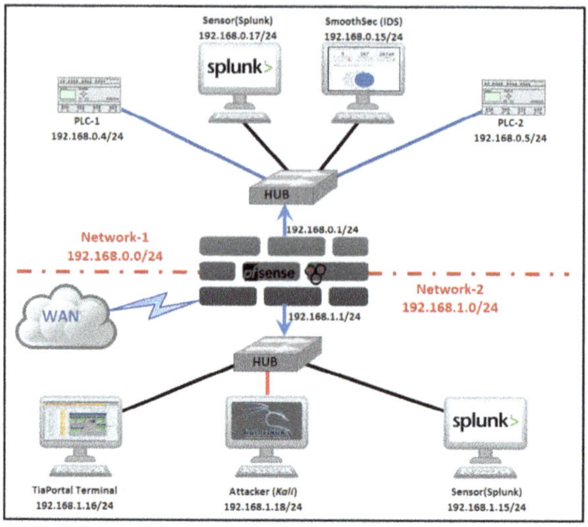

Fig. 3. Segmented network topology (diagnose).

3.2 Attack, Observation and Diagnose

In the second phase of the analyzes, it was aimed to monitor the attack packages carried out in the first stage, to make the package analysis, to create the attack patterns and to identify the similar attacks to be made at a subsequent time by entering the attack rules related to these patterns to detection systems (smoothsec and splunk).

In this context, all of the intrusion and detection systems are located in the 192.168.0.0/24 network in the unsegmented network topology shown in Fig. 2. SmoothSec IDS software (sensor) to diagnose network-related attacks, especially denial of service (DoS) and Start/Stop and Splunk log management software for detecting the MitM attack method have been added to network.

On the other hand, in the segmented network topology shown in Fig. 3, 2 PLCs, Splunk log monitoring and management computer and SmoothSec IDS software computer were located in network-1 (192.168.0.0/24), while TIA Portal management computer, Kali Linux computer and the second Splunk log monitoring and management computer (sensor) were located in network-2 (192.168.10.0/24).

Attack detection is an activity used to monitor and report suspicious activity of a system or network. An attack diagnosis system can be a software or hardware that identifies interventions to a system or a network, or a combination of both [21]. Active IDS attempts to prevent attacks or countermeasures, or at least warn the security administrators or operators.

Passive IDS logs only attack details or generates tracks for control. Currently, IDS systems for ICS are limited and have not reached the level of performance in computer systems. The reasons for this are;

- The lack of a well-known threat model,
- High probability of false alarm or false-negative,
- The development of customized IDS systems in ICS environments has not yet proven itself for real systems.

An IPS is a network security or threat prevention technology that scans network traffic and prevents abuse in situations such as unauthorized access to an application or system. IPS is located behind the firewall to analyze risky content. While IDS scanning and reporting; IPS can actively analyze and process in network traffic [22].

There are two basic approaches to identify attacks: signature-based and abnormality-based. Signature-based diagnostics functionally the same as a traditional antivirus scanner.

For each attack event, the system searches the event library for a known signature. To implement a more systematic and secure solution, an anomaly-based diagnosis should be considered. In the diagnosis based on anomalies, firstly the incoming traffic of a network is scanned. All legal (normal) traffic is filtered and traffic packets that display abnormal behavior are monitored. Both approaches have pros and cons. With the database of known signatures, the signature-based IDS is more reliable and the system works better when it receives matching patterns, but it cannot identify new attacks. On the other hand, in spite of its disadvantages of increasing the number of false alarms, the IDS based on anomalies can identify unknown (especially zero-day) attacks.

Firstly, Start/Stop and DoS attacks were carried out on PLC devices and in order to be able to diagnose the attacks, and obtain the rule according to the steps shown in Fig. 1, intrusion packets were captured through the Wireshark packet analyzer. Subsequently, the signature of the attack was revealed by continuous monitoring of port 102 (Siemens s7-1200) and port 506 (Schneider M241). With the result of the signature, two different rules (see Figs. 4 and 5) that can be identified in the industrial control systems' Start/Stop and DoS attacks were entered into the Snort IDS which is included in the Smoothsec intrusion detection system. The attacks were repeated and the rules were successfully tested to check the diagnostic capability of the rules entered. In this way, the norms against internal and external threats were determined and work was carried out to diagnose deviations (anomalies).

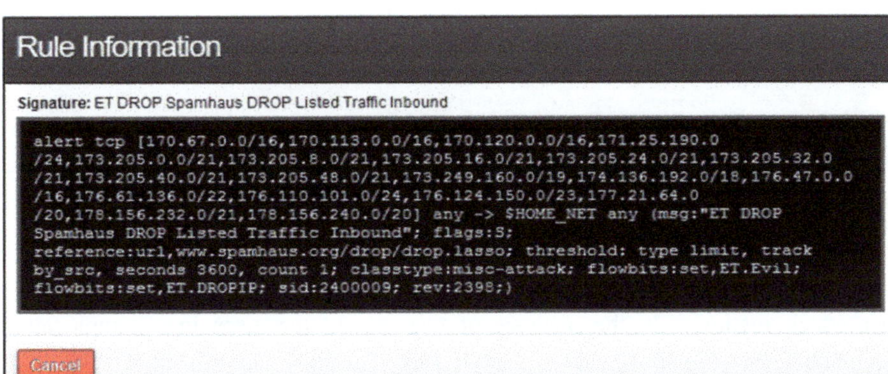

Fig. 4. Rule entered to Snort for detecting (diagnosing) DoS attack.

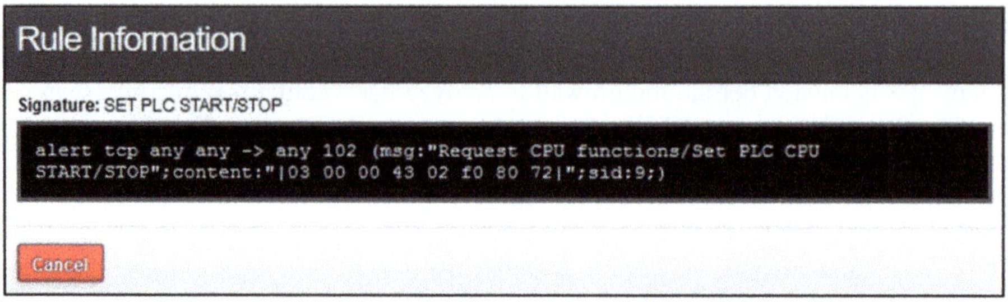

Fig. 5. Rule entered to Snort for detecting (diagnosing) start/stop attack.

In order to get detailed information about the attacks, the list sessions under the events tab on the Snorby monitoring screen are examined. When the screen image presented in Fig. 6 is examined, it is seen that the events identified were sorted by the time they occurred.

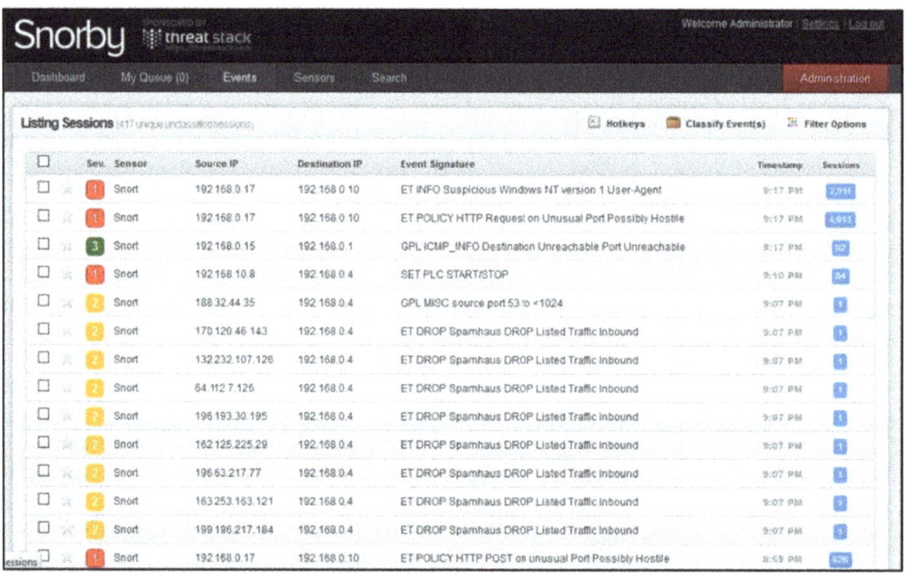

Fig. 6. Transforming packets captured by IDS sensors into events

Another attack on the PLC devices, MitM attack, includes different stages from other attack methods. In the analysis of this attack, the steps in a sample attack diagnosing process as shown in Fig. 7 were followed. In this context, first of all, target PLC and the command system, in which PLC communicate, were identified in order to realize MitM attack with ARP poisoning by scanning the network. Secondly, the attack was carried out and some sensitive information about the system was intercepted by intervening the communication of the PLC. Finally, it was aimed that the warning system based on the continuous control of the ARP table and the visual log display of the Splunk Intrusion Detection System would prevent the system from being damaged in case of similar attack.

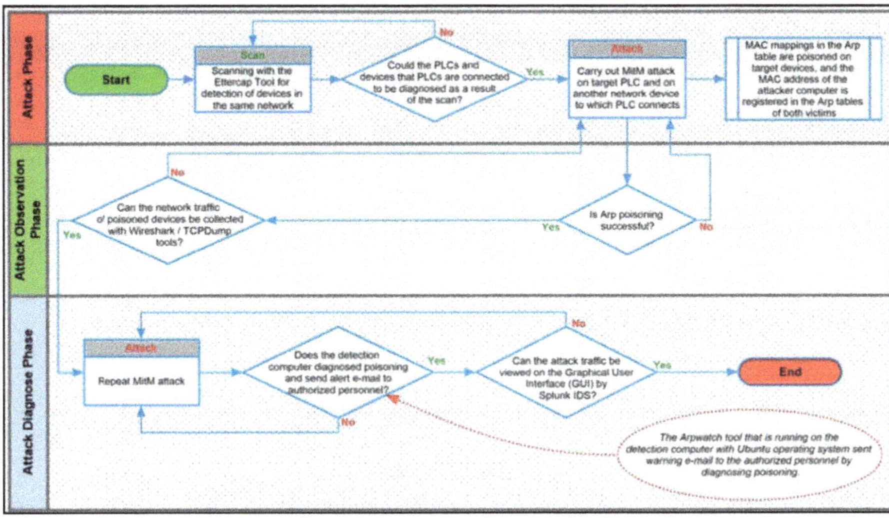

Fig. 7. Phases of reconnaissance, attack and diagnosing of MitM attack.

During the diagnosis of MitM attack, ARP table was created by gathering the MAC addresses of the devices on the network with Arpwatch application installed on Ubuntu (Linux) operating system. In case of any changes (add, change, etc.) in ARP table, it was provided to send email by the application (Arpwatch) to the desired user (root) via SMTP Server (send mail feature). The received e-mails were displayed with the console-based working email client (Alpine) in Fig. 8.

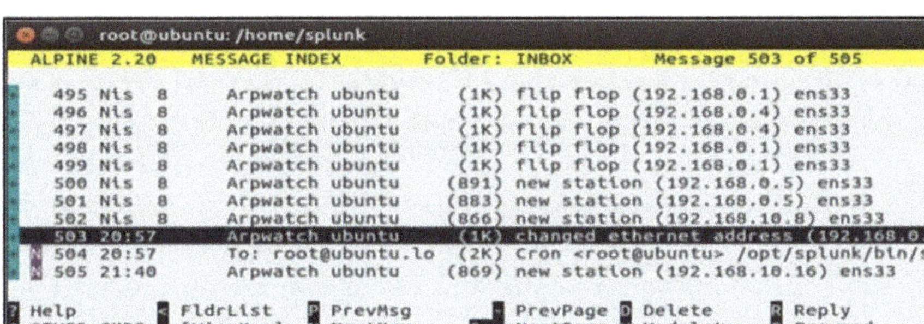

Fig. 8. Alert emails for the diagnosis of MitM attack

The content of the warning package numbered 503 from the attack alert e-mails is shown in Fig. 9. When the packet is evaluated in general, it has been found that the attacker replaced the MAC address (00:e0:4c:53:44:58) of network device, which had 192.168.0.1 IP address, with his MAC address (00:04:1:14:04:36).

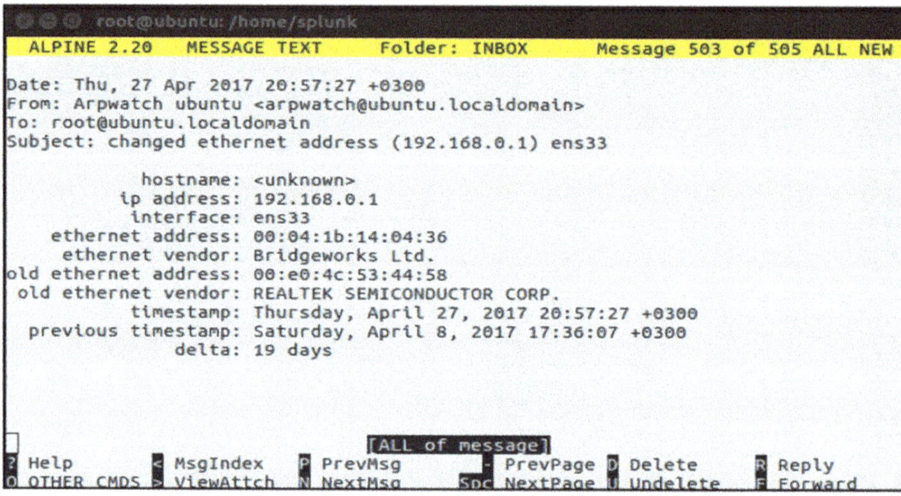

Fig. 9. Alert email contents for the diagnosis of MitM attack.

The warnings created for the diagnosis of MitM attack can be listed on Linux console screen as shown in Fig. 10, and it can be monitored more visually and functionally by transferring MitM warning logs and adapting the sensors of Splunk IDS system in Fig. 11 with Ubuntu operating system.

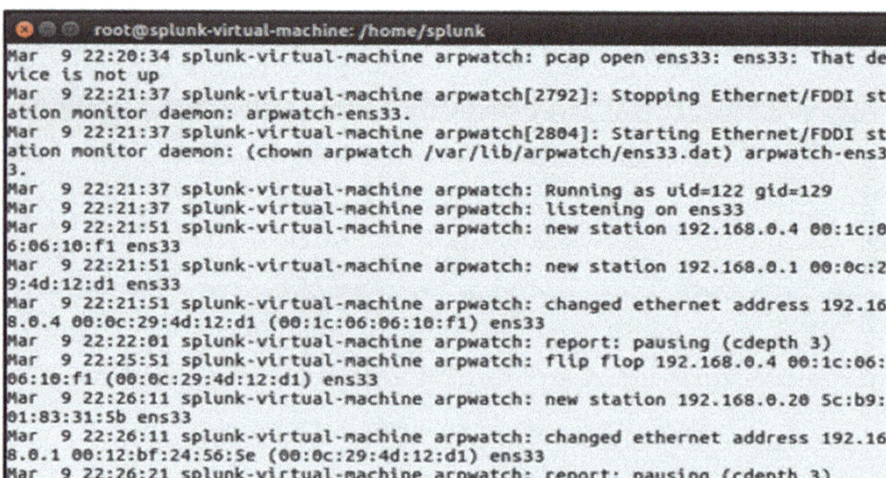

Fig. 10. Console screen of changes in ARP table after MitM attack.

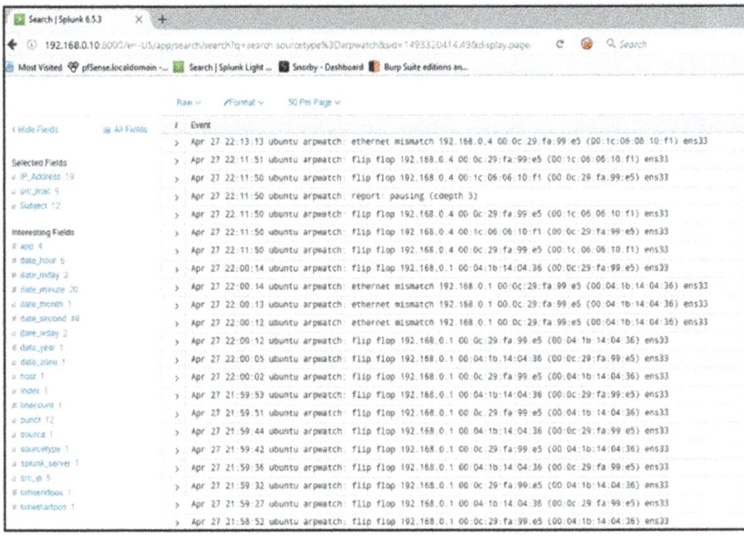

Fig. 11. Splunk log management screen of changes in ARP table after MitM attack.

4 Analyzes and Inferences

The attacks and their successes are analyzed in Table 1. When Table 1 is evaluated, the attack application can be divided into three groups. The first two groups are related to segmentation of the network topology (segmented and unsegmented) and the third is to make the read/write password active or inactive.

In the testbed the attacks firstly carried out on unsegmented network and all the attacks were successful and it was seen that an important component of Industrial Control systems such as PLC had significant vulnerabilities. In the second phase, attacks were carried out on segmented network topology and the result of attacks showed that other attacks were successfully carried out except for the MitM attack. As a result of the MitM attack, the importance of network segmentation is better understood, given the clear transmission of data transmitted in the network and the adverse effects that may occur in the system.

Table 1. Attack/diagnostic results (✓ successful, **X** unsuccessful)

	Attack type		
	DoS	MitM	Start/stop
Unsegmented network topology	✓	✓	✓
Segmented network topology	✓	X	✓
Read/Write protection password	✓	✓	X
SmoothSec	✓	–	✓
Splunk	–	✓	–

In the start/stop attack scenario performed to the PLC, it was seen that the Start/Stop attack was not carried out when the read/write password of PLC was activated by the TIA Portal program but other attacks were successful. As a result of this test, the importance of activating the read/write password and authorizing only the required quantity personnel (least to know) and consequently the effective implementation of password management has been seen.

System administrators can more easily access control computers with the TIA Portal application installed, and therefore, attacks on the TIA Portal can be more easily and quickly identified and necessary measures can be taken. However, it can take much longer to diagnose the attacks on PLCs. As PLCs are field devices such as RTU and IED which are far away, it is very difficult for system/security administrators to notice the symptoms of attacks on existing infrastructures. Cyber-attacks on PLCs are often noticed when vital results occur after the attack, as in the case of STUXNET. For this purpose, PLCs were chosen as targets in attack scenarios because of the more destructive results of the attack.

In the analyzes of the attacks, the mirroring method was used to analyze the entire traffic flow in the network and the contents of these traffic packets were analyzed for anomalies by packet analysis software. Within the scope of the diagnosis of active attacks, The Start/Stop attack was successfully identified by entering the attack pattern as a rule to the Smoothsec system. The DoS attack carried out from the Bogon IP addresses was also successfully identified with the Smoothsec system.

For MitM attack, which is described as a passive attack, Arpwatch application installed on Ubuntu operating system, so in case arp poisoning was performed in order to listen the packages related to any device on the network, e-mail was sent to the system administrator via the application. In addition to warning e-mails, security administrators/operators could see the changes in arp table via the console or via the splunk log management interface.

4.1 Analysis Results

In the security analysis section of the study, important information of a critical device could be obtained by using free accessible analysis programs such as Kali Linux operating system and Wireshark, as generally encryption is not used to transmit or receive packets in communication protocols of ICS. Serious physical hazards can emerge from targeted attacks on vulnerabilities of unprotected industrial protocols. Therefore, MitM, DoS and Start/Stop attacks were performed in the test environment to draw attention to the security vulnerabilities of the PLC. The results of the attacks have shown that PLCs are vulnerable to these attacks and that network segmentation and password protection are critical for critical devices such as PLCs. On the other hand, the results of the analysis for the diagnosis phase of the attacks showed that the continuous monitoring of the communication traffic to the PLCs could provide significant contributions to the identification of possible attacks and to take measures.

5 Conclusion

Attacks have now shifted from physical to cyber environment, as very effective and comprehensive attacks can be carried out with very low cost and criminal sanctions of cyber-crimes is also very low. With the Internet of Things (IoT), the necessity of ICS sensors and devices connected to the Internet, ICS is transformed from an isolated system into a remotely accessible system connected to the Internet. With this process, ICS have become an attractive target for hackers. Therefore, in this study, the vulnerabilities of ICS, which is used in the management of critical infrastructures and very important for social life.

The average time period between the first infiltration of the attackers and the identification of the leakage is 146 days [23]. If the presence of leaks is detected by an external source, the average is 320 days, when it is diagnosed within the organization it takes 56 days. When the figures mentioned above are examined, it is very clear that the existence of the aggressors must be identified by the organization in order to minimize the damage caused by cyber-attacks. Diagnosing the presence of an attacker who is able to intrude the systems can be regarded as a very important step in preventing the attacker from spreading to the systems (lateral movement) and removing the data. After that, it is important to make the recovery operations that will make the systems secure.

Although IDS systems are currently limited to ICS, they become a must for systems protection when they are combined with continuous monitoring, especially compared to IPS. As IPS may cause instability in the operation of the system in case of incorrect interferences. IPS systems are considered to be superior to IDS because of their autonomous operation but it is strictly undesirable to interrupt access to a system used in critical infrastructures in case of false alarms. Therefore, the use of IDS instead of IPS and the 24/7 continuous monitoring of alarms generated by the IDS system is a must essential.

Siemens s-7 1200 and Schneider M241PLC devices were used in the test environment. However, similar applications can be applied to other brands and models. During the diagnostic phase of the study, SmoothSec system including network security tools like Snort, Snorby, Barnyard and Sagan, was developed. In order to maintain continuity, which is the most important requirement of ICS, the samples were transferred to the diagnostic system by using the mirroring technique without adding additional load to the system. The diagnostic phase of the analysis has shown that continuous monitoring of network traffic can contribute significantly to the identification of possible attacks and subsequent measures

References

1. Li, S., Tryfonas, T., Li, H.: The Internet of Things: a security point of view. Internet Res. **26**(2), 337–359 (2016)
2. Mallouhi, M., Al-Nashif, Y., Cox, D., Chadaga, T., Hariri, S.: A testbed for analyzing security of SCADA control systems (TASSCS). In: ISGT, pp. 1–7. IEEE, Anaheim (2011)

3. Lechner, U.: IT-security in critical infrastructures experiences, results and research directions. In: Fahrnberger, G., Gopinathan, S., Parida, L. (eds.) 15th International Conference on Distributed Computing and Internet Technology. LNCS, vol. 11329, pp. 42–59. Springer, Bhubaneswar (2019)
4. Genge, B., Siaterlis, C., Fovino, I.N., Masera, M.: A cyber-physical experimentation environment for the security analysis of networked industrial control systems. Comput. Electri. Eng. **38**(5), 1146–1161 (2012)
5. Erol-Kantarci, M., Mouftah, H.T.: Smart grid forensic science: applications, challenges, and open issues. IEEE Commun. Mag. **51**(1), 68–74 (2013)
6. Zhu, B., Joseph, A., Sastry, S.: A taxonomy of cyber attacks on SCADA systems. In: International Conference on Internet of Things and 4th International Conference on Cyber, Physical and Social Computing, pp. 380–388. IEEE, Dalian (2011)
7. Almalawi, A., Yu, X., Tari, Z., Fahad, A., Khalil, I.: An unsupervised anomaly-based detection approach for integrity attacks on SCADA systems. Comput. Secur. **46**, 94–110 (2014)
8. Pidikiti, D.S., Kalluri, R., Kumar, R.K.S., Bindhumadhava, B.S.: SCADA communication protocols: vulnerabilities, attacks and possible mitigations. CSI Trans. ICT **1**(2), 135–141 (2013)
9. Yang, Y., McLaughlin, K., Littler, T., Sezer, S., Wang, H.F.: Rule-based intrusion detection system for SCADA networks. In: 2nd IET Renewable Power Generation Conference (RPG 2013), pp. 1–4. Institution of Engineering and Technology, Beijing (2013)
10. Jain, P., Tripathi, P.: SCADA security: a review and enhancement for DNP3 based systems. CSI Trans. ICT **1**(4), 301–308 (2013)
11. Dell Security: Annual Threat Report. http://www.netthreat.co.uk/assets/assets/dell-security-annual-threat-report-2016-white-paper-197571.pdf. Accessed 22 Dec 2018
12. Fovino, I.N., Carcano, A., Masera, M., Trombetta, A.: An experimental investigation of malware attacks on SCADA systems. Int. J. Crit. Infrastruct. Prot. **2**(4), 139–145 (2009)
13. Kirsch, J., Goose, S., Amir, Y., Wei, D., Skare, P.: Survivable SCADA via intrusion-tolerant replication. IEEE Trans. Smart Grid **5**(1), 60–70 (2014)
14. Pauna, A., Moulinos, K.: Window of exposure… a real problem for SCADA systems? The European Union Agency for Network and Information Security (ENISA). https://www.enisa.europa.eu/publications/window-of-exposure-a-real-problem-for-scada-systems. Accessed 22 Dec 2018
15. Berger, I., Rieke, R., Kolomeets, M., Chechulin, A., Kotenko, I.: Comparative study of machine learning methods for in-vehicle intrusion detection. In: Katsikas, S.K., Cuppens, F., Cuppens, N., et al. (eds.) International Workshop on Security and Privacy Requirements Engineering. LNCS, vol. 11387, pp. 85–101. Springer, Barcelona (2019)
16. Giani, A., Karsai, G., Roosta, T., Shah, A., Sinopoli, B., Wiley, J.: A testbed for secure and robust SCADA systems. SIGBED Rev. **5**(2), 1–4 (2008)
17. Oman, P., Phillips, M.: Intrusion detection and event monitoring in SCADA networks. In: International Conference on Critical Infrastructure Protection (ICCIP), pp. 161–173. Springer, Hanover (2007)
18. Byres, E., Hoffman, D., Kube, N.: On shaky ground – a study of security vulnerabilities in control protocols. In: 5th International Topical Meeting on Nuclear Plant Instrumentation, Controls, and Human Machine Interface Technology (NPIC & HMIT), vol. 1, pp. 782–788. American Nuclear Society - ANS, Albuquerque (2006)
19. Genge, B., Graur, F., Haller, P.: Experimental assessment of network design approaches for protecting industrial control systems. Int. J. Crit. Infrastruct. Prot. **11**, 24–38 (2015)

20. Sayegh, N., Chehab, A., Elhajj, I.H., Kayssi, A.: Internal security attacks on SCADA systems. In: Third International Conference on Communications and Information Technology (ICCIT), pp. 22–27. IEEE, Beirut (2013)
21. Genge, B., Siaterlis, C.: An experimental study on the impact of network segmentation to the resilience of physical processes. In: 11th International IFIP TC 6 Networking Conference, NETWORKING 2012, pp. 121–134. Springer, Prague (2012)
22. Gaddam, R., Nandhini, M.: An analysis of various snort based techniques to detect and prevent intrusions in networks proposal with code refactoring snort tool in Kali Linux environment. In: International Conference on Inventive Communication and Computational Technologies (ICICCT), pp. 10–15. IEEE, Coimbatore (2017)
23. FireEye: M-Trends 2016 Special Report. https://www.fireeye.com/current-threats/annual-threat-report/mtrends/rpt-2016-mtrends-emea.html. Accessed 22 Dec 2018

Investigating the Impact of Code Refactoring Techniques on Energy Consumption in Different Object-Oriented Programming Languages

Ibrahim Sanlialp[1(✉)] and Muhammed Maruf Ozturk[2]

[1] Department of Computer Engineering, Kırsehir Ahi Evran University, Kırsehir, Turkey
ibrahim.sanlialp@ahievran.edu.tr
[2] Department of Computer Engineering, Suleyman Demirel University, Isparta, Turkey
muhammedozturk@sdu.edu.tr

Abstract. Code refactoring techniques that are used to improve the properties of the code such as readability, performance, maintenance are applied to the code depending on the type of coding. However, these techniques could increase energy consumption that this case can be considered as a hint for re-arranging them. This article includes an empirical experiment that investigates the effect of refactoring techniques energy consumption. C#, Java, and C++ are selected as experimental object-oriented languages. The individual effects of the five different code refactoring techniques are examined on similar applications coded with three different languages. The power consumption profiling tool namely Intel Power Gadget is used for measuring energy consumption of original and refactored codes. The findings of the analysis provide new insights into how a refactoring technique affects energy consumption with regard to the type of programming language.

Keywords: Code refactoring techniques · Object-oriented programming · Energy consumption · Open source code

1 Introduction

As population increases around the world, using information technologies becomes widespread. To maintain such cases, robust infrastructures are needed. Employing large infrastructures leads to high energy consumption. This means emitting adverse gases into the atmosphere that carbon footprints are gradually risen. In this respect, technological equipment should be considered in terms of global warming and climate change so that an urgent plan is required to overcome those problems [1].

Generally speaking, a specific awareness of energy consumption has attracted the interests of the researchers across the world. An approach named green software by practitioners is an evolving paradigm [2] and it aims to develop green software to reduce adverse effects on the environment.

D. J. Hemanth and U. Kose (Eds.): ICAIAME 2019, LNDECT 43, pp. 142–152, 2020.
https://doi.org/10.1007/978-3-030-36178-5_12

Green software can be regarded as a sub-field of software engineering [3]. In this context, there are many researchers presenting both hardware and software solutions to reduce energy consumption [4]. Hardware solutions generally comprise the observation of CPU energy consumption and its prediction. In addition to this, they are for configurations of peripheral devices. Software energy consumption studies mainly cope with mobile devices. The reason is that mobile devices need charging frequently. However, the growing speed of software scale and improvements in cloud computing leverage the level of energy consumption in personal computers. Therefore, new methods based either on refactoring or software development process is applied without changing the functional behavior of the software.

Code refactoring is a software activity changing source codes to improve its internal structure without altering the external behaviour [5]. Code refactoring techniques are used in developing software attributes such as software performance and reliability [6, 7]. They are applied depending on the type of programming language. Despite the fact that integrated development environments provide some refactoring techniques, developers can develop specific techniques to apply refactoring techniques. They are also employed for energy productivity which is evaluated as a quality factor [8]. Refactoring techniques can reduce energy consumption, thereby producing new codes from original codes. However, each refactoring technique may not create a favorable effect in terms of energy consumption. Park et al. validated this case by performing an experiment [9]. They investigate whether refactoring supports effective software development. According to their results, designing a refactoring technique without considering energy consumption generally results in ineffective source codes. They also stressed that 33 out of 63 refactoring techniques of M. Fowler are energy effective [9].

Object-oriented programming languages are becoming more important for practitioners and software engineers. Code smell [10, 11], performance [12, 13], and energy saving [9, 14] have been investigated in object-oriented languages.

Code obfuscation has also been analyzed in terms of energy consumption [22]. It was found that obfuscation has a two-side effect on energy consumption. However, it was only tested on mobile devices. Individual effects of refactoring were tested on embedded systems [19]. But the data sets of that experiment consist of C++ projects. Energy consumption works lack a comprehensive study evaluating well-known object-oriented languages together.

The motivation of this paper is to perform a comprehensive comparison, thereby analyzing common object-oriented languages along with some refactoring techniques. To this end, five refactoring techniques are selected to apply on C#, Java, and C++ source codes. An application coded with three different languages is tested to observe the effects of refactoring. To measure energy consumption, Intel Power Gadget [17] is employed. The results of this paper may provide useful information to green software practitioners to better understand energy saving designs.

The remaining of this paper is organized as follows. Section 2 presents the preliminaries of the method. Section 3 describes the applied methodology. The results of this work are reported and discussed in Sect. 4. Finally, Sect. 5 presents conclusions along with future research.

2 Theoretical Background

Before focusing on the applications for the impact of code refactoring techniques on energy consumption in different object-oriented programming languages, it is important to give brief information about theoretical background regarding the applied approaches.

2.1 General Definition of Energy

A program can be regarded as efficient if it meet three requirements: size, speed, and energy consumption [15]. Hardware, software, and program complexity are main elements affecting energy consumption. Energy saving is of great importance due to environmental and cost issues. Further, adverse effects on the environment can be removed via green software systems.

When computer programs execute, they consume energy. Energy is generally measured in joules (J). Joule is an energy measurement unit put by International System of Units (SI). Electric power spent over a specific period of time can explain energy for this study. Power consumed by an electric device can be measured via Eq. 1:

$$P = V \times I(SI\ units : Watt,\ Volt,\ Ampere) \tag{1}$$

Electric potential is measured though *V* and *I* denotes electric current passing across the resistance. The formulas of energy are given in Eqs. 2 and 3:

$$E = P \times T\ (SI\ units : Joule, Watt, second) \tag{2}$$

$$E = P \times\ V \times\ T\ (SI\ units : Joule, Volt, Ampere, second) \tag{3}$$

For varying voltage-v(t) and current-i(t) in an electrical device, the energy consumption between instants t1 and t2 is given by the integrals Eq. 4 [15]:

$$E = \int_{t2}^{t1} V_c\, I(t)dt = V_c \int_{t2}^{t1} I(t)dt = V_c \int_{t2}^{t1} V_s(t)/R_s dt = V_c/Rs \int_{t2}^{t1} V_s(t) \tag{4}$$

Equation 4 is a more general definition of energy consumption in electrical devices.

$T = t_1 - t_2$ is the integral used for measuring consumption and Vc represents the source voltage that is not unstable. *Rs* is a value of resistance. To obtain the value of *I*, Ohm's Law (*I* = *Vs/Rs)* is used, This law utilizes resistance *Rs* for the value of *I* by measuring Vs at the resistor. These operations help estimate energy consumption or power dissipation for computer programs when they are executed.

2.2 Monitoring Energy Consumption

Computer programs need a great number of resources when they are harnessed. Hard disk, graphics card, storage device, memory, and CPU can be regarded as main resources. During the execution of each process of computer programs, some operations such as disk read/write per process, or CPU computation consume power/energy [16]. When a user estimates power/energy usage of the processor, there are some methods to monitor power/energy usage. Some special tools such as Petra, XEEMU, Jalen, Jolinar, pTop, EnergyChecker, PowerApi, Intel Power Gadget enable the user to estimate power/energy without requiring any hardware instrumentation [17]. In this study, we estimate the energy consumed by original and refactored codes. To this end, Intel Power Gadget API [17] is employed. This tool provides a log file for a period of time. It includes elapsed time, package power limit, processor frequency, GT frequency, processor temperature, average, and cumulative power of the processor. To estimate processor energy consumption, the tool utilizes the formula presented in Eq. 5 [17]:

$$\begin{aligned} &\textit{Processor Energy}\ (\textit{Total energy of the processor}) = \textit{IA Energy} + \textit{GT Energy} \\ &(\textit{if applicable}) + \ \textit{Others}\ (\textit{not measured}) \end{aligned} \tag{5}$$

2.3 Code Refactoring Techniques

Some software engineering techniques can directly affect program energy consumption and performance, such as code refactoring techniques [12]. Refactoring on the source code in the correct way does not only increase the quality of the code but also affects the energy consumed by an application [18]. Refactored codes could increase energy consumption compared to the original codes [19]. There are many studies focusing on code refactoring techniques to reduce energy consumption [12, 21, 24].

Software developers, engineers, or researchers want to use more stable code refactoring techniques to maintain open-source software development [9]. Code refactoring techniques can be implemented in many computer programs in the software industry. These programs may be written by object-oriented languages. There are many code refactoring techniques that can be used in object-oriented programming. Some of them are as follows: “extract method”, “inline method”, “inline temp”, “encapsulate field”, “add parameter”, “pull up method”, “simplify nested loop”, “remove parameter”, “split temporary variable”, “decompose conditional”. In this study, we apply five different code refactoring techniques to analyze the individual effects in terms of energy consumption on similar applications written in three different object-oriented languages.

3 Methodology

As in the aforementioned studies, our study focuses on refactoring techniques, thereby analyzing different object-oriented languages. Our conceptual framework of investigation is given in Fig. 1.

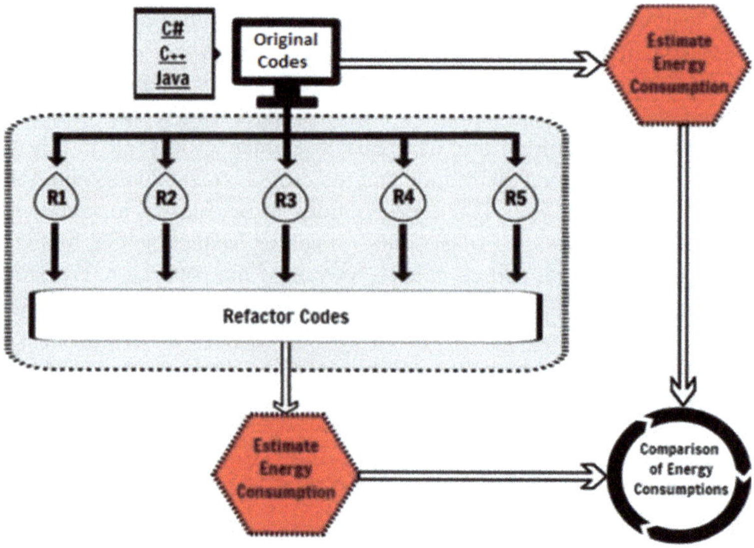

Fig. 1. The conceptual framework of the investigation. In this framework, different source codes taken from object-oriented languages (C#, C++, Java) are given as original codes; code refactoring techniques applied on original codes: R1 (Simplify Nested Loop), R2 (Extract Method), R3 (Encapsulate Field), R4 (Extract Variable), R5 (Consolidate Duplicate Conditional Fragments).

First, three similar applications coded with different programming languages have been retrieved from GitHub [23]. Our framework consists of three main steps: obtaining open-source codes, applying refactoring techniques, and measuring energy consumption. The third step is performed both at the beginning and the end of the experiment. To compare the energy consumption of original codes with refactored ones, Intel Power Gadget [17] is employed. Our system specification that we use Intel Power Gadget is given in Table 1.

Table 1. System specification

CPU	Intel Core i7 6700HQ processor, 2,60 to 3,50 GHz
RAM	16 GB SDRAM, 2133 MHz
Architecture	Skylake, Superpipeline, X86-64
Measurement tap	CPU/processor cores current(IA Energy)

Five different code refactoring techniques applied to original codes are given in Table 2.

Table 2. Applied refactoring techniques

Original Codes	Refactor Codes
```void main() { ... for(i = 0; i <= M; i++){ for(j = 0; j <= N; j++){ final=i*j; ... } } }```	```void main() { ... for(k = 0; k <= M*N; k++){ final = k; ... } }```

Refactoring I: Simplify Nested Loop

May reduction in the number of iterations
Improves energy consumption, performance

Original Codes	Refactor Codes
```int Calculator() { ... //calculate if(x>1) return x*(x-1); else return 1; }```	```int Calculator() { ... return calculate(x); } int calculate(x){ if(x>1) return x*(x-1); else return 1; }```

Refactoring II: Extract Method

Reduces code repetition; improves complexity, maintainability
May increases code size and energy consumption

Original Codes	Refactor Codes
```class Square { public int length; }```	```class Square { private int length; public int Lenght { get { return length; } set { length = value ; } } }```

Refactoring III: Encapsulate Field

Improve modularity, reliability and code quality
May improves performance and energy consumption

```
void Empty()                              void Empty()
    {                                         {
    ...                                       ...
    if(equation.IndexOfAny(symbol, i, 1) > -1)    bool IndexEmpty =
    {                                             equation.IndexOfAny(
    ...                                           symbol,i,1) > -1;
      }                                           if(IndexEmpty)
    }                                             {
                                                  ...
                                                  }
                                              }
```

Refactoring IV: Extract Variable

Improve understandability and code quality
May increases code size and energy consumption

```
        if (IsEmpty())              if (IsEmpty())
    {                                   {
    ...                                 ...
     add();                             }
     }                              else {
else{                                   ...
    ...                                 }
      add();                        add();
}
```

Refactoring V: Consolidate Duplicate Conditional Fragments

Improve maintainability, code size, and readability

4 Experimental Results and Evaluation

Table 3 shows metric values of three programs coded with C#, C++, and Java. LocMetrics [20] is employed to obtain Table 3. According to this table, C# has the highest complexity among them. This case may depend on the type of programming language. Complexity is not directly proportional to energy consumption [10].

Table 3. Metrics of applications used in the investigation

	Calculator (C#)	Calculator (C++)	Calculator (Java)
Source files	9	4	2
Lines of code	3528	1666	532
Physical executable lines of code	3035	1293	431
Logical executable lines of code	2250	993	313
McCabe VG complexity	243	42	4

Refactoring techniques have been selected in accordance with the experimental programming languages as follows: Simplify Nested Loop, Extract Method, Encapsulate Field, Extract Variable, Consolidate Duplicate Conditional Fragments.

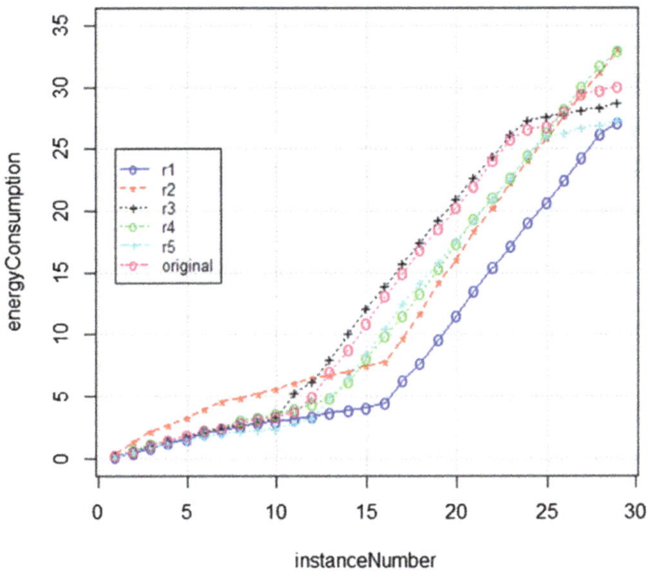

Fig. 2. Energy consumption results of calculator (C#). Code refactoring techniques applied to calculator (C#): r1 (Simplify Nested Loop), r2 (Extract Method), r3 (Encapsulate Field), r4 (Extract Variable), r5 (Consolidate Duplicate Conditional Fragments).

Refactoring techniques are able to reduce energy consumption compared to the original codes except for r3 as seen in Fig. 2. However, the promising one is r1 that it has the largest margin from original codes. On the other hand, r2 requires a specific iteration count to reach a consistent point in terms of energy consumption. For all refactoring techniques, 25 is the threshold value for C# codes. For C#, it has found that r4 and r5 produce similar effects with regard to energy consumption. Figure 3 presents energy consumption rates of refactored and original codes for Java. As in Fig. 2, r1 has achieved remarkable success in reducing energy consumption in Fig. 3. However, Java r3 has been detected as the best technique for Java. For this analysis, r1 and r4 create similar effects on energy consumption. 25 iterations are needed to make stagnant energy consumption rates.

Figure 4 presents the energy consumption rates of refactoring techniques in terms of C++. This language does not create a stagnant model for energy consumption.

In contrast, the increasing number of refactoring operations create a fixed rate effect either positive or negative on energy consumption. Further, energy consumption rates are close to each other compared to Java and C#.

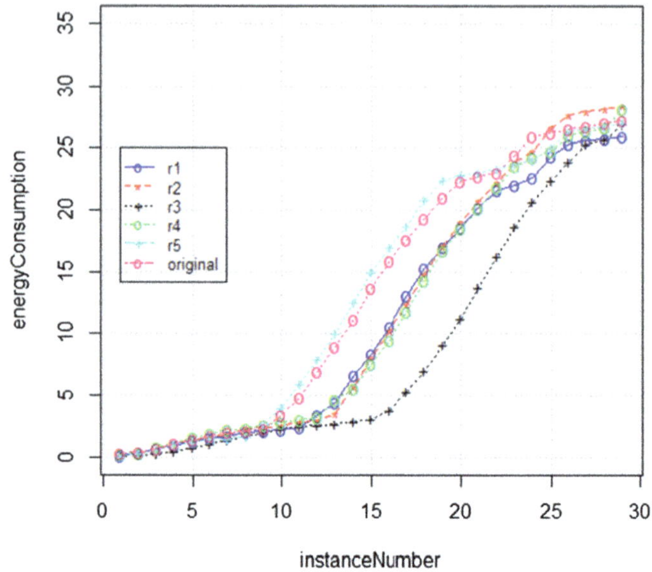

Fig. 3. Energy consumption results of calculator (Java). Code refactoring techniques applied to calculator (Java): r1 (Simplify Nested Loop), r2 (Extract Method), r3 (Encapsulate Field), r4 (Extract Variable), r5 (Consolidate Duplicate Conditional Fragments).

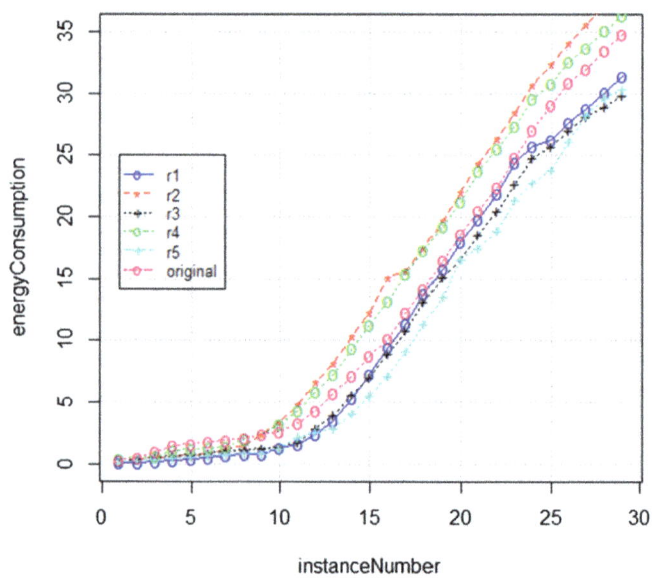

Fig. 4. Energy consumption results of calculator (C++). Code refactoring techniques applied to calculator (C++): r1 (Simplify Nested Loop), r2 (Extract Method), r3 (Encapsulate Field), r4 (Extract Variable), r5 (Consolidate Duplicate Conditional Fragments).

5 Conclusions and Future Work

Code refactoring techniques are applied on the source code to improve various aspects of software engineering. In this work, five refactoring techniques have been applied on similar programs coded with three different programming languages. The experiment shows that Java and C# yield similar energy consumption rates with regard to the consumption curves. This case is clear in the margins of energy consumption figures. The effects on energy consumption of some refactoring techniques such as r3 differ depending on the type of programming language. R4 produces consistent curves with regard to energy consumption. In summary, each refactoring technique has a specific coding design that may create different effects on energy consumption. Therefore the individual effect of a refactoring technique relies on the other techniques which are employed to construct refactoring combination.

In future works, sophisticated combinatorial methods will be investigated to generate energy effective refactoring combinations. Further, refactoring techniques can be analyzed in various platforms such as mobile and embedded devices.

References

1. Agarwal, S., Nath, A., Chowdhury, D.: Sustainable approaches and good practices in green software engineering. Int. J. Res. Rev. Comput. Sci. **3**(1), 1425–1428 (2012)
2. Manotas, I., Bird, C., Zhang, R., Shepherd, D., Jaspan, C., Sadowski, C., Clause, J.: An empirical study of practitioners' perspectives on green software engineering. In: 2016 IEEE/ACM 38th International Conference on Software Engineering (ICSE), pp. 237–248. IEEE (2016)
3. Shenoy, S.S., Eeratta, R.: Green software development model: an approach towards sustainable software development. In: 2011 Annual IEEE India Conference (INDICON), pp. 1–6. IEEE (2011)
4. Hsu, C.H., Kremer, U.: The design, implementation, and evaluation of a compiler algorithm for CPU energy reduction. ACM SIGPLAN Not. **38**(5), 38–48 (2003)
5. Fowler, M.: Refactoring: Improving the Design of Existing Code. Addison-Wesley Professional, Boston (2018)
6. Kwon, Y., Lee, Z., Park, Y.: Performance-based refactoring: identifying & extracting move-method region. J. KIISE: Softw. Appl. **40**(10), 567–574 (2013)
7. Park, J.J., Hong, J.E.: An approach to improve software safety by code refactoring. In: Proceedings of Korea Computer Congress, pp. 532–534 (2013)
8. Gottschalk, M., Jelschen, J., Winter, A.: Saving energy on mobile devices by refactoring. In: EnviroInfo, pp. 437–444 (2014)
9. Park, J.J., Hong, J.E., Lee, S.H.: Investigation for software power consumption of code refactoring techniques. In: SEKE, pp. 717–722 (2014)
10. Gottschalk, M., Josefiok, M., Jelschen, J., Winter, A.: Removing energy code smells with reengineering services. GI-Jahrestagung **208**, 441–455 (2012)
11. Palomba, F., Di Nucci, D., Panichella, A., Zaidman, A., De Lucia, A.: On the impact of code smells on the energy consumption of mobile applications. Inf. Softw. Technol. **105**, 43–55 (2019)
12. da Silva, W.G.P., Brisolara, L., Correa, U.B., Carro, L.: Evaluation of the impact of code refactoring on embedded software efficiency. In: Workshop de Sistemas Embarcados (2010)

13. Sahin, C., Pollock, L., Clause, J.: From benchmarks to real apps: exploring the energy impacts of performance-directed changes. J. Syst. Softw. **117**, 307–316 (2016)
14. Gottschalk, M., Jelschen, J., Winter, A.: Energy-efficient code by refactoring. Softwaretechnik-Trends **33**(2), 23–24 (2013)
15. Bessa, T., Gull, C., Quintão, P., Frank, M., Nacif, J., Pereira, F.M.Q.: JetsonLEAP: a framework to measure power on a heterogeneous system-on-a-chip device. Sci. Comput. Program. **173**, 21–36 (2017)
16. Borghetti, S., Gianfagna, L., Sgro, A.M.: U.S. Patent No. 8,145,918. U.S. Patent and Trademark Office, Washington, DC (2012)
17. Intel power gadget: https://software.intel.com/en-us/articles/intel-power-gadget-20. Accessed 10 Feb 2019
18. Papadopoulos, L., Marantos, C., Digkas, G., Ampatzoglou, A., Chatzigeorgiou, A., Soudris, D.: Interrelations between software quality metrics, performance and energy consumption in embedded applications. In: Proceedings of the 21st International Workshop on Software and Compilers for Embedded Systems, pp. 62–65. ACM (2018)
19. Kim, D., Hong, J.E., Yoon, I., Lee, S.H.: Code refactoring techniques for reducing energy consumption in embedded computing environment. Clust. Comput. **21**(1), 1079–1095 (2018)
20. LocMetrics. http://www.locmetrics.com/index.html. Accessed 15 Dec 2018
21. Banerjee, A., Chong, L.K., Chattopadhyay, S., Roychoudhury, A.: Detecting energy bugs and hotspots in mobile apps. In: Proceedings of the 22nd ACM SIGSOFT International Symposium on Foundations of Software Engineering, pp. 588–598. ACM (2014)
22. Sahin, C., Tornquist, P., Mckenna, R., Pearson, Z., Clause, J.: How does code obfuscation impact energy usage? In: 2014 IEEE International Conference on Software Maintenance and Evolution, pp. 131–140. IEEE (2014)
23. GitHub. https://github.com/postman721/Calculator. Accessed 12 Mar 2019
24. Pinto, G., Castor, F., Liu, Y.D.: Understanding energy behaviors of thread management constructs. ACM SIGPLAN Not. **49**(10), 345–360 (2014)

Determination of Numerical Papillae Distribution Affecting the Taste Sensitivity on the Tongue with Image Processing Techniques

Sefa Çetinkol and İsmail Serkan Üncü(✉)

Faculty of Technology, Isparta University of Applied Sciences, 32100 Isparta, Turkey
sefacetinkol@gmail.com, serkanuncu@isparta.edu.tr

Abstract. In this study, numerical papillae distribution on tongue are determined with image processing and according to the region and papillae number, taste sensitivity is calculated. Aim is to calculate taste sensitivity and examining effect of person's age and nutrition style to the papillae number and taste sensitivity. At the project, codes are written at Python and OpenCV is used for image processing. To detect papillae, tongue image is used and Gaussian Filter is applied to remove the noise. Filtered image is passed from Canny Edge Detection function to detect edges on tongue. Papillae detection is done by using edges. Tongue is examined in three pieces. Each piece is multiplied by coefficient to determine papillae number. Papillae density is calculated by using papillae number and area. Smoking effect is calculated by using tongue color. Taste sensitivity is obtained at the result of polinomial operation of papillae number, papillae density, smoking effect and person's age. Input images are separated to age groups. Relation between age group and papillae number is observed. Also, effect of nutrition to the papillae distribution is examined between the people in same age group. At the trials, papillae is detected thus papillae density and taste sensitivity are calculated.

Keywords: Papillae · Image processing · Taste sensitivity

1 Introduction

Thanks to advancing technology, progress has been made in the field of medical electronics and image processing. It can be ensured that the organs can be observed or diseases can be detected in the medical field thanks to the devices which are created by using image processing techniques with medical electronics. In a study formed by using of medical electronics, it was emphasized that there is a relationship between the physical movements of people and the imagination mechanism that enables the realization of these movements in the brain. Using the person's EEG data, movement imagination mechanism was trained with artificial neural networks according to the physical hand movement performed by the person. As a result of the training, the classification between motion data and imagination mechanism was carried out with a

D. J. Hemanth and U. Kose (Eds.): ICAIAME 2019, LNDECT 43, pp. 153–170, 2020.
https://doi.org/10.1007/978-3-030-36178-5_13

success rate of 99% [1]. In a study in which medical electronics and image processing techniques were used together, it was stated that retinal image can be used to detect diseases such as diabetes and high blood pressure. The retinal image taken with a fundus camera enables the advance observation of future form of diabetes disease in a successful way by using the algorithm generated in the green color channel [2]. In a study created by using image processing techniques and medical electronics, the real-time measurement of the pulse of a person was performed by using the palm image taken by the camera [3]. In a study formed by using image processing techniques in the medical field, a support system was established to detect gastric cancer. Using color changes on the gastric image taken with high-resolution endoscopy allows detection of cancerous cells. It is intended to help surgeons because the cancer cell can be overlooked during the procedures performed by the surgeon [4]. In the studies which are examined, it has become possible to obtain medical data with medical electronics devices created by image processing. In this study, papillae in the tongue are detected and taste sensitivity is calculated with using image processing techniques.

The organs that transmit and interpret the chemical and physical stimuli in the environment are called sensory organs. There are 5 different sensory organs in a human being and these organs are the sight organ, the touch organ, the hearing and balance organ, the olfactory organ and the taste organ [5]. Although each sense is useful to flavor perception, the combination of taste sense and smell sense is mainly considered to the perception of taste [6]. Disturbed foods, useless or harmful nutrients are distinguished by the senses of smell and taste [7]. Living things react to 5 different taste types which are bitter, sweet, sour, salty and umami [8]. It was suggested that fat could also be considered as the main taste in 2011 [9].

The perception of taste sensation is provided by receptors called as taste buds. These buds, mostly found in the tongue, are found in the other organs such as palate and epiglottis. An adult person has between 3 thousand and 10 thousand taste buds. Taste buds are located over the papillae in the tongue. The papillae which are located at special regions shown in Fig. 1 provide the perception of basic flavors such as sour, sweet, bitter and salty. Thanks to the papillae located at the tip of the tongue, salty and sweet flavors are perceived. The sour flavors are perceived by the papillae located on the two sides of the tongue. The papillae located on the root side of the tongue and on the soft palate provide the perception of bitter taste [10].

Taste receptors in the tongue mediate taste sensitivity. The differences in taste receptors and the intensity of the taste papillae affect the taste sensitivity. Also taste sensitivity affects which food we enjoy and which food we choose to eat [12].

There are some researches related to tongue and taste. In one of these studies, it was observed that some people have black hairy tongue, and the existence of black hairy tongue caused by bad oral hygiene, smoking and using alcohol [13]. In a study about development of human tongue, as a result of using the scanning electron microscopy, the development of human tongue papillae was investigated. With examining of papillae in the tongue tissue taken from 3 to 8 month old babies, the emergence of the papillae, the distribution of papillae in the tongue and formation of papillae were examined thus information about formation and development of the papillae is reached [14]. In a study about anorexia nervosa, known as eating disorder, depending on the taste sensitivity of low-body individuals, the taste sensitivity of a person with or

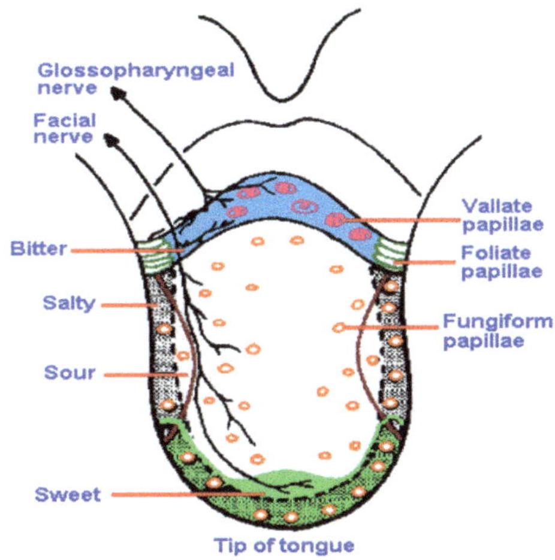

Fig. 1. Structure of tongue [11]

without anorexia disease has been observed to perceive the true taste varieties [15]. In a study on obesity, it has been observed that fat taste has an effect on appetite control and nutrient selection mechanism although it is not among the 5 basic flavors [16].

In this study, image processing techniques are applied to the tongue image and the number of papilla, papilla density and taste sensitivity in the tongue will be calculated. Firstly, the tongue image entered into the system is passed through the Gaussian Filter, thus the noise in the image is eliminated. Canny Edge Detection function is applied to the filtered image and the edges in the tongue are detected. The obtained edge data is used to determine the papillae. The tongue is examined in three parts which are tip of tongue, middle of tongue and top of tongue. Each piece is multiplied by a coefficient to determine the number of papillae. Papilla density is calculated using the number of papillae and area data. The tongue color of the smokers shows changes and the effect of smoking is calculated using the change in tongue color. To calculate the taste sensitivity, the number of papillae, papilla density, smoking effect and age of the person undergo a polynomial procedure. In addition, the input tongue image data is divided into age groups and it is observed that there is a relationship between the age group and the number of papillae. In the examination among people in the same age group, it is observed that nutrition style affects papilla distribution.

2 Materials and Methods

The mechanism of the study about the determination of the numerical distribution of papillae on the tongue affecting the taste sensitivity using Python programming language is shown in Fig. 2 and the OpenCV library is used during the study. According to the mechanism, Gaussian Filter is applied to the tongue image to find the number of papillae and papilla density in the tongue. The Gaussian Filter is used for removing the noise from the tongue image. The Canny Edge Detection function is applied to obtain the edges in the filtered tongue image. It has become possible to detect the papillae using the tongue image which is filtered and obtained the edges. At this stage, the tongue image is divided into three parts such as tip of tongue, middle of tongue and top of tongue. The sections are taken from each piece and the number of papillae is found. Found number of papillae is multiplied by a coefficient of that piece to calculate the desired number of papillae. In order to find the density of papillae, the total area of the papillae is divided by the total area of the sections. It is observed that there is a change in the color of the tongue of smokers. It is masked the changed tongue color via color masking thus smoking effect is obtained. In this study, a relationship is established between age and papillae number and using this relationship, the number of papillae a person should have is calculated according to the age. To calculate the taste sensitivity, a polynomial operation is applied to the number of papillae, papilla density, smoking factor and age factor. In this study, Gaussian Filter, Canny Edge Detection and color masking methods are used.

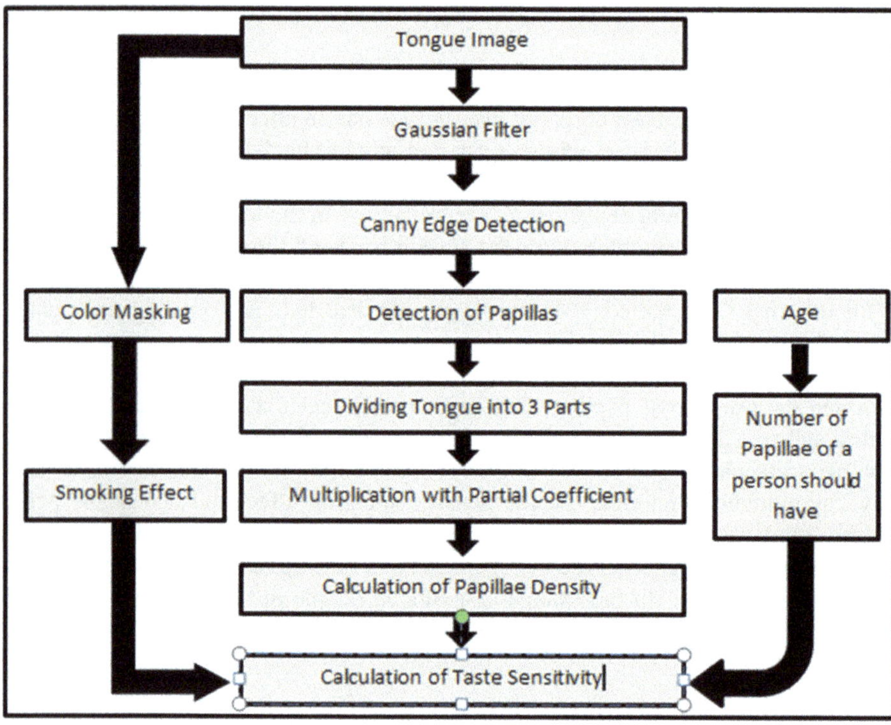

Fig. 2. Working mechanism of the project

2.1 Gaussian Filter

By filtering an image, operations such as smoothing the image, sharpening the image, revealing certain details in the image, finding the edges in the image or sharpening the edges in the image can be performed [17]. The noise is removed by applying Gaussian

Filter to the image [18]. The Gaussian Filter is often used to soften and smooth the image. Equation of Gaussian Filter,

$$H(u, v) = \frac{1}{2\pi\sigma^2} e^{-(u^2 + v^2)/2\sigma^2} \tag{1}$$

horizontal distance to the center is u, vertical distance to the center is v and standard deviation is σ [19].

By applying a gaussian filter to the original image shown at the left of Fig. 3, the sharp spots are eliminated and the image at the right of Fig. 3 is obtained.

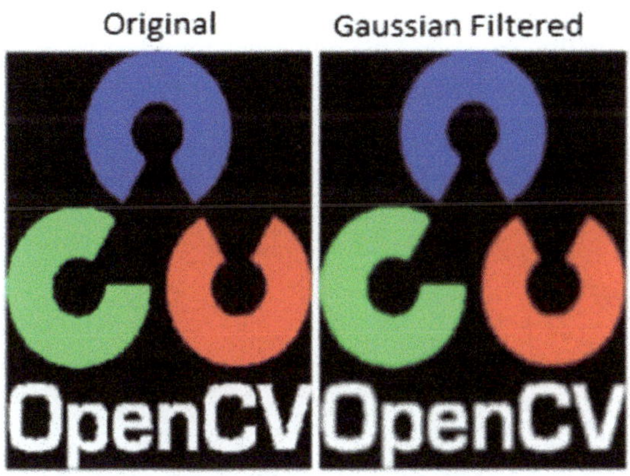

Fig. 3. Gaussian filter [20]

The tongue image is used in this study. However, there are noises in the image taken by the camera and the noises are shown as sharp points at the image. Gaussian filter is applied to the tongue image in order to determine the edges more accurately by eliminating the sharp points from the tongue image. In addition, Canny Edge Detection phase has Gaussian Filter to remove the noise and its size is 5 × 5 units. Applying Gaussian Filter only in the Canny Edge Detection does not produce good results so that a different size of Gauss Filter is applied to remove noise before Canny Edge Detection.

2.2 Canny Edge Detection

The Canny Edge Detection is used to obtain the edges at the input data entering the system and it is sensitive to noise [21]. This method allows the detection of the parts in which there are sudden changes in intensity [22]. Canny Edge Detection is a multi-layer algorithm. In the first stage, the Gaussian Filter is applied to eliminate the noise in the image. The second stage contains the density gradient of the image and the Equation of finding the gradient,

$$\boldsymbol{Gradyan}(\boldsymbol{G}) = \sqrt{G_x^2 + G_y^2} \tag{2}$$

$$\text{Açı}(\vartheta)=\tan\text{-}1\left(\frac{G_y}{G_x}\right) \tag{3}$$

Gx refers to the horizontal direction and Gy refers to the vertical direction. Gradient of the image is calculated with Eqs. 2 and 3. In the third stage, pixels are examined and the local maximum points are perceived as edges and 0 is assigned to the local minimum points and only the edges are left at the output image. At the last stage, a threshold value is assigned and the parts above this threshold value are considered as edges and the remaining parts are indicated as not edges [23].

In order to find the edges in the image at the left part of Fig. 4, Canny Edge Detection is applied to the image and the edges are found as the image at the right side of Fig. 4.

Fig. 4. Canny edge detection [24]

In this study, Canny Edge Detection is applied to the tongue surface to detect the papillae over the tongue. The papillae on the surface of the tongue, whose the edges are found, are detected by the algorithm which is specially prepared for this study.

2.3 Color Masking

Images consist of RGB (red, green, blue) colors. In the Python programming language, images are ordered by BGR (blue, green, red). To mask the image, the image is converted to the HSV (Hue Saturation Value) color format. The HSV color format works more precisely in the terms of color than the BGR color format [25].

Hue (Tone) value in HSV is used for color selection, Saturation indicates color tone according to white color and Value refers to the light illumination [26]. To make color masking, the upper and lower HSV values of the desired color are entered into the system and the limits of the desired color are defined. Image in the HSV color format, the values between these color limits is shown at the output image [27].

The original image is shown at the left side of Fig. 5. The color of the flower in this image is desired to be masked and therefore the upper and lower color limits of the desired color are set. As a result of detecting values between these limits, the pixels that are masked are shown in the central image in Fig. 5. The desired color is masked by the pixels within the limits and the masked image is shown at the right side of Fig. 5.

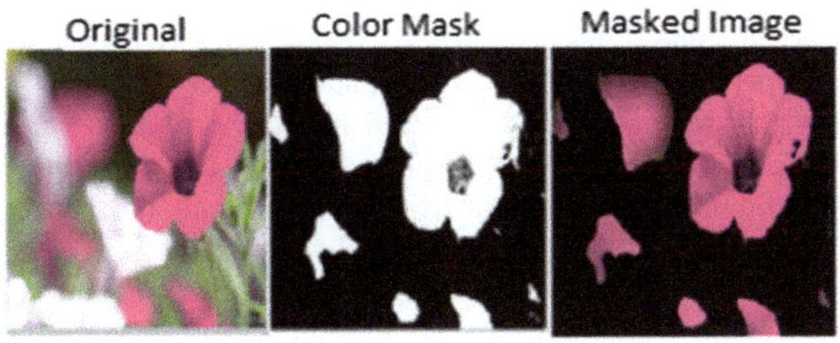

Fig. 5. Color masking [28]

In this study, the color masking process is used to find the effect of smoking which is used in the calculation of taste sensitivity. In the examination, it is observed that there is a change in the tongue color of the smokers. By defining the upper and lower HSV values of the changed tongue color, the changed tongue parts of the smokers are masked. Using the masked tongue image, the masked pixels are obtained and the cigarette effect is calculated using the number of masked pixels.

3 Research Findings

Tongue images are entered into the system to determine the numerical distribution of papillae on the tongue that affects the taste sensitivity with using image processing techniques. Gaussian Filter is applied to the tongue image in order to soften the sharp edges and to eliminate the noises in the images. The edges of the tongue image is obtained in order to detect the papillae in the filtered tongue image. The Canny Edge Detection method is used to obtain the edges on the tongue image. The detection of the papillae is carried out with using the edges on the tongue. At this stage, to examine the effect of the nutrition style of a person on the papillae distribution, the tongue image is examined by dividing into 3 parts such as the tip of tongue, the middle of tongue and the top of tongue. Also, everyone's tongue is not same. Since the tongues may be wide, thin, long or short, sections which are taken from each part are examined. Two sections are taken from the top part of the tongue and the middle part of the tongue. One of these sections is taken from the left side of the tongue and the other is taken from the right side of the tongue. The number of papillae is found as taking the average of the sections' papillae number. The average number of papilla is multiplied by the coefficient of the part to reach the desired number of papillae. At the tip of the tongue, a section is taken and the number of papillae which is found in this section is multiplied by the coefficient of the part and the number of papillae belonging to the tip part of the tongue is reached. Papilla density is calculated by dividing the total area of the papillae in the sections by the total cross-sectional area. In order to find the effect of smoking, color masking is applied to the tongue image. In the examination, it is observed that there is a change in the tongue color of the smokers. The use of this change in tongue color is used to calculate the smoking effect. In this study, the relationship between the number of papillae and the age of a person is obtained. The number of papillae a person should have is calculated according to the person age using this relationship. In order to obtain the numerical value of taste sensitivity, taste sensitivity is calculated as a result of a polynomial operation of papilla number, papilla density, age and smoking effect.

In this study, the tongue image is entered into the system to calculate the taste sensitivity and to find the number of papillae. During the capturing of the tongue images, the light coming to the tongue image is kept constant, the distance between the camera and the tongue is remained constant and the tongue image is taken with the best angle to see the whole tongue surface. The contour of the tongue image is drawn with blue color and blue color masking is applied, thus the tongue image is obtained. In part a of Fig. 6 shows the tongue image with blue color border and in part b of Fig. 6 is obtained by applying blue color masking to the tongue image.

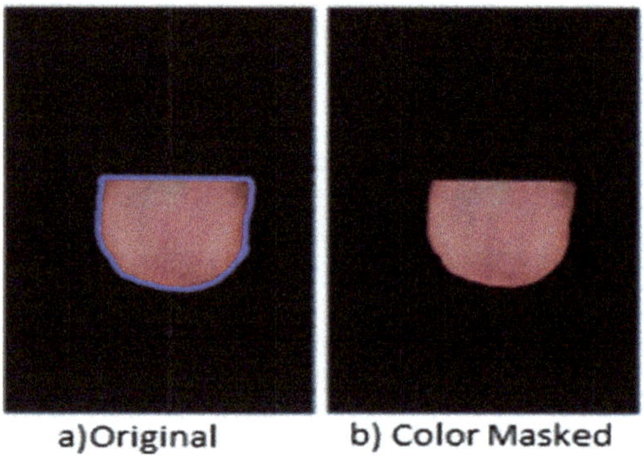

Fig. 6. Masking tongue image

In order to normalize the tongue images obtained as a result of the color masking, the area of the tongue image with 4128 × 2322 resolution is resized to 1000 × 1000 resolution. Tongue images are resized because the tongues are different from each other, even if the tongue images are captured at a constant distance, constant light, and at the most appropriate angle. In part a of Fig. 7 shows the blue color masked tongue image and the tongue image is resized using the coordinates obtained by color masking and it is shown in part b of Fig. 7.

In order to eliminate the sharp spots and noise from the resized tongue image, Gaussian Filter is applied to the tongue image. The filtered image shown in part b of Fig. 8 is obtained by applying Gaussian Filter to the resized tongue image shown in part a of Fig. 8.

In order to detect the papillae on the tongue, Canny Edge Detection method is applied to the Gaussian Filtered tongue image, thus the edges on the tongue image is found as shown in part a of Fig. 9. With the detection of the edges on the tongue, the papillae on the tongue are detected by the algorithm created specifically for this study and detected papillae are shown in part b of Fig. 9.

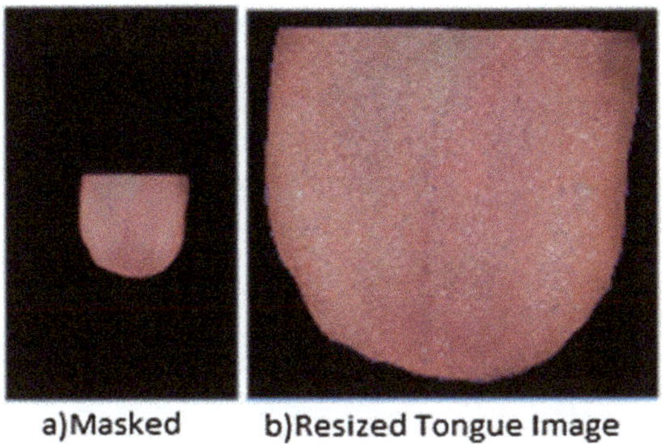

Fig. 7. Normalizing the tongue image

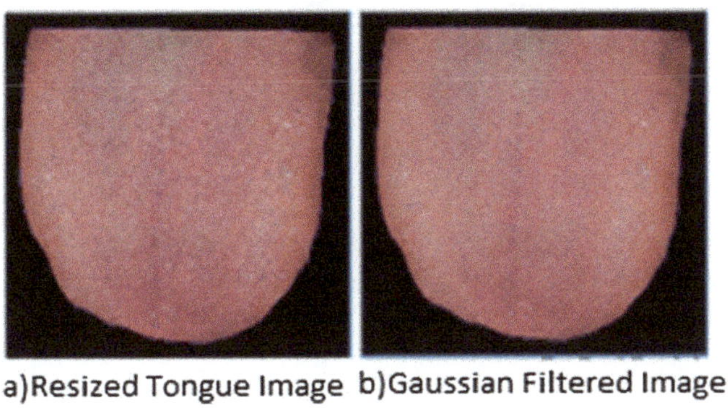

Fig. 8. Application of gaussian filter

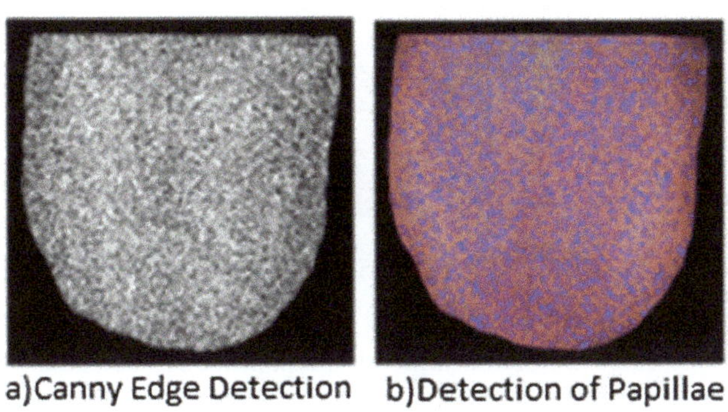

Fig. 9. Canny edge detection and papilla detection

The papillae on the tongue are spread over different areas on the tongue in order to perceive different tastes. Papillae are located at the tip of the tongue in order to perceive sweet and salty flavors. papillae which are found at the middle of the tongue help to perceive sour tastes and the perception of bitter flavors is provided by the papillae at the top of the tongue. For this reason, in order to understand the distribution of papillae according to human nutrition style, the tongue is divided into three parts as tip of tongue, middle of tongue and top of tongue. To examine the distribution of papillae on the tongue image shown in part a of Fig. 10, the tongue image is divided into 3 parts as shown in part b of Fig. 10.

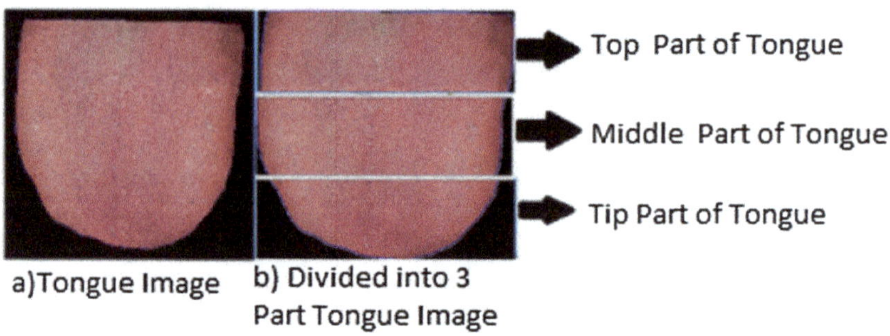

Fig. 10. Separating tongue in 3 parts

Since the tongues of the people are different from each other, sections are taken from each part. In the sectioning process shown in Fig. 11, two sections are taken from the top and middle parts of the tongue, and a section is taken from the tip part of the tongue.

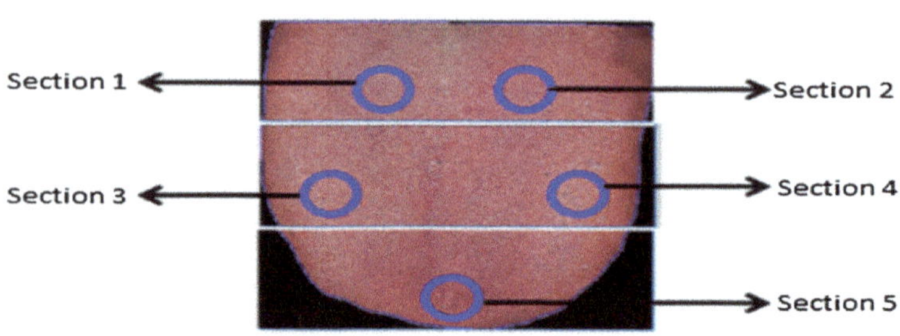

Fig. 11. Sectioning

With the Canny Edge Detection method, the papillae are detected by obtaining the edges on the tongue. At this stage, the number of papillae in the section is obtained and papillae in the section are shown in Fig. 12.

2 sections are taken from the top part of the tongue. For this reason, the papilla rate at the top part of the tongue is found by dividing the total number of papillae found in Sects. 1 and 2 by 2 and average papillae number is multiplied by the coefficient of the top part, thus the number of papillae on the top part is obtained and shown in Eq. 4. The papilla rate at the middle part of the tongue is found by dividing the total number

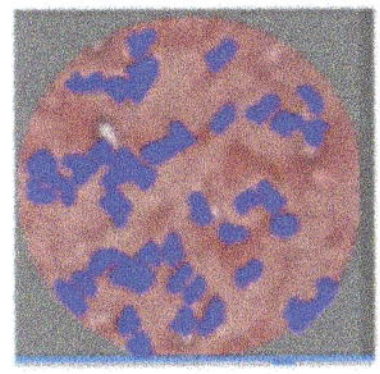

Fig. 12. Papillae in the section

of papillae found in Sects. 3 and 4 by 2 and average papillae number is multiplied by the coefficient of the middle part, thus the number of papillae on the middle part is obtained and shown in Eq. 5. Since there is one section on the tip of the tongue, the number of papillae at the tip part of the tongue is obtained by multiplying the number of the papillae in Sect. 5 with the coefficient of the tip of the tongue and shown in Eq. 6.

$$\mathrm{TOPPN} = \sigma_1 \mathrm{x}(\frac{\mathrm{TOPLPN} + \mathrm{TOPRPN}}{2}) \tag{4}$$

$$\mathrm{MPN} = \sigma_2 \mathrm{x}(\frac{\mathrm{MLPN} + \mathrm{MRPN}}{2}) \tag{5}$$

$$\mathrm{TIPPN} = \sigma_3 \mathrm{xTIPSPN} \tag{6}$$

In order to find the number of papillae in the top part (TOPPN) as Eq. 4, the number of papillae in the left part (Sect. 1) of the top part (TOPLPN) and the number of papillae in the right part (Sect. 2) of the top part (TOPRPN) are added, summation is divided by 2 and multiplied by the coefficient of the top part ($\sigma 1$), so TOPPN is calculated. In order to find the number of papillae in the middle part (MPN) as Eq. 5, the number of papillae in the left part (Sect. 3) of the middle part (MLPN) and the number of papillae in the right part (Sect. 4) of the middle part (MRPN) are added, summation is diveded by 2 and multiplied by the coefficient of the top part($\sigma 2$), thus MPN is calculated. In order to find the number of papillae located in the tip of the tongue (TIPPN) in Eq. 6, the number of papillae section 5 in Fig. 11 in the tip of the tongue (TIPSPN) is multiplied by the coefficient of the top piece ($\sigma 3$). TOPPN, MPN and TIPPN need to be collected to reach the total number of papillae (TPN) in the tongue image and shown in Eq. 7.

$$\mathrm{TPN} = \mathrm{TOPPN} + \mathrm{MPN} + \mathrm{TIPPN} \tag{7}$$

The papilla density on the surface of the tongue is obtained by dividing the total papilla areas by the total sectional area and shown in Eq. 8. The total papilla area (TPA) is obtained by dividing into the total sectional area (TSA) to find the density of the papillae (PD) and is shown in Eq. 8. For the calculation of the TPA shown in Eq. 9, the area of the papillae in Sect. 1 (TOPLPA), the papillae area in the Sect. 2 (TOPRPA), the papillae area in the Sect. 3 (MLPA), the papillae area in the Sect. 4

(MRPA) and the papilla area (TIPPA) contained in Sect. 5 are collected. Since the sections are identical, the area of one of the sections (SA) is multiplied by five in order to obtain TSA and is shown in Eq. 10.

$$PD = TPA/TSA \tag{8}$$

$$TPA = TOPLPA + TOPRPA + MLPA + MRPA + TIPPA \tag{9}$$

$$TSA = 5xSA \tag{10}$$

The change in tongue color is used to calculate the effect of smoking. The effect of smoking is obtained by color masking of changed color of tongue surface because of smoking. A tongue of a smoker is shown in part a of Fig. 13, the coordinations of masked pixels between upper and bottom values of changed color because of smoking is obtained and shown in part b of Fig. 13 and the masked pixels of tongue are shown in part c of Fig. 13.

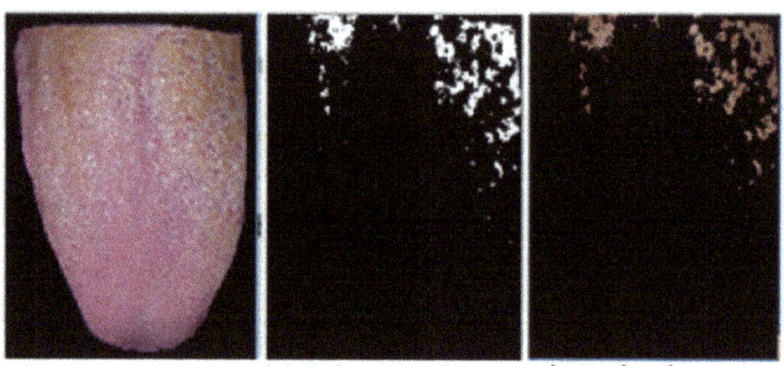

Fig. 13. Smoking effect

As a result of the study, the effect of nutrition style on the papilla distribution of 7 people is shown in Table 1. The distribution of papillae indicated by green is desired papillae distribution and unwanted papillae distribution is indicated by red. The papillae in the top part of the tongue help to perceive the bitter taste, perception of the sour flavors is done by the papillae in the middle part of the tongue and the papillae at the tip of the tongue provide to perceive the sweet and salty flavors. When Table 1 is examined, success rate of 80% (8 green/(8 green + 2 red)) is reached.

In another examination, the approximate number of papillae that a person should have according to the age is calculated. The calculation of the necessary number of papilla (NNP) according to the age (A) is shown at Eq. 11. In Fig. 14, the yellow line indicates the number of papillae that a person should have according to age, the blue line indicates the number of papillae that a person should have based on the classification, and the red dots indicate the age and number of papillae used in the study.

Table 1. Nutrition and papilla distribution

Loved Taste	Number of Papillae at the Top Part of Tongue	Number of Papillae at the Middle Part of Tongue	Number of Papillae at the Tip Part of Tongue
Bitter	27	16	16
Bitter + Sour	14	16	12
Sweet	21	29	35
Sour + Sweet	16	10	15
Salty	29	18	21
Sweet	16	20	21
Bitter	22	19	17

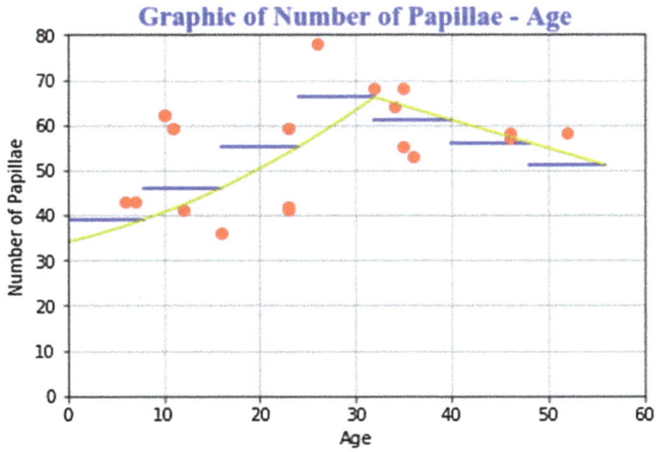

Fig. 14. Graphic of number of papillae-age

$$\text{NNP(A)} = \begin{cases} 30 + (\frac{A}{8} + 2)^2, A < 32 \\ 86 - \frac{5xA}{8}, A \geq 32 \end{cases} \tag{11}$$

Also ages are examined in 8-year age groups. The relationship between age and age group according to the age is shown in Eq. 12. The equation for calculating the number of papilla (NNP) that a person should have according to age group (AG) is shown in Eq. 13. In Fig. 15, the yellow line indicates the number of papillae that a person should have according to age group, the blue line indicates the number of papillae that a person should have based on the classification, and the red dots indicate the age group and number of papillae used in the study.

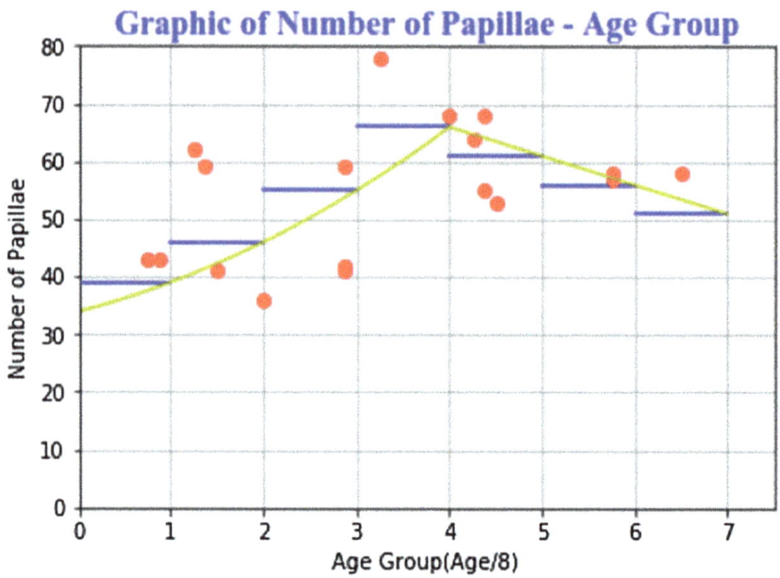

Fig. 15. Graphic of number of papillae-age group

$$\mathrm{AG(A)} = \begin{cases} 1, 0 < A \leq 8 \\ 2, 8 < A \leq 16 \\ 3, 16 < A \leq 24 \\ 4, 24 < A \leq 32 \\ 5, 32 < A \leq 40 \\ 6, 40 < A \leq 48 \\ 7, 48 < A \leq 56 \end{cases} \tag{12}$$

$$\mathrm{NNP(AG)} = \begin{cases} 30 + (\mathrm{AG} + 2)^2, AG < 4 \\ 86 - 5x\mathrm{AG}, AG \geq 4 \end{cases} \tag{13}$$

To calculate the Taste Sensitivity (TS), the number of papillae (TPN) that the person has, number of papillae (NNP) that a person should have according to age, papilla density (PD), age group (AG) and smoking effect (SE) required to be performed by a polynomial function.

$$\mathrm{TS(TPN, PD, SE, AG, NNP)} = \frac{\vartheta_1 \mathrm{x} \sqrt{\frac{\mathrm{TPN}}{\mathrm{NNP}}} + \vartheta_2 \mathrm{xPD} - \vartheta_3 \mathrm{xSE}}{\sqrt[3]{\mathrm{AG}}} \tag{14}$$

The taste sensitivity is calculated as a result of the process of the Eq. 14 and the effect of the number of papillae on taste sensitivity as $\vartheta 1$, the effect of papilla density on taste sensitivity as $\vartheta 2$ and the coefficient of smoking effect on taste sensitivity as $\vartheta 3$ are set. The taste sensitivities in the study are shown in Table 2. According to Table 2, taste sensitivity is generally higher at younger ages, while taste sensitivity decreases as age increases. It is also observed that people who smoke have less taste sensitivity.

Table 2. Calculation of taste sensitivity

Age	Number of Papillae	Papillae density	Smoking effect	Taste sensitivity
7	43	0.737	4	88.69
7	43	0.688	2	85.32
10	62	0.968	79	91.59
10	62	0.93	64	89.53
11	59	0.778	9452	76.91
12	41	0.65	0	66.08
16	36	0.56	1288	58.05
23	85	1.302	338	99.639
23	59	0.889	6716	71.30
23	42	0.67	3	57.56
23	41	0.655	15	56.32
26	78	1.119	20	81.05
32	68	1.158	2543	79.36
34	64	0.964	4	65.429
35	68	1.003	7738	66.369
35	55	0.83	42571	49.23
36	53	0.78	7	55.689
46	57	0.895	32613	52.879
46	58	0.88	63978	46.27
52	58	0.89	3	57.68

In the study, the tongue image has been successfully detected and the detected tongue image is resized. For the removal of the sharp points and noise in the resized tongue image, Gaussian Filter is applied to the resized tongue image. The filtered tongue image is passed through the Canny Edge Detection function and the edges on the tongue surface are obtained. It is provided that the papillae are detected by using the tongue image with the edges. The tongue is divided into 3 parts and the papilla distribution is calculated according to the person's eating habits in other word nutrition style. Papillae density is obtained by dividing total papilla area by total cross-sectional area. Smoking effect is calculated by masking the color change on the tongue because of smoking. The approximate number of papillae that one person should have according to the age is calculated. A polynomial function is applied to the number of papillae, papilla density, age of the person, the number of papillae that a person should have according to age, and the effect of smoking in order to calculate taste sensitivity. Approximately 80% success is achieved in the observation of papilla distribution according to nutrition style. It has been observed that the number of papillae that a person should have increases until a certain age and the number of papillae that a person should have decreases in later ages. It is expected that taste sensitivity will be high in young people and in the study; it is observed that the taste sensitivity is higher in the younger people and the taste sensitivity decreases as the age of the person increases.

4 Discussion and Conclusion

In this study, the effect of papillae on taste sensitivity is investigated by using image processing techniques. Gaussian Filter is applied to the input tongue image in order to remove the sharp points at the image, Canny Edge Detection Method is used to detect the edges on the filtered tongue image. Detection of papillae, papillae distribution and density of papillae is calculated by using these edges. The tongue of smokers is changed and color masking is used to mask these changed color. As a result of color masking, masked piksel is obtained and using these masked piksel count, the smoking effect is calculated. Taste Sensitivity is calculated by the polynomial operation of age, smoking effect, density of papillae, the number of the papillae. In addition, the effect of nutrition style on the papilla distribution and papilla distribution according to age are observed. In further studies, different algorithms can be developed for more successful detection of papillae or a mechanism which the person can put his/her face while the camera is capturing in order to obtain a better result by ensuring that all variables are kept constant.

As a result, for determination of the numerical distribution of papilla on the effect of taste sensitivity with image processing techniques, Gaussian Filter is used to soften sharp points in the tonge image which is entered into the system, Canny Edge Detection function is used to obtain the edges on the tongue. The algorithm which is created specially for this study is applied the edged detected tongue image in order to observe the papilla distribution, the number of papillae and papilla density. In addition, the color change in the tongue color of the smokers masked by color masking and smoking effect is calculated by using the number of masked pixels. The effect of smoking, age, number of papillae, and papilla density are passed through a polynomial function in order to calculate the Taste Sensitivity. The distribution of papillae with respect to the nutrition style of the person is examined and a relationship is established between the age of the person and the number of papillae that a person should have. In the study, the codes are written in the Python programming language and the OpenCV library is used for image processing.

References

1. Tosun, M., Erginli, M., Kasım, Ö., Uğraş, B., Tanrıverdi, Ş., Kavak, T.: EEG verileri kullanılarak fiziksel el hareketleri ve bu hareketlerin hayalinin yapay sinir ağlari ile siniflandirilmasi. Sak. Univ. J. Comput. Inf. Sci. **1**(2), 1–8 (2018)
2. Kara, M., Dımılıler, K.: Bleeding detection in retinal images using image processing. In: 2018 2nd International Symposium on Multidisciplinary Studies and Innovative Technologies (ISMSIT), pp. 1–5 (2018)

3. Gangal, O.B., Öztürk, M.: Real-time pulse counting using palm images. In: 2018 26th Signal Processing and Communications Applications Conference (SIU), pp. 1–4 (2018)
4. Yaşar, A.: Mide Kanserinin Tespiti için Görüntü İşleme Teknikleri Kullanılarak Bir Karar Destek Sisteminin Geliştirilmesi. Unpublished Doctoral Thesis, Selçuk Üniversitesi, Turkey (2018)
5. T.C. Milli Eğitim Bakanlığı: Duyu Organları (39–46). p. 50, Ankara, Turkey (2012)
6. Karadeniz, F.: Lezzet Alma Mekanizması. GIDA **25**(5), 317–324 (2000)
7. Tanalp, R.: Duyu Fizyolojisi. Ankara Üniversitesi Eczacılık Fakültesi Yayınları **32**, 87–101 (1975)
8. Zhao, G.Q., Zhang, Y., Hoon, M.A., Chandrashekar, J., Erlenbach, I., Ryba, N.J.P., Zuker, C.S.: The receptors for mammalian sweet and umami taste. Cell **115**, 255–266 (2003)
9. Mattes, R.D.: Accumulating evidence supports a taste component for free fatty acids in humans. Physiol. Behav. **104**(4), 624–631 (2011)
10. Kurtuldu, E., Miloğlu, Ö., Derindağ, G., Özdoğan, A.: Tat duyu bozukluklarina genel bakış. Atatürk Üniversitesi Diş Hekimliği Fakültesi Derneği. **28**(2), 277–283 (2016)
11. Anonymous: Gustatory and Olfactorysenses. https://michaeldmann.net/mann10.html. Accessed 30 Feb 2019
12. Grimm, E.R., Steinle, N.I.: Genetics of eating behavior: established and emerging concepts. Nutr. Rev. **69**(1), 52–60 (2011)
13. Kutlu, Ö., Özdemir, P., Karadeniz, T.B., Vahaboğlu, G., Ekşioğlu, H.M.: Oral nistatin ve b vitamin kompleksi tedavisine yanit veren siyah killi dil olgusu. Turkderm-Arch Turk Dermatol Venerology **49**, 291–294 (2015)
14. Yıldız, H.T., Özdamar, S.: İnsan fetuslarinda dil papillalarinin gelişiminin taramali elektron mikroskobunda incelenmesi. Sağlık Bilimleri Dergisi (Journal of Health Sciences) **18**(3), 129–137 (2009)
15. Kinnaird, E., Stewart, C., Tchanturia, K.: Taste sensitivity in anorexia nervosa: a systematic review. Int. J. Eat. Disord. **51**, 771–784 (2018)
16. Öztürk-Duran, E.E., Dikmen, D.: Obezitede tat duyusunun etkisi: yağ algisi. Türk Tarım – Gıda Bilim ve Teknoloji Dergisi. **6**(5), 550–556 (2018)
17. Celik, K., Sayan, H.H., Demirci, R.: Gradient adaptive gaussian image filter. In: 2015 23nd Signal Processing and Communications Applications Conference (SIU), pp. 879–882 (2015)
18. Santur, Y., Dilmen, H., Makinist, S., Talu, M.F.: Mean shift ve gauss filtre ile gölge tespiti. Eleco 2014 Elektrik – Elektronik – Bilgisayar ve Biyomedikal Mühendisliği Sempozyumu, 27–29 Kasım, Bursa, Turkey, pp. 738–742 (2014)
19. Huang, M.L., Fu, C.C.: Applying image processing to the textile grading of fleece based on pilling assessment. Fibers 2018 **6**(4), Article No. 73 (2018)
20. Mordvintsev, A., K, A.: Smooting images. https://opencv24-python-tutorials.readthedocs.io/en/stable/py_tutorials/py_imgproc/py_filtering/py_filtering.html. Accessed 30 Feb 2019
21. Rong, W., Li, Z., Zhang, W., Sun, L.: An improved canny edge detection algorithm. In: 2009 Second International Workshop on Computer Science and Engineering, pp. 497–500 (2010)
22. Manasa, N., Mounica, G., Tejaswi, B.D.: Brain tumor detection based on canny edge detection algorithm and it's area calculation. Int. J. Comput. Math. Sci. **5**(3), 10–13 (2016)
23. Mordvintsev, A., Abid, K.: OpenCV- Python tutorials documentation release 1. https://media.readthedocs.org/pdf/opencv-python-tutroals/latest/opencv-python-tutroals.pdf. Accessed 18 Feb 2019
24. Anonymous: The OpenCV tutorials. https://docs.opencv.org/2.4/opencv_tutorials.pdf. Accessed 18 Feb 2019
25. Kolkur, S., Kalbande, D., Shimpi, P., Bapat, C., Jatakia, J.: Human skin detection using RGB, HSV and YCbCr Color Models. Adv. Intell. Syst. Res. **137**, 324–332 (2016)

26. Bora, D.J., Gupta, A.K., Khan, F.A.: Comparing the performance of l*a*b* and hsv color spaces with respect to color image segmentation. Int. J. Emerg. Technol. Adv. Eng. **5**(2), 192–203 (2015)
27. Shukor, A.Z., Natrah, N.A., Tarmizi, A.L., Afiq, A.A., Jamaluddin, M.H., Ghani, Z.A., Shah, H.N.M., Rashid, M.Z.A.: Object tracking and following robot using color-based vision recognition for library environment. J. Telecommun. Electron. Comput. Eng. **10**(2), 79–83 (2018)
28. Prabhakar, A., Devi, N., Devi, R.: Different color detection in an RGB image. Int. J. Dev. Res. **7**(8), 14503–14506 (2017)

Comparison of Image Quality Measurements in Threshold Determination of Most Popular Gradient Based Edge Detection Algorithms Based on Particle Swarm Optimization

Nurgül Özmen Süzme(✉) and Gür Emre Güraksın

Department of Biomedical Engineering, Afyon Kocatepe University, Afyonkarahisar, Turkey
{nozmen, emreguraksin}@aku.edu.tr

Abstract. Determination of the threshold value is one of the challenging processes for edge detection in image processing. In this study, the threshold values of the gradient based edge detection algorithms for Roberts, Sobel, Prewitt were determined using the Particle Swarm Optimization (PSO) algorithm, based on the image quality measurements, Mean Squared Error (MSE), Peak Signal-to-Noise Ratio (PSNR), Structural Similarity Index Metrics (SSIM) and Correlation Coefficients (CC). The threshold values determined by the PSO algorithm and the quality values obtained for the default value of the threshold are compared. In addition the output images obtained by the algorithm were evaluated visually.

Keywords: Image processing · Determining threshold · Quality measurements · Gradient based methods

1 Introduction

Image processing is a kind of signal processing method, in which an input image is converted into an image or a series of features. Density indicates discontinuity or significant variation in gray levels, indicating the edge in the image. Edge detectors detect edges using pixels and edge detection is a local image processing method (Gonzales and Wintz 1987). Edge detection is very urgent for image processing since the boundaries need to be defined for identifying objects.

A gray-scale image transforms a binary edge image with edge detection technique and with this transformation, the most of the useful information is kept. In other operations of image processing; the process is conducted using a binary image, which is a simpler form, instead being concerned with the gray scale image. As a usual way to the problem of edge detection, high spatial frequency development/thresholding algorithms is used. These algorithms use spatial operators that make up the edge improvement power map. A threshold value is then enforced to the edge power map to decide whether there is an edge (Lee et al. 1987).

There are many ways to determine the edge. However, it can be divided into two categories as Gradient Based methods, which determines the edges by looking at the

D. J. Hemanth and U. Kose (Eds.): ICAIAME 2019, LNDECT 43, pp. 171–181, 2020.
https://doi.org/10.1007/978-3-030-36178-5_14

maximum and minimum in the first derivative of the image and Laplacian based methods, which uses zero transitions in the second derivative of the image to find edges of the image (Shinde 2015).

Intuitive selection of threshold values is often a problem encountered by detectors. Operators of Prewitt's (Gonzales and Wintz 1987), Roberts' (Gonzales and Wintz 1987) Sobel's and zero-crossing edge detectors (Marr and Hildreth 1980) generally use the selected threshold values without a precise objective evaluation (Rakesh et al. 2004).

Types of edge detection established on threshold can be divided into two classes: (a) local techniques that use local image neighborhoods, (b) global techniques that use global knowledge and filtering methods for the edge detection. Two of them have strengths and weaknesses. Almost all of the edge detectors use thresholding to find the edge. The pixel values in the image are compared with the threshold value. Higher threshold value than pixel intensity is shown as white in output image, otherwise it is shown as black. Optimizing the threshold value is one of the major and challenging processes in image processing (Ahmad and Choi 1999).

The main problem in determining different edge detectors is that there is no single evaluation method for this. Every person has a different way of analyzing and dividing an image. When the results are different from the same image, a single and special technique will be insufficient. For example root mean square error (ERMS) between the input image and the output image is used to measure the consistency of, edge detectors. This result is used as a criterion for evaluating the different outputs produced by the edge detectors (Kaur et al. 2012a and b).

Digital images are subject to distortion during image acquisition, processing, storage, compression, and reproduction, which can result in reduced image quality (Wang et al. 2004). There basically two approaches to determine image quality which are: Subjective measurements is that expert people give their opinions on image quality; Objective measurements are realized by applying mathematical algorithms (Kumar and Rattan 2012).

In practice, subjective methods are usually too annoying, tedious and expensive. The objective image quality measurement is classified according to whether the original image is used or not. In most of the available approaches, the input picture is fully known and this is called full reference. In some embodiments, the reference image is absent and the non-reference or "blind" quality evaluation approach is applied. A third type of reference image is either moderately present or has attributes of the image, used to evaluate the image quality (Kumar and Rattan 2012).

In this study, the threshold values of Sobel, Roberts and Prewitt algorithms, which are the traditional edge detection operators, were determined by the Particle Swarm Optimization method which is one of the heuristic optimization algorithms. In the process of determining the threshold value by particle swarm optimization, the quality determination metrics were used in image processing are as based on: Peak Signal-to-Noise Ratio (PSNR), Mean Square Error (MSE), Correlation Coefficient (CC) and Structural Similarity Index (SSIM) methods. In this way, the most ideal quality measurement was determined by using PSO algorithm for edge determination by making visual comparisons according to quality measurements used in this study.

2 Edge Detection

The success in finding meaningful edges indicates the effectiveness of computer vision and image processing. Edge detection is the challenge of low-level image processing (Lakshmi and Sankaranarayanan 2010). Gradient besed methods are often used in image processing because they are not complex. In this study, Sobel, Roberts and Prewitt were considered as gradient based image processing algorithms.

Gradient is first order derivative of the image corresponding to two-dimensional function. Thus, in the sampling point group, image is generated as a continuous derivative of image density (Bin and Yeganeh 2012). The frequently used gradient based techniques in edge detection techniques in image are Prewitt Sobel, and Roberts Edge Detection algorithms.

To obtain the gradient of an image, the partial derivatives $\frac{\partial f}{\partial x}$ and $\frac{\partial f}{\partial y}$ must be calculated for each pixel value in the image. Digital numbers are needed, so partial derivatives need a numerical approach to a point in a neighborhood.

$$\mathrm{Gx} = \frac{\partial f(x,y)}{\partial x} = f(x+1,y) - f(x,y) \quad (1)$$

$$\mathrm{Gy} = \frac{\partial f(x,y)}{\partial y} = f(x,y+1) - f(x,y) \quad (2)$$

These two equations can be used with all values of the following masks.

Considering the crossed edges, a two-dimensional mask is needed. The Roberts operator is based on diagonal differences of image.

Table 1. Gradient based edge detection operators

Edge detection operator	Mask for Gx	Mask for Gy
Roberts	1 0 0 −1	0 1 −1 0
Sobel	−1 0 1 −2 0 2 −1 0 1	1 2 1 0 0 0 −1 −2 −1
Prewitt	−1 0 1 −1 0 1 −1 0 1	1 1 1 0 0 0 −1 −1 −1

Table 1 shows the convolution masks of the gradient based edge detection operators.

The magnitude of the gradient is calculated approximately by using a total of absolute values (Seif et al. 2010).

$$M(x, y) = |Gx| + |Gy| \quad (3)$$

In the first derivative-based edge detection process, the gradient of the image must be thresholded to eliminate the wrong edges generated by noise. If the threshold value is too small, the wrong edges may be found, otherwise the actual edges may be missed (Bao et al. 2005).

And the best way to separate objects from the background is to specify a threshold (T) that separates this background and object. Then, the points in the image (x, y) are named as f (x, y) > T in the case of the object and otherwise the background (Ahmad and Choi 1999). Segmented image g(x, y) shown as

$$g(x,y) = \begin{cases} 1, \mathit{iff}(x,y) \geq T \\ 0, \mathit{otherwise} \end{cases} \quad (4)$$

When T is a valid constant on an entire image, this is called global thresholding. When the value of T changes for an image, we use the concept of variable thresholding. At any point, if the value of T is determined by looking at the neighborly relations, the term local or regional thresholds are used (Gonzales and Wintz 1987).

2.1 Partical Swarm Optimisation

The partical swarm has emerged as a simplified simulation of a social system. Its main purpose is to graphically show the elegant but unpredictable movements of the flock (Shi 2001). Due to its simple structure and efficiency, PSO, which is a flock and stochastic approach, is used in a wide area (Güraksın et al. 2014). Optimization is started with a population of random candidate solutions, called particles. A random velocity assignment is made for each particle and it is passed through the problem space as repetitive. Until then the particle itself and the best fit up to the entire population are pulled towards the best fit (Trelea 2003).

The description of PSO algorithm is shown below

$$v_{id} = v_{id} + c_1 rand()(p_{id} - x_{id}) + c_2 Rand()\left(p_{gd} - x_{id}\right) \quad (5)$$

$$x_{id} = x_{id} + v_{id} \quad (6)$$

where c_1 and c_2 are positive constants, and rand() and Rand() are random functions in the [0,1] range; $X_i = (x_{i1}, x_{i2}, \ldots, x_{iD})$ represents the ith particle; $Pi = (p_{i1}, p_{i2}, \ldots, p_{iD})$ represents the best previous position of the ith particle; the symbol g represents the index of the best particle; $V_i = (v_{i1}, v_{i2}, \ldots, v_{iD})$ represents the rate of the position change for particle i.

Equation 5 shows updating velocity dynamically and Eq. 6 shows updating position of particles (Shi 2004).

The procedure for performing PSO is shown as (Shi 2004):

1. Initialize a population of particles with random positions and velocities.
2. Estimate the desired optimization fitness function for each particle in variables.

3. Compare particle's fitness estimation with its p_{best}. If valid value is preferable than p_{best}, then set p_{best} adequate to the valid value, and P_i adequate to the valid position X_i in D-dimensional space.
4. The most successful neighboring particle has been identified so far and the index of the particle is assigned to the g variable.
5. The speed and position of the particle are changed depending on the equation 1a and 1b.
6. Until criterion is met go step 2, usually this criterion is a good fitness or the number of iteration.

Advantages of PSO algorithm matched with other heuristic algorithms for detecting edges is listed below (Setayesh 2013).

(1) Higher Convergence: There are two basic reasons that make the PSO algorithm faster than other heuristic algorithms: The first is the sharing of information among each other by a topology. When a particle is better located by a particle, this information is transmitted immediately to the other particles from the same topology, so that all particles are rapidly approaching the local optimum. Second, it is the use of speed in determining the position of the particle, so it has a better convergence rate than other heuristic algorithms.
(2) Due to the ease of operation, the PSO algorithm is easy to implement.
(3) The only operator in PSO is the calculation of speed, other algorithms contain more operators.
(4) The PSO algorithm has two types of memory, namely cognitive and social, that determine the particle movements. While the cognitive memory saves the best previous positions and the social memory keeps the position of the best point in search space. Although the particles are periodically updated, these two types of memory ensure that the information is acquired.

In addition, the reason why PSO algorithm is used in edge determination is that it works fast and works stably and effectively even in pictures with noise (Setayesh 2013).

3 Image Quality Measurements

Any operation on the image may cause loss of image quality and the information contained in it. Methods for evaluating image quality are divided into two groups as subjective and objective. Subjective methods depend on people's decisions and do not have any criteria. In objective methods, comparison is made by using mathematical criteria (Hore 2010).

Given a source image a and a test image b, both of size M × N, the PSNR between a and b is defined by:

$$\mathrm{PSNR}(\mathrm{a},\mathrm{b}) = 10\ \log_{10}(255^2/MSE(a,b) \tag{7}$$

Where

$$\mathrm{MSE(a,b)} = 1/MN \sum\nolimits_{i=1}^{M} \sum\nolimits_{j=1}^{N} \left(a_{ij} - b_{ij}\right)^2 \tag{8}$$

This equation shows that a higher PSNR value provides a higher image quality.

SSIM is a popular quality measure used to measure similarity between two images. SSIM models the distortion in any image as a combination of three factors as correlation loss, brightness degradation and contrast degradation (Hore 2010).

SSIM includes three parts: Comparison of Luminance l(f, g), Comparison of Contrast c(f, g) and Comparison of Structure s(f, g).

SSIM is described as:

$$\mathrm{SSIM(f,g)} = \mathrm{l(f,g)c(f,g)s(f,g)} \tag{9}$$

Where

$$\mathrm{l(f,g)} = \frac{2\mu_f \mu_g + c_1}{\mu_f^2 + \mu_g^2 + c_1} \tag{10}$$

$$\mathrm{c(f,g)} = \frac{2\sigma_f \sigma_g + c_2}{\sigma_f^2 + \sigma_g^2 + c_2} \tag{11}$$

$$\mathrm{s(f,g)} = \frac{\sigma_{fg} + c_3}{\sigma_f \sigma_g + c_3} \tag{12}$$

Where μ_f, mean of the f, σ_f^2 the variance of the f, σ_{fg} the covariance of the f, g. μ_f and σ_f can be viewed estimates of luminance and contrast of f, σ_{fg}, measures the tendency of f and g to vary together. σ_{fg} is an indication of structural similarity (Wang et al. 2003).

Correlation is the examination of the possibility of a linear relationship between the two measured characteristics. The product-moment correlation coefficient r was defined as Pearson in 1895 by Karl Pearson. Pearson Correlation Coefficient, which is widely used in statistical analysis, pattern recognition and image processing, was the first formal correlation measurement (Kaur et al. 2012a and b).

$$\mathrm{r} = \frac{\sum_i (x_i - x_m)(y_i - y_m)}{\sqrt{\sum_i (x_i - x_m)^2}\sqrt{\sum_i (y_i - y_m)^2}} \tag{13}$$

4 Determination of Threshold Value via PSO Algorithm

Analyzing the effectiveness of PSO based edge detection technique we compared the other mostly used edge detection techniques. Firstly the images are processed with different edge detection techniques as Roberts, Sobel, Prewitt with default threshold values. And calculated MSE, PSNR, CC and SSIM values for this methods. Then new threshold value with PSO algorithm calculated then new MSE, PSNR, CC and SSIM values calculated for images. And the output images compared with human visual system.

Evaluating the PSO algorithm particles created with a random uniform distribution from −1000 to 1000. Iteration number was evaluated as 200. The algorithm is tested on 1 dimensional because one feature is used. The particles of swarms were initialized in the search range [0, 1] where Xmax and Xmin were the maximum and minimum value in search range. The criteria for stopping is maximum iteration number 200 is selected.

As the PSNR rate accesses infinity, the MSE rate accesses zero, indicating a high PSNR value for high image quality (Shi 2004) (Tables 2 and 3).

Table 2. MSE1 MSE for default threshold value and MSE2 MSE for calculated threshold value due to PSO algorithm

	Roberts	Prewitt	Sobel
MSE1	13583,145067749	13602,0074661229	13569,8042140904
MSE2	11427,2617344158	11477,6294173425	11465,7708807572

Table 3. Visual determination for calculated threshold value due to PSO algorithm using MSE

Robert MSE	Prewitt MSE	Sobel MSE
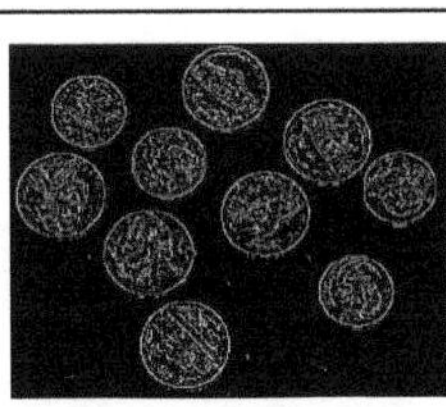	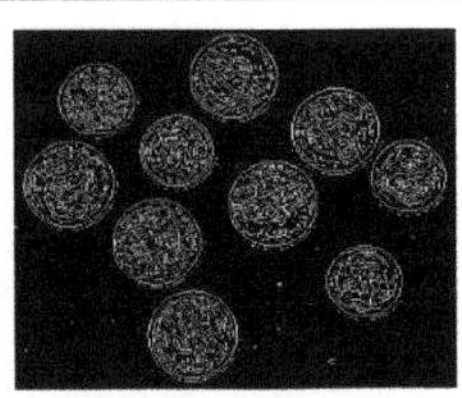	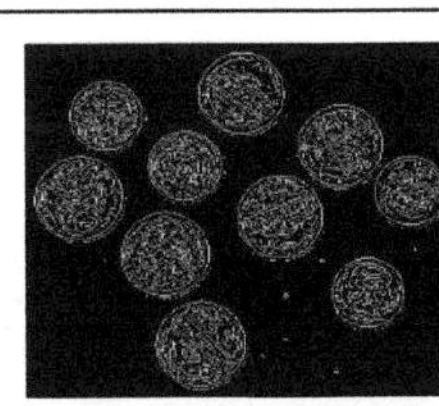

Compared to the default threshold value and the threshold value calculated with PSO, without visual comparison; The MSE value is low in the proposed algorithm and the PSNR value is high (Tables 4 and 5).

Table 4. PSNR1 Psnr for default threshold value and PSNR2 Psnr for calculated threshold value due to PSO algorithm

	Roberts	Prewitt	Sobel
PSNR1	6,800800219	6,79477352	6,805067792
PSNR2	7,5513818612	7,532281625	7,5367710187

Table 5. Visual determination for calculated threshold value due to PSO algorithm using PSNR

Robert PSNR	Prewitt PSNR	Sobel PSNR
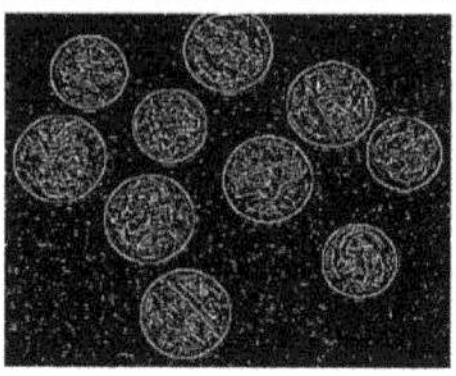	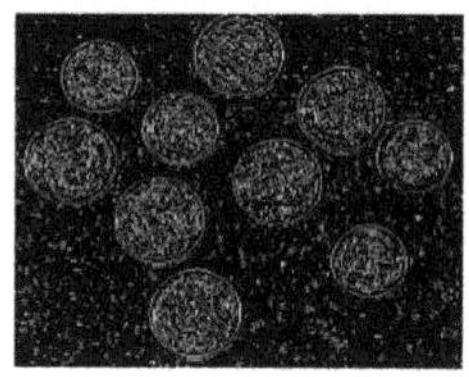	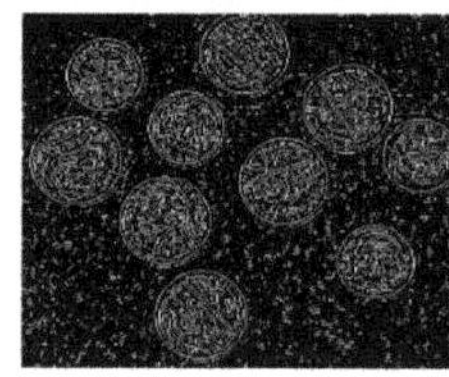

Table 6. SSIM1 SSIM value for default threshold value and SSIM2 SSIM value for calculated threshold value due to PSO algorithm

	Roberts	Prewitt	Sobel
SSIM1	0,013233475	0,008446819	0,009150445
SSIM2	0,016213193	0,011470759	0,012452242

Table 7. Visual determination for calculated threshold value due to PSO algorithm using SSIM

Robert SSIM	Prewitt SSIM	Sobel SSIM
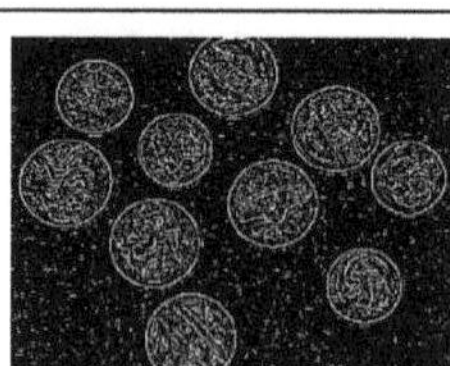	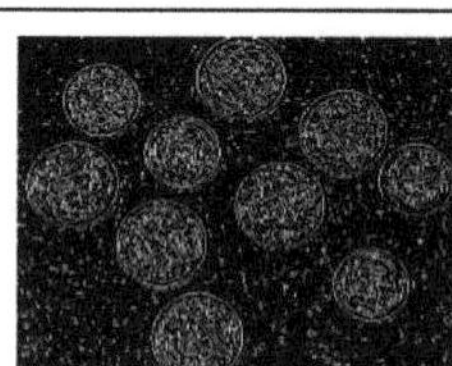	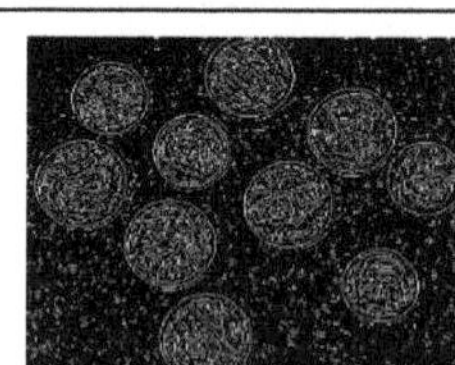

SSIM, the higher the rate of SSIM(x, y) is, the more alike the original images and compared. Compared to the default threshold value and the threshold value calculated with PSO, without visual comparison the SSIM values are higher for proposed algorithm. Higher SSIM value shows higher similarity (Tables 6 and 7).

Table 8. CC1, CC value for default threshold value and CC2, CC value for calculated threshold value due to PSO algorithm

	Roberts	Prewitt	Sobel
CC1	0,110200175	0,101095526	0,108522567
CC2	0,410718587	0,405781671	0,407197929

Table 9. Visual determination for calculated threshold value due to PSO algorithm using SSIM

Robert CC	Prewitt CC	Sobel CC
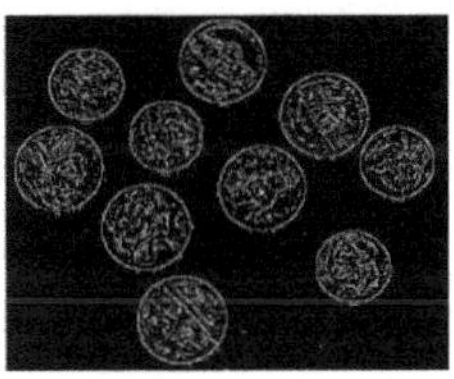	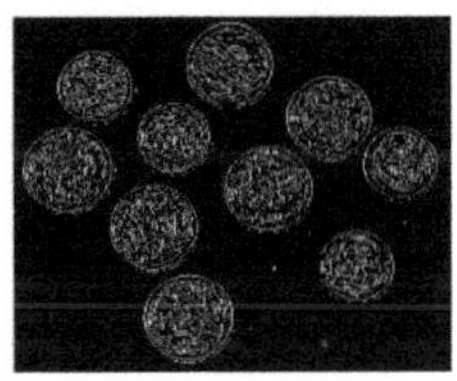	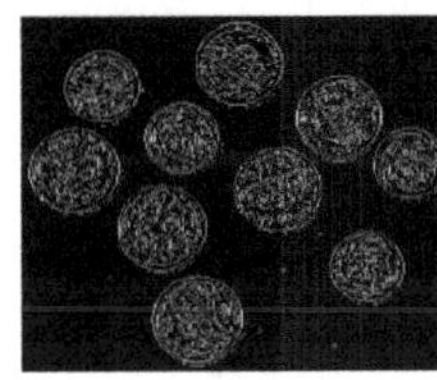

The threshold values for coins image calculated depend on MSE, PSNR, SSIM; CC with PSO algorithm are shown below (Tables 8 and 9).

Table 10. Threshold values due to PSO algorithm

Filter (quality measurement)	Calculated threshold value
Roberts (MSE)	0.011097571847915
Roberts (PSNR)	0.011097571847915
Prewitt (MSE)	0.008574686924883
Prewitt (PSNR)	0.008574686924883
Sobel (MSE)	0.009342735092613
Sobel (PSNR)	0.009342735092613
Roberts (SSIM)	0.006826954629409
Prewitt (SSIM)	0.004798352270326
Sobel (SSIM)	0.005011111986058
Sobel (CC)	0.009506789250893
Roberts (CC)	0.010629198795497
Prewitt (CC)	0.008744090865921

The threshold values calculated according to image quality measurements with the help of PSO algorithm are presented in the table (Table 10).

5 Conclusion

The goal of this paper was to develop a new PSO-based approach to determine threshold values for edge detection. The threshold values determined by using the PSO algorithm according to the image quality determination criteria were applied to the images and the obtained images were evaluated subjectively and objectively. In this study, Sobel, Prewitt, Robert edge determination algorithms which are commonly used in the gradient based edge detection algorithms are used.

Quality indexes are frequently used in image enhancement studies. When we considered at the quality measurements used in the threshold value determination, it was seen that the results of PSNR, MSE and CC based image quality measurements were close to each other. However, according to SSIM quality measurement value as a result of the edge detection process, the threshold value of the result of the algorithm is very low as a result of the determination was more noisy.

References

Ahmad, M.B., Choi, T.S.: Local threshold and boolean function based edge detection. IEEE Trans. Consum. Electron. **45**(3), 674–679 (1999)

Bao, P., Zhang, L., Wu, X.: Canny edge detection enhancement by scale multiplication. IEEE Trans. Pattern Anal. Mach. Intell. **27**(9), 1485–1490 (2005)

Bin, L., Yeganeh, M.S.: Comparison for image edge detection algorithms. IOSR J. Comput. Eng. **2**(6), 1–4 (2012)

Gonzales, R.C., Wintz, P.: Digital Image Processing. Addison-Wesley, Reading (1987)

Güraksın, G.E., Haklı, H., Uğuz, H.: Support vector machines classification based on particle swarm optimization for bone age determination. Appl. Soft Comput. **24**, 597–602 (2014)

Hore, A., Ziou, D.: Image quality metrics: PSNR vs. SSIM. In: 2010 20th International Conference on Pattern Recognition (ICPR), pp. 2366–2369. IEEE, August 2010

Kaur, A., Kaur, L., Gupta, S.: Image recognition using coefficient of correlation and structural similarity index in uncontrolled environment. Int. J. Comput. Appl. **59**(5) (2012)

Kaur, J., Agrawal, S., Vig, R.: A comparative analysis of thresholding and edge detection segmentation techniques. Image **7**(8), 9 (2012)

Kumar, R., Rattan, M.: Analysis of various quality metrics for medical image processing. Int. J. Adv. Res. Comput. Sci. Softw. Eng. **2**(11) (2012)

Lakshmi, S., Sankaranarayanan, D.V.: A study of edge detection techniques for segmentation computing approaches. IJCA Special Issue on "Computer Aided Soft Computing Techniques for Imaging and Biomedical Applications" CASCT, 35–40 (2010)

Lee, J., Haralick, R., Shapiro, L.: Morphologic edge detection. IEEE J. Robot. Autom. **3**(2), 142–156 (1987)

Marr, D., Hildreth, E.: Theory of edge detection. Proc. R. Soc. Lond. B **207**(1167), 187–217 (1980)

Rakesh, R.R., Chaudhuri, P., Murthy, C.A.: Thresholding in edge detection: a statistical approach (2004)

Setayesh, M.: Particle Swarm Optimisation for Edge Detection in Noisy Images (2013)

Seif, A., Salut, M.M., Marsono, M.N.: A hardware architecture of Prewitt edge detection. In: 2010 IEEE Conference on Sustainable Utilization and Development in Engineering and Technology (STUDENT), pp. 99–101. IEEE, November 2010

Shinde, S.G.: Novel hardware unit for edge detection with comparative analysis of different edge detection approaches. Int. J. Sci. Eng. Res. **6**(4) (2015)

Shi, Y. (2001). Particle swarm optimization: developments, applications and resources. In *evolutionary computation, 2001. Proceedings of the 2001 Congress on* (Vol. 1, pp. 81–86). IEEE

Shi, Y.: Particle swarm optimization. IEEE Connect. **2**(1), 8–13 (2004)

Trelea, I.C.: The particle swarm optimization algorithm: convergence analysis and parameter selection. Inf. Process. Lett. **85**(6), 317–325 (2003)

Wang, Z., Bovik, A.C., Sheikh, H.R., Simoncelli, E.P.: Image quality assessment: from error visibility to structural similarity. IEEE Trans. Image Process. **13**(4), 600–612 (2004)

Wang, Z., Simoncelli, E.P., Bovik, A.C.: Multiscale structural similarity for image quality assessment. In: The Thrity-Seventh Asilomar Conference on Signals, Systems & Computers, vol. 2, pp. 1398–1402. IEEE, November 2003

A Hybrid Approach for the Sentiment Analysis of Turkish Twitter Data

H. A. Shehu[1] and S. Tokat[2(✉)]

[1] Institute of Science and Engineering, Computer Engineering Section, Pamukkale University, Denizli, Turkey
harisushehu@gmail.com
[2] Department of Computer Engineering, Pamukkale University, Denizli, Turkey
stokat@pau.edu.tr

Abstract. Social media is now playing an important role in influencing people's sentiments. It also helps analyze how people, particularly consumers, feel about a particular topic, product or an idea. One of the recent social media platforms that people use to express their thoughts is Twitter. Due to the fact that Turkish is an agglutinative language, its complexity makes it difficult for people to perform sentiment analysis. In this study, a sum of 13K Turkish tweets has been collected from Twitter using the Twitter API and their sentiments are being analyzed using machine learning classifiers. Random forests and support vector machines are the two kinds of classifiers that are adopted. Preprocessing methods were applied on the obtained data to remove links, numbers, punctuations and un-meaningful characters. After the preprocessing phase, unsuitable data have been removed and 10,500 out of the 13K downloaded dataset are taken as the main dataset. The datasets are classified to be either positive, negative or neutral based on their contents. The main dataset was converted to a stemmed dataset by removing stopwords, applying tokenization and also applying stemming on the dataset, respectively. A portion of 3,000 and 10,500 of the stemmed data with equal distribution from each class has been identified as the first dataset and second dataset to be used in the testing phase. Experimental results have shown that while support vector machines perform better when it comes to classifying negative and neutral stemmed data, random forests algorithm perform better in classifying positive stemmed data and thus a hybrid approach which consists of the hierarchical combination of random forest and support vector machines has also been developed and used to find the result of the data. Finally, the applied methodologies have been tested on both the first and the second dataset. It has been observed that while both support vector machines and random forest algorithms could not achieve an accuracy of up to 77% on the first and 72% on the second dataset, the developed hybrid approach achieve an accuracy of up to 86.4% and 82.8% on the first and second dataset, respectively.

Keywords: Social media · Sentiment analysis · Artificial intelligence · Twitter · Turkish

D. J. Hemanth and U. Kose (Eds.): ICAIAME 2019, LNDECT 43, pp. 182–190, 2020.
https://doi.org/10.1007/978-3-030-36178-5_15

1 Introduction

Social media is a term used for the Internet applications that allow users to create and share information, ideas, opinion, etc., through virtual communities and social networking. Over the years, social media systems on the web have provided outstanding platforms to facilitate and enable audience participation and engagement that have resulted a new community with a participatory culture [1]. With the growth of Internet-based social media and the increase of the social media users, it became conceivable for a person to communicate with hundreds of other people and this capacity keeps on increasing progressively. Currently, social media is playing a vital role in modern life. Without any restrictions and hesitation, users are allowed to post their view and thoughts on social media. In the last two decades, the social networking sites were seen as a medium for only the purposes of friendship or dating. But with time, the characteristics of the social media platforms has now been changing. While some of the social media platforms allow users to share their thoughts with an easy level of privacy and to interact only with their friends, now users are migrated from traditional means of communication to microblogging sites such as Facebook, Twitter etc. [2].

There are different kinds of social media platforms that are used for different purposes. For instance; there're dating apps such as Bumble, Lovoo and Tinder, multi-purpose messaging apps such as Facebook Messenger, Google Allo, Hangouts, Viber, WeChat, WhatsApp, online news and social networking apps such as Google + , Facebook and Twitter, online news apps such as Flipboard, Google News and Microsoft News, random video chatting apps such as Azar, CamSurf and Chatroulette, microblogging apps such as FriendFeed, Twitter and Tumblr, photo and video sharing apps such as Flickr, Instagram and Pinterest and also video chatting apps such as Google Duo, Imo and Skype.

Twitter is one of the most popular social media platform that is being used nowadays [3]. It is a social networking website that allows users to send and read messages consisting of up to 140 characters initially. This number has been doubled in 2017 for all the languages except Chinese, Japanese and Korean. In Twitter, the messages send are called tweets. Twitter is said to have 310 million active users monthly [4]. Approximately 500 million tweets get tweeted everyday on Twitter [5].

Billions of data on social media make it an attractive tool for researchers to perform research on data analysis. Twitter is widely used for expressing opinion-rich views on certain topics [6]. Users of Twitter (tweeple) are allowed to post their opinion on certain topics based on what they want to post about, by using the hashtag topic "#topic". For example, #technology, #peace, #industry etc. to discuss on technology, peace and industry, respectively.

Many companies and business organizations effectively use social media to gain benefit for their business. Many researches are performed using tweets to make predictions about an event, product, industry, stock market etc. using various classification methods [6]. This has become possible due to the huge amount of data that can be found on Twitter.

Sentiment analysis which is also referred to as opinion mining, emotion mining, attitude mining or subjectivity mining, is the method of computing and identifying a person's view in a given piece of text particularly to identify a person's idea towards a specific topic, product, situation, person or a thing [7]. Sentiments have been expressed via social media through text based messages and images [8]. Currently, some of the social media platforms that allow users to post their views publicly includes Twitter, Facebook, Flickr, LinkedIn, etc.

Users from various locations in the world send their tweets in different languages. The Turkic languages are a large language family extending from Turkey to China that show close similarity to each other in phonology, morphology, and syntax. Turkish language, a member of the Turkic family, is by far the most commonly spoken of the Turkic languages [9].

At the moment, the existing sentiment analysis methods developed for English rarely have productive outcome when it comes to Turkish due to the fact that Turkish is an agglutinative language [10]. The examples below show the difference between English and Turkish and why it's hard to perform sentiment analysis on Turkish texts with extensive agglutination and vowel harmony [11]. With agglutination property, the root words of Turkish language can be extended by many suffixes to produce new meanings. An example is given in Table 1.

Table 1. Example Turkish root words extended to produce new meaning

Word	Suffixes	English meaning
Yap		Do
Yapma	Yap-ma	Don't do
Yaptım	Yap-tı-m	I did
Yapıyorum	Yap-ıyor-um	I'm doing
Yapabilirim	Yap-abilir-im	I can do
Yapabilirdim	Yap-abilir-dim	I could have done
Yapamayabilirdim	Yap-amayabilir-di-m	I might not have been able to do

The added suffix may change the polarity of a root word. An example is given in Table 2. A word that appear to be negative used in a sentence may have a different meaning. An example is given in Table 3.

Table 2. Example of changing polarity of a root word

Word	Suffixes	English meaning	Sentiment polarity
Merhametli	Merhamet-li	Merciful	Positive polarity
Merhametsiz	Merhamet-siz	Unmerciful	Negative polarity

In this paper, we have proposed a hybrid approach that involves the hierarchical combination of random forests (RF) and support vector machines (SVM) to analyze the sentiment of a Turkish Twitter data after the data has been stemmed.

Table 3. Example of negative words that change their meaning when used in a sentence

Sentence	English meaning
Boya **yapma** makinesi kullanarak boya yapabilirsiniz	You can paint using the painting machine
Buradan slayt **yapma** ve video düzenleme programını indirebilirsiniz	You can download slide and video editing program from here

The rest of the paper is organized as: Sect. 2 illustrates related works; Sect. 3, explains the stemming method; Sect. 4 explains the hybrid approach; Sect. 5 compares our results with existing methodology; Sect. 5 concludes the paper and hints future works.

2 Related Work

While there are many studies on sentiment analysis based on text written on English language [12–14], there are relatively few studies currently done on other languages such as Turkish. Some of the influential work that have been performed on sentiment analysis on Turkish texts are reviewed in this chapter.

The study of sentiment analysis to produce a Turkish sentiment dictionary was carried out in a thesis [15]. The Turkish sentiment dictionary was produced by translating an English sentiment dictionary to Turkish. SVM was used with a lexicon that has 27,000 Turkish words with assigned polarity on some movie corpora to determine its performance. Another thesis studied the sentiment analysis for movie review where movie reviews are gathered from various websites such as rec.arts.movies.reviews, rottentomatoes.com and beyazperde.com [16]. Then, SVM was used to perform the analysis. While this study didn't develop any comprehensive Turkish sentiment lexicon, the effects of part of speech (POS), information of words and negation suffix on the sentiment of the reviews were analyzed, respectively. In [17], RF and SVM are used separately in order to classify the tweets in three different forms: raw form, the form in which the tweets are converted into tokens and stop-words are being removed, and the form in which stemming is performed.

The research carried out in the above two mentioned theses [15, 16] were combined in a single research [18] where a comparison between lexicon based and machine learning based sentiment analysis is presented. The performance of the Turkish informal texts was tested on both short (Twitter) and long (movie) dataset. The lexicon is obtained by manually translating words from English to Turkish and the best result obtained using the lexicon based method is 75.2% on the Twitter dataset whereas a result of 79% was obtained using the movie dataset. On the other hand, naïve-Bayes, SVM and J48 Decision Trees are being used as the machine learning techniques to classify the data. SVM outperforms the other classifiers on the Twitter dataset with an accuracy of 85% and both SVM and NB outperformed J48 classifier on the movie dataset with both appearing to have an accuracy of 89.5% [18].

While there are studies that concentrate on analyzing the sentiments in Turkish text, there are also some other studies that concentrated on developing lexicon or building dataset that can be used to perform the sentiment analysis. For instance, a research was carried out to build the first polarity lexicon for Turkish and also propose a semi-automatic approach to do this for other languages as well [19]. The developed lexicon contains polarity score, triplets (positive, negative and neutral/objective) for all Synsets (set of synonyms) in the Turkish WordNet which consists of almost 15,000 Synsets. The combination of three English and one Turkish resources were used for the construction of the developed polarity lexicon called SentiTurkNet which consist of around 27,000. The three English resources used are English WordNet [20], SentiWordNet [21] and SenticNet [22] and the Turkish resource used is the Turkish WordNet [23]. A classifier implemented in Weka using three different algorithms is then used to determine the performance of the developed lexicon. The best accuracy obtained using all features and classifier combination of the three classifiers; nearest neighbor, sequential minimal optimization and logistic regression is 91.11%.

In [10], the polarity of a word or sentences as the sum of the polarity of individual words or phrases are used. They start with a large database of Turkish news pages whose URLs are taken from the GDELT database. They moved ahead to obtain a raw news text and root of words seen in the text by parsing these HTML pages and using Zemberek respectively. A score is then assign to each word using the polarity values obtain from the GDELT database. The results is what is called as SWNetTR-GDELT and it consist of around 14000 unique Turkish words. The data used in the experiment is called SWNetTR-PLUS and is formed by adding almost 10K unique words that exists in SWNetTR-GDELT but not in SWNetTR. The new lexicon was tested using the data and the results were reported. The result shows that the accuracy of determining the polarity of news in Turkish has been increased from 60.6% to 72.2%.

In [24], data are gathered from individuals to form a new dataset. Then the formed dataset was divided into two namely; raw and validated dataset. Furthermore, two different stemming methods which are the fixed prefix stemming (FPS) which is proven to give more accuracy after the fifth character [25] and Zemberek or the dictionary based Turkish stemmer [26] are applied to each dataset which makes a total of four different datasets. Also, several machine learning algorithms such as naïve-Bayes, decision tree, random forest and an updated SVM were run on the formed dataset and it was concluded that SVM classifier yielded a better result and that the model trained with validated dataset yielded a better result than the model trained with non-validated dataset.

In our study, the research carried out in [17] is one-step improved by designing a hybrid and hierarchical structure based on both SVM and RF at the same time.

3 The Stemming Method

Zemberek is an open source natural language framework for Turkic language especially Turkish [26]. The library is used to operate on the data to be used in the analysis. The data operations involve cleaning, tokenization, applying stemming and also removing stopwords from the data. The Zemberek method used in this research doesn't only find the stem of the words but also all their possible stems in cases that the word has more than one stem [26].

The so called stems of the word that are found can be rewritten more than once in the stem result according to how it's used in the context. For example, some words ending with the suffix "ler" which indicates a plural in Turkish language are written three times in the stem result whereas some words ending with the suffix "lik" or "luk" are written two times in the stem result depending on the context.

Tables 4 and 5 show some example of words that have more than one stem and words that are rewritten more than once in the stem result.

Table 4. Example of words having more than one stem.

Word	Meaning	Stem	Meaning
Gözlükçü	Optician	Gözlük Göz	Glasses Eye
Kötülük	Wickedness	Kötülük Kötü	Wickedness Bad

Table 5. Example of words that are written more than once.

Word	Meaning	Number of times rewritten	Meaning
Güzellik	Beauty	2 * güzel	Beautiful
Güzeller	Beauties	3 * güzel	Beautiful
Çalışmalar	Studies	3 * çalış; 3 * çal	Work, Steal

4 The Proposed Hybrid Approach

The hybrid approach consists of a hierarchical combination of RF and SVM algorithms. The developed hybrid approach used in this study is shown in Fig. 1.

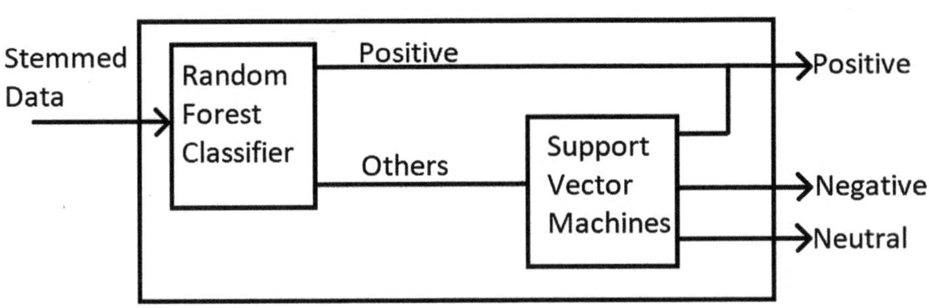

Fig. 1. The proposed hybrid approach

Figure 1 illustrates how the hybrid approach is formed. Firstly, the classifier is fed with stemmed data and the classifier uses RF algorithm to classify the data based on two classes; positive and others. The others class is not only classified into negative and neutral classes but also to positive class using SVM so as to rectify the real positives in others class.

We have used the stemmed data to analyze the sentiment of the tweets in two different cases; both on small dataset which consist of 3,000 tweets and on large dataset that consist of up to 10,500 tweets data with equal distribution from each class. Some of the dataset used for validation consist of a benchmark dataset collected from Başkent University, Turkey [27].

After using RF and SVM to classify the tweets, experimental results have shown that SVM performs better than RF when classifying negative and neutral stemmed data in most cases whereas RF algorithm performs better than SVM in classifying positive stemmed data in all the cases and as a result, the hybrid approach has been developed.

As it can be seen from Fig. 2(a) and (b), the hybrid approach has outperformed SVM in all the two cases and although the hybrid approach didn't outperform RF in some cases, it appears to give a better accuracy result than RF in the overall case.

On the small dataset, on the other hand, while SVM achieved an accuracy of 76.4% and RF achieved an accuracy of 75.9%, the method using hybrid approach has achieved an accuracy of up to 86.4% on the same dataset. And while SVM achieved an accuracy of 67.6% and RF achieved an accuracy of 71.2% on the large dataset, the method using hybrid approach achieved an accuracy of 82.8% on the same dataset.

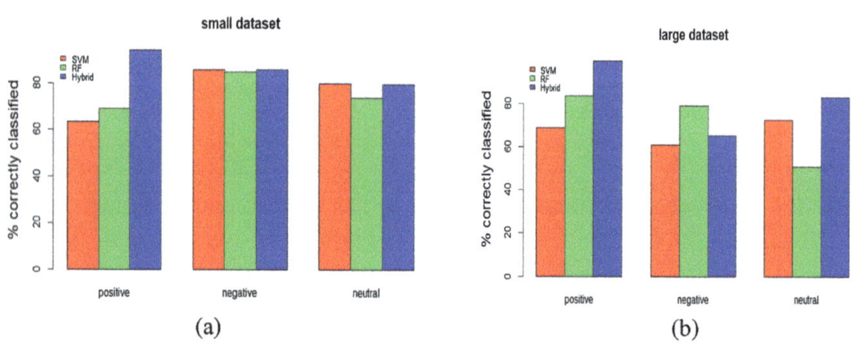

Fig. 2. Bar chart showing the performance comparison of SVM, RF and the hybrid approach on the (a) small dataset (b) large dataset.

5 Conclusions and Future Work

In this research, we have proposed a hybrid approach to perform sentiment analysis on Turkish Twitter data. Our approach work on a stemmed data and is proven to give results with better accuracy than an already existing methodology. We have tested our approach in two different phases; both on small and large dataset and obtained an accuracy of up to 86.4% and 82.8% respectively. As a further study, the proposed methodology can be improved to give a more accurate result, and also test set can be expanded by using different kinds of datasets [28]. Also to handle the imprecise and ambiguous information, or to cope with the linguistic terms fuzzy logic systems can be used [29]. A complete statistical analysis [30] could also be performed in the long run to compare the performance of miscellaneous state-of-the-art approaches.

References

1. Liu, B.: Sentiment Analysis: Mining Opinions, Sentiments and Emotions. Cambridge University Press, Cambridge (2015)
2. Pak, A., Paroubek, P.: Twitter as a corpus for sentiment analysis and opinion mining. In: Proceedings of the 7th International Conference on Language Resources and Evaluation (LREC 10), 1320–1326 (2010)
3. Murthy, D.: Twitter: Social Communication in the Twitter Age. Polity Press, Cambridge (2012)
4. Anastasia, S., Budi, I.: Twitter sentiment analysis of online transportation service provider. In: Proceedings of the International Conference on Advanced Computer Science and Information Systems (ICACSIS), pp. 359–365 (2016)
5. Borgaonkar, P., Sharma, H., Sharma, N., Sharma, A.K.: Social big data analysis: techniques, issues and future research perspective. In: Rathore, V.S., Worring, M., Mishra, D.K., Joshi, A., Maheshwari, S. (eds.) Emerging Trends in Expert Applications and Security, pp. 625–632. Springer, Singapore (2018). https://doi.org/10.1007/978-981-13-2285-3_73
6. Jain, A.P., Katkar, V.D.: Sentiment analysis of Twitter data using data mining. In: International Conference on Information Processing (ICIP) Vishwakarma Institute of Technology, pp. 807–810 (2015)
7. Kiprono, K.W., Abade, E.O.: Comparative Twitter sentiment analysis based on linear and probabilistic models. Int. J. Data Sci. Technol. **2**, 41–45 (2016)
8. Anjaria, M., Guddeti, R.M.: Influence factor based opinion mining of Twitter data using supervised learning. In: Proceedings of the 6th International Conference on Communication Systems and Networks (COMSNETS), pp. 1–8 (2014)
9. Kornfilt, J.: Turkish and the Turkic languages. In: Comrie, B. (ed.) The World's Major Languages, 2nd edn. Routledge, Oxford (1990)
10. Saglam, F., Sever, H., Genc, B.: Developing Turkish sentiment lexicon for sentiment analysis using online news media. In: IEEE/ACS 13th International Conference of Computer Systems and Applications, Agadir, Morocco (2016)
11. Vural, A.G., Cambazoglu, B.B., Senkul, P., Tokgoz, Z.O.: A framework for sentiment analysis in Turkish: application to polarity detection of movie reviews in Turkish. In: Gelenbe, E., Lent, R. (eds.) Computer and Information Sciences III, pp. 437–445. Springer, London (2013). https://doi.org/10.1007/978-1-4471-4594-3_45
12. Pang, B., Lee, L.: Opinion mining and sentiment analysis. Found. Trends Inf. Retr. **2**(1–2), 1–135 (2008)
13. Etter, M., Colleoni, E., Illia, L., Meggiorin, K., D'Eugenio, A.: Measuring organizational legitimacy in social media: assessing citizens' judgments with sentiment analysis. Bus. Soc. **57**(1), 60–97 (2016)
14. Cummins, N., Amiriparian, S., Ottl, S., Gerczuk, M., Schmitt, M., Schuller, B.: Multimodal bag of words for cross domains sentiment analysis. In: IEEE International Conference on Acoustics, Speech, and Signal Processing (ICASSP), Calgary, AB, Canada (2018)

15. Ucan, A.: Automatic sentiment dictionary translation and using in sentiment analysis. M.Sc. thesis, Hacettepe University, Ankara, Turkey (2014)
16. Erogul, U.: Sentiment analysis in Turkish. M.Sc. thesis, Middle East Technical University, Ankara, Turkey (2009)
17. Shehu, H.A.: Sentiment analysis of Turkish Twitter data using polarity lexicon and artificial intelligence. M.Sc. thesis, Pamukkale University, Institute of Science, Computer Engineering, Denizli, Turkey (2019)
18. Turkmenoglu, C., Tantug, C.A.: Sentiment analysis in Turkish media. In: International Conference on Machine Learning, Beijing, China, 21–26 June (2014)
19. Dehkharghani, R., Yanikoglu, B., Saygin, Y., Oflazer, K.: SentiTurkNet: a Turkish plarity lexicon for sentiment analysis. Lang. Resour. Eval. **50**(3), 667–685 (2015)
20. Miller, G.A.: WordNet: a lexical database for English. Commun. Assoc. Comput. Mach. **38**(11), 39–41 (1995)
21. Baccianella, S., Esuli, A., Sebastiani, F.: SentiWordNet 3.0: an enhanced lexical resource for sentiment analysis and opinion mining. In: Proceedings of the 7th Conference on Language Resources and Evaluation (LREC), Valletta, Malta, vol. 10, pp. 2200–2204 (2010)
22. Cambria, E., Olsher, D., Rajagopal, D.: SenticNet 3: a common and common-sense knowledge base for cognition-driven sentiment analysis. In: Proceedings of the 28th AAAI Conference on Artificial Intelligence, pp. 1515–1521 (2014)
23. Bilgin, O., Cetinoglu, O., Oflazer, K.: Building a WordNet for Turkish. Rom. J. Inf. Sci. Technol. **7**(1–2), 163–172 (2004)
24. Tocoglu, M.A., Alpkocak, A.: TREMO: a dataset for emotion analysis in Turkish. J. Inf. Sci. **44**(6), 848–860 (2018)
25. Can, F., Kocberber, S., Balcik, E., Kaynak, C., Ocalan, H.C., Vursavas, O.M.: Information retrieval on Turkish texts. J. Am. Soc. Inf. Sci. Technol. **59**(3), 407–421 (2008)
26. Akın, A.A., Akın, M.D.: Zemberek, an open source NLP framework for Turkic languages (2007). https://github.com/ahmetaa/zemberek-nlp. Accessed 1 Mar 2019
27. Hayran, A., Sert, M.: Sentiment analysis on microblog data based on word embedding and fusion techniques. In: 2017 25th Signal Processing and Communications Applications Conference, pp. 1–4 (2017)
28. Sharif, M.H., Shehu, H.A., Galip, F., Ince, I.F., Kusetogullari, H.: Object tracking from laser scanned dataset. Int. J. Comput. Sci. Eng. Tech. **3**(6), 19–27 (2019)
29. Zadeh, L.A., Abbasov, A.M., Shahbazova, S.N.: Fuzzy-based techniques in human-like processing of social network data. Int. J. Uncertain. Fuzziness Knowl.-Based Syst. **23**(Suppl. 1), 1–14 (2015)
30. Kusetogullari, H., Sharif, M.H., Leeson, M.S., Celik, T.: A reduced uncertainty-based hybrid evolutionary algorithm for solving dynamic shortest-path routing problem. J. Circ. Syst. Comput. **24**(5), 1550067 (2015)

Text Mining and Statistical Learning for the Analysis of the Voice of the Customer

Rosalía Andrade Gonzalez[1], Roman Rodriguez-Aguilar[2], and Jose A. Marmolejo-Saucedo[3](✉)

[1] Facultad de Ciencias, Universidad Nacional Autónoma de México, Mexico City, Mexico
[2] Escuela de Ciencias Económicas y Empresariales, Universidad Panamericana, Augusto Rodin 498, 03920 Mexico City, Mexico
rrodrigueza@up.edu.mx
[3] Facultad de Ingeniería, Universidad Panamericana, Augusto Rodin 498, 03920 Mexico City, Mexico
jmarmolejo@up.edu.mx

Abstract. This paper analyzes the content of texts through a Text Mining classification model for the particular case of the Tweets made about the Miniso brand in Mexico during the period from November 17 to 24, 2018. The analysis involves the extraction of the data, the cleaning of the text and supervised support models for high-dimensional data, obtaining as a result the classification of the tweets in the topics: Positive, Negative, Advertising or Requirements of new Branches. As well as the use of resampling techniques to measure the variability of the performance of the model and to improve the accuracy of the parameters. This practice allows to reduce time spent reading texts, especially in Social Networks, finding faster and more efficient trends that help decision-making and respond quickly to customer demand.

Keywords: Text mining · Sentimental analysis · Supervised models · Ensemble · Bagging · Boosting

1 Introduction

Miniso is a Japanese brand that has revolutionized fast fashion worldwide, has 7,000 stores in the world, making available to its customers more than 5,000 high quality products and Japanese design at affordable prices when producing in China [4]. In December 2016, Miniso opened its first location in Galerías Coapa in Mexico and the opening of new points of sale has not ended since that day. It has 58 stores in 19 cities of the Mexican Republic, while the goal is to close this year (2018) with 100 points of sales and double this number during 2019. However, Miniso has competition since other stores are reaching Mexico. Chinese items such as Mumuso, and although it might seem like a copy of Miniso, what sets Mumuso apart is that the whole store has a Korean theme, which is why it has become the favorite of many people. Other stores such as Minigood, Yoyoso and Moamoa [5], with the same Asian proposal, make direct competition to Miniso.

D. J. Hemanth and U. Kose (Eds.): ICAIAME 2019, LNDECT 43, pp. 191–199, 2020.
https://doi.org/10.1007/978-3-030-36178-5_16

A Voice of the Customer program is an inexhaustible source of feedback to detect strengths, improvement opportunities and feed Miniso's business initiatives, allowing it to continue to have a competitive advantage over its competitors, allowing:

- Listening to customers and understanding their needs
- Capture and interpret the expectations, preferences and experiences of the client with the products and services regularly.
- Regularly measure the impact of your Customer Experience initiatives.
- Retain the clients that you have already created and design ways to attract new ones.
- Solve the problems of your current clients with greater efficiency.
- Maximize the efficiency of your operation by clearly identifying the processes and policies that affect your customers.
- Prioritize improvement initiatives based on the impact they will have.
- Provide you with ideas to innovate in your offer according to what your customers really want.

As part of the proposed Voice of the Customer program, a strategy will be developed to know the opinion of clients about the Miniso brand on Social Networks, particularly on Twitter, through Text Mining [1].

Text Mining is a method to extract useful and important information by identifying patterns within texts, such as trends in the use of words, syntactic structure, etc. This practice allows companies to reduce the time spent reading large texts, which means that key resources can be found more quickly and effectively, helping in making decisions and responding quickly to customer queries. This technique will help the interpretation of the content of the tweets about Miniso and its classification in order to detect, understand and take actions that benefit the satisfaction of its clients.

2 Applications of the Proposed Approach

2.1 Proposed Model

We propose a statistical learning model that combines text-mining methods with SVM (Support Vector Machine) classifier, in addition to resampling methods such as Bagging and Boosting [1].

In the case of the SVM classifier, three types of models were considered:

(a) Linear
(b) Radial
(c) Polynomial

Similarly, to evaluate the efficiency of the classifier Random Forest models were considered to select the best classifier. Finally, two ensemble meta-algorithms were estimated to improve the estimates made:

(a) Bagging
(b) Boosting

Figure 1 shows the proposed model, which combines data mining techniques, classification by means of supervised support and ensemble meta-algorithms.

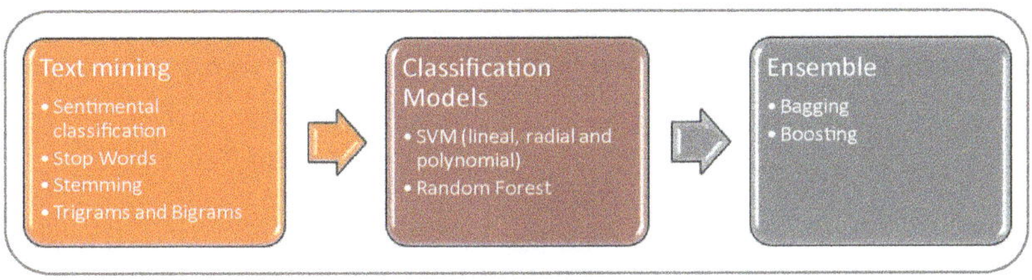

Fig. 1. Proposed model.

2.2 Database

Twitter comments were downloaded into a spreadsheet in Google Drive using an add-on called: Twitter Archiver, which allows you to search for tweets containing a word or hashtag, importing the results into the spreadsheet and adding new tweets every hour.

In this case the search for Tweets was done with the word "Miniso" of comments in Mexico in the period from November 17 to 24, 2018, obtaining 700 Tweets (Fig. 2).

Fig. 2. Removing Tweets in Google Drive spreadsheet with the Twitter Archiver plug-in.

A manual classification of the Tweets was made, classifying them into the following categories:

- POS: Tweets with positive comments towards the Miniso brand
- NEG: Tweets with negative comments towards the Miniso brand
- PUB: Tweets referring to Miniso advertising
- SUC: Tweets referring to customer requests to open new branches.

2.3 Descriptive Analysis

Of the total of 700 tweets in the database, there are 356 tweets (51%) corresponding to Positive comments about Miniso [4], 217 tweets (31%) referring to Negative comments, 71 tweets (10%) of comments about requirements for the opening of new branches and finally 56 tweets (8%) of brand advertising. The above is shown in Fig. 3.

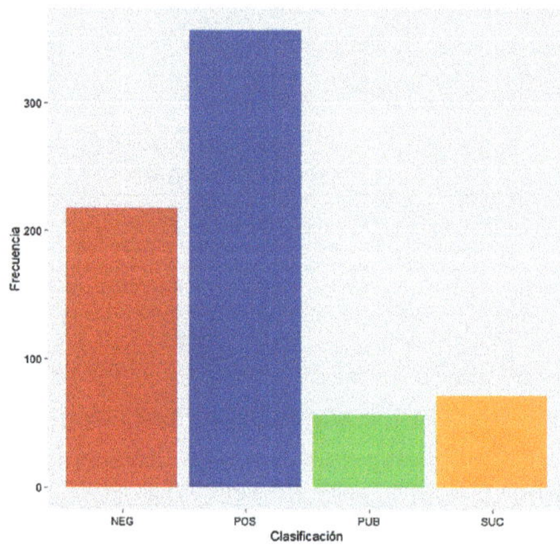

Fig. 3. Classification of Tweets by feeling

2.4 Data Cleaning

Tweets were cleaned in order to obtain only the relevant information each one. Initially there are 2,229 unique words in all the Tweets, this number of unique words is reduced with the following steps:

- Conversion of the text to lowercase.
- Elimination of numbers, punctuation, special characters and accents.
- Elimination of repeated letters repeated together.

It is known that in the Spanish language you cannot write more than one or two identical letters together in the same word. For example there is no word that has more than 1 letter "o" together in the same word, then in the cases in those who have Tweets with this characteristic as the tweet: "Miniso has all the things I need", the modification is made that there is only 1 letter: "Miniso has everything I need" [4].

Elimination of StopWords
By counting the words in the Tweets, it was found that most of the most frequent words correspond to very common words in the Spanish language, called Stop Words, as can be seen in Fig. 4. Even though the library of R: tm has a StopWords dictionary for the Spanish language, it was decided to create one from the most frequent words in the text since the dictionary of the library was not suitable for this case [1].

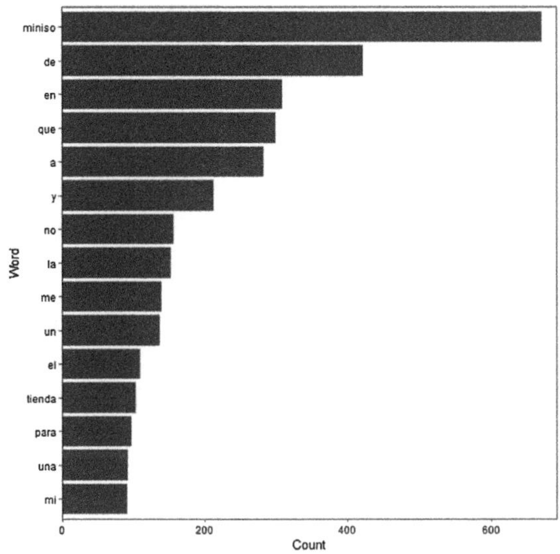

Fig. 4. Most frequent words in Tweets

Stemming (Unification of Words)

Frequently verbs and root words were identified in the tweets (e.g., need, open, buy, inaugurate,…) that are combined and written differently according to the context of the tweets (e.g. needed, opened, bought, inaugurated,..). Thus, from each of these root words, the words that most resembled each other were found through Levenshein's distance, as shown in Fig. 5.

```
$necesitar
[1] "necesitas"    "necesitamos" "necesitaba"  "necesitan"    "necesita"     "necesitabas"

$abrir
[1] "abrir"    "abrira"    "abrieran" "abrieron" "abrio"    "abriran"

$comprar
 [1] "comprando" "comprar"    "compras"    "compraron" "compra"     "comprare"   "comprarme"
 [8] "comparte"  "compradas" "comprado"

$inaugurar
[1] "inauguracion" "inaugurar"    "inaugura"

$encantar
[1] "encantaron" "encanta"    "encantan"

$barato
[1] "baratos"    "baratas"    "barato"     "baratijas"

$bonito
[1] "bonito"   "bonitas" "bonitos" "bonita"
```

Fig. 5. Example of frequent root words and search results for similar words across Levenshein distance

Tri-grams and Bi-grams

We identified the most frequent bi-grams and tri-grams in each category (POS, NEG, PUB and SUC). For example: I want to go, buy things, favorite store, Chinese things, lots of lines, useless things, store three floors, open store, urge open, etc.... forming a catalog and replacing these bi-grams and tri-grams in the text of the tweets concatenated with a "-" (e.g. I want to go, buy-things,...).

Word Cloud

Word clouds were made for each category to visually identify the most common content of the Tweets. As shown in Fig. 6, the Negative Tweets refer mainly to the competition that is Mumuso and Yoyoso [6], as well as to the purchase of unnecessary and useless things in the store, many people in the ranks, the lack of an online store, Chinese and cheap products.

In the cloud of words of the Positive Tweets (Fig. 7) it can be seen that the tweets are referring mainly to the purchase of stuffed animals, to the desire of the customers to go and buy everything in the store, the products are cheap; they love the store and think that the products are necessary.

On the other hand, Fig. 8 shows the most frequent word cloud in the Advertising Tweets category, where it is noted that these refer to the opening of the 3-story store in Mexico City and the Premium line that will have in addition to the discounts that will be in the stores of Miniso for the Good End. Regarding the Tweets of requests of opening of new branches (Fig. 9) are observed the places where the clients consider necessary to open stores, as in Veracruz, Tapachula, Tampico, Tuxtla and Culiacan, etc.

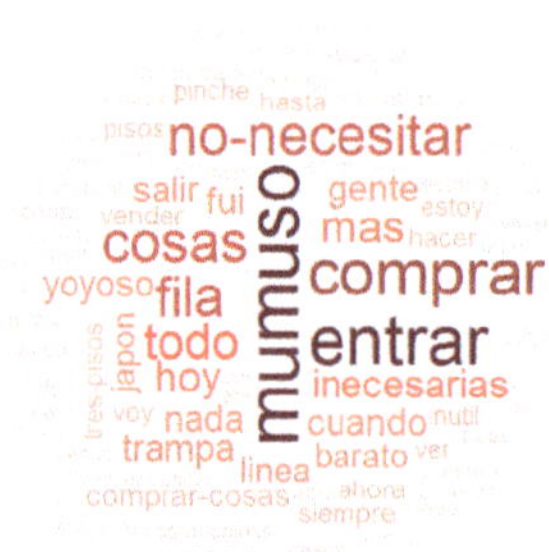

Fig. 6. Negative Tweets

Fig. 7. Positive Tweets

Fig. 8. Advertising Tweets

Fig. 9. Tweets requesting new branches

Elimination of Words Infrequently
The words that appeared only once in all Tweets analyzed were eliminated, which implies the elimination of 782 words. Finally, after having applied the cleaning process, the unique word number was reduced to 1,511 words.

After this, a database transformation was carried out such that each record is a tweet and each variable is one of the unique words and for each tweet the number of times each word is repeated is counted.

Selection of Variables
The most frequent words and terms in each category were selected, so that their frequency is greater than the 80th percentile. So finally, we have 206 variables for the model.

2.5 Classification Models

Different classification models were made for high-dimensional data to determine the category of the tweets. For each of the models, a division of the base was carried out in Training and Validation and it was tested with three different seeds. The efficiency of the models (Misclasification Rate) is shown in Table 1 for the Training set and in Table 2 for the Validation set.

As shown in the Tables 1 and 2, the results of these tests imply that the Linear Support Vector Machine Model turns out to be the model with the lowest misclassified rate, with an average of 4.5% for Training Sets and 18.0% for Validation Sets.

It should be mentioned that the Bagging and Random Forest models were created based on the Classification Tree model in order to improve their performance, an objective that was achieved since the error was reduced from 27% to 7% with the Bagging Model. This model could also be considered as a good option for the classification of tweets.

Table 1. Rate of wrong classifieds by model and seed in the Training Sets

Model		Support Vector Machine			Random forest	Random forest pruned	Bagging	Random Forest	Boosting
	%	Lineal	Radial	Polinomial (2)					
Semilla 1	50%	6.3%	28.6%	51.4%	25.1%	25.4%	6.0%	13.1%	28.9%
Semilla 2	50%	3.4%	28.6%	48.6%	25.7%	25.7%	2.3%	8.6%	26.0%
Semilla 3	50%	6.3%	28.6%	48.0%	28.6%	30.0%	5.1%	9.7%	23.7%
Semilla 1	60%	5.5%	27.9%	50.2%	26.7%	26.7%	5.2%	10.7%	24.0%
Semilla 2	60%	4.3%	25.2%	48.6%	28.1%	28.1%	2.9%	8.1%	19.3%
Semilla 3	60%	4.8%	23.8%	47.9%	28.1%	28.3%	4.0%	8.6%	21.0%
Semilla 1	70%	1.6%	2.5%	3.7%	24.1%	24.3%	5.5%	10.0%	20.0%
Semilla 2	70%	5.1%	24.9%	47.9%	28.4%	28.4%	3.3%	8.4%	18.0%
Semilla 3	70%	5.9%	24.1%	48.1%	29.0%	29.0%	5.1%	7.8%	16.6%
Semilla 1	80%	2.0%	2.1%	3.8%	26.4%	25.9%	5.0%	10.5%	16.6%
Semilla 2	80%	5.7%	23.2%	48.8%	25.0%	25.0%	39.0%	9.3%	15.0%
Semilla 3	80%	2.3%	2.0%	4.1%	28.2%	28.9%	5.2%	8.6%	14.8%
Semilla 1	90%	3.5%	1.9%	5.2%	25.9%	25.7%	5.7%	9.2%	14.1%
Semilla 2	90%	3.3%	1.9%	4.9%	27.0%	27.0%	4.9%	9.0%	13.5%
Semilla 3	90%	7.1%	1.7%	3.8%	25.9%	25.9%	4.9%	8.4%	13.8%
Average		4.5%	16.5%	31.0%	26.8%	27.0%	6.9%	9.3%	19.0%

Table 2. Rate of wrong classifieds by model and seed in the Validation Sets

Model	%	Support Vector Machine			Random forest	Random forest pruned	Bagging	Random Forest	Boosting
		Lineal	Radial	Polinomial (2)					
Semilla 1	50%	19.7%	33.7%	46.9%	30.0%	29.7%	20.6%	24.0%	36.0%
Semilla 2	50%	22.6%	36.9%	49.7%	34.3%	34.3%	27.7%	26.0%	36.6%
Semilla 3	50%	19.4%	36.3%	50.3%	34.9%	35.7%	20.3%	22.6%	38.9%
Semilla 1	40%	18.2%	32.5%	47.5%	30.0%	30.0%	16.8%	18.2%	32.9%
Semilla 2	40%	18.9%	33.6%	50.0%	33.2%	33.2%	21.4%	24.3%	32.1%
Semilla 3	40%	19.3%	32.5%	51.1%	31.8%	30.7%	20.0%	22.5%	34.3%
Semilla 1	30%	19.4%	16.1%	17.5%	30.3%	29.9%	17.1%	19.4%	37.9%
Semilla 2	30%	18.5%	34.6%	52.1%	35.5%	35.5%	19.0%	21.8%	34.1%
Semilla 3	30%	17.1%	32.7%	51.7%	28.9%	28.9%	17.1%	19.4%	27.0%
Semilla 1	20%	17.1%	15.7%	16.4%	27.1%	27.9%	16.4%	17.9%	28.6%
Semilla 2	20%	15.7%	32.1%	50.7%	31.4%	31.4%	17.9%	20.0%	29.3%
Semilla 3	20%	22.9%	12.9%	17.9%	27.1%	27.1%	19.3%	21.4%	30.0%
Semilla 1	10%	7.1%	14.3%	17.1%	21.4%	21.4%	10.0%	14.3%	22.9%
Semilla 2	10%	21.4%	20.0%	20.0%	27.1%	27.1%	17.1%	15.7%	24.3%
Semilla 3	10%	12.9%	11.4%	15.7%	28.6%	28.6%	14.3%	17.1%	34.3%
Average		**18.0%**	**26.4%**	**37.0%**	**30.1%**	**30.1%**	**18.3%**	**20.3%**	**31.9%**

3 Conclusions

A model of Linear Support Vector Machine was found that classifies the tweets with an acceptable error and can be useful to detect the subsequent tweets in the different categories and take actions from them and measure their impact.

With the previous analysis we know the strengths of the Miniso brand are the teddies and their accessible prices, also that there are many people who feel happy with the purchased products and the importance of this model is to continue monitoring that satisfaction. On the other hand, it was also discovered that the main complaint of customers are the long lines to enter or pay and the absence of an online store so it is recommended to take actions in these aspects as more boxes and accelerate the process of creation of the virtual store.

It is also recommended to meet requests for new branches in the most requested places such as Tampico, Monterrey, Tuxtla, Culiacan, Veracruz, Tapachula, Chihuahua and Saltillo, which guarantee the success of the same.

References

1. Kumar, A., Paul, A.: Mastering Text Mining with R. Packt Publishing, Birmingham (2016)
2. Gordillo, J.D.M.: Primer Taller de Análisis de Sentimiento en Twitter con R. de Youtube Sitio web (2016). https://www.youtube.com/watch?v=nOIZnYLlPBo
3. Openminted Communications. ¿Qué es la minería de textos, cómo funciona y por qué es útil? de Universo Abierto Sitio web (2018). https://universoabierto.org/2018/02/22/que-es-la-mineria-de-textos-como-funciona-y-por-que-es-util/

4. José Roberto Arteaga. La estrategia de Miniso que cautivó a los consumidores mexicanos, de Alto Nivel Sitio web (2018). https://www.altonivel.com.mx/empresas/estrategia-miniso-consumidores-mexicanos/
5. Mundotkm. ¡Tiembla, Miniso! Llegó nueva tienda oriental a la CDMX y todos quieren ir,de Mundotkm Sitio web (2018). https://www.mundotkm.com/mx/actualidad/266375/tiembla-miniso-llego-nueva-tienda-oriental-a-la-cdmx-todos-quieren-ir
6. Mariana García. Yoyoso: más que un estilo de vida. 2018, de Roastbrief Sitio web (2018). https://www.roastbrief.com.mx/2018/09/yoyoso-mas-que-un-estilo-de-vida/

A Decision Support System for Role Assignment in Software Project Management with Evaluation of Personality Types

Azer Celikten[1(✉)], Eda Kurt[2(✉)], and Aydin Cetin[2(✉)]

[1] Celal Bayar University, Manisa, Turkey
azer.celikten@cbu.edu.tr
[2] Gazi University, 06500 Ankara, Turkey
edaakurtt@outlook.com, acetin@gazi.edu.tr

Abstract. Recent studies show that personal factors in software engineering are effective on team performance, motivation and job quality. Forming team members or incorporation of a new member into the team that currently working on a software project directly affects the project team's work performance and hence the progress of the project. In this study, a decision support system was developed to provide the ability to select team members according to personal characteristics in order to improve the performance of software project teams. The developed decision support system, determines the project roles that can be appropriate by analyzing the personality type of the project team members. The fuzzy c-means method, one of the fuzzy clustering methods, was used when the IPI personality inventory and personality type team members were determined to be appropriate to their personality type.

Keywords: Decision support system · Fuzzy clustering · Software project management

1 Introduction

With the new industrial revolution called Industry 4.0, significant developments have occurred in the field of software for human life. The systems such as smart systems, artificial intelligence, internet of the things, 3D printers, big data, cyber systems and cloud computing have emerged as a result of improvements in software technologies. With these developments, countries budgets for software sector and costs for employment of software engineers has increased. According to TUBISAD Information Technology Sector 2016 market data, total employment in the information and communication technology sector for Turkey is 120000. At the end of 2016, Software category has the highest employment growth in the information technology sector in Turkey [1]. According to the latest report released by the US Economist Intelligence Unit (EIU) in September 2017, the total number of employment related to software activities in the United States directly or indirectly is 10.5 million [2]. If human resource management in software projects cannot done correctly -for example, if the staff selected for the project are not eligible for the assigned task- projects may fail.

D. J. Hemanth and U. Kose (Eds.): ICAIAME 2019, LNDECT 43, pp. 200–210, 2020.
https://doi.org/10.1007/978-3-030-36178-5_17

Failure of software projects will affect the project costs prejudicial. In the studies conducted that the reasons for the failure of software projects classified as technical, managerial, social and other reasons [3]. Some of the social reasons of failure are as follows:

- Poor communication between team members,
- Inability to retrieve entries due to poor connection with the customer,
- Over conflict between the customer and other stakeholders,
- Being unaware of technological advancement
- Ineffective communication due to management's careless behavior,
- Working under continuous pressure about meeting deadlines,
- Lack of communication and support between departments and units,
- Enforcing projects in areas other than the general skills of technical personnel.

Lack of interpersonal communication, poor time management, relationships between team members and failing to keep abreast of the latest developments related to the personal characteristics of project employees. For this reason, besides the technical qualities of software engineers, their personality types and personal characteristics should be suitable for the task.

The five task definitions defined in the software project teams are as follows:

System Analyst: They conduct research, analyze and evaluate information technology requests, procedures or problems. They develop and implement proposals, suggestions and plans to develop existing or future information systems [6].

Software Designer: Creating interface designs by modeling of the system, creative and innovative direction comes to the fore. They make use of flow diagrams, UML (Unified Modeling Language) and other modeling methods for making the design visual [6].

Software Developer: They research, analyze and evaluate the demands for existing or new software applications and operating systems. In addition, they can design, develop, test and maintain the software solutions to meet these demands [6].

Software Tester: They test that the project coded under the business requirements identified in the requirement analysis phase. Designs the tests and applies these designs effectively after the coding phase. They share with the software developer their failures that do not meet the project requirements or the errors in the software.

Software Maintainer: Software maintenance activities related to all other activities as it is the last activity of the software life cycle. It consists of complex activities that involve a variety of areas of expertise and require a team to work with the client [7]. The organization or persons who performing the maintenance activities are Software Maintainer.

In this study, the degree of conformity of the personnel involved in software projects for the roles of system analyst, software designer, software developer, software tester and software maintainer determined according to personality types. The short Turkish version of the IPI (International Personality Inventory), based on the five-factor personality model, was used for identify personality type [4]. The reason for using this inventory is that the sub-dimensions of the five factor personality types are similar to those required in the software project processes. For example, communication skills related to extraversion personality type. Analytical thinking and problem-solving skills related to openness to intellect personality type. Time management related to conscientiousness personality type and interpersonal skills and the ability to work in-team related to agreeableness personality type. Software project roles requires similar personal characteristics. Team members can be compatible with multiple roles according to personality types. For example, the personality type of extraversion is necessary for all role definitions. But, it may be more important for some roles. So, it is not possible to assign each team member to a single role according to personality type. For that reason, a fuzzy clustering method is used instead of sharp clustering. Using fuzzy c-means algorithm, which is one of the fuzzy clustering methods, a decision support system has been prepared that decides the role of the project staff to work according to the type of personality. With the help of this decision support system, software engineers were determined to be in compliance with their roles.

In the second part of the study, literature studies on role assignment and personality types in software projects are included. In the third chapter, clustering methods and personality type determination inventory used in the study were introduced. In the fourth chapter, the application of fuzzy c-means algorithm for role distribution is explained. The fifth and last section contains the results.

2 Literature Study

Since software engineering is a prominent area with its technical and mathematical aspects, there is not much tendency to human and human characteristics [5]. Cruz et al. found that between 1970 and 2010, 19,000 studies in software engineering were published and only 90 studies were related to personal characteristics in software engineering. 72% of these studies occurred after 2002. This situation shows that personal characteristics and personality types are very important for software projects but there is not enough study. 88% of the 90 studies, reports paired programming, training, team performance, software process allocation, personal characteristics of the software engineer, and individual performance. The remaining part includes teamwork, behavior and leadership performance [5].

In another study, Capretz et al. examined the relationship between the role of the task in software projects and MBTI personality types. While analyst, designer and programmer roles are the most preferred roles in all personality types, it has been determined that tester and maintainer roles are less preferred. [8].

Ahmed et al. [9] reviewed 500 IT jobs in North America, Europe, Asia and Australia and identified soft skills demanded by companies for software project roles. They classified as high, medium and low according to demand level. Highly demanded roles are as follows:

- System Analyst: Communication skills, analytical thinking and problem solving, being a team player.
- Software Designer: Communication skills, interpersonal skills.
- Software Developer: Communication skills.
- Software Tester: Communication skills.

According to the regions, the demanded skills have been changed.

Capretz and Ahmed [10] have matched the MBTI personality types by identifying the technical and soft skills required for the system analyst, designer, programmer, test specialist and maintenance worker from the software project role definitions. Suitable MBTI personality types for project roles have been identified. They concluded that extraversion and feeling for system analysts; intuition and thinking for software designers; introversion, sensing and thinking for programmers; s sensing and judging for software tester; and sensing and perceiving for software maintainers were the appropriate personality types.

Rahman et al. [11] matched the software project roles with five-factor personality types (extraversion, agreeableness, openness to experience, neuroticism, conscientiousness). For the role of system analyst and designer, openness to experience and agreeableness personality types, for the software developer role, openness to experience, agreeableness and extroversion personality types, for software tester and software maintainer roles, openness to experience and conscientiousness personality types have determined. In a study by Salleh et al. [12] with undergraduate students, they investigated conscientiousness personality type's effect on the cooperative study. According to the result, the dimension of the conscientiousness is not effective on the performance of paired work but they stated that the dimension of the openness to experience found effective. Bell et al. [13] used the five-factor personality test in their study on the effects of personal characteristics and design studies on the design studied over second-year undergraduate students. However, unlike the work of Acuna et al., they didn't find a strong relationship between performance and personality traits. They reported a weak relationship between the five-factor dimensions of neuroticism and designing.

Acuna et al. [14] examined the effects of personal characteristics on job satisfaction and software quality in software project team. As a result of the measurement of five-factor personality dimensions on 35 software development teams (105 persons), it was found that job satisfaction was more common among individuals who had agreeableness personality and conscientiousness personality types and who has organizing skills high. In addition, there is a positive correlation between the extraversion personality type and software product quality. In the study of Cunha and Greathead [15], the ability to find the evidence in the codes was associated with MBTI personality types. In an experimental study conducted with 64 undergraduate students, it was found out that individuals who have NT (Intuition, Thinking) personality type were more successful in detecting the code evidence. Acuna and Juristo [16] conducted a study about the appointment of suitable people to project roles. They have determined the weight of the personal skills required for 20 different role definitions responsible for performing various tasks in software projects. Using the 16PF personality test, they performed the personality role matching with the experimental study. At the end of the study, they showed the capabilities-oriented software process model. Martinez et al. conducted a study on IPI personality testing and assigning a task in software projects using ANFIS (Adaptive Network Based Fuzzy Inference System) learning method. 72 software engineers have been assigned to one of the 6 role definitions that are suitable for them [17].

3 Method

The aim of the study is to determine the degree of conformity of the personnel involved in software projects for their roles as analyst, software designer, software developer (programmer), software tester and software maintainer according to their personality types. IPI test based on five factor personality types was used for determine personality types. In order to determine the weights of five factor personality dimensions for software roles, expert project managers' opinions were taken. The fuzzy c-means algorithm was used as the fuzzy clustering method for the role distribution process.

3.1 Personality Types Identification Tools

Personality has been one of the most studied subjects in the areas examining human behavior. The main reason for this is that personality is one of the most important predictors of the quality of interpersonal relationships, adaptation to difficult living conditions, professional achievements, social participation, happiness and health. Various approaches (such as psychoanalytic, behavioral, property approach) have been introduced to examine personality. However, the Five Factor Personality Model which suggesting that personality traits can be evaluated under five factors for the last two decades has begun to come to the forefront [9].

3.1.1 Five-Factor Model Personality Types

The five-factor personality model, which has a long history but has a new perspective, is a hierarchical classification of universal and complete personality traits, which suggests that personality has fundamentally been composed of five factors. It evaluates five dimensions as extraversion, neuroticism, openness to experience, agreeableness, and conscientiousness (Table 1).

Table 1. Properties of five-factor personality types

	Personality type properties
Extraversion	Being full of life, excited, cheerful, talkative and social
Agreeableness	Gentle, elegant, respectful, reliable, flexible, open-hearted, compassionate
Conscientiousness	Formal, meticulous, responsible, determined, self-discipline and prudent
Neuroticism	Anxiety, tendency to depression, irritable, distressed
Openness to experience	Analytical, complex, curious, independent, creative, liberal, non-traditional, original, imaginative, much brave, adaptable to changes, artistic, open-minded

3.1.2 International Personality Inventory

IPI is a personality inventory structured on a five-factor personality model using an international inventory of personalities. In the first version of this inventory, published in 1978, neuroticism, extraversion, and openness to experience consisted of personality types. In 1992, Costa and McCrae added conscientiousness and agreeableness factors. In order to determine the personality types of the project team, test on Turkish software engineers and obtain correct information, Güneri's shortened Turkish adaptation of validity and reliability analysis of IPI was used in this study. The short version of English consists of 50 questions. But in the Turkish version, during the validity and reliability analyses 10 questions were omitted.

3.2 Fuzzy Clustering Analysis

Clustering analysis, a multivariate statistical method, is a technique that helps to distinguish the number of observations from a group of unknown groups or clusters, while all observations in the group are similar [18]. In our study, clustering was performed in which the degree of conformity to roles was determined according to personality types. While the degree of membership of the individual to the cluster is 1 in sharp clustering, it varies between 1 and 0 in fuzzy clustering. In personality type measurements, a person may belong to different personality types at specific rates. For example, a person may have a 0.7-percent extraversion, a 0.8-percent agreeableness, a 0.5-percent conscientiousness, a 0.7-percent, openness to experience and a 0.4-percent neuroticism personality type. At the same time, different personality types may have different effects for project roles. As a result, a person will belong in different rates to a specified set of 5 role sets. Therefore, the problem of problem solving should be solved by fuzzy clustering. Fuzzy c-means clustering method, one of the most commonly used fuzzy clustering methods, was used to determine the role ratio of a person. Fuzzy c-means clustering was performed in 1973 by Proposed by Dunn [19] and developed by Bezdek in 1981 [20]. In the fuzzy c-means method, a membership degree is determined according to the Euclidean distance between the cluster center and the data point.

In the fuzzy c-means algorithm,

X is a set of n elements $X = \{X_1, X_2, X_3, \ldots, X_n\}$, where each X_i is d sized point. $C_1, C_2, C_3, \ldots, C_k$ k is the number of fuzzy set.

i = 1…n and j = 1….k, W = $W = w_{i,j} \in [0, 1]$ is a matrix of membership rates. Membership rate of each i object to the c_j set is expressed with $w_{i,j}$. Steps of algorithm are as follows:

1. Initial values of the membership degrees matrix are determined. Each x_i point membership degrees additions which belongs to the set k, must be equal to 1.

$$\sum_{j=1}^{k} w_{i,j} = 1 \tag{1}$$

2. Set center points are determined for each set with using the Eq. (2). p is the fuzzy coefficient and it must be between 1.25 and 2.

$$c_j = \frac{\sum_{i=1}^{n} w_{i,j}^{p} x_i}{\sum_{i=1}^{n} w_{i,j}^{p}} \tag{2}$$

3. $dist(x_i, c_j)$ refers to the Euclidian Distance between the cluster center points and x_i point. Equation (3) calculates the new membership values of the data points to clusters based on distances between the points.

$$w_{i,j} = \frac{\left(\frac{1}{dist(x_i, c_j)^2}\right)^{\frac{1}{p-1}}}{\sum_{q=1}^{k} \left(\frac{1}{dist(x_i, c_q)^2}\right)^{\frac{1}{p-1}}} \tag{3}$$

4. Sum of the squared error (SSE) must be repeated until it reaches the smallest value.

$$SSE = \sum_{j=1}^{k} \sum_{i=1}^{n} w_{i,j}^{p} dist(x_i, c_j)^2 \tag{4}$$

4 Role Distribution by Fuzzy C-Means Algorithm

The following steps were taken in establishing the system which determines the degree of compliance of the persons working in software projects for their roles.

1. The importance of the five-factor personality types on the roles of analyst, software designer, software developer, software tester and software maintainer was determined. At this stage, conducted a questionnaire to expert project managers. The sub-dimensions of personality types are available in Table 2. Project managers rated the sub-dimensions with a score between 1 and 5 according to their severity for the relevant role.
2. After the normalization process is done by taking the average of the given points, the values in Table 3 are obtained.
3. 5 sets to be used for fuzzy clustering method are determined as analyst, software designer, software developer, software tester and software maintainer.
4. If the personality type ratios of a person are equal to the rate of importance determined for the roles, it is assumed that the person belongs 100% to the role in question. Therefore, the rate of personality type determined for each role can be expressed as the center of gravity of the cluster. Personality type scores have been established for each role by using the personality type sub-dimensions that priorities are determined by the project managers. Table 4 shows the personality type scores of the role sets.
5. Personality type ratios of software engineers are a set of 5-dimensional dots whose distances to cluster centers will be calculated.

6. The membership ratings for the 5 sets were calculated by using the Euclidean distances of the personality type ratios to the cluster centers in 3rd equality.
7. A web-based decision support software has been developed which determines the personality types of the project employees by IPI testing and calculates the membership degrees of the people by using the fuzzy c-means method.
8. Using fuzzy clustering-based decision support software, the suitability levels of software engineers in a software company were calculated. Personality analysis and role membership ratings of a person determined by decision support system are as in Fig. 1.

Table 2. Sub-dimensions of five-factor personality types

	Sub-dimensions of Personality Types
Extraversion	Thruster, Lively, Introversion
Agreeableness	Calmness, Softhearted/Altruism, Reactiveness
Conscientiousness	Responsibility/Stability, Rule-bound, Systematic
Neuroticism	Tendency to Anxiety, Self Confidence
Openness to experience	Analytical thinking and problem solving, Innovative, Sensitive

Table 3. Priority ratio of the five-factor personality types sub-dimensions on software project roles

	Analyst	Designer	Developer	Tester	Maintainer
Thruster	0.78	0.62	0.59	0.67	0.56
Vitality	0.72	0.63	0.59	0.65	0.57
Introversion	0.45	0.48	0.46	0.49	0.53
Calmness	0.66	0.69	0.65	0.65	0.66
Softhearted/altruism	0.62	0.52	0.58	0.50	0.55
Reactiveness	0.58	0.49	0.50	0.64	0.54
Responsibility/stability	0.89	0.82	0.78	0.82	0.78
Rule-bound	0.59	0.61	0.63	0.64	0.60
Systematic	0.73	0.78	0.78	0.85	0.78
Analytical thinking and problem solving	0.97	0.93	0.91	0.85	0.80
Innovative	0.75	0.90	0.87	0.71	0.69
Sensitive	0.68	0.64	0.62	0.60	0.58
Tendency to anxiety	0.55	0.51	0.49	0.65	0.49
Self-confidence	0.79	0.77	0.75	0.75	0.66

Table 4. Personality type values of role cluster centers

Clusters	Extraversion	Openness to experience	Conscientiousness	Agreeableness	Neuroticism
System analyst	0.75	0.80	0.74	0.64	0.55
Software designer	0.62	0.82	0.73	0.60	0.51
Software developer	0.59	0.80	0.73	0.61	0.49
Software tester	0.66	0.72	0.77	0.57	0.65
Software maintainer	0.56	0.69	0.72	0.60	0.49

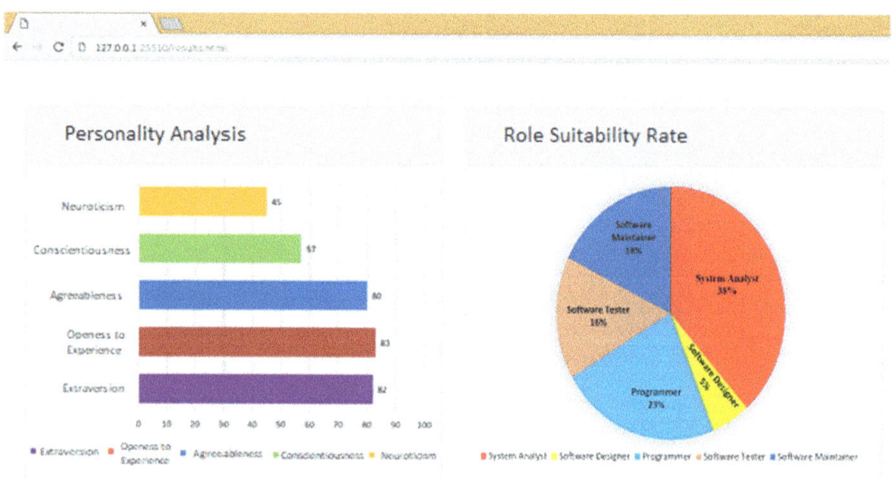

Fig. 1. Personality types and role suitability ratio.

5 Conclusion

In this study, a decision support system has been developed to match the personality types and appropriate roles in order to identify more appropriate people for roles in software project teams. A questionnaire was applied to project managers in order to determine the importance of five factor model personality types sub-dimensions for project roles. According to the results, it was found out that openness to experience is more important for system analyst, software designer and software developer roles, and conscientiousness personality type is more important for software tester and software maintenance roles. The decision support system will help project managers in selecting the right person when creating project teams or selecting new staff for project teams. This system does not include the technical knowledge and skill conditions, but only performs the role determination according to personal characteristics. In future studies, it is planned to develop a decision support system in which technical qualifications and personal characteristics can be measured together.

References

1. Bilgi ve Iletisim Teknolojileri Sektoru 2016 Pazar Verileri (2017). https://www2.deloitte.com/content/dam/Deloitte/tr/Documents/technology-media-telecommunications/TUBISAD-2017-bit-pazar-verileri.pdf. Accessed 27 May 2018
2. The Growing $1 Trillion Economic Impact of Software (2017). https://software.org/wp-content/uploads/2017_Software_Economic_Impact_Report.pdf. Accessed 27 May 2018
3. Rehber, D.: Yazilim Projelerinde Başarisizlik. http://www.emo.org.tr/ekler/d2129f927262c5b_ek.pdf
4. Yöyen, E.G.: Uluslararasi Kişilik Envanteri (IPI) kisa versiyonunun Türkçeye uyarlanmasi: Güvenilirlik ve geçerlilik analizi. Int. J. Soc. Sci. Educ. Res. **2**(4), 1058–1069 (2016)
5. Cruz, S., da Silva, F.Q., Capretz, L.F.: Forty years of research on personality in software engineering: a mapping study. Comput. Hum. Behav. **46**, 94–113 (2015)
6. "Uluslararası Standart Meslek Siniflamasi - ISCO 08". https://biruni.tuik.gov.tr/DIESS/SiniflamaSurumDetayAction.do?surumId=210&turId=41&turAdi=%209.%20Meslek%20S%C4%B1n%C4%B1flamalar%C4%B1. Accessed 27 May 2018
7. Kurtel, K.: Yazilim Bakim Personelinin Görev ve Sorumluluklar Açısından Đncelenmesi. 4. Ulusal Yazilim Mühendisliği Sempozyumu – UYMS 2009 (2009)
8. Capretz, L.F., Varona, D., Raza, A.: Influence of personality types in software tasks choices (2015). https://www.doi.org/10.1016/j.chb.2015.05050
9. Ahmed, F., Capretz, L.F., Campbell, P.: Evaluating the demand for soft skills in software development. IEEE IT Prof. **14**(1), 44–49 (2012). https://doi.org/10.1109/MITP.2012.7
10. Capretz, L.F., Ahmed, F.: Why do we need personality diversity in software engineering? ACM SIGSOFT Softw. Eng. Notes **35**(2), 1–11 (2010)
11. Rehman, M., Mahmood, A.K., Salleh, R., Amin, A.: Mapping job requirements of software engineers to Big Five Personality Traits. In: 2012 International Conference on Computer & Information Science (ICCIS), vol. 2, pp. 1115–1122. IEEE (2012)
12. Salleh, N., Mendes, E., Grundy, J., Burch, G.S.J.: An empirical study of the effects of conscientiousness in pair programming using the five-factor personality model. In: Proceedings of the 32nd ACM/IEEE International Conference on Software Engineering, vol. 1, pp. 577–586. ACM (2010)
13. Bell, D., Hall, T., Hannay, J.E., Pfahl, D., Acuna, S.T.: Software engineering group work: personality, patterns and performance. In: Proceedings of the 2010 Special Interest Group on Management Information System's 48th Annual Conference on Computer Personnel Research on Computer Personnel Research, pp. 43–47. ACM (2010)
14. Acuña, S.T., Gómez, M., Juristo, N.: How do personality, team processes and task characteristics relate to job satisfaction and software quality? Inf. Softw. Technol. **51**(3), 627–639 (2009)
15. Cunha, A.D.D., Greathead, D.: Does personality matter? An analysis of code-review ability. Commun. ACM **50**(5), 109–112 (2007)
16. Acuña, S.T., Juristo, N.: Assigning people to roles in software projects. Softw.: Pract. Exp. **34**(7), 675–696 (2004)

17. Martínez, L.G., Rodríguez-Díaz, A., Licea, G., Castro, J.R.: Big five patterns for software engineering roles using an ANFIS learning approach with RAMSET. In: Advances in Soft Computing, pp. 428–439. Springer, Heidelberg (2010)
18. Ada, A.: Kümeleme Analizi ile AB Ülkeleri ve Türkiye'nin sürdürülebilir kalkınma açısından değerlendirilmesi. Communities (2001)
19. Dunn, J.C.: A fuzzy relative of the ISODATA process and its use in detecting compact well-separated clusters. J. Cybern. **3**(3), 32–57 (1973). https://doi.org/10.1080/01969727308546046. ISSN 0022-0280
20. Bezdek, J.C.: Pattern recognition with fuzzy objective function algorithms (1981). ISBN 0-306-40671-3

A Survey of Methods for the Construction of an Intrusion Detection System

Abdel Karim Kassem[1(✉)], Shymaa Abo Arkoub[2], Bassam Daya[2], and Pierre Chauvet[3]

[1] University of Angers, Angers, France
abedelkarim_kassem@hotmail.com
[2] Lebanese University, Saida, Lebanon
shymaa_arkoub@hotmail.com, b_daya@ul.edu.lb
[3] Université Catholique de l'Ouest, Angers, France
pierre.chauvet@uco.fr

Abstract. Cybercrimes committed using computer networks lead to billions of dollars lose, the illegal access into computer system, stealing valuable data and destroying organization networks which in turn affect the cyber resources. Because of the expansion of attacks or threats on the networks infrastructure, which is nothing but can be consider as an illegitimate intrusion, based on the machine learning methodology, the intrusion detection system (IDS) can consider as one of the most used cyber security mechanisms, thus to detect the promiscuous activities against sensitive and private data. In this paper our target is to provide a guide lines for researchers and developers discussing the IDS construction phases and their latest techniques, we will clarify the most applied data sources employed in the proposition of a model that will be built for the purpose of creating an intelligent detection system. Furthermore, this survey presents the most commons and latest methods employed and used for designing an IDS based on the data mining techniques and discusses the artifacts removal by summarizing the advantages with the disadvantages of the currents methods and addressing the last novel steps into this field of research.

Keywords: Cyber security · Networks · Attacks · IDS · Data mining · Threat · Intrusion

1 Introduction

With the enormous amount of transferred data from different network to another, the fast development of network technologies, the ease of accessing various services and information materials, the rapid use of network and the embedded of internet into many traditional services of health, education, business, finance, government and banking transactions, internet becomes an essential part in our daily regular operations.

The progression in network and its availability for numerous people and machine all the way through the internet makes it a dangerous place for organization and individual's sensitive data and provides an opportunity for intruders to gain illegal access to the system or downgrading the performance of the network using various

D. J. Hemanth and U. Kose (Eds.): ICAIAME 2019, LNDECT 43, pp. 211–225, 2020.
https://doi.org/10.1007/978-3-030-36178-5_18

attacking techniques. Therefore, the presence of secure network is a growing concern to keep data under strict surveillance. Network security is an essential part in the system infrastructure because of the numerous ways that internet has introduce to threat the stability, availability and security of the network systems. Network security includes both hardware such as firewall, proxy servers and software such as malicious code detection, firewall, antivirus and Intrusion detection system (IDS).

IDS act as a network security mechanism for a computer network to prevent data material from intruders and it decreases the massive harm to the system. It is useful to track, detect, capture and classify attacks which will result in a protected network computer system. IDS monitor the traffic coming in, the vulnerabilities and system configuration so that it can guarantee that the system is able to provide high service quality with low response time. IDS is classified as hosted or networks based intrusion detection systems. A host based IDS can monitor single machine, application resources such as log file, file system and disk resources, whereas the network based IDS monitors and analyze the network traffic without affecting the system performance.

Depending on what is being monitored by IDS on the detection techniques used to find abnormal pattern in the monitored data and any analysis type, IDS can be divided into misused detection system, anomaly detection system and hybrid detection:

- Misused detection system is based on the signature of attacks like antivirus soft-ware where the incoming traffic to the network are compared to the signature database of the familiar intrusions that have been defined by security experts to judge whether the monitored data is normal or an abnormal, this kind of detection give high detection rate of known attack without generating an overwhelming number of false positive alarms. The drawbacks to this type of IDS is that it need a daily manual updates to the signature rules sit into database and its unable to detect novel attacks.
- Anomaly detection system is based on a model that describe the normal behavior and usage pattern of the examined system, where the incoming traffic to the network is compared with the reference model and flag any deviation from the normal activities, making this type of IDS able to detect novel attacks coming into the system. The main disadvantage of IDS's used behavior is the fact that it produces a high false positive rates due to unseen legitimated traffic coming into the system, which will be categorize as attacks.
- Hybrid detection system is a combination of both misuse and anomaly detection techniques. They are employed to increase the detection strength of known attacks and decrease the FP rate of unknown attacks.

The currently used IDS requires human intervention by either create signature data base or build the model of normal behavior, which makes it far from intelligent. Therefore, a more advanced IDS that support learning algorithms is highly desired to provide a prospect alternative to human intervention. It should be capable of detecting both of known and unknown hacking attacks activities with the minimum human effort input. The main task of these algorithms is to discover usable pattern from the selected training dataset in order to characterize the normal and attack behavior, moreover to make a prediction on unseen data. Therefore, various data mining algorithms like clustering, classification and association can be proposed as solution for IDS draw-backs which listed previously.

2 Related Works

The literature of using machine learning for cyber security issues has been a widespread research domain, various techniques and methods absorbed by researchers are to provide cyber security solutions.

Bhuyan, Bhattacharyya, Kalita [1] presents an overview of network intrusion detection architecture in general, then discusses and compares various methods, tools, systems and techniques used by researchers to build anomaly detection system. The author reports criteria for performance evaluation of IDS and emphasis the issues and challenges faces by researchers.

Neethu [2] represents and compares three machine learning methods (Naive Bayes, neural network, decision tree) on KDDCup 99 intrusion detection benchmark dataset, the study use PCA to select features from dataset under study and shows that Naive Bayes achieve a higher accuracy with lower cost and consuming time.

Bhuyan et al. [3] discusses the steps taken to build an affective network intrusion dataset with unbiased rule time features that compensate the lack of intrusion dataset. This study explains in details the requirements needed to create an efficient benchmark dataset.

Ibrahim et al. [4] discusses and compares the performance for each of dataset, KDD Cup99 and NSLKDD, this comparison was based on self-organization map (SOM) with artificial neural network, where KDD 99 dataset detection rate (92.37%) greater than NSL-KDD dataset detection rate (75.49%).

Ektela et al. [5], classify intrusions in network using two ML methods, Support Vector Machine(SVM) and classification tree Data mining. The results from these two techniques was compared and analyze in term of detection rate and false alarm rate then they found that the C4.5 algorithm has a better performance than the SVM technique, but with respect to U2R attack SVM performs better.

According to Kayacık et al. in [6] the Authors focus on the individual feature analysis in order to substantiate the machine learning technique performance based on the detectors that are trained on KDD-99 training dataset; This study investigate the relevance related to the data set using analysis measures such as dependency ratio, information gain or correlation coefficient.

The Researchers in [7] present the role of the data mining algorithms concerning the intrusion detection system within the using of log file as a data set, thus to found out patterns to detect the intrusion attacks in threated networks. Moreover, this paper achieves the discovering of the DoS Attacks using a clustering technique as well as testing the ability of the Data Mining techniques. Data mining was used also to incorporate both of the signature and anomaly databases schemes.

3 Data Mining Techniques Overview

Data mining known as data analysis and discovery algorithm step of the knowledge discovery in database, it is the process searching for a patterns pattern from large volume of data in order to transform the collected data into business intelligence giving an informational advantage results.

Machine learning in data mining usually is used in implementing the process of finding and searching for useful and hidden information from large databases in a wide range of profiling practices, mainly it is used in marketing, surveillance, fraud detection and scientific discovery [8]. Its algorithm commonly categorized into four main type of classes:

3.1 Classification

Data mining process in this mode will map the data to some predefined categories, when we examine new data, this process provides level of accuracy to which categories the data instances belong to.

The various classification techniques used in the IDS are listed below:

- Support Vector Machine: SVM is a supervised learning method designed for binary classification "either positive or negative class" [9], it is applied for purpose of finding patterns from collection of data. Generally, the pattern classification applied activity involves two main steps, first mapping the input to higher dimension feature space this is done due to the fact that SVM usually depends on geometrics characteristics of inputted data [8] and second finding the most suite hyper plane that classify the mapped features in the higher dimensional space.
- Fuzzy Logic: It is an application of fuzzy set theory, where a set of rule will be generating depending on the domain data under study, this approach used to find patterns basing on approximate rather than precisely deduced from classical predicate logic [10], this method fit well in case of complex real world problems. This approach involves as first step to classify data on the basis of various statistical metrics and then generating appropriate fuzzy logic rules to be applied on the classified portion of data, thus these rules will be responsible in classifying them to their targeted classes.
- Neural Networks: It is designed to mimic the functionality if human brain. The model architecture consists of multiple node imitate biological neurons, each node has a weighted connected to all previous layer nodes where it received its activation by computing, with the help of simple functions, on all node's weights from previous layer. User is responsible for specifying the number of hidden layers as well as the numbers of nodes in each layers, the nodes in the output layer depends on number of classes under study and that for inputs linked to the number of collected features [11]. Generally, it is used to learn complex nonlinear input-output relationships.
- Naïve Bayes Classifier: in order to study the probabilistic relation between random variable, or in other word to check if a set of variables affected by others a probabilistic model known as Naive Bayesian can be used. Naive Bayesian is a probabilistic classifier that provides high accuracy and speed when dealing with huge dataset, it studies the relation between dependent and independent features, to produce a predicting events.
- Genetic Algorithm: This Algorithm was defined and introduced in the computational biology filed. These sets of algorithms are belonging to a larger category of evolutionary algorithms (EA), they provide the generation of problems optimization

solutions based on the techniques that inspired by natural evolution such as crossover, selection, inheritance and mutation. This algorithm begins by specifying a collection of solution generally known as population, solutions from same population are extracted based on their fitness where they have the chance to reproduce and these selected solutions will be used to produce a new population [12].
- Deep Learning: The method of the deep learning aims to learn the feature hierarchies which it is in relation with features that composing by lower level inside higher level features. The features are learnt by the methods independently with several levels of abstraction, then determining the complex functions mapping the input with the output from the raw data directly without taking into consideration the pre customized features suggested by the researches.

Humans can't know the nature of the connection from a lot of hardware like the input of raw sensory in the abstraction higher level. Thus, as the size of data growing rapidly, the need to learn complex features becomes a necessity. Some researchers mention this capacity as feature engineering or extraction. The transformation process from input that contains raw data to features which defines the problem in a proper way to improve the accuracy of the model for unfamiliar data is known as the 'feature engineering'.

3.2 Clustering

By using this technique, the data are divided onto groups that are not defined and we do not know their properties which are known as cluster. Cluster is a group of instances that describe the data it contains, which reveal the natural structure of the studied data. We can say that data instance in same cluster are very similar to each other than the data contained in any other cluster. Some elements do not belong to any cluster which are known as outlier, we can observe structure of the outlier when comparing them to data found in other cluster and thus reveal the nonstandard system behavior.

The various Clustering techniques used in detecting intrusion are as followed:

- K-Means clustering algorithm: This method is the most widely used clustering technique, it aims to classify the studied data into predefined K clusters specified by user. This algorithm starts by randomly specifying the center of each cluster, then depending on the measured distance "Euclidean distance" between data point and the centroid of all drawn cluster, the algorithm assign this data point to the nearest cluster [13]. The execution time of this methods is proportional to the volume of data set under study. It is can consider as a fast iterative algorithm but actually, it will be also classified as sensitive with respect to the outlier and noise.
- K-Medoids: It is a clustering technique which work in comparable manner as K-Means strategy. K-Means procedure is sensitive to outliers because it works by calculating the mean of data points per cluster, and since mean is effectively affected by extraordinary qualities this implies the added "Nosie sensitivity" property to this algorithm. In contrast K-Medoids is robust to noises due to the fact of using an actual data point, called reference point or Medoid, as cluster's center in place of using mean of data points as a center which makes it limits the separation among centroid and data points implies limit the squared error [13], this approach make

K-Medoids perform better in term of execution time when the volume of data increase. A summary K-Medoids is less influenced by outliers but more expensive in processing.

- EM-Clustering: This is an iterative algorithm at which data points get assigned to a specific cluster depending on data's probability membership which is known as weight. Expectation Maximization algorithm start by a random estimation for hidden parameters by computing the mean and standard deviation of each Gaussian distribution, then this algorithm runs iteratively in order to search the maximum likelihood for these parameters. This approach generalizes well on data because of the fact that clusters are far from each other's. Figure 1 shows the various classification and clustering technique for data mining with their advantage and disadvantage [14].

Types	*Methods*	*Disadvantages*	*Advantages*	*concept*
Classification methods	Simple rules	Complex situations are hard to define in simple rules	Simple to use	If…then … like rules
	Neural Networks	Black box	Can handle complex relations	Mathematical model calculating output based on inputs
	Support Vector Machine	Slow in large sample size	Small chance at over fitting and possible to use dynamically	Classes are separated by a hyperplane by calculating support vectors to the closest points from each class
	Bayesian Networks	Cannot handle missing data well	Efficient with causal relations	Probability of events
	Decision Trees	Very hard to find optimal solutions	Can handle numerical and text data types	Classification by If… then… like tree structure
	k-Nearest Neighbor	Storage intensive and susceptible to noise	Simple to implement	Distance/density calculation between case and classes
	Hidden Markov Models	Events must be independent. (The events may not provide a probability of a following event.)	Can analyze sequences of events in which the events are not independent	The probability of a sequence of observed events is used to calculate the probability of a sequence of non-visible events.
Clustering methods	K-Means	Initial choice of parameter values Hard to find optimal solution and sensitive to cluster shape	Insensitive to noise and cluster shape Pre-classification not necessary	Clustering based on equality Clusters data into a given number of k clusters by minimizing the mean distance to a cluster center
	Self-organizing maps	Resulting model is a black box and creating a model is computational intensive	Good reduction of data feature dimensionality while maintaining relationships between the features	Neural network where output neurons are pixels of a density map and similar cases are mapped close to each other

Fig. 1. The various classification and clustering technique for data mining with their advantage and disadvantage

3.3 Regression

Regression and classification are used to assist in solving similar problems, but regression is used to predicate continues values. Most applications take advantages of regression for purpose of predicating and forecasting, it helps in identifying the behavior of variables under study.

3.4 Association Rule

This technique search for the most occurring item set from a large data volume, used in order to identify the relationship between feature variables and thus indicate the most effect of this relationship on outcome of future values.

4 Data Mining and Cyber Security Approaches

Cyber-attack detection via data mining techniques illustrated in five important steps listed as follow: data acquisition through different tools like sniffing, logging and sensors hardware/software tools), data preprocessing (such as regularization, data cleaning and normalization), feature extraction and dimension reduction, data mining (e.g. classification, clustering) and finally visualizations and interpretation of outputted results.

The intrusion detection system (IDS) based on the data mining techniques used perfectly and can become the capital of the networks. It provides enormous and useful function. For instance, it derives a real time monitoring process as well as incident management in order to provide a barrier security related to events collected from networks, security devices, system and applications. Nevertheless, it provides a workflow which is a priceless tool that track and escalate the incident. Moreover, it could be used in different log manners such as log management, log consolidation and generates reporting for compliance purpose. Actually, the intelligent IDS is an important technique that has many comprehensive benefits for respective components. These benefits can be summarizing into three stages: namely, processing, analysis and visualization, with processing covering the first two stages and analysis the middle two stages.

Mainly, to provide any intelligent system, it should have 100% DR (Detection Rate) and 0% FPR (False Positive Rate). However, in the real world this is really hard to attain this accuracy of prediction. Following are false positive rate (FPR) and detection rate (DR) that are used to estimate the IDS's performance.

The machine learning and the data mining technique uses many algorithms by employing metrics for evaluating the classification of an intelligent models. In the case of proposing an intelligent IDS, the researchers apply different performance measures that are used to assess the model. We will highlight this two mensuration; the false alarm rate and the attack detection rate.

- Attack Detection Rate (ADR): this is the proportion between attacks total number detected by a selected system to the attacks total number present in a dataset.
- Attack Detection Rate = Total detected attacks/Total attacks * 100

- False Alarm Rate (FAR): It is the proportion between the misclassified total number instances to the total number of the normal instances.
- False Alarm Rate = Total misclassified instances/Total normal instances * 100

By applying these two measures, we calculate them by using the confusion matrix technique which is an identifier for the matrix representation with respect to the classification results. The Table 1 below clarifies each phase of the alarm rate; the lower right cell that indicates the number of connections classified as an attack were they are really an attack (TN) and the upper left cell that indicates the number of connections classified as a normal pattern and they are were really Normal (TP). However, the other cells denoted by the number of misclassified connections, the upper right cell indicates the number of connections classified as an attack but they are being really normal (FP), on other hand the lower left indicates the number of connections that classified as normal but they were really attack (FN).

Table 1. The detection alarm rate

	Classified as normal	Classified as attack
Normal	TP	FP
Attack	FN	TN

5 Data Acquisition

This phase in the Intrusion detection system consists of collecting the data from systems under attacks. There are two sources that play an important role to acquire the collection of data: the first consists of records collected from log files which contain user activities. The second source consists of data collected from network traffics at which potential threats get recorded. Attacks can be detected at different zone and network infrastructure, each zone have its own view of data traffic and has its own advantages and disadvantages.

5.1 Log File

Log files record is considering one of the most data rich source to detect attacks on systems, hence all user activates get recorded by the log utility, thus this data source is count as an indicator of serious problems occur on the system, however log files contains thousands if not millions which make it impossible for human to analyze those logs, what we need is a solution that can automatically detect intrusions recorded by logs without human interventions. Each record in log file is a representation of an event occurs in the system, in the form of binary format or plain text where the message content represented by natural text or numerical data or a combination of both [16], for example any hit to web site is associated with a line in web log file. A basic record contains information about who, where, what resource they target from web site.

- Layer 3/4 Firewall logs: these logs contain layer 3 (network layer) and layer 4 (transport layer) information like TCP, UDP, ICMP related data. Firewall logs can be used to detect anomalies carried out to the system network through these protocols, but these logs are not familiar with application layer traffic like HTTP data, so they are not very helpful sources in detecting malicious on higher layers.
- Application Layer Firewall logs: these logs contain layer 7 (application layer) information like HTTP and SOAP and can analyze those requests in great details making it a good place in detecting anomalies targeting then application layer of system network.
- Web server logs: web server like IIS and Apaches counted as end device of HTTP request which by default log activities occur on the system using Common Log Format specifications (CLF). Web server logs do not record data sent in HTTP body of the request such as POST parameters and since many server's forms submit their parameters by HTTP POST requests, web server logs cannot represent a complete view of traffic passing through web server, thus this source will be missing a valuable data that could be useful in detecting malicious activities.

Attacks detection using log files has some advantages and disadvantages over using network traffic benchmarks datasets which are listed as following [17]:

Advantages: The greatest advantage of log records over system traffic are that log documents are effectively accessible and there is no need for costly equipment for investigation. Also, logs may give effective identification particularly to encrypted data carried by protocols such as Secure Sockets Layer (SSL) and Secure Shell Daemon (SSHD).

Disadvantages: logs contain a partial set of the full traffic that is actually occurring in the network for example logs do not contain payload carried out via HTTP body or we may have the case where serious malicious activities initiate by attackers get deleted by the attackers themselves. Many attacks carried out on upper layer protocol such as: DNS Misuse, Finger Daemon Attack, Routing Infrastructure Intrusions, NFS that cannot be detected using log files as a data source, finally anomalies activities detected via log files are not done in real time manner, these activities initiated and executed before we even know about it.

5.2 Networks Traffic

Monitoring network traffic at selected network points is done by IDS which collect traffic packet at real time, IDS may collect activities done on different protocol layer such as network, transport and application layer. Below is a list of IDS data set created for the purpose of intrusion detection base on network traffic:

- DARPA (Lincoln Laboratory 1998, 1999): this data set was created for security analysis purpose [18], this dataset consists of 7 weeks of data acquisition, each day was formed of BRM audit and TCP dump data and TCP dump list file which label each record by attack or not. DARPA consist of various network activities such as FTP, email, browser, SNMP, telnet, IRC actions, which in result will contain attacks such as DOS, buffer overflow, guess password, remote FTP, syn flood, Nmap, rookit. However, the disadvantages from using this data set is that it does not represent real network traffic and it is outdated in term of attack types and it contain irregularities [19, 20].

- KDD'99 (University of California, Irvine 1998, 99): KDD 99 [21] which was formed from DARPA 98, by converting the network traffic recorded by DARPA 98 to network connection with number of 41 features per connection. each connection is considering as a sequence of TCP packets that flow from IP address source to destination IP address under a defined protocol for a specific amount of time, this dataset contains 22 attack type in training set with attack distribution higher than normal flow distribution and the attack types is not equally distributed having DOS accounts the majority of dataset each connection has 41 features.
- NS-KDD: due to the complicated redundant records, about 78% and 75% of records are duplicated in KDD99, NS-KDD predecessor of KDD 99 was presented, the advantages over KDD 99 is that it does not contains redundant records in training set and there is no duplication record in test set, consist of the needed at-tributes from KDD 99 with sufficient amount of records presented in test and training set. Which will cause the learning algorithm to biased toward the frequent records and thus preventing it from unlearning the infrequent records. The duplicate records in the test set will cause the evaluation to biased to method that give better detection rate on the frequent records. Another problem may exist in the labeled records in the train and test set which makes the comparison between IDS difficult due to the variation in classification rate [22].
- DEFCON (The Shmoo Group, 2000): 2 data set version were created DEFCON-8 data set which contain buffer overflow and port scanning attacks, DEFCON-10 data set which contain bad packet, admin privilege, FTP, Telnet port scanning. The network traffic captured by this data set does not reveal real world network traffic [23, 24].
- LBNL (Lawrence Berkeley National Laboratory and ICSI – 2004/2005): this data set consist of recording network traffic header only without including the payload in capture packet, which it is data insufficient set [25].
- CDX (United States Military Academy 2009): CDK contain attacks traffic carried automatically by attackers with the help of common attack tool such as Web Scarab, it also contains normal traffic that carries out to system DNS, Web, Email services. The volume and diversity of data in this set cannot be used to train ML model, however it could be used to evaluate the model [26].
- UMASS (University of Massachusetts – 2011): contain attacks on network traffic due to single TCP download request, which make neither useful in training nor in testing [27–29].
- ISCX2012 (University of New Brunswick – 2012): this data set was created based on 2 profile "alpha and beta", where alpha is the creation of attacks process at which multistage attacks scenario, while the beta profile represents the creation of normal traffic that capture all packet's payload carried by different protocols such as HTTP, SNMP, POP3, FTP. However, this data set does not contain nowadays most used protocols such as HTTPS, SFTP.
- ADFA (University of New South Wales – 2013): [16, 30, 31].

- CAIDA (Center of Applied Internet Data Analysis – 2002/2016): this data set consist of three different data repository where each one of them contains data specific to an event, these data set are listed as follow:
  - CAIDA OC48: provide different type of observed data at OC48 link [32].
  - CAIDA DDOS: provide DDOS attack traffics [33].
  - CAIDA internet trace: provide traffic observed at high speed internet link [34].
- CICIDS2017: This data set is developed in the Canadian institute for cyber-security (CIC) which contains benign and incoming network attacks, it contains normal and most up to date network attack, which represent real world traffic, it also includes the network traffic resulted as an analysis using CIC Flow Meter with a labeled flows based on many networks traffics features such as the time stamp, source and destination IPs, source and destination ports, protocols and attack (CSV files) [35]. This data set use Beta profile used by ISCX2012 data set to create a real background traffic that mimic 25 user's interactions over several protocols as HTTP, HTTPS, SSH, etc. and thus generate a natural normal traffic. The implemented attacks in this data set has the ability of including the latest attacks type listed as follow: DoS, DDoS, Brute Force FTP, Web Attack, Brute Force SSH Heartbleed, Infiltration and Botnet.

There are numbers of advantages and disadvantages of using IDS over networks which are listed as follow:

Advantages: The essential benefit of the networks traffics analysis is the ability to detect and close blind spots in communication as well as the security gaps on end devices by performing successful attacks in a specific period time [36], moreover, the performed data is usually extracted from the networks traffics which contains a lot of information that is not even recorded in a data set extracted from the web server log file.

Disadvantages: if a traffic is encrypted, such as the encryption of HTTP traffic by SSL (HTTPS), IDS may not be able to decrypt it, making the captured data useless for further preprocessing. Most IDS work on TCP/IP level, thus traffics from higher layer may not be captured by IDS. Therefore, attackers will aim to target these higher levels protocols in order to initiate the attacks for example using techniques such as (fragmentation, encoding).

6 Features Extraction

Due to the tremendous measure of features caught from system traffic and logging, feature extraction process become the most imperative factors that influences the adequacy of an IDS, this process depends on reducing the features by removing the predictors that have no impact on IDS classification process, which will improve the overall IDS process with an increase in both computation speed and detection accuracy. There are a lot of feature extraction techniques which helps in extracting useful predictors from the gathered data, such as Latent Semantic Indexing (LSI) [37] which used to capture knowledge from unstructured data by filtering information inside records and recognize connections between the ideas contained in that. Basically it's a strategy that search for covered up connections between words to improve knowledge extraction. Another technique used for feature extraction is Principal Component Analysis (PCA) which used as a feature reduction technique, that is, it decreases the number of the selected feature by developing a new dataset that hold the same quality of data as the original source but with smaller number of predictors. According to available literatures related to the construction of an intelligent IDS, we will focus on various features extraction methods. According to [38] the statistical results properties of an extracted packet sequence are used as the input features that can be used in the learning step. With respect to [39] the resulting features represent the signatures gathered from the preprocessing phase. The researchers in [40] highlight on a technique to extract the features that depend on an empirical probability of a token presented in a sample.

The Fig. 2. below shows an example of the features about the NSL-KDD data set [37].

No.	Features	Types	No.	Features	Types
1	duration	Continuous	22	is_guest_login	Symbolic
2	protocol_type	Symbolic	23	count	Continuous
3	service	Symbolic	24	srv_count	Continuous
4	flag	Symbolic	25	serror_rate	Continuous
5	src_bytes	Continuous	26	srv_serror_rate	Continuous
6	dst_bytes	Continuous	27	rerror_rate	Continuous
7	land	Symbolic	28	srv_rerror_rate	Continuous
8	wrong_fragment	Continuous	29	same_srv_rate	Continuous
9	urgent	Continuous	30	diff_srv_rate	Continuous
10	hot	Continuous	31	srv_diff_host_rate	Continuous
11	num_failed_logins	Continuous	32	dst_host_count	Continuous
12	logged_in	Symbolic	33	dst_host_srv_count	Continuous
13	num_compromised	Continuous	34	dst_host_same_srv_rate	Continuous
14	root_shell	Continuous	35	dst_host_diff_srv_rate	Continuous
15	su_attempted	Continuous	36	dst_host_same_src_port_ra	Continuous
16	num_root	Continuous	37	dst_host_srv_diff_host_rat	Continuous
17	num_file_creations	Continuous	38	dst_host_serror_rate	Continuous
18	num_shells	Continuous	39	dst_host_srv_serror_rate	Continuous
19	num_access_files	Continuous	40	dst_host_rerror_rate	Continuous
20	num_outbound_cmds	Continuous	41	dst_host_srv_rerror_rate	Continuous
21	is_host_login	Symbolic			

Fig. 2. Features extraction about the NSL-KDD data set

7 Conclusion

The big issue faced in the international security is concerning the protection of the computer systems from any cyber-attacks. As a solution to solve this issue, many researchers put their efforts using different datasets to apply machine learning algorithms and artificial intelligent techniques for cyber-attacks protection target.

In this paper, we have discussed the state of the art in the latest network intrusion detection system and their current methods of data mining as well as the proposed data set which it is common in this field, after comparing the related work's results, we can find that the classification of cyber-attack requires much research studies. Furthermore, we highlight many techniques in data mining that used in the cyber-attack approach for the literature containing classification, ensemble techniques and clustering thus we observe the evaluation measures for testing the attack detection rate or system performance.

As our perspective work, we are going to generate a new dataset for the purpose of making it public to be available for the researchers, moreover, by employing this proposed data set we will highlight on developing an intelligent model based on the last machine learning techniques for detecting the illegal activities as well as the common updated of cyber-attacks.

References

1. Bhuyan, M.H., Bhattacharyya, D.K., Kalita, J.K.: Network anomaly detection: methods, systems and tools. IEEE Commun. Surv. Tutorials **16**(1), 303–336 (2014)
2. Neethu, B.: Adaptive intrusion detection using machine learning. Int. J. Comput. Sci. Netw. **13**(3), 118–124 (2013)
3. Bhuyan, M.H., Bhattacharyya, D.K., Kalita, J.K.: Towards generating real life datasets for network intrusion detection. Int. J. Netw. Secur. **17**(6), 683–701 (2015)
4. Mahmood, S., Ibrahim, L.M., Basheer, D.: A comparison study for intrusion database (KDD99, NSL-KDD) based on self-organization map (SOM) artificial neural network. J. Eng. Sci. Technol. **8**(1), 107–119 (2013)
5. Ektefa, M., Memar, S., Sidi, F., Affendey, L.S.: Intrusion detection using data mining techniques, pp. 200–203. IEEE (2010)
6. Kayacik, H.G., et al.: Selecting features for intrusion detection: a feature relevance analysis on KDD 99. In: PST (2005)
7. Ng, J., Joshi, D., Banik, S.M: Applying data mining techniques to intrusion detection. In: 12th International Conference on Information Technology New Generations, Las Vegas, NV, pp. 800–801 (2015)
8. Chauhan, A., Mishra, G., Kumar, G.: Survey on data mining techniques in intrusion detection. Int. J. Sci. Eng. Res. **2**(7), 1–4 (2011)
9. Harshna, Navneet, K.: Survey paper on data mining techniques of intrusion detection. Int. J. Sci. Eng. Technol. Res. **2**(4) (2013)
10. Sharma, Y., Sharma, S.: Intrusion detection system: a survey using data mining and learning methods. Comput. Eng. Intell. Syst. **8**(7) (2017)
11. Agrawal, S., Agrawal, J.: Survey on anomaly detection using data mining techniques. Procedia Comput. Sci. **60**, 708–713 (2015)
12. Sahasrabuddhe, A., Naikade, S., Ramaswamy, A., Sadliwala, B., Futane, P.: Survey on intrusion detection system using data mining techniques. Int. Res. J. Eng. Technol. **4**(5), 1780–1784 (2017)
13. Bharti, K.K., Shukla, S., Jain, S.: Intrusion detection using clustering. IJCCT **1**(2), 158–165 (2010)
14. Dharamkar, B., Singh, R.R.: A review of cyber attack classification technique based on data mining and neural network approach. Int. J. Comput. Trends Technol. **7**(2), 100–105 (2014)
15. Baseman, E., Blanchard, S., Li, Z., Fu, S.: Relational synthesis of text and numeric data for anomaly detection on computing system logs. In: 15th IEEE International Conference on Machine Learning and Applications (ICMLA), Anaheim, CA, pp. 882–885. IEEE (2016)

16. Xie, M., Hu, J., Slay, J.: Evaluating host based anomaly detection systems: application of the one class SVM algorithm to ADFA-LD. In: 11th International Conference on Fuzzy Systems and Knowledge Discovery (FSKD), Xiamen, pp. 978–982. IEEE (2014)
17. Meyer, R., Cid, C.: Detecting attacks on web applications from log files. SANS Institute InfoSec, Reading Room (2008)
18. Lee, J.-H. et al.: Effective value of decision tree with KDD 99 intrusion detection datasets for intrusion detection system. In: 10th International Conference on Advanced Communication Technology, vol. 2, pp. 1170–1175 (2008)
19. McHugh, J.: Testing intrusion detection systems: a critique of the 1998 and 1999 DARPA intrusion detection system evaluations as performed by Lincoln laboratory. ACM Trans. Inform. Syst. Secur. **3**, 262–294 (2000)
20. Brown, C., Cowperthwaite, A., Hijazi, A., Somayaji, A.: Analysis of the 1999 DARPA/Lincoln laboratory IDS evaluation data with Net ADHICT. In: Proceedings of the IEEE Symposium on Computational Intelligence for Security and Defense Applications, Piscataway, NJ, pp. 1–7 (2009)
21. Kayacik, H.G., Zincir-Heywood, A.N., Heywood, M.I: Selecting features for intrusion detection: a feature relevance analysis on KDD 99 intrusion detection datasets. In: Third Annual Conference on Privacy, Security and Trust, Canada (2005)
22. Solanki, M., Dhamdhere, V.: Intrusion detection technique using data mining approach: survey. Int. J. Innovative Res. Comput. Commun. Eng. **2**(11), 6235–6359 (2014)
23. DEFCON 8, 10 and 11, The Shmoo Group (2000). http://cctf.shmoo.com
24. Nehinbe, J.O., Weerasinghe, D.: A simple method for improving intrusion detections in corporate networks. In: Information Security and Digital Forensics. First International Conference ISDF, pp. 111–122. Springer, Berlin (2010)
25. Lawrence Berkeley National Laboratory (LBNL)/ICSI Enterprise Tracing Project. https://www.icir.org/enterprisetracing/download.html. Accessed 30 July 2013
26. Sangster, B., O'Connor, T. J., Cook, T., Fanelli, R., Dean, E., Adams, W.J., Morrell, C., et al.: Toward instrumenting network warfare competitions to generate labeled datasets. In: CSET 2009 Proceedings of the 2nd Conference on Cyber Security Experimentation and Test, Canada, p. 9 (2009)
27. Nehinbe, J.O.: Acritical evaluation of datasets for investigating IDSs and IPSs researches. In: Proceedings of the IEEE 10th International Conference on Cybernetic Intelligent Systems (CIS), New York, pp. 92–97 (2011)
28. Prusty, S., Levine, B.N., Liberatore, M.: Forensic investigation of the one swarm anonymous FileSharing system. In: Proceedings of the ACM Conference on Computer and Communications Security (CCS), New York (2011)
29. UMass Trace Repository. Optimistic TCP ACKing, University of Massachusetts Amherst (2011). http://traces.cs.umass.edu
30. Creech, G., Hu, J.: Generation of a new IDS test dataset: time to retire the KDD collection. In: Proceedings of the IEEE Wireless Communications and Networking Conference (WCNC), New York, NY, pp. 4487–4492 (2013)
31. Xie, M., Hu, J.: Evaluating host based anomaly detection systems: a preliminary analysis of ADFALD. In: Proceedings of the 6th International Congress on Image and Signal Processing (CISP), vol. 03, pp. 1711–1716. Springer, Berlin (2013)
32. CAIDA: CAIDA data set OC48 Link A (2002).https://www.caida.org/data/passive/passiveoc48dataset.xml
33. CAIDA: CAIDA DDoS Attack Dataset (2007). https://www.caida.org/data/passive/ddos-20070804dataset.xml
34. CAIDA. CAIDA Anonymized Internet Traces 2016 Dataset (2016).https://www.caida.org/data/passive/passive2016dataset.xml

35. UNB. https://www.unb.ca/cic/datasets/ids-2017.html
36. Finally Safe News Page. https://www.finally-safe.com/news-detail/network-traffic-analysis-vs-siem/. Accessed 31 July 2018
37. Chen, L.-S., Syu, J.-S.: Feature extraction based approaches for improving the performance of intrusion detection systems. In: Proceedings of the International MultiConference of Engineers and Computer Scientists, IMECS 2015, Hong Kong, vol. I, (2015)
38. Sekar, R., Gupta, A., Frullo, J., Shanbhag, T., Tiwari, A., Yang, H., Zhou, S.: Specification based anomaly detection: a new approach for detecting network intrusions. In: Proceedings of the 9th ACM Conference on Computer and Communications Security, CCS 2002, pp. 265–274. ACM, New York (2002)
39. Rieck, K., Schwenk, G., Limmer, T., Holz, T., Laskov, P.: Botzilla: detecting the phoning home of malicious software. In: Proceedings of the 2010 ACM Symposium on Applied Computing, Sierre, Switzerland, pp. 1978–1984. ACM (2010)
40. Newsome, J., Karp, B., Song D.: Polygraph: automatically generating signatures for polymorphic worms. In: IEEE Symposium on Security and Privacy (S&P 2005), USA, pp. 226– 241. IEEE (2005)

A Novel Hybrid Model for Vendor Selection in a Supply Chain by Using Artificial Intelligence Techniques Case Study: Petroleum Companies

Mohsen Jafari Nodeh[1], M. Hanefi Calp[2(✉)], and İsmail Şahin[3]

[1] Department of Informatics Institute, Gazi University, Ankara, Turkey
mohsen@gazi.edu.tr
[2] Department of Management Information Systems, Karadeniz Technical University, Trabzon, Turkey
mhcalp@ktu.edu.tr
[3] Department of Industrial Design Engineering, Gazi University, Ankara, Turkey
isahin@gazi.edu.tr

Abstract. Oil is an important strategic material which is associated with vital and major components of national security and economy of each country. In this context, supplier selection in a supply chain of oil companies has a direct effect on immunization and optimization of the production cycle, refining and distribution of petroleum, gas and petroleum products in oil producer and exporter countries. Therefore, creating and owning a purposeful and intelligent process for evaluation and analysis of suppliers is one of the inevitable needs and concerns of these countries. Many of the methods which are currently widely applied in the management of oil companies utilize traditional supplier selection methods which are unfortunately limited to individual and subjective evaluation in weighing decision maker's criteria, incorrect assessment rules, and inefficient decision-making methods. In this paper, with an in-depth look at the supplier selection in supply chain management of oil companies project, a novel model has been proposed based on an object-oriented framework. This model which finally leads to optimal selection and ranking of suppliers, reducing the time and cost in the selection process and also reduced human errors by using data mining techniques and neural networks in the reasoning method cycle based on the case. The proposed model was implemented on data bank information of the Oil Company. Finally, the results of the proposed model are compared with several other models. Results show that using reduced errors, improved accuracy, and efficiency the proposed model has been able to have a good performance in the supplier selection.

Keywords: Vendor selection · Supply chain · Artificial intelligence · Supplier selection

D. J. Hemanth and U. Kose (Eds.): ICAIAME 2019, LNDECT 43, pp. 226–251, 2020.
https://doi.org/10.1007/978-3-030-36178-5_19

1 Introduction

One of the main goals of data mining is finding models, patterns and repetitive rules through large amounts of data. In comparison with other branches of artificial intelligence, data mining significantly increases logical reasoning and learning of mathematical models. The results of Supplier Selection Based Data Mining represent a repeating pattern of "if … then …" and understanding and interpretation of the rules is very simple. The decision makers realize suppliers' problems by analyzing these rules. Common learning methods data include: Rough set, Genetic algorithms, Decision tree, neural network, case based reasoning (CBR).

Due to the product's uniqueness, frequency of purchase, and the ability to experience the historic decision of oil companies, this paper uses supplier selection method based on data mining techniques.

In the selection of suppliers using the same sample is very practical and has shown its effectiveness as a supplier selection strategy. In addition, the high volume of data with techniques of data mining is fully compatible with the spirit of the industry.

Due to the complexity of reasoning, CBR must solve several key problems:

- Case warehouse which is under the management and support of CBR, should resolve time and space and complexity problems.
- Character distilling and weight setting in CBR should create the possibility for comparing the current and main case existing in the memory.
- CBR must have a good appeal system power, the ability to maintain high reservoir and can support users to extract knowledge effectively.

1.1 Supplier Selection Methods

Scientific research to find a systematic approach in decision making, purchasing and supplier selection commerce is developed in more than three decades. In a review of methods for supplier selection by de Boer et al. [1] writers the trade Supplier Selection Process is divided into four stages. (1) Defining the problem, (2) defining the criteria, (3) preliminary evaluation of suppliers and (4) making a final choice [2]. Preliminary evaluation reduces "all" suppliers to a smaller set of acceptable 135 suppliers, ways of doing this step can be classified into four categories: Categorical [3], data envelopment analysis (DEA) [4–8], Clustering analysis (CA) [9] and case based reasoning method [10–13]. The suggested methods for the final selection stage are also classified in a linear weighting models [14], Total Cost of Ownership models [15] mathematical programming models [4, 5, 16], statistical models [1] and artificial intelligence models form [17–22]. Most of the suggested methods at this stage are dependent on mathematical programming and weighting models [10]. Mathematical Programming allows the decision maker to decide the issue using variables in the objective function to formulate for maximizing revenue and minimizing costs. Weber et al. [4] presented a multi-objective planning In order to determine the environmental characteristics of a number of conflicting provisions in the form of criteria for selecting providers. They also developed models using non-harmonized trading strategies to allow Transactions with suppliers selected by the multi-objective programming model. They showed that the values of a chosen supplier alters based on the number of suppliers [2]. In linear weighting models, weight allocates to the criteria, the highest weight indicates the highest importance.

Degrees of criteria are multiplied by their weight and the sum is shown as a number for each supplier. Trade supplier with the highest total degree is suggested for selection. Lee et al. [14] provided the supplier select and management system which it uses a linear weighting model to calculate the weight of sensible and insensible criteria and categorizing the suppliers' performance. This system can identify the chosen supplier's weaknesses and compare them with other suppliers. The supplier select and management system provide ways to improve a trade supplier's performance [2].

Suppliers Analysis and Selection Techniques

In order to provide supplier selection processes, De Boer et al. [1] have done a wide examination on reported decision making approaches in literature from the time when Weber et al. [23] reviewed it. Unlike other reviews, De Boer et al. covered all stages of the selection process includes primary definition the formulation of criteria, preliminary evaluation and final selection of the best suppliers. Figure 1 has summarized some of these popular presented methods by De Boer et al. and conceptual studies after them including works of Das et al. [24], Kumar et al. [20, 22] and Chai et al. [25].

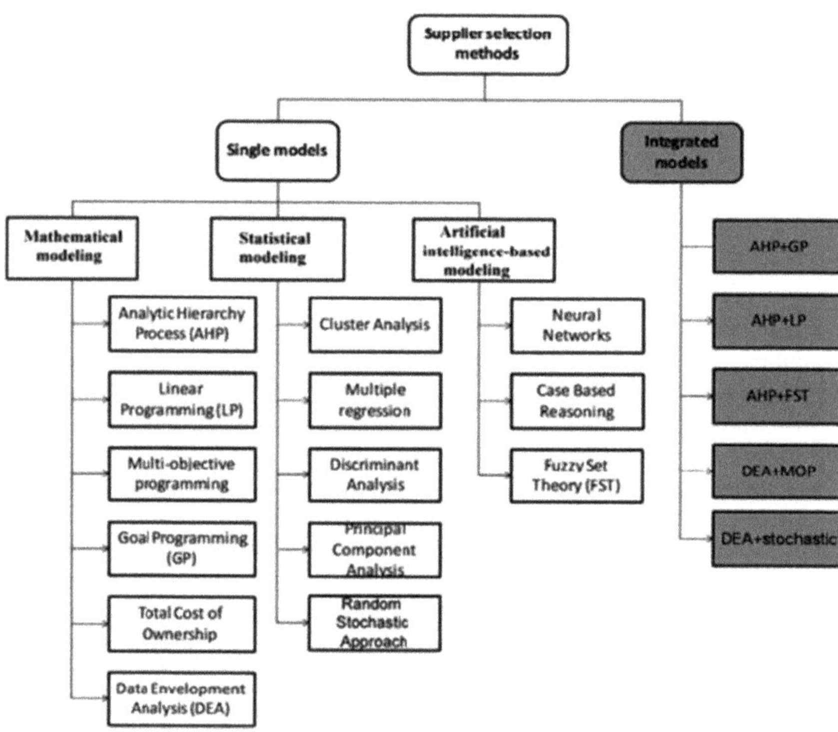

Fig. 1. Supplier selection methods.

Linear Weighting Patterns

This is a method that solves the single-source problems. It is described by a weight to each rule, whatever the weight the higher it is more important. The outcome of any supplier, summary score of every law and product of every law, is the supplier's weight. So every person who obtained the highest outcome will be the best supplier.

The Analytical Hierarchy Process (AHP) is the most widespread method of a linear weighting model [26]. This simple method is applicable and effective which it is used in choosing a supplier [14] and evaluating suppliers [27]. The problem with AHP is, it depends on experts and the role of elements and personal cases in it are more. In addition, sometimes using selected features will be incorrect and will cause diversion in the result of the decision.

Transaction Cost Approach (TCA)

Transaction cost approach (TCA) is a normal approach to solve single purchase problems. Its main idea is calculating purchase cost which includes selling price, purchase costs, transportation costs of a supplier that can meet the needs. Compare to different purchases of suppliers, the one who has the lowest price will be selected. Timmerman invented the cost rate approach in 1986 which calculates product's, transportation's and services' final price that are related to related price to select a supplier [3]. This method can calculate the certain ration of each law in the final price. Roodhoof et al. [28] followed the activity method based on value (Activity based on casting) or ABC for choosing suppliers in 1996. Monczka et al. [29] in 1998 suggested that the supplier method in oil companies chooses using direct and indirect prices and analyzing these prices in activity purchases. Degraeve et al. [30] before 2000, compared 13 selected patterns and calculated the price of each pattern with real purchase data of the Belgian steel company and the consequence is that mathematical programming approach is better than the evaluation approach and multi-products are better than singe products.

Mathematical Programming Method

Mathematical programming method is an important method to solve problems in one or more sources, including programs for single or multi product, linear and nonlinear Programs combined. Since the program is easily explained with mathematical methods and software Supplier selection is a good way to solve problems. In 1974 Gaballa for the first time used the linear system for his problems [31]. Benton set A non-linear program features with the ability to select multi-product, multi-source, multi-client, limited supply, limited amount, Storage costs and the order of prices for its purpose [23]. Ghodsypour and O'Brien invented a combined non-linear pattern which solves multi sources with the ability to limit selection and determines the number of purchases. The aim of this project is to reduce and minimize the cost of buying, Store, transfer and order that the quality and limit its production capacity to itself [32]. According to the purpose of minimizing costs, the single-purpose model doesn't have the ability to solve multi-rule problems. The solution is the multi-purpose linear programming model that can solve the supplier selection criteria challenges. In 1993, Weber & Current used a multi-functional linear model for supplier selection, the price, quality and supplying them as their goals which can choose to monopolize demands, poli tics, and the chosen amount [33]. It should be noted that there is no sufficient scientific and historical data about the complexities of society, and production systems and therefore it is difficult to have problems to abstract mathematical models and precise methods of mathematics for solving it. In addition, to obtain an analytical solution, it is very difficult when problems are large and complex [34].

Traditional selection methods based on analysis of statistics from suppliers which are at high levels have major flaws which is mentioned in the following:

- Decision-making which Lacks a dynamic status and learning capability, remains suspended in the nature of these patterns. The range is specified and its adaptability is poor. So we can't function effectively with inaccurate information [35].
- The decision making standard is based on people's taste and arranging lists and weight does not matter.
- The result for supplier selection is shown as a score or a degree that it is not easy to understand. Moreover, it doesn't define the quality of the supplier's operation. So, monitoring the supplier for improvement is difficult [34].

Since the results of this choice cannot be explained, finding hidden rules and its covered patterns is hard. Thus, decision making results are not predictable. There are two ways: from one hand a large number of mental patterns exist and on the other, is the action taken on a limited number of them. So it is necessary to invent ways which with their powerful transmission, decrease the gap between theory and reality.

CBR Approaches for Suppliers Selection Based on Data Mining
CBR is a set of three thinking-logical, innovation and inner sense of human beings. In 1982, Schank writes a book named dynamic memory which was the start of recent studies about the details of the case-based reasoning method and presented the CBR creation method in computers [36].

In some ways, data mining and CBR technology and results, both used to make decisions, and both began Similar. In the CBR system, a lot of machines and information exists such as: The structure and content of the item, examining the similarities, survival of the item Storage, review, the list's features, item Points and etc. Also there are some problems to obtain these information [37].

In 2006, Li and Shiu invented the information classification system with CBR [38]. In 2007, Huang & Chen made a personal network learning system based on hereditary algorithm with CBR [39]. In recent years, Liu Choy and Lo used CBR in supplier selection and have done extensive research in this field [13]. Chan and Harris have done a research on the environmental management of supplier performance evaluation [11]. In 2001, Kim and Han suggested the combined data mining and case-based reasoning methods based on support vector machine (SVM) in order to predict the classification of bonds. This research Showed that the combination of CBR and SVM, is an effective method in transition index and can increase the accuracy of prediction significantly [40]. In 2011, Zhao et al. [34] presented a new model using CBR based on clustering data in choosing the suppliers of China's National Petroleum Corporation. In the context of using and combining data mining methods with the case-based reasoning method.

It is important to note that these studies actually suggest or provide a systematic model to improve the ability of trust management decisions. Thanks to advances in science, particularly statistics, nowadays researchers have more facilities to build more reliable models. In this paper, the novel intelligent model to predict the optimum supply achieving these objectives is presented.

2 Material and Method

2.1 The CRISPs-DM Process

This process includes the environmental data monitoring and analysis functions. In the DM-CRISP model, three steps are taken to monitor and evaluate the data. These steps are Business Understanding, Data Understanding (Analyzing and examining data) and Data Preparation. In business understanding phase, since the scope of the reviewed activities is related to project supply chain management and in particular the project supply management, the basic familiarity was studied with the supply chain and processes of providing project management was taken place. In data understanding phase, data mining process requires some understanding of the data that it's investigating. This process has got two important functions such as "Collecting initial data" and "Explaining and describing data". In final phase, data was prepared to process data mining [41, 42].

In this research, 1430 records were collected which after data cleaning was reduced to 1118 records. To overcome these problems, various methods and solutions were used which some of them will be mentioned in the following. After preparing the data, 36 attributes for each item were obtained and finally 29 attributes were selected which can be seen in Table 1.

Table 1. Proposed model attribute names.

No	Attribute	No	Attribute	No	Attribute
1	Project name	11	Responsiveness and accountability	21	The number of payment in contract
2	Order name	12	Country code	22	Presentation and answering capabilities in meetings
3	Supplier name	13	Time objectives	23	The company AVL
4	Supplier's nationality	14	Flexibility in production	24	Percentage of payment
5	Producer or dealer	15	The number of proposals for contract	25	Total score in the technical criteria
6	Price forecast	16	Cooperation in specific technical demands	26	The main part of the contract
7	The suggested retail price	17	Company status near the employer	27	Number of different products presented in the proposal
8	Ability to write and recommend proposal	18	The original offer of contract	28	The main part of the contract payment
9	Delivery time	19	Communications	29	Moon and year of proposal
10	Delivery type	20	Win or reject		

2.2 The Proposed CBR Framework

This framework is the operational environment is the case-based reasoning (CBR). Operating modules that put the 4R process in the CBR cycle are embedded in this structure. Embedded triple module and figure of the model (Fig. 2) are as follows:

- CBR case development module
- CBR solution provide module
- CBR advisor module

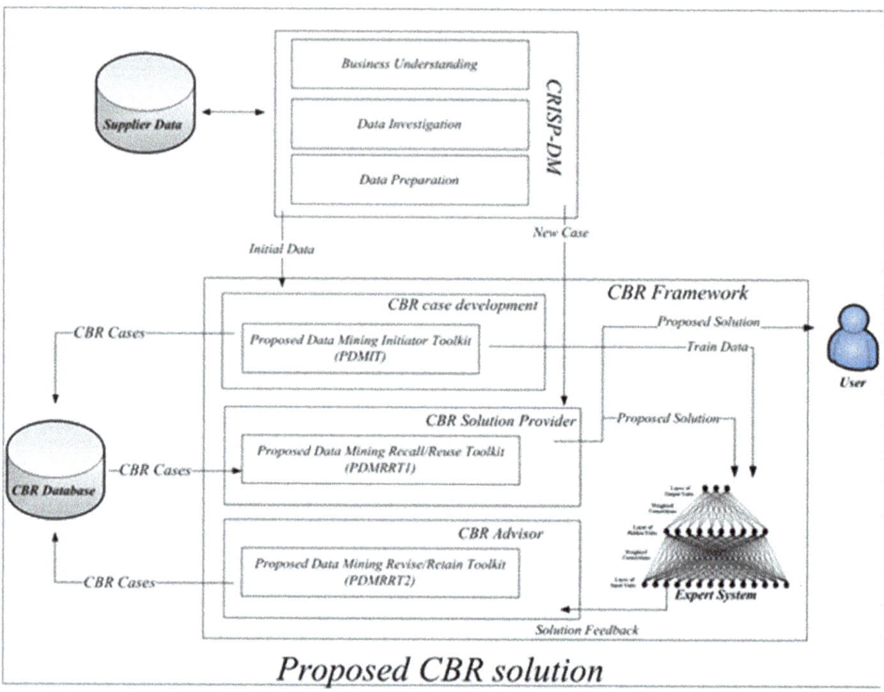

Fig. 2. Proposed model architecture

CBR Case Development Module

Theoretical Assumptions

It is assumed that the primary training base cases an n member set of our suppliers: Supplier Initiation Matrix:

$$SIM = \begin{bmatrix} SI_1 \\ \vdots \\ SI_n \end{bmatrix}; SI_i\text{: } i^{th} \text{ Supplier for case initiation } i = 1\ldots n$$

It is assumed that we have the following after passing by the CRISP-DM modules containing m descriptive variables from the suppliers:

Attributes Initiation Matrix:

$$AIM = \begin{bmatrix} AI_1 \\ \vdots \\ AI_m \end{bmatrix}; AI_j{:}\, j^{th} \textit{ Attributes for case initiation } j = 1\ldots m$$

Using the SAIM Matrix described as follows, a relationship is placed between the suppliers and the descriptive variables:

$$\text{Supplier/Attribute Initiation Matrix:}\, SAIM = \begin{bmatrix} SAI_{11} & \ldots & SAI_{1m} \\ \vdots & \ddots & \vdots \\ SAI_{n1} & \ldots & SAI_{nm} \end{bmatrix};$$

SAI_{ij}: *the decriptor for* j^{th} *attribute of* i^{th} *Supplier in initiation* $i = 1\ldots n,\, j = 1\ldots m$

According to the above definition we have:

$$\text{SIM} = \text{SAIM} \;\S\; \text{AIM}$$

Which § in it is as in which the elements of the matrix and describing them next to each other. The present assumption is that after the transition from modules DM-CRISP, in the decision variable p module earns about suppliers be:

$$\text{Decision Matrix:}\, DM = \begin{bmatrix} D_1 \\ \vdots \\ D_p \end{bmatrix}; D_l{:}\, l^{th} \textit{ Decision Value for } l = 1\ldots p$$

Using SDM as follows matrix varies between suppliers and decision-making about suppliers' relationship is in place:

$$\text{Supplier/Decision Matrix:}\, SDM = \begin{bmatrix} SD_{11} & \ldots & SD_{1p} \\ \vdots & \ddots & \vdots \\ SD_{n1} & \ldots & SD_{np} \end{bmatrix};$$

SD_{il}: *the decriptor for* l^{th} *decision of* i^{th} *Supplier* $i = 1\ldots n,\, l = 1\ldots p$

According to the above definition we have:

$$\text{SIM} = \text{SDM} \;\S\; \text{DM}$$

Which § in it is as in which the elements of the matrix and describing them next to each other.

If all the information about the suppliers in terms of descriptive variables and decision variables are considered, we have:

$$\text{SIM} = \text{SM} \ \S \ \text{M}$$

Which in it:

$$\text{SM} = \left\{ \begin{array}{ccccccc} SAI_{11} & \ldots & SAI_{1m} & SD_{11} & \ldots & SD_{1p} \\ \vdots & \ddots & \vdots \mid & \vdots & \ddots & \vdots \\ SAI_{n1} & \ldots & SAI_{nm} & SD_{n1} & \ldots & SD_{np} \end{array} \right\}$$

$$M = \left\{ \begin{array}{c} AI_1 \\ \vdots \\ AI_m \\ \hline D_1 \\ \vdots \\ D_p \end{array} \right\};$$

Implementation of the Module

As also shown in Fig. 2, the module in addition to the training data set for training the neural network as an expert, is responsible to develop the case base for the proposed model. Before taking any action necessary to the importance of each aspect of which is specified in the data base. For this purpose, the technique gain ratio is used. The weights obtained in this way is used for the extraction of decision variables 8 variables that have the most weight as a decision variable and the remaining features are considered as an explanatory variable.

Also, as indicated in Fig. 3 entire database should be three sets of training data, test data collection and validation data set are divided. For this purpose, 70% of data as training data, test data and 20% as the remaining 10% were considered as validation data set. As mentioned, the other measures contained in the training data set for neural network training modules are created as connoisseurs. This dataset contains data on the educational database for all descriptive characteristics and decision. Also, the training data set of decision variables used to train the decision tree mining rules. Mining rules have been used to produce case base.

Proposed Data Mining Initiator Toolkit (PDMIT)

Data mining techniques used in the PDMI toolkit described early in the development of the case base:

– *Weighting using Gain ratio*

Since the database provided by the suppliers include numerical variables, class variables of interest rate method for weighting characteristics is used. How to calculate interest rates are as follows:

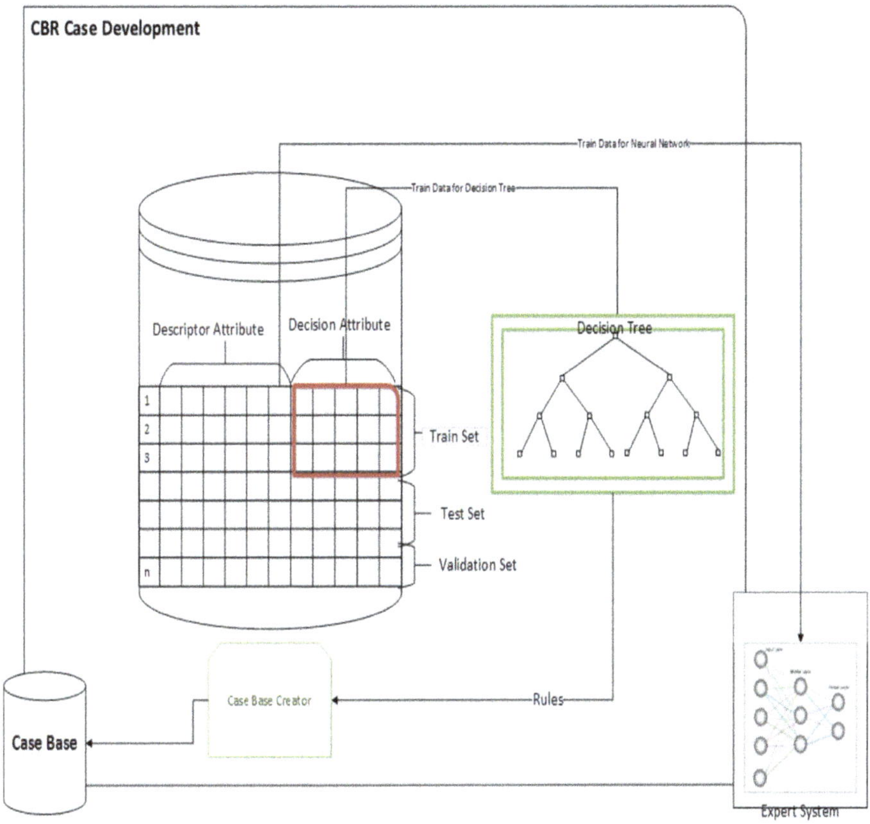

Fig. 3. CBR Case development module architecture

Step 1: searching for the means of the main data. Since the characteristics of each feature on the original data are different, so first of all we mean the quality of each index declining oil companies' equation is computed. (After obtaining the means list of each variable). Finally, the degree and significance of each variable is categorized based on score producer. Calculating the singular coefficient: $Y_{in} = a_i X_{in}^{bi}$ which i is an integer, according to a numerical value to each of which we look back, bi and ai are decline or recursive parameters. N means the number of hosts. x_{in} means the value of any feature that is provided by the Department of Management evaluation.

Calculating b_i and a_i: $Y_{imax} = a_i X_{imax} b_i = 1.30$. $Y_{imax} = a_i X_{imax} b_i = 1.30$

These are based on the valuation of regular management department.

$$a_i = 1.3/power(X_{imax}, b_i)b_i = ln(1.3)/ln(X_{imax}/X_{imin})$$

Step 2: Calculate information gain: based on the ID3 decision tree, we can use instead of the classification decision of A (instead of obtaining information).

Obtaining information as mentioned function is defined as follows: we assume that A has different values different to v number. We can use A to divide S to v number of sets: $\{s_1, s_2, \ldots, s_v\}$. We can use A to divide S to a subset of v s. If A is selected as a test example (which covers the clearest example). So those subunits indicate the expanded and separate branches of S. we assume that S_{ij} is the record number of group C_i in the subcategory of S_j. So the intended entropy or information is ramified from A:

$$E(A) = \sum_{j=1}^{r} \frac{S_{ij} + \cdots + S_{mj}}{S} I(S_{ij} + \cdots + S_{mj})$$

The information that will be achieved in branch A is as follows:

$$Gain(A) = Info(s_1, s_2, \ldots, s_m) - E(A)$$

After selecting the properties influencing the choice of supplier, these characteristics need to be assigned weights and the weights are calculated in a way to reflect the effect of each of the following characteristics. Since the Selected features that include batch and numerical variables are variables, and on the other hand because the distance calculation in CBR technique requires weight values that are in the range of zero to one, the gain ratio technique in the orange Canvas software is used. Also, the more of these rhythms will be used for the selection of decision variables in the proposed framework. Figure 4 shows the weights obtained ReliefF techniques and the Gain ratio.

	Attribute	#	ReliefF	Gain ratio
1	No. of suggesti...	3	0.037	0.070
2	TOTAL SCORE	C	-0.001	0.059
3	Fulfillment / Technical Spec Compliance	C	-0.009	0.056
4	delivery location	C	-0.008	0.055
5	status code	8	0.069	0.050
6	No. of peyment step	5	0.008	0.047
7	delivery type	C	0.097	0.046
8	Writing ability & Appearance	C	-0.005	0.038
9	main suggestion	C	-0.010	0.035
10	Proposal Quality / Preciseness	C	-0.008	0.035
11	AVL code	3	-0.030	0.034
12	Flexibility	C	-0.003	0.029
13	(actual/ estima...	C	0.001	0.020
14	Presentation / Meeting	C	-0.036	0.020
15	Communication	C	0.004	0.019
16	Response	C	-0.020	0.018
17	Total delivery time (day)	C	0.002	0.009
18	actual price ()	C	-0.008	0.007
19	supplier code	C	-0.025	0.004
20	country code	C	-0.032	0.004
21	project code	5	-0.018	0.002
22	DEVIATION No.	C	0.008	0.002
23	proposal year	C	0.002	0.001
24	Pre Peyment	C	-0.021	0.001

Fig. 4. Weighting attributes by ReliefF and Gain ratio.

– *Training neural networks as the expert*

In this paper, Multilayer Perceptron Neural Network (MLP) is used as the expert. For training this network, all features including the training of the decision variables and descriptive variables are used. Therefore, the input layer of the network will have 29 neurons, hidden layer also has 20 neurons and also the output layer contains a neuron that identifies the input supplier will be selected or not selected. Parameters set for MLP neural network training are as follows:

- Number of repeats: 500 times
- Learning rate: 0.1
- Activation function: Sigmoid

MLP neural network has been trained in the testing phase with new examples. In addition to its opinion to select or deselect suppliers based on thresholds defined for the decision variables, sends its feedback to the CBR advisor module till it improves the performance of the database. This expert opinion is applied for every test. In the proposed model validation phase, the expert does not interfere with the operation of this phase. In the proposed model validation phase, the expert does not interfere with the operation of this phase.

MLP neural network implementation is done as an expert in MATLAB with the write function as MLPexpert.ml. As indicated in Fig. 5 the number of neurons in the input layer is 29 neurons that includes all the decision variables, and descriptive variables problem; Moreover, 20 neurons are in the hidden layer and a layer neurons output indicates the supplier is selected or not selected.

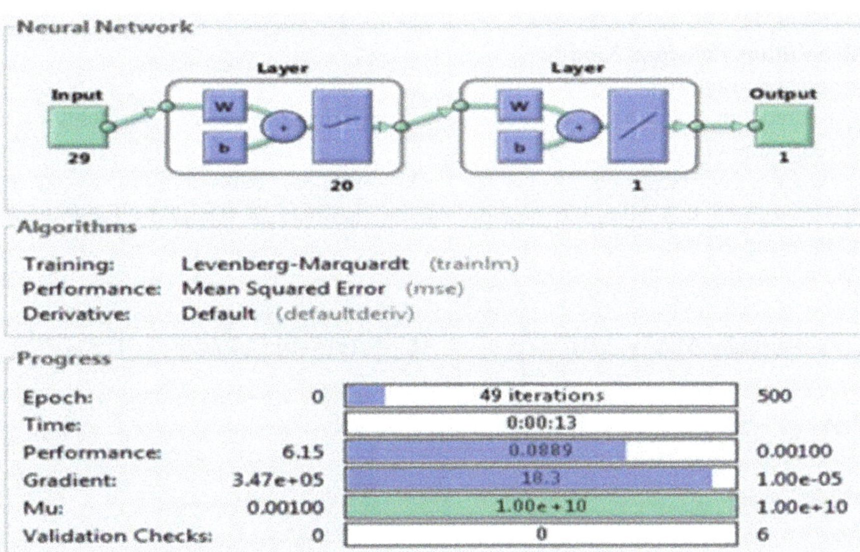

Fig. 5. Training MLP in MATLAB.

– *Establishing a case-base based on extracted rules from the decision-tree*

The main objective of the module of creating and developing cases, is providing a case base based on rules. Decision Tree algorithm rules by using the training data set and only decision variables are done. The output of the decision tree will be if-then rules that the foremost step includes thresholds for 8 variables and the step after is the determination of selection and/or skipping suppliers. After extracting data, in the case base creator, these rules are used and all cases that approve at least one of these rules have been placed in the database. In other words, each instance of database values in their decision variables, which are accepted by one of the extracted rules, are placed in the intended case-base. The noteworthy point is that in establishing case-base, only existing samples in the training data set are used and sample testing phase of the neural network as the expert are added or revised.

In the PDMIT Module, the numbers of features as decision variables were used to train the decision tree mining rules. To identify and select these characteristics of the weights obtained from the section above and 8 features the highest weights are used as decision variables selected. In other words, the first 8 features as decision variables are considered in Fig. 4.

With regard to database mining and decision variables related to suppliers, decision tree training data set can be created in such a way that only the relevant part of the decision variables from the database. In other words, the dataset used to train the decision tree model training data set includes all samples and only features of the proposed decision variables have been identified. As shown in Figs. 6 and 7, Clementine SPSS software and Orange has been used to train the decision tree. Decision tree rules if-then taught two classes for each supplier is selected and not selected on the basis of the decision variables creates.

CBR Solution Provider Module

Theoretical Assumptions

It is assumed that a new case to assess and provide the desired variables by the proposed framework:

New Case Matrix: NCM

It is assumed that after the passage of the modules of the DM-CRISP with m variable in our NCM:

$$\text{Attributes Matrix:} AIM = \begin{bmatrix} AI_1 \\ \vdots \\ AI_m \end{bmatrix}; AI_j : j^{th}\, Attributes\, for\, case\, j = 1...m$$

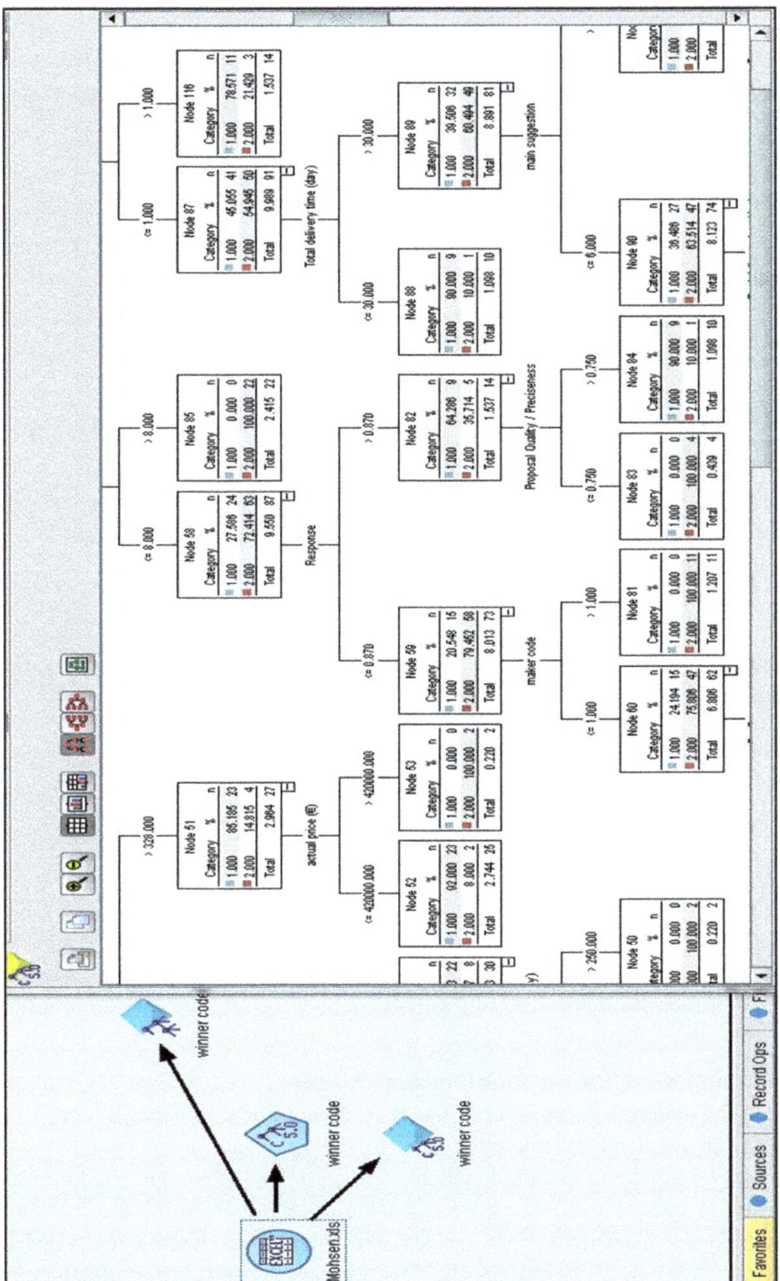

Fig. 6. Training decision tree.

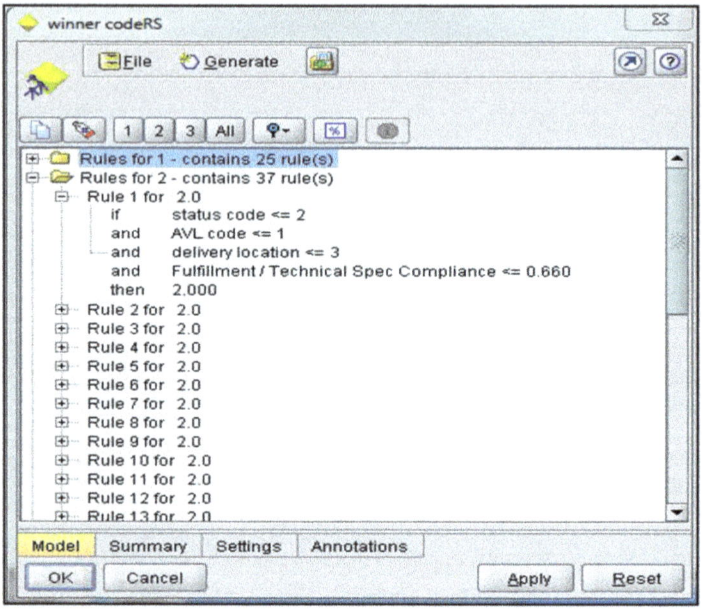

Fig. 7. Extracted rules from SPSS.

Using the Matrix CAM, the following of the new case and descriptive variables related to that relationship:

$$\text{Case/Attribute Matrix: CA}M = [\,CAM_{11} \quad \ldots \quad CAM_{1m}\,];$$

$$CAM_{1j}\text{: } \textit{the decriptor for } j^{th} \textit{ attribute of New Case}, j = 1\ldots m$$

According to the above equations:

$$\text{NCM} = \text{CAM} \; \S \; \text{AIM}$$

Which § in it is as the multiplier of elements of such Matrix and descriptive next to each other.

Implementation of the CBR Solution Provider Module

After creating the database from cases in the case development module, the model with new examples to follow is the nearest case to the new sample in case-base. Case solution strategy proposed new case will be retrieved. So, what this module will be most interested in how to find the nearest is using a new case. Since the cased in the database include the numerical variables and the categorical parameters, SOM clustering algorithm was used based on the Prototype-K for clustering databases and improve the efficiency of the recovery. After clustering, retrieving the nearest case in search for a cluster that its center has the least distance to the new sample, is enough.

Table 2. Samples of extracted rules from decision tree.

No	If	then
1	if status code <= 2 and AVL code <= 1 and delivery location <= 3 and Fulfillment / Technical Spec Compliance > 0.660 and supplier code <= 328 and country code <= 7 and No. of payment step <= 1 and Total delivery time (day) <= 72 and (actual/ estimate) <= 0.986	1
2	if status code <= 2 and AVL code <= 1 and delivery location <= 3 and Fulfillment / Technical Spec Compliance > 0.660 and supplier code > 328 and actual price (€) <= 420,000	1
3	if status code > 2 and Fulfillment / Technical Spec Compliance >0.875 and AVL code <= 1	1
4	if status code <= 2 and AVL code > 1 and (actual/ estimate) <= 1.022 and Proposal Quality / Preciseness <= 0.820 and DEVIATION No. <= 8 and Response <= 0.870 and maker code <= 1 and Pre Payment<= 70 and Flexibility <= 0.300	1
5	if status code <= 2 and AVL code <= 1 and delivery location <= 3 and Fulfillment / Technical Spec Compliance <= 0.660	2
6	if status code <= 2 & AVL code <= 1 & delivery location > 3	2
7	if status code <= 2 and AVL code > 1 and (actual/ estimate) <= 1.022 and Proposal Quality / Preciseness <= 0.820 and DEVIATION No. > 8	2
8	if status code <= 2 and AVL code <= 1 and delivery location <= 3 and Fulfillment / Technical Spec Compliance > 0.660 and supplier code >328 and actual price (€) > 420,000	2

Proposed Data Mining Recall/Reuse Toolkit (PDMRRT1)

The SOM Algorithm Based on K-Prototype

The SOM technique which was presented by Kohonen is one of the most distinctive unsupervised neural network models for applying clustering. A notable feature is that maintaining SOME similarities between input data and generate a map of the High-dimensional input, they largely on a two-dimensional space usually reflects as much as possible while trying to maintain the topological relationship between the input data. Similar to the same input data nodes on the map assigned. The SOM technique have been impaired in calculating the hands variables are similarities in the face of data; In this thesis to overcome this defect, the scale used for similarities in the techniques used Prototype-K. As stated in the second quarter of this technique by taking a mixture that includes both data and numerical variables are variables hands, the scale used to measure a representative sample. SOME proposed model-based algorithm K-Prototype are as follows:

1. All weights and learning rate α and neighborhood radius are initialized.
2. An input x of the input data set is presented to the network.

3. The winner neurons are selected based on criteria (The scale used in the Prototype-K technique):

$$d(X, Y) = \sum_{j=1}^{p} (X_i - Y_j)^2 + \gamma \sum_{j=p+1}^{m} \delta(X_i, Y_j)$$

4. Updating the winning neuron weight and neighboring neurons from k repetition to k + 1:

For numerical variables:

$$w_{ij}(k+1) = \begin{cases} w_{ij}(k) + \propto (k)\left[x - w_{ij}(k)\right] & \text{if } (i, j) \in N_c(k) \\ w_{ij}(k) & \text{otherwise} \end{cases}$$

For categorical variables:

$$w_{ij}(k+1) = \underset{j}{\mathrm{mode}}([x, w_{i,j}(k)])$$

5. Repeating steps 2 to 4 for the whole input vectors.
6. Reducing the learning rate and neighborhood radius determined based on an approach for the next period.

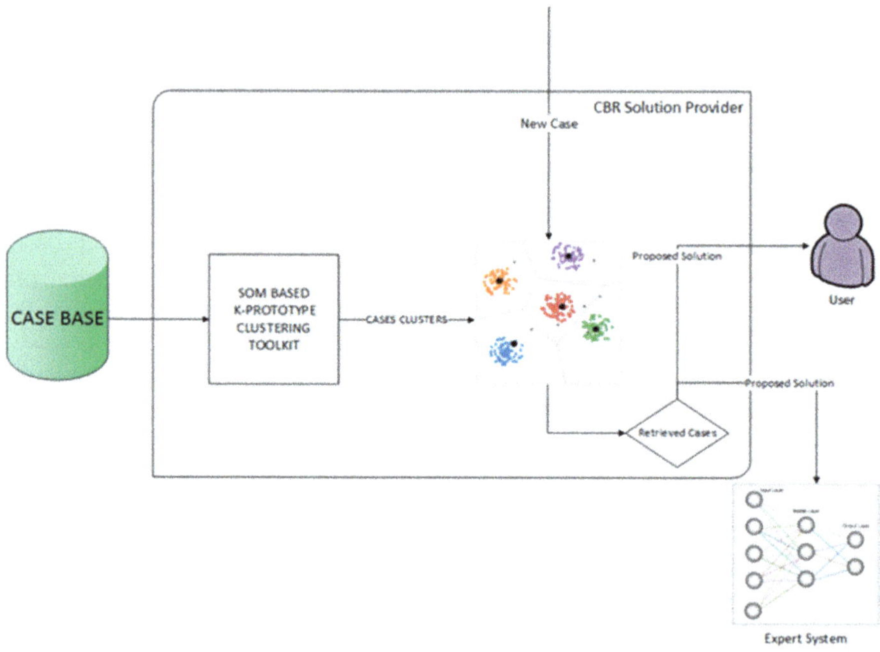

Fig. 8. CBR solution provider module architecture.

Repeating steps 2 to 6 until time realization termination condition (usually a certain number of repetitions).

Database Clustering
In the PDMRRT module, to improve the retrieval performance, nearest to the version of the base case, the SOM-based clustering technique is used K-Prototype. By implementing these techniques, the CBR database is partitioned into K clusters in such a way that the samples are similar to each other in a cluster. Davies – Bouldin index is used to calculate the optimal K that will be explained later.

Calculating the Davies-Bouldin Index
Clustering of clusters is aimed to achieve the highest minimum distance from each other and are within their cluster. In other words, clustering samples must be such that there is the highest concentration in each cluster and the cluster centers farthest from each other. Davis-Bouldin criteria for evaluating clustering is used. This index is a measure of clustering is that the compression and resolution into consideration and is calculated as follows:

$$index = \frac{1}{k}\sum_{i=1}^{k} \max_{i \neq j}\left(\frac{a_i + a_j}{d(c_i, c_j)}\right)$$

Where k is the number of clusters. a_i is distance Inside cluster correspond to cluster i. $d(c_i, c_j)$ equals to the distance between cluster i and j. Clustering is more appropriate to have the lowest index. To select the optimal k, for every k = 2, 3,..., 12 in the MATLAB software, the Davies-Bouldin index is calculated in a function called Davies-Bouldin.m which in Table 3, its results are shown. If specified in the table for k = 3 index is the lowest.

Table 3. Davies-Bouldin index for variant k.

K	2	3	4	5	6	7	8	9	10	11	12
Index	0.5203	0.3525	0.5135	0.5122	0.5199	0.5065	0.5134	0.5530	0.5339	0.5300	0.5346

Clustering Using SOM Based on K-Prototype
As previously mentioned, samples in the database are also characterized by a set of numerical and categorical characteristics; Therefore, it is necessary to choose K in parts of the algorithm SOM such as primary centers, distance calculation and update centers, this point should be considered. Centers for initialization vectors, the number of K samples of the base case selected randomly and were considered as centers of K clusters. Thus, even the values assigned to the properties of the cluster centers will be valid values. In calculating the distance as well as the data together with numerical variables and parameters are set. The purpose of using the software MATLAB function as distance.m was written to calculate this distance. SOM algorithm to calculate the distance in samples of cluster centers to cluster sample allocation is closer than the scale used.

Update centers separate functions to update numerical variables and class variables were written cluster centers. Numericupdate.m function has the task of updating numerical variables; this algorithm update function in SOM is standard practice. Nominal update.m function also updates the variables hands and with the addition of each cluster to cluster presented its fashion value and the variable batch assignment of the cluster center. After the implementation of SCM based on K- Prototype, for clustering on the base created in PDMIT module has been implemented. This implementation is done using MATLAB software and by function as CBclustering.m.

Since the cluster members during the testing phase CBR model by adding new models to the database of the proposed changes. It is necessary that in this phase update cluster centers are done using numericupdate.m functions and nominalupdate.m. Table 4 shows the number of each cluster after finishing the test phase.

Table 4. The number of each cluster after finishing the test phase.

Cluster	First	Second	Third
Number	212	134	416

In retain and reuse process, after identifying the nearest recycle the closest cluster, a new sample interval is calculated with all members of the cluster. Table 5 shows an example of the similarity of arranged new cluster members selected.

Table 5. Similarity between new case.

Index	78	91	105	371	42	8	296	511
Similarity	79.5%	74.12%	64%	48.23%	43.59%	41.82%	35.16%	32.91%

After clustering the cases to the nearest retain to new cases it is sufficient to search only the center of the cluster is the smallest distance using a new sample by reaching the nearest solution as the solution proposed by the expert presented.

CBR Advisor Module

Theoretical Assumptions

It is assumed that when a new case, with the last part of the NCM definition provided for evaluation and decision variables are analyzed by the proposed framework, having regard to the transition from modules DM-CRISP using AIM and CAM matrix for describing variable m describing the matrix NDM proposed as a solution is proposed. The CAM Matrix is provided as input to the expert system and the feedback returned by the expert system will be in the form of a matrix of variables decision which it is called PDM, we'll have:

$$\text{Proposed Decision Matrix: } PDM = \begin{bmatrix} PDM_{11} & \ldots & PDM_{1p} \end{bmatrix};$$

$$PDM_{1l}\text{: the decriptor for } l^{th} \text{ decision } \; l = 1\ldots p$$

According to the above equations:

$$\text{SIM} = \text{PDM} \; \S \; \text{DM}$$

Which in it, § is as the multiplication of such Matrix values and their description is beside each other.

The proposed solution module creator CBR strategy that was called NDM was studied next to the PDM Matrix and based on the Performance evaluation similarities of the cases in the database, update rate is determined. If you need to add new records to the CBR database, the new case will be added by this module. Finding similar operations using the structure of object-oriented cases occurs. Embedded behaviors in cases on receipt of decision variables in the valuation matrix of NDM and PDM, uses data mining to investigate the finding similarities in the result of the matrix of NDM deals and based on this modification of cases and added new cases or removing the existing case occurs. The action mechanism is like this that the output values of expert system is investigated and compared in a PDM Matrix form with an output value of NDM Matrix. If the variables are the same aim in both matrixes, variables are evaluated based on different thresholds for each decision, if the specified limit difference is less than framework threshold for each decision variable, the creation module output of the CBR solution is confirmed and is established in the case database and changes in the existing cases structure and consequently related to its laws. In contrast, if the difference between decision variables in NDM and PDM matrix outputs are greater than the specified threshold for each of the variables decision, Output module, causing the solution is reviewed. In this case, the module of the existing amendments to the CBR corrects and modifies the existing cases. For this purpose, the following process is done.

The new training case is again considered and creating platform cases of n + 1 sets of suppliers are also considered:

Supplier Revised Matrix:

$$SIM = \begin{bmatrix} SI_1 \\ \vdots \\ SI_{n+1} \end{bmatrix}; SI_i\text{: } i^{th} \text{ Supplier for case initiation } \; i = 1\ldots n+1$$

In this case, the supplier has the same definition for m describes the former:

Attributes Initiation Matrix:

$$AIM = \begin{bmatrix} AI_1 \\ \vdots \\ AI_m \end{bmatrix}; AI_j\text{: } j^{th} \text{ Attributes for case initiation } \; j = 1\ldots m$$

The SARM Matrix can be stated as follows between suppliers and descriptive variables:

$$\text{Supplier/Attribute Revise Matrix}: SARM = \begin{bmatrix} SAI_{11} & \cdots & SAI_{1m} \\ \vdots & \ddots & \vdots \\ SAI_{n+1,1} & \cdots & SAI_{n+1,m} \end{bmatrix};$$

$$SAI_{ij}: \textit{the decriptor for } j^{th} \textit{ attribute of } i^{th} \textit{Supplier in initiation } i = 1\ldots n,\ j = 1\ldots m$$

According to the above we have:

$$\text{SRM} = \text{SARM} \ \S\ \text{AIM}$$

In which § is as multiplication of values of such matrixes and describing them next to each other.

Taking into consideration the decision variable p about suppliers as follows:

$$\text{Decision Matrix}: DM = \begin{bmatrix} D_1 \\ \vdots \\ D_p \end{bmatrix}; D_l: l^{th}\ \textit{Decision Value for } l = 1\ldots p$$

SDM matrix are as follows and redefine between suppliers and variables communicate decisions about suppliers:

$$\text{Supplier/Decision Revised Matrix}: SDRM = \begin{bmatrix} SD_{11} & \cdots & SD_{1p} \\ \vdots & \ddots & \vdots \\ SD_{n+1,1} & \cdots & SD_{n+1,p} \end{bmatrix};$$

$$SD_{il}: \textit{the decriptor for } l^{th} \textit{ decision of } i^{th} \textit{Supplier } i = 1\ldots n,\ l = 1\ldots p$$

According to the above:

$$\text{SRM} = \text{SDRM} \ \S\ \text{DM}$$

In which § is as multiplication of values of such matrixes and describing them next to each other.

If using the information Amendments to the suppliers of the explanatory variables and variables be considered in terms of decision-making in aggregate prosper, we have:

$$\text{SRM} = \text{SRRM} \ \S\ \text{M}$$

$$\text{SRRM} = \left\{ \begin{array}{ccc|ccc} SAI_{11} & \cdots & SAI_{1m} & SD_{11} & \cdots & SD_{1p} \\ \vdots & \ddots & \vdots & \vdots & \ddots & \vdots \\ SAI_{n+1,1} & \cdots & SAI_{n+1,m} & SD_{n+1,1} & \cdots & SD_{n+1,p} \end{array} \right\}$$

$$M = \left\{ \begin{array}{c} AI_1 \\ \vdots \\ AI_m \\ \hline D_1 \\ \vdots \\ D_p \end{array} \right\};$$

Implementation of CBR Advisor Module

CBR model proposed in this module original cycle is completed. In other words, corrective action and storage of CBR cycle is done in this module.

As seen in Fig. 6, this module receives feedback neural network, as well as the expert Solution Module strategy underlying the proposed CBR, comparing these two definitions on the eve of the similarity of each decision variables and the difference between targets, one of the following approaches taken:

Insert approach: This approach is taken when the proposed solution by the same supplier is certified on select or deselect as well as decision variables like new sample with the most similar example of threshold be more determined. In other words, it confirms Certified Solution proposed by the CBR. In this new approach with the proposed solution as the solution, it is added in the database.

Modify approach: This approach is taken when the proposed solution by the same supplier is certified on select or deselect but variables' similarity of new decision sample be less than the determined threshold. In this new approach with expert opinion as a solution, it is replaced by the most similar case in the case database.

Deleting approach: If the solution proposed by the expert selected or not, is a different supplier selection, this approach is adopted. In this approach the most similar case is removed from the case database.

It is noteworthy to mention that after the election of each of the above-mentioned approaches, it is necessary to update the cluster and rues centers and clustering should also apply modifications.

Proposed Data Mining Revise/Retain Toolkit (PDMRRT2)

For advisor module implement MATLAB software is used. For each of the approaches in this module, special functions were written that according to the proposed approach, expert opinion and threshold of decision variables like one of these functions is executed. Also, since the samples centers to update the database includes numeric variables as well as variables hands are also needed to separate functions for updating values of these variables to be written in cluster centers.

For the first solution, a function is written as addcase.m and for the second function, modify.m. The third strategy was implemented by deleting.m function. In addition to the above measures, cluster members and cluster centers are also updated. What is noteworthy to note that the changes imposed by the expert opinion and add case to the base in the testing phase to a database model to achieve the optimum done and updated expert opinion in validation phase and don't perform.

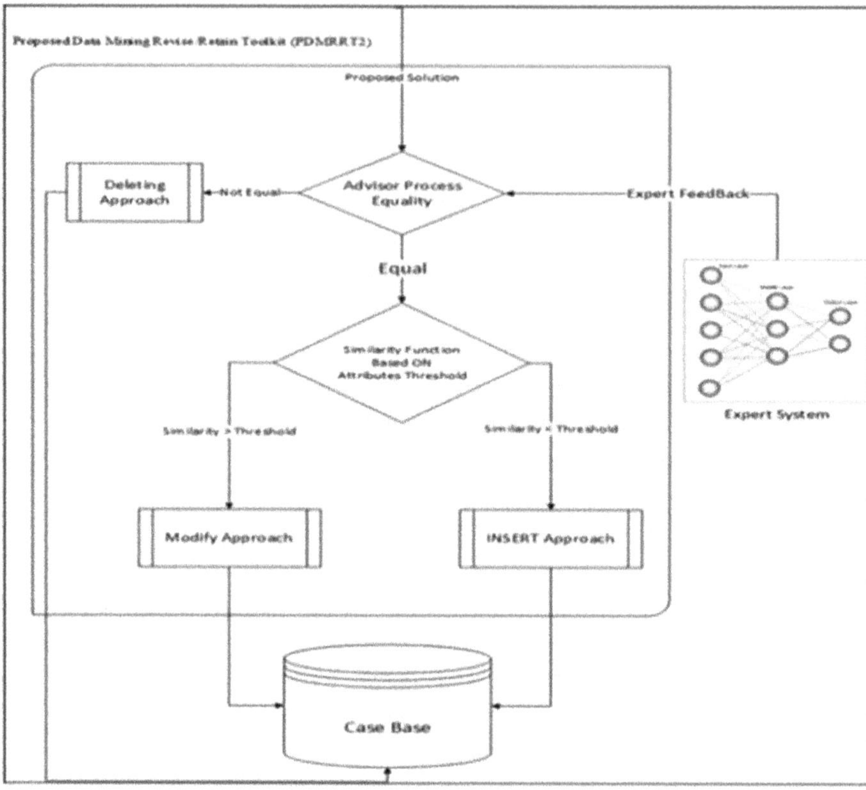

Fig. 9. CBR advisor module architecture.

3 Conclusion

In this paper, a novel model has been proposed based on an object-oriented framework with the supplier selection. This model which leads to optimal selection and ranking of suppliers, reduced the time and cost in the selection process and also reduced human errors by using data mining techniques and neural networks in the reasoning method cycle based on the case. The results of the proposed model are compared with several other models. Results show that using reduced errors, improved accuracy, and efficiency the proposed model has been able to have a good performance in the supplier selection. Also the proposed model has a CBR framework that includes modules for creating and developing the database, the module is responsible for strategy and module amendments which the tools and data mining techniques have been used in each of these modules.

The creation and development of base module about the rules extracted from decision tree trained by the training data set based on the decision variables to develop the database of metadata to suppliers is used. In addition, MLP neural network module is taught as an expert. In the nearest recovery solution for recovery module creates a new sample and to provide the base were proposed approach to clustering. In clustering, SOM combined technique is used based on K-Prototype and also according to the Davies-Bouldin index, the number of clusters were selected. Finally, the Corrective module according to the proposed strategy, expert opinion and decision variables one of the approaches. Add the similarity threshold, modification and removal is taken. According to the obtained results it can be stated that:

- Reducing the impact of human error by replacing it with expert system based on neural network is another useful outcome of this study to explore it.
- The decision tree is created based on the decision variables based on the weight and characteristics are obtained dependency. As a result, rules of GPA increases the reliability of the base and create more efficiency.

To evaluate the result of the operation model with the aim of practical application in the oil and gas industry is prepared and implemented, the two tangible results of the model include the discovery of an effective decision variable and also retrieve an ideal supplier of the Chartered Company, were examined by the experts.

The results achieved in the first, showing the 8 decision variables extracted from the model to the experts, they number six variables of the decision were approved. Experts have disagreement among them about the remaining variables in the model proposed code status and role in creating rules. With further study on mining legislation have shown that most of the rules of its front have the code status, In Tully Law, the provider has been selected as the winner. The importance of this intention to experts was offered.

References

1. De Boer, L., Labro, E., Morlacchi, P.: A review of methods supporting supplier selection. Eur. J. Purchasing Supply Manage. **7**(2), 75–89 (2001)
2. Hong, G.H., Ha, S.H.: Evaluating supply partner's capability for seasonal products using machine learning techniques. Comput. Ind. Eng. **54**(4), 721–736 (2008)
3. Timmerman, E.: An approach to vendor performance evaluation. IEEE Eng. Manage. Rev. **15**(3), 14–20 (1987)
4. Weber, C.A., Current, J.R., Desai, A.: Non-cooperative negotiation strategies for vendor selection. Eur. J. Oper. Res. **108**(1), 208–223 (1998)
5. Weber, C.A., Desai, A.: Determination of paths to vendor market efficiency using parallel coordinates representation: a negotiation tool for buyers. Eur. J. Oper. Res. **90**(1), 142–155 (1996)
6. Weber, C.A., Current, J., Desai, A.: An optimization approach to determining the number of vendors to employ. Supply Chain Manage. Int. J. **5**(2), 90–98 (2000)
7. Liu, J., Ding, F.Y., Lall, V.: Using data envelopment analysis to compare suppliers for supplier selection and performance improvement. Supply Chain Manage. Int. J. **5**(3), 143–150 (2000)
8. Parthiban, P., Zubar, H.A., Katakar, P.: Vendor selection problem: a multi-criteria approach based on strategic decisions. Int. J. Prod. Res. **51**(5), 1535–1548 (2013)
9. Holt, G.D.: Which contractor selection methodology? Int. J. Project Manage. **16**(3), 153–164 (1998)

10. Lin, R.H., Chuang, C.L., Liou, J.J., Wu, G.D.: An integrated method for finding key suppliers in SCM. Expert Syst. Appl. **36**(3), 6461–6465 (2009)
11. Humphreys, P., McIvor, R., Chan, F.: Using case-based reasoning to evaluate supplier environmental management performance. Expert Syst. Appl. **25**(2), 141–153 (2003)
12. Choy, K.L., Lee, W.B., Lau, H., Lu, D., Lo, V.: Design of an intelligent supplier relationship management system for new product development. Int. J. Comput. Integr. Manuf. **17**(8), 692–715 (2004)
13. Choy, K.L., Lee, W., Lo, V.: Design of a case based intelligent supplier relationship management system—the integration of supplier rating system and product coding system. Expert Syst. Appl. **25**(1), 87–100 (2003)
14. Lee, E.K., Ha, S., Kim, S.K.: Supplier selection and management system considering relationships in supply chain management. IEEE Trans. Eng. Manage. **48**(3), 307–318 (2001)
15. Liu, Y., Yu, F., Su, S.Y., Lam, H.: A cost–benefit evaluation server for decision support in e-business. Decis. Support Syst. **36**(1), 81–97 (2003)
16. Wang, G., Huang, S.H., Dismukes, J.P.: Product-driven supply chain selection using integrated multi-criteria decision-making methodology. Int. J. Prod. Econ. **91**(1), 1–15 (2004)
17. Çebi, F., Bayraktar, D.: An integrated approach for supplier selection. Logistics Inf. Manage. **16**(6), 395–400 (2003)
18. Kilic, H.S.: An integrated approach for supplier selection in multi-item/multi-supplier environment. Appl. Math. Model. **37**(14–15), 7752–7763 (2013)
19. Kumar, S.K., Tiwari, M.K., Babiceanu, R.F.: Minimisation of supply chain cost with embedded risk using computational intelligence approaches. Int. J. Prod. Res. **48**(13), 3717–3739 (2010)
20. Kumar, M., Vrat, P., Shankar, R.: A fuzzy goal programming approach for vendor selection problem in a supply chain. Comput. Ind. Eng. **46**(1), 69–85 (2004)
21. Kumar, M., Vrat, P., Shankar, R.: A fuzzy programming approach for vendor selection problem in a supply chain. Int. J. Prod. Econ. **101**(2), 273–285 (2006)
22. Kumar, A., Jain, V., Kumar, S.: A comprehensive environment friendly approach for supplier selection. Omega **42**(1), 109–123 (2014)
23. Weber, C.A., Current, J.R., Benton, W.C.: Vendor selection criteria and methods. Eur. J. Oper. Res. **50**(1), 2–18 (1991)
24. Das, A., Narasimhan, R., Talluri, S.: Supplier integration—finding an optimal configuration. J. Oper. Manage. **24**(5), 563–582 (2006)
25. Chai, J., Liu, J.N., Ngai, E.W.: Application of decision-making techniques in supplier selection: a systematic review of literature. Expert Syst. Appl. **40**(10), 3872–3885 (2013)
26. Saaty, T.L.: What is the analytic hierarchy process? Springer (1988)
27. De Boer, L., van der Wegen, L., Telgen, J.: Outranking methods in support of supplier selection. Eur. J. Purchasing Supply Manage. **4**(2–3), 109–118 (1998)
28. Roodhooft, F., Konings, J.: Vendor selection and evaluation an activity based costing approach. Eur. J. Oper. Res. **96**(1), 97–102 (1997)
29. Monczka, R.M., Trecha, S.J.: Cost-based supplier performance evaluation. J. Purchasing Mater. Manag. **24**(1), 2–7 (1988)
30. Degraeve, Z., Labro, E., Roodhooft, F.: An evaluation of vendor selection models from a total cost of ownership perspective. Eur. J. Oper. Res. **125**(1), 34–58 (2000)
31. Gaballa, A.A.: Minimum cost allocation of tenders. J. Oper. Res. Soc. **25**(3), 389–398 (1974)
32. Ghodsypour, S.H., O'brien, C.: The total cost of logistics in supplier selection, under conditions of multiple sourcing, multiple criteria and capacity constraint. Int. J. Prod. Econ. **73**(1), 15–27 (2001)

33. Weber, C.A., Current, J.R.: A multiobjective approach to vendor selection. Eur. J. Oper. Res. **68**(2), 173–184 (1993)
34. Zhao, K., Yu, X.: A case based reasoning approach on supplier selection in petroleum enterprises. Expert Syst. Appl. **38**(6), 6839–6847 (2011)
35. Li, H., Sun, J.: Case-based reasoning ensemble and business application: a computational approach from multiple case representations driven by randomness. Expert Syst. Appl. **39**(3), 3298–3310 (2012)
36. Schank, R.C.: Dynamic Memory: A Theory of Reminding and Learning in Computers and People, vol. 240. University Press, Cambridge (1982)
37. Aamodt, A., Sandtorv, H.A., Winnem, O.M.: Combining case based reasoning and data mining-a way of revealing and reusing rams experience. In: Safety and Reliability; Proceedings of ESREL, vol. 98, pp. 16–19 (1998)
38. Li, S., Ragu-Nathan, B., Ragu-Nathan, T.S., Rao, S.S.: The impact of supply chain management practices on competitive advantage and organizational performance. Omega **34**(2), 107–124 (2006)
39. Huang, G., Li, X., He, J., Li, X.: Data mining via minimal spanning tree clustering for prolonging lifetime of wireless sensor networks. Int. J. Inf. Technol. Decis. Making **6**(02), 235–251 (2007)
40. Kim, K.S., Han, I.: The cluster-indexing method for case-based reasoning using self-organizing maps and learning vector quantization for bond rating cases. Expert Syst. Appl. **21**(3), 147–156 (2001)
41. Mosaddar, D., Shojaie, A.A.: A data mining model to identify inefficient maintenance activities. Int. J. Syst. Assur. Eng. Manage. **4**(2), 182–192 (2013)
42. Shearer, C.: The CRISP-DM model: the new blueprint for data mining. J. Data Warehousing **5**(4), 13–22 (2000)

Effect the Number of Reservations on Implementation of Operating Room Scheduling with Genetic Algorithm

Tunahan Timuçin(✉) and Serdar Biroğul

Department of Computer Engineering, Duzce University, 81620 Düzce, Turkey
{tunahantimucin, serdarbirogul}@duzce.edu.tr

Abstract. In this paper, the problem of the most efficient use of the Operating Rooms (ORs) which one of the most important departments of hospitals, was tackled. Efficient use of operating rooms is a scheduling problem with many constraints. This type of problem is defined as NP-Hard. Complex problems involving multiple constraints are defined as NP-Hard type problems. As the NP-Hard type problem does not consist of polynomial values, the solution of such problems becomes complicated. Such problems cannot be solved by classical mathematical methods. For the solution of NP-Hard type problems which have high level of complexity and many constraints, heuristic and meta-heuristic algorithms such as Genetic Algorithm (GA), tabu search algorithm, simulated annealing algorithm and partical swarm optimization algorithm have emerged. In this paper, the operating room scheduling problem is solved by the genetic algorithm. When coding the program, the C# programming language was preferred because of the visual advantages and user-friendliness of the language.

Keywords: Operating Room Scheduling · Genetic Algorithm · Constrained optimization

1 Introduction

Hospitals are one of the most important enterprises of the changing world balance and increasing world population. Hospitals and other health care organizations are experiencing a serious pressure on providing the highest quality service at the lowest possible cost, with the increasing world population, the growing average age and the number of patients in the hospitals, and the increasing number of patients on the waiting list. The most allocated department in hospitals is Operating Rooms. The number of surgeries per year is increasing due to the increase in world population.

The number of surgeries performed annually in Turkey from 2002 until 2014 in the calculations made has already increased by about 8 times [1]. Hospitals use a variety of methods to reduce waiting time of patients, to determine the number of technical team and equipment accurately and to minimize the cost in case of surplus number of patients. One of these ways is to solve this problem by applying the scheduling method

D. J. Hemanth and U. Kose (Eds.): ICAIAME 2019, LNDECT 43, pp. 252–265, 2020.
https://doi.org/10.1007/978-3-030-36178-5_20

with the help of software programs. In this century, where technology and electronic tools are so developed, the importance of Operating Room Scheduling is increasing.

Scheduling is the timing of all activities and steps required to produce a product or to deliver a service. Some of the optimization problems in the literature can be solved by classical mathematical methods and algorithms. Increasing the size and the complexity of the problem, make it impossible to solve the problems with classical methods. Therefore, various heuristic and meta-heuristic algorithms have emerged to solve these complex problems. Meta-heuristic is the general-purpose heuristic method designed to guide problem-specific intuitions to perform searches in areas of high-quality solutions to the search space [2]. Meta-heuristic cover a wide range from simple local search algorithms to sophisticated and complex learning processes. These algorithms are; Genetic Algorithm, Simulated Annealing, Tabu Search, Ant Colony, Variable Neighbor Search algorithms such as [3].

2 Operating Room Scheduling and Optimization

It was mentioned that the operating rooms are one of the most important and most resource-allocated units of hospitals. In this section, contains how to find the solutions to the operating room scheduling problems in the literature, and which parameters are important for this problem.

Marques et al. [4] proposed an integer linear programming model for the timing of separation of weekly surgery and elective surgery to maximize the operating room profession, ensure the most efficient use of operating rooms, minimize hospital costs and increase profit. In their article, the similar study by Guido et al. [5], for each surgeon, they performed the daily and weekly limitation of the work schedule limitation, and they applied this as a constraint to the surgeons not working on this time and formed their work accordingly. The same authors who proposed the linear programming model then developed a solution approach based on genetic algorithms, but no details were reported about the solved samples [6]. Molina-Pariente et al., tried to make a weekly operating room comparison with 17 meta-heuristic algorithms by scheduling and tried to present the most efficient algorithm. In doing so, they applied the clinical prioritization method, which was tried only 1 time before and tried to minimize the waiting time [7]. Saadouli et al. evaluated the operation room planning on two sources as operating room and rescue beds. For the solution of the problem, a backpack model was used for the selection of intraday operations, and a mixed integer programming model was used to optimize the timing of the operating rooms [8]. Riise et al. developed their own different Adaptive Structure and Improvement and search algorithm and applied them in three different situations as daily, weekly and patient acceptance [9].

3 Genetic Algorithm

Genetic Algorithm, in short definition, is defined as an optimization algorithm based on the parameter coding principle and aims to reach a solution with random search techniques. Used to find the most specific and best solution within a data set. Genetic Algorithm was first mentioned by Bagley (1967)). Although this was the beginning of the genetic algorithm, the real evolution of the genetic algorithm was started after John H. Holland's proposed machine. Holland explained how the principles of Genetic Algorithm can be applied to artificial intelligence problems in his book "Adaptation in natural and artificial system" [10]. Holland considered GA as an abstract model of evolution in this study. In 1989, Goldberg, a Dutch student, took GA to the Academic aspect by applying the Genetic Algorithm to the gas pipeline control in his book "Genetic algorithms in Search Optimization and Machine Learning" [11]. Today, GA is actively used in both the theoretical and practical areas in the following areas and in the sub-areas of these areas; Scheduling Problems, Routing Problems, Game Programming, Mechanical Learning, Automatic Programming and Information Systems vs.

3.1 Basic Genetic Algorithm

When coding the Genetic Algorithm, the basic steps of the algorithm as well as the genetic operators that are specific to the problem should be implemented in a loop. Before starting the coding, all the steps, criteria, pre-planning the entire plan, such as which operators will be used, increase the quality of the algorithm. It shows the pseudocode of the genetic algorithm (see Fig. 1).

```
Genetic Algorithm
{
    Create Initial Population
    Set Criterias
        while(!(end of iteration) or !(healing stopped))
            {
                Evaluate chromosomes according to fitness function
                Send chromosomes to match pool
                Get the best chromosome with elitism
                Apply crossover operator to chromosomes
                Apply mutation operator to chromosomes
                Apply repair operator to each chromosomes
                Apply the Roulett Wheel method according to the fitness value of each chromosome
                Create new population from previous population
            }
                Get the best chromosome as a result
}
```

Fig. 1. Genetic algorithm pseudo code

Genetic Algorithm generally uses 4 operators.

1. Operator used to encode parameters
2. Reproductive Operator
3. Crossover Operator
4. Mutation Operator

In addition to these 4 operators, there is also a repair operator as an operator who should be applied in constrained optimization problems such as operating room scheduling in this article.

3.2 Repair Operator

In constrained optimization problems such as the problem in this paper, individuals in the population pool, after implementing crossover and mutation operators, the gene information of offsprings is lost or they have different sequences from what they should be. Therefore, the algorithm moves away from the solution space. This problem makes it necessary to pass the individuals through the Repair operator when each generation is completed. For this reason, in addition to exist operators, the Repair operator is also used. The Repair Operator corrects the deteriorating genes of chromosomes that have lost their usual sequence. Therefore, the Repair Operator is also called the Correction Operator. For example; let's assume that the parameters to be encoded are integers from 0 to 9 and 2 chromosomes are generated from these integers. It represents the genes lost as a result of crossover (see Fig. 2).

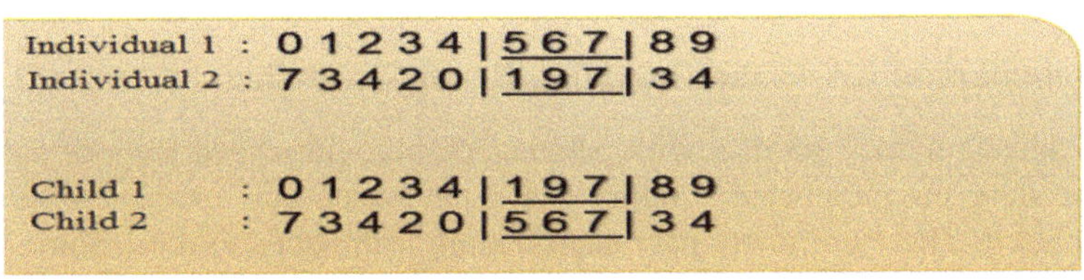

Fig. 2. Without using repair operator

After the crossover operator is applied, genes 5 and 6 are lost when the genes are transferred to the first child (see Fig. 3). In order for these genes to be replaced, excess genes removed and healthy recovery may continue, the repair operator must be implemented.

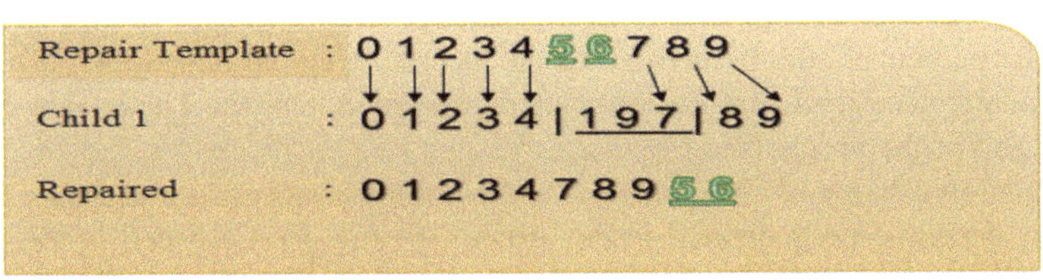

Fig. 3. Using repair operator

4 Use of GA in Constrained Optimization Problems

Optimization is the most efficient use of all resources such as labor, raw materials, capacity, equipment, money in a system. In addition, it is a technology that enables to reach the objectives such as minimizing cost, maximizing profit and maximizing capacity utilization. The constraints in optimization problems make the problem difficult. This is a new area of limited optimization problems, these problems have led to the emergence of new constrained optimization techniques.

Examples of limited optimization problems include; various examples such as Operating Room Scheduling, Nurse Scheduling, and Course Scheduling can be given. When applying Genetic Algorithm to the constrained optimization problems, the occurrence of unconstrained sequences after the classical GA operators such as crossover, mutation, makes the problem difficult.

In such problems, various techniques are reveal to overcome the problems. One of them is the advancement of the solution through the penalty function. In this method, genes or chromosomes that do not comply with the specified constraints are penalized per constraint (penalties may vary depending on the type of constraint). These penalties then appear as evaluation parameters.

4.1 Adaptation of GA to the Operating Room Scheduling

Genetic Algorithm does not deal with parameters, but with coded form of parameters. When encoding the parameters, coding technique which best expresses the solution should be used. There are many parameter coding techniques to date. Some of those; binary encoding, Permutation coding, Value coding, Tree encoding etc.

4.2 Adaptation of GA to the Operating Room Scheduling

Value Encoding

As problems such as real numbers are complicated, methods such as binary coding will be incompatible, so value coding is used for such problems. Value coding in itself; is divided into sub-branches such as Integer Coding, Fraction Coding. In this article, Integer Coding, which is one of the sub-branches of Value Coding, is used as the most appropriate parameter coding method for the Operating Room Scheduling problem. In integer coding, genes and chromosomes are expressed in integers. The integer encoding method used in the article (see Fig. 4).

The population consists of chromosomes and chromosomes consists of genes. It shows a random chromosome generated in the article (see Fig. 4). Each cell in the chromosome is called a gene. It shows 4 genes. 1 gene consists of 8 integers. The first 3 digits of these 8 integers; represents the Institution code of doctors in the hospital. The second 3 digits are; represents the types of surgery in the hospital. In this article, 6 types of surgery are included. These; pediatrics, endoscopy and radiology, general surgery, eye diseases, otolaryngology and orthopedics. Duration of surgery; can change according to the type of surgery, according to the complications that may occur at the moment, the patient's response to surgery and many factors, the type of surgery. Therefore, in

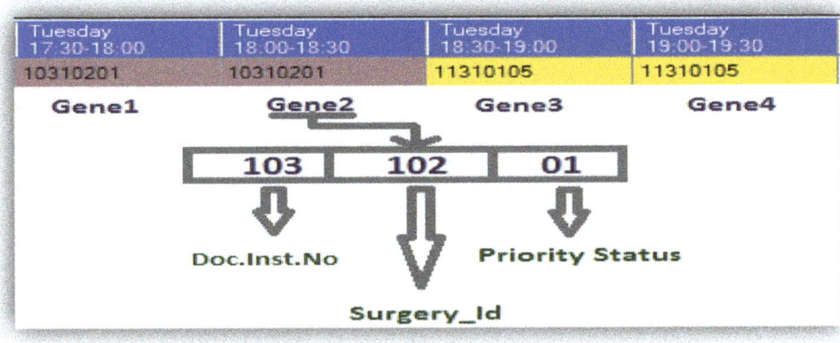

Fig. 4. Chromosome structure

this article, duration of average surgeries was determined and assigned to type of surgeries. In addition to surgery times; the total operating time was obtained by calculating the time to prepare the operating room. These values are calculated and used in this article. Table 1 shows these values.

Table 1. Duration of operation and operating room preparation times

Surgery _Name	Surgery _time	Surgery_Preparation_Time
Pediatrics	t	t
Endoscopy and radiology	t	t
General surgery	6t	2t
Eye diseases	2t	t
Otolaryngology	2t	t
Orthopedics	4t	2t

The last 2 digits of the 8 integers in the gene structure indicate the priority of the 6 surgeries used. Priority status is one of the constraints of the algorithm that has been developed and must be observed. The last 2 digits of the surgery with the highest ranking are 06 and the lowest with the operation is 01. Priorities of Operations Types are determined as follows;

General Surgery $\rightarrow$ 06 Endoscopy and Radiology $\rightarrow$ 05 Orthopedics $\rightarrow$ 04

Eye Diseases $\rightarrow$ 03 Otolaryngology $\rightarrow$ 02 Pediatrics $\rightarrow$ 01

When the operations are placed in the chart, first of all, the general surgical operation, then the endoscopy and radiology and in this order should be settled. Each gene that does not comply with this order is punished to try to ensure that the algorithm takes the most appropriate result.

4.3 Examination of the Software Program

In this article, Genetic Algorithm is used to solve the Operating Room Scheduling problem, and the Algorithm is developed using C# programming language.

Before running the program, the number of operating rooms, the number of initial chromosomes to be included in the population pool, the crossover rate to be used for the Crossover Operator and the mutation rate values to be applied for the Mutation Operator determine. Program is run by using the "Run" Algorithm button.

The number of operating rooms is between 1 and 3, the number of chromosomes is between 2 and 50, the crossover rate is between 50% and 90% and the mutation rate is between 0.01% and 0.15%. It is aimed to find the best solution by using different combinations. In addition to these parameters, the program also has a reservation section. It shows an image from the reservation screen (see Fig. 5).

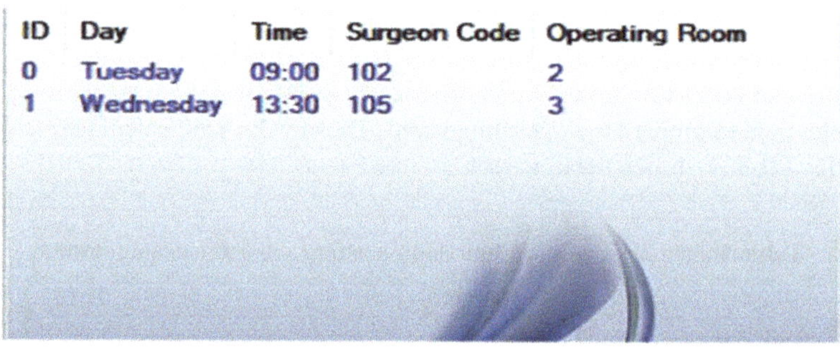

Fig. 5. Reservation screen

In this section, reservations that can be specified in advance are recorded. The doctor who will make the reservation will make a reservation on which day and hour of surgery to register. In this case, even if the number of iterations increases and the populations change, no other surgery is placed on the day and time of the reservations. The reservation section is one of the algorithm constraint parameters.

Program Constraints

The biggest difficulty in the optimization problems, which have a high number of constraints such as the Operating Room Scheduling problem, is to obtain structures that do not provide constraints and try to minimize these structures. For each gene that does not comply with the constraints, pre-determined Hard and Soft constraints have been applied, for each of the constraints, different and appropriate penalty scores were determined and the genes that did not comply with these constraints were punished by the constraint penalty coefficient. The goal is to achieve the best solution by minimizing the penalty point by using genetic algorithm.

Hard Constraints

I. The surgeries must follow the priority order (the last 2 digits of the genes indicate priority status (for example, because 06 > 05 has a priority of 06 and the operation must be performed first)).
II. A doctor cannot be present in more than one operating room at the same time.
III. In one operating room, only one patient may undergo surgery.
IV. Surgery cannot be interrupted, should keep in block.
V. The operating time cannot exceed 20.00, which is the end of the day.
VI. No doctor can be assigned to the time of reservation except for the doctor who made the reservation.
VII. To a doctor, cannot be assigned more patients than the specified number of patients.

Soft Constraints

VIII. Leisure times should be reduced as much as possible.

For example, in Figure, a part of the chromosome is shown while the program is running (see Fig. 6). It was stated that the first 3 digits of the 8 digit number in the gene indicated the doctor institution code, the next 3 digits showed the surgery code and the last 2 digits showed the surgery priority. At this point, "01, is the last 2 digits of the "10410201" gene on Wednesday. The priority value of "10010006" surgery that started after this operation is end is "06". This is a breach of the priority constraint because" "06" is high priority (01 < 06). This gene will be penalized by the number of penalties determined for the priority criterion.

	08:00	08:30	09:00	09:30	10:00	10:30
▸ Monday		10210105	10210105	10210105	10210105	10210105
Tuesday		10110006	10110006	10110006	10110006	10110006
Wed.day	10310201	10410201	10410201	10010006	10010006	10010006

Fig. 6. Examination of the constraint.

It also appears leisure time on Monday at 08.00 a.m. and Tuesday at 08.00 a.m. Leisure time has been tried to be reduced. This constraint has lower weight from the other constraints. The surgery hours were accepted as 12 h per day and between 08.00 and 20.00. As mentioned in article V. of the hard constraint clauses, failure to perform surgery outside than working hours is one of the problem constraints.

5 Experimental Results

In this section, the effect of the number of reservations on the solution of the problem has been examined in the program developed for the implementation of the algorithm. The results are shown on the graph and the representation of the scheduling structure is given.

5.1 The Effect of Number of Reservations on the Problem Solution

In C# application program, while the operating room scheduling problem is solved, it is also possible for the surgeon to make reservations easily by the user. As applied to one of the constraint of the problem, the day and hour of the operation of the doctor who has been reserved cannot be assigned to another doctor's surgery unless the reservation is canceled. The effect of the number of this feature on problem solving is shown in this section (Table 2).

Table 2. Number of reservations is equal 3.

# Operating rooms	Size of population	Crossover ration (1/100)	Mutation ration (1/1000)	#Number of reservation
3	40	85	10	3

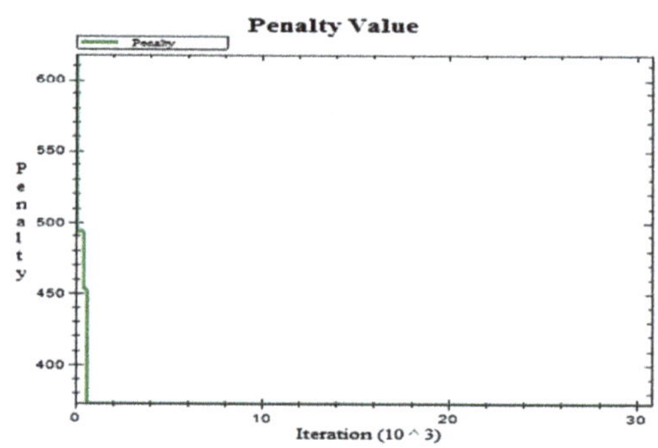

Fig. 7. Penalty value

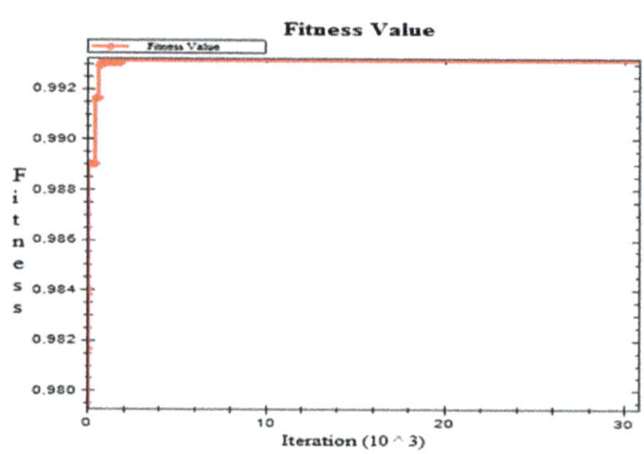

Fig. 8. Fitness value

In the first experiment, 3 surgeon made reservation and algorithm parameters, operating room number 3, crossover rate 85%, mutation rate was determined as 0.1%. As a result of 30000 iterations, as shown in Figures, the penalty value decreased to 370 value and the value of fitness increased to 0.993 (see Figs. 7 and 8).

In case of reservation; as the area of displacement of other chromosomes will decrease, it will be difficult to reduce the penalty value. The other factors affecting the trial are the number of patients assigned to doctors. In this experiment, the number of surgeon was fixed and 8, while the number of patients ranging from 2–8 was assigned to surgeons (Table 3).

The day of reservations, hours and doctors are as follows;

- Tuesday - 08.00–104 coded Doctor (Operating Room 1)
- Tuesday - 14.00–107 coded Doctor (Operating Room 1)
- Wednesday - 08.00–101 coded Doctor (Operating Room 1)

The results of these reservations and the status of the reservations are shown below (see Figs. 9 and 10).

	08:00	08:30	09:00	09:30	10:00	10:30	11:00	11:30	12:00	12:30	13:00	13:30
Monday	10110006	10110006	10110006	10110006	10110006	10110006	10110006	10110006	10010006	10010006	10010006	10010006
Tuesday	10410201	10410201	10010006	10010006	10010006	10010006	10010006	10010006	10210105	10210105	10610503	10610503
Wednesday	10110006	10110006	10110006	10110006	10110006	10110006	10110006	10110006	10010006	10010006	10010006	10010006
Thursday	10110006	10110006	10110006	10110006	10110006	10110006	10110006	10110006	10510304	10510304	10510304	10510304
Friday	10210105		10310201	10310201		10310201	10310201	10610503	10610503	10610503	10110006	10110006

Fig. 9. Solution chromosome operating room 1 (first 6 h)

	14:00	14:30	15:00	15:30	16:00	16:30	17:00	17:30	18:00	18:30	19:00	19:30
Monday	10010006	10010006	10010006	10010006	10710402	10710402	10710402	10410201	10410201	10310201	10310201	
Tuesday	10710402	10710402	10710402	10010006	10010006	10010006	10010006	10010006	10010006	10010006	10010006	
Wednesday	10010006	10010006	10010006	10210105	10210105	10610503	10610503	10610503		10310201	10310201	
Thursday	10510304	10510304	10710402	10710402	10710402	10310201	10310201	10210105	10210105			10210105
Friday	10110006	10110006	10110006	10110006	10110006	10110006	10510304	10510304	10510304	10510304	10510304	10510304

Fig. 10. Solution chromosome operating room 1 (last 6 h)

Table 3. Number of reservations is equal 8.

# Operating rooms	Size of population	Crossover ration (1/100)	Mutation ration (1/1000)	# Number of reservation
3	40	85	10	8

In the second experiment, the results of 8 surgeons were reserved and the number of operating rooms is 3, the crossing over rate is 85% and the mutation rate is 0.1%. As a result of 30000 iterations, the penalty value decreased to 450 and the value of fitness increased to 0,991 (see Figs. 11 and 12). In this experiment where the other values were kept the same as the first experiment, the decrease in the penalty value was due to the decrease of the gene change movements of the algorithm.

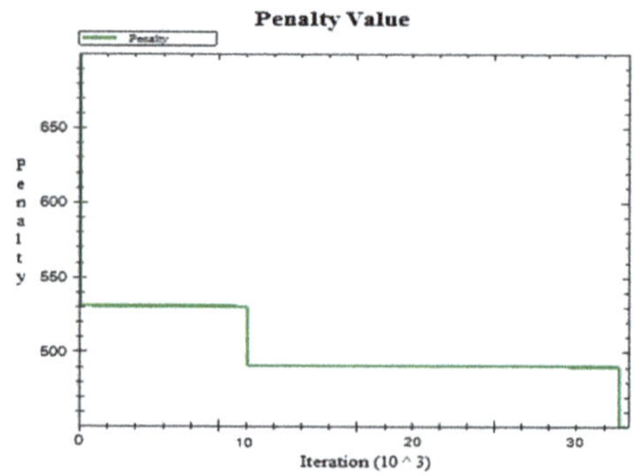

Fig. 11. Penalty value

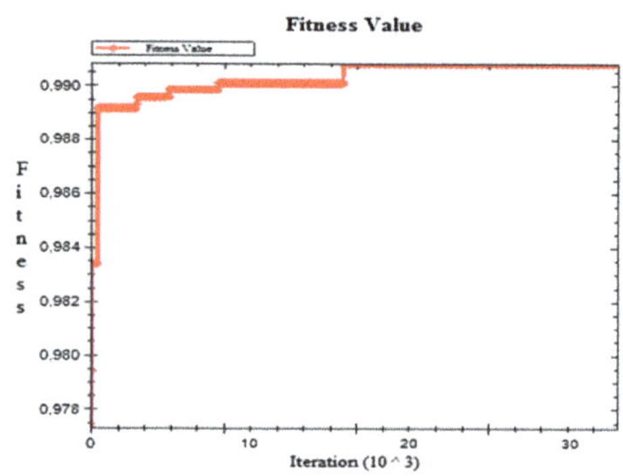

Fig. 12. Fitness value

The day of reservation, hours and doctors are as follows;

- Tuesday -08.00–104 coded Doctor (Operating Room 1)
- Tuesday - 14.00–107 coded Doctor (Operating Room 1)
- Wednesday - 08.00–101 coded Doctor (Operating Room 1)
- Friday - 12.00–100 coded Doctor (Operating Room 1)
- Monday - 14.00–107 coded Doctor (Operating Room 2)
- Tuesday - 09.00–106 coded Doctor (Operating Room 2)
- Friday - 09.30–105 coded Doctor (Operating Room 2)

The results of these reservations and the status of the reservations are shown below (see Figs. 13, 14, 15, 16 and 17).

	08:00	08:30	09:00	09:30	10:00	10:30	11:00	11:30	12:00	12:30	13:00	13:30
Monday	10510304	10510304	10510304	10510304	10510304	10510304	10610503	10610503	10610503	10610503	10610503	10610503
Tuesday	10410201	10410201	10010006	10010006	10010006	10010006	10010006	10010006	10310201	10310201	10210105	10210105
Wednesday	10110006	10110006	10110006	10110006	10110006	10110006	10110006	10110006	10010006	10010006	10010006	10010006
Thursday	10310201	10310201	10010006	10010006	10010006	10010006	10010006	10010006	10010006	10010006	10510304	10510304
Friday	10010006	10010006	10010006	10010006	10010006	10010006	10010006	10010006	10010006	10010006	10010006	10010006

Fig. 13. Solution chromosome operating room 1 (first 6 h)

	14:00	14:30	15:00	15:30	16:00	16:30	17:00	17:30	18:00	18:30	19:00	19:30
Monday	10610503	10610503	10610503	10610503	10610503	10610503					10410201	10410201
Tuesday	10710402	10710402	10710402	10410201	10410201	10310201	10310201	10210105	10210105		10210105	10210105
Wednesday	10010006	10010006	10010006	10010006	10410201	10410201	10310201	10310201	10210105	10210105	10210105	10210105
Thursday	10510304	10510304	10510304	10510304	10110006	10110006	10110006	10110006	10110006	10110006	10110006	10110006
Friday	10010006	10010006	10010006	10010006	10110006	10110006	10110006	10110006	10110006	10110006		

Fig. 14. Solution chromosome operating room 1 (last 6 h)

	08:00	08:30	09:00	09:30	10:00	10:30	11:00	11:30	12:00	12:30	13:00	13:30
Monday			10110006	10110006	10110006	10110006	10110006	10110006	10110006	10110006		
Tuesday			10610503	10610503	10610503	10510304	10510304	10510304	10510304	10510304	10510304	
Wednesday	10510304	10510304	10510304	10510304	10510304	10510304						
Thursday												
Friday				10510304	10510304	10510304	10510304	10510304	10510304			

Fig. 15. Solution chromosome operating room 2 (first 6 h)

	14:00	14:30	15:00	15:30	16:00	16:30	17:00	17:30	18:00	18:30	19:00	19:30
Monday	10710402	10710402	10710402			10510304	10510304	10510304	10510304	10510304	10510304	
Tuesday	10510304	10510304	10510304	10510304	10510304	10510304						
Wednesday												
Thursday												
Friday												

Fig. 16. Solution chromosome operating room 2 (last 6 h)

	08:00	08:30	09:00	09:30	10:00	10:30	11:00	11:30	12:00	12:30	13:00	13:30
Monday												
Tuesday												
Wednesday	10610503	10610503	10610503									
Thursday												
Friday												

Fig. 17. Solution chromosome operating room 3 (first 6 h)

6 Conclusion and Evaluation

Operating Room Scheduling problem should be evaluated according to the level of complexity. There are significant constraints in determining complexity levels. These features can be listed as follows; Total Number of Constraints, Number of Operating Rooms, Number of Surgeons, size and creation of Initial Population, Crossing and Mutation Ratios, Parameter Encoding Type, Selection Mechanism, Function of Repair Operator.

In this article, Genetic Algorithm is used to solve the operating room scheduling problem or find the closest solution. GA, one of the optimization algorithms; trying to reach the exact solution, even if it cannot find the best solution can offer close solutions. In this article, a population of 2–50 was tested and it was observed that the most suitable solution was the initial population of 40. Another important factor is the crossover and mutation rates. In the experiments, it was found that it is best to keep the crossover rate between 50% and 90%. In this article, experiments were conducted in the range specified in the literature and the most appropriate value was 85%. Another feature, the mutation rate is generally between 0.01% and 0.15%. The best result in this interval was 0.10%.

In the constrained NP-hard scheduling problems, such as the operating room scheduling, the emergence of genes that are lost, degenerated, or defective in the application of genetic operators suggests the necessity to correct these genes. The repair operator directly affects the solution of the problem. If the repair operator is not applied, it is shown that the solution situation has become and the necessity to apply correction.

Duration of surgery; it can vary for many reasons. These; complications during surgery, personal abilities of doctors, etc. In order to make optimization, the average duration of these operations was determined as fixed value and the problem was included in this form. Another neglected issue is emergency surgery. Emergency surgery was resolved by referral to another hospital.

The structure of this study will be tried to be applied to the Epigenetic Algorithm which has recently started to take place in the literature. The results of this study and the results of the Epigenetic Algorithm will be compared and examined which one will give better results.

References

1. A-b-c Grubu toplam ameliyatlar. http://rapor.saglik.gov.tr/istatistik/rapor/index.php. Accessed 27 Oct 2018
2. Dorigo, M., Stutzle, T.: Ant colony optimization for NP-Hard problems. In: Ant Colony Optimization, 1st ed. ch. 5, pp. 167–181. Springer, Boston (2004)
3. Engin, O., Fığlalı, A.: Akış tipi çizelgeleme problemlerinin genetik algoritma yardımı ile çözümünde uygun çaprazlama operatörünün belirlenmesi. Doğuş Üniversitesi Dergisi, c. 3, s. 2, pp. 27–35 (2002)
4. Marques, I., Captivo, M., Vaz Pato, M.: An integer programming approach to elective surgery scheduling. Oper. Res. Spectrum **34**(2), 407–27(2012)

5. Conforti, D., Guerriero, F., Guido, R.: A multi-objective block scheduling model for the management of surgical operating rooms: New solution approaches via genetic algorithms. In: Proceedings of IEEE Workshop on Health Care Management (WHCM), Venice, Italy (2010)
6. Marques, I., Captivo, M., Vaz Pato, M.: Planning elective surgeries in a portuguese hospital: study of different mutation rules for a genetic heuristic. In: Lecture Notes Management Science, Netherlands (2012)
7. Molina-Pariente, J.M., Hans, W.E., Framinan, J.M., Gomez-Cia, T.: New heuristics for planning operating rooms. Comput. Ind. Eng. **90**, 429–443 (2015)
8. Hadhemi, S., Badreddine, J., Abdelaziz, D., Lotfi, M, Abir, B.: A stochastic optimization and simulation approach for scheduling operating rooms and recovery beds in an orthopedic surgery department. Comput. Ind. Eng. **80**, 72–79 (2015)
9. Riise, A., Mannino, C., Burke, E.K.: Modelling and solving generalised operational surgery scheduling problems. Comput. Oper. Res. **66**, 1–11 (2016)
10. Holland, J.H.: Adaptation in Natural and Artificial Systems: An Introductory Analysis with Applications to Biology, Control, and Artificial Intelligence. England, Oxford (1975)
11. Golberg, D.E.: Genetic Algorithms in Search, Optimization, and Machine Learning. Boston, MA (1989)

Identifying Driver Behaviour Through Onboard Diagnostic Using CAN Bus Signals

Gül Fatma Türker[1(✉)] and Fatih Kürşad Gündüz[2]

[1] Department of Computer Engineering, Suleyman Demirel University, Isparta, Turkey
gulturker@sdu.edu.tr

[2] Department of Computer Engineering, Bartın University, Bartin, Turkey
fatihgunduz@bartin.edu.tr

Abstract. Nowadays, traffic accidents occur due to the increasing number of vehicles. In the researches, it was determined that most of the accidents were caused by the driver. Audible and visual warnings of drivers against possible situations in traffic will reduce the risk of errors and accidents. it was observed that the traffic signs were not enough stimuli for the drivers. For this reason, stimulating electronic applications are developed for drivers in Intelligent Transport Systems. The selection of the correct stimulators by measuring the response of the drivers to different situations in different road conditions will provide a more efficient driving. For this purpose, in order to evaluate the driving behavior of the driver in this study, the speed and RPM information received by means of OBD (Onboard Diagnostic) access to the ECU (Electronic Control Unite) data of the vehicle was evaluated instantaneously. Thus driving information provides aggressive driver detection and warns of traffic hazard situations. For this purpose, an experimental system was created by using machine learning algorithms. The vehicle's speed and RPM data have been used to determine the acceleration of the vehicle and drive. Four different types of drivers have been identified in this designed system. In this way, the driver will be able to detect their own driving. Research will be carried out on how to influence traffic flow by identifying aggressive driver behaviors. It is foreseen that some of the accidents caused by the driver can be prevented.

Keywords: Aggressive driver · OBD · CAN bus · Vehicle RPM · Vehicle speed · Diagnostic

1 Introduction

Intelligent transportation system (ITS) applications are supported with different technologies and produce solutions to many traffic problems. Intelligent Transportation Systems supports drivers about vehicle information and drivers in case of dangerous situation [1]. Smart phones, which plays an important role in intelligent transportation system, are able to make data sharing, vehicle safety, driving safety, road safety and traffic control because they support the web. Thanks to the development of smartphones, drivers are warned of possible traffic accident accidents, instant traffic tracking and monitoring of forecast reports [2]. Vehicles are equipped with many sensors and

D. J. Hemanth and U. Kose (Eds.): ICAIAME 2019, LNDECT 43, pp. 266–275, 2020.
https://doi.org/10.1007/978-3-030-36178-5_21

ECUs (Embedded Control Units) [3]. CAN-bus (Controller Area Network) provides in handling the entire communication between the sensors, ECUs and actuators. Smart phones can be accessed via CAN On-Board Diagnostic (OBD-II) systems [4, 5].

Evaluation of drivers and vehicles that affect road traffic is very important [6]. Investigations are conducted to reduce errors and accidents caused by driver behavior [7, 8]. Drivers can make driving dangerous. More than 50% of traffic accidents are due to the use of aggressive driving [9]. Numerous studies are being developed, especially for aggressive driver detection [10–12].

Shinar and Compton for aggressive driver detection over 2000 behaviors were observed over a total of 72 h at six different sites. The behaviors were observed like cutting across one or more lanes in front of other vehicles. It was found that driver behaviour changes caused congestion. A link between congestion and the frequency of aggressive behaviour was detected [13].

In-vehicle communication technology, Controller Area Network signals are used for studies in traffic [14]. Zardosht et al. were categorized driver behavior in Preturning Maneuvers using in vehicle CAN bus signals. CAN bus data streams such as vehicle speed, gas pedal pressure, brake pedal pressure, steering wheel angle, and acceleration were collected and analysed. They considered all turns for each driver and extract statistical features from the signals and used cluster analysis to categorize drivers into groups reflecting different driving styles [15]. Karaduman et al. have developed an algorithm for determining the most appropriate variables of the CAN bus data for the aggressive/calm driving detection problem [16]. Taha And Nasser detected road conditions (potholes, speedbumps, slowdowns, etc.) via CAN bus and smartphones to enforce safe and responsible driving [17].

Driver behavior creates differences in vehicle usage. It directly affects the vehicle's fuel consumption and vehicle transportation safety [18, 19]. Thanks to the OBD II technology in the vehicle, many data such as speed and RPM data can be obtained. Thus, studies are carried out with data that reflects vehicle status to assess driver behavior. [20–22].

In this study, the speed and RPM data received from the CAN bus of the vehicle were processed to evaluate the driver behavior. In this system, the driving style data of the vehicle is taken over the ECU through OBD. OBD socket device which established a bluetooth connection with smart phone has been developed and time of entering the service of the driver was informed by email via the internet. Unsafe driving behaviours are being improved to minimize the number of the accidents. This study is organized; Sect. 2 presents on-board diagnostic systems. In Sect. 3, driving classification methods. In Sect. 4, the simulation results have presented clearly. In section conclusion, K-Neighbors algorithm provided 93% accuracy in the experiments.

2 On-Board Diagnostic Systems

On-Board Diagnostic systems provide access to the vehicle's sub-systems, sensor information, and various data about the engine. With OBD technology, data collection, driver information about the vehicle, monitoring of the vehicle's error codes are easily performed [23]. Data are collected by the communication of the vehicle Electronic

Control Unit (ECU) with the OBD [24] (Chen et al., 2009). Some software and hardware supporting OBD is still being developed for various applications [25–28].

Control Area Network (CAN) protocol is designed to generate a powerful serial data transmission in automotive applications. The CAN nodes can be connected with a data transmission speed of 1 Mbit/s and a bus with a data transmission speed of 40 m and a data rate of 40 Kbit/s over a 1000 m bus. The CAN bus has been the only authorized interface for the OBD since 2008 and is used by nearly all personal and commercial vehicle manufacturers [29].

The hardware part uses the ELM327 diagnostic device, which provides wireless communication to support the OBD-II standard for CAN bus access. The ELM327 OBD-II diagnostic device allows you to receive information from vehicle sensors and display them on the PC and mobile devices. Various protocols are used for ELM327 OBD-II. Wi-Fi protocols supported by the ELM327 OBD-II diagnostic device: ISO15765-4 (CAN), ISO14230-4 (KWP2000), ISO9141-2, J1850 VPW, J1850 PWM. In this study, drivers were monitored instantaneously in order to be able to distinguish the driving styles. A dataset with the speed and RPM of the vehicle is created.

3 Material and Method

3.1 Artificial Neural Networks

Artificial Neural Networks (ANN) inspired by the brain function of people, can learn by trial. It can reveal the unknown and discernible relations between the data. ANN is an ideal algorithm for problems where the solution is not a linear problem. However, non-linear models are not suitable if there is no linear relationship between the problem and the solution [30].

ANN does not need any relationship between input and output [31]. ANN enables modeling without any hypothesis. First, input and output information is given to the system, the relationship between them is learned and the network is trained. This model, which is instructional learning, is generally preferred [32].

Multilayer artificial neural network model was used in the study. The system consists of three layers, the input layer, the hidden layer and the output layer.

- Input Layer: The required layer for entering data.
- Hidden Layer: This is the section where the entered data is processed. Many problems can be solved with one intermediate layer. If the relationship between the input and output of the problem is not linear, more than one hidden layer can be used.
- Output Layer: The layer that gives out the information coming from the interface [33].

The number of machining elements in the input and output layers is determined according to the problem. The handlers in the input layer depend on the handlers in the hidden layer, and the handlers in the output layer depend on it. The information flow is forward (Fig. 1).

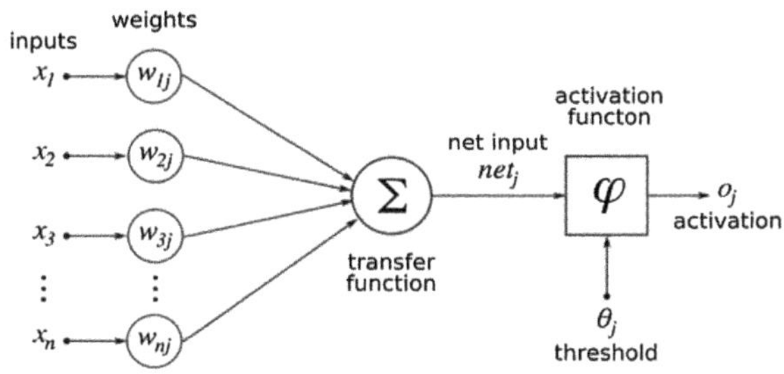

Fig. 1. Artificial neural cell [34]

$$y_i = \mathrm{f}\left(\sum\nolimits_{j=1}^{n} x_j * w_{ji}\right) \quad (1)$$

$$E(w) = 1/2\Sigma_k(t_k - 0_k)^2 \quad (2)$$

Equation (1) formulates feed forward, equality (2) formulates feedback.
Equality 1, Xj; j. the value of the neuron,
yi; i. value transferred to the neuron.

3.2 Naive Bayes

The Naive-Bayes algorithm calculates a probability set by Thomas Bayes and makes the classification. It works with a defined calculation according to the principles of probability. For this purpose, data from a certain rate is presented to the system. The algorithm calculates the probability values according to this data and determines the category of the data. The more data, the more categories can be identified.

Qualifications are independent from each other in this classifier. All data samples are accepted at the same level. Whenever he wants to classify any sample, he calculates which class belongs to which class he uses by using Eq. 1. The highest probability of these values is considered an instance of that class.

$$P(S_i/X) = \mathrm{P}(X/S_i) * P(S_i)/(P(X)) \quad (3)$$

$P(S_i/X)$: Probability of event Si when X event occurs
$P(X/S_i)$: Probability of event X when Si event occurs
$p(S_i), \mathrm{P}(X)$: $S_i$veX: The possibility of events

Because each X sample is evaluated at the same level, P (X) is the same for each sample.

$$P(S_i/X) = \mathrm{P}(X/S_i) * P(S_i)$$

For each class, Eq. 3 is applied and each class belongs to which class [35].

3.3 Decision Tree

Decision tree learning algorithm is a frequently used algorithm for classification and prediction. Decision trees are very suitable for classification in terms of easy interpretation and intelligibility. The decision tree technique consists of 2 steps: classification of data, learning and classification. In the learning step, the training data is analyzed by the algorithm to perform modeling data in advance. Rules are created for the data to be classified. The classification data are classified according to the rules for classification [36].

Decision trees is a controlled machine learning algorithm. Continuously divides data up to a certain point. The tree consists of two concepts; leaves and decision nodes. The decision decides the nodes and the leaves are the results. From this aspect, the decision trees are similar to flowcharts. The hit rate is much better than other methods.

Methods of creating decision trees consist of 2 stages.

Creating a tree: Examples in the learning set are divided recursively depending on the properties in the root

Tree pruning: To get rid of incorrect data in the learning set.

The decision tree can deepen each branch when classifying its data. This method creates a memorization by allowing too much division between data. As a result, a great decision tree is formed and memorizes.

Pruning
Pruning is performed to reduce errors in the data. Branches of low importance are pruned. Thus, memorization of trees is prevented and predictive power increases.

There are 2 methods of pruning.

1. Early Pruning: An approach that stops early growth of the tree.
2. Late Pruning: The process is carried out after the completion of the tree.

3.4 K-Nearest Neighbors (KNN) Algorithm

The KNN algorithm is an example-based classification algorithm. In sample based learning algorithms, the learning process classifies based on the data held in the training set. A new example is compared to the examples in the training set and classified according to the nearest example. In the algorithm, when a new sample is introduced into the system, the distance closest to K is calculated and the new value is added to a set. In the calculation of distance, distance calculation methods such as euclidean distance, distance from manhattan can be used which are used k-means and hierarchical clustering. Euclidean distance will be used in this example [37].

This algorithm consists of five steps.

1. Determine the K value.
2. Calculate distances from other objects to the target object.

3. Distance is sorted and the nearest neighbors are located depending on the minimum distance.
4. The nearest neighbor categories are collected.
5. The most appropriate neighboring category is selected.

4 Aggressive Driver Classification Results

4.1 Assessment Criteria

The classification algorithms used in the study will be evaluated according to accuracy, precision and Recall criteria. This concept called F-measure is used as a measure of performance in data sciences.

Accuracy: Ratio of correctly predicted data to all data.
Precision: Ratio of correct estimates
Recall: Finding correct estimates
F-1 Score: The harmonic mean of the precision and recall values.

Complexity matrix: The complexity matrix is a measurement tool that provides information about the accuracy of the estimates.

Sklearn Library python-supported, large-scale machine learning algorithms can be used in all problems as medium-scale consultant and consultant. The library offers general use for those who are not experts in machine learning. The results were evaluated with Accuracy, Recall, Precision and F1-Score.

Artificial Neural Networks (Tables 1 and 2):

Table 1. Evaluation criteria of ANN

		Precission	Recall	F1-Score
Quiet Driver	1	0.72	0.96	0.82
Mid Level Quiet Driver	2	0.81	0.93	0.77
No Quiet Driver	3	0.63	0.91	0.74
Aggresive Driver	4	0.96	0.99	0.98

Table 2. ANN rate of hit and complexity matrix

Accuracy Score %81					
Confusion Matrix Predicted		1	2	3	4
Quiet Driver	1	[160	0	4	3]
Mid Level Quiet Driver	2	[51	0	88	8]
No Quiet Driver	3	[10	0	160	6]
Aggresive Driver	4	[1	0	3	456]

Naive Bayes (Tables 3 and 4):

Table 3. Evaluation criteria of Naive Bayes

		Precission	Recall	F1-Score
Quiet Driver	1	0.71	0.96	0.82
Mid Level Quiet Driver	2	0.93	0.44	0.60
No Quiet Driver	3	0.46	0.89	0.60
Aggresive Driver	4	0.96	0.65	0.77

Table 4. Naive Bayes rate of hit and complexity matrix

Accuracy Score %71		
Confusion Matrix Predicted		1 2 3 4
Quiet Driver	1	[160 2 2 3]
Mid Level Quiet Driver	2	[53 65 25 4]
No Quiet Driver	3	[11 3 157 5]
Aggresive Driver	4	[1 0 161 298]

Decision Trees (Tables 5 and 6):

Table 5. Evaluation criteria of decision trees

			Precission	Recall	F1-Score
Quiet Driver	1	1	0.85	0.89	0.87
Mid Level Quiet Driver	2	2	0.68	0.67	0.68
No Quiet Driver	3	3	0.77	0.84	0.80
Aggresive Driver	4	4	0.96	0.92	0.94

Table 6. Decision trees rate of hit and complexity matrix

Accuracy Score %85		
Confusion Matrix Predicted		1 2 3 4
Quiet Driver	1	[149 11 4 3]
Mid Level Quiet Driver	2	[13 99 26 9]
No Quiet Driver	3	[9 14 147 6]
Aggresive Driver	4	[4 21 14 421]

KNN Algorithm (Tables 7 and 8):

Table 7. Evaluation criteria of KNN

		Precission	Recall	F1-Score
Quiet Driver	1	0.93	0.93	0.93
Mid Level Quiet Driver	2	0.85	0.84	0.84
No Quiet Driver	3	0.94	0.86	0.90
Aggresive Driver	4	0.96	1.00	0.98

Table 8. KNN rate of hit and complexity matrix

Accuracy Score %93		
Confusion Matrix Predicted		1 2 3 4
Quiet Driver	1	[156 7 1 3]
Mid Level Quiet Driver	2	[8 123 8 8]
No Quiet Driver	3	[3 14 152 7]
Aggresive Driver	4	[0 1 1 458]

Experiments were performed on 1356 samples. Artificial Neural Networks, Naive Bayes, Decision Tree and KNN algorithms are used for classification algorithms. Criteria for evaluating algorithms are Accuracy, Recall, Precision and F1-Score.

5 Conclusion

In this study, 881 OBD records were tested. Artificial Neural Networks, Naive Bayes, Decision Tree and KNN algorithms algorithms were used for classification algorithms. Algorithms have been shown how to classify performance and how many drivers are classified. Only Speed and RPM were used to help in the classification of the data. 25% of this data was used as an aid in the classification of other data. The experiments were conducted with Python software language. In this experiment, KNN has become prominent in terms of performance compared to Decision trees, Neural Networks and Naive Bayes algorithms. The KNN algorithm provided 93% accuracy in the experiments.

References

1. Hadiwardoyo, S.A., Patra, S., Calafate, C.T., Cano, J.C., Manzoni, P.: An intelligent transportation system application for smartphones based on vehicle position advertising and route sharing in vehicular ad-hoc networks. J. Comput. Sci. Technol. **33**(2), 249–262 (2018)

2. Yang, Y., Chen, B., Su, L., Qin, D.: Research and development of hybrid electric vehicles CAN-Bus data monitor and diagnostic system through OBD-II and android-based smartphones. Adv. Mech. Eng. **5**, 741240 (2013)
3. Sathyanarayana, A., Boyraz, P., Purohit, Z., Lubag, R., Hansen, J.H.: Driver adaptive and context aware active safety systems using CAN-bus signals. In: 2010 IEEE Intelligent Vehicles Symposium (IV), pp. 1236–1241. IEEE (2010)
4. Avatefipour, O., Malik, H.: State-of-the-art survey on in-vehicle network communication (CAN-Bus) security and vulnerabilities. arXiv preprint arXiv:1802.01725 (2018)
5. Zaldivar, J., Calafate, C.T., Cano, J.C., Manzoni, P.: Providing accident detection in vehicular networks through OBD-II devices and android-based smartphones. In: 2011 IEEE 36th Conference on Local Computer Networks (LCN), pp. 813–819. IEEE (2011)
6. Liu, M., Chen, Y., Lu, G., Wang, Y.: Modeling crossing behavior of drivers at unsignalized intersections with consideration of risk perception. Transp. Res. Part F Traffic Psychol. Behav. **45**, 14–26 (2017)
7. Săucan, D.Ş., Micle, M.I., Popa, C., Oancea, G.: Violence and aggressiveness in traffic. Procedia Soc. Behav. Sci. **33**, 343–347 (2012)
8. McCall, J.C., Trivedi, M.M.: Driver behavior and situation aware brake assistance for intelligent vehicles. Proc. IEEE **95**(2), 374–387 (2007)
9. Bushman, B.J., Steffgen, G., Kerwin, T., Whitlock, T., Weisenberger, J.M.: "Don't you know I own the road?" The link between narcissism and aggressive driving. Transp. Res. Part F Traffic Psychol. Behav. **52**, 14–20 (2018)
10. Kovácsová, N., Rošková, E., Lajunen, T.: Forgivingness, anger, and hostility in aggressive driving. Accid. Anal. Prev. **62**, 303–308 (2014)
11. Mackey, J.J., Mackey, D.M.: U.S. Patent No. 6,392,564. Washington, DC: U.S. Patent and Trademark Office (2002)
12. Kaysi, I.A., Abbany, A.S.: Modeling aggressive driver behavior at unsignalized intersections. Accid. Anal. Prev. **39**(4), 671–678 (2007)
13. Shinar, D., Compton, R.: Aggressive driving: an observational study of driver, vehicle, and situational variables. Accid. Anal. Prev. **36**(3), 429–437 (2004)
14. Raz, O., Fleishman, H., Mulchadsky, I.: U.S. Patent No. 7,389,178. Washington, DC: U.S. Patent and Trademark Office (2008)
15. Zardosht, M., Beauchemin, S.S., Bauer, M.A.: Identifying driver behavior in preturning maneuvers using in-vehicle CANbus signals. J. Adv. Transp. **2018**, 10 (2018). (Article ID 5020648)
16. Karaduman, O., Eren, H., Kurum, H., Celenk, M.: An effective variable selection algorithm for aggressive/calm driving detection via CAN bus. In: 2013 International Conference on Connected Vehicles and Expo (ICCVE), pp. 586–591. IEEE (2013)
17. Taha, A.E.M., Nasser, N.: Utilizing CAN-Bus and smartphones to enforce safe and responsible driving. In: 2015 IEEE Symposium on Computers and Communication (ISCC), pp. 111–115. IEEE (2015)
18. Jakobsen, K., Mouritsen, S.C., Torp, K.: Evaluating eco-driving advice using GPS/CANBus data. In: Proceedings of the 21st ACM SIGSPATIAL International Conference on Advances in Geographic Information Systems, pp. 44–53. ACM (2013)
19. Yeh, C.F., Lin, L.T., Wu, P.J., Huang, C.C.: Using on-board diagnostics data to analyze driving behavior and fuel consumption. In: International Conference on Smart Vehicular Technology, Transportation, Communication and Applications, pp. 343-351. Springer, Cham (2018)
20. Zhang, M., Chen, C., Wo, T., Xie, T., Bhuiyan, M.Z.A., Lin, X.: SafeDrive: online driving anomaly detection from large-scale vehicle data. IEEE Trans. Industr. Inf. **13**(4), 2087–2096 (2017)

21. El Masri, A.E.B., Artail, H., Akkary, H.: Toward self-policing: detecting drunk driving behaviors through sampling CAN bus data. In: 2017 International Conference on Electrical and Computing Technologies and Applications (ICECTA), pp. 1–5. IEEE (2017)
22. Tu, Y., Zhang, F., Wang, Z.: Based on hidden Markov model to identify the driver lane-changing behavior of automobile OBD internet of vehicles research and design. Advances in Computer and Computational Sciences, pp. 257–263. Springer, Singapore (2018)
23. Cabala, M., Gamec, J.: Wireless real-time vehicle monitoring based on android mobile device. Acta Electrotechnica et Informatica **12**(4), 7–11 (2012)
24. Chen, Y., Xiang, Z., Jian, W., Jiang, W.: Design and implementation of multi-source vehicular information monitoring system in real time. In: IEEE International Conference on Automation and Logistics 2009. ICAL 2009, pp. 1771–1775. IEEE (2009)
25. Rahmani, M., Koutsopoulos, H.N., Ranganathan, A.: Requirements and potential of GPS-based floating car data for traffic management: Stockholm case study. In: 2010 13th International IEEE Conference on Intelligent Transportation Systems (ITSC), pp. 730– 735. IEEE (2010)
26. Jhou, J.S., Chen, S.H., Tsay, W.D., Lai, M.C.: The implementation of OBD-II vehicle diagnosis system integrated with cloud computation technology. In: 2013 Second International Conference on Robot, Vision and Signal Processing (RVSP), pp. 9–12. IEEE (2013)
27. Sourav, H., Ali, M., Mary, G.I.: Ethernet in embedded automotive electronics for OBD-II diagnostics. Int. J. Appl. Eng. Res. **8**(19), 2417–2421 (2013)
28. Baek, S.-H., Jang, J.-W.: Implementation of integrated OBD-II connector with external network. Inf. Syst. **50**, 69–75 (2015)
29. Society Automotive Engineering (SAE). http://www.sae.org. Accessed 23 Dec 2018
30. Zhang, G., Patuwo, B.E., Hu, M.Y.: Forecasting with artificial neural networks: the state of the art. Int. J. Forecast. **14**, 35–62 (1998)
31. Kaastra, I., Boyd, M.: Designing a neural network for forecasting financial and economic time series. Neurocomputing **10**, 215–236 (1996)
32. Haykin, S.: Neural Networks: A Comprehensive Foundation. Prentice Hall PTR, Upper Saddle River (1994)
33. Öztemel, E.: Yapay sinir ağlari. PapatyaYayincilik, Istanbul (2003)
34. Artificial Neural Networks. https://en.wikibooks.org/wiki/Artificial_Neural_Networks/Activation_Functions. Accessed 23 Dec 2018
35. Bermejo, P., Gámez, J.A., Puerta, J.M.: Improving the performance of Naive Bayes multinomial in e-mail foldering by introducing distribution-based balance of datasets. Expert Syst. Appl. **38**(3), 2072–2080 (2011)
36. Albayrak, A.S.: Türkiye'de Yerli ve Yabancı Ticaret Bakanlıklarının Finansal Etkinliğ Göre Sınıflandırılması: Karar Ağacı, Lojistik Regresyon ve Diskriminant Analizi Modellerinin Bir Karşılaştırılması. Süleyman Demirel Üniversitesi İktisadi ve İdari Bilimler Fakültesi Dergisi **14**(2), 113–139 (2009)
37. Mitchell, T.: Machine Learning. McGraw Hill, New York (1997)

Statistical Learning Applied to Malware Detection

Roman Rodriguez-Aguilar[1] and Jose A. Marmolejo-Saucedo[2(✉)]

[1] Escuela de Ciencias Económicas y Empresariales, Universidad Panamericana, Augusto Rodin 498, 03920 Mexico City, Mexico
rrodrigueza@up.edu.mx

[2] Facultad de Ingeniería, Universidad Panamericana, Augusto Rodin 498, 03920 Mexico City, Mexico
jmarmolejo@up.edu.mx

Abstract. This work shows an application of statistical learning methodologies in order to determine the important factors for malware detection. Support Vector Machines and Lasso Regression performed Malware classification with additional re-sampling methods. The results show that the Lasso Regression allows an efficient selection of relevant variables for the construction of the classifier, also the integration of support vector machines improves the efficiency of the classifier through the application of resampling methods. The model presented in this paper uses a statistical learning approach through the selection of variables, non-linear classification, and resampling methods.

Keywords: Malware detection · Statistical learning · Support Vector Machine · Lasso Regression · Resampling

1 Introduction

The accelerated evolution of Information and Communication Technologies has allowed the use of mobile communication devices has increased in the last two decades. Becoming a luxury to a need, allowing the diversification of its costs to access people of any kind social, opening unlimited communication channels (voice, video, and data), constant access to various social networks, joining the daily life of the vast majority of people in the world, moving from a specific operating team to one that can be used as a camera, for the management of bank accounts, such as GPS, personal agenda, email, e-book reader, music and video player, videogames, texting, etc.

The foregoing allows serious questions to be established regarding whether a team so directly involved in a person's life can also be used as a tool to affect the privacy of an individual's personal information. In addition, it is considered the Android operating system, considered today as the most used for mobile devices such as cell phones and tablets, car computers, televisions, etc. Malware detection in the antivirus industry manly was based on heuristic features that identify malware files by code fragments, hashes of code fragments, file properties and combination of these issues. However, malware programmers invented techniques like server-side polymorphism that implies

D. J. Hemanth and U. Kose (Eds.): ICAIAME 2019, LNDECT 43, pp. 276–294, 2020.
https://doi.org/10.1007/978-3-030-36178-5_22

hundreds of thousands of malicious samples being discovered every day and therefore its detection is complex [1].

It has turned towards the integration of machine learning models for the detection of malware as a promising option. [2] shows a study to compare different algorithms to classify malware and clean files. [3] developed a model of machine learning comparing structural features of the software and machine learning techniques to detect malicious executables. Recently, [4] applied techniques such as deep learning to detect malware in the framework of Internet of Things applications, the model presented is a combination of behavior graphs and neural network stacked autoencoders. [5] focus his research in evaluating classifiers to identify malware for mobile devices, using the anomaly-based approach with Bayes network and Random Forest. [6] using a set of variables to learning and classification malware behavior using as target variable malware labeled by antivirus scanner.

[7] proposes a hybrid model (semi-supervised) of malware protection detect unknown malware using a set of labeled and unlabeled instances. [8] shows a performance of the machine learning model using classification algorithms generally used (supervised and unsupervised) for the detection of malicious web content. [9] compares three supervised and two unsupervised models to detect malicious web pages and found that performance is quite equal. [10] shows an application of machine learning in headers of executable files and found the utility of analyzing the metadata of executable files to classify clean software and infected software. To date, several studies have been carried out; however, two sources were identified for the data. One is the employee for this work considered because of the dynamic analysis of the applications considered malicious [11].

The present work addresses malware detection through a statistical learning approach, combining Lasso Regression for the selection of variables, SVM (Support Vector Machines) and RF (Random Forest) as classifiers. Additionally, a resampling method was considered to improve the accuracy of the estimations. It is a proposal that consumes few computing resources and allows to dynamically updating the models over time.

2 Theoretical Background

2.1 Lasso Regression

Regression Shrinkage and selection with Lasso methodology were developed by Tibshirani (1996) is a linear model that objective is to minimize the square errors subject to a constraint (the sum of the absolute value of parameters less than or equal to a constant). The main result is that after this optimization process generates some parameters equal cero, hence give interpretable models, and as consequence allows selecting relevant variables in the model.

Consider data $(x_i, y_i), i = 1, 2, .., N$, were x_i are the predictor variables and y_i are the dependent variable. Assume that the observations x_{ij} are independent and standardized, then Lasso estimate $\left(\hat{\alpha}, \hat{\beta}\right)$ is defined by,

$$\left(\hat{\alpha}, \hat{\beta}\right) = min\left\{\sum\nolimits_{i=1}^{N} \left(y_i - \alpha - \sum\nolimits_j \beta_j x_{ij}\right)^2\right\} \text{subject to} \sum\nolimits_j |\beta_j| \leq t. \quad (1)$$

So simplifying, the estimators would be as follows:

$$\left(\hat{\beta}\right) = min\left\{\sum\nolimits_{i=1}^{N} \left(y_i - \alpha - \sum\nolimits_j \beta_j x_{ij}\right)^2\right\} + \lambda \sum\nolimits_j |\beta_j| \quad (2)$$

With $t \geq 0$ as a tuning, parameter controls the amount of shrinkage, Eq. (1) represent an optimization problem with linear constraints and quadratic objective function. Lasso regression will often generate coefficients that are exactly zero. Lasso also shrinks coefficients to reduce variance, but it also shares similarities with better subset regression, where for each number of variables we seek to choose the variables that give the best model. However, the Lasso approach is more scalable and can be calculated more simply [5].

2.2 Support Vectors Machines

The SVM is considered as part of the Supervised Learning models. Involves statistical learning theory, optimization theory, reduction of spaces by kernel functions and efficient numerical algorithms. Its development begins with a linear regression problem, where the least-squares method on a hyperplane having the minimum square of the distance between two points. Leaving the solution of a system of linear equations, which can have many solutions becoming in a more robust method against values out of range.

An optimal hyperplane for linear separate observations in the direction of the minimum point distance from the plane that begins to maximize was constructed as a generalized portrait algorithm. The idea of combining a linear algorithm with an inspired approximation core by examining specific cores for applications. Finally, an important approach is that an important factor in its implementation is the use of Quadratic Optimization. [12] The most applications have nonlinear behavior and are not linearly separable. [13] Did the analysis of hard margin classifiers to the no separable cases as a constrained linear model, soft margin classification model:

$$\frac{1}{2}\|S\|^2 + K\sum\nolimits_{i=1}^{m} \varepsilon_i \rightarrow min_{w,b,\varepsilon} \quad (3)$$

$$subject\ to\ y_i(\langle s, x_i\rangle + d) \geq 1 - \varepsilon_i, \varepsilon_i \geq 0, i = 1, \ldots, n$$

In the case of no linear learning the problem, use a kernel function and in the soft margin classification, the optimization problem is modified in the following way.

$$\frac{1}{2}\|f\|_{HK}^2 + C\sum\nolimits_{i=1}^{m} L_h(y_i, f(x_i) + b) + \rightarrow min_{f \in HK} \quad (4)$$

2.3 Bagging

The bootstrap aggregating (bagging) is a method of improving the precision of estimation. Bagging averages model predictions over bootstrap samples, thus reducing its variance. In the classification problem in place to prediction problem, the meta-estimation is defined by the frequency of all models estimated. The analysis of relevant variables in the classification problem is performed using a concentration coefficient such as the Gini coefficient [14]. For each sample, the model is fitted, obtaining an estimation or response variable. The bagging meta-estimator is:

$$\hat{f}_b(x) = \frac{1}{B}\sum\nolimits_{b=1}^{B} \hat{f}^{*b}(x) \tag{5}$$

The proposed model integrates three approaches to statistical learning, variable selection, classification and resampling for validation of estimates taking into account the feedback of the model when changing data or temporality of the same (Fig. 1).

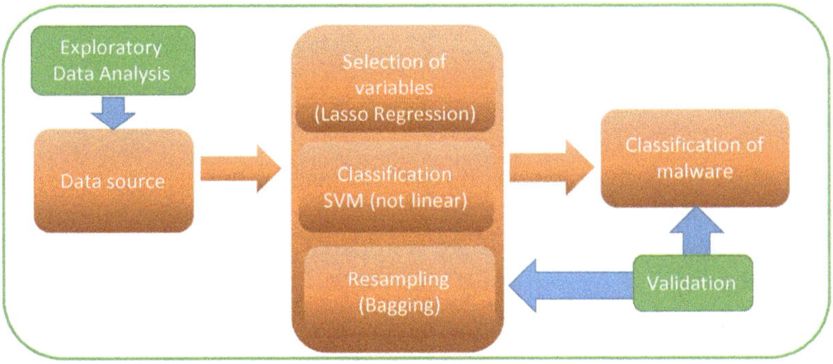

Fig. 1. Model of statistical learning proposed.

3 Application of the Proposed Model

The data used containing 215 attributes extracted from 15,036 applications (37% malware and 63% benign apps). This dataset has been used to develop a multilevel classifier for Android malware detection [11]. The database includes numerical values in all the variables, with values of 0 and 1, referring to the execution of the operation in 1 and 0 when not, also, the transformation of B and S was made, by 0 and 1, respectively to be able to perform the corresponding mathematical operations. Subsequently, with the SAS University tool, the task of descriptive analysis was carried out to identify the behavior of the variables based on the main class classification (Annex 1).

Subsequently, the data was loaded for manipulation using the R language, so it was processed and loaded into a matrix for handling in the Lasso regression functions, with which once executed, the values that should be discarded due to their high correlation with the rest of the data. Thus, the following variables are those that were discarded, after performing a Lasso regression (Table 1):

Table 1. Variables discarded through the Lasso Regression

1. Service Connection	15. Read_User_Dictionary
2. Android.content.pm.Signature	16. MessengerService
3. Get Calling Uid	17. Android.intent.action.Call
4. Runtime.getRuntime	18. Bind_Text_Service
5. Write Sync Settings	19. Mount Format FilesSystems
6. Android.intent.action.Send_Multiple	20. Runtime.loadLibrary
7. Bind_Remoteviews	21. Expand_Status_Bar
8. Receive Boot Completed	22. Internal System Window
9. Read_Calend	23.Android.intent.action.Action_Power_Connected
10. TelephonyManager.getSimCountryIso0	24. Bind VPN_Service
11. Vibrate	25. Device_Power
12. SendDataMessage	26. Access_Fine_Location
13. Delete_Packages	27. Set_Preferred Applications
14. Reorder Tasks	

The selection of the best lambda that allows making the best predictions is done through cross-validation, in this case, the dependent variable is the malware classifier, and the total variables of the database are considered as explanatory variables. Taking into account the tradeoff between bias and variance of the model estimators Fig. 2 shows the best value for lambda.

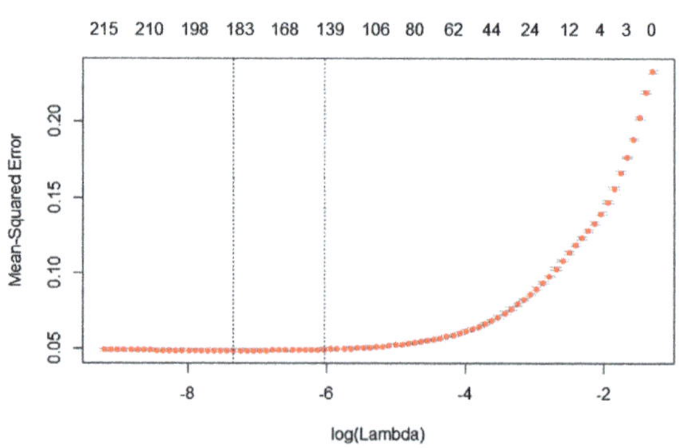

Fig. 2. Best lambda for Lasso Regression

The values of the coefficients obtained after the evaluation of the regression were ordered, where it is determined that those that are highlighted (Annex 2), are the significant variables. So by means of an interpretation of the problem, the result is logical, due that the function createSubprocess. For example, is an operation that is usually generated during an attack of malicious code, as well as sendMultipartTextMessage, is something that does not happen in the opposite way during attacks. The main results of Lasso Regression are in Table 2.

Table 2. Lasso Regression

Parameter	Value
Best λ	0.0006520529
Average square error	0.0453905412
Coefficients	189

The results of the estimation of SVM shows a total of 4,086 support vectors and a mean square error of 0.2172, it is the main classifier used in the model but it will be combined with decision trees to validate the results (Table 3).

Table 3. SVM

Parameter	Value
Kernel	Kernel
Optimal cost	0.01
Gamma Value	0.005319149
Number of Support Vectors	4086
Mean square error	0.2172682

The proposed model is an ensemble between Lasso Regression and SVM that allows replicating the classification of malware through the set of explanatory variables. Was estimated the Receiver Operating Characteristic curve (ROC) with an adjustment of 71% in terms of the confusion matrix (Fig. 3).

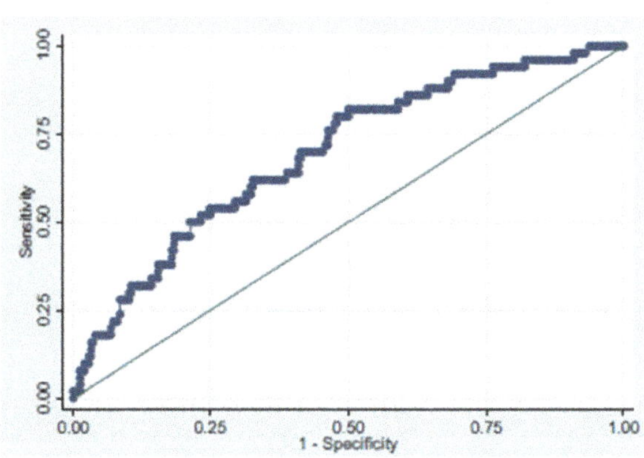

Fig. 3. ROC for SVM model (Area under ROC curve 0.71450)

Meta-Estimator Using Bootstrap Aggregating (Bagging)
Finally, in order to improve the performance of the estimates, bootstrap sampling was performed to estimate a meta-estimator of the SVM classification predictions using one-out validation criterion, 500 samples were used for bagging estimation. As the number of samples increases, the classification error decreases (Fig. 4).

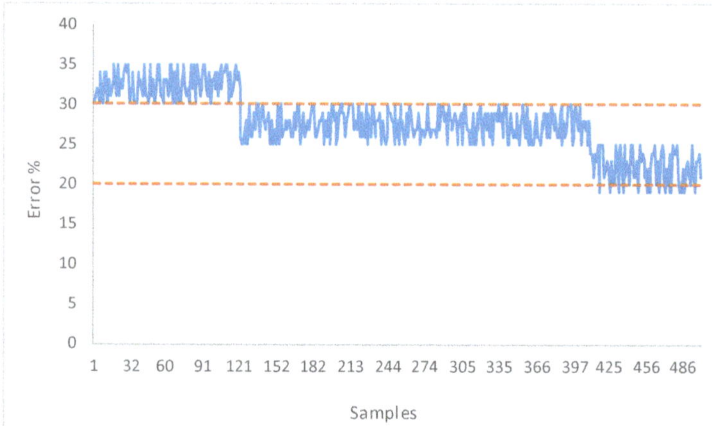

Fig. 4. Bagging and error estimate

The results of the meta-estimator are generated based on each sample taken, in this case, it is relevant to know which variables were the most important in terms of their appearance in a larger number of models. For this, the Gini concentration index is used, which allows us to evaluate the importance of the variables according to their concentration or presence in a larger number of estimated models. The Gini Index shows the relative importance of variables with respect to the most important variable (value of 100 means that it is the most important variable) (Fig. 5).

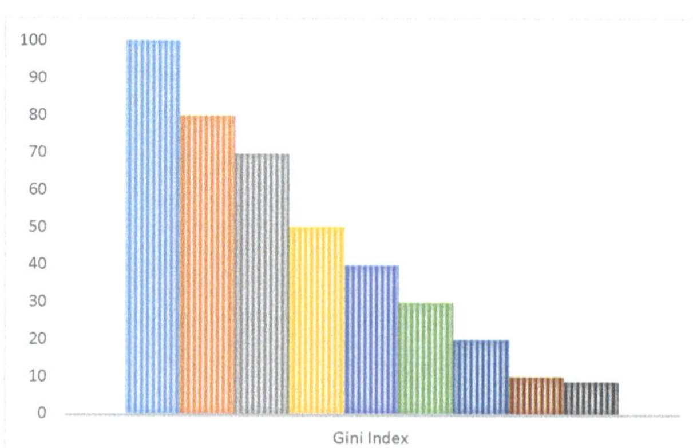

Fig. 5. Variable importance (first 10 variables)

The Gini coefficient makes it possible to identify the set of variables that are most relevant and that have a presence in most of the estimated models (Table 4).

Table 4. First ten variables in order of relevance according to the Gini index

Variable	Name
V1	createSubprocess
V2	SEND_SMS
V4	INTERNET
V5	android.telephony.gsm.SmsManager
V6	android.telephony.SmsManager
V7	CONTROL_LOCATION_UPDATES
V8	DELETE_CACHE_FILES
V9	READ_HISTORY_BOOKMARKS
V10	ACCESS_LOCATION_EXTRA_COMMANDS

4 Conclusions

The detection of malware means of machine learning techniques supervised and unsupervised is a job that requires a time of the research. Also, the number of variables that should be considered because of vary according to the version of the Operating System and the Hardware that contains it. Currently, the system is associated according to the controllers that correspond specifically to the communication equipment, its processor, communication protocols that enable it, among other aspects that must be considered. Likewise, the version controls used are a mechanism that generates complexity in its analysis for all possible variations.

The present work seeks to contribute to the state of the art, to what was done, with other algorithms and using a statistical learning approach. Considering Lasso Regression Lasso to select relevant variables and later it was evaluated a classification model by SVM and bagging resampling method.

The proposed model is a case of application of a statistical learning ensemble, combining Lasso Regression, SVM and Bagging. The results show a ROC curve with an area under the curve of 71%, this implies a percentage of error in the classification of acceptable malware, same that when implementing the resampling decreases up to 20% considering 500 samples.

In addition, in the case of the android operating system despite being a system that inherits the security and robustness of a Linux environment, its applications inherit the vulnerabilities of the Java programming language, which currently has presented several APIs that have different vulnerabilities that are exploited by attackers. There are efforts to improve the operating system itself, as well as the channel signaling system on which it is used.

Annex I. Variables

Variable	Description
transact	API call signature
onServiceConnected	API call signature
bindService	API call signature
attachInterface	API call signature
ServiceConnection	API call signature
android.os.Binder	API call signature
SEND_SMS	Manifest Permission
Ljava.lang.Class.getCanonicalName	API call signature
Ljava.lang.Class.getMethods	API call signature
Ljava.lang.Class.cast	API call signature
Ljava.net.URLDecoder	API call signature
android.content.pm.Signature	API call signature
android.telephony.SmsManager	API call signature
READ_PHONE_STATE	Manifest Permission
getBinder	API call signature
ClassLoader	API call signature
Landroid.content.Context.registerReceiver	API call signature
Ljava.lang.Class.getField	API call signature
Landroid.content.Context.unregisterReceiver	API call signature
GET_ACCOUNTS	Manifest Permission
RECEIVE_SMS	Manifest Permission
Ljava.lang.Class.getDeclaredField	API call signature
READ_SMS	Manifest Permission
getCallingUid	API call signature
Ljavax.crypto.spec.SecretKeySpec	API call signature
android.intent.action.BOOT_COMPLETED	Intent
USE_CREDENTIALS	Manifest Permission
MANAGE_ACCOUNTS	Manifest Permission
android.content.pm.PackageInfo	API call signature
KeySpec	API call signature
TelephonyManager.getLine1Number	API call signature
DexClassLoader	API call signature
HttpGet.init	API call signature
SecretKey	API call signature
Ljava.lang.Class.getMethod	API call signature
System.loadLibrary	API call signature
android.intent.action.SEND	API call signature
Ljavax.crypto.Cipher	API call signature
WRITE_SMS	Manifest Permission

(continued)

(*continued*)

Variable	Description
READ_SYNC_SETTINGS	Manifest Permission
AUTHENTICATE_ACCOUNTS	Manifest Permission
android.telephony.gsm.SmsManager	API call signature
WRITE_HISTORY_BOOKMARKS	Manifest Permission
TelephonyManager.getSubscriberId	API call signature
mount	Commands signature
INSTALL_PACKAGES	Manifest Permission
Runtime.getRuntime	API call signature
CAMERA	Manifest Permission
Ljava.lang.Object.getClass	API call signature
WRITE_SYNC_SETTINGS	Manifest Permission
READ_HISTORY_BOOKMARKS	Manifest Permission
Ljava.lang.Class.forName	API call signature
INTERNET	Manifest Permission
android.intent.action.PACKAGE_REPLACED	Intent
Binder	API call signature
android.intent.action.SEND_MULTIPLE	Intent
RECORD_AUDIO	Manifest Permission
IBinder	API call signature
android.os.IBinder	API call signature
createSubprocess	API call signature
NFC	Manifest Permission
ACCESS_LOCATION_EXTRA_COMMANDS	Manifest Permission
URLClassLoader	API call signature
WRITE_APN_SETTINGS	Manifest Permission
abortBroadcast	API call signature
BIND_REMOTEVIEWS	Manifest Permission
android.intent.action.TIME_SET	Intent
READ_PROFILE	Manifest Permission
TelephonyManager.getDeviceId	API call signature
MODIFY_AUDIO_SETTINGS	Manifest Permission
getCallingPid	API call signature
READ_SYNC_STATS	Manifest Permission
BROADCAST_STICKY	Manifest Permission
android.intent.action.PACKAGE_REMOVED	Intent
android.intent.action.TIMEZONE_CHANGED	Intent
WAKE_LOCK	Manifest Permission
RECEIVE_BOOT_COMPLETED	Manifest Permission
RESTART_PACKAGES	Manifest Permission
Ljava.lang.Class.getPackage	API call signature

(*continued*)

(*continued*)

Variable	Description
chmod	Commands signature
Ljava.lang.Class.getDeclaredClasses	API call signature
android.intent.action.ACTION_POWER_DISCONNECTED	Intent
android.intent.action.PACKAGE_ADDED	Intent
PathClassLoader	API call signature
TelephonyManager.getSimSerialNumber	API call signature
Runtime.load	API call signature
TelephonyManager.getCallState	API call signature
BLUETOOTH	Manifest Permission
READ_CALENDAR	Manifest Permission
READ_CALL_LOG	Manifest Permission
SUBSCRIBED_FEEDS_WRITE	Manifest Permission
READ_EXTERNAL_STORAGE	Manifest Permission
TelephonyManager.getSimCountryIso	API call signature
sendMultipartTextMessage	API call signature
PackageInstaller	API call signature
VIBRATE	Manifest Permission
remount	Commands signature
android.intent.action.ACTION_SHUTDOWN	Intent
sendDataMessage	API call signature
ACCESS_NETWORK_STATE	Manifest Permission
chown	Commands signature
HttpPost.init	API call signature
Ljava.lang.Class.getClasses	API call signature
SUBSCRIBED_FEEDS_READ	Manifest Permission
TelephonyManager.isNetworkRoaming	API call signature
CHANGE_WIFI_MULTICAST_STATE	Manifest Permission
WRITE_CALENDAR	Manifest Permission
android.intent.action.PACKAGE_DATA_CLEARED	Intent
MASTER_CLEAR	Manifest Permission
HttpUriRequest	API call signature
UPDATE_DEVICE_STATS	Manifest Permission
WRITE_CALL_LOG	Manifest Permission
DELETE_PACKAGES	Manifest Permission
GET_TASKS	Manifest Permission
GLOBAL_SEARCH	Manifest Permission
DELETE_CACHE_FILES	Manifest Permission
WRITE_USER_DICTIONARY	Manifest Permission
android.intent.action.PACKAGE_CHANGED	Intent
android.intent.action.NEW_OUTGOING_CALL	Intent

(*continued*)

(*continued*)

Variable	Description
REORDER_TASKS	Manifest Permission
WRITE_PROFILE	Manifest Permission
SET_WALLPAPER	Manifest Permission
BIND_INPUT_METHOD	Manifest Permission
divideMessage	API call signature
READ_SOCIAL_STREAM	Manifest Permission
READ_USER_DICTIONARY	Manifest Permission
PROCESS_OUTGOING_CALLS	Manifest Permission
CALL_PRIVILEGED	Manifest Permission
Runtime.exec	API call signature
BIND_WALLPAPER	Manifest Permission
RECEIVE_WAP_PUSH	Manifest Permission
DUMP	Manifest Permission
BATTERY_STATS	Manifest Permission
ACCESS_COARSE_LOCATION	Manifest Permission
SET_TIME	Manifest Permission
android.intent.action.SENDTO	Intent
WRITE_SOCIAL_STREAM	Manifest Permission
WRITE_SETTINGS	Manifest Permission
REBOOT	Manifest Permission
BLUETOOTH_ADMIN	Manifest Permission
TelephonyManager.getNetworkOperator	API call signature
/system/bin	Commands signature
MessengerService	API call signature
BIND_DEVICE_ADMIN	Manifest Permission
WRITE_GSERVICES	Manifest Permission
IRemoteService	API call signature
KILL_BACKGROUND_PROCESSES	Manifest Permission
SET_ALARM	API call signature
ACCOUNT_MANAGER	API call signature
/system/app	Commands signature
android.intent.action.CALL	Intent
STATUS_BAR	Manifest Permission
TelephonyManager.getSimOperator	API call signature
PERSISTENT_ACTIVITY	Manifest Permission
CHANGE_NETWORK_STATE	Manifest Permission
onBind	API call signature
Process.start	API call signature
android.intent.action.SCREEN_ON	Intent
Context.bindService	API call signature

(*continued*)

(*continued*)

Variable	Description
RECEIVE_MMS	Manifest Permission
SET_TIME_ZONE	Manifest Permission
android.intent.action.BATTERY_OKAY	Intent
CONTROL_LOCATION_UPDATES	Manifest Permission
BROADCAST_WAP_PUSH	Manifest Permission
BIND_ACCESSIBILITY_SERVICE	Manifest Permission
ADD_VOICEMAIL	Manifest Permission
CALL_PHONE	Manifest Permission
ProcessBuilder	API call signature
BIND_APPWIDGET	Manifest Permission
FLASHLIGHT	Manifest Permission
READ_LOGS	Manifest Permission
Ljava.lang.Class.getResource	API call signature
defineClass	API call signature
SET_PROCESS_LIMIT	Manifest Permission
android.intent.action.PACKAGE_RESTARTED	Intent
MOUNT_UNMOUNT_FILESYSTEMS	Manifest Permission
BIND_TEXT_SERVICE	Manifest Permission
INSTALL_LOCATION_PROVIDER	Manifest Permission
android.intent.action.CALL_BUTTON	Intent
android.intent.action.SCREEN_OFF	Intent
findClass	API call signature
SYSTEM_ALERT_WINDOW	Manifest Permission
MOUNT_FORMAT_FILESYSTEMS	Manifest Permission
CHANGE_CONFIGURATION	Manifest Permission
CLEAR_APP_USER_DATA	Manifest Permission
intent.action.RUN	Intent
android.intent.action.SET_WALLPAPER	Intent
CHANGE_WIFI_STATE	Manifest Permission
READ_FRAME_BUFFER	Manifest Permission
ACCESS_SURFACE_FLINGER	Manifest Permission
Runtime.loadLibrary	API call signature
BROADCAST_SMS	Manifest Permission
EXPAND_STATUS_BAR	Manifest Permission
INTERNAL_SYSTEM_WINDOW	Manifest Permission
android.intent.action.BATTERY_LOW	Intent
SET_ACTIVITY_WATCHER	Manifest Permission
WRITE_CONTACTS	Manifest Permission
android.intent.action.ACTION_POWER_CONNECTED	Intent
BIND_VPN_SERVICE	Manifest Permission

(*continued*)

(*continued*)

Variable	Description
DISABLE_KEYGUARD	Manifest Permission
ACCESS_MOCK_LOCATION	Manifest Permission
GET_PACKAGE_SIZE	Manifest Permission
MODIFY_PHONE_STATE	Manifest Permission
CHANGE_COMPONENT_ENABLED_STATE	Manifest Permission
CLEAR_APP_CACHE	Manifest Permission
SET_ORIENTATION	Manifest Permission
READ_CONTACTS	Manifest Permission
DEVICE_POWER	Manifest Permission
HARDWARE_TEST	Manifest Permission
ACCESS_WIFI_STATE	Manifest Permission
WRITE_EXTERNAL_STORAGE	Manifest Permission
ACCESS_FINE_LOCATION	Manifest Permission
SET_WALLPAPER_HINTS	Manifest Permission
SET_PREFERRED_APPLICATIONS	Manifest Permission
WRITE_SECURE_SETTINGS	Manifest Permission
class	B = Benign

Annex II. Coefficients of Lasso Regression

Variable	Coefficient
createSubprocess	0.264244253
SEND_SMS	0.258899979
INTERNET	0.207230368
android.telephony.gsm.SmsManager	0.202311101
android.telephony.SmsManager	0.170603554
CONTROL_LOCATION_UPDATES	0.167463464
chmod	0.129396768
DELETE_CACHE_FILES	0.128579622
SET_TIME	0.126521959
READ_SMS	0.124454019
READ_HISTORY_BOOKMARKS	0.12137216
HttpUriRequest	0.118544076
ACCESS_LOCATION_EXTRA_COMMANDS	0.113866999
MODIFY_PHONE_STATE	0.105379828
WRITE_SOCIAL_STREAM	0.097755195
SUBSCRIBED_FEEDS_READ	0.096845453
WRITE_PROFILE	0.094601544

(*continued*)

(*continued*)

Variable	Coefficient
Runtime.exec	0.086924025
DUMP	0.083824425
READ_PHONE_STATE	0.083395724
UPDATE_DEVICE_STATS	0.083345931
TelephonyManager.getLine1Number	0.079018974
SecretKey	0.074446082
Ljava.lang.Class.getResource	0.074278545
android.intent.action.PACKAGE_RESTARTED	0.072230441
PERSISTENT_ACTIVITY	0.070560713
android.intent.action.BOOT_COMPLETED	0.066059447
TelephonyManager.getDeviceId	0.065452464
CLEAR_APP_CACHE	0.065119831
TelephonyManager.getSubscriberId	0.064089987
WRITE_HISTORY_BOOKMARKS	0.063187614
SET_ALARM	0.062644562
BIND_WALLPAPER	0.061718151
SET_ORIENTATION	0.061584459
onBind	0.057236584
remount	0.052002868
SET_WALLPAPER	0.050447888
GLOBAL_SEARCH	0.050315561
divideMessage	0.042985093
android.intent.action.BATTERY_LOW	0.039620621
DexClassLoader	0.038455601
X.system.bin	0.037747743
mount	0.037456259
android.intent.action.NEW_OUTGOING_CALL	0.036929947
ACCESS_COARSE_LOCATION	0.036554244
findClass	0.036472269
ACCESS_WIFI_STATE	0.035110756
SYSTEM_ALERT_WINDOW	0.034679765
SET_WALLPAPER_HINTS	0.032870177
PROCESS_OUTGOING_CALLS	0.031816527
MANAGE_ACCOUNTS	0.030118538
Ljava.lang.Class.getMethod	0.029238659
AUTHENTICATE_ACCOUNTS	0.029026365
android.intent.action.CALL_BUTTON	0.028889754
defineClass	0.028221046
READ_FRAME_BUFFER	0.023514533
PackageInstaller	0.018827645

(*continued*)

(*continued*)

Variable	Coefficient
KeySpec	0.018161017
Ljava.lang.Class.getDeclaredField	0.016860114
WRITE_APN_SETTINGS	0.016567216
intent.action.RUN	0.015230216
BLUETOOTH	0.013431058
android.intent.action.BATTERY_OKAY	0.013242983
android.intent.action.PACKAGE_ADDED	0.010809145
HARDWARE_TEST	0.010537673
Ljava.lang.Class.getPackage	0.010420027
getCallingPid	0.009211549
android.intent.action.PACKAGE_CHANGED	0.008287811
Ljava.lang.Class.getClasses	0.008147411
WRITE_EXTERNAL_STORAGE	0.006609881
RECEIVE_WAP_PUSH	0.006325588
PathClassLoader	0.005271246
CHANGE_WIFI_STATE	0.004566972
FLASHLIGHT	0.004039508
Ljava.lang.Class.forName	0.003478003
android.intent.action.TIMEZONE_CHANGED	0.001800096
READ_SOCIAL_STREAM	0.001715007
DISABLE_KEYGUARD	0.000932628
BIND_INPUT_METHOD	0.000500278
GET_TASKS	0.000274389
BIND_ACCESSIBILITY_SERVICE	−0.00026438
RESTART_PACKAGES	−0.00148047
android.os.IBinder	−0.00279385
TelephonyManager.getSimCountryIso1	−0.00288437
Ljava.lang.Class.getField	−0.00296154
IBinder	−0.00342076
BROADCAST_SMS	−0.00398095
SUBSCRIBED_FEEDS_WRITE	−0.00454201
ACCESS_NETWORK_STATE	−0.00459926
bindService	−0.00471922
READ_EXTERNAL_STORAGE	−0.00519205
android.intent.action.SENDTO	−0.00563882
Ljava.lang.Class.cast	−0.00718701
STATUS_BAR	−0.0093266
REBOOT	−0.00970475
NFC	−0.00980768
android.os.Binder	−0.01051688

(*continued*)

(*continued*)

Variable	Coefficient
CHANGE_CONFIGURATION	−0.01179464
RECEIVE_SMS	−0.01204579
BLUETOOTH_ADMIN	−0.0139069
android.intent.action.PACKAGE_DATA_CLEARED	−0.01435001
X.system.app	−0.01479621
BIND_APPWIDGET	−0.01481123
getBinder	−0.01611587
WAKE_LOCK	−0.01650988
TelephonyManager.getCallState	−0.01720881
BROADCAST_STICKY	−0.01764756
TelephonyManager.getNetworkOperator	−0.01889666
WRITE_SECURE_SETTINGS	−0.01918746
WRITE_USER_DICTIONARY	−0.01993922
KILL_BACKGROUND_PROCESSES	−0.02030457
android.intent.action.SEND	−0.02039022
READ_SYNC_SETTINGS	−0.0208841
android.content.pm.PackageInfo	−0.02119871
Landroid.content.Context.registerReceiver	−0.02174869
READ_CONTACTS	−0.02274993
CAMERA	−0.02403828
RECORD_AUDIO	−0.02466568
TelephonyManager.getSimSerialNumber	−0.02478753
chown	−0.02591173
abortBroadcast	−0.02719413
Ljava.lang.Class.getCanonicalName	−0.0271965
Ljava.lang.Class.getMethods	−0.02771707
IRemoteService	−0.02791817
CALL_PHONE	−0.03090892
android.intent.action.SCREEN_ON	−0.03210006
BROADCAST_WAP_PUSH	−0.03226206
HttpPost.init	−0.03234517
WRITE_SMS	−0.0324788
TelephonyManager.getSimOperator	−0.03469507
android.intent.action.ACTION_POWER_DISCONNECTED	−0.03616017
BATTERY_STATS	−0.0364922
android.intent.action.SCREEN_OFF	−0.03728614
INSTALL_LOCATION_PROVIDER	−0.03791462
READ_SYNC_STATS	−0.03810809
attachInterface	−0.03829324
Ljavax.crypto.Cipher	−0.03949882

(*continued*)

(*continued*)

Variable	Coefficient
INSTALL_PACKAGES	−0.03961002
android.intent.action.SET_WALLPAPER	−0.04164843
MASTER_CLEAR	−0.04187666
onServiceConnected	−0.04191473
Process.start	−0.04247342
transact	−0.04265317
ProcessBuilder	−0.04321857
GET_PACKAGE_SIZE	−0.04354681
USE_CREDENTIALS	−0.04364622
android.intent.action.PACKAGE_REMOVED	−0.04367305
Landroid.content.Context.unregisterReceiver	−0.04397466
WRITE_CALL_LOG	−0.04446323
CALL_PRIVILEGED	−0.04503315
READ_LOGS	−0.04560355
WRITE_CONTACTS	−0.04869014
ACCOUNT_MANAGER	−0.05009465
Ljava.lang.Class.getDeclaredClasses	−0.05091621
Ljava.lang.Object.getClass	−0.0509497
android.intent.action.PACKAGE_REPLACED	−0.05162474
WRITE_SETTINGS	−0.05171949
CLEAR_APP_USER_DATA	−0.05230378
GET_ACCOUNTS	−0.05414448
WRITE_CALENDAR	−0.05974735
System.loadLibrary	−0.06144877
MODIFY_AUDIO_SETTINGS	−0.06205213
ADD_VOICEMAIL	−0.06566482
TelephonyManager.isNetworkRoaming	−0.06653053
READ_PROFILE	−0.06791394
ACCESS_MOCK_LOCATION	−0.06795926
Binder	−0.06942009
CHANGE_NETWORK_STATE	−0.07220123
RECEIVE_MMS	−0.0792903
URLClassLoader	−0.07967425
Ljavax.crypto.spec.SecretKeySpec	−0.08006777
android.intent.action.ACTION_SHUTDOWN	−0.08283051
WRITE_GSERVICES	−0.08450157
MOUNT_UNMOUNT_FILESYSTEMS	−0.08603744
android.intent.action.TIME_SET	−0.08665047
Context.bindService	−0.08838194
CHANGE_WIFI_MULTICAST_STATE	−0.08942738

(*continued*)

(*continued*)

Variable	Coefficient
HttpGet.init	−0.09067209
ClassLoader	−0.09084002
CHANGE_COMPONENT_ENABLED_STATE	−0.10764022
Ljava.net.URLDecoder	−0.10774214
Runtime.load	−0.10884032
ACCESS_SURFACE_FLINGER	−0.11486367
READ_CALL_LOG	−0.12663504
BIND_DEVICE_ADMIN	−0.13304341
SET_ACTIVITY_WATCHER	−0.15191197
SET_TIME_ZONE	−0.18878256
SET_PROCESS_LIMIT	−0.19163698
sendMultipartTextMessage	−0.34691918

References

1. Kaspersky Lab: Machine learning for malware detection. Kaspersky for Bussines (2019)
2. Gavrilut D., Cimpoesu M., Anton D., Ciortuz, L.: Malware detection using machine learning. In: Proceedings of the International Multiconference on Computer Science and Information Technology, pp. 735–741 (2009)
3. Aydogan, E., Sen, S.: Analysis of machine learning methods on malware detection. In: 22nd Signal Processing and Communications Applications Conference (2014)
4. Xiao, F., Lin, Z., Sun, Y., Ma, Y.: Malware detection based on deep learning of behavior graphs. Math. Prob. Eng. **2019**, 10 (2019)
5. Amalina, F., Feizollah, A., Bradul, N., Gani, A.: Evaluation of machine learning classifiers for mobile malware detection. Soft Comput. **20**(1), 343–357 (2016)
6. Rieck, K., Holz, T., Willems, C., Dussel, P., Laskov, P.: Learning and classification of malware behavior. In: International Conference on Detection of Intrusions and Malware, and Vulnerability Assesment (2008)
7. Santos, I., Nieves, J., Bringas, P.: Semi-supervised learning for unknown malware detection. In: Abraham, A., Corchado, J.M., González, S.R., De Paz Santana, J.F. (eds.) International Symposium on Distributed Computing and Artificial Intelligence. Advances in Intelligent and Soft Computing, vol. 91. Springer, Heidelberg (2011)
8. Hou, Y., Chang, Y., Chen, T., Laih, C., Chen, C.: Malicious web content detection by machine learning. Expert Syst. Appl. **37**(1), 55–60 (2010)
9. Markel, Z., Bilzor, M.: Building a machine learning classifier for malware detection. In: Second Workshop on Anti-malware Testing Research (WATeR) (2014)
10. Martín, I., et al.: Android malware characterization using metadata and machine learning techniques. Secur. Commun. Networks **2018**, 11 (2018)
11. Yerima, S., et al.: DroidFusion: A novel multilevel classifier fusion approach for android malware detection. IEEE Trans. Cybern. **49**, 453–466 (2018)
12. Tibshirani, R.: Regression shrinkage and selection via the lasso. J. Roy. Stat. Soc. **58**, 267–288 (1996)
13. Cortes, C., Vapnik, V.: Support vector networks. Mach. Learn. **20**, 273 (1995)
14. Hastie, T., Tibshirani, R., Friedman, J.: The Elements of Statistical Learning. Springer Series in Statistics: EU (2008)

A Novel Model for Risk Estimation in Software Projects Using Artificial Neural Network

M. Hanefi Calp[1(✉)] and M. Ali Akcayol[2]

[1] Department of Management Information Systems, Karadeniz Technical University, Trabzon, Turkey
mhcalp@ktu.edu.tr

[2] Department of Computer Engineering, Gazi University, Ankara, Turkey
akcayol@gazi.edu.tr

Abstract. Software projects generally involve more risks due to unexpected negative results. Therefore, the risks encountered in software projects should be detected and analyzed on time, and effective precautions should be taken in order to complete the projects successfully. The aim of this study was to estimate the deviations that may occur in the software project outputs according to risk factors by using artificial neural networks (ANNs). Thus we aimed to minimize loses that may occur in project processes with the developed model. Firstly, a comprehensive and effective list of risk factors was created. Later, a checklist form was prepared for Team Members and Managers. The data collected include general project data and risk factors, and these are the inputs of the model. The outputs of the model are the deviations in the project outputs. MATLAB package program was utilized to develop the model. The performance of the model was measured according to Regression Values and Mean-Squared Error. The model obtained has forty-five inputs, one hidden layer that has fifteen neurons, and five outputs (45-15-5). In addition, the training-R, testing-R, and MSE values of the model were found as 0.9978, 0.9935, and 0.001, respectively. It is seen that the estimation results obtained with the model using the real project data coincide with the actual results largely and the error rates were also very low (close to zero). The experimental results clearly revealed that model performance is high, and it is very effective to use ANNs in risk estimation processes for software projects.

Keywords: Software project · Risk factor · Risk estimation · Risk management · Artificial neural networks

1 Introduction

Nowadays, developed software projects are used in almost every area of the software industry including production and consumption. Many software projects are carried out in order to fulfil the needs in these areas. The scope of these projects is growing day by day and as a result of this, software is becoming complicated. This situation negatively affects the completion of software projects successfully on time and with estimated resources [1–3]. Software projects generally contain more risks due to unexpected negative results. The concept of risk means loss and probability. Loss describes the

D. J. Hemanth and U. Kose (Eds.): ICAIAME 2019, LNDECT 43, pp. 295–319, 2020.
https://doi.org/10.1007/978-3-030-36178-5_23

hazards that might be encountered such as time, money, reputation, quality; probability describes the potential of occurrence of this dangers. In the literature, the software risks should be detected in advance and analyzed in software development projects. However, when these are not realized, the risks are defined as the potential dangers that may cause losses to such as effort, time, money, prestige, etc. [4–7]. According to another definition, risk is defined as lack of reliability in the system life cycle [8]. It is also defined as events that occur in a software project and threaten the success of the project. [9] Risks are found in all projects and can not be removed altogether. However, it is possible to reduce the adverse effects on the project by using effective management methods. Risk also refers to economic or financial loss caused by uncertainty and physical injury or delay [10]. Risk is an uncertain condition that is determined, controlled and eliminated by risk management through selection and analysis of appropriate strategies that affect the project in terms of quality, time and cost [11]. Therefore, considering the risk definitions, the risks encountered in software projects should be detected and analyzed on time, and effective precautions should be taken to improve the project process. Otherwise, significant resources such as money, time and personnel that are planned at the beginning of the project cannot be successfully managed and therefore the projects become unsuccessful. In this context, in the studies of the Standish Group International at 2004, it was demonstrated that 53% of the projects was delayed or overspent, 18% of these projects was abandoned or modified. Only 29% of these projects were completed on time and on budget [1, 12, 13]. These results indicate that some measures should be taken by estimating the deviations that may occur in order to the complete the projects successfully.

The purpose of this study is to estimate the deviations that may occur in the software project outputs according to risk factors by using ANNs that is one of the quantitative estimate methods, and thus, is to minimize losses that may occur in project processes with the developed model. The differences of this study from other studies are (1) extensive preparation of risk factors that are considered very important and represent inputs; (2) use of many real project data in different areas and (3) model's capability to estimate the deviations on the issues that have great importance such as project time, cost, number of personnel, targets, and general success of the project. In the study, review of literature on the subject; the ANNs subject; all details of the experimental study; the results and discussion, the conclusions, suggestions and future work have been mentioned respectively.

2 Literature Review

In the literature, some techniques such as ANNs [14–17], Support Vector Machines (SVM) [18], Bayesian Belief Networks (BBNs) [19], Decision Trees [20], Discriminant Analysis [21, 22] Principal Component Analysis (PCA), Inference Diagrams, Monte Carlo Analysis, Classification and Regression Tree, Genetic Algorithms (GA) and Fuzzy Logic (FL) are used to solve complex problems [1, 12, 18, 23, 24]. These techniques have also been used in risk prediction and analysis in software projects. For example, Neumann developed a model to categorize risks in his study. Developed model has capability to discriminate high-risk software by using PCA and

ANN together. PCA was used to normalize the input data, thus remove the ill effects. ANN was used for the determination and classification of the risks [25]. Xu et al. developed an application by using fuzzy expert system in order to assess the operational risks of the software. In the study, they studied the assessment of risks that may occur in the software development process. In addition, they demonstrated the positive impact of the fuzzy expert system method on risk assessment activity [26]. Fan and Yu proposed a theoretical model to provide a better understanding of software risk management. In addition, they developed a BBN-based method to estimate potential risks, identify the sources of risks, and to make decisions dynamically by using a feedback loop. Analytical and simulation results are reported [27]. Yang et al. developed a model by using GA during the integration of software to minimize risks. It was a study based on software integration project implemented in the Derbyshire Fire Rescue Service [28]. Yong et al. detected the software risk factors that is important for successful outcome by forming a model to minimize the risks by using a neural network. The inputs of the model are software risk factors. They obtained the inputs via interviewing. Outputs are the outcome of the project. The real software projects data were used for analysis. In order to increase performance of the model, PCA and GA were utilized. The experimental results indicated that the risk analysis model was effective [1]. Hu et al. established a model to assess risks using ANN and SVM in project development process. In the suggested model, the inputs are software risk factors and the output is the final outcome. The experimental results showed the model was valid. The experimental results in their study indicated that increase model surpassed SVM in performance after optimizing the ANN model with GA [12]. Hui showed a new mathematical approach for risk management. Hui demonstrated how to form the model optimization by using linear programming and algorithm method. In addition, he analyzed an approval of the BBN optimal values and their comparison to the output from a optimal solution [29]. Li-ying, and Xin-zheng, established the risk assessment index system, based on the analysis of the project's internal and external environment, by BP neural network and MATLAB neural network toolbox. The results revealed that the model had more accuracy and practicability [30]. Sharma and Khan conducted a study that provides risk-based software testing by using FL. In the study, they showed risk management in software testing by using FL. They concluded that the fuzzy model they developed was effective for risk management [31]. Hu et al. suggested a model by using BNs with causality constraints for risk analysis in software development projects. Suggested model can explore causalities based on expert knowledge, and perform better in prediction compared to algorithms such as logistic regression, C4.5, and Naïve Bayes [32].

When the studies in this area were examined, results obtained from models that especially formed by ANN is more successful than other methods, and it seems that ANN is an effective solution in some conditions that are difficult to relate when detecting the relationship between risk factors and project output. Therefore, this situation has also been particularly important in choosing ANN in the study.

Besides, it was seen in the studies that analysis were made on the specific topics such as only operational risks, high risk factors, risks that could occur during early stages of software life cycle or during software integration. At the same time, it was seen that the development of models predicted only project outcome (successful, failed, etc.). Whereas in this study, a model has been developed that can enable an inclusive estimate on the matters like project time, budget, number of staff, target, general success of the project by using a total of 40 risk factors in various topics under eight section and many real project data. Before getting into the details of the application, the features of ANN method has been briefly discussed in the third chapter.

3 Artificial Neural Networks

An ANN has an algorithm for which modeling inspiration was biological neuron cell structures and that can learn by itself in process. ANNs are utilized in control applications, robotics, pattern recognition, medical science, power systems, signal processing, estimation, and particularly in system modeling [33–36]. An example of neural network model is shown in Fig. 1.

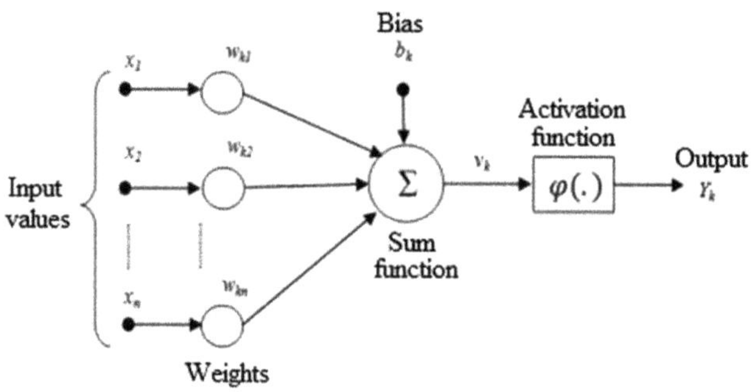

Fig. 1. An ANN model [37].

The ANN model seen in Fig. 1 is defined mathematically as Eq. 1.

$$u_k = \sum_{j=1}^{n} w_{kj}x_j$$

$$y_k = \varphi(u_k + b_k)$$

$$v_k = u_k + b_k$$

$$y_k = \varphi(v_k) \qquad (\text{w : weight, x: input, y: output, b: bias}) \tag{1}$$

An ANN is formed through a group of neurons establishing a connection with each other. ANNs can derive new information as a result of learning by generalizations from examples and provide solutions for nonlinear problems. An ANN is divided into two parts as training and testing sets. The purpose of training is to minimize the error value by adjusting the weights in the network. The training process continues until the output value targeted is obtained. The performance of training is assessed by testing unused data (that are not used during the training) in the network [35, 38–41]. For the training of the network, feed-forward back-propagation network architecture is effective. Feed-forward networks allow the progress in one direction from input to output. A typical feed-forward neural network consists of an input layer, usually one or two hidden layer(s), and an output layer (Fig. 2) [42, 43].

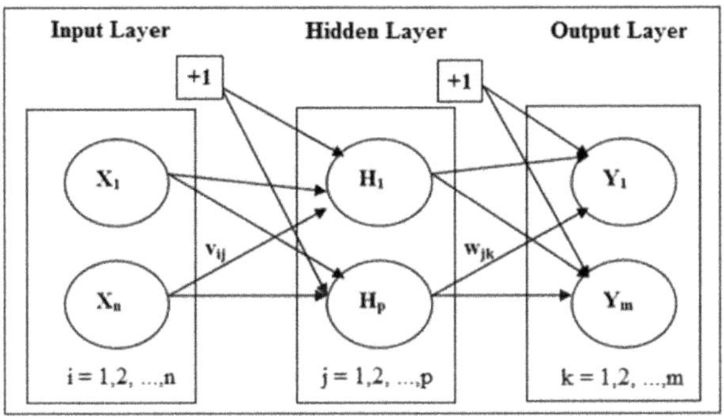

Fig. 2. A feed-forward ANN models.

The training of the network is carried out with the distribution of the weights of the connections between neurons in each layer. The distribution of the weights is performed with the error function presented in Eq. 2 [44]. (d_j: targeted result, o_j: obtained result).

$$E_p = \frac{1}{2}\sum_j \left(d_j^p - o_j^p\right)^2 \quad (2)$$

The difference of the error function is used to adjust the weights Eq. 3.

$$\Delta^p w_{ji} = -\eta.\left(\frac{\partial E^p}{\partial w_{ji}}\right) \quad (3)$$

In this equation, any value can be assigned for the η constant (learning rate). Equation 4 is used to re-adjust the weights.

$$w_{ij}(t+1) \cong w_{ij}(t) + \eta.\delta_j.i_i \quad (4)$$

In this equation (Eq. 4), $w_{ij}(t)$: weight, i_i may be the result value of the node i or δ_j may be the error term of the node j. The error term for an output node (δ_j) (Eq. 5),

$$\delta_j \cong o_j.(1-o_j).(d_j-o_j) \quad (5)$$

is obtained. For a hidden node j node, the error term $\left(\delta_j\right)$ (Eq. 6):

$$\delta_j \cong o_j.\left(1 - o_j\right).\sum\nolimits_k \delta_k.w_{jk} \tag{6}$$

With the addition of a "moment" term (α), it can be effected on weight changes (Eq. 7),

$$w_{ij}(t+1) \cong w_{ij}(t) + \eta.\left(d_j - o_j\right).i_i + \alpha.\left(w_{ij}(t) - w_{ij}(t-1)\right) \tag{7}$$

4 Experimental Study

This section includes the material and method of the study, data collection process, details of the model, and the experimental results obtained.

4.1 Material and Method

In the present study, the structure of the model was determined first. Model has input, hidden, and output layers. The most important process at this point is to determine the inputs of the model. The inputs are software risk factors. These software risk factors suggested by researchers have been explored in detail in the selection process of the factors. Some of the risk factors commonly used in the literature are provided in Table 1 [2, 45–47].

In addition, in the literature, the risks encountered in software projects are collected under eight domains, namely time, budget (cost), management, technical, program, contract/legal risks, personnel, and other risks [51, 52].

A comprehensive and effective list of risk factors was created for use in generating the model based on the risk factors accepted in the literature. To this end, academic staff in Software Engineering Department at different universities in Ankara, experts, software project managers, and project customers were interviewed. As a result of these interviews, for data collection, a risk identification form (see Appendix) was prepared that concerns Team Members (Developer and Testing Personnel) and Managers (Analyst, Expert, Team Leader, Project Manager and General Manager).

4.2 Data Collection

In the scope of the research, the data collection form was uploaded to an online system (Google Drive) and was made available to the participants. In addition, approximately 3000 e-mail were sent to Software Companies in Technoparks and elsewhere in Turkey to fill the risk identification form. However, the response rate was quite low, and they were mostly incomplete. Therefore, face-to-face interviews were conducted in the area.

Table 1. Software risk factors according to researchers.

Boehm [4]	SEI [48]	Conrow and Shishido [49]	Kansala [50]
Personnel shortfalls Unrealistic schedules and budgets Developing the wrong software functions Developing the wrong user interfacing Gold plating Continuing stream of requirements changes Shortfalls in externally furnished components Shortfalls in externally performed tasks Real-time performance shortfalls Straining computer science capabilities	Product Engineering Requirements Design Engineering specialties Code Development Environment Management Process Development System Management Methods Development Process Work Environment Program Constraints Resources, Customer Program Interface Contact	Project level Excessive, immature, unrealistic, or unstable requirements Lack of user involvement Underestimation of project complexity or dynamic nature Project attributes Performance shortfalls (includes errors and quality) Unrealistic cost or schedule (estimates and/or allocated amounts) Management Ineffective project management (multiple levels possible) Engineering Ineffective integration, assembly and test, quality control, specialty engineering, or systems engineering (multiple levels possible) Unanticipated difficulties associated with the user interface Work environment Immature or untried design, process, or technologies selected Inadequate work plans or configuration control Inappropriate methods or tool selection or inaccurate metrics Poor training Other Inadequate or excessive documentation or review process Legal or contractual issues (such as litigation, malpractice, ownership) Obsolescence (includes excessive schedule length) Unanticipated difficulties with subcontracted items Unanticipated maintenance and/or support costs	Volatility of requirements Availability of key staff (=developers) Dependence on key staff Interfaces to other systems Unnecessary features ('gold plating') Commitment of customer Capability of contact person Analysis skills of staff Delivery reliability of subprojects Complexity of functional model Commitment of staff Logical complexity of software Maintainability of software Availability of project manager Complexity of data model

However, we still encountered some problems in the process of data collection in the area. Problems encountered are listed below:

- time constraints,
- reluctance of the staff working at companies due to their busy work schedule,
- reluctance of the staff due to the concerns for releasing confidential company information,
- unable to reach out the relevant staff due to their being off-site,
- reluctance of the staff due to the lack their managers' permission,
- company executives' prejudiced approach against surveys and interviews,
- reluctance of the staff due to large number of questions in the form.

Despite all the difficulties, this form was filled by software engineers, personnel, and managers employed in software development processes. Most of the data were collected from companies that develop software projects in Technoparks in Ankara. In this process, 467 real project data in total were collected: 373 from 774 different companies and 94 from online forms. Data were collected in a period of about 7 months and the summary information about data collected is given in Table 2.

Table 2. Collected data.

University name	Technopark name	Number of companies	Web address	Number of collected data
Gazi University	Gazi Technopark	117	http://www.gaziteknopark.com.tr	59
Hacettepe University	Hacettepe Technopark	147	http://www.hacettepeteknokent.com.tr	109
ODTÜ	ODTÜ Technopark	285	http://odtuteknokent.com.tr	104
Bilkent University	Bilkent Cyberpark	225	http://www.cyberpark.com.tr	101
Data obtained by online form		–	https://docs.google.com/forms/d/1RbjOGRtOHSzU2bxO1w_pJXtf3ymrWRKCavkCiKUOwmw/viewform	94

4.3 Reliability Analysis of the Collected Data

The reliability analysis of the variables (risk factors) used in this study, has an importance in terms of being scientific. According to the analysis result, Cronbach's Alpha value of the research scale (reliability value) was calculated as 0.927. So, the calculated value is greater than 0.7. This result shows that variables are "reliable" in terms of measurement.

4.4 Model

In the study, a model was developed to estimate the deviations that might occur in the software projects using ANN, and thus to minimize the losses to be encountered. The risk factors encountered in the software projects comprise the model's inputs. The model's outputs are the deviations in time, budget and the number of personnel, target and success of the project. Matlab 2016a package program was utilized in the creation of the model, and the Regression values (R) and the Mean Squared Error (MSE) were taken into account. Firstly, the data obtained were normalized with min-max method to express with the finite numbers. Here, v_R indicates the actual value of input, v_{min} indicates the minimum value of input, and $v_{\max}$ indicates the maximum value of input (Eq. 8) [34].

$$v_n = \frac{v_R - v_{min}}{v_{max} - v_{min}} \tag{8}$$

Then the network training process was started. In the training process, the aim was to reach to minimum errors with the few numbers of hidden layers and neurons. In the training of the network, the feed-forward back propagation network architecture was selected because more effective results were obtained. A great number of trials were made in different ways in order to form the network model. These trials were performed by determining the training and testing data at different rates (70%–30%, 75%–25%, 80%–20%) and using different normalization methods (D_min_max, min_max, median), the training functions (trainlm, trainscg, trainbr), the learning functions (traingd, traingdm), the performance functions (mse, msereg, sse), the activation functions (tansig, logsig, purelin), the iterations (through 100–1000) and the number of hidden layers (through 1–3) having different neuron numbers (through 1–25). Some of the successful results obtained by trials are given in Tables 3, 4, 5 and 6 as samples. Tables 3, 4, 5 and 6 show that the best results for the problem were obtained from the "45-15-5" models and from the training and testing masses determined at the rates of 70%–30%.

Table 3. Neural network models and obtained results-1.

Dataset	Performance criteria	Neural network model (single hidden layer)								
		Training set: 70% Testing set: 30%								
		Training function: trainlm								
		Activation functions Hidden layer: tansig Output layer: tansig			Activation functions Hidden layer: logsig Output layer: tansig			Activation functions Hidden layer: tansig Output layer: purelin		
		45-10-5	**45-15-5**	45-20-5	45-10-5	**45-15-5**	45-20-5	45-10-5	**45-15-5**	45-20-5
Training	R	0.9814	**0.9942**	0.9216	0.9891	**0.9941**	0.9875	0.9822	**0.9985**	0.9811
Testing	R	0.9558	**0.9896**	0.9060	0.9881	**0.9862**	0.9794	0.9716	**0.9861**	0.9808
	MSE	0.006	**0.006**	0.008	0.005	**0.003**	0.004	0.007	**0.001**	0.017

Table 4. Neural network models and obtained results-2.

Dataset	Performance criteria	Neural network model (single hidden layer)								
		Training set: 70% Testing set: 30%								
		Training function: trainscg								
		Activation functions Hidden layer: tansig Output layer: tansig			Activation functions Hidden layer: logsig Output layer: tansig			Activation functions Hidden layer: tansig Output layer: purelin		
		45-10-5	**45-15-5**	45-20-5	**45-10-5**	45-15-5	45-20-5	45-10-5	**45-15-5**	45-20-5
Training	R	0.9770	**0.9860**	0.9489	**0.9910**	0.9797	0.9884	0.9664	**0.9978**	0.9866
Testing	R	0.9693	**0.9809**	0.9093	**0.9828**	0.9658	0.9691	0.9171	**0.9935**	0.9430
	MSE	0.010	**0.010**	0.012	**0.004**	0.008	0.004	0.008	**0.001**	0.010

Table 5. Neural network models and obtained results-3.

Dataset	Performance criteria	Neural network model (single hidden layer)								
		Training set: 75% Testing set: 25%								
		Training Function: trainlm								
		Activation functions Hidden layer: tansig Output layer: tansig			Activation functions Hidden layer: logsig Output layer: tansig			Activation functions Hidden layer: tansig Output layer: purelin		
		45-10-5	**45-15-5**	45-20-5	45-10-5	**45-15-5**	45-20-5	45-10-5	**45-15-5**	45-20-5
Training	R	0.9685	**0.9682**	0.9675	0.8427	**0.9726**	0.9635	0.9649	**0.9807**	0.9311
Testing	R	0.8417	**0.8963**	0.8900	0.8696	**0.8942**	0.8634	0.8056	**0.8249**	0.8734
	MSE	0.02	**0.02**	0.04	0.004	**0.03**	0.03	0.03	**0.03**	0.04

Table 6. Neural network models and obtained results-4.

Dataset	Performance criteria	Neural network model (single hidden layer)								
		Training set: 80% Testing set: 20%								
		Training function: trainlm								
		Activation functions Hidden layer: tansig Output layer: tansig			Activation functions Hidden layer: logsig Output layer: tansig			Activation functions Hidden layer: tansig Output layer: purelin		
		45-10-5	**45-15-5**	45-20-5	45-10-5	45-15-5	**45-20-5**	45-10-5	**45-15-5**	45-20-5
Training	R	0.9787	**0.9470**	0.9760	0.9613	0.9687	**0.9714**	0.9724	**0.9740**	0.9658
Testing	R	0.8599	**0.9112**	0.8993	0.9087	0.8819	**0.8673**	0.8735	**0.8882**	0.8625
	MSE	0.02	**0.02**	0.02	0.02	0.02	**0.01**	0.02	**0.02**	0.03

During the trials, a few subjects that considered very important in the network's training process attracted attention. The first one is the process of determination of the neuron's activation functions; the number of neurons in layers and the hidden layers' number. The number of layers and neurons is increased according to the problem's difficulty, however; small numbers caused the training not to be provided well whereas being big numbers led to extension in the process. Another one is that the network's performance increased particularly for the models having the hidden layers more than

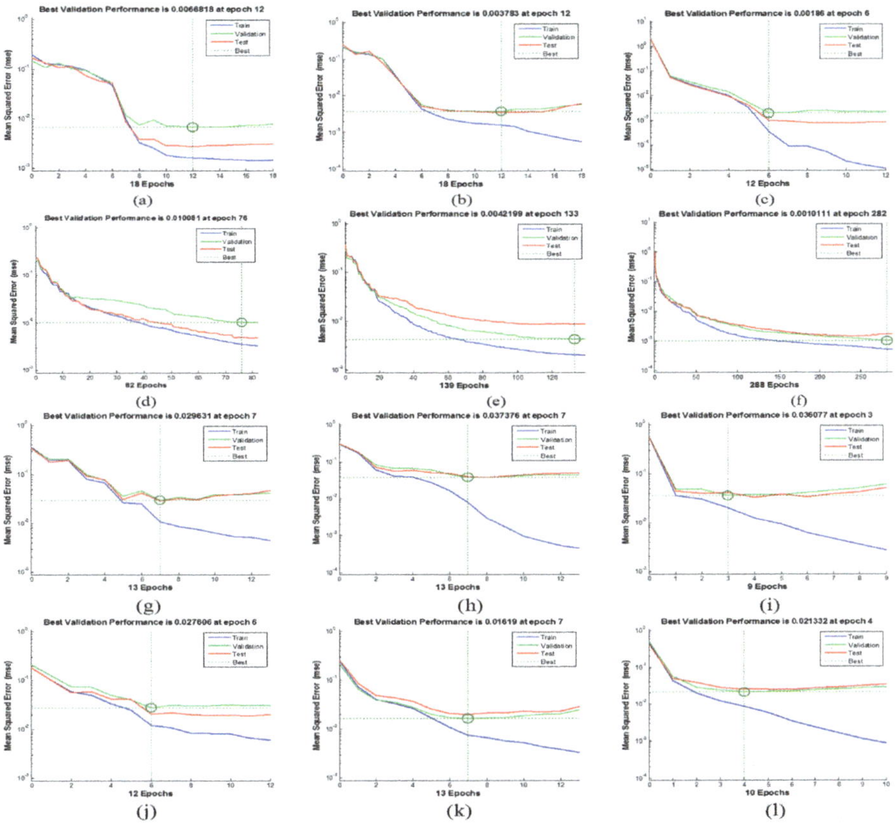

Fig. 3. Performances according to ANN models and functions.

one when the numbers of neurons in the hidden layer are equal. In addition, it was seen that the value R obtained from the training set increased whenever the number of the neurons in the hidden layers increases. However, the R-values decreased in the networks including particularly 25 or more neurons in the hidden layers when the network formed were tested with the test set,. This case showed that the network forgot what it learned or memorized.

The graphs given in Fig. 3 can also be presented as a proof for this case. In Fig. 3, the error graphs of the best results in each group among the results given in Tables 3, 4, 5 and 6 are presented. The striking subject in the error graphs is that the training and testing error curves are close to zero and each other as much as possible. Otherwise, as stated before, when the training error curve approaches to zero and the testing error curve goes away from zero after the certain step, it shows that the model forgot what it learned or memorized. It is clearly seen that particularly Fig. 3-(d, e and f) training-testing and validation error curves are closer to zero than the others are and the best-trained model of them is Fig. 3-(f) when Fig. 3 is examined in details.

Then, network models were made from data kept for testing and various number of neurons and hidden neurons and these network models were tested with the testing data never seen by the network. The estimation accuracies of ANN models with different architectures were evaluated by comparing estimation values found in testing process with actual values. As a result, after determining 70% of all data collected as training set and 30% as testing set, the best network model was obtained by selecting the normalized method "min_max", the activation function "purelin", the training function "trainscg" and performance function "MSE". The best model obtained has forty-five inputs, the single hidden layer (fifteen neurons) and five outputs (45-15-5) (Fig. 4).

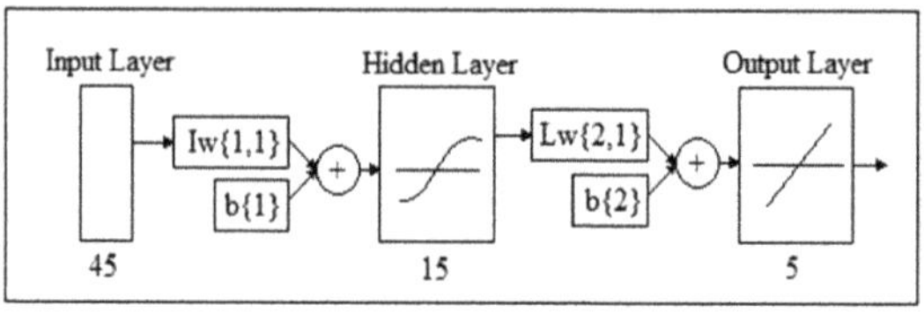

Fig. 4. The obtained ANN model (45-15-5).

In Eq. 9, the ANN model estimating the project's outputs according to the risk factors is formulized.

$$f_{x,ANN} = Tansing\left(b + \sum\nolimits_{h=1}^{n_h} I_h\, Purelin\left(b_h + \sum\nolimits_{i=1}^{n_i} w_{ih} x_i\right)\right) \quad (9)$$

As seen in Fig. 4 and Eq. 9, the activation functions "purelin" were used in the output layer; the activation functions "tansig" were used in the model's hidden layer.

In Table 7, the statistical parameters of the training and testing data sets (the values of R and the rate of errors) were presented.

Table 7. Features of the ANN model.

Dataset	Performance criteria	Neural Network Model (Single Hidden Layer) 45-15-5
Training	R	0.99782
Testing	R	0.99352
Validation	R	0.99604
All	R	0.99686
	MSE	0.0010111

Here, it is remarkable that the R-values of training and testing are very high (close to 1); the rate of errors is very low (close to 0). It is understood by looking at the R-values here that the actual and estimated outputs are very close to each other, the relationship between these outputs is very high and the model's training is quite successful. In Fig. 5, the indication of neural network model obtained from the data is available as nodes.

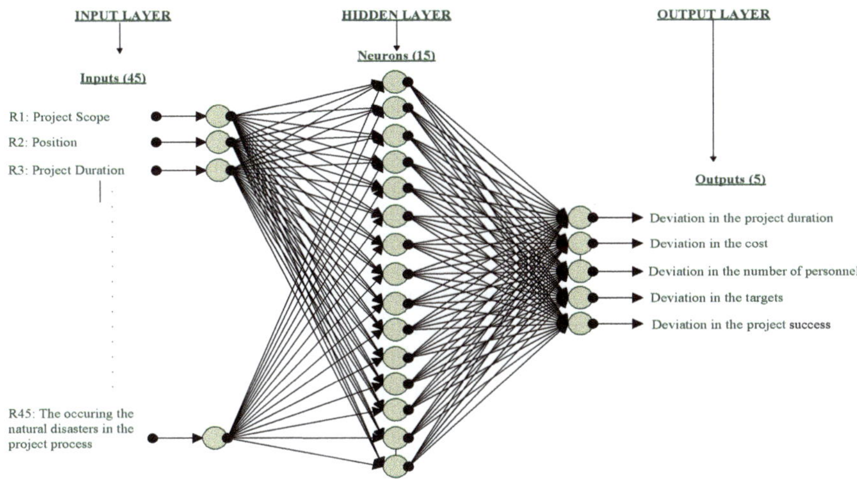

Fig. 5. The ANN model based on nodes (45-15-5).

In Fig. 3(f), the error curve of the network model is presented. It is seen that the training reached to the best training performance with error rate of 0,0010111 in 282nd step, and the error levels are very close to zero (quite low) when the training-testing-validation error curves of the network model are examined carefully here. This is an important sign in determination of model as the estimation values will be reached with the high accuracy.

5 Results and Discussion

After the model was determined and its details were given, the experimental results were analyzed by measuring its performance, and it was discussed. Firstly, the model was embedded in the prototype console application developed using .Net program after being converted to the .dll library in MATLAB. The pseudo-codes developed using MATLAB and .Net are given in below.

The pseudocode of the prediction function prepared using MATLAB package program

```
function [ output ] = PY_Prediction_Function (input)
load C:\PY_TahminNet\net.mat;
x1=mapminmax('apply',giris,network11.inputs{1,1}.processSettings{1});
y1 = tansig(network11.IW{1} * x1 + network11.b{1});
y2 = purelin(network11.LW{2} * y1 + network11.b{2});
annoutput=mapminmax('reverse',y2,network11.outputs{1,2}.processSettings{1});
output=annoutput;
end
```

The pseudocode that required to run the prediction function using .Net program

```
PY_Prediction_Class rm = new PY_Prediction_Class();
 double[,] input = new double[45, 1];
  for (int i=0;i<45;i++)       {
      input[i,0]= 0;       }
        MWNumericArray matinput = input;
```

Later, the console application was demonstrated to experts on the subject. The experts consist of project managers (3 person) in the software companies in Techno-Park, and academicians in Computer Engineering (3 person) and Industrial Engineering (3 person) departments. The performance of the model was tested and analyzed by experts. In literature, the values of the Root Mean Squared Error-RMSE (Eq. 10), the Mean Absolute Percentage Error-MAPE (Eq. 11), the Mean Squared Error-MSE (Eq. 12) and the Linear Regression-R2 (Eq. 13) are used in order to measure the estimation performance of the suggested network structure. Here, it is preferred to obtain low value from the Eqs. 10–12; high value from the Eq. 13 [53–56].

$$RMSE = \sqrt{\frac{\sum (y_t - \hat{y}_t)^2}{T}} \tag{10}$$

$$MAPE = \frac{1}{T} \cdot \sum \left| \frac{y_t - \hat{y}_t}{y_t} \right| \times 100 \tag{11}$$

$$MSE = \frac{\sum (y_t - \hat{y}_t)^2}{T} \tag{12}$$

$$R^2 = 1 - \left(\frac{\sum_t (y_t - \hat{y}_t)^2}{\sum_t (\hat{y}_t)^2} \right) \tag{13}$$

(y_t = Actual values, ^y = Estimated values, T = Number of estimate)

The performance of the network model obtained in the study was measured with the Eqs. 10–13 by using ten (10) pieces of real project data that have different areas and are selected randomly. The experimental results obtained from the projects according to their areas are given in Tables 8 and 9. Here, lines specified according to areas shows the project outputs (the deviations in time, budget and the number of personnel, target and success of the project).

According to Table 8, the values of MSE, MAPE and RMSE were reached respectively with a rate of 0.0008, 1.48, 0.0174 in average through the model formed by using the real project data at the result of the example model application, and a high success rate (98.51%) was obtained.

Table 8. The experimental results for different project areas.

Neural Network Model (45-15-5)								
Project No	Project Scope	Project Outputs (Deviations)	Actual Values	Estimated Values	RMSE	MAPE	MSE	R^2
1	Network App	The project duration	0.0000	0.0030	0.0030	0.3000	0.0000090000	99,70000
		The budget	0.5000	0.4670	0.0330	2.2000	0.0010890000	97,80000
		The number of personnel	0.0000	0.0520	0.0520	5.2000	0.0027040000	94,80000
		The target	0.0000	0.0500	0.0500	5.0000	0.0025000000	95,00000
		The project success	0.2500	0.2070	0.0430	3.4400	0.0018490000	96,56000
2	Mobile App	The project duration	0.2500	0.2460	0.0040	0.3200	0.0000160000	99,68000
		The budget	0.7500	0.7720	0.0220	1.2571	0.0004840000	98,74286
		The number of personnel	0.2500	0.2560	0.0060	0.4800	0.0000360000	99,52000
		The target	0.2500	0.2890	0.0390	3.1200	0.0015210000	96,88000
		The project success	0.2500	0.2490	0.0010	0.0800	0.0000010000	99,92000
3	Game App	The project duration	0.0000	0.0060	0.0060	0.6000	0.0000360000	99,40000
		The budget	0.0000	0.0130	0.0130	1.3000	0.0001690000	98,70000
		The number of personnel	0.0000	0.0050	0.0050	0.5000	0.0000250000	99,50000
		The target	0.0000	0.0070	0.0070	0.7000	0.0000490000	99,30000
		The project success	0.2500	0.2610	0.0110	0.8800	0.0001210000	99,12000
4	Communication App	The project duration	0.5000	0.5060	0.0060	0.4000	0.0000360000	99,60000
		The budget	0.2500	0.2480	0.0020	0.1600	0.0000040000	99,84000
		The number of personnel	0.2500	0.2330	0.0170	1.3600	0.0002890000	98,64000
		The target	0.0000	0.0140	0.0140	1.4000	0.0001960000	98,60000
		The project success	0.2500	0.2470	0.0030	0.2400	0.0000090000	99,76000
5	ERP App	The project duration	0.0000	0.0020	0.0020	0.2000	0.0000040000	99,80000
		The budget	1.0000	1.0150	0.0150	0.7500	0.0002250000	99,25000
		The number of personnel	0.2500	0.2380	0.0120	0.9600	0.0001440000	99,04000
		The target	0.0000	0.0420	0.0420	4.2000	0.0017640000	95,80000
		The project success	0.2500	0.2670	0.0170	1.3600	0.0002890000	98,64000
6	Education App	The project duration	0.0000	0.0023	0.0023	0.2300	0.0000052900	99,77000
		The budget	0.2500	0.2612	0.0112	0.8960	0.0001254400	99,10400
		The number of personnel	0.0000	0.0189	0.0189	1.8900	0.0003572100	98,11000
		The target	0.0000	0.1317	0.1317	13.1700	0.0173448900	86,83000
		The project success	0.2500	0.2430	0.0070	0.5600	0.0000490000	99,44000

(*continued*)

Table 8. (*continued*)

Neural Network Model (45-15-5)								
Project No	Project Scope	Project Outputs (Deviations)	Actual Values	Estimated Values	RMSE	MAPE	MSE	R^2
7	Embedded Systems App	The project duration	1.0000	0.9664	0.0336	1.6800	0.0011289600	98,32000
		The budget	1.0000	0.9640	0.0360	1.8000	0.0012960000	98,20000
		The number of personnel	0.5000	0.5082	0.0082	0.5467	0.0000672400	99,45333
		The target	0.0000	0.0206	0.0206	2.0600	0.0004243600	97,94000
		The project success	0.2500	0.2598	0.0098	0.7840	0.0000960400	99,21600
8	Database App	The project duration	0.5000	0.4975	0.0025	0.1667	0.0000062500	99,83333
		The budget	0.0000	0.0044	0.0044	0.4400	0.0000193600	99,56000
		The number of personnel	0.7500	0.7373	0.0127	0.7257	0.0001612900	99,27429
		The target	0.0000	0.0377	0.0377	3.7700	0.0014212900	96,23000
		The project success	0.0000	0.0087	0.0087	0.8700	0.0000756900	99,13000
9	Defense Industry App	The project duration	0.5000	0.4840	0.0160	1.0667	0.0002560000	98,93333
		The budget	0.2500	0.2569	0.0069	0.5520	0.0000476100	99,44800
		The number of personnel	1.0000	1.0037	0.0037	0.1850	0.0000136900	99,81500
		The target	0.0000	0.0054	0.0054	0.5400	0.0000291600	99,46000
		The project success	0.2500	0.2656	0.0156	1.2480	0.0002433600	98,75200
10	Cloud Computing App	The project duration	0.2500	0.2572	0.0072	0.5760	0.0000518400	99,42400
		The budget	0.7500	0.7461	0.0039	0.2229	0.0000152100	99,77714
		The number of personnel	0.5000	0.5019	0.0019	0.1267	0.0000036100	99,87333
		The target	0.0000	0.0375	0.0375	3.7500	0.0014062500	96,25000
		The project success	0.2500	0.2496	0.0004	0.0320	0.0000001600	99,96800

The comparative indication of actual and estimation values obtained from the model according to their areas and the rates of errors are given in Fig. 6, the results' regression curve is given in Fig. 7. It is seen that the estimation results obtained with the model by using the real project data coincide with the actual results largely and the errors' rates are also very low (close to zero) when Fig. 7 is examined.

According to the regression curve in Fig. 7, it is observed that there is a high relationship with the rate of 0.9933 between the actual values and the estimation values obtained from the model for different project areas.

The experimental results for the same project areas are given in Table 9. According to Table 9, the values of MSE, MAPE and RMSE were reached respectively with a rate of 0.0007, 1.40, 0.0180 in average through the model formed by using the real project data at the result of the example model application, and a high success rate (98.59%) was obtained.

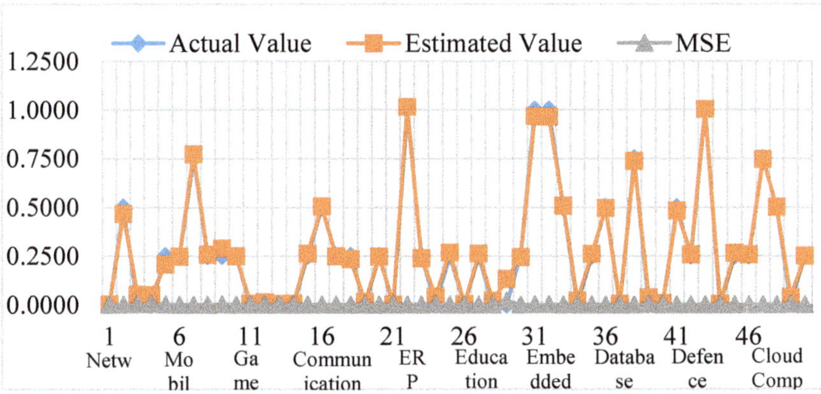

Fig. 6. The comparison of actual and estimated data for different project areas.

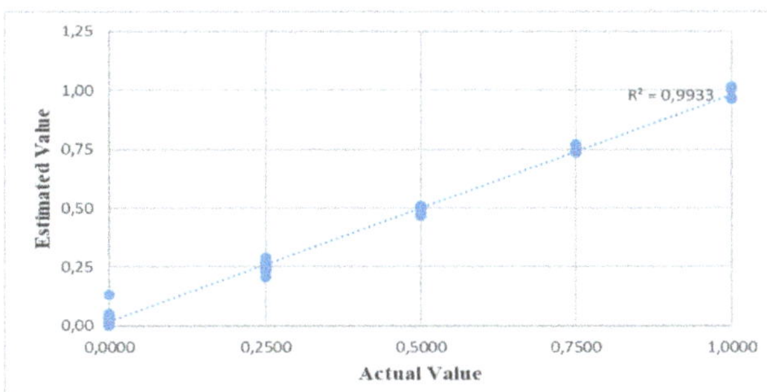

Fig. 7. The regression graph of the actual and estimated values for different project areas.

Table 9. The experimental results for the same project areas.

Neural network model (45-15-5)								
Project no	Project scope	Project outputs (deviations)	Actual values	Estimated values	RMSE	MAPE	MSE	R^2
1	Network App.-1	The project duration	0,2500	0,2435	0,0065	0,5200	0,0000422500	99,48000
		The budget	1,0000	0,9749	0,0251	1,2550	0,0006300100	98,74500
		The number of personnel	0,2500	0,2318	0,0182	1,4560	0,0003312400	98,54400
		The target	0,0000	0,0113	0,0113	1,1300	0,0001276900	98,87000
		The project success	0,2500	0,2296	0,0204	1,6320	0,0004161600	98,36800
2	Network App.-2	The project duration	0,2500	0,2359	0,0141	1,1280	0,0001988100	98,87200
		The budget	1,0000	1,0288	0,0288	1,4400	0,0008294400	98,56000
		The number of personnel	0,5000	0,4992	0,0008	0,0533	0,0000006400	99,94670
		The target	0,0000	0,0233	0,0233	2,3300	0,0005428900	97,67000
		The project success	0,2500	0,2717	0,0217	1,7360	0,0004708900	98,26400

(*continued*)

Table 9. (*continued*)

Neural network model (45-15-5)								
Project no	Project scope	Project outputs (deviations)	Actual values	Estimated values	RMSE	MAPE	MSE	R^2
3	Network App.-3	The project duration	0,5000	0,4966	0,0034	0,2267	0,0000115600	99,77330
		The budget	1,0000	1,0092	0,0092	0,4600	0,0000846400	99,54000
		The number of personnel	1,0000	0,9913	0,0087	0,4350	0,0000756900	99,56500
		The target	0,7500	0,7512	0,0012	0,0686	0,0000014400	99,93140
		The project success	0,2500	0,2643	0,0143	1,1440	0,0002044900	98,85600
4	Network App.-4	The project duration	0,2500	0,2520	0,0020	0,1600	0,0000040000	99,84000
		The budget	0,5000	0,4824	0,0176	1,1733	0,0003097600	98,82670
		The number of personnel	0,0000	0,0083	0,0083	0,8300	0,0000688900	99,17000
		The target	0,0000	0,0222	0,0222	2,2200	0,0004928400	97,78000
		The project success	0,2500	0,2857	0,0357	2,8560	0,0012744900	97,14400
5	Network App.-5	The project duration	0,5000	0,5041	0,0041	0,2733	0,0000168100	99,72670
		The budget	1,0000	1,0128	0,0128	0,6400	0,0001638400	99,36000
		The number of personnel	0,5000	0,5011	0,0011	0,0733	0,0000012100	99,92670
		The target	0,0000	0,0144	0,0144	1,4400	0,0002073600	98,56000
		The project success	0,2500	0,2533	0,0033	0,2640	0,0000108900	99,73600
6	Embedded Systems App.-1	The project duration	0,0000	0,0063	0,0063	0,6300	0,0000396900	99,37000
		The budget	0,2500	0,2557	0,0057	0,4560	0,0000324900	99,54400
		The number of personnel	0,2500	0,2573	0,0073	0,5840	0,0000532900	99,41600
		The target	0,0000	0,0041	0,0041	0,4100	0,0000168100	99,59000
		The project success	0,2500	0,2379	0,0121	0,9680	0,0001464100	99,03200
7	Embedded Systems App.-2	The project duration	0,2500	0,3031	0,0531	4,2480	0,0028196100	95,75200
		The budget	0,2500	0,1900	0,0600	4,8000	0,0036000000	95,20000
		The number of personnel	0,5000	0,4747	0,0253	1,6867	0,0006400900	98,31330
		The target	0,2500	0,2330	0,0170	1,3600	0,0002890000	98,64000
		The project success	0,2500	0,1623	0,0877	7,0160	0,0076912900	92,98400
8	Embedded Systems App.-3	The project duration	1,0000	1,0033	0,0033	0,1650	0,0000108900	99,83500
		The budget	0,2500	0,2843	0,0343	2,7440	0,0011764900	97,25600
		The number of personnel	1,0000	0,9862	0,0138	0,6900	0,0001904400	99,31000
		The target	0,2500	0,2008	0,0492	3,9360	0,0024206400	96,06400
		The project success	0,2500	0,2480	0,0020	0,1600	0,0000040000	99,84000

(*continued*)

Table 9. *(continued)*

Neural network model (45-15-5)								
Project no	Project scope	Project outputs (deviations)	Actual values	Estimated values	RMSE	MAPE	MSE	R^2
9	Embedded Systems App.-4	The project duration	0,5000	0,5153	0,0153	1,0200	0,0002340900	98,98000
		The budget	1,0000	1,0204	0,0204	1,0200	0,0004161600	98,98000
		The number of personnel	0,5000	0,5032	0,0032	0,2133	0,0000102400	99,78670
		The target	0,0000	0,0188	0,0188	1,8800	0,0003534400	98,12000
		The project success	0,2500	0,2470	0,0030	0,2400	0,0000090000	99,76000
10	Embedded Systems App.-5	The project duration	1,0000	1,0025	0,0025	0,1250	0,0000062500	99,87500
		The budget	0,2500	0,2554	0,0054	0,4320	0,0000291600	99,56800
		The number of personnel	1,0000	1,0245	0,0245	1,2250	0,0006002500	98,77500
		The target	0,0000	0,0752	0,0752	7,5200	0,0056550400	92,48000
		The project success	0,2500	0,2713	0,0213	1,7040	0,0004536900	98,29600

The comparative indication of actual and estimation values obtained from the model according to their areas and the rates of errors are given in Fig. 8, the results' regression curve is given in Fig. 9. It is also seen that the estimation results obtained with the model by using the real project data coincide with the actual results largely and the errors' rates are also very low (close to zero) when Fig. 8 is examined.

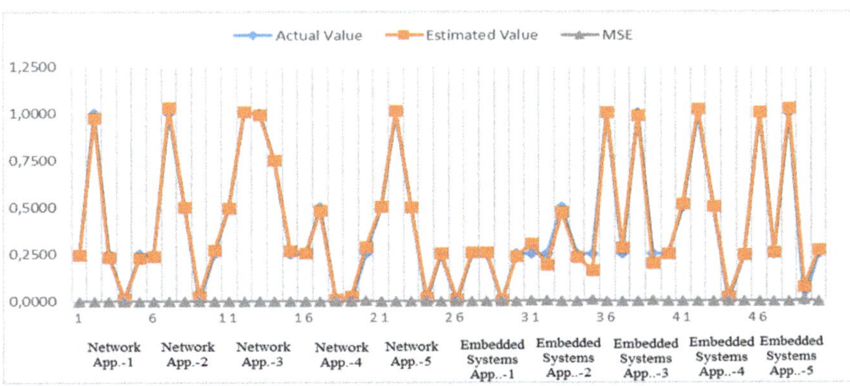

Fig. 8. The comparison of actual and estimated data for the same project areas.

According to the regression curve in Fig. 9, it is observed that there is a very high relationship with the rate of 0.9942 between the actual values and the estimation values obtained from the model for the same project areas.

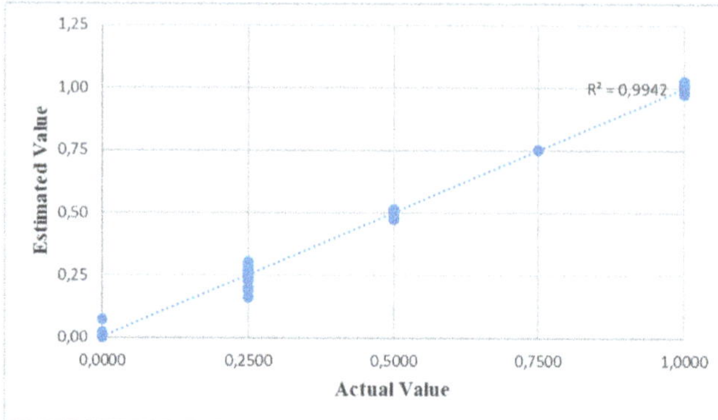

Fig. 9. The regression graph of the actual and estimated values for the same project areas.

The experimental results, the analyses and the comparisons for the same and different project areas clearly set forth that the suggested ANN model could perform the estimations with high accuracy, consequently the model was quite well in the performance and the usage of ANN was very effective in estimation of the software project risks as the different project areas.

In addition, experts stated that the prediction performance is quite high and they suggested that the model could be realized early risk assessment by embedding in project management package programs such as MS Project. However, they said that the proposed estimation model could be used not only in software area but also in other risk management areas.

6 Conclusions, Suggestions and Future Work

In this study, an ANN based model estimating the project risks in the software projects was developed. As the model has forty-five inputs, single hidden layer (fifteen neurons) and five outputs (45-15-5), the estimation performance is very high. The experimental results showed that the suggested estimation model gave better results as the number of data used in the training was large and various. The proposed model is strong enough to predict the risk factors that can be encountered in software projects in different areas and fit the structure of the problem, especially since it takes into account the scope of the project.

In addition to that, when the other models suggested in similar studies were examined and compared with the proposed model, it was found that the results were both more accurate and the training time is faster. At this point, the experimental results found with the traditional methods can be used as contributor in supporting ANN although the estimation with ANN gives effective results. Furthermore, the network structures in conformity with the problem's structure should be researched for the formed model to give good results. Also, the number of hidden layers, the number of neurons in layers and the neuron activation functions should be determined properly.

When this number is large, this case should not be ignored as this large number might extend the process and cause the network to memorize.

In the next stage of the study, it is planned to compare the estimation model performance in details on the basis of the areas, to increase the number and the variety of the real data used, to use the hybrid approaches particularly while forming the model and finally to develop a web-based system to suggest the risk control methods by considering the estimation results obtained from the model (model's outputs). All details of the planned web-based system will be submitted for publication in another article after completion of the development phase.

Acknowledgments. The authors would like to thank experts in technoparks at Gazi University, at Hacettepe University, at Middle East Technical University, and at Ankara University for their very helpful suggestions during obtaining data about their software projects.

Appendix

Risk Identification Form

Project Scope Mobile App. () Network App. () Game App. () Communication App. () Other (…)

Position Analyst () Experts () Team Leader () Project Manager () General Manager () Developer () Testing Personnel ()

Project Time 0-6 months () 6 - 12 months () 12-24 months () 24-48 months () more than 48 months ()

Project Budget (TL) Under 50000 () 50000 TL -100000 TL () 100 000-250 000 () 250000-500000 () 500000 over ()

Personnel Number Less than 10 () 10-30 () 30-50 () 50-100 () more than 100 ()

Is there a deviation in the project duration? No () Yes () …%

(The time difference between specified delivery date in the beginning with actual date)

Is there a deviation in the budget? No Yes () …%

(The difference between specified budget in the beginning with budget at the project delivery date)

Is there a deviation in the number of personnel? No Yes () …%

(The difference between specified personnel number in the beginning with personnel number at the project delivery date)

Is there a deviation from the target? No () Yes () …%

(The completion status of the project work packages specified in the beginning)

Is there a deviation in project success? No () Yes () …%

(The completion of project on time, with determined budget, number of personnel and target)

Risk Factors	Risk identification	Risk Level: Very Low 1	Low 2	Inter mediate 3	High 4	Very High 5	Occurrence Frequency of Risk (Enter a number from 1-5)
Time Risks	Difficulties experienced in the preparation of the schedule plan						
	Occurrence of the problems in the hardware distribution						
	Occurrence of the delay in project delivery duration						
	Difficulties experienced in the implementation of the prepared and approved schedule plan						
	Unexpected changes in project structure						
Budget (Cost) Risks	Encountering with budget changes						
	Being not able to be completed project with the targeted and approved budget						
	Occurrence of difficulties in determining of the resource costs (human, tool and material)						
	Occurrence of the difference between real and planned cost						
	Investment restrictions for the project						
Management Risks	Unfulfilled the responsibilities of the personnel						
	Not being used resources effectively						
	Occurrence of the problems in the management according to the plan of the project						
	Challenging priorities in aligned correctly						
	Experiencing difficulties in providing the communication among the team						
Technical Risks	The lack of conformity of the physical aspects (Safety, work environment, radiation, material) of a software development environment						
	The lack of used tools and equipment's						
	Occurrence of the equipment changes in the project process						
	Presence of software activities that challenge the personnel in the project						
	Not being able to benefit from advanced technology						
Planning and Program Risks	The lack of clearity of the project scope						
	Showing ambivalent attitudes of the contractor's.						
	Changing of the customer's priorities in the project process						
	Experiencing of the difficulties in finding the materials						
	Being large the size of project						
Contract and Legal Risks	Being severe of the contract conditions						
	Experiencing problems related to data copyright						
	Occurrence of changes in the contract text						
	Being used of unlicensed software						
	Breach of contract made by stakeholders						
Personnel Risks	Being insufficient number of personnel						
	Experiencing of the personnel changes						
	Being inadequate of the education level of personnel						
	Experiencing health problems of personnel in the project process						
	The lack of team spirit among team members						

	Experiencing some delays in the delivery of equipment and supplies						
	Being inadequate of the project in terms of goals and plants						
Other Risks	Being problematic of the personnel in terms of the ethical and moral						
	Experiencing political instability in the project process						
	Experiencing natural disasters in the project process						

References

1. Yong, H., Juhua, C., Zhenbang, R., Liu, M., Kang, X.: A neural networks approach for software risk analysis, In: Sixth IEEE International Conference on Data Mining Workshops, ICDM Workshops, pp. 722–725. IEEE, December 2006
2. Calp, M.H., Akcayol, M.A.: Risk factors and risk management process encountered in software project. Marmara Univ. J. Sci. **1**, 1–20 (2015). https://doi.org/10.7240/mufbed.72223
3. Tavares, B.G., da Silva, C.E.S., de Souza, A.D.: Risk management analysis in Scrum software projects. Int. Trans. Oper. Res. **26**, 1884–1905 (2017)
4. Gülebaglan, S.: Risk management in software development, Master thesis, Canakkale Onsekiz Mart University, Institute of Science (1991)
5. Boehm, B.W.: Software risk management: principles and practices. IEEE Softw. **8**(1), 32–41 (1991)
6. Islam, S., Mouratidis, H., Weippl, E.R.: An empirical study on the implementation and evaluation of a goal-driven software development risk management model. Inf. Softw. Technol. **56**(2), 117–133 (2014)
7. Teklemariam, M.A., Mnkandla, E.: Software project risk management practice in Ethiopia. Electron. J. Inf. Syst. Dev. Ctries. **79**(1), 1–14 (2017)
8. Cole, S., Giné, X., Tobacman, J., Topalova, P., Townsend, R., Vickery, J.: Barriers to household risk management: evidence from India. Am. Econ. J.: Appl. Econ. **5**(1), 104–135 (2013)
9. Bannerman, P.L.: Risk and risk management in software projects: a reassessment. J. Syst. Softw. **81**(12), 2118–2133 (2008)
10. Zavvar, M., Yavari, A., Mirhassannia, S.M., Nehi, M.R., Yanpi, A., Zavvar, M.H.:. Classification of risk in software development projects using support vector machine. J. Telecommun., Electron. Comput. Eng. (JTEC) **9**(1), 1–5 (2017)
11. Soofifard, R., Bafruei, M.: An optimal model for Project Risk Response Portfolio Selection (P2RPS) (2017)
12. Hu, Y., Huang, J., Chen, J., Liu, M., Xie, K.: Software project risk management modeling with neural network and support vector machine approaches. In: Third International Conference on Natural Computation, ICNC 2007, vol. 3, pp. 358–362. IEEE, August 2007
13. Elzamly, A., Hussin, B.: Quantitative and intelligent risk models in risk management for constructing software development projects: a review. Int. J. Softw. Eng. Its Appl. **10**(2), 9–20 (2016)
14. Es, H.A., Kalender, F.Y., Hamzacebi, C.: Forecasting the net energy demand of Turkey by artificial neural networks. J. Fac. Eng. Arch. Gazi Univ. **29**(3), 495–504 (2014)

15. Hamzaçebi, C.: Artificial Neural Network: Uses for Estimation: Applied MATLAB and Neurosolutio. Ekin Publishing, Bursa (2011)
16. Khoshgoftaar, T.M., Lanning, D.L.: A neural network approach for early detection of program modules having high risk in the maintenance phase. J. Syst. Softw. **29**(1), 85–91 (1995)
17. Khoshgoftaar, T.M., Allen, E.B., Hudepohl, J.P., Aud, S.J.: Application of neural networks to software quality modeling of a very large telecommunications system. IEEE Trans. Neural Netw. **8**(4), 902–909 (1997)
18. Tang, A.G., Wang, R.L.: Software project risk assessment model based on fuzzy theory. In: 2010 International Conference on Computer and Communication Technologies in Agriculture Engineering, vol. 2, pp. 328–330. IEEE (2010)
19. Chulani, S., Boehm, B., Steece, B.: Bayesian analysis of empirical software engineering cost models. IEEE Trans. Softw. Eng. **25**(4), 573–583 (1999)
20. Selby, R.W., Porter, A.A.: Learning from examples: generation and evaluation of decision trees for software resource analysis. IEEE Trans. Softw. Eng. **14**(12), 1743–1757 (1988)
21. Munson, J.C., Khoshgoftaar, T.M.: The detection of fault-prone programs. IEEE Trans. Softw. Eng. **18**(5), 423–433 (1992)
22. Rodriguez, V., Tsai, W.T.: Evaluation of software metrics using discriminant analysis. In: Proceedings-IEEE Computer Society's International Computer Software & Applications Conference. IEEE (1987)
23. Chua, D.K.H., Loh, P.K., Kog, Y.C., Jaselskis, E.J.: Neural network for construction project success. Expert Syst. Appl. **13**(4), 317–328 (1997)
24. Roiger, R., Geatz, M.: Data Mining: A Tutorial-Based Primer. Addison Wesley, New York (2003)
25. Neumann, D.E.: An enhanced neural network technique for software risk analysis. IEEE Trans. Softw. Eng. **28**(9), 904–912 (2002)
26. Xu, Z., Khoshgoftaar, T.M., Allen, E.B.: Application of fuzzy expert systems in assessing operational risk of software. Inf. Softw. Technol. **45**(7), 373–388 (2003)
27. Fan, C.F., Yu, Y.C.: BBN-based software project risk management. J. Syst. Softw. **73**(2), 193–203 (2004). https://doi.org/10.1016/j.jss.2003.12.032
28. Yang, L., Jones, B.F., Yang, S.H.: Genetic algorithm based software integration with minimum software risk. Inf. Softw. Technol. **48**(3), 133–141 (2006)
29. Hui, A.K.T.: Risks analysis of software development using bayesian belief network and non-linear programming methods. The Claremont Graduate University and California State University, Long Beach (2009)
30. Li-ying, X., Xin-zheng, W.: Risk evaluation the construction project based on BP Neural Network. In: International Conference on Management and Service Science (MASS), pp. 1–4, 24–26 August 2010 (2010)
31. Sharma, O., Khan, A.Y.: Risk based testing: a fuzzy approach. Int. J. Comput. Sci. Eng. **3** (10), 3346 (2011)
32. Hu, Y., Zhang, X., Ngai, E.W.T., Cai, R., Liu, M.: Software project risk analysis using Bayesian networks with causality constraints. Decis. Support Syst. **56**, 439–449 (2013)
33. Kalogirou, S.A., Bojic, M.: Artificial neural networks for the prediction of the energy consumption of a passive solar building. Energy **25**(5), 479–491 (2000)
34. Bayram, S., Kaplan, K., Kuncan, M., Ertunc, H.M.: Ball bearings space of time statistical feature extraction and neural networks with error estimation method size. In: Automatic Control National Meeting, TOK 2013, Malatya, 26–28 September 2013 (2013)
35. Sahin, I.: Prediction of surface roughness of Al/SiC composite material with artificial neural networks. J. Fac. Eng. Arch. Gazi Univ. **29**(1), 209–216 (2014)

36. Ozdemir, V.: Determination of Turkey's carbonization index based on basic energy indicators by artificial neural networks. J. Fac. Eng. Arch. Gazi Univ. **26**(1), 9–15 (2011)
37. Efendigil, T., Önüt, S., Kahraman, C.: A decision support system for demand forecasting with artificial neural networks and neuro-fuzzy models: a comparative analysis. Expert Syst. Appl. **36**(3), 6697–6707 (2009)
38. Kalogirou, S.A.: Artificial intelligence for the modeling and control of combustion processes: a review. Prog. Energy Combust. Sci. **29**(6), 515–566 (2003)
39. Fındık, T., Taşdemir, Ş., Şahin, İ.: The use of artificial neural network for prediction of grain size of 17-4 pH stainless steel powders. Sci. Res. Essays **5**(11), 1274–1283 (2010)
40. Askin, D., Iskender, I., Mamizadeh, A.: Dry type transformer winding thermal analysis using different neural network methods. J. Fac. Eng. Arch. Gazi Univ. **26**(4), 905–913 (2011)
41. Karataş, C., Sozen, A., Dulek, E.: Modelling of residual stresses in the shot peened material C-1020 by artificial neural network. Expert Syst. Appl. **36**(2), 3514–3521 (2009)
42. Hamzacebi, C., Kutay, F.: Electric consumption forecasting of Turkey using artificial neural networks up to year 2000. J. Fac. Eng. Arch. Gazi Univ **19**(3), 227–233 (2004)
43. Huang, J., Li, Y.F., Xie, M.: An empirical analysis of data preprocessing for machine learning-based software cost estimation. Inf. Softw. Technol. **67**, 108–127 (2015)
44. Erdal, H.: Contribution of machine learning methods to the construction industry: prediction of compressive strength. Pamukkale Üniversitesi Mühendislik Bilimleri Dergisi **21**(3), 109–114 (2015)
45. Elzamly, A., Hussin, B.: Managing software project risks with proposed regression model techniques and effect size technique. Int. Rev. Comput. Softw. **6**(2), 250–263 (2011)
46. Hijazi, H., Alqrainy, S., Muaidi, H., Khdour, T.: Risk factors in software development phases. Eur. Sci. J. **10**(3), 213–232 (2014)
47. Verner, J.M., Brereton, O.P., Kitchenham, B.A., Turner, M., Niazi, M.: Risks and risk mitigation in global software development: a tertiary study. Inf. Softw. Technol. **56**(1), 54–78 (2014)
48. Higuera, R.P., Haimes, Y.Y.: Software Risk Management (No. CMU/SEI-96-TR-012). Carnegie-Mellon Univ Pittsburgh Pa Software Engineering Institute (1996)
49. Conrow, E.H., Shishido, P.S.: Implementing risk management on software intensive projects. IEEE Softw. **14**(3), 83–89 (1997). https://doi.org/10.1109/52.589242
50. Kansala, K.: Integrating risk assessment with cost estimation. IEEE Softw. **14**(3), 61–67 (1997)
51. Bodea, C.N., Dascalu, M.I.: Modeling research project risks with fuzzy maps. J. Appl. Quant. Methods **4**(1), 17–30 (2009)
52. Hoodat, H., Rashidi, H.: Classification and analysis of risks in software engineering. World Acad. Sci. Eng. Technol. **56**(32), 446–452 (2009)
53. Karaatlı, M., Helvacıoglu, O.C., Omurbek, N., Tokgoz, G.: An artificial neural network based automobile sales forecasting. Int. J. Manag. Econ. Bus. **8**(17), 87–100 (2012). https://doi.org/10.11122/ijmeb.2012.8.17.290
54. Ozkan, M.T., Eldem, C., Köksal, E., Metem, Ş.S.U.A.V.: Notch sensitivity factor determination with artificial neural network for shafts under the bending stress. Pamukkale Univ. J. Eng. Sci. **19**(1), 24–32 (2013). https://doi.org/10.5505/pajes.2013.88598
55. Gandomi, A.H., Roke, D.A.: Assessment of artificial neural network and genetic programming as predictive tools. Adv. Eng. Softw. **88**, 63–72 (2015)
56. Calp, M.H.: An estimation of personnel food demand quantity for businesses by using artificial neural networks. J. Polytech. **22**(3), 675–686 (2019). https://doi.org/10.2339/politeknik.444380

Routing of Maintenance and Repair Operations of Mobile Based Fault Notifications of Municipal Services

Tuncay Yiğit[1] and Huseyin Coskun[2(✉)]

[1] Department of Computer Engineering, Süleyman Demirel University, Isparta, Turkey
tuncayyigit@sdu.edu.tr

[2] Department of Computer Technologies, Uşak University, Uşak, Turkey
huseyin.coskun@usak.edu.tr

Abstract. The aim of this study is to solve routing problem of maintenance and repair service of municipal administration within the scope of municipal services. The fault locations have been consisted of mobile notifications which are reported by citizens. In this context, a mobile application has been developed to able to notify the faults by citizens in the scope of municipal services. The citizens are able to notify the fault of a municipal service via multimedia content based on their location by using the mobile application. The fault notifications that reach to municipal administration needs to be routed and scheduled as to location and time. In this study, the routing problems of transportation of maintenance and repair service have been solved with the aim of repair the fault reported by citizens in different places in city. These fault notifications are graded by three level as very urgent, urgent and non-urgent and grouped by service type. The number of maintenance staff and vehicles are defined for each service type. The Integer Linear Programing, Simulated Annealing and Genetic Algorithm optimization methods have been utilized to solve the routing problem. The best improving value of total routing distance of maintenance operation and best optimization rate has been found as 11.9%, 13.34% respectively.

Keywords: Mobile software · Artificial intelligence · Genetic Algorithm · Integer Linear Programing · Simulated Annealing · Applied mathematics · Intelligent systems · Big data · Optimization · Municipal services

1 Introduction

The municipality services include that executing and management urban society services by municipal employees to aim of best maintaining of life of society. The large number of municipal tasks are associated with numerous similar responsibilities like local traffic and transportation, car parking control, management and disposal of waste, maintenance of public parks, roads and sidewalks, water supply systems, lighting of streets, environmental planning and cleaning. It's important that these services are carried out efficiently and affordably for sustainability of these services. Also, some of these services requires unplanned maintenance and repairs. The performing maintenance and repair of some services are as important as sustainability of these services.

D. J. Hemanth and U. Kose (Eds.): ICAIAME 2019, LNDECT 43, pp. 320–334, 2020.
https://doi.org/10.1007/978-3-030-36178-5_24

The timely performing municipal maintenance and repair services affects citizens' pleasantness directly. The maintenance and repair affairs usually are carried out in a plan by municipal employees. Sometimes, maintenance and repair services which emerged with result of a fault in somewhere in the city are hard to determine and follow by the municipal employees. In such a case, the citizens notify or complain to municipality about the faults. The maintenance and repair of the faults that notified by citizens mostly are included this plan or carried out in another plan by other employees. It is interfered directly in emergency case like fire, flooding. The citizens are able to report the faults by phone or over web site form. Recently, the mobile application of some municipalities has been being used by citizens to notify the faults too. When the studies on this topic is researched, it's seen that the citizens don't play active role to improvement municipal services in municipalities' mobile application, contrarily most of these mobile applications are only used to give information to citizens or tourist. On the contrary, if the users aren't active role as a matter of course in application model which will be developed toward aimed process as to reason of mobile applications' using, the designed and developed mobile applications remain incapable. If the studies on either mobile application used by citizens to report fault in city or classification of faults and hazards, routing, scheduling solves to maintenance has been analyzed, it is seen that there aren't studies which assembles these processes together or in one mobile application. The Atlanta Municipality has developed an application for cycling users to determine routes and identify potholes on these roads with the application of Cycle Atlanta [1]. With this application, the roads have been determined, improved and the faults of the roads have been eliminated. Firstly, The Johannesburg Roads Agency [2] introduced the freely downloadable Find it and Fix it application for smartphones and then Seattle Municipality [3] provide the citizens to report pothole on the roads, nonlegal graffiti paintings, abandoned vehicles and faults of street lights through the mobile application named as Find It, Fix It. Kenteris et al. [4] has designed a mobile application for tourist guide on multiplatform. They offer mobile application based on customizable tourist guide model which provide a specific travel plan to travel touristic places of the city. Horton et al. [5] have taken out a patent to monitoring municipal operations and develop an alarm system. The methods and systems for efficient network and device-based monitoring and responses to municipal functions, operations, infrastructure and public institutions with the proposed system use automatic notifications of municipal agencies or municipal operation information employees through offered system in this study. Gutiérrez et al. [6] have developed a mobile application to utilize more efficiency municipal services using infrastructure of IOT and virtual reality through participant of the citizens. With this application, citizens can find out where the services of municipalities are, how they are carried out and what services are in their location. In this context, a mobile application has been designed to make the citizens play active role in improvement and sustainability of municipal services in this study. The novel model has been presented through the mobile application toward report efficiently notify and compliant about municipal services, classification of these faults automatically and offer maintenance route to municipal employees. In this model, the notification which reported by citizens' mobile device in different place of city has been classified with artificial intelligence as to its subject and a routing and scheduling method has been presented to determine a working route with the aim of resolving and maintenance of these fault notifications by municipal employees. The working diagram of the model as mentioned is given in Fig. 1.

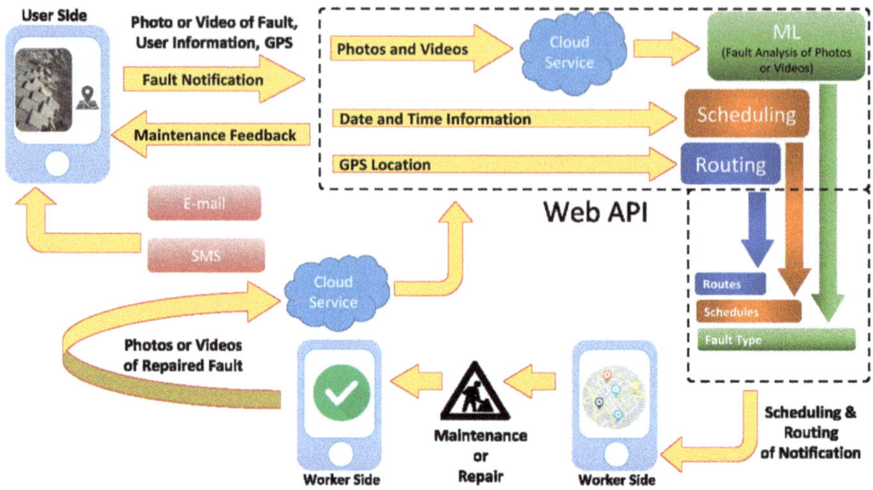

Fig. 1. Working diagram of mobile based fault notification system

Through the mobile application in the model, the citizens are able to notify a municipal fault location based with multimedia content to municipality. The mobile application provides following advantages;

- It's able to determined easily where the fault is without define its address, directions or street name due to system has location-based notification.
- The municipal service content of fault is able to expressed explicitly, easily and more understandable due to multimedia based without the need to describe by citizens. The type of fault in the photo or video is detected through image processing and artificial intelligence techniques. The image-net database of Deng et al. [13] has been used to make this process easier. The object in the image or video is detected, thereby it's obtained that image is belong to which fault type through a rule-based method. The rule-based method has not been explained due to its issue of another study.
- The citizens' personal information, e-mail, phone number is able to send with the notify, as may be wanted by citizens or when needed. Thus, the citizens will be able to follow own notify or will be informed about the fault with feedback by the municipal employee.
- It's specified automatically that which crew will repair or maintenance and which working equipment will be selected to repair with the mobile application. For that purpose, municipal maintenance employees have been recorded to system as to their working areas and the equipment or machines that are used in maintenance or repair process are recorded to system as rule-based list. Thus, the equipment or machines that is needed to maintenance is checked before go to fault location and it has been prevented to forget.

- The optimum cost and time of maintenance or repair is computed, by scheduling and routing of maintenance process.
- It will be contributed to sustainability of municipal services, by storing the statistical data of maintenance or repair process.
- The important faults in terms of municipal services which are difficult to notice or follow by municipal employees will have been detected by citizens.
- The citizens will have contributed significantly and more productive to make municipal services more sustainably.

The faults reported by mobile app are graded by three level as very urgent, urgent and non-urgent and grouped by service type. The number of maintenance staff are described according to service type. For example, the number of staffs has been defined as 5 and 4 persons for pothole and sidewalk fault respectively and 2 persons for park and garden arrangement. The detailed definitions about the number of required staffs are given in material section.

2 Material

The faults at following scopes have been determined to use in the study,

(1) Park and Garden Arrangement,
(2) Traffic Control,
(3) Road and Sidewalk Structure Analysis and Cleaning,
(4) Monitoring Pollution,
(5) Management and Disposal of Waste.

The example fault photos have been taken in different place of city and are given in Table 1.

Table 1. Sample fault photos of municipal services.

1	2	3	4
			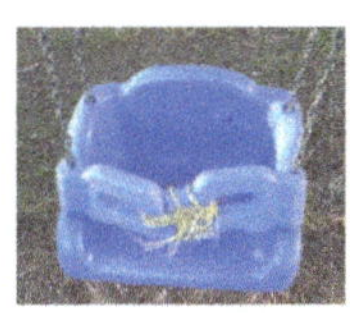
5	**6**	**7**	**8**
	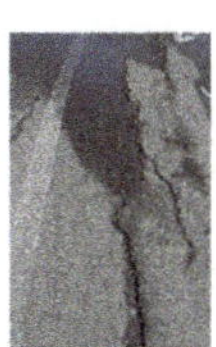		

The scopes, fault type, importance level, GPS location (latitude, longitude) and maintenance duration of municipal services belong to the sample fault photo are given in Table 2.

Table 2. Detail information of sample fault photos in Table 1.

ID	Service type	Fault type	IL	Location (latitude, longitude)	Team members
1	3	Sidewalk crack cleaning	1	38.667578, 29.397540	4
2	5	Deficiency of protection circle or warning sign	2	38.660136, 29.391063	2
3	1	Sport equipment breakage	1	38.668488, 29.400149	2
4	1	Park equipment breakage	1	38.655847, 29.408571	2
5	3	Sidewalk crack	2	38.679531, 29.366186	4
6	3	Pothole	2	38.676180, 29.332701	5
7	5	Deficiency of rubbish bin	1	38.665250, 29.404028	2
8	2	Deficiency of towaway zone sign	1	38.674727, 29.391682	2

The numbered bookmarks of fault photos location are shown in Fig. 2. The number zero bookmark represents municipal maintenance and repair facility and technical works center.

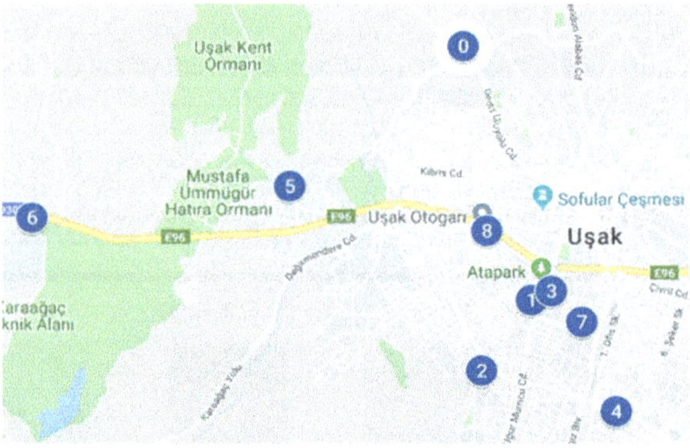

Fig. 2. The map of fault photos location

The municipal maintenance facility location is accepted as starting point because in order that the required distances of fault locations have been calculated. The distances of maintenance location are given in Table 3 as kilometer.

Table 3. Distance matrix.

Location	0	1	2	3	4	5	6	7	8
0	0	4,97	6	4	8,30	3	6,50	5	2,87
1	4,97	0	0,900	0,467	2,82	4	8,10	0,854	2
2	6	0,900	0	0,930	4	5	8	2	2,10
3	4	0,467	0,930	0	2	3,95	7	0,674	1,30
4	8,30	2,82	4	2	0	6	8,90	2	3,93
5	3	4	5	3,95	6	0	4	6	4
6	6,50	8,10	8	7	8,90	4	0	7,20	5
7	5	0,854	2	0,674	2	6	7,20	0	2
8	2,87	2	2,10	1,30	3,93	4	5	2	0

3 Method

In this study, transportation services which is required to carry out municipal maintenance and repair operations are assumed as a transportation model. Basically, it is found to a solution that to minimize the following three issues in transportation problems.

- cost of transportation,
- duration of transportation,
- total number of vehicles in transportation.

The distances between each other of locations which are going to be belong to the model are given in Table 3. It is seen that the model is not complex, if the data are analyzed. Therefore, the transportation model is able to be formulated as Linear Programing using specific notations. Accordingly, integer linear programming (ILP) in class of exact solution methods has been utilized to determine routes with minimum total distance. Besides, the problem has been adapted and solved to capacity vehicle routing problem (CVRP) algorithm which is one of the basic routing problems. Besides, Simulated Annealing (SA) algorithm which is meta-heuristic routing method has been utilized alongside ILP and all results of these methods and municipal services manually defined route are compared.

3.1 Integer Linear Programming

The vehicle routing optimization problems aim to make the vehicle arrive location over specific nodes with minimum cost. These nodes define a graph connected to each other. It is assumed that the CVRP is defined with graph $G = (V, E)$ with this context. $V = \{0,1, \ldots, N, N + 1\}$, $E = \{(i, j): i, j \in V, i \neq j\}$ are defined as node cluster and edge cluster respectively in graph G. The whole route is toured by one vehicle, so only one route is going to be created at the result of solving of model. The definition of parameter, indexes and variables used in the model are given in Table 4.

Table 4. The definition of parameter, variables and indexes.

Parameter	Definition
"0"	Center of municipal maintenance and repair service
M	The number of vehicles
N	The number of fault locations
C	Vehicle capacity
D_{ij}	Distance from i location to j location
q_i	The number of staffs who will go to i locations
Index	**Definition**
i,j,p	$\{0,1, \ldots.. N, N+1\}$ The fault nodes index
k	$\{1, \ldots.., M\}$ Vehicle index
Variables	**Definition**
X_{ijk}	$\begin{cases} 1, \textit{if vehicle k goes to i location to j location} \\ 0, \textit{otherwise} \end{cases}$
Y_j	The variable that prevent to creation sub tour

The objective function of the model is given in Eq. 1 [7].

$$Z_{min} = \sum_{i=0}^{N+1} \sum_{\substack{j=0 \\ j \neq i}}^{N+1} \sum_{k=1}^{M} D_{ij} X_{ijk} \tag{1}$$

There are standard constraints with distinctive values of every problem for solving vehicle routing through linear programing. The constraints for this model are consists of joining of each other fault nodes connected equal number ways, the number of vehicles, location of maintenance and repair center, distance between this center and fault nodes, staff capacity of vehicles and preventing of sub tours. The definitions of these constraints are given in Table 5.

Table 5. The definitions of constraints for the ILP model used in this study [14].

Definition of constraint	Equation	No
The number of vehicles	$\sum_{k=1}^{M} \sum_{j=1}^{N+1} X_{ijk} = M, \quad i = 0,$	(2)
The joining of each other fault nodes connected equal number ways	$\sum_{i=0}^{N+1} X_{ipk} - \sum_{j=0}^{N+1} X_{pjk} = 0, k = 1, \ldots M, \quad p = 1, \ldots N,$	(3)
The using of only one way from one fault location	$\sum_{k=1}^{M} \sum_{\substack{j=0 \\ j \neq 1}}^{N+1} X_{ijk} = 1, \quad i = 1, \ldots.N,$	(4)

(*continued*)

Table 5. *(continued)*

Definition of constraint	Equation	No
Ensuring going out of every vehicle from center and return of every vehicle to center.	$\sum_{j=1}^{N} X_{0jk} = 1, k = 1, \ldots., M,$	(5)
	$\sum_{i=1}^{N} X_{i,N+1,k} = 1, k = 1, \ldots., M,$	(6)
Ensuring not exceeding of staff capacity of vehicle	$\sum_{i=0}^{N+1} \sum_{\substack{j=0 \\ j \neq 1}}^{N+1} q_i X_{ijk} \leq C, k = 1, \ldots., M,$	(7)
Prevention of creation of sub tours	$Y_j \geq Y_i + 1 - N\left(1 - \sum_{k=1}^{M} X_{ijk}\right), i \neq j, i = 1, \ldots..N, j = 1, \ldots, N$	(8)
	$Y_j \geq 0, j = 1, \ldots, N$	(9)

The objective function of the model is defined using data in Tables 2 and 3 according to these definitions and equations following;

$$\begin{aligned} Z_{min} = {} & 4.97X_{01} + 6X_{02} + 4X_{03} + 8.3X_{04} + 3X_{05} + 6.5X_{06} + 5X_{07} + 2.87X_{08} \\ & + 4.97X_{10} + 0.9X_{12} + 0.467X_{13} + 2.82X_{14} + 4X_{15} + 8.10X_{16} \\ & + 0.854X_{17} + 2X_{18} + 6X_{20} + 0.9X_{21} + 0.93X_{23} + 4X_{24} + 5X_{25} \\ & + 8X_{26} + 2X_{27} + 2.10X_{28} + 4X_{30} + 0.467X_{31} + 0.93X_{32} + 2X_{34} \\ & + 3.95X_{35} + 7X_{36} + 0.674X_{37} + 1.3X_{38} + 8.3X_{40} + 2.82X_{41} \\ & + 4X_{42} + 2X_{43} + 6X_{45} + 8.9X_{46} + 2X_{47} + 3.93X_{48} + 3X_{50} \\ & + 4X_{51} + 5X_{52} + 3.95X_{53} + 6X_{54} + 4X_{56} + 6X_{57} + 4X_{58} \\ & + 6.5X_{60} + 8.10X_{61} + 8X_{62} + 7X_{63} + 8.9X_{64} + 4X_{65} + 7.2X_{67} \\ & + 5X_{68} + 5X_{70} + 0.854X_{71} + 2X_{72} + 0.674X_{73} + 2X_{74} + 6X_{75} \\ & + 7.2X_{76} + 2X_{78} + 2.87X_{80} + 2X_{81} + 2.10X_{82} + 1.3X_{83} \\ & + 3.93X_{84} + 4X_{85} + 5X_{86} + 2X_{87} \end{aligned}$$

Equation 2 is utilized to calculate the number of vehicles which go out from municipal maintenance operation facility. In this model, it is represented by M and value is 1. Besides, the number of ways which come and go from one node to another node are equal according to Eq. 3. It is defined that the vehicle needs to go from only one fault node to another fault node according to Eq. 4;

$$X_{10} + X_{12} + X_{13} + X_{14} + X_{15} + X_{16} + X_{17} + X_{18} = 1$$
$$X_{20} + X_{21} + X_{23} + X_{24} + X_{25} + X_{26} + X_{27} + X_{28} = 1$$
$$X_{30} + X_{31} + X_{32} + X_{34} + X_{35} + X_{36} + X_{37} + X_{38} = 1$$
$$X_{40} + X_{41} + X_{32} + X_{34} + X_{35} + X_{36} + X_{37} + X_{38} = 1$$
$$X_{50} + X_{51} + X_{52} + X_{53} + X_{54} + X_{56} + X_{57} + X_{58} = 1$$
$$X_{60} + X_{61} + X_{62} + X_{63} + X_{64} + X_{65} + X_{67} + X_{68} = 1$$
$$X_{70} + X_{71} + X_{72} + X_{73} + X_{74} + X_{75} + X_{76} + X_{78} = 1$$
$$X_{80} + X_{81} + X_{82} + X_{83} + X_{84} + X_{85} + X_{86} + X_{87} = 1$$

The constraint equations that express the number of vehicles which going out from center and return to center is equal 1 and these are defined according to Eqs. 5 and 6;

$$X_{01} + X_{02} + X_{03} + X_{04} + X_{05} + X_{06} + X_{07} + X_{08} = 1$$
$$X_{10} + X_{20} + X_{30} + X_{40} + X_{50} + X_{60} + X_{70} + X_{80} = 1$$

The constraint that the number of maintenance staff need to not exceed capacity of vehicle is defined according to Eq. 7 following;

$$2X_{21} + 2X_{31} + 2X_{41} + 4X_{51} + 5X_{61} + 2X_{71} + 2X_{81} \leq 6$$
$$4X_{12} + 2X_{32} + 2X_{42} + 4X_{52} + 5X_{62} + 2X_{72} + 2X_{82} \leq 6$$
$$4X_{13} + 2X_{23} + 2X_{43} + 4X_{53} + 5X_{63} + 2X_{73} + 2X_{83} \leq 6$$
$$4X_{14} + 2X_{24} + 2X_{34} + 4X_{54} + 5X_{64} + 2X_{74} + 2X_{84} \leq 6$$
$$4X_{15} + 2X_{25} + 2X_{35} + 2X_{45} + 5X_{65} + 2X_{75} + 2X_{85} \leq 6$$
$$4X_{16} + 2X_{26} + 2X_{36} + 2X_{46} + 4X_{56} + 2X_{76} + 2X_{86} \leq 6$$
$$4X_{17} + 2X_{27} + 2X_{37} + 2X_{47} + 4X_{57} + 5X_{67} + 2X_{87} \leq 6$$
$$4X_{18} + 2X_{28} + 2X_{38} + 2X_{48} + 4X_{58} + 5X_{68} + 2X_{78} \leq 6$$

The constraints in Eqs. 8 and 9 prevent to arise of sub route. These constrains haven't been utilized due to there are no sub route in this model. The created model in according to these definitions has been solved using 0−1 ILP equations.

3.2 Simulated Annealing (SA)

The SA is among meta-heuristics optimization algorithm. The SA has presented a solution to escape from local optima for the first time and it has been applied to solve a combinatorial optimization problem by Kirkpatrick et al. [8]. The basic principle of algorithm is to decrease dynamically p value which is possibility of acceptance of bad solution instead of good solution by increasing iteration. When an arrangement is made in this way, in the first parts of the problem solution there is a lot of leap between the solution zones, and when the number of iterations increases and the solutions, we get are near to 0, the search zone becomes narrower. The possibility of selection of bad solution decrease systematically with temperature. The temperature is a definition which decrease uniformly or logarithmically in general dependent on iteration. The unconstrained and bound-constrained optimization problems is able to be solved with SA. The parameters of SA are given in Table 6.

Table 6. The parameter of SA.

Parameter	Definition
T_0	Initial temperature
T	Current temperature
s	Initial solution obtained by the nearest neighbor algorithm
$N(s)$	Current solution
S	Best solution
f(s)	Total Routing Costs (Objective function value of Vehicle Routing Subproblem) for solution s
p	Uniform random number between 0 and 1
alpha	Rate of decrease temperature when update T
s'	Neighboring solution of S in each iteration

The general working of a SA algorithm can be seen from the Fig. 3 with parameters in Table 6 [7].

```
s ← GenerateInitialSolution()
T ← T0
while termination conditions not met do
  s' ← PickAtRandom(N(s))
  if (f(s') < f(s)) then
     s ← s'                % s' replaces s
  else
     Accept s' as new solution with probability p(T, s', s)
  endif
  Update(T)
endwhile
```

Fig. 3. Pseudo code of SA

The control parameters of SA in this study have been selected as study of Miao et al. [17] and are given in Table 7.

Table 7. Control parameters of SA

Parameter	Value
T_0	100
Maximum iteration	120
Maximum number of inner iterations	10
Alpha	0.9

3.3 Genetic Algorithm (GA)

GA is a general search and optimization algorithm developed by influencing the functioning of the natural world [9]. GA is a coincidental research technique that works with an evolutionary approach based on Darwin's best surviving principle [10]. GA was developed by John Holland in 1975 and Goldberg, who was then a student, continued his studies. Goldberg describes GA as a search algorithm that randomly scans the solution space [11]. A basic flowchart of GA algorithm is given in Fig. 4.

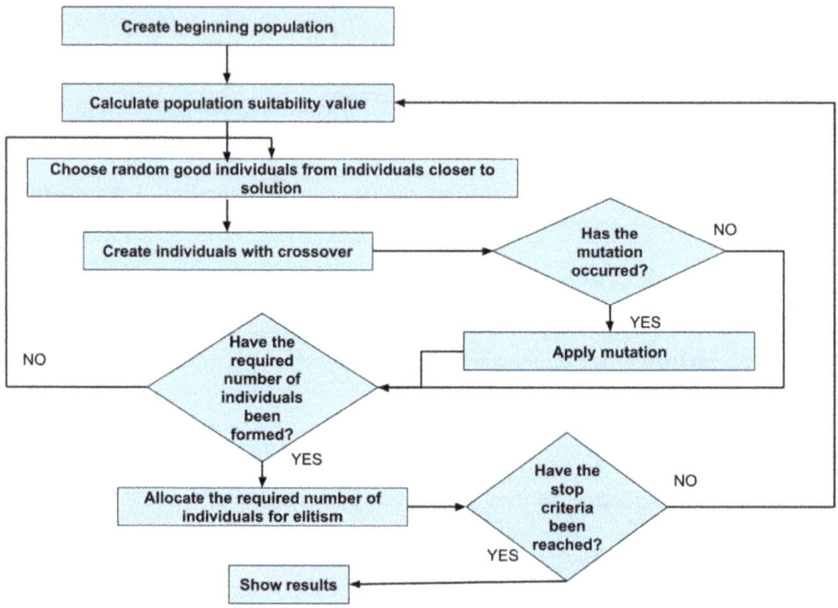

Fig. 4. Flowchart of GA [12]

If the problem in this study is tackled as a permutation optimization (PO) problem, route space for this problem consists of every possible permutation of routes that visit each node once. As to there are 8 maintenance location to be visited, and because once a maintenance location has been visited that it is not eligible to be visited to again, the size of the route space stands as 8! and it means there will be 40320 possible permutations. In order to explain how GA can be applied to PO problems, similarities between them are given in Table 8 [15].

Table 8. Similarities between PO and GA

PO	GA
Encoded solution	Chromosome
Solution	Decoded chromosome
Set of solution	Population
Objective function	Fitness Function (FF)

At the beginning in GA, it is able to start with a heuristic solution which is defined as starting population. It is assumed that each parameter of the heuristic solution is a gene and a whole solution is a chromosome. An example structure of the offspring and chromosome are given in Fig. 5.

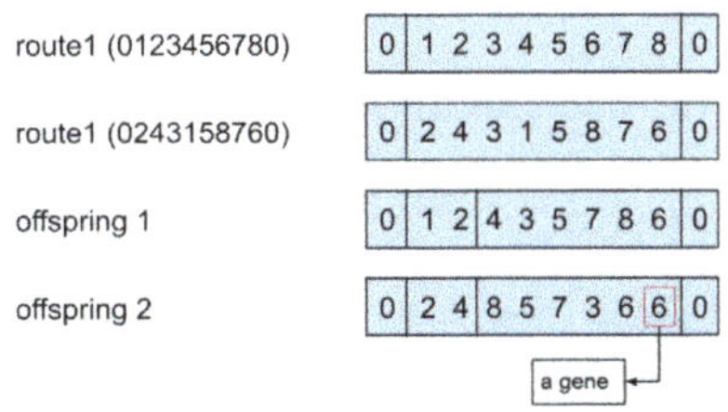

Fig. 5. Structure of offspring and chromosome.

The new generation chromosomes called as offspring are created by crossing over two chromosome or mutating a chromosome, after first chromosomes are created. The chromosomes which has furthest distance to FF are eliminated and it is aimed to reach optimal solution which means FF value is to minimize within each iteration algorithm. The control parameters of GA in this study have been selected as study of Schaffer et al. [16] and are given in Table 9.

Table 9. Control parameters of GA.

Parameter	Value
Population size	200
Location size	8
Crossover size	20
Iteration number	100
Mutation rate	0.1
Crossing over rate	0.9

4 Results and Discussion

If municipal staff member visits the maintenance location over the route which determined by manually, route distance has been calculated as 30.471 km. In this study, capacity value of vehicle and repair team have been restricted with 6. Therefore, it's assumed that one vehicle is allocated for one repair team to maintenance and repair operation of faults. When the model that is applied with ILP have been solved, and results analyzed with these assumptions, it is going to be seen that Z_{min} found as 26.991. Besides, the route solution has been found as 0-8-3-1-7-2-4-5-6-0 with evaluation of following equation. The optimization rate of ILP has been found as 12, 9.

$$X_{08} + X_{83} + X_{31} + X_{17} + X_{72} + X_{24} + X_{45} + X_{56} + X_{60} = 1$$

The distance of routes, and routes which determined by municipal workers using by maps and the results obtained by the optimization methods utilized in this study are given Table 10 and Fig. 6.

Table 10. The manual and experimental solutions.

Route no	Route	Method	Iteration	Distance	Optimization rate (%)
1	0-5-3-1-7-2-8-4-6-0	SA	20	35.904	−15,13
2	0-5-3-1-7-4-8-2-6-0	SA	24	35.004	−12,95
3	0-6-5-8-3-2-1-7-4-0	SA	83	28.784	5,86
4	0-8-1-3-2-4-7-5-6-0	SA	88	28.767	5,92
5	0-8-1-3-7-2-4-5-6-0	SA	99	28.511	6,87
6	**0-2-1-3-7-4-8-5-6-0**	**SA**	**112**	**28.471**	7,02
7	**0-5-6-1-2-4-7-3-8-0**	**GA**	**61**	**26.844**	13,34

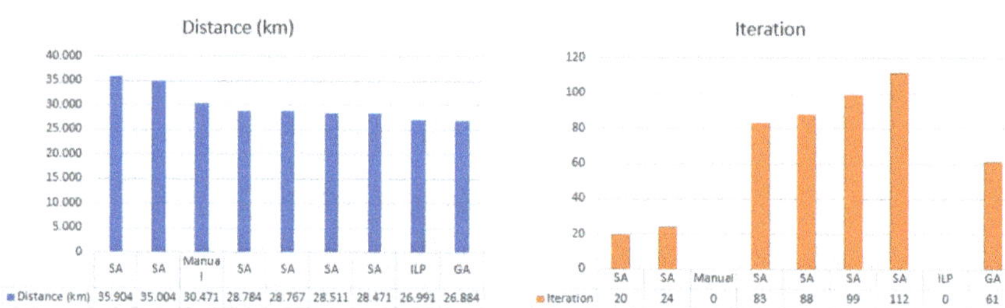

Fig. 6. Distance solutions and iteration results.

ILP and GA methods has found two best distance solution according to experimental results. The solution of GA has obtained best route solutions when lowest number of iterations is considered. It is observed that SA method has better performance when the number of iterations has been increased.

5 Conclusions and Future Works

If those which is carried out in this study are summarized, on behalf of reflection of social municipalism perspective and performing municipality services effectively, a system has been proposed to make citizen to notify municipal service faults. If proposed system is analyzed, it is going to be seen that there are many issues which are be able to studied in machine learning, computer vision and natural language processing with the aim of development and improvement of the system. In this study, routing problem of municipal maintenance and repair operation which involve a specific part of the proposed system has been tackled. Therefore, the distances between one location to other and each location to municipal maintenance and repair facility have been determined by using locations of faults that is notified by citizens. It is seen that problem of the study is a simple CVR problem, when this data is analyzed. ILP, SA and GA optimization methods which are most common using in routing problems have been utilized to calculate shortest distance of the route of maintenance locations. GA has calculated the shortest distance of the route with lowest the number of iterations. It has been assumed in solving of problem that there is one repair team in a vehicle and just one operation facility to maintenance and repair. The routing issue of this study can be expanded on by increasing the number of staffs, vehicles and maintenance facility. Besides, control parameters of the methods need to be adjusted to make methods obtain better solutions. Especially for GA, the studies on issue of adjusting size of population, mutation and crossing over rate can be carried out. The changing of selection operator based on the number of locations and its effect on GA can be researched in future works. Nature-inspired or meta-heuristic optimization algorithms like ant colony, artificial bee colony and artificial immune, particle swarm and tabu search can be experimented on this problem and the results can be compared each other. Dynamic routing problems can be adopted to proposed system. The maintenance and repair operations can be scheduled according to urgencies of faults or duration of maintenances and repairs. The optimal repair routing of fault locations can be calculated by using different optimization parameter with the scope of the scheduling.

References

1. Le Dantec, C.A., Watkins, K.E., Clark, R., Mynatt, E.: Cycle Atlanta and OneBusAway: driving innovation through the data ecosystems of civic computing. In: International Conference on Human-Computer Interaction, pp. 327–338. Springer, Cham, August 2015
2. Find it, Fix it mobile application (2014). https://www.jra.org.za/documents/pressrelease/Med_Rel_mobile_app_launch.pdf
3. Find it, Fix it mobile application (2016). http://www.seattle.gov/customer-service-bureau/find-it-fix-it-mobile-app
4. Kenteris, M., Gavalas, D., Economou, D.: Mytilene E-guide: a multiplatform mobile application tourist guide exemplar. Multimedia Tools Appl. **54**(2), 241–262 (2011)
5. Horton, C., Novotney, K., Herraghty, J.: U.S. Patent No. 8,138,940. U.S. Patent and Trademark Office, Washington, DC (2012)
6. Gutiérrez, V., Galache, J.A., Sánchez, L., Muñoz, L., Hernández-Muñoz, J.M., Fernandes, J., Presser, M.: SmartSantander: Internet of things research and innovation through citizen participation. In: The Future Internet, pp. 173–186. Springer, Heidelberg (2013)
7. Blum, C., Roli, A.: Metaheuristics in combinatorial optimization: overview and conceptual comparison. ACM Comput. Surv. (CSUR) **35**(3), 268–308 (2003)
8. Kirkpatrick, S., Gelatt, C.D., Vecchi, M.P.: Optimization by simulated annealing. Science **220**(4598), 671–680 (1983)
9. Coley, D.A.: An introduction to Genetic Algorithms for Scientists and Engineers. World Scientific Publishing Company, Hackensack (1999)
10. Yiğit, T.: Meslek liseleri haftalık ders çizelgelerinin genetik algoritmalar yardımıyla oluşturulması. Gazi Üniversitesi Endüstriyel Sanatlar Eğitim Fakültesi Dergisi **19**, 25–39 (2006)

11. Goldberg, D.E., Holland, J.H.: Genetic algorithms and machine learning. Mach. Learn. **3**(2), 95–99 (1988)
12. Çolak, R., Yığıt, T.: The effect of genetic algorithm parameters in the solution of the course timetable problem. In: International Conference on Computer Science and Engineering (UBMK), pp. 1090–1094. IEEE, October 2017
13. Deng, J., Dong, W., Socher, R., Li, L.J., Li, K., Fei-Fei, L.: ImageNet: a large-scale hierarchical image database. In: IEEE Conference on Computer Vision and Pattern Recognition, pp. 248–255. IEEE (2009)
14. Şahın, Y., Eroğlu, A.: Kapasite Kısıtlı Araç Rotalama Problemi İçin Metasezgisel Yöntemler: Bilimsel Yazın Taraması. Süleyman Demirel Üniversitesi İktisadi ve İdari Bilimler Fakültesi Dergisi **19**(4) (2014)
15. Potvin, J.Y.: Genetic algorithms for the traveling salesman problem. Ann. Oper. Res. **63**(3), 337–370 (1996)
16. Schaffer, J.D.A.R.: A study of control parameters affecting online performance of genetic algorithms for function optimization. San Meteo, California (1989)
17. Miao, H., Tian, Y.C.: Robot path planning in dynamic environments using a simulated annealing-based approach. In: 10th International Conference on Control, Automation, Robotics and Vision, pp. 1253–1258. IEEE (2008)

Safe Map Routing Using Heuristic Algorithm Based on Regional Crime Rates

Atakan Alpkoçak(✉) and Aydin Cetin(✉)

Gazi University, 06500 Ankara, Turkey
{atakan.alpkocak, acetin}@gazi.edu.tr

Abstract. The general structure of the map routing systems produced for ease of access to the destination point determined by the users in the route planning and guidance is based on the temporal cost improvement. Existing systems are intended to deliver users to the destination as quickly as possible. However, the route, which was created to provide temporal improvement, allows transitions from unsafe areas, and therefore users may face a security problem when following the created route. In this article, it is aimed to develop a new route suggestion method based on safety parameters in order to ensure the safety of the users while creating a transportation route to the destination. In addition to the estimated distance information used by the current applications, crime rates of the regions on the designated route are defined as a parameter and then used in heuristic function.

Keywords: Safe path planning · A* algorithm · Map routing

1 Introduction

Maps and routing systems undertake various tasks between the coordinates created according to the users' input. These tasks include the most appropriate route routing between coordinates and provide information about the region. Users aim to reach the target coordinates they want according to the information given on the map. In order to reach the target coordinate, route suggestion system and method of routing change according to users' needs. This capability can be displayed to the user by using different methods, such as the shortest route or the road suggestion with the least traffic density, depending on the user's preference.

The general structure of the map routing systems produced for the ease of access to the coordinate points determined by the users is based on the temporally cost improvement. The structure of existing systems provides users to reach the target in the fastest way. The route created to ensure temporal improvement allows transitions from unsafe areas. The route created by the existing systems does not include the route alternatives preventing users pass thru from areas with high-risk of crime levels. The lack of such alternative within the methods used may be resulted with distressful driving when using the routes that users do not know. In this article, it is aimed to develop a new route suggestion method in order drivers to have an alternative routing avoiding the routes with safety concern while creating the transportation route to the destination.

D. J. Hemanth and U. Kose (Eds.): ICAIAME 2019, LNDECT 43, pp. 335–346, 2020.
https://doi.org/10.1007/978-3-030-36178-5_25

In order for route suggestion to be performed in regions that can be considered safer, the security rating must be established by using the information obtained from past incidents in all the regions to be passed thru to reach to the user's destination. Each of the nodes on route must have a specific tangible degree of security rating and be processed as input data to the algorithms. While achieving this security rating, the crime rates used for social comparison can be simulated. The nodes of the road network specified in the simulation phase must receive the regional population density and the number of dangerous events. According to historical data, the specific reliability of each network node can be calculated. The most important constraints are the reliability and quality of the data obtained while providing the regional security rating as input to the algorithm. In order for the safe route proposition to be reliable, various information about events considered as dangerously should be included in the algorithm. The reason for this is that the solution set is wide. The comparison of the results of the relevant studies depends to a large extent on the reliability of the route. Route planning in the related studies is based on the number of crime incidents in the regions [1]. The high number of criminal incidents in a region allows the avoidance of that region during the route determination phase. This approach causes a certain level of deficiencies in route planning and decreases the accuracy of planning. The surplus of crime in the region depends to a certain extent on the population of the region [2]. Therefore, the numerical excess of the crime incidents arising from the population density prevents the final decision on the region. Due to the regression relationship between the number of crime incidents and the population density, the high number of crime incidents in the region does not permit classification in the name of the region's credibility [3]. According to the algorithm and heuristic function used in this article, clustering of crime events in a particular region does not affect route planning. The effect of the cluster of crime in a region on the heuristic function is only proportional to the amount of population. For this reason, the use of the regional crime rate value was preferred for a more precise classification of the region.

2 Materials and Methods

In order to develop and deploy a route planning system based on crime rate, the information obtained must be stored on a cloud-based database system. There is a need to create location-based functions on the cloud database to process the stored information and produce the desired result. In the study, The obtained data related to population d, various criminal events and geographic locations are stored on the problem-specific PostgreSQL database server at the cloud service provider. With the open source PostGIS database plug-in, various functions have been created for ease of spatial query as a result of the interaction with the existing database server. OSM data were used for city map information and pgRouting plugin used in the route routing phase. The data set of the algorithm to be used to create the planning should be prepared correctly. In the proposed system A* algorithm used and then A* algorithm is modified in conjunction with regional crime rates.

2.1 A* Algorithm

The A* algorithm is one of the heuristic search algorithms. A* algorithm has heuristic function. It provides the convenience of solving multidimensional problems with the help of heuristic function based on predictive function. In its most basic form, the result is the production function

$$f(x) = g(x) + h(x) \tag{1}$$

The function *g(x)* shown in (Eq. 1) is the actual cost that occurs until the desired search is performed during the search phase. The heuristic function *h(x)* is the problem-specific predictive cost. The estimated cost of a suggestion in the classical route planning is the heuristic distance of the coordinates between the extremes [4]. Some of the heuristic function relations used in the A* Algorithm are Manhattan, Euclidean and Chebyshev distances.

In this article, it is aimed to create the effect of the crime rate of each coordinate point on the function of *h(x)* besides the Euclidean distance relation [5] which is used to generate the predictive distance values of the heuristic function.

2.2 Modified A* Algorithm

A* Algorithm, which is one of the most popular algorithms used for route guidance, cannot solve the local minimum problem [6, 7]. In the route determination phase, the distance between the target coordinate point determined by the user and the current coordinate point is compared heuristicly between all the nodes in the map road network. The route determined during comparison may not be the best route suggestion. The local minimum problem must be exceeded to determine the best route suggestion [8].

In this article, determining the best route suggestion as a result of the effect of the formula of safety degree determined for overcoming the problem with the heuristic function was studied. The heuristic function included in the Modified A* Algorithm was modified using the equation proposed (Eqs. 2 and 4) by Shah et al. [1].

$$d(s,t) = \sqrt{(t_x - s_x)^2 + (t_y - s_y)^2} \tag{2}$$

Equation (2), s, the source coordinate information, t, the target coordinate information, d(s, t), Euclidean the relation between the points s and t points to each other.

$$\alpha(s,t) = t_{crimerate} - s_{crimerate} \tag{3}$$

Equation (3), $\alpha(s, t)$, It refers to the difference in crime rate between target and source location coordinates.

$$\min_{\underset{\rightarrow}{sEt}} \sum_{e \in E} \{\{(1-\beta).d(e)\} + \{\beta.\alpha(e)\}\} \tag{4}$$

Equation (4), β, refers to the degree of safety coefficient. If the crime rate difference is positive $[\alpha(e)]$, i.e. if the crime rate of the target location is higher than the crime rate of the source location, the crime rate to be used in the location should be added with the overall function of the result production function $[\{\beta.\alpha(e)\}]$. The reason for this is to decrease the priority and increase the heuristic distance value obtained with the Euclidean distance relation in the general function.

If the crime rate difference is negative $[\alpha(e)]$, i.e. if the crime rate of the target location is lower than the crime rate of the source location, the crime rate to be used in the location should be subtracted with the overall function of the result production function $[\{\beta.\alpha(e)\}]$. The reason for this is to increase the priority and decrease the heuristic distance value obtained with the Euclidean distance relation in the general function.

2.3 Crime Rate Function

Behaviors that are considered to be dangerous by the community and that are against the law are defined as a crime. The Oxford dictionary defines the safety as "the condition of being protected from or unlikely to cause danger, risk, or injury". Crime rate is the one of the parameters for measuring the safety of a region.

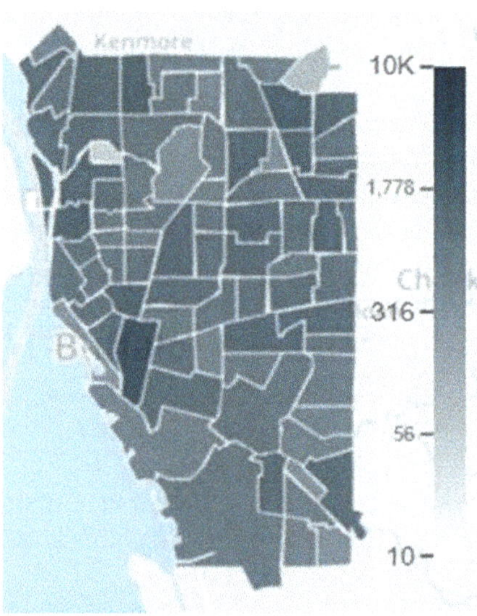

Fig. 1. City of buffalo crime events map

Crime rate is also a simple way to compare the safety of the regions. Figure 1 shows the crime events map pf the city of Buffalo divided into regions. The crime rate is defined as the ratio of the number of incidents that are considered as crimes to the number of population in the region. The increase in the crime rate in a certain region heuristicly constitutes an approach to the high probability of a new crime. Crime rate function is defined as [9]:

$$CrimeRate_S = \frac{n(S)}{p(S)} x b(S) \tag{5}$$

Equation (5), $n(S)$ refers to the number of incidents occurred in the region S, $p(S)$ refers the number of population of the region S, and $b(S)$ refers to the desired population ratio coefficient in the region S (general validity: 100,000).

2.4 Cloud Database

The process of keeping the structural or non-structural data on the virtual server accessible on the internet is called cloud computing, by using cloud technology for data storage, the load of personal devices can be reduced and the speed of data access can be increased. Cloud database can be created by means of the infrastructure and services provided by cloud technology. Cloud databases allow more effective access to data or data processing on the cloud server [10].

2.5 PostgreSQL and PostGIS

Database is a system where data is stored and managed by means of data storage templates created according to data properties [11]. Databases that are systems where various operations are facilitated on interrelated data; There are four different types of databases hierarchical, network, relational, and object-oriented. The use of species varies according to the data model. The SQL query language can be used to process data on databases according to various definitions, processes, or criteria. PostgreSQL, it is an open source relational-object database management system in accordance with SQL standards.

PostGIS is a spatial database extender for PostgreSQL object-relational database [12]. Spatial data management and spatial analysis facilitate the processing of queries on spatial data. It is aimed to facilitate complex spatial analysis using an SQL-like language. It can provide effective solutions by using specific spatial references and standards in the management of GIS and storage of data.

2.6 OpenStreetMap and PgRouting

OSM is a free, editable map of the whole world that is being built by volunteers largely from scratch and released with an open-content license [13]. Thanks to the GPS receivers, user information is obtained.

PgRouting, extends the PostGIS/PostgreSQL geospatial database to provide geospatial routing functionality [14]. It contains various algorithms of path routing.

3 Application

In this article, population information, knowledge of crime events and geographic information of the region should be obtained in order to establish navigation according to the rate of security. The data of the region was determined for the applicability of the study was obtained from the data which is presented and allowed to be used. The obtained data related to population, various criminal incidents and geographic data are stored on the problem-specific PostgreSQL database server at the cloud service provider. Various operations on the stored data are also performed on the cloud database. With the PostgreSQL database service created on the cloud service provider, it is possible to interrogate various geographic, crime and population information to associate data with each other and to process the data according to the desired situation. With the open source PostGIS database plug-in, various functions have been created for ease of spatial query as a result of the interaction with the existing database server. Figure 2 shows the components of the developed application.

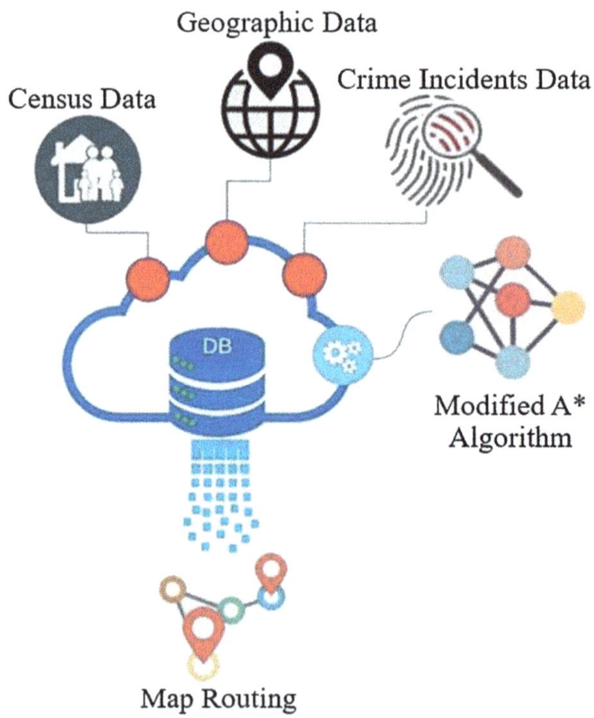

Fig. 2. Components of developed application

Thanks to the spatial functions obtained, the ability to associate the information between the population information and crime information is introduced into the system. The association process is defined by the regional boundary determinants with the spatial object structure of the multi-polygon within the population information, which is one of the data sets obtained [15]. The spatial structure of the Multi-polygon was used to determine the limits of regional validity of the current population information. The number of criminal events occurring within each border was established through a linkage over the regional population data set and the crime rate was obtained. The crime rate of the regions within each multi-polygon obtained by means of the crime rate production function was calculated and the crime rates of all the nodes in the region were appointed.

The region was chosen to be limited to understand the impact of the study results. OSM data were used for City of Buffalo city map information, which was designated as restricted area [16]. The crime data set [17] was used for the information of the crime incidents in the City of Buffalo area to establish the regional safety rate of the designated area. The crime rate calculation was made for each route node by the population information [18] together with the information of the obtained crime incidents. Thanks to the pgRouting plugin used in the route routing phase, various shortest path finding algorithms can be used. PgRouting has open source structure and allows problem-specific changes. The current structure of A* Algorithm does not operate according to crime rate. The structure of the algorithm has been changed for the creation of the problem solution space and the applicability of the A* Algorithm in problem. Through this change, it has been studied to overcome the local minimum problem, which is one of the limitations of the techniques used by the current systems.

Route guidance is provided to the target coordinate point that users want to reach with the effect of the safety rating parameter. The calculation of the security rating provides the result-based production function of the crime rate, which is a combination of the information of the crime events in the region and the population information. The system architecture created is supported by Modified A* Algorithm. The Modified A* Algorithm helps to plan users'route from their current location to the destination they want to reach. The Modified A* Algorithm finds the actual distance between each node on the route using the Euclidean equation [5].

The crime rates between each node on the route are calculated by the obtained data sets. Depending on the result of the heuristic function, the Modified A* Algorithm affects the result of the Euclidean relationship with a requirement.

The Modified A* Algorithm plans a route with the new node weight produced as a result of the effect. The most reliable route information, determined after the completion of the route planning phase, provides the route information data transmission with the interaction of the cloud service provider with the android mobile devices. Obtained the route information by means of the GPS satellite-based navigation system on the OpenStreet Map based map used can be displayed on Android mobile devices by users as seen in Figs. 3 and 4.

Fig. 3. Route planning with A* algorithm on Android device

Fig. 4. Route planning with A* algorithm on Android device

4 Results

In this article, the heuristic security function is used in the realization of the safe route suggestion. With the heuristic function used, users can determine their own safety factors and intervene in the current route planning. In order to understand the effect of the intervention, algorithms developed over classical approaches and the results of the proposed method were compared. Comparison of the distance of route between the A* algorithm and the modified A* algorithm is given in Table 1.

Table 1. Comparison of the distance of route between the A* algorithm and the modified A* algorithm

Source coordinate degrees		Target coordinate degrees		Route distance value (km)	
Latitude	Longitude	Latitude	Longitude	A*	Modified A*
42.912332	−78.83911	42.905421	−78.899764	44.361562	42.693787
42.912332	−78.83911	42.912090	−78.832399	50.357428	47.432536
42.912332	−78.83911	42.912367	−78.854937	55.390113	46.496656
42.912332	−78.83911	42.920840	−78.898551	49.529583	49.900937
42.912332	−78.83911	42.853633	−78.834812	50.804338	49.913266

Route planning is shown through map orientation application on android mobile devices. The coordinate degree determined on the map is determined as the fixed user position[1]

The costs of the different route directions arising from the differences in algorithm and heuristic function vary. The route obtained as a result of the intuitive function produced by the proposed method is costly as seen in Table 1. However, in accordance with the heuristic function approach used in this study, the cost is acceptable at certain scales.

In Table 2 comparison of between the A* algorithm and the Modified A* algorithm is given. As seen in Table 2, the results of the routes formed from the current and target location points determined by the modified heuristic function have established superiority to the classical heuristic function at various stages. Due to the lack of population information in certain regions, deficiencies occur in the study of the function of the modified heuristic result production function. The reason for this deficiency is to obtain population information in certain blocks. As a result of this lack of information, the crime rate values of the adjacent nodes in the blocks are equal. This equality does not affect the production of the heuristic function. In order for the heuristic function to make an accurate route planning, the population information should be obtained as small parts (multi-polygon) as possible on the map.

[1] Latitude: 42.912332, Longitude: −78.839111

Table 2. Comparison of between the A* algorithm and the modified A* algorithm

Source coordinate degrees		Crime rate of source coordinate	Target coordinate degrees		Crime rate of target coordinate	Average of crime rate	
Latitude	Longitude		Latitude	Longitude		A*	Modified A*
42.912332	−78.83911	59	42.936378	−78.879822	52.01	44.361562	42.693787
42.912332	−78.83911	59	42.853633	−78.834812	52	50.357428	47.432536
42.912332	−78.83911	59	42.916580	−78.903661	24	55.390113	46.496656
42.912332	−78.83911	59	42.92084	−78.898551	24	49.529583	49.900937
42.912332	−78.83911	59	42.930066	−78.895476	15	50.804338	49.913266

In Figs. 5 and 6, comparison of route map suggestions of Modified A* algorithm and the A* algorithm for different targets are shown. The route created for the reach of the target points with close source points due to the lack of population information of various regions is similar to the classical approaches.

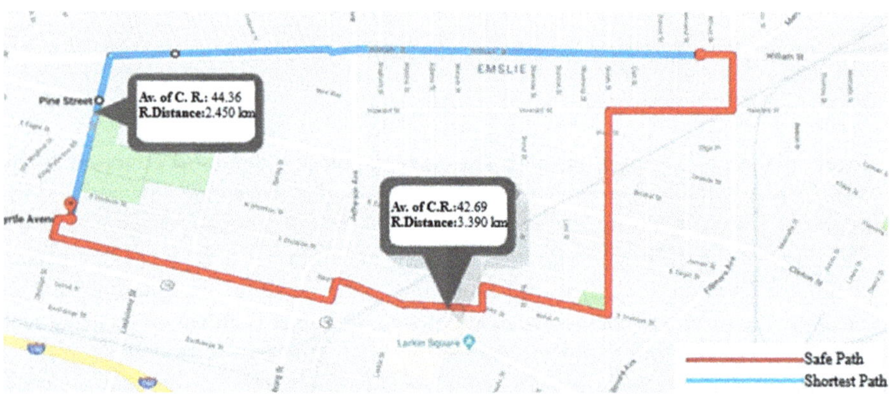

Fig. 5. Route map of comparison between the modified A* algorithm and the A* algorithm

The value of β used in the general function of Modified A* Algorithm is 0.99. Thanks to the β value, the effect of the crime rate value on the overall function is maximized. As a result of the effect achieved at the maximum level, the route distance increases and the route average crime rate decreases. A significant decrease in the β value, which can vary according to the preference of the user, was observed in the route distance cost decrease and the route average of crime rate increase. The limitation of the distance cost increase can be achieved by changing the β value by the users.

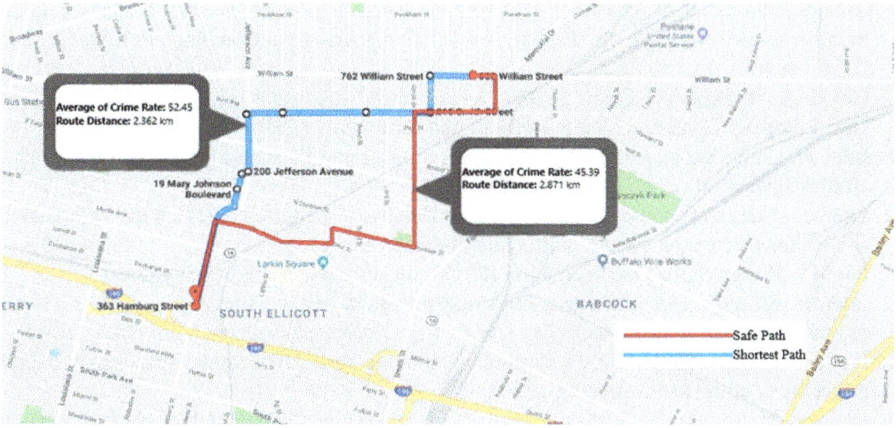

Fig. 6. Route map of comparison between the modified A* algorithm and the A* algorithm

5 Conclusions

In this article, it is aimed to develop a new route suggestion method based on safety parameters in order to ensure the safety of the users while creating a transportation route to the destination. In addition to the estimated distance information used by the current applications, crime rates of the regions on the designated route are defined as a parameter and then used in heuristic function. The degree of safety determined by users in regions supports the route determination phase. Route is offered to users according to their preferences enabling users to prevent from the routes they do not know in regions where crime rates are high. According to comparison of crime rates of regions, the developed heuristic result production function did not become evident in every situation in the regions where the regional population data are missing.

References

1. Shah, S., Bao, F., Lu, C.T., Chen, I.R.: Crowdsafe: crowd sourcing of crime incidents and safe routing on mobile devices. In: Proceedings of the 19th ACM SIGSPATIAL International Conference on Advances in Geographic Information Systems, pp. 521–524. ACM, November 2011
2. Chang, Y.S., Choi, S.B., Lee, J., Jin, W.C.: Population Size vs. Number of Crime-Is the Relationship Super-Linear? (2013)
3. Nolan, J.J.: Establishing the statistical relationship between population size and UCR crime rate: its impact and implications. J. Crim. Justice **32**(6), 547–555 (2004)
4. Nakisa, B., Rastgoo, M.N., Norodin, M.J.: Balancing exploration and exploitation in particle swarm optimization on search tasking. Res. J. Appl. Sci. Eng. Technol. **8**(12), 1429–1434 (2014)
5. Internet: Euclidean Distance. https://en.wikipedia.org/wiki/Euclidean_distance
6. Internet: Local and global maxima and minima. https://en.wikipedia.org/wiki/Maxima_and_minima#/media/File:Extrema_example_original.svg

7. Aine, S., Sharma, C., Likhachev, M.: Learning to search more efficiently from experience: A multi-heuristic approach. In: Eighth Annual Symposium on Combinatorial Search, May 2015
8. Goyal, A., Thakral, A., Sharma, G.K.: Improved A* algorithm for query optimization. In: 10th European Conference on Modelling and Simulation (2006)
9. Internet: California Department of Justice,Computational Formulas. https://oag.ca.gov/sites/all/files/agweb/pdfs/cjsc/prof10/formulas.pdf
10. Internet: Margaret Rouse, What is Cloud Database? February 2017. https://searchcloudapplications.techtarget.com/definition/cloud-database
11. Internet: PostgreSQL: Documentation (2018). https://www.postgresql.org/docs/
12. Internet: PostGIS: Documentation https://postgis.net/documentation/
13. Internet: OSM:OpenStreetMap Wiki (2014). https://wiki.openstreetmap.org/wiki/Main_Page
14. Internet: PGROUTING: User's Documentation, 01 June 2018. https://docs.pgrouting.org/pdf/en/pgRoutingDocumentation-2.6.0.pdf
15. Internet: Multipolygon Illustration. https://wiki.openstreetmap.org/wiki/File:Multipolygon_Illustration_6.svg
16. Internet: City of Buffalo Geographic Data (2018). https://download.geofabrik.de/north-america/us/new-york.html
17. Internet: City of Buffalo Crime Data (2018). https://data.buffalony.gov/Public-Safety/Crime-Incidents-Data-Lens-/vhp3-62vz
18. Internet: City of Buffalo Census Data (2010). https://data.buffalony.gov/Government/Block-Groups/i76q-u5zs

Image Spam Detection Using FENOMAA Technique

Aziz Barbar[1] and Anis Ismail[2(✉)]

[1] Department of Computer Science, Faculty of Arts and Sciences, American University of Science Technology, Beirut, Lebanon
abarbar@aust.edu.lb

[2] Department of Computer Engineering, University Institute of Technology, Lebanese University, Sidon, Lebanon
anismaiil@ul.edu.lb

Abstract. Today's most important mean of communication is the e-mail that allows people all over the world to communicate, share data, and perform business. Yet there is nothing worse than an inbox full of spam; i.e., information crafted to be delivered to a large number of recipients against their wishes. Numerous anti-spam methods and solutions have been proposed and deployed, but they are not effective as most mail servers rely on blacklists and rules engine leaving a big part on the user to identify the spam, while others rely on filters that might carry high false positive rate. The proposed solution is "Image Spam Detection using FENOMAA Technique". FENOMAA which stands for Feature Extraction Neural network with OCR enhanced by Mail Authentication and Analyzer of context is an aggregation of methods that detects spam and specifically image spam and is composed of four essential parts that include authentication, OCR enhanced by Neural Network, Feature Extraction enhanced by Neural Network, and Context Analyzer Spell Checker. The authentication method in which the SPF record is encapsulated within the Domain Keys Identified Mail (DKIM) header along with the use of trusted third party in DKIM as an alternative to DNS, thus allowing additional verification and strength to the proposed method. Also, The OCR and the Feature Extraction methods are enhanced by the addition of Neural Network where the neural net is composed and trained according to the criteria mentioned in this paper in order to generate the desired results. The Context Analyzer Spell Checker is present in order to correct, verify, and optimize the results taken from both neural nets.

Keywords: Image spam · Neural network · OCR · DKIM

1 Introduction

The Internet population has been growing and spreading in a way that not only connects each of its users in a virtual world, but also affects them. As the Internet is constantly evolving, claims that this virtual community of people (users) is increasing and that this growth is generating an inestimable value, a value of people connected together in a certain period of time. Now, imagine what this could be as a target for marketing and advertising. At the same time, it could also be detrimental to these users

D. J. Hemanth and U. Kose (Eds.): ICAIAME 2019, LNDECT 43, pp. 347–364, 2020.
https://doi.org/10.1007/978-3-030-36178-5_26

as this type of marketing and advertising changes, affecting the world's resource structure and users. Consider a table whose four wooden legs are the resource structure and capable of holding a load of 50 kg. Now bring a load of 70 kg. You will notice that the table would be damaged and broken. Apply it now to the Internet population whose re-source structure is its communication capable of containing a certain level of bandwidth. If we abuse and upgrade this level, the Internet population will be blocked and will be affected by itself and its users, which will be costly for the entire population a bur-den that starts from spam.

Email spam targets individual users with direct mail messages and junk mail lists: they are often produced by scanning Usenet messages, pinching Internet mailing lists, or probing for addresses on the web. Email spam usually costs users currency out of pocket. Many people whose phones are billed for Internet use read or receive their mail while billed for every minute they spend, spam costs them extra money. In addition, the transmission of spam by Internet Service Providers (ISPs) and online services is expensive, and these costs are directly passed on to subscribers.

In addition to wasting time and money on unwanted e-mails, spam also absorbs a lot of network bandwidth. As a result, many institutions, as well as individuals, are devoted to fighting spam through a variety of techniques. But since the Internet is public, there is really nothing to do to prevent spam, just as it is impossible to avoid spam.

Spam is still available until now because it does not cost the spammer anything other than managing mailing lists and it is difficult to keep the spammer responsible for mass mailings. Therefore, given the aforementioned feasibility, spammers are many and the volume of unsolicited mail has become very important. Costs incurred, such as lost efficiency and fraud, have become a handicap for the public and ISPs, who have been forced to increase their capabilities to cope with the increase in spam.

Email spam has progressively increased, even exponentially, since the early 1990s, getting billions of messages per day. Spam has aggravated, baffled and bothered email users. Anti-spam laws have been frequently implemented and some have been avoided. Spam represents on average 94% of all emails sent [1]. ISPs have attempted to recover the price of spam by filing lawsuits against spammers, although they have usually not been able to repair the damage regardless of winning in court [2].

In just a few years, spam has paying attention primarily on email, where it still exists today. In addition, the violent spamming of several major spammers, such as Cyber Promotions' Sanford Wallace, in the mid-1990s, has contributed to making spam an essentially public phenomenon in the minds of the public [3].

The problem that is discussed in this paper is the impact of e-mail spam and the various kinds of threats that we face in addition to describing the way spammers operate by gathering e-mail addresses and obfuscating the message content in order to prepare them for delivery. In this paper, we propose a solution to contribute in the struggle against e-mail spam that uses images as a way to overcome current spam filters, even though some people believe that sending multiple messages for the case of advertisement (spam) is a right for them; however, this causes more damage to the internet than they think because they are only looking at the issue from their own perspective and not from the other side where spam drains out the resources of the internet and put on a large cost yearly, thus we aim that the solution proposed would contribute to the limitation of spam through filtering the message content and image based content and by introducing the Domain Authentication approach (DKIM) which in addition to the filtering mechanism will be able to identify and verify whether this e-mail is a spam or not.

2 Related Works

This section lists a variety of solutions for tackle spam and picture based spam, where the luminosity is shacked on the process and procedure used to skirmish spam and the diverse features each solution contains. Also, the filtering steps that each solution requires to detect and avoid spam are offered.

Even though Internet Security suites typically include as standard e-mail spam filter, spammers keep on to find ways around the checks that are made. One of those workarounds is the use of images with text buried in the image data. This kind of spam can be checked for, but at present it is done using machine identification. Spammers can rise above those checks by creation the text fuzzy and adding alteration or rotation to an image. Kaspersky Lab has a statistics-based process for detecting image-based spam that is used to bypass conventional text-based filters. The machinery analyses whether text is controlled in images based on the graphic sample of words and lines, said developer Eugene Smirnov. Spam is anticipated to continue to be a problem in 2009, particularly with the rise in the number and popularity of websites that allow user-generated content. Kaspersky Anti-Spam 3.0 and above provides careful and correct protection from spam for users of corporate mail systems and public e-mail services.

Zhe Wang, William Josephson, Qin Lv, Moses Charikar, and Kai Li [4] in Filter Image Spam with Near-Duplicate Detection provide an image-based spam detection method that uses near-duplicate detection to detect spam images. Methods of detecting a division of spam images, then using multiple image spam filters to identify all spam images that "look like" spam captured by traditional methods. Battista Biggio, Giorgio Fumera, Ignazio Pillai and Fabio Roli [5] in Filtering Image Spam by Hidden Content Detection suggest an approach based on low-level image dealing out techniques to detect one of the main individuality of most of the image spam, namely the use of incomprehensible content techniques to overcome OCR tools by finding the noise level of a given image spam. Jason R. Bowling, Priscilla Hope, Kathy J. Liszka [6] in Spam Image Identification Using an Artificial Neural Network, Provides a Method for Identifying Image Spam Using the Fast Artificial Neural (FANN) Library Model Network and forming the network of artificial neurons. A detailed preprocessing process for spam image files is obtainable, followed by a description of how to form an artificial neural network to differentiate between harm and spam. Mr. Muztaba Fuad, Debzani Deb, Mr. Shahriar Hossain [7] in A Shapeable Spam Detection System, presents the plan and achievement of a fuzzy logic-based email classification system that can be trained who learns the most effective blur rules during the training phase. Then apply the fuzzy control model to classify invisible messages. M. Soranam-ageswari, C. Meena [8] in Extracting Statistical Characteristics for Spam Classification of Images Using Artificial Neural Networks presents an experimental system for the classification of image spam by considering a histogram of statistical characteristics and the average value of a picture block. A comparative study of the image classification based on the color histogram and the average value is presented.

Image spam hunter [9] proposed a detection method based on a probabilistic reinforcement tree that takes into account global image characteristics such as color and gradient histogram. They suggested a prototype system based on learning to differentiate spam from normal images. The system begins by grouping spam into groups based on a similarity measure of histograms in colors and gradients. Next, create a probabilistic reinforcement tree on the training set (chosen from clustered groups) to distinguish the images. The results obtained from the SVM classifier are considered as a basis for the comparison of their system. The proposed technique achieved a precision rate of 89.44% with a false positive rate of 0.86%.

Detection of image spam based on file properties, Hough transformation, and histogram [10] allowed focused detection to be focused by specific image attributes. The authors of this article extracted features related to three different domains and compared their individual performance. The first approach takes into account the characteristics related to the properties of the file, such as format, width, height and aspect ratio, which are calculated at a low cost of calculation. The second method uses a classification based on color histograms of the image. The final approach uses Hough transformations for classification. The Hough transformation is a method of searching for different shapes in images using edge detection. According to their experimental results, an approach using file properties eliminates about 80% of image spam and the method using the histogram to implement a measurement of distance reduces by 84% the spam while the method of transformation of Hough reaches 88% accuracy and also minimizes the false positive rate.

The image-based fuzzy inference (F-ISDS) image-sensing technique [11] has introduced a server-side solution that can extract multiple domain features and further classify it based on the fuzzy inference system (F-ISDS) [11]. Fuzzy inference. F-ISDS uses Principal Components Analysis (PCA) dimensionality reduction to reduce the number of functions by mapping them to their main components. In addition, linear regression analysis is used to model the relationship between the main components and the extracted characteristics. Based on the model, the membership functions are designed for a Fuzzy Inference System classifier. Although this work covered a wide range of features, we achieved 85% accuracy.

3 Proposed Solution

This section describes the needs that led to the development of the new approach. We start by describing the needs and the current solutions effectiveness along with their effect. Then, we continue with an overview on the authentication using DKIM also the filtering image spam using OCR and Feature Extraction and the proposed enhancement on them in order to provide the optimal recognition and then listing the layers of our method and then describe its process.

3.1 Effect of Image Spam E-Mails

The effect of image spam on e-mail spam is huge since most undetected kind of spam that finds its way to the users is image spam. The first reason is that it needs a more complex approach to tackle it than those regular spam e-mails that only contains text. The second reason is that spammers take advantage of this difficulty and use image spams as a mean for generating money when advertising spam. The third reason is that millions of spam messages find their way to the user's e-mail inbox which will generate more traffic that is useless, costly, and consuming more bandwidth. This will have a big burden on ISPs who will have to constantly upgrade their network infrastructure to meet the ongoing large bandwidth demand and thus incurring more cost on the ISPs which as well will raise the cost on the end-users.

3.2 DKIM and SPF Proposed Enhancement

Domain authentication is a protocol introduced by the published Domain Keys Identified Mail (DKIM) document. It provides authentication, verification and traceability of messages to help you determine if a message is legitimate.

The draft standard adopted by the Internet Engineering Working Group is more promising because it exploits the power of cryptographically secure digital signatures. The preliminary approval of the Internet Engineering Task Force makes DKIM an official proposed standard.

The way it works is simple: If PayPal sends an e-mail notification to customers about their accounts; the company's outgoing mail server will discreetly insert a digital signature into the legitimate message. (Since the signature is embedded in the message headers, it is usually not visible to readers.)

Let's say the recipient has a Yahoo email address, Yahoo's mail servers can automatically check the list of PayPal's Internet domain names to verify that the digital signature is valid and that the message really comes from Paypal.com.

If the signature does not reach the recipient, the message is probably spam - or a phishing attack designed to try to trick end-users by asking them to divulge the details of their PayPal account. Although the DKIM standard does not actually specify that messages with invalid signatures should be marked as spam, ISPs will probably do so [12].

DKIM is more promising than existing anti-spam and anti-phishing technologies, which rely on techniques such as establishing a "blacklist" of known fraudsters or detecting such messages by trying to identify common features. Plenty of spam that gets sent out is spoofed, so it may appear to come from a particular e-mail address but it actually comes from somewhere else. IP validation goes out and validates that the e-mail address is really from the sender.

DKIM's weakness is that it does not check if the user sending the e-mail is allowed to do so on the domain that is being used. Therefore, this opens some holes that can be exploited within DKIM in order to overcome its authentication. The proposed method includes adding a Sender Policy Framework (SPF) record to identify whether a user of a certain domain is allowed to send e-mails from it. We encapsulate this SPF record within the DKIM header, when received, we unpack the SPF record from the DKIM header and we continue with the DKIM process of authentication and then we analyze both the SPF record and the DKIM authentication to come out with a result if the e-mail is authenticated or not. Also, another enhancement would target the public key sharing which in DKIM is stored on the user's DNS by default. This could make the user a target for attacks by spammer in order to obtain such keys and use them to exploit the DKIM authentication and in order to prevent it, a third party should handle public key sharing since they would be more equipped and ready to battle any attacks on them.

3.3 OCR and Feature Extraction Proposed Enhancement

The proposed detection method includes having two neural networks working in parallel; one of these neural networks is composed of an OCR and the other of Feature Extraction.

The first stage encompasses the training of the OCR by obtaining obfuscated and non-obfuscated templates of letters that are injected in the neural network, and the Feature Extraction to obtain letter feature templates extracted from the letters and will also feed the neural network for learning.

In the second stage after the learning is complete, both neural networks are now able to take input and give the needed output. The input will be the sub-image based letters that are extracted from the entire image based e-mail into 8x8 sub-image based letters. They are given as input for both neural networks where we will obtain the needed output letter, afterward; both results from the neural nets are compared. If both are the same then we discard one and continue with the other, else if one is different from the other we take one and store the other in a temporarily variable to be used in the final stage within the Context based Analyzer Spell Checker if the one chosen is considered to be incorrect.

The third stage holds the Context based Analyzer Spell Checker in which after all the image is processed. We take the resulted text and analyze its words to check whether it is spelled correctly or not and also if the word is correct within the context of the sentence. This is done within the analyzer by the use of features including probabilistic values and identifying the context based Part of Speech (POS) taggers and see if they are grammatically valid and if not to take the most probable suggestion of the word according to the sentence context.

3.4 Complete Solution Algorithm

The proposed algorithm is based on a layered architecture method where the e-mail received will have to pass through several layers by which this e-mail is checked to reach a decision whether to be regarded as spam or not. Below are the layers in which the e-mail passes through.

The first layer presents the authentication mechanism that is derived from the proposed enhanced authentication method where we have to identify that the e-mail received is verified and authenticated.

The second layer presents the distributor where a decision is made after the authentication procedure, also the distributor identifies if the e-mail contains imaged based spam or a text based spam and then forwards it to the right process.

The third layer presents the enhanced image based spam detection engine that is based on OCR and Feature Extraction using neural network along with the Context analyzer Spell Checker where the image based spam is processed into text based spam.

The fourth layer presents the text based spam detection engine, which follows the known way for detecting such spam using dictionaries.

The fifth layer presents the list engine which consists of a black or white list where the spam is added to the blacklist or white list depending on the results of the authentication and the detection of spam using the enhanced proposed method.

The following is the process the e-mail message passes through each layers and what will happen in each layer by which we detect if the e-mail contains spam or not.

The e-mail sent from the sender will go through the list engine layer to check the e-mail within the white list and the black list, if it is in the black list the e-mail is directly discarded, however, if it is in the white list it will depend on the configuration, if it is configured to deliver them directly then they will be delivered directly to the recipient. Moreover, if the e-mail in the white list is configured to always go through the filter check, or the e-mail is neither in the blacklist nor the white list, then the e-mail will be forwarded to the authentication layer to go through the proposed authentication procedure which includes the verification that the received e-mail comes from the domain it says its coming from through the use of the DKIM header and that the one sending the e-mail is authorized to do so by the use of SPF record, then if authentication is unsuccessful in both ways the e-mail will be discarded and its address will be sent to the list engine layer to be added to the blacklist. Also, in case one of the DKIM record or SPF record return false the e-mail will be discarded unless indicated otherwise through the configuration via the user. However, if the authentication is successful and the SPF records returns true then the e-mail will go to the distributor layer where it is checked whether it's an image based or text based e-mail, then if it's an image based it will be forwarded to the image based spam detection layer where the algorithm will check whether this image contains spam or not and then a decision is made according to the results obtained, otherwise if the e-mail does not contain attachment it is sent to the text based spam detection layer where the algorithm will check if the text contains spam. The decision of previous layers would be in case there is spam detected whether it is from the image based spam detection engine or the text based spam detection engine the e-mail will be discarded and the address is given to the list engine to be added to the blacklist, else if there is no spam detected the e-mail is delivered to the recipient and then the e-mail address is sent to the list engine to be added to the white list. In addition, it could be configured to check the e-mails for spam even if they are within the white list or to have them delivered directly to the recipient without checking them.

4 Design Specifications and Implementation

This section explains the design specifications and implementation in details for each part of the proposed method which includes enhancement of authentication using SPF embedded within DKIM and use of trusted third party instead of DNS in DKIM, also filtering image spam with OCR and Feature Extraction via neural networks along with the introduction of Context based Spell Checker into the filtering method.

4.1 Mail Authentication

Mail authentication has become a vital part of a spam solution with its ability to detect the presence of address spoofing in which the spammers mostly use in their process of sending spam to users in order to hide their true identity.

The proposed method includes the above described header however with the change of the "q" to the use of a trusted third party to provide the encryption public key instead of the DNS usage. This is due to several security concerns and reliability issues that will be addressed and those include the possibility of being a target to intrusion, hacks, exploits and Dos attacks which will lead to denial of the service and will hurt the availability and reliability of providing keys. Another proposal would be the introduction to an additional record within the DKIM header that is the SPF (Sender Policy Framework) record. This record would be a valuable introduction to the DKIM because SPF avoids abuses of domain names. However, unlike DKIM it does not validate that a given e-mail actually comes from the claimed user. Users within a given domain can forge each others' address ("cross-user forgery" or even "cross-domain" forgery where outbound mail servers are shared) and by combining both they would complete each other. The process goes by having the mail server acquiring the SPF DNS record of the outgoing e-mail which looks like |spf|example.com. In SPF "v = spf1 a mx -all", note that |spf| is not originally part of the record but it is added over here by our proposed method as an indicator of the start of an SPF record. Afterward, the result is added into the DKIM header to be concatenated with the hashed "bh" and "h" and then be encrypted, also to have an "spf" component added to the header to knows the spf is present and providing the method in which SPF is provided within the header and then the e-mail is sent to the destination mail server, once the e-mail arrive and the DKIM header extracted, the mail server acquires the address of the trusted third party from the "q" record and requests a public key in order to decrypt the "bh" and after it is decrypted the mail server will validate if the e-mail comes from the domain it says it comes from by comparing the DKIM hash by the hash constructed by him according to the DKIM specification and then it take the SPF record and validate if the user sending the e-mail is authorized to send e-mail from this domain. Moreover, any one of these two turns out to be invalid will determine the e-mail as un authenticated. In Fig. 1 the proposed enhanced DKIM header is shown as described above.

```
DKIM-Signature: a=rsa-sha1; q=keys.verisign.com; c=simple/simple;
d=example.com; s=appliances; i=@store.example.com;
t=1117574938; x=1118006938; h=from:to:subject:date;spf=concatenated;
bh=allzndU2Nzg5jsEypzQ1njc4OTAxejr0NTY3ODkwdTI=;
b=AEwzdV`yvt6OfABHkl16B785RjKCdLXdyuU4zGee+ruD00lszZVoG4L
KjNiYzRBfh2DDqwCCvb2f55BJOc9SG2q8LoXSBlEniSba8vMo
```

Fig. 1. Proposed DomainKeys embedded with SPF header

"q" is the address of the trusted third party to retrieve the encryption public key
"b" is the encryption of hashes "bh" and "h" concatenated with the SPF record "|spf|example.com. IN SPF "v = spf1 a mx -all"
"spf" shows the presence of SPF record and its embedded method

4.2 Image Spam Detection

The image spam extracted from the e-mail being received will undergo some image pre-processing done by the OCR before feeding the acquired sub images into the two neural networks. The pre-processing the image includes the following. Figure 2 shows an image based spam received from e-mail.

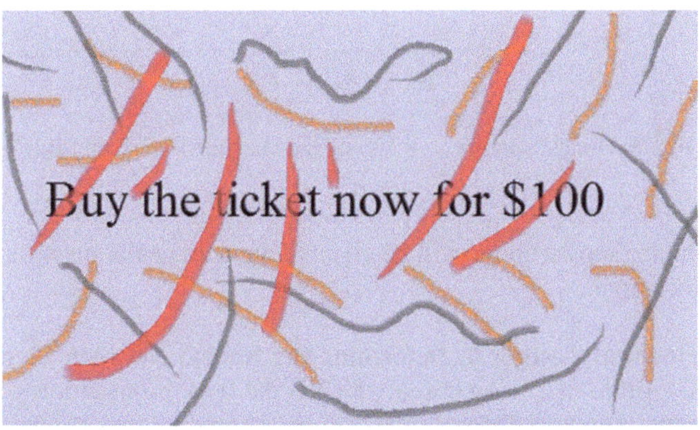

Fig. 2. Image spam sample

In Fig. 3 the image is cleared of noise and went through binarization.

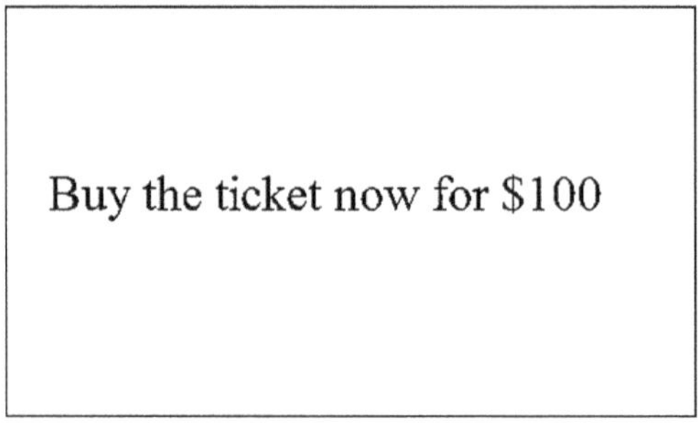

Fig. 3. Image spam cleaned and binarized

In Fig. 4 below the image is cropped to the least size to take off unneeded pixels.

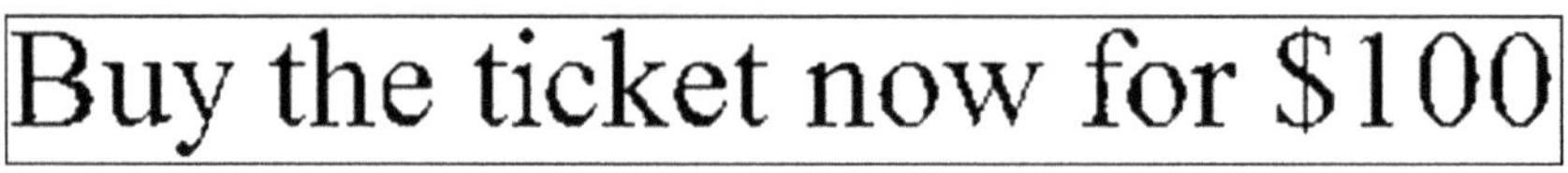

Fig. 4. Image spam cropped

Image is broken into 8 × 8 sub-images each containing a single character, the given samples in Fig. 5 below shows only the image word ‘Buy’ broken into sub-images.

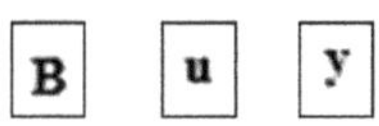

Fig. 5. Image spam 8 × 8 sub-image characters of word “Buy”

Each sub-image shown above will be given to the both neural network after training is done for processing.

4.2.1 Experimental Testing to Determine the Neural Network Structure

The training algorithm that was chosen which is the back propagation was because of the testing results shown in Table 1.

Table 1. Neural network algorithm testing result

Algorithm	MSE	Recognition rate (Training set %)
CBPN	0.0211	92.6
RBPN	0.1517	83.1
CGBPN-CGF	0.1478	82.3
QNBPN	0.1382	87.5

The Sigmoid Activation function was also selected from Table 2 after tests were done and the following results were obtained (Fig. 6).

Table 2. Neural network activation function testing result

Activation function	MSE	Recognition rate (Training set %)
Sigmoid	0.0211	92.6
Satlin	0.0864	86.3
Hardlim	0.2589	74.2
Purelin	0.2432	77.4

The number of the hidden neurons that was chosen was due to the following test results obtained and shown in Table 3 and Fig. 7.

Table 3. Neural network no. of hidden neurons testing result

No. of hidden neurons	MSE	Recognition rate (Training set %)
20	0.2114	57.9
40	0.1432	59.8
60	0.0988	62.1
80	0.0831	64.3
100	0.0726	71.1
120	0.0667	77.2
140	0.0525	80.6
160	0.0453	83.4
180	0.0368	87.1
200	0.0211	92.6

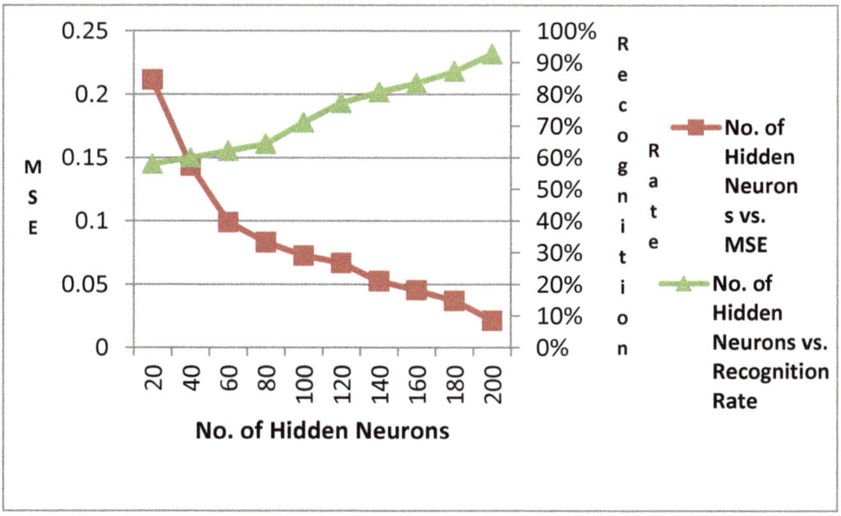

Fig. 6. Neural network no. of hidden neurons vs. MSE and recognition rate

4.2.2 OCR Enhanced via Neural Network

The OCR neural network type that will be used is the Multilayer Perceptron Feedforward Neural Network and the algorithm being used for training the neural net is BackPropagation. This neural network will consist of 3-layer Perceptron with one hidden layer, the three layers are the input layer consisting of 64 inputs which are the 8×8 sub-image character binary representation where the white pixels are encoded as 0 and the black pixels are encoded as 1 and are given to the 64 inputs, the hidden layer which is composed of 200 neuron and one hidden layer is used because it has been tested and shown that having more than one hidden layer wont effect the accuracy of the output whereas it will make it slower in the learning process, moreover we used 200 neuron because its being shown as tested above that the more neurons we have in the hidden layer the more accurate the result is and the higher the recognition rate is however this is to a certain limit, the output layer will be composed of 6 neurons which can give us 26 = 64 choices that represent the English alphabet in small and capital letters in addition to numbers from 0 until 9 and \$ sign which will amount to $26 + 26 + 10 + 1 = 63$. Figure 8 shows OCR MLP Feedforward Neural Network Model that is used.

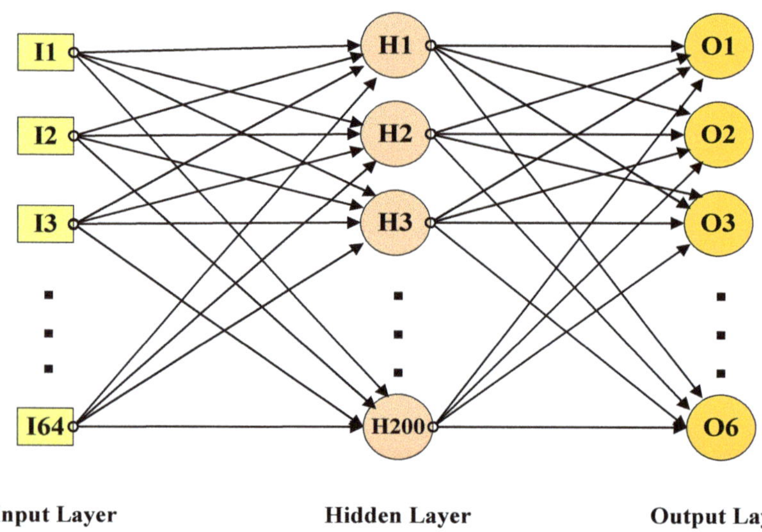

Fig. 7. OCR MLP Neural Network Model

The above neural network needs to be trained on a training set or templates before being able to function. Therefore, the training set covers the English alphabet of small and capital letters including numerals from 0 to 9 and \$ and this training set was chosen to cover the given example of "Buy the ticket for \$100". If we want to add more to this training set we need to increase the number of output in order to be able to present the several kinds of characters. Moreover the alphabet will be given numbers from 1 to 63 to present each different character starting from a → z, A → Z, 0 → 9, and $. These templates will be feed to the input as we train the neural network the weights between the input layer and the hidden layer. The weights between the hidden layer and the output layer will be assigned random numbers between −0.5 and +0.5 at the beginning and a random value for the bias between −0.5 and +0.5 and this is used in order to maintain stability to the network. An activation function, which is the sigmoid function, will be used on the values received in the hidden layer in order to limit the value between 0 and 1. Afterward the network start training on the templates input into it where the Error is calculated from the difference between desired output and the acquired output and then this value is back propagated in order to modify the weight using the delta function and we keep iterating on the training set until we achieve a certain error threshold that we can accept example 0.01 or as our testing experiment above showed the MSE to reach 0.0211 before we stopped or when there is no more improvement to the error rate.

Finally the neural network would be trained and able to function and receive input in order to provide an optimal output which should be the text character of the image character being input.

The use of neural network within the OCR is to enhance its ability to recognize several variations of the characters and these variations include having the characters being subjected to certain obfuscation methods and the ability of the neural network to learn to recognize these characters after being obfuscated.

4.2.3 Feature Extraction Enhanced via Neural Network

The Feature Extraction neural network type that will be used is the Multilayer Perceptron Feedforward Neural Network and the algorithm being used for training the neural net is BackPropagation. This neural network will consist of 3-layer Perceptron with one hidden layer, the three layers are the input layer consisting of 64 inputs which are the numbers representation of the nature and feature of the adjacent pixel's neighborhood since each pixel which are equal to 64 pixels because it is an 8 × 8 sub-image have an 8 adjacent neighbor pixels therefore the neighbors are checked whether they are white or black and depending on the adjacent black neighbor pixel a certain number is assigned to that pixel which represent the feature of the adjacent pixels for example as shown below.

```
***
*+*   →  where the '+' represent the pixel checked and '*' represent the 8
***         adjacent pixels
```

If all the '*' are white pixels the '+' will remain alone and it would have the numerical equivalent of '0' else the numerical equivalent of the presence of each '*' is shown below.

128 1 2
64 + 4 → the numerical equivalent of each '*' present
32 16 8

Example

 *
+ → the numerical equivalent of this feature is 2 + 4 + 16 + 64 = 86
 *

The hidden layer which is composed of 132 neuron and one hidden layer is used because it has been tested and shown that this number is sufficient according to the given nature of the values given in the input, the output layer will be composed of 6 neurons which can give us 26 = 64 choices that represent the English alphabet in small and capital letters in addition to numbers from 0 until 9 and $ sign which will amount to 26 + 26 + 10 + 1 = 63.

The Feature Extraction MLP Feedforward Neural Network Model used is shown in Fig. 8.

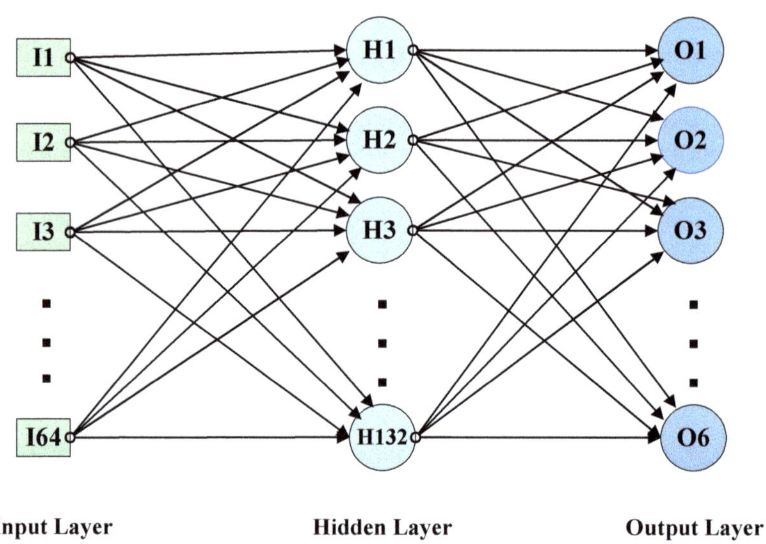

Fig. 8. Feature extraction MLP Neural Network Model

The above neural network needs to be trained on a training set or templates before put into function. Therefore, the training set includes a numerical representation of the pixel's neighbor feature and these set of numerical series would constitute a certain English alphabet of small and capital letters or numerals from 0 to 9 and $ and this training set was chosen to cover the given example of "Buy the ticket for $100. Note that the series of numerical feature for each character are taken from a dictionary having each character and its equivalent set of numerical features. Moreover the alphabet will be given numbers from 1 to 63 to present each different character starting from a → z, A → Z, 0 → 9, and $ in order to deduce the character from the neural net's output. These templates will be feed to the input as we train the neural network the weights between the input layer and the hidden layer, the weights between the hidden layer and the output layer will be assigned random numbers between −0.5 and +0.5 at the beginning and a random value for the bias between −0.5 and +0.5 and this is used in order to maintain stability to the network, also an activation function which is the sigmoid function will be used on the values received in the hidden layer in order to limit the value between 0 and 1. Afterward the network start training on the templates input into it where the Error is calculated from the difference between desired output and the acquired output and then this value is back propagated in order to modify the weight using the delta function. We keep iterating on the training set until we achieve a certain error threshold that we can accept example 0.01 or as we reached in the testing to 0.0245 before we stopped or when there is no more improvement to the error rate.

Finally after the neural network is trained it will be able to function and receive input in order to provide an optimal output which should be the text character of the image character being input.

The first part of acquiring the feature set of a certain character is done by iterating each pixel within the 8×8 sub-image and storing the numerical equivalent of each pixel's neighbor feature and these set of features are then feed to the neural network in order to deduce the character from the input feature set. In addition, the neural network was introduced to this part of the Feature Extraction to enhance the speed of the character recognition.

4.2.4 Context Analyzer Spell Checker

The Context Analyzer Spell Checker is a spell checker that checks the spelling of a word given and if it is misspelled it would correct the spelling by choosing the right word with the highest probability in the list of correct words however it would choose this probability according to context the word belongs to. In addition, the sentence the words belong to is subjected to a grammatical structure analysis within the phase of the context analysis in which each word is assigned with a Part Of Speech Tag (POS) which is for example if the word is a subject, verb, or object.

The domain that the Context Analyzer Spell Checker belongs to is that of Natural Language Processing (NLP) which is a field of computer science and linguistics concerned with the interactions between computers and human (natural) languages. The Context Analyzer in the NLP is based on Lexical analysis that is defined in computer science, as the process of converting a sequence of characters into a sequence of tokens. A program or function which performs lexical analysis is called a lexical analyzer, lexer or scanner.

This feature was proposed and added at the end of the neural network recognition phase in order to provide an additional acknowledgement and enhancement to the end result and in order to overcome any minimal error that might come out from both neural nets.

The process begins when the neural nets output their result. Therefore, if identical character results between both neural nets it will be considered as an additional confirmation of the correctness of the output, however in case of two different output one is taken and the other is stored in a temporarily variable and then a spell check process in undertaken on each word to determine if it is misspelled or not according to context evaluation and also a grammatical structure inspection, in case of a misspelled word or a grammatical structure error the character stored in temporarily variable if it exist for this certain word is taken and then the process of reevaluation is done again, if the results were positive than we continue else we take the most probable word suggestion according to the context.

4.3 Complete Solution Process

The Fig. 9 below shows the complete solution model which includes several components such as the proposed OCR Neural Network with Feature Extraction Neural Network in addition to the Context Analyzer Spell Checker and the proposed Mail Authentication which include DKIM and SPF, moreover, a list engine was included which holds white list and a black list.

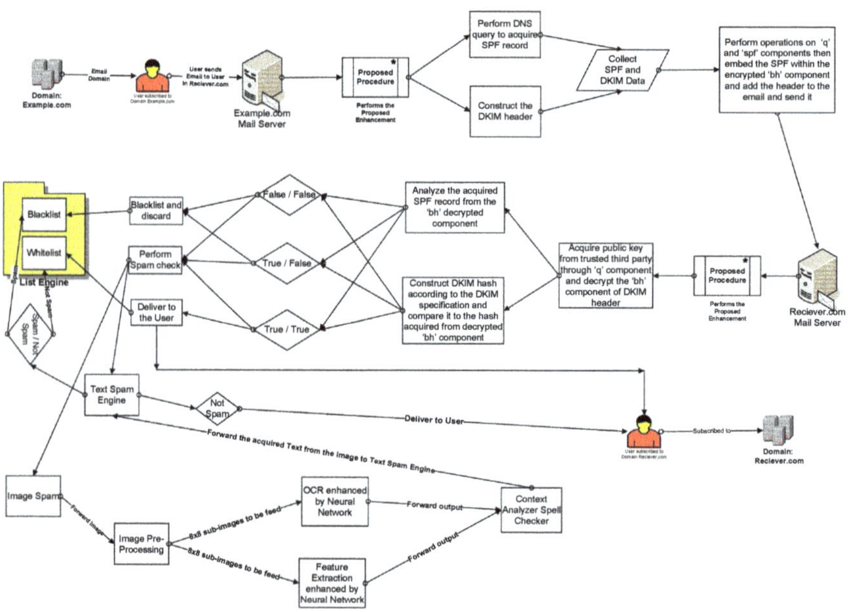

Fig. 9. Complete solution model

5 Conclusions and Future Work

The proposed solution holds two main proposed methods, the first proposed method includes the authentication of the e-mail where DKIM along with SPF were used and SPF was embedded within DKIM to have make sure that the e-mail isn't a result of spoofing because a high percentage of spam e-mails are sent through the use of that method. Moreover, the second proposed method encompasses the filtering procedure that is composed of two neural networks working in parallel: one for the OCR and the other is for the Feature Extraction. In this method both will start to attain results after the end of the learning/training process, and these results will be compared with each other. The appropriate result is chosen according to the algorithm and then the final results will be injected into a context analyzer spell checker. The context analyzer spell checker plays an important role in correcting the mistakes or misrecognized characters from both neural networks. The context analyzer spell checker was introduced to act as a reviewer on the results that are forwarded to it. Finally, having the above neural nets in addition to the context analyzer spell checker will give more accuracy to the end result and lower the false positive rate and thus achieving a higher recognition rate.

This solution gives users the ability to choose the suitable configuration regarding the actions on the e-mail being received. Sample e-mails with DKIM header encapsulating SPF record and spam based images were used to test the application.

Ongoing improvements and enhancements on anti-spam solutions are still taking place to lead to higher recognition rate that would help in reducing the number of spam.

References

1. Stone, B.: Spam back to 94% of all Email New York Times, March 2009. http://bits.blogs.nytimes.com/2009/03/31/spam-back-to-94-of-all-e-mail/
2. Schryen, G.: Anti-Spam Measures: Analysis and Design, Springer (2007). ISBN: 978-3540717485
3. Cormack, G.V.: Email Spam Filtering: A Systematic Review. Now Publishers Inc., Hanover (2008). ISBN: 978-1601981462
4. Wang, Z., Josephson, W., Lv, Q., Charikar, M., Li, K.: Filtering image spam with near-duplicate detection. In: Conference on E-mail and Anti-Spam (2008)
5. Biggio, B., Fumera, G., Pillai, I., Roli, F.: Image spam filtering by content obscuring detection. In: International Conference on Image Analysis and Processing (2008)
6. Bowling, J.R., Hope, P., Liszka, K.J.: Spam image identification using an artificial neural network. In International Conference on Artificial Neural Networks (2009)
7. Muztaba Fuad, M., Deb, D., Shahriar Hossain, M.: A trainable fuzzy spam detection system. In: International Conference on Computer and Information Technology (2005)
8. Soranamageswari, M., Meena, C.: Statistical feature extraction for classification of image spam using artificial neural networks. In: International Conference on Machine Learning and Cybernetics (2010)
9. Gao, Y., Yang, M., Zhao, X., Pardo, B., Wu, Y., Pappas, T.N., Choudhary, A.: Image spam hunter. In: 2008 IEEE International Conference on Acoustics, Speech and Signal Processing, ICASSP 2008, pp. 1765–1768 (2008)

10. Win, Z.M., Aye, N.: Detecting image spam based on file properties, histogram and hough transform. J. Adv. Comput. Netw. **2**(4), 287–292 (2014)
11. Jain, U., Dhavale, S.: Image Spam Detection Technique based on Fuzzy Inference system. Masters report, Department of Computer Engineering, Defense Institute of Advanced Technology (2015)
12. Leiba, B., Fenton, J.: DomainKeys Identified Mail (DKIM): using digital signature for domain verification. In: Conference on E-mail and Anti-Spam, August 2007

Fault Detection of CNC Machines from Vibration Signals Using Machine Learning Methods

Huseyin Canbaz and Kemal Polat(✉)

Department of Electrical and Electronics Engineering, Bolu Abant Izzet Baysal University, Bolu, Turkey
huseyincanbaz7@gmail.com, kpolat@ibu.edu.tr

Abstract. Every machine used in industry has become indispensable in today's production process. As a result, malfunctions of these machines have devastating consequences. In this context, it is very important to diagnose faults before or during the start-up phase. In this paper, the vibration signals obtained from the computerized CNC machine drill bit whiledrilling the metal plate. For fault diagnosis, feature extraction was first performed. A total of 21 properties were extracted from the vibration data in the time domain, frequency domain, and time-frequency domain. The features that were taken at certain intervals from the dataset were doubled to get more accurate results. After the feature extraction process, the data was normalized. The normalization process was determined as min-max and z-score. After the normalization process, classification was carried out by Support Vector Machine (SVM) and K-nearest neighbors (kNN) methods. As a result of these processes, %96,45 and %94,37 accuracy rates for SVM and kNN were obtained respectively.

Keywords: Fault diagnosis · kNN · SVM · Machine Learning · CNC drill bit faults

1 Introduction

With the advances in technology and science, the mechanical systems used in the modern industry have become more functional and more complex than ever. Rotating machines are an indispensable part of this system. The failure of the rotating machinery used in almost every area, the decrease in efficiency of which may or may not stop for a short period can cause serious damage to the production process. Therefore, error analysis in rotating machines has been one of the most important issues in system design and maintenance.

The error analysis of rotating machines has basically three types of tasks: (1) to diagnose whether the system is faulty or normal, (2) to diagnose the fault and the cause of the error when it is in the initial phase, (3) to predict in which direction the error will proceed [1]. Thus, error analysis can be considered as the identification of problems in the working conditions of rotating machines.

When the studies in the literature are examined, it is seen that most of the error analysis is performed using signal processing techniques [2, 3]. In recent years, the use

D. J. Hemanth and U. Kose (Eds.): ICAIAME 2019, LNDECT 43, pp. 365–374, 2020.
https://doi.org/10.1007/978-3-030-36178-5_27

of more current methods such as Machine Learning (ML), Artificial Intelligence (AI) and Artificial Neural Network (ANN) has made it possible to make more effective studies in this field and make it possible to obtain more accurate results [4, 5].

In this study, different structures have been tested for diagnosing faulty and normal situations at the drill bit of computerized CNC machine. In such studies, it is possible to obtain the data in the electrical laboratory or to work on the data previously obtained. Verna et al. created a dataset in their study [6]. In this study, this dataset is used as the main data source. Since these data include a CNC device and drill bit that is currently used in industry, the results are well suited to the actual conditions.

The method used to obtain the dataset is as follows: the vibration signals are recorded during the drill bit drilling of the metal plate and after the drilling. The uni-axis accelerometer PCB 63001 was used to obtain vibration signals. The analog data obtained using this sensor have been transferred to the computer environment and converted to digital data.

There are a total of 3 error conditions in this dataset.

Chisel wear: Drilling process starts as soon as chisel point penetrate work piece. Due to high shear and stress, the temperature at chisel point gets raised and results into a blunted chisel edge

Flank Wear: Due to friction between the work piece and flank of the drill bit, abrasive wear occurs.

Outer corner wear: Due to high impact forces and friction between drill bit and hole's inner head wall, the outer corner of the drill erodes or gets chipped off. This defect causes wear on one or both outer corners of the drill point (Fig. 1).

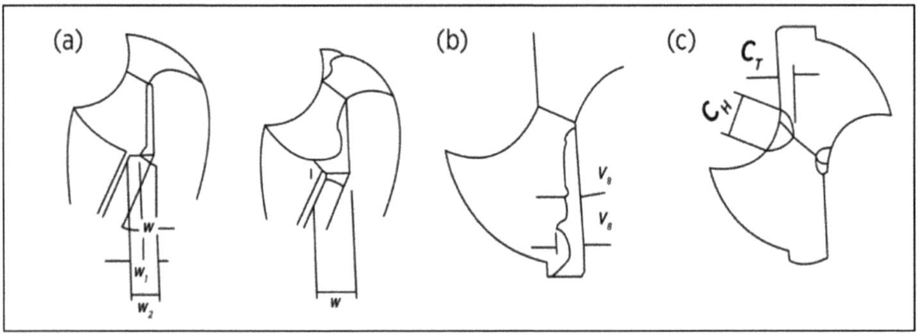

Fig. 1. Different types of wear (a) Outer corner wear, (b) Flank wear, (c) Chisel wear.

These are frequently encountered error conditions, and also a health status data is presented. The data obtained were optimized using pre-processing methods to prevent undesirable noise. The pre-processing methods of the data are 4, and they are respectively; filtering the data using an appropriate filter, trimming the excess parts, smoothing the data and scaling the data. In addition to these methods, various pre-processing methods were used to compare the data before use. These additional pre-processing methods are; the data is divided into appropriate lengths, increasing the number of data gradually, min-max normalization and z-score normalization methods to normalize the data.

After the data were made suitable for the process, feature extraction was performed. Eight properties were obtained in the time domain. These; absolute mean, maximum peak, RMS value, variance, kurtosis, peak factor, shape factor, and skewness. Afterward, 8 features were obtained through Fast Fourier Transform (FFT) and Discrete Cosine Transform (DCT). Then, through Morlet Wavelet Transform (MWT), five features were obtained: standard deviation, wavelet entropy, kurtosis, skewness, and variance. Thus, a total of 21 properties were obtained. After that, the Support Vector Machine (SVM) application has been implemented [7]. The SVM method is of interest for the high accuracy rate in this field. In addition to the SVM method, k nearest neighbor (kNN) algorithm was applied [8]. As a result, algorithms with very high accuracy have been developed.

2 Material

The data set used in this study was performed by Verna et al. [6]. The single axis PCB 63001 accelerometer sensor was used to obtain the data. The sensor is placed on the plate to be drilled. The data is at 32.768 Hz Sampling frequency and transferred to the computer via NI data acquisition system. The NI data acquisition system consists of the NI 9234 signal processing unit and the NI 9172 computer connection. After the acquisition of the data, the data preprocessing step was applied.

3 Method

The methods used to classification the data are given in the block diagram below (Fig. 2).

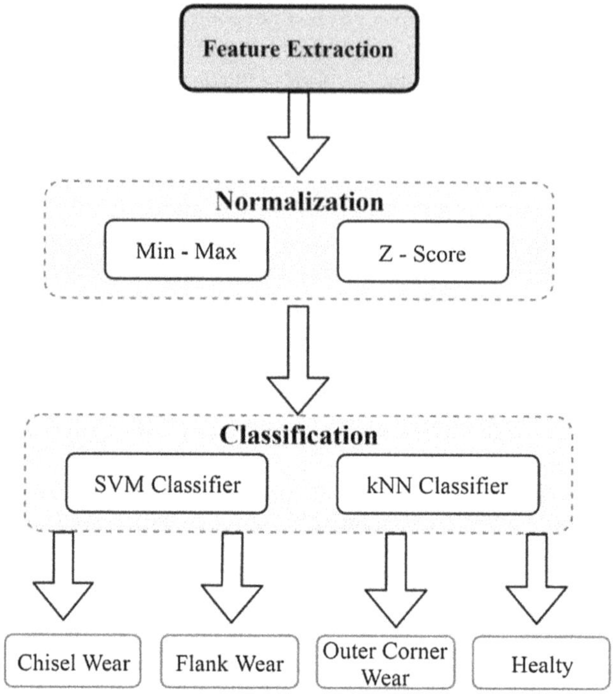

Fig. 2. Block diagram of methods

3.1 Feature Extraction

Matlab program was used for the feature extraction. A total of 21 properties were obtained, including 8 in the time domain, 8 in the frequency domain and 5 in the time-frequency domain, using the codes written. The features obtained are explained in the introduction. All properties are used, and size reduction is not applied. As the distribution of data is balanced, size reduction algorithms reduce accuracy. After feature extraction, the data were normalized.

3.2 Normalization

The raw data were normalized to get better results than data. For the normalization process, 2 methods including min-max normalization and z-score normalization were determined. When the studies in this area are examined, it is observed that min-max normalization gives the best results. In this paper, z-score normalization and min-max normalization were applied to compare. In addition to these two methods, the non-normalization data was named as the standard state and included in the comparison. After the normalization process, the data classification process was started.

3.3 Classification Methods

To classify the data, k nearest neighbors (kNN) and Support Vector Machine (SVM) methods were used. K Fold Cross Validation method was used in the classification process [9]. Here k is selected as 5.

3.3.1 K Nearest Neighbors Method

The kNN algorithm is an easy-to-use monitoring algorithm. Although it is used to solve both classification and regression problems, it is often used in the industry to solve classification problems [10].

The algorithm works on points where the classes are specific within the dataset. The algorithm locates the neighbors that are closest to each point in the dataset. It uses 3 basic methods to measure the distance of points to each other. The most commonly used Euclidean, Minkowsky and Chi-square methods are shown below.

$$\sqrt{\frac{\sum_{i=1}^{m}(x_i - y_i)^2}{m}} \tag{1}$$

Euclidean distance method can be calculated with Eq. (1) [11].

$$\sum_{i=1}^{m}\frac{1}{sum_i}\left(\frac{x_i}{size_Q} - \frac{y_i}{size_I}\right)^2 \tag{2}$$

Chi-square distance method can be calculated with Eq. (2) [12].

$$\left(\sum_{i=1}^{m}|x_i - y_i|^r\right)^{1/r} \tag{3}$$

Minkowsky distance method can be calculated with Eq. (3) [12].

In this study, the k parameter is set to 5.

3.3.2 Support Vector Machine Method

Support Vector Machines are mainly used to distinguish between two classes of data in the most appropriate way. For this purpose, the decision limits or hyper planes are determined. Nowadays, SVMs are used in many classification problems ranging from facial recognition systems to voice analysis[13].

The most accurate hyperplane steps are given below (Fig. 3).

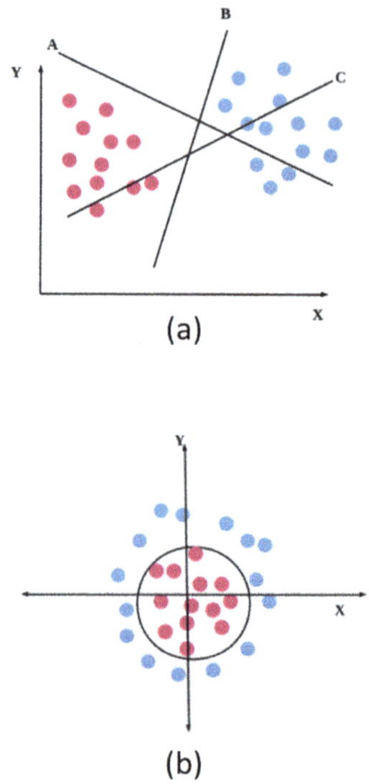

Fig. 3. Hyperplane methods at Support Vector Machine (a) Linear separation (b) Non-Linear separation

4 Experimental Result

4.1 Performance Criteria

When comparing the results, some performance criteria were determined in order to be able to evaluate equally. Cross-validation value is set to 5 for both classification processes. This value can be increased as much as desired, but the processing time increases significantly if it is too high.

The other performance criterion was applied to the number of data. In order to observe the changes in the accuracy rates obtained as a result of the experiments, the number of data was doubled and increased to 4 times. In the case of no changes in the number of data, the number of data is doubled, and the number of data is increased to 4 times the tables and graphs will be shown in the result section.

4.2 Results

The data obtained as a result of the SVM method is shown in Table 1. The graphical information showing the change of these data according to the increase in the number of data is shown in Fig. 4.

Table 1. Accuracy ratios from the SVM method

Amount of data	Min-Max norm.	Z-Score norm.	Standard
Raw data	71,67%	16,67%	73,33%
Doubled data	80,41%	20,42%	82,92%
Four times data	96,45%	21,25%	85,20%

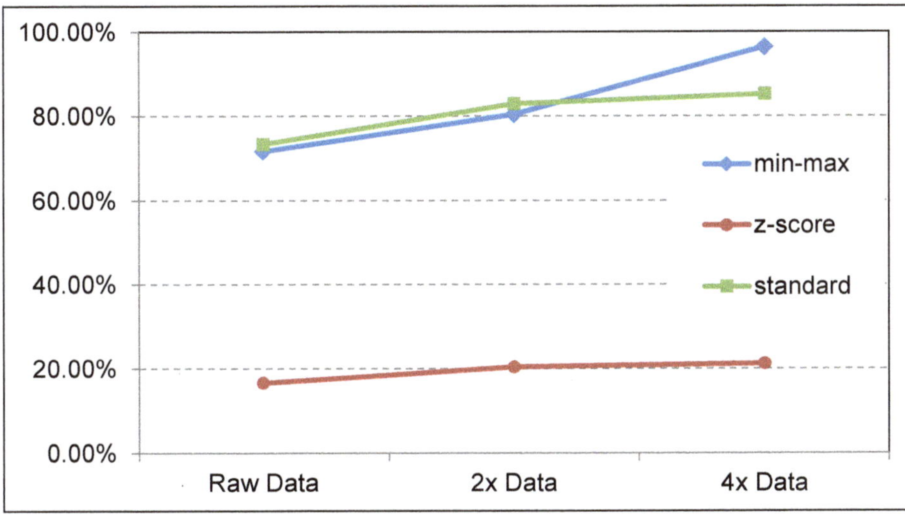

Fig. 4. SVM method graphical results

The Confusion Matrix for the best result is min-max normalization, and the number of data has been increased to 4 times in Table 2 [14]. The average values for the best results are given in Table 3.

Table 2. Confusion matrix of the best result with SVM

Class ->	a	b	c	d
a = 0	117	2	0	1
b = 1	0	119	0	1
c = 2	1	0	113	6
d = 3	3	0	3	114

Table 3. Average values of the best result with SVM

The name of the value	Weighted avg.
TP rate	0,965
FP rate	0,012
Precision	0,965
Recall	0,965
F-Measure	0,965
MCC	0,953
ROC area	0,984
PRC area	0,948

The results so far have been obtained by SVM. The results obtained with kNN will be given hereafter. Accuracy rates obtained by kNN method are shown in Table 4. The graphical information showing the change of these data according to the increase in the number of data is shown in Fig. 5.

Table 4. Accuracy ratios from the kNN method

Amount of data	Min-Max norm.	Z-Score norm.	Standard
Raw data	60,83%	29,17%	69,17%
Doubled data	77,50%	57,92%	75,42%
Four Times data	94,37%	59,37%	82,08%

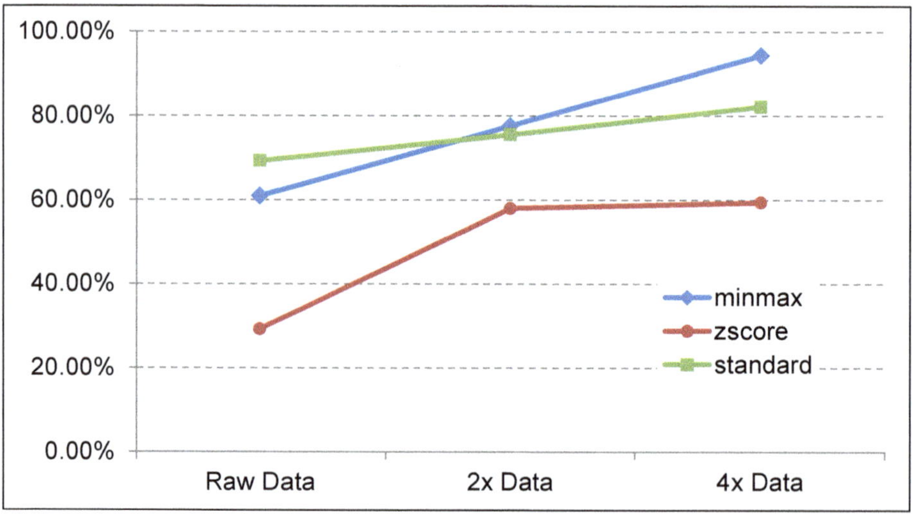

Fig. 5. kNN method graphical results

The Confusion Matrix for the best result is min-max normalization and the number of data has been increased to 4 times in Table 5 [10]. The average values for the best results are given in Table 6.

Table 5. Confusion matrix of the best result with kNN

Class ->	a	b	c	d
a = 0	115	3	2	0
b = 1	2	116	1	1
c = 2	2	1	114	3
d = 3	4	0	8	108

Table 6. Average values of the best result with kNN

The name of the value	Weighted avg.
TP rate	0,944
FP rate	0,019
Precision	0,944
Recall	0,944
F-Measure	0,944
MCC	0,925
ROC area	0,995
PRC area	0,985

5 Conclusions and Future Direction

As a result, it is possible to obtain high accuracy rates with both methods. With SVM method, better results are obtained with min-max normalized data and standard data, while better results are obtained with kNN method and z-score normalized data. When the studies performed in this area are examined, it is seen that the accuracy rates provided by z-score normalized data are not acceptable. For this reason, the best result is the SVM method, and the data type which gives the best result is the min-max normalization, and the data is quadrupled. More advanced machine learning methods will be added to the experiments for further studies.

References

1. Liu, R., Yang, B., Zio, E., Chen, X.: Artificial intelligence for fault diagnosis of rotating machinery: a review. Mech. Syst. Sig. Process. **108**(2018), 33–47 (2017)
2. Ayaz, E., Şeker, S.: İleri işaret işleme yöntemleri ile elektrik motorlarında rulman arıza tanısı. J. İTÜ Eng. **1**(1), 1–12 (2002)
3. Ertekin, Z., Özkurt, N., Yılmaz, C.: Disk Fren Sistemlerinde Dalgacık Tepeleri Yöntemi ile Ses Analizi. J. Fac. Eng. Architect. Cukurova Univ. **32**(4), 193–200 (2017)

4. Janssens, O., Loccufier, M., Slavkovikj, V., Vervisch, B., Stockman, K., Verstockt, S., Walle, R.V.D., Hoecke, S.V.: Convolutional neural network based fault detection for rotating machinery. J. Sound Vibr. **377**(2016), 331–345 (2016)
5. Zhao, R., Yan, R., Chen, Z., Mao, K., Wang, P., Gao, R.X.: Deep learning and its applications to machine health monitoring. Mech. Syst. Sig. Process. **115**(2019), 213–237 (2019)
6. Verna, N.K., Sevakula R.K., Dixit S., Salour, A.: Data driven approach for drill bit monitoring. Reliab. Digest (2015)
7. https://towardsdatascience.com/support-vector-machine-introduction-to-machine-learning-algorithms-934a444fca47. Accessed Mar 2019
8. https://www.analyticsvidhya.com/blog/2018/03/introduction-k-neighbours-algorithm-clustering/. Accessed Mar 2019
9. https://machinelearningmastery.com/k-fold-cross-validation/. Accessed Mar 2019
10. Michalski, R.S., Stepp, R.E., Diday, E.: A recent advance in data analysis: clustering objects into classes characterized by conjunctive concepts. In: Progress in Pattern Recognition, pp. 33–56. North-Holland (1981)
11. Manning, C., Raghavan, P., Schütze, H.: Introduction to information retrieval. Nat. Lang. Eng. **16**(1), 100–103 (2010)
12. Jain, A.K., Duin, R.P.W., Mao, J.: Statistical pattern recognition: a review. IEEE Trans. Pattern Anal. Mach. Intell. **22**(1), 4–37 (2000)
13. https://towardsdatascience.com/support-vector-machine-introduction-to-machine-learning-algorithms-934a444fca47. Accessed May 2019
14. https://towardsdatascience.com/understanding-confusion-matrix-a9ad42dcfd62. Accessed May 2019

Energy Hub Economic Dispatch by Symbiotic Organisms Search Algorithm

Uğur Güvenç[1(✉)], Burçin Özkaya[2], Hüseyin Bakir[1], Serhat Duman[1], and Okan Bingöl[2]

[1] Department of Electrical and Electronics Engineering, Düzce University, Düzce, Turkey
{ugurguvenc, huseyinbakir, serhatduman}@duzce.edu.tr
[2] Department of Electrical and Electronics Engineering, Isparta University of Applied Sciences, Isparta, Turkey
{burcinozkaya, okanbingol}@isparta.edu.tr

Abstract. Energy hub receives various energy carriers such as gas, electricity, and heat in its input and then converts them into required demands such as gas, cool, heat, compressed air, and electricity. The energy hub economic dispatch problem is a non-smooth, high-dimension, non-convex, and non-differential problem, it should be solved subject to equality and inequality constraints. In this study, symbiotic organisms search algorithm is carried out for energy hub economic dispatch problem to minimize the energy cost of the system. In an attempt to show the efficiency of the proposed algorithm, an energy hub system, which has 7 hubs and 17 energy production units, has been used. Simulation results of the symbiotic organisms search algorithm have been compared with some heuristic algorithms to show the ability of the proposed algorithm.

Keywords: Energy hub · Energy hub economic dispatch · Symbiotic organisms search algorithm

1 Introduction

In the world, traditional energy sources such as fossil fuel and coal are getting depleted and also their damages to the environment are increasing day by day. Moreover, the global energy demand is increasing due to population growth. Considering all these situations, the importance of the energy issues has doubled [1, 2]. Recently, a new approach, which is called energy hub, has developed in the energy optimization field. Energy hub is broaden as a robust solution for optimum operation of multi carrier energy substructures such as electrical, cool, and heat demands. In other saying, energy hubs are widely used in various applications for several purposes to meet demand of gas, heat, and electricity. Instead of the individual management of energy carriers, the energy hub investigates the all of the energy systems such as heat, natural gas, electricity, and etc. In energy hubs, the combination of several converters provides the necessary demands by combining more than one energy carrier [3, 4].

In the literature, several studies have been conducted about energy hub approach. In [1], the author presents a general structure in order to model of energy systems

D. J. Hemanth and U. Kose (Eds.): ICAIAME 2019, LNDECT 43, pp. 375–385, 2020.
https://doi.org/10.1007/978-3-030-36178-5_28

consisting of electricity, heat, gas, biomass, etc. The optimum operation of the system including the storage systems, energy markets, wind farm, system uncertainties, and demand response programs, are also examined. Moreover, for the modeling of the system, uncertainties such as wind speed, demand, and market prices, stochastic programming are used. In [3], a self-adaptive learning with time varying acceleration coefficient-gravitational search algorithm (SAL-TVAC-GSA) is proposed to solve single and multiple objective energy hub economic dispatch (EHED) problems. In order to test the algorithm, an energy hub system, which has 39 hubs and 76 energy production units, is used. The aim of the study is that hub losses and energy cost are minimized. The simulation results of the proposed algorithm are compared with different algorithms according to quality solution and computational performance. Reference [5] presents an approach for combined optimization of coupled power flows of different energy carriers. In the study, the authors use the energy hub concept to model the system by introducing the dispatch factors and coupling matrix. Using the energy hubs, input power of natural gas, electricity, and heat are converted to heat and electricity through a coupling matrix. In [6], a varied number of interdependent energy hubs are realized to assess the impacts of energy hub integration. The system is optimized for several scenarios to examine the potential reduction of greenhouse gas emissions and the financial viability. The simulation results show that the environmental and economic benefits are created with the increase in the number of interdependent hubs. Reference [7] presents an energy hub including wind turbines, biomass reactors, solar panels, nuclear plants, fuel cells, and electrolyzes. The hub meets the electricity and hydrogen demands. Three different scenarios have been assessed to specify the optimal technology using economic and technical considerations. In [8], an energy hub, whose inputs are natural gas and electrical energy, have proposed for meeting the heating, cooling, and electrical demands. The authors have presented a mixed integer nonlinear programming model for scheduling CHP based energy hub to maximize the efficiency.

In the study, symbiotic organism search algorithm (SOS) based energy hub system has presented. The aim of the study is to minimize the energy cost. Inputs of the energy hubs are electricity, heat, and natural gas and they are converted into the requested demands such as gas, electricity, cool, heat, and compressed air via different energy infrastructures. In order to test the proposed algorithm, 7 hubs and 17 energy production units have been used. To show the efficiency of the SOS algorithm, simulation results have been compared to results of the particle swarm optimization (PSO), genetic algorithm (GA), and moth swarm algorithm (MSA).

In the following sections, the paper is organized as follows: the energy hub and energy hub system used in the study are introduced and then, the objective function and constraints of the system which is used for the case studies are given in Sect. 2. Following this, main structure of SOS algorithm is explained. In Sect. 4, the efficiency of the SOS algorithm is confirmed by comparisons with different optimization algorithms such as GA, PSO, and MSA. Finally, concluding remarks are given.

2 Formulation of the Problem

2.1 Energy Hub

An energy hub consumes power at their input ports such as compressed air, natural gas, electricity, and etc. and delivers certain required energy services such as heating, electricity, compressed air, cooling, etc. at the output ports. Energy hubs consist of three basic elements: direct connections which is used for delivering an input carrier to the output, converter which converts the energy into a different form, and storage devices used for different forms of energy [3, 5, 6, 9–13]. In Fig. 1, a general energy hub architecture is given. In the Fig. 1, the input energy carriers of hubs are natural gas, heat, and electricity, and the output of the system are gas, electricity, cool, heat, and compressed air. The energy production units are the compressor (C) which products compressed air from electricity, electrical transformer (T) which use and supply the electrical energy, combined heat and power (CHP) producing the heat and electricity from the natural gas, combined heat, cool, and power (CHCP) consuming the natural gas for producing electricity, cooling, and heating, heater exchanger (HE) that uses and supplies heating, and gas furnace (GF) which produces heat from the natural gas. Moreover, the carriers are electricity (e), heat (h), cool (c), compressed air (a), and natural gas (g).

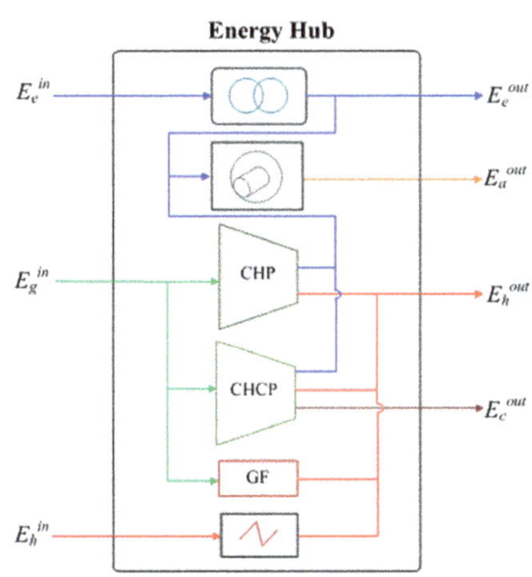

Fig. 1. A general energy hub architecture

In an energy hub, the conversions from input carriers to output carriers are provided with coupling matrix. The matrix model of energy hub is given in Eq. (1) [3, 5, 6, 9–13].

$$\begin{bmatrix} E_{\alpha}^{out} \\ E_{\beta}^{out} \\ \cdot \\ \cdot \\ \cdot \\ E_{\omega}^{out} \end{bmatrix} = \begin{bmatrix} C_{\alpha\alpha} & C_{\beta\alpha} & \ldots & C_{\omega\alpha} \\ C_{\alpha\beta} & C_{\beta\beta} & \ldots & C_{\omega\beta} \\ \cdot & \cdot & \cdots & \cdot \\ \cdot & \cdot & \cdot & \cdot \\ \cdot & \cdot & & \cdot \\ C_{\alpha\omega} & C_{\beta\omega} & \ldots & C_{\omega\omega} \end{bmatrix} \begin{bmatrix} E_{\alpha}^{in} \\ E_{\beta}^{in} \\ \cdot \\ \cdot \\ \cdot \\ E_{\omega}^{in} \end{bmatrix} \quad (1)$$

Here, $\left[E_{\alpha}^{out}, E_{\beta}^{out}, \ldots, E_{\omega}^{out}\right]^{T}$ and $\left[E_{\alpha}^{in}, E_{\beta}^{in}, \ldots, E_{\omega}^{in}\right]^{T}$ respectively denotes the output and input vector, and C is the coupling matrix, where $\alpha, \beta, \ldots, \omega$ represents the different energy carriers. In a SISO system, the coupling factor is suitable to the efficiency of the converter. An energy carrier can be transformed into different energy forms. In this case, the dispatch factors need to be stated. The dispatch factor (v_i) defines how much of each energy carrier is flowing into each converter.

In this study, based on the compressor, transformer, CHP, CHCP, heater exchanger, and gas furnace, the five energy carriers, which are natural gas, compressed air, cool, heat, and electricity, are used. Energy conversions of the energy hubs are given in Fig. 2. These hub structures are taken from Ref. [3]. They can be stated as below:

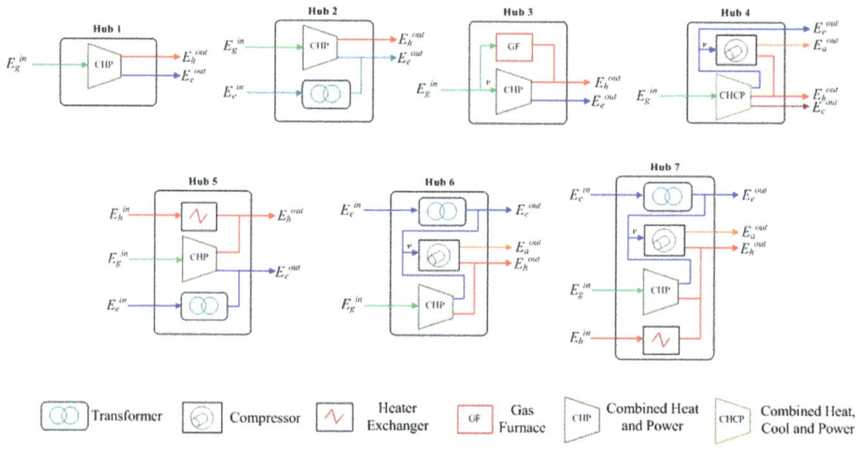

Fig. 2. The hub structures

***Hub 1*:** The hub includes a CHP unit.

$$\begin{bmatrix} E_{e}^{out} \\ E_{h}^{out} \end{bmatrix} = \begin{bmatrix} \eta_{CHP_e} \\ \eta_{CHP_h} \end{bmatrix} E_{g}^{in} \quad (2)$$

Here η_{CHP_e} and η_{CHP_h} respectively represents the efficiency of electrical and heat.

***Hub 2*:** Electricity and heat can be supplied by this hub via using the natural gas and electrical as follows:

$$\begin{bmatrix} E_e^{out} \\ E_h^{out} \end{bmatrix} = \begin{bmatrix} \eta_T & \eta_{CHP_e} \\ 0 & \eta_{CHP_h} \end{bmatrix} \begin{bmatrix} E_e^{in} \\ E_g^{in} \end{bmatrix} \tag{3}$$

where, η_T denotes the transformer efficiency.

***Hub 3*:** In this hub, delivering electricity and heat energy by using the natural gas is performed:

$$\begin{bmatrix} E_e^{out} \\ E_h^{out} \end{bmatrix} = \begin{bmatrix} v_3\eta_{CHP_e} \\ v_3\eta_{CHP_h} + (1 - v_3)\eta_{GF} \end{bmatrix} E_g^{in} \tag{4}$$

where η_{GF} represents the efficiency of the gas furnace.

***Hub 4*:** Compressed air, cooling, heating, and electricity are obtained by using the natural gas.

$$\begin{bmatrix} E_e^{out} \\ E_h^{out} \\ E_c^{out} \\ E_a^{out} \end{bmatrix} = \begin{bmatrix} (1 - v_4)\eta_{CHCP_e} \\ \eta_{CHCP_h} + v_4\eta_{CHCP_e}\eta_{C_h} \\ \eta_{CHCP_c} \\ v_4\eta_{CHCP_e}\eta_{C_a} \end{bmatrix} E_g^{in} \tag{5}$$

Here, η_{C_a} compressor air's efficiencies of the air, η_{C_h} efficiency of the heat of the compressed air, η_{CHCP_h}, η_{CHCP_e}, and η_{CHCP_c} show the cooling, electrical, and heating efficiencies of the CHCP unit, respectively.

***Hub 5*:** In this hub, CHP, transformer, and gas furnace transform electricity and heat into gas, heating, and electricity as below:

$$\begin{bmatrix} E_e^{out} \\ E_h^{out} \end{bmatrix} = \begin{bmatrix} \eta_T & \eta_{CHP_e} & 0 \\ 0 & \eta_{CHP_h} & \eta_{HE} \end{bmatrix} \begin{bmatrix} E_e^{in} \\ E_g^{in} \\ E_h^{in} \end{bmatrix} \tag{6}$$

where η_{HE} is the efficiency of the heater exchanger.

***Hub 6*:** The hub employs CHP, compressor, and transformer via the following equation.

$$\begin{bmatrix} E_e^{out} \\ E_h^{out} \\ E_a^{out} \end{bmatrix} = \begin{bmatrix} (1 - v_6)\eta_T & (1 - v_6)\eta_{CHP_e} \\ v_6\eta_T\eta_{C_h} & \eta_{CHP_h} + v_6\eta_{CHP_e}\eta_{C_h} \\ v_6\eta_T\eta_{C_a} & v_6\eta_{CHP_e}\eta_{C_a} \end{bmatrix} \begin{bmatrix} E_e^{in} \\ E_g^{in} \end{bmatrix} \tag{7}$$

***Hub 7*:** HE, compressor air, transformer, and CHP convert the gas, heating, and electricity into heating, compressed air, and electricity as below:

$$\begin{bmatrix} E_e^{out} \\ E_h^{out} \\ E_a^{out} \end{bmatrix} = \begin{bmatrix} (1 - v_7)\eta_T & (1 - v_7)\eta_{CHP_e} & 0 \\ v_7\eta_T\eta_{C_h} & \eta_{CHP_h} + v_7\eta_{CHP_e}\eta_{C_h} & \eta_{HE} \\ v_7\eta_T\eta_{C_a} & v_7\eta_{CHP_e}\eta_{C_a} & 0 \end{bmatrix} \begin{bmatrix} E_e^{in} \\ E_g^{in} \\ E_h^{in} \end{bmatrix} \tag{8}$$

2.2 Objective Function

In this paper, the aim is minimized the energy cost expose to satisfy inequality and equality constraints. The objective function is calculated with Eq. (9) [3].

$$
\begin{aligned}
Min \quad EC = \sum_{i \in \{g,h\}} \sum_{j=1}^{n_i} (c_{j,i} + b_{j,i} E_{j,i}^{in} + a_{j,i} \left(E_{j,i}^{in} \right)^2) + \sum_{j=1}^{n_e} (c_{j,e} + b_{j,e} E_{j,e}^{in} \\
+ a_{j,e} \left(E_{j,e}^{in} \right)^2 + \left| d_{j,e} \sin \left(e_{j,e} \left(E_{j,e}^{in,min} - E_{j,e}^{in} \right) \right) \right|)
\end{aligned} \tag{9}
$$

In Eq. (9), EC represents the energy cost, $(a_{j,i}, b_{j,i}, c_{j,i})$ are the cost coefficients of jth source relate to i^{th} input carrier, $d_{j,e}$ and $e_{j,e}$ are the cost coefficients for modeling the valve point effect of electrical carrier, n_i represents the total number of energy sources related with i^{th} input carrier, $E_{j,i}^{in}$ represents the energy production and i^{th} input carrier. In the optimization problem, there is three constraint.

1. Energy flow equation is given in Eq. (10). Moreover, all output carrier of hubs must be fulfilled the Eq. (11). In equations, N_h represents the number of hub.

$$
E_{out,i} = E_{in,i} C_i, \qquad i = 1, 2, \ldots, N_h \tag{10}
$$

$$
\sum_{j=1}^{N_h} E_{j,i}^{out} = E_i^{demand}, \qquad i \in \{e, h, c, a\} \tag{11}
$$

2. For all energy units, Eq. (12) should be evaluated.

$$
E_{j,i}^{in,min} \le E_{j,i}^{in} \le E_{j,i}^{in,max}, \qquad i = \in \{e, g, h\} and j = 1, 2, \ldots, n_i \tag{12}
$$

3. Value of all dispatch factors should be in [0, 1].

3 Symbiotic Organisms Search Algorithm

The Symbiotic Organisms Search (SOS) algorithm was developed by Cheng and Prayogo in 2014. SOS is a new and effective meta-heuristic algorithm which is based on symbiotic interaction strategies. The SOS algorithm works on the interactive behavior among organisms found in nature. Organisms have symbiotic close relationships that emerge as the long-term interplay between biological species [14, 15].

SOS is a population based algorithm and begins with the first population that is called ecosystem. Every solution in the ecosystem is called organism. In the first ecosystem, a group of organisms is randomly generated in the search space. Each living organism, as a suggestion for the solution of the problem, shows the degree of fitness for the given objective. The SOS algorithm is improved by using the three most widespread symbiotic relationships which are mutualism, commensalism, and parasitism in nature. Each stage specifies the movement of the organism and whether the other organism will replace it [16].

Mutualism Phase: Mutualism is a symbiotic relationship between two different organisms which can work together to benefit each other. In this phase, X_i is called to be the ith organism in the ecosystem. The other organism X_j, which will interact with X_i is randomly selected from the ecosystem. Both organisms are in a reciprocal relationship to enhance the mutual survival advantage in the ecosystem. New candidate solutions for X_i and X_j organisms are calculated based on the mutualistic symbiosis shown in Eqs. (13) and (14) [14].

$$X_{inew} = X_i + rand(0,1) * (X_{best} - Mutual\_Vector * BF_1) \tag{13}$$

$$X_{jnew} = X_j + rand(0,1) * (X_{best} - Mutual\_Vector * BF_2) \tag{14}$$

$$Mutual\_Vector = \frac{X_i + X_j}{2} \tag{15}$$

The parameters BF_1 and BF_2 are the benefit factors determined as 1 or 2 in the algorithm. The benefit factors symbolize the level of utility to each organism. In other words, they show whether an organism partially or fully benefits from the interaction. The (X_{best}) expresses the highest degree of adaptation. *Mutual_Vector* is a vector representing the characteristic of the relation between organisms X_i and X_j and calculated as shown in Eq. (15). Finally, if the new fitness value of organisms is better than their pre-interaction fitness, they are updated [14, 17].

Commensalism Phase: In the commensalism phase, an individual organism develops a self-beneficial relationship alone, while the other organism remains unharmed. As in the mutualism phase, the organism called X_j is randomly selected from the ecosystem to interact with the X_i organism. In this phase, organism X_i attempts to utility from the interaction. However, organism X_j itself neither benefits nor suffers from the relationship. The new candidate solution of X_i is calculated using Eq. (16) [14]. Eventually, X_i is updated only if its new fitness is better than previous fitness value.

$$X_{inew} = X_i + rand(-1,1) * (X_{best} - X_j) \tag{16}$$

Parasitism Phase: Parasitism is a relationship between two organisms in which one organism provides benefits and the other organism is damaged. At this stage, X_i is used to composing an artificial parasite named *Parasite_Vector*. When creating the *Parasite_Vector*, X_i is copied first and then randomly modified in the search field. As in the previous steps, X_j is randomly selected from the ecosystem and serves as a host to the *Parasite_Vector*. *Parasite_Vector* tries to change X_j in the ecosystem. The fitness value of both organisms is appreciated. If *Parasite_Vector* has a better fitness value, it will kill organism X_j and stand its position in the ecosystem. If the fitness value of X_j is better than the others, X_j will have immunity from the parasite and the *Parasite_Vector* will no longer be able to live in that ecosystem [14, 17].

After all, phases are completed, the next iteration of the algorithm starts. This process continues until the final criteria are met. The pseudo code of the energy hub system with SOS algorithm is given in Fig. 3.

```
Start
Initialize : Generate an ecosystem (check eco_size)
While stopping conditions are not satisfied do
    For i = eco_size
        Calculate the fitness value of each organism in ecosystem
        Find the best organism, X_best
// Mutualism phase
        Select one organism randomly, X_j (j != i)
        [X_inew, X_jnew] = mutualism(X_i, X_j, X_best)
        If f([X_inew, X_jnew]) < f([X_i, X_j])
        Replace X_inew and X_jnew with X_i and X_j
        End if
// Commensalism phase
        Select one organism randomly, X_j (j != i)
        X_inew = commensalism(X_i, X_j, X_best)
        If f(X_inew) < f(X_i)
        Replace X_inew with X_i
        End if
// Parasitism phase
        Select one organism randomly, X_j (j != i)
        X_parasite = parasitism(X_i)
        If f(X_parasite) < f(X_j)
        Replace X_parasite with X_j
        End if
    End for
End while
Stop – Save the most suitable organism X_best in the ecosystem
```

Fig. 3. The pseudo code of the energy hub system with SOS algorithm

4 Simulation Results

The SOS algorithm has been tested on energy hub system which has 7 hub and 17 control variables. 13 variables of them are for energy sources and the remaining of them are for the dispatch factors. The total demand of the system is given in Table 1. In Appendix A, the system data is given. For the simulation of the system, MATLAB is used and the tests are implemented on an Intel(R) Core(TM) i5 CPU, 4 GB RAM, and 64 bit operating system PC. In the study, the results of the algorithms have obtained from 30 independent run.

Table 1. Total demands

Carriers	Demand (pu)
Electricity	1.5
Heat	1.6
Cool	0.1
Compressed air	0.2

The obtained optimal solution of energy cost minimization of system obtained by SOS algorithm is compared with the PSO, GA, and MSA in Table 2. In this table, the objective value of SOS is 2336.2166 mu, while objective value of GA, PSO, and MSA are 2388.0599, 2355.2932, and 2345.8561 mu, respectively. Moreover, the total energy losses in the hubs that is attained by the GA, PSO, MSA, and SOS are 1.1546, 1.2275, 0.5949, and 0.5653 pu. In Fig. 4, the optimum values of dispatch factors of hubs obtained by SOS are given.

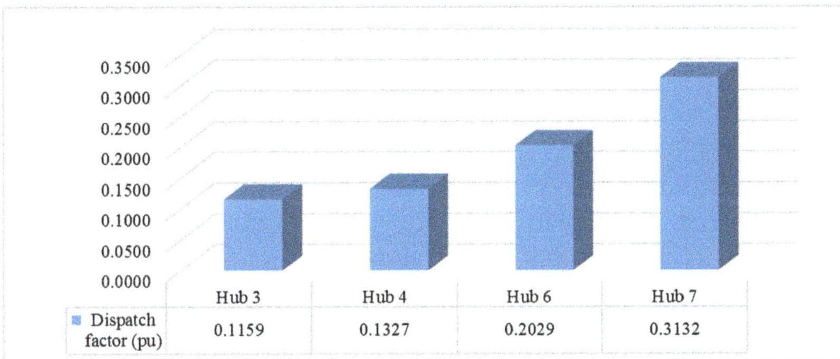

Fig. 4. The optimum values of dispatch factors of hubs attained by SOS

Table 2. Comparative results of energy cost minimization using different techniques.

Hub no	Energy type	GA	PSO	MSA	SOS
1	Gas	0.7897	0.6388	0.5012	0.5400
2	Electricity	0.2146	0.2000	0.2000	0.2000
	Gas	0.4125	0.1542	0.1000	0.1051
3	Gas	0.8141	0.9484	0.3017	0.1507
4	Gas	0.3448	0.3448	0.3448	0.3448
5	Electricity	0.2001	0.2000	0.2000	0.2000
	Gas	0.2057	0.2000	0.3397	0.2001
	Heat	0.1330	0.1000	0.5599	0.6999
6	Electricity	0.1405	0.1000	0.1000	0.1000
	Gas	0.7896	1.2157	0.2000	0.2331
7	Electricity	0.2068	0.2001	0.8280	0.8411
	Gas	0.2033	0.2000	0.2000	0.2007
	Heat	0.1000	0.1256	0.1195	0.1496
Total production (pu)	Electricity	0.7620	0.7001	1.328	1.3411
	Gas	3.5597	3.7019	1.9874	1.7745
	Heat	0.2330	0.2256	0.6794	0.8495
CF (mu)		2388.0599	2355.2932	2345.8561	**2336.2166**
Total losses (pu)		1.1546	1.2275	0.5949	**0.5653**
Total penalty (mu)		5.95e−10	1.16e−5	1.0016e−9	4.01e−4

5 Conclusion

In this paper, SOS algorithm is used to optimize highly high-dimensional, non-convex, nonlinear, non-differential, and non-smooth energy hub economic dispatch problem. The aim of the algorithm minimizes the energy cost of the energy hub system. In order to test the efficiency of the SOS algorithm, the system has been also applied to GA, PSO, and MSA algorithms. According to test results, SOS has better quality solution for giving the good convergence while satisfying the all constraints.

Appendix

See Tables A.1 and A.2.

Table A.1. Hub data

Hub no	Efficiency
1	$\eta_{CHP_e} = 0.3$, $\eta_{CHP_h} = 0.4$
2	$\eta_T = 1$, $\eta_{CHP_e} = 0.27$, $\eta_{CHP_h} = 0.41$
3	$\eta_{CHP_e} = 0.31$, $\eta_{CHP_h} = 0.38$, $\eta_{GF} = 0.8$
4	$\eta_{CHCP_e} = 0.3$, $\eta_{CHCP_h} = 0.31$, $\eta_{CHCP_c} = 0.29$, $\eta_{C_a} = 0.7$, $\eta_{C_h} = 0.2$
5	$\eta_T = 0.97$, $\eta_{CHP_e} = 0.32$, $\eta_{CHP_h} = 0.44$, $\eta_{HE} = 1$
6	$\eta_T = 0.99$, $\eta_{CHP_e} = 0.3$, $\eta_{CHP_h} = 0.32$, $\eta_{C_a} = 0.59$, $\eta_{C_h} = 0.21$
7	$\eta_T = 1$, $\eta_{CHP_e} = 0.32$, $\eta_{CHP_h} = 0.41$, $\eta_{C_a} = 0.6$, $\eta_{C_h} = 0.2$

Table A.2. Data of energy sources

Hub no	Entire energy	Cost coefficients of entire energy					Energy production limits (pu)	
		a (*mu*)	b (*mu/pu*)	c (*mu/pu*^{2})	d (*rad/pu*)	e (*mu*)	E_{min}	E_{max}
1	Gas	20	150	65	–	–	0.5	3.4
2	Electricity	30	180	60	140	4	0.2	1.25
	Gas	20	170	90	–	–	0.1	1
3	Gas	25	120	50	–	–	0.15	1
4	Gas	10	220	60	–	–	0.1	3.2
5	Electricity	10	220	160	190	3.6	0.2	1.1
	Gas	20	200	100	–	–	0.2	1.8
	Heat	12	170	210	–	–	0.1	0.7
6	Electricity	80	200	25	100	4.2	0.1	0.75
	Gas	25	100	40	–	–	0.2	1.9
7	Electricity	95	130	300	90	4.9	0.2	1.9
	Gas	29	220	330	–	–	0.2	1
	Heat	32	135	110	–	–	0.1	0.5

References

1. Vahid-Pakdel, M.J., Nojavan, S., Mohammadi-Ivatloo, B., Zare, K.: Stochastic optimization of energy hub operation with consideration of thermal energy market and demand response. Energy Convers. Manag. **145**, 117–128 (2017)
2. Krause, T., Andersson, G., Frohlich, K., Vaccaro, A.: Multiple-energy carriers: modeling of production, delivery, and consumption. Proc. IEEE **99**(1), 15–27 (2011)
3. Beigvand, S.D., Abdi, H., La Scala, M.: A general model for energy hub economic dispatch. Appl. Energy **190**, 1090–1111 (2017)
4. Maniyali, Y., Almansoori, A., Fowler, M., Elkamel, A.: Energy hub based on nuclear energy and hydrogen energy storage. Ind. Eng. Chem. Res. **52**(22), 7470–7481 (2013)
5. Geidl, M., Andersson, G.: Optimal power flow of multiple energy carriers. IEEE Trans. Power Syst. **22**(1), 145–155 (2007)
6. Maroufmashat, A., Elkamel, A., Fowler, M., Sattari, S., Roshandel, R., Hajimiragha, A., Entchev, E.: Modeling and optimization of a network of energy hubs to improve economic and emission considerations. Energy **93**, 2546–2558 (2015)
7. Maniyali, Y., Almansoori, A., Fowler, M., Elkamel, A.: Energy hub based on nuclear energy and hydrogen energy storage. Ind. Eng. Chem. Res. **52**(22), 7470–7481 (2013)
8. Moghaddam, I.G., Saniei, M., Mashhour, E.: A comprehensive model for self-scheduling an energy hub to supply cooling, heating and electrical demands of a building. Energy **94**, 157–170 (2016)
9. Mohammadi, M., Noorollahi, Y., Mohammadi-ivatloo, B., Hosseinzadeh, M., Yousefi, H., Khorasani, S.T.: Optimal management of energy hubs and smart energy hubs–a review. Renew. Sustain. Energy Rev. **89**, 33–50 (2018)
10. Geidl, M., Koeppel, G., Favre-Perrod, P., Klöckl, B., Andersson, G., Fröhlich, K.: The energy hub–a powerful concept for future energy systems. In: Third Annual Carnegie Mellon Conference on the Electricity Industry, vol. 13, p. 14 (2007)
11. Orehounig, K., Evins, R., Dorer, V.: Integration of decentralized energy systems in neighbourhoods using the energy hub approach. Appl. Energy **154**, 277–289 (2015)
12. Carradore, L., Bignucolo, F.: Distributed multi-generation and application of the energy hub concept in future networks. In: 2008 43rd International IEEE Universities Power Engineering Conference, UPEC 2008, pp. 1–5 (2008)
13. Aghamohamadi, M., Samadi, M., Rahmati, I.: Energy generation cost in multi-energy systems; an application to a non-merchant energy hub in supplying price responsive loads. Energy **161**, 878–891 (2018)
14. Cheng, M.Y., Prayogo, D.: Symbiotic organisms search: a new metaheuristic optimization algorithm. Comput. Struct. **139**, 98–112 (2014)
15. Sonmez, Y., Kahraman, H.T., Dosoglu, M.K., Guvenc, U., Duman, S.: Symbiotic organisms search algorithm for dynamic economic dispatch with valve-point effects. J. Exp. Theor. Artif. Intell. **29**(3), 495–515 (2017)
16. Banerjee, S., Chattopadhyay, S.: Power optimization of three dimensional turbo code using a novel modified symbiotic organism search (MSOS) algorithm. Wirel. Pers. Commun. **92**(3), 941–968 (2017)
17. Baysal, Y.A., Altas, I.H.: Power quality ımprovement via optimal capacitor placement in electrical distribution systems using symbiotic organisms search algorithm. Mugla J. Sci. Technol. **3**(1), 64–68 (2017)

An Extended Business Process Representation for Integrating IoT Based on SWRL/OWL

Lynda Djakhdjakha[1,3(✉)], Djehina Boukara[1], Mounir Hemam[2], and Zizette Boufaida[3]

[1] Department of Computer Science, University of 08 May 1945, BP 401, Guelma, Algeria
ldjakhdjakha@yahoo.fr, djakhdjakha.lynda@univ-guelma.dz, jouhaynajinyl4@gmail.com

[2] Department of Computer Science, University of Khenchela, 40000 Khenchela, Algeria
Mounir.hemam@gmail.com

[3] LIRE Laboratory, Department of Computer Science, Mentouri University of Constantine, 25000 Constantine, Algeria
zboufaida@gmail.com

Abstract. Nowadays, the increasing growth of the combination of Internet of Things (IoT) and Semantic Web as a strong paradigm for business management has led to the proposal of new research themes. This article examines the integration of sensor data transmitted by Internet of Things (IoT) into the company's business processes based on Semantic Web technologies. Our main objective is to use the efficiency of IoT, the power of Business Process Execution Language (BPEL), the expressiveness of Web Ontologies Language (OWL) and Semantic Web Rule Language (SWRL) to represent business processes and make them react to changes in the environment.

Keywords: Internet of things · Sensor data · Semantic web · Ontology · Business process · BPEL · WSDL · OWL · SWRL

1 Introduction

The need to solve organisational problems leads major research efforts to develop tools to manage and control their processes. Nowadays, the Semantic Web paradigm is used in organisations for a wide range of their applications. Since its inception, with its various technologies, the focus has been on providing tools for the interoperability and flexible integration of applications within and across organisations' borders.

This article discusses the use of ontologies to represent the organisation's business processes. Ontologies are used to represent knowledge in a formal and reusable way. Ontology is defined as an explicit specification of a conceptualisation, where a conceptualisation is an abstract view of the real world that we represent for a specific objective [1]. In description logic [1], an ontology composes of two parts (i) the

D. J. Hemanth and U. Kose (Eds.): ICAIAME 2019, LNDECT 43, pp. 386–405, 2020.
https://doi.org/10.1007/978-3-030-36178-5_29

TBox defines the terminological part, it provides the data of an ontology (ii) the ABox defines the assertion part, it provides the schema, with their restricted resources.

Today, Organisations are looking to use real-time data detected from IoT to improve product quality and decision-making[1]. In a dynamic context, within Internet of Things (IoT), business process can gain competitive advantage by using real data provided by sensors during their runtime [2]. IoT is a network of sensors and smart devices that communicate over the internet to share information and to complete different tasks. IoT use has grown exponentially and has become more and more an essential part of organisation's life. In the few years, the impact of IoT on both individuals and businesses will be very significant [3]. In a highly dynamic environment, where organisations evolve very rapidly, they need solutions to be able to adapt to on-going changes. In [3], author states that the real challenge for IoT is not technological, but rather to integrate change in organisations.

This paper introduces a new representation, it aims to combine the efficiency of IoT and a power of business process to give organisations the ability to integrate IoT in their processes. To meet the necessary requirement, this representation must save the change data history, able to give a comparison between the different changes to give the best decision and able to include information about whom is making the change, why and when. In this case, ontologies can be the key to resolve this problem. Ontologies can be used in which the evolution plays an important part. They can evolve from one consistent state to another one [4]. Ontology evolution is considered as the process of modifying and adapting ontology in a consistent way and timely [4].

Our work is based on Web Services Business Process Executable Language (WS-BPEL 2.0[2]) the most popular language for formally describing business process. Nevertheless, BPEL lacks a mechanism (i) to represent semantic information and (ii) to monitor variable changes, so it makes it difficult to define processes that react to environment changes [2]. To address this problem we use Ontology Web Language (OWL 2[3]) an ontology language for the Semantic Web with formally defined meaning. OWL 2 ontology provides classes, properties, individuals, and data values. It provides different facilities and makes it possible to describe complex elements of executable business processes and different concepts related to sensor data. OWL 2 allows an automatic inference [5].

However, in our context, where activities are considered as the main building blocks of BPEL and the elements most influenced by changes of environment, OWL 2 is not sufficient for representing different activities after receiving new sensor data values. So, it should be noted that the representation of activities plays an important role in the semantics of business process. In order to solve this problem, we will combine Semantic Web Rule Language (SWRL[4]) to OWL 2. SWRL rules are used for describing classes and data types, binary and some special predicates for properties. Their form is: a body => a head. Their meaning is: when the conditions specified in the

[1] https://www.slideshare.net/CiscoBusinessInsights/journey-to-iot-value-76163389.

[2] https://www.oasis-open.org/committees/download.php/23964/wsbpel-v2.0-primer.htm.

[3] https://www.w3.org/TR/owl2-syntax/.

[4] https://www.w3.org/Submission/SWRL/.

body are verified, the conditions specified in the head must also be verified. With SWRL rules, we can extend the set of our representation axioms to make it more complete.

The remainder of the paper is organised as follows: Sect. 2 presents a background on business process. Section 3 introduces a motivating example, while Sect. 4 presents an overview on related works. In Sect. 5, we present the set of transformation rules from BPEL to OWL 2 ontology extended by SWRL rules. A sample example is analysed in Sect. 6. Finally, Sect. 7 summarises the results of this work and draws conclusions.

2 Background

Business process is a collection of structured and related activities that accomplish a specific goal. In the last years, business process modelling has become an effective way of managing organisation's processes. Business processes are modelled as a series of activities connected together using data and control dependencies, that can accomplish a specific product or service [6].

In literature, there are many tools for modelling and executing the organisation's processes. Business Process Modelling Notation (BPMN) offers a standard notation for executable BPEL processes. It provides a formal mapping between BPMN to Business Process Execution Language (BPEL) [7]. BPEL [8], identified also as BPELWS or BPEL4WS, provides an XML-based language to describe both the interface between the participants in a process and the full operational logic of the process and its execution flow [9]. BPEL process definition identifies the technical details of a workflow that provides a complex Web Service build from a set of elementary Web Services [7].

The basic elements of a BPEL process are variables, PartnerLinks, basic and structured activities, and handlers. Where, variables store the process data and messages that are exchanged with web services, PartnerLinks define the required port types of a message exchange by declaring which partner acts according to which role, defined in a partner link. Basic Activities identify the operations which are executed in a process. These include basic Web Service operations: invoke, receive, or reply.

Other activities allow to assign data values to the variables (assign) or to wait for the process to stop for a certain time interval. Structured Activities are used for the definition of control flow, and to specify concurrency of activities (flow), alternative branches (switch), or loops (while). While structured activities can be nested, links can be used to express synchronisation constraints between activities. Handlers can be defined in order to respond to the occurrence of a fault, an event, or if a compensation has been triggered.

BPEL is used for required orchestration and choreography between diverse web services [7]. Web service composition can integrate applications and produce intra-organisational business processes. Web Service Description Language (WSDL) has become a standard for defining web service interfaces, it describes the interfaces of all services [7].

3 Motivating Example

Nowadays, IoT technologies can be used in various environments. In enterprise applications, the supervision, control and management of different values measured by sensors can be used to offer new services, change or improve other existing services. In this section, we provide a motivating example to demonstrate the necessity of an extended representation for integrating IoT into business process. This example represents a business process with a set of sub-processes (sale, production, supply) who fulfil a sales order and achieve a business objective of an industrial company. Each sub-process is a set of activities. The activities are related together and executed in a predefined order under normal conditions as shown in the Fig. 1:

Fig. 1. A part of BPEL activities

But, under certain conditions (for example: under very high or very low temperature), the process will not be executed in the same way. In order to achieve our objective, it is necessary to consider a series of questions:

- How to integrate values measured by the sensor data in business process?
- How to describe different blocks of activities without the values sent by sensor devices?
- How do we represent the new block after receiving a new value from a sensor?

Furthermore:

- How to respond to intelligent queries to improve managers' decisions?
- How to make this business process semantically accessible to answer these requests?

Therefore, it is necessary to define a new representation of business processes that allows to describe the semantics of the different BPEL elements within the values of the detection devices and to identify what will happen after receiving the signals from the latter.

4 Related Work

This section presents a brief overview of the current state of the art of related work, and we will describe this work according to different field categories. The first issue we address is related to the integration of sensor data transmitted from the IoT into business process execution language. This issue has been addressed in a few research papers extending WS-BPEL with context variables [7]. In [2], authors extend WS-BPEL with context variables that are aware of changes to sensor reported values. They proposed a new language construct to handle expected exceptions. However, their solution's deficiency is the impossibility to represent semantic information.

The second issue we consider is the introduction of semantics in the business process area. A considerable research effort has been carried out in this field. The interest in using ontologies to represent static and dynamic aspects of an organisation [10] and for compliance management [11, 12] has been increased. In [6], authors presented an approach for describing, storing, and discovering business processes. The proposed approach extended the concept of similarity between process models defined in [13] and based on concepts of inheritance between business processes proposed in [14].

A very interesting work was developed in [15], where authors proposed a visual query language to query and discover business process models using BPEL. Several researches have already been carried out to address the semantic limitations of BPEL, WSDL and BPMN languages by extending or mapping them to OWL [6] or OWL-S [16]. Based on our representation, we note that this domain-based ontology can significantly improve the semantic level of the dynamic aspects of the business process.

In addition, a further issue concerns the development of IoT ontologies. LOV4IoT[5] referenced more than 330 ontologies based on IoT projects. The existing ontologies are restricted to a certain domain [17]. Successful ontology studies in the field of IoT have been presented in [17, 18] and they are based on the classification presented in [19]. Semantic Sensor Network ontology (SSN) is considered as a standard ontology for describing the sensor resources and the data provided as observations proposed by W3C[6] [20].

[5] http://ci.emse.fr/ic2017/tutoriel-swot/diapo/04_Ontologies.pdf.

[6] World Wide Web Consortium.

In [17], authors provide an overview of the most recent IoT ontologies to highlight the fundamental ontological concepts required for an IoT-based application. In our work, the list of fundamental concepts could be very useful to support observations of the sensor devices to enable business processes to respond to environmental changes.

According to OWL reasoning based on SWRL rules state of the art, the authors presented in [21] the application of SWRL/OWL for the representation of knowledge relevant to a supply logistics scenario. They used the SWRL rules to demonstrate their situation awareness application SAWA. In [22], Zhai et al. proposed a rule-based reasoner for inferring and providing query services based on OWL and SWRL. We support a similar solution to support SWRL rules, and we have applied them in our business process representation.

This paper aims to illustrate the use of OWL 2 ontology and SWRL for the representation of a company's dynamic aspects. Therefore, we present a semantic representation for describing and storing business process. This representation extends the concepts of business processes modelled using BPEL by fundamental ontological concepts required for an IoT-based application and a set of SWRL rules to extend the set of OWL 2 axioms in order to make business process representation more expressive and more complete.

5 Business Process Representation Based on SWRL/OWL

In this section, we design a BPEL OWL-based ontology. The main objects of this ontology include the BPEL elements, in particular the activity blocks and concepts that are aware of the changes in the values reported by the sensors. In a first step, we have derived three subtasks: (A) Transformation rules from BPEL elements in terms of OWL 2 ontology, (B) extension of the OWL 2 ontology by adding new concepts for representing sensor data values and (C) instantiation and population of the OWL 2 ontology. In a second step, we combine SWRL rules to OWL 2 declarations to complete the representation.

5.1 Transformation Rules from BPEL Elements Based on OWL 2 and SWRL

Our aim is to produce a consistent semantic OWL 2 (DL) representation based on SROIQ [23]. To simplify the formal structure of the proposal representation, we use the functional-style syntax that can be used for serialization.

The conceptualization of OWL 2 ontology is based on the XML schema defined by OWL specification, so, first we identify the different IRI:

```
<Prefix name="rdf" IRI="http://www.w3.org/1999/02/22-rdf-syntax-ns#"/>
<Prefix name="rdfs" IRI="http://www.w3.org/2000/01/rdf-schema#"/>
<Prefix name="xsd" IRI="http://www.w3.org/2001/XMLSchema#"/>
<Prefix name="owl" IRI="http://www.w3.org/2002/07/owl#"/>
```

Then each BPEL element is represented as a Declaration (Class). A BPEL process is specified as a class and its corresponding attributes as data properties, as follows:

Declaration (Class(:Process))
Declaration (DataProperty (:hasName))
DataPropertyDomain(:hasName :Process)
DataPropertyRange (:hasName xsd :string)
...

BPEL processes exchange functionalities by using web services defined in WSDL definition declared as PartnerLinks and characterised by a PartnerLinkType. We can declare a partner as Declaration (Class(:PartnerLink)) that is related to its specific process by Declaration (ObjectProperty (:hasPartnerLink)) with ObjectPropertyDomain (:hasPartnerLink :Process) and ObjectPropertyRange (:hasPartnerLink : PartnerLink) and has a functional attribute Declaration (DataProperty (:hasName)) where DataPropertyDomain (:hasName PartnerLink), DataPropertyRange (:hasName xsd :string). BPEL elements related to a partner link are represented as follows:

Declaration (Class (:PartnerLinkType))
Declaration (Class (:MyRole))
Declaration (ObjectProperty (:hasPartnerLinkType))
ObjectPropertyDomain (:hasPartnerLinkType :PartnerLink)
ObjectPropertyRange (:hasPartnerLinkType :PartnerLinkType)
Declaration (ObjectProperty (:hasMyRole))
ObjectPropertyDomain (:hasMyRole :PartnerLink)
ObjectPropertyRange (:hasMyRole :MyRole)

In BPEL, variables declaration appears directly under the process element, in our work, variables have a great importance, because they hold the data that constitute the state of a BPEL business process during its execution. We declare a variable by specifying his name as Declaration (Class(:Variable)) and Declaration (DataProperty (:hasName)) where DataPropertyDomain (:hasName :Variable) and DataPropertyRange (:hasName xsd :string). A process is related with its variables by Declaration (ObjectProperty (:hasVariable)) where ObjectPropertyDomain(:hasVariable : Process) ObjectPropertyRange (:hasVariable :Variable). The values contained in such variables come from message exchanged with a partner or it is intermediate data that is private to the process. So, we can add to the variable specification one of the three attributes (type, messageType or element), as follows:

```
Declaration ( DataProperty ( :hasMessageType))
DataPropertyDomain ( :hasMessageType :Variable)
DataPropertyRange ( :hasMessageType xsd :string)
Declaration ( DataProperty ( :hasElement)
DataPropertyDomain ( :hasElement :Variable)
DataPropertyRange ( :hasElement xsd :string)
Declaration ( DataProperty ( :hasType))
DataPropertyDomain ( :hasType:Variable)
DataPropertyRange ( :hasType xsd :string)
```

Another BPEL element which seems very important in this work is the activity. Activities are presented as the very important building blocks of BPEL processes. We can distinguish two types of activities: (Declaration (Class :BasicActivity), Declaration (Class : StructuredActivity)) specified as SubClassOf (Class(: Activity)). Basic activities (receive activity, the replay activity and the invoke activity) consume messages from and providing messages to web service partners. They declared as disjoint SubClassOF(:InvokeActivity :BasicActivity), SubClassOf (:ReceiveActivity : BasicActivity) and SubClassOf(:ReplayActivity :BasicActivity). Based on this representation, if an individual <invoke name="supplyService"…/> is in class BasicActivity, and the class BasicActivity is a subclass of class Activity a reasoner will infer a new fact : <invoke name="supplyService"…/> is an activity which is implicitly contained in the OWL 2 ontology. Each activity specifies the partner link and operation of the partners as object properties where ObjectPropertyDomain (:hasPartnerLink : BasicActivity), ObjectPropertyRange (:hasPartnerLink :PartnerLink) and ObjectPropertyDomain (:hasOperation :BasicActivity), ObjectPropertyRange (: hasOperation :Operation) respectively. The input and output variables are represented as follows:

```
Decalaration (ObjectProperty ( :hasInPutVariable))
ObjectPropertyDomain (:hasInPutVariable :BasicActivity)
ObjectPropertyRange (:hasInPutVariable :Variable)
Declaration (ObjectProperty ( :hasOutPutVariable))
ObjectPropertyDomain (:hasOutPutVariable :BasicActivity)
ObjectPropertyRange (:hasOutPutVariable :Variable)
```

A sequence defines a collection of activities which are executed sequentially. We declare Declaration (Class (:Sequence)) as SubClassOf (:Sequence :Structured Activity).

The if-then-else activity is defined as SubClassOf (:IfThenElseActivity :StructuredActivity), it allows to execute exactly one path of the activity from a given set (True path or False path). The behaviour is to check a condition and if that condition evaluates to true, the associated branch is taken, otherwise an alternative branch is executed. Condition is an important part of if-then-else activity. In order to make our representation useful, we define Declaration (Class(: IfCondition)), SubClassOf(: Condition) and Declaration (DataProperty (:hasValue)) with as DataPropertyDomain (:hasValue :IfCondition) and DataPropertyRange (:hasValue xsd :boolean). Object ExactCardinality(1 :hasBranchTrue :BrachTrue) and ObjectExactCardinality(1 :has-BranchFalse :BrachFalse) are defined as exact cardinality expression contain those individuals that are connected by (:hasBrabchTrue) or (:hasBranchFalse) to exactly one instance of (:BranchTrue) or (:BranchFalse) defined as SubClassOf(:BranchTrue :Activity), SubClassOf(:BranchFalse :Activity). Therefore, we extend our representation by the following SWRL rules:

```
Prefix (var :=<urn :swrl#>)
DLSafeRule(
      Body(
        ClassAtom( :IfThenElseActivity Variable (var :x))
        ClassAtom( :IfCondition Variable (var :y))
        ObjectPropertyAtom( :hasCondition Variable (var :x) Variable(var :y))
        DataPropertyAtom( :hasValue Variable (var :y) "true"^^xsd:boolean))
        ClassAtom( :BranchTrue Variable (var :z))
        ObjectPropertyAtom( :hasBranchTrue Variable (var :x) Variable(var :z))
      Head(
       ClassAtom( :BranchTrue Variable (var :z))
      )
)
```

This rule indicates that if that condition(y) evaluates to true, the BranchTrue is executed. The next rule indicates that if that condition(y) evaluates to false, the BranchFalse is executed.

```
Prefix (var :=<urn :swrl#>)
DLSafeRule(
      Body(
        ClassAtom( :IfThenElseActivity Variable (var :x))
        ClassAtom( :IfCondition Variable (var :y))
        ObjectPropertyAtom( :hasCondition Variable (var :x) Variable(var :y))
        DataPropertyAtom( :hasValue Variable (var :y) "false"^^xsd:boolean))
        ClassAtom( :BranchTrue Variable (var :z))
        ObjectPropertyAtom( :hasBranchTrue Variable (var :x) Variable(var :z))
      Head(
       ClassAtom( :BrachFalse Variable (var :z))
      )
)
```

In addition, BPEL provides three repetitive activities: The while activity, repeatUntil and forEach activity. They are declared as subclasses of StructuredActivity class. For the three activities the condition verified by each activity is very important, we add three subclasses of condition class (SubClass (:WhileCondition :Condition), SubClass (:RepeatCondition :Condition), SubClass (:ForEachCondition :Condition)), and we propose to combine each activity representation with a SWRL rule as follows:

The while activity allows to repeat execute the ChildWhileActivity defined as SubClassOf(:ChildWhileActivity : Activity) as long as a given WhileCondition evaluates to true. This one is checked at the beginning of each iteration, if the Whilecondition does not evaluate to true, the WhileActivity might not be executed at all.

```
Prefix (var :=<urn :swrl#>)
DLSafeRule(
        Body(
          ClassAtom( :WhileActivity Variable (var :x))
          ClassAtom( :WhileCondition Variable (var :y))
          ObjectPropertyAtom( :hasWhileCondition Variable (var :x) Variable(var :y))
          DataPropertyAtom( :hasValue Variable (var :y) "true"^^xsd:boolean))
          ClassAtom( :ChildWhileActivity Variable (var :z))
          ObjectPropertyAtom( :hasChildWhileActivity Variable (var :x) Variable(var :z))
        Head(
          ClassAtom( :ChildWhileActivity) Variable (var :x))
        )
)
```

For the repeatUntil activity, the repeat condition is evaluated at the end of each iteration, in this case the repeatUntil body is executed at least once. We add Declaration (Class(: RepeatCondition)), SubClassOf(:Condition) and Declaration (DataProperty (:hasValue)) with as DataPropertyDomain (:hasValue :RepeatCondition) and DataPropertyRange (:hasValue xsd :boolean). ObjectMinCardinality(1 :hasRepeatUntilBody :RepeatUntilBody) is defined as minimum cardinality expression contain those individuals that are connected by (:hasRepeatUntilBody) to at least one instance of (:RepeatUntilBody) defined as SubClassOf(:RepeatUntilBody :Activity). We extend the repeatUntil representation by the follows SWRL rule, which means that if the repeatUntil condition evaluates to false, the repeatUntilBody is executed:

```
Prefix (var :=<urn :swrl#>)
DLSafeRule(
    Body(
      ClassAtom( :RepeatUntilActivity Variable (var :x))
      ClassAtom( :RepeatCondition Variable (var :y))
      ObjectPropertyAtom( :hasCondition Variable (var :x) Variable(var :y))
      DataPropertyAtom( :hasValue Variable (var :y) "false"^^xsd:boolean))
      ClassAtom( :RepeatUntilBody Variable (var :z))
      ObjectPropertyAtom( :hasRepeatUntilBody Variable (var :x) Variable(var :z))
    Head(
      ClassAtom( : RepeatUntilBod Variable (var :x))
    )
)
```

The third one is the forEach activity, it iterates sequentially N times over a given set of activities. For the forEachActivity, we add Declaration (DataProperty (:hasValue)) with as DataPropertyDomain (:hasValue :ForEachCondition) and DataPropertyRange (:hasValue xsd :integer). In this case, the value of (:hasValue) is defined as datatype restriction (:hasValue DataTypeRestriction(xsd :integer xsd :minExclusive “1”^^xsd : integer maxExclusive “N”^^xsd :integer)). ObjectMinCardinality(1 :hasForEach SetActivity :ForEachSetActivity), ObjectMaxCardinality(N : hasForEachSetActivity : ForEachSetActivity) are defined as minimum and maximum cardinality expression contain those individuals that are connected by (:hasForEachSetActivity) to at least one instance of (:ForEachSetActivity) defined as SubClassOf(:ForEachSetActivity : Activity). We extend the forEach activity representation by the follows SWRL rules, which means that if the value of ForEachCondition is less than N and greater than 1, the ForEachSetActivity is executed:

```
Prefix (var :=<urn :swrl#>)
DLSafeRule(
    Body(
      ClassAtom( :ForEachActivity Variable (var :x))
      ClassAtom( :ForEachCondition Variable (var :y))
      ObjectPropertyAtom( :hasForEachCondition Variable (var :x) Variable(var :y))
      ClassAtom( :ForEachSetActivity Variable (var :z))
      ObjectPropertyAtom( :hasForEachSetActivity Variable (var :x) Variable(var :z))
      DataPropertyAtom ( :hasValue Variable(y) Variable(var :value)
      DatatypeRestriction(xsd :integer          xsd :minInclusive          "1"xsd :integer
xsd :maxInclusive"N"xsd :integer Variable(var :value)
    Head(
      ClassAtom( :ForEachSetActivity  Variable (var :z))
    )
)
```

5.2 Extension of the Semantic Business Process Representation

In this section, we used the research efforts in the field of IoT-ontologies development to extend our semantic representation of business processes. We focus on extending the BPEL OWL 2 ontology with concepts related to sensor data used in existing IoT-ontologies with provision to use, to adopt, or to add new concepts and/or relationships for building a simple hierarchy that can respond to our goal. We define SensorDevice as the most important concept in our extension. It is defined as subClassOf (:SensorDevice :Thing). Sensors may be characterized by type (internal or external), state (static or dynamic) and may be associated with an observation and location concepts. Thus, the concept Observation represents the observation provided by the sensor device, it is associated with time:

```
Declaration (Class (:SensorDevice))
Declaration (DataProperty ( :hasId))
FunctionalDataProperty( :hasID)
DataTypeRange ( :hasId xsd :integer)
Declaration (DataProperty ( :hasType))
DataOneOf ( "dynamic"^^xsd:string "static"^^xsd:string)
Declaration (DataType ( :hasState))
DataOneOf ( "internal"^^xsd:string "external"^^xsd:string)
Declaration (DataProperty ( :hasUnit))
DataTypeRange ( :hasUnit xsd :string)
Declaration (Class (:Location))
Declaration (DataProperty ( :hasIdLocation))
functionalDataProperty( :hasIdLocation)
DataTypeRange ( :hasId xsd :integer)
Declaration (DataProperty ( :hasName))
DataTypeRange ( :hasName xsd :string)
Declaration (Class ( :Observation))
Declaration (DataProperty ( :hasValue))
DataTypeRange ( :hasValues xsd :float)
Declaration (DataProperty ( :hasTime))
DataTypeRange ( :hasTime xsd :date time)
Declaration (ObjectProperty ( :hasObservation))
ObjectPropertyDomain ( :hasObservation :SensorDevice)
ObjectPropertyDomain ( :hasObservation :Observation)
Declaration (ObjectProperty ( :hasLocation))
ObjectPropertyDomain ( :hasLocation :SensorDevice)
ObjectPropertyDomain ( :hasLocation :Location)
```

In BPEL, data is written to and read from variables that are declared under a business process[7]. According to this affirmation, we can identify an object property as bridge rule between business process concepts and concepts related of sensor data as: Declaration (ObjectProperty (:hasImpactOnVariable)) where ObjectPropertyDomain (: hasImpactOnVariable :Observation) and ObjectPropertyRange (: hasImpactOnVariable :Variable). Therefore, we propose the following set of SWRL rules:

```
Prefix (var :=<urn :swrl#>)
    Declaration (ObjectProperty( :hasImpactOnProcess :Observation :Variable)
    DLSafeRule(
    Body(
        ClassAtom ( :Process Variable (var :x))
        ClassAtom ( :Observation Variable (var :y))
        ClassAtom ( :Variable Variable (var :z))
        ObjectPropertyAtom ( :hasVariable Variable (var :x) Variable(var :z))
        ObjectPropertyAtom ( :hasImpactOnVariable Variable (var :y) Variable(var :z)))
    Head(
        ObjectPropertyAtom ( :hasImpactOnProcess Variable (var :y) Variable(var :x))
        )
)
```

This rule indicates that if a process has a variable, and this one is impacted by an observation, we can identify that this process is also impacted by this observation. The follow rule indicates that if an activity has as an impacted InPutVariable, this one is also impacted by this observation.

```
Prefix (var :=<urn :swrl#>)
    Declaration (ObjectProperty( :hasImpactOnActivity :Observation :Variable)
    DLSafeRule(
    Body(
        ClassAtom ( :Variable Variable (var :x))
        ClassAtom ( :Observation Variable (var :y))
        ClassAtom ( :Activity Variable (var :z))
        ObjectPropertyAtom ( :hasInPutVariable Variable (var :z) Variable(var :y))
        ObjectPropertyAtom ( :hasImpactOnVariable Variable (var :y) Variable(var :x)))
    Head(
        ObjectPropertyAtom ( :hasImpactOnActivity Variable (var :y) Variable(var :z))
        )
)
```

7 https://www.oasis-open.org/committees/download.php/23964/wsbpel-v2.0-primer.htm#_Toc166509687.

The impact on an activity can produce several changes on different types of activities, depending on the domain of the discourse and observations measured by the IoT, for example, a condition of an activity can also change his value, based on the last SWRL rule, we can declare the object property atom (:hasImpactOnCondition), which indicates that if an activity While is impacted by an observation, and if this one has a WhileCondition then we can identify that the latter observation can modify the value of this condition, it is also impacted by this observation:

```
Prefix (var :=<urn :swrl#>)
    Declaration (ObjectProperty( :hasImpactOnCondition :Observation :Variable)
    DLSafeRule(
    Body(
        ClassAtom ( :WhileActivity Variable (var :x))
        ClassAtom ( :Observation Variable (var :y))
        ObjectPropertyAtom ( :haImactOnActivity Variable(var :y) Variable(var :x))
        ClassAtom ( :WhileCondition Variable (var :z))
        DataPropertyAtom( :hasValue Variable (var :z) "true"^^xsd:boolean))
    Head(
        ObjectPropertyAtom ( :hasImpactOnCondition Variable (var :y) Variable(var :z))
        DataPropertyAtom( :hasValue Variable (var :z) "false"^^xsd:boolean))
        )
)
```

5.3 Instanciation of the Semantic Business Process Representation

Our representation is developed and it seems useful to describe a business domain influenced by sensor data. To instantiate the BPEL Ontology in an efficient way, we accompany the proposal representation by an API that allows loading BPEL files directly from Java code, creating instances and store results as OWL file. The API uploads all WSDL files imported by the BPEL process. It creates the following set of named individuals using the following algorithm:

```
Algorithm instantiation;
Input
BPEL and WSDL files;
Output
OWL file;
Begin
1 For each BPEL Process
2 Create ClassAssertion (:Process :ProcessName);
3 For each PartnerLink
4 Create ClassAssertion (:PartnerLink :PartnerLinkName);
5 Create ObjectPropertyAssertion(:hasPartnerLink
:ProcesName :PartnerLinkName);
6 Create ObjectPropertyAssertion(:hasPartnerLinkType
:PartnerLinkName :PartnerLinkTypeName);
7 Create ObjectPropertyAssertion(:hasMyRole
:PartnerLinkName :MyRoleName);
8 End for;
9 For each Variable
  10  Create ClassAssertion(:Variable :VariableName);
  11  Create ObjectPropertyAssertion(:hasVariable:ProcesName
:VariableName);
  12  End for;
13  For each Activity
14  For each BasicActivity
15  If ReceiveActivity then
16  Create ClassAssertion(:ReceiveActivity
:ReceiveActivityName);
17  Create ObjectPropertyAssertion (:hasPartnerLink
:ReceiveActivityName :PartnerLinkName);
18  Create ObjectPropertyAssertion (:hasPortType
:ReceiveActivityName :PortTypeName);
19  Create ObjectPropertyAssertion (:hasInputVariable
:ReceiveActivityName :VariableName);
20  Create ObjectPropertyAssertion (:hasOutPutVariable
:ReceiveActivityName :VariableName)
21  Else if InvokeActivity then
22  Create ClassAssertion(:InvokeActivity
:InvokeActivityName);
23  Create ObjectPropertyAssertion (:hasPartnerLink
:InvokeActivityName :PartnerLinkName);
24  Create ObjectPropertyAssertion (:hasPortType
:InvokeActivityName :PortTypeName);
```

```
Create ObjectPropertyAssertion (:hasVariable :InvokeActivityName : VariableName)
Else if ReplyActivity then
Create ClassAssertion(:ReplayActivity :ReplayActivityName);
Create ObjectPropertyAssertion (:hasPartnerLink :ReplyActivityName :PartnerLinkName);
Create ObjectPropertyAssertion (:hasOperation :ReplyActivityName :OperationName);
Create ObjectPropertyAssertion (:hasPortType :ReplyActivityName :PortTypeName);
Create ObjectPropertyAssertion (:hasVariable :ReplyActivityName : VariableName);
End if;
End for;
For each StructuredActivity
If sequenceActivity then
Create ClassAssertion(:SequenceActivity :SequenceActivityName);
For each Activity from sequence
Create ObjectPropertyAssertion (:hasActivity : SequenceActivityName :ActivityName);
Create DataPropertyAssertion (:hasValue :ActivityName "order""^^xsd:integer);
End for
Else if IfThenElseActivity then
Create ClassAssertion(:IfThenElseActivity :IfThenElseActivityName);
Create ObjectPropertyAssertion(:hasIfCondition :IfThenElseActivityName :IfConditionName);
Create DataPropertyAssertion (:hasValue :IfConditionName "value"^^xsd:boolean);
Create ObjectPropertyAssertion(:hasBranchTrue :IfThenElseActivityName :BranchTrueName);
Create ObjectPropertyAssertion(:hasBranchFalse :IfThenElseActivityName :BranchFalseName)
Else if WhileActivity then
Create ClassAssertion(:WhileActivity :WhileActivityName);
Create ObjectPropertyAssertion(:hasWhileCondition :WhileActivityName :WhileConditionName);
Create DataPropertyAssertion (:hasValue :WhileConditionName "value" "^^xsd:boolean);
Create ObjectPropertyAssertion(:hasChildWhileActivity :WhileActivityName :ChildWhileActivityName)
```

```
Else if RepeatUntilActivity then
Create ClassAssertion(:RepeatUntilActivity :
RepeatUntilActivityName);
Create ObjectPropertyAssertion(:hasRepeatCondition :
RepeatUntilActivityName : Repeat1ConditionName);
Create DataPropertyAssertion (:hasValue
:RepeatConditionName "value""^^xsd:boolean)
Create ObjectPropertyAssertion(:hasRepeatUntilBody :
RepeatUntilActivityName : RepeatUntilBodyName);
Else if ForEachActivity then
Create ClassAssertion(:ForEachActivity :
ForEachActivityName);
Create ObjectPropertyAssertion(:hasForEachCondition :
ForEachActivityName : ForEachConditionName);
Create DataPropertyAssertion (:hasValue
:ForEachConditionName "N""^^xsd:integer);
Create ObjectPropertyAssertion(:hasForEachSetActivity
:WhileActivityName : ForEachSetActivityName);
End if;
End for;
End for;
End For;
End.
```

After receiving signals from sensor devices, the API can create new named individuals of each SensorDevice with its different proprieties (identifier, type (dynamic, static), state (internal or external)), location and observation.

6 Sample Example

The above motivating example demonstrates the applicability of this work. Figure 2 presents a part of the production_BPEL.bpel file, it imports production_WSDL.wsdl, deby_Ble.wsdl, Achat_WSDL.wsdl, etc. A process named production_BPEL has several Partnerlinks, several variables and a set of activities.

```
<process name="Production_BPEL" targetNamespace="http://enterprise.netbeans.org/bpel/BonDeCommandeBPEL/Production_BPEL">
  <import namespace="http://j2ee.netbeans.org/wsdl/BonDeCommandeBPEL/Production_WSDL" location="Production_WSDL.wsdl" importType="http://schemas.xmlsoap.org/wsdl/"/>
  <import namespace="http://j2ee.netbeans.org/wsdl/Derby_Ble" location="Derby_Ble.wsdl" importType="http://schemas.xmlsoap.org/wsdl/"/>
  <import namespace="http://j2ee.netbeans.org/wsdl/Derby_StockProduit" location="Derby_StockProduit.wsdl" importType="http://schemas.xmlsoap.org/wsdl/"/>
  <import namespace="http://j2ee.netbeans.org/wsdl/BonDeCommandeBPEL/Achat_WSDL" location="Achat_WSDL.wsdl" importType="http://schemas.xmlsoap.org/wsdl/"/>
  <import namespace="http://j2ee.netbeans.org/wsdl/Derby_Production_Historique" location="Derby_Production_Historique.wsdl" importType="http://schemas.xmlsoap.org/wsdl/"/>
  <import namespace="http://j2ee.netbeans.org/wsdl/DerbyRisque" location="DerbyRisque.wsdl" importType="http://schemas.xmlsoap.org/wsdl/"/>
 <partnerLinks>
   <partnerLink name="EtatRisque" partnerLinkType="tns:jdbcpartner" partnerRole="jdbcPortTypeRole"/>
   <partnerLink name="BleStock" partnerLinkType="tns:jdbcpartner" partnerRole="jdbcPortTypeRole"/>
   <partnerLink name="Achat_Client" partnerLinkType="tns:Achat_WSDL" partnerRole="Achat_WSDLPortTypeRole"/>
   <partnerLink name="ProductionHisto" partnerLinkType="tns:jdbcpartner" partnerRole="jdbcPortTypeRole"/>
   <partnerLink name="UpdateVente" partnerLinkType="tns:jdbcpartner" partnerRole="jdbcPortTypeRole"/>
   <partnerLink name="ServiceVente" partnerLinkType="tns:Production_WSDL" myRole="Production_WSDLPortTypeRole"/>
  </partnerLinks>
 <variables>
   <variable name="UpdateInRisque2" messageType="tns:inputMsg"/>
   <variable name="FindInRisc2" messageType="tns:inputMsg"/>
```

Fig. 2. Part of production BPEL file

If we apply the algorithm described in Sect. 5.3, we obtain an OWL file as shown in Fig. 3.

```
<DataPropertyAssertion>
   <DataProperty IRI="#partnerLinkName"/>
   <NamedIndividual IRI="#Appro_BPEL"/>
   <Literal datatypeIRI="http://www.w3.org/1999/02/22-rdf-syntax-ns#PlainLiteral">Achat_Client</Literal>
 </DataPropertyAssertion>
<DataPropertyAssertion>
   <DataProperty IRI="#partnerLinkName"/>
   <NamedIndividual IRI="#Appro_BPEL"/>
   <Literal datatypeIRI="http://www.w3.org/1999/02/22-rdf-syntax-ns#PlainLiteral">UpdateVente</Literal>
 </DataPropertyAssertion>
<DataPropertyAssertion>
   <DataProperty IRI="#partnerLinkName"/>
   <NamedIndividual IRI="#Appro_BPEL"/>
   <Literal datatypeIRI="http://www.w3.org/1999/02/22-rdf-syntax-ns#PlainLiteral">ProductionHisto</Literal>
 </DataPropertyAssertion>
<DataPropertyAssertion>
   <DataProperty IRI="#partnerLinkName"/>
   <NamedIndividual IRI="#Appro_BPEL"/>
   <Literal datatypeIRI="http://www.w3.org/1999/02/22-rdf-syntax-ns#PlainLiteral">BleStock</Literal>
 </DataPropertyAssertion>
<DataPropertyAssertion>
   <DataProperty IRI="#partnerLinkName"/>
   <NamedIndividual IRI="#Appro_BPEL"/>
   <Literal datatypeIRI="http://www.w3.org/1999/02/22-rdf-syntax-ns#PlainLiteral">EtatRisque</Literal>
 </DataPropertyAssertion>
<ObjectPropertyRange>
   <ObjectProperty IRI="#hasActivity"/>
   <Class IRI="#Activity"/>
 </ObjectPropertyRange>
```

Fig. 3. Part of the result OWL file

The results showed that the API can successfully be giving all instances in an automatic manner. The result representation is enable to treat impact of values measured by sensor devices on business processes. It can respond to the follow list of simple queries:

- Which variables are impacted by sensor data values?
- Which activities are impacted?

7 Conclusion

This paper presents a simple representation of business processes, which is specifically designed to integrate IoT concepts. In particular, we have shown that we can represent different BPEL elements within values transmitted by the IoT using SWRL rules.

Our representation has the advantage of being based on the standards and tools proposed for the Semantic Web and recommended by the W3C. It allows semantic information to be added to the business process elements. It is sufficient to represent the knowledge necessary to identify what is happening in an evolving business process. We can also restore the original version of the ontology at any time using the instantiation algorithm.

Our future research activities will focus on analyzing the data obtained from the sensors and ensuring that reasoning is always possible on the modified representation.

References

1. Baader, F., Calvanese, D., McGuinness, D.L., Nardi, D., Patel-Schneider, P.F. (eds.) The Description Logic Handbook: Theory, Implementation, and Applications (2003)
2. Domingos, D., Martins, F., Cândido, C., Martinho, R.: Internet of things aware WS-BPEL business processes context variables and expected exceptions. J. UCS **20**(8), 1109–1129 (2014)
3. van Schalkwyk, P., XMPro: A Practical Framework to Turn IoT Technology into Operational Capability, March 2018 EDITION (2018)
4. Haase, P., Stojanovic, L.: Consistent evolution of OWL ontologies. In: The Semantic Web: Research and Applications, Second European Semantic Web Conference, ESWC 2005, Heraklion, Crete, Greece, 29 May–1 June 2005, Proceedings, pp. 182–197 (2005)
5. Lavbic, D., Vasilecas, O., Rupnik, R.: Ontology-based multi-agent system to support business users and management, CoRR, vol. abs/1807.0 (2018)
6. Belhajjame, K., Brambilla, M.: Ontology-based description and discovery of business processes. In: Enterprise, Business-Process and Information Systems Modeling, 10th International Workshop, BPMDS 2009, and 14th International Conference, EMMSAD 2009, held at CAiSE 2009, Amsterdam, The Netherlands, 8–9 June 2009, Proceedings, vol. 29, pp. 85–98 (2009)
7. Recker, J.C., Mendling, J.: On the translation between BPMN and BPEL: conceptual mismatch between process modeling languages. In: 18th International Conference on Advanced Information Systems Engineering. Proceedings of Workshops and Doctoral Consortiums, pp. 521–532 (2006)
8. Diane Jordan, I., John Evdemon, M.: Web Services Business Process Execution Language Version 2.0 (2007)
9. Beeri, C., Eyal, A., Kamenkovich, S., Milo, T.: Querying business processes with BP-QL. Inf. Syst. **33**(6), 477–507 (2008)
10. Filipowska, A., Kaczmarek, M., Kowalkiewicz, M., Zhou, X., Born, M.: Procedure and guidelines for evaluation of BPM methodologies. Bus. Proc. Manag. J. **15**(3), 336–357 (2009)
11. El Kharbili, M., Stein, S., Markovic, I., Pulvermüller, E.: Towards a Framework for Semantic Business Process Compliance Management (2008)
12. Namiri, K., Stojanovic, N.: Towards a formal framework for business process compliance. In: Multikonferenz Wirtschaftsinformatik, MKWI 2008, München, 26.2.2008–28.2.2008, Proceedings (2008)
13. Khoshkbarforoushha, A., Jamshidi, P., Gholami, M.F., Wang, L., Ranjan, R.: Metrics for BPEL Process Reusability Analysis in a Workflow System, CoRR, vol. abs/1405.6 (2014)
14. Christensen, E., Curbera, F., Meredith, G., Weerawarana, S.: Web Services Description Language (WSDL) 1.1 (2001). https://www.w3.org/TR/wsdl/#_http:operation

15. Beeri, C., Eyal, A., Kamenkovich, S., Milo, T.: Querying business processes. In: Proceedings of the 32nd International Conference on Very Large Data Bases, Seoul, Korea, 12–15 September 2006, pp. 343–354 (2006)
16. Aslam, M.A.: Towards integration of business processes and semantic web services. University of Leipzig (2008)
17. Bajaj, G., Agarwal, R., Singh, P., Georgantas, N., Issarny, V.: A study of existing Ontologies in the IoT-domain, CoRR, vol. abs/1707.0 (2017)
18. Agarwal, R., et al.: Unified IoT ontology to enable interoperability and federation of testbeds. In: WF-IoT, pp. 70–75 (2016)
19. Ye, P.N.J., Coyle, L., Dobson, S.: Ontology-based models in pervasive computing systems. Knowl. Eng. Rev. **22**, 315–347 (2007)
20. Compton, O.C.M., Barnaghi, P., Bermudez, L., Garcia-Castro, R., Cox, S., Graybeal, J., Hauswirth, M., Henson, C., Herzog, A.: The SSN ontology of the W3C semantic sensor network incubator group. Web Semant. Sci. Serv. Agents World Wide Web **17**, 25–32 (2012)
21. Matheus, C.J., Baclawski, K., Kokar, M.M., Letkowski, J.J.: Using SWRL and OWL to capture domain knowledge for a situation awareness application applied to a supply logistics scenario. In: Proceedings of the 1st International Conference on Rules and Rule Markup Languages for the Semantic Web; Galway, Ireland (2005)
22. Zhai, Z., Martínez Ortega, J.F., Lucas Martínez, N., Castillejo, P.: A rule-based reasoner for underwater robots using OWL and SWRL. Sensors **18**(10), 3481 (2018)
23. Horrocks, I., Kutz, O., Sattler, U.: The even more irresistible SROIQ. In: Proceedings of the 10th International Conference on Principles of Knowledge Representation and Reasoning (KR 2006), pp. 57–67 (2006)

A Review on Watermarking Techniques for Multimedia Security

Hüseyin Bilal Macit[1(✉)] and Arif Koyun[2]

[1] Mehmet Akif Ersoy University, Burdur, Turkey
hbmacit@gmail.com
[2] Süleyman Demirel University, Isparta, Turkey

Abstract. To ensure the safety of the analog data, it is enough to store it physically. With the transfer of data to the digital environment, the concept of data security has been moved to a different dimension. The emergence of the Internet brought data security among the most discussed issues. Higher data transfer speed and higher storage capacity of data make more difficult the protection and authentication of digital data including personal and commercial audio, video and image. Concepts such as cryptology, steganography, digital signature and watermarking have been used for data security. In this study, the concept of watermarking which is used for data security is explained. History of watermarking, watermarking methods, watermarking types and watermarking media are mentioned. Watermarking methods are introduced in the spatial and frequency domain. Comparison of methods and performance metrics are described.

Keywords: Digital watermarking · Spatial domain · Frequency domain

1 Introduction

Data production is getting easier every day with the development of electronic systems. It is estimated that 1.2 trillion in 2017 and 2.2 trillion digital photographs created in 2018 in the world. Figure 1 shows last 5 years of these statistics [1].

The Internet has become a good distribution tool for digital media due to its low cost, fastness and no storage needs. Numerical data can be copied in this communication and storage environment without loss and can be distributed very quickly. This situation increased privacy and security problems. Every personal data can be followed by easily accessible communication channels. It has become a big problem to protect the rights of digital data manufacturers, to protect data against malicious interventions and to ensure data reliability [2]. Cryptography tools are widely used in the secure communication of personal and corporate data. In the transmission channel, each data is encrypted and transported by algorithms and keys created by modern mathematical methods. However, obtaining the unencrypted data once is sufficient to reproduce it. This has led to the search for new data security methods.

Most used methods of ensuring data security are labeling, digital signature, cryptology, watermarking and steganography in the 21st century digital environment [3]. Labeling, also known as fingerprint insertion is the process of inserting an imperceptible

D. J. Hemanth and U. Kose (Eds.): ICAIAME 2019, LNDECT 43, pp. 406–417, 2020.
https://doi.org/10.1007/978-3-030-36178-5_30

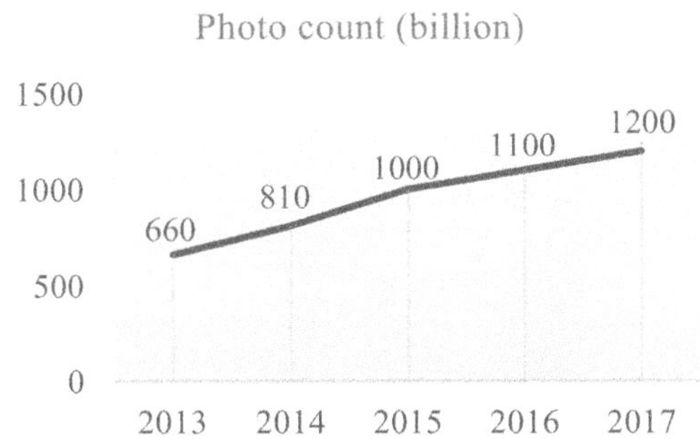

Fig. 1. Count of digital photos taken worldwide between years 2013–2017

confidential information which only represents that consumer to protect digital media copyright before being distributed [4]. Digital signing is the document owner's signing of the document with his personal key which is his encryption. Digital signing can be defined as watermarking using a personal and public key [5]. Cryptology is the technique used to encrypt data as it is shown in Fig. 2. Cryptology, which is closely associated with digital watermarking, is not sufficient to protect some digital information. Because, once the data is encrypted with cryptological methods, hereafter there is no security for the data [6].

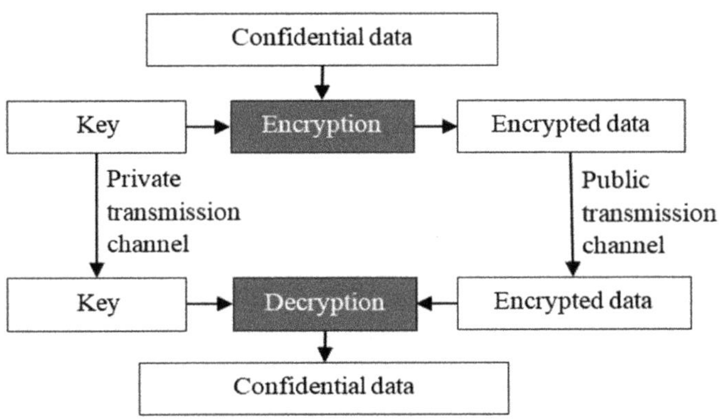

Fig. 2. Cryptology application

In some cases, sending data without encryption is safer than cryptography. In this case steganographic methods are used. A steganographic method hides the transmitting data into another data which is not suspicious. Steganography is a good alternative when cryptography is illegal or inadequate. Steganography consists of three basic elements: hidden data, concealment function and cover data as shown in Fig. 3. Confidential data includes the data that is sent to the recipient by the sender in a secure manner. The hide function includes the secret method of hiding data and the key on the sender side. Cover data contains any data that will not be suspicious in the transmission environment where confidential data will be embedded [7].

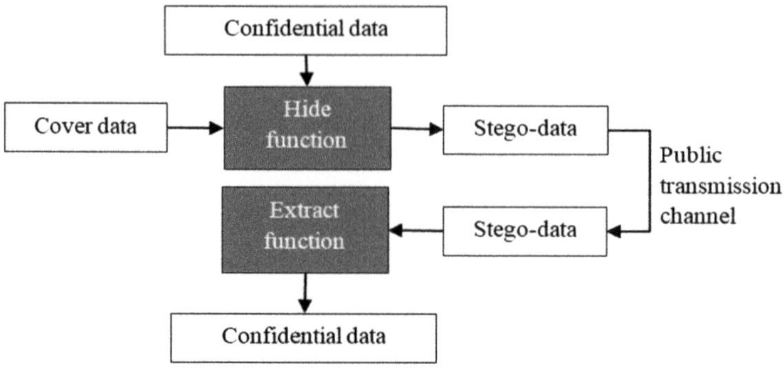

Fig. 3. Steganographic system

In cases such as copyright protection, it is necessary to use confidential data to protect the cover data. In these cases, watermarking techniques that are evolved from steganography are used. Watermarking technique can be used to embed a confidential data called a watermark into a digital signal and then extract the data by performing the opposite operation. Digital signal can be text, audio, picture or video. As shown in Fig. 4, watermarking system generally consists of watermark encoder, decoder, watermark and key [8].

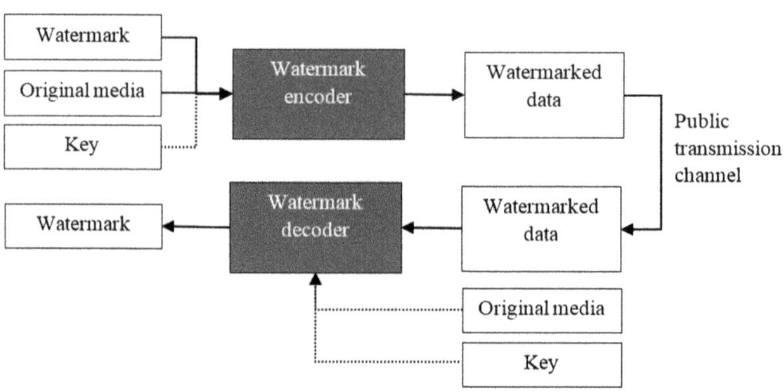

Fig. 4. Watermarking process

The use of watermarking techniques goes back to ancient civilizations. The first known applications were made on raw papers in Italy in 1282. It was used to report the brand of paper and the factory producing paper. It is known that watermarks are used against forgery in money and other documents in the 18th century [9]. It is estimated that the term "watermark" emerged in the late 18th century as the German term "wassermarke" [10]. In 1954, Muzak Corporation has requested a patent for watermarking their musical works. Komatsu and Tominaga were the first to use the term "digital watermark" [11]. The increase in watermarking studies has progressed in parallel with the increase in copyright protection studies.

1.1 Watermarking Properties

The performance of a watermarking method can be measured by its performance in properties. The effectiveness of digital watermarking is evaluated by many different performance criteria. Different criteria can be prioritized for each application, and some criteria can be completely ignored. Most important properties of a watermarking system are; imperceptibility, security, data load, cost, false positive rate (FPR), efficiency and process complexity.

Imperceptibility. It can be defined as the maximum level of similarity between the original and the watermarked data.

Security. The security of a watermark refers to its ability to resist enemy attacks. An enemy attack is a process that is specially designed to completely destroy the watermark, make it unreadable, or prevent its purpose.

Data Load. The number of bits of the watermark that can be embedded in the original media is called the data load or capacity of the watermarking method. The speed of the watermark detector must also increase as the data load of the watermarking media increases. The lower the data load, the more powerful and secure the watermark media.

Cost. The main issue that affects the cost of watermarking is the expected speed in the watermarking and detecting process.

False Positive Rate (FPR). A watermark detector software or hardware may detect a false watermark on a media that does not contain any watermarks. This possibility is called as FPR.

Efficiency. It is the success of the watermark coder embedding watermark to the digital media by the method. Immediately after the watermark embedding process is completed, the watermark detector will perform a scan on the digital media and the embedding activity is expected to be 100%.

Process Complexity. Watermark embedding and extracting should have low process complexity but it generally depends on watermarking type. For example, process complexity of the watermarking process of a copyright protection application can be high [12] unlike the watermarking types which require real time extraction.

1.2 Watermarking Applications

Digital watermarking techniques have different and wide area applications. Mostly used ones are copyright protection, proof of ownership, transaction control, verification, copy control, device control, content archiving and broadcast monitoring.

Copyright Protection. It is to prevent unauthorized duplication or reproduction of the media. The process of making a second copy of the original media or recording the media to any known device is called duplication. In a media that uses digital watermarking, the watermark automatically reproduces within the media as the media is duplicated. It is also not easy to remove the watermark from the media, as the watermark can only be detected by a special software.

Proof of Ownership. It is used to prove who the manufacturer of the media is. The best method developed for proof of ownership is to create a central database for watermarks. The media owner encodes the watermark and sends the watermarked media to a copyright office. This is an efficient but high cost method.

Transaction Control. It is used to identify who copied the media as a pirate. For example, separate watermarks are produced for each of the authorized distributors of the media, and the media is watermarked with the watermark of the distributor before it is delivered to these distributors. If the pirate distribution of the media is then detected, a pirated copy is captured and the watermark of the hacked copy is extracted to determine which distributor distributed the pirate.

Verification. It is the way of determining whether the original of the media has been modified. As digital technologies evolve, it becomes difficult to distinguish whether a work is real or not. Figure 5 shows an example of verification problem. Here; a credit card number and validation date is corrupted by a simple image editing software. If this image is part of a legal process, corruption here may be a serious problem.

Fig. 5. Original and corrupted images of a credit card

Copy Control. Unlike other watermarking techniques, copy control does not follow the illegal copy generator after copying. On the contrary, it aims to protect the rights of the owner before copying. Rather than catching and punishing the person who did an illegal act, it is better to stop him before the action is taken. In the copy control, the product is delivered encrypted even to the person who purchased the original product. In fact, the product can be found in the market already. However, the key to decrypt the media is given only to the person who purchased the original copy of the product. The most common example of this method is encrypted TV broadcasts. The person who has bought the broadcast media has a smart card with a crypto key and can decode the publicly available encrypted broadcast with this key.

Device Control. It is basically embedding a special information about the media processed by a device with a method that only the same device can detect it.

Content Archiving. Watermarking techniques used for content archiving are used to avoid confusion of the identity or characteristics of media. This type of watermarking is the mostly used in medical media. Because the results may be fatal if medical image or other digital media of patients are confused. Therefore, when a media is produced, it is a good solution to embed information of the patient, doctor, the disease and similar data to medical media as an invisible watermark.

Broadcast Monitoring. It is a digital watermarking application especially used in television and radio stations. A digital watermark is embedded to the media to be published and it is detected whether the correct broadcast has been reached to the audience by taking these signals on the monitoring device [13].

1.3 Watermarking Media

Watermarking is examined in different categories according to the type of original media. The types of media to watermark are usually text, image, audio and video media.

Text Watermarking. It is often used in applications for protecting and distributing the copyright of the text owner. It is disadvantageous in comparison with other types of media, due to the fact that the data contain less redundancy and people are very sensitive to changes in a text. There are different methods for text watermarking as shown in Table 1.

Table 1. Text watermarking examples

Original sentence	He drivers the car
Format based method	He drivers the car
Word based method	He drivers
Syntactic based	The car is driven by him
Semantic based	Drive the car

Image Watermarking. Natural or artificial identification of objects is called image. The digital picture is represented by a two-dimensional array in N-line and M-column or it represented as a function. Digital images are divided into three categories as binary, grayscale and RGB. RGB is the optimal image type for watermarking because of its data load.

Audio Watermarking. Audio watermarking methods are examined in four categories. Phase coding, spectrum modulation, patchwork and echo hiding. The ideal amount of data load is 1kps for each 1 kHz [14]. The distortion in the cover audio should be too low to be detected by Human auditory system (HAS). Audio watermarking methods are frequently used for copy protection in audio. The biggest problem of audio watermarking is data quality losses in original audio [15].

Video Watermarking. Videos can be considered as a combination of pictures shown one after the other. Therefore, most of the video watermarking methods are based on image watermarking techniques and can directly apply to raw video or compressed video.

1.4 Perceptual Classification

Human perception is one of the important issues of watermarking. Watermark must be strong against attacks and invisible to human perception. In this manner, the digital watermarks can be classified into 3 categories as visible, invisible and semi-transparent.

Visible Watermarks. Watermark can be easily detected by human visual system (HVS) or HAS. For example, a TV channel can place its own logo in a corner of the image. Visible watermarks should not be easily removed from the original data.

Invisible Watermarks. It is the type of watermarking that the watermark cannot be detected by HVS or HAS. The aim is that the receiver is not aware of the presence of the watermark.

Semi-transparent Watermarks. In semi-transparent watermark applications, changes that can be perceived as limited on the original media are acceptable. Usually a logo is placed on the background of an image without reducing the image's similarity to the original.

2 Watermarking Methods

Watermarking methods are examined in two categories according to the analog or digital processing of the media where the watermark will be embedded. Watermarking methods which applied in digital environment are generally pixel-based methods and these are called spatial domain methods. If the watermarked media is examined in the frequency domain and the watermarking is applied here, these methods are referred to frequency domain methods. This classification of methods is shown in Fig. 6.

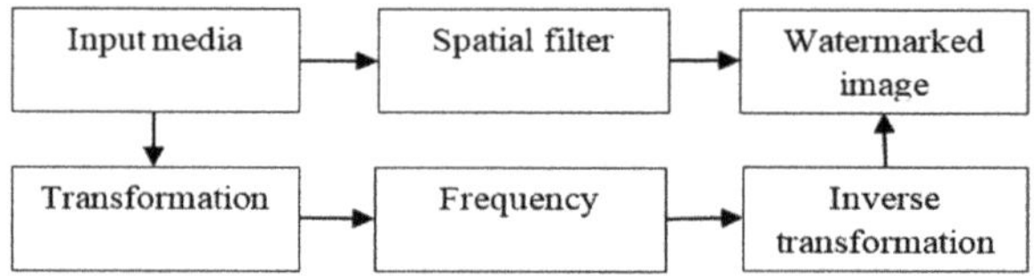

Fig. 6. Application watermarking according to domain

2.1 Spatial Domain Watermarking

Basically, the watermark is embedded to the pixels of the still images, the sound samples, or the pixel values of the video frames. No conversion is applied to the main signal during watermark embedding. Most used methods of spatial domain watermarking are least significant bit, spread spectrum method and echo hiding.

Least Significant Bit (LSB). It is the most common used method of watermarking. It is based on embedding the watermark to the most insignificant bits of the original media. The watermark is converted to a bit sequence and each bit is replaced with the LSB value of the pixel values of the original media as it is shown in Fig. 7. This method can embed watermark on media such as pictures, videos and audios.

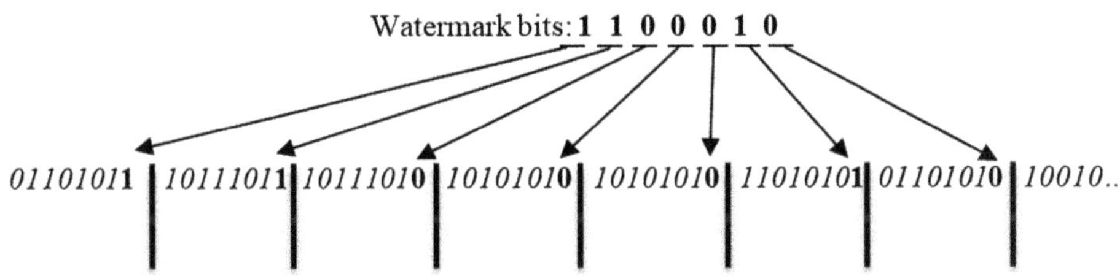

Fig. 7. Sample of LSB watermarking

Spread Spectrum Method. Spread spectrum is the data communication method used in many areas, especially military communication. The data is not continuously transmitted at the same frequency, but embedded in continuously changing frequency bands according to a specific algorithm.

Echo Hiding. This watermarking method is only acceptable on audio signals where data is embedded into cover audio by adding up delayed versions of audio signal on itself. In conventional method, data bits are represented by single echoes with known delays for each bit.

2.2 Frequency Domain Watermarking

In this watermarking methods; the original media and watermark conerts to frequency domain and the coefficients are embedded in the desired frequency level [16]. Usually, different frequency bands are separated and watermarked with mathematical functions. The watermark has spread over the entire media when the media is converted back to the spatial domain from the frequency domain and this makes it difficult for the attackers to perceive and extract them. Most used methods of frequency domain watermarking are discrete cosine transform.

Discrete Cosine Transform (DCT). The DCT maps the digital data to the frequency domain with the cosine waveform and transfers the frequency domain data back to spatial domain by inverse discrete cosine transformation (IDCT) [17]. If two dimensional DCT is applied to a two dimensional N1xN2 pixel media, N1xN2 size DCT coefficients are obtained. These coefficients can be used in compression, watermarking or similar applications. This method divides original data to blocks. Block count is selected by algorithm developer. After the signal is divided into blocks, DCT is applied to each block independently. After the original media and watermark data are converted to frequency domain, the sequence of medium frequency coefficients is calculated and the quantization matrix is generated. Two high-energy blocks are selected in the quantization matrix. The original media is cropped evenly to the size of the watermark object, including the selected frequency coefficients. The watermark is mapped to the resulting cropped object as shown in Fig. 8.

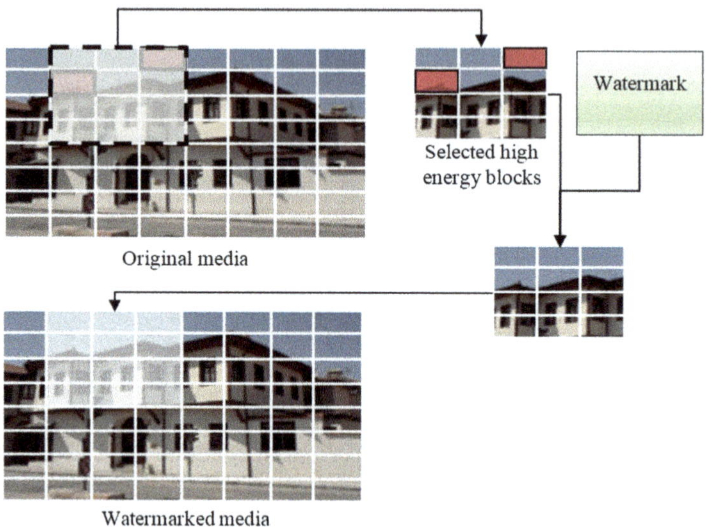

Fig. 8. Sample DCT image watermarking process

Discrete Wavelet Transform (DWT). DWT transforms a function into frequency components. Thus, each component of functions or images can be studied separately. The main purpose is to separate the high and low frequency fields inside the original media and to clear the image from the noise. This means that the watermark can be embedded in the area where no noise is present and which actually belongs to the image. DWT separates the input signal into four separate frequency sub-bands each time for watermarking as shown in Fig. 9. These bands are Low-Low (LL), which is the same as the original cover data, High-Low (HL): which are horizontal high frequencies, Low-High (LH) which are original signal details with vertical high frequencies and High-High (HH) which are highest frequencies [18].

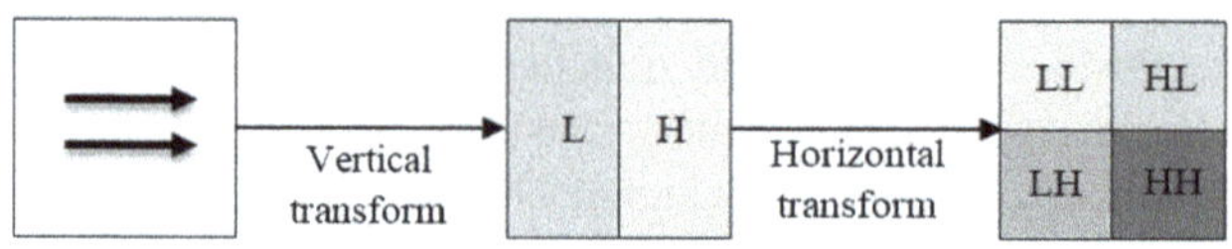

Fig. 9. DWT Transform of a signal

Watermarking is examined in different categories according to the type of original media. The types of media to watermark are usually text, image, audio and video media.

3 Performance Metrics

There are some mathematical measures that evaluate the performance of a watermarking method. These measures are mostly depended on perceptual and structural similarity of cover data and watermarked data. Most used measures are mean squared error (MSE), peak signal to noise ratio (PSNR) and structural similarity index (SSIM).

3.1 Mean Squared Error

It is a fast and simple way to check for distortions between cover and watermarked image. It is calculated with Eq. 1. Here, I is the original image and I_w is the watermarked image and x_i and y_i are i. Samples of these images. MSE is usually transformed to PSNR for a better result.

$$MSE(I, I_w) = \frac{1}{N}\sum\nolimits_{i=1}^{N} (x_i - y_i)^2 \tag{1}$$

3.2 Peak Signal to Noise Ratio

It examines the noise between two different images using MSE. Despite MSE, higher PSNR value means higher similarity between images. Equation 2 shows the calculation of PSNR. Here; L is the dynamic range of the allowed image pixel or audio sample densities.

$$PSNR = 10log_{10}\frac{L^2}{MSE} \tag{2}$$

3.3 Structured Similarity Index Measure

It measures similarity between two images, close to human perception. For example, blurred images are perceived as low quality by HVS, and this cannot evaluate by the MSE method, but can evaluate in SSIM metric measurements. SSIM first calculates three parameters; luminosity, degradation, and degradation. These factors calculate as in Eqs. 3, 4 and 5, respectively.

$$l(I, I_w) = \left(\frac{2\mu_I\mu_{I_w} + k_1}{\mu_I^2 + \mu_{I_w}^2 + k_1}\right) \tag{3}$$

$$c(I, I_w) = \left(\frac{2\sigma_I \sigma_{I_w} + k_2}{\sigma_I^2 + \sigma_{I_w}^2 + k_2} \right) \tag{4}$$

$$s(I, I_w) = \left(\frac{2\sigma_{II_w} + k_3}{\sigma_I + \sigma_{I_w} + k_3} \right) \tag{5}$$

SSIM is calculated in Eq. 6 after calculating l, c and s.

$$SSIM(I, I_w) = l(I, I_w)^{\alpha}.c(I, I_w)^{\beta}.s(I, I_w)^{\gamma} \tag{6}$$

4 Conclusion

This paper is prepared as a reference for studies on watermarking. A brief history and definition of watermarking are mentioned. Properties of watermarking are explained. Watermarking methods are classified according to application domain, media type and human perception. Table 2 shows a comparison of these watermarking methods over media types.

Table 2. Comparison of watermarking methods

Domain	Media	Imperceptibility	Data payload	Security	Process complexity
Spatial	Image	High	High	Low	Low
	Audio	Medium	Medium	Low	Low
	Video	High	Medium	Medium	Medium
Frequency	Image	Medium	Medium	High	High
	Audio	Medium	High	Medium	Medium
	Video	Medium	High	Medium	High

References

1. Cakebread, C.: People will take 1.2 trillion digital photos this year - thanks to smartphones. Erişim tarihi, 05 July 2018 (2017). https://www.businessinsider.com/12-trillion-photos-to-be-taken-in-2017-thanks-to-smartphones-chart-2017-8
2. Boyacı, O.: Doğal Dilde Steganografi, İstanbul Technical University, Graduate School of Natural and Applied Sciences, Master's thesis, 73 p (2017)
3. Yalman, Y., Ertürk, İ.: Gerçek Zamanlı Video Kayıtlarına Veri Gizleme Uygulaması, XI. Akademik Bilişim Konferansı Bildirileri, Harran University, Şanlıurfa, Turkey, pp. 545–552 (2009)
4. Arnold, M., Schmucker, M., Wolthusen, S.D.: Techniques and Applications of Digital Watermarking and Content Protection, 273 p. Artech House, London (2003)
5. Fridrich, J.: Minimizing the embedding impact in steganography. In: Proceeding of the 8th Workshop on Multimedia and Security, Geneva-Switzerland, pp. 2–10 (2006)

6. Yıldırım, İ.: Şifreli ve Şifresiz Videolar İçin Yinelemeli Histogram Değiştirme Tabanlı Tersinir Video Damgalama, Sakarya University, Graduate School of Natural and Applied Sciences, Ph.d. Thesis, 86 p, Sakarya, Turkey (2017)
7. Kutucu, H., Dişli, A., Akça, M.: Çok Katmanlı Steganografi Tekniği Kullanılarak Mobil Cihazlara Haberleşme Uygulaması. Akademik Bilişim Konferansı, Eskişehir, Turkey (2015)
8. Kahalkar, C.: Digital audio watermarking for copyright protection. Int. J. Comput. Sci. Inf. Technol. **3**, 4185–4188 (2012)
9. Shih, F.Y.: Digital Watermarking and Steganography Fundamentals and Techniques, p. 200. CRC Press, London (2005)
10. Simpson, J., Weiner, E.: Oxford English Dictionary. Oxford University Press, Oxford (1989). 22000s
11. Komatsu, N., Tominaga, H.: Authentication system using concealed image in telematics. Mem. Sch. Sci. Eng. Waseda Univ. **52**, 45–60 (1988)
12. Abbasfard, M.: Digital Image Watermarking Robustness: A Comparative Study, Computer Engineering Division, Master's thesis, Delft University of Technology, Netherlands (2009). 74s
13. Fındık, O.: Yapay Zeka Teknikleri Kullanarak Sabit Görüntüler İçin Sayısal Damgalama, Selçuk University, Graduate School of Natural and Applied Sciences, Ph.d. thesis, 120 p (2010)
14. Bhattacharyya, S., Kundu, A., Sanyal, G.: A novel audio steganography technique by M16MA. Int. J. Comput. Appl. **30**(8), 26–34 (2011)
15. Rashid, R.S.: Binary Image Watermarking on Audio Signal Using Wavelet Transform, Çankaya University, Mathematic-Computer Division, Master's thesis, 42 p, Ankara, Turkey (2014)
16. Mahmoud, K., Datta, S., Flint, J.: Frequency domain watermarking: an overview. Int. Arab J. Inf. Technol. **2**(1), 33–47 (2005)
17. Tsai, S.E., Yang, S.M.: An effective watermarking method based on energy averaging in audio signals. Math. Probl. Eng. **2018**, 1–8 (2018). Article ID: 6420314
18. Abdülkhaev, A.: A New Approach for Video Watermarking, Gaziantep University, Graduate School of Natural and Applied Sciences, Master's thesis, 66 p (2016)

Realization of Artificial Neural Networks on FPGA

Mevlut Ersoy(✉) and Cem Deniz Kumral

Department of Computer Engineering, Suleyman Demirel University, Isparta, Turkey
mevlutersoy@sdu.edu.tr, cemkumral2434@gmail.com

Abstract. Artificial Neural Networks (ANNs) are generally modeled and used as software based. Software models are insufficient in real time applications where ANN output needs to be calculated. ANN has an architecture that can operate in parallel to calculate hidden layers. The fact that ANN has such an architecture makes it potentially fast in calculating certain transactions. However, the speed of these operations in real-time systems depends on the specification of the hardware. Therefore, ANN design has been realized on FPGA which is capable of parallel processing. In this way, the ANN structure was realized in a hardware structure and it was provided to be used on real-time structures.

Keywords: Artificial Neural Networks · Field Programmable Gate Array · Very High Speed Integrated Circuit Hardware Description Language · Real time systems

1 Introduction

Artificial Neural Networks (ANN) is a popular machine learning method used in many applications developed today because of its high performance in nonlinear operations. ANNs can produce solutions in many areas such as pattern recognition, signal processing and control systems. When the studies in this area are considered, it is performed by using software based simulation results [6]. In recent years, it has become necessary to use ANN structure in real time and hardware based applications [5]. For the implementation of ANN structures, the proposed system architectures are analog, digital and both. Analog architectures give clearer results, but are more difficult to apply and also suffer from storage problems for neuron weights. Digital designs are more accurate and have no storage problems [6]. In applications using ANN architecture, operations are performed in two stages as learning and testing. In these stages, the learning stage is quite complex and long [3]. For these reasons, ANN provides efficiency in applying these technologies on working principles. With microprocessors and graphic processing units, structures and parallel designs can be made via software. Recently, demand for ANN in hardware and real-time systems has increased. Field Programmable Gate Array (FPGA) integrated circuits are also popular for such structures. FPGAs provide a portable hardware infrastructure and allow parallel software design. Because of these achievements, the use of ANNs in hardware applications

D. J. Hemanth and U. Kose (Eds.): ICAIAME 2019, LNDECT 43, pp. 418–428, 2020.
https://doi.org/10.1007/978-3-030-36178-5_31

improves performance. Application Specific Integrated Circuit (ASIC) and Very Large Scale Integration (VLSI) technologies are used to design integrated modules that operate entirely in parallel. However, the development of these integrations is both costly and time-consuming. Furthermore, these integrated designs are only suitable for one target application. Thanks to FPGAs, flexible designs can be made as well as cost and design cycle savings.

In this study, the hardware design of an artificial neural network testing stage applied on FPGA is presented. The ANN algorithm was trained through the Artificial Neural Network Library in MATLAB before being transferred to FPGA. During the training phase of ANN software, channel signal strength measurements with 2.4 GHz operating frequency and 100 MHz bandwidth were used as sample data sets in closed spaces with different structures [12, 13]. According to the outputs of the trained ANN model, a classification was performed as weak signal, medium signal and strong signal. For the ANN model, the most common hardware definition language, Very High Speed Integrated Circuit Hardware Description Language (VHDL) was applied using hardware identification process. The application of Xilinx VIVADO software was performed and the operating speed and memory usage rates were examined in the simulation environment.

Regarding the subject of the article and the research done, the remaining content is arranged as follows: In the second part, machine learning algorithm applications made with FPGA are mentioned and general literature review is explained. Following this, general information about Artificial Neural Networks and FPGAs is given and the findings obtained from the applications are given and a general discussion about them is given in the third section. Subsequently, the final chapter and the article ended with explanations of the results and some possible future studies.

2 Related Works

Chang et al. in their work, they presented a hardware application of Long Term Memory (LSTM) repetitive artificial neural network on programmable logic Zynq 7020 FPGA from Xilinx. They applied an RNN with 2 layers and 128 hidden units in the hardware and performed the testing using a character level language model. They concluded that the application was 21 times faster than the ARM Cortex-A9 CPU built into the Zynq 7020 FPGA. They concluded that this study could potentially become an RNN co-processor for future mobile devices [1].

Abrol and Mahajan presented a hardware application of an artificial neural network with Field Programmable Gate Arrays (FPGA). The parallel architecture of a neural network reduces the time it takes to calculate certain tasks and makes the neural network suitable for implementation on the VLSI. Implementation of an Artificial Neural Network (ANN) is largely dependent on the effective implementation of a single neuron. They have argued that FPGA-based architectures are suitable for artificial neural networks applications. As a result of their work, they presented the work of different researchers to assist young researchers in their research studies [2].

Yilmaz in his study, realized a trainable artificial neural network (ANN) structure on the chip with Altera FPGA circuits. He worked with the XOR problem and a sensor linearization problem and used a ANN structure based on a fixed point number system. His used neural network trained by the error's back propagation algorithm. As the learning rule was delta bar delta has chosen rule. He designed and simulated these applications with Altera's QUARTUS II FPGA design program and MATLAB. In addition, the XOR problem with the simplified ANN structure was realized with the Altera Cyclone EP1C6Q240C8 FPGA based UP3 development board. As a result of his study, he showed that FPGA is a more suitable solution for some ANN based systems in terms of cost, time saving, rearrange ability and parallel design capability [3].

Ali and Mohammed in his work, they presented a hardware design of a machine learning algorithm on Field Programmable Gate Arrays (FPGA). The digital system architecture is designed to perform a multi-layered neural network. They defined the architecture they designed using the Very High Speed İntegrated Circuits Hardware İdentification Language (VHDL). As a result of the study, they presented their analysis on the memory usage and the operating speed of the system, which is formed by the realization of artificial neural network on FPGA [4].

Şahin and Becerikli in their work, presented hardware implementation of a machine learning algorithm using FPGAs. The digital system architecture was introduced using the Very High Speed Integrated Circuits Hardware Description Language (VHDL) and implemented in the FPGA chip. They tested their design on an FPGA demo card. They argued that the choice of dynamic range, precision, and arithmetic hardware architecture used had a direct effect on the processing density obtained in neural network applications. Using the appropriate precision arithmetic design, they analyzed the memory and speed results. Of the study, they reported that a sufficiently high speed can be achieved and that small size memory usage can be provided for real time ANN applications [5].

Savran and Ünsal in his work, they presented an artificial neural network hardware application using Field Programmable Gate Arrays (FPGA). They designed a multilayer neural network with forward feed to perform a digital system architecture. Designed architecture, were implemented using very high speed integrated circuits hardware identification language (VHDL) and FPGA demo card. As a result of the study, thanks to the parallel working principle of FPGAs, artificial neural network's working time is shortened and the performance is greatly increased [6].

Muthuramalingam et al. in their work, discussed the issues involved in the realization of a multi-input neuron using FPGA. They have proposed an application method with a source/speed transition to process signed decimal numbers. The developed VHDL encoding was tested using the Xilinx XC V50hq240 Chip. They discussed the problems of using a lookup table (LUT) for a nonlinear function using a lookup table method to increase processing speed. Percentage savings at source and rate improvement with a LUT for a neuron have been reported. They also attempted to derive a generalized formula for a multiple-entry neuron to make it easier to estimate the total resource requirement and speed that can be obtained for a given multilayer neural network. They using the proposed method of application, they presented a Space vector modulator for a neural network-based application, i.e. a vector-controlled driver [7].

Botros and Abduz-Aziz in their work, they presented the hardware implementation a machine learning algorithm using Xilinx FPGAs. The network was composed of an input layer consisting of five nodes, a single hidden layer consisting of four nodes, and an output layer consisting of two nodes. Training was done offline on a traditional digital computer, where the final values of weights are obtained we they implemented each node with two XC3042 FPGAs and 1k $\times$ 8 EPROM. They obtained phase datas by simulating the neural network a computer and successfully tested the node values of the outputs by compare with this simulation datas [8].

Omondi and Rajapakse in their work, they worked on a book containing the applications of the realization of artificial neural networks on FPGA. In the result of working they presented obtained scientific datas and the results obtained for each application in book [9].

Chen and Plessis in their work, they implemented a forward-feeding Artificial Neural Network (ANN) on Field Programmable Gate Arrays (FPGA). They conducted a study to find the minimum certainty required to maintain a minimum 95% recognition rate for two characters in an optical character recognition application. In order to reduce the circuit size, they enabled the bit series architecture to perform arithmetic, and as a result of their work they demonstrated the optimum use of FPGA resources [10].

3 Method and Material

3.1 Overview of Artificial Neural Networks

Artificial Neural Networks (ANN) are logical software developed to perform basic functions of the brain such as learning, remembering, generalizing and generating new information by mimicking the working mechanism of human brain. In short, ANNs are synthetic structures that mimic biological neural networks.

ANNs are composed of artificial nerve cells (neurons) formed on the basis of human nerve cells. It is a structure that collects artificial nerve cell input data and produces a value according to the threshold function. In the neuron structure, the inputs are collected and the cell create an effect when the inputs exceed the threshold. This process is the basic working logic of the artificial nerve cell. The basic structure of the artificial nerve cell is given in Fig. 1.

ANN consists of artificial neural cells connected to each other and is usually arranged in layers. ANN is an architecture scattered in parallel to the ability to collect, postcellular weights, and to store and generalize information after learning. In the learning process, the process is about renewing the weights to achieve the desired result. Although ANNs can perform transactions very quickly, they are far from being able to compete with the human brain. However, they are successful in complex mappings and data classification. The basic structure of a multi-layered ANN architecture is given in Fig. 2.

The basic features of ANN architecture are given below.

Learning: ANNs can be trained according to desired outputs for the application. In training process, it is necessary to give weights between connections. ANN does not

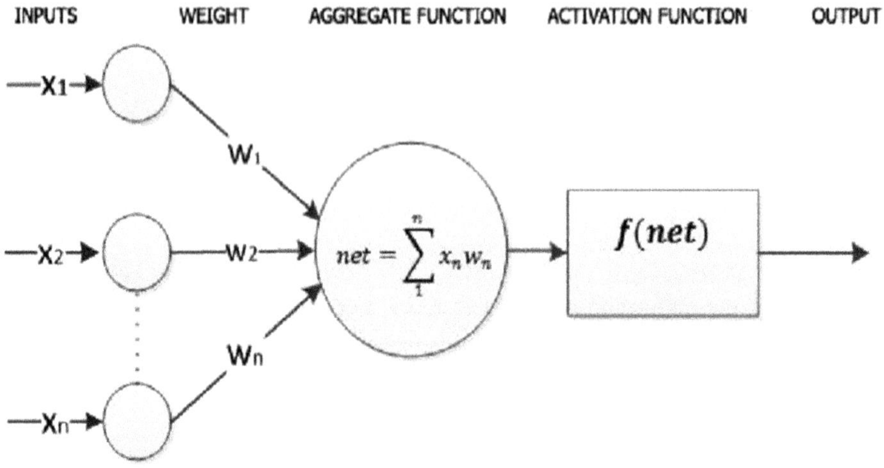

Fig. 1. Basic structure of artificial nerve cell [11]

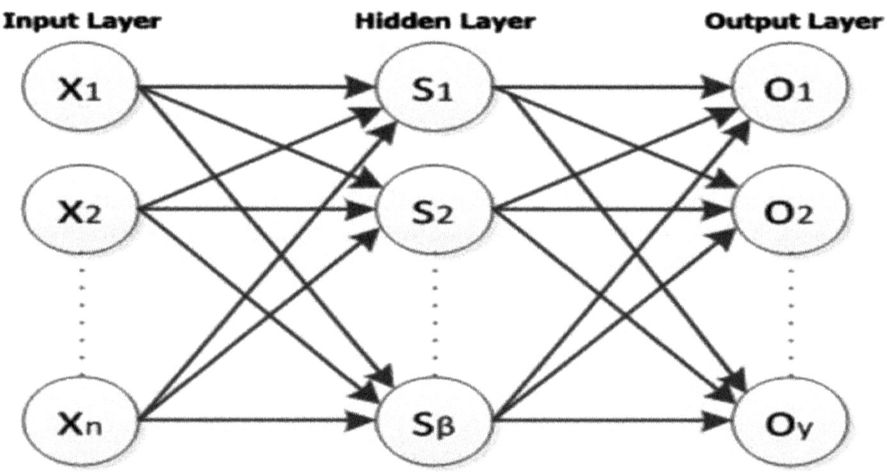

Fig. 2. Multi-layered ANN architecture [11]

work with connections that have a certain weight after the learning process. For this reason, ANN performs the self-learning process according to the data that it was previously trained.

Generalization: Generalization is that the ANN generates appropriate responses in itself for entries not encountered during the learning period. ANN is therefore used in many complex applications. After learning a certain problem, ANN can produce the desired output for the test samples that it does not encounter during the training. With ANNs, for character recognition application can performed process to identify a damaged or missing character.

Hardware and Speed: ANN can be designed as a circuit since it has a parallel structure. This feature speeds up the ANN's ability to process information and makes it available for real-time applications.

Nonlinearity: Considering the basic structure of ANN, a linear structure can be seen. But, since the basic unit cells are not in a linear structure, ANN works by producing nonlinear results. Therefore, it has a high utilization rate in solving complex problems.

Adaptability: ANN that is trained to solve a specific problem can also be trained according to a new problem by making changes to on the problem. A ANN trained for pattern recognition can also be trained to be used on the signal processing application.

The most important feature of ANN is its distributed parallel structure, learning and generalization.

3.2 FPGA Definition and General Features

FPGAs are abbreviation of "Field Programmable Gate Array" phrases. Depending on the amount of transistors on the FPGA, the designer can design and all the works that integrals can do with a single FPGA. FPGAs are integrated circuits that allow the designer to easily design digital circuits in digital design and to perform and test and validate designs in a physical environment on a prototype device.

FPGAs allow the designer to change the hardware structure after the production stage according to the desired function. This feature makes FPGAs one of the most basic features that distinguish it from the commonly used microprocessors. FPGAs are integrated circuits capable of parallel processing. Multiple processes can be written to an entity defined in the VHDL design and these processes can work simultaneously with each other. For example, all programs written on a microprocessor run sequentially, that is, processing in a main, interrupts to perform different operations, operations can be executed from the main by using timers, but this means a slowdown in the program. But there is no such problem in FPGA design.

For example, when processing images on a microcontroller, the image is received and processed, then the second image segment is retrieved, so the second image is not received until the first image is processed and output to the output. If these operations are not performed quickly enough, there is a possibility that the second image will be lost. In FPGA, the process is much faster. The second image can be imported while the first image is received and processed. When the first image is output, the second image can be processed and the third image can be uploaded to the FPGA. This makes FPGA indispensable for complex operations at very high speeds process. FPGAs are used in many different fields from consumer electronics, especially aerospace and defense industry to automotive industry.

FPGAs are composed of three main parts that can be arranged: programmable logic blocks, input-output blocks surrounding this block array, and interconnections. The visual representation of the basic structure of FPGA is given in Fig. 3.

The programmable logic blocks are embedded in the interconnections. Their configuration and communication between them is achieved through interconnections. The input-output blocks establish the connection between the intermediate connections and the package legs of the integrated circuit.

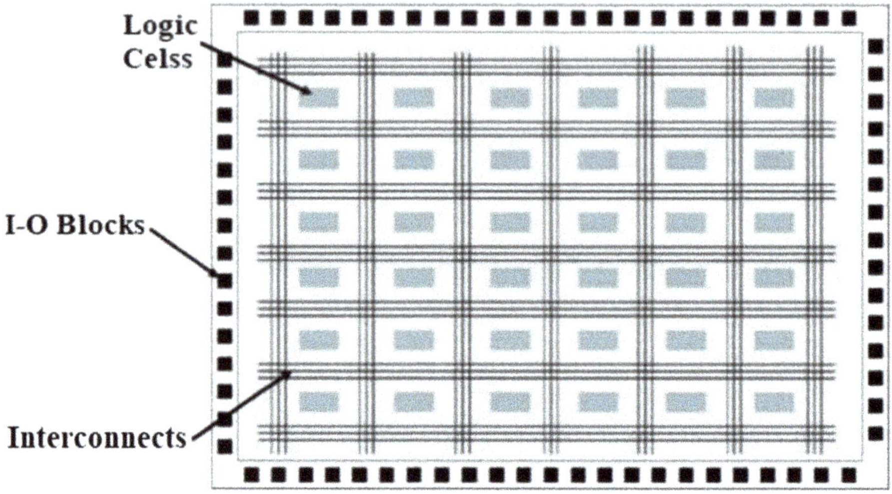

Fig. 3. Basic FPGA architecture

To create the FPGA design, must either create the HDL language or the schematic design. Today, the most common method for designing FPGA applications is VHDL (Very High Speed Integrated Circuit Hardware Description Language). Different to this, the Verilog language is also used.

VHDL is a hardware description language used to describe digital electronic systems. Basically, the structure of a component in VHDL consists of an interface specification and an architecture specification. Interface definitions start with the ENTITY keyword and contain the input and output ports of the component. The name of the component is written after the ENTITY keyword. Then comes the architectural part where the application performs the actual logic operations and is identified using the ARCHITECTURE keyword. For alternative applications that perform the same function a defined interface can have several different architectural bodies [4]. An example of a piece of code written in the VHDL language in the VIVADO environment is given in Fig. 4.

Once a digital system is identified via VHDL, the written code needs to be simulated. The reason for the simulation is to verify whether the VHDL code correctly applies the intended design and conforms to the design specifications. The defined VHDL model is transferred to synthesis tools such as FPGA and implemented on hardware.

3.3 ANN Application on FPGA

In this study, Xilinx Artix-7 FPGA based BASYS3 board was used. The board has 16 switches, 16 LEDs, 5 push-buttons, 4 7-segment displays, USB-based serial ports, VGA output for display connection, serial flash memory and 4 general-purpose 8-bit digital input-output connectors. The card is coding via a separate micro-USB port. No extra power connection is required as the card receives operating power through this

```
library IEEE;
use IEEE.STD_LOGIC_1164.ALL;

-- Uncomment the following library declaration if using...

entity lineCoding is
    Port ( clk : in STD_LOGIC;
           data_in : in STD_LOGIC;
           data_out_Man : out STD_LOGIC;
           A : inout STD_LOGIC;
           B : inout STD_LOGIC;
           C : inout STD_LOGIC;
           D : inout STD_LOGIC;
           Q : inout STD_LOGIC := '0';
           data_out_difMan : inout STD_LOGIC;
           data_out_NRZ : out STD_LOGIC;
           data_out_NRZ_L : out STD_LOGIC;
           data_out_NRZ_I : inout STD_LOGIC := '0'
          );
end lineCoding;

architecture Behavioral of lineCoding is

begin

-----------MANCHESTER-------------
data_out_Man <= data_in xor clk;
-----------------------------------
```

Fig. 4. VHDL code piece

input. The BASYS3 card supports the Xlinx VIVADO Design environment for design, programming and debugging. The applications performed in this study were made in this design environment [14].

In this study, hardware design of an artificial neural network applied on FPGA is presented. The ANN software was trained in Artificial Neural Networks processing libraries in MATLAB before being transferred to FPGA. While the ANN software was trained, a data set including with different structures indoor channel measurements performed over radio frequency (RF) signals at 2.4 GHz operating frequency and 100 MHz bandwidth was used. The weight outputs calculated as a result of the trained ANN model were subjected to hardware identification using the Very High Speed Integrated Circuit Hardware Description Language (VHDL) and written to the VHDL package file using the Xilinx VIVADO development environment. After the codes written in this file are compiled and synthesized, the data outputs of ANN are transferred to the BASYS3 FPGA card through the ports and the hardware realization process is performed. As a result of these processes, the signals of the FPGA card were checked and the signals were controlled according to a predetermined threshold value. Finally a classification was performed as weak signal, medium signal and strong signal. After the implementation is completed, the completion time of the system operations was compared with the operation time of the same application which is run only in MATLAB environment. The values obtained by examining the resource usage rates are given as explanations and tables in the results section. The block diagram of the ANN application on FPGA is given in Fig. 5.

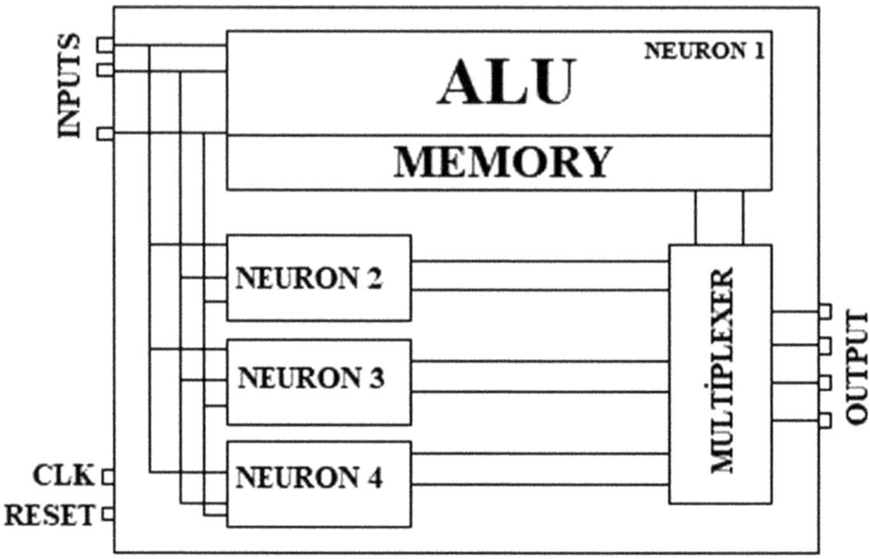

Fig. 5. ANN block diagram on FPGA

4 Discussion and Results

In this study, ANN test stage was performed by choosing VHDL hardware design and definition language on FPGA. Although the syntax structure of the VHDL language differs from that of high-level languages, it can be easily learned and used by someone who does not have much hardware knowledge. With VHDL, it is possible to design very complex designs with little hardware knowledge. ANN test phase was found to be accurate in a short time using FPGA. It has been shown that the use of FPGA in hardware based ANN designs can be used in real time systems and produce accurate results. In addition, the VHDL language can be used with the FPGA, thus demonstrating that flexible design of real-time systems is possible and can be modified by editing if necessary. ANN application on FPGA was compared with MATLAB in terms of working time. The ANN design in FPGA uses its own adder and multiplier circuits, while the MATLAB program uses its own processor. FPGA operations are done in 16 bits to save memory. However, due to the flexibility of the VHDL language, it is possible to change operations quickly and perform more operations with more bits and more precision. MATLAB program run by computer processor and VHDL based program run by FPGA processor were compared according to the number of iterations. The comparison results are shown in Table 1.

Table 1. Comparison of processing times

Iteration number	Matlab process time	FPGA process time
2121	1950000 μs	850 μs

As shown in Table 1, it was seen that the application performed on FPGA completed the operations much faster than the application run with MATLAB. It is possible to achieve faster results by increasing the operating frequency of the FPGA card.

Table 2. Comparison of error rates

Iteration number	Matlab (Min. – Max.)	FPGA (Min. – Max.)
2121	0.01–0.03	0.02–0.04

In this study, the ANN application on FPGA is compared with MATLAB in terms of accuracy and error control. The comparison results are shown in Table 2. According to Table 2, the error rates of the application on MATLAB were lower than the FPGA processor. However, while 32-bit and 64-bit applications are performed in the computer processor, 16-bit operations are performed in the FPGA application. Therefore, rounding errors occurred in the decimal part of the numerical value of the weights of ANN neurons during the test phase.

Table 3. Comparison of error rates

Source type	Number of resource uses
Flip – Flop	81
LUT	121
Block RAM	3

When the resource usage of FPGA based ANN application is examined, the source types used by the encoded hardware circuit and the number of types used are given in Table 3. As a result of running the application on FPGA card, 81 Flip-Flop, 121 LUT and 3 Block RAM were used.

Table 4. Signal classification results

	Signal Level		
	Weak	Medium	Strong
Matlab	55	506	239
FPGA	56	505	239

In this study, a sample data set consisting of 2000 values including indoor channel measurements performed at 2.4 GHz operating frequency and 100 MHz bandwidth over radio frequency (RF) signals were used in indoor spaces with different structures. The 70% value in this dataset was used as training data and the rest was used as test data. After the application was started, the signals of the FPGA card were checked and the signals were checked according to a predetermined threshold and a classification process was performed as a weak level signal, a middle level signal and a strong level signal. The results of the signal classification are shown in Table 4.

5 Conclusions and Future Work

As a result of the study, it has been shown that using VHDL language and ANN model on FPGA has significant positive effects on computational speed and memory usage. Thanks to the infinite FPGA circuit used and the great flexibility with VHDL, it has been found that a design can be used in many real-time applications in a short time. It has been shown that FPGAs can work with artificial neural networks because of their speed, security and parallel processing capabilities that traditional processors do not have, and it is appropriate to use FPGA processor in ANN style architectures. At the same time, more successful results can be achieved by increasing the number of logic gates and flip flops used in the FPGA. FPGA has been shown to be a very suitable solution for ANN based systems in terms of cost, time saving, re-configurability and parallel design capability.

References

1. Chang, A.X.M., Martini, B., Culurciello, E.: Recurrent neural networks hardware implementation on FPGA. arXiv preprint arXiv:1511.05552 (2015)
2. Abrol, S., Mahajan, R.: Artificial neural network implementation on FPGA chip. Int. J. Comput. Sci. Inform. Technol. Res. **3**(1), 11–18 (2015)
3. Yilmaz, N.: Alan programlamalı kapı dizileri (FPGA) üzerinde bir yapay sinir ağları (YSA)'nın tasarlanması ve donanım olarak gerçekleştirilmesi. Doctoral dissertation, Selçuk Üniversitesi Fen Bilimleri Enstitüsü (2008)
4. Ali, H.K., Mohammed, E.Z.: Design artificial neural network using FPGA. IJCSNS **10**(8), 88 (2010)
5. Sahin, S., Becerikli, Y., Yazici, S.: Neural network implementation in hardware using FPGAs. In: International Conference on Neural Information Processing. Springer, Heidelberg (2006)
6. Savran, A., Ünsal, S.: Hardware implementation of a feed forward neural network using FPGAs. In: The third International Conference on Electrical and Electronics Engineering (ELECO 2003) (2003)
7. Muthuramalingam, A., Himavathi, S., Srinivasan, E.: Neural network implementation using FPGA: issues and application. Int. J. İnform. Technol. **4**(2), 86–92 (2008)
8. Botros, N.M., Abdul-Aziz, M.: Hardware implementation of an artificial neural network. In: IEEE International Conference on Neural Networks. IEEE (1993)
9. Omondi, A.R., Rajapakse, J.C. (eds.): FPGA Implementations of Neural Networks, vol. 365. Springer, Dordrecht (2006)
10. Chen, Y., du Plessis, W.: Neural network implementation on a FPGA. In: IEEE AFRICON 6th Africon Conference in Africa, vol. 1. IEEE (2002)
11. Yigit, T., Ersoy, M.: The testing of wireless local area network with ANN (2013)
12. AlHajri, M.I., et al.: Classification of indoor environments based on spatial correlation of RF channel fingerprints. In: 2016 IEEE İnternational Symposium on Antennas and Propagation (APSURSI). IEEE (2016)
13. AlHajri, M.I., Ali, N.T., Shubair, R.M.: Classification of indoor environments for IoT applications: a machine learning approach. IEEE Antennas Wirel. Propag. Lett. **17**(12), 2164–2168 (2018)
14. Kelleci, B.: VHDL ve Verilog ile Sayısal Tasarım, Doç

Estimation of Heart Rate and Respiratory Rate from Photoplethysmography Signal for the Detection of Obstructive Sleep Apnea

E. Smily Jeya Jothi and J. Anitha(✉)

Karunya University, Coimbatore, India
smilyedwin@gmail.com, anithaj@karunya.edu

Abstract. In this proposed work, a new technique has been devised for recognizing patients with OSAS aided by extracting their respiratory rate and heart rate from photoplethysmogram (PPG) signal as the PPG signal contains respiratory induced intensity variations in addition to cardiac synchronous pulsations. Principle Component Analysis (PCA), a data reduction technique is utilized to extract the respiratory activity from the PPG signal. Singular Value Ratio (SVR) trend is equipped to estimate the periodicity and heart rate. Experimental results depict a strong correlation between the normal and extracted respiratory signals. The possibility to reliably estimate respiratory rate from the PPG signal is particularly appealing due to the simplicity and the ease with which it can be measured. Respiratory rate extracted from the PPG signal taken from both normal and abnormal subjects are monitored and compared. Contrary to the conventional polysomnographic method of detecting sleep apnea, the proposed method can be carried out during day time.

Keywords: Obstructive Sleep Apnea Syndrome (OSAS) · Principal component analysis · Singular Value Decomposition (SVD) · PPG Derived Respiratory Rate (PDR)

1 Introduction

Sleep is a physiological process that which performs restorative functions for the brain and the body and is indeed very essential for humans [15]. Obstructive Sleep Apnea Syndrome (OSAS), the most common breathing disorder in adults and in children, is a condition where the upper airway gets blocked up repeatedly during sleep which is due to various reasons, while the respiratory effort continues during sleep for at least 10 s. Sleep apnea is a naturally recurring condition characterized by reduced or absent consciousness, relatively suspended sensory activity and inactivity of all the voluntary muscles. It can be easily distinguished from wakefulness by decreased ability to react to external stimuli, and it can be reversed more easily than being in coma. This condition, so called sleap apnea could affect the quality of human health and sleep when occurs very often and may even cause death in severe cases. Statistics shows that sleep apnea affects up to 25 million people across the globe and those who already possess it are unaware of it [7, 10, 16]. Researchers depicts the availability of various biosignals and the extraction of several features from them using sophisticated signal processing

D. J. Hemanth and U. Kose (Eds.): ICAIAME 2019, LNDECT 43, pp. 429–436, 2020.
https://doi.org/10.1007/978-3-030-36178-5_32

algorithms which would aid in early detection of sleep apnea syndrome [2, 5, 11, 12, 14]. Estimation of respiration rate and heart rate from Photoplethysmogram signal is an alternative approach to Polysomnography [1, 3, 4, 6], which is an overnight sleep study and it involves too many sensors connected to the patients which reduces ease and cause discomfort. Photoplethysmography is an optical technique which continuously monitors and records the light intensity from a source, that which is scattered by the tissues and collected by a suitable photodetector. It is non-invasive and assist in monitoring the arterial oxygen saturation (SaO_2) continuously during anesthesia and in intensive care units [1]. The PPG signal is composed of five frequency components ranging from 0.007 to 2.5 Hz. These frequency components relate to respiration, cardiac activity, blood pressure, thermoregulation, and autonomous nervous system. Two of the five frequency components of the PPG signal are dominating. One frequency component due to the pulsatile motion of blood flow in vessels i.e. the arterial pulse which given an alternating signal, AC component and the other relates to respiration i.e. a large quasi-DC component. PPG signal consists of a periodic respiratory signal due to respiratory induced intensity variations, with a frequency range of 0.2 to 0.33 Hz [13]. The photodetector output of the PPG sensor is proportional to the intensity of the light transmitted through the finger and it keeps oscillating in response to the cardiac cycle due to cardiac induced increase in the blood volume during every systole. The respiratory fluctuations are shown in the PPG baseline and is inversely proportional to the changes in the blood volume. The main objective of this work is to find the Principal Components of the PPG signal using Singular Value Decomposition and Covariance.

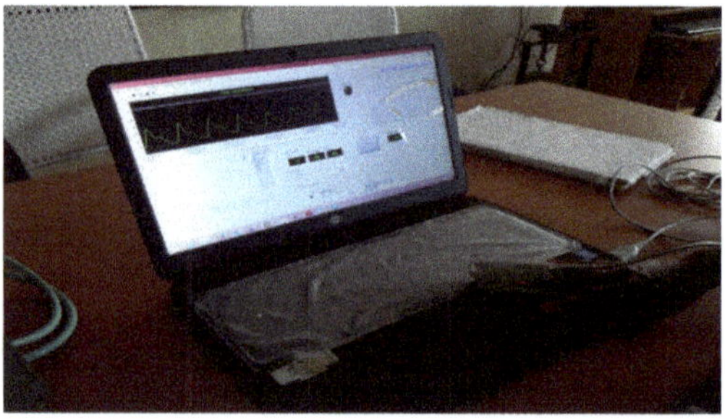

Fig. 1. Experimental Setup for heart rate and respiration rate measurement using Finger-tip pulse oximeter

2 Methodology

2.1 Database

The PPG signals used for the analysis purpose were collected in two forms. Real time PPG signals were acquired from the PPG sensor and the signals of Sleep Apnea patients were collected from the Physionet database. The MIT-BIH polysomnography database consists of 16 recordings monitored continuously from males who suffer from sleep apnea. The recordings are of 2 to 7 h duration and sampled at 250 Hz.

2.2 Signal Preprocessing

The PPG signal recorded using the finger-tip pulse oximeter sensor is prone to artifacts due to movement of limbs and cannot be processed without preprocessing the signal. Hence, the signal is applied with a smoothening filter to remove the low frequency trend. The filter employed is a moving average filter with a smoothening factor of 4 and is usually employed in one-dimensional signal processing [17]. The running average slides over the entire data set and takes the average of specified number of datapoints prescribed by the smoothing factor. The averaged signal is reconstructed without any loss of information (Fig. 2).

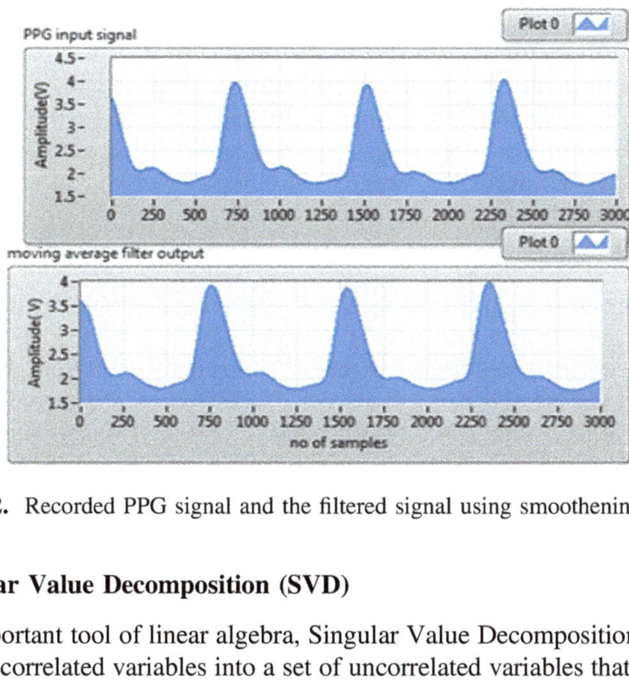

Fig. 2. Recorded PPG signal and the filtered signal using smoothening filter

2.3 Singular Value Decomposition (SVD)

Being an important tool of linear algebra, Singular Value Decomposition is capable of transforming correlated variables into a set of uncorrelated variables that would in turn expose the various relations among the original data set. SVD is also a method of identifying the dimensions and ordering the dimensions along which the data points show much variations. Once the variations are identified and their dimensions are known, SVD helps in finding the best approximation of the original data points using fewer dimensions. Thus, SVD is a better method which can be utilized for data reduction.

Singular Value decomposition in linear algebra states that a rectangular matrix say M, can be split into the product of three matrices: an orthogonal matrix N, a diagonal matrix D and the transpose of the orthonormal matrix P.

$Mmn = Nmn \ \ Dmn \ \ Pmn^T$ Where $N^TN = I$, $P^TP = I$, the columns of N are orthonormal eigen vectors of MM^T, the columns of P are orthonormal eigen vectors of M^TM, and D is a diagonal matrix containing the square roots of eigen values from N or P in descending order. Understanding that the singular values of a given data matrix carries information regarding the noise level, energy and rank of the matrix, it is being used in signal processing to perform data compression, noise removal and pattern extraction. Hence in this hypothesis, SVD is used for removing motion artifacts from corrupted PPG signals. PPG signals are in synchronous with the electrocardiogram signal (ECG) and it can provide the heart rate as it is a periodic signal which changes continuously with changes in heart rate.

2.4 Singular Value Ratio (SVR)

The matrix formation is done with reference to the frequency of the PPG signal and the expected range of heart rate. Singular value decomposition is applied to the data matrix thus formed. The ratio of first two singular values also called as the Sigular Value Ratio (SVR) is computed for the entire length of the PPG signal. It is then plotted against the time period to obtain the SVR spectrum of the PPG signal. The time period for which the value of SVR spectrum is maximum is considered to be the heart rate and the heart rate per second denotes the periodicity. Thus heart rate is extracted from the SVR trend of PPG signal.

2.5 Principal Component Analysis (PCA)

One of the standard technique used in signal processing for feature extraction and for data reduction is Principal Component Analysis (PCA) [14, 15]. It performs data reduction by extracting the desired number of principal components from the data. PCA aims at reducing the large dimensional data space (observed variables) into smaller dimensional feature space (independent variables). It intensionally permits to reduce a set of observed variables into a smaller set of independent variables called principal components which are merely linear combinations of optimally-weighted observed variables and are used in the subsequent analysis.

The principal component that is extracted first elucidates a maximal amount of total variance in the observed variables. The component that is extracted second, elucidates a maximal amount of variance in the data set that was not accounted by the first component. Thus the first component and the second component remains uncorrelated with each other. Principal component analysis can either be performend by covariance matrix or by SVD. This work involves PCA using covariance matrix. The ratio of the first two singular values also called as singular value ratio accounts for periodicity which is denoted as n. The data matrix X is of size m x n, where n is the periodicity and m denotes the number of periods. Hereby, using the periodicity of the PPG signal a data set is formed as given.

$$X(t) = [x_1(t),\ x_2(t),\ x_3(t),\ x_4(t),\ldots\ldots x_m(t)] \tag{1}$$

By removing the mean trend from each of the xi, the covariance matrix can be computed. The covariance matrix is a square matrix of size m x m, for which the eigen values (aj) and the eigen vectors (λj) are calculated. In order to set the order of significance of the components, the eigen vectors are arranged corresponding to the eigen values in descending order. This arrangement signifies in signal compression by ignoring the components with less eigen values. Now, depending on the eigen vectors of the covariance matrix, the principal components are ordered as $Z_j = a_j x$, where j = 1, 2, 3, 4…n. The eigen vectors account for the principal component analysis in extraction of respiratory signal from the PPG signal. The correlation with the respiratory component is exhibited only by the first few principal components in every cycle.

3 Experimental Results

The implementation of the hypothesis has been carried out using LabVIEW platform which is the most significant software in providing a graphical representation of the signal, thereby enabling the visualization of the signal being processed. Two different database has been used in this study. The first one is the apnea database available in Physionet, which includes the PPG data of 16 apnea patients. The second database has been created in real time by recording the PPG signal of 15 normal individuals using the finger tip pulse oximeter as shown in Fig. 1.

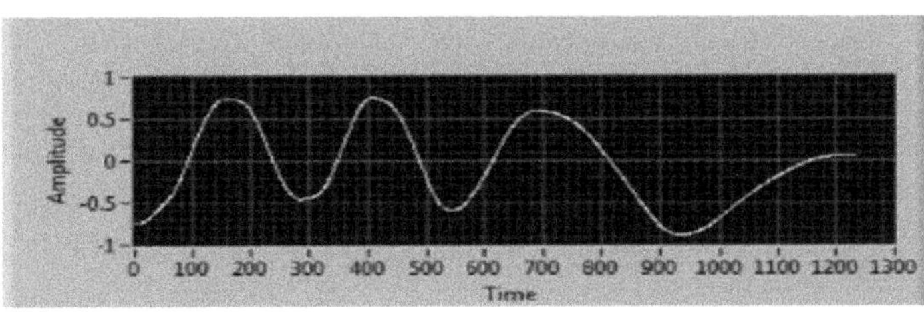

Fig. 3. Respiratory rate extracted for 10 s

The time period for which the value of SVR spectrum is maximum is considered to be the heart rate. The heart rate thus extracted from the SVR trend of the PPG signal is compared with the number of PPG peaks/min. The LabVIEW results for the respiratory rate extracted using PCA is shown in Fig. 3. It acquires 1000 samples per 10 s with a sampling frequency of 100 Hz. The number of peaks shown per 1000 samples will determine the respiratory rate per 10 s.At the end of each minute, the total number of respiration counts are summed up and displayed as respiration rate per minute. Table 1 depicts the comparison of the respiration rate obtained using PCA and the reference respiration rate measured using the standard thermistor based respiration sensor.

Table 1. Comparison of respiratory rate obtained from the principal component analysis of normal PPG signals and the reference standard.

Data	PPG derived Resp. Rate/min	Reference Resp. Rate/min	% Error
Normal	14	16	12.5
Normal	17	18	5.5
Normal	19	18	5.5
Apnea	9.5	11	13.63
Apnea	9	10	5.5

The highest and lowest error percentage were 13.63% and 5.5% respectively. The visual inspection of the respiratory rate derived from the PPG signal shows a clear match with that of the reference respiratory rate. Yet, quantification of the degree of similarity in time domain and frequency domain is determined in terms of relative correlation coefficient and magnitude squared coherence respectively. Relative correlation coefficient is measured as the "maximum value of cross correlation obtained between derived and original signal", upon the "maximum value of auto correlation of original respiratory signal" (Table 2).

Table 2. PPG derived respiratory rate from the first principal components

Data	Relative correlation coeff.	Magnitude squared coeff.
Normal	0.69	0.99
Normal	0.67	0.95
Normal	0.62	0.93
Apnea	0.70	0.90
Apnea	0.66	0.96

4 Sleep Apnea Detection Using PDR

Test results illustrates the efficiency of the principal component analysis algorithm in estimating the respiratory rate from the PPG signal. Monitoring of heart rate and respiratory rate during sleep manifests the degree of severity of sleep apnea, as it is a sleep related breathing disorder. Generally, respiratory rate is being measured either by the abdominal respiratory movement sensor or a thermistor based sensor fixed near the nostrils of the apnea patients. It causes a discomfort to the patients and this results in repeated visits to the hospitals so as to determine the presence or absence of apnea. The proposed methodology provides a solution to the above problem of discomfort and repeated visits, as it involves only a single sensor fitted to the finger from which multiple nocturnal parameters viz., respiration rate, heart rate, blood pressure, blood glucose, blood hemoglobin, temperature, heart rate variability etc., can be measured even when the patient is asleep.

5 Conclusions

This work sought to determine the effectiveness of PDR based on principal component analysis in distinguishing the apnea in breathing signals.In our previous published research, we used the empirical mode of decomposing the signals into 4 intrinsic mode functions to derive the respiratory rate from the PPG signal. Empirical mode decomposition is a non-linear and an adaptive technique that empirically extracts the oscillatory tones embedded in a signal, but when applied to cardiovascular signals, the intrinsic modes expressing cardiac related informations widely vary with the signal, the measurement methods and the patient [8, 9, 16]. Therefore reconstruction of a noiseless signal from the intrinsic modes becomes a tedious process. Principal component analysis is a signal processing technique applied for data reduction, feature extraction and data compression. It helps in deriving a small number of uncorrelated principal components thereby retaining the maximum amount of information of the original signal.

In order to improve the efficiency of sleep apnea detection in real time, the proposed algorithm can be applied to the cardiovascular signals measured from the patients using a single lead ECG sensor. Since the cardiovascular signals exhibit non-linear and non-stationary behaviour, the previously discussed empirical mode decomposition technique can be applied to extract the intrinsic mode functions. To recompose the results of decomposition, the proposed PCA can be used on the intrinsic modes by removing the correlations between them and thereby generating smaller number of orthogonal components.

References

1. Lazaro, J., Gil, E., Bailon, R., Laguna, P.: Deriving respiration from the pulse photoplethysmographic signal. In: Computing in Cardiology, pp. 713–716. IEEE Computer Society Press (2011)
2. Gutta, S., Cheng, Q., Nguyen, H., Benjamin, B.: Cardiorespiratory model-based data-driven approach for sleep apnea detection. IEEE J. Biomed. Health Inform. **99**, 1–5 (2018)
3. Cheng, M., Sori, W.J., Jiang, F., Khan, A., Liu, S.: Recurrent neural network based classification of ECG signal features for obstruction of sleep apnea detection. IEEE Trans. Biomed. Eng. **2**, 199–202 (2017)
4. Chen, L., Zhang, X., Song, C.: An automatic screening approach for obstructive sleep apnea diagnosis based on single-lead electrocardiogram. IEEE Trans. Autom. Sci. Eng. **12**, 106–115 (2015)
5. Varon, C., Caicedo, A., Testelmans, D., Buyse, B., Huffel, S.V.: A novel algorithm for the automatic detection of sleep apnea from single lead ECG. IEEE Trans. Biomed. Eng. **9**(62), 2269–2278 (2015)
6. Bsoul, M., Minn, M., Tamil, L.: Apnea MedAssist: real-time sleep apnea monitor using single-lead ECG. IEEE Trans. Inform. Technol. Biomed. **3**(15), 416–427 (2011)
7. Yilmaz, B., Asyali, M., Arikan, E., Yektin, S., Ozgen, F.: Sleep stage and obstructive apneaic epoch classification using single-lead ECG. Biomed. Eng. Online **9**, 39 (2010)
8. Faust, O., Bairy, M.G.: Nonlinear analysis of physiological signals: a review. IEEE J. Mech. Med. Biol. **12**(04), 1240015 (2012)

9. Koley, B.L., Dey, D.: Automatic detection of sleep apnea and hypopnea events from single channel measurement of respiration signal employing ensemble binary svm classifiers. Measurement **46**, 2082–2092 (2013)
10. Jafari, A.: Sleep apnoea detection from ECG using features extracted from reconstructed phase space and frequency domain. Biomed. Signal Process. **8**, 551–558 (2013)
11. Al-Angari, H.M., Sahakian, A.V.: Automated recognition of obstructive sleep apnea syndrome using support vector machine classifier. IEEE Trans. Inf Technol. Biomed. **3**(16), 463–468 (2012)
12. Almazaydeh, L., Elleithy, K., Faezipour, M.: Obstructive sleep apnea detection using SVM-based classification of ECG signal features. In: Proceedings of 2012 Annual International Conference on IEEE EMBC, pp. 4938–4941 (2012)
13. Eiseman, N.A., Westover, M.B., Mietus, J.E., Thomas, R.J., Bianchi, M.T.: Classification algorithms for predicting sleepiness and sleep apnea severity. J. Sleep Res. **1**(21), 101–112 (2012)
14. Alickovic, E., Kevric, J., Subasi, A.: Performance evaluation of empirical mode decomposition, discrete wavelet transform, and wavelet packed decomposition for automated epileptic seizure detection and prediction. Biomed. Signal Process. Control **39**, 94–102 (2018)
15. Araslanova, R., Paradis, J., Rotenberg, B.W.: Publication trends in obstructive sleep apnea: evidence of need for more evidence. World J. Otorhinolaryngol.-Head Neck Surg. **3**, 72–78 (2017)
16. Pinheiro, E., Postolache, O., Girao, P.: Empirical mode decomposition and principal component analysis implementation in processing non-invasive cardiovascular signals. Measurement **45**, 175–181 (2012)
17. Jothi, E.S.J., Anitha, J.: Screening of sleep apnea syndrome with decomposition of photoplethysmogram to extract respiratory rate. In: Proceedings of IEEE International Conference on Signal Processing and Communication, pp. 328–331 (2017)

Improved Social Spider Algorithm via Differential Evolution

Fatih Ahmet Şenel(✉), Fatih Gökçe, and Tuncay Yiğit

Department of Computer Engineering, Süleyman Demirel University, 32260 Isparta, Turkey
{fatihsenel,fatihgokce,tuncayyigit}@sdu.edu.tr

Abstract. In this study, Social Spider Algorithm (SSA) is improved by combining it with the mutation strategies of Differential Evolution (DE) algorithm. Without altering the main operation of the SSA, we utilized the DE mutation strategies to diversify the individuals in the population. We selected two different mutation strategies to obtain two different hybrids and evaluated them. These mutation strategies are the most basic strategy and currently the best strategy of the DE algorithm proposed in the literature. Two different hybrids are evaluated via several tests performed on five different real-world problems. According to our results, the SSA is improved successfully. The hybrid utilizing the most basic strategy of the DE algorithm has given better results than the currently best strategy with an improvement 15.56% on average in all tests. It has been found that the currently best strategy resulted in worse performance probably since it provided more diversity on the population.

Keywords: Improvement · Differential Evolution · Metaheuristics · Mutation · Optimization · Social Spider Algorithm

1 Introduction

Today, swarm based metaheuristic optimization algorithms which take place among the artificial intelligence techniques are utilized to solve many real–world optimization problems. Generally, metaheuristic optimization algorithms are developed by inspiring from the living beings in nature. When the literature is examined, algorithms such as Artificial Bee Colony [11], Particle Swarm Optimization [12], Genetic Algorithm [4], Social Spider Algorithm (SSA) [21,22], Social Spider Optimization (SSO) [3], Differential Evolution (DE) algorithm [13,16,17] fit into this class of metaheuristic algorithms. Despite the fact that new optimization algorithms are continuing to be added to the literature, there are also studies to improve the already proposed optimization algorithms. Researchers have been investigating the existing optimization algorithms and studying to improve them sometimes by modifying them standalone and sometimes by hybridizing them with other algorithms.

The SSO and SSA algorithms are both inspired from the spiders in nature. However, they are developed on completely different phenomenons. While the

D. J. Hemanth and U. Kose (Eds.): ICAIAME 2019, LNDECT 43, pp. 437–445, 2020.
https://doi.org/10.1007/978-3-030-36178-5_33

SSO is motivated from the mating strategies of the spiders, the SSA is adapted from the hunting strategies of them. When we consider the literature, there are many studies related to improving the SSO algorithm and its hybridization with other algorithms [2,15,18,20]. Although there are many studies about using the SSA to solve optimization problems [6,19,24], there are few studies about improving the SSA algorithm or its hybridization with other algorithms [8,10,23].

In this study, our aim is to improve the SSA algorithm by utilizing the mutation strategies of the DE algorithm. Similar to our approach, Elsayed et al. [8] aimed to improve the SSA algorithm by combining it with a strategy similar to the mutation strategy of the DE algorithm. However, they modified the general operation structure of the SSA and deactivated its random walking characteristic. In our study, we make use of the mutation strategies of the DE algorithm without altering the main operation of the SSA. Our approach depends on mutating the individuals in the SSA population via the mutation strategies of the DE algorithm so that the diversity of the individuals are increased providing a better dispersion over the search space.

The DE algorithm, which fits into the field of Evolutionary Computation, is an optimization algorithm improving a candidate solution by iteratively applying mutation, recombination or crossover, and selection. The DE algorithm continues to be improved with the proposals of different mutation strategies mainly through the competitions organized in IEEE Congress on Evolutionary Computation [1]. In our study, we employ two different strategies of the DE algorithm which are proposed in [9,17] and [13], respectively. The first strategy [9,17] is known as the most basic strategy in the literature. The second one is a recently proposed and claimed to be the currently best strategy for the DE algorithm. In the rest of this paper, we will call these two strategies as SSABSC and SSABST, respectively, for the most basic and currently best strategies.

The rest of the paper is organized as follows. In the next section, the SSA and the mutation strategies of the DE algorithm are explained in detail. Section 3 presents our experimental results. We conclude our paper in Sect. 4.

2 Optimization Methods

2.1 Social Spider Algorithm

The SSA is metaheuristic optimization algorithm developed by inspiring from hunting strategies of the spiders on a virtual web [21,22]. The spiders which are also called as agents are assumed to be on a common web which acts as a common communication medium. The spiders keep their locations and fitness values in their memories and share these information with other spiders on the web via vibrations that they generated. These vibrations provide a social communication system among all the spiders on the web. The spiders are assumed to identify directions and intensity of the vibrations on the web well. Whenever the spiders change their locations, they share their new locations and fitness values.

The most important process of the SSA is the generation of the vibrations and the reading of them. The vibrations are calculated with Eq. 1, and get values in a range of $[0, +\infty)$. The vibration, while they are going far from the source, get the values in the range of decreasing value $[0, +\infty)$. In Eq. 1, the calculation of the intensity of vibrations is shown.

$$I = \log\left(\frac{1}{f(s) - C} + 1\right) \tag{1}$$

Here I denotes the intensity of vibrations, f(s) represents fitness value, C is a confidently small constant such that all possible fitness values are larger than C [22]. Intensity of vibrations in the SSA can be decreased by using two different methods. These are increasing the number of iteration and lengthening the distance. Equations 2 and 3 show the reduction of vibration intensity with the increase of iteration and distance, respectively.

$$I(t+1) = I(t) \times r_a \tag{2}$$

$$I(p) = I(s) \times \exp\left(-\frac{D(s,p)}{D_{\max} \times r_a}\right) \tag{3}$$

Here t represents the iteration step, r_a stands for user defined vibrations decreasing coefficient, s expresses the position of vibrations' source and p expresses the position of the point that the vibrations are got. D(s,p), represents the distance between two points, I(s) expresses the intensity in the source of vibrations and D_{max} refers to the farthest distance between two points from each other on web (searching space) [21].

The SSA consists of three phases. These are initialization, iteration and termination phases. In initialization phase, defining cost functions and algorithm parameters are completed. Initially placed at random positions, spiders' vibrations are set to zero. In iteration phase, all of the spiders share their new locations' fitness values on web whenever they change their positions. Each spider evaluates all the vibrations that they get and they pay attention to the vibration that has the best value. As comparing the previous best intensity value (V_{prev}) in its memory and the existing best intensity value (V_{best}), each spider updates their positions. In Eq. 2, process of updating positions is shown.

$$P_s(t+1) = P_s(t) + (P_{best}(t) - P_s(t)) \odot (1 - R \odot R) \tag{4}$$

Here; P_s(t), denotes the position of s spider in t iteration step, $\odot$ process represents element-wise multiplication of matrix and vectors, P_{best} represents the position of the best vibration source and R is a vector of random float-point numbers generated from range [0–1] uniformly. When the number of user-defined iterations is reached or a user-defined stop criterion is provided, the iteration phase is completed and the results obtained by passing to the termination phase are given to the user as the output.

2.2 Mutation Strategies of the DE Algorithm

In this study, one of the most basic mutation strategy of the DE algorithm (DE/rand/1/bin) and the newest version of the mutation operator developed by Mohamed and Suganthan [13] has been used to improve the SSA. These are named respectively as SSA1 and SSA2. DE/rand/1/bin mutation operator is shown in Eq. 5 and the mutation operator which is developed by Mohamed and Suganthan is shown in Eqs. 6, 7 and 8.

$$v_i^{G+1} = x_{r_1}^G + F \times \left(x_{r_2}^G - x_{r_3}^G\right) \tag{5}$$

Here G denotes the iteration step and F expresses a coefficient which is in the range of [0–2].

$$v_i^{G+1} = x_c^{-G} + F_1 \times \left(x_{best}^G - x_{better}^G\right) + F_2 \times \left(x_{best}^G - x_{worst}^G\right) + F_3 \times \left(x_{better}^G - x_{worst}^G\right) \tag{6}$$

Here x_c^{-6} represents the combination of three individuals that are chosen and in Eq. 7, this combination process is shown.

$$x_c^{-G} = w_1^* \times x_{best} + w_2^* \times x_{better} + w_3^* \times x_{worst} \tag{7}$$

The W_i^* values provide the $w_i^* \geq 0$ and $\sum_{i=1}^{3} w_i^* = 1$ conditions and are calculated as shown in Eq. 8.

$$w_i = 1 - \left| \frac{f\left(x_{best}\right) - f\left(x_i\right)}{f\left(x_{best}\right) - f\left(x_{max}\right)} \right|, i = 1, 2, 3 \tag{8}$$

Here; $f\left(x_{best}\right) = f\left(x_{\min}\right)$ value represents $\min\left\{f\left(x_i\right)\right\}$ value and $f\left(x_{\max}\right)$ denotes the maximum value which has been observed in each iterations. F_1, F_2 and F_3 coefficients are calculated as it is shown in Eq. 9.

$$
\begin{gathered}
F_i = rand\left(0, k_i\right), i = 1, 2, 3 \\
k_1 = \begin{cases} \left|\frac{f(x_{better})}{f(x_{best})}\right| + \varepsilon \; if \left|\frac{f(x_{better})}{f(x_{best})}\right| < 1 \\ \left|\frac{f(x_{best})}{f(x_{better})}\right| + \varepsilon \text{ otherwise} \end{cases} \\
k_2 = \begin{cases} \left|\frac{f(x_{worst})}{f(x_{best})}\right| + \varepsilon \; if \left|\frac{f(x_{worst})}{f(x_{best})}\right| < 1 \\ \left|\frac{f(x_{best})}{f(x_{worst})}\right| + \varepsilon \text{ otherwise} \end{cases} \\
k_3 = \begin{cases} \left|\frac{f(x_{worst})}{f(x_{better})}\right| + \varepsilon \; if \left|\frac{f(x_{worst})}{f(x_{better})}\right| < 1 \\ \left|\frac{f(x_{better})}{f(x_{worst})}\right| + \varepsilon \text{ otherwise} \end{cases}
\end{gathered} \tag{9}
$$

Here ε represents a number which is close to zero to protect the coefficients from zero.

In Eq. 10, a formula is applied to increase the likelihood of this mutation strategy as iteration progresses and to apply the mutation strategy of the standard DE algorithm in initial iterations.

$$\begin{aligned}
&\boldsymbol{if}\left(u(0,1) \geq \left(1-\frac{G}{GEN}\right)^2\right) \boldsymbol{then} \\
&v_i^{G+1} = \begin{aligned}[t] &x_c^{-G} + F_1 \times \left(x_{best}^G - x_{better}^G\right) + F_2 \times \left(x_{best}^G - x_{worst}^G\right) \\ &+ F_3 \times \left(x_{better}^G - x_{worst}^G\right)\end{aligned} \\
&\boldsymbol{else} \\
&v_i^{G+1} = x_{r_1}^G + F \times \left(x_{r_2}^G - x_{r_3}^G\right) \\
&\boldsymbol{end}
\end{aligned} \tag{10}$$

GEN is total iteration numbers.

2.3 Proposed Method

In this study, it is aimed to increase the variety in population with SSA1 and SSA2 algorithms in addition to SSA's own standard updates in each iterations. For this process, by using them separately, both one of the most basic mutation strategy of the DE algorithm's (DE/rand/1/bin, In Eq. 5) and one of newest strategy (Eq. 10) effects on results are performed. In the SSA, results have been obtained by applying this mutation operators to all individuals separately at the end of each iterations.

3 Experimental Results

In this study, all the processes are coded with Microsoft Visual Studio 2017 C# programming language. Five different real-world problems are chosen from Literature to conduct test processes. The detailed information about these problems are given in Table 1.

In all experiments the population number of the SSA algorithm is chosen as 100. Three different algorithms as SSA, SSA1 and SSA2 have been operated for each problem. Every single test process is repeated 100 times. In Table 2, there are mean, minimum, maximum and standard deviation results of data which are obtained after 100 times.

As it is seen in Table 2, in all tests SSA falls behind the SSA1 algorithm. It is understood that SSA needs exploration ability with these results. Better results have been obtained by increasing the variety of the population in SSA with the mutation strategy of DE algorithm. When SSA1 and SSA2 algorithms are compared, it can be seen that SSA1 has better results in all problems except the Problem 4. And in Problem 4, they generated very close values to each other. But when all the test are taken into consideration, it is seen that SSA1

Table 1. Real-world problems that have been used.

Problem	Name	Description	Constraints	Bounds
1	Design of Gear Train [7]	**Min** $f(x) = \left(\frac{1}{6.931} - \frac{x_1 x_2}{x_3 x_4}\right)^2$	–	$12 \leq x_i \leq 60$ $i = 1, 2, 3, 4$ and $x_i$should be integer
2	Design of Pressure Vessel [14]	**Min** $f(x) = 0.6224x_1x_3x_4 + 1.7781x_2x_3^2$ $+3.1661x_1^2x_4 + 19.84x_1^2x_3$	$g_1(X) = -x_1 + 0.0193x_3 \leq 0$ $g_2(X) = -x_2 + 0.00954x_3 \leq 0$ $g_3(X) =$ $-\pi x_3^2 x_4 - \frac{4}{3}\pi x_3^3 + 1296000 \leq 0$ $g_4(X) = x_4 - 240 \leq 0$	$0 \leq x_1, x_2 \leq 100$ $10 \leq x_3, x_4 \leq 200$
3	Process Flowsheeting Problem [5]	**Min** $f(x) = -0.7x_3 + 5(x_1 - 0.5)^2 + 0.8$	$g_1(X) = \exp(x_1 - 0.2) - x_2 \leq 0$ $g_2(X) = x_1 - 1.2x_3 - 0.2 \leq 0$ $g_3(X) = x_2 + 1.1x_3 + 1.0 \leq 0$	$0.2 \leq x_1 \leq 1$ $-2.22554 \leq x_2 \leq -1$ $x_3 \in 0, 1$
4	Optimal thermohydraulic performance of an artificially roughened air heater [7]	**Max** $f(x) = 2.51 \ln e^+ + 5.5 - 0.1R_M - G_H$ $where: R_M = 0.95x_2^{0.53}, G_H =$ $4.5\left(e^+\right)^{0.28}(0.7)^{0.57}$ $e^+ = x_1x_3\left(\frac{f^-}{2}\right)^{\frac{1}{2}}, f^- =$ $\frac{f_S + f_r}{2}, f_s = 0.079x_3^{-0.25}$ $f_r =$ $2\left(0.95x_3^{0.53} + 2.5\ln\left(\frac{1}{2x_1}\right)^2 - 3.75\right)^{-2}$	–	$0.02 \leq x_1 \leq 0.8$ $10 \leq x_2 \leq 40$ $3000 \leq x_3 \leq 20000$
5	Tension/compression spring design optimization problem [5]	**Min** $f(x) = (x_3 + 2)x_2x_1^2$	$g_1(X) = 1 - \frac{x_2^3 x_3}{71785x_1^4} \leq 0$ $g_2(X) =$ $\frac{4x_2^2 - x_1x_2}{12566\left(x_2x_1^3 - x_1^4\right)} + \frac{1}{5108x_1^2} - 1 \leq 0$ $g_3(X) = 1 - \frac{140.45x_1}{x_2^2x^3} \leq 0$ $g_4(X) = \frac{x_2 + x_1}{1.5} - 1 \leq 0$	$0.05 \leq x_1 \leq 2$ $0.25 \leq x_2 \leq 1.3$ $2 \leq x_3 \leq 15$

Table 2. Experimental results. As stated in Table 1, problems 1,2,3 and 5 are minimization problems, whereas 4 is a maximization problem. Therefore, higher values are better for problem 4.

Problem	Metrics	SSA	SSA1	SSA2
1	Mean	0,000027	0,000010	0,000049
	Min	0,000002	0,000002	0,000002
	Max	0,000049	0,000049	0,000648
	Std	0,000014	0,000011	0,000084
2	Mean	6395,465	5998,370	13856,818
	Min	5925,272	5894,427	6644,887
	Max	7291,143	6566,748	102575,593
	Std	282,880	126,326	15920,264
3	Mean	1,147688	1,119886	1,177891
	Min	1,077541	1,076546	1,080886
	Max	1,350000	1,316611	1,350000
	Std	0,073284	0,043656	0,076387
4	Mean	4,193155	4,214160	4,214161
	Min	4,091498	4,213262	4,213202
	Max	4,214219	4,214220	4,21422
	Std	0,022700	0,000135	0,000159
5	Mean	0,015003	0,014146	0,022099
	Min	0,012740	0,012697	0,013027
	Max	0,020580	0,021157	0,058158
	Std	0,001699	0,001685	0,007735

algorithm is better than the other two algorithms. If the mutation strategy of SSA2 algorithm is taken into consideration, it can be understood that it has a lot of variety in population. But having much more variety in population effects SSA in a negative direction as it is understood from the results. Therefore, it is understood that exploration ability of SSA needs to be improved but not much. It can be concluded that SSA has an advanced level of exploration ability and it just needs partial improvement.

The best case in which the SSA1 algorithm improves SSA is Problem 1. Improvement ratio is 62,96% in this case. It is clear that there is an improvement with a very high ratio. The least improvement has been obtained in Problem 4 and success ratio is 0,5% for this problem. When all the problems are taken into consideration, the average ratio of improvement is 15,56%.

4 Conclusions

Metaheuristic optimization algorithms are often inspired by nature. However, it is not possible to purpose a new algorithm. Instead of this, researches are conducted to improve the algorithms which have already been available. Generally, hybrid approaches are aimed with combining the good sides of more than one algorithms.

In this study, an improved SSA has been aimed by adding the mutation strategy of the DE algorithm to the SSA. The originality of the SSA has been preserved during this process.

Although SSA is an algorithm which generates successful results, with this study it is revealed that its exploration ability is not enough. Mutation strategy is a useful step which has a process to increase variety in populations. By adding mutation strategy to the SSA, the variety in population has been increased and the results have been observed. Mutation strategy improved the results averagely 15,56%. In future studies, it is aimed to test the improved SSA1 algorithm on many more problems. It is also among the targets to identify the strategy with the most improvement by integrating many existing mutation strategies in the DE algorithm into the SSA.

References

1. IEEE Congress on Evolutionary Computation (2018). https://ewh.ieee.org/conf/cec/. Accessed 21 Dec 2018
2. Abd El Aziz, M., Hassanien, A.E.: An improved social spider optimization algorithm based on rough sets for solving minimum number attribute reduction problem. Neural Comput. Appl. **30**(8), 2441–2452 (2018). https://doi.org/10.1007/s00521-016-2804-8
3. Cuevas, E., Cienfuegos, M., Zaldívar, D., Pérez-Cisneros, M.: A swarm optimization algorithm inspired in the behavior of the social-spider. Expert Syst. Appl. **40**(16), 6374–6384 (2013). https://www.sciencedirect.com/science/article/pii/S0957417413003394
4. Davis, L.: Handbook of Genetic Algorithms, vol. 115 (1991)
5. Dong, M., Wang, N., Cheng, X., Jiang, C.: Composite differential evolution with modified oracle penalty method for constrained optimization problems. Math. Probl. Eng. **2014**, 1–15 (2014). http://www.hindawi.com/journals/mpe/2014/617905/
6. El-bages, M., Elsayed, W.: Social spider algorithm for solving the transmission expansion planning problem. Electr. Power Syst. Res. **143**, 235–243 (2017). https://www.sciencedirect.com/science/article/pii/S0378779616303510
7. El Dor, A., Clerc, M., Siarry, P.: Hybridization of differential evolution and particle swarm optimization in a new algorithm: DEPSO-2S, pp. 57–65. Springer, Heidelberg (2012). https://doi.org/10.1007/978-3-642-29353-5_7
8. Elsayed, W., Hegazy, Y., Bendary, F., El-bages, M.: Modified social spider algorithm for solving the economic dispatch problem. Eng. Sci. Technol. Int. J. **19**(4), 1672–1681 (2016). https://www.sciencedirect.com/science/article/pii/S2215098616305006#b0180

9. Fan, H.Y., Lampinen, J.: A trigonometric mutation operation to differential evolution. J. Glob. Optim. **27**(1), 105–129 (2003). https://doi.org/10.1023/A:1024653025686
10. Gupta, S., Arora, S.: A hybrid firefly algorithm and social spider algorithm for multimodal function, pp. 17–30. Springer, Cham (2016). https://doi.org/10.1007/978-3-319-23036-8_2
11. Karaboga, D., Basturk, B.: A powerful and efficient algorithm for numerical function optimization: artificial bee colony (ABC) algorithm. J. Glob. Optim. **39**(3), 459–471 (2007). https://doi.org/10.1007/s10898-007-9149-x
12. Kennedy, J., Eberhart, R.: Particle swarm optimization. In: Proceedings of ICNN 1995 - International Conference on Neural Networks, vol. 4, pp. 1942–1948. IEEE (1995). http://ieeexplore.ieee.org/document/488968/
13. Mohamed, A.W., Suganthan, P.N.: Real-parameter unconstrained optimization based on enhanced fitness-adaptive differential evolution algorithm with novel mutation. Soft Comput. **22**(10), 3215–3235 (2018). https://doi.org/10.1007/s00500-017-2777-2
14. Mortazavi, A., Toğan, V., Nuhoğlu, A.: Interactive search algorithm: a new hybrid metaheuristic optimization algorithm. Eng. Appl. Artif. Intell. **71**, 275–292 (2018). https://www.sciencedirect.com/science/article/pii/S0952197618300514
15. Ouadfel, S., Taleb-Ahmed, A.: Social spiders optimization and flower pollination algorithm for multilevel image thresholding: a performance study. Expert Syst. Appl. **55**, 566–584 (2016). https://www.sciencedirect.com/science/article/pii/S0957417416300550
16. Storn, R.: On the usage of differential evolution for function optimization. In: Proceedings of North American Fuzzy Information Processing, pp. 519–523, June 1996. https://doi.org/10.1109/NAFIPS.1996.534789
17. Storn, R., Price, K.: Differential evolution - a simple and efficient heuristic for global optimization over continuous spaces. J. Glob. Optim. **11**(4), 341–359 (1997). https://doi.org/10.1023/A:1008202821328
18. Sun, S.C., Qi, H., Ren, Y.T., Yu, X.Y., Ruan, L.M.: Improved social spider optimization algorithms for solving inverse radiation and coupled radiation-conduction heat transfer problems. Int. Commun. Heat Mass Transf. **87**, 132–146 (2017). https://www.sciencedirect.com/science/article/pii/S0735193317301793
19. Tawhid, M.A., Ali, A.F.: A simplex social spider algorithm for solving integer programming and minimax problems. Memet. Comput. **8**(3), 169–188 (2016). https://doi.org/10.1007/s12293-016-0180-7
20. Tawhid, M.A., Ali, A.F.: A hybrid social spider optimization and genetic algorithm for minimizing molecular potential energy function. Soft Comput. **21**(21), 6499–6514 (2017). https://doi.org/10.1007/s00500-016-2208-9
21. Yu, J.J.Q., Li, V.O.K.: Base station switching problem for green cellular networks with Social Spider Algorithm. In: 2014 IEEE Congress on Evolutionary Computation (CEC), pp. 2338–2344. IEEE, July 2014. http://ieeexplore.ieee.org/document/6900235/
22. Yu, J.J.Q., Li, V.O.K.: A social spider algorithm for global optimization. Appl. Soft Comput. **30**, 614–627 (2015). http://arxiv.org/abs/1502.02407
23. Yu, J.J.Q., Li, V.O.K.: Parameter sensitivity analysis of social spider algorithm. In: 2015 IEEE Congress on Evolutionary Computation (CEC), pp. 3200–3205. IEEE, May 2015. http://ieeexplore.ieee.org/document/7257289/
24. Yu, J.J., Li, V.O.: A social spider algorithm for solving the non-convex economic load dispatch problem. Neurocomputing **171**, 955–965 (2016). https://www.sciencedirect.com/science/article/pii/S0925231215010188

Gender Determination from Teeth Images via Hybrid Feature Extraction Method

Betül Uzbaş[1(✉)], Ahmet Arslan[2], Hatice Kök[3], and Ayşe Merve Acılar[4]

[1] Department of Computer Engineering, Faculty of Engineering and Natural Sciences, Konya Technical University, Konya, Turkey
buzbas@ktun.edu.tr

[2] Enelsis Industrial Electronic Systems Research and Development Co. Ltd., Konya, Turkey
ahmetarslan@enelsis.com

[3] Faculty of Dentistry, Department of Orthodontics, Selcuk University [SÜ], Konya, Türkiye
hatice.kok@selcuk.edu.tr

[4] Department of Computer Engineering, Faculty of Engineering and Architecture, Necmettin Erbakan University, Konya, Turkey
m.acilar@erbakan.edu.tr

Abstract. Teeth are a significant resource for determining the features of an unknown person, and gender is one of the important pieces of demographic information. For this reason, gender analysis from teeth is a current topic of research. Previous literature on gender determination have generally used values obtained through manual measurements of the teeth, gingiva, and lip area. However, such methods require extra effort and time. Furthermore, since sexual dimorphism varies among populations, it is necessary to know the optimum values for each population. This study uses a hybrid feature extraction method and a Support Vector Machine (SVM) for gender determination from teeth images. The study group was composed of 60 Turkish individuals (30 female, 30 male) between the ages of 19 and 27. Features were automatically extracted from the intraoral images through a hybrid method that combines two-dimensional Discrete Wavelet Transformation (DWT) and Principle Component Analysis (PCA). Classification was performed from these features through SVM. The system can be easily used on any population and can perform fast and low-cost gender determination without requiring any extra effort.

1 Introduction

Determining the features of an unidentified person is highly important in archeology and forensic science [1,2]. In archeology, it is crucial to determine demographic information when examining findings obtained from historical settlements. In the case of mass graves, if bones are deformed or pieces of the bones cannot be separated from one another, information regarding age and gender

D. J. Hemanth and U. Kose (Eds.): ICAIAME 2019, LNDECT 43, pp. 446–456, 2020.
https://doi.org/10.1007/978-3-030-36178-5_34

distribution can be obtained from the teeth. In forensic science, it is important to determine the identities of people, especially those who have died in accidents, have been murdered, and in the event of mass disasters, among others. In such cases, teeth are generally used for identification because they are more durable than other tissues. Teeth are an important source for identification in forensic cases because they can remain intact even after the most serious fires other than cremation. Gender determination is an important step in identification. For this reason, the present study suggests a new intelligent system for gender determination from teeth images.

Several studies have been conducted to determine the gender of an individual using information from the teeth. The techniques include visual/clinical methods, microscopic methods, and advanced methods. The features used for gender determination through visual/clinical methods are tooth size, root length, crown diameter, canine dimorphism, tooth morphology, dental index, and odontometric differences [1]. Kolte et al. [3] showed that gingival thickness and width vary with gender. Anand et al. [4] studied a sample of 50 periodontally healthy volunteers and found that there was a significant difference between females and males in terms of gingival thickness. Suazo et al. [5] found that differences in buccolingual diameters were statistically significant between sexes.

Hasanreisoglu et al. [6] observed that the dimensions of the maxillary central incisor and canine are greater in men than in women in a Turkish population. Al-Sehaibany [7] found a statistically significant difference in the width-to-height ratios of the lateral incisors between men and women. Parekh et al. [8] studied 368 students (216 males, 152 females) aged 18–24 years old and found sexual dimorphism in the mesiodistal width of the maxillary canine teeth. Lakhanpal et al. [9] determined that mesiodistal measurements are more suitable than buccolingual dimensions for gender discrimination when used independently. Kaushal et al. [10] found that the mandibular canines measured directly or from casts showed a statistically significant sexual dimorphism in a North Indian population between the ages of 17 and 21. In another study on a North Indian population, Rai and Anand [11] discovered that the mandibular canine had significant mean differences in all measurements. Nagesh et al. [12] showed that there was a statistically significant sexual dimorphism in mandibular canines. In a study on a population of 60 individuals (30 males and 30 females) between the ages of 15 and 34, Sai Kiran et al. [13] found that mandibular canines were a valuable source for gender determination.

Using the right mandibular canine index, Bakkannavar et al. [14] achieved 73.2% accuracy in gender prediction for males and 75.6% for females with an overall accuracy of 74.2%. Ahuja and Manchanda [15] showed that the use of mandibular canine index and upper lip length for gender determination provided statistically significant results. Anuthama et al. [16] computed a new formula to differentiate male and female teeth using discriminant function analysis for a South Indian population. Shin [17] extracted features from model images of plaster figures of teeth using PCA and performed gender determination using the k nearest neighbor algorithm. Akkoc et al. [18] performed Gray Level

Co-occurrence Matrix (GLCM) and Random Forest (RF) based gender determination from the maxillary tooth plaster model images. In our another work [19], we extracted features from the 3D maxillary tooth plaster model images using Discrete Cosine Transform (DCT) and classified by RF.

Gender determination using face images, gait, and body features is a popular research area in computer science. Manually measured values and statistical approaches are generally used to determine gender using information from the teeth. Relevant studies have been performed in dentistry, anatomy and forensic science research. The state of the art and recent studies are given in Table 1. There are inadequate studies on gender determination from teeth images using automatic feature extraction in the computer science literature.

Table 1. State of the art and recent studies

Authors	Ref	Year	Research area of study	Feature extraction process	Evaluation method
Hasanreisoğlu et al.	[6]	2005	Dentistry	Digital caliper Adobe PhotoShop	Statical Tests
Garn et al.	[20]	1977	Dentistry	Optical digitizing device (OPTOCOM)	Statical Tests
Kaushal et al.	[10]	2003	Anatomy	Vernier caliper, Divider	Statical Tests
Sai kiran et al.	[13]	2014	Forensic Science	Digital vernier caliper, Divider	Statical Tests
Bakkannavar et al.	[14]	2015	Forensic Science	Digital vernier caliper, Divider	Statical Tests
Anuthama et al.	[16]	2011	Forensic Science	Digital vernier caliper	Statical Tests
Rao et al.	[21]	1989	Forensic Science	Vernier caliper, Divider	Statical Tests
Muller et al.	[22]	2001	Forensic Science	Vernier caliper	Statical Tests
Iscan and Kedici	[23]	2003	Forensic Science	Vernier caliper	Statical Tests
Karaman	[24]	2006	Forensic Science	Digital caliper	Statical Tests
Shin	[17]	2006	Computer Science	PCA	KNN
Akkoç et al.	[18]	2016	Computer Science	GLCM	RF
Akkoç et al.	[19]	2017	Computer Science	DCT	RF

Manually measured values and statistical approaches are generally used for gender determination from teeth. These values are measured intraorally or from dental casts using digital vernier calipers or from radiographic images using drawing software like AutoCad. Examples from of current measuring techniques are given in Fig. 1.

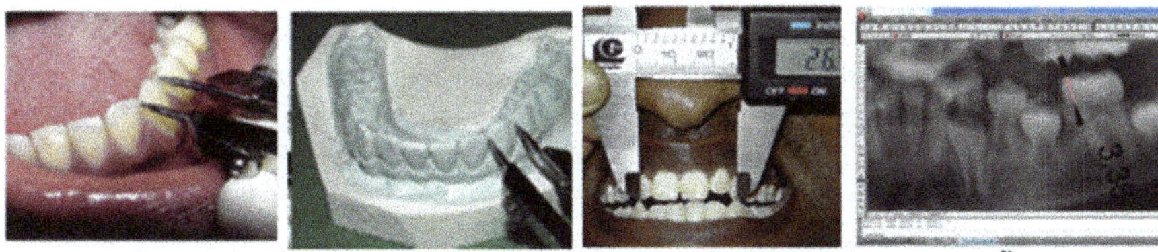

Fig. 1. Current measurement techniques: (a) intraoral measurement [13], (b) measurement on dental casts [13], (c) intraoral intercanine width measurement [14], (d) measurement on radiographies through drawing software programs [25]

This study uses a hybrid feature extraction method and a SVM for gender determination from intraoral images. This paper is organized as follows. Section 2 provides background information about feature extraction and classification methods used in the study. Section 3 describes the proposed intelligent system. Section 4 gives the experimental framework and results. Finally, we present our conclusions and perspectives in Sect. 5.

2 Methods

2.1 Principle Component Analysis

PCA is a popular method that is used for reducing the size of data. It is an eigenvector-based multivariate analysis technique that generally selects the best variance in the data [26]. The pixels in an image have a large degree of correlation include unnecessary information. The basic components of the image give uncorrelated coefficients. A reasonable solution is to use the basic components as features through PCA [27]. PCA also has low computation time and is an effective method for extracting feature sets by creating a feature space [28].

Feature extraction from teeth images was performed using a hybrid technique that involves PCA. Before applying PCA, the rows of a two-dimensional image were combined to create a vector. A $1 \times P$ vector is obtained from each image matrix with dimensions of $M \times N$. The P value is the product of M and N. This procedure is repeated for each image in the dataset and a $K \times P$ matrix is created for K number of images. In PCA, a covariance matrix is first created from the dataset, and its eigenvectors and eigenvalues are determined. The eigenvectors are sorted based on the size of their eigenvalues. This procedure also gives the order of importance of the obtained components. The most meaningful first R components are selected for the size reduction process ($R < P$).

2.2 Discrete Wavelet Transform

Wavelets are mathematical functions that divide data into different frequency components. Each component is then studied with a resolution that matches its scale [29]. Wavelet transformations are classified into continuous and DWT. DWT is a popular and very useful feature extraction method for biomedical signals [30]. Sub-band coding involves various steps of filtering and sampling. In the first stage, the input stream is filtered by a low pass filter and a high pass filter. The approximation coefficients are output by the low pass filter, and the detail coefficients are output by the high pass filter. For a stream of length N, at the end of stage one, there are two output streams corresponding to low and high frequency with a length of N/2. At the second stage, the output of the low pass filter in the first stage is passed through the same low pass and high pass filters, and two outputs with lengths of N/4 are obtained. This process is repeated M times [31].

In a one-dimensional DWT, starting from signal s with length N, the approximation coefficient CA1 and detail coefficient CD1 are calculated with the high pass and low pass filters. The two-dimensional DWT (2D-DWT) decomposes a digital image at level j into four components: the approximation coefficient at level j + 1 and three detail coefficients (horizontal, vertical, and diagonal). We used approximation coefficients and obtained dominant features for use in classification.

2.3 Support Vector Machine

SVM was developed by Cortes and Vapnik for solving two-group classification problems [32]. SVM was first designed for the classification of two-class linear data and later generalized for multi-class classification and non-linear data [33]. The purpose of the SVM is to predict the optimal separating plane to classify the data as well as possible.

The goal of SVM is to estimate the function that will separate the two classes in an optimal way. SVM is thus a good choice for gender determination because it has an appropriate structure for binary classification. The SVM algorithm gave the best results in level studies using the face [34], human gait [35], and fingerprints [36]. Moghaddam [34] also showed that even in at low-resolution, SVM produced good results in a study about gender determination using the face.

3 Proposed Intelligent System

The intelligent system has two phases: feature extraction and classification. The features are obtained from the anterior teeth through a hybrid feature extraction method. These are input to the classification algorithm to perform gender determination. In the first stage of the study, six maxillary anterior teeth were manually clipped from images of the anterior teeth. The color of the teeth is not necessary for the proposed system, so the obtained images were converted from the RGB color space into a greyscale image. In the feature extraction phase, 2D-DWT is first applied to the image. The rows of the two-dimensional image are

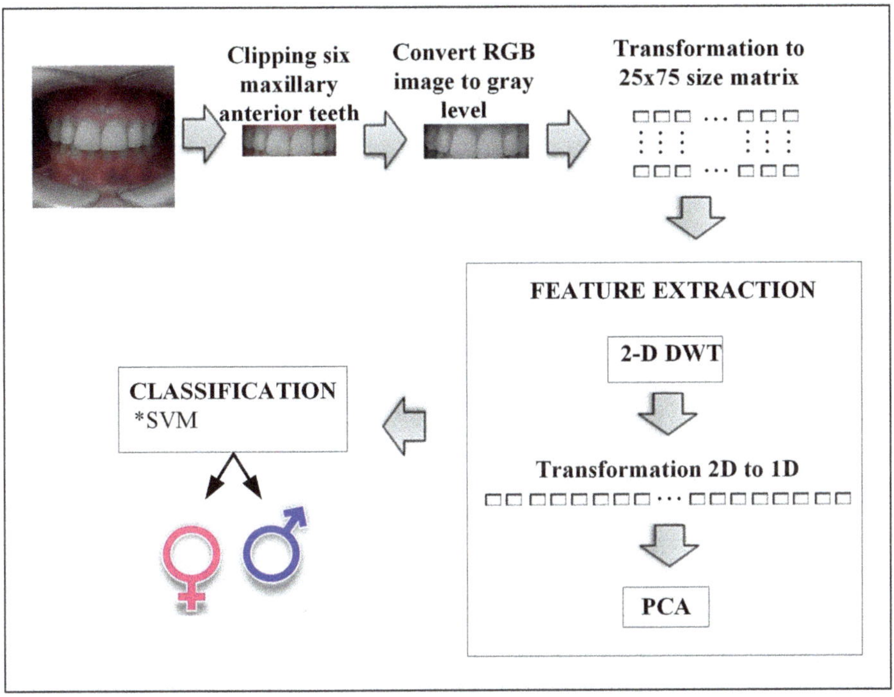

Fig. 2. Steps of automatic gender determination

combined to create a vector, which is subjected to PCA. The steps are presented in Fig. 2.

The Daubechies (db) mother wavelet and haar functions were compared, and db3 gave a better result for the DWT. Using approximation coefficients of level 1, a two-dimensional matrix with a size of 13×38 was obtained. After applying 2D-DWT, the two-dimensional images were converted into one-dimensional $1 \times N$ vectors. The rows of the two-dimensional image were again combined to create a vector, as shown in Fig. 3.

The N value was obtained by multiplication of the number of row pixels (13) and the number of column pixels (38). As a result of the multiplication, each image was represented by a vector with a size of 1×494. Afterwards, the principal components were obtained from this vector. As a result of the experiments, the highest success rate was achieved by using 22 principal components.

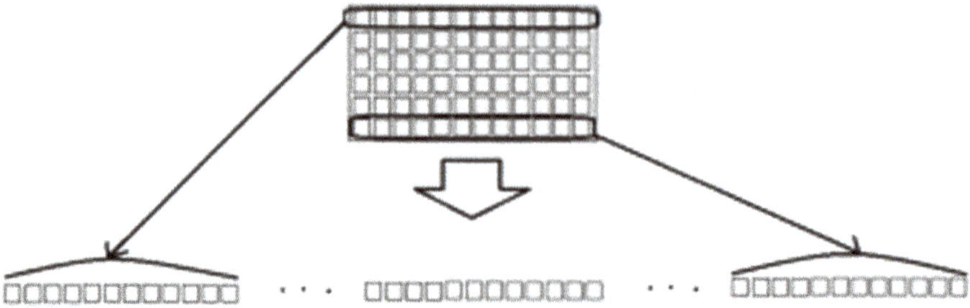

Fig. 3. Process for converting two-dimensional matrix into one-dimensional vector

4 Experimental Framework and Results

4.1 Dataset

The dataset was obtained from a group of 60 Turkish individuals (30 female, 30 male) between the ages of 19 and 27. Approval was obtained from the ethical committee of the Faculty of Dentistry of Necmettin Erbakan University (decision no: 2015/002). The dataset was generated from an anterior teeth image archive collected by expert dentist Hatice KÖK. The data set included individuals with normal teeth combinations (i.e., without missing and/or redundant anterior teeth or crowding). Those with extreme crowding or double row sequences were excluded since the teeth could not be fully seen.

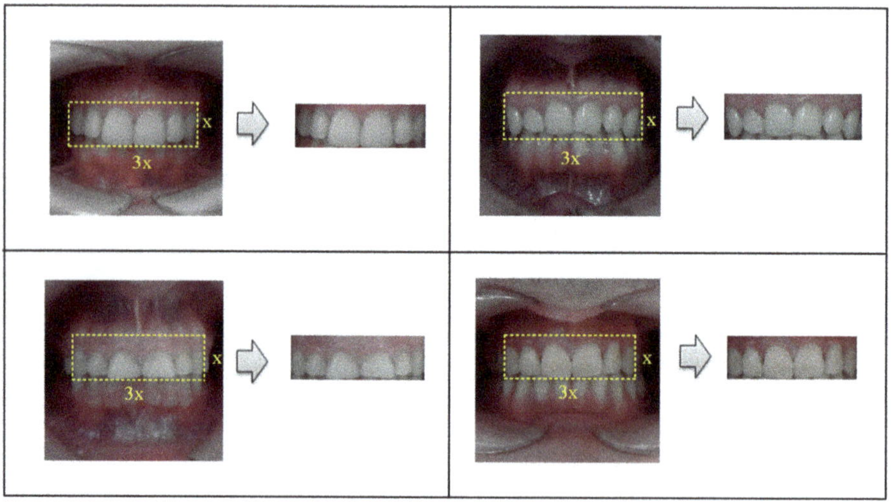

Fig. 4. Sample dataset for six maxillary anterior teeth

The heads of individuals were aligned vertically, dental images were captured from the front, and manual segmentation was performed. Considering the width limit of the six maxillary anterior teeth, starting from the bottom limit where the teeth appear, the six maxillary anterior teeth were cropped in the upward direction at 1/3 of their width. The teeth with smaller width were cropped with the gingival margin, while wider teeth were cropped so that the gingiva would be more visible. This process is shown in Fig. 4. The images were resized to 25×75 while maintaining the actual width/height ratio.

4.2 Experimental Results

In this study, the gender of an individual was determined by using a hybrid feature extraction method and SVM. The feature extraction process was performed in Matlab, and Orange 2.7 software was used for classification. The data were separated randomly 5 times (50 samples for training, 10 samples for testing) through 5 different random seed values (1, 2, 3, 4 and 5) in the Orange random generator. Training data samples were used for determining the optimal SVM parameters using 10-fold cross validation for each run. The optimum SVM was determined as c-SVM, the cost was 6.9, and the Linear kernel was used. The training results are shown in Table 2.

Table 2. Performance measurement results obtained from the training data

Seed	CA (%)	AUC (%)	F-Score (%)
1	92	95	92
2	82	88.33	80.85
3	88	90	87.5
4	88	95	88
5	82	90	82.35

After finding the optimum model parameters, the performance was measured using test data. The test results are shown in Table 3. The average Classification Accuracy (CA), the Area Under the ROC Curve (AUC), and F-Score values were 88%, 98.4%, and 86.56%, respectively.

Table 3. Performance measurement results obtained from the test data

Seed	CA (%)	AUC (%)	F-Score (%)
1	80	96	80
2	90	96	88.89
3	100	100	100
4	80	100	75
5	90	100	88.89

Table 4 shows the sensitivity, specificity, and correct classifications (CC)/incorrect classifications (IC) for females and males. The sensitivity of the system is defined as the percentage of females correctly predicted, and the specificity of the system is defined as the percentage of males correctly predicted.

Table 4. Sensitivity and specificity success rates for the female class

Seed	Sensitivity (%)	Specificity (%)	Female (CC)/(IC)	Male (CC)/(IC)
1	80	80	4/1	4/1
2	80	100	4/1	5/0
3	100	100	5/0	5/0
4	60	100	3/2	5/0
5	80	100	4/1	5/0

5 Conclusion

Teeth are excellent materials that contain information about an individual. Particularly in archeology and forensic science, teeth are valuable for determining the identity of an unknown individual. In this study, the gender of an individual was determined by using a hybrid feature extraction method and SVM. The primary contributions as follows:

- A system that performs gender determination from teeth images via hybrid feature extraction method with SVM was proposed
- The features used for classification are automatically extracted from the images, which minimizes effort for measurement
- The loss of time spent for the measuring process is eliminated
- This method is an intelligent system with a low implementation cost that can easily be used, especially in the fields of archeology and forensic science
- The proposed system can easily be adapted to any population with various teeth sizes.

In future work, we aim to design a system that performs the segmentation process automatically. Also, automatic gender determination will be performed using different perspectives of dental plaster cast models and intraoral images.

References

1. Monali, C., Pritam, P., Tapan, M., Kajal, D.: Gender determination: a view of forensic odontologist. Indian J. Forensic Med. Pathol. **4**(4), 147–151 (2011)
2. Vodanovic, M., Demo, Z., Njemirovskij, V., Keros, J., Brkic, H.: Odontometrics: a useful method for sex determination in an archaeological skeletal population? J. Archaeol. Sci. **34**(6), 905–913 (2007). https://doi.org/10.1016/j.jas.2006.09.004

3. Kolte, R., Kolte, A., Mahajan, A.: Assessment of gingival thickness with regards to age, gender and arch location. J. Indian Soc. Periodontol. **18**(4), 478–481 (2014)
4. Anand, V., Govila, V., Gulati, M.: Correlation of gingival tissue biotypes with gender and tooth morphology: a randomized clinical study. Indian J. Dent. **3**(4), 190–195 (2012). https://doi.org/10.1016/j.ijd.2012.05.006
5. Suazo, G.I., Cantín, L.M., López, F.B., Sandoval, M.C., Torres, M.S., Gajardo, R.P., Gajardo, R.M.: Sexual dimorphism in mesiodistal and bucolingual tooth dimensions in Chilean people. Int. J. Morphol. **26**(3), 609–614 (2008)
6. Hasanreisoglu, U., Berksun, S., Aras, K., Arslan, I.: An analysis of maxillary anterior teeth: facial and dental proportions. J. Prosthet. Dent. **94**(6), 530–538 (2005). https://doi.org/10.1016/j.prosdent.2005.10.007
7. Al-Sehaibany, F.: Analysis of maxillary anterior teeth and various facial dimensions among adolescents in Rıyadh, Saudi Arabia. J. Pak. Dent. Assoc. **20**(2), 67–72 (2011)
8. Parekh, D.H., Patel, S.V., Zalawadia, A.Z., Patel, S.M.: Odontometric study of maxillary canine teeth to establish sexual dimorphism in Gujarat population. Int. J. Biol. Med. Res. **3**(3), 1935–1937 (2012)
9. Lakhanpal, M., Gupta, N., Rao, N.C., Vashisth, S.: Tooth dimension variations as a gender determinant in permanent maxillary teeth. JSM Dent. **1**(1), 1014 (2013)
10. Kaushal, S., Patnaik, V.V.G., Agnihotri, G.: Mandibular canines in sex determination. J. Anat. Soc. India **52**(2), 119–124 (2003)
11. Rai, B., Anand, S.: Gender determination by diagonal distances of teeth. Internet J. Biol. Anthropol. **1**(1), 1–4 (2007)
12. Nagesh, K.S., Iyengar, A.S., Kapila, R., Mehkri, S.: Sexual Dimorphism in human mandibular canine teeth: a radiometric study. J. Indian Acad. Oral Med. Radiol. **23**(1), 33–35 (2011)
13. Sai Kiran, C., Khaitan, T., Ramaswamy, P., Sudhakar, S., Smitha, B., Uday, G.: Role of mandibular canines in establishment of gender. Egypt. J. Forensic Sci. **4**(3), 71–74 (2014). https://doi.org/10.1016/j.ejfs.2014.05.003
14. Bakkannavar, S.M., Manjunath, S., Nayak, V.C., Pradeep Kumar, G.: Canine index - a tool for sex determination. Egypt. J. Forensic Sci. **5**, 157–161 (2015). https://doi.org/10.1016/j.ejfs.2014.08.008
15. Ahuja, P., Manchanda, A.: Application of oral hard and soft tissue structures in sex determination. Internet J. Forensic Sci. **4**(2), 1–7 (2009)
16. Anuthama, K., Shankar, S., Ilayaraja, V., Kumar, G.S., Rajmohan, M.: Vignesh: determining dental sex dimorphism in South Indians using discriminant function analysis. Forensic Sci. Int. **212**, 86–89 (2011). https://doi.org/10.1016/j.forsciint.2011.05.018
17. Shin, Y.: Gender identification on the teeth based on principal component analysis representation. In: LNCS, vol. 4069, pp. 300–304 (2006)
18. Akkoc, B., Arslan, A., Kok, H.: Gray level co-occurrence and random forest algorithm-based gender determination with maxillary tooth plaster images. Comput. Biol. Med. **73**, 102–107 (2016). https://doi.org/10.1016/j.compbiomed.2016.04.003
19. Akkoc, B., Arslan, A., Kok, H.: Automatic gender determination from 3D digital maxillary tooth plaster models based on the random forest algorithm and discrete cosine transform. Comput. Methods Programs Biomed. **143**, 59–65 (2017). https://doi.org/10.1016/j.cmpb.2017.03.001
20. Garn, S.M., Cole, P.E., Wainwright, R.L., Guire, K.E.: Sex discriminatory effectiveness using combinations of permanent teeth. J. Dent. Res. **56**(6), 697 (1977)

21. Rao, N.G., Rao, N.N., Pai, L., Kotian, M.S.: Mandibular canine index - a clue for establishing sex identity. Forensic Sci. Int. **42**, 249–254 (1989)
22. Muller, M., Lepi-Pegurier, L., Quatrehomme, G., Bolla, M.: Odontometrical method useful in determining gender and dental alignment. Forensic Sci. Int. **121**, 194–197 (2001)
23. Iscan, M.Y., Kedici, P.S.: Sexual variation in bucco-lingual dimensions in Turkish dentition. Forensic Sci. Int. **137**, 160–164 (2003). https://doi.org/10.1016/s0379-0738(03)00349-9
24. Karaman, F.: Use of diagonal teeth measurements in predicting gender in a Turkish population. J. Forensic Sci. **51**(3), 630–635 (2006). https://doi.org/10.1111/j.1556-4029.2006.00133.x
25. Toprak, O.K.: Availability of tooth development in the gender determination. Graduation thesis, Ege University (2013)
26. Das, N., Reddy, J.M., Sarkar, R., Basu, S., Kundu, M., Nasipuri, M., Basu, D.K.: A statistical-topological feature combination for recognition of handwritten numerals. Appl. Soft Comput. **12**(8), 2486–2495 (2012)
27. Sinha, C.: Gender classification from facial images using PCA and SVM. National Institute of Technology Rourkela (2013)
28. Gumus, E., Kilic, N., Sertbas, A., Ucan, O.N.: Evaluation of face recognition techniques using PCA, wavelets and SVM. Expert Syst. Appl. **37**, 6404–6408 (2010)
29. Ceylan, R.: A tele-cardiology system design using feature extraction techniques and artificial neural networks. Selçuk University (2009)
30. Karlik, B.: Machine learning algorithms for characterization of EMG signals. Int. J. Inf. Electron. Eng **4**(3), 189–194 (2014)
31. Thyagarajan, K.S.: Still Image and Video Compression with MATLAB. Wiley-IEEE Press, Hoboken (2011)
32. Cortes, C., Vapnik, V.: Support vector machine. Mach. Learn. **20**, 273–297 (1995)
33. Canbay, Y.: Classification of Diabetes Data Using Support Vector Machines. Erciyes University, Kayseri (2013)
34. Moghaddam, B., Yang, M.-H.: Gender classification with support vector machines. In: IEEE International Conference on Automatic Face and Gesture Recognition Grenoble 2000, pp. 306–311. IEEE (2000)
35. Begg, R.K., Palaniswami, M., Owen, B.: Support vector machines for automated gait classification. IEEE Trans. Biomed. Eng. **52**(5), 828–838 (2005)
36. Arun, K.S., Sarath, K.S.: A machine learning approach for fingerprint based gender identification In: IEEE Recent Advances in Intelligent Computational Systems, Trivandrum 2011, pp. 163–167. IEEE (2011)

Simulated Annealing Algorithm for a Medium-Sized TSP Data

Mehmet Fatih Demiral[1(✉)] and Ali Hakan Işik[2]

[1] Department of Industrial Engineering, Burdur Mehmet Akif Ersoy University, Burdur, Turkey
mfdemiral@mehmetakif.edu.tr

[2] Department of Computer Engineering, Burdur Mehmet Akif Ersoy University, Burdur, Turkey
ahakan@mehmetakif.edu.tr

Abstract. Traveling Salesman Problem (TSP) is among the most popular and widely studied NP-hard problems in the literature. There exist many mathematical models, applications and proposed techniques for TSP. In a classic TSP, the problem consists of dispersed locations in a space; so salesman aims to visit all of the locations constructing the best minimal tour. The problem has great attention by scientists in the field of operations research and other scientific areas since it was put forth. There exists quite a lot exact, heuristic and meta-heuristic technique for TSP. In this study, Simulated Annealing (SA) algorithm has been applied on a group of randomly generated medium-sized TSP problems. Besides, as a neighborhood structure, two well-known operators, which are reverse and swap-reverse, were implemented through SA. At last, the findings and algorithm performance were given by a comparison with operators and randomly generated TSP data.

Keywords: Travelling salesman problem · Simulated annealing algorithm · Meta-heuristics · Neighborhood structure · Artificial intelligence

1 Introduction

One of the recently studied research area is the meta-heuristic search. A meta-heuristic algorithm is an iterative or population-based technique that searches near optimal solutions within a reasonable computation time. It has bio-inspired or nature-inspired capabilities to solve complex combinatorial problems. Hybrid meta-heuristics are also among the powerful techniques for NP-hard problems. There has been many proposed meta-heuristics since last decades. For instance, SA applications [1–4], genetic algorithms (GA) [5–8], cuckoo search optimization (CSO) [9, 10] and hybrid meta-heuristics [11–14] are well-known examples of optimization algorithms.

This study analyzes the application of the SA algorithm on symmetric travelling salesman problem (TSP) in terms of using two distinct neighborhood operators. Travelling salesman problem is a popular benchmark problem for scientists to evaluate the performance of meta-heuristic algorithms. In a classical TSP, salesman aims to find the best minimal tour while visiting each city dispersed in a space only once.

The rest of the paper is organized as follows: In Sect. 2, the travelling salesman problem is under consideration. Some descriptive information is given in Sect. 3.

D. J. Hemanth and U. Kose (Eds.): ICAIAME 2019, LNDECT 43, pp. 457–465, 2020.
https://doi.org/10.1007/978-3-030-36178-5_35

In Sect. 4, an application with SA is done and experimental results are given, and finally, last section includes conclusion and some advices related to this work.

2 Travelling Salesman Problem

Travelling salesman problem (TSP) is a well-known and widely studied combinatorial problem in operations research. The problem has great attention by scientists since last decades. There have been many exact, heuristic and meta-heuristic algorithms for TSP. However, there exists no exact method to find optimal TSP solutions in polynomial time. Thus, it is reasonable to find near optimal solutions to that problem within acceptable times.

The fundamental formula for travelling salesman problem is given below [15]:

$$Min. \quad \sum_{i=1}^{n} \sum_{j \neq i, j=1}^{n} c_{ij} x_{ij} \tag{1}$$

$$\sum_{i=1, i \neq j}^{n} x_{ij} = 1, \qquad j = 1, \ldots, n \tag{2}$$

$$\sum_{j=1, i \neq j}^{n} x_{ij} = 1, \qquad i = 1, \ldots, n \tag{3}$$

$$u_i - u_j + n x_{ij} \leq n - 1, \qquad 2 \leq i \neq j \leq n \tag{4}$$

$$x_{ij} \in \{0, 1\} \tag{5}$$

In the above formula,
n: Number of cities
c_{ij}: Cost (distance) of assignment between city i and city j
x_{ij}: Decision variable, which equals 1 wherever a connection between city i and city j occurs.

The constraint (4) ensures that each cycle is a form of complete tour, not a subset of whole cities. For symmetrical TSP, the number of possible solutions for n cities is the $(n-1)!/2$. Therefore, the number of possible solutions can be very large for problems, which have medium and large number of cities [16]. There are several cases of TSP. Fundamental cases of TSP are symmetric, asymmetric and time-windows instances. TSPLIB [17] gives special variants and test data sets on travelling salesman problem. A wide range of articles have been worked upon these topics in recent years.

3 Simulated Annealing Algorithm

The Simulated Annealing Algorithm (SA) is an artificial intelligence based optimization algorithm introduced by Kirkpatrick, Gelatt and Vecchi in 1983 [18]. The SA is a stochastic search technique based on a single solution for solving global optimization problems. This algorithm is rather simple and effective. As many other algorithms require, it uses neighborhood search. The algorithm with many neighborhood structures were applied and proven in many different studies [2, 3, 13, 19].

The possibility of accepting worse solutions in TSP is controlled by the following formula:

$$P(x^* = x_2) = \begin{cases} e^{\frac{Z_1 - Z_2}{T}}, & if \quad Z_2 \geq Z_1 \\ 1, & if \quad Z_2 < Z_1 \end{cases} \tag{6}$$

Z_1 : *Objective value of* x_1 *solution*
Z_2 : *Objective value of* x_2 *solution*
T : *Temperature parameter of SA*

At first, the SA starts with the initial solution x and initial temperature T. Temperature is chosen relatively high at initial state. As algorithm iterates, it decreases with a relatively high ratio ρ. ρ is the cooling constant that can be set in range $0.7 - 0.99$ [20].

$$T = T * \rho \tag{7}$$

The steps of the standard SA algorithm are as follows (Fig. 1):

Algorithm 1: SA

```
Input: initial temperature T0, minimum temperature Tmin, maximum number of iteration Kmax,
probability of temperature drop ρ;
Output: the optimal solution xbest;
Generating an initial solution x0;
xbest ← x0;
Computing the value of the objective function f(x0) and f(xbest);
Ti ← T0;
while Ti>Tmin do
  Δf ← f(xnew)−f(xbest);
  if Δf < 0 then
     xbest ← xnew;
  end if
  if Δf ≥ 0 then
    p ← e^(Δf/T);
    if c ← random[0, 1] ≥ p then
       xbest ← xnew;
    else
       xbest ← xbest;
    end if
  end if
  i ← i + 1;
  Ti ← ρ × Ti;
end while
Return xbest;
```

Fig. 1. The steps of the standard SA algorithm [21].

Figure 2 presents a flow chart of the standard SA algorithm.

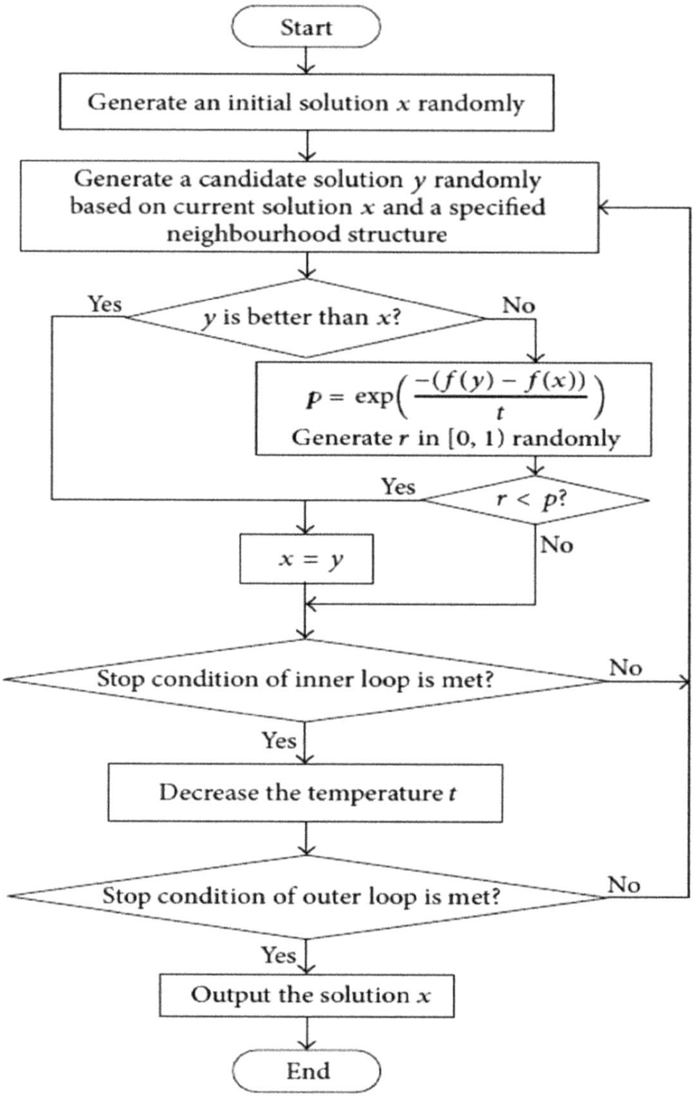

Fig. 2. A flow chart of the standard SA algorithm [3].

3.1 Data and Parameter Settings

In this application, randomly generated medium size problems have been handled with 8 different sizes (N10, N15, N20, N25, N30, N35, N40, N45). Contrary to known, in order to converge near optimal solutions quickly, a relatively low initial temperature T_0 has been taken ($T_0 = 40000$). Cooling rate is also kept relatively low because of escaping from poor solutions ($\rho = 0.50$). L = 50, the iteration limit for temperature change is sufficient to get best results for the medium size data. K = 3000, the global iteration limit is also sufficient to get best results for those group of problems (N10–N45).

4 Experimental Results

As we mentioned before, to evaluate the performance of the simulated annealing (SA) algorithm, we have compared its results with two well-known operators on a randomly generated test dataset. All computations have been implemented in MATLAB and run on an Intel® Core™ i5 3210-M CPU 2.5 GHz speed with 8 GB RAM.

As seen in Table 1, SA algorithm with Reverse-Operator gives the best results in terms of Average-Best solution and CPU time. Besides that, the differences between operators are getting larger while increasing number of cities. SA with Swap-Reverse has a negative effect on the solution quality, especially on the relatively large data sizes ($N \geq 30$).

Table 1. The problem ID, average and best solution values, CPU Time values for the randomly generated TSP problems

Problem-ID	SA-Reverse Operator			SA-Swap Reverse Operator		
	Average solution	Best solution	CPU Time (secs)	Average solution	Best solution	CPU time (secs)
N10	**293.81**	**293.08**	0.127	304.59	293.08	0.186
N15	**379.62**	**374.69**	0.133	402.28	375.78	0.186
N20	**394.36**	**371.91**	0.155	485.94	419.06	0.249
N25	**457.92**	**432.22**	0.174	537.28	494.14	0.235
N30	**430.42**	**414.47**	0.189	551.34	504.58	0.251
N35	**570.25**	**531.49**	0.239	754.88	626.19	0.288
N40	**566.17**	**494.38**	0.222	791.16	713.55	0.297
N45	**601.58**	**536.58**	0.251	950.85	818.13	0.304

Table 2. % deviation from the best known results

Problem-ID	SA-Reverse Operator			SA-Swap Reverse Operator		
	Average %Dev.	Best % Dev.	CPU time (secs)	Average %Dev.	Best % Dev.	CPU time (secs)
N10	–	–	0.127	%3,67	–	0.186
N15	–	–	0.133	%5,97	%0,3	0.186
N20	–	–	0.155	%23,2	%12,7	0.249
N25	–	–	0.174	%17,3	%14,3	0.235
N30	–	–	0.189	%28,1	%21,7	0.251
N35	–	–	0.239	%32,4	%17,8	0.288
N40	–	–	0.222	%39,7	%44,3	0.297
N45	–	–	0.251	%58,1	%52,5	0.304

As seen from Table 2, although the Reverse-Operator CPU times are slightly better than the Swap-Reverse Operator times, the obviously large extent differences can be observed in Average and Best solutions. In particular, in N40 and N45, the deviations from best-known solutions have been simply detected.

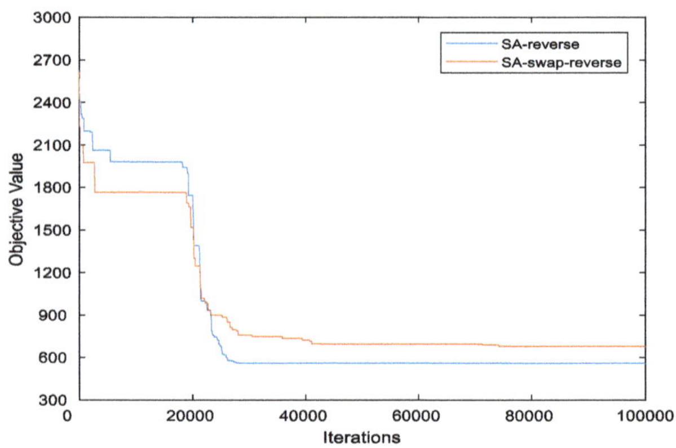

Fig. 3. Convergence curves for SA in terms of Reverse and Swap-Reverse Operators (one run for N45 city problem).

As seen in Fig. 3, SA with reverse-operator converges faster than the other method with Swap-reverse operator. If we examine the results, the reverse operator would find better results in earlier iterations. The reason of that would be the searching direction of swap-reverse is too diversified and may not lead to promising regions of the solution space [22].

Figure 4 demonstrates a set of obtained solutions by the simulated annealing algorithm with reverse operator on the randomly generated test data sets.

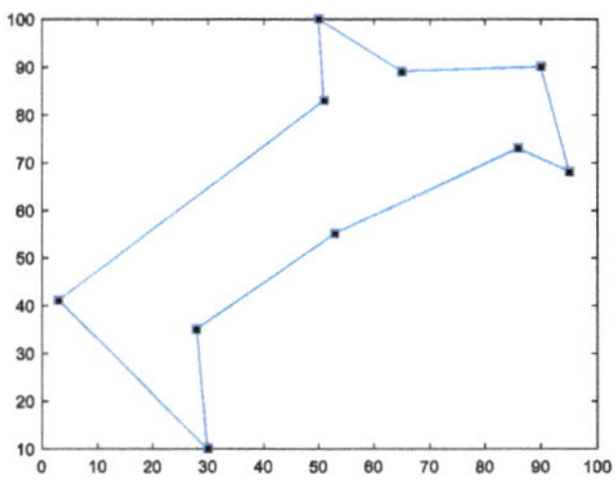

Tour Path of N10 City Problem

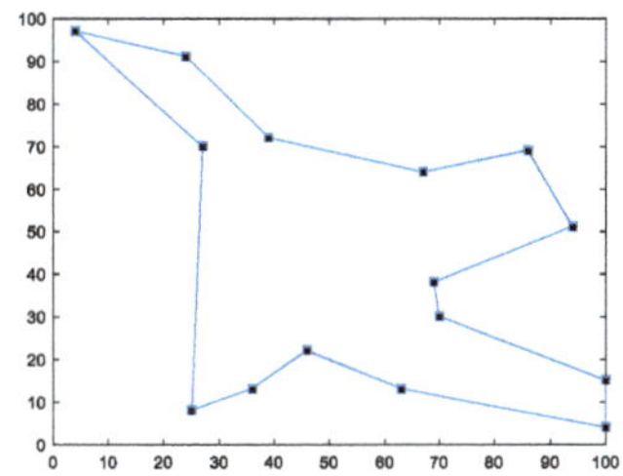

Tour Path of N15 City Problem

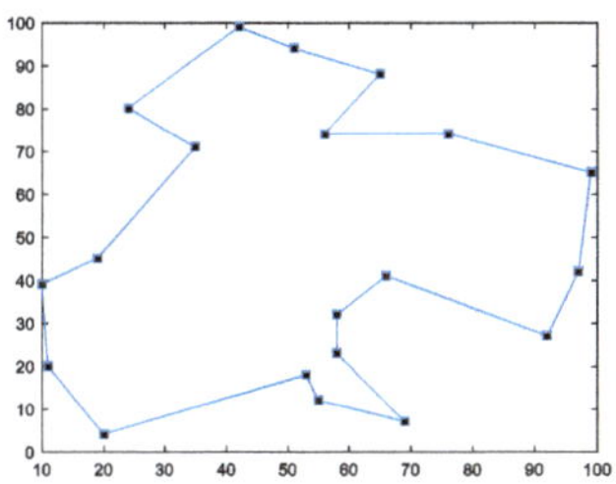

Tour Path of N20 City Problem

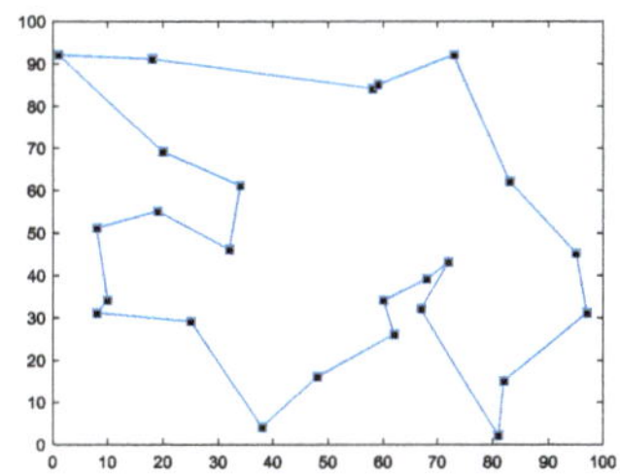

Tour Path of N25 City Problem

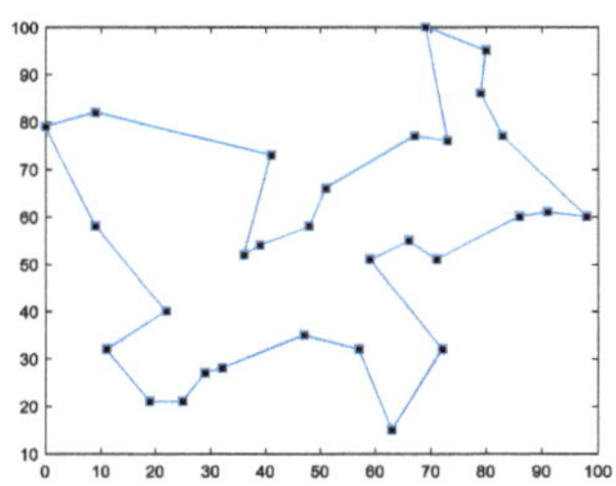

Tour Path of N30 City Problem

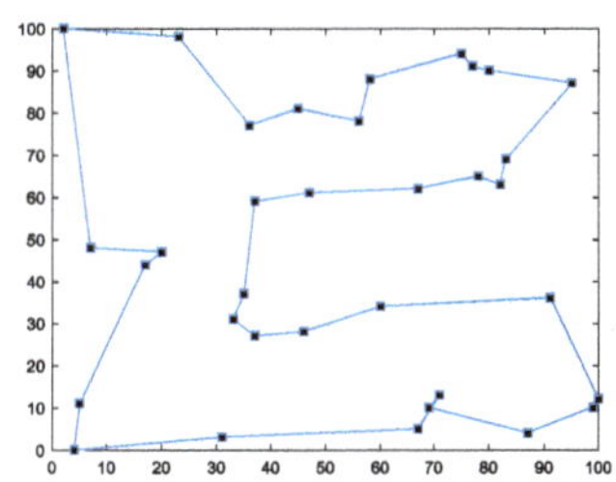

Tour Path of N35 City Problem

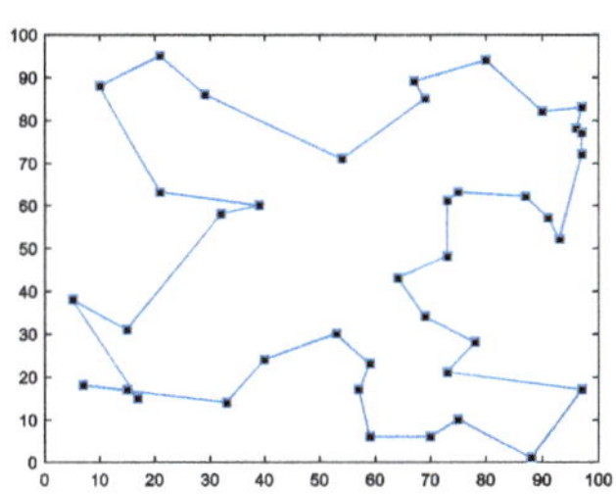

Tour Path of N40 City Problem

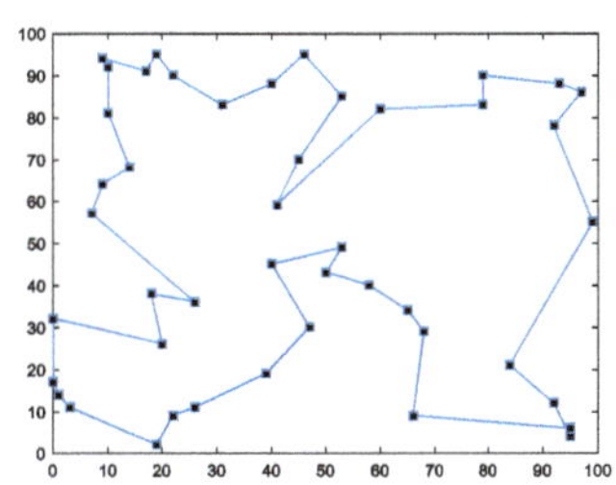

Tour Path of N45 City Problem

Fig. 4. TSP solutions found by the SA algorithm with Reverse-Operator on the test data sets

5 Conclusions and Future Work

In many publications, the importance of neighborhood operators on performance of meta-heuristics has been strongly emphasized. The diversification and intensification phase of algorithm directly depends on the type of the operator being used. Generally speaking, different neighborhood operators contribute differently. In this study, in view of average results, best results and CPU time, reverse operator is superior to swap-reverse operator with SA. In addition, the reverse operator with SA converges best solution better than the swap-reverse operator with SA. However, the computational results are valid for medium sized TSP problems and a few number of iterations in the application.

In future studies, the convergence of both operators might be investigated on random and benchmark TSP problems with other meta-heuristic and heuristic algorithms. Additionally, more experimental study can be done with larger set of random, benchmark and real data with meta-heuristics. Both operators and many others can also be implemented on different types of combinatorial problems.

References

1. Geng, X., Chen, Z., Yang, W., Shi, D., Zhao, K.: Solving the travelling salesman problem based on an adaptive simulated annealing algorithm with greedy search. Appl. Soft Comput. **11**(4), 3680–3689 (2011)
2. Xinchao, Z.: Simulated annealing algorithm with adaptive neighborhood. Appl. Soft Comput. **11**(2), 1827–1836 (2011)
3. Zhan, S.-H., Lin, J., Zhang, Z.-J., Zhong, Y.-W.: List-based simulated annealing algorithm for travelling salesman problem. Comput. Intell. Neurosci. **2016**, 12 (2016). Article ID 1712630
4. Wang, Z., Geng, X., Shao, Z.: An effective simulated annealing algorithm for solving the travelling salesman problem. J. Comput. Theoret. Nanosci. **6**(7), 1680–1686 (2009)
5. Tsai, H.-K., Yang, J.-M., Tsai, Y.-F., Kao, C.-Y.: Some issues of designing genetic algorithms for traveling salesman problems. Soft. Comput. **8**(10), 689–697 (2004)
6. Winter, G., Galvan, B., Alonso, S., Gonzales, B., Jimenez, J.I., Greiner, D.: A flexible evolutionary agent: cooperation and competition among real-coded evolutionary operators. Soft. Comput. **9**(4), 299–323 (2005)
7. Sathya, S.S., Kuppuswami, S.: Gene silencing-a genetic operator for constrained optimization. Appl. Soft Comput. **11**(8), 5801–5808 (2011)
8. Ter-Sarkisov, A., Marsland, S.: K-Bit-Swap: a new operator for real-coded evolutionary algorithms. Soft. Comput. **21**(20), 6133–6142 (2017)
9. Ouaaraba, A., Ahioda, B., Yangb, X.-S.: Discrete cuckoo search algorithm for the traveling salesman problem. Neural Comput. Appl. **24**(7–8), 1659–1669 (2014)
10. Rajabioun, R.: Cuckoo optimization algorithm. Appl. Soft Comput. **11**(8), 5508–5518 (2011)
11. Chen, S.-M., Chien, C.-Y.: Solving the traveling salesman problem based on the genetic simulated annealing ant colony system with particle swarm optimization techniques. Expert Syst. Appl. **38**(12), 14439–14450 (2011)

12. Elhaddad, Y., Sallabi, O.: A new hybrid genetic and simulated annealing algorithm to solve the traveling salesman problem. In: Ao, S.I., Gelman, L., Hukins, D.W.L., Hunter, A., Korsunsky, A.M. (eds.). World Congress on Engineering 2010, WCE, vol. I, pp. 11–14. Newswood Limited, London (2010)
13. Lin, Y., Bian, Z., Liu, X.: Developing a dynamic neighborhood structure for an adaptive hybrid simulated annealing – tabu search algorithm to solve the symmetrical traveling salesman problem. Appl. Soft Comput. **49**, 937–952 (2016)
14. Ezugwu, A.E.-S., Adewumi, A.O., Frîncu, M.E.: Simulated annealing based symbiotic organisms search optimization algorithm for traveling salesman problem. Expert Syst. Appl. **77**(1), 189–210 (2017)
15. Miller, C.E., Tucker, A.W., Zemlin, R.A.: integer programming formulation of traveling salesman problems. J. ACM **7**(4), 326–329 (1960)
16. Hatamlou, A.: Solving travelling salesman problem using black hole algorithm. Soft. Comput. **22**(24), 8167–8175 (2018)
17. TSPLIB. Ruprecht-Karls-Universität Heidelberg. https://wwwproxy.iwr.uni-heidelberg.de/groups/comopt/software/TSPLIB95/
18. Kirkpatrick, S., Gelatt Jr., C.D., Vecchi, M.P.: optimization by simulated annealing. Science **220**(4598), 671–680 (1983)
19. Miki, M., Hiroyasu, T., Ono, K.: Simulated annealing with advanced adaptive neighborhood. In: Second International Workshop on Intelligent Systems Design and Application Proceedings, pp. 113–118. Dynamic Publishers, Atlanta (2002)
20. Halim, A.H., Ismail, I.: Combinatorial optimization: comparison of heuristic algorithms in travelling salesman problem. Arch. Comput. Methods Eng. **26**(2), 367–380 (2019)
21. Zhou, A.-H., Zhu, L.-P., Hu, B., Deng, S., Song, Y., Qui, H., Pan, S.: Traveling-salesman-problem algorithm based on simulated annealing and gene-expression programming. Information **10**(1), 7 (2019)
22. Szeto, W.Y., Yongzhong, W., Ho, S.C.: An artificial bee colony algorithm for the capacitated vehicle routing problem. Eur. J. Oper. Res. **215**(1), 126–135 (2011)

Gene Selection in Microarray Data Using an Improved Approach of CLONALG

Ezgi Deniz Ülker(✉)

European University of Lefke, Mersin-10, Turkey
eulker@eul.edu.tr

Abstract. Expression of a gene at a molecular level can lead to the diagnosis and prediction of diseases especially cancer. In microarray datasets, large amount of genes exists and this makes the classification process complex. Since the selected genes must contain enough number of features for classification, selecting suitable features for classification is crucial. Due to the complexity of the problem, it is accepted as a NP-hard problem and an evolutionary approach; Clonal Selection Algorithm (CLONALG) is chosen to produce solution for this problem. In this paper, an Improved Clonal Selection Algorithm (ICSAT) with K-nearest neighbor (K-NN) method are used together to select the relevant features of genes for an accurate gene classification. The proposed method ICSAT-KNN is tested on three gene datasets and compared with two existing algorithms. The obtained classification accuracy values are quite competitive or even better than the values of the compared algorithms. Experimental results show that ICSAT-KNN method can serve as a reliable tool for feature selection and accurate data classification.

Keywords: Clonal Selection Algorithm · Gene selection · Gene classification · K-NN method · Microarray data

1 Introduction

The selection of relevant genes contributes substantially for genetic nature of diseases, since the genes are the atomic representation of a cell. However, the challenge arises when the selection of relevant features among the large amount of genes is necessary. When the large number of genes is investigated for selection and classification, an accurate and a reliable selection method is required. Gene selection is preferred in the classification of cancer cells or diagnosis and prediction of diseases. Selecting suitable genes that represent almost all necessary features is crucial and this process requires lots of time. Due to these difficulties, researchers have preferred artificial intelligence techniques in microarray studies such as Genetic Algorithm, Particle Swarm Optimization, Simulated Annealing, Tabu Search and Bee Colony Optimization [1–8].

In this paper, an improved version of CLONALG is selected to obtain an accurate gene selection from microarray datasets. There are some motivations to use CLONALG as a gene selection method. Clonal selection algorithm (CLONALG) is one of the proved, powerful optimization algorithms proposed by De Castro and Von Zuben to solve pattern recognition tasks [9], and after its first appearance, it has been adapted

D. J. Hemanth and U. Kose (Eds.): ICAIAME 2019, LNDECT 43, pp. 466–472, 2020.
https://doi.org/10.1007/978-3-030-36178-5_36

to many complex problems from different fields [10–13]. The improved clonal selection algorithm (ICSAT) was proposed in one of our recently published works for solving the complex problem in the field of microwave engineering [14].

CLONALG imitates the antibody-antigen reaction to obtain the global optimum solution without trapping into many local optimum points. This makes the algorithm suitable to study with the large amount of genes. Furthermore, in this study the performance of CLONALG is improved by the use of Tournament Selection operator in the crucial steps; genetic variation and natural selection. Tournament Selection (TS) operator has been generally used in Genetic Algorithm [15–17]. However, because of its easy implementation and satisfactory performance, TS operator is preferred by other algorithms as well [18–20]. TS operator gives chance to all of the candidates in the population to be chosen. While this attribute allows for an increase in diversity, on the other hand, the decrease in convergence speed can be observed because of the lack of elitism feature.

In this work, the algorithm ICSAT is combined with K-Nearest Neighbor (K-NN) method to serve as an accurate tool for gene classification. K-NN is one of the preferred techniques because of its effectiveness and simplicity [21]. K-NN classifies the data according to the distance between training/test data and the number of the nearest neighbors.

The rest of the paper is organized as follows: Sect. 2 describes the steps of ICSAT with the K-NN method. Section 3 discusses the experimental results of ICSAT-KNN with the comparison between two algorithms for the three cancer datasets. Section 4 gives the concluding remarks of the paper.

2 Implementation of ICSAT and K-NN Method

Clonal selection was originated from the theory of evolution which has diversity, genetic variation and natural selection. The CLONALG was first proposed by De Castro and Von Zuben to solve pattern recognition tasks [9]. It is based on the behavior of antibodies, when they encounter an antigen. The antibodies bind to the antigen to avoid it from the organism. When the same antigen enters to the same organism, the antibodies with higher affinity values are created. In CLONALG, selection of antibodies, cloning them according to their affinity values and mutating them are the core steps that directly affect the performance. In the improved clonal selection algorithm, these core steps are controlled by a Tournament Selection (TS) operator which is generally used in Genetic Algorithm.

The use of TS operator is quite efficient in terms of having high diversity by giving a chance to all antibodies to be selected. In the performance of TS operator, selecting the group size (G_{size}) plays an important role. It determines the number of antibodies to be selected for a competition. In our experiments, G_{size} is used as 15 among 100 members in the population [14]. The winners of the competition are used for the next generation and the ones that they lost the competition are not passed to the next generation. TS operator is used in the selection, cloning and mutating steps.

In ICSAT, the elimination of worst antibodies is removed from the algorithm. The antibodies which have worst affinity values are kept in the population even in the first level of optimization to spread their desired features to the next generation. TS operator is used to have an increased diversity in the ICSAT, because of the necessity of finding the meaningful features among the large number of genes. The improvements done in the original CLONALG are emphasized as italic in Fig. 1.

Step 1: Initialization of a set of antibodies (Ab) by B lymphocyte cells.
Step 2: Definition of the objective function, antigen, (Ag) that is needed to be optimized.
Step 3: Calculation of affinity values for each Ab.
Step 4: *Competition of randomly selected G_{size} number of Abs from the population and keeping the winner for the next generation.*
Step 5: *Selecting G_{size} number of Abs randomly from the population and applying a mutation to the winners of the tournament.*
Step 6: Recalculation of affinity values for new mutated antibodies.
Step 7: *Selecting G_{size} number of Abs randomly among the n number of the worst affinity values and applying an equal mutation rate to the winners of the tournament.*
Step 8: *Replacing the Abs, if the affinity value of the mutated Ab is better than the affinity value of the same Ab that is not mutated.*
Step 9: Repeat steps 3-8 until a stopping criterion is met.

Fig. 1. The steps of ICSAT.

After the selection of genes by ICSAT, K-Nearest Neighbor (K-NN) method is used to classify data according to the training and validation data. This method is mainly based on a supervised learning algorithm. The result of classification depends on the highest K-nearest neighbor categories. It was proposed by Fix and Hodges [21] and it is preferred by many areas due to its easy implementation. However, K-NN method has some variables that are needed to be adjusted such as K (the number of nearest neighbors), D (the distance between the classify data and all of the training data), C (the number of categories of the nearest neighbors). The aim is to use K-NN method with ICSAT is not only to select the minimum number of features but to select the ones with the relevant information for classification. The steps of K-NN method are given in Fig. 2.

Step 1: Calculation the distance (Euclidean) between the training and validation data.
Step 2: Sorting the distances and determining the K-nearest neighbors.
Step 3: Assigning the data to the most common class among its K-nearest neighbors.
Step 4: Repeating the steps 1-3, until all of the samples are selected once as a validation data.
Step 5: Calculating fitness by using LOOCV.

Fig. 2. The steps of K-NN method.

The algorithm performs by initializing the randomly generated antibodies in binary string form as selected {1} and non-selected {0} features. In this study, K is assigned as 1 and 1-NN method is used to classify the selected data. The affinity value of 1-NN is calculated according to leave-one-out cross validation (LOOCV) method.

LOOCV method is used to determine the training data sets among the selected gene features by selecting only one sample from the original data set as the validation data and by selecting the rest of the samples as the training data. This process repeats, until all of the samples are selected once as a validation data. The following figure Fig. 3 shows the ICSAT-KNN method applied as 1-NN in the form of a flowchart.

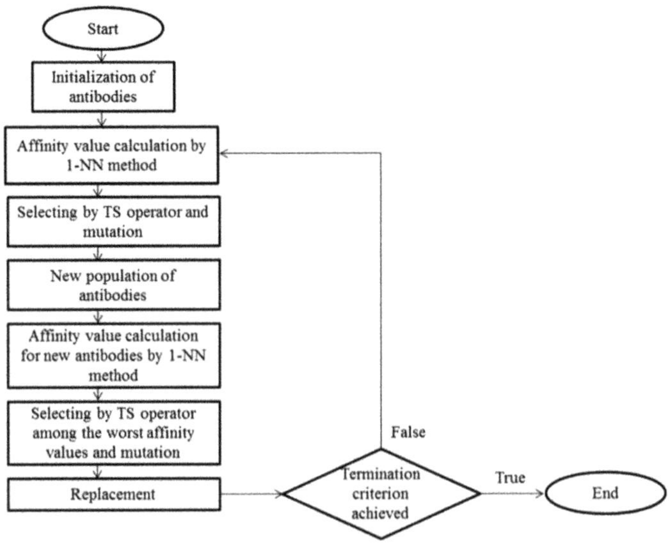

Fig. 3. The flowchart of ICSAT-KNN method.

3 Implementation of ICSAT and K-NN Method

In this study, 3 gene datasets are taken from http://www.gems-system.org. The selected gene datasets belong to different type of samples; brain tumor, leukemia and prostate tumor. Table 1 shows the information of the selected samples and the format of the used datasets is given in Table 2. It includes the number of samples, categories, genes and the selected number of genes with their percentage values. Table 3 demonstrates the comparative classification accuracy values by two different methods from the literature; Multi-class support vector machines (MC-SVM) [22], improved binary PSO for feature selection (IBPSO-KNN) [7] and our proposed method ICSAT-KNN. The results obtained by MC-SVM and IBPSO-KNN are taken as they appeared in the original studies [7, 22].

The average percentage value for the selected genes is 0.16 and our approach improved clonal selection algorithm (ICSAT) with K-NN method has achieved an accurate classification. This justifies the argument of the small portion of genes achieves high accuracy rather than the larger ones and all of the features are not required to have a good accuracy rate.

Table 1. Information of selected samples.

Gene dataset	Description
Brain_Tumor1	DNA microarray gene expression profiles derived from 99 patient samples. The medulloblastomas included primitive neuroectodermal tumors (PNETs), a typical teratoid/rhabdoid tumors (AT/RTs), malignant gliomas and the medulloblastomas activated by the Sonic Hedgehog (SHH) pathway
Leukemia1	DNA microarray gene expression profiles of acute myelogeneous leukemia (AML), acute lymphoblastic leukemia (ALL) B-cell and T-cell
Prostate_Tumor	cDNA microarray gene expression profiles of prostate tumors. Based on MUC1 and AZGP1 gene expression, the prostate cancer can be distinguished as a subtype associated with an elevated risk of recurrence or with a decreased risk of occurrence

Table 2. Format of selected datasets for classification.

Gene dataset	No. of samples	No. of categories	No. of genes	No. of selected genes	Percentage values of selected genes
Brain_Tumor1	90	5	5920	867	0.15
Leukemia1	72	3	5327	1128	0.21
Prostate_Tumor	102	2	10509	1352	0.13

Table 3. Comparative classification accuracy values.

Gene dataset	ICSAT-KNN	MC-SVM	IBPSO-KNN
Brain_Tumor1	95.68%	91.67%	94.44%
Leukemia1	98.29%	97.50%	100.00%
Prostate_Tumor	92.12%	92.00%	92.16%
Average	96.36%	93.72%	95.53%

In general, classification accuracy values depend on having the smallest gene dataset that contains similar and meaningful features. According to the overall average classification values, ICSAT-KNN has achieved the maximum average classification value (96.36%) for the selected samples.

When the algorithms are compared individually, the proposed method ICSAT-KNN gives better classification accuracy values than MC-SVM in all instances. The classification accuracy values for Brain_Tumor, Leukemia1 and Prostate_Tumor achieved by ICSAT-KNN are 95.68%, 98.29% and 92.12%, respectively. In the comparison with the MC-SVM method, there is an improvement of 4.01%, 0.79% and 0.12% in classification accuracy values for all datasets. In the comparison of IBPSO-KNN, ICSAT-KNN performs better for the Brain_Tumor1 with the improvement of 1.24%. Although the expected performance improvement is not achieved for the datasets Leukemia1 and Prostate_Tumor by ICSAT-KNN, it produces good quality classification values. The obtained values by the proposed algorithm are not as good as IBPSO-KNN, but quite competitive. The classification accuracy can be improved by the selection of the genes that contain sufficient feature information. In this study the average percentage value for gene selection is 0.16. However, for the Leukemia and Prostate_Tumor datasets, the expected performance is not achieved by ICSAT-KNN in the comparison with IBPSO-KNN. This shows that it is not enough to select the less number of genes, but must carry the enough quality of information for the classification.

The algorithm ICSAT is not a highly parameter dependent algorithm like the other evolutionary techniques. However, the number of candidates in the competition (G_{size}) and the number of nearest neighbors (K) values are needed to be optimized for ICSAT-KNN. This may yield the algorithm to have a high quality in different sizes of feature selection and classification problems.

4 Conclusion

In this study, an improved version of CLONALG with TS selection operator and K-nearest neighbor method are used to have an efficient feature selection and an accurate data classification. The original algorithm CLONALG was strengthened with the use of TS operator. It gives chance to be used to all of the candidates in the population and by this all of the features have a chance to be analyzed for an accurate classification.

Three different datasets are studied and compared with the MC-SVM and IBPSO-KNN methods. The classification accuracy achieved by ICSAT-KNN is better than the values obtained by MC-SVM. The enhancement percentages for the datasets in the classification by ICSAT-KNN are 4.01%, 0.79% and 0.12%, when it is compared with MC-SVM. However, the obtained values except one of the datasets are not as superior as the values of IBPSO-KNN, but quite competitive. This can be a reason of the selection small amount of genes, but not including enough number of features without applying parameter tuning in G_{size} and K values.

As a future work, different values in feature selection can be reached with the use of G_{size} and K values. However, even without having parameter tuning in the algorithm ICSAT-KNN, the selection and classification performances are quite promising, when the algorithm is compared with the powerful algorithms. After applying some fine-adjustments, the algorithm's performance may be increased and it can be used for such kind of problems in the future.

References

1. Ghosh, M., Begum, S., Sarkar, R., Chakraborty, D., Maulik, U.: Recursive memetic algorithm for gene selection in microarray data. Expert Syst. Appl. **116**, 172–185 (2019)
2. Ving, D., Lam, C.: Gene selection using a hybrid memetic and nearest shrunken centroid algorithm. In: Proceedings of the 9th International Joint Conference on Biomedical Engineering Systems and Technologies, pp. 190–197, Rome (2016)

3. Cai, R., Hao, Z., Yang, X., Wen, W.: An efficient gene selection algorithm based on mutual information. Neurocomputing **72**(4–6), 991–999 (2009)
4. Alshamlan, H.M., Badr, G.H., Alohali, Y.A.: Genetic bee colony (GBC) algorithm: a new gene selection method for microarray cancer classification. Comput. Biol. Chem. **56**, 49–60 (2015)
5. Alshamlan, H.M., Badr, G.H., Alohali, Y.A.: mRMR-ABC: a hybrid gene selection algorithm for cancer classification using microarray gene expression profiling. Biomed. Res. Int. **2015**, 15 (2015)
6. Prasad, Y., Biswas, K.K., Hanmandlu, M.: A recursive PSO scheme for gene selection in microarray data. Appl. Soft Comput. **71**, 213–225 (2018)
7. Chuang, L.Y., Chang, H.W., Tu, C.J., Yang, C.H.: Improved binary PSO for feature selection using gene expression data. Comput. Biol. Chem. **32**(1), 29–38 (2008)
8. Li, S., Wu, X., Tan, M.: Gene selection using hybrid particle swarm optimization and genetic algorithm. Soft. Comput. **12**(11), 1039–1048 (2008)
9. De Castro, L.N., Von Zuben, F.J.: The clonal selection algorithm with engineering applications. In: Proceedings of GECCO, pp. 36–39, Nevada (2000)
10. Babayigit, B., Akdagli, A., Guney, K.: A clonal selection algorithm for null synthesizing of linear antenna arrays by amplitude control. J. Electromagn. Wave **20**, 1007–1020 (2002)
11. Gao, S., Dai, H., Yang, G., Tang, Z.: A novel clonal selection algorithm and its application to traveling salesman problem. IEICE Trans. Fundam. Electron. Commun. Comput. Sci. 90 (10), 2318–2325 (2007)
12. Deniz, E., Ülker, S.: Clonal selection algorithm application to simple microwave matching network. Microw. Opt. Technol. Lett. **53**(5), 991–993 (2011)
13. De Castro, L.N., Von Zuben, F.J.: An evolutionary immune network for data clustering. In: Neural Networks Proceedings of Sixth Brazilian Symposium, pp. 84–89, Rio De Janeiro (2000)
14. Ülker, E.D.: An improved clonal selection algorithm using a tournament selection operator and its application to microstrip coupler design. Turkish J. Electr. Eng. Comput. Sci. **25**(3), 1751–1761 (2017)
15. Blickle, T., Thiele, L.: A mathematical analysis of tournament selection. In: Proceedings of the 6th International Conference on Genetic Algorithms, pp. 9–16, Pittsburgh (1995)
16. Blickle, T.: Tournament selection. Evol. Comput. **1**, 181–186 (2000)
17. Miller, B.L., Goldberg, D.E.: Genetic algorithms, tournament selection, and the effects of noise. Complex Syst. **9**, 193–212 (1995)
18. Li, G., Wang, Q., Du, Q.: Improved harmony search algorithms by tournament selection operator. In: IEEE Congress on Evolutionary Computations, pp. 3116–3123, Sendai (2015)
19. Qu, B.Y., Suganthan, P.N.: Novel multimodal problems and differential evolution with ensemble of restricted tournament selection. In: IEEE Congress on Evolutionary Computation, pp. 1–7, Barcelona (2010)
20. Angeline, P.J.: Using selection to improve particle swarm optimization. In: IEEE International Conference on Evolutionary Computation Proceedings, pp. 84–89, Alaska (1998)
21. Fix, E., Hodges, J.L.: Discriminatory analysis-nonparametric discrimination: consistency properties. California University Berkeley (1951)
22. Statnikov, A., Aliferis, C.F., Tsamardinos, I., Hardin, D., Levy, S.: A comprehensive evaluation of multicategory classification methods for microarray gene expression cancer diagnosis. Bioinformatics **21**(5), 631–643 (2004)

Improvement for Traditional Genetic Algorithm to Use in Optimized Path Finding

Hasan Alp Zengin and Ali Hakan Işik(✉)

Burdur Mehmet Akif Ersoy University, Burdur, Turkey
hasalp38@gmail.com, ahakan@mehmetakif.edu.tr

Abstract. Genetic algorithm tries to find the optimized solution with different process stages. All stages are inspired by the natural mechanisms with the genes as individuals. Modelling that natural loop in Computer systems to find the optimized populations which is various combinations of genes, provide a good method to find a solution for problems that can't solve with any mathematical definition. Today, genetic algorithm is using for diverse fields like path finding, robotic, medical, network, big data and so more. In this work, genetic algorithm improved for path finding methods. All stages are examined and discussed to find possible improvements. A new step which is called as "Fate Decide Operator" is implemented and compared with traditional genetic algorithm. Fate decide algorithm's tests shows that the fate decide operator has some advantages for path finding algorithms. Improved genetic algorithm can be used in various problems.

Keywords: Genetic algorithm · Evolutionary · TSP

1 Introduction

Genetic algorithms are a diversity of search and optimization techniques that make use of fundamentals from Darwin's evolution theory to progress towards a solution. Although they just an approximation to real biological evolutionary process, the have been proven to solve problems powerfully. For example there isn't most usable methods to solve the optimal walking animation for a particular robot except genetic algorithms [3]. The prominent characteristic of GA is that it tests and manipulates a set of possible solutions simultaneously which assures that GA finds the optimal solution that cannot be found by "hill-climbing" search algorithms or "gradient descent" techniques in some cases like finding optimized route [4]. GAs apply the possible solution to various processes to find better solution. Algorithm imitates the process of natural selection where the optimal individuals are selected for reproduction of next generation [6]. Reaching to optimal result, actualized by dominating old generation with new generation. Populations are possible solution set that contains pre-defined

D. J. Hemanth and U. Kose (Eds.): ICAIAME 2019, LNDECT 43, pp. 473–483, 2020.
https://doi.org/10.1007/978-3-030-36178-5_37

chromosome count. Chromosomes represents different solutions of the problem. The main aim of genetic algorithm is finding the finest chromosome with trying variant gene sequences or values. Each individual represents a possible solution in a search space. The individuals in the population are then made to go through a process of evolution. Genetic algorithms are resembling of genetic structure and behavior of chromosomes within a population of individuals using the following basics:

- Individuals competes to be best in population
- Those individuals most successful in each competition will produce more efficient seeds than those individuals that perform poorly.
- Good individuals' genes, spread throughout the population so that two good parents will sometimes produce seeds that are better than themselves.
- With this way, each successive generation will become more proper to their environment.

Traditional genetic algorithm can be examined in 6 steps. Figure 1 shows the steps of traditional genetic algorithm mechanism.

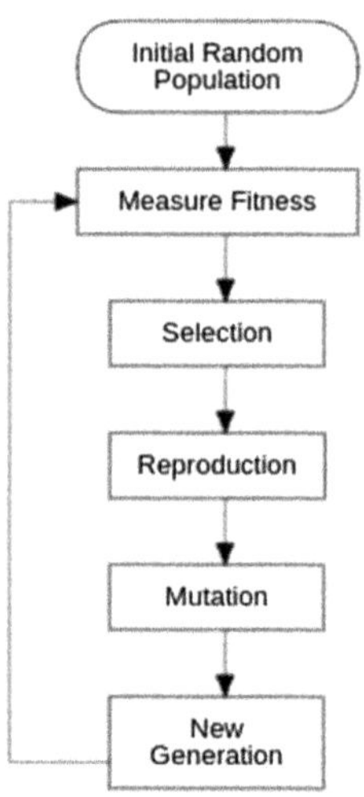

Fig. 1. Traditional genetic algorithm

- Initial Random Population: Populations contains chromosomes and chromosomes contains genes. To create random population, genetic algorithm uses different codding methods [7]. Sequenced permutation coding, value coding, tree coding and the most famous binary coding. Coding methods depends the problem.

- Measure Fitness: Finding a measurement for chromosomes is the mathematical part of genetic algorithm [5]. This measurement usually selects as a function. Each problem has own fitness function to find most optimized chromosome.
- Selection: Genetic algorithm has selection methods to match chromosomes each other with the purpose of getting ready for crossover operator. This operator mainly aims to increasing the change of gathering finest next generation.
- Crossover: Selected chromosomes has a contribution ratio for new generation. This process known as crossover rate in genetic algorithm. Mostly CR (Crossover Rate) selects between 0.6 and 0.95. that means %60 of selected chromosome can transfer its genes to next generation [1].
- Mutation: Mutation is a genetic operator used to ensure genetic variety from one generation of a population to the next with an probability. This probability should be set low otherwise the search will turn to a basic random search.

2 Methodology

That study aims implementing genetic algorithm for finding most optimized path. Finding most optimized path requires ad-hoc genetic algorithm parameters for coding, measurement, selection and crossover.

2.1 Initial Random Population

First step of genetic algorithm is Initialing random population. That step starts with preparing genes. Generally, genes have 4 different types [2].

- Sequenced Permutation Coding: Sequenced permutation is a rearrangement of the elements of an ordered list. That makes the diversity of possible order for each individual that has n element is n!.
- Value Coding: Each element of array represents with an fixed value.
- Tree Coding: In this method it is expressed as a tree structure consisting of each series of objects.
- Binary Coding: This method represents elements of array as binary digits. This is most optimized method if problem is relevant to apply.

Coordinate based applications are mostly compatible with sequenced permutation coding so chosen for the study. Each coordinate in path represents genes in genetic algorithm. In TSP case each chromosome has to include all genes. Because main objective is trying to find closest route to travel without missing any point.

Chosen solution for creating permutation with selected coordinates exposed in Fig. 2.

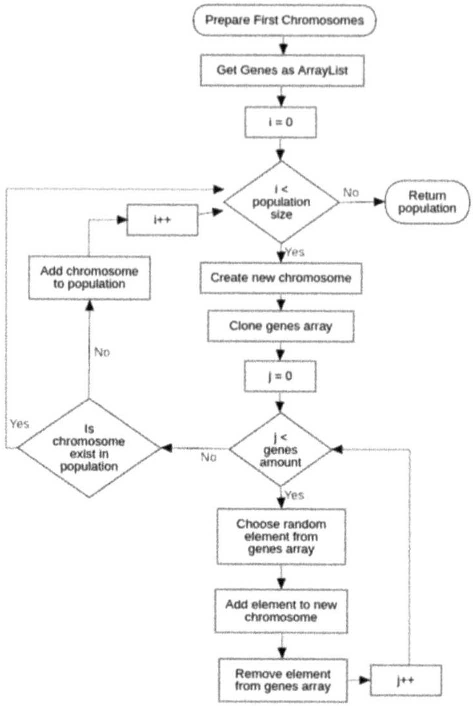

Fig. 2. Preparing first chromosomes algorithm

2.2 Measure Fitness

This study aims finding best route for selected coordinates. Actual focus is finding the shortest path between these coordinates. To find shortest path there is various type of distance calculation methods [9,10].

- Euclidean Distance: This simple metric is the straight-line distance between two points in space. This metric is commonly used in problems that distance is not rigorous parameter.
- Manhattan Distance: Based on the grid like street geography of the New York borough of Manhattan. Uses only vertical and horizontal distances to draw line between 2 point.
- Minkowski Distance: Minkowski is a metric in a normed vector space which can be considered as a combination of both the Euclidean and the Manhattan distance.
- Chebyshev Distance: Chebyshev distance is distance between vectors defined by the maximum difference between coordinate axes. It's also known as chessboard distance.

These 4 distance calculation methods are most generic metrics in coordinate systems. In this work euclidean metric is used to keep it simple. Measuring fitness value of chromosomes algorithm shown in Algorithm 1.

Set measurement to 0
for each gene in chromosome **do**
get first 2 genes from the chromosome
$x_{distance} = gene1[0] - gene2[0]$
$y_{distance} = gene1[1] - gene2[1]$
$distance = \sqrt{x^2_{distance} + y^2_{distance}}$
measurement = measurement + distance
remove first gene from chromosome
end for

Algorithm 1: Measure fitness value algorithm

2.3 Selection

Selection operator can be defined as selecting the best chromosome. This is simply tries to select 2 best chromosome to pass next operation "crossover". Selection operator can be considered as the imitation of natural selection. There is 6 genetic types of selection in genetic algorithm.

1. Natural (Sequenced) Selection: Random selection of chromosomes according to their fitness values.
2. Weighted Selection: After sorting operation according to their fitness values. N percentage of population selects without random variables.
3. Threshold Selection: A threshold value selects by user. The chromosomes that are less than this value die and the others continue to live.
4. Random Selection: Chromosomes select randomly without sorting.
5. Roulette Wheel Selection: Roulette wheel is one of most popular selection method. In this method chromosomes will have an percentage on wheel according to their fitness value. After that random value selects to find selected chromosome [11].
6. Tournament Selection: Chromosomes are grouped and eliminates it's opponent. This operation repeats until N chromosome selects.

In this work new selection method that can be called as evolved roulette wheel selection has used. Evolved roulette is not killing any chromosome. Instead of killing them it matches 2 chromosomes with each other. After selecting chromosome with roulette wheel method. It removes selected mother and co chromosome from next population and crosses them with the fair rate. That way increases the change to find optimized chromosome without taking risk to lost any good chromosome in population. Algorithm 2 shows our algorithm to prepare roulette wheel. After that operation a random value between 0 and 100 defining to choose a slice from roulette wheel.

Selection and crossover operation are consecutive in GAs. After selection done algorithm takes next chromosome as co-chromosome to crossover with selected mother crossover. This process loops until there is no left chromosome pair for crossover.

Set sum to 0
for each chromosome in population **do**
  $sum+ = chromosome'sfitnessvalue$
end for
set degree to 0
for each chromosome in population **do**
  $slice\_start\_degree[chromosome] = degree$
  $degree = (chromosome'sfitnessvalue/sum) * 100$
end for
return slice_start_degree

Algorithm 2: Prepare roulette wheel algorithm

2.4 Crossover

The aim is produce the child chromosomes by altering the ancestor chromosome to produce child chromosomes with higher suitability than parent chromosomes. Crossover operation can be categorized in 2 branch.

- Position based crossover
- Sequence based crossover

In position based crossover, a group of crossover point (one or more points) is randomly appointed on selected chromosomes to be crossed. The part of the second chromosome in these positions is placed in the respective positions of the first chromosome, and then the remaining genes are filled by taking the first chromosome genes respectively, which are not exist in the new chromosome. As different way in sequenced based crossover, a group of gene amount is constantly selected. The selected gene amount of mother chromosome directly includes to new generated chromosome. New chromosome's empty genes are filled by the co-chromosome with the same sequence. This study uses sequence based crossover to produce new generation. Selected chromosome and co-chromosome which comes from end of population are using to create new chromosome. %70 of selected chromosome directly passes its genes to next generation, rest of genes which is not included in chromosome comes from co-chromosome with the sequence. This algorithm exposed in Algorithm 3.

Crossover operator provides a chance to generate more optimized chromosome without losing selected chromosome's gene sequence. After child chromosomes generated with chromosome pairs' crossover, child chromosomes exposed to mutation.

2.5 Mutation

Mutation operation is exist in traditional genetic algorithm. Child chromosomes that comes after crossover operation are subjected to change with a certain chance [8]. Mutation operation has different application types depends the coding method of chromosome like swaps, inversion and scrambles. Improved algorithm utilizes swapping mutation. 2 random selected genes in chromosome swaps

```
for i=0; gene_amout * %70; i++ do
  append parentChromosomeGenes[i] to childChromosomeGenes
end for
for j=0; gene_amount; j++ do
  if childChromosomeGenes not contains coChromosomeGenes[j] then
    append coChromosomeGenes[j] to childChromosomeGenes
  end if
end for
return childChromosomeGenes
```

Algorithm 3: Crossover Algorithm

with each other with the chance of %1. Mutation operator algorithm exposed in Algorithm 4. Mutation operator visualized in Fig. 3. Afterwards mutation operation applied, algorithm applies new operation which decides fate of old chromosomes.

```
if mutationChange equals 0 then
  pos1 = random value between 0 and gene amount
  pos2 = random value between 0 and gene amount
  while pos1 equals pos2 do
    pos2 = random value between 0 and gene amount
  end while
  swap childChromosomeGenes[pos1] and childChromosomeGenes[pos2]
end if
return childChromosome
```

Algorithm 4: Mutation Algorithm

Mutation operator visualized in Fig. 3. Afterwards mutation operation applied, algorithm applies new operation which decides fate of old chromosomes.

2.6 Fate Decide

That operator is not exist in traditional genetic algorithm. In traditional algorithm new generated chromosomes has rights to swap because of elitism principle. In this algorithm additional fate decide operator decides who survives and dies. Most important thing about this operator is changing chromosome for next generation is not obligatory. If child chromosome('s) has worst fitness value than old chromosomes. Child chromosomes' can't stay alive for next generation. This operator allows not to loose possible good chromosomes. That means fitness values always go to good. Additionally fate decide operator needs extra fitness value measurement of child chromosomes to make that decide as seen in Fig. 4.

Fate Decide is the last operation of loop which is returns new generation. After first cycle completed, loop continues until stop condition provided.

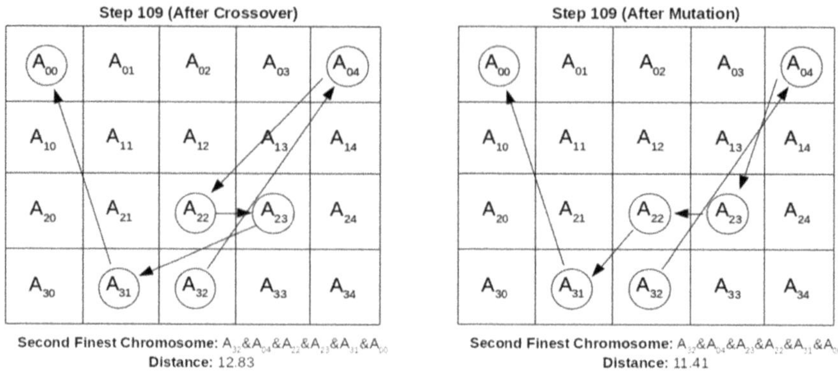

Fig. 3. Mutation operator for path finding algorithm

3 Results

Fate decide operation is the new approach in the genetic algorithm. That mechanism is alternative of elitism principle of genetic algorithm. Table 1 shows the comparing between elitism and fate decide operation.

Table 1. Fate decide and elitism operation comparison with step count

Gen count	Fate decide operation	Elitism
6	1388	4008
6	5229	1254
6	657	2557
7	607	1305
7	3136	6743
7	9501	12409
8	6935	6898
8	3180	7475
8	1747	7493
9	2051	5047
9	9979	11305
9	6229	2276

This test results obtained to solve and path finding problem. Results shows that the face decide algorithm is faster to find optimized solution than elitism. Randomized factors such as initializing first population also has effect over step count but tests done for multiple times and average values taken to eliminate random factors. In 12 result, fate decide operator won 9 of 12 against elitism

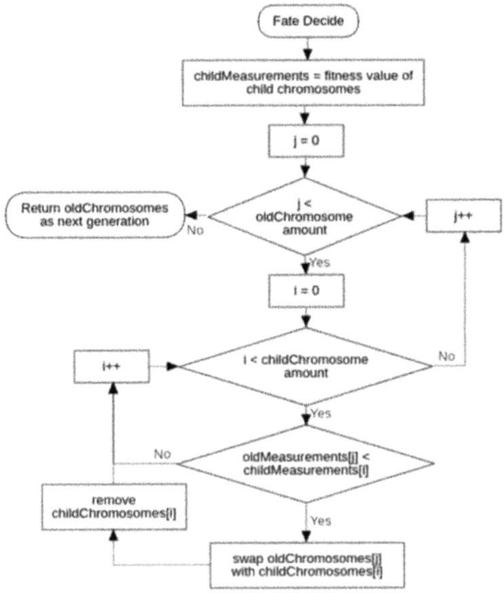

Fig. 4. Fate decide algorithm

Table 2. Calculation time and working time examples

Gen count	Optimized step	Max step
5	23	3124
5	151	3124
5	378	3124
6	750	7776
6	397	7776
6	694	7776
7	890	16808
7	2055	16808
7	3541	16808
8	1391	32768
8	4262	32768
8	5039	32768

operator. Improved algorithm loops until reach to $LoopCount = GenCount^5$ loop amount which is defined dynamically with genes amount to increase chance to reach optimum value. Reason of that is the test results which shows complexity of problem depends to quantity of genes. Disadvantage of this stop value is calculation time increasing exponential as seen in Table 2.

This algorithm also applied to a cleaner robot study that serves an GUI as seen in Fig. 5 to choose coordinates of room for simulating.

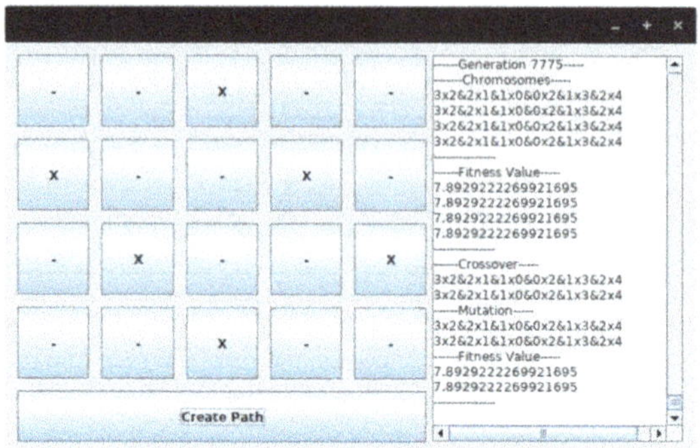

Fig. 5. Cleaner robot simulation

4 Conclusion

This study aimed modernizing genetic algorithm for optimized path finding algorithms. Traditional genetic algorithm was not enough to work in optimized path finding algorithm so traditional algorithm improved for coordinate based genetic algorithms. Traditional algorithm could kill the successful chromosome to continue with child chromosomes. Improved algorithm uses tolerable high mutation chance to prevent wasting step. Another advantage of this algorithm is possibility of finding alternative paths which has same fitness value.

Results shows optimized path finding in early steps. Code of that test published as open source.

To conclude up the research, comparison test shows the fate decide operation can be more preferable against elitism because of it finds the optimum value faster than %75 of tries. Especially Traveling salesman problems are fully compatible to this improvement.

References

1. Xu, J., Pei, L., Zhu, R.: Application of a genetic algorithm with random crossover and dynamic mutation on the travelling salesman problem. In: ICICT2018 (2018). https://doi.org/10.1016/j.procs.2018.04.230
2. Kumar, M.: Write a program to print all permutations of a given string. GeeksforGeeks (2016). https://www.geeksforgeeks.org/write-a-c-program-to-print-all-permutations-of-a-given-string/. Accessed 15 Aug 2018
3. Bräunl, T.: Embedded Robotics: Mobile Robot Design and Applications with Embedded Systems, 2nd edn. Springer, Heidelberg (2006). https://doi.org/10.1007/3-540-34319-9

4. Lin, F., Yang, Q.: Improved genetic algorithm operator for genetic algorithm. J. Zhejiang Univ. Sci. (2018). https://doi.org/10.1016/j.proenv.2011.12.055
5. McCall, J.: Genetic algorithm for modelling and optimisation. J. Comput. Appl. Math. (2005). https://doi.org/10.1016/j.cam.2004.07.034
6. Haldurai, L., Madhubala, T., Rajalakshmi, R.: A study on genetic algorithm and its applications. Int. J. Comput. Sci. Eng. **4**(10), 139–143 (2016)
7. Maaranen, H., Miettinen, K., Penttinen, A.: On initial populations of a genetic algorithm for continuous optimization problems. J. Glob. Optim. (2017). https://doi.org/10.1007/s10898-006-9056-6
8. Greenwell, R.N., Angus, J.E., Finck, M.: Optimal mutation probability for genetic algorithms. Math. Comput. Model. (1995). https://doi.org/10.1016/0895-7177(95)00035-Z
9. Shahid, R., Bertazzon, S., Ghali, W.A.: Comparison of distance measures in spatial analytical modeling for health service planning. BMC Health Serv. Res. (2009). https://doi.org/10.1186%2F1472-6963-9-200
10. Kim, Y., Moon, B.: Distance measures in genetic algorithms. Gen. Evol. Comput. GECCO (2004). https://doi.org/10.1007/978-3-540-24855-2_43
11. Saini, N.: Review of selection methods in genetic algorithms. Int. J. Eng. Comput. Sci. (2016) https://doi.org/10.18535/ijecs/v6i12.04

Investigation of the Most Effective Meta-Heuristic Optimization Technique for Constrained Engineering Problems

Hamdi Tolga Kahraman(✉) and Sefa Aras

Software Engineering Department, Karadeniz Technical University, Trabzon, Turkey
{htolgakahraman, sefaaras}@ktu.edu.tr

Abstract. One of the most common areas of meta-heuristic search (MHS) algorithms is optimization problems. In addition, the performance of only a few of the hundreds of MHS algorithms in the literature is known for constrained engineering design problems. The reason for this is that in most of the studies in which MHS algorithms have been developed, only classical benchmark problems are used to test the performance of the algorithms. Besides, applying MHS techniques to engineering problems is a costly and difficult process. This clearly demonstrates the importance of investigating the performance of new and powerful MHS techniques in engineering problems. In this paper, we investigate the search performance of the most recent and powerful MHS techniques in the literature on constrained engineering problems. In experimental studies, 20 different MHS techniques and five constrained engineering problems most commonly used in the literature have been used. Wilcoxon Runk Sum Test was used to compare the performance of the algorithms. The results show that the performance of MHS algorithms in classical benchmark problems and their performance in constrained engineering problems do not exactly match.

Keywords: Meta-heuristic search · Constrained engineering problems · Optimization

1 Introduction

In the last decade, great progress has been made in meta-heuristic search (MHS) algorithms. Hundreds of new algorithms have been developed and these algorithms have been tested in dozens of different benchmark problems. As a result of studies on meta-heuristic search algorithms, more successful solutions can be obtained for many optimization problems. The benchmark problems used to test search performance of MHS algorithms are usually classified into four categories according to their types. These are unimodal, multimodal, hybrid and composition respectively. The search performance of the algorithms is evaluated depending on the type of the problems [1–4]. Unimodal problems are used to test the exploitation capability of algorithms. Multimodal problems are used to measure the exploration capability of algorithms. Hybrid problems are used to test the balanced search capabilities of algorithms and

D. J. Hemanth and U. Kose (Eds.): ICAIAME 2019, LNDECT 43, pp. 484–501, 2020.
https://doi.org/10.1007/978-3-030-36178-5_38

composition problems to test the stability of algorithms in complex search spaces. These four problem types are continuous and unconstrained optimization examples. One of the optimization problems often encountered in real world applications is constrained engineering problems. Although many MHS algorithms have been developed in the literature, the number of algorithms tested in constrained optimization problems is low. This indicates the need to conduct research on constrained engineering optimization problems. In this paper, a comprehensive study is conducted to investigate the performance of MHS algorithms on constrained engineering real-world optimization problems. For this purpose, experimental studies were carried out by using 4 engineering problems and 20 MHS algorithm which is current and powerful developed in recent years. Wilcoxon runk sum test was used to analyze the results obtained from experimental studies [5–7]. The sections of the article are arranged as follows. Studies on constrained optimization problems are given in the second section. Engineering problems and MHS algorithms used in the paper are given in the third section. The comparison of competing algorithms and the performance of algorithms on engineering problems are given in fourth section.

2 Related Study

This section provides a literature review on engineering design problems optimized by meta-heuristic search algorithms. For this purpose, a review table was prepared with up-to-date studies. The engineering design problems commonly used in the literature and the features of them are given in Table 1. The MHS algorithms used in the optimization of the problems, the optimum solutions obtained and the best fitness values are given in the fourth, fifth and sixth columns of Table 1, respectively. Butterfly optimization algorithm (BOA) was used to solve three classical engineering problems spring design, welded beam design, and gear train design, respectively. The optimization results of BOA for the problem spring design are compared with seven MHS techniques. These are Grey Wolf Optimization (GWO) [8], Gravitational Search Algorithm (GSA) [9], PSO [10], ES [11], GA [12], HS [13], and DE [14]. Additionally, the mathematical approaches (numerical optimization technique) were used to solve same problems. For the welded beam design problem, the BOA [15] algorithm was compared with the GWO [8], GSA [9], HS [13] and 7 other competing methods. For the gear train design problem, the BOA algorithm was compared with the CAPSO [16], CS [17], PSO [18], GeneAS [19], SA [20] and 6 other competing methods. The optimum results on the three problems have also obtained the BOA algorithm. The main handicaps in this study are the lack of comparisons with current and strong algorithms and the selection of different algorithms as competitors in all three problems. The method of obtaining the results of competing algorithms is also inappropriate. Performing benchmarking on the results of experimental studies not performed under the same conditions can be misleading. Another study on constrained engineering optimization problems is weighted differential evolution algorithm

(WDE) [21]. In the solution of design problems pressure-vessel, speed-reducer, tension/compression string, and welded-beam, PSO2011 [22], CPI-JDE [23], Advanced Artificial Cooperative Search Algorithm [24], CS [17], ABC [25], JADE [26], BSA [27], and WDE were compared. As a result of experimental study, the mean-solution and standard deviation of mean-solution values were obtained. Table 1 shows the average values of the best results of WDE from 30 independent runs for four design problems. Another study applied constrained engineering problems is the Spider Monkey Optimization (SMO) with Nelder–Mead (NM) method [28]. It includes three design problems. These are welded beam, pressure vessel, and tension/compression spring. The proposed SMONM algorithm was compared its results with thirteen algorithms that are from previous studies. It showed a more successful performance than the competing algorithms in which two of the three problems were compared. In the spring design problem, it was the second after the SC-PSO [29] algorithm. ISA [30] algorithm was also tested on five constrained mechanical design problems. These problems are design of a gear system, pressure vessel, welded beam, tension/compression spring, and speed reducer. In Gear design problem, ISA method is compared with 6 competitor algorithm. The ISA performed similarly with 4 competing algorithms and found a better solution than the TLPSO [31] and PSOTC [32] algorithms. In the pressure vessel problem, ISA method was compared with 10 competing algorithms. ISA has found a better solution than competing algorithms. In the welded beam problem, ISA method was compared with 9 competing algorithms. Competing algorithms ABC [25], ES [11], UPSO [33], PSO-DE [34], MBA [35] and Elitist TLBO [36] found better solutions than ISA. In the speed reducer problem, ISA has established superiority to all of its competitors. In the speed reducer problem, ISA has established superiority to all of its competitors. In the tension/compression spring problem, ISA obtained similar results with competitors Elitist TLBO, MBA, ABC, PSO-DE, and MDE [37]. IAFOA [38] is an improved fruit fly optimization algorithm. It was used to optimize three engineering design problems, which are oil compression spring, welded beam and speed reducer. IAOFA was compared the five variants of FOA and the other five algorithms IASFA [39], CLPSO [40], DLHS [41], ALO [42], and GWO [8]. For the three problems, IAOFA performed better than other comparative algorithms. HFPSO [43] is a hybrid algorithm of firefly and particle swarm optimization methods. It was used to optimize three engineering design problems, which are the pressure vessel, the welded beam and the tension-compression spring. It was compared with sixteen competitors. For the pressure vessel design problem, EPSO [44] and HFPSO obtained best solution. For welded beam and spring design problems, it has obtained worse results than competing algorithms. EPO [45] is a bio-inspired algorithm, which mimics the huddling behavior of emperor penguins. It was compared eight competing algorithms which are SHO [46], GWO [8], PSO [18], MVO [47], SCA [48], GSA [9], GA [12], and HS [13]. It obtained best solution for the design problems, which are pressure vessel and speed reducer, welded beam, and tension/compression spring. MBHS (mine blast harmony search) algorithm [49] was developed to improve the exploration and exploitation capabilities of the HS [13] and MBA [50]. It was tested on constrained design problems, which are pressure vessel, welded beam, and speed reducer and compared with algorithms, which are PSO-DE [51], ABC [25], WCA [52], TLBO [53], MBA [35], and HS [13]. MBHS was reportedly outperformed its competitors in all problems. VPL (Volleyball Premier League) algorithm [54] was applied to solve three classical engineering design optimization problems. It performed better results from its competitors for two design problems. These are compression spring and pressure vessel. For the welded beam design problem, the competitors of it performed better performance. iDEaSM [55] is a modified version of differential evaluation algorithm. Two engineering optimization problems were used to test the it which are welded beam and pressure vessel. For the welded beam design problem it achieved better performance its five competitors and worse performance from one competitor. For pressure vessel, it achieved better performance than all competitors (Table 2).

Table 1. Engineering optimization problems.

No.	Design problems	Variables	Variable range
P_1	Spring (Tension/compression string)	x_1 = wire diameter (d), x_2 = mean coil diameter (D), x_3 = the number of active coils (P)	$0.05 \leq x_1 \leq 2.00$, $0.25 \leq x_2 \leq 1.30$, $2.00 \leq x_3 \leq 15.00$
P_2	Pressure Vessel	T_s (x_1, thickness of the shell), T_h (x_2, thickness of the head), R (x_3, inner radius) and L (x_4, length of the cylindrical section of the vessel	$0 \leq (x_1, x_2) \leq 100$, $10 \leq (x_3, x_4) \leq 200$
P_3	Speed-reducer	x_1: face width, x_2: module of teeth, x_3: number of teeth on pinion, x_4: length of the first shaft between bearings, x_5: length of the second shaft between bearings, x_6: diameter of the first shaft, x_7: diameter of the first shaft	$2.6 \leq x_1 \leq 3.6$, $0.7 \leq x_2 \leq 0.8$, $17 \leq x_3 \leq 28$, $7.3 \leq x_4 \leq 8.3$, $7.8 \leq x_5 \leq 8.3$, $2.9 \leq x_6 \leq 3.9$, $5.0 \leq x_7 \leq 5.5$
P_4	Welded Beam	h: the thickness of weld l: the length of attached part of the bar t: the height of the bar b: thickness of the bar	$0.1 \leq (h, b) \leq 2$, $0.1 \leq (l, t) \leq 10$
P_5	Gear Train	T_a, T_b, T_d, T_f	$12 \leq T_{a,b,d,f} \leq 60$

Table 2. Engineering optimization problems' results.

No.	Algorithms	Optimum solution	Best fitness
P_1	BOA [15]	*d*: 0.0513, *D*: 0.3348, *P*: 12.9227	0.0119
	VPL [54]	*d*: 0.0501, *D*: 0.3316, *P*: 12.8342	0.0123
	SMONM [28]	*d*: 0.0519, *D*: 0.3622, *P*: 11.4386	0.0126
	ISA [30]	*d*: 0.0516, *D*: 0.3555, *P*: 11.3584	0.0126
	EPO [45]	*d*: 0.0510, *D*: 0.3429, *P*: 12.0898	0.0126
	WDE [21]	–	0.0126
	IAFOA [38]	–	0.0126
	HFPSO [43]	–	0.0128
P_2	EPO	x_1: 0.7780, x_2: 0.3832, x_3: 40.3151, x_4: 200.0000	5580.0700
	ISA	x_1: 0.7782, x_2: 0.3846, x_3: 40.3221, x_4: 199.9651	5884.8400
	WDE	–	5885.3327
	HFPSO	–	5885.3328
	SMONM	x_1: 0.7783, x_2: 0.3847, x_3: 40.3275, x_4: 199.8889	5885.5950
	MBHS [49]	–	5885.8640
	iDEaSm [55]	x_1: 0.7781, x_2: 0.3846, x_3: 40.3210, x_4: 199.9802	5887.1083
	VPL	x_1: 0.8152, x_2: 0.4265, x_3: 42.0912, x_4: 176.7423	6044.9565
P_3	EPO	x_1: 3.5012, x_2: 0.7000, x_3: 17.0000, x_4: 7.3000 x_5: 7.8000, x_6: 3.3342, x_7: 5.2653	2994.2472
	ISA	x_1: 3.4999, x_2: 0.7000, x_3: 17.0000, x_4: 7.3000 x_5: 7.7153, x_6: 3.3502, x_7: 5.2866	2994.4685
	MBHS	–	2994.4746
	WDE	–	2994.9252
	AFOA	–	2996.3480
P_4	BOA	*h*: 0.1736, *l*: 2.9690, *t*: 8.7637, *b*: 0.2188	1.6644
	EPO	*h*: 0.2054, *l*: 3.4723, *t*: 9.0352, *b*: 0.2011	1.7235
	SMONM	*h*: 0.2057, *l*: 3.4704, *t*: 9.0366, *b*: 0.2057	1.7248
	ISA	*h*: 0.2057, *l*: 3.4704, *t*: 9.0366, *b*: 0.2057	1.7248
	iDEaSm	*h*: 0.2057, *l*: 3.4704, *t*: 9.0366, *b*: 0.2057	1.7248
	IAFOA	–	1.7248
	HFPSO	–	1.7248
	WDE	–	1.7248
	MBHS	–	1.7248
	VPL	*h*: 0.2152, *l*: 6.8989, *t*: 8.8150, *b*: 0.2162	2.2647
P_5	BOA	T_a: 43, T_b: 16, T_d: 19, T_f: 49	2.7×10^{-12}
	ISA	T_a: 43, T_b: 16, T_d: 19, T_f: 49	2.7×10^{-12}

3 Method

In order to obtain a detailed view on the performance of MHS algorithms in constrained engineering design problems, we chose 20 algorithms that are up-to-date and powerful. These problems are spring design (tension/compression string), pressure vessel design, speed-reducer design, and welded beam design.

3.1 Constrained Engineering Design Problems

Problem 1. Tension/Compression Spring Design: The main objective of this problem is to minimize the weight of spring. There are three variables in the definition of problem. These are the wire diameter (*d*), mean coil diameter (*D*), and the number of active coils (*P*). In the optimization process, the problem constraints surge frequency, minimum deflection, and shear stress are considered.

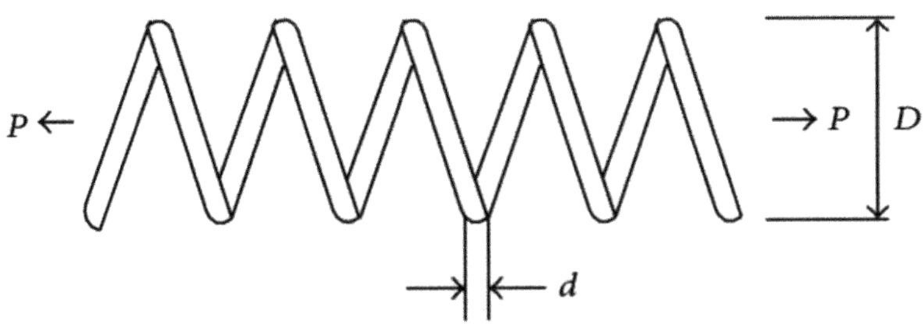

Fig. 1. Tension/compression spring design problem [56].

Figure 1 shows the schematic illustration of the problem. The ranges of variables are given in Table 1. The mathematical formulation of objective function and constraints of this problem are described as follows:

$$\begin{aligned}
&\textit{Consider } \vec{x} = [x_1 x_2 x_3] = [dDP], \qquad \textit{Minimize } f(\vec{x}) = (x_3 + 2)x_2 x_1^2, \\
&\textit{Subject to}: \\
&g_1(\vec{x}) = 1 - \frac{x_2^2 x_3}{71785 x_1^4} \le 0, \qquad g_2(\vec{x}) = \frac{4x_2^2 - x_1 x_2}{12566(x_2 x_1^3 - x_1^4)} + \frac{1}{5108 x_1^2} \le 0, \\
&g_3(\vec{x}) = 1 - \frac{140.45 x_1}{x_2^2 x_3} \le 0, \qquad g_4(\vec{x}) = \frac{x_1 + x_2}{1.5} - 1 \le 0, \\
&\textit{where}: 0.05 \le x_1 \le 2.0, 0.25 \le x_2 \le 1.3, 2.0 \le x_3 \le 15.0
\end{aligned} \tag{1}$$

Problem 2. Pressure Vessel Design: The pressure vessel problem is a structural engineering optimization problem including the cost of the materials, forming and welding. Please see the Fig. 2. for schematic illustration of it and its features. There are four design variables: thickness of pressure vessel (T_s), thickness of head (T_h), inner radius of the vessel (R) and length of the vessel (L) without the heads. The ranges of design variables are given in Table 1.

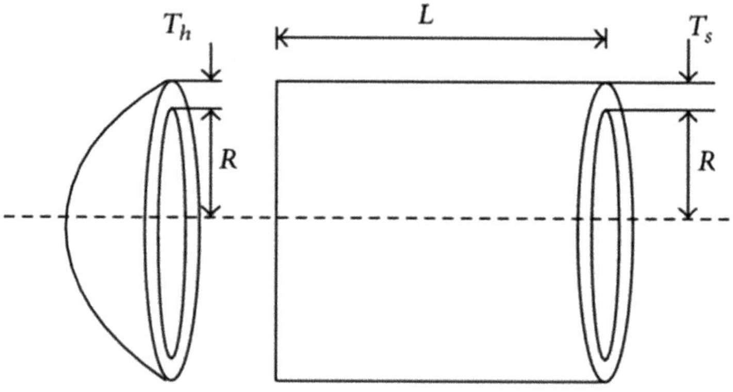

Fig. 2. Pressure vessel design problem [57]

The mathematical formulation of objective function and constraints of this problem are described as follows:

$$
\begin{aligned}
&Consider\ \vec{x} = [x_1 x_2 x_3 x_4] = [T_s T_h R L],\\
&Minimize\ f(\vec{x}) = 0.6224 x_1 x_3 x_4 + 1.7781 x_2 x_3^2 + 3.1661 x_1^2 x_4 + 19.84 x_1^2 x_3,\\
&Subject\ to:\\
&g_1(\vec{x}) = 0.0193 x_3 - x_1 \le 0, \qquad g_2(\vec{x}) = 0.00954 x_3 - x_2 \le 0,\\
&g_3(\vec{x}) = 1296000 - \pi x_3^2 x_4 - \frac{4}{3}\pi x_3^3 \le 0, \qquad g_4(\vec{x}) = x_4 - 240 \le 0,\\
&where: 1 * 0.0625 \le x_1, x_2 \le 99 * 0.0625,\ 10.0 \le x_3 x_4 \le 200.0
\end{aligned}
\tag{2}
$$

Problem 3. Welded Beam Design: The welded beam problem is a structural engineering optimization problem. The objective of this problem is to find the best dimensions for design variables b, t, h and l which are used to move a P load and support the minimum production cost. Please see the Fig. 3. for schematic illustration of it and its features. There are four design variables: the thickness of weld (h), the length of attached part of the bar (l), the height of the bar (t), and thickness of the bar (b). The ranges of design variables are given in Table 1.

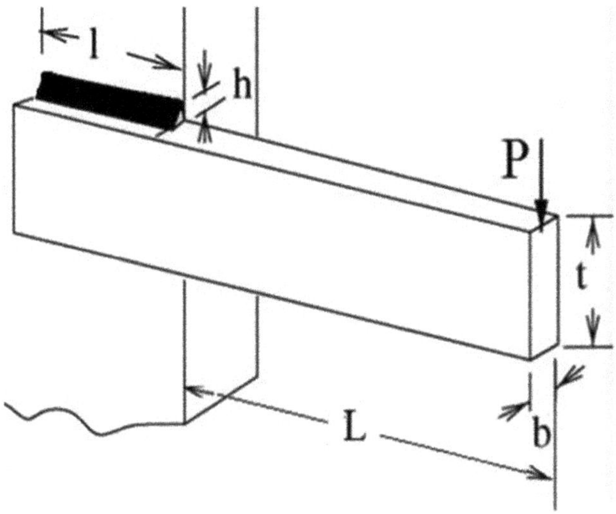

Fig. 3. Welded beam design problem [58]

The mathematical formulation of objective function and constraints of this problem are described as follows:

$$
\begin{aligned}
& \textit{Consider } \vec{x} = [x_1 x_2 x_3 x_4] = [hltb], \\
& \textit{Minimize } f(\vec{x}) = 1.10471 x_1^2 x_2 + 0.04811 x_3 x_4 (14.0 + x_2), \\
& \textit{Subject to}: \\
& g_1(\vec{x}) = \tau(\vec{x}) - 13600 \le 0, \quad g_2(\vec{x}) = \sigma(\vec{x}) - 30000 \le 0, \\
& g_3(\vec{x}) = \delta(\vec{x}) - 0.25 \le 0, \quad g_4(\vec{x}) = x_1 - x_4 \le 0, \\
& g_5(\vec{x}) = 6000 - P_c(\vec{x}) \le 0, \quad g_6(\vec{x}) = 0.125 - x_1 \le 0, \\
& g_7(\vec{x}) = 1.10471 x_1^2 + 0.04811 x_3 x_4 (14.0 + x_2) - 5.0 \le 0, \\
& \textit{where}: 0.1 \le x_1, 0.1 \le x_2, x_3 \le 10.0, x_4 \le 2.0, \\
& \tau(\vec{x}) = \sqrt{(\tau')^2 + (\tau'')^2 + \frac{(x_2 \tau' \tau'')}{\sqrt{0.25\left(x_2^2 + (x_1 + x_3)^2\right)}}}, \\
& \tau' = \frac{6000}{\sqrt{2} x_1 x_2}, \sigma(\vec{x}) = \frac{504000}{x_3^2 x_4}, \delta(\vec{x}) = \frac{65856000}{(30 * 10^6) x_4 x_3^3}, \\
& \tau'' = \frac{6000(14 + 0.5 x_2)\sqrt{0.25\left(x_2^2 + (x_1 + x_3)^2\right)}}{2\left[0.707 x_1 x_2 \left(\frac{x_2^2}{12} + 0.25 (x_1 + x_3)^2\right)\right]}, \\
& P_c(\vec{x}) = 64746.022(1 - 0.0282346 x_3) x_3 x_4^3
\end{aligned}
\tag{3}
$$

Problem 4. Speed Reducer Design: The speed reducer design is a minimization problem. The objective of this problem is to find the minimum weight of speed reducer. It has seven design variables. These are the face width (x_1), module of teeth (x_2), number of teeth in the pinion (x_3), length of the first shaft between bearings (x_4), length

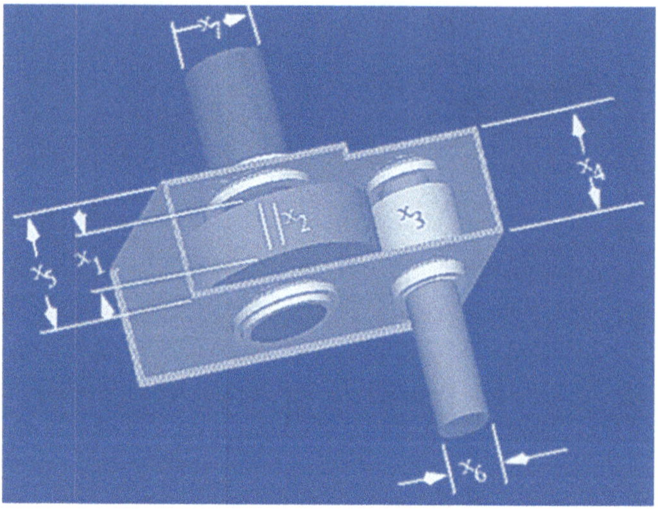

Fig. 4. Speed reducer design problem [59]

of the second shaft between bearings (x_5), and the diameters of the two shafts (x_6, x_7). Please see the Fig. 4. for schematic illustration of it and its features.

The mathematical formulation of objective function and constraints of this problem are described as follows:

$$\begin{aligned}
&\textit{Minimize}\, f(\vec{x}) = 0.7854x_1x_2^2\left(3.3333x_3^2 + 14.9334x_3 - 43.0934\right) \\
&\qquad - 1.508x_1\left(x_6^2 + x_7^2\right) + 7.4777\left(x_6^3 + x_7^3\right) \\
&\qquad + 0.7854\left(x_4x_6^2 + x_5x_7^2\right), \\
&\textit{Subject to}: \\
&g_1(\vec{x}) = \frac{27}{x_1x_2^2x_3} - 1 \leq 0, g_2(\vec{x}) = \frac{397.5}{x_1x_2^2x_3^2} - 1 \leq 0, \\
&g_3(\vec{x}) = \frac{1.93x_4^3}{x_2x_6^4x_3} - 1 \leq 0, g_4(\vec{x}) = \frac{1.93x_5^3}{x_2x_7^4x_3} - 1 \leq 0, \\
&g_5(\vec{x}) = \frac{\sqrt{\left[(745(x_4/x_2x_3))^2 + 16.9 * 10^6\right]}}{110x_6^3} - 1 \leq 0, \\
&g_6(\vec{x}) = \frac{\sqrt{\left[(745(x_5/x_2x_3))^2 + 157.5 * 10^6\right]}}{85x_7^3} - 1 \leq 0, \\
&g_7(\vec{x}) = \frac{x_2x_3}{40} - 1 \leq 0, g_8(\vec{x}) = \frac{5x_2}{x_1} - 1 \leq 0, g_9(\vec{x}) = \frac{x_1}{12x_2} - 1 \leq 0, \\
&g_{10}(\vec{x}) = \frac{1.5x_6 + 1.9}{x_4} - 1 \leq 0, g_{11}(\vec{x}) = \frac{1.1x_7 + 1.9}{x_5} - 1 \leq 0, \\
&\textit{where}: 2.6 \leq x_1 \leq 3.6, 0.7 \leq x_2 \leq 0.8, 17 \leq x_3 \leq 28, 7.3 \leq x_4 \leq 8.3, \\
&7.3 \leq x_5 \leq 8.3, 2.9 \leq x_6 \leq 3.9, 5.0 \leq x_7 \leq 5.5
\end{aligned} \tag{4}$$

Problem 5. Gear Train Design: The objective of this design problem is to explore the optimal number of tooth for four gears (T_a, T_b, T_d, T_f) of a train to minimize the gear ratio. The mathematical formulation of objective function and constraints of this problem are described as follows:

$$\begin{aligned} &Consider\ \vec{x} = [x_1 x_2 x_3 x_4] = \left[T_a T_b T_d T_f\right], \\ &Minimize\ f(\vec{x}) = \left(\frac{1}{6.931} - \frac{x_2 x_3}{x_1 x_4}\right)^2, where: 12 \le x_1, x_2, x_3, x_4 \le 60 \end{aligned} \tag{5}$$

For detailed information about design problems, please follow the reference studies [15, 21, 28, 30, 38, 43, 49, 54–59].

3.2 Competing Algorithms

The competing algorithms are the PSO (Particle Swarm Optimization, 1993) [10], ABC (Artificial Bee Colony, 2007) [25], GSA (Gravitational Search Algorithm, 2009) [9], DSA (Differential Search Algorithm, 2012) [64], BSA (Backtracking Search Algorithm, 2013) [27], SOS (Symbiotic Search Algorithm, 2014) [62], GWO (Grey Wolf Optimizer, 2014) [8], MFO (Moth-flame Optimization, 2015) [65], LSA (Lightning Search Algorithm, 2015) [60], CSA (Crow Search Algorithm, 2016) [61], SCA (Sine Cosine Algorithm, 2016) [48], WOA (Whale Optimization Algorithm, 2016) [66], CKGSA (Chaotic kbest gravitational search algorithm, 2016) [69], SSA (Salp Swarm Algorithm, 2017) [67], CGSA (Chaotic gravitational constants for the gravitational search algorithm, 2017) [72], TLABC (Teaching Learning Based Artificial Bee Colony, 2018) [63], MS (Moth Search, 2018) [68], COA (Coyote Optimization Algorithm, 2018) [70], BOA (Butterfly optimization algorithm, 2019) [15], ASO (Atom Search Algorithm, 2019) [71].

4 Experimental Study

In this section, parameter settings and experimental study results of algorithms are given.

4.1 Settings

See Table 3.

Table 3. Parameter settings for algorithms.

Algorithm	Parameter settings
SOS	Ecosystem size = 50
GWO	Number of search agents = 30
PSO	Swarm size = 30, cognitive constant = 2, social constant = 2
COA	$N_p = 20$, $N_c = 5$, number of coyotes = $N_p * N_c$
CSA	Flock size = 20, awareness probability = 0.1, flight length = 2
BSA	Population size = 30, mixrate = 1, F = 3 * randn
DSA	Superorganism size = 30, $p_1 = 0.3 * \text{rand}$, $p_2 = 0.3 * \text{rand}$
GSA	Population size = 50, initial gravitational constant = 100, decreasing coefficient = 20
CKGSA	Population size = 50, initial gravitational constant = 100, decreasing coefficient = 20, biotic potential = 4
CGSA	Population size = 50, initial gravitational constant = 100, decreasing coefficient = 20, chaos index = 2
TLABC	Colony size = 50, limit = 200, scale factor = rand
SSA	Salp population size = 30
WOA	Number of search agents = 30
SCA	Number of search agents = 30, a = 2
MFO	Number of moths = 30
LSA	Population size = 50, channel time = 10
MS	Population size = 50, number of kept moths at each generation = 2, $\beta = 2.5$, max walk step = 1, acceleration factor = $5^{1/2} - 1$
BOA	Population size = 50, modular modality = 0.01, p = 0.8
ASO	Number of atom population = 50, $\alpha = 50$, $\beta = 0.2$
ABC	Colony size = 50, limit = 100, number of food source = colony size/2

4.2 Results of Experimental Study

This section presents the performance of 20 competing MHS algorithms on spring, pressure vessel, speed reducer, welded beam, and gear train design problems, respectively. The values of design variables, numerical results, algorithm rankings, and algorithms' optimal solutions for problems are presented in Tables 4, 5, 6, 7 and 8.

Table 4. Comparison results for tension/compression spring design problem.

Algorithm	*d*	*D*	*P*	Cost	Rank
TLABC	0,0500	0,4036	6,8220	0,0089	1
MFO	0,0500	0,4036	6,8220	0,0089	1
PSO	0,0500	0,4036	6,8220	0,0089	2
COA	0,0500	0,4036	6,8220	0,0089	3
SOS	0,0500	0,4036	6,8220	0,0089	4
BSA	0,0500	0,4036	6,8220	0,0089	5

(continued)

Table 4. (*continued*)

Algorithm	d	D	P	Cost	Rank
SSA	0,0500	0,4036	6,8220	0,0089	6
LSA	0,0500	0,4036	6,8220	0,0089	7
CSA	0,0500	0,4036	6,8220	0,0089	8
MS	0,0500	0,4036	6,8220	0,0089	9
GWO	0,0500	0,4036	6,8227	0,0089	10
DSA	0,0500	0,4036	6,8235	0,0089	11
SCA	0,0500	0,4036	6,8305	0,0089	12
ABC	0,0500	0,4036	6,8332	0,0089	13
BOA	0,0500	0,4012	6,9521	0,0090	14
WOA	0,0504	0,4134	6,5599	0,0090	15
ASO	0,0513	0,4351	6,2078	0,0094	16
GSA	0,0529	0,4711	5,3923	0,0098	17
CKGSA	0,0520	0,4481	6,0908	0,0098	18
CGSA	0,0500	0,4035	7,9553	0,0100	19

Table 5. Comparison results for pressure vessel design problem.

Algorithm	T_s	T_h	R	L	Cost	Rank
MFO	0,7782	0,3846	40,3196	200,0000	5.885,3328	1
COA	0,7782	0,3846	40,3196	200,0000	5.885,3328	2
PSO	0,7782	0,3846	40,3196	200,0000	5.885,3328	3
SOS	0,7782	0,3846	40,3196	200,0000	5.885,3328	4
TLABC	0,7782	0,3846	40,3196	200,0000	5.885,3328	5
CSA	0,7782	0,3846	40,3196	200,0000	5.885,3337	6
BSA	0,7782	0,3847	40,3197	199,9995	5.885,3529	7
MS	0,7792	0,3852	40,3746	199,2363	5.887,1486	8
GWO	0,7790	0,3857	40,3616	199,4234	5.889,1658	9
ABC	0,7792	0,3854	40,3735	199,8786	5.901,4671	10
DSA	0,7900	0,3907	40,9208	191,7977	5.907,6935	11
SSA	0,7913	0,3912	41,0011	190,7255	5.908,2021	12
ASO	0,8022	0,3965	41,5662	183,3456	5.927,7419	13
LSA	0,8101	0,4004	41,9755	178,1664	5.942,2335	14
WOA	0,7938	0,4051	40,6795	195,0500	6.009,6241	15
SCA	0,8148	0,4003	41,8395	181,5057	6.029,5581	16
CKGSA	0,9016	0,4456	46,7136	127,2162	6.144,5625	17
GSA	0,9300	0,4597	48,1867	123,6532	6.512,3960	18
BOA	0,8238	0,6801	42,7364	178,4996	7.182,8614	19
CGSA	0,9954	1,7998	45,9159	136,4149	11.958,1658	20

Table 6. Comparison results for speed reducer design problem.

Algorithm	x_1	x_2	x_3	x_4	x_5	x_6	x_7	Cost	Rank
MFO	2,6000	0,7000	17,0000	7,3000	7,8000	2,9000	5,0000	2376,2280	1
COA	2,6000	0,7000	17,0000	7,3000	7,8000	2,9000	5,0000	2376,2280	1
PSO	2,6000	0,7000	17,0000	7,3000	7,8000	2,9000	5,0000	2376,2280	1
SOS	2,6000	0,7000	17,0000	7,3000	7,8000	2,9000	5,0000	2376,2280	1
CSA	2,6000	0,7000	17,0000	7,3000	7,8000	2,9000	5,0000	2376,2280	1
BSA	2,6000	0,7000	17,0000	7,3000	7,8000	2,9000	5,0000	2376,2280	1
ABC	2,6000	0,7000	17,0000	7,3000	7,8000	2,9000	5,0000	2376,2280	1
DSA	2,6000	0,7000	17,0000	7,3000	7,8000	2,9000	5,0000	2376,2280	1
SSA	2,6000	0,7000	17,0000	7,3000	7,8000	2,9000	5,0000	2376,2280	1
ASO	2,6000	0,7000	17,0000	7,3000	7,8000	2,9000	5,0000	2376,2280	1
LSA	2,6000	0,7000	17,0000	7,3000	7,8000	2,9000	5,0000	2376,2280	1
WOA	2,6000	0,7000	17,0000	7,3000	7,8000	2,9000	5,0000	2376,2280	1
SCA	2,6000	0,7000	17,0000	7,3000	7,8000	2,9000	5,0000	2376,2280	1
CKGSA	2,6000	0,7000	17,0000	7,3000	7,8000	2,9000	5,0000	2376,2280	1
GSA	2,6000	0,7000	17,0000	7,3000	7,8000	2,9000	5,0000	2376,2280	1
CGSA	2,6000	0,7000	17,0000	7,3000	7,8000	2,9000	5,0000	2376,2280	1
TLABC	2,6000	0,7005	17,0000	7,3942	7,9708	2,9000	5,0000	2381,9217	2
BOA	2,6071	0,7000	17,0540	7,7394	7,8986	2,9275	5,0019	2397,1255	3
GWO	2,6127	0,7002	17,0923	7,7936	8,1423	2,9392	5,0178	2403,1767	4
MS	2,6250	0,7000	17,1006	7,9145	8,0304	2,9572	5,0109	2425,0374	5

Table 7. Comparison results for welded beam design problem.

Algorithm	h	l	t	b	Cost	Rank
DSA	0,2444	6,2186	8,2915	0,2444	2,3811	1
PSO	0,2444	6,2186	8,2915	0,2444	2,3811	2
TLABC	0,2444	6,2186	8,2915	0,2444	2,3811	3
WOA	0,2444	6,2186	8,2915	0,2444	2,3811	4
MFO	0,2444	6,2186	8,2915	0,2444	2,3811	5
SOS	0,2444	6,2186	8,2915	0,2444	2,3811	6
CSA	0,2444	6,2186	8,2917	0,2444	2,3812	7
SCA	0,2444	6,2178	8,2927	0,2444	2,3812	8
COA	0,2443	6,2167	8,2948	0,2444	2,3815	9
SSA	0,2415	6,2900	8,3236	0,2442	2,3891	10
BSA	0,2418	6,2821	8,3254	0,2442	2,3893	11
CKGSA	0,2429	6,0857	8,4895	0,2430	2,3906	12
BOA	0,2374	6,4907	8,2760	0,2453	2,4052	13
CGSA	0,2428	5,9298	8,6762	0,2436	2,4129	14
LSA	0,2426	6,1617	8,5827	0,2432	2,4252	15

(continued)

Table 7. (*continued*)

Algorithm	h	l	t	b	Cost	Rank
ASO	0,2323	6,6408	8,3231	0,2470	2,4373	16
MS	0,2454	6,3801	8,0856	0,2570	2,4618	17
GWO	0,2830	5,9448	7,8776	0,2854	2,6837	18
GSA	0,2271	8,5607	8,0873	0,2569	2,7425	19
ABC	0,2678	5,9864	8,7444	0,3925	3,7745	20

Table 8. Comparison results for gear train design problem.

Algorithm	T_a	T_b	T_d	T_f	Cost	Rank
MFO	49	19	16	43	2,70E−12	1
COA	49	16	19	43	2,70E−12	1
PSO	49	16	19	43	2,70E−12	1
SOS	49	16	19	43	2,70E−12	1
TLABC	43	16	19	49	2,70E−12	1
CSA	49	16	19	43	2,70E−12	1
MS	43	16	19	49	2,70E−12	1
GWO	49	16	19	43	2,70E−12	1
DSA	49	16	19	43	2,70E−12	1
SSA	43	19	16	49	2,70E−12	1
ASO	49	19	16	43	2,70E−12	1
LSA	49	19	16	43	2,70E−12	1
SCA	49	19	16	43	2,70E−12	1
BOA	49	19	16	43	2,70E−12	1
CGSA	43	19	16	49	2,70E−12	1
BSA	51	15	26	53	2,31E−11	2
WOA	51	15	26	53	2,31E−11	2
GSA	57	31	13	49	9,94E−11	3
CKGSA	56	23	13	37	6,60E−10	4
ABC	57	23	14	39	3,25E−07	5

5 Conclusions

This paper study is one of the most comprehensive studies in the field of constrained engineering optimization problems. In this respect, significant information is provided for researchers. Considering the experimental study results, it is seen that MFO, COA, PSO and SOS algorithms are more successful than competing algorithms in engineering design problems. In most of the problems, algorithms have similar performances. As an interesting result, it can be said that the current algorithms are not very successful. For example, the ASO algorithm has been published in a highly effective scientific journal in 2019 but has performed worse than its competing algorithms.

References

1. Han, X., Liu, Q., Wang, H., Wang, L.: Novel fruit fly optimization algorithm with trend search and co-evolution. Knowl.-Based Syst. **141**, 1–17 (2018)
2. Tian, D., Shi, Z.: MPSO: modified particle swarm optimization and its applications. Swarm Evol. Comput. **41**, 49–68 (2018)
3. Sun, G., Ma, P., Ren, J., Zhang, A., Jia, X.: A stability constrained adaptive alpha for gravitational search algorithm. Knowl.-Based Syst. **139**, 200–213 (2018)
4. Tang, D., Liu, Z., Yang, J., Zhao, J.: Memetic frog leaping algorithm for global optimization. Soft. Comput. **23**, 1–29 (2018)
5. Derrac, J., García, S., Molina, D., Herrera, F.: A practical tutorial on the use of nonparametric statistical tests as a methodology for comparing evolutionary and swarm intelligence algorithms. Swarm Evol. Comput. **1**(1), 3–18 (2011)
6. Martin, L., Leblanc, R., Toan, N.K.: Tables for the Friedman rank test. Can. J. Stat. **21**(1), 39–43 (1993)
7. Hooker, J.N.: Testing heuristics: we have it all wrong. J. Heuristics **1**(1), 33–42 (1995)
8. Mirjalili, S., Mirjalili, S.M., Lewis, A.: Grey wolf optimizer. Adv. Eng. Softw. **69**, 46–61 (2014)
9. Rashedi, E., Nezamabadi-Pour, H., Saryazdi, S.: GSA: a gravitational search algorithm. Inf. Sci. **179**(13), 2232–2248 (2009)
10. He, Q., Wang, L.: An effective co-evolutionary particle swarm optimization for constrained engineering design problems. Eng. Appl. Artif. Intell. **20**(1), 89–99 (2007)
11. Mezura, M.E., Coello, C.: An empirical study about the usefulness of evolution strategies to solve constrained optimization problems. Int. J. Gen Syst **37**(4), 443–473 (2008)
12. Coello, C.A.C.: Use of a self-adaptive penalty approach for engineering optimization problems. Comput. Ind. **41**(2), 113–127 (2000)
13. Mahdavi, M., Fesanghary, M., Damangir, E.: An improved harmony search algorithm for solving optimization problems. Appl. Math. Comput. **188**(2), 1567–1579 (2007)
14. Huang, F.Z., Wang, L., He, Q.: An effective co-evolutionary differential evolution for constrained optimization. Appl. Math. Comput. **186**(1), 340–356 (2007)
15. Arora, S., Singh, S.: Butterfly optimization algorithm: a novel approach for global optimization. Soft. Comput. **23**(3), 715–734 (2019)
16. Gandomi, A.H., Yun, G.J., Yang, X.S., Talatahari, S.: Chaos-enhanced accelerated particle swarm optimization. Commun. Nonlinear Sci. Numer. Simul. **18**(2), 327–340 (2013)
17. Gandomi, A.H., Yang, X.S., Alavi, A.H.: Cuckoo search algorithm: a metaheuristic approach to solve structural optimization problems. Eng. Comput. **29**(1), 17–35 (2013)
18. Parsopoulos, K.E., Vrahatis, M.N.: Unified particle swarm optimization for solving constrained engineering optimization problems. In: Wang, L., Chen, K., Ong, Y.S. (eds.) Advances in Natural Computation, pp. 582–591. Springer, Berlin (2005)
19. Deb, K., Goyal, M.: A combined genetic adaptive search (GeneAS) for engineering design. Comput. Sci. Inf. **45**, 26–30 (1996)
20. Zhang, C., Wang, H.: Mixed-discrete nonlinear optimization with simulated annealing. Eng. Optim. **21**(277), 91 (1993)
21. Çivicioğlu Beşdok, P., Beşdok, E., Günen, M.A., Atasever, Ü.H.: Weighted differential evolution algorithm for numerical function optimization: a comparative study with cuckoo search, artificial bee colony, adaptive differential evolution, and backtracking search optimization algorithms. Neural Comput. Appl. 1–15 (2018)
22. Civicioglu, P.: Artificial cooperative search algorithm for numerical optimization problems. Inf. Sci. **229**, 58–76 (2013)

23. Wang, Y., Liu, Z.Z., Li, J.: Utilizing cumulative population distribution information in differential evolution. Appl. Soft Comput. **48**, 329–346 (2016)
24. Civicioglu, P., Besdok, E.: A+ evolutionary search algorithm and QR decomposition based rotation invariant crossover operator. Exp. Syst. Appl. **103**, 49–62 (2018)
25. Karaboga, D., Basturk, B.: A powerful and efficient algorithm for numerical function optimization: artificial bee colony (ABC) algorithm. J. Glob. Optim. **39**(3), 459–471 (2007)
26. Zhang, J., Sanderson, A.C.: JADE: adaptive differential evolution with optional external archive. IEEE Trans. Evol. Comput. **13**(5), 945–958 (2009)
27. Civicioglu, P.: Backtracking search optimization algorithm for numerical optimization problems. Appl. Math. Comput. **219**, 8121–8144 (2013)
28. Singh, P.R., Elaziz, M.A., Xiong, S.: Modified Spider Monkey Optimization based on Nelder-Mead method for global optimization. Exp. Syst. Appl. **110**, 264–289 (2018)
29. Cagnina, L.C., Esquivel, S.C., Coello, C.A.C.: Solving engineering optimization problems with the simple constrained particle swarm optimizer. Informatica **32**(3), 319–326 (2008)
30. Mortazavi, A., Toğan, V., Nuhoğlu, A.: Interactive search algorithm: a new hybrid metaheuristic optimization algorithm. Eng. Appl. Artif. Intell. **71**, 275–292 (2018)
31. Lim, W.H., Isa, N.A.M.: Two-layer particle swarm optimization with intelligent division of labor. Eng. Appl. Artif. Intell. **26**(10), 2327–2348 (2013)
32. Lim, W.H., Isa, N.A.M.: Particle swarm optimization with increasing topology connectivity. Eng. Appl. Artif. Intell. **27**, 80–102 (2014)
33. Parsopoulos, K.E., Vrahatis, M.N.: Unified particle swarm optimization for solving constrained engineering optimization problems. In: International Conference on Natural Computation, pp. 582–591. Springer, Heidelberg (2005)
34. Mortazavi, A., Toğan, V., Nuhoğlu, A.: Interactive search algorithm: a new hybrid metaheuristic optimization algorithm. Eng. Appl. Artif. Intell. **71**, 275–292 (2018)
35. Sadollah, A., Bahreininejad, A., Eskandar, H., Hamdi, M.: Mine blast algorithm for optimization of truss structures with discrete variables. Comput. Struct. **102**, 49–63 (2012)
36. Rao, R.V., Waghmare, G.G.: Complex constrained design optimisation using an elitist teaching-learning-based optimisation algorithm. IJMHeur **3**(1), 81–102 (2014)
37. Montes Montes, E., Coello, C.A.C., Velazquez Reyes, J.: Increasing successful offspring and diversity in differential evolution for engineering design. In: Seventh International Conference on Adaptive Computing in Design and Manufacture, pp. 131–139 (2006)
38. Wu, L., Liu, Q., Tian, X., Zhang, J., Xiao, W.: A new improved fruit fly optimization algorithm IAFOA and its application to solve engineering optimization problems. Knowl.-Based Syst. **144**, 153–173 (2018)
39. Jiang, M., Yuan, D., Cheng, Y.: Improved artificial fish swarm algorithm. In: IEEE Fifth International Conference on Natural Computation, ICNC, vol. 4, pp. 281–285 (2009)
40. Liang, J.J., Qin, A.K., Suganthan, P.N., Baskar, S.: Comprehensive learning particle swarm optimizer for global optimization of multimodal functions. IEEE Trans. Evol. Comput. **10** (3), 281–295 (2006)
41. Pan, Q.K., Suganthan, P.N., Liang, J.J., Tasgetiren, M.F.: A local-best harmony search algorithm with dynamic subpopulations. Eng. Optim. **42**(2), 101–117 (2010)
42. Mirjalili, S.: The ant lion optimizer. Adv. Eng. Softw. **83**, 80–98 (2015)
43. Aydilek, İ.B.: A hybrid firefly and particle swarm optimization algorithm for computationally expensive numerical problems. Appl. Soft Comput. **66**, 232–249 (2018)
44. Ngo, T.T., Sadollah, A., Kim, J.H.: A cooperative particle swarm optimizer with stochastic movements for computationally expensive numerical optimization problems. J. Comput. Sci. **13**, 68–82 (2016)
45. Dhiman, G., Kumar, V.: Emperor penguin optimizer: a bio-inspired algorithm for engineering problems. Knowl.-Based Syst. **159**, 20–50 (2018)

46. Dhiman, G., Kumar, V.: Spotted hyena optimizer: a novel bio-inspired based metaheuristic technique for engineering applications. Adv. Eng. Softw. **114**, 48–70 (2017)
47. Mirjalili, S., Mirjalili, S.M., Hatamlou, A.: Multi-Verse Optimizer: a nature-inspired algorithm for global optimization. Neural Comput. Appl. **27**(2), 495–513 (2016)
48. Mirjalili, S.: SCA: a sine cosine algorithm for solving optimization problems. Knowl.-Based Syst. **96**, 120–133 (2016)
49. Sadollah, A., Sayyaadi, H., Yoo, D.G., Lee, H.M., Kim, J.H.: Mine blast harmony search: a new hybrid optimization method for improving exploration and exploitation capabilities. Appl. Soft Comput. **68**, 548–564 (2018)
50. Sadollah, A., Bahreininejad, A., Eskandar, H., Hamdi, M.: Mine blast algorithm: a new population based algorithm for solving constrained engineering optimization problems. Appl. Soft Comput. **13**(5), 2592–2612 (2013)
51. Liu, H., Cai, Z., Wang, Y.: Hybridizing particle swarm optimization with differential evolution for constrained numerical and engineering optimization. Appl. Soft Comput. **10** (2), 629–640 (2010)
52. Eskandar, H., Sadollah, A., Bahreininejad, A., Hamdi, M.: Water cycle algorithm–a novel metaheuristic optimization method for solving constrained engineering optimization problems. Comput. Struct. **110**, 151–166 (2012)
53. Rao, R.V., Savsani, V.J., Vakharia, D.P.: Teaching-learning-based optimization: a novel mechanical design optimization problems. Comput. Aided Des. **43**, 303–315 (2011)
54. Moghdani, R., Salimifard, K.: Volleyball premier league algorithm. Appl. Soft Comput. **64**, 161–185 (2018)
55. Awad, N.H., Ali, M.Z., Mallipeddi, R., Suganthan, P.N.: An improved differential evolution algorithm using efficient adapted surrogate model for numerical optimization. Inf. Sci. **451**, 326–347 (2018)
56. Long, W., Wu, T., Liang, X., Xu, S.: Solving high-dimensional global optimization problems using an improved sine cosine algorithm. Exp. Syst. Appl. **123**, 108–126 (2019)
57. Dong, M., Wang, N., Cheng, X., Jiang, C.: Composite differential evolution with modified oracle penalty method for constrained optimization problems. Math. Probl. Eng. 1–15 (2014). https://doi.org/10.1155/2014/617905
58. Amir M.: Towards an approach for effectively using intuition in large-scale decision-making problems. Ph.D. thesis, University of Debrecen (2013)
59. Lin, X., Zhang, F., Xu, L.: Design of gear reducer based on FOA optimization algorithm. In: International Conference on Smart Vehicular Technology, Transportation, Communication and Applications, pp. 240–247. Springer, Cham (2017)
60. Shareef, H., Ibrahim, A.A., Mutlag, A.H.: Lightning search algorithm. Appl. Soft Comput. **36**, 315–333 (2015)
61. Askarzadeh, A.: A novel metaheuristic method for solving constrained engineering optimization problems: crow search algorithm. Comput. Struct. **169**, 1–12 (2016)
62. Cheng, M.Y., Prayogo, D.: Symbiotic organisms search: a new metaheuristic optimization algorithm. Comput. Struct. **139**, 98–112 (2014)
63. Chen, X., Xu, B.: Teaching-learning-based artificial bee colony. In: International Conference on Swarm Intelligence, pp. 166–178. Springer, Cham (2018)
64. Civicioglu, P.: Transforming geocentric cartesian coordinates to geodetic coordinates by using differential search algorithm. Comput. Geosci. **46**, 229–247 (2012)
65. Mirjalili, S.: Moth-flame optimization algorithm: a novel nature-inspired heuristic paradigm. Knowl.-Based Syst. **89**, 228–249 (2015)
66. Mirjalili, S., Lewis, A.: The whale optimization algorithm. Adv. Eng. Softw. **95**, 51–67 (2016)

67. Mirjalili, S., Gandomi, A.H., Mirjalili, S.Z., Saremi, S., Faris, H., Mirjalili, S.M.: Salp swarm algorithm: a bio-inspired optimizer for engineering design problems. Adv. Eng. Softw. **114**, 163–191 (2017)
68. Wang, G.G.: Moth search algorithm: a bio-inspired metaheuristic algorithm for global optimization problems. Memetic Comput. **10**, 151–164 (2018)
69. Mittal, H., Pal, R., Kulhari, A., Saraswat, M.: Chaotic Kbest gravitational search algorithm (CKGSA). In: 2016 Ninth International Conference on Contemporary Computing (IC3), pp. 1–6. IEEE, August 2016
70. Pierezan, J., Coelho, L.S.: Coyote optimization algorithm: a new metaheuristic for global optimization problems. In: Proceedings of the IEEE Congress on Evolutionary Computation (CEC), Rio de Janeiro, Brazil, pp. 2633–2640 (2018)
71. Zhao, W., Wang, L., Zhang, Z.: Atom search optimization and its application to solve a hydrogeologic parameter estimation problem. Knowl.-Based Syst. **163**, 283–304 (2019)
72. Mirjalili, S., Gandomi, A.H.: Chaotic gravitational constants for the gravitational search algorithm. Appl. Soft Comput. **53**, 407–419 (2017)

The Development of Artificial Intelligence-Based Web Application to Determine the Visibility Level of the Objects on the Road

Mehmet Kayakuş[1,2(✉)] and Ismail Serkan Üncü[1,2]

[1] Akdeniz University, 07600 Antalya, Turkey
mehmetkayakus@akdeniz.edu.tr
[2] Faculty of Technology, Isparta University of Applied Sciences, 32100 Isparta, Turkey
serkanuncu@isparta.edu.tr

Abstract. The smallest objects that can lead to accidents on the road are called critical objects. The timely recognition of critical objects by drivers increases the safety of people driving. With a mathematical model called "visibility level" developed by Adrian in 1982, the visibility of critical objects by drivers is expressed numerically. The visibility level is defined as the ratio of the difference between the luminance of the object and the background luminance to the minimum luminance difference (luminance difference threshold) required for the object to be seen. In this model, there are many factors such as luminance of the object, background luminance, contrast polarity, observer age and the duration of the observation. Using the web application developed in this study, the visibility level of the critical object is expressed mathematically. The fundamental logic of the software consists of two stages. In the first stage, the luminance of the road and the object on the road is learned. An artificial intelligence-based algorithm has been developed to learn these luminance values. In the second stage, the observer's features are entered into the system as the level of visibility may vary according to the features of the observer. In the second stage, the features of the observer are entered into the system, since the level of visibility will change according to the features of the observer. Once this information has been entered into the system, the software can calculate how well the object is visible to the observer.

Keywords: Visibility level · Web application · Artificial intelligence

1 Introduction

Road lighting helps drivers and pedestrians get visual information about the road at night. Road lighting encourages the use of the road at night and ensures efficient use of the road. Road lighting is an important factor in preventing accidents at night [1]. Past studies show the importance of road lighting in preventing road accidents [2]. In addition, artificial road lighting facilitates road finding and improves driving time by using this information [3–5].

D. J. Hemanth and U. Kose (Eds.): ICAIAME 2019, LNDECT 43, pp. 502–508, 2020.
https://doi.org/10.1007/978-3-030-36178-5_39

There are three different methods to assess road lighting measurements: illuminance, luminance, and small target visibility [3]. Firstly, the illumination level criterion was used for road lighting. Today, road lighting measurements are based on luminance. It is required to provide minimum average luminance values according to the road class. A new method of calculation, including the human factor, was needed in road lighting measurements. As a result of scientific studies, a calculation method named as visibility level (VL) was proposed.

VL is recommended as a standard for road lighting measurements in the United States [6], but is not yet accepted as a standard in Europe [7]. Adrian [8–10], to calculate the visibility of small objects on the road, psychophysical data were collected and suggested that the visibility level could be a reference value for road lighting design. His work followed Blackwell's definition of the VL [11] as the ratio between a target luminance contrast L/L_b and a threshold luminance contrast L_t/L_b (L_b is background luminance). Compared to indexes such as illuminance and luminance, VL is a more convenient method. Because it is more related to visual performance, which is part of the driving event [12].

2 Materials and Methods

2.1 Visibility Level of the Critical Object

According to Adrian [9], the visibility level is the best theoretical method for evaluating road lighting. The main hypothesis will be to reduce the accident rates when the lighting system is optimized to increase the visibility of the objects in the road: road lighting reveals objects before automotive lighting can do, and improves anticipation. The VL model creates a scenario that constructs and collects information on how well a small object at a certain distance in the road can be seen by drivers. It can be adjusted in terms of driving safety. The lighting system can be optimized to enhance the visual task for critical safety hazards. (collision with an obstacle) [3].

In this study, the VL developed by Adrian [9] will be used to measure the visibility of the critical object on the road. According to Adrian's visibility equations, visibility level of the critical object which is on the road surface defined by Adrian [9].

$$VL = \frac{\Delta L}{\Delta L_e} \tag{1}$$

VL: Visibility level, ΔL: luminance difference between the object and its background, ΔL_e: The luminance difference needed for minimal visibility, between a target of certain angular size and its background [13].

$$\Delta L = L_c - L_f \tag{2}$$

L_c: luminance of the target (cd/m^2), L_f: luminance of the background (cd/m^2).

$$\Delta L_f = k.\left(\frac{\sqrt{\Phi}}{\alpha} + \sqrt{L}\right)^2 * \frac{a(\alpha, L_f) + t_g}{t_g} * F_{CP} * YF \tag{3}$$

Where, k: factor for the probability of perception (k = 2.6 for 100% probability), Φ: Luminous flux function (lm), L: luminance function (cd/m^2), F_{cp}: contrast polarity factor, AF: age factor a(a; Lb): Parameter depending on size of target and background luminance, t_g: observation time (s).

For safe driving, lighting engineers have a certain visibility equal to what they call a "field factor". It has been implicated that this value should be higher than the threshold values in laboratory conditions in order to be observed in complex and different driving conditions. The French recommends that the level of visibility for road lighting should be greater than seven [14, 15].

2.2 Working Principle of the System

The basic operating logic of the software consists of two stages. In the first stage, the luminance of the road and the critical object are measured. An artificial intelligence-based algorithm has been developed to learn these luminance values. Thanks to this developed algorithm, the luminance values of the desired points on the road are learned without the need for hardware measurement device. This step was performed in the previous study.

For the analysis of the visibility of critical objects on the road, the system in Fig. 1 was prepared. In this system, the observer is located 60 m from the first two luminaires as in the international standards. They are placed between the first two luminaires in critical objects by leaving 1 m of space between them. After measurements were made, measurements were repeated until the last armature was shifted by 1 m.

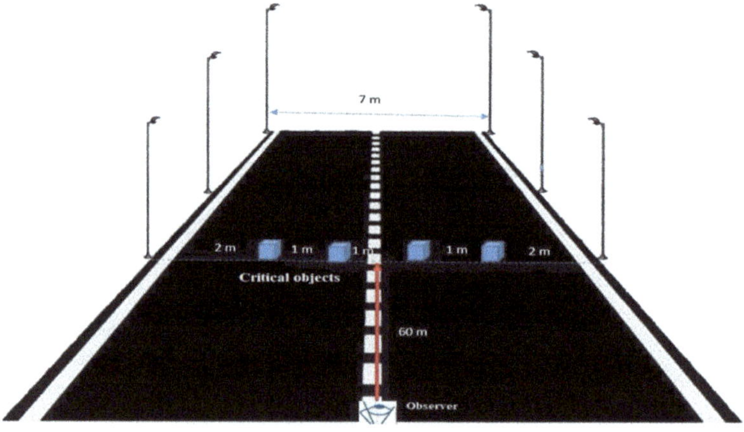

Fig. 1. Visibility level measurement system

Artificial intelligence-based measurement of the study was made in [16] publication. The user will first upload a photo of the road to the software. Then click on the critical object on the road and the road will determine the measurement points. The developed artificial intelligence algorithms will measure the luminance of the determined measurement points. Three different artificial intelligence algorithms were used. Artificial neural networks (ANN), fuzzy logic (FL) and Anfis. Figure 2 shows the interface of the software developed.

Fig. 2. VL software interface

Then the user's age and the visual angle information is entered into the system. Adrian [9, 17] calculates the visibility level according to the visibility model by clicking the calculate visibility factor button.

3 Experimental Study

Test road is in M2 road class and LED lighting is used.

Different tests were carried out by changing the position of the critical body on the road. Thus, it has been tested in factors such as the structure of the road and the location of the luminaires. To determine the visibility threshold value for critical objects, 96 tests were performed with three different observers. Figure 3 shows the critical objects used in the calculation of visibility level.

Fig. 3. Critical objects used in calculation of visibility level

Calculated luminance levels of the critical object and road; age of the observer and the visual angle of the object were entered into the system. Calculated visibility values of the critical body on the road are shown in Table 1.

Table 1. Visibility level of critical objects

Critical objects	Road luminance			Critical object luminance			Visibility level				Result
	ANN	FL	ANFIS	ANN	FL	ANFIS	Age	ANN	FL	ANFIS	
Object-1	15,33	14,21	15,89	15,37	14,17	16,03	25	0,24	0,07	1,26	Invisible
							45	0,18	0,06	0,97	Invisible
							65	0,12	0,03	0,62	Invisible
Object -2	3,93	4,09	3,85	4,86	4,75	4,90	25	15,77	10,90	17,98	Visible
							45	12,19	8,43	13,89	Visible
							65	7,80	5,39	8,89	Invisible
Object -3	3,34	3,78	3,42	4,20	4,22	3,92	25	16,56	7,99	9,59	Visible
							45	12,79	6,17	7,41	Invisible
							65	8,18	3,95	4,74	Invisible
Object -4	2,78	3,26	3,33	4,66	4,49	4,66	25	41,61	24,11	25,31	Visible
							45	32,15	18,62	19,55	Visible
							65	20,57	11,92	12,50	Visible

4 Conclusion

In this study, artificial intelligence-based software has been developed to measure the visibility of the critical object called the smallest object that can lead to accidents on the road. When you use the web based measurement system you will save time and reduce costs.

This measurement system was developed by combining Adrian visibility level calculation method and artificial intelligence based software. This study shows how well the observer can see the critical object. Using the developed software, the lower limit value was determined in order to be able to see the critical object in a large number of tests performed on different observers. As a result of this study, it is concluded that the critical object can be seen clearly when the visibility level is 9,5 and above.

This work was supported by The Scientific Research Projects Coordination Unit of Akdeniz University. Project Number: FBA-2018-3899.

References

1. Li, F., Chen, D., Song, X., Chen, Y.: LEDs: a promising energy-saving light source for road lighting. In: 2009 Asia-Pacific Power and Energy Engineering Conference, Wuhan, China, 27–31 March 2009, pp. 1–3 (2009)
2. Zhou, H., Pirinccioglu, F., Hsu, P.: A new roadway lighting measurement system. Transp. Re. Part C: Emerg. Technol. **17**(3), 274–284 (2009)
3. Mayeur, A., Bremond, R., Bastien, J.M.C.: The effect of the driving activity on target detection as a function of the visibility level: implications for road lighting. Transp. Res. Part F: Traffic Psychol. Behav. **13**(2), 115–128 (2010)
4. Hills, B.L.: Vision, visibility, and perception in driving. Perception **9**(2), 183–216 (1980)

5. Sivak, M.: The information that drivers use: is it indeed 90% visual? Perception **25**(9), 1081–1089 (1996)
6. Illuminating Engineering Society of North America (IESNA): American National Standard Practice for Roadway Lighting, IESNA Publication RP-8-00, New York, USA (2000)
7. (CEN) EN 13201: Road Lighting. European Committee for Standardization, Brussels, Belgium (2016)
8. Adrian, W.: Visibility levels under night-time driving conditions. J. Illum. Eng. Soc. **16**, 3–12 (1987)
9. Adrian, W.: Visibility of targets: model for calculation. Lighting Res. Technol. **2**, 181–188 (1989)
10. Adrian, W.: Fundamentals of roadway lighting. Light Eng. **12**, 57–71 (2004)
11. Commission Internationale de l'Eclairage (CIE) 19/2: An Analytic Model for Describing the Influence of Lighting Parameters upon Visual Performance, Vienna, Austria (1981)
12. Brémond, R., Dumont, E., Ledoux, V., Mayeur, A.: Photometric measurements for visibility level computations. Lighting Res. Technol. **43**, 119–128 (2011)
13. Güler, O., Onaygil, S.: A new criterion for road lighting: average visibility level uniformity. J. Light Vis. Environ. **27**(1), 39–46 (2003)
14. Association Française de l'Eclairage: Recommendations Relatives à l'Eclairage des Voies Publiques (2002)
15. Brémond, R., Bodard, V., Dumont, E., Nouailles-Mayeur, A.: Target visibility level and detection distance on a driving simulator. Lighting Res. Technol. **45**(1), 76–89 (2013)
16. Kayakuş, M., Üncü, I.S.: Research note: the measurement of road lighting with developed artificial intelligence software. Lighting Res. Technol. **51**(6), 969–977 (2019)
17. Adrian, W.: Visibility levels in street lighting: an analysis of different experiments. J. Illum. Eng. Soc. **22**, 49–52 (1993)

A Study on the Performance of Base-m Polynomial Selection Algorithm Using GPU

Oğuzhan Durmuş[1(✉)], Umut Can Çabuk[2], and Feriştah Dalkılıç[1]

[1] Computer Engineering Department, Dokuz Eylül University, 35160 Buca, Izmir, Turkey
oguzhan.durmus@ceng.deu.edu.tr, feristah@cs.deu.edu.tr
[2] International Computer Institute, Ege University, 35100 Izmir, Turkey
umut.can.cabuk@ege.edu.tr

Abstract. Factorization of large integers has been being considered as a challenging problem in computer science and engineering since the earliest times of the computer technology. Despite the comprehensive efforts, there is still no reported deterministic polynomial-time algorithm; however, its complexity class is in fact not yet decided. A fast and robust polynomial-time algorithm for this problem is required to increase the processing capabilities of current systems. Yet, there are also hesitations at the same time within the community, due to the potential security threats that may appear in such a case. The (asymptotically) fastest algorithm ever found so far to factor large integers is the general number field sieve. Its performance depends on selection of "good" polynomials, which requires a specific procedure for such a selection. Another significant performance factor surely is the power of the processing hardware and their peripherals. This article unveils and discusses the impacts of heterogeneous computing using a graphics processor units (GPU) instead of a central processing unit (CPU) on the performance of polynomial selection and so of factoring large integers. Further, the article presents implementation details and a comparative performance evaluation of the Base-m polynomial selection method to select "good" polynomials. Accordingly, the GPU is found to be more effective over larger numbers with more roots, while the CPU appeared more effective over smaller numbers with less roots, possibly due to the excessive overheads in the GPU processing procedures.

Keywords: Polynomial selection · General number field sieve · Graphical processing unit (GPU) · Heterogeneous computing

1 Introduction

Computer technology is reshaping the life of the entire humankind, as well as the commerce and industries. In the 21st century, however, this impact is explicitly increasing with the advances in the computer technology. These advances provide cheaper, smaller, faster, more robust and more energy efficient computers and/or computer parts. Nevertheless, the needs are also changing by the time. Two major

D. J. Hemanth and U. Kose (Eds.): ICAIAME 2019, LNDECT 43, pp. 509–517, 2020.
https://doi.org/10.1007/978-3-030-36178-5_40

needs can be listed as reducing the energy consumption (which is obviously out of the scope of this paper) and increasing the computational power.

Knowing the shortcomings, engineers and developers push the limits in order to avoid bottlenecks in the technology. For instance, conventional central processing units (CPUs), also called general purpose processors or briefly processors; despite all the recent advances on their speed and efficiency, still lack of power when it comes to make vast arithmetic and geometric operations on very large amounts of input data. What they lack is not actually the clock speeds, but the parallelization aspects. Hence, engineers and developers nowadays prefer to use graphics processor units (GPUs), which are originally created for graphic and signal processing, also for computationally intense processes that contain straightforward arithmetic operations, excluding complex logical operations. These operations include, but not limited to matrix multiplication, factorization of large numbers, determination of prime numbers, and many other operations involving matrices and very large numbers.

Cryptology, consisting of cryptography and cryptanalysis, is essentially a computationally heavy discipline, whose techniques require many arithmetic and geometric operations in order to provide or outreach the security. Further applications and implementations of cryptography, such as the blockchain or the public key infrastructure, also require operations on very large numbers. A more precise example would be the Rivest-Shamir-Adleman (RSA) algorithm [1], which is used for public key encryption and digital certificates that verify parties of transactions (especially in e-commerce) [2]. Factoring very large integers is one of the most popular problems in the cryptography, as RSA (and many other algorithm) takes advantage of difficulty in factoring large integer numbers [3]. Because it contains the step:

$$n = p \times q \tag{1}$$

wherein n is the number in Eq. 1, which is produced from production of p and q, where p and q are both prime numbers; but those are very large prime integers. In the RSA algorithm [1], n is public, which means everyone can see and access n, while p and q are private, which means no one can access these numbers except the owner/issuer. This privacy of the p and q variables provides some level of difficulty against breaking RSA.

To break RSA, very large prime numbers must be produced. However, producing large prime numbers requires (primality) testing in ($10^{\text{number of digits}}/2$) numbers [3] before a prime number is guaranteed to be found. General number field sieve (GNFS), on the other hand, is a well-known algorithm for integer factorization. But, it has an arguably large time complexity, which implies a slow working rate, especially for very large numbers. The time complexity of the method can be represented in the big-O notation as in Eq. 2 [4]:

$$O(\exp(1.923 + O(1))(\log N)1/3(\log \log N)2/3)) \tag{2}$$

Considering the complexity given in Eq. 2, the growth of time is exponential in asymptotical analysis method. So, generation of very large prime integers presumably takes too long time [5]. Additionally, the polynomial selection is only the 3% of the total workload [5].

In this article, to speed up of the GNFS algorithm, the polynomial selection step that is the step of the GNFS algorithm, is studied comprehensively. This step requires too many arithmetic processes. These processes are simple, but they repeat many times. To achieve this difficulty, GPU is used. A modern GPU can perform simple (especially arithmetic) instructions with very high performance. Recently, especially in the last decade, the processing power of GPUs (in terms of gigaflops) had surpassed the general-purpose CPUs by far, at least for particular means of processes or operations [6].

GPUs have much more arithmetic logic units (ALUs), when compared to multi-core CPUs. ALUs are the sub-parts of processors that can perform arithmetic (and logical) operations. In the experimental parts of this work, Nvidia GeForce GTX 860M is used as the GPU. This GPU contains 640 processor cores, which support compute unified device architecture (CUDA) runtime library for parallel computing. So that, this GPU can perform parallel execution of given instructions, but parallel execution requires parallel code, too. So, the efficiency of any algorithm depends on its parallel program code.

In this study, 3 polynomial selection methods were intended to be implemented and compared to find the best method for generating "good" polynomials. These algorithms are, namely the Base-m method, the Murphy method [7] and the Kleinjung method [8]. The Base-m method constructs polynomials using the expansion of m. Murphy method generates polynomials according to their roots' properties and the Kleinjung method implies restrictions to improve the performance. Nevertheless, due to the time constraints, only the Base-m method was implemented and analyzed.

In the following section (Sect. 2) related works are detailed, in Sect. 3 the Base-m method is explained, in Sect. 4 the implementation of Base-m method is given followed by Sect. 5, where the results are given and discussed. The last section concludes the study.

2 Related Works

Most prior researches had been made using the single-core processor architecture [5] and/or multi-core CPU processors. Nevertheless, in this work, a many-core GPU is preferred, while the prior works did not take advantage of the power of the GPU, which is its vast parallel processing capability. Many works in the literature, focused on algorithmic developments and algorithmic (time, process etc.) complexity of the algorithms. Hence, implementational aspects and comparison have not been comprehensively studied.

Zhu and Su, in their conference paper [9], focused on mathematical foundations of the mentioned problem and compared three existing methods. They have investigated Base-m, Murphy and Kleinjung methods, and proposed improvements on the Kleinjung method. Since these are all linear methods, it takes too much time to generate good polynomials. This study constitutes an origin for our implementational work, even though their efforts were limited to theoretical analysis, consisting size and root properties of the polynomials.

Kleinjung et al. [5] have worked on factoring 768-bit RSA modulus via GNFS algorithm utilizing 80 single core processors running in parallel, and up to 1 TB of RAM memory. Their implementational work that presents outstanding results, was another motivation for our study. But, in contradistinction to ours, they have preferred multiple single core processors and a special memory setup, instead of multicore processors and/or GPUs.

Murphy, in his prominent thesis study [10], has made efforts on building a novel and effective polynomial model, for 140-digit (465-bit) and 155-digit (512-bit) RSA modulus factoring using number field sieve. However, he has not applied nor mentioned any means of parallelization and potential computational speedups. Before choosing Base-m for our implementational study, we also considered his method and left it as a future work because of our project's time limitations.

Lenstra et al. [8] made estimations regarding hardness and time complexity of factoring large numbers if TWIRL, a hypothetical high-performance parallel computing device, was used as the processing platform. Though the obtained results are notable, TWIRL is still hypothetical, and no functional implementation has been reported.

Shamir, the co-inventor of the TWIRL device, had previously invented the TWINKLE device, the predecessor of the TWIRL device, having the same purpose of factoring large integers using GNFS. In his work [11], he has explained how large integers can be factored using that device and presented a remarkable performance analysis. Shamir has noted that this device might cost ca. $5000 in case of bulk production. Yet, similar to the TWIRL device, there is no known implementation of this device. Even if it were realized, using standard powerful GPUs would be easier and/or cheaper in most cases as long as the same performance is promised.

In this article, parallelization is the main focus, unlike many other works. Because, nowadays parallel processing power of GPUs is a major trending topic in computer science. Heterogeneous computing is realized in the implementation phase, where we have chosen to analyze parallel implementation of the Base-m method on a GPU among two other algorithms, because of our hardware and time restrictions. Our results include direct observations regarding this parallelization, as well as the extrapolation attempts.

3 Base-m Method

The Base-m method is the main linear method for the polynomial selection problem. This method linearly searches polynomials in a given interval. Following Eqs. 3 to 7 define the Base-m method [12].

Let

$$\mathrm{N}^{1/(\mathrm{d}+1)} < \mathrm{m} < \mathrm{N}^{1/\mathrm{d}} \tag{3}$$

$$\mathrm{N} = \sum_{0}^{d} a_i m^i (0 \leq ai < m) \tag{4}$$

$$f(x) = \sum\nolimits_{i=0}^{d} a_i x^i \tag{5}$$

$$\mathrm{g(x) = x - m} \tag{6}$$

$$0 \equiv m\ mod(N) \tag{7}$$

The Base-m method essentially requires 2 inputs, which are called as N and the degree d. These values are used in Eq. 3 for restricting the m value. N is an integer, but it must be a big number since RSA uses big integers and the polynomial requires big coefficients within. In Eq. 4, N is found using m, and a_i values. a_i values are the polynomial coefficients. The most critical part of algorithm is Eq. 5. Because, this function has got a degree and it will be used to find the RSA key. Its growth is obviously not linear. Both of Eqs. 5 and 6 have a root of m. The root must fit as in Eq. 7, which contains a modulo operation.

4 Base-m Method Implementation

This section includes our pseudo codes of the Base-m method for single core, multiple cores and GPU processor versions. The essential difference between these versions can be seen in our implementation, which is omitted from this paper, yet source codes are given online (http://people.cs.deu.edu.tr/dalkilic/basemcs.html).

4.1 Single Core CPU Software

The single core implementation is pretty straightforward.

```
N = MAXIMUM_INTEGER;
d = 5; //In the related works that was used as 5

int m[] = GenerateMValues(N, d);
int a[d];
for p = 0 to numberOfMValues
a = findAValues(m[p], d, N);
if(a != null)
  print("Some a values are found.");
  createGFunction(m);
  bool isCorrect = testPolynomials(m[p], N, a[]);

if(isCorrect)
  print("Polynomials that are produced are approved.")
    break;
  else
    print("Polynomials are not approved, the process will be
executed again.")
```

4.2 Multiple Cores CPU Software

The multi-core implementation is slightly more complicated since each core should be assigned to a specific task.

```
N = MAXIMUM_INTEGER;
d = 5; //In the related works that was used as 5

int m[] = GenerateMValues(N, d);
int a[d];
int **foundValues = (int)malloc(sizeof(m)*sizeof(a))

#pragma omp parallel for default(none) shared(p, m, numberOf-
MValues, d, foundValues) private (a, N)
for p = 0 to numberOfMValues
a = findAValues(m[p], d, N);
if(a != null)
  print("Some a values are found.");
  createGFunction(m);
  bool isCorrect = testPolynomials(m[p], N, a[]);

  if(isCorrect)print("A polynomial is found");

#pragma omp critical
    foundValues[count] = a;
```

4.3 GPU Software

The GPU implementation is easier assuming task assignments are handled in a lower layer.

```
__global__ void
vectorPow(const float *A, float *B, int numberOfElements, int
degree)
{
  int i = blockDim.x * blockIdx.x + threadIdx.x;
  if (i < numberOfElements)
  {
    B[i] = powf(A[i], i % 4);
  }
}
```

5 Results and Discussion

Throughout this article, some advantages of making factorization and related computations on a GPU are discussed. There is another point worth mentioning, which is the application launch times and their resource usages. GPU usually occupies more RAM and in some cases its launch time is more than the CPU, at least according to our tests.

First things first, the GPU requires memory allocation in its own RAM and also in CPU's (general purpose) RAM. Because, the main host of a personal computer is always the CPU. The CUDA framework uses pointer to pointer at this point. It points to the CPU's RAM. This operation takes relatively long time and doubles the memory usage, also the framework itself requires some additional RAM resource. The turn-around time of the GPU is greater than the CPU's time, because of the memory usage behaviors. So, the GPU is appropriate for long operations, because the cost of copying a process is too high. And according to the architecture, the CPU may spend its time on idle state. Results of our experiments, shown in Table 1 confirm these claims, too.

Table 1. Experiment results.

For range of N 0 to 2,147,483,647					
	# of roots	Turnaround time (s)	Processing time (s)	Memory usage (MB)	# of Blocks in GPU (w/512 threads)
CPU	268	2.08	2.08	2	
	644,176	4400.774	4400.774	225	
GPU	268	0.948	0.002	90	9
	644,176	1.17	0.164	132	20,131
For range of N 0 to 18,446,744,073,709,551,615					
CPU	268	2.08	2.08	2	
	644,176	4400.774	4400.774	225	
GPU	268	0.948	0.002	90	9
	644,176	1.17	0.164	132	20,131

The statistics given in Table 1 are all obtained from the computations containing equations of the third-order. During the experiments, the computer was plugged in to the power line and had reliable operational conditions. In the source codes of the subject program, N is first defined as a "long int" and later as an "unsigned int64" (equivalent to "unsigned long long int") consecutively, as shown in the first and second rows.

When all these advantages and disadvantages are considered together, the GPU requires additional time at the start and the end of the applications and additional power resources, plus it may cause overheating, but it gives a high performance and shortens the calculations, especially in CPU intensive calculations. This is presented in Fig. 1. It shows 3 measurements; the CPU and GPU turnaround times, and the GPU calculation time. In the graph, each value in the x axis show the average values of 10 measurements, wherein 100 measurements were made in total.

As an important note regarding the difficulty of the given problem; it should be stated that when the interval values determined by N is chosen larger, the values of m to be examined becomes larger, too. Since all values of m (in the calculated interval depending on N) are tried one at a time, like brute-force, the difficulty of the problem increases saliently by N. Further, especially for large numbers, many of the roots found via Base-m are either negative or zero, which are both useless to continue running the method. Though some positive roots are also found often (but flatly not always), this fact increases the difficulty, too. Hence, the problem becomes cushier for GPUs, which can handle these repeating tasks faster, thanks to their vast parallelism. Conversely, they pose an excessive overhead, mostly caused by the need of copying the relevant data to the GPU's own RAM memory, which overly limits the potential performance gains. To state that, the mere GPU calculation times are shown separately in Fig. 1.

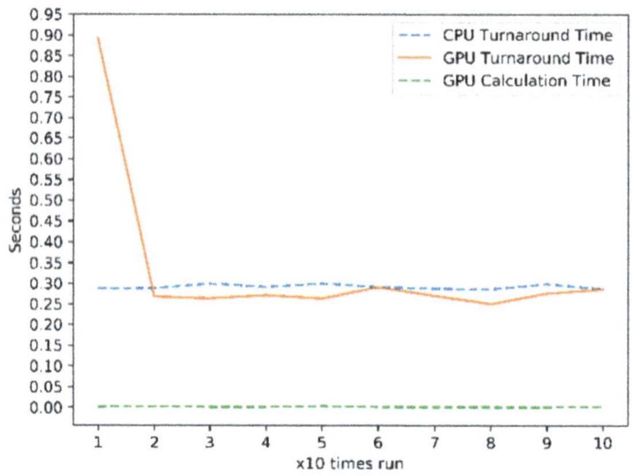

Fig. 1. Comparison of CPU and GPU time stats.

6 Conclusions

As a clear observation, when the number of the roots to be searched are increased, also means N is increased, the CPU's time and (main) memory consumption increases, too. But, if the number of the roots (N) so the number of the equations are rather small, then the CPU works faster and more effectively. If the assigned job is a long-term job, then the launch and end time can be neglectable. Additionally, the GPU uses the float data type, so data loss can be possible. Also, the GPU requires some additional power requirements, especially in power saving computers and laptops.

Per to the experiments, there is no rule of thumb regarding the CPU/GPU choice within this problem. Hence; depending on the job, more likely, its difficulty, complexity and duration, the more advantageous option changes. In simpler jobs, i.e. smaller numbers with less roots, the CPU performs better, owing to its compact architecture and lower overhead. The GPU on the other hand, performs better, when larger numbers with more roots are to be used, thanks to its higher processing capabilities (provided by vast processor cores).

As mentioned earlier; only the Base-m method (among 3 selected algorithms) could be implemented and investigated, due the time limitations of this study. Thus, our conclusions should be examined by using other methods, too. Implementation and analysis of the two other algorithms, namely the Murphy and Kleinjung methods, as well as the numeric comparison of all three, are left for future works.

References

1. Rivest, R.L., Shamir, A., Adleman, L.: A method for obtaining digital signatures and public-key cryptosystems. Commun. ACM **21**(2), 120–126 (1978)
2. Cavallar, S., et al.: Factorization of a 512-bit RSA modulus. In: Preneel, B. (ed.) Advances in Cryptology EUROCRYPT 2000, LNCS, vol. 1807, pp. 1–18. Springer, Heidelberg (2000)
3. Niven, I., Zuckerman, H.S., Montgomery, H.L.: An Introduction to the Theory of Numbers, 5th edn. Wiley, New York (1991)
4. Sun, H.M., Yang, W.C., Laih, C.S.: On the design of RSA with short secret exponent. J. Inf. Sci. Eng. **18**(1), 1–18 (2002)
5. Kleinjung, T., et al.: Factorization of a 768-bit RSA modulus. In: Annual Cryptology Conference, pp. 333–350. Springer, Heidelberg (2010)
6. Nvidia: CUDA C Programming Guide, PG-02829-001_v10.0, October 2018 Design Guide. https://docs.nvidia.com/cuda/pdf/CUDA_C_Programming_Guide.pdf. Accessed 27 Dec 2018
7. Kleinjung, T.: On polynomial selection for the general number field sieve. Math. Comput. **75** (256), 2037–2047 (2006)
8. Lenstra, A., et al.: Factoring estimates for a 1024-Bit RSA modulus. In: Laih, C.S. (ed.) Advances in Cryptology ASIACRYPT 2003, LNCS, vol. 2894, pp. 55–74. Springer, Heidelberg (2003)
9. Zhu, H., Su, S.: The improvement of the commonly used linear polynomial selection methods. In: 9th International Conference on Computational Intelligence and Security, Leshan, China, pp. 459–463 (2013)
10. Murphy, B.A.: Polynomial selection for the number field sieve integer factorization algorithm. Australian National University (1999)
11. Shamir, A.: Factoring large numbers with the TWINKLE device. In: Cryptographic Hardware and Embedded Systems, pp. 2–12. Springer, Heidelberg (1999)
12. Buhler, J.P., Lenstra, H.W., Pomerance, C.: Factoring integers with the number field sieve. In: Lenstra, A.K, Lenstra, H.W. (eds.) The Development of the Number Field Sieve. Lecture Notes in Mathematics, vol. 1554. Springer, Heidelberg (1993)

Analysis of Permanent Magnet Synchronous Motor by Different Control Methods with Ansys Maxwell and Simplorer Co-simulation

Huseyin Kocabiyik[1], Yusuf Oner[2(✉)], Metin Ersoz[3], Selami Kesler[2], and Mustafa Tumbek[2]

[1] Ustunel Submercible Pump A.S., Izmir, Turkey
[2] Department of Electrical and Electronics Engineering, Pamukkale University, Denizli, Turkey
yoner@pau.edu.tr
[3] Senkron Ar-ge Engineering A.S., Denizli, Turkey

Abstract. In this study, permanent magnet synchronous motor control methods were investigated. In the literature, permanent magnet synchronous motor control methods are applied with different methods. In order to make the analysis of permanent magnet synchronous motor designed with Ansys Maxwell program to be realistic, control algorithms must be applicable to motor during magnetic analysis. In this context, different control algorithms were written to the previously designed synchronous motor and the real performance of the motor during the analyzes was tried to be reached.

As known, FOC (field oriented control control) and DTC (direct torque control) methods are used as the control method of permanent magnet synchronous motors. These methods are linked by magnetic analysis by creating Ansys Simplorer program. Thus, the magnetic analysis of the motor under control with a control method was carried out in Ansys Maxwell program. In this way, the actual behavior of the motor has been an idea. In the study, without control algorithm, analyses were made using FOC and DTC methods and motor behavior was compared.

Keywords: Permanent magnet synchronous motor · FOC and DTC controller · Ansys Maxwell and Simplorer co-simulation

1 Introduction

Permanent magnet synchronous motors are of the class AC motors, similar to discrete pole motors, but with strong magnets on the rotor to form a field winding. The permanent magnet causes the excitation voltage not to be altered. PMSMs have a greater overturning moment than motors of the same size and with field winding. In PMSMs, the absence of field winding, excitation winding and assembly, and the ease of ventilation systems make the motor more efficient.

D. J. Hemanth and U. Kose (Eds.): ICAIAME 2019, LNDECT 43, pp. 518–525, 2020.
https://doi.org/10.1007/978-3-030-36178-5_41

In the vector control method, the current component of the AC motor is replaced by independent flux and torque components and controlled in a similar manner to a DC motor. In AC motors, all the fluxes rotate at synchronous speed, forming a magneto motor force in 3 phase current. The vector control coincides with the magneto motor force so that the axes of a flux are perpendicular to each other. In particular, it is easier to place magneto motor force of the stator current so that it is orthogonal to the rotor flux. This vector control method is called Field Oriented Control (FOC) due to the fact that the current is directed to the rotor current.

In 1986, a control method called Direct Torque Control (DTC) was developed by Depenbrock and Takashi and applied to asynchronous machines. In the DTC method, it is base on the creation of a switching sequence which directly removes errors in the moment with the flux tolerance value [1, 2].

Takahashi and Noguchi presented the DTC control method, which pro-vides quick response for the asynchronous motor, as an alternative to the vector control method. This method was based on selecting the most suitable voltage vector for the inverter using flux and torque. The accuracy of the study has been ensured with simulations [2].

Depenbrock presented Direct Self-Control (DSC) as a result of his studies. In this method, where they developed for an asynchronous motor fed by a voltage sourced inverter, the moment and three phase stator fluxes are estimated with the measured current and voltage information and stator resistance, and compared with reference values. The inverter switching signals are obtained by selecting the output information of the flux hysteresis controllers according to the moment state. The difference from direct torque control is that there is no switching table. This creates the basis for direct torque control [3]. Zhu et al. Presented a sensorless control system scheme based solely on motor line measurements. The proposed scheme combined the linearization-based controller with a non-linear condition monitor which estimates the rotor position and speed [4].

Morimoto et al. presented a new sensorless control strategy for the synchronous motor with salient poles. The system using expanded electromotor force is used in estimation of position and speed. The proposed system is simple and has been proven by the experimental results [5].

Casadei et al. have made a thorough research on DTC and FOC methods and compared their advantages and disadvantages. They focused on torque response and torque fluctuations [6].

Moon et al. have developed an estimated current system for the calculation of inverter output voltages in current and voltage control systems for SMSM drives [7].

Tang et al. have presented a DTC approach that provides low torque and flux fluctuations and a fixed switching frequency. A significant reduction in flux and torque fluctuations is demonstrated with the approach presented in this study by comparing it with the classical DTC method. The speed estimation is given integrated with the DTC scheme [8].

Kim et al. have developed a permanent magnet synchronous motor PWM VSI inverter for controlling voltage disturbances for the first time [9].

Tripathi et al. emphasized that the field weakening method was insufficient at high speeds and developed the DTC-SVM method [10].

Considering the torque of inertia in real applications, they proposed an adaptive control scheme suitable for the speed control system in SMSM drivers [11].

2 Control Methods of the Permanent Magnet Synchronous Motor

Control methods of alternating current motors are generally divided into two groups as scalar and vector control. The vector control is also divided into two parts as Indirect (Field Oriented) and Direct Control [12].

2.1 Scalar Control (V/F Control)

The purpose of the V/f control is to achieve a constant magnetic flux by increasing the voltage and frequency ratio at the same rate. It is a simple and costless control method. It is often used in applications running at constant speed. It is open-loop controlled, and does not require any position sensors. One of the biggest disadvantages is the inability to intervene in the direct moment. When the ratio of voltage to frequency is kept constant, the magnetic flux remains constant. However, at high frequencies, the magnetic flux in the air gap decreases due to the voltage drop in the stator impedance and the voltage must be increased in order to maintain the torque level.

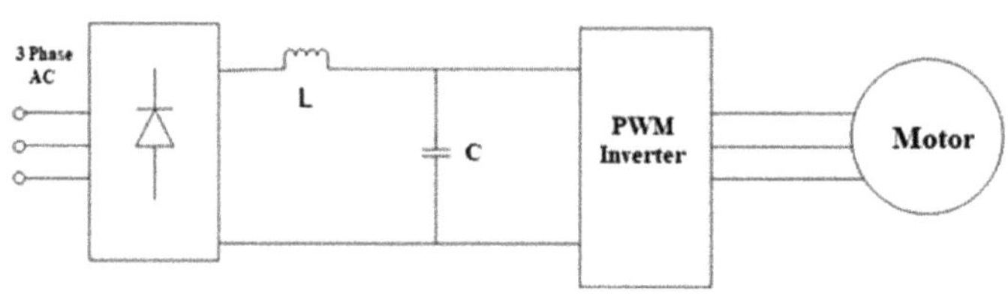

Fig. 1. Fixed DC and PWM converter

A circuit diagram is shown in Fig. 1 to obtain variable frequency and voltage. In Fig. 1, PWM methods are applied to change both the voltage and the frequency at the inverter output while the DC voltage is constant. Because it is not possible to return the energy to the source due to the diode rectifier, the inverter will produce harmonics in the AC supply.

2.2 Field Oriented Control (FOC)

With the FOC technique, the stator current can be divided into two different current components, one providing the flux and the other generating moment. Thanks to this distinction, the magnetic flux and torque control are separated from each other, leading the control characteristic to be linear. The stator current is coupled to the magnetic flux vector and converted into a synchronous rotating reference axis. In these two components of the stator, current the d-axis current is I_d similar to the excitation current of a DC motor, and the q-axis current I_q similar to the inductor current. The FOC technique allows the flux and moment to be in-dependent from each other and to control them in a similar way with a DC Motor [13].

In Field Oriented Control, position information is required to obtain these currents because the currents are converted into vectors in the axis of the d-q in the rotor reference plane. Figure 2 shows the block diagram of the FOC control.

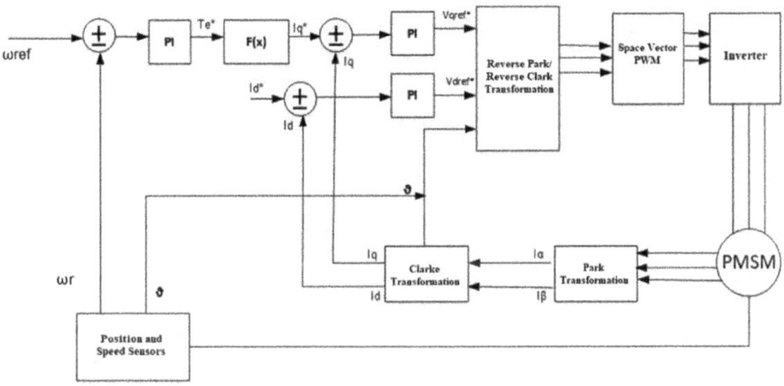

Fig. 2. Fixed DC and PWM converter

2.3 Direct Torque Control (DTC)

The DTC method was first developed by Takahashi, Nagochi and Depenbrock [1–3]. The method of these developers was to control the flux and torque with the appropriate switching elements in hysteresis control. It was produced and marketed by ABB company in 1996. After the developments in the DTC method, this control method has become more common in industry and applications. The DTC method in AC Motors, as the name suggests, is to control the direct moment and flux components with 6 or 8 suitable voltage space vectors selected from the appropriate switching table. It is also defined as the application of a switching series that directly removes the error margin in the moment via the reference and calculated flux to the power electronics elements in the inverter [14]. In the traditional DTC method, two hysteresis bands are used, one for the moment error and one for the flux error. The purpose of the flux hysteresis controller is to maintain the stator flux in the desired reference orbit, and the purpose of the moment hysteresis controller is to maintain the value of the torque in the desired bandwidth [15].

3 Design and Fem Analysis of the Permanent Magnet Synchronous Motor

Permanent magnet synchronous motor is designed for 0.55 kw, 4 terminals, 220 V, 50 Hz, with internal rotor and surface magnet. Figure 3 shows the designed permanent magnet synchronous motor.

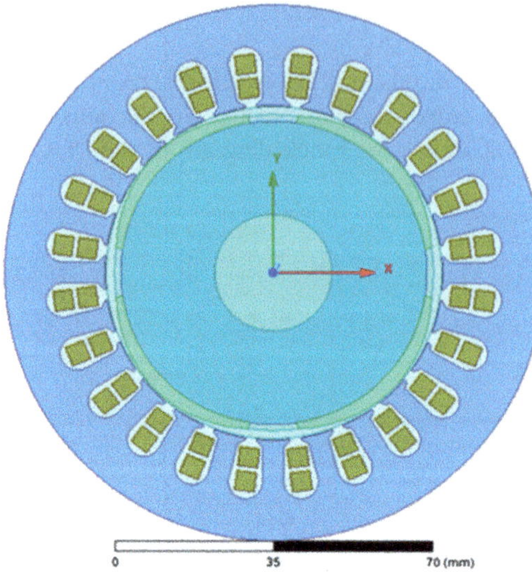

Fig. 3. The designed permanent magnet synchronous motor

The permanent magnet synchronous motor was analyzed according to the finite element method in Ansys Maxwell program and the speed time charts in Fig. 3 and the torque-velocity charts in Fig. 4 were obtained.

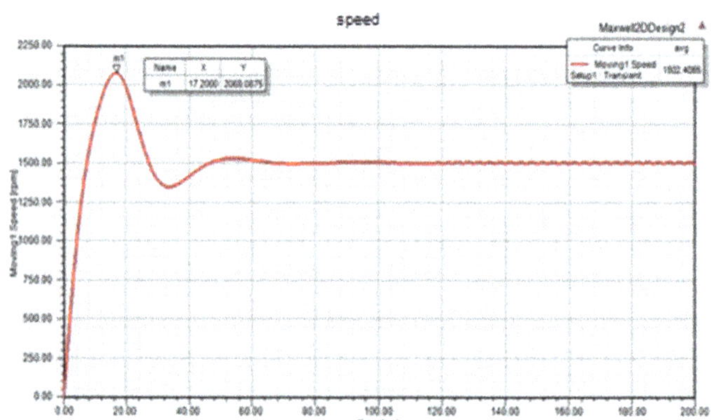

Fig. 4. Speed-time chart

When Fig. 4 is examined, the motor synchronous speed is 1500 rpm ac-cording to the frequency of 50 Hz. The motor reaches this speed at 80 ms. At the same time, the motor speed has increased to 2000 rpm making a peak, and then fluctuated to reach the nominal speed of 1500 rpm. During this analysis, no control algorithm was used, the direct solution of Ansys Maxwell program was utilized.

The moment-time chart is given in Fig. 5. When the graph is examined, it is seen that the motor is in an unstable operating zone for up to 80 ms and stabilized with the speed settling at the nominal value, and has the average moment of 3.66 Nm at the rated speed.

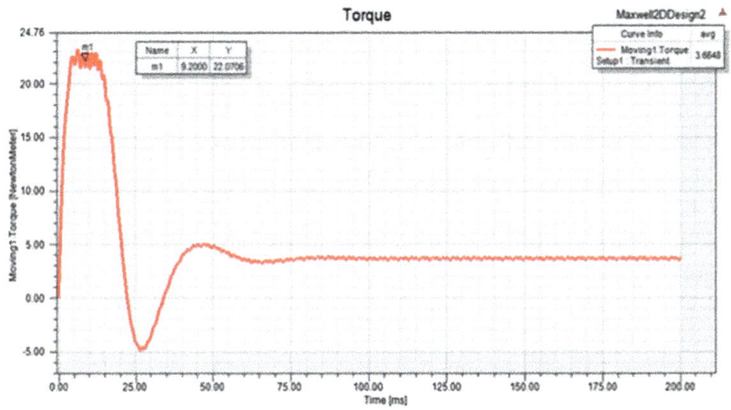

Fig. 5. Moment-time chart

According to two different control methods (Field oriented FOC and direct torque control DTC) designed in the motor simulator, and the speed-time and torque-time charts shown in Fig. 6 are obtained.

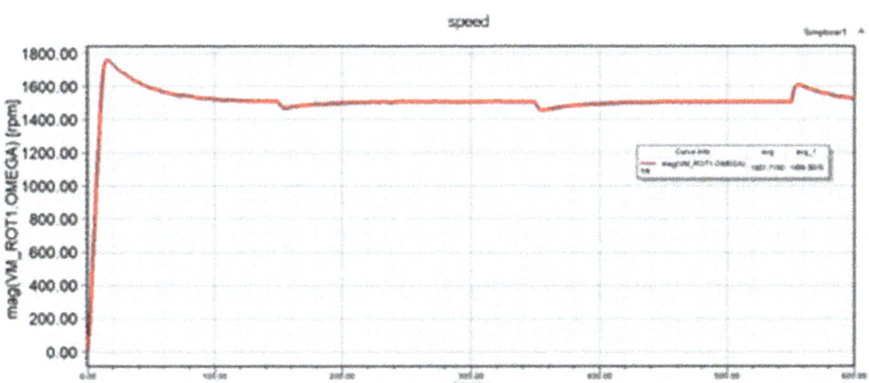

a

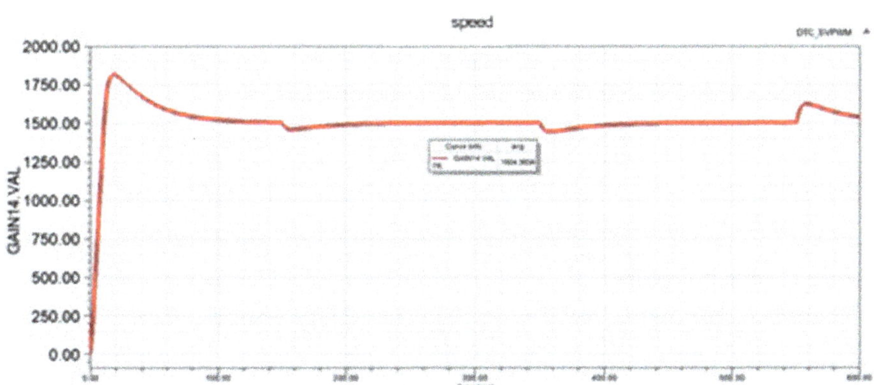

b

Fig. 6. (a) Field orientated FOC control speed-time graph, (b) Direct torque control DTC Moment-time graph

4 Conclusion and Recommendations

Permanent Magnet Synchronous Motors have started to take its place in the industry in recent years, thanks to magnet topologies with increased efficiency and high magnetic density and researches. Developments in the fields of power electronics and microprocessors also facilitated the control of PMSMs. Although scalar control is a simple system and has open-loop control, important issues such as complex mathematical calculations in vector control and switching times have been overcome by power electronics topologies.

ANSYS was connected to the Simplorer program of an PMSMs designed in Maxwell program to conduct the Field Oriented and Direct Torque Control. The control block diagrams were carried out simultaneously with the Maxwell 2D interface and the motor was driven in two control methods.

In both methods, phase transformations were provided, speed and angle information were used. In the FOC method, the I_d current was kept at 0, the PI was obtained by comparing with the I_d current component, the read speed value was compared with the reference speed value and the PI was tested and the I_q value was obtained, and finally it was again compared with the read I_q value, PI was tested again and the U_q value was obtained. In the DTC method, the estimated value of the flux in the I_q current was also calculated from the speed reading to the torque, then the U_d and U_q values were obtained by the PI controllers and it was used to generate the pulse voltages in the SVPWM block. Thanks to the PI controllers, the hysteresis band was not used, so that moment and flux could be controlled separately. The voltages in the d-q axis were used to generate the required pulse voltages in the Space Vector Modulated PVM part according to the equations. When two control methods are compared, one of the biggest ad-vantages of Direct Torque Control is that it allows to directly control the moment and flux, as its name suggests. As seen in the moment graphs, it is observed that the fast and dynamic moment response provides the desired load values faster than the Field Oriented Control and the moment fluctuations are less.

ANSYS Simplorer 2015.2 version was used as the simulation program. Although there are PWM and Sine PWM blocks in the ready blocks of the program, there is no SVPWM block. In the following versions, this block can be added to reduce the number of SVPWM blocks and timing periods. In addition, the operation can be transferred to other programming languages and to a real motor driver, and the actual motor can be controlled with the ANSYS Simplorer interface.

References

1. Depenbrock, M.: Method and Device for Controlling of a Rotating Field Machine. EPO Patent EP0179356 (1986)
2. Takahashi, I., Noguchi, T.: A new quick-response and high-efficiency control strategy of an induction motor. IEEE Trans. Ind. Appl. **22**, 820–827 (1986)
3. Depenbrock, M.: Direct self control of inverter-fed induction machines. IEEE Trans. Power Electron. **3**, 420–429 (1988)

4. Zhu, Z.Q., Shen, J.X., Howe, D.: Improved speed estimation in sensorless PM brushless AC drives. In: IEMDC 2001. IEEE International Electric Machines and Drives Conference (Cat. No. 01EX485), Cambridge, MA, USA, pp. 960–966 (2001)
5. Morimoto, S., Kawamoto, K., Sanada, M., Takeda, Y.: Sensorless control strategy for salient-pole PMSM based on extended EMF in rotating reference frame. IEEE Trans. Ind. Appl. **38**(4), 1054–1061 (2002)
6. Casadei, D., Profumo, F., Serra, G., Tani, A.: FOC and DTC: two viable schemes for induction motors torque control. IEEE Trans. Power Electron. **17**(5), 779–787 (2002)
7. Moon, H.T., Kim, H.S., Youn, M.J.: A discrete-time predictive current control for PMSM. IEEE Trans. Power Electron. **18**(1), 464–472 (2003)
8. Tang, L., Zhong, L., Rahman, M.F., Hu, Y.: A novel direct torque control for interior permanent-magnet synchronous machine drive with low ripple in torque and flux-a speed-sensorless approach. IEEE Trans. Ind. Appl. **39**(6), 1748–1756 (2003)
9. Kim, H.W., Youn, M.J., Cho, K.Y.: New voltage distortion observer of PWM VSI for PMSM. IEEE Trans. Industr. Electron. **52**(4), 1188–1192 (2005)
10. Tripathi, A., Ashwin, M.K., Sanjib, K.P.: Dynamic control of torque in overmodulation and in the field weakening region. IEEE Trans. Power Electron. **21**(4), 1091–1098 (2006)
11. Li, S., Liu, Z.: Adaptive speed control for permanent-magnet synchronous motor system with variations of load inertia. IEEE Trans. Industr. Electron. **56**(8), 3050–3059 (2009)
12. Kronberg, A.: Design and simulation of field oriented control and direct torque control for a permanent magnet synchronous motor with positive saliency. Msc thesis, Uppsala Universitet, Sweden (2012)
13. Merzoug, M.S., Naceri, F.: Comparison of field-oriented control and direct torque control for permanent magnet synchronous motor (PMSM). World Acad. Sci. Eng. Technol. **45**, 299–304 (2008)
14. Menlibar, O.: Asenkron motorda moment dalgalanmalarının ve gürültünün azaltılması. Ph. D. thesis, Yildiz Technic University, (2009)
15. Tang, L., Rahman, M.F.: A new direct torque control strategy for flux and torque ripple reduction for induction motors drive by using space vector modulation. In: 2001 IEEE 32nd Annual Power Electronics Specialists Conference, PESC, 3 (2001)

A Comparison of Data Mining Tools and Classification Algorithms: Content Producers on the Video Sharing Platform

Ercan Atagün[1(✉)] and İrem Düzdar Argun[2]

[1] Department of Computer Engineering, Duzce University, Duzce, Turkey
ercan-atagun@hotmail.com
[2] Industrial Engineering Department, Duzce University, Duzce, Turkey
iremduzdar@duzce.edu.tr

Abstract. With the development of internet technologies, the use of video sharing sites has increased. Video sharing sites allow users to watch videos of others. In addition, users can create an account to upload content and upload videos. These platforms stand out as the places where individuals are both producers and consumers. In this study, data about YouTube which is a video sharing site was used. The content of the content, which is also called as a channel on YouTube, was made by using a set of producers. The data set with 5000 samples on YouTube channels is taken from Kaggle. The data were classified using 4 different data mining tools such as Weka, RapidMiner, Knime and Orange using Naive Bayes and Random Forest algorithms. The parameters are requested from the user in order to obtain a more efficient result in the application of data mining algorithms and in the data preprocessing steps and in the data mining steps. Although these parameters are common in some data mining software, they are not included in all data mining software. Data mining software provides management of some parameters while other parameters cannot be managed. These changes affect the accuracy value in the study and affect the accuracy value in different ratios. Changing the values of the parameters revealed differences in the accuracy rates obtained. A data mining software model has been proposed by emphasizing to what extent the management of the parameters of the study and the extent of the management of the parameters should be connected to the data mining software developer.

Keywords: Data mining · Classification · Youtube

1 Introduction

1.1 Video Sharing Platforms

The increase in Internet bandwidth is changing the habits of users on internet usage. Users use video sharing web sites more.

YouTube ranks second in the most visited sites on the world and ranks first in the video sharing category [1]. According to the June 2018 report on YouTube, 300 h of video per minute is uploaded on YouTube [2]. This video sharing platform, which has a monthly active 1.9 billion users, accounts for about one third of the internet user [3].

D. J. Hemanth and U. Kose (Eds.): ICAIAME 2019, LNDECT 43, pp. 526–538, 2020.
https://doi.org/10.1007/978-3-030-36178-5_42

When the literature is examined, it is possible to find studies in many different fields using this platform. Ata and Atik in their study, video sharing sites on the impact of learning in a study on the classroom education, e-learning, social media learning and mobile learning on YouTube are included in the training applications are mentioned. YouTube and other similar video-sharing platforms are expected to become a part of the education process by becoming widespread [4]. In another study on the impact of video sharing site on education, it is expected that the availability of internet and mobile applications can be easily accessible from anywhere, and this platform is expected to become increasingly widespread [5]. Jovic et al. comparing data mining software, a researcher presented the advantages and disadvantages of the tools. It has been discussed whether the algorithms related to classification, clustering, association rules and regression are present or not, and the areas of application of these algorithms are supported or not supported. In addition, programming languages have been compared with other features such as distribution licenses, interface and command line screens, general purposes [6]. Chen et al. conducted a survey study to compare open source data mining software. Problems such as elimination of deficiencies in data entry sources, low performance and scalability have been discussed [7]. Wahbeh et al. he tested different data mining software such as Weka, Knime, Tanagra, and Orange for performance by applying classification algorithms [8]. King and Elder compared fourteen data mining software. The software's capabilities, flexibility, usability have been evaluated with features such as [9]. Rangra and Bansal have addressed Weka, Keel, R, Knime, RapidMiner and Orange software. Weka and Knime are featured in the forefront with the introduction of RapidMiner and Orange programming support and limited visual interface [10]. When the data mining tools are compared, it is determined that the properties of the vehicles are generally or not. In the literature, a study on the situation where the same algorithms produce different results in different software has hardly been studied. RapidMiner, Microsoft Azure Learning Studio, Weka, H2O and Spark compared to a study in the architecture, compared, compared to the data preprocessing, except Spark Graphical User Interface, except the H2O can be used Oracle database, such as the file types they have been compared in general. The differences in the classification process revealed that the tools differed by an in-memory data processing, parallel programming and iterative programming tasks [11]. In this study, it has been studied to determine the class which measures the channel value by evaluating the video loading, number of subscribers and the number of views of the channels that produce content on YouTube. It was also determined that content producers will have high quality channel class.

2 Data Mining and Open Source Code Data Mining Programs

With the development of information technologies, the number of devices that produce and collect data in technology is constantly increasing. In this case, large amounts of data occur in many areas. In addition to hardware factors and data formation, computer networks, scientific calculations and commercial trends have also been instrumental in increasing the importance of data mining. Data mining is based on a large amount of data to produce meaningful results and to make predictions about the future, meaningful and acceptable relationship rules are determined by computer programming [12]. According to Kitler and Wang, data mining has been defined as the ability to predict important variables that can be predicted from a large number of potential variables [13]. In general, data mining can be defined as the process by which data obtained for a given topic in a given time period can be used to obtain information that can be used and decided on by using certain algorithms.

2.1 Data Preprocessing

In order to achieve consistent and optimum results in data mining, the deficiencies of the data to be used should be eliminated. The data obtained can cause many problems due to various reasons. The data is now a group of data with unnecessary attributes in the data set to be used to obtain the result [14]. Data preprocessing steps: Data Cleanup, Data Aggregation, Data Transformation, Data Reduction. Data Cleanup includes completing or removing missing data, and eliminating deviating (outlier) values. Data Merge, combining data from different sources; Data Transformation, normalization processes for effective implementation of algorithms; Data reduction is the removal of one or more sets of attributes from the data set [15].

2.2 Weka

Weka is an open-source data mining software developed with Java language in the University of Waikato, New Zealand. It has basic data mining algorithms. Data entry is done with its own .arff file extension [16]. Classification, clustering, association rules and regression can be applied [17].

2.3 Knime

Knime is an open-source data mining software developed by the Java language in the University of Konstanz [18]. It can read data in basic format such as.tff file type which can be read by Weka. It can also read XML-based common data read files [19].

2.4 RapidMiner

RapidMiner is a Java language developed data mining software in Dortmund University of Technology in Germany [20]. This software, which provides easy model creation with drag and drop technique, contains more than 1500 data mining and machine learning algorithms. RapidMiner also supports databases such as Access, MSSQL, PostgreSQL, and Sybase [16].

2.5 Orange

Orange is a data mining software developed by the University of Slovenia and using Python and C++ programming languages. It can be used as an interactive software with the addition of the developed code to the software [21]. This software uses data formats such as .csv, .svm, .xml, .arff, .tab, .txt [22].

3 Material and Method

3.1 Naïve Bayes

Naive Bayes is one of the supervised classification algorithms. Requires data to be tagged. In general, it calculates the probability of each data to the result class. The concept of conditional probability stands out here. By calculating the probability of each data group, the class of data to be determined is determined by multiplying each conditional probability [23].

$$P(X/Y) = (P(Y/X) * P(X))/P(Y) \tag{1}$$

P(X) refers to the probability of the occurrence of the X event, the probability of the P (Y) Y event. P(Y|X) shows the probability of the occurrence of the event Y even though X event occurs. In the same way, P(X|Y) indicates the probability of occurrence of X event when Y is occurring [24].

3.2 Random Forest

Random Forest is a supervised learning algorithm. Creates a forest structure by creating and expanding a random tree from the trained trees. Multiple decision trees are created and merged into those trees to be a consistent structure (Table 1).

Table 1. Two-class confusion matrix.

Actual class	Predicted class	Predicted class
	A	B
A	TP	FN
B	FP	TN
Total	P	N

$$\text{Accuracy} = (TP + TN)/(P + N) \tag{2}$$

$$\text{Recall} = TP/(TP + FN) \tag{3}$$

$$\text{Precision} = TP/(TP + FP) \tag{4}$$

$$\text{Error rate} = 1 - \text{accuracy} \tag{5}$$

$$\text{Error} = (FN + FP)/(P + N) \tag{6}$$

Accuracy is the measure of how accurate the classification is estimated. The most commonly used measure of success is the parameters. Recall is the measure of avoidance of class A in which class A is actually estimated. Precision indicates how many of the A estimates are correct.

3.3 Data Set

The data set was taken from Kaggle [25]. The data consists of 6 attributes in the first step. Because 2 attributes were not used for classification, the data was removed in the preprocessing step. One of them is "Rank" and the other one is Channel_Name. Rank is the information that allows the data to be different from each other. The first attribute of the 4 attributes used in the classification shows which quality channel the data has, the second attribute indicates the number of video uploads, the third attribute indicates the number of subscribers owned by the channel, the fourth attribute shows the total number of views of the channel. In all classification processes, 80% of 5000 data was used for the training and 20% for the test (Table 2).

Table 2. Weka-random forest complexity matrix

Actual	Predicted A++	Predicted A+	Predicted A	Predicted Aa	Predicted A−	Predicted B+
A++	1	0	0	0	0	0
A+	0	3	4	0	1	1
A	0	1	75	0	38	79
Aa	0	0	0	0	0	0
A−	0	0	27	0	27	156
B+	0	0	41	0	44	502

The classification of the decision tree obtained by using the J48 algorithm in Weka is listed below. The results of A+ and B+ channels are listed according to this result:

```
Video views > 1260606050
| Video views <= 3866472711
| | Video views <= 2535153504
| | | Video Uploads <= 130
| | | | Subscribers <= 2628950
| | | | | Subscribers <= 565955: A+
| | | | | Subscribers > 565955: B+
```

The properties of the channel with the label A+ and B+ are listed in the above decision tree. If the number of video views is greater than 1260606050, less than 3866472711 and 2535153504, and the number of video uploads is less than or equal to 130, the number of subscribers is labeled as A+ if less than or equal to 565955. If the number of subscribers is greater than 565955, it becomes a channel with the B+ tag.

Video views <= 1260606050
| Video views <= 510079162
| | | Subscribers <= 10523367
| | | | Video Uploads <= 308: A
| | | | Video Uploads > 308: A-

If the number of video views is smaller than 1260606050 and 5100791623, and the number of subscribers is smaller than 10523367, it is labeled Class A if the number of video uploads is less than or equal to 308. If the number of video uploads is greater than 308, it is labeled as class.

Table 3. Classification accuracy values

Data mining tools	Naïve Bayes	Random Forest
Weka	55.10	60.80
RapidMiner	61.80	62.90
Orange	63.10	60.60
Knime	61.70	60.80

In Table 3, There are accuracy values. Naïve Bayes and Random Forest algorithms is applied for Weka, RapidMiner, Orange, Knime.

4 Application

Different algorithms can be used for different data sets for classification. Algorithms used in this study The algorithms determined in 4 different data mining software were selected according to the data. Naive Bayes and Random Forest algorithms have been used in 4 different data mining such as Weka RaipdMiner, Knime and Orange and some parameter differences have been revealed. There is an interactive hierarchical model made over clustering. However, this study is expected to constitute a new model based on classification [26].

Table 4. Parameters and software

Parameters	Weka	RapidMiner	Knime	Orange
useKernelEstimator	✓	✗	✗	✗
MaxDept	✓	✗	✗	✗
Percentage split	✓	✓	✓	✓
linear sampling	✗	✓	✓	✗
Stratified sampling	✗	✓	✓	✓
Shuffled sampling	✗	✓	✗	✗

The parameters used in the study for data mining operations are shown in Table 4. The useKernelEstimator in the Weka software is a parameter that uses a kernel estimator for numeric attributes instead of normal distribution. By default in Weka, a Gauss distribution is assumed for each numeric property. The algorithm can be modified to use a kernel estimator with the useKernelEstimator argument, which can better match the actual distribution of attributes in the dataset. It can also automatically convert numeric properties to nominal properties with the useSupervisedDiscretization parameter.

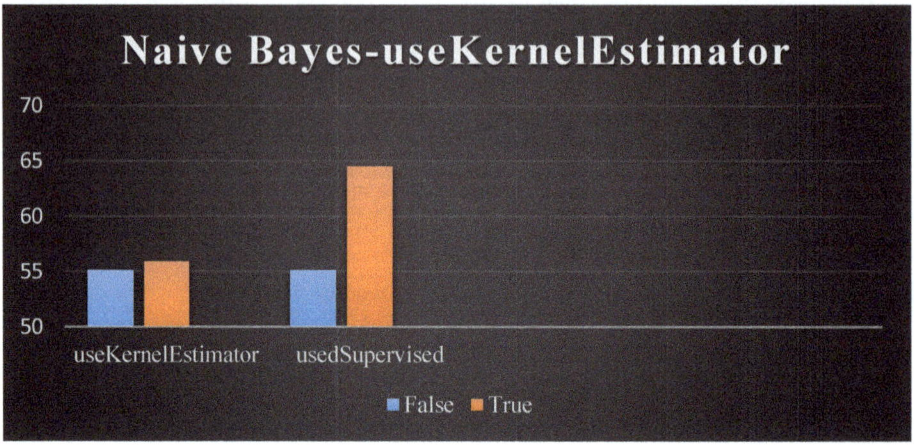

Fig. 1. This figure is about UseKernelEstimator and usedSupervised. When UseKernelEstimator is false and true, accuracy is change. Similarly, usedSupervised is false, accuracy is 55.1%, and when it is true, accuracy is 64.5%.

UseKernelEstimator, the value of this value in the Weka Naive Bayes algorithm is shown in Fig. 1 is shown. When the value of useKernelEstimator from the Weka interface is changed from True to False, Naive Bayes has a success rate of 55.1% to 55.9%. The success rate was also increased by 64.5% when usedSupervised True was performed.

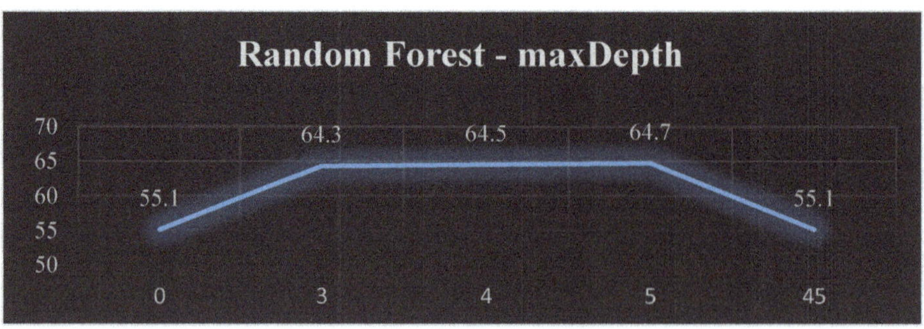

Fig. 2. This figure is about accuracy and maxDepth. MaxDepth is given 0, 3, 4, 5 and 45.

MaxDepth refers to the maximum tree depth for Random Forest, a parameter based on a treebased algorithm. This parameter indicates that the maximum tree depth is infinite. Figure 2 shows the effect of the maxDepth parameter on the success rate. The maxdept property found under the Weka Random Forest algorithm was 64.3% when it was set to 0 by default, 64.5% when it was 4, 64.7% when it was 5, and the success rate was the same as the default value when it was maxdept 45.

Fig. 3. This figure is about bagSizePercent. Horizontal values is about number of bagSizePercent and vertical values is about accuracy.

BagSizePercent, the size of the data set that is divided into pieces in the learning set and defined as a bag. As shown in Fig. 3, bagSizePercent works with 60.8% success with a value of 100 by default. When this value is 10, the success rate is 64.6%; when this value is 4, the success rate is 65.1%; when this value is 200 and 100, the success rate is 60.8%.

StoreOutOfBagPredictions includes information on whether to store out-of-bag estimates, outputOutOfBagComplexityStatistics, or whether complexity-based statistics will be issued when evaluating outside the bag. However, the storeOutOfBagPredictions value has not changed when True and outputOutOfBagComplexityStatistics is True.

NumDecimalPlaces is the number of decimal places to use for output of numbers in the model; numExecutionSlots refers to the number of sections to be used to create the set in the data. When NumDecimalPlaces was 2, the success rate did not change when changed to 4 and 1, and NumExecutionSlots did not change the result when the default value was 1 with a value of 0, 5, 8.

NumFeatures is the parameter that sets the number of randomly selected properties. In Fig. 4, the success rate of the numFeatures from 0 to 3, 12, 16, 80 was 61.4.

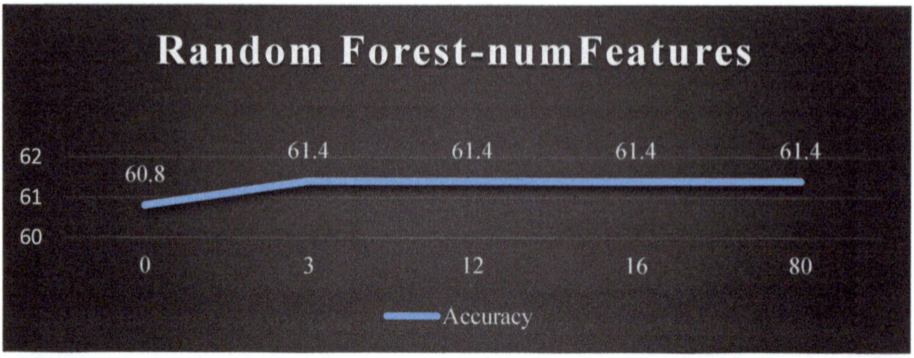

Fig. 4. This figure is about numFeatures and accuracy. NumFeatures is 0, 3, 12, 16, 80 and accuracy changes.

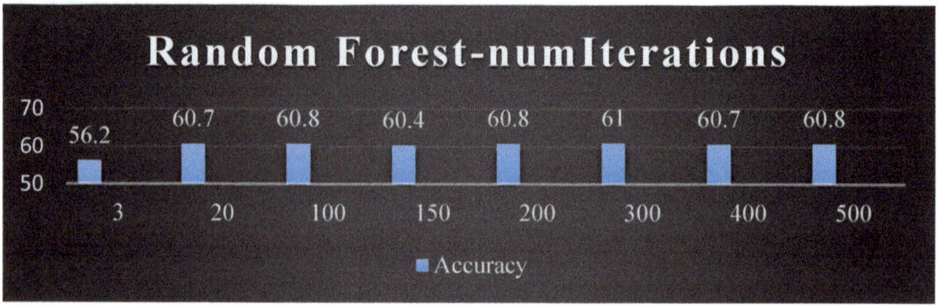

Fig. 5. This figure is about numIterations and accuracy. Accuracy changes depending on the value of numIterations.

In Fig. 5, The effect of the numIterations parameter on the success rate is shown. NumIterations indicate the number of iterations to be performed. When the default value of the numIterations parameter is 100, the success rate is 60.8%. When the parameter is changed to 150, the success rate is 60.4%. When the parameter is changed to 200, the success rate is 60.8%. When the parameter is changed to 300, the success rate is 61%. When the parameter is changed to 400, the success rate is 60.7%. When the parameter is changed to 500, the success rate is 60.3%. When the parameter is changed to 20, the success rate is 60.7%. When the parameter is changed to 3, the success rate is 56.2%.

Table 5. Cross validation parameters and accuracy

Parameters	Weka	RapidMiner	Knime	Orange
Stratified sampling	×	%61.80	%61.80	%49.80
Shuffled sampling	×	%61.80	×	×
automatic	×	%61.80	×	×
Linear sampling	×	×	%91.00	×
Draw randomly	×	×	%61.20	×
Randomly stratified	×	×	×	%50.7

In Table 5, cross validation techniques were compared. Percentage split indicates how much of the learning will be the test data for the classification according to the percentage to be given. There are differences in how to choose data for learning and testing. For example, RapidMiner software includes stratified sampling, shuffled sampling, automatic and linear sampling. The results obtained using the Naïve Bayes algorithm are shown on Table 5. RapidMiner also shows a value of 61.80% for stratified sampling, shuffled sampling, and automatic, 95.70% for linear sampling. These values show a success rate of 61.2% for linear sampling in Knime software and 61.8% for stratified sampling randomly.

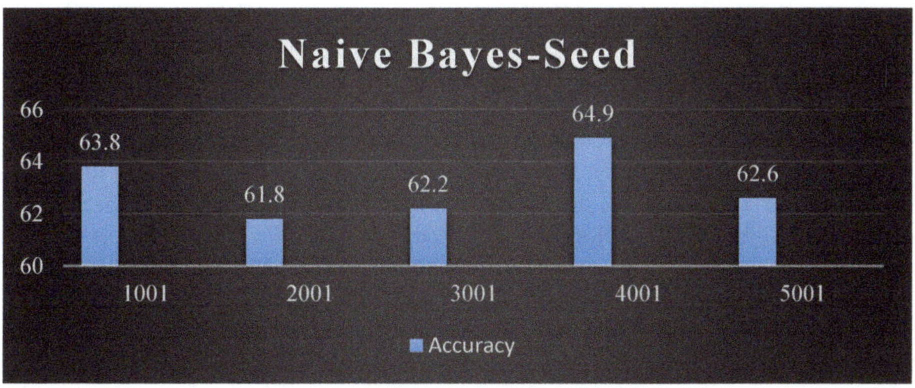

Fig. 6. This figure is about Seed. Accuracy changes depending on the value of Seed. Seed is 1001, 2001, 3001, 4001 and 5001. Accuracy is 63.8%, 61.8%, 62.2%, 64.9%, 62.6%.

Figure 6 shows the effect of seed value on the success rate. The Seed value showed success rates in RapidMiner software for the value of 1001, 63.70%, 61.80% for 2001, 621% for 3001, 64.90% for 4001 and 62.60% for 5001. Changes made to the default value in 2001 affected the success rate. The values that affect the result with the help of the user interface in the Weka tool: useKernelEstimator (for Naive Bayes), usedSupervised (for Naive Bayes, the most increase), bagSizePercent (Random Forest, maximize), maxdept, numfeatures, numIteration. Since it is not possible to manage these values with the Orange interface, success rates are predicted to change. With Knime, the maximum number of unique setpoints per attribute can be changed with the user interface. Thus, a solution can be found with the most optimal category number. Choosing the type of cross validation procedures in RapidMiner software can affect the result. Linear sampling, shuffled sampling, changed the result of stratified sampling. The Seed value is the value of the random number generator in Java. Random Forest is slightly more affected than random. The random separation of data in the form of training and testing is an important issue in data groups that have poor consistency and are not distributed evenly. The number generator used with Java in Weka is the same number of times each time. This number is determined by the seed value. The default value is 1, but a random number sequence is generated using values such as 2,3 or 34.

5 Conclusion and Discussion

In this study, according to the features of YouTube content producers, two different algorithms were used for classification. Different results were observed when tested in different data mining software. The software can give different results in different same algorithms depending on development purposes. For example, software that brings speed to the foreground and software that brings the memory to the foreground can encode the same algorithm differently, but it will also affect the result. The coincidence used when selecting training and test data can affect the result. In another point, when the complexity matrices of the software were examined, it was determined that 2 of the data allocated for testing by Orange were deleted. The identification of the data preprocessing by the user by the software in the software and the differences in the methods used in the data preprocessing process can differentiate the results. Another difference will be the differences in the success rates of the model when the data with different values are the same in the algorithms that make up the model and the results are different. In addition, there are differences arising from the programming languages in which the software is developed. For example, in Java, the type of data is partially modifiable; In this study, parameter information at user level in data mining tools and the effect of these parameters on classification success were investigated. Some parameters affected the success rate of some parameters are determined by the comparison of some affected parameters. With this study, it is concluded that the parameters affecting the success rate of more than 5% at the user level and the parameters that affect below 5% should be under the management of the developers who convert the data mining algorithms into code. In this case, the parameter step that the data analyst needs to interpret and must manage is included in the last step, and erroneous results can be obtained and the wrong measures can be applied. For example, these results may bring in the loss of individual investments or institutional investments that will invest. Developers contribute to data mining software by writing a common code for all data groups based on general algorithm knowledge. However, this situation is not accepted by the data analyst and subject specific experts. After applying the algorithms in the last part of the steps of data mining, the results are interpreted by the experts. This study emphasizes the active inclusion of all data mining processes in the process, not after the results of specific experts. A new data mining software to be developed reveals that experts should be aware of the structure of data and have pre-intuitions related to the subject.

References

1. ALEXA Homepage. https://www.alexa.com. Accessed 15 Dec 2018
2. OMNICOREAGENCY Homepage. https://www.omnicoreagency.com. Accessed 16 Dec 2018
3. PRESS Homepage. https://www.youtube.com/yt/about/press. Accessed 16 Dec 2018
4. Ata, A., Atik, A.: ALTERNATİF BİR EĞİTİM-ÖĞRETİM ORTAMI OLARAK VİDEO PAYLAŞIM SİTELERİ: ÜNİVERSİTELERDEKİ YOUTUBE UYGULAMALARI. Soc. Sci. **11**(4), 312–325 (2016)
5. Yıldırım, N., Özmen, B.: Video paylaşim sitelerinin eğitsel amaçli kullanimi. Educ. Sci. **7** (1), 288–295 (2012)
6. Jovic, A., Brkic, K., Bogunovic, N.: An overview of free software tools for general data mining. In: 2014 37th International Convention on Information and Communication Technology, Electronics and Microelectronics (MIPRO), pp. 1112–1117. IEEE, Croatia (2014)
7. Chen, X., Ye, Y., Williams, G., Xu, X.: A survey of open source data mining systems. In: Pacific-Asia Conference on Knowledge Discovery and Data Mining, pp. 3–14. Springer, Heidelberg (2014)
8. Wahbeh, A.H., Al-Radaideh, Q.A., Al-Kabi, M.N., Al-Shawakfa, E.M.: A comparison study between data mining tools over some classification methods. Int. J. Adv. Comput. Sci. Appl. **8**(2), 18–26 (2011)
9. King, M.A., Elder, J.F., Gomolka, B., Schmidt, E., Summers, M., Toop, K.: Evaluation of fourteen desktop data mining tools. In: 1998 IEEE International Conference on Systems, Man, and Cybernetics, vol. 3, pp. 2927–2932. IEEE, Beijing (1998)
10. Sharma, N., Bansal, K.L.: Comparative study of data mining tools. J. Adv. Database Manag. Syst. **2**(2), 35–41 (2015)
11. Dusanka, D., Darko, S., Srdjan, S., Marko, A., Teodora, L.: A comparison of contemporary data mining tools. In: XVII International Scientific Conference on Industrial Systems, IS 2017, pp. 150–155, Novi Sad (2017)
12. Köktürk, F., Ankarali, H., Sümbüloğlu, V.: Veri Madenciliği Yöntemlerine Genel Bakış. Turkiye Klinikleri J. Biostat. **1**(1), 20–25 (2009)
13. Kittler, R., Wang, W.: The emerging role for data mining. Solid State Technol. **42**(11), 45 (1999)
14. Ozekes, S.: Veri Madenciligi Modelleri ve Uygulama Alanlari. Istanb. Ticaret Univ. J. **3**, 65–82 (2003)
15. Oğuzlar, A.: Veri ön işleme. Erciyes Üniversitesi İktisadi ve İdari Bilimler Fakültesi Dergisi **21**, 67–76 (2003)
16. Kaya, M., Özel, S.A.: Açık Kaynak Kodlu Veri Madenciliği Yazılımlarının Karşılaştırılması. In: 14th Academy Information Conference, pp. 47–53, Mersin (2014)
17. WEKA Homepage. https://www.cs.waikato.ac.nz/ml/weka. Accessed 17 Dec 2018
18. KNIME Homepage, https://www.knime.com. Accessed 17 Dec 2018

19. Akpolat, O., Odabaş, S.Ç.: KNIME Yazilimi Ile Kimyasal Analiz Verilerinin Kümelenmesi. In: 16XVI. Akademik Bilişim Konferansı, Aydın (2016)
20. RAPIDMINER, Homepage. https://rapidminer.com. Accessed 18 Dec 2018
21. Demšar, J., Curk, T., Erjavec, A., Gorup, Č., Hočevar, T., Milutinovič, M., Polajnar, M., Toplak, M., Starič, A., Štajdohar, M.: Orange: data mining toolbox in Python. J. Mach. Learn. Res. **14**(1), 2349–2353 (2013)
22. Düzdar, İ., TemüR, G.: Comparing of results and implementation of clustering methods of data mining software with a data set. In: Electric Electronics, Computer Science, Biomedical Engineering's Meeting (EBBT), pp. 1–5. IEEE, Istanbul (2017)
23. Cooper, G.F., Herskovits, E.: A Bayesian method for the induction of probabilistic networks from data. Mach. Learn. **9**(4), 309–347 (1992)
24. Çaliş, K., Gazdaği, O., Yildiz, O.: Reklam İçerikli Epostaların Metin Madenciliği Yöntemleri ile Otomatik Tespiti. Bilişim Teknolojileri Dergisi **6**(1), 1–7 (2013)
25. KAGGLE Homepage. https://www.kaggle.com. Accessed 18 Dec 2018
26. Yeşilbudak, M., Kahraman, H. T., Karacan, H.: Veri madenciliğinde nesne yönelimli birleştirici hiyerarşik kümeleme modeli. Gazi Üniversitesi Mühendislik-Mimarlık Fakültesi Dergisi **26**(1), 27–39 (2011)

Normal Mixture Model-Based Clustering of Data Using Genetic Algorithm

Maruf Gogebakan[1(✉)] and Hamza Erol[2]

[1] Department of Applied Mathematics, Faculty of Computer Science, Abdullah GUL University, Kayseri, Turkey
maruf.gogebakan@agu.edu.tr
[2] Department of Computer Engineering, Faculty of Engineering, Mersin University, Mersin, Turkey
herol@mersin.edu.tr

Abstract. In this study, a new algorithm was developed for clustering multivariate big data. Normal mixture distributions are used to determine the partitions of variables. Normal mixture models obtained from the partitions of variables are generated using Genetic Algorithms (GA). Each partition in the variables corresponds to a clustering center in the normal mixture model. The best model that fits the data structure from normal mixture models is obtained by using the information criteria obtained from normal mixture distributions.

Keywords: Genetic Algorithm · Gaussian mixture models · Model based clustering · Information criteria

1 Introduction

Components in multivariate normal mixture distributions match a fragmentation in heterogeneous data [2]. Mixture model of normal distributions is defined as

$$f\left(x_j;\theta\right) = \sum\nolimits_{i=1}^{g} \pi_i f_i\left(x_j;\psi_i\right) \tag{1}$$

where π_i represents probability weights for i. cluster centers $0 < \pi_i < 1$ and $\sum_{i=1}^{g} \pi_i = 1$ and $\psi_i = (\mu_i, \Sigma_i)$ denotes μ_i mean vectors and Σ_i covariance matrix. $f_i\left(x_j;\mu_i,\Sigma_i\right)$ multivariate Gaussian densities shown as,

$$f_i\left(x_j;\mu_i,\Sigma_i\right) = \frac{1}{(2\pi)^{\frac{g}{2}}|\Sigma_i|^{\frac{1}{2}}} exp\left\{-\frac{1}{2}\left(x_j-\mu_i\right)^T\Sigma_i^{-1}\left(x_j-\mu_i\right)\right\} \tag{2}$$

thus, in the equation of (1), $\theta = \pi_1,\ldots,\pi_g,\psi_1,\ldots,\psi_g$ vectors are the multivariate parameter vector of normal mixture distributions in the Ω space vector [1].

D. J. Hemanth and U. Kose (Eds.): ICAIAME 2019, LNDECT 43, pp. 539–543, 2020.
https://doi.org/10.1007/978-3-030-36178-5_43

2 Data Set and Method

The use of Genetic Algorithms for model-based clustering by using fragmentation of variables in multivariable data is described on synthetic big data.

2.1 Variable Data Fragmentation in Multivariable Data

In multivariable data, the fragmentation of the variable is determined using univariate normal mixture distributions [7]. Univariate normal mixture distribution define as

$$f(x;\theta) = \sum\nolimits_{i=1}^{g} \pi_i f_i(x;\mu_i,\sigma_i) \tag{3}$$

where $(x;\theta)$, g and π_i represents normal mixture distribution density function, number of component in distribution and mixed probability weight respectively.

The number of fragmentation of variables is obtained by optimization with information criteria using Gaussian mixture distributions [7]. In multivariable data, fragmentation numbers can be estimated approximately by looking at the graphs of the variables (Histogram and Normal Probability Plots) [6].

2.2 Determining the Observations of Fragmentation in Variable by k-Means Algorithm

In multivariable data, the fragmentation of the variables is determined with the help of normal mixed models, and the values of the observations falling into the fragmentation in the heterogeneous variables are assigned to the subgroups in the data. The determination of the observations to the subgroups is calculated using the Euclid distance with the help of the following equation (Table 1).

Table 1. Determining the fragmentations of variables in the multivariate data set based on normal mixture models.

Variables	X_1	X_2	$X_3X_4X_5\ldots X_{15}$
Fragments	2	2	1 1 1 … 1

$$argmin_s \sum\nolimits_{i=1}^{k} \sum\nolimits_{j\in s_i} \| x_j - \mu_i \|^2 \tag{4}$$

2.3 Determining the Number of Cluster and Location of Normal Mixture Models Using Genetic Algorithms

When creating normal mixed models, the number of sets, the location of the cluster centers, the structure of the sets and the volume are determined by heterogeneous (fragmented) variables [3]. Genetic Algorithms are used to determine the number of clusters and mixture models. Lower and upper limit of the numbers cluster depending on the fragmentation of variables in the mixture model, $s = 1, \ldots, p$ and k_s are obtained as follows to show disintegration on the variable.

$$C_{max} = \prod_{s=1}^{p} k_s \; ve \; C_{min} = max\{k_s\} \tag{5}$$

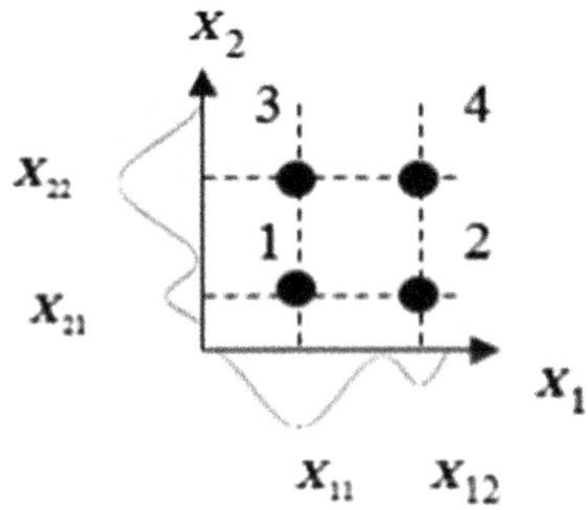

Fig. 1. Cluster centers corresponding to heterogeneous variables in multivariate data

It is obtained in the form of $C_{max} = 2.2.1\ldots1.1 = 4$ and $C_{min} = max\{2, 2, 1, \ldots 1, , 1\} = 2$. corresponding to the fragmentation in the data set (Fig. 1).

2.4 Number and Structure of Normal Mixed Models

In the case of multivariate data, the fragmentation of the heterogeneous variables X_1 and X_2 the homogeneous structure of the other variables, the number of models indicated by the total number of M_{Total} is calculated as follows.

$$M_{Toplam} = 2^{C_{Max}} - 1 = 2^4 - 1 = 15 \tag{6}$$

2.5 Number of Candidate Normal Mixture Models and Genetic Algorithms

In multivariable data, the number of normal mixture models is determined depend on the number of cluster centers and subgroups of heterogeneous variables. In normal mixture models, at least one subset (cluster) corresponds to the fragmentation of heterogeneous variables, whereas models with centers that do not correspond to subgroups are invalid models. As a result of the calculation, the locations and model numbers of the centers are given in Table 2.

Genetic Algorithm determines the components of normal mixed models calculated by fragmentation. While mixture models are represented by DNA sequence, centers with clustering in the sequence are shown as 1, non-clustering centers are 0 and models are generated with Genetic Algorithm. Gene sequences and model numbers corresponding to the assumption mixture models are given in Table 3.

Table 2. Cluster numbers, clusters' positions, and model numbers for candidate models matching the hypothesis among mixture models.

Cluster numbers	Location of centers	# of model
2	1st Column 1 and 2nd Column 1 Cluster	2
3	1st Column 2 and 2nd Column 1 Cluster	4
4	1st Column 2 and 2nd Column 2 Cluster	1
# of model		15
# of candidate model		7

Table 3. The number of normal mixture models, the number of suitable models and the DNA sequence in data with fifteen variables.

# of cluster	# of models	# of suitable models	DNA sequence
			1 2 3 4 …14 15
1 Cluster model	4	–	–
2 Cluster models	6	2	100100…0
			011000…0
3 Cluster models	4	4	111000…0
			110100…0
			101100…0
			011100…0
4 Cluster models	1	1	111100…0

2.6 Parameter Estimation of Normal Mixture Models Using Genetic Algorithm

Component weights, mean vectors and covariance matrices are calculated from the sample for normal mixture distribution models in multivariable data. Fragmentations in heterogeneous variables constitute mixed models using the number of clusters and the positions of the genes. While applying the genetic algorithm on the model, the 0 and 1 genes in the DNA sequence representing the model determine the presence or absence of clustering in the position corresponding to the mixture distribution.

3 Conclusion

In this study, the method for calculating the normal mixed models on the big data for clustering using genetic algorithm has been proposed. In the method, fragmentation in the appropriate variables of the data was found based on univariate normal mixture distributions. Cluster centers corresponding to subgroups of variables were determined and calculation method using Genetic Algorithms was proposed to determine the number of models. Normal mixture models have been created by applying genetic algorithms and DNA sequences corresponding to the obtained models are shown. The calculation of the parameters in the mixture model was used to calculate the information criteria for each model using the genetic algorithm and to obtain the best normal mixture model.

Acknowledgment. The authors also would like to thank to editor(s) for his support to adapt the study content to Springer chapter requirements.

References

1. Fraley, C., Raftery, A.E.: How many clusters? Which clustering method? Answers via model-based cluster analysis. Comput. J. **41**, 578–588 (1998)
2. McLachlan, G.J., Chang, S.U.: Mixture modelling for cluster analysis. Stat. Methods Med. Res. **13**, 347–361 (2004)
3. Erol, H.: A model selection algorithm for mixture model clustering of heterogeneous multivariate data. In: 2013 IEEE International Symposium on Innovations in Intelligent Systems and Applications Innovations in Intelligent Systems and Applications, Albena, pp. 1–7 (2013). https://doi.org/10.1109/inista.2013.6577617
4. Servi, T., Erol, H.: On total number of candidate component cluster centers and total number of candidate mixture models in model based clustering. Selçuk J. Appl. Math. **8**(2), 57–69 (2007)
5. Akaike, H.: A new look at the statistical model identification. IEEE Trans. Autom. Control **19** (6), 716–723 (1974)
6. Schwarz, G.: Estimating the dimension of a model. Ann. Stat. **6**, 461–464 (1978)
7. Gogebakan, M., Erol, H.: A new semi-supervised classification method based on mixture model clustering for classification of multispectral data. J. Indian Soc. Remote Sens. **46**(8), 1323–1331 (2018)
8. Erol, H., Gogebakan, M., Erol, R.: Grid structures and orientations of clusters using discretization of variables in big data. In: Proceedings of International Conference on Engineering, Technology, and Applied Science ICETA 2017, pp. 16–31 (2017). ISSN 2411-9318

Analyzing and Processing of Supplier Database Based on the Cross-Industry Standard Process for Data Mining (CRISP-DM) Algorithm

Mohsen Jafari Nodeh[1], M. Hanefi Calp[2(✉)], and İsmail Şahin[3]

[1] Department of Informatics Institute, Gazi University, Ankara, Turkey
mohsen@gazi.edu.tr

[2] Department of Management Information Systems, Karadeniz Technical University, Trabzon, Turkey
mhcalp@ktu.edu.tr

[3] Department of Industrial Design Engineering, Gazi University, Ankara, Turkey
isahin@gazi.edu.tr

Abstract. In recent years, Data Mining has grown significantly in almost every field. Sectors such as banking, insurance, pharmaceuticals and retailing utilize data mining techniques widely to reduce costs, improve research and increase sales. However, large projects are being carried out on this issue and standards on data mining technique are required. In order to respond to this request, it can be said that CRoss-Industry Standard Process for Data Mining (CRISP-DM) is the most important effort. CRISP-DM is used in many studies, grew as an industry standard, and is defined as a series of sequential steps that guide the application of data mining technique. The CRISP-DM reference model for data mining provides an overview of the life cycle of a data mining project and includes the phases, related tasks and outputs of a project. CRISP-DM is an effort to provide industrial standards for DM applications, including business understanding, data understanding, data preparation, modeling, evaluation and deployment steps. The quality and accuracy of each of the CRISP-DM steps related to DM applications used in different fields is very important for the success of the whole project. In this study, firstly the CRISP-DM algorithm and steps are investigated, later, the data related stages of the CRISP-DM usage in a project (data monitoring and evaluation) are examined, and explained by using an example application. In this process, the supplier database based on the CRISP-DM algorithm for Data Mining has been analyzed and processed. The aim of this study is to explain the role of the steps of CRISP-DM, and specially to process, understand and prepare information in the process of data discovery. In this way, it is expected that create awareness about CRISP-DM, and the impact of this process on projects is expected to be clearly understood.

Keywords: Supplier selection · Data mining · CRISP-DM

1 Introduction

Data mining is an issue that can be applied in almost every field and important results can be obtained. At this point, data is the most basic element and the high amount of data is means that some important conclusions can be made. Significant data can be

D. J. Hemanth and U. Kose (Eds.): ICAIAME 2019, LNDECT 43, pp. 544–558, 2020.
https://doi.org/10.1007/978-3-030-36178-5_44

produced even in the slightest process. Therefore, it is important for enterprises to make these data meaningful and to use them in making right decisions [1]. Data mining are used in many different areas such as machine learning, database and data warehousing technology, statistics, information retrieval, neural networks, pattern recognition, information-based systems, data visualization, artificial intelligence, image and signal processing. It is therefore a very useful tool for the growth of a business, and it has problem definition, iterative process of six stages, such as data discovery, data preparation, modeling, evaluation and implementation [2]. Data mining methods and techniques have become an important area of research, because being involved in a data analysis process can reveal secret relationships, behavior patterns, asset profiles, and similar schemes in data stored in large databases or repositories. Information discovered through intelligent methods may not be obtained by conventional methods, such as statistical analysis, data queries, or other analytical methods, due to the vague idea of the large amount of collected data and information [3]. Data mining can be defined as the process of finding previously unknown patterns and trends in the database and using it to create predictive models. In other words, it can be defined as data selection and discovery using large data stores to reveal previously unknown patterns [4].

Data mining requires a standardized approach that will help to translate enterprises issues, and propose appropriate data transformations and data mining techniques, and provide tools to assess the effectiveness of results and document the experience. CRISP-DM solves these problems by defining a process model for carrying out data mining projects, regardless of the industry sector and the technology used. CRISP-DM process was started in 1996 by Daimler Chrysler. This process (CRISP-DM) is an iterative process model and has been the most preferred methodology ever since. CRISP-DM is an effort to provide industrial standards for the implementation of DM and consists of six stages; business understanding, data understanding and preparation, modeling, evaluation, and deployment [3, 5–8].

This study purposed to explain the role of the steps of CRISP-DM, and especially to process, to understand, and to prepare information in the process of data discovery. Thus, it is hoped that create awareness about CRISP-DM, and the impact of this process on projects is expected to be openly understood. In this study, analyzing and processing of the supplier database with the CRISP-DM algorithm was explained by a sample application. Firstly, the literature reviewed; later, the material and method of the study were explained, and finally, the results and recommendations obtained from the study were given.

2 Literature Research

In the literature, it is possible to find many studies on CRISP-DM, its processes and applications in different fields. For example, Mosaddar and Shojaie proposed a model based on data mining to define inefficient maintenance activities in a maintenance system. They identified certain inefficient maintenance activities and stated that the model they proposed gave maintenance decision makers the opportunity to make changes in the maintenance systems under supervision. As a result, organizations have demonstrated that they focus on correcting these unproductive activities and thus reducing costs associated with their realization [9].

Bošnjak et al., have studied the limitations and challenges of discovering the knowledge of analyzing the data of small and medium-sized enterprises. Many useful findings were obtained during data modeling, but some additional irregularities were observed in the data subsets. In order to avoid the exclusion of large amounts of input data from intelligent analyzes, more attention should be paid to the data collection process. As a result, if the data analyst in the later stages of the CRISP-DM methodology had overseen the data collection process, they would have a view that the KDD (Knowledge in Data Discovery) process would be less troublesome [3].

Nadali and et al., evaluated the success level of CRISP-DM methodology based data mining projects by using blurred expert systems. It was aimed to design a fuzzy expert system to evaluate the success level of data mining projects based on the quality of CRISP-DM methodology stages as one of the data mining methodologies. In the study, the CRISP-DM stages were reported as inputs of the Fuzzy Inference System (FIS) model and the output was taken as the success level of the data mining project. By using the proposed system, as the evaluation system, the success level of data mining projects can be determined by considering the quality level of the main stages of data mining projects [8].

Catley et al., examined the current CRISP-DM approach for modeling using an intensive care unit case study. They emphasized the need for an extended CRISP-DM approach while modeling clinical systems that apply techniques such as Case Study, Data Mining, and Temporal Temporal Abstraction. They proposed an extended CRISP-DM methodology to support the TDM known as CRISP-TDM. They stated that this approach supports clinical research in multidimensional time series [10].

Rocha and Junior aimed to assess the use of decision tree techniques with the CRISP-DM management model to help prevent bank fraud. This article presents a study on decision trees, which is an important concept in the field of artificial intelligence. The study focused on discussing how these trees are. They showed that they helped with the analysis of the information related to the bank transactions and the fraud determination decision process. This was proved by the CRISP-DM management model of data mining in large operational databases that are logged from internet bank transactions [11].

Cespivova and et al., the introduction of medical ontology and other historical information in the process of association mining has been studied. The inventory used consists of the LISp-Miner tool, the UMLS ontology, the STULONG data set on cardiovascular risk, and a set of simple qualitative rules. By the experiment, they demonstrated that an ontology could bring benefits to all stages of the KDD cycle as described in CRISP-DM [12].

When the studies are examined, it is seen that each study has its pros and cons compared to the other and the importance of the study is obvious. In the third chapter, all the details of the study are given.

3 Material and Method

This framework is responsible for environmental data monitoring and analysis functions.

3.1 The CRISPs-DM Framework Used in the Scope of Study

The model DM-CRISP is considered to have three executive steps for monitoring and assessing the data. These steps include (Fig. 1):

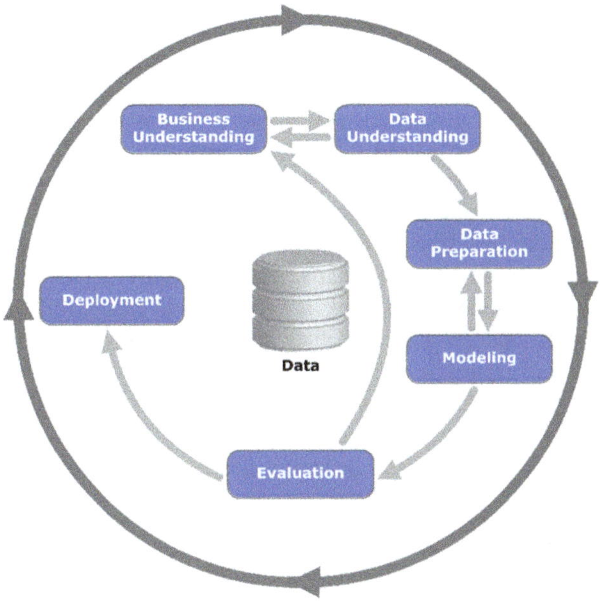

Fig. 1. CRISP-DM Process Diagram [13].

- Business Understanding
- Data Understanding (Analyzing and examining data)
- Data Preparation [9, 13].

Business Understanding
Since the scope of the reviewed activities is related to project supply chain management and in particular the project supply management, in the first step, the basic familiarity should've been studied with the supply chain and processes of providing project management should've taken place.

In the procedure of company dealing with manufacturers (employers), the company, by receiving the basic list of trusted supplier of employers (AVL), seeks and negotiates with those suppliers and other powerful ones, also finally, the list of selected providers for support is presented to them. In this approach, contracts with suppliers are done after the approval of the employer.

That defined operational tasks for the ACL section in company is selecting better and appropriate suppliers and thus, proper evaluation of suppliers is one of the important business goals of this section. Therefore, several steps are considered for evaluating powerful suppliers' reviews. In this approach, evaluations are done on information obtained from received inquiries and suppliers' submitted documents. After initial inquiries, in order to know whether suppliers are ready to cooperate with the company or not, suppliers are examined into 2 groups of technical and financial criteria. In this process at every stage approved companies in the previous step are only studied, eventually leading to the approval of the great Company by the employer.

Data Understanding

Data mining process requires some understanding of the data that it's investigating.

Collecting Initial *Data*

There was no integrated database of information including information received from suppliers and assessments carried out in this project. The only existing information bank in the company was the access file of the list of production companies and/or companies that import materials and equipment related to the gas and oil industry which contains information about the name of the company, national, products, locations and the company's site. Therefore, it was decided that information of intended companies to be collected from various files and databases. For each project, a set of several files and forms exist that it should be integrated with other information in order to do data mining projects. Thus, samples were collected from the information contained in the projects.

Files and forms in each project can be divided into the following categories:

- Trusted provider list form (ACL)
- Project procurement plan
- Project cash flow
- Initial inquiry forms of Suppliers' cooperation
- Submitted documents and proposals of suppliers
- Technical evaluations of suppliers
- Suppliers' bids comparing table
- Suppliers' contract

Explaining and Describing *Data*

After preliminary exploration in relation to quality and reliability of data, useful subsets for the proposed model were selected. At this stage, all contents of the AVL files, technical evaluations of suppliers, project cash flow and compeer table of suppliers' suggested prices is considered and is selected for the continuation of the project. Profile standardization of database providers for the company name and national identification in various files and business (producer/intermediary) companies were used. The files of the procurement planning projects also were used for determining the schedule and resource requirements to determine the winning companies.

Data Preparation
After choosing optimal subsets of data to process data mining, data preparation phase, which is the most time-consuming part of the cycle, began. Due to the lack of a comprehensive database and distribution subsidiaries in different parts of selective information overload, too much time was spent doing this step of the algorithm. In addition to the distribution of information in this section, there were also other problems in this section Such as lack of uniformity in the name of the supplier in different files, various technical evaluation criteria and incomplete information in various projects of suppliers. Also, although in this research, projects which were done in the last 4 years were studied which due to long production cycle, 1240 records were collected which after data cleaning was reduced to 911 records. To overcome these problems, various methods and solutions were used which some of them will be mentioned in the following.

Selecting and Combining Attributes
After preparing the data, 36 attributes for each item were obtained. It was necessary that selection and combination of attributes take place because a number of these attributes weren't complete. As will be described in detail in the next section, the Software Clementine SPSS was used for feature selection and finally 29 attributes were selected which can be seen in Table 1.

Among the features that the combination of features or lack of sufficient data have been removed, the suggested retail price Suppliers (combination of features), provided by the supplier warranty (lack of information), Inspection (loss Sufficient data) and technical advantages (lack of information) can be mentioned. In this part, the preparation required for each of the attributes is presented.

Standardizing Data
One of the problems in preparing the data collection for the suppliers' database, lack of standard in company and request names in various files. The lack of standard criteria for technical evaluation of different projects (or even in a project) was also among the other problems facing the project. Therefore, at first referring to the supplier profile database project, all the company names were standardized. The file name and custom code with reference to the planning and preparation of the project was revised and standardized. About the lack of standard of technical criteria, combination of criteria was also considered based on common sense. As a result of the evaluation criteria set forth in each checklist (there were three checklists for evaluating suppliers which each had more than 10 criteria), seven standard criteria were selected by assigning a weight based on previous weight. Also in order to not let the technical points of suppliers by removing some of the criteria, their former total scores were also among the features.

Data Cleaning
At the clean-up stage of the set of data, there were several unusual types of data. This unusual data can be divided into data and missing documentation. The purpose of noisy data, the data that due to typographical errors or miscalculation are inserted into the data set. To identify and remove this type of data, the SPSS Clementine software is used.

Table 1. Proposed model attribute names.

No.	Attribute	No.	Attribute	No.	Attribute
1	Project name	11	Responsiveness and accountability	21	The number of payment in contract
2	Order name	12	Country code	22	Presentation and answering capabilities in meetings
3	Supplier name	13	Time objectives	23	The company AVL
4	Supplier's nationality	14	Flexibility in production	24	Percentage of payment
5	Producer or dealer	15	The number of proposals for contract	25	Total score in the technical criteria
6	Price forecast	16	Cooperation in specific technical demands	26	The main part of the contract
7	The suggested retail price	17	Company status near the employer	27	Number of different products presented in the proposal
8	Ability to write and recommend proposal	18	The original offer of contract	28	The main part of the contract payment
9	Delivery time	19	Communications	29	Moon and year of proposal
10	Delivery type	20	Win or reject		

The most important type of malformed data collection are data records with missing values. The appearance of this type of data can be due to a lack of integrated databases, incomplete some files for some data records and Implementation of evaluation process stages by the external team. The priority of data cleaning, modifying and maintaining Data record was considered. For this purpose, two usage methods of the same information of the supplier or remove the sample was placed on the agenda. In this approach, first the scores and conditions of the intended company in other requests and past co-operations were examined. If multiple data record was found of those companies, in relation to the supplier's approval status, the nearest conditions for blank fields were inserted; also, if the sample contains a large amount of missing data, it will be eliminated.

Data Normalization

Due to the nature and value of the various orders, some of the fields of information are heterogeneous values. In order to compare the values of these fields of information of providers in different orders with each other, the values of Data records in those certain fields should be normalized. Therefore, the basis compared to the overall evaluation of these properties suppliers should be reviewed. In this data set, fields of information related to topics commodity prices and the delivery of features were included which they needed to change on a compare basis. In order to normalize the bid price providers in different orders, the basis for comparing the proposed price with the predicted price was considered for each order. Therefore, all bids were presented as a multiplier of the predicted price, which this way of showing, created the possibility for comparing different orders. Also the homogenization of time of delivering goods interval between the time of delivering goods (time announced by the supplier) and Time Objective (scheduled in supply) were used as the basis for comparison.

Adjusting Data Structure

To take the necessary measures for data preparation, the least operation necessary for data set for modeling a proposed model was created. But since modeling data mining is done with the help of specialized software, other measures to set the data format models to achieve a simpler and more efficient solutions are needed. Thus, because of lack of data mining models with numerical values better solutions offer qualitative transformation of values and character to numeric values on the agenda. For this conversion, different amounts of qualitative characteristics and character were assigned in a specific code and to index data for each record in a numerical data set was created. With this set of measures, data collection was prepared for modeling.

3.2 The CRISP-DM Implementation

In order to achieve better results more interpretable data mining model of the data, some of the criteria and combined criteria of Table 2 combination is used at all stages of the proposed model. Among these synthetic benchmarks can be as follows:

Table 2. The standard deviation values.

Initial value	Changed value
Time deviation < -90	−3
$-90 <$ Time deviation < -30	−2
$-30 <$ Time deviation < 0	−1
Time deviation $= 0$	0
$0 \leq$ Time deviation < 30	1
$30 <$ Time deviation < 90	2
$90 <$ Time deviation	3

- The combined ratio measures of the proposed price to price ratio instead of predicted price criteria and supplier's proposed price.
- The combined standard deviation of time between the scheduled time (target) and supplier's time-variation instead of the supplier's proposed time and scheduled time criteria.

Also, because the most used techniques with limited data sets and discrete (set) present better solutions compare to provide continuous data collection, it was decided that some of the recorded values be converted to a set from a continuous mode for Data records. In this first part of the conditions of continuous values certain criteria set values are expressed and then the corresponding variables used in this model is expressed.

Continuous variable change criteria to the set values Name and criteria set criteria that has become their amounts, are as follows.

1. Time deviation between scheduled time and proposed time from the supplier.

Since the scheduled time and the proposed deviation and fragmented supplier has greatly expanded the range of 350 days to 312 days beyond the target time earlier than the target time [−350, 312]. It was decided that based on the amount of time deviation, new values be assigned to data records. So based on the standard deviation values Table 2 count was limited to a set time.

2 The ratio of the proposed price to the predicted price

The criterion has scattered values [0–6.61]. Since the default bid suppliers is equal to the price provided, this ratio was placed based on dividing, which was one. According to the classification criteria of the top suppliers with very little chance of winning, the bid price ratios have been placed in one class.

Table 3 shows the change criteria of continuous values to the set values.

Table 3. The change criteria of continuous values to the set values.

Initial value	Changed value
Price ratio < 0.5	−3
0.5 < Price ratio < 0.75	−2
0.75 < price ratio < 1	−1
Price ratio = 1	0
1 < price ratio < 1.25	1
1.25 < price ratio < 1.5	2
1.5 < price ratio < 1.75	3
1.75 < price ratio < 2	4
2 < price ratio	5

3. Seven criteria for technical evaluation stage

Among the set of 29 characteristics considered for the project, seven criteria evaluate the technical and quality suppliers. These criteria are expressed as percentage points, but the fact that most of the techniques are not suitable responses to the wide range, the decision to change the standardized value from continuous sets to a more limited set. Table 4 shows the Criteria continuous change of the seven criteria set values.

Since the values of the seven quarters qualitative approach used models percentage, the changed values are saved from tech-set- to tech-set-7.

Table 4. The criteria continuous change of the seven criteria set values.

Initial value	Changed value
Percentage < 50	1
50 < percentage < 60	2
60 < percentage < 70	3
70 < percentage < 80	4
80 < percentage < 90	5
90 < percentage < 100	6
Percentage = 100	7

4. Total technical score

The criterion is expressed as a percentage of total technical points. All the reasons for changing the mentioned values for the above qualitative criteria is also applied in this criterion. To make the values of these premier supplier selection criteria is more effective, this criterion is divided in more categories. Table 5 shows the continuous values of the standard set values.

Table 5. The continuous values of the standard set values.

Initial value	Changed value
Percentage < 50	1
50 < percentage < 60	2
60 < percentage < 65	3
65 < percentage < 70	4
70 < percentage < 75	5
75 < percentage < 80	6
80 < percentage < 85	7
85 < percentage < 90	8
90 < percentage < 95	9
95 < percentage < 100	10
Percentage = 100	11

5. The number of product variations proposal

When issuing a request for proposals from the company, the order type of product or commodity Supplier Company is announced. By submitting proposals firms Inquiry, the number of departures products of the company are discussed with the Order Request and according to the conditions of the inquiries received and flexibility in product it is then decided in relation to doing changes in the order. The data set collected in the database of this project, consists of the number of deviations which are much dispersed and wide range [50–0]. Thus, the model used to improve performance and results, decided to change the standardized data set was limited to the amount distributed to a set of values. Table 6 shows the criterion change of dispersed values of this criterion to the value of a categorical.

Table 6. The criterion change of dispersed values of this criterion to the value of a categorical.

Initial value	Changed value
Number of deviations = 0	0
0 < number of deviations < 5	1
5 < number of deviations < 10	2
10 < number of deviations < 15	3
15 < number of deviations < 20	4
20 < number of deviations < 30	5
30 < number of deviations < 40	6
40 < number of deviations	7

It should be noted that in all the above transforms, in some cases due to withdrawal or rejection supplier of information on the supplier be available by the company, the word NULL is used for showing the lack of information in the considered criterion. Obviously, data records that are shown with this word when using these criteria are removed before calculation. In the next part, the variables in the model are implemented.

The Proposed Model *Variables*

Now the following this section, the topic and concept of other usable variables are presented.

Winner-code:

This variable is a binary variable that shows the supplier's selection or rejection. If the supplier is selected by the company (to win), the variable shows the value 1 and in refusal of the supplier, shows the value 2. This is used as the target variable.

Country-code:

In the data collection for the project, there are information suppliers from 28 different countries. In order to simplify the transfer of data to the model, an equivalent code for each of the countries included is considered in the data set. Therefore, this variable determines the country supplier using the amount that contains.

Marker-code:

This criterion indicates in which area the company is operating (attribute 5). This variable measures the three options, each providing one of these three values 1, 2 and 3 allocated. A value of 1 indicates that the supplier is the manufacturer. Value 2 indicates that the supplier is a trading company (medium) and the product is produced by another company. 3 also represents the amount that your supplier is the producer, but some of the products provided by other companies.

Status-code:

As well as technical and quality evaluation of suppliers, completion of a checklist in which the terms and conditions for each candidate suppliers depending on demand, is inserted. With the completion of the technical evaluation process and quality suppliers, this checklist and final negotiated agreement in its initial acceptance is specified. In this project, eight current situations in the early adoption of selected providers is defined in the form of this variable. In this variable, the best condition shown with a value of 1 and the worst with a value of 8.

AVL-code
This criterion studies the presence of the supplier's standard filters employer. The variable of this criterion has also 3 collections and in each order, the supplier only can allocate the three values, 1, 2, and 3. Value 1 indicates that the provider is in the list of employers. Value 2 also indicates that the supplier is not in the list of employers and value 3 also indicates that the employer hasn't presented a list for approved contractors in the intended order.

Proposal-month
This criterion represents the month submit proposals to the provider. The corresponding variable of this criterion also includes 12 values, these values represent the months of the year.

Proposal-year
This measure represents the year submitted a proposal to the supplier. This criterion also offer the corresponding variable indicates the time of year.

No-suggestion
When conducting financial negotiations between the company and suppliers, financial requests in the form of one or more supplier's offer. In this offer, Suppliers discuss methods of contract, delivery, time delivery and other items related to financial topics. This variable indicate the number of supplier's proposals.

Main-suggestion-code
Usually when announcing the proposal from the vendor, supplier Priority Payment for the contract will be announced. So to show this priority, 11 mentioned methods were used in corporate data banks and equivalent code was assigned to each.

As a result of this coding suggest that this variable shows, the main supplier by value is determined. Table 7 shows the specified payment methods in the data set.

Table 7. The specified payment methods in the data set.

Code	Payment method	Code	Payment method
1	Cash	7	Irrevocable L/C at sight
2	L/C	8	Cash, irrevocable L/C at sight
3	L/C at sight	9	L/C, cash
4	CAD	10	L/C, T/T
5	T/T	11	L/C at sight, T/T
6	Irrevocable		

Codes 8 and after indicate the combined payment of the contract price.

No-payment-step
The contract price is usually paid in several steps. This variable represents the number of payments that have been raised in the negotiations corporate finance and supplier.

Prepayment
During the financial negotiations to determine the percentage of the contract as prepaid. The prepaid variable rate shows determination in negotiations.

Main-part-code
When conducting financial negotiations between the company and the suppliers, on time payments negotiated agreement on the main part of the payment time is agreed (attribute19). Through this variable, changing time or times that the largest part of the payment made is ascertained.

Delivery-type
In international trade, there are several contracts to deliver the goods. Each of the types of contracts is shown with certain words. By defining this variable, it is proposed that contracts register suppliers to deliver every order. Also the values of these variables to simplify analysis by the application for each vocabulary how the contract was considered equivalent code. Table 8 shows the various delivery contracts in the data set and their means code.

Table 8. The various delivery contracts in the data set and their means code

Code	Contract type	Code	Contract type	Code	Contract type	Code	Contract type
1	CFR	3	CIF	5	FOB	7	FCA
2	EXW	4	CIP	6	CPT		

Feature selection
The collected data forms, documentation suppliers of Oil and Gas Company 40 features were extracted for each supplier; but all extracted features are helpful and necessary for the implementation of the proposed model is due to feature selection issue to consider. For choosing features, the SPSS Clementine software is used that the selected features lack the following:

- Features more than 70% of missing values.
- Distribution characteristics above 90% in one category.
- Features with diverse values above 95%.
- Features with similar concepts.

After selecting a feature, the number of properties decreased to 29 properties. Figure 2 shows the output in selecting features using the SPSS Clementine software:

Missing Data
Although the feature selection was trying to selected features which are no missing data, but it is necessary to prevent errors and reduce the accuracy of the proposed model, missing data should more carefully be examined and be eliminated if they are.

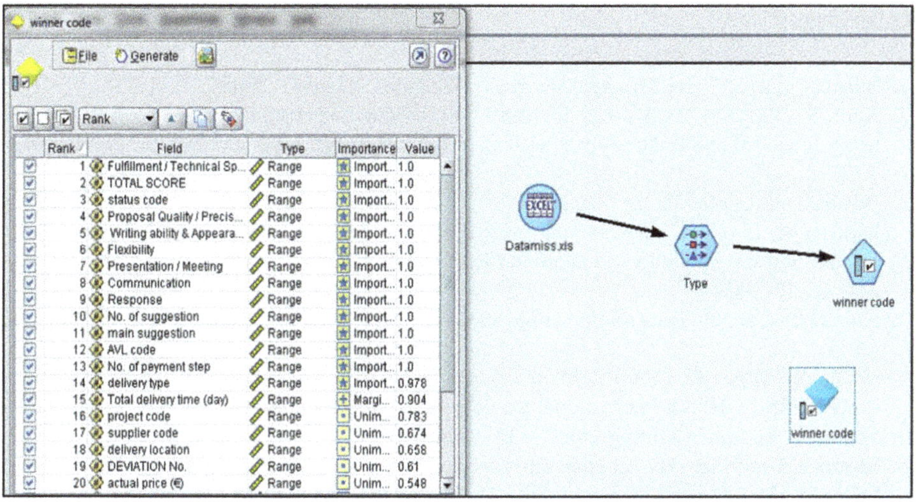

Fig. 2. Feature selection result with SPSS Clementine

For this purpose, by using the MATLAB function as findNaN.m for 7 identification of missing data was written which is the output value of missing data in the database. For imputation of these missing data, the follomodewing approaches are considered:

- Removing samples that have lots of missing values.
- A function that recognizes the nearest sample to sample a missing value attribute value to replace the amount lost. This is done by a function named missing.m in the MATLAB environment.

4 Conclusion

In this study, the CRISP-DM algorithm and steps were investigated, later, the data related stages of the CRISP-DM usage in a project (data monitoring and evaluation) was examined, and explained by using an example application. In this process, the supplier database based on the CRISP-DM algorithm for Data Mining was analyzed and processed. In addition, it has presented the results of applying the proposed model and the general model of data preparation step by step in the process of DM-CRISP until the final accuracy is described. Data preparation, data mining software and different techniques of measures including feature selection, weighting and resolve missing data was done. Actually, it was purposed to explain the role of the steps of CRISP-DM, and especially to process, understand and prepare information in the process of data discovery. In this way, it is expected that create awareness about CRISP-DM, and the impact of this process on projects is expected to be clearly understood.

References

1. Silahtaroglu, G.: Data Mining. Papatya Publications, Istanbul (2008)
2. Kaur, K., Vashisht, S.: A novel approach for providing the customer churn prediction model using enhanced boosted trees technique in cloud computing. Int. J. Comput. Appl. **114**(7), 1–7 (2015)
3. Bosnjak, Z., Grljevic, O., Bosnjak, S.: CRISP-DM as a framework for discovering knowledge in small and medium sized enterprises' data. In: 2009 5th International Symposium on Applied Computational Intelligence and Informatics. SACI 2009, pp. 509–514. IEEE (2009)
4. Koh, H.C., Tan, G.: Data mining applications in healthcare. J. Healthc. Inf. Manag. **19**(2), 65 (2011)
5. Wirth, R., Hipp, J.: CRISP-DM: towards a standard process model for data mining. In: Proceedings of the 4th International Conference on the Practical Applications of Knowledge Discovery and Data Mining, pp. 29–39. Citeseer (2000)
6. Shafique, U., Qaiser, H.: A comparative study of data mining process models (KDD, CRISP-DM and SEMMA). Int. J. Innov. Sci. Res. **12**(1), 217–222 (2014)
7. Chapman, E., Clinton, J., Kerber, R., Khabaza, T., Reinartz, T., Shearer, C., Wirth, R.: CRISP-DM 1.0 Step-by-Step Data Mining Guide. SPSS (2000). http://www.crispdm.org/CRISPWP-0800.pdf
8. Nadali, A., Kakhky, E.N., Nosratabadi, H.E.: Evaluating the success level of data mining projects based on CRISP-DM methodology by a Fuzzy expert system. In: 2011 3rd International Conference on Electronics Computer Technology (ICECT), vol. 6, pp. 161–165. IEEE (2011)
9. Mosaddar, D., Shojaie, A.A.: A data mining model to identify inefficient maintenance activities. Int. J. Syst. Assur. Eng. Manag. **4**(2), 182–192 (2013)
10. Catley, C., Smith, K., McGregor, C., Tracy, M.: Extending CRISP-DM to incorporate temporal data mining of multidimensional medical data streams: a neonatal intensive care unit case study. In: 2009 22nd IEEE International Symposium on Computer-Based Medical Systems. CBMS 2009, pp. 1–5. IEEE (2009)
11. da Rocha, B.C., de Sousa Junior, R.T.: Identifying bank frauds using CRISP-DM and decision trees. Int. J. Comput. Sci. Inf. Technol. **2**(5), 162–169 (2010)
12. Cešpivová, H., Rauch, J., Svatek, V., Kejkula, M., Tomeckova, M.: Roles of medical ontology in association mining CRISP-DM cycle. In: ECML/PKDD04 Workshop on Knowledge Discovery and Ontologies (KDO 2004), Pisa, vol. 220 (2004)
13. Shearer, C.: The CRISP-DM model: the new blueprint for data mining. J. Data Warehous. **5**(4), 13–22 (2000)

On the Prediction of Possibly Forgotten Shopping Basket Items

Anderson Singh(✉) and Patrick Hosein

The University of the West Indies, Saint Augustine, Trinidad and Tobago
{anderson,patrick}@lab.tt

Abstract. With the easy availability of shopping transaction data, it is now possible to use historical information to enhance the shopping experience of shoppers. In particular, we focus on the issue of forgotten items. We have all had the experience of shopping at a supermarket and either forgetting to buy certain items or not realizing that certain items are needed. In this paper, we focus on predicting such items during a shopper's visit and providing hints to the shopper that they may in fact need certain items that were not purchased. Note that, by providing hints for forgotten items, the shopper benefits by avoiding a return trip to the supermarket and the supermarket benefits because the shopper purchases additional items. We provide examples to illustrate the utility of the proposed approach.

Keywords: Machine learning · Prediction · Market basket analysis · Artificial intelligence

1 Introduction

In the retail industry, market basket analysis [8,12] is a popular technique used to identify patterns and trends among customers and their purchased items. It is this data that can be used to influence major corporate decisions. In real-world situations, a customer's shopping history can be considered as a sequence of baskets, where each basket consists of the set of purchased items at a particular visit. We can use this information to predict what the customer should purchase at a future visit through the use of recommender systems [1,9]. The problem of predicting the customer's next basket has been addressed in different contexts, such as e-commerce websites, which recommend to their customers relevant items, related to their individual lifestyles. It is important to note that we can increase customer satisfaction and total sales by recommending appropriate items to customers.

In this paper, we propose a new method known as Interval Based Predictor (IBP), which solves the next basket recommendation problem in the context of a supermarket environment. IBP provides the customer with an appropriate list of items that they may have forgotten to purchase at their present supermarket visit. IBP is able to learn the consumption time of a particular item and hence

D. J. Hemanth and U. Kose (Eds.): ICAIAME 2019, LNDECT 43, pp. 559–571, 2020.
https://doi.org/10.1007/978-3-030-36178-5_45

be able to predict when next it would be needed. It is worth mentioning that this approach only uses data for the customer under consideration, unlike traditional approaches which require data from multiple customers, that is, the method is customer-centric.

It should be noted that both the customer and the supermarket benefit from our approach since the customer avoids a return visit to the supermarket, and the supermarket benefits from the potential increase in revenue. To this end, we propose a metric to estimate the potential increase in revenue from each customer, and consequently, for all shoppers of the supermarket. In the evaluation section of our paper, we compare the proposed next basket recommendation method against multiple state-of-the-art methods.

2 Related Work and Contributions

To the best of the authors' knowledge, there has been no previous work on recommending potentially forgotten items to customers. Instead, the literature is focused on next basket recommendation, which provides the customer with a reasonable shopping list of items that they may purchase at their next visit based on historical transactions. On the other hand, our objective is to remind the customer about potentially forgotten items just before the start of their checkout procedure. Since our work is closely related to next basket recommendation, we adopt a similar approach to [7,13,17] in analyzing the existing literature. Thus, we classify recommender systems based on their type, which can be either general, sequential, or hybrid.

A general recommender system provides recommendations based on the general taste of the customer, that is, it does not consider factors such as the sequence in which the items were purchased or how recent the items were purchased. A commonly used technique in general recommender systems is collaborative filtering. The term collaborative filtering implies that individuals can assist each other in locating content which is of interest to them. Tapestry [6] is one of the first collaborative filtering implementations that recommends documents posted in newsgroups to users. A survey of collaborative filtering including model-based and memory-based techniques are discussed in [15]. Furthermore, [11] discussed the state of research with regard to collaborative filtering and implicit feedback.

A sequential recommender utilizes both historical transactions as well as its accompanying sequential properties to provide a list of recommendations to the customer. A customer's historical transactions can be treated as an ordered sequence of baskets, in which, a basket consists of a set of items purchased at a particular point in time. It is this sequential data that is exploited to provide recommendations. A recommender based on the Markov decision process is proposed in [14], where the generation of recommendations is treated as a sequential decision problem because the recommendations are based on the customer's past behaviour. It should be noted that sequential recommenders are not only applicable to predicting a customer's next basket, for example, [10] investigated real website usage data to compare the effects of contiguous sequential patterns and less constrained sequential patterns. Another example of sequential recommender applicability is discussed in [3] which proposes a method for generating music playlists based on Markov Chains. In [4], an external memory network that uses hidden states was proposed to capture historical data. Furthermore, a comprehensive overview of sequential mining algorithms based on their efficiency and applicability is provided in [2].

A hybrid recommender system is based on a combination of sequential and general recommendation techniques which leverages the advantages from each method. For example, FPMC [13] is a hybrid recommendation system which combines a general recommendation technique, matrix factorization and a sequential recommendation technique, Markov Chains. HRM [17] is another novel hybrid recommendation method aimed at improving the results of FPMC. The authors of HRM noted that a drawback of FPMC is that the factors that affect a customer's purchasing decisions are combined linearly, however, this assumption is not true. HRM overcame this issue through the introduction of nonlinear aggregation operations, which provided better results. DREAM [18] is yet another hybrid recommendation technique based on recurrent neural networks. The authors of DREAM stated that HRM presents a partial solution to this issue through the inclusion of nonlinear aggregation operations. The authors of DREAM further state that FPMC and HRM do not model the global sequential behavior of a customer, that is, those methods only consider the sequential properties between adjacent baskets. In [7], a novel hybrid recommendation method, Temporal Based Predictor (TBP), which combines both sequential and pattern-based recommender techniques was presented. TBP exploits recurring sequential purchases known as Temporal Annotated Recurring Sequence (TARS) in the customer's historical transactional data. TARS is able to capture multiple factors which affect the purchasing decisions of a customer such as the co-occurrency, sequentiality, periodicity, and recurrency of items in the basket. In [5], the task of predicting a customer's next basket was considered as a machine learning problem, but more specifically, a classification problem. The authors proposed learning separate binary classifiers for every item for each customer. In other words, a single classifier is associated with every customer and item pair which determines whether or not the given customer will purchase the particular item. Let us consider a supermarket that has X customers, where each customer purchases approximately Y items. In this situation, the recommendation system would consist of approximately XY distinct binary classifiers. The authors also explored the application of various machine learning classification methods such as, decision trees, perceptron, Naive Bayes, and Winnow.

3 Problem Definition

We have all had the experience of going grocery (or any) shopping and forgetting to buy one or more items. When this happens we have to make an additional trip to the grocery store for the missed items. In this paper we address this problem. We consider one such visit and attempt to predict any items that may have been missed. Just before the customer checks out, we determine potentially missed items and inform the customer who can then go back for the items before checking out. The metric we use is the percentage increase in the total number of basket items and percentage increase in revenue because of these additional purchases. This reflects, in some way, the increased revenue achieved by the store and hence the utility of the approach.

We model the problem as follows. Let B represent the set of basket items picked up prior to providing hints to the user. Let F represent the set of forgotten items and finally, let P denote the set of predicted basket items. The set $P \cap B$ represents the set of items that were picked up and were correctly predicted to be purchased. The set $P - (P \cap B)$ represents the set of items that were predicted to be purchased but were not. This is the potential set of forgotten items.

The probability of successfully predicting a basket item is given by $s = \frac{|P \cap B|}{|B|}$. Now consider a forgotten item. If this item had not been forgotten then the probability that it would have been successfully predicted would have also been s. The fact that it was forgotten does not change this probability. Therefore the probability that a forgotten item is correctly predicted should also be s. Therefore the expected number of predicted items that were not bought because they were forgotten is given by $s|F|$. If the customer purchases all forgotten but predicted items then the expected fractional increase in purchased items, ρ, is given by

$$\rho = \frac{s|F|}{|B|} = \frac{|P \cap B||F|}{|B|^2} \tag{1}$$

Let us illustrate with a simple example. Suppose that our predictor has an accuracy of 60% so that $s = 0.6$. Assume that if one goes into a store to purchase 20 items that they tend to forget at least one of the items. We can then assume that $|F| = 0.05|B|$. Therefore we obtain $\rho = 0.03$. In other words there is an expected increase of 3% of purchased items and, if we assume any item is equally likely to be forgotten, this translates into a 3% increase in revenue.

Note that this is an upper bound on revenue increase since the customer may return to the same grocery for the items anyway rather than going somewhere else in which case there is no revenue increase. However, there is still the benefit of convenience since these hints would avoid the customer making a second trip either to the same grocery or another one.

The prediction success probability s can be obtained for each customer based on their prior purchases. A prediction algorithm is used to predict the basket and then this can be compared with the actual basket to determine the success probability s. We assume that the basket items in set B are known. However the set F, the set of forgotten items, is typically not known. One approximation that can be used is the set $P - (P \cap B)$ which can be obtained from the present visit. This is the set of predicted items that were not purchased and is the list of items which will then be provided as hints to the user. If this set only contained forgotten items then it is what we would need. However, in some cases it may also contain items that are neither in the bought or forgotten sets. By using feedback from the customer (who would reject such items) we can identify the truly forgotten items and use this set to obtain $s|F|$.

4 The Proposed Approach

4.1 Basket Prediction

In this section, we explain the algorithm used to predict a customer's basket at the supermarket. We consider visit i for a single customer which occurs at time t_i, and compute the predicted basket for that visit. We look at a single item j, and determine whether or not it should be included in the predicted basket for visit i. Let T_i represent the average time between visits for the customer, and let G_j represent the average time between visits where the customer purchased item j. Also, we use L_j to represent the last time the customer purchased item j such that $L_j < t_i$. Using this data, $L_j + G_j$ provides an estimated time when the customer would exhaust their stock of item j at home. Hence, $L_j + G_j$ is the ideal time for the customer to repurchase item j. However, the customer may not be shopping at the supermarket at that exact moment. Therefore, a reasonable repurchase time must be computed.

Let us consider the trivial case where $L_j + G_j \leq t_i$. In such a situation, item j should be repurchased by the customer since their stock at home has been exhausted. When $L_j + G_j \geq t_i + T_i$, item j should not be purchased because the customer has sufficient stock at home which will not be exhausted at least until the next visit. In the last case where $t_i < L_j + G_j < t_i + T_i$, item j will be exhausted by the customer some time between the current visit and the next visit. Since the customer uses item j with the time period G_j, then it must be also repurchased with this same time period. If item j is always bought at the current visit, then the customer will continuously increase their stock at home. Furthermore, if item j is always bought at the next visit, the customer will continuously exhaust their stock at home. Therefore, a random decision must be made to purchase item j at the current visit. A random number r is generated between 0 and 1. If $r \geq \frac{L_j + G_j - t_i}{T_i}$, then item j should be purchased at the current visit, otherwise, it should be purchased at the next visit. In this approach, the time period with which item j is repurchased is given by

$$\frac{L_j + G_j - t_i}{T_i}(t_i - L_j + T_i) + \left(1 - \frac{L_j + G_j - t_i}{T_i}\right)(t_i - L_j)$$

which is simply G_j, the desired repurchase time period. If item j is recommended to the customer, a suggested purchase quantity Q_j is also provided.

It is important to note that the above procedure is repeated for each item the customer has purchased in the past to construct the predicted basket, however, we may want to place a constraint on the size of the predicted basket to a given value, say k, to avoid an unreasonably large predicted basket. To achieve this, we select the top k items ranked in descending order by the number of unique visits where each item was purchased by the given customer. We let U_j represent the number of unique visits where the customer purchased item j. The corresponding pseudo-code is included in Algorithm 1. Note that S is computed in Algorithm 2.

Algorithm 1. *predict_basket*(i(visit), T_i, t_i, S, k(limit))

```
P ← ∅ (predicted basket)
for ((G_j, L_j, U_j, Q_j, j(item)) ∈ S) do
    if L_j + G_j ≤ t_i then
        P ← P ∪ {(j, U_j, Q_j)}
    else
        if L_j + G_j > t_i and L_j + G_j < t_i + T_i then
            r = rand(0, 1)
            if r ≥ (L_j + G_j − t_i)/T_i then
                P ← P ∪ {(j, U_j, Q_j)}
P ← top_k_items(P,k)
return P
```

4.2 Training the Basket Prediction Algorithm

At the start of the customer's next supermarket visit, for some forgetting factor $0 < \alpha < 1$, we perform an update as follows

$$T_{i+1} = \alpha T_i + (1 - \alpha)(t_{i+1} - t_i). \tag{2}$$

If item j is purchased by the customer at visit i, then we compute

$$G_{j+1} = \alpha G_j + (1 - \alpha)(t_i - L_j). \tag{3}$$

However, if item j is not purchased by the customer at visit i, then we simply set $G_{j+1} = G_j$. In this case, item j will then be purchased at a future visit and the variable will be updated at that time. If the customer purchases quantity q of item j at visit i, where $q > 0$, then we update

$$Q_{j+1} = \beta Q_j + (1 - \beta)(q). \tag{4}$$

where β is a factor $0 < \beta < 1$ used to exponentially smooth the quantity purchased. Also, if item j is purchased at visit i, we update the number of unique visits where the customer purchased this particular item.

$$U_{j+1} = U_j + 1 \tag{5}$$

The updates above are performed for each item the customer has purchased at visit i. Given both the predicted and actual baskets, the success probability and resulting performance metric ρ can be calculated. Algorithm 2 provides the pseudo-code for learning both the time intervals for item repurchases and the customer's time interval for revisiting the supermarket. The update is performed just before the customer begins their shopping at visit $i + 1$.

5 Numerical Results

In this section we provide numerical results for real customer transactions at a major supermarket to show the effectiveness of our approach.

Algorithm 2. *learn_time_intervals(S, B (basket),* t_{i+1}, t_i, α, β*)*

$T_{i+1} \leftarrow \alpha T_i + (1-\alpha)(t_{i+1} - t_i)$
for $(j \in B)$ **do**
  if $\nexists\, x : x \in S$ *and* $x = (G_j, L_j, U_j, Q_j, j))$ **then**
    $L_{j+1} \leftarrow t_i$
    $G_{j+1} \leftarrow 0$
    $U_{j+1} \leftarrow 1$
    $Q_{j+1} \leftarrow item\_count(B, item)$
    $S \leftarrow S \cup \{(G_j, L_j, U_j, Q_j, j)\}$
  else
    $G_{j+1} \leftarrow \alpha G_j + (1-\alpha)(t_i - L_j)$
    $Q_{j+1} \leftarrow \beta Q_j + (1-\beta) item\_count(B, j)$
    $U_{j+1} \leftarrow U_j + 1$
    $L_{j+1} \leftarrow t_i$
return S

5.1 Dataset and Experimental Approach

The data was supplied by a major supermarket chain. It contains customer transactions for a branch of the supermarket chain and includes data on items from various categories, such as electronics, food, health, and pharmaceuticals. The dataset spans purchases made during 2015, a one-year period. In the data pre-processing phase only those customers who made at least one purchase every month were selected. Furthermore, the items in the dataset were grouped into general classes as opposed to being used as individual items. For instance, *artisan romaine lettuce* and *artisan Boston lettuce* are grouped under the general class of *lettuce*. The rationale for grouping the items under general classes is that the customer may purchase product substitutes in different brands or sizes. Thus, to understand the level of customer stock, it is better to generalize the items. We obtained data for a total of 433 customers and focused on 615 basket items. These customers made an average of 36 visits (baskets) and the average basket size over all customers and visits was 28.

In conducting our experiments, we adopt the *leave one out* strategy for the purpose of model evaluation. In this strategy, the dataset is split into a training set and test set for each customer such that each test set consists of only the last transaction for a given customer. The remaining transactions for each customer are placed into their respective training set. In other words, for each customer, their historical data is used to train their individual predictive model and their last basket is used to check the accuracy of the approach. In terms of the predicted basket's length, we use the average length of the customer's basket, k^*, based on their historical transactions. However, we also perform experiments where the length of the predicted basket is based on a fixed value $k \in [2, 20]$. With regard to parameters, we set $\alpha = 0.4$ and $\beta = 0.5$, however, we varied these values and found the results to be fairly consistent, which suggests that our method is reasonably stable. We evaluate our model based on the F1-score, which is the harmonic mean of precision and recall [16].

In terms of increase in revenue, the size of the forgotten set, $|F|$, is estimated as $\gamma|B|$, where $|B|$ is the size of the customer's basket and γ is a factor, $0 < \gamma < 1$, which represents the customer's forgetfulness. Let us consider a customer that must purchase 20 items at the supermarket. If the customer forgets to purchase a single item, then 5% of the basket has been forgotten. Therefore, we use a value of $\gamma = 0.05$ which can be interpreted as the customer will forget 5% of their basket. We believe $\gamma = 0.05$ is a reasonable value to use for our experiments.

5.2 Results

In Table 1 we demonstrate the intuition in the item selection process used in our method. It should be noted that the items in Table 1 we both predicted by our algorithm and purchased by the customer. The customer visited the supermarket on the 22-12-2015, and given that the time period between visits is approximately 9 days, their next visit is estimated to be on the 31-12-2015.

Table 1. Examples of correctly predicted items

Item	Past visits			Visit	Current	Visit	Estimated
	Date 1	Date 2	Date 3	Interval	Date	Interval	Next visit
Flour	Oct 15	Nov 6	Dec 4	24 days	Dec 22	9 days	Dec 31
Milk	Nov 12	Nov 27	Dec 10	14 days			
Tuna	Nov 19	Nov 27	Dec 4	11 days			

First, let us consider the item, *flour*, which has a consumption time of approximately 24 days for this customer. Thus, given that the customer last purchased this item on the 04-12-2015, we expect that it will be exhausted on 28-12-2015. Therefore, at the current visit date, 22-12-2015, the customer does not need *flour*, since there is still a fraction remaining at home. However, given that the estimated next visit is on 31-12-2015, there would be a 3-day window where the customer will not have *flour*. Hence, the algorithm makes a random decision to recommend the *flour* at the current visit as opposed to the customer foregoing the item for 3 days.

In a similar manner, *milk*, which has a consumption time of approximately 14 days, was last purchased by the customer on 10-12-2015, and at the current visit date, 22-12-2015, would not yet have been exhausted. However, in 2 days, the *milk* would have been fully consumed, and the customer would be forced to forgo the item for 7 days, that is, until the estimated next visit on 31-12-2015. Consequently, as with *flour*, a random decision is made to recommend *milk* at the current visit, as opposed to the next visit.

Finally, *tuna*, has a consumption time of approximately 11 days for this customer and thus, based on the item's last purchase date, 04-12-2015, the customer's stock is expected to become exhausted on the 15-12-2015. Therefore, at the current visit date, 22-12-2015, the customer has not purchased *tuna* in 18 day and has been deprived of the item for 7 days. As a result, the algorithm recommends that the customer should purchase *tuna* at the current visit.

Intuitively, from Table 1 and the above explanation, the reader should have a clearer understanding of the item selection criteria. Therefore, based on these criteria, we believe the items that were both predicted by our method and were not purchased by the customer are good candidates for forgotten items. However, we do not truly know if these items are forgotten since hints were never supplied to the customers used for training the model.

In Table 2, we show such examples of probable forgotten items for a chosen customer. The customer visited the supermarket on the 14-12-2015, and their next visit is estimated to be on the 27-12-2015, given that the interval between visits is approximately 13 days.

Table 2. Examples of potentially forgotten items

Item	Past visit dates			Visit	Current	Visit	Estimated
	Date 1	Date 2	Date 3	Interval	Date	Interval	Next Visit
Pepper	Oct 23	Nov 6	Nov 27	25 days	Dec 14	13 days	Dec 27
Bread	Nov 6	Nov 20	Nov 27	10 days			

First, let us consider the item, *pepper*, that has an approximate consumption time of 25 days for the given customer. Thus, based on the customer's last purchase date, 27-11-2015, we expect that their stock of *pepper* will be exhausted on the 22-12-2015. Now, as at the current visit date, 14-12-2015, the stock of *pepper* is not yet depleted. However, if the customer does not purchase the item at this time, they must forgo it for approximately 5 days, that is, until their next visit. Hence, the algorithm makes a choice to recommend the item to the customer in order to avoid depriving them of it for the 5-day period. Similarly, *bread* has a consumption time of 10 days for the selected customer, and given that the customer last purchased bread on 27-11-2015, we expect that it would be exhausted on 07-12-2015. Thus, on 14-12-2015, the current visit date, the customer has already been deprived of *bread* for 7 days and thus the algorithm recommends that bread be purchased by the customer.

We also computed the average expected increase in revenue for a number of visits. This is plotted in Fig. 1. The overall average increase in revenue is approximately 1.74%.

5.3 Comparison with Prior Methods

Our objective is not to predict the customer's next basket, but rather to provide the customer with a list of items that may have been forgotten. Nevertheless, it

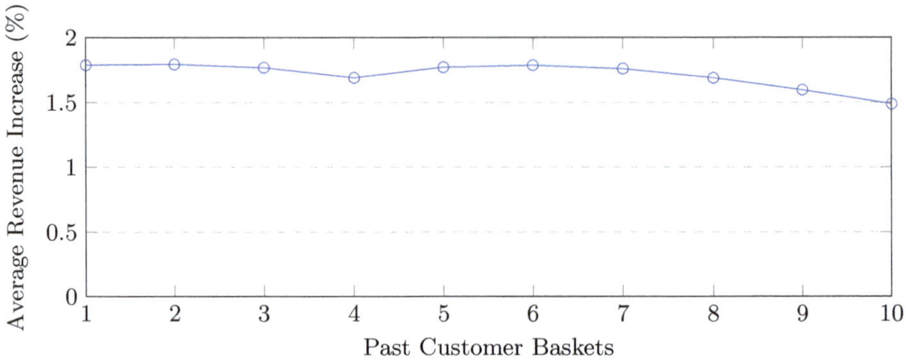

Fig. 1. Expected revenue increase for 10 visits

is useful to compare the performance of our algorithm against several baseline methods used for next basket prediction. The baseline methods chosen share a central characteristic to our method in that they are customer-centric in nature, which means that each customer's predictive model is based solely on their historical data. The following is a brief description of each of the chosen baseline methods as well as ours.

LST: [5,7] The predicted basket consists of the same set of items as the last basket bought by the given customer.
TOP: [5,7] The predicted basket consists of the most frequent items bought by the given customer.
TBP: [7] The predicted basket consists of items that are obtained from recurring patterns or sequences in the given customer's past transactions.
IBP: Our proposed algorithm.

We computed the F1-Scores for each algorithm and obtained values of 0.28 for IBP, 0.30 for TBP, 0.33 for TOP and 0.22 for LST. It should be noted that the size of the predicted basket, k, is equal to the average size of the customer's basket based on their historical data, k^*.

In comparing the methods, we adopt the strategy in [7], where the training time of the methods are reported. We found that for IBP, TOP and LST, the training time was about 0.01 s. However, for TBP, the training time was about 55 s. The difference in running time is a clear benefit of using the proposed method over TBP, despite the 2% under performance in the F1-score metric. Also, TBP requires that the model for each customer be retrained as time progresses because the quality of the recommendations will begin to decay. However, the proposed method and TOP both allow a given customer's model to be updated as they make future visits to the supermarket without entirely rebuilding the model.

Although TOP outperforms the proposed method, we believe for the task of recommending potentially forgotten items, our algorithm is more appropriate. In our work, the set of potentially forgotten items is a subset of the items that were recommended but not bought because the customer is not aware that either their stock at home will finish before the next visit, or has already finished. The proposed method takes into account the approximate time interval the given customer takes to return to the supermarket, and also the approximate time it takes for each item to be consumed. Using this data, the method can determine when a customer's stock for an item is about to be exhausted or has already been exhausted, and then, make a recommendation accordingly. However, the TOP algorithm simply recommends the most popular set of items bought by a customer, and this does not reflect items which may have been forgotten. In fact, if an item is popular with a given customer, it is less likely to be forgotten.

Also, TOP is not able to predict items which are bought infrequently, whether because they have a long consumption time or because they are bought in bulk. We believe that these items will generally incur a higher cost to the customer and thus it is in the supermarket's interest to recommend such items to achieve a revenue gain. It was found that the average price of an item that was both recommended and bought by the proposed method is \$23.82, as opposed to TOP, where the average price is \$20.18. Therefore the proposed method recommends items which cost approximately \$3.64 more than those recommended by TOP.

We also report on the F1-scores of the proposed method and the baseline methods aforementioned, where the size of the predicted basket, k, is fixed for all customers in the dataset. We perform these experiments for different values of

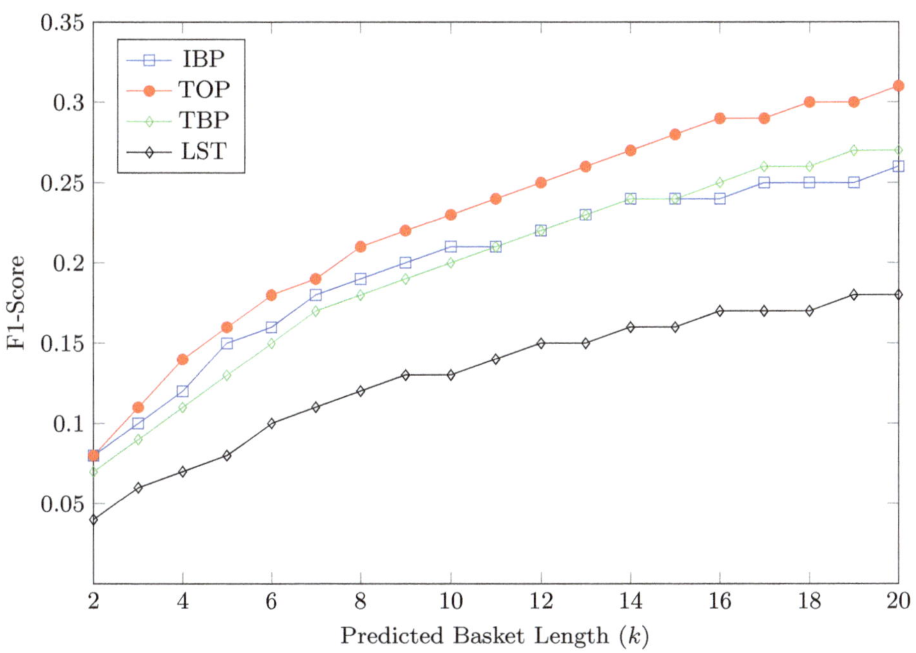

Fig. 2. Performance comparison

$k \in [2, 20]$ which is then plotted in Fig. 2. It can be seen that TOP outperforms both the proposed method and TBP. However, the proposed method outperforms TBP in the range $k \in [2, 11]$ which suggests it is better able to predict smaller baskets. It is also worth mentioning that the proposed method and TBP has approximately the same F1-scores when the size of the predicted basket is in the range $k \in [12, 15]$ after which TBP begins to outperform the proposed method.

6 Conclusions and Future Work

The contributions of this paper are two-fold. Firstly, we introduced a time-based method to recommend items that a customer may have potentially forgotten at their visit to the supermarket. Then, we show the advantages of our work and how it compares to existing next basket recommendation methods. Secondly, we demonstrate the effectiveness of our method by estimating its increase in revenue on a real-world dataset. Although the proposed algorithm did not yield the best results in the performance comparison with other algorithms, it took into account individual item costs and therefore, resulted in an increase in revenue stream unlike the cost of items predicted by the other algorithms.

Future work will include cultural habit changes where the consumer starts or stops purchasing a product based on a lifestyle change for example. We also plan to develop and deploy a mobile application that customers may use to obtain hints on potentially missed items when they shop. We believe that the revenue increase, even if it is 2%, will prove to be significant to the supermarket especially since the only cost to them is the development of the application. Consumers also benefit by avoiding return trips to the supermarket for forgotten items.

References

1. Bobadilla, J., Ortega, F., Hernando, A., Gutérrez, A.: Recommender systems survey. Knowl.-Based Syst. **46**, 109–132 (2013). https://doi.org/10.1016/j.knosys.2013.03.012, http://www.sciencedirect.com/science/article/pii/S0950705113001044
2. Chand, C., Thakkar, A., Ganatra, A.: Sequential pattern mining: survey and current research challenges. Int. J. Soft Comput. Eng. **2**(1), 185–193 (2012)
3. Chen, S., Moore, J.L., Turnbull, D., Joachims, T.: Playlist prediction via metric embedding. In: Proceedings of the 18th ACM SIGKDD International Conference on Knowledge Discovery and Data Mining, pp. 714–722. ACM (2012)
4. Chen, X., Xu, H., Zhang, Y., Tang, J., Cao, Y., Qin, Z., Zha, H.: Sequential recommendation with user memory networks. In: Proceedings of the Eleventh ACM International Conference on Web Search and Data Mining, pp. 108–116. ACM (2018)
5. Cumby, C., Fano, A., Ghani, R., Krema, M.: Predicting customer shopping lists from point-of-sale purchase data. In: Proceedings of the Tenth ACM SIGKDD International Conference on Knowledge Discovery and Data Mining, KDD 2004, pp. 402–409. ACM, New York (2004). https://doi.org/10.1145/1014052.1014098
6. Goldberg, D., Nichols, D., Oki, B.M., Terry, D.: Using collaborative filtering to weave an information tapestry. Commun. ACM **35**(12), 61–70 (1992)

7. Guidotti, R., Rossetti, G., Pappalardo, L., Giannotti, F., Pedreschi, D.: Market basket prediction using user-centric temporal annotated recurring sequences. In: 2017 IEEE International Conference on Data Mining (ICDM), pp. 895–900. IEEE (2017)
8. Ledolter, J.: Data Mining and Business Analytics with R. Wiley, Hoboken (2013)
9. Lu, J., Wu, D., Mao, M., Wang, W., Zhang, G.: Recommender system application developments: a survey. Decis. Support Syst. **74**, 12–32 (2015). https://doi.org/10.1016/j.dss.2015.03.008, http://www.sciencedirect.com/science/article/pii/S0167923615000627
10. Mobasher, B., Dai, H., Luo, T., Nakagawa, M.: Using sequential and non-sequential patterns in predictive web usage mining tasks. In: 2002 IEEE International Conference on Data Mining, ICDM 2003, Proceedings, pp. 669–672. IEEE (2002)
11. Najafabadi, M.K., Mahrin, M.N.: A systematic literature review on the state of research and practice of collaborative filtering technique and implicit feedback. Artif. Intell. Rev. **45**(2), 167–201 (2016). https://doi.org/10.1007/s10462-015-9443-9, https://doi.org/10.1007/s10462-015-9443-9
12. Olson, D.L.: Market basket analysis. In: Descriptive Data Mining, pp. 29–41. Springer (2017)
13. Rendle, S., Freudenthaler, C., Schmidt-Thieme, L.: Factorizing personalized Markov chains for next-basket recommendation. In: Proceedings of the 19th International Conference on World Wide Web, pp. 811–820. ACM (2010)
14. Shani, G., Heckerman, D., Brafman, R.I.: An MDP-based recommender system. J. Mach. Learn. Res. **6**(Sep), 1265–1295 (2005)
15. Su, X., Khoshgoftaar, T.M.: A survey of collaborative filtering techniques. Adv. Artif. Intell. **2009**, 4 (2009)
16. Tan, P.N.: Introduction to Data Mining. What's New in Computer Science, 2nd edn. Pearson (2018). https://www.xarg.org/ref/a/0133128903/
17. Wang, P., Guo, J., Lan, Y., Xu, J., Wan, S., Cheng, X.: Learning hierarchical representation model for nextbasket recommendation. In: Proceedings of the 38th International ACM SIGIR Conference on Research and Development in Information Retrieval, pp. 403–412. ACM (2015)
18. Yu, F., Liu, Q., Wu, S., Wang, L., Tan, T.: A dynamic recurrent model for next basket recommendation. In: Proceedings of the 39th International ACM SIGIR conference on Research and Development in Information Retrieval, pp. 729–732. ACM (2016)

Consensus Approaches of High-Value Crypto Currencies and Application in SHA-3

Murat Emeç[1], Melike Karatay[2](✉), Gökhan Dalkılıç[1], and Erdem Alkım[3]

[1] Dokuz Eylül University, İzmir, Turkey
murat.emec@ceng.deu.edu.tr, dalkilic@cs.deu.edu.tr
[2] Ege University, İzmir, Turkey
melike.karatay@ceng.deu.edu.tr
[3] Ondokuz Mayıs University, Samsun, Turkey
erdem.alkim@omu.edu.tr

Abstract. In view of the widespread use of information technologies, the security of data against third parties should be maintained. Blockchain technology is used for this data and especially for transactions in finance. However, although bitcoin is well known as the crypto currency, the use of the blockchain technology is gaining importance in different areas as well. Thus, consensus algorithms have been developed in order to increase the security and especially the integrity of the crypto currencies. These consensus algorithms indirectly influence the market value of crypto currencies. In this study, consensus algorithms of some crypto currencies have been investigated. In addition to the SHA256 algorithm, the advantages and disadvantages of using Shake and Keccak algorithms as a part of the consensus algorithms, have been analyzed. SHA256, Keccak and Shake algorithms are compared regarding to their performance. As a result, we recommend that the cryptographic hash function can be replaced with the Shake algorithm can be an extendable output function for new crypto currencies.

Keywords: Security · Blockchain · Consensus · Cryptocurrency · High-value crypto · Hash functions · Shake · Keccak

1 Introduction

Blockchain is a system that allows users to connect to different networks and establish a connection through this network. There are two types of blockchain system which are called as unauthorized systems and granted systems [1]. In recent years, cryptocurrencies such as Bitcoin and Ethereum, which are the subject of curiosity by everyone, are some examples of unauthorized systems. In such systems, participants can restrict the transactions and participate in the blockchain consensus. Examples of granted systems include Hyperledger and multichain. In such systems, participants must obtain permission for their operations, while at the same time there are several restrictions for the blockchain to take the lead [2]. Since unauthorized systems are open to the public, any user can easily join the blockchain, so security problems would be arisen because

D. J. Hemanth and U. Kose (Eds.): ICAIAME 2019, LNDECT 43, pp. 572–583, 2020.
https://doi.org/10.1007/978-3-030-36178-5_46

of the size of the blockchain. Consensus protocols are developed in order to maintain the system safely. For example, a participant in the system may join the blockchain with more than one identity at the same time and cause the system to fail during the consensus process.

It is important to develop consensus protocols, to reveal positive and negative aspects of existing approaches or to build alternative approaches to inadequate systems in order to minimize the problems to be experienced by the participants, network or other reasons. The security of the consensus protocol should be taken into account in the beginning when selecting the blockchain platform. The negative aspects of the selected protocol also affect the usability and effectiveness of the blockchain system. The security of a blockchain technology can be expressed with the size of the distributed database [3].

In this study, the advantages of the algorithms used by Crypto currencies, i.e. Proof of Work, Ripple, Byzantine Fault Tolerance, Proof of Stake consensus protocols are investigated. Hash functions suggested to be used in blockchain technology, generate fixed-length output by definition. On the other hand the Shake algorithm, upon request of the user, takes an arbitrary length string and produces an arbitrary length output. We also analyzed the advantages and disadvantages of using extendable output function that gives arbitrary output length for the blockchain structure.

2 Related Works

Baliga [4] has analyzed and gave an overview of the popular blockchain consensus models. In his study, he summarized the history of consensus, bitcoin proof of work invention, consensus models used in blockchain and comparative analysis of these consensus models. The metrics used by the author in comparisons of the protocol are given in detail. In the conclusion section, before emphasizing the correct platform and system, he emphasized that consensus protocols should be paid attention according to the functional and non-functional aspects.

In [5], the authors analyzed the practicability and efficiency of the features proposed by the consensus protocols. In addition to the blockchain types, the comparison of the consensus algorithms, the stellar consensus protocol and Hyperledger fabric are given. As a result, they concluded that Hyperledger cannot provide as much freedom as Stellar to users. Rather, it is a limitation that Hyperledger supports Java as the programming language. It is also mentioned that blockchain technology will be widely used in various Internet of things (IoT) applications as well as contributing to the growth of the economy.

In [6], the authors discussed the scalability of consensus algorithms, and the importance of time spent verifying blocks and security risks. The general problems of consensus algorithms, such as the Byzantine general's problem, Byzantine fault tolerance (BFT), delegated BFT (dBFT) are discussed. The comparison of high profile consensus algorithms, such as proof of work (PoW), Ripple protocol consensus algorithm (RPCA), proof of stake (PoS), Stellar consensus protocol (SCP) are given. In addition, some metrics like energy saving and tolerated power of adversary are given in the framework of the comparison. In conclusion, the authors conclude that PoW algorithm, which is widely used in blockchain platforms, will be replaced by more efficient algorithms. Finally, a comprehensive analysis of the RPCA and SCP has been proposed.

In [7], the authors discussed common blockchain architecture, consensus algorithms, blockchain applications, blockchain technology challenges and recent developments and also possible solutions. It is also mentioned that blockchain technology is not only bitcoin, but can be implemented in many areas (IoT, finance, traditional industry, etc.). In the study, the importance of rapidly developing smart contract applications is also mentioned. However, due to many shortcomings in smart contracting languages, it is stated that there are many new studies that can be applied to the problem. Finally, they mentioned that they are planning to conduct an in-depth research on smart contracts.

In [8], the authors discussed important blockchain consensus protocols in both permissioned blocking and permissionless blocking systems. In this study, authors proposed an attack to some important consensus protocols, by comparing the possible counter measures. In this study, because of the similarity of consensus protocols with cryptographic systems, it has been mentioned that the security system developers of blockchain need to have some distributed system experience. Finally, it is noteworthy that the widespread use of blokchain systems and performance comparisons and their resistance to real attacks will be remarkable.

3 High Value Crypto Currencies and Consensus Algorithms

3.1 Litecoin (LTC)

Litecoin (LTC) is an open source distributed crypto currency. It was launched in October 2011. The Litecoin consensus is based on the Scrypt PoW algorithm using the SHA-256 PoW algorithm used by bitcoin. Usually, the gold of the bitcoin is defined as "silver" [9].

It differs from Bitcoin in two ways. The use of the Scrypt PoW algorithm in Litecoin is much faster than Bitcoin. A blockchain in bitcoin can be generated in an average of 10 min. In the Litecoin, it takes an average of 2.5 min to produce a blockchain block. Another different aspect is the supply limit. It is 84 million in Litecoin and a 21 million supply limit in Bitcoin [10]. When a block is created in Litecoin, a different block header is obtained each time. A block header contains the following fields: Version, hashPrevBlock, hashMerkleRoot, Time, Bit, Nonce. By combining a single block at the Litecoin, the same amount of effort is made to combine a block of 10,000 operations.

The Bits field represents the target using a floating-point encoding. Three bytes for this encoding and one base-256 bytes are used. Only 5 lowest bits are used. Nonce starts at 0 and increases for each call to hash function. As the overflow occurs, the extraNonce portion of the production process is increased, which causes the Merkle root to change.

Just looking at these areas, users frequently sort out the mess one after another, and the fastest miners will always win. Scrypt (N = 1024, r = 1, p = 1 parameters) is used to calculate the work-goal mixes and all other purposes to check the SHA-256d (SHA-256 twice). When calculating hash operations, particular attention is paid to the order of bytes [11].

3.2 Ethereum (ETC)

Ethereum is a decentralized platform called smart contracts released in July 2015. Smart contracts are "self-governing" contracts that are programmed, such as fraud, censorship or interventions by third parties. Unlike Bitcoin, which has a capacity that goes beyond the Ethereum currency, a process that a software application can do is similar to the process of executing a smart phone operating system. The local currency of the Ethereum is called "ether". Like Bitcoin, Ethereum is already using a PoW consensus, but in the future, studies are under construction to adopt a PoS consensus mechanism known as the Casper Protocol [12]. The development of Ethereum is supported by the Ethereum Foundation, a Swiss non-profit organization founded by the inventors of Ethereum.

It uses Ethash, a Hashimoto hybrid, to replace Scrypt or SHA256 as evidence of Ethereum's PoW algorithm. In Hashimoto, using the blockchain as a dataset, selecting the index N from the blockchain, collecting the operations in these indices, returns the result after performing XOR operation on these data. Hashimoto is based on 256-bit arithmetic, which has a significant computational load. To solve this problem, Hashimoto uses least significant 64 bits when adding the data set to the directory.

KECCAK-256 is used in Ethereum for hash operations. Because NIST changed the padding to SHA3-256 (M) = FALSE [512], it was different from the original padding proposed by the Keccak team. Users refer to the "original" SHA3 Keccak as "Keccak" and the new NIST SHA3 standard as "SHA-3" [13].

3.3 Ripple (XRP)

Ripple has been released in 2012 by Ripple (Labs) Inc. It provides transfers close to the currency, regardless of forms. The crypto currency is XRP. In the Ripple platform, the XRP can process more than 1500 operations per second. Ripple uses its own private consensus to validate transactions instead of PoW or PoS. The question of how the XRP is or how it will be distributed in the future remains unclear.

In the XRP, the digital signature proves that a transaction is authorized to set a specific set of actions. Only signed transactions are sent to the network and approved. Each digital signature is based on an encrypted key pair associated with the sending account. A key pair can be created using one of XRP Ledger's supported encryption signing algorithms.

The objects of the ledger have a unique ID in the XRP. The identity is derived by mixing the important content of the object with a namespace identifier. The ledger object type determines the namespace identifier to use and the content to be included in the mixed content. In this way, each identity becomes unique. The SHA-512 is used to calculate the hash value and the result is added to the first 256 bytes. This algorithm called SHA-512 half provides security results similar to SHA-256, but works faster on 64-bit processors.

3.4 Dash

Dash (DASH), also known as Darkcoin, was first released in January 2014. Consistently, X11 PoW algorithm is used. In addition to PoW, Dash also uses Masternode infrastructure. Masternode is a server connected to the Dash network, which guarantees certain minimum performance to perform some tasks with PrivateSend and InstantSend calls can be made to establish the authority to vote by checking the validity of a transaction sent by using the Dash masternode network, and if this exists, the information is transferred to the network. This means that Dash is capable of processing almost instantly, such as credit card transactions.

X11 is designed and developed to make the creation of application-specific integrated circuits (ASICs) much more difficult. This approach is largely driven; At the beginning of 2016, the ASICs for X11 are now a major part of the current network complexity but cannot yet lead to the level of centralization that is present in Bitcoin.

Inspired by Quark's chain-mixed approach, more complexity are added, increasing the number of hashes; however, unlike Quark, instead of selecting some hashes randomly, it is different from the determination of mixed tours as a priori. In the X11 algorithm, 11 different hashes use more than one round, thus becoming one of the safest and most complex encryption mixes used today.

3.5 Monero (XMR)

Monero (XMR) was released in April 2014. Monero is cryptographically developed to maintain operations in complete security. The sending and receiving addresses of the data and the amount of transactions performed are encoded in the process for security purposes [14].

The Monero consensus is based on the CryptoNote PoW algorithm. Unconfirmed crypto coins, such as Bitcoin and Litecoin, are blacklisted. If they have been used for illegal operations before, this is forever in the blockline. Unlike other currencies on the market, Monero (XMR) has not been previously released.

Monero, unlike the 256-bit secure hash algorithm (SHA-256) used by Bitcoin, prevents ASIC development by using a memory hard algorithm (CryptoNight) which makes ASIC development more difficult than. Mixed functions are encryption tools that can generate a unique number for each entry. These algorithms are designed to cause a completely different output with any changes to the input. The term "hash" is used to refer to both the function itself and its output for a particular input.

As a result, CPU and GPU mining is suitable for Monero even in 2018. Monero has billions of available devices with mining capability. It is even possible to mine Monero from any phone or computer with a web browser.

3.6 Bitcoin Cash (BTC)

Bitcoin Cash (BCH) was released on August 1, 2017. Bitcoin is based on the SHA-256 PoW algorithm. But there are some changes different from Bitcoin. Some bitcoin developers have the paradigm that arose when they increased the block size from 1 MB to 8 MB to reduce transaction costs and improve approval times. Bitcoin Cash

(BCH) is known as "hard fork" in the crypto community. The consensus, as in Bitcoin, uses the PoW mechanism.

Hashcash system is found for e-mail users in 1997 by Adam Back [15]. For e-mail users, it is anti-spam tool that uses cryptographic hash functions. It ignores low waiting period but cares high waiting period, and the e-mails are accepted as spam-mail, while sending e-mail among e-mail users. Actually, HashCash is Proof-of-Work by nature. Because, HashCash proves that cryptographic hash functions do that specific work. In Bitcoin, this the process is so-called "mining".

In Bitcoin, the definition of mining is as follows;

Bitcoin uses SHA-256 algorithm as the hash function. The hash functions generate predefined formatted hash value. This format is the number of leading zeros in hash values. This property is named "difficulty" in mining. The difficulty value is set so that one block can be produced in ten minutes. More and more number of zero in initial bits increase the difficulty. Because, if number of zeros in initial bits are wanted to be many, then these zero-bits are found so much difficult by brute-force approach. Miners use random numbers to find specified number of zeros in initial bits of the output, which is by brute-force.

3.7 NEO

NEO was first launched as an Antshares project in February 2014. It was later released as NEO in June 2017. NEO, sometimes referred to as the "Chinese Ethereum" [16]. NEO, like Ethereum and Cardano, is a platform in which intelligent contracts and decentralized transactions are carried out. NEO is a project created for the purpose of a "smart economy" which is technically not a crypto currency. NEO can support up to 10,000 transactions per second. As the consensus mechanism, dBFT is used.

Problem of dBFT is as follows;

Let's assume that there are 9 generals in the Byzantine Empire and they surround the city of Rome. Each general has his own army. In order to capture Rome, the generals must act strategically. If any general does not implement the consensus decision, his army will be defeated. There are two options of attack and withdrawal, and they are offered to vote every day. If the option exceeds 50%, the decision is applied. Each general is in a different region. Therefore, the generals communicate their decisions to each other by courier. This system has some problems. First problem, Romans can be bribe ant of the Byzantine couriers and can learn the plan. Second problem, there may be generals who do not have a consensus. Third problem, bribes can be given by the Romans to change votes to any of the couriers. The fourth problem, couriers may deliver the message incorrectly or may not deliver the message [6]. The problem is similar to the problem in distributed computer systems.

There are many protocols to solve this the problem. The NEO creators chose dBFT. Because, this method has good scalability and scalability is a very important issue.

Underlying logic of dBFT can be explained as; Assume that there are delegates and citizens in a country. Citizens vote by electing the delegate as their president. The selected delegate do what citizens want and saves them. If the citizens are not satisfied, they may change the delegate unanimously. When a new proposal is made, a random speaker from each group tells the happiness number in the community and the result is presented to the delegates. If the result is greater than 66% or equal to 66%, and if the delegates verify the result then they accept new proposal [17]. This analogy is the consensus of the NEO blockchain. In here, happiness number is the hash value in NEO blockchain.

3.8 Cardano (ADA)

Cardano (ADA) was released in September 2015. It was developed as a platform for smart contracts and decentralized applications such as Ethereum and NEO. Cardano took many lessons from the Bitcoin and Ethereum communities and aimed for security, scalability and interoperability. Cardanon's currency is ADA. With ADA, sending and receiving transactions from digital funds can be realized [18].

Cardano is the first blockchain project designed by a team led by academicians and engineers, differing from Ethereum and other crypto coins. The basis of the Cardano consensus is based on the PoS algorithm. More special name is Ouroboros. In the Ouroboros protocol, each section contains a set of approvers. Blocks are produced by those called leaders. The procedures to be included in these blocks must be approved by the appropriate number of approvers. To create equilibrium, approvers are rewarded by leaders in certain periods.

3.9 Stellar (XLM)

Stellar is an open-source, distributed system like Ripple. In 2014, it was launched by Ripple's founders. Stellar has been developed to combat poverty and perform cost-effective financial transactions individually. It can also be used in intelligent contracts. Stellar has its own special consensus protocol. Currency used is called XLM. The development of Stellar is supported by stellar.org [8].

Actually, Stellar uses the BFT consensus algorithm which is resistant to Sybil attacks. This is called Federated Byzantine Fault Tolerance. The consensus that is used by Stellar has some differences compared to other BFT consensus algorithms. It works on values without considering the majority of votes. The biggest difference from other BFT consensus algorithm is that the nodes determine the combination of quorums. Each node selects some nodes that are important, and the kernel must contain such an array for each node.

3.10 IOTA (MIOTA)

IOTA (MIOTA) was released in 2016. It was designed and developed so that machines and people can transfer their data in a safe and decentralized environment at no cost. IOTA also uses distributed ledger technology, such as other crypto currencies. Transactions are also referred to as "circulating" because of intermixing [19]. IOTA has a consensus algorithm as PoW or PoS, instead of the participants of the network to perform a new transaction, each time two consensus is established by confirming the old process.

IOTA does not use blockchain structure. IOTA uses hash directed acyclic graph (Hash-DAG) instead of blockchain, this is called the Tangle. All nodes effect each other. In the same time, all transactions include proof-of-work. Each transaction involves two or more hash values of the previous transaction. This creates a structure like blockchain hash which is called as DAG. Actually, the structure is a graph and DAG is the edge. Node value is the weight of difficulty of the puzzle. When a new node is added, all nodes verify other nodes. The oldest nodes are approved by new ones to prevent cheating. The previous nodes of the new node are related to how much weight they should approve. The nodes decide which side to trust. This is decided by statistical tests.

4 Using SHA-3 Algorithm in Block Mining

Keccak has been accepted as the new standard by NIST in 2012 as the SHA-3 hash function [20]. This hash function is based on sponge constructions [21]. Sponge construction consists of permutation function and includes parameters rate and capacity. Rate is the message block being processed and security is provided with the capacity. The permutation's length is the sum of the rate and the capacity. The vector used in Keccak is expressed by $\mathbf{v} = \mathbf{5} \times \mathbf{5} \times \mathbf{w}$ and $\mathbf{w} = \mathbf{2}^{\mathbf{l}}, \mathbf{l} = \mathbf{1}, \mathbf{2}, \ldots, \mathbf{6}$. Therefore, vector size is [25, 50, 100, 200, 400, 800, 1600]. The number of rounds varies by vector size and indicated by $\mathbf{12 + 2l}$. The function f is the most important part of the Keccak structure. F function takes place in five steps. These steps are θ, ρ, π, χ, ι [22]. Each step provides different security property in cryptanalysis. SHA-3 types are defined as Keccak-512, Keccak-384, Keccak-256, Keccak-224, which are fixed-length output hash functions, together with Shake-128 and Shake-256 which are extendable output functions with the same security definitions as SHA-3. The security claim of SHA-3 is high because it is related to the capacity instead of the length of the output.

Hash functions are functions that take input, which its size can vary, and generate fixed-length output. A cryptographic hash function is a hash function with collision resistance and preimage resistance. Blockchain is functionally used with the outputs the cryptographic hash values of the entries of the previous block of the chain. With this repetitive operation, each block can be returned to the first starting block of each block by verifying the integrity of the previous block. A blockchain database uses peer-to-peer network and has distributed structure. Namely, there is no central authority. Peer-to-peer networks do not contain a security vulnerability for attackers. Information and data manipulation is possible in case of a central authority. But thanks to the distributed database, it works like a consensus mechanism, thus the blockchain makes this impossible. Cryptocurrency is a blockchain technology. A block is a list of payments, where each one referring back to the one that went before. Cryptocurrency can be transferred peer-to-peer. Approvals from other users are received when a new block is included in the chain. All created chain can be displayed but unchangeable. Control and security are provided with symmetric and asymmetric encryption algorithms.

The SHA-1 and SHA-2 hash function are in the same category. Hash functions should be able to provide the desired n-bit security against attacks. Yet length-extension attack on SHA-2 is not known for SHA-3, which is the new standard by NIST, because SHA-1 and SHA-2 use the similar algorithms.

Blockchain is built on hash functions. Usually, SHA-256 algorithm is used in the blockchain. That is, blockchain uses hash function with 256 bit fixed-length output. The first hash value consists of zeros in the chain. The hash functions of other blocks take previous hash value as the input. In this way, the hash value of the new blocks is produced by doing mixed operations between the nonce and the previous hash function. Nonce is a random alpha-numeric string. There are certain features that stand out in the hash value of the new block such that presence of leading zeros in hash values [23]. There are two problems here for miners. First, the hash value of the same input does not have a different output. Therefore, trial and error method used to find the leading number of zeros. The second and the most important problem in here, blockchain system wants short output and decides how many bits of the 256 bits will be zero. Leading zeros determine the difficulty of the work.

For example, in SHA-256, we want the first 10 bits to be zero. But, it took a long time to find a value of 10 bits zero. In this case, the first 9 bits of the next block should be zero. In this way, the number of zero bits is decided. In our work, instead of searching for zero bits within 256 bits, we have searched for what changes in the chain system to look for these zero bits in longer hash values. For our study, we used Keccak algorithm which is the SHA-3 hash function and Shake algorithm which is an arbitrary length version of Keccak algorithm. Then, the use of these three algorithms in mining was compared with the duration of hash power and the time to find the leading zero bits.

5 Results and Analysis

This study examines ten sub-coins which currently have the highest market capacity (see Table 1). The selection of the sub-coin is performed not only by considering the popularity of sub-currencies in the crypto community, but also considering various different characteristics. Some sub-coins use the PoW mechanism as the consensus, while others use different consensus mechanisms.

Table 1. Top ten crypto-currencies by Market Cap as of 2018-12-12.

Coins	Consensus	Algorithm	TPS	Market Cap [24]
Litecoin (LTC)	*PoW*	*Scrypt PoW*	*56*	*$ 1.4 B*
Ethereum (ETC)	*PoW*	*EtHash PoW*	*15*	*$ 8.7 B*
Ripple (XRP)	*XRP*	*RPCA*	*1500*	*$ 11.6 B*
Dash	*PoW*	*X11 PoW*	*10*	*$ 0.5 B*
Monero (XMNR)	*PoW*	*CryptoNight Pow*	*1700*	*$ 0.6 B*
Bitcoin (BTC)	*PoW*	*HashCash*	*60*	*$ 1.3 B*
NEO	*BFT*	*DBFT*	*10000*	*$ 0.3 B*
Cordano (ADA)	*PoS*	*Ouroboros*	*7*	*$ 0.7 B*
Stellar (XLM)	*BFT*	*FBFT*	*1000*	*$ 1.8 B*
IOTA (MIOTA)	*Specific*	*Tangle*	*182*	*$ 0.6 B*

In this work, the results are investigated by using SHA-3 which gives arbitrary length output instead of SHA-256 algorithm is used in blockchain. We wrote a mining software for Shake, Keccak and SHA-256 algorithms by using Python3 and Python programming language. This software searches the number of consecutive zeros up to twenty-eight within the nonce value. If it does not find any number of zeros, it produces a new hash value. The test computer has 4 GB RAM, AMD A8-7410 APU CPU, and AMD Radeon R5 GPU.

In Fig. 1, according to the output of our software, SHA-256, Keccak and Shake algorithms were compared with the number of hashing per second and number of zero bits for the first 28 bits. As it can be seen, the number of hash values that were produced per second by Shake algorithm is less than other algorithms. The reason for this is that the length of the hash is very long. Another name of the number of hash produced per second is the hash power. We already know that SHA-3 hash functions are slow. However, our study is not based on the slowness of SHA algorithms. In Figs. 2, 3 and 4, this can be clearly seen.

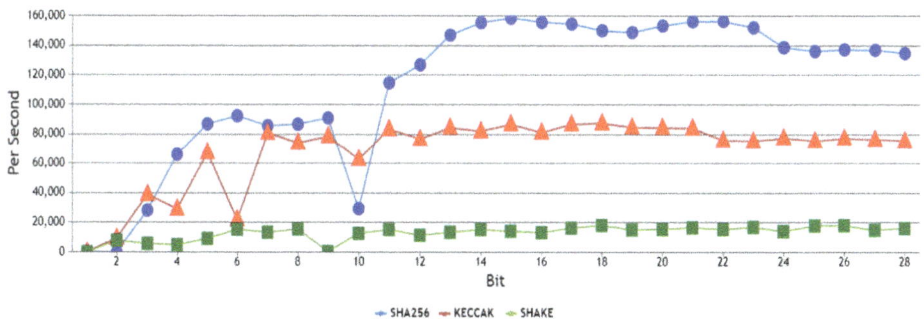

Fig. 1. Comparison of hash power of SHA256, Keccak, and Shake algorithms.

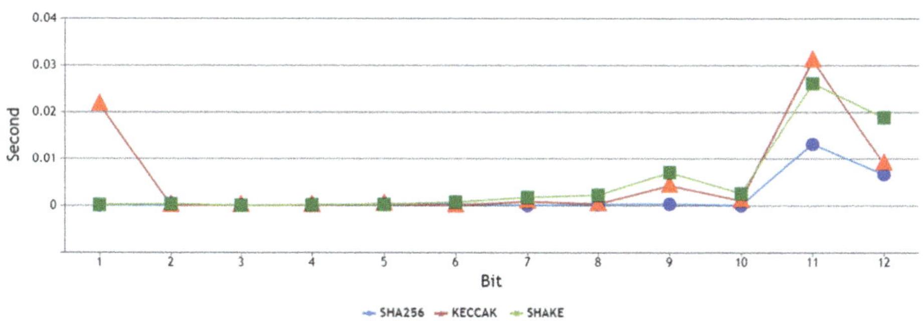

Fig. 2. Time-dependent comparison of SHA256, Keccak, Shake algorithms (1 to 12 bits)

The difficulty in the block chain is an increasing value over time. This situation poses a threat to the new cryptocurrencies. In Fig. 2, according to the output of our software, SHA-256, Keccak and Shake algorithms were compared by searching time in seconds and number of zero bits for the first twelve-bits.

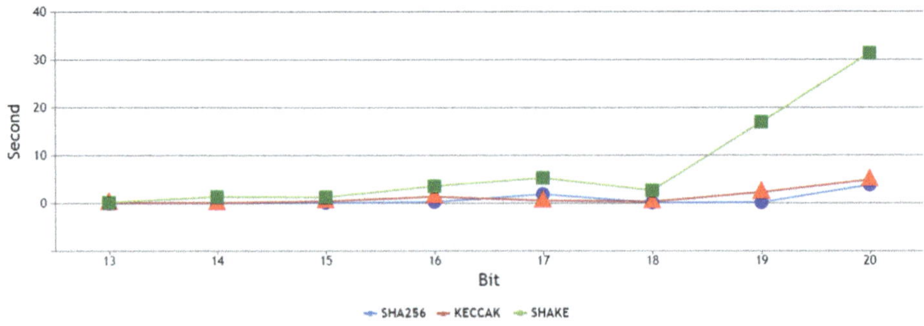

Fig. 3. Time-dependent comparison of SHA256, Keccak, Shake Algorithms (13 to 20 bits)

Shake gives similar results with other algorithms up to thirteen bits. However, after the thirteenth bit, time to find zero bits gets longer than the other algorithms. Although the Shake algorithm produces short length hash value, it is very difficult to find the leading zero bits in the hash value produced, compared to other algorithms. In Figs. 3 and 4, we can see this situation more clearly after the sixteenth bit.

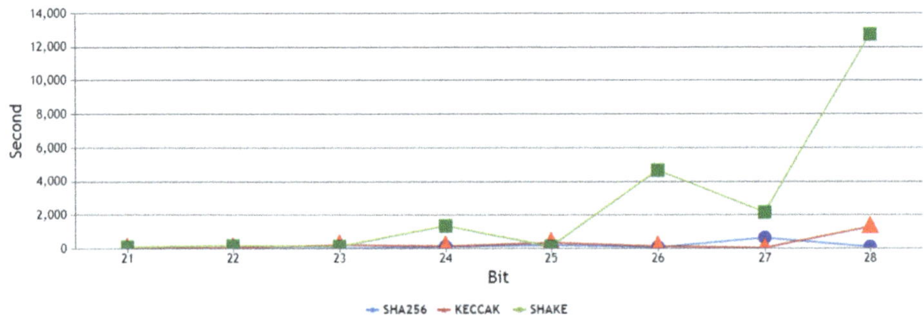

Fig. 4. Time-dependent comparison of SHA256, Keccak, and Shake algorithms (21 to 28 bits)

6 Conclusion

All of the hash values produced by the Shake algorithm are in different sizes. Although this situation slows down the blockchain structure, we recommend that there is a system that can be used for new cryptocurrencies. As we know that the blockchain structure will grow very much in the future, we will need the Shake hash function together with the advancing technology in the blockchain structure.

In little-known cryptocurrencies, even with the consensus protocols, there are still some problems. For example, two different results are obtained by adding two different data to the same hash value. In this case, there is an open problem: 'which hash output is used for producing next hash value?'. This can cause the problem that little-known cryptocurrency systems, which its users have small computing power, can be broken by users of big cryptocurrencies with powerful computers. In the future works, after the creation of a block, the possibility of creating another alternative block and the probability of creating the first block with Shake will be investigated.

References

1. Narayanan, A., Bonneau, J., Felten, E., Miller, A., Goldfeder, S.: Bitcoin and Cryptocurrency Technologies: A Comprehensive Introduction. Princeton University Press, Princeton (2016)
2. Androulaki, E., Barger, A., Bortnikov, V., Cachin, C., Christidis, K., De Caro, A., Muralidharan, S.: Hyperledger fabric: a distributed operating system for permissioned blockchains. In: Proceedings of the Thirteenth EuroSys Conference, p. 30. ACM (2016)
3. Li, Z., Ren, W., Liu, X., Xie, L.: Distributed consensus of linear multi-agent systems with adaptive dynamic protocols. Automatica **49**(7), 1986–1995 (2013)
4. Baliga, A.: Understanding blockchain consensus models. In: Persistent (2017)
5. Sankar, L.S., Sindhu, M., Sethumadhavan, M.: Survey of consensus protocols on blockchain applications. In: 2017 4th International Conference on Advanced Computing and Communication Systems (ICACCS), pp. 1–5. IEEE (2017)
6. Bach, L.M., Mihaljevic, B., Zagar, M.: Comparative analysis of blockchain consensus algorithms. In: 2018 41st International Convention on Information and Communication Technology, Electronics and Microelectronics (MIPRO), pp. 1545–1550. IEEE (2018)
7. Zheng, Z., Xie, S., Dai, H., Chen, X., Wang, H.: An overview of blockchain technology: architecture, consensus, and future trends. In: 2017 IEEE International Congress on Big Data (BigData Congress), pp. 557–564. IEEE (2017)
8. Blockchains consensus protocols in the wild. arXiv pre-print arXiv:1707.01873 (2017)
9. What is Litecoin? A basic beginners guide. https://blockgeeks.com/guides/litecoin. Accessed 21 Dec 2018
10. What is Litecoin? Here's everything you need to know. https://www.digitaltrends.com/computing/what-is-litecoin. Accessed 26 Dec 2018
11. Block Hashing Algorithm—Litecoin. https://litecoin.info/index.php/Block_hashing_algorithm. Accessed 28 Jan 2019
12. What is Ethereum Casper Protocol? Crash Course. https://blockgeeks.com/guides/Ethereum-casper. Accessed 22 Dec 2018
13. Nist. https://csrc.nist.gov/projects/hash-functions/sha-3-project. Accessed 28 Jan 2019
14. What to Know Before Trading Monero. https://www.coindesk.com/what-to-know-before-trading-monero. Accessed 24 Dec 2018
15. Nakamoto, S.: Bitcoin: a peer-to-peer electronic cash system (2017)
16. Cryptocurrency Industry Spotlight: Who is NEO's Da Hongfei? https://coincentral.com/cryptocurrency-industry-spotlight-neos-da-hongfei. Accessed 25 Dec 2018
17. Vukolić, M.: The quest for scalable blockchain fabric: proof-of-work vs. BFT replication. In: International Workshop on Open Problems in Network Security, pp. 112–125. Springer, Cham (2015)
18. On the Origin of Cardano. https://medium.com/on-the-origin-of-smart-contract-platforms/on-the-origin-of-cardano-a6ce4033985c. Accessed 24 Dec 2018
19. Tennant, L.: Improving the Anonymity of the IOTA Cryptocurrency (2017)
20. Bertoni, G., Daemen, J., Peeters, M., Van Assche, G.: Keccak sponge function family main document. Submission to NIST (Round 2), vol. 3, no. 30 (2009)
21. Cryptographic sponges. http://sponge.noekeon.org. Accessed 21 Dec 2018
22. Bertoni, G., Daemen, J., Peeters, M., Van Assche, G.: Keccak specifications, Submission to NIST (round 2), pp. 320–337 (2009)
23. Badev, A., Chen, M.: Bitcoin: technical background and data analysis (2014)
24. Coin Market Cap. https://coinmarketcap.com/coins/views/all/. Accessed 21 Dec 2018

Estimation of Foam Concrete Mixture Rate with Randomed Forest Algorithm

Şemsettin Kilinçarslan[1(✉)], Emine Yasemin Erkan[1,2], and Murat Ince[1,2]

[1] Civil Engineering Department, Enginering Faculty, Suleyman Demirel University, 32600 Isparta, Turkey
semsettinkilincarslan@sdu.edu.tr

[2] Isparta Vocational School of Technical Sciences, Isparta University of Applied Sciences, 32600 Isparta, Turkey

Abstract. Concrete is one of the most widely used construction materials in the world. When the property of concrete is developed over time and they find new application areas, their usage areas are increasing. Foam concrete is a light-weight concrete type and is also known as porous or cellular concrete. In this study, foam concrete series having densities of 450, 600 and 750 kg/m^3 were obtained by using different aggregate types such as pumice, fly ash, barite and limestone. In addition to density, compressive strength and thermal conductivity coefficient, the radiation absorption coefficient was the most important factor for the use of foam concrete. The aim of this study is to estimate the amount of parameters such as density, radiation absorption coefficient, cement, water, barite, pumice, fly ash and limestone in the concrete mixture with randomed forest algorithm. Trial mixes are prepared and tested to determine the parameters to reach the desired mixture. Continuous experimentation is a time consuming and costly process. Several studies have been conducted to estimate the mixing ratios based on existing experimental data. In this study, using the randomed forest algorithm and the limited experimental data available, the desired mixing ratios were estimated. Ten concrete mixes were made to train the algorithm. Data such as density, radiation absorption coefficient, compressive strength and thermal conductivity coefficient are estimated to be mixed in parameters such as cement, water, barite, pumice, fly ash and limestone.

Keywords: Foam concrete · Forest algorithm · Mixture rate

1 Introduction

Concrete is a building material which is widely used in construction works. The concrete mixture mainly contains cement, water, aggregate and additives suitable for the purpose of the concrete to be produced [1, 2].

Some of the disadvantages of concrete have been produced by using different aggregates, adding additives, changing the composition, using only normal aggregate and cement, making changes in the construction technique or using a few of these methods together [3]. While looking for solutions to the disadvantages of conventional concrete, one of the ways used was to reduce the volume of the concrete by using

D. J. Hemanth and U. Kose (Eds.): ICAIAME 2019, LNDECT 43, pp. 584–588, 2020.
https://doi.org/10.1007/978-3-030-36178-5_47

different methods. By reducing the unit volume mass of the normal concrete used in the buildings, the self-weight of the reinforced concrete element can be reduced and thus the structure can be mitigated. Reducing the structure and reducing the sections of the structural system can be achieved economy [4].

The use of light concrete instead of normal concrete in life and work places provides significant advantages in terms of sound insulation in terms of sound insulation as well as providing more positive results in terms of sound insulation. Lightweight concrete allows the reduction of the total static load, thus facilitating the design of more reliable and economical structures. Due to all these advantages, researches have started to increase with the use of chemical and mineral additives and to produce high performance light concrete with natural or artificial aggregates [5, 6]. Lightweight concrete is defined as the weight (density) of the unit having a dry weight, the density of which is greater than 0.8 g/cm^3 and less than 2.0 g/cm^3. Lightweight concrete is produced by the use of light aggregates having different types and specialties [7].

Random forest algorithm is a controlled learning algorithm [8]. This algorithm consists of many randomly generated decision trees. This algorithm is a method known as bagging, which can perform both regression and classification tasks with the use of multiple decision trees [9]. In the bagging method, each decision tree includes training on a different data sample that is made by changing the sampling [10]. The aim of this study is to estimate the amount of parameters such as density radiation absorption coefficient, cement, water, barite, pumice, fly ash and limestone in the concrete mixture with randomed forest algorithm.

2 Materials and Method

In this study, micronized limestone, barite and pumice were used as aggregate. For the production of foam concrete, aggregates were used after being micronized. For foam concrete produced with fly ash, UK, FC for foam concrete produced with micronized limestone, BFC for foam concrete produced with barite aggregate, and LC abbreviations were used for concretes produced with Kayseri pumice.

Foam with a density of 80 g/L in average was added to the mixtures until theoretical fresh density value calculated was reached according to the mix designs. For each set, six 150 mm cube specimens were produced. All specimens were cured at 20 ± 2°C under 95% of relative humidity (RH). The compressive strength and splitting tensile strength of specimens were tested under dried air conditions (~%6 RH). The dry densities of foam concrete specimens were determined according to TS EN 678. Compressive strength and splitting tensile tests were applied on three 150 mm cube specimens in accordance with TS EN 1354 and TS EN 12390-6, respectively. Foam concretes were produced according to the amount of mixture given in Table 1.

28 days compressive strength test on 6 × 10 × 10 cm cube samples was carried out at 2400 N/hr loading speed conditions using a uniaxial pressure test device within the principles specified in TS EN 1354 (TSE, 2007). For thermal conductivity coefficient determination, Lasercomp Fox 50 thermal conductivity device which is working with heat flow measurement method is used. Foam concretes were estimated to be 10 × 10 × 2 and 10 × 10 × 4 cm with stone cutting tool (Fig. 1).

Table 1. Foam concrete series and mixture amounts.

Experimental group	Cement (kg)	Water (kg)	Barite (kg)	Pumice (kg)	Fly Ash (kg)	Limestone (kg)
UK 450	340	198,9	0	0	102	0
UK 600	530	286,3	0	0	106	0
FC 450	300	180	0	0	0	150
FC 600	300	240	0	0	0	300
FC 750	300	300	0	0	0	450
BFC 450	300	180	150	0	0	0
BFC 600	300	240	300	0	0	0
BFC 750	300	300	450	0	0	0
LC 600	300	135	0	300	0	0
LC 750	300	135	0	450	0	0

Fig. 1. Samples used in experiments.

In order to determine the radiation absorption properties of 10 × 10 × 2 cm foam concretes, it was tested with Gamma Spectroscopy System with Cs-137 and Co-60 radiation sources and 662 keV, 1173 keV and 1332 keV gamma energy values.

In this study, foam concrete series having densities of 450, 600 and 750 kg/m^3 were obtained by using different aggregate types such as pumice, fly ash, barite and limestone. In addition to density, compressive strength and thermal conductivity coefficient, the radiation absorption coefficient was the most important factor for the use of foam concrete. Table 2 gives the test results of the foam concrete samples.

Performance and accuracy values such as Root Mean Square Error (RMSE), RSquare (R^2) and Mean Absolute Error (MAE) are given in Table 3.

The low RMSE and MAE values and the high R^2 values indicate that the mix ratios of the random forest algorithm are estimated successfully.

Table 2. Test results of foam concrete samples.

Specimen of Foam Concrete	Density (kg/m^3)	Radiation Absorption Coefficient μ 1173 keV	Compressive strength (MPa)	Thermal Conductivity Coefficient (w/mK)
UK 450	443,1	0,04352	1,345	0,117
UK 600	636,85	0,047332	2,727	0,173
FC 450	452,34	0,044955	0,113	0,176
FC 600	602,14	0,045431	1,775	0,213
FC 750	751,94	0,048051	1,986	0,225
BFC 450	452,36	0,042044	0,495	0,125
BFC 600	602,2	0,045655	1,256	0,175
BFC 750	752,03	0,047233	1,423	0,184
LC 600	601,83	0,043666	1,8	0,151
LC 750	751,49	0,050338	2,496	0,174

Table 3. Performance and accuracy values of foam concrete material mixing ratios.

	Cement	Water	Barite	Pumice	Fly Ash	Limestone
RMSE	0,21372	0,22437	0,30000	0,56324	0,23220	0,32863
R^2	0,90324	0,86049	0,96153	0,86571	0,68858	0,95384
MAE	0,12	0,15384	0,18	0,345	0,1456	0,27

3 Conclusions

In this study, it was aimed to determine the mixing ratio of the materials such as cement, water, barite, fly ash, pumice and limestone which should be present in the mixture from the limited foam concrete parameters obtained from the real experiment results using the random forest algorithm. Using the random forest algorithm, the predicted mixing ratios were estimated with high RMSE and MAE values with high R^2. Thus, problems such as time-consuming and costly experimentation are eliminated. In addition, different concrete mixes which will be formed by using different test parameters besides the parameters used in the study of the foam concretes can also be successfully estimated with random forest algorithm without any further experiments.

References

1. Neville, A.M.: Properties of concrete, vol. 4. Longman, London (1995)
2. Kilincarslan, Ş., Davraz, M., Akça, M.: The effect of pumice as aggregate on the mechanical and thermal properties of foam concrete. Arabian J. Geosci. **11**(11), 289 (2018)
3. Baradan, B.: Construction Materials z II. Dokuz Eylul University Faculty of Engineering and Architecture, s 1–3, İzmir (1991)
4. Konuk, H., Özyurt, N., Taşdemir, C., Yüceer, Z., Sönmez, R.: Carrier properties of lightweight concrete. In: 1st National Building Material Congress, İstanbul (2002)

5. Kilinçarslan, Ş., Davraz, M., Koru, M., Ekıztaş, F.: Investigation of Properties of Foam Concretes Produced Using Pumice at Different Ratios. Acta Phys. Pol., A **132**(3), 708–709 (2017)
6. Davraz, M., Kilinçarslan, Ş., Koru, M., Tuzlak, F.: Investigation of relationships between ultrasonic pulse velocity and thermal conductivity coefficient in foam concretes. Acta Phys. Pol., A **130**(1), 469–470 (2016)
7. TS EN 206-1, "Concrete-Section 1: Feature, Performance, Manufacturing and Conformity, Turkish Standards Institute, Ankara (2002)
8. Breiman, L.: Random forests. Mach. Learn. **45**(1), 5–32 (2001)
9. Svetnik, V., Liaw, A., Tong, C., Culberson, J.C., Sheridan, R.P., Feuston, B.P.: Random forest: a classification and regression tool for compound classification and QSAR modeling. J. Chem. Inf. Comput. Sci. **43**(6), 1947–1958 (2003)
10. Strobl, C., Malley, J., Tutz, G.: An introduction to recursive partitioning: rationale, application, and characteristics of classification and regression trees, bagging, and random forests. Psychol. Methods **14**(4), 323 (2009)

Pre-processing Effects of the Tuberculosis Chest X-Ray Images on Pre-trained CNNs: An Investigation

Erdal Tasci(✉)

Computer Engineering Department, Ege University, Izmir, Turkey
arif.erdal.tasci@ege.edu.tr

Abstract. Tuberculosis (TB) is a serious infectious disease which is one of the top causes of death worldwide. In 2017, 1.6 million people died from the disease according to the World Health Organization (WHO). The earlier identification and treatment of the TB is critical for preventing death and decreasing risk of transmitting the disease to others. Computer-aided diagnosis (CADx) systems are essential tools to speed up the decision-making process of experts and provide more efficient, accurate and systematic solutions. Chest radiography (CXR) is one of the most common and effective imaging technique for the detection of thoracic diseases such as TB and lung cancer. In this study, three different region of interests (ROIs) based pre-processing methods are applied to two CXR image datasets (namely, Montgomery and Shenzhen). We used three pre-trained convolutional neural networks (CNNs) (namely, AlexNet, VGG16, VGG19) as deep learning models and deep feature extractors for automatic classification of TB disease. We investigate the pre-processing effects of TB CXR images on the classifier whether ROI is selected and remaining regions of images are set pixel values to white, black and same pixel values in the original images. Experimental results indicate that proposed methods contribute to the classifier performance gain considerably in terms of accuracy rate.

Keywords: Tuberculosis · Diagnosis · Deep learning · Convolutional neural networks · Region of interest · Classification · Machine learning

1 Introduction

Tuberculosis is a potentially infectious disease caused by Mycobacterium tuberculosis bacteria (MTB). TB affects the lungs primarily and other parts of the body. This disease also spreads through the air from person to person [3]. According to the WHO, 10 million people fell ill with TB in 2017 [21]. The largest number of the cases and deaths of TB disease occur in developing countries such as South-East Asia, Western Pacific and African regions [21]. It is curable and preventable if necessary diagnosing are done and medicines are taken properly, otherwise, it can be fatal.

D. J. Hemanth and U. Kose (Eds.): ICAIAME 2019, LNDECT 43, pp. 589–596, 2020.
https://doi.org/10.1007/978-3-030-36178-5_48

With the use of CADx systems by means of machine learning and deep learning techniques, the diagnosis process of the TB disease becomes easier, faster and more accurate. Additionally, CXR is a primary and rapid medical imaging technique for TB detection. This technique has low cost and high sensitivity for TB and allows lung abnormalities to be identified [20].

In recent years, convolutional neural networks have become a popular deep learning approach for image recognition due to their excellent results in artificial intelligence and computer vision applications [12]. Moreover, pre-trained CNNs use transfer learning process and eliminate the need for training the CNNs from scratch. Thanks to the transfer learning, deep and distinctive features can be extracted from images directly and efficiently. However, the classification performance of the system can be affected with respect to the effects of the image pre-processing techniques applied before using CNNs. Various pre-processing techniques consist of image enhancement and image segmentation methods such as filtering, thresholding, etc. [16].

In this study, we aim to investigate the effects of ROI based pre-processing techniques on two CXR image datasets (i.e., Montgomery and Shenzhen) by using three different pre-trained CNN models (i.e, AlexNet, VGG16, VGG19) as the feature extractors. These pre-processing techniques include white background, black background and the same pixel values in the original image.

The rest of the paper is organized as follows: Sect. 2 summarizes the previous works in the field of TB detection. Section 3 introduces the proposed methods for TB detection. Afterward, the experimental results are given in Sect. 4. Finally, Sect. 5 concludes the paper with future perspectives.

2 Related Work

Researchers in this field applied various methods including deep learning based classification and traditional machine learning based classification due to the fact that diagnosis of TB is a challenging task. Additionally, contrast enhancement, bone suppression, and lung boundary detection are commonly used for X-ray images [8].

An automated segmentation technique which provides automatic TB cavity detection from CXRs was proposed in [17]. To detect TB cavities, a hybrid knowledge-based Bayesian classification was taken and various diverse coefficients and measures like circularity were applied in this approach.

In [15], wavelet transform was proposed to identify TB from CXRs. To this end, thirty line profiles were obtained and these profiles are represented with Daubechies coefficients. The coefficients were used as features for the detection of TB disease.

In [4], various features including Gabor, Gist, the histogram of oriented gradients (HOG), and pyramid histogram of oriented gradients (PHOG) were extracted from the images without segmentation to recognize between TB and non-TB CXRs automatically. These features were evaluated on image datasets by using five-fold cross-validation technique in terms of ROC analysis.

In [1], a mobile device-based CADx software was developed for TB detection. In this study, global and local features were extracted and a deep convolutional neural network-based X-ray images were analyzed. They obtained promising results for their approaches.

In [13], pre-trained convolutional neural networks were used as the feature extractors for TB detection. In the pre-processing stage, the lung regions are segmented with SIFT Flow algorithm and the segmentation masks were utilized for CXR datasets (namely, Montgomery and Shenzhen). They presented three different proposals including simple CNN feature extraction, bag of CNN features and ensemble scheme for deep learning based feature extraction and classification.

In [5], a deep CNN-based method was proposed for TB detection. They used a lightweight architecture which has a number of layers in between that of LeNet and Alexnet architectures. So, it required less time and fewer parameters to be trained. In addition, they compared the performance of three different optimizers including Adam, Momentum and SGD optimizer in terms of accuracy.

In [22], in order to recognize TB bacillus, a CNN based model, named tuberculosis AI (TB-AI), was developed by using deep learning and machine learning algorithms. Their model had high sensitivity for the recognition of bacilli. However, TB-AI was weak for discriminating a contaminant bacillus from a pathogenic bacilli.

In [11], two different CNNs (namely, AlexNet and GoogleNet) were used to classify various images into TB or healthy on four different CXR datasets. They also applied a data augmentation technique with rotations and ensemble technique to improve performance results. They evaluated their system in terms of AUC values.

3 Methods

In this paper, we apply three different ROI based pre-processing techniques (namely, Proposal 2–4) on the CXR images. We also utilize the pre-trained CNNs (namely, AlexNet, VGG16, and VGG19) for feature extraction. These techniques are explained in the following subsections.

3.1 Pre-processing

Proposal 1: Original Images Without Segmentation. In this approach, no segmentation technique is applied. So, the original images are used with pre-trained CNNs to extract deep features before the classification stage (Fig. 1). Then, these features are given into the classifier (i.e., Random Forest).

Proposal 2: Original Images with Segmentation. In proposal 2, the CXR images are segmented with the pre-defined lung masks on two datasets. The rectangular regions of lung masks (i.e. minimum bounding box) are obtained and used with the pre-trained CNNs (Fig. 1). The pixel values in the rectangular

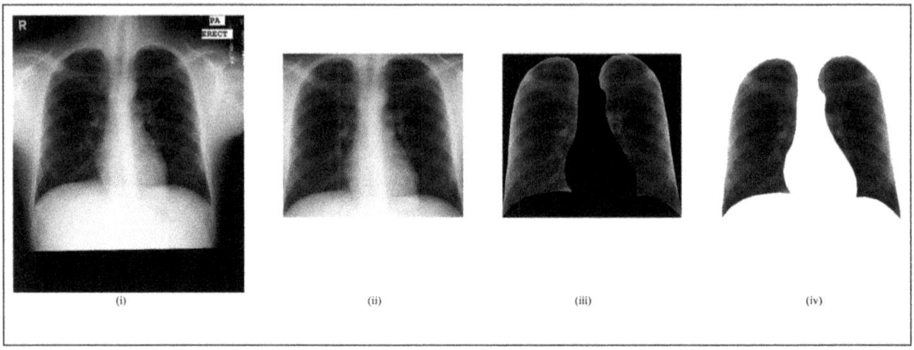

Fig. 1. Sample CXR images for proposals [7] (i) Proposal 1 (ii) Proposal 2 (iii) Proposal 3 (iv) Proposal 4

region but not including the lung mask are set to the pixel values of original input images.

Proposal 3: Segmented Images with Black Background. In proposal 3, the CXR images are segmented in a similar manner to proposal 2 (i.e. predefined lung masks). However, the pixel values in the rectangular region but not including the lung mask are set to the black pixel values (Fig. 1).

Proposal 4: Segmented Images with White Background. In proposal 4, the CXR images are segmented in a similar manner to proposal 2 (i.e. predefined lung masks). However, the pixel values in the rectangular region but not including the lung mask are set to the white pixel values (Fig. 1).

3.2 Feature Extraction

Feature extraction process can be carried out manually with hand-crafted features or automatically with CNN models. In this study, in order to extract features, we utilized three different pre-trained CNNs namely AlexNet, VGG16 and VGG19. These CNN models are trained on more than a million images from the ImageNet database and are used for image recognition tasks [6,10,18]. A different number of features can be extracted from different layers of CNN. The number of features extracted from these CNNs is 4096. In addition, we used 'FC6' and 'FC7' (fully connected) layers of AlexNet, VGG16, and VGG19 CNN models.

3.3 Classification

Classification stage categorizes data based on their characteristics. To categorize deep features, we used Random Forest classifier. Random Forest, an ensemble classification approach, includes a combination of tree predictors [2].

4 Experimental Results

In this section, we present the experimental setup, image datasets used in this study. Then, we give experimental results in detail by using three different pre-trained CNNs and the classifier in the following subsections.

4.1 Experimental Process

In this study, we implemented the experiments on a desktop PC with 8 GB NVIDIA GTX1080 GPU memory. We repeated these experiments five times with 80–20 train-test ratio for extracting features and classifying randomly. We also obtained the mean and standard deviation of accuracy values for the results. Besides, we set the number of trees to 150 for Random Forest classifier.

4.2 Image Datasets

There are two image datasets namely Montgomery and Shenzhen for tuberculosis diagnosis. These datasets can be accessed in [14]. Montgomery dataset consists of 138 posterior-anterior x-rays, of which 80 x-rays are normal and 58 x-rays are abnormal with tuberculosis disease. In addition, Shenzhen Hospital X-ray set consists of 340 normal x-rays and 275 abnormal x-rays having diverse manifestations of tuberculosis disease.

Apart from these datasets, pre-defined lung masks are used for pre-processing and segmentation stage. For Montgomery dataset, this process is easy due to the fact that there are also segmentation masks. However, there is no segmentation mask for Shenzhen dataset. Therefore, manually segmented lungs masks are utilized in [9,19]

4.3 Results

The accuracy and AUC (Area Under the ROC Curve) values are illustrated in Tables 1 and 2. The best results for each proposal method and dataset are shown in bold. According to Table 1, the best accuracy values are obtained with Proposal 2, VGG19 model and FC6 layer for the Montgomery dataset. Besides, Proposal 4 with AlexNet model and FC7 layer gives best accuracy values for Shenzhen dataset.

As it can be observed from Table 2, the best AUC values are provided with Proposal 1, VGG16 model and FC6 layer for Montgomery dataset. Additionally, Proposal 3 with AlexNet and FC6 layer gives the highest AUC values for Shenzhen dataset. Standard deviation values can be varied with respect to the used method, CNN models and layers.

Table 1. The accuracy values for the proposals with Random Forest classifier

ACC (%)	AlexNet		VGG16		VGG19	
Proposal 1	fc6	fc7	fc6	fc7	fc6	fc7
Montgomery	76.43 ± 3.19	70.00 ± 8.22	77.14 ± 7.41	76.43 ± 9.31	**77.86 ± 6.87**	74.29 ± 7.74
Shenzhen	**86.02 ± 2.29**	83.54 ± 3.29	84.07 ± 4.20	84.60 ± 1.01	83.54 ± 4.36	84.42 ± 1.84
ACC (%)	AlexNet		VGG16		VGG19	
Proposal 2	fc6	fc7	fc6	fc7	fc6	fc7
Montgomery	80.00 ± 4.07	80.00 ± 9.31	77.86 ± 5.30	78.57 ± 6.68	**80.71 ± 5.42**	72.14 ± 6.87
Shenzhen	82.83 ± 4.54	**84.60 ± 1.34**	83.89 ± 2.11	81.77 ± 2.84	81.24 ± 3.51	81.77 ± 4.18
ACC (%)	AlexNet		VGG16		VGG19	
Proposal 3	fc6	fc7	fc6	fc7	fc6	fc7
Montgomery	78.57 ± 5.05	**80.00 ± 5.42**	70.71 ± 4.66	75.00 ± 4.37	69.29 ± 11.46	80.00 ± 9.65
Shenzhen	**85.84 ± 2.58**	84.60 ± 2.97	85.84 ± 3.31	83.89 ± 3.88	83.89 ± 3.28	83.36 ± 3.51
ACC(%)	AlexNet		VGG16		VGG19	
Proposal 4	fc6	fc7	fc6	fc7	fc6	fc7
Montgomery	75.71 ± 8.14	70.71 ± 1.60	75.71 ± 8.53	**77.14** ± 8.22	73.57 ± 7.41	67.14 ± 6.39
Shenzhen	86.02 ± 2.53	**88.32 ± 1.92**	84.78 ± 3.15	83.00 ± 2.68	85.66 ± 2.11	86.19 ± 4.67

Table 2. The AUC values for the proposals with Random Forest classifier

AUC	AlexNet		VGG16		VGG19	
Proposal 1	fc6	fc7	fc6	fc7	fc6	fc7
Montgomery	0.85 ± 0.03	0.77 ± 0.10	**0.88 ± 0.06**	0.86 ± 0.08	0.85 ± 0.06	0.81 ± 0.08
Shenzhen	**0.90 ± 0.04**	**0.90 ± 0.03**	0.88 ± 0.03	0.89 ± 0.02	0.88 ± 0.04	0.90 ± 0.02
AUC	AlexNet		VGG16		VGG19	
Proposal 2	fc6	fc7	fc6	fc7	fc6	fc7
Montgomery	0.87 ± 0.07	0.85 ± 0.05	0.84 ± 0.06	0.84 ± 0.05	**0.87 ± 0.03**	0.78 ± 0.07
Shenzhen	0.90 ± 0.03	**0.91 ± 0.02**	0.90 ± 0.02	0.87 ± 0.02	0.89 ± 0.02	0.90 ± 0.03
AUC	AlexNet		VGG16		VGG19	
Proposal 3	fc6	fc7	fc6	fc7	fc6	fc7
Montgomery	0.78 ± 0.05	**0.84 ± 0.03**	0.77 ± 0.09	0.80 ± 0.03	0.75 ± 0.12	0.81 ± 0.11
Shenzhen	**0.93 ± 0.00**	0.91 ± 0.01	0.91 ± 0.03	0.90 ± 0.04	0.90 ± 0.03	0.90 ± 0.02
AUC	AlexNet		VGG16		VGG19	
Proposal 4	fc6	fc7	fc6	fc7	fc6	fc7
Montgomery	**0.82 ± 0.07**	0.79 ± 0.10	0.75 ± 0.04	0.81 ± 0.08	0.76 ± 0.07	0.74 ± 0.06
Shenzhen	**0.92 ± 0.01**	**0.92 ± 0.01**	0.92 ± 0.03	0.90 ± 0.02	0.90 ± 0.01	0.91 ± 0.03

5 Conclusion

In this paper, the effects of the three pre-processing methods using deep learning models are investigated. Deep learning models such as CNNs are useful for facilitating the feature extraction process and obtaining distinctive information about the data. Pre-processing methods are also important for improving recognition performance. According to the experimental results, Proposals (2–4) provided performance gain in accuracy rate compared to Proposal 1 for Montgomery and

Shenzhen datasets. In addition, different methods contribute to the performance of the system in various proportions.

Apart from the used pre-processing approaches, diverse approaches including contrast enhancement and bone suppression can be applied. Furthermore, other classifiers can be tried as future work. Another direction of this study, the combination of these deep features can be considered.

References

1. Alcantara, M.F., Cao, Y., Liu, C., Liu, B., Brunette, M., Zhang, N., Sun, T., Zhang, P., Chen, Q., Li, Y., et al.: Improving tuberculosis diagnostics using deep learning and mobile health technologies among resource-poor communities in Peru. Smart Health **1**, 66–76 (2017)
2. Breiman, L.: Random forests. Mach. Learn. **45**(1), 5–32 (2001)
3. CDC: Tuberculosis (TB) disease: Symptoms and risk factors (2019). https://www.cdc.gov/features/tbsymptoms/index.html
4. Chauhan, A., Chauhan, D., Rout, C.: Role of Gist and PHOG features in computer-aided diagnosis of tuberculosis without segmentation. PloS One **9**(11), e112980 (2014)
5. Hooda, R., Sofat, S., Kaur, S., Mittal, A., Meriaudeau, F.: Deep-learning: a potential method for tuberculosis detection using chest radiography. In: 2017 IEEE International Conference on Signal and Image Processing Applications (ICSIPA), pp. 497–502. IEEE (2017)
6. ImageNet: Imagenet database (2016). http://www.image-net.org
7. Jaeger, S., Candemir, S., Antani, S., Wáng, Y.X.J., Lu, P.X., Thoma, G.: Two public chest x-ray datasets for computer-aided screening of pulmonary diseases. Quant. Imaging Med. Surg. **4**(6), 475 (2014)
8. Jaeger, S., Karargyris, A., Candemir, S., Siegelman, J., Folio, L., Antani, S., Thoma, G.: Automatic screening for tuberculosis in chest radiographs: a survey. Quant. Imaging Med. Surg. **3**(2), 89 (2013)
9. Kaggle: communications engineering branch (2018). https://www.kaggle.com/eduardomineo/u-net-lung-segmentation-montgomery-shenzhen/data
10. Krizhevsky, A., Sutskever, I., Hinton, G.E.: ImageNet classification with deep convolutional neural networks. In: Advances in Neural Information Processing Systems, pp. 1097–1105 (2012)
11. Lakhani, P., Sundaram, B.: Deep learning at chest radiography: automated classification of pulmonary tuberculosis by using convolutional neural networks. Radiology **284**(2), 574–582 (2017)
12. LeCun, Y., Bengio, Y., Hinton, G.: Deep learning. Nature **521**(7553), 436 (2015)
13. Lopes, U., Valiati, J.F.: Pre-trained convolutional neural networks as feature extractors for tuberculosis detection. Comput. Biol. Med. **89**, 135–143 (2017)
14. NIH: Communications engineering branch (2014). https://ceb.nlm.nih.gov/repositories/tuberculosis-chest-x-ray-image-data-sets/
15. Noor, N.M., Rijal, O., Fah, C.Y.: Wavelet as features for tuberculosis (MTB) using standard X-ray film images. In: 6th International Conference on Signal Processing, vol. 2, pp. 1138–1141. IEEE (2002)
16. Qin, C., Yao, D., Shi, Y., Song, Z.: Computer-aided detection in chest radiography based on artificial intelligence: a survey. Biomed. Eng. Online **17**(1), 113 (2018)

17. Shen, R., Cheng, I., Basu, A.: A hybrid knowledge-guided detection technique for screening of infectious pulmonary tuberculosis from chest radiographs. IEEE Trans. Biomed. Eng. **57**(11), 2646–2656 (2010)
18. Simonyan, K., Zisserman, A.: Very deep convolutional networks for large-scale image recognition. arXiv preprint arXiv:1409.1556 (2014)
19. Stirenko, S., Kochura, Y., Alienin, O., Rokovyi, O., Gordienko, Y., Gang, P., Zeng, W.: Chest x-ray analysis of tuberculosis by deep learning with segmentation and augmentation. In: 2018 IEEE 38th International Conference on Electronics and Nanotechnology (ELNANO), pp. 422–428. IEEE (2018)
20. WHO: Chest radiography in tuberculosis detection (2019). https://apps.who.int/iris/bitstream/handle/10665/252424/9789241511506-eng.pdf?sequence=1
21. WHO: Tuberculosis (2019). https://www.who.int/news-room/fact-sheets/detail/tuberculosis
22. Xiong, Y., Ba, X., Hou, A., Zhang, K., Chen, L., Li, T.: Automatic detection of mycobacterium tuberculosis using artificial intelligence. J. Thorac. Dis. **10**(3), 1936 (2018)

A Comparison of Neural Network Approaches for Network Intrusion Detection

Mehmet Uğur Öney[1,3](✉) and Serhat Peker[2,3]

[1] May Cyber Technology Inc., Ankara, Turkey
ugur.oney@maysiber.com
[2] Bakırçay University, İzmir, Turkey
serhat.peker@bakircay.edu.tr
[3] Atılım University, Ankara, Turkey

Abstract. Nowadays, network intrusion detection is an important area of research in computer network security, and the use of artificial neural networks (ANNs) have become increasingly popular in this field. Despite this, the research concerning comparison of artificial neural network architectures in the network intrusion detection is a relatively insufficient. To make up for this lack, this study aims to examine the neural network architectures in network intrusion detection to determine which architecture performs best, and to examine the effects of the architectural components, such as optimization functions, activation functions, learning momentum on the performance. For this purpose, 6480 neural networks were generated, their performances were evaluated by conducting a series of experiments on KDD99 dataset, and the results were reported. This study will be a useful reference to researchers and practitioners hoping to use ANNs in network intrusion detection.

Keywords: Network intrusion detection · Data mining · Data classification · Machine learning · ANNs

1 Introduction

Today, with rapid growth of internet and network technologies, companies are exposed to an increased risk of malicious attacks and threats on their networks. A single case of an intrusion can steal, damage or delete the valuable data of an organization. As a result of these, such intrusions and attacks may cause huge loses and decrease the efficiency of the organization as well. Monitoring and protecting the computer network against possible suspicious attacks are thus essential to keep the data confidential and secure in the organizations.

Among the ways of protecting computer systems and networks from attacks, network intrusion detection system (NIDS) is an effective tool and they have been widely deployed as a core part of network architectures. NIDS scans network traffic for abnormalities, and when it identifies an attack or an anomalous behavior,

D. J. Hemanth and U. Kose (Eds.): ICAIAME 2019, LNDECT 43, pp. 597–608, 2020.
https://doi.org/10.1007/978-3-030-36178-5_49

raises alarm and responds to such undesirable and malicious activity. Different machine learning techniques such as fuzzy sets [1–4], artificial immune system [5–7], support vector machines (SVMs) [8–11], or k-nearest neighbour algorithm (k-NN) [12–15] have been employed to perform network intrusion detection.

Recent surveys [16,17] have shown that with the rising trend of deep learning and their ability in high dimensional datasets, artificial neural networks (ANNs) are remarkably popular and effectively used in the network intrusion detection by outperforming traditional machine learning techniques such as decision trees, naïve bayes and k-nearest neighbour algorithms. However, the performance of ANNs depends on many critical components including the network topology, optimization algorithms, activation functions, and the other parameters. Finding the best-suited ANN topology and adjusting the parameters of these models is a challenging problem. It is also an important issue to analyse the accuracy and speed of ANN models before they are implemented in highly sensitive applications like network intrusion detection.

The current paper aims to evaluate the performance of well-known ANNs and the effects of ANN components in network intrusion detection. For this purpose, several experiments are conducted on a popular benchmark dataset and the performances are measured using various evaluation metrics used in the field of network intrusion detection.

The remainder of the paper is organized as follows. The related work is presented in Sect. 2. Section 3 presents an overview of the methodology. The set-up of our experiments is reported in Sect. 4 and the results are discussed in Sect. 5. The paper is concluded in Sect. 6, which provides a summary and directions for future work.

2 Related Works

One of the earliest research on intrusion detection was done by Denning [18] in which a model was introduced for a real-time intrusion detection expert system to detect a wide range of security violations. Afterwards, several research efforts on anomaly detection have been carried out by employing a variety of techniques. The use of neural networks (ANNs) has become increasingly popular in many NIDS model and there are a significant number of studies in the related literature.

One of the pioneer study in NIDS with ANN was conducted by Debar et al. [19]. Cahyo et al. [20] compared the detection accuracy of ANN and SVM on the anomaly-based IDS and used all the features in the dataset KDD99. In this study, ANN outperformed the SVM. In another study, Akhilesh Kumar Shrivas et al. [21] proposed a ensemble classifier of artificial neural networks (ANNs) and a Bayesian net. In recent studies, Tang et al. [22] built a Deep Neural Network (DNN) model for an IDS, while Dong et al. [23] utilized a DBN (deep belief network) model to classify network traffic. Further, Aminanto et al. [25] employed a Stacked Auto Encoder (SAE) as a classifier to classify TCP/IP network traffic. Niyaz et al. [25] also used a deep learning based approach for an efficient and flexible NIDS.

Chiba et al. [26] proposed an important study about the comparison of neural network architectures. In this study, they generated 48 neural networks with different the parameters and compared them in a KDD99 environment. However, their research was limited with the feedforward neural networks. Another comprehensive comparison study has been done by Kwon et al. [17] in 2017. In this study, they compared the structure of the AEN with RNNs. This research pointed out that the AEN detection accuracy was higher than RNN network in a static test dataset. On the other hand, construction parameters of the created networks were not detailed in this study. Because of this reason, replicating the results is not possible.

As seen in the previous studies, while ANNs have been widely employed in the field of network intrusion detection, such attention is distinctly lacking when it comes to performance comparison of ANNs in that area. Therefore, this research attempts to fill this gap through proposing a comparative study.

3 Methodology

In this research, we have adopted a methodology consisting of five main phases, which are determination of the neural network architectures, estimating parameters for the neural network architectures, generation of the all possible combinations of the architectures with these parameters, building each architecture, and collecting and evaluating the results respectively. We systematically reviewed the literature of the NIDS domain, and we determined the following frequently used neural network architectures which is shown in Table 1.

Table 1. List of Neural Network architectures in the NIDS domain.

#	Neural network architecture	Abbreviation
1	Convolutional Neural Network	CNN
2	Recurrent Neural Network	RNN
3	Feed Forward Neural Network	FFNN
4	AutoEncoder Networks	AEN
5	Self-Organizing Map	SOM
6	Evolutionary Neural Network	EEN

Similarly, we determined the significant parameters of the neural networks according to the architectures. These components are as follow:

- **Optimizers** are specific algorithms which determine the distance between neural networks output and the labeled value. As an optimizer algorithm, we decide to choose the most suitable algorithms are Stochastic Gradient Descent and Adam optimizer in the domain. Adam and SGD optimizers are the most popular optimization algorithms and are a common way to optimize neural networks [27].

- **Activation functions** calculate the output value of a neuron. On activation function side, we choose the most popular [28] nine function for the NIDS domain. These functions are; elu, relu, selu, linear, sigmoid, softsign, softmax, softplus, tanh.
- **Learning rate** of the optimizer algorithms determines the size of the steps that needed to reach the local minimum. In this research, learning rates of the networks are determined as commonly accepted rage of 0.01 to 0.001 [29,30].
- **Momentum** defines oscillation of the learning rate in each epoch. We decay the network with the rate of 0.0 and 1×10^{-6} [29].
- **Dropout rate** is a regulation technique that prevents overfitting problem in the neural network. Neurons are shoot down with the rate of 0.0, 0.25 and 0.50 [31].
- **Depth of the network** is defined the number of hidden layer of the network. We determined the depth of the network as 1, 2, 5, 10, 25 hidden layer depth. We limited the neural network depth with 25 layer because of the limitation of the hardware.

After the determination of the topologies and the parameters, we created the all possible combinations of the neural networks according to network architectures and parameters. The count of all these combinations is $6 * 2 * 9 * 2 * 2 * 3 * 5 = 6480$.

4 Experiments

This section is divided into three subsections. Section 4.1 describes the dataset used and data preprocessing procedures applied to the dataset. Section 4.2 presents an overview of our experimental set-up. Finally, the metrics used to compare the performances of different ANN topologies in NIDSs are introduced in Sect. 4.3.

4.1 Dataset Description and Data Preprocessing

In this study, to examine the performances of different ANN topologies, we used KDD99 benchmark dataset, which is a popular and has been used as a benchmark in many studies evaluating the network-based intrusion detection systems. The KDD99 is a popular dataset prepared by Stolfo et al. [32] and is built based on the data captured from DARPA'98 IDS evaluation program [33]. It is 4 gigabytes of compressed raw tcpdump data pertaining to 7 weeks of network traffic, which can be processed into about 5 million connection records, each with about 100 bytes.

The dataset consists of approximately 4,900,000 single connection instances and there are 42 features (including the attack type) in each instance. The 41 features were used to classify each instance as normal or attack. Before building models, a set of standard pre-processing steps were applied to the dataset. For example, non-numeric features were transformed into numerical format. Further, input data were normalized to the interval [0, 1], before running ANN models.

4.2 Experimental Setup

The pictorial depiction of our experiments is shown in Fig. 1.

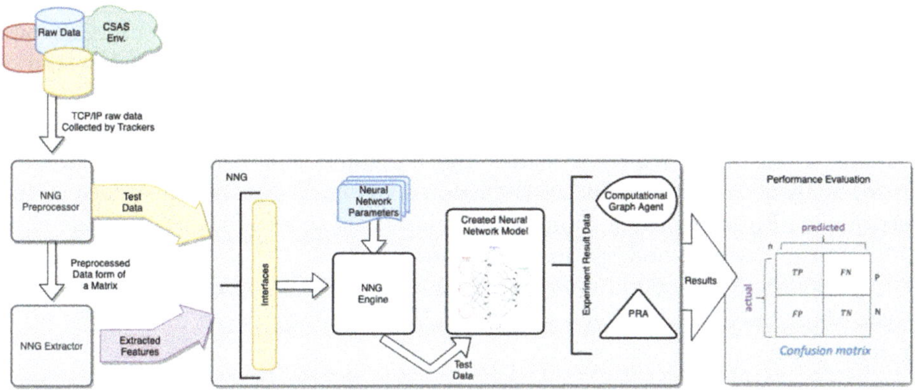

Fig. 1. Overview of the experiments.

First, Neural Network Generator (NNG) Preprocessor reads raw data with tracker interfaces. These interfaces specified for each source type which are big data tracker, Rest API tracker, and file tracker. KDD99 data stored in the big data environment for this study because of data size and accessibility. Each tracker reads data and transfers them to a matrix. NNG Preprocessor application was developed on C++ v17 programming language with R-Interpreter and Python-Interpreter.

In second phase, preprocessor starts to decode data. The computational graph which will be created by NNG needs numerical data and collected data contains protocol-specific values in it. As a result of that, decoding data is a crucial step for the rest of the experiment.

NNG Extractor application extracts the basic features. The matrix, which is passed from NNG Preprocessor, is created from raw TCP/IP dump data. Extractor gathers each packages' duration information, protocol type, port information, connection flag, and sends/receives byte counts. Extractor also labels each feature for the type. If data feature is extracted from a time frame, extractor application marks this feature as a "Cont" otherwise data mark as "Disc". NNG Extractor application was developed on Python 3.6 programming language with Pandas v0.23.4 as a data analysis library, NumPy v1.15.4 as a scientific computing library, and SciKit v0.20.1 for regression calculation library.

After the extraction processes, data arrives to the NNG Engine. The NNG Engine creates neural network computational graphs according to given architecture and then evaluates the networks with the test data. Each generation of the network changes parameters stored in a matrix of activation functions, depth of the network, optimization algorithm, learning rate, learning momentum, and dropout rate. NNG Engine assures that all generated neural networks used the same preprocessed dataset as an input and the entire dataset was firstly split into two equal partitions. Then, the first half of the dataset was split into two parts (first 80% for training and the remaining 20% for validation). The remaining half part of the entire dataset (50%) is used for testing. NNG Engine was developed on Python v3.6 programming language with TensorFlow v1.12 as low-level computational graph API, and Keras v2.2.4 as high-level neural networks API.

Finally, NNG collects the performance and statistical data for the tested network. Each statistical data is stored in a comma separated value file for the future analysis with the name of the created network. Performance data of the network is also stored as a binary. This data contains information about each epoch of the network and the metric values.

All the experiments were performed on the Intel x64 based i7 3.5 Ghz CPU, 32 GB of memory, two NVIDIA 1080x graphic card connected with SLI and 5 TB of disk space.

4.3 Performance Measurements

In order to evaluate performance of different ANN topologies in NIDS, we used the following widely known metrics: accuracy, recall, precision, F-measure, specificity, false positive rate (FPR) as defined in Eqs. 1–6. All these metrics are formulated based on the entries of the confusion matrix, shown in Table 2.

Four possible outcomes of the confusion matrix for a record in this study are given as:

- True Positive (TP) is the number of attack records classified correctly;
- True Negative (TN) is the number of normal records classified correctly;
- False Positive (FP) is number of normal records classified incorrectly and
- False Negative (FN) is the number of attack records classified incorrectly.

Table 2. Confusion matrix.

		Predicted class	
		Normal	Attack
Actual class	Normal	True negative (TN)	False positive (FP)
	Attack	False negative (FN)	True positive (TP)

True positive rate (TPR), also known as recall or sensitivity, is the proportion of positive cases that are correctly identified and is calculated as follows:

$$TPR = \frac{TP}{TP + FN} \tag{1}$$

Precision is the proportion of the classified positive cases that are correct and is calculated as follows:

$$Precision = \frac{TP}{TP + FP} \tag{2}$$

The **F-measure**, also known as **F-score** is the harmonic mean of precision and recall and is calculated as follows:

$$F - score = 2 \times \frac{precision \times recall}{precision + recall} \tag{3}$$

True negative rate (TNR), also known as specificity, is the proportion of negative cases that are correctly identified and is calculated as follows:

$$TNR = \frac{TN}{TN + FP} \tag{4}$$

False positive rate (FPR), also known as false alarm rate (FAR), is the proportion of positive cases that are incorrectly identified and is calculated as follows:

$$FPR = \frac{FP}{TN + FP} = 1 - Specificity \tag{5}$$

Accuracy is the proportion of the total number of predictions that were correct and is calculated as follows:

$$Accuracy = \frac{TN + TP}{TN + TP + FN + FP} \tag{6}$$

The closer the value of these measures (except for FPR) to one is, the better the network topology is. The values of FPR should be close to zero for better topologies. We also used the time needed to build a model. It is an important factor for the evaluation of NIDS and the lower its value is, the better the network topology is.

5 Results

The detection time statistics of the architectures is shown in Fig. 2. According to this figure, ENN architecture has the worst detection time performance with 7.94 s on average, whereas the CNN architecture has the best detection time performance with 0.23 s.

Figure 3 demonstrates the accuracy performance results of the ANN architectures. In terms of accuracy, EEN architecture performs better than the other architectures with a value of 98.23%. AEN follows this architecture with an accuracy value of 94.71%. However, when the detection times are considered, we can conclude that AEN architecture is better than EEN architecture.

Figure 4 shows the effect of optimizers on the performances of ANNs. According to this figure, when Adam optimizer is used, all performance measures show

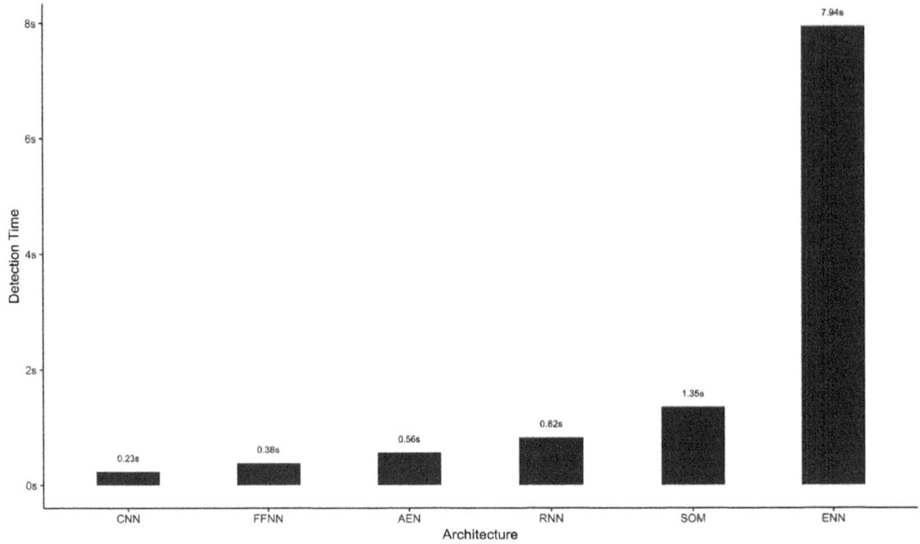

Fig. 2. Architecture average detection time.

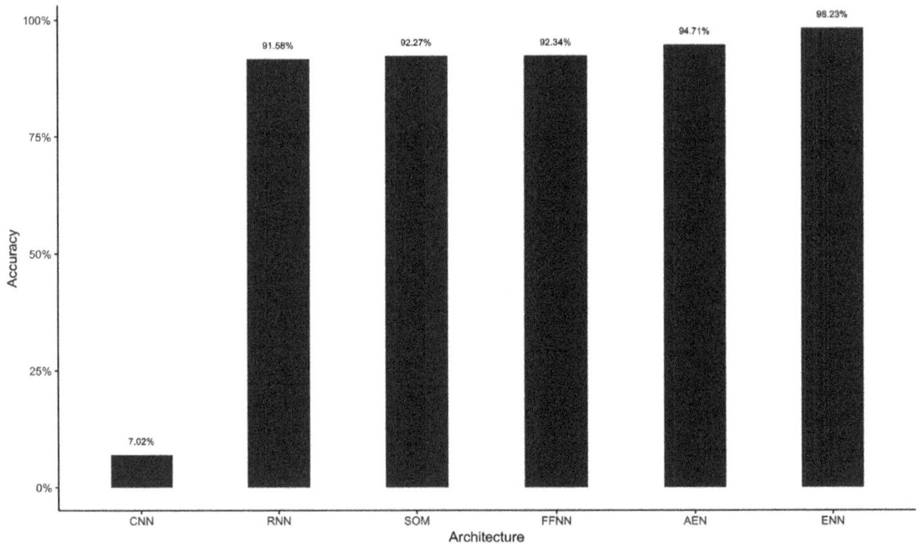

Fig. 3. Architecture average Accuracy.

better results. Thus, it can be concluded that neural networks with Adam optimizer performs better than the ones with SGD optimizer.

Figure 5 shows the effect of activation functions on the performances of ANNs. Softplus activation function has the lowest false positive rate on average with a value of 10.8%. However, in term of the other measures, Softmax,

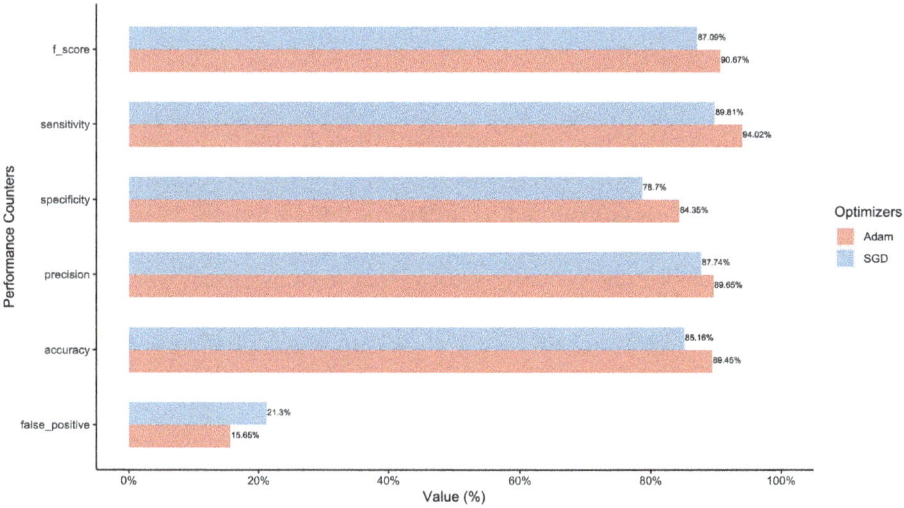

Fig. 4. Compression of optimizers.

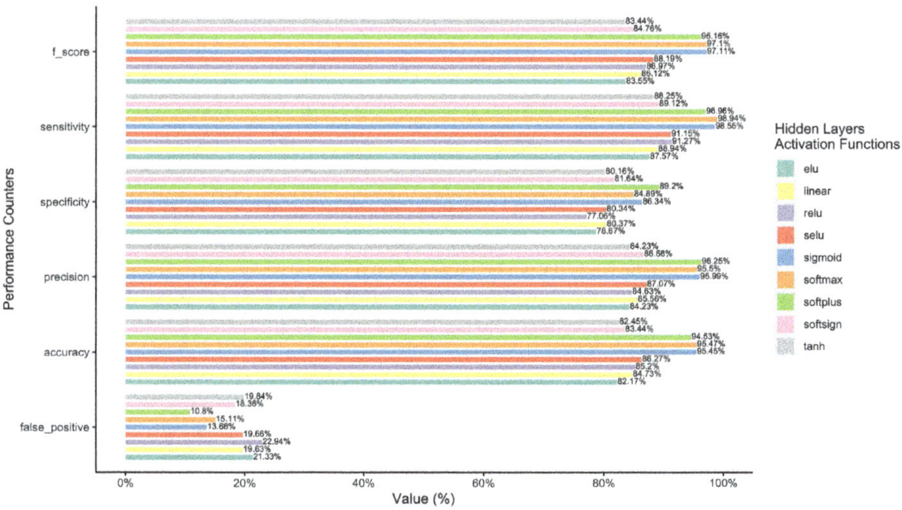

Fig. 5. Hidden layer activation function performance counters.

Softplus and Sigmoid activation functions have similar results by significantly outperforming the others. These three activation functions have the same characteristics and all of them belongs to the logistic function family. Therefore, neural networks, which employed logistic function as an activation function, tend to give better results than the ones with other activation functions.

6 Conclusion

ANNs have been widely and successfully used in network intrusion detection. In this manner, we have presented an empirical comparison of well-known ANN architectures for network intrusion detection. The components of ANN architectures have been also investigated by conducting many experiments on the KDD99 dataset which is considered as a benchmark within NIDS research area. The results have demonstrated that AEN neural network topology is the most suitable architecture for the NIDSs by providing high accuracy with reasonable time margins. As an activation function, logistic based functions like sigmoid and softplus are the best suitable for the neuron step activators. Further, the usage of the Adam optimizer brings significant performance benefit for the ANN topologies in NIDS field.

The findings can be utilized to design efficient network intrusion detection systems which can be used for security analysis and management in organizations. Moreover, this research may provide researchers and practitioners with a guideline for effectively employing ANNs in NIDS field. A possible direction in the future work might be to increase the variance of the testing parameters like optimizer function and depth of the networks by using a more capable, modern GPU supported experimental environment.

References

1. Kuok, C.M., Fu, A., Wong, M.H.: Mining fuzzy association rules in databases. ACM SIGMOD Rec. **27**(1), 41–46 (1998). https://doi.org/10.1145/273244.273257. 00653
2. Dickerson, J.E., Dickerson, J.A.: Fuzzy network profiling for intrusion detection. In: PeachFuzz 2000. 19th International Conference of the North American Fuzzy Information Processing Society - NAFIPS (Cat. No.00TH8500). 00330, Atlanta, GA, USA, pp. 301–306. IEEE (2000). https://doi.org/10.1109/NAFIPS.2000.877441. ISBN: 978-0-7803-6274-1
3. Tajbakhsh, A., Rahmati, M., Mirzaei, A.: Intrusion detection using fuzzy association rules. Appl. Soft Comput. **9**(2), 462–469 (2009). https://doi.org/10.1016/j.asoc.2008.06.001. 00238
4. Saniee Abadeh, M., Habibi, J., Lucas, C.: Intrusion detection using a fuzzy genetics-based learning algorithm. J. Netw. Comput. Appl. **30**(1), 414–428 (2007). https://doi.org/10.1016/j.jnca.2005.05.002. 00200
5. Kim, J., Bentley, P.J.: Towards an artificial immune system for network intrusion detection: an investigation of clonal selection with a negative selection operator. In: Proceedings of the 2001 Congress on Evolutionary Computation (IEEE Cat. No.01TH8546), vol. 2. 00361, Seoul, South Korea, pp. 1244–1252. IEEE (2001). https://doi.org/10.1109/CEC.2001.934333. ISBN: 978-0-7803-6657-2
6. Toosi, A.N., Kahani, M.: A new approach to intrusion detection based on an evolutionary soft computing model using neuro-fuzzy classifiers. Comput. Commun. **30**(10), 2201–2212 (2007). https://doi.org/10.1016/j.comcom.2007.05.002. 00251
7. Dasgupta, D., Gonzalez, F.: An immunity-based technique to characterize intrusions in computer networks. IEEE Trans. Evol. Comput. **6**(3), 281–291 (2002). https://doi.org/10.1109/TEVC.2002.1011541. 00505

8. Mukkamala, S., Janoski, G., Sung, A.: Intrusion detection using neural networks and support vector machines. In: Proceedings of the 2002 International Joint Conference on Neural Networks, IJCNN 2002 (Cat. No.02CH37290). 00802, Honolulu, HI, USA, pp. 1702–1707. IEEE (2002). https://doi.org/10.1109/IJCNN.2002.1007774. ISBN: 978-0-7803-7278-8
9. Sung, A.H., Mukkamala, S.: Identifying important features for intrusion detection using support vector machines and neural networks. In: Proceedings of the 2003 Symposium on Applications and the Internet (SAINT 2003). 00483, Orlando, FL, USA, pp. 209–216. IEEE Computer Society (2003). https://doi.org/10.1109/SAINT.2003.1183050. ISBN: 978-0-7695-1872-5
10. Chen, W.-H., Hsu, S.-H., Shen, H.-P.: Application of SVM and ANN for intrusion detection. Comput. Oper. Res. **32**(10), 2617–2634 (2005). https://doi.org/10.1016/j.cor.2004.03.019. 00308
11. Deng, H., Zeng, Q.-A., Agrawal, D.P.: SVM-based intrusion detection system for wireless ad hoc networks. In: 2003 IEEE 58th Vehicular Technology Conference, VTC 2003-Fall (IEEE Cat. No.03CH37484), vol. 3. 00111, Orlando, FL, USA, pp. 2147–2151. IEEE (2003). https://doi.org/10.1109/VETECF.2003.1285404. ISBN: 978-0-7803-7954-1
12. Liao, Y., Vemuri, V.R.: Use of K-nearest neighbor classifier for intrusion detection. Comput. Secur. **21**(5), 439–448 (2002). https://doi.org/10.1016/S0167-4048(02)00514-X. 00000
13. Li, Y., Guo, L.: An active learning based TCM-KNN algorithm for supervised network intrusion detection. Comput. Secur. **26**(7–8), 459–467 (2007). https://doi.org/10.1016/j.cose.2007.10.002. 00123
14. Stein, G., et al.: Decision tree classifier for network intrusion detection with GA-based feature selection. In: Proceedings of the 43rd Annual Southeast Regional Conference on - ACM-SE 43, vol. 2. 00232, Kennesaw, Georgia, p. 136. ACM Press (2005). https://doi.org/10.1145/1167253.1167288. ISBN: 978-1-59593-059-0
15. Aburomman, A.A., Reaz, M.B.I.: A novel SVMkNN- PSO ensemble method for intrusion detection system. Appl. Soft Comput. **38**, 360–372 (2016). https://doi.org/10.1016/j.asoc.2015.10.011. 00121
16. Öney, M.U., Peker, S.: The use of artificial neural networks in network intrusion detection: a systematic review. In: 2018 International Conference on Artificial Intelligence and Data Processing (IDAP). 00000, pp. 1–6. IEEE (2018)
17. Kwon, D., et al.: A survey of deep learning-based network anomaly detection. Clust. Comput., 1–13 (2017)
18. Denning, D.E.: An intrusion-detection model. IEEE Trans. Softw. Eng. **2**, 222–232 (1987). https://doi.org/10.1109/TSE.1987.232894. 04895
19. Debar, H., Becker, M., Siboni, D.: A neural network component for an intrusion detection system. In: IEEE Symposium on Security and Privacy. 00727, pp. 240–250 (1992)
20. Cahyo, A.N., Hidayat, R., Adhipta, D.: Performance comparison of intrusion detection system based anomaly detection using artificial neural network and support vector machine. In: Proceedings of the 12th International Conference on Synchrotron Radiation Instrumentation – SRI 2015. 00002, New York, NY, USA, p. 070011 (2016). https://doi.org/10.1063/1.4958506
21. Shrivas, A.K., Dewangan, A.K.: An ensemble model for classification of attacks with feature selection based on KDD99 and NSL-KDD data set. Int. J. Comput. Appl. **99**(15), 8–13 (2014). 00026

22. Tang, T.A., et al.: Deep learning approach for network intrusion detection in software defined networking. In: 2016 International Conference on Wireless Networks and Mobile Communications (WINCOM). 00082, Fez, Morocco, pp. 258–263. IEEE, October 2016. https://doi.org/10.1109/WINCOM.2016.7777224. ISBN: 978-1-5090-3837-4
23. Dong, B., Wang, X.: Comparison deep learning method to traditional methods using for network intrusion detection. In: Proceedings of the IEEE ICCSN. 00031, pp. 581–585 (2016)
24. Aminanto, M.E., Kim, K.: Deep learning-based feature selection for intrusion detection system in transport layer (2016). 00002
25. Javaid, A., et al.: A deep learning approach for network intrusion detection system. In: Proceedings of the 9th EAI International Conference on Bio-Inspired Information and Communications Technologies (Formerly BIONETICS). 00118, New York City, United States. ACM (2016). https://doi.org/10.4108/eai.3-12-2015.2262516. ISBN: 978-1-63190-100-3
26. Chiba, Z., et al.: A novel architecture combined with optimal parameters for back propagation neural networks applied to anomaly network intrusion detection. Comput. Secur. **75**(00001), 36–58 (2018). https://doi.org/10.1016/j.cose.2018.01.023. 00001
27. Kim, J., Le, T.-T.-H., Kim, H.: An effective intrusion detection classifier using long short-term memory with gradient descent optimization. In: International Conference on Platform Technology and Service (2017)
28. Sibi, P., Allwyn Jones, S., Siddarth, P.: Analysis of different activation functions using back propagation neural networks. J. Theor. Appl. Inf. Technol. **47**, 5 (2005). 00105
29. Chandra, P., Singh, Y.: An activation function adapting training algorithm for sigmoidal feedforward networks. Neurocomputing **61**, 429–437 (2004). https://doi.org/10.1016/j.neucom.2004.04.001. 00070
30. Singh, Y., Chandra, P.: A class +1 sigmoidal activation functions for FFANNs. J. Econ. Dyn. Control. **28**(1), 183–187 (2003). https://doi.org/10.1016/S0165-1889(02)00157-4. 00000
31. Srivastava, N., et al.: Dropout: a simple way to prevent neural networks from overfitting. J. Mach. Learn. Res. **15**, 1929–1958 (2014)
32. Stolfo, S.J., et al.: Cost-based modeling for fraud and intrusion detection: results from the JAM project. In: Proceedings DARPA Information Survivability Conference and Exposition, DISCEX 2000, vol. 2. 00532, Hilton Head, SC, USA, pp. 130–144. IEEE Computer Society (1999). https://doi.org/10.1109/DISCEX.2000.821515. ISBN: 978- 0-7695-0490-2
33. Tavallaee, M., et al.: A detailed analysis of the KDD CUP 99 data set. In: IEEE Symposium On Computational Intelligence for Security and Defense Applications, CISDA 2009, pp. 1–6. IEEE (2009). ISBN: 1- 4244-3763-6

A Case Study: Comparison of Software Cost Estimation of Smart Shopping List Application

Tuncay Yiğit[1] and Huseyin Coskun[2(✉)]

[1] Department of Computer Engineering, Süleyman Demirel University, Isparta, Turkey
tuncayyigit@sdu.edu.tr

[2] Department of Computer Technologies, Uşak University, Uşak, Turkey
huseyin.coskun@usak.edu.tr

Abstract. The aim of this study is to evaluate development cost of smart shopping software which is going to be planned to developed using cost estimation methods of software development in software engineering. The smart shopping software aims that faster and easier preparing of shopping lists which is prepared by consumer with the aims of buying product from supermarket with lower cost. The most of consumers demand that a products price which they are going to buy or total amount of shopping price are lowest while shopping. Therefore, product prices of supermarkets need to be known. They try to find out which supermarket lowest price of products is in by researching of product price before shopping and they go shopping with the aim of lowest total cost of shopping. The smart shopping software planned to develop in this study aims to make consumers to shop with lower total cost within the help of artificial intelligence and applied mathematics. It is going to provided that to be found out supermarkets where optimal product prices are. The cost of best shopping route for shopping from these supermarkets is going to be determined. In this study, it has been made an estimate of development cost of smart shopping software. Therefore, the attributes of the software were firstly defined. These attributes are classified to determine complexity of them in following categories; interface design, database, controller classes, web API, artificial intelligence. So, the difficulties of performing these tasks have been obtained. After that, the tasks required to develop these attributes were determined within the scope of the functional points. In conclusion, the required workforce and time costs and project size were estimated by the methods using different software development cost estimation approaches and the results of the methods were evaluated.

Keywords: Software development cost estimation · Software development effort estimation · Software engineering · Function point · Parametric estimation model

1 Introduction

The people generally prepare a list before supermarket shopping. This list is created shortly by writing product quantity next the product name. It is necessary that preparing a shopping list and not purchasing another product except this list for economically

D. J. Hemanth and U. Kose (Eds.): ICAIAME 2019, LNDECT 43, pp. 609–620, 2020.
https://doi.org/10.1007/978-3-030-36178-5_50

purchasing [1, 2]. Also, it's aimed that most affordable product is purchased with the lowest price as for that shopping list in supermarket shopping. Therefore, the customers make a search of product prices [3, 4]. The customers look at supermarkets website or mobile application while making a search of products prices. Hence it considerably takes time that searching prices of products which is needed to be purchased.

The pricing out for one product takes 12 min on average based on local supermarket in experimental researches. This search has been done based upon web site of supermarket and product information bulletin. If there is no price information of the product which is going to be purchased on website or product information bulletin, the most economically price of the product is looked for by visiting for other supermarkets. Taking into consideration that, the numbers of the product which is going to be purchased is too many, the cost of timing and transportation to other supermarkets are going to have increased.

The smart shopping software aims that faster and easier preparing of shopping lists which is prepared by consumer with the aims of buying product from supermarket with lower cost. The most of consumers demand that a products price which they are going to buy or total amount of shopping price are lowest while shopping. Therefore, product prices of supermarkets need to be known. They try to find out which supermarket lowest price of products is in by researching of product price before shopping and they go shopping with the aim of lowest total cost of shopping with the prepared shopping list. The smart shopping software planned to develop in this study aims to make consumers to shop with lower total cost within the help of artificial intelligence and applied mathematics. It is going to provided that to be found out supermarkets where optimal product prices are. The cost of best shopping route for shopping from these supermarkets is going to be determined. In smart shopping application, the price information which is obtained from shopping receipts are going to be entered by consumer after supermarket shopping. Thus, it's planned with the smart shopping application that displaying most current product price which is going to be purchased by consumer with aim of price information which is obtained from shopping receipts. Accordingly, the smart shopping application is termed as *Social Price (SOP)* due to the product price information are collected by the consumers. Before SOP application is developed, the literature on this scope has been researched. when these studies are analyzed, there is no application toward smart shopping list application developed in this article [5–19].

Software project cost estimation is to be predicted essential workforce and time cost without be carried out the project after calculation of project size. It's difficult but possible to know that how many workforces is going to be used and how long project is going to be taken time. For this reason, various methods and techniques are utilized based mathematical calculation, analysis previous project experience, classification and disintegrate to meaningful part of the problem solving. Therefore, software project cost estimation is carried out with the aim of developing the SOP in this study.

1.1 Social Price (SOP)

The product prices need to be obtained primarily with intent to be displayed to consumer which supermarket has most economically prices. The SOP provides that supermarket shopping receipts are uploaded to application with mobile device camera using image processing techniques. Then it reads the receipt to obtain following information with optical character recognition (OCR).

- Product name,
- Price,
- Quantity,
- Supermarket name and address,
- Date-time information.

This information is inserted to database. Also, different quantities and prices are inserted to database for similar products. These steps are followed to create smart shopping list using SOP;

1. The consumer who makes supermarket shopping creates the list consist of products to be purchased.
2. The SOP search the information about the product on shopping list and represent to consumer which product is most economically price on which supermarket.
3. The price list is displayed as to nearest supermarket to consumer location based or the consumer add the products to shopping list as to product prices and distance of supermarket to consumer location and create final shopping list.
4. The consumer teaches where are the most economically price on which supermarket. Then goes to supermarket and purchases the product by checking the shopping list.

The working diagram of the SOP is shown in Fig. 1.

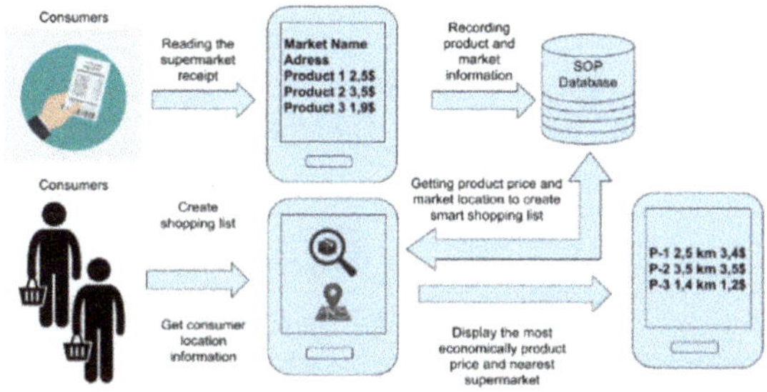

Fig. 1. Working diagram of the SOP.

2 Methods

The experiences of previous project are accepted as the most important tool for project cost estimation. The duration, staff member, spent resources and properties of developed software in previous project are going to be guided for next projects. The various modelling methods have been developed about project cost estimation. The size of project, development progress and development environment are effective on choosing of the modelling methods. These models don't make certain correct result due to using of complex and many variables in cost estimation models and estimation process involving uncertainty. The previous experiences are able to give more correct result in some situations. For this reason, if possible, three different cost estimation methods need to be utilized at least. Accordingly, in this study, the cost estimation of developing SOP mobile application has been implemented by using many cost estimation methods. While the cost estimation of the project is going to be carried out;

- Basic Code Line Number (KLOC) Based,
- Function Point (FP) Based

parametric estimation models are going to be used [20]. The following methods have been utilized for project cost estimation.

2.1 Code Line (KLOC) Based Parametric Estimation Models

The code line based parametric estimation models have been obtained from many completed software project and application in various establishment [21]. The cost estimation models have been derived from regression analysis on data obtained from past software project experience. As the size of software, cost estimation model equations base on one thousand code line value (KLOC). The models based on code line predict only the workforce and effort unit is person-month. Effort represent the main cost factor for projects. The general structure of cost estimation models takes the form Eq. 1:

$$E = A + B x S^{C} \tag{1}$$

In Eq. 1; A, B, and C are experimentally acquired constants, E is effort in person months, and S is the basic input and is able to be either KLOC or FP [24].

The models based KLOC are given in Table 1 [22, 23].

Table 1. The based KLOC models.

No	Model name	Effort (E) equation	Time (T) equation
1	Walston – Felix Model	$E = 5.2 \times (KLOC)^{0.91}$	$2.5 \times E^{0.35}$
2	Bailey – Basili Model	$E = 5.5 + 0.73 \times (KLOC)^{1.16}$	$2.5 \times E^{0.40}$
3	Halstead Model	$E = 5.2 \times (KLOC)^{1.5}$	$2.5 \times E^{0.25}$
4	Doty Model	$E = 5.288 \times (KLOC)^{1.047}$	$2.5 \times E^{0.20}$
5	Barry Boehm Model	$E = 3.2 \times (KLOC)^{1.05}$	$2.5 \times E^{0.38}$
6	COCOMO Organic Model	$E = 2.4 \times (KLOC)^{1.05}$	$2.5 \times E^{0.28}$
7	COCOMO Semidetac Model	$E = 3.0 \times (KLOC)^{1.12}$	$2.5 \times E^{0.35}$
8	COCOMO Embedded Model	$E = 3.6 \times (KLOC)^{1.2}$	$2.5 \times E^{0.32}$

2.2 Function Point (FP) Based Parametric Estimation Models

Function point-based estimation model is more usable than code line-based estimation model. FP prioritize functionality of software to be delivered to software's users. FP is calculated to define complexity and size of software as to functionality. While KLOC based estimation evaluate size of software as to developer's perspective, FP based estimation evaluate as to user's perspective. The many factors effect on functionality toward software are weighted as to their effects on software by being analyzed them due to they not be able to measured directly. Ultimately, a number is obtained and this number is be able to be utilized to evaluate various software project relatively. FP based cost estimation models are given Table 2 [22].

Table 2. The FP based models.

No	Model name	Effort (E) equation	Time (T) equation
1	Albrecht–Gaffney Model	$E = 13.39 + 0.0545 \times (FP)$	$2 \times E^{0.12}$
2	Kemerer Model	$E = 60.62 + 7.728 \times (10^{-8}) \times FP^3$	$2 \times E^{0.18}$
3	Matson–Barret–Meltichamp Model	$E = 585.7 + 15.12 \times (FP)$	$2 \times E^{0.2}$

When the FP based model effort equations are analyzed, it's clearly shown that the enormous differentiations of each model arise in effort estimation even if same FP value is used as size of software. The basic reason of arising these differentiations is to be utilized restricted data in proposed models and is that the project metrics of data came from project which has different characteristic.

2.3 Calculation of Function Point and Total Code Line

The weights of factors affect complexity of project are found by trying their effects on the complexity for calculation of FP. Some methods are applied on arranging of these weights. Two operation has been carried out to calculate FP. First, Unadjusted FP (UFPs) based on objective measurements is calculated by analyzing of the project information environment component with the FP model. FP model diagram is given in Fig. 2.

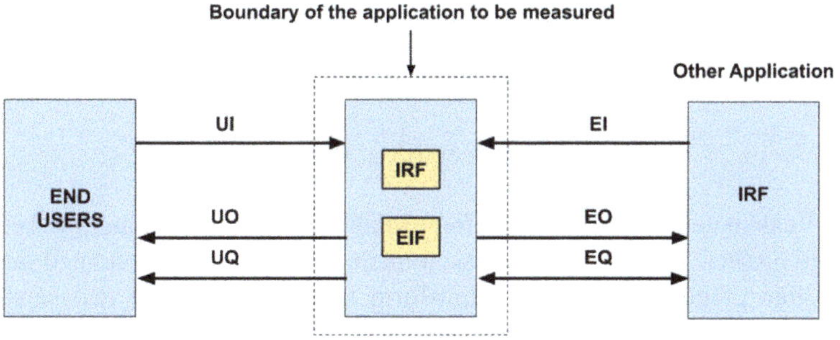

Fig. 2. Function point model.

The functionality of software has been analyzed with five different measurement parameters to calculate UFPs. The functionality parameters have been classified as to three complexity level which are simple, normal or medium and complex. Then the weighting factors are obtained. The weighting factors of smart shopping list mobile application are given in Table 3.

Table 3. The weighting factors of the project information environment component.

Measurement parameter	Weight factor			Total
	Simple	Normal	Complex	
User Input (UI) - External Input (EI)	6	10	9	25
User Out (UO) - External Out (EO)	4	6	5	15
User Query (UQ) - External Query (EQ)	10	9	9	28
Internal Registry Files (IRF)	2	3	1	6
External Interface Files (EIF)	8	5	5	18
Total	30	33	29	92

The measured values are acquired by summing of all [25]. The IRF and EIF identify the data functionalities. The value of EIF and IRF are the number of information points which transferred by all types of computer or system. The UI and EI are defined as data input during using of the software by end user and other application respectively and mustn't be confused as querying. EI, EO and EQ are identified as transaction functionality with other applications. The UFPs value of the smart shopping list application has been calculated as 92. Secondly, the technical complexity factor (TCF) is need to be calculated to calculate FP of software after calculation of UFPs. TCF is summation of weighted answers which are between 0 and 5 points from 15 questions in Table 4. There are six weight values in Table 4. The terms which define level of general system characteristics like *no effect, very little, normal, average, significant* and *inevitable* are represented by zero, one, two, three, four, five points respectively.

TCF has been calculated following as;

$$TCF = \sum\nolimits_{i=0}^{15} G(i) \tag{2}$$

TCF has been calculated as 39 via Eq. 2. After UFPs and TCF have been calculated, FP is going to be calculated by using formula in Eq. 3 [26].

$$FP = UFPs * (0.65 + 0.01 * TCF) \tag{3}$$

The FP calculated as 95,68. Therefore, totally 96 function points has been calculated for the project. The indirectly measurements are able to be calculated take FP for instead of line of codes. There are transform table that join the two separate cost

Table 4. Analysis of technical complexity factors.

General system characteristics (G)	0	1	2	3	4	5
1. Does the application require reliable backup and recovery?		X				
2. Does data communication be needed?				X		
3. Does the project have distributed operation functions?				X		
4. Is the performance critical or vital?		X				
5. Will the system operate in a hardworking operating environment?		X				
6. Does the system require online data entry?					X	
7. Does online data input require multiple screens for an interim process?				X		
8. Are the main registry being updated online?					X	
9. Are inputs, outputs, registries, or queries complex?				X		
10. Are internal transactions complex?						X
11. Will the code be designed to be reusable?				X		
12. Will conversion and installation be considered in design?			X			
13. Is the system being developed for different institutions located in multiple locations?		X				
14. Designed application, easy to be used and can be easily changed by the user?				X		
15. Will different algorithms work intertwined?			X			

estimation approach. It's possible that switching between LOC based and FP based measurements. This operation is be able to be carried out with transform values in Table 5.

Table 5. Transform values from FP to LOC [24]

No	Programming language	LOC/FP
1	C	128
2	COBOL	105
3	FORTAN	105
4	PASCAL	90
5	ADA	70
6	C++	53
7	Object Oriented Languages	30
8	4. Generation Languages	20
9	Code Generators	15
10	SQL	13

Since the project will be developed using object-origin languages, it is predicted that the software will consist of a total of $LOC = 96x30 = 28704$ (2.88 KLOC) line codes.

2.4 Comparison of Models

The differences between parametric effort estimation models results and real effort results have been evaluated to compare models. Root Mean Square Error (RMSE) and Magnitude of Relative Error (MRE) has been utilized as error criterion. RMSE is used to evaluate differences among values which estimate by a model or method [27]. RMSE is represented by the Eq. 4 presented below;

$$RMSE = \sqrt{\frac{1}{N}\sum\nolimits_{i=1}^{N}\left(E_{estimation} - E_{real}\right)^2} \tag{4}$$

RMSE is more sensitive against greater prediction error than mean error due to consider error squares. MRE is an evaluation metric that is usually utilized to evaluate performance of software project cost estimation models. MRE is represented by the Eq. 5 presented below;

$$MRE = \frac{|E_{real} - E_{estimation}|}{E_{real}} \tag{5}$$

MRE is a type of error that shows how close the actual value is. In most problems, MRE means more determinative than RMSE [28].

3 Results and Discussion

The proposed estimation model data and actual data have been compared and the effort and time estimation values has been calculated. Also, number of full-time programmers to develop software and number of code lines per programmer have been found. The RMSE values of each models have been obtained. The actual data have been given on Table 6.

Table 6. Actual data of software development process

Effort (person/month)	Time (month)	Full-time programmer	Number of code line (KLOC)	Number of code lines per programmer (KLOC)
8.66	4.33	2	3.854	1.927

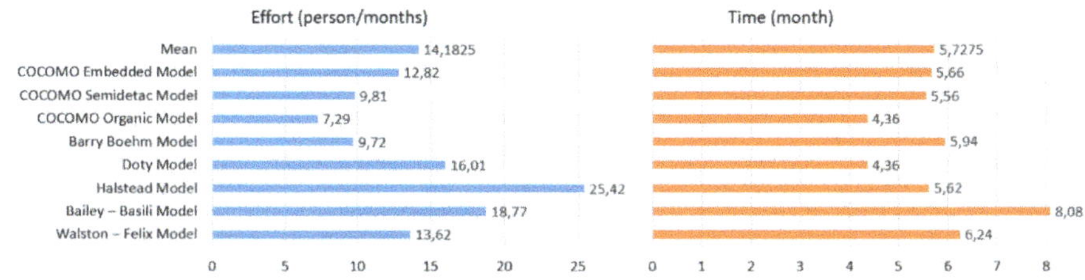

Fig. 3. Effort and time values of code line number-based models

If results of code line number-based estimation models have been analyzed, it's seen that while effort values are different each other, a few of time values are the same and some of them are so close each other. As seen in Fig. 3, COCOMO Organic Model has the best effort value and also best time value with Doty model. As seen in Fig. 4, COCOMO Semidetac, Organic and Barry Boehm models have the best number of full programmers.

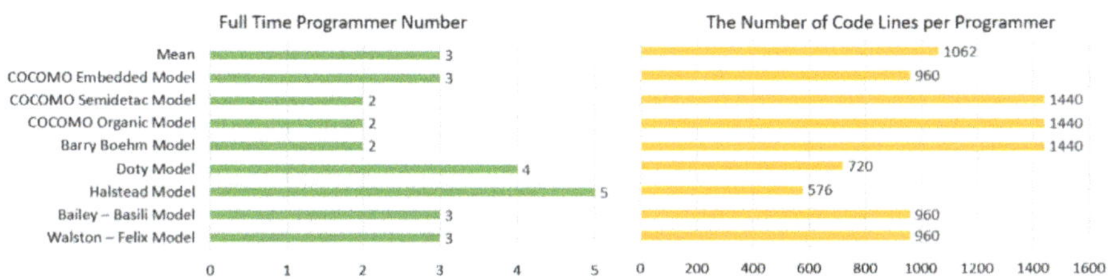

Fig. 4. Number of full-time programmer and code lines per programmer

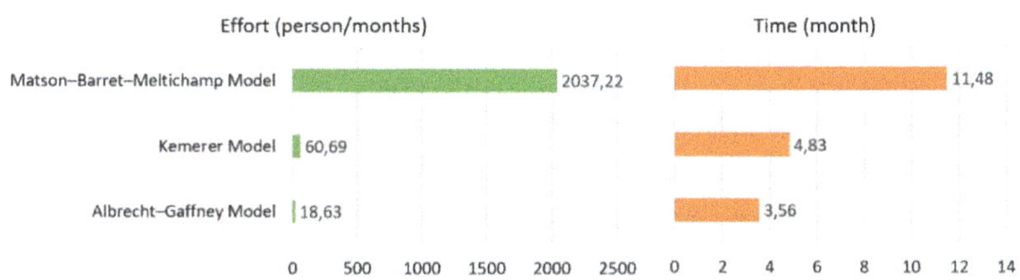

Fig. 5. Effort and time values of FP based models

If results of FP-based estimation models in Fig. 5 are analyzed, it's seen that while effort values are quite different each other, these differences have been decreased for time values. During the evaluation process of RMSE and MRE values, if results of RMSE and MRE values in Fig. 6 are analyzed, it's seen that Barry Boehm Model has best RMSE and MRE value for effort and COCOMO Organic and Doty Model has the best RMSE value and COCOMO Embedded Model has the best MRE value for time data.

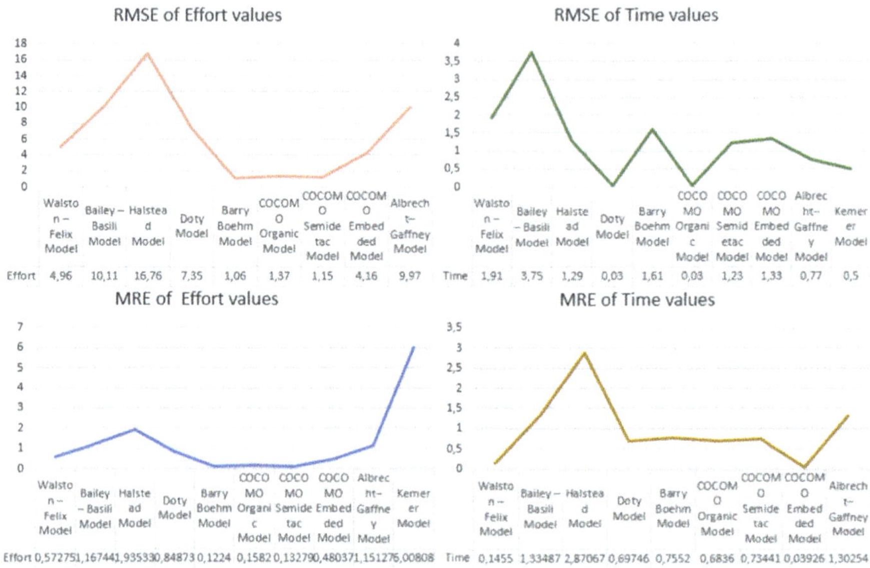

Fig. 6. RMSE and MRE for time and effort values

4 Conclusions and Future Works

The main objective of a successful software project is to produce software that will meet the customer's expectations at a predetermined time and budget. The failure of many software projects stems from the inaccuracy of the initial estimates of the planning phase. Software components that are not meticulously revealed result in time-outs and budget overruns, which result in unsuccessful results.

In this study, effort and time cost estimation of smart shopping mobile application has been carried out with function point and code line number-based estimation models with the aim of researching reasons of similar situation. The weighted factors have been calculated by analysis of project information environment component with the aim of obtain the number of project's function point. Thereafter, technical complexity factor of project has been calculated by weighting the answer of complexity questions to find function point of project. Software cost has been estimated before development of software. Project time and effort value have been determined as to COCOMO Semidetac estimation model. By new technical complexities which arises during development of software, the effort and time costs of software has been increased and accuracy of cost estimation models has been affected. Therefore, if real and estimated effort and time values are wanted to be close, information environment component and technical complexity factor of project are need to be calculated meticulously before development of software.

Consequently, it is seen that a calibration of information environment component is required in order to provide a successful model in different software development. It was found that the success of the obtained results increased in proportion to the number of selected objects and the correct selection of their complexity. In order to achieve more successful results, it is recommended that calibration be carried out on a periodic basis, with more homogeneous data on a large scale. The traditional metrics were utilized in this study. Results can be compared using object-oriented metrics.

References

1. Huang, Y., Yang, Z.: Write or type? How a paper versus a digital shopping list influences the way consumers plan and shop. J. Assoc. Consum. Res. **3**(3), 396–409 (2018)
2. Thomas, A., Garland, R.: Supermarket shopping lists: their effect on consumer expenditure. Int. J. Retail Distrib. Manage. **21**(2) (1993)
3. Arnaud, A., Kollman, A., Berndt, A.: Generation Y: the development and use of shopping lists. Adv. Soc. Sci. Res. **2**(9), 1–13 (2015)
4. Bellini, S., Cardinali, M.G., Grandi, B.: Does shopping preparation influence consumer buying decisions. Int. Bus. Res. **9**(10), 201–211 (2016)
5. Keller, M.B.: U.S. Patent Application No. 12/775,329 (2011)
6. Chu, T.H., Hui, F.C., Chan, H.C.: Smart shopping system using social vectors and RFID. In: International Multi Conference of Engineers and Computer Scientists, vol. 1, pp. 239–244, March 2013
7. Gopalakrishnan, P., Roberts, R., Rathi, S., Lortz, V.: U.S. Patent Application No. 13/976,026 (2013)
8. Heinrichs, F., Schreiber, D., Schöning, J.: The hybrid shopping list: bridging the gap between physical and digital shopping lists. In: Proceedings of the 13th International Conference on Human Computer Interaction with Mobile Devices and Services, pp. 251–254. ACM, August 2011
9. Groß, M.: Mobile shopping: a classification framework and literature review. Int. J. Retail Distrib. Manage. **43**(3), 221–241 (2015)
10. Voropanova, E.: Conceptualizing smart shopping with a smartphone: implications of the use of mobile devices for shopping productivity and value. Int. Rev. Retail Distrib. Consum. Res. **25**(5), 529–550 (2015)
11. Wenxing, O., Lei, W., Zhipeng, J., Changhong, Y.: Implementation of smart shopping system based on NFC Technology. In: 2015 Seventh International Conference on Measuring Technology and Mechatronics Automation, pp. 553–555. IEEE, June 2015
12. Nehrenz, K., Singleton, V.C., Stanley, R.: U.S. Patent and Trademark Office, U.S. Patent No. 9,307,355, Washington, DC (2016)
13. Liu, J., You, J.: Smart shopper: an agent-based web-mining approach to Internet shopping. IEEE Trans. Fuzzy Syst. **11**(2), 226–237 (2003)
14. Dacko, S.G.: Enabling smart retail settings via mobile augmented reality shopping apps. Technol. Forecasting Soc. Change **124**, 243–256 (2017)
15. Jayawilal, W.H., Premeratne, S.: The smart shopping list: an effective mobile solution for grocery list-creation process. In: 2017 IEEE 13th Malaysia International Conference on Communications (MICC), pp. 124–129. IEEE, November 2017
16. Katuk, N., Jayasangar, T., Yusof, Y.: Design and development of smart list: a mobile app for creating and managing grocery lists. Baghdad Sci. J. **16**, 462–476 (2019)
17. Gültekin, G., Bayat, O.: Smart location-based mobile shopping Android application. J. Comput. Commun. **2**(08), 54 (2014)

18. Willems, K., Smolders, A., Brengman, M., Luyten, K., Schöning, J.: The path-to-purchase is paved with digital opportunities: an inventory of shopper-oriented retail technologies. Technol. Forecast. Soc. Change **124**, 228–242 (2017)
19. You, C.W., Wei, C.C., Chen, Y.L., Chu, H.H., Chen, M.S.: Using mobile phones to monitor shopping time at physical stores. IEEE Pervasive Comput. **10**(2), 37–43 (2011)
20. Boehm, B., Abts, C., Chulani, S.: Software development cost estimation approaches—a survey. Ann. Softw. Eng. **10**(1–4), 177–205 (2000)
21. Kaur, J., Singh, S., Kahlon, K.S.: Comparative analysis of the software effort estimation models. World Acad. Sci. Eng. Technol. **46**, 485–487 (2008)
22. Laird, L.M., Brennan, M.C.: Software Measurement and Estimation: A Practical Approach, vol. 2. Wiley, New York (2006)
23. Soner, T.: Parametrik tahmin modellerin yazılım projelerine uygulanmasına yönelik bir yazılım paketinin geliştirilmesi. Master Thesis. Ankara University (2014)
24. Pressman, R.S.: Software Engineering: a Practitioner's Approach. Palgrave Macmillan, New York (2005)
25. Schach, S.R.: Object-Oriented and Classical Software Engineering, vol. 6. McGraw-Hill, New York (2007)
26. Sommerville, I.: Software Engineering, 8th edn. Addison Wesley, Boston (2006)
27. Barnston, A.G.: Correspondence among the correlation, RMSE, and Heidke forecast verification measures; refinement of the Heidke score. Weather Forecast. **7**(4), 699–709 (1992)
28. Foss, T., Stensrud, E., Kitchenham, B., Myrtveit, I.: A simulation study of the model evaluation criterion MMRE. IEEE Trans. Softw. Eng. **29**(11), 985–995 (2003)

Forecasting Housing Prices by Using Artificial Neural Networks

Tolga Yesil[1(✉)], Fatma Akyuz[2], and Utku Kose[2]

[1] Uşak University, Ankara-İzmir Highway 8.Km., Uşak, Turkey
tolga.yesil@usak.edu.tr
[2] Süleyman Demirel University, Çünür, 32260 Isparta, Turkey
fatma.akyuz@usak.edu.tr, utkukose@sdu.edu.tr

Abstract. Artificial Intelligence techniques have penetrated all different areas of life since their first appearances and the use of intelligence systems has had a big potential for multi-disciplinary perspective by offering successful solutions to real world based problems. As there are many different techniques that can be examined under Artificial Intelligence research area, some of these techniques can become more efficient due to their wide solution scope. Artificial neural network (ANN) models are known remarkable solution ways among these techniques. The aim of this paper is to forecast housing prices by using ANN. Research data were collected from the 'sahibinden.com' web site, where houses are listed for sale advertisements in Turkey (particularly Uşak for this research), and these data were used for training and testing period of the ANN model designed. As findings of this study, the ANN model estimated prices of houses for sale successfully, according to analyzes over the ANN model. Eventually, a decision support/adaptive suggestion mechanism is provided in this study, as an intelligent solution about optimum sales price of houses, which is needed by a large mass of society.

Keywords: Housing prices · Forecasting sale price · Artificial intelligence · Artificial neural networks · Machine learning

1 Introduction

In time, practical and effective processing of information has become a necessity for dynamically changing conditions of today's world. Nearly all areas benefiting from computer based systems have required use of many different techniques because of different issues appeared while processing the information from real world. Undoubtedly, conversion of information to data thanks to computer based systems has always played important role for getting effective solutions to real based problems in different areas. That situation has affected the humankind's life directly and caused rise of different developments. Nowadays, the humankind tries to give a high level intelligence to the machines, as besides the owned, real intelligence. The humankind aims to make machines, which are intelligent and effective to take right decisions. As a result of these efforts, a notion called as artificial intelligence has appeared in time [1]. Nearly all active research by the field of artificial intelligence has played top roles in solutions

D. J. Hemanth and U. Kose (Eds.): ICAIAME 2019, LNDECT 43, pp. 621–632, 2020.
https://doi.org/10.1007/978-3-030-36178-5_51

for real based problems, as causing it to become one of the strongest scientific areas. By means of artificial intelligence, there are many positive achievements far away from findings by traditional solutions so that field is currently accepted as a scientific area for both nowadays and the future.

McCarthy, the father of the concept of artificial intelligence, defines it as the science or engineering of making intelligent machines [2]. On the other hand, Nabiyev sees the artificial intelligence as a computer or a computer-controlled machine, which is generally assumed to be providing human oriented reasoning and generalization [3]. Finally, Luger expresses that the artificial intelligence is a branch of Computer Science for dealing with the automation of intelligent behavior [4]. Artificial intelligence is often employed in different fields and one of its main solution approaches is forecasting, which is a need for real world problems [5, 6].

Moving from the explanations, aim of this research is to forecast house prices by using Artificial Neural Networks (ANN), which is especially a machine learning technique in the context of artificial intelligence. ANN is briefly a mathematical computational model, which can be trained with different algorithms, in order to learn from known samples for providing accurate solutions against new cases of the considered problem. In the study, research data for housing prices were collected from the 'sahibinden.com' web site, where houses are listed for sale advertisements in Turkey (particularly Uşak for this research). The collected data were used for training and testing period of the ANN model designed within this research. With the study, it is aimed to provide a decision support/adaptive suggestion mechanism, which is somehow an intelligent solution about optimum sales price of houses, which is needed by a large mass of society.

Considering the subject of the study, the remaining sections of the paper are organized as follows: The next section provides a review of alternative studies-works based on using intelligent systems for price forecasting. In this way, it is aimed to inform the target readers about alternative applications of artificial intelligence on the problem of price estimation. Following to that section, the third section provides information regarding the ANN technique used in this study. In this regard, the background of the ANN technique is briefly explained. Next, the fourth section is devoted to the application of ANN for forecasting housing prices and the findings are provided in the fifth section. Finally, the paper is ended with discussions about conclusions and some future work ideas.

2 Literature Review

Before giving details of the performed research-study, it is also necessary to have sufficient information about the associated literature. According to the subject of the study, the literature has been generally reviewed and some foremost efforts by researchers have been briefly explained under the following paragraphs.

In their study, Yuksel and Akkoc has employed an ANN model for forecasting gold price, considering silver price, Brent Petrol, US dollars/Euro parity, the rate of USD bond interest, Euro Next100 index, Dow Jones index and USD consumer price index, respectively [7]. The study provided positive enough results in terms of forecasting gold price, thanks to the ANN technique. In another study, Tabales et al. forecasted the housing price by using an ANN model, which was thought as an alternative to the hedonic model [8]. They collected the data considering some features of a house, such as number of bedroom, number of shower, exist of a garage, state of air condition systems…etc. As realized in Spain, a total of 135.920 data was included for the analyses. At the end of this study, the ANN model was successful enough with an accuracy rate changing between 81% and 91%, for the housing price forecasting. Ecer used ANN and hedonic model in order to forecast second hand automobile prices, considering Turkey [9]. Ecer compared the related two techniques and used 42 different variables of 640 automobiles as collected from the web site: sahibinden.com. By using 60% of data for training, 15% for validation, and 25% for testing respectively, the ANN model was found to be better (with Mean Square Error-MSE: 0,006) than the hedonic model (with MSE: 0,018) for the related price forecasting problem. In another study, Asilkan and Irmak performed the same research of second hand automobile price forecasting by using the ANN [10]. The data collected from 40 different web sites as including also advertisements of Volkswagen Golf V 1.9 Tdi Trendline, BMW 525 D DPF and Mercedes A150 Classic for the time period from January 2005 to December 2007 were used accordingly. These data were analyzed by considering the means of months. According to evaluation works done over the ANN model, it was found to be better for forecasting second hand automobiles prices. Another similar research study was done by Karaatli et al., by performing the forecasting for automobile quantity this time [11]. Selim and Demirbilek provided a remarkable research, in which they tried to determine home rental prices in Turkey [12]. In the study, research data was collected from household budget survey of 2004, which was prepared by the Turkish Statistical Institute. Household budget survey included 7514 information of household. In the study, the ANN technique was used for analyzing the period because of nonlinear nature of the data. In addition to the ANN technique, the solution approach of this study consisted of also hedonic model. In terms of housing prices, Khalafallah provided a study in which a forecasting solution was introduced by running an ANN model [13]. The study was for the people working in the context of the associated sector in USA and the research employed some data such as interest rate, percentage change of real estate, time and trading size for training the ANN model. Eventually, forecasted housing prices were validated for a period of 6–12 months and it was observed that the ANN model had a forecasting performance with a 2% rate of error. Din et al. compared different real estate valuation models by considering environmental factors through linear regression model and the ANN technique [14]. In the reference model, they employed a standard linear regression model in order to measure environmental quality. In the study, a geographic information system (GIS) developed in Switzerland was compared with linear models containing environmental quality grades extracted from the urban habitat database. The study briefly concludes that the formed price indices were very similar to the traditional hedonic model. Additionally, it was found that the price indices of nonlinear ANN models showed a generalized form and detailed price behaviors of different models included significant differences depending on the input selection from environmental factors. This research was supported by Nghiep and Al for assessing real-estate for forecasting housing value [15]. The study included both ANN technique and regression model and obtained the findings showed that the ANN model was more successful than the regression model for the target problem. For again assessing housing process in Selcuklu area of Konya, Turkey, Ozkan et al. employed ANN and regression models, by considering 8 different parameters of 115 houses [16]. The study reveals a success rate of 84,54% for the ANN model, and 83% for the regression model. In the context of forecasting housing prices, Limsombunchai considered the region of Christchurch, New Zealand [17]. In that study, the author used 200 samples, which were collected from Harcourt web sites. The forecasting problem was tried to be solved by using ANN and hedonic models, by using a set of data with the parameters such as size of the house, age of the house, total number of bedrooms, the number of bathrooms, and geographic position. According to the obtained findings, the ANN model is more successful than the hedonic model. In another study, Worzala et al. used a total of 288 house sale data for forecasting prices by using an ANN model [18]. The forecasting was briefly for the Fort Collins, Colorado (USA) and one of the most remarkable findings of the study was long analyzing time by the ANN, which means a negative result for the ANN. By using the ANN technique, Do and Grudnitski examined the effect of housing age on housing value [19]. They formed the data set with the information from the year of 1991, for 242 houses located in San Diego, California (USA). The data consisted of attributes such as housing sale price, housing age, the number of bedrooms, and the size. According to the findings obtained via an ANN model with single hidden layer, there was a negative relation between the age and the price of the house.

In the associated literature including use of ANN technique, there are many alternative studies in order to forecast and evaluate housing price. In this sense, interested readers are referred to also [20–23]. The main point that can be derived from already done studies is that the ANN oriented solutions can provide effective solutions for forecasting prices of different components including also houses. So, that has become an encouraging factor for the authors of this study and an ANN model has been employed by considering a different area and data scope for the problem of forecasting housing prices.

3 Artificial Neural Network

Before giving information regarding the research done in this study, it is important to learn some about ANN. In this way, readers having not too much idea about ANN were aimed to be enabled for having enough knowledge. In this context, the next subsections are briefly devoted to general structure of ANN models and the techniques used for training them, respectively.

3.1 The Structure of Artificial Neural Network

With the help of algorithmic structures designed to be based on various logical and mathematical foundations, Artificial Intelligence can produce effective and fast solutions to complex and difficult problems, by simulating human thinking styles/behaviors, actions by living organisms, or occurred routine events in the nature [3, 24, 25]. During the past period, many artificial methods and techniques have been developed and even sub-research topics have been formed under the field of Artificial Intelligence. In general, Artificial Neural Networks is one of the most preferred Artificial Intelligence techniques in scientific researches, thanks to its very wide application scope and flexibility.

Artificial Neural Networks (ANNs), which are formed by combining parallel distributed processing elements and developed based on the functions of the human brain, are known as an effective Artificial Intelligence technique that can produce effective Machine Learning oriented solutions for almost every kind of problems [3, 26, 27]. ANN models are generally used for forecasting, optimization, pattern recognition and control problems [26, 27]. In order to understand how a default ANN runs, it is possible to take a look at to the human nervous system. In detail, the human body can communicate with the real world via electrical signals transmitted within an environment where brain and neurons take place. Here, the human brain can process and store necessary electrical signals (information) simultaneously by benefiting from some chemical reactions, which are used during also instant, very rapid responses. Neurons in the human brain are connected with the real world depending on the intensity, repetition and condition of the stimuli [28]. ANN technique is briefly inspired from the (human) brain and working mechanisms of neurons. As a result of developments in modeling of ANN, it has become a popular technique used in all fields.

Although there are many different ANN models in the literature (with different working mechanisms and structures) [29, 30], the most frequently used ANN model has a feedforward, multilayered structure [30, 31]. In a typical multilayered, feedforward ANN structure, artificial neurons collect data and uses special activation functions to derive new values from them, just before transmitting these new values to other artificial neurons coming after them. Each artificial neuron here is connected to others with weights and included in one of three layer types as input layer (where input data received), hidden layer(s) (where data has processed more and more), and output layer (where the resulting data is obtained) [29–32]. At this point, the objective is to optimize all weights connecting artificial neurons so that the ANN model can give correct outputs to the known, sample input data [30, 31].

3.2 Training Artificial Neural Networks

Effective and efficient training of ANN models is extremely important to produce correct solutions for the target problem. Traditionally, algorithms such as Back-Propagation Algorithm [26, 30, 33] or Marquardt Algorithm [34] are used for training ANN models. At this point, especially Back-Propagation Algorithm is widely preferred in training ANN models with multilayered, feedforward architecture. In addition to the Back-Propagation Algorithm, different algorithms or swarm intelligence oriented optimization techniques are also employed for training ANN because they can provide better results [35–40]. As a result, a trained ANN model can produce accurate outputs - results in response to input (test and/or application) data that has not been shown to it before and can thus be used long times for the target problem. It is also important that ANN models can encounter with overfitting problem (giving high results for training data but being unsuccessful for test/application data) so some solutions such as cross-validation or alternative training algorithms are preferred [41, 42]. In the study considered here, ANN was training with the Bayesian Regularization algorithm, which is robust against overfitting problem [43, 44].

4 Using Artificial Neural Networks for Forecasting Housing Prices

In this section, details of the ANN based home sale price forecasting process, which was carried out in accordance with the subject and aim of the study, are explained briefly. In this context, the forecasting process is generally discussed first, and after that, the data structure used for the forecasting process and the application with the ANN are explained accordingly.

4.1 General Framework of the Research

In this study, the city of Uşak in Turkey was selected within the scope of the general forecasting process for housing prices. Uşak, which is located in western Anatolia area of Turkey become an official province on July 15, 1953 [45]. Nowadays, Uşak hosts around a total of 31.113 students studying at Uşak University established on 17.03.2006 [46]. The real estate market in Uşak is very active as a result the increasing number of students and also immigration caused by the war and other problems in the Middle East [47]. Additionally, another important factor for the housing market in Uşak is the urban transformation project with a budget of approximately 450 million Turkish Liras (which is around 80.500.000 USD during 2019 June). With the completion of the urban transformation project, 1051 new houses were added to the real estate market of Uşak, as a positive contribution [48].

According to the Turkey Statistical Institute (TÜİK-Turkstat), the number of housing sales in Uşak is increasing over the years. By December 2008, the number of housing sales had been 1143 and that value was announced as 6681 in 2018. Total housing sales between 2008 and 2018 in Uşak are provided in Table 1 [49].

Table 1. Total housing sales in Uşak, according to years [49].

Years	2008	2009	2010	2011	2012	2013	2014	2015	2016	2017	2018
Number of sales	1.143	1.919	2.104	2.583	2.256	2.865	3.354	4.241	5.492	6.320	6.681

As it can be seen from Table 1, there is an increasing rate in the number of housing sales in Uşak. The Table 1 also shows a quantitative view for the development in the real estate market of Uşak. At this point, Uşak was chosen as the sample of this study, because of the increasing population and the dynamic real estate market of this city.

4.2 Data and the Solution Process

In this study, the data as the research sample included information from housing ads obtained from the 'sahibinden.com' web site. For training the ANN model, a total of 700 ads from the related web site (not considering the city) were used. As it was mentioned before, the exact forecasting was done by using the data for the city of Uşak. At the time of this study-research, there were a total of 696 ads in Uşak city center. While preparing the exact data for training, the advertisements that were not suitable for being used as sample (i.e. ads with incomplete information, different component sales) were removed accordingly. So, the remaining 636 real estate data were included in the analysis-forecasting application.

In the forecasting, NNTool interface of the MATLAB software system was used. Details regarding the formed ANN model is provided in Table 2.

Table 2. Details of the formed ANN model for forecasting housing sales.

Input data (for each artificial neuron)	Hidden layer(s)	Output data for the output layer
Location of the house	A total of 3 hiden layers (Each of them employs 37 artificial neurons)	Price for the sale (1 artificial neuron)
Type of the house (flat, villa...etc.)		
Total area of the house (square metre)		
Total number of rooms		
The age of the house-building		
Total floor number of the house		
Heating system		
Total number of baths		
Being located in a site		
Sale from owner or real estate agent		

As it can be understood from Table 2 that a total of 10 attributes were used for the process of forecasting sale price. The ANN model was based on the structure of multilayered feedforward mechanism. On the other hand, the model was trained via Bayesian Regularization (TrainBr) algorithm through a total of 1000 iterations. In order to evaluate the result-success, Mean Squad Error (MSE) was selected as the performance evaluation criteria.

5 Forecasting Application and Findings

After performing the related training phase for the formed ANN model, forecasting application was done over the related test data from Uşak city center. Figure 1 shows MAE and regression graphics viewed at the end of the training phase done over the ANN model.

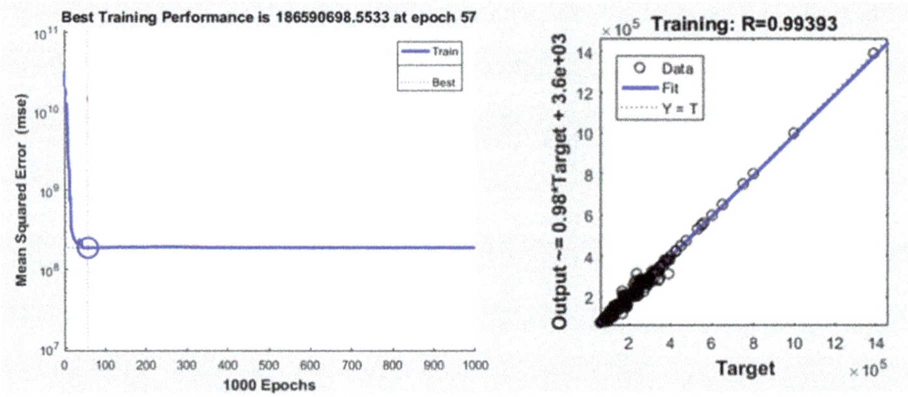

Fig. 1. MAE and regression graphics for the training phase of the ANN model

When the graphics for MAE and regression changes (Fig. 1) are examined, it can be seen that the best training performance value was 186590698,5533. The ANN model achieved that value in the 57th iteration. Additionally, the regression value was calculated as 0.99393 at the time of that training-ending point. Finally, the output error value was detected as −2.9595.

5.1 Findings for the Test Data

Table 3 shows findings for the forecasting application done with the trained ANN model. Based on the training data of Uşak city, each two columns from left to side show forecasted prices and the real prices (desired prices), respectively.

Table 3 shows that the trained ANN model in this study provided near forecasting findings for the data of Uşak city. Those findings have been obtained via multilayered, feedforward ANN model trained by Bayesian Regularization algorithm. Of course, findings may be improved by trying different combinations of ANN models, input attributes and training algorithms.

Table 3. Findings for the forecasting application.

Forecasted price	Real price	Forecasted price	Real price	Forecasted price	Real price	Forecasted price	Real price
141760.7866	135000	105854.4736	100000	123347.6771	130500	165943.9661	155000
113326.2921	98000	86499.959	65000	117999.3448	88000	168801.1678	215000
257107.2777	340000	162618.4975	190000	200114.0903	217000	251146.743	310000
134204.9949	175000	99149.2791	147000	107274.6776	155000	228442.2947	148000
123716.7904	165000	389621.7843	265000	402956.8107	240000	411334.7153	395000
246916.6307	215000	189156.1377	194000	157959.8927	295000	91600.2058	120000
183226.4863	140000	368542.7144	260000	177010.0429	380000	96794.2945	115000
285559.6589	280000	251146.743	210000	132935.0181	220000	69482.9443	175000
846199.3848	500000	204418.9513	163000	301123.8951	125000	92500.9323	260000
188131.0645	235000	391558.6184	259000	115821.7791	115000	161964.1852	260000
65867.8452	350000	81067.628	100000	108691.088	117000	332910.7527	195000
214604.1686	220000	189156.1377	175000	165047.9587	189000	83950.5481	90000
113349.6414	135000	134035.6361	140000	263600.4836	350000	80810.6049	195000
992036.8937	375000	281308.6906	400000	143398.6251	135000	83950.5481	90000
113349.6414	140000	138019.9225	330000	89025.5212	105000	80810.6049	195000
146139.2167	190000	171300.5542	170000	250826.8196	230000	200114.0903	280000
93592.0641	78000	138639.9252	143000	421398.5839	270000	243858.0464	280000
155395.6909	180000	146139.2167	182000	127052.9109	135000	344284.4281	440000
130663.5155	150000	180447.07	185000	132375.4498	175000	203958.0157	215000
145535.8559	156000	90591.6473	85000	508690.4887	365000	94511.2177	87000
392540.7831	345000	219808.4283	205000	99913.0105	77000	254075.3116	245000
155395.6909	155000	204303.5772	197000	86712.2182	125000	92895.3681	177500
191185.0207	265000	193239.9686	345000	251146.743	295000	65857.8867	195000
337006.8821	158000	89025.5212	110000	89459.8441	85000	109030.6902	185000
188110.4237	145000	117740.5616	210000	112051.8814	110000	273858.4038	170000
275040.7358	235000	164475.3219	195000	191372.3433	160000	546533.6157	230000
159207.9263	150000	176026.8643	350000	79290.6831	88000	273858.4038	144000
121019.7957	135000	201908.2319	175000	293624.2084	165900	551979.4543	260000
82218.5406	255000	167138.1271	172000	119038.2814	95000	148624.3861	260000
140564.7401	120000	243858.0464	315000	541694.7739	420000	111944.7639	360000
233811.4373	175000	258339.0231	115000	96507.8895	97000	389222.2542	65000
103374.6541	249000	250826.8196	230000	89459.8441	78000	106586.2637	82500
239743.0275	165000	173213.1895	235000	643325.5989	408000	232905.6304	279000
148590.1454	150000	255937.3476	220000	99913.0105	105000	93358.3927	147000
281826.9747	350000	538763.0812	310000	168801.1678	370000	105854.4736	92000
163591.6942	180000	266362.5317	300000	117362.3512	98000	144006.2485	300000
120288.3741	144000	538763.0812	310000	273674.014	225000	118084.5772	380000
				202831.3357	225000	599134.6439	340000

6 Conclusions and Future Work

This study provided a research on forecasting housing prices via ANN model. In detail, the formed ANN model was trained by Bayesian Regularization technique in order to eliminate overfitting and fed with data obtained from the 'sahibinden.com' web site, where lots of ads for housing sales in Turkey are listed widely. For the scope of this research, the city of Uşak was chosen for the forecasting application. After training the ANN model with training data from different cities, the forecasting performance was evaluated by considering the data for Uşak. According to the findings, the ANN model achieve well enough forecasting regarding sale prices for houses with different attributes. Moving from the findings, it can be indicated that the ANN model can be used as a decision support/adaptive suggestion mechanism, which is somehow an intelligent solution about optimum sales price of houses, which is needed by a large mass of society.

There are also some future works planned as associated with the research here. There are some future works in order to train the ANN model with different algorithms and evaluate the findings over the same data included in this study. On the other hand, the same data will be evaluated with also alternative Machine Learning techniques. Finally, forecasting of housing prices will be analyzed with different Machine Learning solutions, by using more data from different sources.

References

1. Ergezer, H., Dikmen, M., Özdemir, E.: Recognition systems with artificial neural networks. PiVOLKA **2**(6), 14–17 (2003). (in Turkish)
2. McCarthy, J.: What is Artificial Intelligence? (2007). http://www-formal.stanford.edu/jmc/whatisai/whatisai.html. Accessed 18 Jan 2019
3. Nabiyev, V.V.: Artificial Intelligence (Problems-Methods-Algorithm). Seçkin Press, Ankara (2014). (in Turkish)
4. Luger, G.F.: Inteligência Artificial: Estruturas e estratégiaspara a solução de problemas complexos. Bookman (2004)
5. Kose, U., Arslan, A.: Forecasting chaotic time series via anfis supported by vortex optimization algorithm: applications on electroencephalogram time series. Arab. J. Sci. Eng. **42**(8), 3103–3114 (2017)
6. Raza, M.Q., Khosravi, A.: A review on artificial intelligence based load demand forecasting techniques for smart grid and buildings. Renew. Sustain. Energy Rev. **50**, 1352–1372 (2015)
7. Yüksel, R., Akkoç, S.: Forecasting gold prices with artificial neural networks and an application. Dogus Univ. J. **17**(1) (2016). (in Turkish)
8. Tabales, J.M.N., Caridad, J.M., Carmona, F.J.R.: Artificial neural networks for predicting real estate price. Revista de Métodos Cuantitativos para la Economía y la Empresa **15**, 29–44 (2013)
9. Ecer, F.: Second hand car prices estimation and identification of determinants for prices in Turkey. Anadolu Univ. Soc. Sci. J. **13**(4), 101–112 (2013). (in Turkish)
10. Asilkan, Ö., Irmak, S.: Forecasting future prices of second hand cars by using artificial neural networks. Suleyman Demirel Univ. Fac. Econ. Adm. Sci. J. **14**(2), 375–391 (2009). (in Turkish)

11. Karaatli, M., Helvacioglu, O.C., Omurbek, N., Tokgoz, G.: Forecasting of car sales with the method of artificial neural networks. Int. J. Manage. Econ. Bus. Adm. **8**(17), 87–100 (2012). (in Turkish)
12. Selim, S., Demirbilek, A.: Analysis of the rental value of housing in Turkey: an approach of hedonic model and artificial neural network. J. Aksaray Univ. Fac. Econ. Adm. Sci. **1**(1), 73–90 (2009). (in Turkish)
13. Khalafallah, A.: Neural network based model for predicting housing market performance. Tsinghua Sci. Technol. **13**(S1), 325–328 (2008)
14. Din, A., Hoesli, M., Bender, A.: Environmental variables and real estate prices. Urban Stud. **38**(11), 1989–2000 (2001)
15. Nghiep, N., Al, C.: Predicting housing value: a comparison of multiple regression analysis and artificial neural networks. J. Real Estate Res. **22**(3), 313–336 (2001)
16. Özkan, G., Yalpır, Ş., Oygunol, O.: An investigation on the price estimation of residable real-estates by using artificial neural network and regression methods. In: XIIth Applied Stochastic Models and Data Analysis International conference (ASMDA), Chania, Crete, Greece (2007)
17. Limsombunchai, V.: House price prediction using artificial neural network: a comparative study with hedonic price model. In: New Zealand Agricultural and Resource Economics Society Conference, Blenheim, New Zealand, 25–26 June 2004
18. Worzala, E., Lenk, M., Silva, A.: An exploration of neural networks and its application to real estate valuation. J. Real Estate Res. **10**(2), 185–201 (1995)
19. Do, Q., Grudnitski, G.: A neural network analysis of the effect of age on housing values. J. Real Estate Res. **8**, 253–264 (2009)
20. Çalışkan, M., Tuncer, M., Deniz, D.: Forecasting prices and directions of stocks with artificial neural networks. J. Eskişehir Osmangazi Univ. Fac. Econ. Adm. Sci. **10**(3), 177–194 (2015). (in Turkish)
21. Erdoğan, E., Özyürek, H.: Price forecasting with artificial neural networks. J. Soc. Sci. Humanit. **4**(1), 85–92 (2012). (in Turkish)
22. Özçalıcı, M.: Multistage price forecasting with artificial neural networks: a research on BIST30 stocks. Dokuz Eylül Univ. J. Fac. Econ. Adm. Sci. **31**(2), 209–227 (2016). (in Turkish)
23. Ertuğrul, I., Bekin, A.: Comparative analysis of artificial neural networks and time-series forecasting models for some essential food prices in Turkey. Kafka Univ. J. Fac. Econ. Adm. Sci. **7**(13), 253 (2016). (in Turkish)
24. Michalski, R.S., Carbonell, J.G., Mitchell, T.M. (eds.): Machine Learning: An Artificial Intelligence Approach. Springer, Heidelberg (2013)
25. Nilsson, N.J.: Principles of Artificial Intelligence. Morgan Kaufmann, San Mateo (2014)
26. Elmas, Ç.: Artificial Neural Networks. Seçkin Press, Ankara (2003). (in Turkish)
27. Hassoun, M.H.: Fundamentals of Artificial Neural Networks. MIT Press, Cambridge (1995)
28. Duman, B.: Why Brain Based Learning?. Pegem Academy Press, Ankara (2015). (in Turkish)
29. Krenker, A., Kos, A., Bešter, J.: Introduction to the Artificial Neural Networks. INTECH Open Access Publisher (2011)
30. Yegnanarayana, B.: Artificial Neural Networks. PHI Learning Pvt. Ltd. (2009)
31. Hornik, K., Stinchcombe, M., White, H.: Multilayer feedforward networks are universal approximators. Neural Networks **2**(5), 359–366 (1989)
32. Ugur, A., Kinaci, A.C.: A web-based tool for teaching neural network concepts. Comput. Appl. Eng. Educ. **18**(3), 449–457 (2010)
33. Phansalkar, V.V., Sastry, P.S.: Analysis of the back-propagation algorithm with momentum. IEEE Trans. Neural Networks **5**(3), 505–506 (1994)

34. Hagan, M.T., Menhaj, M.B.: Training feed forward networks with the Marquardt algorithm. IEEE Trans. Neural Networks **5**(6), 989–993 (1994)
35. Gudise, V.G., Venayagamoorthy, G.K.: Comparison of particles warm optimization and backpropagation as training algorithms for neural networks. In: Proceedings of the 2003 IEEE Swarm Intelligence Symposium, SIS 2003, pp. 110–117. IEEE (2003)
36. Ilonen, J., Kamarainen, J.K., Lampinen, J.: Differential evolution training algorithm for feed-forward neural networks. Neural Process. Lett. **17**(1), 93–105 (2003)
37. Karaboga, D., Akay, B., Ozturk, C.: Artificial bee colony (ABC) optimization algorithm for training feed-forward neural networks. In: International Conference on Modeling Decisions for Artificial Intelligence, pp. 318–329. Springer, Heidelberg (2007)
38. Kose, U., Arslan, A.: Optimization of self-learning in Computer Engineering courses: an intelligent software system supported by artificial neural network and vortex optimization algorithm. Comput. Appl. Eng. Educ. **25**(1), 142–156 (2017)
39. Montana, D.J., Davis, L.: Training feed forward neural networks using genetic algorithms. In: IJCAI, vol. 89, pp. 762–767 (1989)
40. Valian, E., Mohanna, S., Tavakoli, S.: Improved cuckoo search algorithm for feedforward neural network training. Int. J. Artif. Intell. Appl. **2**(3), 36–43 (2011)
41. Tetko, I.V., Livingstone, D.J., Luik, A.I.: Neural network studies. 1. Comparison of overfitting and overtraining. J. Chem. Inf. Comput. Sci. **35**(5), 826–833 (1995)
42. Zhang, G., Hu, M.Y., Patuwo, B.E., Indro, D.C.: Artificial neural networks in bankruptcy prediction: general framework and cross-validation analysis. Eur. J. Oper. Res. **116**(1), 16–32 (1999)
43. Burden, F., Winkler, D.: Bayesian regularization of neural networks. In: Artificial Neural Networks, pp. 23–42. Humana Press (2008)
44. Ticknor, J.L.: A Bayesian regularized artificial neural network for stock market forecasting. Expert Syst. Appl. **40**(14), 5501–5506 (2013)
45. Bilgen, N.: Urban functions and use of urban areas in Uşak. In: Urban Economic Research Symposium, March 2004, pp. 58–69 (2004). (in Turkish)
46. Uşak University Student Affairs: Uşak University Student Guide (2017). https://ogrencirehberi.usak.edu.tr/3031. Accessed 12 Feb 2019. (in Turkish)
47. Hürriyet Newspaper-Local News: Aparts are Investors' Favorite in Uşak (2015). http://www.hurriyet.com.tr/usak-ta-yatirimcilarin-gozdesi-apart-daireler-37141503. Accessed 13 Feb 2019. (in Turkish)
48. Republic of Turkey Prime Ministry Housing Development Administration of Turkey: Uşak Urban Transformation Project is an Exemplary Project. http://www.toki.gov.tr/haber/usak-kentsel-donusum-projesi-ornek-bir-projedir. Accessed 12 Feb 2019
49. Turkey Statistical Institute (TÜİK-Turkstat), Number of Home Sales by Province and Year. http://www.tuik.gov.tr/PreTablo.do?alt_id=1056. Accessed 15 Mar 2019

High Power Density and High Speed Permanent Magnet Synchronous Generator Design

Benhar Aydogan[1], Yusuf Öner[2(✉)], Metin Ersoz[3], Selami Kesler[2], and Mustafa Tumbek[2]

[1] Technical Sciences Vocational High School, Pamukkale University, Denizli, Turkey
[2] Department of Electrical and Electronics Engineering, Pamukkale University, Denizli, Turkey
yoner@pau.edu.tr
[3] Senkron Ar-ge Engineering A.S., Denizli, Turkey

Abstract. In this paper, it is realized by using numerical and FEM methods of a high power permanent magnet synchronous generator design which can produce high frequency voltage which can be used in space vehicles, rockets and aircraft. The permanent magnet synchronous generator designed to be used in the mentioned vehicles is designed according to the space constraints in the vehicles and according to the boundary conditions which can give the highest power. Numerical and FEM analysis solutions have been developed for both size and high power. Generator sizes obtained as a result of numerical analysis in the Ansys Maxwell Program, optimization studies were performed to obtain the most appropriate geometric size and power density. With the results of FEM analysis, experimental studies have been started by making a permanent magnet synchronous generator prototype. In experimental studies, results obtained as a result of analysis were compared with experimental results. At the end of the experimental results, high-power density, high-speed synchronous generator has also been shown to have high efficiency.

Keywords: Permanent magnet synchronous generator · Generator design

1 Introduction

A synchronous machine has two main sections as a static part (stator) and a movable part (rotor). Stators and rotors of synchronous machines have windings. They are preferred in places where constant speed is desired regardless of load. In order to rotate the machine, two fluxes are needed, one from the stator and the other from the rotor. It is possible to make different machine designs based on this principle. When the direct current excitation winding in the rotor of the synchronous machine is replaced by a permanent magnet, and the magnetic flux generated in the rotor is supplied by a magnet, a permanent magnet synchronous machine (SMSM) is obtained [1].

Rapid developments in technology are observed in electrical machines as in all areas. While these developments aim to increase the performance of machines, they

D. J. Hemanth and U. Kose (Eds.): ICAIAME 2019, LNDECT 43, pp. 633–642, 2020.
https://doi.org/10.1007/978-3-030-36178-5_52

also aim to contribute to reducing the size of these machines. One of the most important materials affecting the performance and size of the machine is the magnet.

In recent years, the invention of new magnets with better magnetic characteristics and the reduction in cost have increased the applications of permanent magnet synchronous machines [1]. The use of magnets is very convenient for synchronous machines in terms of their working principle and design. Its dynamic performance is further increased compared to a synchronous machine with winding in its rotor. The stator of the synchronous machines has a three-phase winding as in the induction machine. By using a permanent magnet instead of winding in the rotor, the drawbacks caused by brush and collector are also eliminated. In addition, since the excitation losses are eliminated, thermal limits are extended and more power can be received from a machine with the same volume. The use of a permanent magnet with better magnetic characteristics, such as NdFeB, in the rotor, keeps the air gap induction at a higher value than the winding machines, eliminates the copper losses in the rotor windings, and ensures higher efficiency compared to a synchronous motor with the same power. In addition, the size of the motor are considerably smaller [2].

Ionel has presented a booklet with principles of permanent magnet synchronous machine design for field attenuation applications [3]. This booklet contains optimum machine parameters, optimum machine topology, machine sizing and numerical studies to improve design performance. Slemon has carried out studies on the preferability of standard surface mounted permanent magnet machines in many applications [4].

Many authors have presented several different control algorithms to improve the field attenuation capacity for surface-mounted permanent magnet machines [5, 6].

Schneider has compared surface mounted permanent magnet machines with 5 different rotor structures, which are used for forklifts or small hand-held vehicle applications where limited field attenuation is required, considering rated speed and field attenuation areas. As a result of the comparison of pole types in Fig. 1 he found that the best structure in terms of the best torque/current ratio and simpler structure was the b rotor structure in Fig. 1 [7].

There are many parameters in the design of synchronous generators. Yield, one of these parameters, was investigated in this study. The weight of the used materials, copper and iron losses, power and torque per unit volume, magnet material effect, air gap, unit air gap are of factors in permanent magnet synchronous generators [8]. Generally, the design methods of synchronous generators can be made by the computer software with finite element analysis, electromagnetic thermal and mechanical stress analyzes [9].

The finite element analysis (FEA) is a numerical method for solving partial differential equations in many engineering fields. This method can be used to solve problems such as magnetic field distribution or temperature distribution in the design of electric machines [10]. The Finite Element Analyses is currently the most popular numerical method used for electromagnetic field problems. Because the finite elements method is a powerful method for a very wide range of application areas [11].

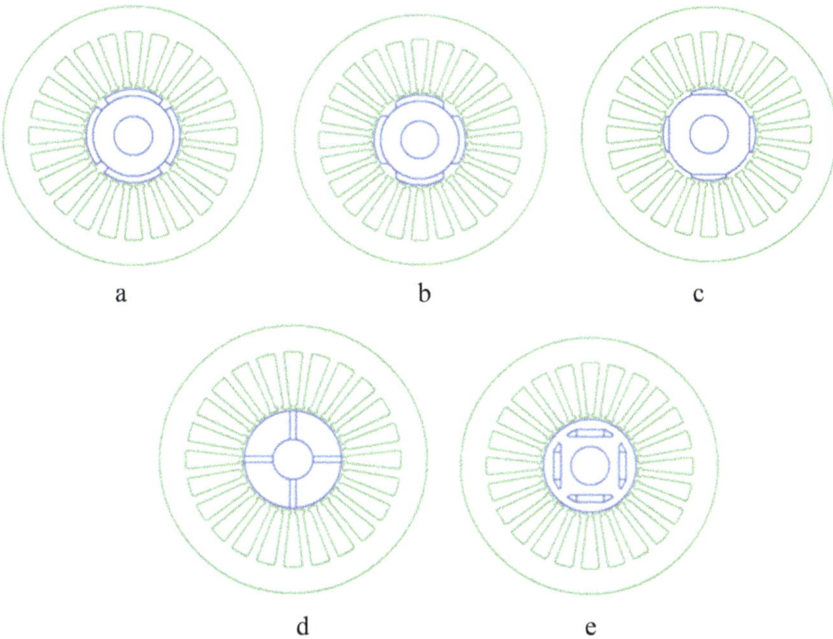

Fig. 1. a-b-c-d-e shows different rotor types of a permanent magnet synchronous motor.

2 Design and Analysis of High-Speed Permanent Magnet Synchronous Generator with Ansoft Maxwell

By using Ansys Maxwell program, we tried to make a permanent magnet synchronous generator (SMSG) design which can operate at 12000 rpm. As known, the high-speed turbo-generators are manufactured with longer and narrower diameters compared to conventional generators. In this study, a high power density synchronous generator was designed, which can generate 5 kw output power. Initially, parametric analysis of SMSG was made numerically in the Ansys Rmxprt module.

In order to create the SMSG structure, it is necessary to create the geometric structure of the generator and to select the appropriate magnet geometry and thickness. Therefore, the first geometric structure of the SMSG shown in Fig. 2 was designed. On this geometric structure, parametric analysis of the generator parameters were performed by taking into account the output power and efficiency of the generator.

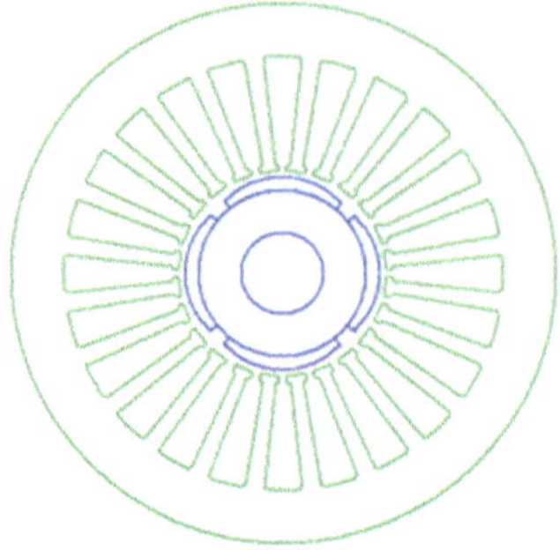

Fig. 2. The geometric design of the permanent magnet synchronous generator

Certain initial conditions need to be determined at the beginning of the generator design. The initial design conditions of the generator are given in Table 1.

Table 1. The initial design conditions of the generator

Parameter	Value
Output power	5000 W
Power factor	0.85
Terminal voltage	220 V
Number of poles	4
Frequency	400 Hz
Number of grooves	24
Stator outside diameter	130 mm
Stator inside diameter	50 mm
Stator length	90 mm
Number of conductors per groove	13
Air gap	1 mm
Rotor inside diameter	20 mm
Rotor outside diameter	47.6 mm
Rotor length	90 mm
Magnet width	1–5 mm
Magnet type	NdFe35

According to these initial parameters, numerical calculations were made in Ansys Rmxprt program and parametric analysis of parameters to obtain the best efficiency and output power were made. The most appropriate values of magnet thickness, generator lamination length, magnet offset and embrace were determined as a result of parametric analysis.

As a result of the parametric analyses, the ideal values for the generator lamination length was found to be 90 mm, magnet thickness was 3.6 mm, magnet offset was 5 and the embrace value was 0.86. After these values were found, the output power was obtained as 5 Kw and the yield as 95.37%.

Regarding the results obtained in parametric analysis, the generator's cogging torque was shown in Fig. 3.

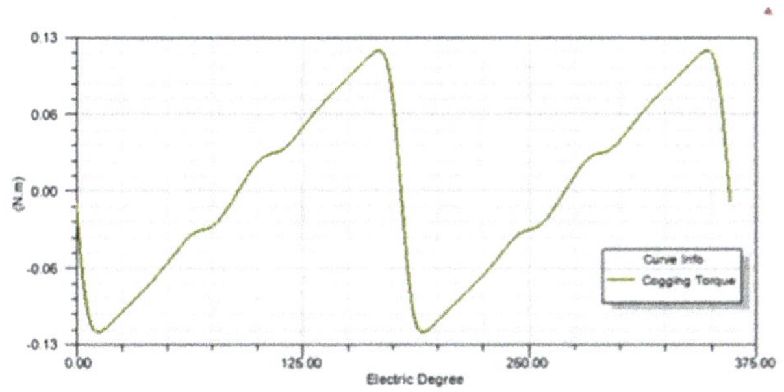

Fig. 3. The cogging torque of the permanent magnet synchronous generator

Figure 3 shows that the designed generator's cogging torque is around 0.11 Nm. This value can be considered suitable for a permanent magnet generator structure.

Figure 4 shows the induced line and phase voltages in the generator windings.

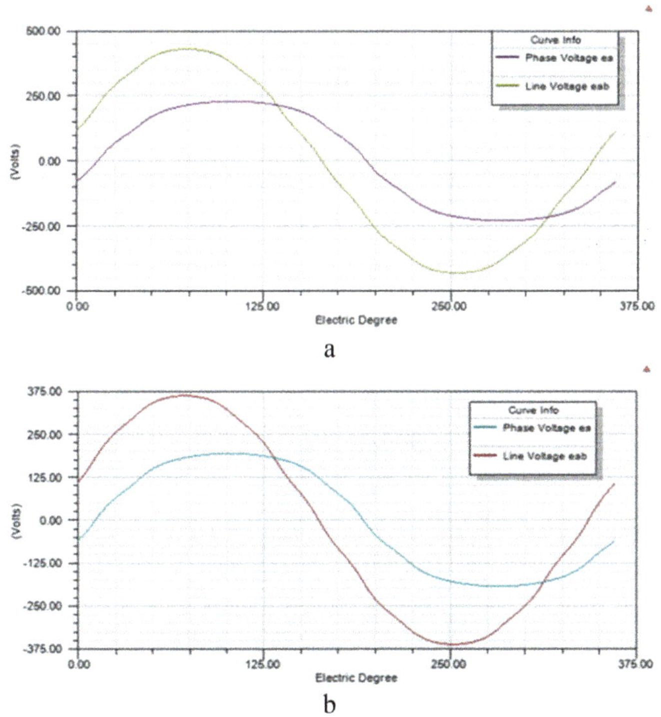

Fig. 4. (a) Induced phase and line voltages at idle, (b) Induced line and phase voltages under load, in the windings of the permanent magnet synchronous generator

Figure 4, which shows the voltage values induced at idle and loaded states in the winding of the synchronous generator, shows the phase-to-phase voltage induced at the g windings as 12000 rpm.

Figure 5 shows the 3 phase currents of the synchronous generator under load.

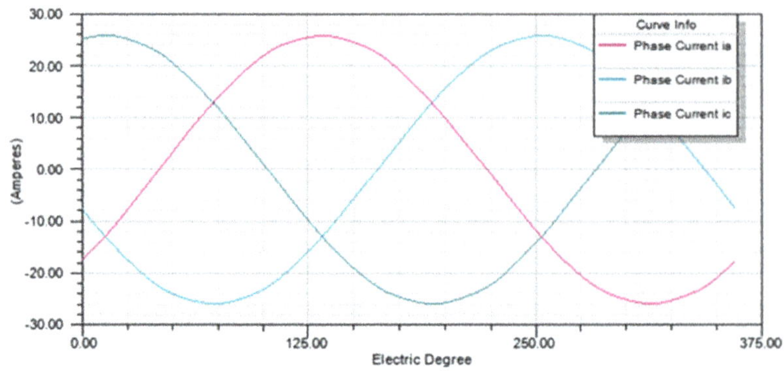

Fig. 5. Load currents of permanent magnet synchronous generator

3 2D and Simplorer Linked Magnetic Analysis of Permanent Magnet Synchronous Generators

The two-dimensional model of synchronous generator was formed according to the values obtained as a result of parametric analyses in the Rmxprt program. On the 2D model, the load was connected to the generator with the simplorer circuit and the performance of the generator in various situations was examined. Figure 6 shows the 2D model of the generator.

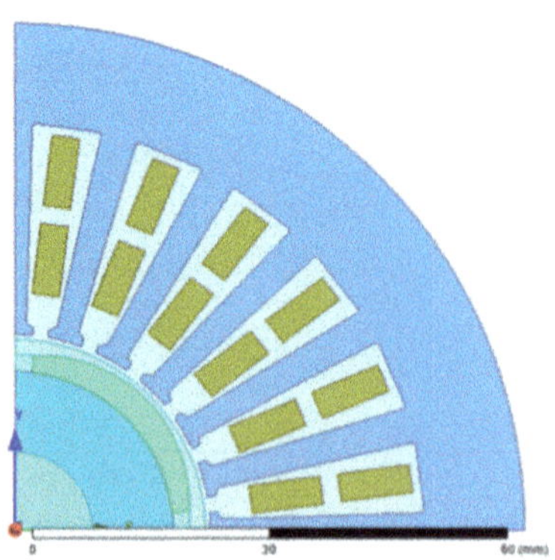

Fig. 6. The 2D model of the synchronous generator

The 2D model of the generator was connected with the simplorer program and a common solution was made to investigate the behavior of the generator in idle, loaded and short circuit situations. For this study, the simplorer circuit diagram in Fig. 7 was installed. In this circuit, the model created in 2D was ensured to work by being embedded in simplorer. As seen in the diagram, ammeters and voltmeters were connected to generator outputs. Then a controlled switch was placed on each phase and the switch was left open in the initial position to make the generator run at idle for the first 15 ms. During this period, the generator's idle current and the terminal voltage were measured. Then, the controlled switch was closed at 15 ms and the generator was connected to a star connected ohmic load. This ohmic load generator was switched to operate at full load. The generator run at full load between 15 ms and 30 ms and the voltage and current values in the generator terminals were read.

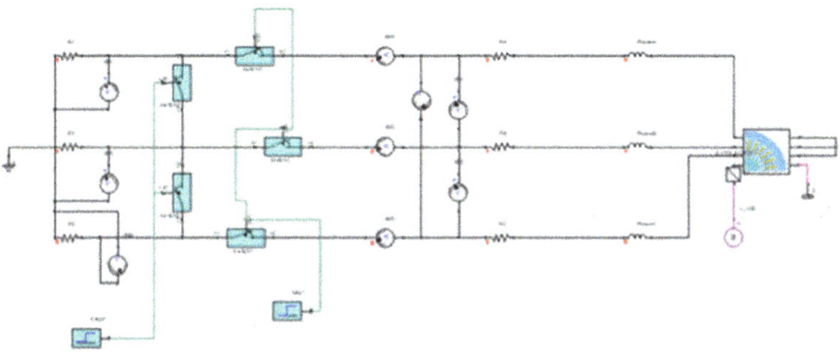

Fig. 7. Simplorer and Maxwell 2D model common connection

Two controlled switches were placed between the generator phases and the generator terminals were short-circuited at 30 ms. The short circuit status of the generator was examined up to 55 ms (Fig. 8).

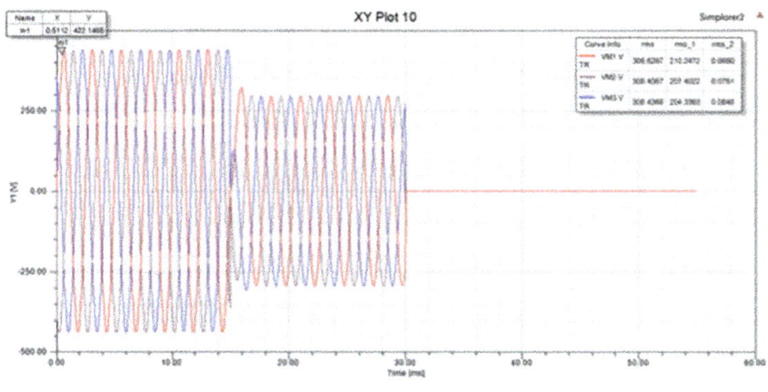

Fig. 8. Voltage status according to load on generator terminals

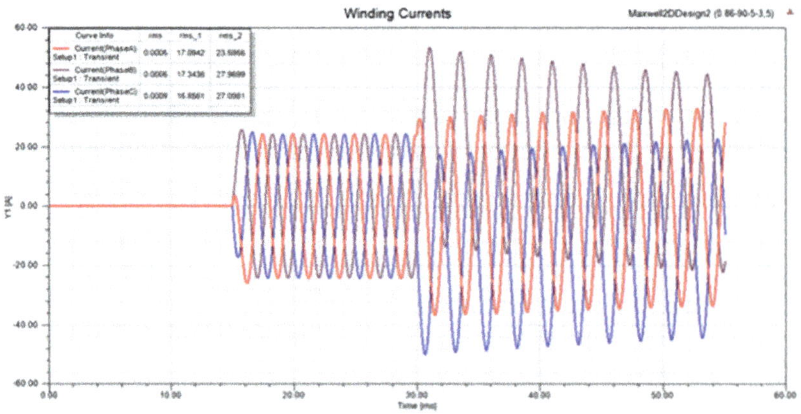

Fig. 9. Load-dependent generator phase currents curves

As shown in Fig. 9, the generator was run empty between 0 and 15 ms, in full load between 15 ms–30 ms, and in short circuit condition between 30 ms–55 ms. Considering the marked points and calculated rms values, it is seen that the circuit and analysis are consistent.

In Fig. 10, it is seen that the generator is empty in 0–15 ms. Therefore, it does not generate power. The generator has full load in 15–30 ms. In this case, the generator supplies power to the circuit around 5 Kw. Between 30 ms and 55 ms, the generator terminals are short-circuited. As a result, the generator does not generate power. The negative torque indicates that the generator generates power.

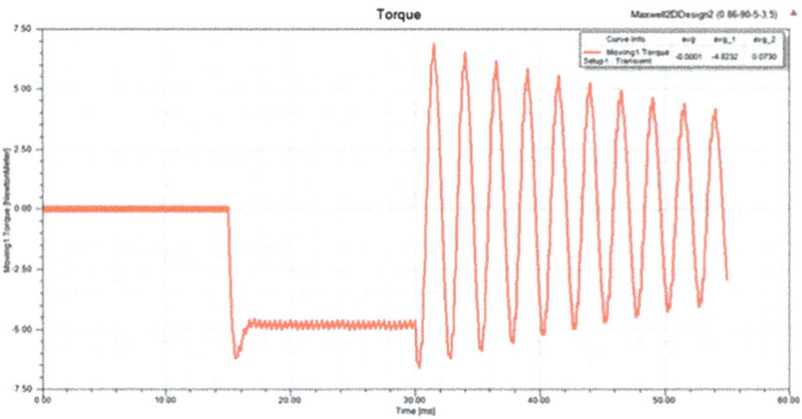

Fig. 10. Load-dependent generator load charts

Figure 11 shows the magnetic analysis of the generator at empty. In the figure, it is seen that the fluxes are at normal level in the stator grooves and yoke of the generator.

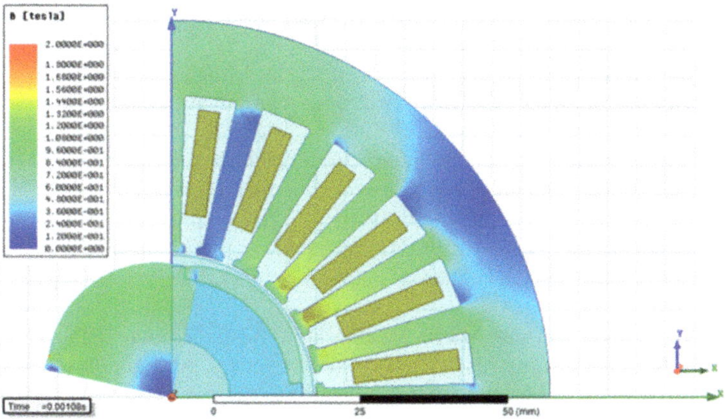

Fig. 11. Load-dependent generator load charts

4 Conclusions

A parametric analysis was performed to determine the magnet sizes to be attached to the surface of the generator whose stator and rotor sizes were determined. 11,067 results were obtained as a result of the parametric analyzes. The magnet sizes that are most suitable for us have been determined according to the method described above and other selected parameters. As a result of these analyzes, the magnet parameters which are ideal for us are were found to be Embrace: 0.86 Offset: 5 mm. Thickness: 3.6 mm.

Transient analyzes were performed in 2D analyzes of the High Speed Generator whose analytical design was completed and parametric analyzes were performed, and the results of this analysis were evaluated to determine the dimensions of the generator prototype to be produced. Based on these results, the required magnet dimensions and generator parameters have been determined and it has been proved that it generates the desired 5 kW power and it can easily be able to produce voltage at 12000 rpm without the need for extra cooling, considering the current density value.

Acknowledgments. This study was supported by Pamukkale University Scientific Research Projects Coordination Unit with Project no. 2019KKP007 and 2016FEBE044. We would like to thank PAUBAP for their contribution.

References

1. Özçıra, S.: Sabit Mıknatıslı Senkron Motorun Kontrol Yöntemleri ve Endüstriyel Uygulamaları. Msc. Thesis, Yıldız Technic University, Istanbul (2007, in Turkish)
2. Texas Instruments: Digital Signal Processing Solution for Permanent Magnet Synchronous Motor. Texas Instruments Application Note Literature Number: 044 (1997)
3. Bianchi, N., Bolognani, S.: Unified approach to the analysis and design of an AC motor drive for flux-weakening operations. In: IEEE Industry Applications Conference on Thirty-Third IAS Annual Meeting, vol. 1, pp. 95–102 (1998

4. Slemon, G.R.: Achieving a constant power speed range for PM drives. IEEE Trans. Ind. Appl. **31**(2), 368–372 (1995)
5. Ionel, D.M., Eastham, J.F., Miller, T.J.E., Demeter, E.: Design considerations for permanent magnet synchronous motors for flux weakening applications. IEE Proc. – Electr. Power Appl. **145**(5), 435 (1998)
6. Dhaouadi, R., Mohan, N.: Analysis of current-regulated voltage-source inverters for permanent magnet synchronous motor drives in normal and extended speed ranges. IEEE Trans. Energy Convers. **5**(1), 137–144 (1990)
7. Schneider, T., Koch, T., Binder, A.: Comparative analysis of limited field weakening capability of surface mounted permanent magnet machines. IEE Proc. – Electr. Power Appl. **151**(1), 76 (2004)
8. Yüzer, Ö.E.: Düşük Düşülü Mikro Hidroelektrik Santralleri için Sabit Mıknatıslı Bir Senkron Generatör Tasarımı. Msc. Thesis, Gazi University, Ankara (2011)
9. Boldea, I.: Synchronous generators. In: The Electric Generators Handbook. Taylor & Francis (2005)
10. Bastos, J.P.A., Sadowski, N.: Electromagnetic modeling by finite element methods. In Electrical and Computer Engineering. CRC Press (2003)
11. Ocak, C.: Mikro-Hidroelektrik ve Rüzgar Santralleri için Düşük Devirli Eksenel Akılı Kalıcı Mıknatıslı Generatör Tasarımı ve Analizi. Msc. Thesis, Gazi University, Ankara (2009)

Selection and Training of School Administrators in Different Countries

Fatma Köprülü(✉), Behcet Öznacar, and Nevriye Yilmaz

Near East University, Yakındoğu Blv, Lefkoşa, Nicosia, Cyprus
{fatma.koprulu, behcet.oznacar, nevriye.yilmaz}@neu.edu.tr

Abstract. This study presents the results of the content analysis of articles published in Scopus "Scientific Article Database" Web of Science, Ulakbilim, Yök and Google Scholar from 2010 to 2018. The research contains 277 articles. These articles were examined and subsequently screened as based on the keywords "education administration", "school administrators", "selection of administrators", "training of administrators" and "content analysis". As a result of the screening, 54 articles were reached. Content analysis is a method providing necessary indicators and a glimpse of multiple articles for authors, readers and others in the relevant field. In the content analysis research, the years of publication of the articles, journals they were published in, the number of authors, the number of references, the number of keywords, the country the research was conducted in, the method used, the number of samples, data collection methods and analysis methods were examined.

As a result of the research, it was concluded that in the United States, England and many foreign countries, a master's degree is obligatory to be an educational administrator. However, the related degrees are not adequately taken into consideration by the Ministry of Education in Turkey.

Keywords: Education administration · School administrators · Selection of administrators · Training of administrators · Content analysis

1 Introduction

Today, the school headmaster is not only seen as the director of the institution but also as the leader in the institution he manages. Therefore, the appointment and training of school administrators is seen (accepted) as an important factor in improving the quality of education systems. The selection of candidates who can perform leadership duties in educational institutions will facilitate the realization of determined goals regarding education at both institutional and national levels (Süngü 2013). The United States is a pioneer in the education of school administrators around the world. Referring to Turkey, it is clear that a serious policy has emerged in this direction (Steel 2002). In recent years, efforts to educate school administrators have been given great importance to ensure the effectiveness of the schools in developed countries (Özmen 2002). In some countries, national leadership institutes have been established to examine the new approaches that have emerged under the leadership of education, to train qualified

D. J. Hemanth and U. Kose (Eds.): ICAIAME 2019, LNDECT 43, pp. 643–653, 2020.
https://doi.org/10.1007/978-3-030-36178-5_53

leaders in schools, to meet the educational needs of these leaders and to allocate resources for training programs for leaders. The content of standards and training programs in institutes and the way in which they are evaluated are taken into consideration by many countries. Research shows that in many countries, "*it is difficult to find qualified candidates for school directorate, the training programs for school directorate are usually inadequate, the work to improve vocational development and knowledge is not continuous and can not be carried out in coordination, barriers to vocational saturation are increasing day by day, and many trainers dedicated to their profession are leaving their jobs at an early age* (Valentine et al. 2004)."

Educational institutions work to educate individuals according to the set objectives and qualifications. In order to determine the degree of success of the applications, institutional evaluation and supervision are required (Bozkurt 1995). As Taymaz (2003) stated, the evaluation of the operation of a system for the purpose of the system is of great importance for the integrity of the system and the product that comes out of the system. In Turkey, 1962, according to the decisions of the Central government and the 7th National Board of Education, it can be said that it was the first serious emphasis on the growing needs of educating school administrators. The education programs of University of Ankara Faculty of Education and Hacettepe University Department of Education, which were established after this date, are important resources for school administrators. One of the most important of these programs, the Educational Administration and Supervision undergraduate programs were closed in 1997 during the restructuring of the Council of Higher Education. Education at the undergraduate level was considered as the choice of being a school administrator (Çelik 2002). Şimşek (2004) defined our administrator selection policy which is the scene of the complexity of the regulations in three periods; as the apprenticeship model, the Educational Sciences model that emerged in the 1970s and the exam model that emerged in the 1998 regulation. The general conditions to be sought as managers are as follows: "To have a higher education, teaching and the civil service nomination to be removed, including administrative positions, to have at least three years of teaching experience, to have the conditions to be a teacher in the school to be appointed, the last three years not to be removed from the administrative position as a result of investigation, not to have any problem related to the obligatory work and to be successful in the selection exam. There are many exams to be school principals in Japan. These exams are held for each level of management and courses are arranged to prepare candidates for the exam (Bartell and Willis 1987). In Japan, school principals are trained through in-service training. This training is provided in regional and local education centres (Hara 1994). The National Teacher Development Center in Japan organizes training programs on urgent and important issues for managers and assistant managers in each region (MEXT 2010). In England, the concepts of "in-service training" and "continuous professional development" can be used interchangeably regarding the education of school principals. The content of these programs is based on the roles and responsibilities of school principals and modules based on management theory and practical management skills (Balcı and Çınkır 2002). In addition, in England, prospective principals are observed when giving a lesson or directing a meeting. Activities such as working with an experienced principal and taking part in school development projects are carried out during the training process (Balcı et al. 2007; Thody et al. 2007; Day et al. 2000). Educational management in Germany is carried out locally by states. Teachers are appointed by the government and accepted as officers. Köprülü et al. (2017) stated that in North Cyprus, the managers and lecturers should attend seminars in and out of the school in order to improve their performances in their professional education life. Owing to designated strategies, while managers could make inspections in a more sound way, lecturers could present the curriculum in a more disciplined and organized way. The students, on the other hand, could benefit from the stability which emerged as a result.

In Germany, school principals are chosen from teachers who have teaching qualifications at a certain level and who have teaching and management experience for a certain period of time. The French education system is managed by the Minister of Education. There are 28 local academies in France and each region is built around a university. The rectors, who are the principals of the University, are responsible for regional education and appoint candidates to administrative positions in their region. It is the rector's responsibility to appoint the members of the review commission, to supervise the audits and to chair the academic council serving in the cabinet (Erdoğan 2003).

2 The Aim of This Study

The aim of this study is to analyze the content of 54 articles published in social sciences between 2010–2018 in Scopus, Web of science, Ulakolim, Yök tez and Google scholar data bases by content analysis method, to examine the distribution of related articles by science, journals, author number, the number of sources, the country used in the research, the method used, the years of publication, the number of keywords, the methods of analysis, the method of data collection, the distribution according to the types of publications and the distribution of the criteria for selection and training of school administrators in different countries. In addition, to examine the selection and training of school administrators in different countries, and in this way to reveal the weaknesses and strengths of the practices among the countries and to make recommendations to improve these practices. To compare and contrast education in different countries, comparison method is used regarding educational applications. With this method, it is possible to see the applications in other countries on currently and to draw conclusions. In this study, some countries selected to make comparisons among world countries on the selection and training of school administrators were examined and then the similarities and differences in the dimensions were discussed. The countries selected for this study are Turkey, the United States, Japan, the United Kingdom, Germany, France, Finland, Belgium, Denmark.

3 Methodology

This study contains the content analysis of quantitative research in the field of Educational Sciences and Social Sciences. Büyüköztürk et al. (2012) claim that content analysis is a systematic and reproducible method in which a text is summarized into smaller categories than the whole, based on predetermined rules. "The basic process in content analysis is to gather data that are similar to each other within the framework of specific concepts and themes and to interpret them in a way that the reader can understand." (Yıldırım and Şimşek 2011, p. 227).

In this study, 54 articles from 277 articles containing education management, motivation and school keywords, which were published in Scopus, Web of Science, Ulakbilim, YÖK and Google Scholar databases between 2010–2018, were examined.

3.1 Data Collection Tool

In the research, a classification form was used as a data collection tool created by the researcher. The classification of the articles included in the study prepared in the classification form consists of the journals, the number of authors, the number of sources, the country of research, the method used, the years of publication, the number of keywords, the methods of analysis, the method of data collection, distribution by publication types and schools in different countries in terms of criteria for selection and training of managers.

3.2 Statistical Analysis of Data

In the research frequency analysis of descriptive test statistics was used. In the research, the distribution of articles, the journals, the number of authors, the number of sources, the country of research, the method used, the years of publication, the number of keywords, the methods of analysis, the method of data collection, the distribution by type of publications and the selection and the training of school administrators in different countries are determined by frequency analysis and the results are illustrated with frequency distribution tables in terms of the distribution criteria. While analyzing the articles, the information obtained from the classification form is classified within them. According to the continents where the research is conducted, it is classified according to the types of research, including quantitative, qualitative and mixed and classified according to survey, interview and literature review in terms of data collection method. In addition, the test statistics which are used for reviewed articles were also classified.

4 Results

The distribution of the study according to the disciplines of science is given in Table 1 below. 54 (19.49%) of the studies were in social sciences, 38 (13.71%) in the field of business administration, 55 (19.85%) in the field of medicine, 42 (15.16) in the field of dentistry, 38 (13,71%) in the field of computer science, 15 (5,41%) in the field of psychology, 14 (5,05%) in the field of engineering, and 21 (7,58%) in economics. In the light of the findings, most of the studies are in the field of medicine, social sciences and dentistry, respectively.

Table 1. Distribution of publications in scientific disciplines

Distribution of publications in scientific disciplines	Number (n)	Percentage (%)
Social sciences	54	19,49
Field of business administration	38	13,71
Field of medicine	55	19,85
Field of dentistry	42	15,16
Field of computer science	38	13,71
Field of psychology	15	5,41
Field of engineering	14	5,05
Field of economics	21	7,58
Total	277	100,00

When Table 1 is examined, it is observed that 11.1% of the research articles were published in 2010, 9.25% in 2011, 11.1% in 2012, 11.1% in 2013, 14.8% in 2014, 7.4% in 2015, 11.1% in 2016, 13% in 2017 and 9.25% in 2018 (Table 2).

Table 2. Distribution of articles published according to the date of publication

Year of publication	Number (n)	Percentage (%)
2010	6	11,1
2011	5	9,25
2012	7	13
2013	6	11,1
2014	8	14,8
2015	4	7,4
2016	6	11,1
2017	7	13
2018	5	9,25
Total	54	100,00

When Table 3 is examined, the distribution of the articles included in the research according to the journals they published in, it is determined that 18,52% of them are published in the Journal of Educational Administration.

When Table 4 is examined, it can be seen that 33,33% of the articles included in the study have one author, 25,93% have two authors, 16,67% have three authors, 14,82% have four authors and 9,25% have five authors or more.

Table 5 shows the distribution of the articles included in the research given according to the countries where the articles written, and it was determined that 31.49% written only in Turkey, 25.93% written only in Turkey and European countries, 14.82% written in Turkey and Asian countries, 11.11% written in African countries, 9.25% written in The Republic of South Africa and 7.40% in Australia.

Table 3. Distribution of articles published in the field of social sciences according to the journals published

Name of the journal	No (n)	Percentage (%)
Journal of Educational Administration	10	18,52
Turkish Journal of Educational Studies	6	11,11
Development and Learning in Organizations an International Journal	4	7,40
Education and Science TEDMEM	4	7,40
International Journal of Educational Management	3	5,55
Mustafa Kemal ÜniversitesiSosyalBilimlerEnstitüsüDergisi	3	5,55
TÜBİTAK DestekliAraştırmaProjesi	3	5,55
EğitimveBilim	2	3,71
Ahi EğitimÜniversitesiKırşehirEğitimFakültesiDergisi	2	3,71
Mersin ÜniversitesiEğitimFakültesiDergisi	2	3,71
Gazi EğitimFakültesiDergisi	2	3,71
SakaryaÜniversitesi Journal of Education	2	3,71
Mehmet Akif ErsoyÜniversitesiEğitimFakültesiDergisi	2	3,71
AİBÜ SosyalBilimlerEnstitüsüDergisi	2	3,71
KuramsalEğitimbilim	1	1,85
European Journal of Education	1	1,85
History of Educational Review	1	1,85
International Journal of Educational Research ELSEVIER	1	1,85
International Journal of Leadership in Education ROUTLEDGE	1	1,85
International Journal of Public Administration ROUTLEDGE	1	1,85
Journal of Education Policy	1	1,85
Total	54	100,00

Table 4. Distribution of articles published in the field of social sciences by the number of authors

Number of authors	Number (n)	Percentage (%)
One author	18	33,33
Two authors	14	25,93
Three authors	9	16,67
Four authors	8	14,82
Five authors and more	5	9,25
Total	54	100,00

Table 6 shows the distribution of articles included in the study according to the number of sources used. It was found that 46.3% of the articles were used in 30 and below sources, 18.52% in between 31–50, 33.33% in between 51–100, 1.85% in 101 and over.

Table 5. The distribution of the articles in social sciences by countries

Country	Number (n)	Percentage (%)
Turkey	17	31,49
Turkey and European Countries	14	25,93
Turkey and Asian Countries	8	14,82
African Countries	6	11,11
Republic of South Africa	5	9,25
Australia	4	7,40
Total	54	100,00

Table 6. Distribution of articles by source number

Source number	Number (n)	Percentage (%)
30 and below	25	46,3
Between 31–50	10	18,52
Between 51–100	18	33,33
101 and more	1	1,85
Total	54	100,00

Table 7 provides the distribution of articles according to the number of keywords used. 66.67% of the articles 3 or less, 22.23% of articles 4–5, 7.30% of articles between 6–10 and 3.70% of them is not specified in the table.

Table 7. Distribution of articles by keyword number

Keyword number	Number (n)	Percentage (%)
3 and below	36	66,67
In 4–5	12	22,23
Between 6–10	4	7,40
Not specified	2	3,70
Total	54	100,00

When Table 8 is examined, it is determined that 44,45% of the articles included in the research are qualitative, 29,62% are comparative, 18,52% of them are quantitative and 7,41% of them are document analysis method.

In Table 9, it was found that 44.45% of the articles included in the study used semi-structured interview form method, 37,03% of them literature review and 18,52% of them questionnaire method as data collection tools.

Table 10 presents the distribution of articles according to data analysis methods, content analysis in 33.33%, comparative analysis in 25.93%, descriptive analysis in 12.97%, correlation and factor analysis 11.11%, compilation in 9,26% and in 7,40% analysis method is not specified.

Table 8. Distibution of articles by types of research

Type of research	Number (n)	Percentage (%)
Qualitative	24	44,45
Comparative	16	29,62
Quantitative	10	18,52
Document analysis	4	7,41
Total	54	100,00

Table 9. Distribution of articles by data collection methods

Data collection method	Number (n)	Percentage (%)
Semi-structured interview	24	44,45
Literature review	20	37,03
Questionnaire method	10	18,52
Total	54	100,00

Table 10. Distribution of articles by analysis methods

Analysis method	Number (n)	Percentage (%)
Content analysis	18	33,33
Comparative analysis	14	25,93
Descriptive analysis	7	12,97
Correlation analysis-factor analysis	6	11,11
Compilation	5	9,26
Not specified	4	7,40
Total	54	100,00

Table 11 indicates that choosing and educating school administrators included in this study are needed to be teachers. While the necessity of in-service training in the Turkish Republic of Northern Cyprus is not sought, it is seen that it is necessary in all other countries covered by the research. As can be seen from Table 11, in most of the countries involved in the study, at least a higher education requirement is needed to be a school administrator. However, a master's degree is needed to be a school administrator in the UK, USA, Japan and Finland. In Turkey, France, North Cyprus and South Africa central government is responsible from choosing and appointing school administrators. In Germany, North Cyprus, Finland and South Africa, there is no need for administrative experience in choosing and school administrators.

Table 11. Criteria for selection and training of school administrators in different countries

Countries	Predivted education	Obligation of exams	Selection and assignment authorization	Necessity of in-service training	Necessity of administrative experience	Condition for teaching experience	Requirement of administrative certificate
England	Y.Y.L.	G	Y	G	G	G	G
Turkey	Y.Ö	G	M	G	G	G	G.M
Germany	Y.Ö.	G	Y	G	G.M	G	G.M
The USA	Y.Y.L.	G	Y	G	G	G	G
France	Y.Ö.	G	M	G	G	G	G.M
Japan	Y.Y.L.	G	Y	G	G	G	G
Belgium	Y.O	G.M	Y	G	G	G	G
North Cyprus	Y.Ö	G	M	G.M.	G.M.	G	G.M
Belgium	Y.Ö.	G.M	Y	G	G	G	G
Finland	Y.Y.L.	G.M	Y	G	G.M.	G	G.M
Denmark	Y.O.	G	Y	G	G	G	G
Republic of South Africa	Y.Ö.	G.M	M	G	G.M	G	G.M

Needed: G
Not needed: G.M
Central: M
Local: Y
Master degree in Administration: Y.Y.L.
Bachelor degree: Y.Ö

5 Conclusion and Discussion

The survey carried out with keywords "education management", "school", "manager", "manager selection", "educating administrator" and "content analysis", there are different categories from 277 articles published between 2010 and 2018.

According to the findings, it is seen that the Journal of Education Administration is in the first place. The distribution of publications according to the disciplines is examined and 54 publications are determined to be in the field of social sciences. When the distribution of the research studies examined according to the country/region, it can be seen that 17 of them are in Turkey and 14 of them are Turkey and Europe. No studies have been conducted on this subject in Northern Cyprus.

When the research approaches of the publications were examined, it was concluded that qualitative research method was mostly used, but the comparative research method was also extensively used. Based on this result, it can be said that researchers prefer to work with people they can easily access. Driscoll (1995) stated that research in the field of education should be open to different research techniques, and studies should not be conducted with a single research technique (Göktaş and Diğ 2012).

As a result of the researches, it was emphasized that the educational status envisaged in choosing school administrators in England, USA, Japan and Finland is at least a master's degree to be a school administrator. Akın et al. (2012)'s study shows that in the United States of America, Japan and the UK, a master's degree is required to be a school administrator, while in Turkey, a master's degree and doctorate degrees from related fields have not been taken into consideration by the ministry.

According to Bush and Jackson (2002), it is generally accepted that in developed countries, school administrator education should be given at post graduate level. In Turkey, Germany, France, Northern Cyprus, Finland and South Africa, it has been concluded that no administrative certificate is required in order to be a school administrator. According to Bursalıoğlu (2000), problems related with educating school administrators are not only faced in Turkey but in developed countries as well. In the United States, England and many foreign countries, it has been concluded that a master's degree is required to be a school administrator, but it is not taken into consideration by the Ministry in Turkey.

6 Recommendations

The following can be recommended according to the results of the research:

1. The process of selecting and appointing administrators should be reviewed; administration should be considered as a profession and the people who have received pre-service training of this work should be included.
2. In addition to a central examination that measures the knowledge of management and leadership of education, a measurement instrument technique that measures the characteristics of experience, education and interest should be developed and implemented in a fair manner.
3. A new regulation, which considers the expectations of school administrators and general satisfaction should be studied.

References

Akın, U.: Okul Yöneticilerinin Seçimi ve Yetiştirilmesi: Türkiye ve Seçilmiş Ülkelerden Farklı Uygulamalar, Karşılaştırmalar. AİBÜ Sosyal Bilimler Enstitüsü Dergisi, Güz 2012, Cilt:12, Yıl:12, Sayı:2 **12**, 1–30 (2012)

Balcı, A.: Etkili okul. PegemA Yayıncılık, Ankara (2007)

Balcı, A., Çınkır, Ş.: 21.yüzyıl eğitim yöneticilerinin yetiştirilmesi sempozyumu kitapçığı (16–17 Mayıs 2002). Ankara: Ankara Üniversitesi Eğitim Bilimleri Fakültesi Yayınları No: 191 (2002)

Bartell, C.A., Willis, D.B.: American and Japanese Principals: A Comparative Analysis of Excellence in Instructional Leadership. Annual Meeting of the American Educational Research Association, Washington, D.C. (1987)

Bozkurt, E.: Necessity for evaluation in education. J. Educ. Manag. (1995). Pegem Publishing House, Ankara

Bursalıoğlu, Z.: Eğitimde Yönetimi Anlamak Sistemi Çözümlemek. Pegem A Yayıncılık, Ankara (2000)

Bush, T., Jackson, D.: A preparation for school leadership: international perspectives. Educ. Manage. Adm. Leadersh. **30**, 417–429 (2002)

Büyüköztürk, Ş., Kılıç Çakmak, E., Akgün, Ö.E., Karadeniz, Ş., Demirel, F.: Bilimsel Araştırma Yöntemleri. 11. Baskı. Ankara: Pegem Akademi. s. 177, 265 (2012)
Çelik, V.: Eğitim Yöneticisi Yetiştirme Politikasına Yön Veren Temel Eğilimler. 21. Yüzyıl Eğitim Yöneticilerinin Yetiştirilmesi Sempozyumu. Ankara (2002)
Day, C., Harris, A., Hadfield, M.: Grounding knowledge of schools in stakeholder realities: a multi-perspective study of effective school leaders. Sch. Leadersh. Manag. **21**(1), 19–42 (2000)
Driscoll, M.E.: Thinking like a fish: the implications of community for parents and schools. In: Schneider, B., Cookson, P. (eds.) Transforming Schools, pp. 209–236. Garland Publishing, New York, London (1995)
Göktaş, Y., Hasançebi, F., Varışoğlu, B., Akçay, A., Bayrak, N., Baran, M., Sözbilir, M.: Türkiye'deki eğitim araştırmalarındaki eğilimler: bir içerik analizi [Trends in Educational Research in Turkey: A Content Analysis]. Kuram ve Uygulamada Eğitim Bilimleri 12(1), 443–460 (2012)
Erdoğan, İ.: Çağdaş Eğitim Sistemleri. Sistem Yayıncılık, İstanbul (2003)
Köprülü, F., Aksoy, Y., Öznacar, B.: An analysis of preparatory school managers' strategies for managing change. Int. J. Econ. Perspect. **11**(1) (2017)
MEXT: Improvement of the Quality of Teachers (2010). http://www.mext.go.jp/english/org/struct/015.htm
O'Hara, D.G.: The Superintendent's first contract. Sch. Adm. **51**(7), 19–21 (1994)
Özmen, F.: Okul Müdürlerinin YetiştirilmesiGelişmiş Ülkelerdeki Uygulamalardan Örnekler. 21. Yüzyıl Eğitim Yöneticilerinin Yetiştirilmesi Sempozyumu. Ankara (2002)
Süngü, H.: Recruiting and preparing school principals in Turkey, Germany, France and England. Sakarya Univ. J. Educ. **2**(1), 33–48 (2013)
Şimşek, H.: Eğitim yöneticilerinin yetiştirilmesi: Karşılaştırmalı örnekler ve Türkiye için öneriler (2004). http://www.hasansimsek.net
Steel, L., Levine, R.: Educational Innovation in Multiracial Contexts: The Growth of Magnet Schools in American Education. U.S. Department of Education, Washington, D.C. (1994)
Taymaz, H.: Okul Yönetimi. Pegem A Yayıncılık, Ankara (2003)
Thody, A., Papanaoum, Z., Johansson, O., Pashiardis, P.: School principal preparation in Europe. Int. J. Educ. Manage. **21**(1), 37–53 (2007)
Valentine, J., Clark, D., Hackmann, D., Petzko, V.: A National Study of Leadership in Middle Level Schools, Volume II: Leadership for Highly Successful Middle Level Schools. National Association of Secondary School Principals, Reston, VA (2004)
Yıldım, A., Şimşek, H.: Sosyal bilimlerde nitel araştırma yöntemleri. Seçkin Yayıncılık, Ankara (2011)

Security on Cloud Computing Using Pseudo-random Number Generator Along with Steganography

Moolchand Sharma[1(✉)], Suman Deswal[2], Jigyasa Sachdeva[1], Varun Maheshwari[1], and Mayank Arora[1]

[1] Maharaja Agrasen Institute of Technology, Delhi, India
sharma.cs06@gmail.com, sumandeswal.cse@dcrustm.org, jigyasa.sachdeva24@gmail.com, varunmaheshwari27@gmail.com, arora.mayank7@gmail.com
[2] Deenbandhu Chhotu Ram University of Science and Technology, Murthal, Sonepat, Haryana, India

Abstract. Cloud computing is a convenient and flexible method to use computing assets such as storage, database and networks while "on the cloud", i.e. to access over the internet. Data communicated is not secure and hence a technique was formulated which included- standard encryption techniques is used to encrypt the data, followed by applying steganography and then finally, splitting it across two or more hosts to transfer it. Steganography refers to encrypting information in the pixels of an image. In the spatial format, an alternative to embedding information sequentially in the picture elements of a cover image, a randomized method- PseudoRandom Number Generator (PRNG) is implemented to improve the overall security and efficacy of the system.

Keywords: Cloud computing · Steganography · AES (Advanced encryption standard) · PRNG (Pseudo-Random Number Generator) · PSNR (Peak Signal-to-Noise Ratio) · Splitting algorithm

1 Introduction

The word 'Steganography' has its origins from the Greek words- "stegos" and "grafia" meaning 'cover' and 'writing' respectively. Hence, steganography refers to coveted writing. It is a field dealing with hiding secret information in a media [18]. One of the techniques used in Steganography is Image compression. This further has two types- lossy compression and lossless compression. Lossy compression does not conserve the righteousness of the image and distorts it where as Lossless compression is able to conserve the authenticity of the image data correctly. Graphic Interchange Format and BitMaP are examples of lossless compression formats [19]. JPEG is an example for lossy compression format.

Steganography embeds a message inside a cover image. The following criteria define the requirements of an algorithm-

D. J. Hemanth and U. Kose (Eds.): ICAIAME 2019, LNDECT 43, pp. 654–665, 2020.
https://doi.org/10.1007/978-3-030-36178-5_54

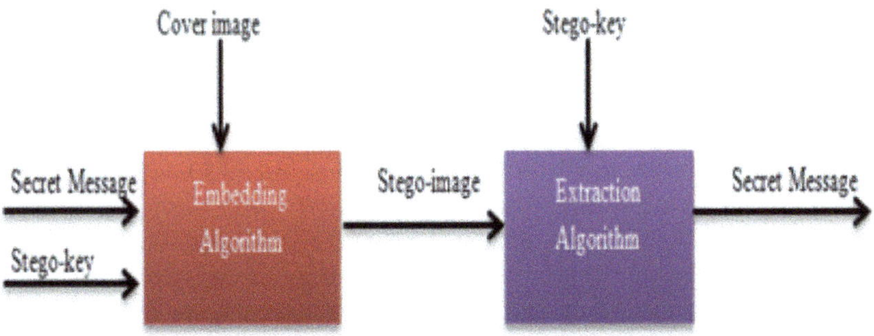

Fig. 1. The basic model of steganography

- Invisibility- The cover image embedding the image should look exactly the same as it was without the message. Hence, the invisibility quotient should be high.
- Payload Capacity- In traditional methods, less information was capable of being encrypted. Payload capacity refers to capacity of a technique to hold information. This should be high.
- Hiding Capacity- The amount of data that could be covered in proportion with the amount of the cover object is known as the Concealing Capacity. More concealing capacity leads to the utilisations of smaller cover images and hence, the data communication needed to publicise the stego image is abated.
- Perceptual Transparency- Not being able to detect a hidden message is termed as perceptual transparency.

Figure 1 depicts a basic model of steganography. Steganography process has three main components- carrier, message, and password. Plain text, cipher text, or image is the types of data in which the message can be sent by the sender. The password makes sure that only the prescribed receiver can decode the received message by using the unique Stego-key. Stego-image is the cover-image which is formed with the secretly hidden message.

2 Literature Review

Image Steganography can be done in two types of domains- spatial and transform. Spatial domain deals with changing the pixel's binary values to message bits' ASCII values. Transform domain deals with changing the frequency coefficients of the image in which the message is to be embedded.

One of the very common methods used in the spatial domain is Least Significant Bit Technique. In this the rightmost bit of every byte of a pixel is sequentially alternated with the message bits. It is a convenient and easy technique. Any image is a 2-D matrix with scales of colors in 1 or 3 bytes. 1 byte use black and white scales whereas 3 bytes employ RGB models. For both these models, there exist LSB techniques with respective pros and cons. A 3- byte insertion would increase the size upto 5–6 Mb, increasing suspicion. A 1-byte insertion would lead to severe distortion as the changes are very easily highlighted. But since the cover page pixels are sequentially manipulated in the medium image, the encryption and decryption of information becomes very easy.

The following are the drawbacks of LSB Technique-

- The most basic of LSBs insertion for 24-bit pictures inserts 3 bits/pixel. Since every pixel is 24 bits, we can hide 3 hidden bits/pixel/24 data bits/pixel = 1/8 hidden bits/data bits. So, for this case we hide 1 bit of the embedded message for every 8 bits of the cover image. Hence, the payload capacity of the system is less.
- This method is very insecure and prone to distortions and changing the authentic data meaning. This is because even a slight amount of noise causes huge distortion. Hence, the system is not robust.

Hence, because of the above disadvantages of LSB; a new technique is implemented for security while sending information over the cloud. The technique implemented is PRNG- Pseudo-Random Number Generator.

3 Preliminary Work

In the previous paper, the prime objective was to create a protected data storing technique that overpowers the traditional approach used for securing information transmission in cloud computing. Services such as- applications, storage servers and networks is provided by cloud computing [3]. Nowadays most organisations are employing cloud services because of the convenience and on the spot services it provides [2]. On using the cloud, a workstation interacts with a wide variety of servers connected in the network. Cloud computing can be managed because it is convenient and effortless to the service provider interaction [1, 4]. Steganography and the proposed splitting algorithm were important methodologies used in the system design.

In this extended paper, we have cited a more efficient technique using the previous version to make the entire system more efficient.

3.1 Existing System

The capacity to safeguard and protect information is crucial to the growth of electronic commerce and data security. Cryptography refers to hiding the message in a stego media like images, audio, video and thus creates a encrypted data format. The information is converted into an unreadable code during encryption. AES- Advanced Encryption Standard is a symmetric key encryption standard which is used throughout the world to secure data and maintain the data integrity.

AES (Advanced Encryption Standard)

AES (Advanced Encryption Standard) is a symmetric-key encryption standard for secret key encryption. In other words, AES is based on a design principle known as Substitution permutation network [11]. It is comparably fast in both software and hardware. The algorithm uses a combination of XOR Operation (Exclusive-OR), octet substitution with an S-box, row and column rotations and a Mix Column. The basic algorithms of AES include Sub Bytes, Shift Rows, Columns, Add Round Key.

Algorithm:

- Step 1: Produce the key from the inputted message.
- Step 2: Generate the cipher message using AES.
- Step 3: Transmit this message to the recipient.

This process is reversed at the receiving end to fetch the original message.

Steganography

Steganography is a technique used to hide useful information as stego-text in images called a stego-image. A user at the receiving end having the key is able to fetch the concealed text using reverse phenomenon [20]. An encrypted message depicts some useful information is being sent but a message concealed within an image is difficult to be attacked because of its perceptual transparency. This is where Steganography has an undue advantage over cryptography [14, 15]. In simple words, we could say that image Steganography exploits the human visual system (HVS). Even though it is slightly difficult to implement, with proper methodologies we can hide informative text in an image [13, 22].

Introduction to terms:

- Cover-media: the medium which acts as a carrier to hide the message
- Secret Message: data or information which is to be hidden
- Stego-media: carrier along with the hidden message

Least Significant bit:

The earlier paper used image Steganography, i.e. an image is used as a cover image. The most important concern is to create a identical stego image and the original image. Any visible changes would violate the basic fundamental of image steganography. The image steganography used in the previous paper used sequential Least Significant Bit method (LSB method). In LSB method changes are made to the least significant bit of each pixel [12, 23]. The value of pixel is replaced with the bit value of the message to create a stego image. All pixel are converted into their respective equivalent binary representation [21]. Consider the following illustration-

Pixel: (10101110 11001010 10111011 00010101 11010010 10001000 10111001 00010101)
Message: (10101111)
Result: (10101111 11001010 10111011 00010100 11010011 10001001 10111001 00010101)

Algorithm:

Step 1: Input the data and change it into its binary representation.
Step 2: Select the carrier image and repeat the same.
Step 3: Fetch the least significant bit of each pixel.
Step 4: substitute lsb value of all pixels with respective message bits.

An image will be generated. This is termed as the stego-image. At the receiving end, this is processed to get back the original information.

Split Algorithm
The splitting algorithm divides the data over multiple hosts which are not connected directly to similar network. The original message can only be merged by the owner who has access to all the hosts and thus can gather the splitted data. In the earlier paper splitting algorithm is applied after applying sequential lsb technique in image steg-nography [5]. The stego image is taken as an input by splitting algorithm to split the data in 'n' numbers and stored in randomly chosen server and sent by different hosts.

System Design

1. Begin.
2. Input the data or information to be transmitted on the cloud.
3. Apply AES Encryption to the data or information
4. Perform steganography using LSB algorithm, i.e. hide the message in a cover image.
5. Using split algorithm divides the stego-image into n parts and send over
6. Recover the splitted n parts of stego- image and combine them to create the stego image
7. Recover the message using de-steganography.
8. Recover and generate the original message using decryption technique.
9. End.

4 Proposed System

Because of the drawbacks mentioned; Least Significant Bit is a traditionally verified less efficient method to perform steganography. LSB has a huge Mean Squared Error which causes distortion to the image. Hence, a new method is required to increase the Peak Signal to Noise Ratio. Hence, any method proposed should have the following characteristics-

- If the data in encrypted before passing it into the image, even if the data is detected; it has to be decrypted first. This encryption technique should be employed in a manner such that the transformation does not cause blurring of the image. The encrypted data should not cause distortion to the image.
- In this paper, we are using randomized placements of bits of the message in the pixels of the image. The technique would use a prime number and select a specific area to embed the information.
- Post image steganography, the image is splitted and sent to two or more hosts.

4.1 Pseudo Random Number Generator

Pseudo random number generator is a technique which generates a series of unique numbers with respect to a set of numbers. Let's consider a prime number- 'p'. The function used to generate these values are- **'x^2 mod p'** for all values of x such that- $2x < p$;

'**p – x^2mod p**' for all other values of x in the range.

For example- Let us take a prime number '11'. For all the integral values of x less than 5.5 (13/2), generate unique numbers by employing the function- x^2 mod p; i.e. {0, 1, 2, 3, 4, 5}. For rest of the integral values {6, 7, 8, 9, 10}, unique numbers are generated by the function- p- x^2 mod p.

For x = 0: x^2 mod p = 0 mod 11 = 0.
For x = 4: x^2 mod p = 16 mod 11 = 5.
For x = 7: p- x^2 mod p = 11 − (49 mod 11) = 11 − 5 = 6.

All the numbers get unique random numbers as shown in Figs. 2 and 3.
Consider the pixels in grey and blue scale of an image, given in the Table 1 below.

$$x * x \% p \quad (3)$$

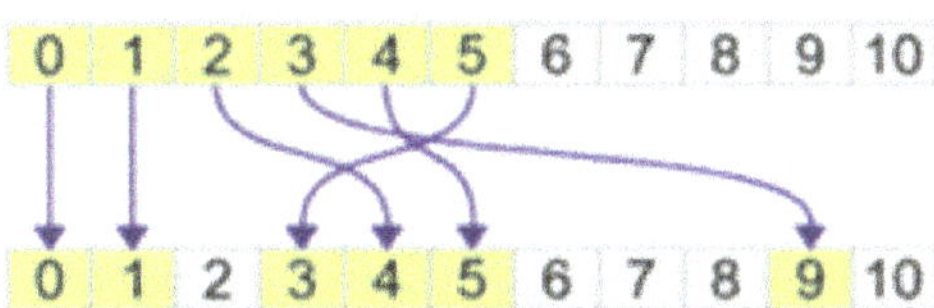

Fig. 2. Quadratic residues of 0, 1, 2, 3, 4, 5

$$p - x * x \% p \quad (4)$$

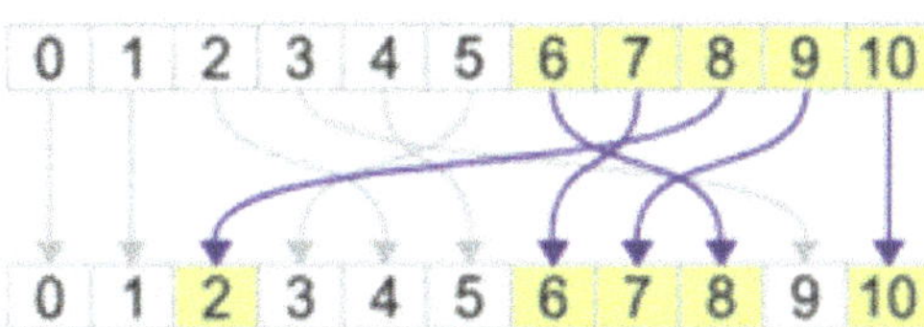

Fig. 3. Quadratic residues of 6, 7, 8, 9 and 10

Table 1. Pixels in grey and blue scale of an image.

Pixel no. in sequence manner	Pixels before embedding				Pixels after embedding	
	Blue pixel	Green pixel	Blue pixel bits	Green pixel bits	Blue pixel bits	Green pixel bits
1	131	137	10000011	10001001	10000001	10001001
2	120	135	01111000	10000111	01111001	10000111
3	106	132	01101010	10000100	01101001	10000101
4	113	131	01110001	10000011	01110000	10000011
5	109	129	01101101	10000001	01101100	10000001
6	111	131	01101111	10000011	01101101	10000010
7	103	127	01100111	01111111	01100100	01111111
8	112	133	01110000	10000101	01110000	10000101
9	108	122	01101100	10000100	01101101	10000101

Let's take an example of embedding the message "abc" in the image. The ASCII code's binary equivalents of the letters are- 01100001, 01100010, and 01100011. We are considering 3-2-3 substitution. 8 bits are divided in slots of 3, 2 and 3 respectively. The random numbered positions generated are- {1, 4, 5, 3, 8, 6, 2, 7, 9}. This means the 1st 3 bits go to the 1st pixel's blue and green bits. The next two bits go to the 4th pixels' blue scale. The last 3 go to the 5th pixel's blue and green scale. The substitution happens in from right to left- Least significant to most significant bits order to decrease the Mean Squared Error value. Amongst the 3 bits substituted in blue and green are- 2 in blue and 1 in green respectively. Hence, the order is: {(2 blue, 1 green), (2 blue), (2 blue, 1 green), (2 blue, 1 green), (2 blue), (2 blue, 1 green) …}.

Consider the pseudo code of the proposed method is as follows:

Input: Cover Image C, Secret Message n, Permutation P
Output: Stego Image S.
Steps:
1. Split the Cover Image C into three layers- Red (R), Green (G), Blue (B).
2. Convert H, and I into blocks; H= {H1,H2,H3...bn},I= { I1,I2,I3...gn... } where each block represents only one pixel.
3. Transform all block from H and I to its ASCII format.
4. Split n into characters, n= {n1, n2, n3….n...}.
5. Take n_i from N, and change it to its binary form
6. Take an array Y of integer numbers.
7. For i =1 to length (Y)
8. Generate RN (i) using PRNG generator by using equations 3rd, and 4th. Every number of these random numbers will represent the position of the pixels in C.
9. End for.
10. Select 8 numbers from RN (i) which located between
{1 to 8}, transfer these numbers in different array called Y.
11. for v=1 to len (M)*3
12. RS (v) =Y (i) +8
13. End for
14. For messageCnt=1 to len(M)
15. L=1
16. Take from B, and from G.
17. H(RS(L),7:8)=(messageCnt ,1:2)
18. I(RS(L),end)= (messageCnt,3)
19. L++
20. H(RS(L),7:8)= (messageCnt ,4:5)
21. L++
22. H(RS(l),7:8)=(messageCnt ,6:7)
23. I(RS(l),end)= (messageCnt,8)
24. L++
25. end for
26. Convert h, and I from binary to decimal system
27. Construct the stego Image S Merge the three layers R, G, B

Implementation Figures

(1) Creating stego image (encoding and decoding)- embedding the message in the image as shown in Fig. 4.
(2) Using Split algorithm - The stego image is split into multiple pieces as shown in Fig. 5.
(3) Transfer of the splitted stego-image – The multiple parts are shared to the desired location to random servers as shown in Fig. 6.
(4) Combining the n-parts – At the receiver end the splitted n parts are combined into a single image as shown in Fig. 7.

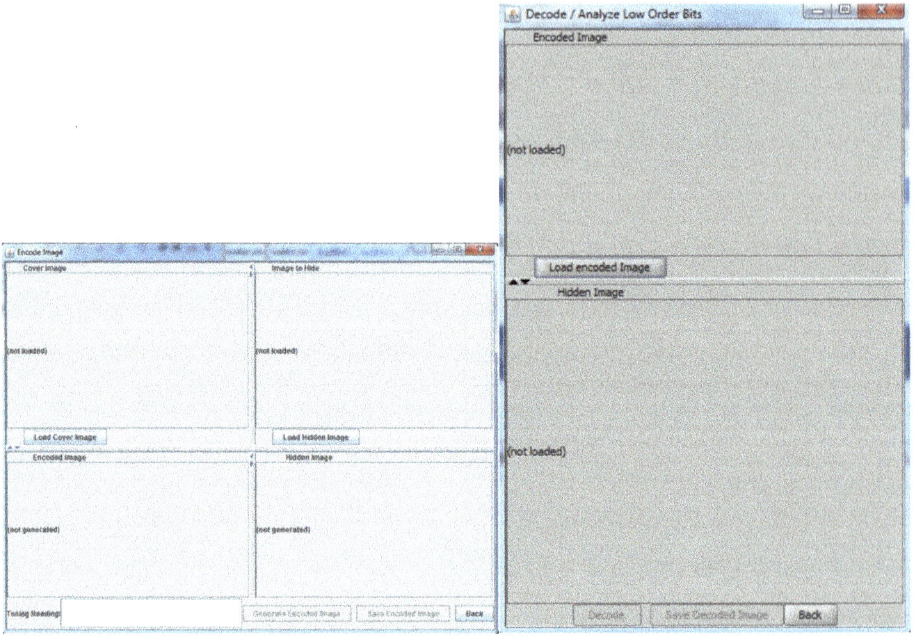

Fig. 4. Creating stego image (encoding and decoding)

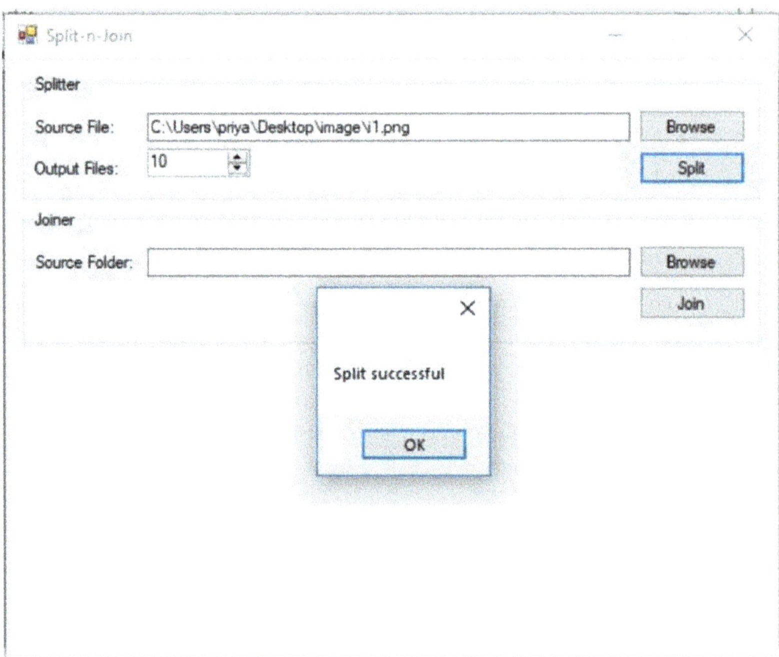

Fig. 5. Stego image is split into multiple pieces

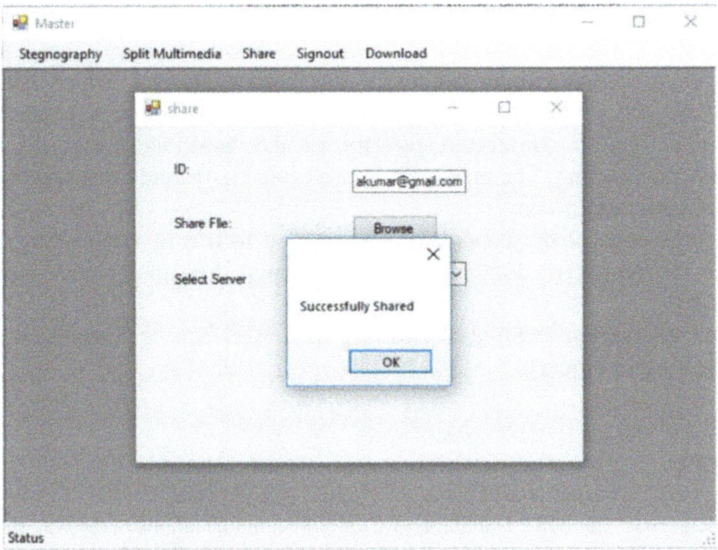

Fig. 6. Transfer of the splitted stego-image

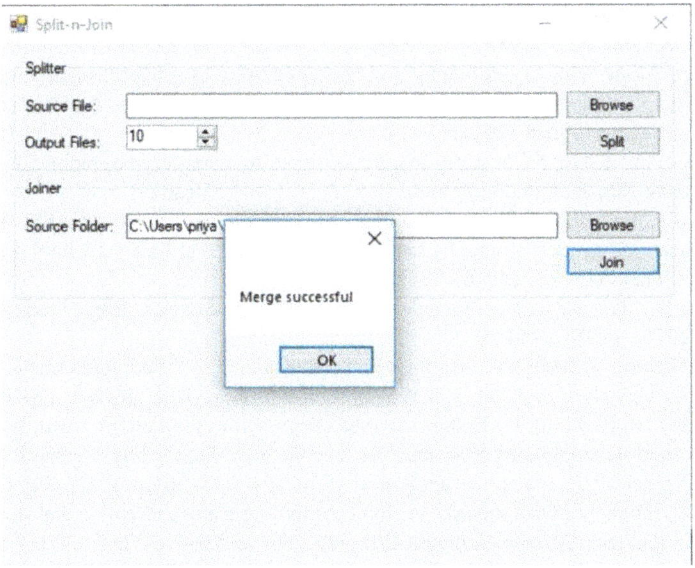

Fig. 7. Combining the n-parts

5 Future Scope and Conclusion

In this paper, a Steganographic technique has been suggested, which has a higher embedding capacity and a improved PSNR as compared to trivial sequential LSB, by using Pseudo Random Number Generator in the least significant bit (lsb), the encryption is more strong. The method is implemented in the previous system design to increase the system efficiency.

The future scope of the paper can be extended to use of various steganography methods in the transform domain as well as spatial domain includes audio, video steganography.

Different encryption techniques stronger than AES can be introduced to further improve the system efficacy.

References

1. Peter Mell, T.G.: The NIST Definition of Cloud, Recommendations of the National Institute, September 2011
2. Columbus, L.: Roundup of Cloud Computing Forecasts, Forbes (2017)
3. Ryan, M.D.: Cloud Computing Privacy Concerns on Our Doorstep, pp. 36–38, January 2011
4. Suruchee, P.A.B.R., Nandgaonkar, V.: A comprehensive study on cloud computing. Int. J. Comput. Sci. Mobile Comput. **3**(4), 733–738 (2014)
5. Garg, P., Sharma, M., Agrawal, S., Kumar, Y.: Security on Cloud Computing Using Split Algorithm Along with Cryptography and Steganography. Springer, Singapore (2019)
6. Daugelaite, J., Sleator, R.D., O'Driscoll, A.: Big data', hadoop and cloud computing in genomics. Science Direct (2013)
7. Shakeeba, P.T., Khan, S.: Security in cloud computing using cryptographic algorithms. Int. J. Innovative Res. Comput. Commun. Eng. **3**(1), 148–155 (2015)
8. Sateesh, G., Lakshmi, E.S., Ramanamma, M., Jairam, K., Yeswanth, A.: Assured data communication using cryptography and steganography. Int. J. Latest Technol. Eng. Manage. Appl. Sci. **5**(3), 545–551 (2016). ISSN 2278 – 2540
9. Radhika, D.U.G., Bajaj, D., Gokhale, U.M.: Int. J. Latest Res. Eng. Technol. (IJLRET) **02**(05) (2016). ISSN: 2454-5031
10. Muthulakshmi, P., Shathvi, K., Aarthi, M., Seethalakshmi, V.: Encrypted image with hidden data using AES algorithm. Int. J. Sci. Eng. Technol. Res. (IJSETR) **5**(4) (2016)
11. Yogeswari, G., Eswaran, P.: Enhancing data security for cloud environment based on AES algorithm and steganography technique. Int. J. Adv. Res. Trends Eng. Technol. **3**(20) (2016)
12. Munish Katoch, R.J.: Image steganography: a review **5**(4) (2016). ISSN (Print) 2319 5940
13. Handa, K., Singh, U.: Data security in cloud computing using encryption and steganography. Int. J. Comput. Sci. Mobile Comput. **4**(5), 786–791 (2015)
14. Alangar, V.: Cloud computing security and encryption. Int. J. Adv. Res. Comput. Sci. Manage. Stud. **1**(5) (2013)
15. Chen, D., He, Y.: A study on secure data storage strategy in cloud computing. J. Convergence Inf. Technol. **5**(7), 175–179 (2010)
16. Sugumar, R., Arul Marie Joycee, K.: Ensure and secure data confidentiality in cloud computing environment using data obfuscation technique. Int. J. Adv. Stud. Comput. Sci. Eng. **6**, 16–21 (2017)

17. Almohammad, A.: Steganography-based Secret and Reliable Communications: Improving Steganographic Capacity and Imperceptibility, Doctor of Philosophy Thesis, Department of Information System and Computing, Brunel University, August 2010
18. Moudgil, S.K., Goel, A.K., Sharma, M.: Steganography on audio wave tenth layer by using signal to noise ratio test and spectrogram analyses. Int. J. Appl. Eng. Res. **13**(4), 1931–1935 (2018). ISSN 0973-4562
19. Ng, K.-H., Liew, S.-C., Ernawan, F.: Colour image steganography using SHA-512 and lossless composition. Int. J. Imaging Robot. **18**. (ISSN 0974-0627)
20. Palathingal, A.G., George, A., Thomas, B.A., Paul, A.R.: Enhanced cloud data security using combined encryption and steganography. Int. Res. J. Eng. Technol. (IRJET) **05**(03) (2018)
21. Sharabyan, M.T., Ghorbani, H.: A new combined method with high security for digital images steganography based on imperialist competitive algorithm and symmetric encryption algorithm. Int. J. Res. Comput. Appl. Robot. **6**(1), 1–12 (2018)
22. Agath, A., Sidpara, C., Upadhyay, D.: Critical analysis of cryptography and steganography. IJSRSET **4**(2) (2018). ISSN: 2395-1990 | Online ISSN: 2394-4099 National Conference on Advanced Research Trends in Information and Computing Technologies (NCARTICT-2018), Department of IT, L. D. College of Engineering, Ahmedabad, Gujarat, India In association with International Journal of Scientific Research in Science, Engineering and Technology
23. Danny Adiyan, Z., Purboyo, T.W., Nugrahaeni, R.A.: Implementation of secure steganography on jpeg image using LSB method. Int. J. Appl. Eng. Res. **13**(1), 442–448 (2018). ISSN 0973-4562
24. Rashmi, D.D.: Securing software as a service model of cloud computing: issues and solutions. Int. J. Cloud Comput.: Serv. Archit. **3** (2013)

Neural Network Prediction of the Effect of Nanoparticle on Properties of Concrete

Şemsettin Kilinçarslan[1](✉), Metin Davraz[1,2,3], Nanh Ridha Faisal[1,2,3], and Murat Ince[1,2,3]

[1] Civil Engineering Department, Engineering Faculty, Suleyman Demirel University, 32600 Isparta, Turkey
semsettinkilincarslan@sdu.edu.tr

[2] Senirkent Vocational School, Isparta University of Applied Sciences, 32600 Isparta, Turkey

[3] Isparta Vocational School of Technical Sciences, Isparta University of Applied Sciences, 32600 Isparta, Turkey

Abstract. Artificial intelligence techniques are used for optimization in engineering studies and it is observed that it gives better results than other classical methods. Artificial intelligence techniques, which are widely used in the solution of civil engineering systems, are examined with their technical aspects and basic principles. To improve the mechanical and physical properties of the concrete, nano and micro scale materials can be added and their effects on concrete properties are investigated intensively. Nanoparticles added to the cement-based materials, it is possible to accelerate the improvement of the early age characteristics of the cement. In this study, 9 different mixtures were designed with nanoparticles and ultra-fine materials such as cement, water, standard sand, super plasticizer, fumed silica, precipitated silica, F-class fly ash and silica fume. Mechanical and physical tests were performed on these samples. Mechanical properties such as pressure and bending strength and physical properties such as water absorption were found. The effect of mechanical and physical properties on artificial neural networks has been tried to be estimated. Concrete mixes; The accuracy of 7-day estimation data was found as $R^2 = 1$ and the error rate was RMSE = 0.00108, the accuracy of the 28-day forecast data was $R^2 = 0.9888$ and the error rate was RMSE = 0.0887. An experimental study with the results obtained using the neural network of the results obtained have been found to show very close values. So, it will be possible to obtain the desired mixture values with an effortless, economical and shorter time.

Keywords: Nano · Concrete · Mortar · NNP · Differential evolution algorithm

1 Introduction

The construction sector is one of the most important sectors for our national economy. Given the foreign trade volume, employment opportunities, direct relations with other sectors and the current global competitiveness of the sector, the strategic importance of the construction sector for our country is even more evident. When building the future, it is extremely important to consider not only technological changes, but also social and

D. J. Hemanth and U. Kose (Eds.): ICAIAME 2019, LNDECT 43, pp. 666–675, 2020.
https://doi.org/10.1007/978-3-030-36178-5_55

demographic changes. The changes that we will experience in the social structure, cultural life and our priorities in the next 10 years give us important clues about how we need the buildings. Considering the basic social needs of the next 10 years and the housing properties that will be needed in the future, it is seen that the most important innovation areas in the construction sector are in the building materials. Giving new and superior properties to existing materials or developing completely new and superior materials can be considered as the most important strategic area of innovation. We will need construction materials that are lighter, more flexible, have a higher strength, have a longer life, are less harmful to the environment both during their use and during their production, which can benefit from natural energy resources more effectively. In the development of these building materials, nanotechnology, which is one of the most important technology areas of our day, can make very important contributions [1].

Artificial neural networks are a method of counseling that can predict the outputs of new data inputs from existing data. Artificial neural networks, the most widely used and known method of artificial intelligence, are widely used in many fields including optimization in science and engineering [2]. This method is used to create a mathematical model between the inputs and outputs obtained and to give close outputs to the next new inputs that the system does not know [3].

Artificial neural networks are used frequently in scientific studies because of their many benefits such as working with little data, working fast, being easy to study, and being used in many different areas for different purposes [4, 5]. The most basic neural network, the input consists of the three main biride, including hidden and output layers. Input and output layer of neurons located in the problem to be solved as the number of inputs and outputs. In the hidden layer, the determination of the number of neurons can be found in various experimental studies. Figure 1 shows the structure of a typical three-layer artificial neural network.

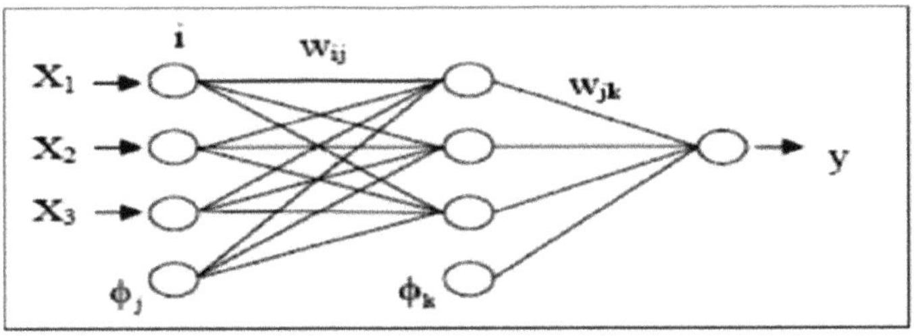

Fig. 1. Structure of artificial neural networks.

2 Material and Method

By using nano materials, C-S-H gels formed in the cement structure of reactive silica and adherence are increased. However, there are limited number of studies on the production of nanoparticles. Using nano minerals to ensure the strengthening of the bond between fine aggregate-cement paste. In this study, Portland cement (EN 197-1 CEM I 42,5 R) obtained from Isparta Göltaş Cement factory was used as a binder in the production of mortar, and the tap water used in the Süleyman Demirel University West campus was used as the mixing water. Standard sand, fumed silica (FS), precipitated silica (PS), Class F fly ash (FA) and silica fume (SF) (>2 mm, *10 nm *<500 nm, ~<1 μm, ~10 μm) in the order given dissimilar dimensions (as in Fig. 2).

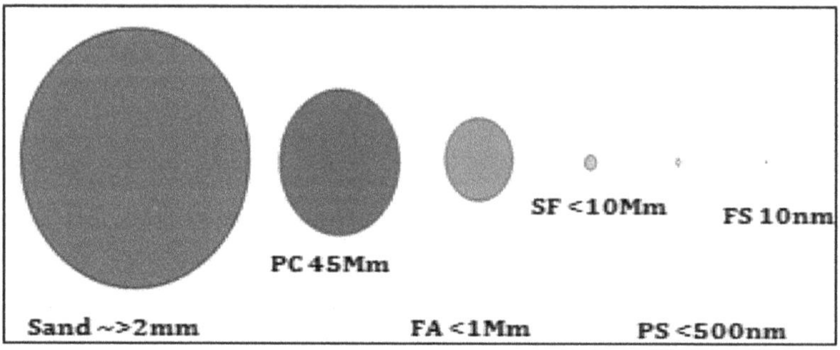

Fig. 2. Different sizes of used working materials.

Different mixes were designed with super plasticizer (SP) and ultra-thin materials and 9 different samples were produced by using these materials in different ratios as in Fig. 3 and Table 1. Water absorption rate, compressive strength and flexural strength tests of the concrete were produced. A total of 27 40 × 40 × 160 mm prismatic specimens were used for these tests (Fig. 4).

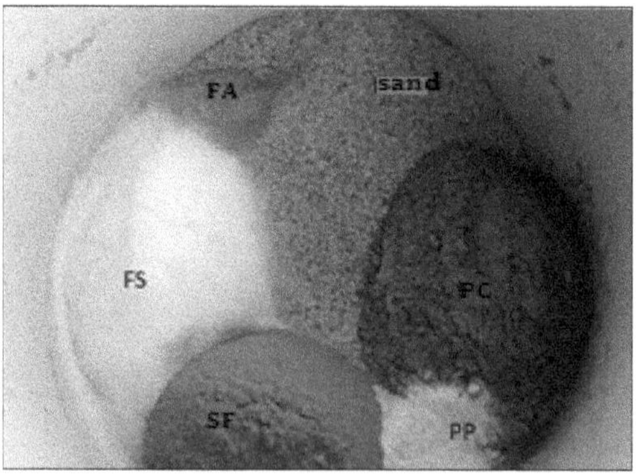

Fig. 3. The image of the used composite mixture.

The prepared mixtures were filled and molded in two stages. Then the shaking table is provided for compressing the mortar is subjected to vibration effects. After 24 ± 2 h, the molds were removed and the samples were cured at 20 °C for 7, 28 and 90 days in saturated water.

Fig. 4. The produced samples.

2.1 Water Absorption

Samples were removed from the pool before the water absorption test according to TS-EN 480-5 and weighed with precision balance and their weights were measured. After that's, the samples were kept in the oven at 105 °C for approximately 12 h and the weight measurements were made every hour. Water absorption, is expressed as a percentage of water absorbed by the unit weight or volume (as in Table 2). The water absorption rate was calculated by the following equation:

$$WA = \frac{W2 - W1}{W1} \%100 \tag{1}$$

where;
WA: water absorption rate, W2: Weight of the saturated sample in the liquid (water) (g), W1: Weight of the saturated sample in air (g).

According to the results of different mixing ratios applied nano and micro mortars 7, 28, 90 days of water absorption values are reduced compared to the control sample (Fig. 5).

Table 1. Amount of material used in mortar mixtures.

w/b = 0.47								
Mixture type	PC (g)	Sand (g)	FA (g)	SF (g)	PS (g)	FS (g)	SP (g)	Water (g)
Control	450	1350					4.5	211.5
%25FA	337.5	1350	112.5				4.5	211.5
%20FA%5SF	337.5	1350	90	22.5			13.5	211.3
%15FA%10SF	337.5	1350	67.5	45			18	211.0
%10FA%15SF	337.5	1350	45	67.5			22.5	211.8
%12FA%10SF% 3PS	337.5	1350	54	45	13.5		24.75	211.4
%12FA%10SF% 2PS%1FS	337.5	1350	54	45	9	4.5	24.75	211.4
%12FA%10SF% 1PS%2FS	337.5	1350	54	45	4.5	9	24.75	211.4
%12FA%10SF% 3FS	337.5	1350	54	45		3.5	24.75	211.4

Table 2. Mixtures of water absorption values.

Mixture type	WA_{7g}	WA_{28g}	WA_{90g}
C	6,85	5,70	3,91
%25 FA	4,44	3,80	3,40
%20FA%5 SF	4,99	3,29	3,20
%15 FA %10 SF	4,73	3,87	2,89
%10 FA %15 SF	4,02	3,15	2,42
%12 FA %10 SF%3PS	4,33	4,04	**1,82**
%12 FA %10 SF%2PS%1FS	**3,29**	**3,28**	2,13
%12 FA %10S%1PS-%2FS	4,00	3,57	2,13
%12 FA %10SD %3FS	4,17	3,66	2,31

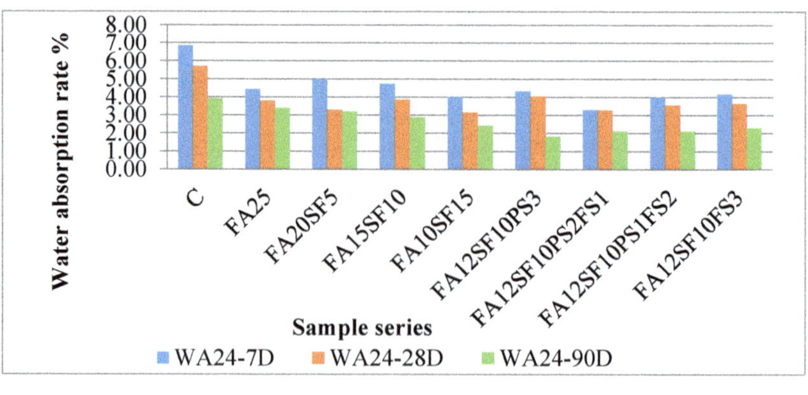

Fig. 5. Results of water absorption test.

2.2 Flexural Strength

Flexural strength determination of cement mortar was performed according to TS EN 196-1 standard. In this test method, 7, 28 and 90 day bending strengths of the samples were measured according to the loading method 3-point testing of them (Fig. 6) and the test results were given (Table 3 and Fig. 7). Flexural strength calculation Rf, equation is given below:

$$R_f = \frac{1,5 \times f_f \times l}{b^3} \tag{2}$$

where;
Rf: Flexural strength (MPa). b: The side length (mm) of the square section of the prism-shaped sample, Ff: The load (N) applied to the center at the moment of the prism-shaped sample, L: the distance between the support rollers (mm).

Fig. 6. Flexural sample image.

Table 3. Values of Flexural strength (f_{cf}) test.

Mixture type	f_{cf}-7D MPa	f_{cf}-28D MPa	f_{cf}-90D MPa
C	7,47	7,23	7,24
%25 FA	5,52	6	6,67
%20 FA-%5SF	3,55	5,49	7,28
%15 FA-%10SF	4,46	6,79	8,07
%10 FA-%15SF	5,09	7,51	7,84
%12 FA-%10SF-%3PS	5,94	7,26	8,61
%12 FA-%10SF-%2PS-%1FS	8,55	8,15	9,83
%12 FA-%10SF-%1PS-%2FS	7,51	8,73	9,60
%12 FA-%10SF-%3FS	7,74	7,89	8,66

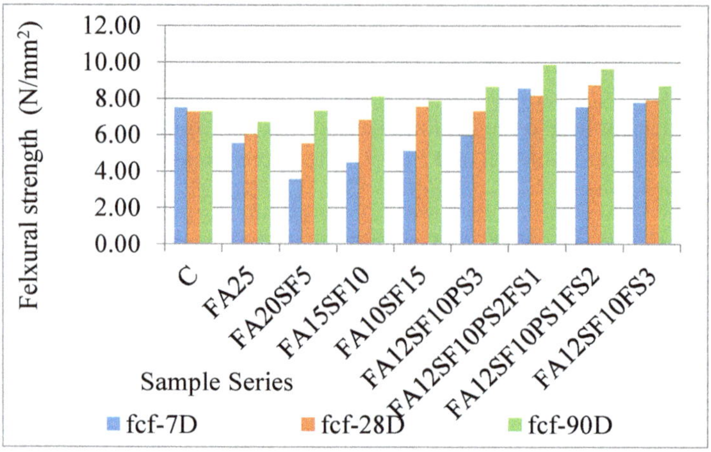

Fig. 7. The results of flexural strength.

2.3 Compressive Strength

The compressive strength of the two parts from the flexural test according to the same standard was used for the test. Prism samples were automatically loaded with load rate of 2400 N/s 7, 28 and 90 day compressive strength value measured was determined (Table 4 and Fig. 8) The experimental results are given. The compressive strength was calculated by the following equation:

$$Rc = \frac{f_c}{1600} \tag{3}$$

f_c: Compressive strength (MPa), f_c: Maximum load in fracture (N), 1600: Area of loading plates (40 mm × 40 mm; mm^2).

Table 4. Compressive strength (f_c) test values.

Mixture type	fc-7D MPa	fc-28D MPa	fc-90D MPa
C	40.90	49.16	51.87
%25 FA	29.99	40.06	45.80
%20 FA-%5SF	25.48	35.44	46.38
%15 FA-%10SF	30.05	42.4	47.84
%10 FA-%15SF	33.65	50.42	55.30
%12 FA-%10SF-%3PS	32.10	42.50	44.81
%12 FA-%10SF-%2PS-%1FS	47.78	49.81	51.21
%12 FA-%10SF-%1PS-%2FS	44.95	53.62	56.61
%12 FA-%10SF-%3FS	44.19	51.06	54.55

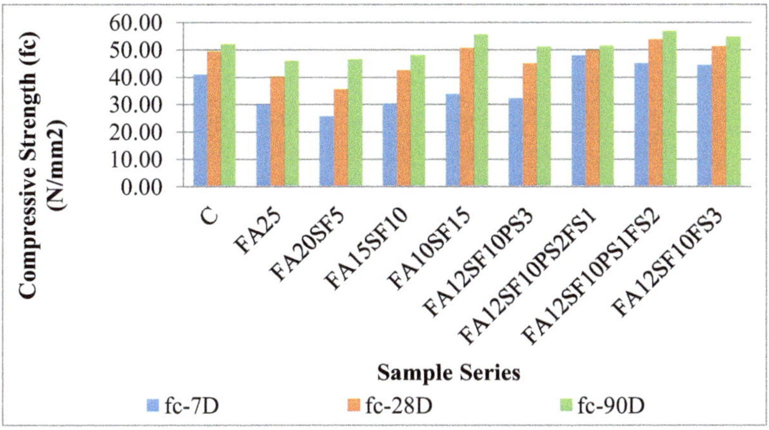

Fig. 8. Results of compressive strength.

In Fig. 9, 28 day pressure (f_c) and bending (f_{cf}) strength values of all samples belonging to 9 groups are given on the graph. In the figure, when the compressive strength values of the samples increased the bending strength values increased.

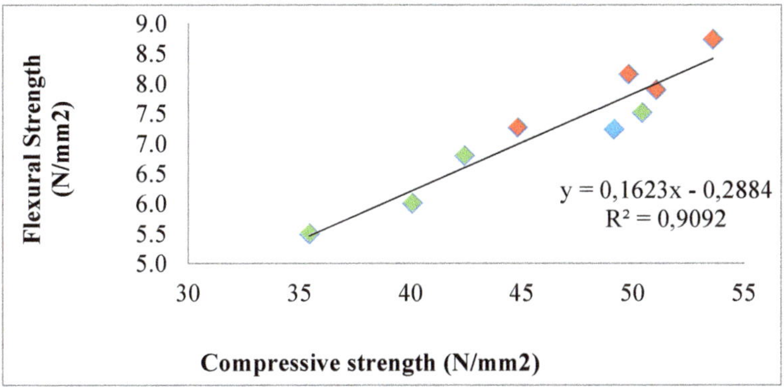

Fig. 9. Pressure (f_c) and bending (f_{cf}) strengths (28 days).

There are 9 mix results from experiments on mortar mixes. 7 of them were used for artificial neural networks, 2 of which were test data. artificial neural networks, water absorption, pressure and flexural strengths are tried to be estimated according to the material mixing ratios specified in Table 1. Actual and estimated values are given in Table 5. Also Root Mean Square Error (RMSE) and rsquare (R^2) is calculated as the performance and accuracy of the results are given in Table 6.

Table 5. Actual and estimated values.

Data	Water absorption rate		Compressive strength		Flexural strength	
	Actual	Estimate	Actual	Estimate	Actual	Estimate
Test1 7D	0.4235	0.426	0.2639	0.2636	0.2456	0.2446
Test1 28D	0.3258	0.3259	0.4062	0.4063	0.4209	0.4210
Test1 90D	0.5095	0.7436	0.3054	0.2485	0.4544	0.4424
Test2 7D	0.1	0.1001	0.9	0.8998	0.9	0.8998
Test2 28D	0.1407	0.1356	0.7323	0.8981	0.7567	0.8970
Test2 90D	0.2186	0.1215	0.5338	0.5787	0.9	0.8353

Table 6. Performance and accuracy values of artificial neural network estimation results.

	7D	28D	90D
RMSE	0.001079	0.08865	0.110897
R^2	1	0.9888	0.8135

3 Conclusions

The improvement in the water absorption property proves that micro additives and especially composite additives have lower porosity to the control sample. Therefore, the durability characteristics of micro and composite additive samples are better than the control sample. Composite (nano + micro) mineral admixtures and superplasticizers were used together with higher compressive strength than the control sample. Improvements in pozzolanic reactions were observed by adding different amounts of nano materials to the mortar. Addition of nano minerals in cement mortar with micro additives in appropriate proportions has caused an increase in mechanical strength. The addition of micro additives generally increases the pozzolanic activity and the filling of the fillers However, cement and displacement ratios of micro and nano-additives are also important. Puzolanic activity at high rate of fly ash + low rate of silica fume with 25% displacement with cement gave lower results than cement. Puzolanic activity increased as the ratio of silica fume increased. All obtained results (strength and toughness properties) when evaluated in a specific amount of nano additive (1–3%) for use with micro additives is only favorable than micro admixture.

The concrete mixture; The accuracy of 7 day estimation data was found as $R^2 = 1$ and the error rate was RMSE = 0.00108, the accuracy of the 28 day forecast data was $R^2 = 0.9888$ and the error rate was RMSE = 0.0887. It was determined that the results obtained with an experimental study were very close to the results obtained by using artificial neural network. So, NNP value effortless mixtures using neural network, it may be possible to achieve economical and less time.

References

1. URL. Innovation and nanotechnology in the future of the construction industry (2019). http://insaatdunyasi.com.tr/uncategorized/78insaat-sektorunun-geleceginde-inovasyon-ve-nanoteknoloji/
2. Dreiseitl, S., Ohno-Machado, L.: Logistic regression and artificial neural network classification models: a methodology review. J. Biomed. Inform. **35**(5–6), 352–359 (2002)
3. Afrand, M., Najafabadi, K.N., Sina, N., Safaei, M.R., Kherbeet, A.S., Wongwises, S., Dahari, M.: Prediction of dynamic viscosity of a hybrid nano-lubricant by an optimal artificial neural network. Int. Commun. Heat Mass Transf. **76**, 209–214 (2016)
4. Das, S., Pal, P., Singh, R.M.: Prediction of concrete mix proportion using ANN technique. Int. Res. J. Eng. Technol. (IRJET) **2**(5), 820–825 (2015)
5. Guresen, E., Kayakutlu, G., Daim, T.U.: Using artificial neural network models in stock market index prediction. Expert Syst. Appl. **38**(8), 10389–10397 (2011)

Tangibility of Fuzzy Approach Risk Assessment in Distributed Software Development Projects

Kökten Ulaş Birant[1], Ali Hakan Işık[2], and Mustafa Batar[3(✉)]

[1] Department of Computer Engineering, Dokuz Eylül University, İzmir, Turkey
ulas@cs.deu.edu.tr
[2] Department of Computer Engineering, Burdur Mehmet Akif Ersoy University, Burdur, Turkey
ahakan@mehmetakif.edu.tr
[3] The Graduate School of Natural and Applied Sciences, Dokuz Eylül University, İzmir, Turkey
mbatar@cs.deu.edu.tr

Abstract. Software is a product or a service or a project whose requirements are captured; specification document is prepared by requirements analysis; in/out architecture is designed; related source codes, variables, methods, classes, modules and relationships among these are written; necessary test and integration operations are done; when needed, maintenance, repair and update operations are executed by software or computer engineers and also that has its own documentation jobs and user manual, and has numeric and textual data, and also multimedia tools in its memory. Software development projects have a large financial burden and need to invest in high volumes. When looked at costs based on the international tangible data on computer software; it was $150 billion in 1985, it was $2 trillion in 2010 and it passed over $5 trillion after 2015. Also, in the year of 2018, the giro of Apple Store is about $250 million. Despite the costs, expenses and investments that are exponentially increasing every year, the rate of successful development of the software projects is not very high. Based on the "CHAOS" report (international size) prepared in 2016 only 17% of the software projects developed were completed in a timely manner, in the allocated budget and in accordance with the requirements. 53% of the projects were completed over time and/or over budget and/or also without fulfilling the requirements exactly. 30% of the software projects cannot have been completed in the development phase and were cancelled. For that software development projects with such high expenses and low success rate can have a better quality structure, a risk assessment approach has to be determined for the risk assessment methodology. So, some problems which may form software risks can be recognized and determined on time before causing trouble and endangering for software development projects. In this paper, some software risk assessment approaches underlying software risk management were introduced and explained in detail in addition to "fuzzy approach". In addition, that this fuzzy technique is more useful and more effective for software risk evaluation in comparison with the others was showed and expressed by giving linguistic rules in this article. This paper tried to explain the terms from the general to the specific.

D. J. Hemanth and U. Kose (Eds.): ICAIAME 2019, LNDECT 43, pp. 676–683, 2020.
https://doi.org/10.1007/978-3-030-36178-5_56

Keywords: Software engineering · Software · Software project management · Software risk management · Software risk assessment · Fuzzy approach

1 Software

Software engineering may be described as a department that performs the engineering tasks in order to develop software (computer software). This computer software is a product or a service or a project whose requirements are captured; specification document is prepared by requirements analysis; in/out architecture is designed; related source codes, variables, methods, classes, modules and relationships among these are written; necessary test and integration operations are done; when needed, maintenance, repair and update operations are executed by software or computer engineers and also that has its own documentation jobs and user manual, and has numeric and textual data, and also multimedia tools in its memory [1]. Computer Software is the item that product engineers plan and assemble. It incorporates programs that execute inside a PC of any size and engineering, archives that envelop printed version and virtual structures and information that join numbers and content yet additionally incorporates portrayals of pictorial, video and sound data [2].

2 Risk in Software Development

Dangers are a piece of activities. Programming advancement ventures are not diverse as task arranging is finished with least data. Be that as it may, the level of hazard differs with multifaceted nature, size and area. Extension creep, absence of comprehension of issues, equivocal prerequisites and absence of assets, equipment, systems administration and security issues are a portion of the basic hazard components in programming advancement ventures. Along these lines, there is a need to oversee hazard in programming advancement. In spite of the fact that specialists and expert have composed on hazard the board in programming improvement, almost no work have been done so as to include all the concerned partners in overseeing danger and incorporating the hazard the executives procedure with an all-encompassing undertaking the board approach [3]. Different structures of programming hazard the board have been proposed; some of which manage just the application part and a few models manage chance comprehensively covering application, association and between association levels. Nonetheless, there is a deficiency of writing on programming advancement chance administration from designers' viewpoint.

3 Software Project Management

A Software Project is the entire strategy of programming advancement from necessity social affair to testing and upkeep, did by the execution procedures in a predefined timeframe to accomplish proposed programming item. Programming is said to be an immaterial item. Programming improvement is a sort of every single new stream in world business and there's next to no involvement in building programming items [4]. Most programming items are customized to accommodate customer's prerequisites. The most vital is that the basic innovation changes and advances so oftentimes and quickly that experience of one item may not be connected to the next one. All such business and ecological limitations get hazard programming improvement thus it is basic to oversee programming ventures proficiently [5]. Programming venture the board is a workmanship and art of arranging and driving programming ventures. It is a sub-order of venture the board in which programming ventures are arranged, executed, checked and controlled.

4 Software Risk Management

Programming hazard the board starts with the thought that product chance is an issue that should be overseen. Programming hazard at its center stems from issues inside the product itself, i.e., the source code that is presented amid advancement. Programming hazard the board should then address two programming sorts of issues: Software disappointment and non-execution task and program the executives and conveyance. Programming hazard the board adopts a proactive strategy programming hazard by giving a methodology and system to search for zones where a product imperfection impacts the ease of use of the product for end clients and the business. For instance, a disastrous disappointment as the consequence of a product bug that does not enable the product to run accurately or at all is a kind of programming hazard that must be overseen [6]. Programming hazard as an effect on undertaking the board, program the board or conveyance is one in which programming imperfections and multifaceted nature affect the capacity to discharge programming on-time or inside spending plan. The effect here is in deferrals and expenses to the business that must be assimilated. For instance, an imperfection found late in the advancement procedure could result in re-work that takes days or weeks to address in this way deferring an undertaking. Both of these issues require solid hazard the board practices to relieve against the hazard. However, do individuals really deal with this hazard? In the first place, they ought to distinguish and comprehend the main driver.

5 Software Risk Assessment

Hazard appraisal is a term used to portray the general procedure or technique where you: One recognizes dangers and hazard factors that can possibly cause hurt (risk distinguishing proof). One breaks down and assesses the hazard related with that danger (chance examination, and hazard assessment). One decides fitting approaches to dispose of the danger, or control the hazard when the peril can't be dispensed with (chance control) [7]. A hazard evaluation is a careful take a gander at your work environment to recognize those things, circumstances, forms, and so forth that may cause hurt, especially to individuals. After ID is made, you investigate and assess how likely and serious the hazard is. At the point when this assurance is made, you can straightaway, choose what measures ought to be set up to successfully take out or control the mischief from occurring. The CSA Standard Z1002 "Word related well-being and security - Hazard ID and end and hazard evaluation and control" utilizes the accompanying terms: Hazard appraisal – the general procedure of peril distinguishing proof, chance examination, and hazard assessment. Risk distinguishing proof – the way toward discovering, posting, and portraying dangers [8]. Hazard investigation – a procedure for appreciating the idea of dangers and deciding the dimension of hazard. Hazard assessment – the way toward looking at an expected hazard against given hazard criteria to decide the hugeness of the hazard. Hazard control – activities executing hazard assessment choices.

6 Software Risk Assessment Approaches

The Fault Tree Analysis (FTA) was first presented by Bell Laboratories and is a standout amongst the most generally utilized strategies in framework unwavering quality, viability and wellbeing examination. It is a deductive methodology used to decide the different blends of equipment and programming disappointments and human blunders that could cause undesired occasions (alluded to as best occasions) at the framework level [9]. The deductive examination starts with a general end, at that point endeavours to decide the explicit reasons for the decision by building a rationale outline called a blame tree. This is otherwise called adopting a best down strategy. The principle motivation behind the blame tree examination is to help recognize potential reasons for framework disappointments before the disappointments really happen. It can likewise be utilized to assess the likelihood of the best occasion utilizing diagnostic or measurable strategies [10]. These counts include framework quantitative dependability and viability data, for example, disappointment likelihood, disappointment rate and fix rate. Subsequent to finishing a FTA, you can concentrate your endeavours on enhancing framework security and unwavering quality.

A Decision Tree is a decision help gadget that uses a tree-like model of decisions and their possible outcomes, including chance event results, resource costs, and utility. It is one way to deal with demonstrate a figuring that just contains prohibitive control explanations [11]. Decision trees are usually used in exercises investigate, expressly in decision examination, to help recognize a technique bound to accomplish a goal, yet then again are an outstanding gadget in AI.

Probabilistic Risk Analysis (PRA) is exact and unquestionable expectations of stream examples and contaminant transport over an extensive variety of ecological streams are famously monotonous due to complexities presented by fierce streams, spatial geometries and normal vulnerabilities [12]. The absence of adequate site portrayal, computational assets, and sometimes, the deficiency of conceptualizations and scientific models of significant physical wonders muddle demonstrating endeavours further. These issues present a principal absence of conviction about any anticipated stream and cast questions on the achievability of acquiring a solitary deterministic forecast of a pollution situation.

An event tree is an inductive expository graph in which an occasion is broke down utilizing Boolean rationale to look at a sequential arrangement of ensuing occasions or outcomes. For instance, occasion tree examination is a noteworthy segment of atomic reactor security engineering. An occasion tree shows succession movement, arrangement end states and grouping explicit conditions crosswise over time [13].

Disappointment Modes Effects Analysis (FMEA) is a well ordered methodology for recognizing every single imaginable disappointment in a plan, an assembling or get together procedure or an item or administration. "Disappointment modes" signifies the ways or modes, in which something may fall flat. Disappointments are any blunders or deformities, particularly ones that influence the client and can be potential or real. "Impacts investigation" alludes to examining the outcomes of those disappointments [14]. Disappointments are organized by how genuine their outcomes are, the way as often as possible they happen and how effectively they can be distinguished. The reason for the FMEA is to take activities to dispose of or decrease disappointments, beginning with the most elevated need ones.

Hazard Analysis Critical Control Point (HACCP) is an administration framework in which sustenance wellbeing is tended to through the investigation and control of organic, synthetic, and physical risks from crude material generation, obtainment and dealing with, to assembling, appropriation and utilization of the completed item [15].

Root Cause Analysis (RCA) is a strategy for critical thinking utilized for recognizing the underlying drivers of issues or problems [16]. A factor is viewed as a main driver if expulsion thereof from the issue blame succession keeps the last unwanted result from repeating; while a causal factor is one that influences an occasion's result, however isn't an underlying driver. In spite of the fact that evacuating a causal factor can profit a result, it doesn't keep its repeat with sureness.

Risk Ranking and Filtering is a standout amongst the most widely recognized assistance strategies utilized for Risk Management. This technique is otherwise called "Relative Risk Ranking", "Hazard Indexing" and "Hazard Matrix and Filtering". Its plan is to give more honed concentration to the basic dangers inside a framework – commonly, from a vast and complex arrangement of hazard situations [17]. Hazard Ranking and Filtering works by separating by and large hazard into hazard segments and assessing those segments and their individual commitments to generally speaking danger.

The Center for Risk Management of Engineering Systems at the University of Virginia built up the method of Hierarchical Holographic Modeling (HHM) in the mid-1980s as a hazard examination device. Intended to catch the many-sided interdependencies of basic frameworks, HHM was created to conquer the difficulty of a solitary model's capacity to speak to all the significant parts of a perplexing framework. In demonstrating substantial scale, complex frameworks, numerous satisfactory scientific and calculated models develop, each speaking to an alternate, yet exact perspective. The execution of HHM permits the improvement of different, substantial models that "catch the quintessence of the numerous measurements, dreams, and points of view of foundation frameworks." Hierarchical implies the need to appreciate risks as they appear at different measurements in a pecking request (e.g., threats at the course of action of systems level, the individual structure level, the sub-structure level and the fragment level) [18]. Holographic suggests the need to get the chance to danger from different perspectives (e.g., monetary, security, geographic, specific, etc.). The displaying approach is dealt with to oversee both holographic and different leveled thoughts and usages a formalized philosophy (Risk Filtering and Ranking Methodology) to sort out threats.

7 Fuzzy Approach in Software Risk Assessment

Fluffy rationale is a way to deal with processing dependent on "degrees of truth" as opposed to the standard thing "genuine or false" (1 or 0) Boolean rationale on which the cutting edge PC is based [19]. The possibility of fluffy rationale was first exceptional by Dr. Lotfi Zadeh of the University of California at Berkeley during the 1960s. Dr. Zadeh was taking a shot at the issue of PC comprehension of regular dialect. Characteristic dialect (like most different exercises throughout everyday life and in reality the universe) isn't actually converted into the outright terms of 0 and 1. (In the case of everything is eventually describable in parallel terms is a philosophical inquiry worth seeking after, however by and by much information we should need to bolster a PC is in some state in the middle of thus, as often as possible, are the aftereffects of registering). It might see fluffy rationale as the manner in which thinking truly works and twofold or Boolean rationale is essentially an exceptional instance of it. Fluffy rationale incorporates 0 and 1 as outrageous instances of truth (or "the condition of issues" or "reality") yet additionally incorporates the different conditions of truth in the middle of so that, for instance, the aftereffect of a correlation between two things could be not "tall" or "short" but rather ".38 of stature." Fluffy rationale appears to be nearer to the manner in which our minds work. We total information and frame various halfway certainties which we total further into higher realities which thusly, when certain edges are surpassed, cause certain further outcomes, for example, engine response [20]. A comparable sort of process is utilized in neural systems, master frameworks and other man-made brainpower applications. Fluffy rationale is fundamental to the advancement of human-like capacities for AI, once in a while alluded to as fake general knowledge: the portrayal of summed up human psychological capacities in programming so that, looked with a new errand, the AI framework could discover an answer.

Some linguistic rules of software risk assessment based on fuzzy approach for distributed software projects [21]:

*PM: Process Maturity TZD: Time Zone Difference F: Formality
*LD: Language Difference CI: Communication Infrastructure
*CE: Common Experiences TC: Task Coupling AK: Application Knowledge
*T: Transparency RS: Requirements Stability NP: Novelty of Product
*CD: Cultural Differences

*If "PM" is high and "TZD" is low then Risk is low.
*If "F" is high and "LD" is low and "CI" is high then Risk is low.
*If "T" is high and "TZD" is low then Risk is low.
*If "RS" is high and "NP" is low and "LD "is low and "CD" is low then Risk is low.

*If “CE” is high and “TC” is low then Risk is low.
*If “AK” is high and “CI” is high then Risk is low.
*If “Productivity” is high and “Communication” is high then Risk is low.

8 Conclusion

Hazard In this paper, software risk assessment which is the basis of software risk management was dwelt and its prominent approaches – Fault Tree Analysis, Decision Tree, Probabilistic Risk Analysis, Event Tree, Failure Modes Effects Analysis, Hazard Analysis Critical Control Point, Root Cause Analysis, Risk Ranking and Filtering and Hierarchical Holographic Modeling – were showed and told. In addition, Fuzzy Approach of software risk evaluation was explained in detail and by its linguistic and logical rules, that it works more effectively in the assessment level of software risk management was expressed in this work.

“You can’t manage the process which you don’t measure.” This statement which is claimed to be said by Peter DRUCKER shows that the software development process has to be measured by clear and objective variables – Process Maturity, Time Zone Difference, Formality, Language Difference, Communication Infrastructure, Common Experiences, Task Coupling, Application Knowledge, Transparency, Requirements Stability, Novelty of Product, Cultural Differences. Thus, some reliable data are specified and determined so that this software risk evaluation process can be managed by benefiting from these trustworthy results. According to the results of this evaluation, manpower, what is the main resource of software development process will be used more effectively. And then, the benefits of the “Software Risk Assessment and Management” based on “Software Engineering” may be seen more tangible.

References

1. Roger, S.: Pressman and Bruce Maxim: Software Engineering: A Practitioner’s Approach, 8th edn. McGraw-Hill Higher Education, New York (2015)
2. Sommerville, I.: Software Engineering, 10th edn. Pearson Education Limited, England (2016)
3. Dey, P.K., Kinch, J., Ogunlana, S.O.: Managing risk in software development projects: a case study. Ind. Manag. Data Syst. **107**(2), 284–303 (2007)
4. OmarElgabry’s Blog, A general introduction to software engineering. https://medium.com/omarelgabrys-blog/software-engineering-introduction-part-1-b79238ec97ee. Accessed 6 May 2019
5. Tutorialspoint, Software Project Management. https://www.tutorialspoint.com/software_engineering/software_project_management.htm. Accessed 7 May 2019
6. National Research Council: The Owner’s Role in Project Risk Management. The National Academies Press, Washington (2005)
7. Sadiq, M., Rahmani, Md.K.I., Ahmad, M.W., Jung, S.: Software risk assessment and evaluation process (SRAEP) using model based approach. In: International Conference on Networking and Information Technology. IEEE (2010)

8. Mohamud, S.A., Basri, S.: Software risk assessment: a review on small and medium software projects. In: Software Engineering and Computer Systems, ICSECS, pp. 214–224 (2011)
9. Quality Progress. http://asq.org/quality-progress/2002/03/problem-solving/what-is-a-fault-tree-analysis.html. Accessed 7 May 2019
10. Weibull. https://www.weibull.com/basics/fault-tree/index.htm. Accessed 7 May 2019
11. Medium Home Page, Decision Trees Algorithms. https://medium.com/deep-math-machine-learning-ai/chapter-4-decision-trees-algorithms-b93975f7a1f1. Accessed 7 May 2019
12. U.S. NRC, Probabilistic Risk Assessment (PRA). https://www.nrc.gov/about-nrc/regulatory/risk-informed/pra.html. Accessed 7 May 2019
13. Rausand, M.: Event Tree Analysis (2005). https://hkarms.org/ASUS_Server/myftp_web_resources/2005_Conference_Presentation/Sys_Any_Event_Tree_Anly.pdf. Accessed 7 May 2019
14. NPD solutions, Failure Modes Effects Analysis. http://www.npd-solutions.com/fmea.html. Accessed 7 May 2019
15. HACCP Principles & Application Guidelines. https://www.fda.gov/food/hazard-analysis-critical-control-point-haccp/haccp-principles-application-guidelines. Accessed 7 May 2019
16. Leaf. http://www.leafexcellenceinauditingforum.com/2018/09/28/root-cause-analysis/. Accessed 7 May 2019
17. Manufacturing Technology Committee – Risk Management Working Group, Risk Ranking and Filtering. http://pqri.org/wp-content/uploads/2015/08/pdf/Risk_Rank_Filter_Training_Guide.pdf. Accessed 7 May 2019
18. Haimes, Y.Y., Lambert, J., Li, D., Schooff, R., Tulsiani, V.: Hierarchical holographic modeling for risk identification in complex systems. In: International Conference on Systems, Man and Cybernetics. IEEE (1995)
19. Garcia-Diaz, N., Lopez-Martin, C., Chavoya, A.: A comparative study of two fuzzy logic models for software development effort estimation. In: Iberoamerican Conference on Electronics Engineering and Computer Science (2013)
20. Garrido, A., Yuste, P.: Looking for oriental fundamentals fuzzy logic. Brain **4**(1–4), 121–125 (2013)
21. Lamersdorf, A., Münch, J., del Viso Torre, A.F., Sánchez, C.R., Heinz, M., Rombach, D.: A rule-based model for customized risk identification in distributed software development projects. In: Proceedings of the International Conference on Global Software Engineering. IEEE (2010)

A Simple Iterative Algorithm for Boolean Knapsack Problem

Fidan Nuriyeva[1,2], Urfat Nuriyev[3], and Onur Ugurlu[3(✉)]

[1] Faculty of Science, Department of Computer Science, Dokuz Eylul University, Izmir, Turkey
nuriyevafidan@gmail.com
[2] Institute of Control Systems of ANAS, Baku, Azerbaijan
[3] Faculty of Science, Department of Mathematics, Ege University, Izmir, Turkey
urfatnuriyev@gmail.com, ugurluonur88@gmail.com

Abstract. The Boolean Knapsack Problem, which has numerous real life applications, is a combinatorial optimization problem. Since the problem belongs to NP-Hard, many researchers have worked on several variants of the problem. In this paper a simple iterative algorithm based on a greedy strategy to solve the classical boolean knapsack problem is proposed. In algorithm, firstly initial solutions are generated, then a greedy criteria is used to improve these solutions. To demonstrate the performance of the proposed method, experiments are carried out with various benchmark instances of boolean Knapsack Problem. The computational results show that the proposed algorithm is functional enough to achieve acceptable results in reasonable times.

Keywords: Combinatorial optimization · Boolean knapsack problem · Greedy method

1 Introduction

Knapsack problem is a well-known combinatorial optimization problem which was first presented by Dantzig in the 1950s and then attracted a great number of researchers [1]. There are many types of knapsack problems. They depend upon the distribution of the items and knapsacks. This paper focus on Boolean Knapsack Problem (KP01). The aim of the Boolean Knapsack problem is to fill the knapsack with items without crossing the weight limit of the knapsack and have a maximum profit. KP01 can be widely applied in capital budgeting [2], cutting stock and loading problems [3], job scheduling [4] and so on.

Solution methods of KP01 can be divided into exact, metaheuristic and heuristic algorithms. Kellerer et al. [5] studied on several exact methods for large instances of the knapsack problem. Jolai et al. [6] present a linear mathematical programming formulation based exact algorithm, another exact algorithm to solve KP01 was proposed by Lin et al. [7]. Figueira et al. [8] improved the

D. J. Hemanth and U. Kose (Eds.): ICAIAME 2019, LNDECT 43, pp. 684–689, 2020.
https://doi.org/10.1007/978-3-030-36178-5_57

dynamic programming approach for KP01 by using many complementary dominance relations.

Exact algorithms can produce optimal solutions for small-scale problems, but real world applications often involve a greater number of variables. Thus, researchers focus their attention towards the heuristic and metaheuristic algorithms to find near-optimal solutions in acceptable times. One of the early heuristic approach "a polynomial approximation schemes" was first proposed for KP by Sahni [9]. Nikitin and Nuriyev [10] proposed a dynamic programming based heuristic for KP01, another effective heuristics were proposed by Csirik [11]. Recently, Lv et al. [12] presented a new hybrid solution method with a greedy strategy to solve KP01.

In recent years, many metaheuristic methods have been employed for solving KP01. Shen et al. [13] develop a genetic algorithm improved by a dual population. Shi [14] proposed a improved ant colony optimization algorithm. Kulkarni and Shabiri [15] proposed Cohort intelligence algorithm for KP01.

Although several KP01 problems have been resolved by existing algorithms, there are still numerous new and difficult real world problems and many of these existing algorithms may lose their efficiency if the size of the problems largely increases. Besides, especially in the real life application, it mostly focuses on finding near-optimal solutions rather than spending too much time for the optimal solutions. For this purpose, heuristic algorithms are more effective than exact and metaheuristic algorithms for solving KP01 [12]. This paper presents an iterative heuristic algorithm for KP01.

This paper is organized as follows: In Sect. 2, the preliminaries of KP01 is briefly described. Section 3 details the proposed iterative algorithm. Computational experiments are presented in Sect. 4 and Sect. 5 deals with conclusion.

2 Boolean Knapsack Problem

For the Boolean Knapsack Problem, we have a set of N items, where ith items owns its value v_i and weight w_i. The capacity of a knapsack is shown as b. The KP01 can be formulated mathematically as follows:

$$max \quad c = f(x_1, x_2, ..., x_n) = \sum_{i=1}^{n} v_i x_i \tag{1}$$

s.t.

$$\sum_{i=1}^{n} w_i x_i \leq b \tag{2}$$

$$x_i \in \{0, 1\}, \quad i = 1, ..., n \tag{3}$$

3 The Proposed Iterative Algorithm

This section presents the heuristic iterative algorithm to solve KP01. Main framework of the proposed solution method was presented in [18]. Since the proposed method based on greedy strategy, we describe the standard greedy algorithm for KP01 firstly. The greedy algorithm is well-known and most straightforward solution approach for KP01. It considers only the current situation and makes the best choice for that situation. For solving KP01, the greedy algorithm selects an item according to some greedy criteria and load it into knapsack at each step. This process continues until no item can be selected for the knapsack.

The proposed method for KP01 consists of two main parts. In the first part, initial solutions are found and priority values of the items are determined by considering these solutions. In accordance with this purpose, the greedy algorithm is used for three times according to u_i, v_i and w_i where u_i is the ratio of v_i and w_i. In the second part, the solutions are tried to improve iteratively. In each iteration, the priority values of the items are updated and the greedy algorithm is used according to these values to find better solutions. Algorithm stops if a certain number of iteration or a termination condition is reached. Main steps of the algorithm are as following:

Let p_i be the priority value for all items $i = 1, ..., n$, k be the number of iteration and f^* represents the best solution, c^* represents the value of the objective function and finally t represents the ratio of c^k/c^*, where c^k is the solution value which is found at iteration k.

Step 1. Set $p_i = u_i$ for $i = 1, ..., n$ and $k = 0$.

Step 2. Run the greedy algorithm according to u_i and calculate the f^k, c^k. Update f^*, c^*, k and p_i. The priority values are updated as below:

$$p_i^0 = \begin{cases} p_i^0 + 1, & if \quad x_i^0 = 1 \\ p_i^0, & otherwise \end{cases}$$

Step 3. Run the greedy algorithm according to v_i and calculate the f^k, c^k, t. Update f^*, c^*, k and p_i. The priority values are updated as below:

$$p_i^k = \begin{cases} p_i^{k-1} + t, & if \quad x_i^k = 1 \\ p_i^{k-1}, & otherwise \end{cases}$$

Step 4. Run the greedy algorithm according to w_i and calculate the f^k, c^k, t. Update f^*, c^*, k and p_i. The priority values are updated as below:

$$p_i^k = \begin{cases} p_i^{k-1} + t, & if \quad x_i^k = 1 \\ p_i^{k-1}, & otherwise \end{cases}$$

Step 5. Run the greedy algorithm according to p_i and calculate the f^k, c^k, t. Update f^*, c^*, k and p_i. The priority values are updated as below:

if $t > 1$ then

$$p_i^k = \begin{cases} p_i^{k-1} + t, & if \quad x_i^k = 1 \; and \; x_i^* = 0 \\ p_i^{k-1}, & otherwise \end{cases}$$

if $t < 1$ then

$$p_i^k = \begin{cases} p_i^{k-1} - t, & if \quad x_i^k = 1 \; and \; x_i^* = 0 \\ p_i^{k-1}, & otherwise \end{cases}$$

if $t = 1$ then

$$p_i^k = \begin{cases} p_i^{k-1} - 1, & if \quad x_i^k = 1 \; or \; x_i^* = 1 \\ p_i^{k-1} + 1, & otherwise \end{cases}$$

Step 6. Go to *Step 5* if a termination condition is not reached.

4 Computational Experiments

This section gives the experimental results of the proposed algorithm. The algorithm is implemented in C and executed on a Intel Core i7 4.00 GHz CPU with 8 GB RAM. Ten low-dimensional and fourteen large-scale KP01 problems are used to test the efficiency of the algorithm.

Standard ten low-dimensional KP01 test problems, which have been studied by several researchers [15–17], are considered here. The results of the proposed algorithm on ten low-dimensional KP01 problems are given in Table 1. The first three columns represent the problem name, number of the items and optimal value of the problem, respectively. The fourth column gives the solution value achieved by the proposed algorithm, last column gives the running time of the algorithm. The iteration number k is set to n for the low-dimensional problems. As it is seen in Table 2, the proposed algorithm find optimal values for all low-dimensional KP01 problems.

To test the performance of the proposed method further, fourteen large-scale KP01 problems are considered. These instances are taken from [19], where sizes of these problems vary from 100 to 5000. In these problems, the knapsack capacity was calculated according to the following formula:

$$b = 3/4 \sum_{i=1}^{n} w_i \tag{4}$$

The results of the proposed algorithm on fourteen large-scale KP01 problems are given in Table 2. The first three columns indicate the problem name, number of the items and optimal value of the problem, respectively. The solution value achieved by the proposed algorithm and its running time (in seconds) are given in last two columns. The iteration number k is set to *100* for large-scale problems. Table 2 demonstrate that the proposed algorithm can find near-optimal solutions in reasonable times for large-scale KP01 problems.

Table 1. The results of the low-dimensional KP01 problems

Problem	n	Opt. value	Proposed alg.	Time(s)
Kp1	10	295	295	0.0
Kp2	20	1024	1024	0.0
Kp3	4	35	35	0.0
Kp4	4	23	23	0.0
Kp5	15	481.07	481.07	0.0
Kp6	10	52	52	0.0
Kp7	7	107	107	0.0
Kp8	23	9767	9767	0.0
Kp9	5	130	130	0.0
Kp10	20	1025	1025	0.0

Table 2. The results of the large-scale KP01 problems

Problem	n	Opt. value	Proposed alg.	Time(s)
Kp11	100	4795	4795	0.000
Kp12	200	10134	10127	0.000
Kp13	300	14843	13838	0.015
Kp14	400	19446	19445	0.031
Kp15	500	23389	23389	0.062
Kp16	600	28490	28489	0.078
Kp17	700	33434	33434	0.109
Kp18	800	38319	38318	0.140
Kp19	900	42881	42881	0.187
Kp20	1000	48126	48125	0.234
Kp21	2000	95647	95645	0.984
Kp22	3000	143361	143361	2.358
Kp23	4000	190228	190226	4.217
Kp24	5000	243225	243225	6.373

5 Conclusions

This paper presents a simple iterative algorithm based on a greedy strategy to solve KP01. A large number of test problems are studied to validate the efficiency of the proposed method. The computational results demonstrate that the proposed algorithm can find high quality solutions in reasonable times. Besides, unlike the many existing algorithms for KP01, the proposed method is easy to implement. Considering its simplicity and its performance on large-scale problems, it is fair to claim that the proposed method can be very useful for real-life applications. In the future, we plan to adapt our iterative method for different forms of knapsack problems.

References

1. Dantzig, G.B.: Discrete-variable extremum problems. Oper. Res. **5**(2), 266–277 (1957)
2. Bas, E.: A capital budgeting problem for preventing workplace mobbing by using analytic hierarchy process and fuzzy 0-1 bidimensional knapsack model. Expert Syst. Appl. **38**(10), 12415–124222 (2011)
3. Wascher, H., Schumann, H., et al.: An improved typology of cutting and packing problems. Eur. J. Oper. Res. **183**(3), 1109–1130 (2007)
4. Martello, S., Toth, P.: Knapsack Problem: Algorithms and Computer Implementations. Wiley, New York (1990)
5. Kellerer, H., Pferschy, U., Pisinger, D.: Knapsack Problems. Wiley, New York (2005)
6. Jolai, F., Rezaee, M.J., et al.: Exact algorithm for biobjective 0-1 knapsack problem. Eur. J. Oper. Res. **194**, 544–551 (2007)
7. Lin, G., Zhu, W., Ali, M.M.: An exact algorithm for the 0-1 linear knapsack problem with a single continuous variable. J. Global Optim. **50**, 657–673 (2011)
8. Figueira, J.R., Paquete, L., Simoes, M.: Algorithm improvements on dynamic programming for the bi-objective 0-1 knapsack problem. Comput. Optim. Appl. **56**, 97–111 (2013)
9. Sahni, S.: Approximate algorithms for the 0-1 knapsack problem. J. ACM **22**, 115–124 (1975)
10. Nikitin, A.I., Nuriev, U.G.: On a method of the solution of the knapsack problem. Kibernetika **2**, 108–110 (1983)
11. Csirik, J.: Heuristics for the 0-1 min-knapsack problem. Acta Cybern. **10**(1–2), 15–20 (1991)
12. Lv, J., Wang, X., et al.: Solving 0-1 knapsack problem by greedy degree and expectation efficiency. Appl. Soft Comput. **41**, 94–103 (2016)
13. Shen, W., Xu, B., Ping Huang, J.: An improved genetic algorithm for 0-1 knapsack problems. In: Second International Conference on Networking and Distributed Computing (ICNDC), China, pp. 32–35. IEEE (2011)
14. Shi, H.: Solution to 0-1 knapsack problem based on improved ant colony algorithm. In: International Conference on Information Acquisition, China, pp. 1062–1066. IEEE (2006)
15. Kulkarni, A.J., Shabir, H.: Solving 0-1 knapsack problem using cohort intelligence algorithm. Int. J. Mach. Learn. Cybern. **7**(3), 427–441 (2016)
16. Zhang, X., Huang, S., et al.: Solving 0-1 knapsack problems based on amoeboid organism algorithm. Appl. Math. Comput. **219**(19), 9959–9970 (2013)
17. Zhou, Y., Chen, X., Zhou, G.: An mproved monkey algorithm for a 0-1 knapsack problem. Appl. Soft Comput. **38**, 817–830 (2016)
18. Nuriyev, U.G., Nuriyeva, F.: Practical aspects of solving combinatorial optimization problems. Adv. Math. Models Appl. **3**(3), 179–191 (2018)
19. Guler, A., Berberler, M.E., Nuriyev, U.G.: A new genetic algorithm for the 0-1 knapsack problem. Acad. Platf. J. Eng. Sci. **4**(3), 9–14 (2016)

A Review of the Solutions for the Container Loading Problem, and the Use of Heuristics

Merve Aydemir(✉) and Tuncay Yigit

Department of Computer Engineering, Suleyman Demirel University, Isparta, Turkey
merve.aydemir90@gmail.com, tuncayyigit@sdu.edu.tr

Abstract. Today, due to the rapid growth in the area of transportation, the need for logistics increases as a result of the growth in the local and global trade volumes. In the current competitive conditions in the national and international markets, each company aims to deliver its products with the lowest cost and high customer satisfaction. One way to reduce costs is to fill the container volume in the best way. Therefore, container loading problems are one of the important problems in terms of logistics. In this paper, studies on container loading problems will be discussed. Issues to be considered when dealing with container loading problems will be identified. It will be analyzed which constraints should be taken into account when placing the boxes in the container, whether boxes should be placed in pallets or individually, which heuristic algorithms should be used to reach the maximum volume in the container after the box placement or which heuristic algorithms give better results.

Keywords: Optimization · Heuristic algorithms · Container loading problems · Logistics

1 Introduction

The efficient use of transportation vehicles in the logistic sectors is of considerable economic importance. The high utilization of the existing carrying capacity provides significant cost savings. Therefore, efficient packaging methods are gaining importance in these sectors.

Container loading problem is one of the problems of cutting and packing. The purpose of this problems:

- Placing a certain number of boxes in a container
- To maximize the container volume by providing some constraints [1].

A box set which will be placed into the container is divided into three groups:

- Homogeneous (only identical boxes)
- Weakly heterogeneous (a few different types of boxes)
- Strongly heterogeneous (many different types of boxes) [3].

D. J. Hemanth and U. Kose (Eds.): ICAIAME 2019, LNDECT 43, pp. 690–700, 2020.
https://doi.org/10.1007/978-3-030-36178-5_58

Container loading problems are classified in two ways; input minimization problem types and output maximization problem types. Input minimization problem types are divided into seven types:

- Single Stock-Size Cutting Stock Problem: Loading a set of weak heterogeneous boxes to the minimum number of identical containers.
- Multiple Stock-Size Cutting Stock Problem: Loading a set of weak heterogeneous boxes to a minimum number of weak heterogeneous containers.
- Residual Cutting Stock Problem: Loading a set of weak heterogeneous boxes to a minimum number of strong heterogeneous containers.
- Single Bin-Size Bin Packing Problem: Loading a set of strong heterogeneous box set to a minimum number of identical containers.
- Multiple Bin-Size Bin Packing Problem: Loading a set of strong heterogeneous box set to a minimum number of weak heterogeneous containers.
- Residual Bin Packing Problem: Loading a set of strong heterogeneous box set to a minimum number of strong heterogeneous containers.
- Open Dimension Problem: Loading a particular set of boxes into a container with one or more variable dimensions to minimize the volume of the container.

Output maximization problem types are the following:

- Identical Item Packing Problem: Loading a single container with the maximum number of identical items.
- Single Large Object Placement Problem: Loading of a single container by selecting a set of weak heterogeneous boxes that will maximize the loading value.
- Multiple Identical Large Object Placement Problem: Loading of multiple identical containers by selecting from the weak heterogeneous box set of items to maximize loading value.
- Multiple Heterogeneous Large Object Placement Problem: Loading of a set of more than one kinds of weak or strong heterogeneous containers by selecting from the weak heterogeneous box set of items to maximize loading value.
- Single Knapsack Problem: Loading of single container by selecting from a set of strong heterogeneous boxes that will maximize the loading value.
- Multiple Identical Knapsack Problem: Loading of more than one identical container by selecting from the strong heterogeneous box set of items to maximize the loading value.
- Multiple Heterogeneous Knapsack Problem: Loading of a set of more than one kinds of weak or strong heterogeneous containers by selecting from the strong heterogeneous box set of items to maximize loading value [4].

Container loading problem is defined as NP-Hard problems and the solution of the problem requires exponential time. Therefore, heuristic and meta-heuristics take an important place in the solution of NP-Hard problems [2].

In this paper, the constraints for container loading problem are mentioned in Sect. 2. In Sect. 3, the approaches used to place boxes in containers are described. In Sect. 4, heuristic methods used in the solution of container loading problem are mentioned. In Sect. 5, the performances of applied heuristic methods were evaluated. Finally, the results of the applied heuristic methods are described in Sect. 6.

2 Constraints for Container Loading Problem

This section includes the constraints used in the container loading problems. There are different types of constraints in the container loading problems [5]. We can divide them as compulsory and non-compulsory constraints. The most basic compulsory constraints for container loading problem are the followings:

- Each box is completely placed inside the container
- Each box should not conflict with another box
- Each box is placed parallel to the boundary surface of the container [6].

Non-compulsory constraints are shown in Fig. 1 [4].

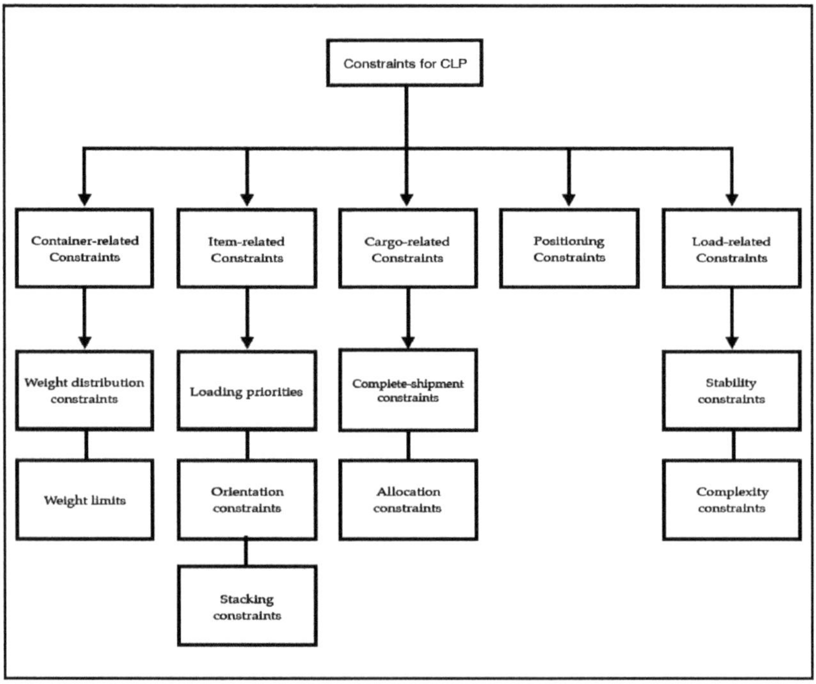

Fig. 1. Non-compulsory constraints for container loading problems

2.1 Container-Related Constraints

Weight Distribution Constraints. This constraint requires the weight to be spread across the container floor. The balanced transport of the items reduces risks during transport. Therefore, heavier items should be placed near the container bottom and lighter items should be placed higher.

Weight Limits. Generally, although the purpose is to maximize the container volume, this constraint is taken into account if the loaded items are too heavy or there is a weight limit that the container can take.

2.2 Item-Related Constraints

Loading Priorities. This constraint can be considered differently in the input minimization and output maximization problems. In case of input minimization, boxes with a high contribution to the objective function may have higher loading priority. Even if all the boxes are to be loaded, there may be constraints on shelf-life, delivery locations, necessarily load together certain boxes or definitely not being loaded them together.

Orientation Constraints. This constraint is the most commonly considered constraint in the literature. It can be positioned in six different ways when a box is allowed to rotate in any size. The boxes may not rotate in height dimension, but only width - length rotation can be allowed, or in any way rotation may not be allowed due to the load in the truck.

Stacking Constraints. It is about how to put a box on the other one. They result from the limited load-bearing strength of the boxes. Some factors are considered such as how much pressure the boxes can withstand or how many boxes can be placed on the box. If it is considered box-based, it may be necessary to discuss this constraint differently in any size as the surfaces of each box may be different.

2.3 Cargo-Related Constraints

Complete-Shipment Constraints. The problems of the output maximization type require that the box placed in the best possible way, but since there is not enough container space, some boxes must be left behind. Depending on whether a box type is loaded, it may need to be loaded in certain boxes. If an element of a subset is not installed at the same time, all other elements of that subset must also not to be installed. So either they are all loaded or none are loaded.

Allocation Constraints. This constraint is usually used for the multiple container loading problems. There may be boxes that should not be loaded into the same container in a single time. For example, food and cleaning products are not carried together. Conversely, all the boxes may need to be placed in the same container in order to take the order at a single time.

2.4 Positioning Constraints

This constraint considered when boxes are delivered to different customers. Some boxes may need to be placed at a specific point or a specific area of the container.

A particular set of boxes may need to be placed together or necessarily within a certain distance. In addition, certain areas can be separated for the boxes according to the loading and unloading conditions.

2.5 Load-Related Constraints

Stability Constraints. This constraint should be taken into account in order to avoid damage to the boxes or vehicles and employees carrying boxes. It ensures that a box that is in vertical position is balanced on the top of others or on the container floor. The bottom surface of the box must be supported to a certain extent when placed on the floor of the container or on the other box. In addition, it is also valid for supporting the side surfaces of the boxes in a horizontal condition with other boxes or container wall at a certain rate.

Complexity Constraints. Complex installation formats can cause a waste of time and difficulties in the case of manual installation of boxes. At the same time, automatic loading of complex structures may require additional work.

3 Box Placement Approaches

Different heuristic placement approaches were used to place boxes in the container loading problems [7].

1. Wall Building Approach: The container is often filled with vertical cuboid layers which run along the longest side of the container. Firstly, it is tried to be placed on the upper side as in Fig. 2, then on the front side as in Fig. 3, and the last on the right side as in Fig. 4.
2. Stack Building Approach: The boxes are packed in stacks, which are themselves arranged on the container floor in a way that saves the most space. Importance of this approach is that the stacks do not themselves form walls.
3. Horizontal Layer Building Approach: The container is filled from bottom to top in the form of horizontal layers that fill the maximum of the load surface underneath.
4. Block Building Approach: This approach is similar to horizontal layer building approach. The difference between them is to fill the bottom of the container with the least possible loss. The container is filled with blocks that have the same type of boxes with the same orientation.
5. Guillotine Cutting Approach: This approach is based on a slicing tree representation of the packing. Each slicing tree corresponds to the division of the container corresponding to the boxes into smaller pieces by guillotine cutting. For this reason, the leaf nodes correspond to the boxes to be placed [8].

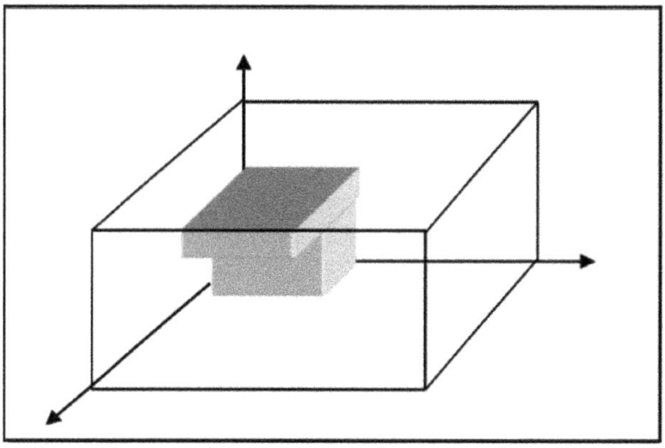

Fig. 2. Placing upper side

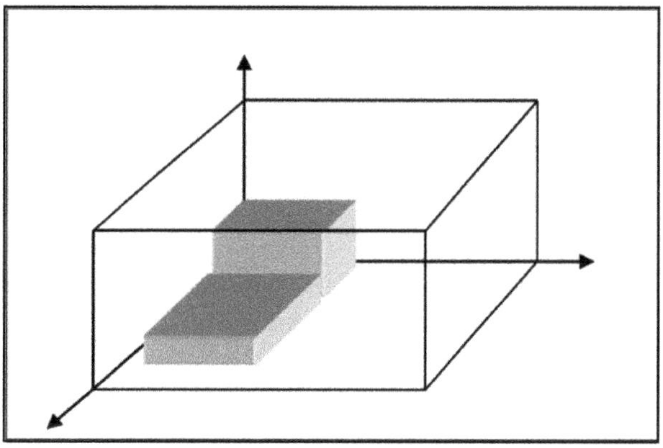

Fig. 3. Placing front side

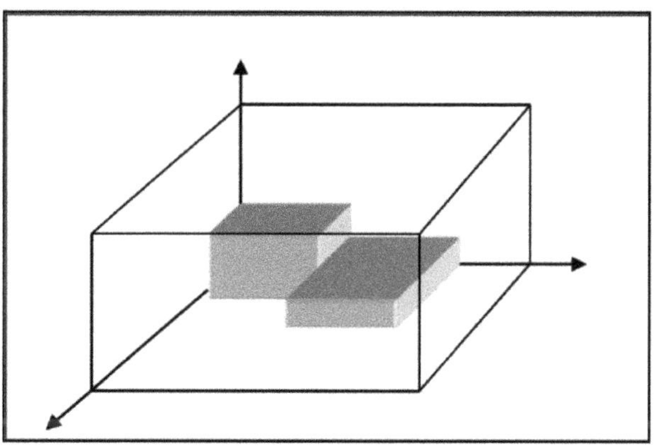

Fig. 4. Placing right side

4 Heuristics for Container Loading Problem

The heuristic word means the search technique that cannot guarantee fitness or the best solution, or in most case, the best-looking solutions without expressing how close a particular appropriate solution is to best [9]. Heuristic algorithms are algorithms inspired by natural phenomena to accomplish any purpose. These algorithms have convergence but cannot the guarantee the exact solution. They can only guarantee a solution close to the exact solution. Heuristic algorithm are needed because of the fact that heuristic algorithms can be much simpler in terms of decision-makers, optimization problems have a structure in which definite solution finding cannot be identified, and its usability as part of finding the exact solution [10]. The heuristic methods for container loading problems can be divided into three groups according to the method type:

- Tree Search Methods
- Heuristic Methods
- Meta-heuristic Methods

Tree Search Methods. Morabito and Araneles suggested AND/OR-graph approach [11]. Eley presented an algorithm that builds homogeneous blocks of identically orientated items [12]. Hifi proposed a tree search method using hill-climbing strategy [13]. Pisinger suggested a new heuristic based on the wall-building approach [7]. Bortfeldt and Mack proposed a heuristic which was derived from branch-and-bound approach [14]. Fanslau and Bortfeldt suggested tree search method using the block-building approach [8]. Zhang, Peng and Leung presented heuristic block-loading algorithm based on multi-layer search for the three-dimensional container loading problem [15]. Sheng, Wei, Zhiyuan and Xiwei proposed a binary tree search algorithm [16]. Araya and Riff used a beam search strategy which is variant of the branch-and-bound search [17]. Sheng et al. suggested a heuristic algorithm consist of a greedy sub-algorithm and a tree search sub-algorithm [28]. Terno et al. used branch-and-bound method with layer-wise loading strategy [30].

Heuristic Methods. Bischoff et al., Bischoff and Ratcliff, Lim et al. presented conventional heuristic methods like greedy algorithm or local search algorithm [18–20].

Meta-heuristic Methods. Gehring and Bortfeldt, Bortfeldt and Gehring suggested a genetic, parallel genetic and hybrid genetic algorithms [3, 5, 21]. Bortfeldt et al., Gendreau et al. presented a tabu search and a parallel tabu search algorithm [22, 29]. Mack et al., Zhang et al. and Sheng et al. proposed simulated annealing and hybrid simulated annealing algorithms [23, 24, 31]. Moura and Oliveira, Parreno et al. suggested a GRASP method for the container loading problems [25, 26]. Zhou and Liu presented a swarm optimization algorithm [27].

5 Experimental Results

The BR data set are generally used for solving the container loading problem. The BR data set consist of a total of 15 sets and there are 100 problem sets for each one. The box types within the sets vary from homogeneous boxes to strongly heterogeneous boxes. Table 1 shows how many different box types are in each set.

Table 1. Number of Box types

Test	Number of boxes
BR1	3
BR2	5
BR3	7
BR4	10
BR5	12
BR6	15
BR7	20
BR8	30
BR9	40
BR10	50
BR11	60
BR12	70
BR13	80
BR14	90
BR15	100

Table 2 shows the result of SPBBL [14], CLTRS [8], HBMLS [15], HBTS [16], BSG-CLP [17] and QTS [28] for BR1-BR15 according to volume utilization. The results range from 87.3% to 95.47%. The best result in each data set was obtained from CLTRS method.

Tables 3 and 4 shows the result of CBGAT [5], CBGAS-P [21], CBGAS [3], TSA [22], HYB [23], HCLPMC [31], GRModGRASP [25], GRASP [26] and SOA [27] for BR1-BR15 according to the volume utilization. The results range from 83.21% to 93.77%. The best result in each data set were obtained from different methods. For BR1 and BR2, the best result belongs to TSA method. For BR3 and BR4 the best result belongs to HYB method. The best result between BR5 and BR15 belongs to SOA method.

When tree search methods and meta-heuristic methods are compared, tree search methods has better results.

Table 2. Results of tree search methods

	SPBBL	CLTRS	HBMLS	HBTS	BSG-CLP	QTS
BR1	87.3	95.05	94.43	90.57	94.50	90.99
BR2	88.6	95.43	94.87	91.46	95.03	91.92
BR3	89.4	95.47	95.06	92.39	95.17	92.84
BR4	90.1	95.18	94.89	92.33	94.97	92.79
BR5	89.3	95.00	94.68	92.42	94.80	92.85
BR6	89.7	94.79	94.53	92.35	94.65	92.86
BR7	89.2	94.24	93.96	92.11	94.09	92.69
BR8	87.9	93.70	93.27	91.93	93.15	92.46
BR9	87.3	93.44	92.60	91.61	92.53	92.13
BR10	87.6	93.09	92.05	91.39	92.04	91.98
BR11	–	92.81	91.46	91.13	91.40	91.74
BR12	–	92.73	90.91	90.96	90.92	91.39
BR13	–	92.46	90.43	90.59	90.51	91.73
BR14	–	92.40	89.80	90.25	89.93	90.39
BR15	–	92.40	89.24	89.79	89.33	90.13

Table 3. Results of meta-heuristic methods

	CBGAT	CBGAS-P	CBGAS	TSA	HYB
BR1	85.80	88.10	87.81	93.52	93.26
BR2	87.26	89.56	89.40	93.77	93.56
BR3	88.10	90.77	90.48	93.58	93.71
BR4	88.04	91.03	90.63	93.05	93.30
BR5	87.86	91.23	90.73	92.34	92.78
BR6	87.85	91.28	90.72	91.72	92.20
BR7	87.68	91.04	90.65	90.55	91.20
BR8	–	90.26	89.73	–	–
BR9	–	89.50	89.06	–	–
BR10	–	88.73	88.40	–	–
BR11	–	87.87	87.53	–	–
BR12	–	87.18	86.94	–	–
BR13	–	86.70	86.25	–	–
BR14	–	85.81	85.55	–	–
BR15	–	85.48	85.23	–	–

Table 4. Results of meta-heuristic methods

	HCLPMC	GRModGRASP	GRASP	SOA
BR1	87.59	89.07	93.27	92.67
BR2	87.73	90.43	93.38	93.19
BR3	88.36	90.86	93.39	93.44
BR4	88.2	90.42	93.16	93.21
BR5	88.17	89.67	92.89	92.94
BR6	87.78	89.71	92.62	92.70
BR7	87.44	88.05	91.86	92.31
BR8	86.55	86.13	91.02	91.92
BR9	85.93	85.08	90.46	91.50
BR10	85.1	84.21	89.87	91.23
BR11	84.74	83.98	89.36	90.85
BR12	84.18	83.64	89.03	90.59
BR13	83.88	83.54	88.56	90.17
BR14	83.47	83.25	88.46	89.70
BR15	83.28	83.21	88.36	89.20

6 Conclusion

In national and international markets, transportation by land, air and sea is important for organizations to carry out transportation at a lower cost. The low cost brings benefits to the organizations as well as the satisfaction among the customers and organizations. One way of keeping costs low has been to maximize the volume of containers.

In this paper, we review that what is container loading problem, what are the heuristic algorithms suggested in the solution of the problem in literature and the results of suggested algorithms.

References

1. Dereli, T., Daş, G.S.: Konteyner Yükleme Problemleri için Karınca Koloni Optimizasyonu Yaklaşımı. Gazi Üniv. Müh. Mim. Fak. Der. **25**(4), 881–894 (2010)
2. Özsüt, Z.: Mathematical Models and Solution Methods for Container Loading Problems. Master thesis, Anadolu University, Eskisehir, Turkey (2015)
3. Bortfeldt, A., Gehring, H.: A hybrid genetic algorithm for the container loading problem. Eur. J. Oper. Res. **131**, 143–161 (2001)
4. Bortfeldt, A., Wäscher, G.: Constraints in container loading - a state-of-the-art review. Eur. J. Oper. Res. **229**, 1–20 (2013)
5. Gehring, H., Bortfeldt, A.: A genetic algorithm for solving the container loading problem. Int. Trans. Oper. Res. 4, 5, 6, 401–418 (1997)
6. Huang, Y., Hwang, F.J., Lu, H.: An effective placement method for the single container loading problem. Comput. Ind. Eng. **97**, 212–221 (2016)
7. Pisinger, D.: Heuristic for the container loading problem. Eur. J. Oper. Res. **141**, 382–392 (2002)

8. Fanslau, T., Bortfeldt, A.: A tree search algorithm for solving the container loading problem. INFORMS J. Comput. **22**(2), 222–235 (2010)
9. Russell, S.J., Norwig, P.: Artificial Intelligence a Modern Approach. Prentice Hall, Upper Saddle River, 07458 (1995)
10. Karaboga, D.: Yapay Zeka Optimizasyon Algoritmaları. Nobel Yayın Dağıtım (2011)
11. Morabito, R., Arenales, M.: An AND/OR-graph approach to the container loading problem. Int. Trans. Oper. Res. **1**(1), 59–73 (1994)
12. Eley, M.: Solving container loading problems by block arrangement. Eur. J. Oper. Res. **141**, 393–409 (2002)
13. Hifi, M.: Approximate algorithms for the container loading problem. Int. Trans. Oper. Res. **9**, 747–774 (2002)
14. Bortfeldt, A., Mack, D.: A heuristic for the three-dimensional strip packing problem. Eur. J. Oper. Res. **183**, 1267–1279 (2007)
15. Zhang, D., Peng, Y., Leung, S.C.H.: A heuristic block-loading algorithm based on multi-layer search for the container loading problem. Comput. Oper. Res. **39**, 2267–2276 (2012)
16. Sheng, L., Wei, T., Zhiyuan, X., Xiwei, L.: A tree search algorithm for the container loading problem. Comput. Ind. Eng. **75**, 20–30 (2014)
17. Araya, I., Riff, M.C.: A beam search approach to the container loading problem. Comput. Oper. Res. **43**, 100–107 (2014)
18. Bischoff, E.E., Janetz, F., Ratcliff, M.S.W.: Loading pallets with non-identical items. Eur. J. Oper. Res. **84**, 681–692 (1995)
19. Bischoff, E.E., Ratcliff, M.S.W.: Issues in the development of approaches to container loading. Omega **23**(4), 377–390 (1995)
20. Lim, A., Rodrigues, B., Wang, Y.: A multi-faced buildup algorithm for three-dimensional packing problems. Omega **31**, 471–481 (2003)
21. Gehring, H., Bortfeldt, A.: A parallel genetic algorithm for solving the container loading problem. Int. Trans. Oper. Res. **9**, 497–511 (2002)
22. Bortfeldt, A., Gehring, H., Mack, D.: A parallel tabu search algorithm for solving the container loading problem. Parallel Comput. **29**, 641–662 (2003)
23. Mack, D., Bortfeldt, A., Gehring, H.: A parallel hybrid local search algorithm for the container loading problem. Int. Trans. Oper. Res. **11**, 511–533 (2004)
24. Zhang, D.F., Peng, Y., Zhu, W.X., Chen, H.W.: A hybrid simulated annealing algorithm for the three-dimensional packing problem. Chin. J. Comput. **32**(11), 2147–2156 (2009)
25. Moura, A., Oliveira, J.F.: A GRASP approach to the container-loading problem. IEEE Intell. Syst. **20**(4), 50–57 (2005)
26. Parreno, F., Alvarez-Valdes, R., Oliveira, J.F., Tamarit, J.M.: A maximal-space algorithm for the container loading problem. INFORMS J. Comput. **20**(3), 412–422 (2007)
27. Zhou, Q., Liu, X.: A swarm optimization algorithm for practical container loading problem. In: IECON 2017 - 43rd Annual Conference of the IEEE Industrial Electronics Society, pp. 5690-5695 (2017)
28. Sheng, L., Hongxia, Z., Xisong, D., Changjian, C.: A heuristic algorithm for the container loading of pallets with infill boxes. Eur. J. Oper. Res. **252**, 728–736 (2016)
29. Gendreau, M., Iori, M., Laporte, G., Martello, S.: A tabu search algorithm for a routing and container loading problem. Transp. Sci. **40**(3), 342–350 (2006)
30. Terno, J., Scheithauer, G., Sommerweiss, U., Riehme, J.: An efficient approach for the multi-pallet loading problem. Eur. J. Oper. Res. **123**, 372–381 (2000)
31. Sheng, L., Xiuqin, S., Changjian, C., Hongxia, Z., Dayong, S., Feiyue, W.: Heuristic algorithm for the container loading problem with multiple constraints. Comput. Ind. Eng. **108**, 149–164 (2017)

A Deep Learning Model Based on Convolutional Neural Networks for Classification of Magnetic Resonance Prostate Images

Fatih Uysal[1(✉)], Fırat Hardalaç[1], and Mustafa Koç[2]

[1] Department of Electrical and Electronics Engineering, Gazi University, Ankara, Turkey
{uysal,firat}@gazi.edu.tr
[2] Department of Radiology, Fırat University, Elazığ, Turkey
mkoc@firat.edu.tr

Abstract. When looking at prostate cancer, it is seen that it is one of the very common types of cancer in men. In literature review, it is understood that there are a lot of studies for the treatment and diagnosis of this type of cancer with various image processing methods on prostate images. On prostate biopsy, secondary haemorrhage areas of T2-weighted magnetic resonance (MR) in prostate images can cause false diagnoses. T1-weighted MR prostate images help diagnose these cases. In such cases, in order to prevent misdiagnosis; A new classification procedure for MR prostate images with convolutional neural networks (CNN) was performed. As a result of this process, a new deep learning model based on CNN which can classify T1-weighted and T2-weighted MR prostate images has been developed.

Keywords: Convolutional neural networks · Deep learning · Magnetic resonance prostate images · Image classification · Artificial intelligence

1 Introduction

Prostate cancer is one of the very common types of cancer in men. Imaging of the prostate organ using computer tomography, transrectal ultrasonography and MRI can be performed. [1] Anatomically, MR prostate images are divided into two types as T1-weighted images and T2-weighted images. T1-weighted prostate images are used as an aid to diagnosis in biopsy relationship bleeding and in such images these areas of hemorrhage appear to be hyperintense. Secondary hemorrhage areas in T2-weighted MR prostate images of prostate biopsy may cause false diagnosis of tumor and in these cases T1-weighted MR prostate images may be used to aid diagnosis [2]. In such cases, in order to help diagnose and prevent misdiagnosis, in this study, a new CNN-based deep learning model was developed for the classification of T1-weighted and T2-weighted MR prostate images.

When the literature is examined, it has been observed that many studies have been performed with various image processing methods by using prostate images for the

D. J. Hemanth and U. Kose (Eds.): ICAIAME 2019, LNDECT 43, pp. 701–708, 2020.
https://doi.org/10.1007/978-3-030-36178-5_59

treatment and diagnosis of prostate cancer. These studies; segmentation and classification of prostate images in general is understood to be in the form. Looking at the classification of prostate images in recent years; by Gummeson, the study of the classification of prostate cancer using CNNs [3]; by Liu et al., a deep learning based classification model for prostate cancer [4]; by Sahran et al., multi-level approach to learning work to classify the images of the prostate [5]; by Arvidsson et al., the study of generalization of prostate cancer classification with deep learning [6]; by Li et al., diagnosis of prostate cancer with path R_CNN [7]; by Reda et al., diagnosis of prostate cancer using a new CNN-based model [8]; by Tian et al., a new image-specific transfer approach for segmentation of prostate in MR images [9] were performed. In this study; In addition to the literature, a new deep learning model based on CNN, which can classify T1 and T2-weighted MR prostate images, has been developed by increasing classification accuracy in the previous study [10].

In the Sect. 2 of this study; CNNs which are one of the deep learning algorithms and used when classifying are elicudated under the title of method. In Sect. 3; the application part of the study is mentioned. Finally; the results and recommendations are explained.

2 Methods

As the method used to classify magnetic resonance prostate images, a deep learning model based on CNN was developed and used. CNN provides successful results when applied to a lot of problems such as image classification, image segmentation and pattern recognition. The network, which has a deepened structure as a result of increasing the number of hidden layers in artificial neural networks, is expressed as CNN. Another important feature that separates these two networks; is DropOut method used in convolutional neural networks to prevent memorization during network training [11].

A CNN is generally a combination with a layer of input and a layer of output in addition to many layers of hidden. Layers of hidden can be layers of fully connected, layers of convolution and layers of pooling. A simple CNN structure is indicated in Fig. 1. This CNN structure consists a layer of input, a layer of output, a layer of convolution and a layer of pooling [4].

Among the layers forming the CNN; in addition to the input, output, convolution, pooling and fully connected layers, DropOut layer, the rectified linear units layer (ReLU) and classification layer can also be included. If the layers that make up the CNN are mentioned.

Layer of Input: It is the first layer of CNN. In this layer, the data (images) are given to the network as raw. In some cases, however, reduction the size of the input images may be required to reduce the need for memory and shorten the duration of training. In such cases, the input images are given to the network after the image size reduction.

Layer of Convolution: In convolution layer, convolution process is performed by using filters of various sizes on the input images and a feature map is created. The filter coefficients change continuously with each learning during training, so that the network can determine which parts of the image are more important [11]. Convolution is

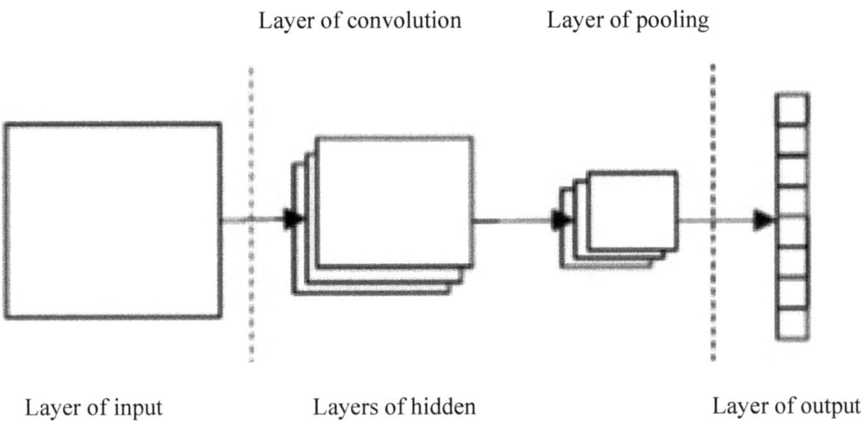

Fig. 1. CNN structure [4].

performed by applying the filter to the input image from left to right and from top to bottom according to a certain stride value. A 3×3 output matrix obtained when applying a 2×2 filter with a number of strides 1 to an input matrix of 4×3 is shown in Fig. 2.

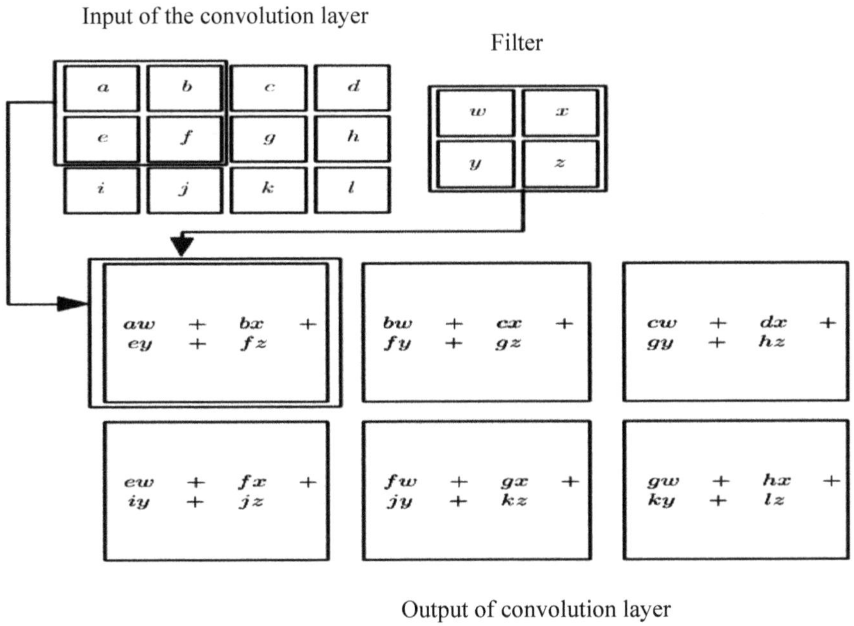

Fig. 2. Convolution process [12].

Rectified Linear Units Layer: It is the layer used after the convolution layer. This layer generates new output by resetting the input data to be negative. CNN learns faster with the use of this layer [11]. The part where this layer is located, can also be called the detector stage in some cases [12].

Layer of Pooling: The layer of pooling is frequently used after the ReLU layer. The basic function of this layer is to decrease the size of the input image of the next convolution layer. The size reduction performed in this layer reduces the computational load and prevents the system from memorizing. As in the layer of convolution, filters of various sizes are also used in this layer [11]. These filters; makes the pooling process determined on the input image of the pooling layer. This pooling process is performed by shifting the filter from left to right on the input image and from top to bottom according to a specific stride value. This can be the pooling process of average or maximum. In pooling of average, the averages of the values in the input matrix are calculated. In pooling of maximum, the maximum value of the input matrix is found.

Fully Connected Layers: In CNN; usually there is a fully connected layer immediately after the pooling layer. This fully connected layer is linked to all areas of the pooling layer that precedes it [11].

DropOut Layer: In case of training on CNN using large data; sometimes it can do network memorization. In such cases, this layer can also be used to remove some nodes of the network and to prevent network memorization [11]. In Fig. 3, from left to right respectively; a normal network and the structure of a CNN after this layer are shown.

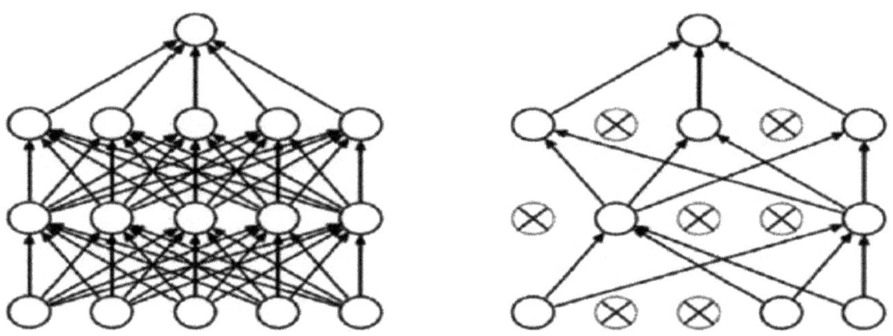

Fig. 3. A CNN after a DropOut layer

Classification Layer: After the fully connected layer, the classification layer is used. The number of objects to be classified and the output of the classification layer must be the same. Although softmax classifier is generally preferred due to its success, it can be used in this layer in different classifiers [11].

3 Implementation

In this study; A new CNN-based deep learning model has been developed that can classify T1-weighted and T2-weighted MR prostate images. MR prostate images used in the study were obtained via a website shared as open data and detailed information about access to data is mentioned in the 'information' section.

The information required for the classification of MR prostate images is shown in Table 1. In this table, image classes and the number of images belonging to these classes are given. In addition, the slice ranges of the images belonging to the classes are also indicated. They have different values for each class.

Table 1. MR prostate image classes

MR prostate images	MR prostate image numbers		Slice ranges
	Test data	Training data	
T1 axial	50	255	28 < S < 48
T1 axial coilout	50	255	44 < S < 56
T2 fse axial	50	255	25 < S < 47
T2 fse sagittal	50	255	26 < S < 37
T2 fse coronal	50	255	29 < S < 36

From the MR prostate images, 5 different classes can be created. These can be T1 axial, T1 axial coilout, T2 fse (fast spin echo) axial, T2 fse sagittal and T2 fse coronal. In this study, 255 images were used as the training data for each class. There are a total of 1275 training data. In the test data, 50 images are used for each class and there are a total of 250 test data.

One image of MR prostate images of each class is shown in Fig. 4.

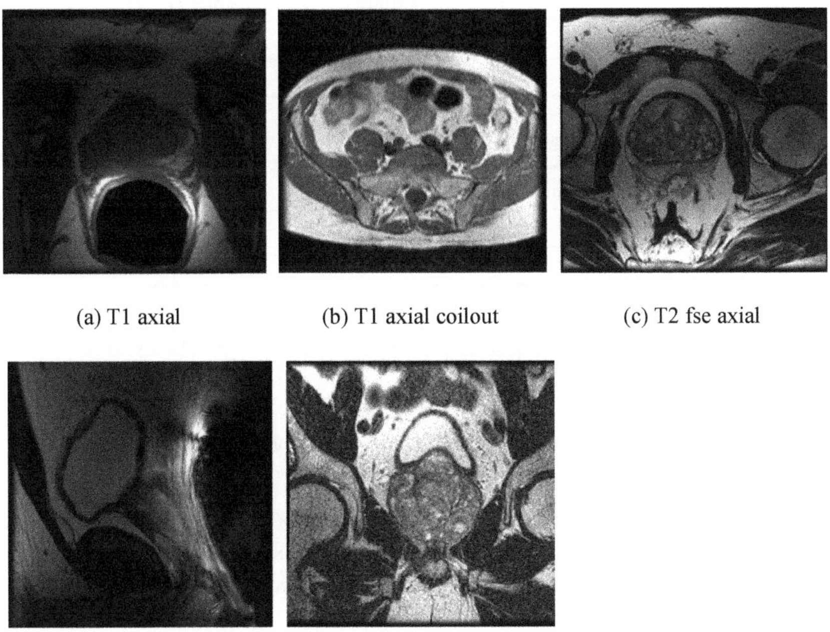

(a) T1 axial (b) T1 axial coilout (c) T2 fse axial

(d) T2 fse sagittal (e) T2 fse coronal

Fig. 4. MR prostate images for each class

Firstly training and test data was created. Then, the implementation stage of the study was started. This step was carried out using MATLAB program and MatConvNet. MatConvNet is the MATLAB toolbox and CNN can be implemented using this toolbox. Thus; image classification, segmentation etc. operations can be performed [13].

Each of the image data used in the study is 256 × 256 pixels in size. First of all, it is aimed to reduce the duration of training, while preparing the image database. For this reason, the size of each image is reduced to 128 × 128 pixels. This reduction leads to a slight loss in the data. But this process is very important in terms of shortening the training period. Due to the classification results obtained, it can be said that this situation does not affect the performance of the system in a negative way. After this reduction, a new CNN model has been developed that can classify MR prostate images.

When developing the CNN model, the convolution layer and pooling layer output matrix values were calculated with the formulation of $(W - F + 2P)/2S + 1$. In this formulation; W is input image size, F is filter value of the corresponding layer, P is zero padding value and S is step value [14].

The new deep learning model based on CNN developed for the classification of T1-weighted and T2-weighted MR prostate images is shown in Fig. 5.

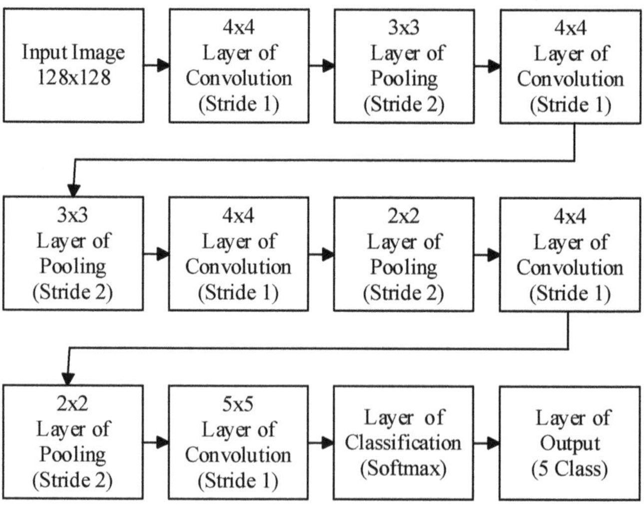

Fig. 5. CNN-based deep learning model for the classification of MR prostate images

An MR prostate image with an input image size of 128 × 128 pixels can be applied to this CNN-based deep learning model. After this process, this input image can be passed through the layers shown in Fig. 5 respectively. As a result, the class of this image can be detected in the output layer.

4 Results and Recommendations

The CNN-based deep learning model is used to classify T1 and T2-weighted MR prostate images. Initially, training is carried out by using the images in the training data. After this process, the weight values of the filters in the layers are determined. Finally, the classification resulting from the verification of this model with the images in the test data is indicated in Fig. 6.

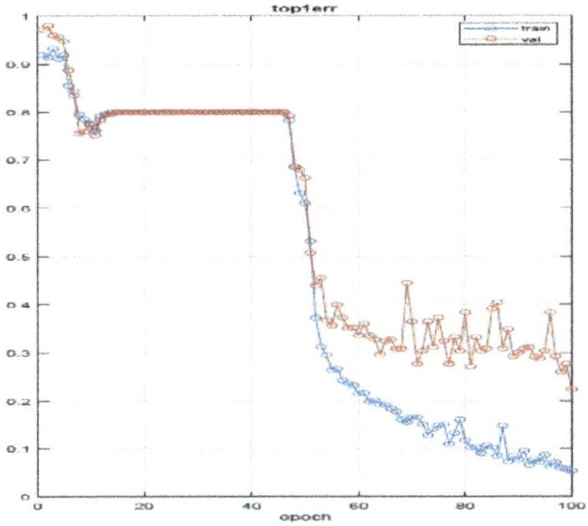

Fig. 6. The result of the classification of MR prostate images

The classification accuracy rates obtained in the first CNN model using the same data are approximately 90% for the training data and 70% for the test data [10]. In the new CNN-based deep learning model, the classification accuracy was increased by using different convolution and pooling layers from the first model.

As a result of the training of the developed new CNN model with MR prostate images, as shown in Fig. 6, the accuracy rates in the classification are %94.7 for the training data and %77.6 in the test data. The process of classifying test data; T1 and T2-weighted (all test data) prostate images were tested.

CNN models can be used in future studies to reduce the error of classification of MR prostate images and thus increase the accuracy of classification. The number and types of layers used in the hidden layer can be increased, the size of the used filters can be changed and/or the number of images used for training can be increased.

Information

The data used in this study has a png extension and a size of 256 × 256 pixels. These data are obtained from the internet address of http://prostatemrimagedatabase.com which is available as of 15.08.2018.

References

1. Gençhellaç, H., Yılmaz, E.: Prostat Görüntüleme. Türk Radyoloji Seminerleri **3**, 138–148 (2015)
2. Bakır, B., Tuğcu, V., Binbay, M., et al.: Prostat Kanserinde Multiparametrik Manyetik Rezonans Görüntüleme ve Hedefe Yönelik Biyopsi Kılavuzu. Türk Üroloji Derneği, İstanbul (2018)
3. Gummeson, A: Prostate cancer classification using convolutional neural networks. Master's thesis, Faculty of Engineering, Lund University (2016)
4. Liu, Y., An, X.: A classification model for the prostate cancer based on deep learning. In: Li, Q., Wang, L., Zhou, M., Sun, L., Qiu, S., Liu, H. (eds.) 10th International Congress on Image and Signal Processing, Biomedical Engineering and Informatics 2017, CISP-BMEI, pp. 1–6. IEEE (2017)
5. Sahran, S., Abdullah, A., Albashish, D., Abu-Taleb, S.J.: Multi-level learning approach for prostate histopathology images classification. In: 8th International Conference on Information Technology 2017, ICIT, pp. 947–955. IEEE (2017)
6. Arvidsson, I., Overgaard, N.C., Marginean, F.E., Krzyzanowska, A., Bjartell, A., Åström, K., Heyden A.: Generalization of prostate cancer classification for multiple sites using deep learning. In: 15th International Symposium on Biomedical Imaging 2018, ISBI, pp. 191–194. IEEE (2018)
7. Li, W., Li, J., Sarma, K.V., Ho, K.C., Shen, S., Knudsen, B.S., Gertych, A., Amold, C.W.: Path R-CNN for prostate cancer diagnosis and gleason grading of histological images. IEEE Trans. Med. Imaging **38**(4), 945–954 (2018)
8. Reda, I., Ayinde, B.O., Elmogy, M., Shalaby, A., El-Melegy, M., et al.: A new CNN-based system for early diagnosis of prostate cancer. In: 15th International Symposium on Biomedical Imaging 2018, ISBI, pp. 207–210. IEEE (2018)
9. Tian, P., Qi, L., Shi, Y., Zhou, L., Gao, Y., Sheri, D.: A novel image-specific transfer approach for prostate segmentation in MR images. In: International Conference on Acoustics, Speech and Signal Processing 2018, ICASSP, pp. 806–810. IEEE (2018)
10. Uysal, F., Hardalaç, F., Koç, M.: Classification of T1 and T2 weighted magnetic resonance prostate images using convolutional neural networks. In: Medical Technologies National Congress 2018, TIPTEKNO, pp. 1–4. IEEE (2018)
11. İnik, Ö., Ülker, E.: Deep learning and deep learning models used in image analysis. Gaziosmanpasa J. Sci. Res. **6**(3), 85–104 (2017)
12. Goodfellow, I., Bengio, Y., Courville, A.: Deep Learning. An MIT Press, Cambridge (2016)
13. MatConvNet Homepage. http://www.vlfeat.org/matconvnet/. Accessed 30 Jan 2019
14. CS231n Homepage. https://cs231n.github.io/convolutional-networks/. Accessed 30 Jan 2019

Effect of Representation of Information in the Input of Deep Learning on Prediction Success

Hikmet Yücel(✉)

Inovasyon Mühendislik Ltd., 26480 Eskişehir, Turkey
hikmet.yucel@inovasyonmuhendislik.com

Abstract. Stock price prediction has been always a hot topic in financial mathematics. There are lots of studies with various techniques on this area. Prediction studies are based on the assumption that the next price has a relation with the past values. On the other hand, artificial neural networks and especially deep learning is one of the best tools for extraction of the relations between inputs and outputs. Naturally, deep learning is becoming a widely used tool day by day. There are some developed neural nets especially for time series like RNN and LSTM. Besides the structure of the neural net, the type of the input is also important for the success of the prediction. Same information can be represented in different formats without any loss. This study investigates the effect of representation of the information in the input on the prediction success. Experiments are done with the USDTRY day close values starting from 1994 till 2017. 4 different representations are tested namely time series itself, Real DFT, DWT and DCT. The transforms used are lossless and time series can be reconstructed exactly so they are just different representations of the information. 64 tests with different time spans are conducted. The same neural net structure with 512 input neurons, 3 hidden layers, 10 output neurons is used for all tests except a slight difference for Real DFT. The results show that DCT gives the best results due to its high energy compaction property. Energy compaction results a sorted array in terms of significance therefore results stable weight calculation in neural nets.

Keywords: Stock price prediction · Deep learning · Representation of information · Real DFT · DWT · DCT

1 Introduction

Time series data can be defined as collection of large number of data values within a uniform time interval. This time interval can be represented as year, month, week, day etc. Over last few decades, time series analysis is an attractive area among researchers. The aim is to discover the characteristics (frequency, amplitude, phase, stationarity level, seasonality length, etc.) of a process by removing the randomness.

Time series modeling is to study the past observations and to developed a model that presents the inherent patterns in the series. This developed model can be used to predict the future values. Therefore time series forecasting can be expressed as

D. J. Hemanth and U. Kose (Eds.): ICAIAME 2019, LNDECT 43, pp. 709–723, 2020.
https://doi.org/10.1007/978-3-030-36178-5_60

predicting the future by understanding the past [1]. For the success of time series forecasting, fitting a proper model to the analysed time series is important.

Because of the particular importance of forecasting in many fields such as science, business, finance [2–4] etc., time series studies have many application areas.

In many application areas, time series forecasting is studied for 3 different periods:

- Short-Term period (less than three months)
- Mid-Term period (three months to one year)
- Long-Term period (more than a year)

Time series forecasting has a particular importance in stock market. The stock market may contain high risks for holders of common shares due to price volatility. Over many years, a lot of works have been done to improve the forecasting accuracy and various important techniques are developed in literature for prediction of future values:

- Autoregressive (AR) and Moving-Average (MA) approaches assumes that a value in a time series is a weighted linear sum of past values and/or residual deviations and time series follows a particular known statistical distribution. Autoregressive Integrated Moving Average (ARIMA) [5–7] model has subclasses of other models, such as the Autoregressive (AR) [5, 8], Moving Average (MA) [9] and Autoregressive Moving Average (ARMA) [7, 8] models. Exponential Smoothing methods are subset of autoregressive approach whose weights decrease exponentially.
- Transfer-function Models tries to link the output by several inputs. The relationship between the inputs and outputs can be estimated using previous time intervals.
- Time-Series Decomposition methods breaks up the series into components. These components may be trend, seasonal and irregular components or low, medium or high frequencies.
- Evolutionary Algorithms (EA) [15, 16] try to optimize (minimize/maximize) the given objective function under the specified constraints.
- Fuzzy Based Techniques considers more than one observations such as low, medium, high, close prices. They composed of the fuzzification stage, defining the fuzzy rules & relations and the defuzzification stage.
- Support Vector Machines (SVM) [10–12] implements structural risk minimization principle to minimize an upper bound of generalization error and overfitting is unlikely to ocur.
- Artificial Neural Networks (ANN) [9, 13, 14] are biologically inspired algorithms that have inherent capability of non-linear modeling, without any presumption about the statistical distribution followed by the observations.

Although, linear models are popular due to their relative simplicity in understanding and implementation; in practical real life, time series mostly have non-linear patterns. Recently, ANNs gained immense popularity in time series forecasting. The basic idea behind the invention of ANNs was to construct a model for mimicking the intelligence of human brain into machine. Like a human brain, ANNs try to identify regularities and patterns in the input, learn from experience and then provide generalized results based on acquired knowledge. The main advantage of the ANNs is their inherent capability of non-linear modeling, without any assumption about the statistical distribution so ANNs can capture complex data patterns. ANNs are data driven and constructs the appropriate model adaptively based on the historical data. ANNs have been successfully applied in especially for forecasting and classification.

There are many studies in literature for time series forcesting based on ANNs [17–23]. Some of them contains hybrid techniques which combines ANNs and other forecasting methods [24, 25].

There are different metrics defined in literature for performance evaluation of predicition. With these metrics a prediction technique can be compared with other techniques. The most used performance evaluation metircs are given below:

The Mean Absolute Error (MAE):

$$\mathrm{MAE} = \frac{1}{n}\sum\nolimits_{t=1}^{n} |e_t| \tag{1}$$

The Mean Absolute Percentage Error (MAPE):

$$\mathrm{MAPE} = \frac{1}{n}\sum\nolimits_{t=1}^{n} \left|\frac{e_t}{y_t}\right| 100 \tag{2}$$

The Mean Percentage Error (MPE):

$$\mathrm{MPE} = \frac{1}{n}\sum\nolimits_{t=1}^{n} \left(\frac{e_t}{y_t}\right) 100 \tag{3}$$

The Mean Squared Error (MSE):

$$\mathrm{MSE} = \frac{1}{n}\sum\nolimits_{t=1}^{n} e_t^2 \tag{4}$$

The Mean Square Root Error (MSRE):

$$\mathrm{MSRE} = \frac{1}{n}\sum\nolimits_{t=1}^{n} \sqrt{e_t^2} \tag{5}$$

The Sum of Squared Error (SSE):

$$\mathrm{SSE} = \sum\nolimits_{t=1}^{n} e_t^2 \tag{6}$$

The Root Mean Squares Error (RMSE):

$$\mathrm{RMSE} = \sqrt{\frac{1}{n}\sum\nolimits_{t=1}^{n} e_t^2} \tag{7}$$

Using machine learning techniques, selecting the right input variables is very important. There must be actually some kind of correlation between input and output variables to make ANNs learn. Most of the studies uses some form on lagged price values. The commonly used parameters are daily opening, high, low and close prices. Some of the studies used technical indicators as input such as simple moving average (SMA), exponential moving average (EMA), relative strength index (RSI), rate of change (ROC), moving average convergence/divergence (MACD).

In this study, the effect of the representation of the information in the input on the success of the prediction is investigated. 4 types of representation are used as input in the experiments namely time series itself, RDFT, DWT, DCT. All representations are lossless so the results can be compared. The theoretical background of these representations and deep learning is given in Sect. 2. The details about the experiments and their results are in Sect. 3. Conclusion and future works are described in Sect. 4.

2 Theoretical Background

In this study, 3 different transforms namely RDFT (real discrete fourier transform), DWT(discrete wavelet transform) and DCT(discrete cosine transform) are used for preprocessing of the input. Before focusing the experiments done, brief information about theory of the mentioned transforms and deep learning is given.

2.1 RDFT – Real Discrete Fourier Transform

DFT is a transformation of finite sequence of equally spaced samples of a function into a same length sequence of equally spaced samples of the Discrete Time Fourier Transform (DTFT).

DFT transforms $\{x_n\} := x_0, x_1, \ldots, x_{N-1}$ into $\{X_k\} := X_0, X_1, \ldots, X_{N-1}$ using Eq. (8) where x_n and X_k are sequence of complex numbers with length N.

$$X_k = \sum_{n=0}^{N-1} x_n e^{-\frac{i2\pi}{N}kn} \tag{8}$$

$$X_k = \sum_{n=0}^{N-1} x_n[\cos(2\pi kn/N) - i\sin(2\pi kn/N)] \tag{9}$$

As mentioned above, the output of the DFT is complex numbers. In stock price series, the values are real and DFT of the real valued sequence results to complex conjugate pairs. Therefore the output of the DFT of stock price series has also some redundancy. RDFT is a variant of fourier transform. RDFT is a special alignment of DFT output. It omits the redundant values and realign the others as in Fig. 1.

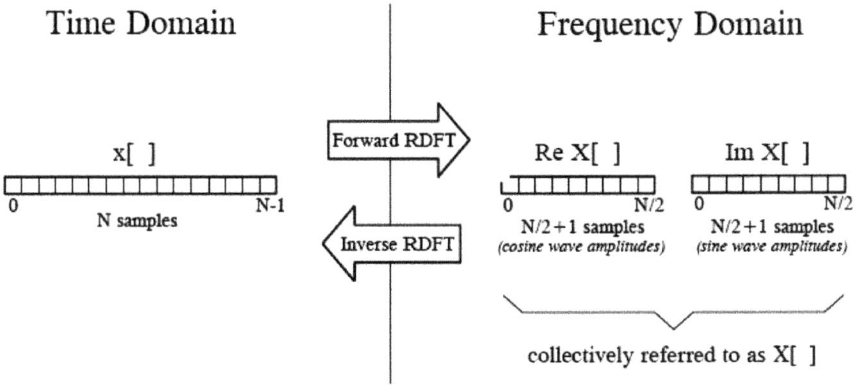

Fig. 1. RDFT output construction [26].

2.2 RDFT – Real Discrete Fourier Transform

Discrete Wavelet Transform (DWT) converts a finite sequence into a group of constituent signals (wavelets). The wavelet components have well defined dominant frequency. Wavelets are transient functions of short limited duration centered around a specific time. Different wavelets shown in Fig. 2 are used in literature for different applications.

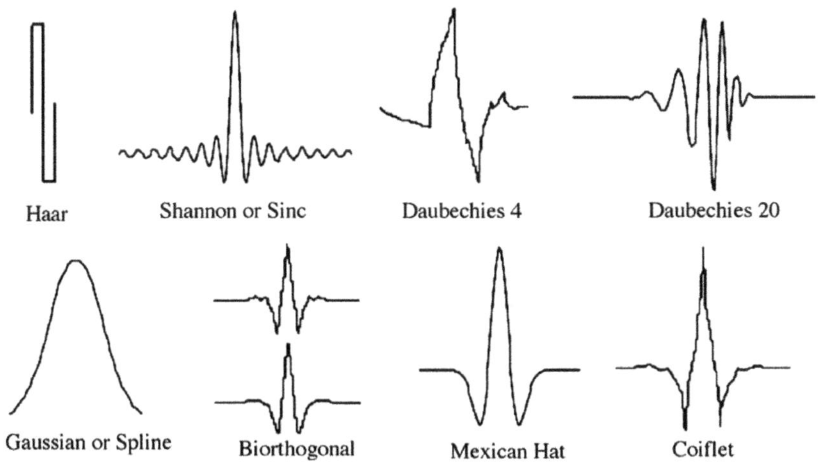

Fig. 2. Wavelet types.

DWT of a signal can be obtained by passing it through a low pass filter (LPF) and a high pass filter (HPF) in parallel as shown in Fig. 3. These two filters must be quadrature mirror filters. LPF and HPF are determined according to the selected wavelet. These filters remove the half of the frequency content, therefore considering the Nyquist rule, the output of the filters can be downsampled with factor of 2 without the loss of information. The output of the LPF and HPF is called approximation and detail respectively.

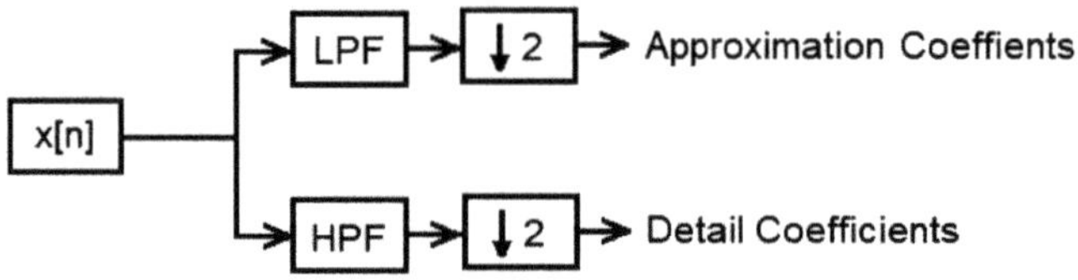

Fig. 3. DWT outputs.

The DWT can be applied in a cascaded manner such as in Fig. 4.

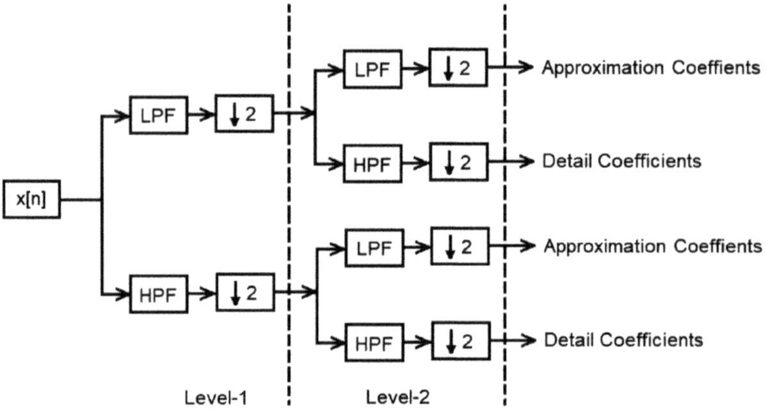

Fig. 4. Cascaded DWT.

In this manner, - with input length N - the output of the DWT after M level is 2^M real valued sequence with length $N/2^M$.

The main advantage of the DWT over DFT is the temporal resolution. DWT allows an analysis in both time and frequency domains, therefore it gives information on the evolution of the frequency content of a signal over time.

2.3 DCT – Discrete Cosine Transform

Discrete cosine transform (DCT) is a well known transformation since the output is real valued and it has wide range of applications. DCT decomposes a finite sequence of data points into sum of cosine functions at different frequencies. The output of the transform is real valued magnitudes of these cosines.

$$X_k = \sum_{n=0}^{N-1} x_n \cos\left[\frac{\pi}{N}\left(n + \frac{1}{2}\right)k\right] \qquad k = 0, \ldots, N-1 \tag{10}$$

DCT uses only cosine functions whereas DFT uses both cosines and sines in the form of complex exponentials. However, DCT implies different boundary conditions from the DFT.

The major advantage of DCT is the energy compaction. It concentrates energy of the sequence on the leading values of the output. The more values kept means the more detailed approximation. Therefore, it is used widely data and image compression. Keeping the leading values and discarding the rest is enough get the approximation of the signal.

2.4 Deep Learning

Deep learning is a subfield of machine learning. It is based on the Artificial Neural Networks (ANN) which deals with algorithms inspired by the structure and function of the brain. ANN, tries to imitate the evaluation of information in human brain using similar software structures to neurons. A neuron cell is given in Fig. 5. Dendrites are input and axon is the output of the cell. Nuerons are connected to each other by synapse connections.

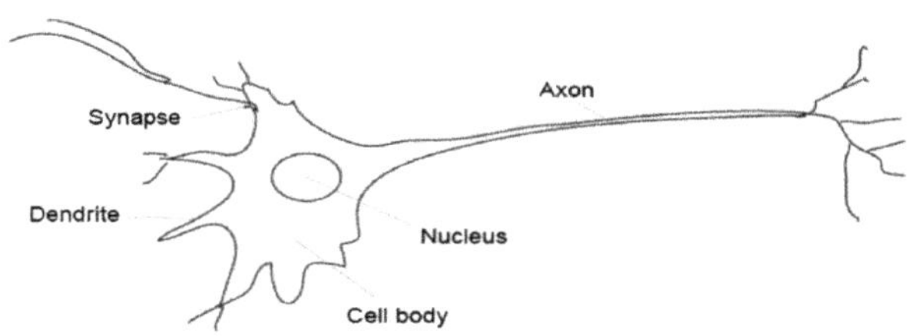

Fig. 5. Structure of a neuron cell.

The perceptron structure which simulates a neuron cell is shown in Fig. 6. Different functions like step, sigmoid etc. can e used as a activation function.

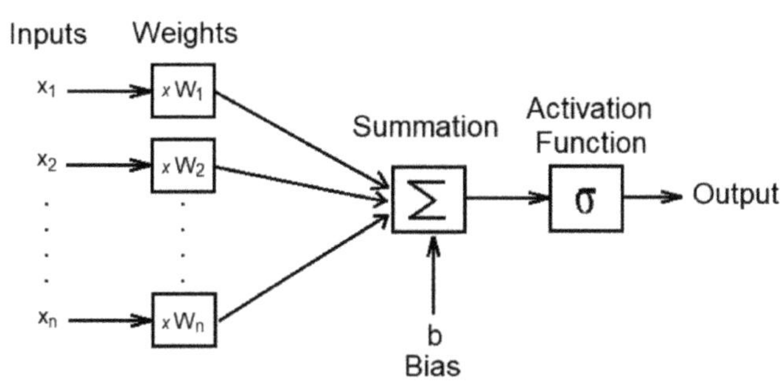

Fig. 6. Structure of a perceptron.

Deep learning networks are similar to how nervous system structured where each neuron connected each other and passing information as shown in Fig. 7.

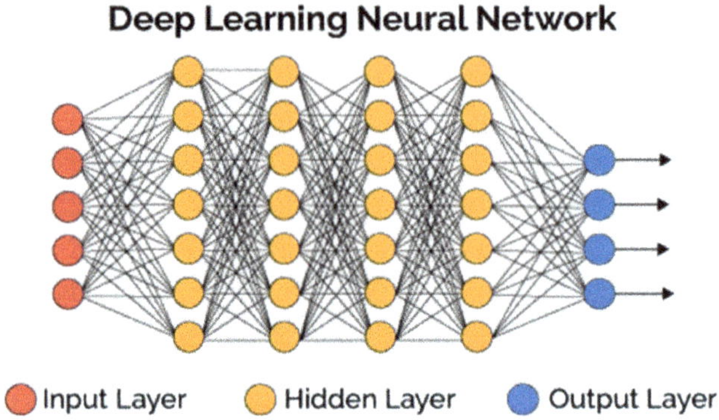

Fig. 7. ANN layers.

There are two basic phases in deep learning algorithms namely training and test. In train phases, the algorithm tries to minimize the error in the output of the given input by updating the weights using train data. In the test phase, the trained net are feeded with new inputs and the outputs of the net are evaluated with an error metric. There are various types of networks developed in the literature for different applications. Convolutional Neural Networks (CNN) is widely used for pattern recognition in images. Recurrent Neural Network (RNN) and a special type Long Short Time Memory Network (LSTM) are successful especially in time series analysis. Generative Adversarial Networks (GAN) are composed of two nets namely Generator and Discriminator. GAN's has a huge potential, because they can learn to mimic any distribution of data. The brief history of ANN and deep learning is given in Fig. 8.

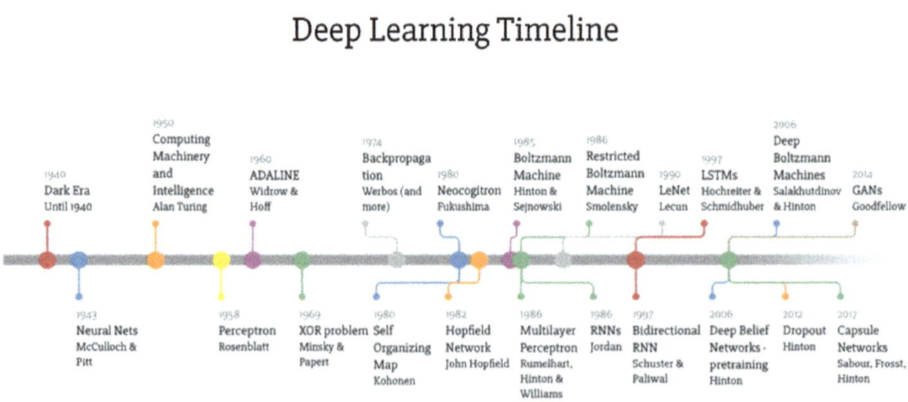

Fig. 8. Deep learning timeline [27].

3 Experiments

Experiments are done with the USDTRY day close values with length 5991, starting from 13.06.1994 till 28.04.2017. Two sliding windows are used to create the inputs and outputs. A sliding window with length 512 is used for inputs. Another sliding window following the first with length 10 for the outputs. 5469 input-output pairs are created with this method. Data creation is process is given in Fig. 9.

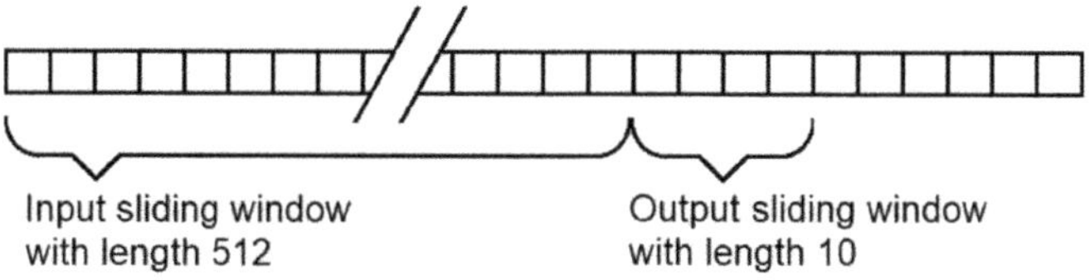

Fig. 9. Creation of input-output pairs.

The structure of the neural network is formed as shown in Fig. 10. The input layer is composed of 512 neurons for each historic day close value input and the output layer has 10 neurons for each future day prediction output. Between them there are 3 fully connected hidden layer with number of neurons 128, 64, 32 respectively. Relu activation function shown in Fig. 11 is used after each hidden layer. A regression layer is used after the output layer. This is a common neural network structure with 3 hidden layers. This classical structure is specially selected to observe the importance of the representation of the information in the input.

1	'sequenceinput'	Sequence Input	Sequence input with 512 dimensions
2	'fc_1'	Fully Connected	128 fully connected layer
3	'relu_1'	ReLU	ReLU
4	'fc_2'	Fully Connected	64 fully connected layer
5	'relu_2'	ReLU	ReLU
6	'fc_3'	Fully Connected	32 fully connected layer
7	'relu_3'	ReLU	ReLU
8	'fc_4'	Fully Connected	10 fully connected layer
9	'regressionoutput'	Regression Output	mean-squared-error with response 'Response'

Fig. 10. Structure of the neural network.

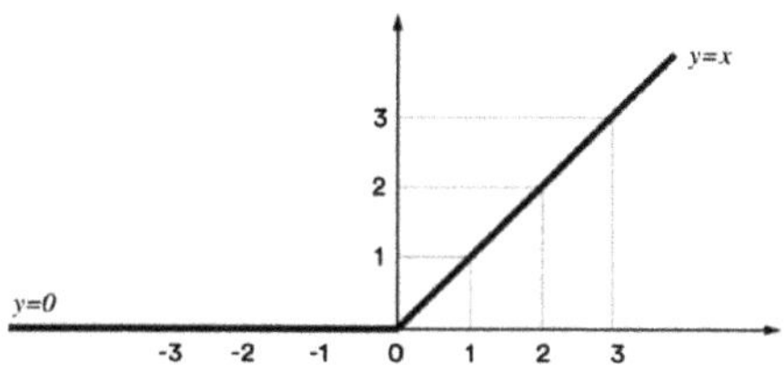

Fig. 11. ReLU activation function.

In real world, with the passing time, the older stock price information becomes less important. Therefore, price info older than 1440 days (more than 48 months) are not used in training. In addition, to have a better test conditions, 16 experiments were implemented with different time slices that are shifted 90 days. For each experiment, 1440 input-output pairs is used for train and next 90 input-output pairs are used for test. The selections of the train and test data is shown in Fig. 12.

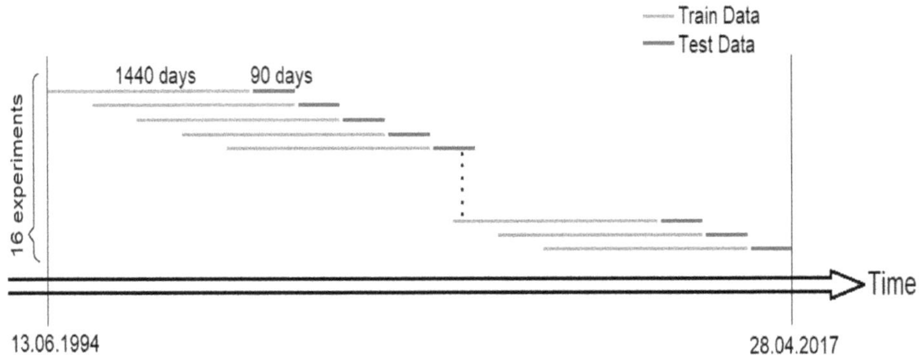

Fig. 12. Selection of train and test data.

There are 16 experiments and 4 different representations, so we have 64 different training and test session. For each training, epoch number is limited to 2000, mini batch size is set to 10, initial leaning rate is set to 1e − 4 and adam solver is used. The Mean Absolute Percentage Error metric is used to evaluate the success.

The first representation of input is used as time series itself. The second representation is obtained using RDFT. Since USDTRY values are real, DFT of them is conjugate symmetric. Using RDFT, for an input with length N, we obtain two outputs with length N/2 + 1 which are the coefficients of real and imaginary parts. We concatenate them and used as input. Since the total length is N + 2, the neural network structure is slightly changed by adding 2 more neurons to the input layer. Third representation is DWT and the fourth is DCT.

The examples from outputs of the experiments are shown in Figs. 13, 14, 15 and 16. The mean of the all outputs of the experiments are calculated and given in Fig. 17 & Table 1.

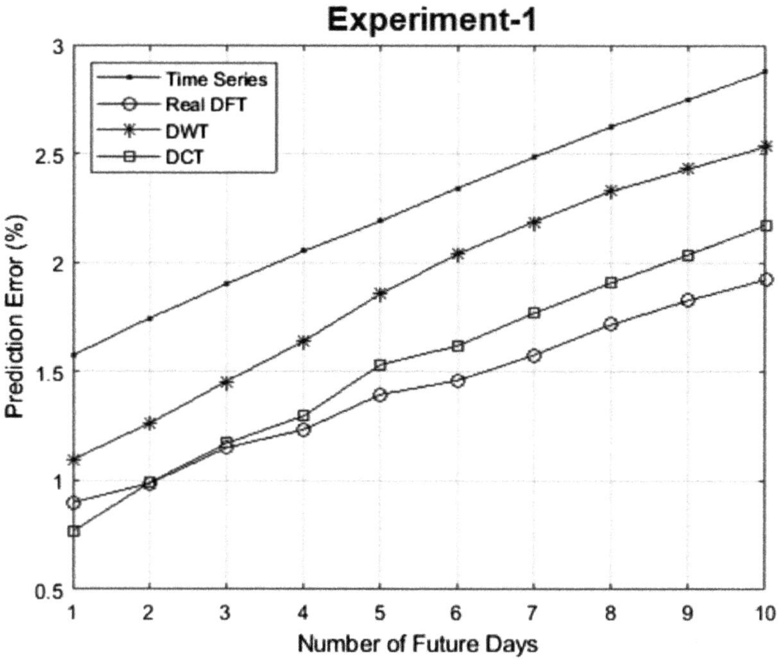

Fig. 13. Example output 1.

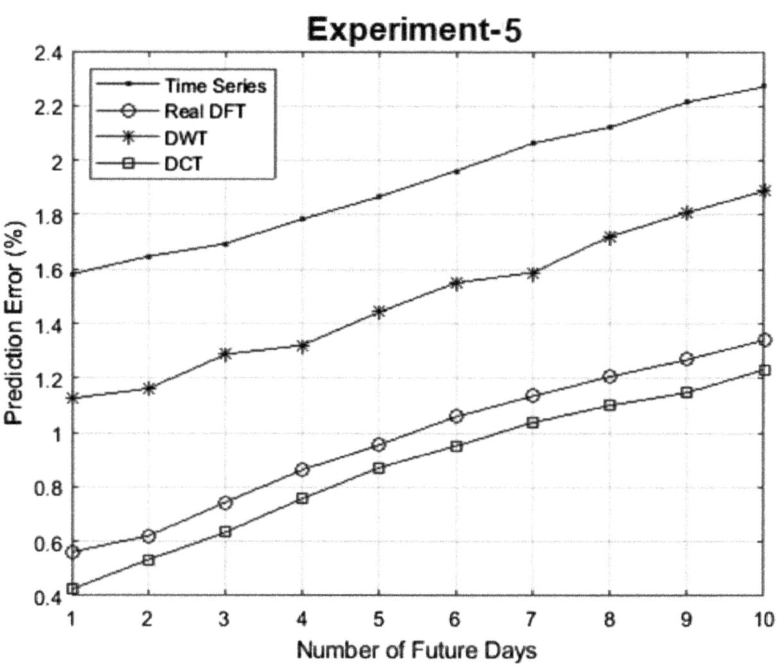

Fig. 14. Example output 2.

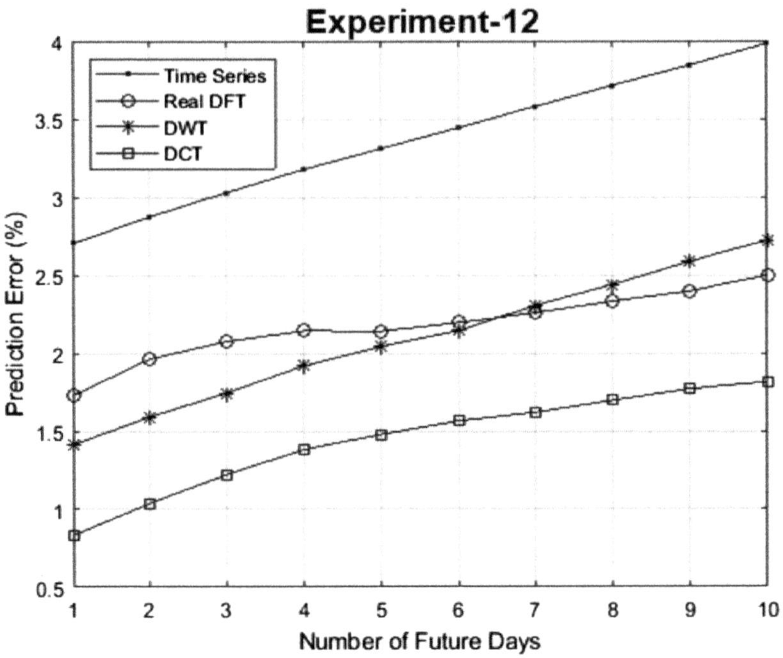

Fig. 15. Example output 3

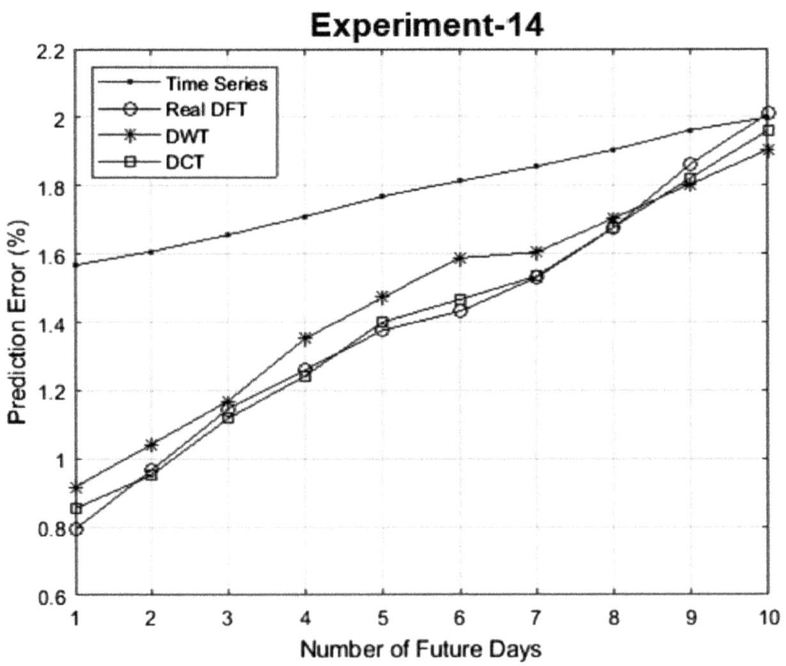

Fig. 16. Example output 4

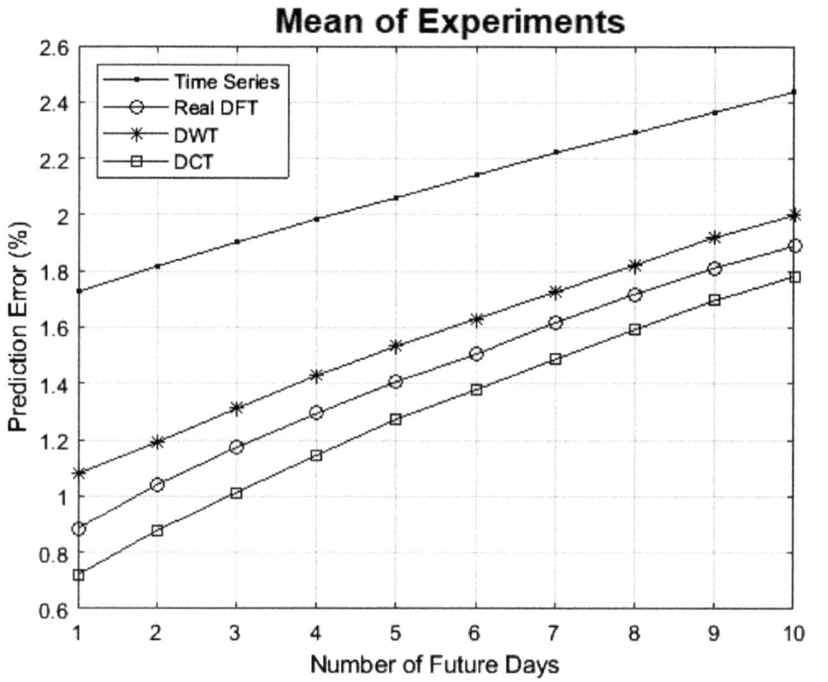

Fig. 17. Mean of all outputs of the experiments.

Table 1. Mean of all outputs of the experiments.

Prediction error (%)	Number of future days									
Representation of the input	1	2	3	4	5	6	7	8	9	10
Time series	1.73	1.82	1.90	1.99	2.06	2.14	2.22	2.29	2.37	2.44
RDFT	0.89	1.04	1.17	1.29	1.41	1.51	1.62	1.72	1.81	1.89
DWT	1.08	1.19	1.31	1.43	1.53	1.63	1.73	1.82	1.92	2.00
DCT	0.72	0.88	1.01	1.15	1.27	1.38	1.49	1.59	1.70	1.78

4 Conclusions and Future Work

Time series forecasting has particular importance in many application areas. There are numerous methods developed in literature. Among researchers, ANNs and deep learning are attracting more attention recently because of its capability to capture complex patterns. However, the success of ANNs considerably depend on the representation form of the input. Changing to representation of the information in the input may leads to increase in prediction success. Test results show that DCT of the USD-TRY currency rate input has significantly decrease the prediction error. The explanation of this phenomena may be energy compaction property of DCT. DCT compacts the energy to the leading values therefore ANN can calculates the input weights more consistently.

In the feature work, these representations of currency rate time series will be studied with the other ANN architectures such as RNN and LSTM which are more successful in time series prediction.

References

1. Raicharoen, T., Lursinsap, C., Sanguanbhokai, P.: Application of critical support vector machine to time series prediction. In: Proceedings of the 2003 International Symposium on Circuits and Systems, ISCAS 2003, vol. 5, pp. V-741–V-744. IEEE, Bangkok (2003)
2. Zhang, G.P.: A neural network ensemble method with jittered training data for time series forecasting. Inf. Sci. **177**(23), 5329–5346 (2007)
3. Zhang, Y., Wu, L.: Stock market prediction of S&P 500 via combination of improved BCO approach and BP neural network. Expert Syst. Appl. **36**(5), 8849–8854 (2009)
4. Adhikari, R., Agrawal, R.K.: An introductory study on time series modeling and forecasting. arXiv preprint arXiv:1302.6613 (2013)
5. Box, G.E., Jenkins, G.M., Reinsel, G.C., Ljung, G.M.: Time Series Analysis: Forecasting and Control. Wiley, Michigan (2015)
6. Zhang, G.P.: Time series forecasting using a hybrid ARIMA and neural network model. Neurocomputing **50**, 159–175 (2003)
7. Cochrane, J.H.: Time series for macroeconomics and finance. Manuscript, University of Chicago (2005)
8. Hipel, K.W., McLeod, A.I.: Time Series Modelling of Water Resources and Environmental Systems. Elsevier, Amsterdam (1994)
9. Kihoro, J., Otieno, R.O., Wafula, C.: Seasonal time series forecasting: a comparative study of ARIMA and ANN models (2004)
10. Suykens, J.A., Vandewalle, J.: Least squares support vector machine classifiers. Neural Process. Lett. **9**(3), 293–300 (1999)
11. Cao, L.J., Tay, F.E.H.: Support vector machine with adaptive parameters in financial time series forecasting. IEEE Trans. Neural Netw. **14**(6), 1506–1518 (2003)
12. Kim, K.J.: Financial time series forecasting using support vector machines. Neurocomputing **55**(1–2), 307–319 (2003)
13. Kamruzzaman, J. (ed.): Artificial Neural Networks in Finance and Manufacturing. IGI Global, Hershey (2006)
14. Zhang, G., Patuwo, B.E., Hu, M.Y.: Forecasting with artificial neural networks: the state of the art. Int. J. Forecast. **14**(1), 35–62 (1998)
15. Sable, S., Porwal, A., Singh, U.: Stock price prediction using genetic algorithms and evolution strategies. In: 2017 International Conference of Electronics, Communication and Aerospace Technology, ICECA, vol. 2, pp. 549–553. IEEE, Coimbatore (2017)
16. Kim, H.J., Shin, K.S.: A hybrid approach based on neural networks and genetic algorithms for detecting temporal patterns in stock markets. Appl. Soft Comput. **7**(2), 569–576 (2007)
17. Bekiros, S.D., Georgoutsos, D.A.: Direction-of-change forecasting using a volatility-based recurrent neural network. J. Forecast. **27**(5), 407–417 (2008)
18. Chen, A.S., Leung, M.T., Daouk, H.: Application of neural networks to an emerging financial market: forecasting and trading the Taiwan Stock Index. Comput. Oper. Res. **30**(6), 901–923 (2003)

19. Hamid, S.A., Iqbal, Z.: Using neural networks for forecasting volatility of S&P 500 Index futures prices. J. Bus. Res. **57**(10), 1116–1125 (2004)
20. Hanias, M., Curtis, P., Thalassinos, J.: Prediction with neural networks: the Athens stock exchange price indicator. Eur. J. Econ. Finance Adm. Sci. **9**, 21–27 (2007)
21. Liao, Z., Wang, J.: Forecasting model of global stock index by stochastic time effective neural network. Expert Syst. Appl. **37**(1), 834–841 (2010)
22. Ning, B., Wu, J., Peng, H., Zhao, J.: Using chaotic neural network to forecast stock index. In: International Symposium on Neural Networks, ISNN, vol. 5551, pp. 870–876. Springer, Heidelberg (2009)
23. Zhu, X., Wang, H., Xu, L., Li, H.: Predicting stock index increments by neural networks: the role of trading volume under different horizons. Expert Syst. Appl. **34**(4), 3043–3054 (2008)
24. De Faria, E.L., Albuquerque, M.P., Gonzalez, J.L., Cavalcante, J.T.P., Albuquerque, M.P.: Predicting the Brazilian stock market through neural networks and adaptive exponential smoothing methods. Expert Syst. Appl. **36**(10), 12506–12509 (2009)
25. Abraham, A., Nath, B., Mahanti, P.K.: Hybrid intelligent systems for stock market analysis. In: International Conference on Computational Science, pp. 337–345. Springer, Heidelberg (2001)
26. Smith, S.W.: The scientist and engineer's guide to digital signal processing (1997)
27. Towards data science. https://towardsdatascience.com/a-weird-introduction-to-deep-learning-7828803693b0. Accessed 15 May 2019

The Applicability of Instructional Leadership in Educational Institution

Behcet Oznacar[1(✉)] and Gulyuz Debes[2]

[1] Near East University, Yakındoğu Blv, Lefkoşa, Nicosia, Cyprus
behcet.oznacar@neu.edu.tr
[2] Akdeniz Karpaz University, Sht. M. Ruso Avenue. No. 79 Belyaka Street, Nicosia, Cyprus
dgulyuz@yahoo.com

Abstract. The aim of this study is to show effective school leadership techniques which would enhance the quality of classroom teaching and learning practices of teachers and administrators. Thus, the pressures to improve the outcomes of education in institutions have directed the focus to the realization level of the leadership of schools. Although, instructional leadership involves learner autonomy, it is found that level of demonstrating leadership and learner autonomy behaviours is higher than it has been considered as the teachers in our institutions and administrators believe such behaviours are necessary in class to foster learning. Those who apply instructional leadership techniques create better learning atmosphere and improve themselves professionally compared to others who are lacking such skills. This helps to the improvement of these institutions and teachers who are willing to work in collaboration. However, this must follow support of peers and evaluation of self and others, in order to see if these are applied correctly. This research is done to reveal the opinions of teachers and managers and their past experiences in such cases. Pre-test and post-test were applied in the study and the results were evaluated. Moreover, the pros and cons of instructional leadership techniques used in educational institutions were identified. In this study, some recommendations are made to create better learning environment, more democratic and comfortable atmosphere supporting all educators and learners as well as their personal growth that would enable them to have a say while giving vital decisions.

Keywords: Instructional leadership · Learner autonomy · Self development · Techniques

1 Introduction

As a result of the changes that are taking place in order to keep up with the conditions of our time, education needs to constantly improve itself. Instructional leadership provides a good teacher training, more desirable learning conditions for teachers, guides the actions of the school to transform its working environment into a satisfactory and productive one. In order to get better results for education the manager, teachers, students, families and the school board can work together to create an organization climate. It is a leadership associated with teaching processes, including teaching leadership,

D. J. Hemanth and U. Kose (Eds.): ICAIAME 2019, LNDECT 43, pp. 724–735, 2020.
https://doi.org/10.1007/978-3-030-36178-5_61

students, teachers, and teaching programs, in order to reveal the hidden power of the academic staff for the realization of the educational objectives of the school and to sustain it in the educational process. Educational organizations are one of the organizations affected by the increasing process of intensity and the speed of the changes (Deniz 2017). Nowadays, new meaning has been given to education, and the purpose, content, methods and tools used in education have begun to change. Therefore, it became very important that school administrators and teachers who play an important role in education can renew themselves and survive in this change (Memduhoğlu and Zengin 2012). In order to achieve certain goals in education, the structure created in the school organization requires good establishment, a good model and a good leadership. In this sense, differentiating leadership types "Teaching Leadership", which is the subject of education management, comes to the fore. Changes in school managers' roles require school administrators to focus more on supervision. Of course, this is the result of seeing school administrators as a teaching leader. According to this understanding, school administrators is seen as instructional leaders, who controls teachers activities in the classroom and focus on making the necessary arrangements to improve the activities and teaching strategies they apply to their students within the aim of helping students to develop themselves and learn better (Aydın 2016). That is also associated with discussions regarding educational management (Debeş, and Öznacar 2018; Öznacar 2019).

As Öznacar (2016) stated, instructional leadership has been available from the beginning of history because for those people who has hierarchical nature, they enjoy having the power in group to control the others around them. Having this power and skills in education helps to the success of students and curriculum and better achievement of all the objectives. According to Gümüşeli (1996), instructional leadership is a set of behaviors that the principal himself/herself shows or shown by others in order to increase student' success in schools and especially focuses on the teaching and learning process. Educational leadership is sometimes used as educational or pedagogical 'leadership.' As Çelik (2015) pointed out instructional leaders ensure that instruction is carried out in accordance with the determined objectives, while trying to overcome the disruptions in teaching. Here it can be said that, instructional leadership is the leadership model to control and assess the quality of teaching. Hence, the most important thing is to develop the effectiveness of teaching. The biggest responsibility is on the school administrators' because while they are trying to improve the teaching and learning process, they should work in collaboration with teachers. Plus, during this process, they should be interacting with students' and show the impact of these on school outcomes (Şişman 2014).

Instructional leadership defines the mission of the school, manages the teaching and training programs and helps to the development of a positive learning climate (Sağır and Memişoğlu 2012). For this reason, an effective instructional leader has the responsibility to develop vision, share it, and organize a learning plan that is appropriate to the development and individual differences of all students Apaydın et al. (2011). Instructional leaders are those who have a balance between their interpersonal relations behaviors and values in and around school, solid bureaucratic adoption of democracy, empowerment and continuity to support control, change, influence, take risk instead of maintaining status, support innovation instead of stability (Gündüz and Balyer 2012). Today, schools, which have the duty of preparing the future of the country as a result of being an information society, have to see the future and determine the required changes for the future, realize them and make them effective continuously by arranging them accordingly. Recently, not only the educational aspect of schools; the society is also open to the emotional side of the students', social diversity, high sensitivity of the technology to protect the spiritual value of the public eye, collaborate with the employee, teach in a democratic way, be ready for the competitive environment of today's world, without deteriorating from the concrete side of life. Therefore, productivity is expected in the organizations that are open to all these changes and they are expected to resist all the negative effects that are likely to happen and question its own structure while performing all these. In order to achieve this, it is one of the prerequisites to display innovative and effective leadership behaviors in schools (Beycioğlu and Aslan 2010). Instructional leader is the person who has the pioneering ability and productive feature that enables the institution to live and develop in accordance with the objectives of the institution, who has a certain vision, has a deep knowledge of the educational processes and learning theories, undertakes the risk, has sufficient humanity skills and an unlimited energy (Derbedek 2008).

2 Objectives of Instructional Leadership

Since instructional leadership is a form of leadership that is fully developed in relation to education, the main starting point is the development of education and training. The main purpose of it is to provide purely educational and productive environment. The main aim in instructional leadership is to lead the teaching process in the most effient way to get the most benefit out of students' and teachers' and to remind them that their job is to ensure that they will reach success and in a qualified way. In this context, the instructional leaders should take the lead in creating and distributing knowledge in the school and gaining the techniques of efficient working and creating an environment that provides these conditions. In order for all operations in schools to be effective, the collaboration of the three powers (student, teacher and society) is needed and how the curriculum would work will depend on the success of their interaction. The main objective of instructional leadership is to improve the quality of these powers for the purposes of the school to make the necessary requirements successfully (Aygün 2014).

2.1 Instructional Leadership Features

According to Aygün (2014), the features expected from instructional leaders to carry their institutions to success and keep up with the recent changes are as follows;

- To have self confidence
- To stay calm
- To be persuasive
- To be able to overcome obstacles on the way reaching to success
- The keep the drive of success high and to show a different approach than people are usually used to hear
- Diligent, applying what they learn and know, being a pioneer to lead the others towards success, taking common decisions with all school staff and students,

helping new innovations and changes to take place in educational institutions and reflecting these to the school atmosphere, creating new visions for the school, trying to apply all the new technological innovations and developments that the time has brought, taking into account the needs of students' and teachers' and demonstrating curriculum flexibility.

They are perceived as doing a hard job by demonstrating a high level of performance in improving educational practices and student success. On the other hand, Çelik and Uğurluoğlu (2009) defined the features of instructional leaders as follows;

- Identifying the priority issues in the education and training program,
- Devoting himself to the realization of the goals and objectives of the school,
- Creating a positive school climate to meet the expectations of the education stakeholders,
- Creating effective communication with teachers for the development of school teaching policies,
- Encouraging educational stakeholders to use new teaching methods and techniques,
- Supplying new teaching materials,
- Organizing and participating in-service trainings,
- Identifying and developing a clear vision that does not conflict with school objectives in the long term,
- When making decisions about the school ensuring the participation of all the relevant units,
- Providing the support that teachers need for their academic development,
- Managing time effectively and minimizing factors that affect negatively.

3 Method

This research is done to find out the instructional leadership of the school administrators in educational institutions and pre-test and post-test questions are being considered. It has a descriptive nature within the aim of searching instructional leadership techniques that are being applied in schools to find the appropriate leadership style that would be the best for all administrators, teachers and students. The qualitative research design is used in this research that puts forward the opinions of administrators working in the educational institutions in connection with leadership styles that are used at schools.

4 Working Group

60 interviews were conducted within the scope of this research with the school administrators, teachers and students that are working in Northern Cyprus. Regarding the participators participating in this research, 37.5% (4) of the individuals who participated in the research were managers, 50% (31) teachers and 12.5% (25) were students.

5 Development of Data Collection Tool

The questionnaire used in this study consists of four (4) sections; Sect. 1 is related to demography factor, Sect. 2 consists of items related to the school leaders' instructional leadership practice; competencies needed for management, features needed for instructional leadership, identification of instructional leadership and the importance of instructional leadership in education. The interview hours were negotiated with the participants and the appropriate day and hours were set upon being informed and was done with written and audio recordings. The comments and suggestions were made as a result of the negotiations obtained from the interviews that were written down. Yıldırım and Şimşek (2008), pointed out that interaction should be based on confidence and empathy which is important between the participant and the researcher.

6 Analysis of Data

6.1 Analysing of Interviews

The content analysis of this research is done with the selection of the questions that are written in this research and developing data collection tools. The records that were taken during the negotiations were recorded and then transferred to computer, analysed and the results was written as a result. Once the data is transferred positive results were obtained. The qualitative data is the data being transformed in numbers and percentages. The objective of the content analysis is carried out within the purpose of shedding light to the conceptualization of the data and the themes that could define the case. The findings have been explained step by step and the data was interpreted accordingly with the results of the qualitative research. Through these results, recommendations and suggestions were made to the school administrators, teachers and students.

7 Results

In this part of the study, the findings of the demographic information belonging to the sample group and the results obtained by the analysis of the data obtained for the sub-problems of the researcher with the appropriate statistical method are included and comments on these findings are given. Below, the pre-test and post-test results that are applied in the seminars and held in different topics within the scope of the research are separately expressed in tables.

8 Personal Information

A: Quantitative Data Analysis

In Table 1, the gender distribution of the participants (67%) was male and eleven (33%) were female.

Table 1. Gender distribution of participants

		Frequency	Percentage
Gender	Women	40	67
	Men	20	33
	Total	60	100

In Table 2, it can be seen that 84.4% of the respondents were married and 16.6% were single.

Table 2. Marital status of participants

		F	%
Marital status	Married	50	84.4
	Single	10	16.6
	Total	60	100

As seen in Table 3; According to the gender distribution of the participants, 24 women (60%) and 36 (40%) were male.

Table 3. Frequency distribution by gender.

		F	%
Gender	Male	36	40
	Female	24	60
	TOTAL	60	100

As seen in Table 4; 37.5% (4) of the individuals who participated in the research were managers, 50% (31) teachers and 12.5% (25) were students.

Table 4. Frequency distribution by position

		F	%
Position	Manager	4	37,5
	Teacher	31	50
	Student	25	12,5
	Total	60	100

8.1 Pre-test and Post-test Findings

A: Qualitative Data Analysis

Question 1. 4 Competencies needed for management

Participants' opinions;

The issue of combining technology and instructional leadership should be kept on the agenda in order to manage the education systems correctly and reliably (P2).

Within the education system, there is a need for technological leadership in order to control and direct the digital media and advancements in technology (D4).

In educational institutions, firstly, technology must be accepted for use and the importance of instructional leadership must be understood (D1).

As it is seen in Table 5, In education, 4 competencies for management stand out. After the adoption of technology, which is the common point of all participants, instructional leadership, then open leadership, effective leadership, technologic leadership and finally combining all with the acceptance of technology are the most common. If we look at the group of administrators among these themes, the technology adoption is the least, while the instructional leadership concept is the first among the 4 competencies. On the other hand, open leadership and effective leadership are the following themes.

Table 5. 4 outstanding competences in education, for management.

Theme	Manager		Teacher		Student		Total	
	N	%	N	%	N	%	N	%
Instructional leadership	11	%72.6	10	%50	–	–	21	%52.5
Open leadership	8	%52.8	18	%90	3	%60	29	%72.5
Effective leadership	11	%72.6	15	%75	4	%80	30	%75
Technologic leadership	12	%79.2	16	%80	5	%100	33	%82.5
Acceptance of technology	14	%92.4	18	%90	5	%100	37	%92.5

Question 2. Features needed for instructional leadership

Participants' opinions;

Creating a positive climate in class within the aim of reaching success and giving the decisions together are the most important features of instructional leaders.

Instructional leaders should be aware of the recent developments and innovations and must feel free to apply them in their teaching environment to take the education level into a better level.

When Table 6 is examined, it can be seen that the establishment of positive climate and working in collaboration to give the decisions together are the most important features of instructional leaders. Besides these, they are expected to have self confidence, to be persuasive and hardworking as well as having good skills to overcome any problems they face with when they are trying to carry the institution into a better place by controlling and leading others into the right way.

Table 6. What kind of features are needed to be instructional leaders.

Theme	Manager		Teacher		Student		Total	
	N	%	N	%	N	%	N	%
Being self confident, persuasive and diligent	6	%39.6	12	%60	2	%40	20	%50
Being able to overcome the obstacles	5	%33.3	8	%40	4	%80	17	%42.5
Being able to control and lead others	8	%52.8	6	%35	3	%60	17	%42.5
Giving the decisions together	10	%66.6	16	%80	5	%100	31	%77.5
Keeping up with the recent developments And innovations and applying them in school	10	%66.6	15	%75	4	%80	29	%72.5
Trying to create positive climate leading to success	5	%33.3	7	%35	5	%100	17	%42.5

Question 3. Identification of instructional leadership

Participants' opinions;

Necessary steps should be taken to strengthen collaborative work and identifying and developing clear vision to cope with the problems that are likely to occur are very important and should be given priority in teaching and training programs.

In Table 7, it can be seen that the individuals participating in the research stated that the individuals were mostly productive in defining the efficiency of the leadership. After this point of view, working collaboratively, identifying and developing a clear vision regarding the needs in education and training programs are stated respectively.

Table 7. Identification of the productivity of instructional leadership by participants.

Theme	Manager		Teacher		Student		Total	
	N	%	N	%	N	%	N	%
All are working in collaboration	5	%33	17	%85	4	%80	26	%65
They are being productive and successful together	13	%85.8	12	%60	3	%60	28	%70
They can identify and develop a clear vision and can figure out the priority issues in their education and training programs	3	%19,8	9	%45	1	%20	13	%32.5

Question 4. The importance of instructional leadership in education

Participant's opinions;

To fulfill the needs of students' and teachers' instructional leaders have a big responsibility.

Not just supplying their needs, but also instructional leaders should encourage all to use the recent developments and innovations in their teaching programs to help students learn better and to improve the instution.

In Table 8, it can be seen that instructional leaders play a big role in shaping the education according to the current and future needs of all stakeholders. Therefore, instructional leaders are needed in education to determine which training programs are necessary for the improvement of teachers and decide on the needed supplements as they are the ones controlling all and considering everyone's opinion when giving decisions to create a successful and healthy learning environment in which everyone is learning.

Table 8. Why instructional leadership is important in education.

Theme	Manage		Teacher		Other	
	N	%	N	%	N	%
To provide the support teachers and students' need	13	%85.8	7	%35	3	%60
To encourage educational stakeholders to use the new teaching methods and techniques and to catch up with the time	8	%52.8	3	%15	2	%40
To organize service training for teachers and to control and lead them in the work they do in and out of the class	3	%19.8	10	%50	2	%40

9 Discussion

This research aimed to find out, whether instructional leadership is applicable in educational institutions or not, and what kind of benefits it will bring to education and educational institutions.

According to the respondents, both teachers' and managers' viewed instructional leadership as an essential value that is needed to be implemented at schools because it generates positive atmosphere, helps effective learning, uses and organizes all activities and resources in such a way that everyone can get the most benefit out of them, and creates a vision that would help to the success of the institutions. This is also supported by Musfah et al. (2015) and in his study he stated that instructional leadership is about managing the school activities and school members. Therefore, instructional leaders are expected to delegate some duties and tasks to the right people, determine the correct time and place for school activities. Similar to our findings Firmaningsih-Kolu (2015) stated that instructional leaders demonstrate positive behaviour and support the learning process with good attitude and good communication skills. So that these actions promote the growth of students' learning and lead to the success and improvement of the institution that is being applied. Most of the researches done on the related topic show that, implication of instructional leadership in educational institutions has reached the conclusion that it is directly effective on the students' and teachers' success.

When we look at the competencies seen as important for management within the education system, it is seen that instructional leadership adoption is first, then effective leadership, technologic leadership and the acceptance of technology respectively. One of the most remarkable points here is the application of instructional leadership which is a noteworthy understanding within the management processes. The vast majority of respondents consider instructional leadership as one of the important competencies that should be in the educational process in the current time. However, Davies (2010) mentions that technology has a large impact on a wide area that cannot be shown as fixed and singular leadership processes of the component of all schools, regardless of leadership. According to him, technology leadership seems to be closely related to a management process that will make positive contributions to the goal of bringing schools together with digitalization in management processes as opposed to our findings (Davies 2010).

When we look at the table that examines the features needed for instructional leadership for the age we live in now, we first see the concept of being self confident, persuasive and diligent. After this, other concepts such as; being able to overcome obstacles, being able to control and lead others, giving the decisions together, keeping up with the recent developments and innovations and creating a positive climate that would lead to success follows. According to Sabancı et al. (2013) their results demonstrated that the most common leadership characteristic was using computer based technology and respondents stated that current leaders had limited skills in using the latest technologies adapted to educational needs. However, our study did not focus on the use of technology. According to our results, establishing positive climate and working in collaboration to give the decisions together are the most important features of instructional leaders. Besides these, they are expected to have self confidence, to be persuasive and hardworking as well as having good skills to overcome any problems they face with when they are trying to carry the institution into a better level by controlling and leading others into the right way.

According to the participants responses given on the identification of instructional leadership, our results indicated that working collaboratively and developing a clear vision regarding the needs in education and training programs are the most important factors that must be found in instructional leaders. On the contrary, Vogel (2018) results differ from our findings because his results indicated that teacher supervision, giving feedback and coaching dialogues, appeared to be the main vehicle of instructional leadership. Promoting professional development of teachers was identified a little more than a third of the principals in this study as part of their instructional leadership role. Thus, it is believed that instructional leaders with what they feel and do are the most directly impacts that take place in the classroom and promote student learning.

10 Conclusion

The findings of this research indicated that the demand and need for instructional leadership in schools are higher. Therefore, the need for instructional leaders and the features of an effective leader pre-test and post-test questions are considered due to these. There are many researches on the subject.

Consequently, the main aim of this study was to get teachers', students' and managers' perceptions about current state of their school leaders' characteristics, identifying and classifying them in weak and strong facets and trying to get a picture of future leadership characteristics depending on their estimations. The results illustrated that instructional leaders are needed in schools and it helps to the improvement of the education system, training of teachers' and developing the education in which students' can get the most benefit of it. The second kind of the data provided us what characteristics teachers estimate to a school leader in the future. Further qualitative analysis can be carried on by comparing current state of leadership in centralized and decentralized education systems and how teachers picture the future of school leadership.

As it has been pointed out in this research, the need for instructional leadership at schools has grown significantly due to the advancements in technology in the last decade. For this reason, if we can identify the best leadership style that would work the most in our classroom regarding our students' can be very beneficial. There would not only be one type fitting to all classrooms. Therefore, the needs', opinions, financial situation of the institution must be considered carefully before deciding on the best one. This would create more advantages to the institutions and it would be less costly, less time consuming and less demotivating for all.

Based on this study and our results, online instructional leadership can be applied at first for trial to see how it will work in institutions and if it works it can be utilized to mentor additional training programs for the development of teachers' and administrators'. Moreover, it can be said that leadership for learning generates an awareness that should be practiced by instructional leaders in adapting the needs and the nature of the schools' environment and atmosphere. So, there is no particular model that would quickly fit to the class and will be available to be used. During this, time to try and evaluate the advantages and disadvantages of it is needed. Therefore, being patient, self confident and persuasive are the most important characteristics that one should have while trying it out. It emphasizes a concern on collaboration, and promotes participative decision making, helps instructional leaders to evaluate and understand how their behavioral style as a leader would affect their relationship with their colleagues and learners and what benefits it would bring to the institution and learners' lives.

This study would help educational leaders to see the pros and cons of application of instructional leadership and fulfill the requirements needed for adaptation to emerging technologies and adoption of many new innovations into their careers and deal with all stakeholders in education effectively while applying all learned leadership styles into their practices.

References

Apaydın, C., Vilkınas, T., Cartan, G.: Effective leadership determinants of secondary school teachers in Turkey. Ank. Univ. Fac. Educ. Sci. J. **44**(1), 107–130 (2011). http://dergipark.gov.tr/auebfd/issue/38386/445157

Aydın, İ.: Audit in Teaching: Assessment, Determination and Development Situation. Pegem Akademi Publishing, Ankara (2016)

Aygün, B.: Examining high school teachers' awareness of instructional leadership roles and their approach to factors limiting instructional leadership. (Unpublished Master's Thesis). Istanbul: Fatih University Institute of Social Sciences (2014)

Beycioğlu, K., Aslan, B.: Teacher leadership scale: validity and reliability study. Prim. Educ. Online **9**(2), 764–775 (2010)

Çelik, F.: Örgütsel Güvenin Is Tatmini ve Örgütsel Baglilikla Iliskisi: Resmi-Özel Ögretim Kurumlarinda Bir Arastirma, Master's thesis, Istanbul Gelisim Üniversitesi Sosyal Bilimler Enstitüsü (2015)

Çelik, Y., Uğurluoğlu, Ö.: Strategic leadership and characteristics in organizations, pp. 121–126 (2009)

Davies, P.M.: On school educational technology leadership. Manag. Educ. **24**(2), 55–61 (2010)

Debeş, G., Öznacar, B.: Evaluation of the opinions of the manager, teacher, employees (secretary and servants) about school management of the digitalization and management processes of the system engineering model in education. Amaz. Investig. **7**(16), 243–253 (2018)

Deniz, Ü.: Ögretimsel liderligin sinif denetimi üzerine yansimasi: ögretimsel denetim, Master's thesis, Balikesir Üniversitesi Sosyal Bilimler Enstitüsü (2017)

Derbedek, H.: The effects of instructional leadership characteristics of primary school principals on teachers' self-efficacy. Unpublished Master thesis, Pamukkale (2008)

Gümüşeli, A.İ.: Elementary school principals leadership behavior in Istanbul. (Published Research), Yildiz Technical University, Institute of Social Sciences, Istanbul (1996)

Gündüz, Y., Balyer, A.: Okul Müdürlerinin Etkili Liderlik Davranışlarının İncelenmesi. Kuramsal Eğitim Bilim Derg. **5**(2), 237–253 (2012). http://dergipark.gov.tr/akukeg/issue/29345/314033

Firmaningsih-Kolu, Y.: The role of the principal's instructional leadership at schools in Indonesia. Master's thesis, December 2015

Memduhoğlu, H.B., Zengin, M.: Applicability of instructional control in Turkish education system as a model of contemporary education supervision. J. Theor. Educ. Sci. **5**(1), 131–142 (2012)

Musfah, J.: Kepemimpinan Kepala Sekolah, pp. 1–14. Fakultas Ilmu Tarbiyah dan Keguruan UIN Syarif Hidayatullah Jakarta, Jakarta (2015). http://www.academia.edu/4105182/Kepemimpinan_Kepala_Sekolah

Öznacar, B.: Determination of instructional leadership administrators. Int. J. Environ. Sci. Educ. **11**, 957–972 (2016)

Öznacar, B.: Identifying the administrator and teacher competencies according to the views of primary school administrators. Rev. San Gregorio **1**(30), 15–23 (2019)

Sabancı, A., Şahin, A., Kasalak, G.: Understanding school leaders' characteristics and estimating the future. Open J. Lead. **23**, 56–62 (2013)

Sağır, M., Memişoğlu, S.P.: Ilkögretim okulu yöneticilerinin ögretimsel liderlik rollerine iliskin ögretmen ve yönetici algilari. Egitim ve Ögretim Arastirmalari Dergisi, **1**(2), 1–12 (2012)

Şişman, M.: Teaching Leadership. Pegem Academy Publishing, Ankara (2014)

Vogel, R.L.: Learning outside the classroom: how principals define and prepare to be instructional leaders. Educ. Res. Int. **2018**, 1–14 (2018)

Yıldırım, A., Şimşek, H.: Sosyal Bilimlerde Nitel Araştırma Yönetmleri. Seçkin Yayıncılık, Ankara (2008)

Production of Myoelectrically Controlled 3D Bionic Hand

Ferdi Alakus[1], Pinar Koc[2(✉)], Orhan Duzenli[3], and Kenan Unlu[4]

[1] Department of Electric and Electronics Engineering, Suleyman Demirel University, Isparta, Turkey
ferdialakus@hotmail.com

[2] Department of Computer Education, Suleyman Demirel University, Isparta, Turkey
pinarkoc1995@gmail.com

[3] Department of Manufacturing Engineering, Suleyman Demirel University, Isparta, Turkey
orhan321532@gmail.com

[4] Department of Electronics and Communications Engineering, Namık Kemal University, Tekirdağ, Turkey
kenanunlu7@gmail.com

Abstract. In this study making of 3D bionic hand for people who was born without hand or who has lost their hand after accidents. Researches are carried out before making the 3D design of bionic hand, upper extremity drawings and the situations which should be considered while transferring to computer environment investigated to create healthy products for disabled individuals. Also thanks to the system, features, software, hardware etc. used in the bionic hand; a functional hand has been created for people to be able to get it cheap and easy. In addition to all these, because the study has been being carried out with local software and hardware people can easily reach technical support. The study because of its reputational extent, alongside having a place among other national and international studies, it will also help employment in sub-industry in Turkey.

Keywords: Bionic hand · EMG sensor · Upper extremity · Prosthetics · 3D prosthetics

1 Introduction

Throughout history there have been many innovations, advancements and changes.

Since the ancient times where humanity first occured even though all the changes that have happened at first were about needs and necessities the changes that came afterwards were about more comfort, beauty, luxury etc. These changes display differences depending on the studies and areas. Among these necessities and needs medicine is one of the areas that needs to be emphasized. From time to time humankind in its lifespan faces health problems and to solve these problems studies has been

D. J. Hemanth and U. Kose (Eds.): ICAIAME 2019, LNDECT 43, pp. 736–743, 2020.
https://doi.org/10.1007/978-3-030-36178-5_62

carried out. As a result of the accidents that people have been through their lives, they might get permanent or long lasting injuries. These injuries are named as handicap or disability based on their degrees. It is possible to come across with various different definition of disability. World health organization (WHO) defines disability as "any restriction or lack (resulting from an impairment) of ability to perform an activity in the manner or within the range considered nonnal for a human being" [2]. Another definition for disability is; a situation where an individual is having physical or mental restrictions, deficiency or limitations while executing life functions [3, 4].

Disability type is decided by disability situation. Disability types are described as; Orthopedic, Visual, Hearing, Motor Speech, Mental, Chronic Illness. For individuals who has health problems, based on the type of their disability, researches have been being carried out in order to find solution or cure to their disabilities and products have been being made to provide them with options which would help them execute their needs as closely as individuals who have not a disability.

Prosthetics/Orthotics studies are being done according to the defect's characteristics in disability type. Based on the reports from Turkish Institute of Statistics 12% of the population in Turkey are disabled and 1,25% of the people are physically challenged [5]. The challenge degree classification in three threshhold values for the five different functionality individuals have is shown below in the Table 1.

Table 1. Challenge degrees for five different functionality

Base fields	Challenge degrees		
	Z1	Z2	Z3
Visual	9,3	2,2	0,3
Hearing	10,5	2,3	0,2
Mobility	14,9	6,0	1,4
Cognition	13,1	2,7	0,3
Personal care	4,7	2,2	1,2

Challenge degrees in five different functionality in Table 1 is stated as;

Z1: little challenged and above;
Z2: very challenged and above;
Z3: not able to perform at all. When challenge degrees investigated all of the challenge degrees have their highest values in mobility functionality [8]. Figure 1 shows the disability population rates according to the type of disability as it was published in Turkey Statistical yearbook (2004) [6, 7].

		Ortopedik Özürlü	Görme Özürlü	İşitme Özürlü	Dil ve Konuşma Özürlü	Zihinsel Özürlü
	Türkiye	1,25	0,6	0,37	0,38	0,48
Cinsiyet	**Erkek**	1,48	0,7	0,41	0,48	0,58
	Kadın	1,02	0,5	0,33	0,28	0,38
Yerleşim Yeri	**Kent**	1,09	0,52	0,32	0,33	0,38
	Kır	1,49	0,73	0,45	0,46	0,64
Bölge	**Marmara**	1,11	0,53	0,37	0,33	0,37
	Ege	1,19	0,61	0,36	0,38	0,53
	Akdeniz	1,22	0,58	0,34	0,38	0,55
	İç Anadolu	1,27	0,63	0,37	0,38	0,5
	Karadeniz	1,6	0,66	0,45	0,46	0,63
	Doğu Anadolu	1,27	0,57	0,31	0,4	0,44
	Güneydoğu Anadolu	1,25	0,7	0,36	0,45	0,45

Fig. 1. Disabled population rate (%) by type of disability

According to the disabled population rate in Fig. 1, it can be seen that 1.25% of the disabled population in Turkey are orthopaedically challenged [1]. In the study, the problems were determined according to the data that was gathere from the complaints and needs of the people who do not have a hand from birth or who have lost it. These problems shows the inadequacy of the existing prosthetic hands (upper extremity) clearly and points out to the fact that they have to be renewed in time or that they are not mobile and thus needs to be renewed. Prosthetics'-orthotics' renewing time differ from country to country. Orthotics' lifespan in Germany and England are not limited, in upper extremity prosthetics it is 2 to 5 years. In countries such as Netherlands, France, Sweden, Hungary while lifespan of orthotics are 1.5 to 5 years, prosthetics' lifespan is 2 to 5 years. In the USA prosthetics' lifespan is 3 years and in Turkey the lifespan for orthotics is 3 years while for the lower extremity prosthetics it is 5 years and for the upper extremity prosthetics it is 10 years [5].

Overall aims in the study of producing bionic hand (upper extremity);

- With the 3D printing technology it has been being planned to build a lower arm (upper extremity) structure which can execute many functions through EMG (Electromyography) signals instead of the existing static prosthetic hand for the people who does not have a hand from birth or who have lost it.
- Having been made by a 3D printer; the upper extremity will have low cost, furthermore it will be possible for it to have a long lifespan thanks to various sensors being used and softwares developed locally.
- Local design, software and hardware will enable the device to be developed constantly as well as reaching the user easily.
- Similar products are being imported from outside the country and thus having a high cost makes it difficult to reach these, and the product suggested in this study will help remove this difficulty.

2 Material and Method

Discussions with the people who are experts on the area have been carried out about what sort of updates could be done to make the prosthetic hand a better and healthier one in the light of literature review done about currently existing upper extremity prosthetics in today's world. In the light of discussions the overall characteristics of the study have been made and thanks to the innovations and advantages that it will bring new ideas are briefly formed about how it can be completely local and how it can have an affordable cost.

The researches have been done for the study whose extend was set to determine what sizes and what structure the upper extremity will be produced and what should be considered in sizes. The important factors have been examined concerning motor movement of joints and the type of material (passive, classic, modular pylon, myo-electrically controlled) which would later be used while measuring [9, 10]. Designs gathered after the examination have been drawn in Solidworks software and have been modelled three dimensionally.

One of the most important points in the bionic hand system is getting the finger designs and joints ready for printing due to the importance from motor movements to skill to grip. Servo motor alongside linear motor in the thumb and only linear motor in other fingers were used. Because there are built in potentiometer feature in the motors used it does not require an additional pressure sensor and the fact that it is affordable makes it superior when compare to other prosthetics in the market.

The palm is designed parabolically to provide an ergonomically comfortable structure. There are holes at the back of the designated palm for the motors and motor drivers to be placed. The final states of the models which was made according to the drawings in Solidworks was saved as .stl (Stereolithography) files and got printed in 3D printer. During printing process it was made possible to reuse any faulty prints thanks to the flament recycling feature. Printed parts were cleaned with the mechanic works done on them and assembling process was carried out carefully. After the assemblying is completed, electronics process was initiated to embed the EMG feature. Visuals regarding the bionic hands ready for the electronics process are shown in Figs. 2, 3, 4, 5 and 6 below.

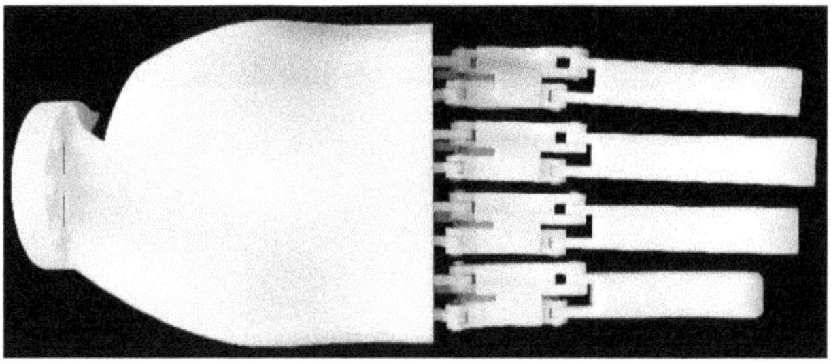

Fig. 2. Palm side view of bionic hand

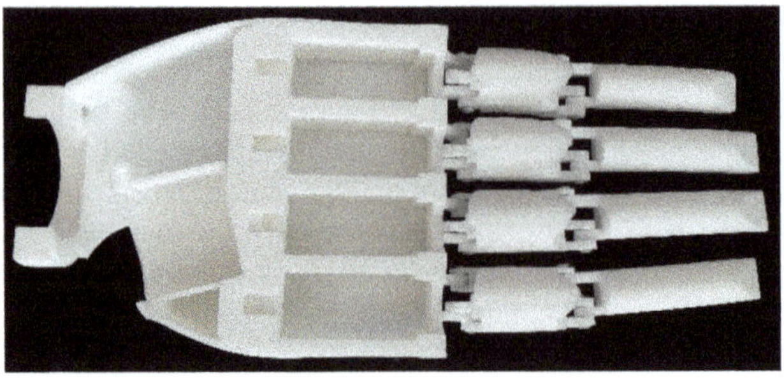

Fig. 3. Back view of the bionic hand

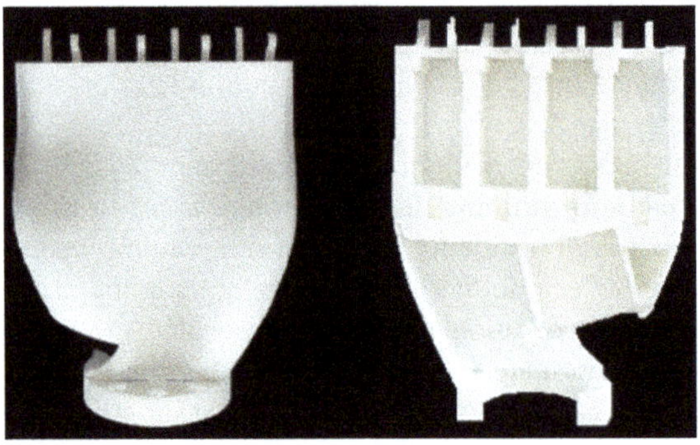

Fig. 4. Palm and back view of bionic hand

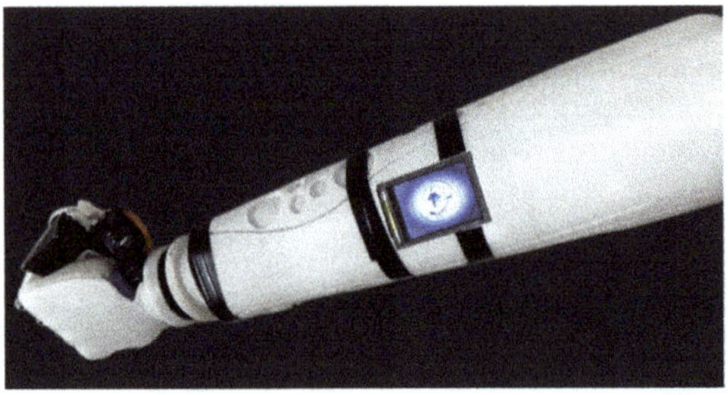

Fig. 5. Bionic arm holding position image

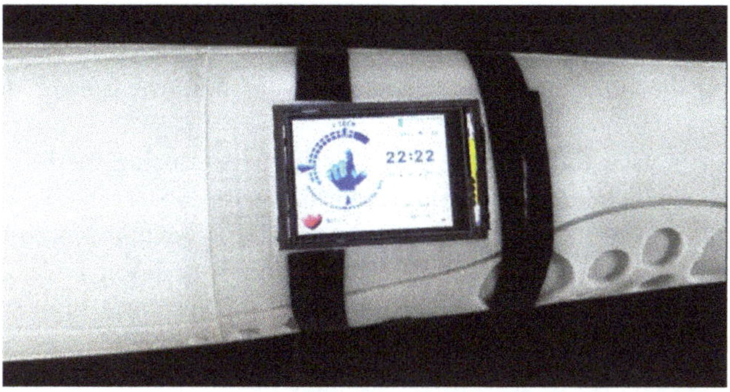

Fig. 6. Touchscreen display of bionic arm

In order for bionic hand to work as a system once orthopaedic systems were taken care of myo armband was first used to be able to build the electronic system. Myo armband is the device which transmits the signals from the arm to the structure via a program beyond the arm for the people who is without a hand or who has lost it. Myo armband detects the signals with standart movement and transmit these signals to pq12 linear motors and servo motors. Pq12 linear motor and servo motor are the motors which are used to provide fingers of the bionic hand with movement power. Fingers' functions made sure to be stable by the linear motors. On the thumb; however, servo motor is also used alongside linear motor. Thanks to the motors used, it was aimed for the bionic hand user to be able to perform all the axis movement that a person with a healthy organic hand can perform. Visuals regarding the bionic hand with the linear motors embedded are shown in Fig. 7.

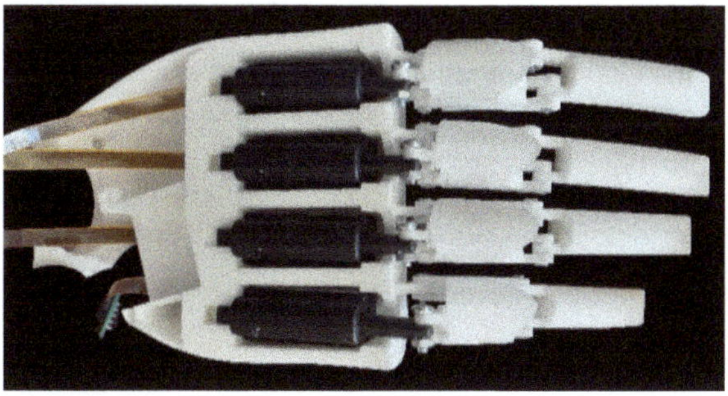

Fig. 7. Bionic hand with embedded linear motors

The communication between myo armband and microcontroller/processor is executed in myo armband's own socket. Python programming language is being used to execute the software tasks while connecting Raspberry pi and myo armband. Because the pins (analog input and PWM outlet) on Rapsberry pi are not at the desired amount an additional communication over UART (Universal Asynchronous Receiver

Transmitter) and microprocessor was made. After the initiation some sensors (heat, pulse and RTS module) are additionally used. The heat sensor measuring the card temperatures on the bionic hand is used to prevent the arm from getting hurt from overheating.

Pulse sensor is used to check the pulse of the person using the bionic hand. RTC module is a system wherein clock information can be observed. To display all the information gather from these sensors and the module a nextion screen is used. Li-ion batteries are used to power the system. The reason why li-ion batteries are used is because mAh (miliampere-hour) and the current of the system is high and these li-ion batteries can handle such transmission. In order to be able to keep the equal usage of the batteries under control BMS (Battery Management System) is used. In the systems with BMS, should the system overdraw current BMS gets activated. BMS is crucially important to prevent the hand from getting injured by interfering in an unpredictable excess current situation and by keeping the input-output balance intact.

All five functions supported by myo armband can be comfortably done once the electronic systems are defined on mechanically and electronically ready bionic hand. Functional processes that can be done by bionic hand are moving right, moving left, grasp and staying stable. Thanks to the bionic hand the disabled person can perform his daily needs which require him to use his hands as well as a person with a healthy organic hand can perform his.

3 Evaluation

The specifications which makes the bionic hand in an advantage to its peers;

- Because the bionic hand is made using 3D printing technology it is lighter, more wallet friendly and easy to carry.
- As it can be controlled by EMG signals, it is ergonomic and thanks to the sensors (heat, pulse) it contains it is functional.
- It is easy to reach technical support and be provided spare parts when in need as it is produced locally both in software and hardware.
- It has high sensitivity.
- Alongside having long battery life and equal battery usage, it has such a system which has countermeasures against overheating and burns.

4 Conclusions and Future Study

It can clearly be seen that despite the fact that there are many projects, conferences, manifestos etc. towards the subject of disabled people these things alone are not enough. Because even though publishing manifestos or having public speeches are methods to make awareness, in order to create a real sort of social integration all the individuals' needs should be provided with regardless of having a disability or not. That is to say that doing studies to make the life of people with disabilities easier is as vitally important as making awareness about these people or defending their rights

[11]. When the study history is examined, it is emphasized that the services provided to people with disabilities are not enough [6]. About the subject of disability it is emphasized that the rate of people with disabilties are lower to that of people without disabilities and that this situation is seen and perceived as normal [12]. In the light of comments and analyses in addition to studies, conferences and etc. about disability; with the thoughts of making the lives of people with disabilties a bit easier and making them feel good about themselves, it was decided to produce bionic hand for people who were born without a hand or who have lost it. The study will not only provide people with disabilities and make them feel more complete and steer their motivation in a more positive direction but also it will have positive feedbacks for national and international studies. The first feedback is that thanks to this study future studies will be guided. It will build a bridge between university and industry and thus increase the academic value of the study in addition to this it will help the development and continuation of such. It will also help local employement in sub-industry area.

Acknowledgment. The study in this paper has been supported by KOSGEB Isparta, Turkey in the context of Research-Development, Innovation and Industrial Application Funding Programme.

References

1. Devlet İstatistik Enstitüsü: Türkiye İstatistik Yıllığı 2004. D.İ.E., Ankara (2004)
2. World Health Organization: International classification of impairments, disabilities, and handicaps: a manual of classification relating to the consequences of disease, published in accordance with resolution WHA29. 35 of the Twenty-ninth World Health Assembly, May 1976 (1980)
3. Coleman, D.: Assisted suicide and disability-another perspective. Hum. Rts. **27**, 6 (2000)
4. Whyte, S.R., Ingstad, B.: Disability and culture: an overview. Disabil. Cult. 3–32 (1995)
5. Alsancak, S., Altınkaynak, H., ve Güner, S.: Sosyal Güvenlik Kurumu verilerine göre Türkiye'de hastaya özel yapılarak uygulanan protez ve ortezlerin sayısal çeşitlilik analizi. Fizyoterapi Rehabil. **24**(1), 99–103 (2013)
6. Cangür, Ş., Sığırlı, D., Bülent, E.Dİ.Z., Ercan, İ., ve Kan, İ.: Türkiye'deki Özürlü Grupların Yapısının Çoklu Uyum Analizi ile Incelenmesi. Uludağ Üniversitesi Tıp Fakültesi Dergisi **31**(3), 153–157 (2005)
7. Türkiye İstatistik Kurumu. Türkiye Sağlık Araştırması Mikro Veri Seti 2012. Türkiye İstatistik Kurumu Matbaası, Ankara (2013)
8. Ekenci, M.T.: Türkiye Sağlık Araştırması'nın Engelliliğe İlişkin Yaygınlık Ölçümü Açısından İncelenmesi. Sosyal Politika Çalışmaları Dergisi (35/2) (2012)
9. Altınkaynak, H.: Alt ve Üst Ekstremite Protezlerinde Ölçü Alma ve Üretim Yöntemleri. Ankara Üniversitesi Dikimevi Sağlık Hizmetleri Meslek Yüksekokulu Yıllığı **2**(1), 19–24 (2001)
10. Altınkaynak, H.: Amputelerde Kullanılan Destekleyici Teknoloji. Ankara Üniversitesi Dikimevi Sağlık Hizmetleri Meslek Yüksekokulu Yıllığı **4**(1), 31–36 (2003)
11. Burcu, E.: Türkiye'deki engelli bireylere ilişkin kültürel tanımlamalar: ankara örneği. Hacettepe Üniversitesi Edebiyat Fakültesi Dergisi **28**(1), 37–54 (2011)
12. Sutherland, A.T.: Disabled We Stand. Souvenir Press, London (1981)

The Arab Students' Needs and Attitudes of Learning English: A Study of Computer Engineering Undergraduates in Cyprus

Fatma Köprülü(✉), Seda Cakmak, and Arhun Ersoy

Near East University Yakındoğu Blv, Lefkoşa, Nicosia, Cyprus
{fatma.koprulu, seda.cakmak, arhun.ersoy}@neu.edu.tr

Abstract. The aim of this study is to identify the Arab students' needs and attitudes of learning English. Quantitative method was used in the research. 108 Arab students who took compulsory English courses in the Computer Engineering Department at a private university in Turkish Republic of Northern Cyprus in 2018–2019 academic year composed the universe of the research. The scales were applied to all the students who continued their education in this department. However, it was observed that the number of scales to be evaluated was 81. Therefore, the sample of the study is composed of 81 computer engineering students. The data was collected with the scale of "positive emotions of computer engineering students towards learning English" and "attitudes of computer engineering students towards learning English" developed by researchers. As a result of the research, it was found out that the students' positive feelings towards English language are beneficial and their attitudes towards language learning provide achievement and easiness. Furthermore, it has been indicated that there is a serious relation among students' needs and attitudes.

Keywords: Positive feelings · English for specific purposes · Attitudes · Computer engineering students

1 Introduction

In today's world the importance of communication in order to establish healthy relationship in all areas of daily life is an undeniable fact. In this respect, communicating in English is of paramount importance in every profession in order to be able to compete successfully. Researches that have been done in the area of engineering indicate that knowing English language is very necessary in engineers' professional and academic lives (Sidek et al. 2006; Pendergrass et al. 2001). Pendergrass et al. (2001) revealed that English has a vital importance in the engineering education so, implementing English into engineering departments helps to enhance the performance of engineering students in English language communication. At universities in all over the world, the medium of instruction at most of the departments is English. Therefore, it is important to determine the needs and attitudes of the students correctly as this will enhance the quality of teaching learning process. The term "attitude" is defined as a psychological construct that defines a particular behavior by Al-Mamun et al. (2012). Brown (2001)

D. J. Hemanth and U. Kose (Eds.): ICAIAME 2019, LNDECT 43, pp. 744–749, 2020.
https://doi.org/10.1007/978-3-030-36178-5_63

indicates that attitude is characterized by relationships, self and feelings in the community. Learning occurs if students have positive feeling towards the target language. Life experiences and attitudes are related to each other. Success and failure which is influenced by attitude has a very important role in the language learning process. Tella, Indoshi, and Othuon (2010) also highlighted that students with poor achievement in English language learning have negative feelings towards the target language. Defining the needs of students learning English makes learning easier and quicker for students. University students have different needs to learn English. Some students want to study their undergraduate program in English because they think they will have more chance in their professional life. Students believe that knowing English helps them to get a better job all over the world because English is the language of communication in the world. For some undergraduate students learning English makes their lives easier because they can easily communicate with international students in the university and people in the community. Students also need to learn English to pass their courses, write essays and write thesis. Students' improvement in English Language depends on special and specific needs of students to learn English. It is clear that from the list above, there are a lot of reasons for studying a foreign language (Harmer 2002; Oznacar et al. 2019).

In terms of education, it is important to manage teaching learning processes well (Debeş and Oznacar 2018; Koprulu and Oznacar 2017; Oznacar et al. 2018; Tezer et al. 2019). When it is examined for language learning, importance of the attitude comes in first rows. Learners' attitude is affected by students' needs of target language. Materials, activities and tasks are selected and created according to students' needs and attitudes. The importance of identifying learners' attitudes and needs towards learning English, this research had been arranged to learn computer engineering students' needs and attitudes towards English Language. Musa (1985) pointed out that Arab students know that English is the language of the world and knowing English language will enable them to continuo their postgraduate studies so, they have positive feelings towards English language and its culture. With this in their mind as undergraduate students they really want to learn English language.

Al-Sobhi et al. (2018) reveal that although Arab students are highly motivated to learn English, because of orthographic difference they can have negative feelings towards English language. Negative feelings towards English language makes learning process more difficult. This is one the reasons to have low proficiency in English language learning. Arab students believe that they have to learn English to be successful in their lives, so they gave high priority to learn English.

Sarjit (1993) carried out a study to find out about the language needs. His study showed that the main purpose for learning English was the instrumental motivation which was followed by personal motivation. Wilkins (1972) mentioned that when students want to learn English to pass the exams, to have a better job and to use English to communicate, they are instrumentally motivated. Cooper and Fishman (1977) described personal motivation as the motivation relating to "personal development or personal satisfaction" watching movies and reading books in English are the activities for the personal needs.

2 Objectives of the Study

The main issue of this study is to explore whether the Arab students' are aware of the needs and attitudes of learning English or not. Therefore, the objectives of this study are as follows:

1. To determine which types of positive feelings towards studying computer engineering in English.
2. To decide the type of attitude that computer engineering students have towards learning English.

3 Methodology

This part of the study presents The Arab students' needs and attitudes of learning English. In order to achieve this objective, interview and questionnaire were used. In the present study, the researchers used triangulation mixed method to obtain more reliability and validity. Creswell (2002) stated that triangulation mixed method design is a method which contains different research methods to examine the same issue.

4 Participants

This study was carried out in a computer engineering faculty, North Cyprus. The students who took part in this study were the second and third year computer engineering students aged around 20–25 years in the academic year 2018–2019. The total number of the students was 81 (male 66 and female 15). The system in the computer engineering department provides four years of instruction that qualifies students to graduate with a BA degree. English language is the medium of instruction in the department. By this way, the students can understand the concepts in the field of engineering clearly (Table 1).

Table 1. Profile of the participants

		Frequency	Percent (%)
Gender	Male	66	81.5
	Female	15	18.5

5 Data Collection Procedures

Before starting the implementation of the experiment, Arab students were informed about the subjects for any clarification they might need. Then they were asked whether they would be participating in the experiment or not since participating should be voluntary. First, the questionnaire was distributed and then the researcher briefly explained the objective and the procedure of the interview session.

6 Data Analysis Procedure

Qualitative and quantitative research methods were used in order to collect the data. Interviews conducted with Arab students in the computer engineering department were analyzed by using content analysis method and the questionnaires were analyzed with the Statistical Package for Social Sciences (SPSS) 20.0, and the mean and the standard deviation values of each item were conducted.

7 Limitations of the Study

The present study is limited to a private university in North Cyprus so it only provides a partial view of the Arab students' needs and attitudes towards learning English (Table 2).

Table 2. The computer engineering students' needs towards learning English

Questionnaire item	Yes		No		Total	
	n	%	N	%	n	%
Would you like to study with computers in English language laboratories that have programs in English language?	65	80.2	16	*19.8*	81	*100*

Table 2 indicates that the most of the students (95%) think that all the lessons should be taught in English in order to improve their language skills. In addition, 55% percent of them agree that English lessons should be compulsory at high schools in all Arabic countries. Only 35.8% of the students think that instead of Online English, 101 and 102 lessons should be taught in class 4 h a week.

8 Discussion and Conclusion

In this part of the study, the findings are discussed. The discussion takes place in accordance with the order of survey objectives. That is, the findings with regard to (1) the computer engineering students' needs towards learning English and (2) attitudes towards learning English are examined separately.

The students' attitudes towards learning English, important findings were obtained. According to the findings, whilst the majority of the students answered positively to the opinion that "all the lessons should be taught in English in order to improve your language skills", most of them did not agree that "English 101 and 102 should not be taught online". According to Koprulu and Oznacar (2019), technology helps students to apply what is taught and learned in the class through online education in a more productive and effective way.

It is quite normal the students agreed that all the lessons should be taught in English in order to improve language skills as this indicates that they are aware of the importance of English language throughout the world. On the other hand, they agree

Table 3. The computer engineering students' attitudes towards learning English

Questionnaire item	Agree		Disagree		Don't know	
	Total (n = 81)	%	Total (n = 81)	%	Total (n = 81)	%
(1) All the lessons should be taught in English in order to improve your language skills	77	*95*	3	*3.7*	1	*1.2*
(2) The syllabus of the English lessons at the department of engineering should be ESP based	67	*82.7*	12	*14.8*	2	*2.4*
(3) English 101 and 102 should not be taught online	10	*12.3*	71	*87.6*		
(4) English lessons should be compulsory at high schools in all Arabic countries	45	*55.5*	23	*28.3*	12	*14.8*
(5) Lecturers, assistants and secretaries should communicate in English with the students	64	*79*	12	*14.8*	5	*6.17*
(6) We should always use technological devices in learning English	50	*61.7*	27	*33.3*	4	*4.9*
(7) Emphasis should be on Speaking	66	*81.4*	12	*14.8*	3	*3.7*
(8) Instead of Online English, 101 and 102 lessons should be taught in class 4 h a week	29	*35.8*	45	*55.5*	7	*8.6*

that teaching English online is not a problem for them. Frohberg et al. (2009) revealed that in the education process at all levels mobile learning has an important place with the increasing use of technology.

This study aimed to examine the computer engineer students' needs and attitudes towards learning English language. For the students' needs the findings revealed that the students would like study with computers in English language laboratories that have programs in English language.

Regarding the students' attitudes, the results brought out that they have positive attitudes with regard to; (1) the lessons should be taught in English, (2) the syllabus of the English lessons at the department of computer engineering should be ESP based, (3) emphasis should be on speaking and (4) lecturers, assistants and secretaries should communicate in English with the students.

To sum up, the present study had been implemented to describe computer engineering students' needs and attitudes towards learning English. In fact, these results showed that there are preferential perceptions among using computer laboratories which give a chance to the students to use English language programs since these positive perceptions might be traced to many reasons like they will not have any fear feeling while using these programs. In addition, the students' confident ideas towards learning English could be used by the authorities to build up more effective and English speaking oriented teaching syllabus and system.

References

Al-Mamun, A., Rahman, M., Rahman, A.R., Hossaim, A.A.: Students' attitudes towards English: the case of Life Science School of Khulna University. Int. Rev. Soc. Sci. Humanit. **3**, 200–209 (2012)

Al-Sobhi, B., Md Rashid, S., Abdullah, A.N.: Arab ESL secondary school students' attitude toward English spelling and writing. SAGE Open **8**(1) (2018). https://doi.org/10.1177/2158244018763477

Brown, H.D.: Teaching by Principle and Interactive Approach to language pedagogy. Longman Inc., New York (2001)

Creswell, J.: Educational Research: Planning, Conducting, and Evaluating Quantitative and Qualitative Research. Merrill, Upper Saddle River (2002)

Cooper, R.L., Fishman, J.A.: A study of language attitudes. In: Fishman, J.L. (ed.) The Spread of English, pp. 239–276. Newbury House, Rowley (1977)

Debeş, G., Oznacar, B.: Evaluation of the opinions of the manager, teacher, employees (secretary and servants) about school management of the digitalization and management processes of the system engineering model in education. Amazonia Investiga **7**(16), 243–253 (2018)

Frohberg, D., Goth, C., Schwabe, G.: Mobile learning projects—a critical analysis of the state of the art. J. Comput. Assist. Learn. **25**(4), 307–331 (2009)

Harmer, J.: How to Teach English. Longman, London (2002)

Koprulu, F., Oznacar, B.: The views of lecturers, students and administrators on the impact of organization in foreign language education to academic achievement. Eurasia J. Math. Sci. Technol. Edu. **13**(12), 7865–7874 (2017)

Koprulu, F., Oznacar, B.: Analysis of university students' attitudes towards distance learning in foreign language education based on various variables. BRAIN - Broad Res. Artif. Intell. Neurosci. **10**(1), 123–128 (2019). ISSN 2067-3957

Musa, M.A.: Why don't our students speak English fluently after eight years of study? J. Educ. (UAE) **38**, 67–72 (1985)

Oznacar, B., Koprulu, F., Çağlar, M.: The success of implementing flipped classroom in teaching foreign language for international students. BRAIN. Broad Res. Artif. Intell. Neurosci. **10**(2), 151–158 (2019)

Oznacar, B., Şensoy, Ş., Satılmış, A.: Learning styles and emotional intelligence levels of university teacher candidates. Eurasia J. Math. Sci. Technol. Educ. **14**(5), 1837–1842 (2018)

Pendergrass, N., Kowalczyk, R., Dowd, J., Laoulache, R.: Improving first-year engineering education. J. Eng. Educ. **90**(1), 33–41 (2001)

Kaur, S.: Analysis of the English language needs of consultants at NCVC. M.A. thesis, University of South Australia (1993)

Sidek, S., Ramachandran, S., Ramakrishan, R.: From students to students: adapting technical reports as classroom materials. In: Mukundan, J. (ed.) Focus on ELT Materials, pp. 152–163, Kuala Lumpur (2006)

Tella, J., Indoshi, F.C., Othuon, L.A.: Relationship between students' perspectives on the secondary school English curriculum and their academic achievement in Kenya. Educ. Res. **1**(9), 390–395 (2010)

Tezer, M., Kan, G.Ş., Oznacar, B., Şensoy, Ş., Çaltıkuşu, Ç.: Content analysis of master's degree and doctorate theses where social skills training is approached. Int. J. Cogn. Res. Sci. Eng. Educ./IJCRSEE **7**(1), 43–49 (2019)

Wilkins, D.: Linguistics in Language Teaching. Cambridge University Press, Cambridge (1972)

Developing a Hybrid Network Architecture for Deep Convolutional Neural Networks

H. Hüseyin Sayan[1], Ö. Faruk Tekgözoğlu[1], Yusuf Sönmez[2(✉)], and Bilal Turan[2]

[1] Technology Faculty, Department of Electrical and Electronics Engineering, Gazi University, Ankara, Turkey
hsayan@gazi.edu.tr, faruktekgozoglu@gmail.com

[2] Technical Science Vocational College, Department of Electrical and Energy, Gazi University, Ankara, Turkey
ysonmez@gazi.edu.tr, bilalturan3@hotmail.com

Abstract. Deep learning is a powerful method, which can provide good solutions in many problems such as image recognition, speech recognition and natural language processing. In recent years researchers have exerted effort in the deep learning area in order to provide new network architectures and advanced results. VGG-16 and GoogleNet architectures which achieve highly accuracies in object recognition competition ILSVRC, have been based on to the convolution operation. In this work, we proposed an architecture which includes VGG-16 network convolution blocks and Inception blocks in GoogleNet architecture. The proposed architecture was evaluated on CIFAR-10 and MNIST datasets. The proposed architecture achieved results as close as results in the literature which have high accuracies. The architecture was created by Keras API and TensorFlow backends.

Keywords: Signature recognition · Keras · Convolutional neural network

1 Introduction

Deep learning concept has become quite popular in recent years and many applications and it offers the best solutions to many problems on image recognition, speech recognition and natural language processing. In particular, achievements in the field of deep learning was provide to the emergence of new network architectures and new solutions. Network architectures generally based on the convolution operations. The new architectures are obtained using to different combination of the convolution operations sequence. Convolution operation have been providing decrease image size and increase number of filters over the all network. Another factor for decreasing to image size is the pooling operation. In the beginning of the deep neural architectures, pooling operation was the average pooling but recently trends in pooling operation is become to the maximum pooling. At the same time activation functions were change to “relu” or “softmax” activation from “sigmoid” or “tanh”. For example, one of the first deep neural network LeNet-5 [1], was using “sigmoid” and “tanh” activation functions. On the other hand recently developed convolutional neural networks such as VGG-16 [2] uses “relu”

D. J. Hemanth and U. Kose (Eds.): ICAIAME 2019, LNDECT 43, pp. 750–757, 2020.
https://doi.org/10.1007/978-3-030-36178-5_64

and "softmax" activation functions. In this work, the architecture is a hybrid model, which is created by using VGG-16 convolution blocks and InceptionNet [3] blocks. Proposed convolutional neural network has been used to "relu" and "softmax" activation functions. And the Adam optimizers is chosen as backpropagation algorithm.

There are some hybrid network studies obtained by combining different network architectures in the literature. Szegedy [4] created a hybrid architecture, which is called as Inception-v4 using by Inception and ResNet architectures. Residual connections, which is idea of the Resnet's have been used to inside Inception blocks. Akhtar [5] used another hybrid architecture which combining SVM (Support Vector Machine) and CNN (Convolutional Neural Network) architecture for the sentiment analysis. Sentiment analysis is one of the sub topic of the Natural Language Processing. Sentiment analysis is the Natural Language Processing research area used to understand emotions using by likes, aggression, criticism etc. word patterns of a text. The researchers have been studied for increasing accuracy combining by different types of architectures and algorithms.

2 The Proposed Architecture

In this study, a new hybrid network model created by using convolutional neural network architectures such as Inception blocks from GoogleNet and convolution layers similar to VGG-16 network is proposed. This architecture is shown in Fig. 1.

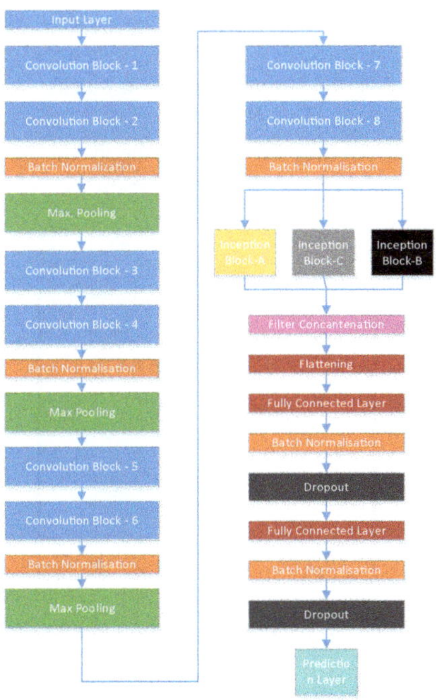

Fig. 1. The proposed architecture

In the proposed architecture all convolution operation blocks have 3 × 3 kernel size. Block – 1 and 2, have 16 and 32 filters, respectively. Convolution block 3 and 4 have 64 and 96 filters respectively, 5 and 6 blocks have 128 and 156 filters respectively, and finally two convolution layers have 156 and 192 filters respectively. The pooling operation in the proposed architecture is the max pooling. Max pooling operation kernel size is 2 × 2 and the stride is 2.

After the convolution operation 8, the Inception-Net blocks from GoogleNet has been placed here. This blocks have the operations shown in Fig. 2. All blocks have 1 × 1 convolution operation and average pooling. Therefore, the blocks operations, which have same operation at the others, has been discarded. Inception-A block was used to directly as same as the original one. In blocks B and C, 1 × 1 convolution and average pooling operations in block A have been removed and only the other operations are applied. Especially in this blocks, convolution operation with different kernel size have been used intentionally. In this way, intended to have different features.

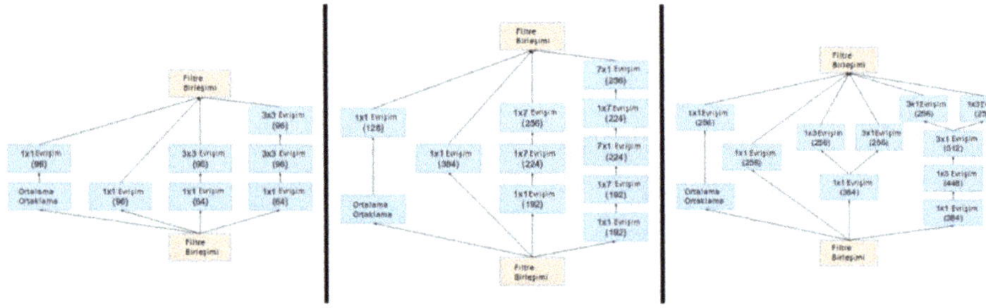

Fig. 2. Original Inception blocks

In the Fig. 2, original Inception block have been shown. On the left one is the Inception A, in the middle is Inception B and on right one is Inception C. This blocks were changed to as shown as in Fig. 3.

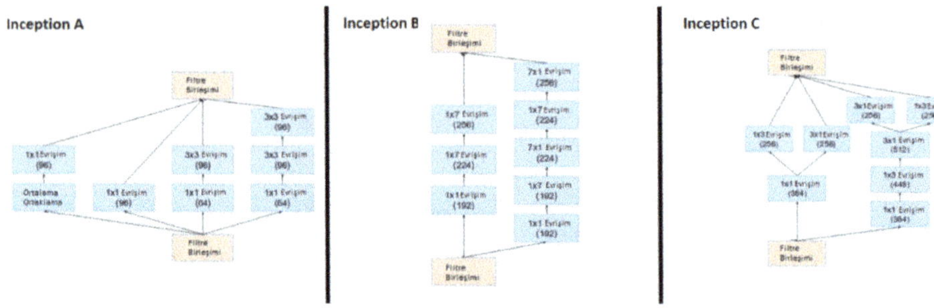

Fig. 3. Modified Inception blocks

Considering the B and C blocks, it is seen that same filter responses are prevented by changes. Also there are 7 × 1 and 1 × 7 convolution kernels in the B block. These operations were changed to 1 × 5 and 5 × 1 convolution kernels.

Inception blocks were increased the number of filters to be sent to the prediction layers. There are two fully connect layers placed at the end of the Inception block outputs. These fully connected layers have 256 neurons. The prediction layer of the network is a softmax layer set to 10 classes.

The Dropout techniques has been used for prevent to over fitting risk between to fully connected layers. Initial dropout values are set to 0.4. For higher epochs, the over fitting still is risk. Therefore, regularization techniques also used in the fully connected layers. Initial regularization value is 1e−5.

3 Experimental Results

MNIST is one of the most popular datasets in the deep learning studies. Dataset includes 60 thousands of training data and 10 thousands of testing data. Datasets image size is 28 × 28. The grayscale handwriting images have been wrote from 0 to 9 [6]. In the Fig. 4, MNIST dataset's examples are shown.

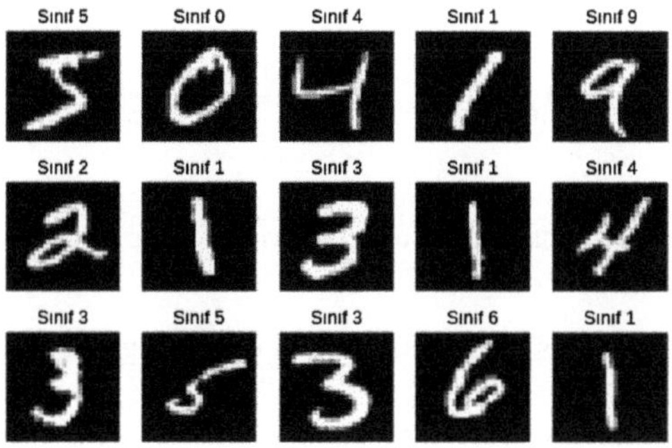

Fig. 4. The MNIST dataset examples

The researchers are frequently studied on the MNIST dataset. Table 1 shows some studies with high performance rates.

Table 1. Some of the researches for MNIST dataset

Name of the study	Loss function	Accuracy (%)
Dynamic Routing Between Capsules [7]	0.25	99.75%
Regularization of Neurol Networks Using DropConnect [8]	0.21	99.79%
Multi-column Deep Neural Networks for Image Classification [9]	0.23	99.79%
APAC: Augmented Pattern Classification with Neural Networks [10]	0.23	99.39%
Batch-normalized Maxout Network in Network [11]	0.24	99.76%
Generalizing Pooling Functions in Convolutional Neural Networks: Mixed, Gated, and Tree [12]	0.29	99.71%
Recurrent Convolutional Neural Network for Object Recognition [13]	0.31	99.69%
On the Importance of Normalisation Layers in Deep Learning with Piecewise Linear Activation Units [14]	0.31	99.69%

Table 2 shows test set result for the proposed architecture with 100 epochs.

Table 2. State of art study and proposed architecture results

Name of the study	Loss function	Accuracy (%)
Dynamic Routing Between Capsules [7]	0.25	99.75%
Proposed Model	0.0268	99.56%

The proposed architecture with only 100 epochs was closed to state of art study for the MNIST dataset. However, the state of art results could not be achieved by this architecture. The proposed architecture is not have data augmentation. Test result curves are shown in the Fig. 5.

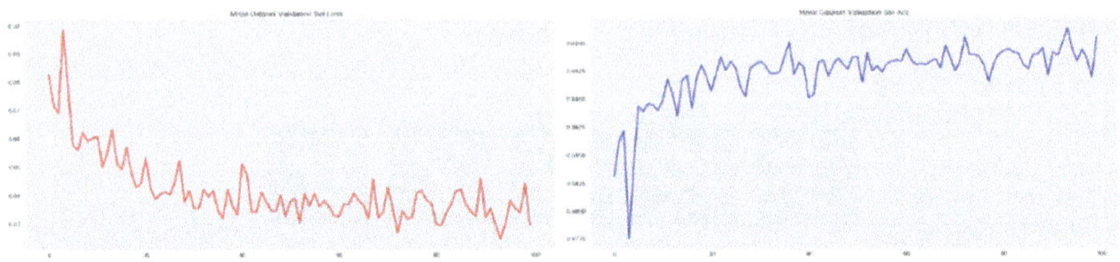

Fig. 5. Loss function and accuracy curve for the proposed architecture

In the Fig. 5, on the left is loss function and on the right is the accuracy for the test dataset.

The proposed architecture have also been evaluated with CIFAR-10 dataset to be sure effectiveness of model performance. CIFAR-10 dataset includes 10 different class with 50 thousands of training data and 10 thousands of testing data. The images are RGB format and have 32 × 32 size [15]. Each class in the dataset has 6000 images. Figure 6 shows the CIFAR-10 dataset examples.

Fig. 6. The CIFAR-10 dataset examples

Some of the academic studies that achieved high success on the CIFAR-10 data set are shown in Table 3.

Table 3. Some of the studies for the CIFAR-10 dataset

Name of the study	Loss function	Accuracy (%)
ShakeDrop Regularization [16]	2.31%	97.69%
Fractional Max Pooling [17]	3.47%	96.53%
Striving for simplicity: the all ConvNet [18]	4.41%	95.59%
All you need is a good init [19]	5.84%	94.16%
Lessons Learned from Manually Classifying CIFAR-10 [20]	6%	94%
Generalizing Pooling Functions in Conv. Neural Networks: Mixed, Gated, and Tree [12]	6.05%	93.95%
Dynamic Routing Between Capsules [7]	10.6%	89.4%

The accuracy value of the proposed model on the CIFAR-10 dataset after 50 epochs are given in Table 4.

Table 4. Proposed model results for CIFAR-10 dataset

Name of the study	Loss function	Accuracy (%)
Proposed Model - 50 epochs	0.5913	83.93

According to these results, it can be said that the proposed model places between high performances studies in the literature. However, proposed model could not have achieved the literature accuracies. The studies in the literature have been evaluate for higher than 500 epochs. But our model was evaluated only 50 epochs because of hardware limitation. Figure 7 shows the proposed model test results for the CIFAR-10 dataset.

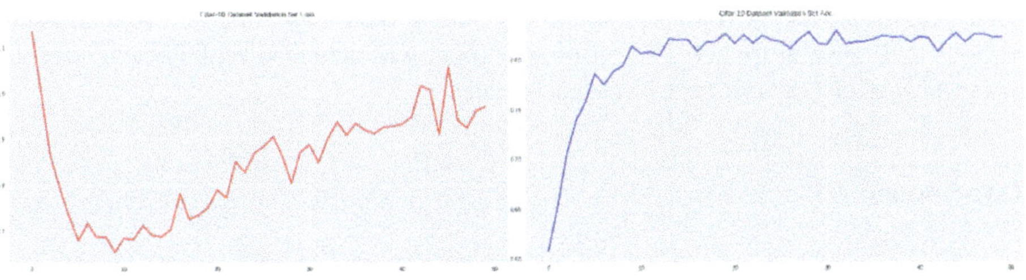

Fig. 7. Loss function and accuracy curve for the proposed architecture

Figure 7 shows the results of the proposed model in the CIFAR-10 data set. In the Fig. 7, shown on the left is loss function. If we look at it carefully, it can be seen that the curve is overfitting after 10 epochs. This is the most common situations in the deep learning. After 10th epochs, the model began to memorize the training dataset. Therefore, the proposed model needs to regularization methods as well as more dropout to. Apart from these, data augmentation can also prevent to overfitting of the model. Shown on the right is the test accuracy of the proposed model. Figure 8 shows the proposed model that have regularization, more dropout and data augmentation for 100 epochs.

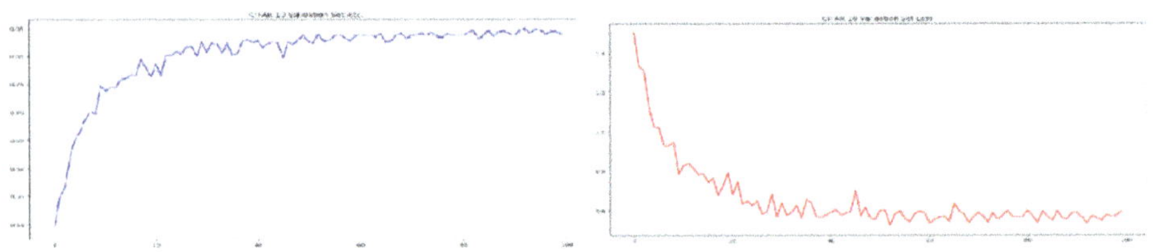

Fig. 8. Accuracy and loss function for the proposed architecture

Results in Fig. 8, the proposed architecture have been evaluated for the CIFAR-10 dataset and 100 epochs. Best accuracy to achieved is 84.86%. For achieving to this result, the model has used data augmentation. The data augmentation which used in this study is the flipping 40°, shifting 0.1 and zoom with 0.1.

4 Conclusions

The proposed architecture is a mixed model where some blocks belong to VGG-16 and GoogleNet. The proposed model was tested with MNIST dataset for 100 epochs. For the MNIST dataset, proposed model was reached to 99.56% test accuracy. These results were achieved without data augmentation. At the same time the proposed model was tested on the CIFAR-10 dataset for 50 and 100 epochs. Results the proposed model was memorizing to training dataset for 50 epochs. Therefore, the proposed model was tested again for 100 epochs with using data augmentation, regularization methods and more dropouts. The proposed model was achieved 84.86% test accuracy for the CIFAR-10 dataset with these techniques.

References

1. LeCun, Y., Bottou, L., Bengio, Y., Haffner, P.: Gradient-based learning applied to document recognition. IEEE **86**(11), 2278–2324 (1998)
2. Simonyan, K., Zisserman, A.: Very deep convolutional networks for large-scale image recognition. arXiv:1409.1556v6

3. Szegedy, C., Liu, W., Jia, Y., Sermanet, P., Reed, S., Anguelov, D., Erhan, D., Vanhoucke, V., Rabinovich, A.: Going deeper with convolutions. arXiv:1409.4842v1
4. Szegedy, C., Joffe, S., Vanhoucke, V., Alemi, A.: Inception-v4, Inception-ResNet and the impact of residual connections on learning. arXiv:1602.07261v2
5. Akhtar, S., Kumar, A., Ekbal, A., Bhattacharyya, P.: A hybrid deep learning architecture for sentiment analysis. In: 26th International Conference on Computational Linguistics: Technical Papers, pp. 482–493, December 2016
6. LeCun, Y., Cortes, C., Burges, C.J.C.: The MNIST Database of Handwritten Digits. http://yann.lecun.com/exdb/mnist/
7. Sabour, S., Frosst, N., Hinton, G.E.: Dynamic Routing Between Capsules. Google Brain, Toronto (2017)
8. Wan, L., Zeiler, M., Zhang, S., LeCun, Y., Fergus, R.: Regularization of neural networks using DropConnect (2013)
9. Cireşan, D., Meier, U., Schmindhuber, J.: Multi-column Deep Neural Networks for Image Classification, IDSIA-USI-SUPSI, Switzerland (2012)
10. Sato, I., Nishimura, H., Yokoi, K.: APAC: Augmented Pattern Classification with Neural Networks, Denso IT. arXiv:1505.03229v1, May 2015
11. Chang, J., Chen, Y.: Batch-Normalized Maxout Network in Network. arXiv:1511.02583v1, November 2015
12. Lee, C., Gallagher, P., Tu, Z.: Generalizing Pooling Functions in Convolutional Neural Networks: Mixed, Gated and Tree. arXiv:1509.08985v2, October 2015
13. Liang, M., Hu, X.: Recurrent convolutional neural network for object recognition. In: CVPR 2015 (2015)
14. Liao, Z., Carneiro, G.: On the importance of normalisation layers in deep learning with piecewise linear activation units, November 2015
15. Krizhevsky, A.: The CIFAR-10 Dataset. https://www.cs.toronto.edu/~kriz/cifar.html
16. Anonymous Authors: ShakeDrop regularization. In: ICLR 2018 Conference (2018)
17. Graham, B.: Fractional Max-Pooling (2015). arXiv:1412.6071v4
18. Springenberg, J.T., Dosovitskiy, A., Brox, T., Riedmiller, M.: Striving for simplicity: the all convolutional net. In: ICLR 2015 (2015). arXiv:1412.6806v3
19. Mishkin, D., Matas, J.: All you need is a good init. In: ICLR 2016 Conference (2016). arXiv:1511.06422v7
20. Karpathy, A.: Lessons Learned from Manually Classifying CIFAR-10. http://karpathy.github.io/2011/04/27/manually-classifying-cifar10/

Blockchain-Based Secure Recognized Air Picture System Proposal for NATO Air C2 Capabilities

Enis Konacakli[1(✉)] and Enis Karaarslan[2]

[1] Department of Computer Engineering, Eskisehir Technical University, Eskişehir, Turkey
enisk@eskisehir.edu.tr

[2] Department of Computer Engineering, Mugla Sitki Kocman University, Muğla, Turkey
enis.karaarslan@mu.edu.tr

Abstract. The command and control (C2) systems are of prime importance for leveraging the superiority of the air forces in combat operations. The key element for these systems is a single, coherent recognized air picture (RAP) which provides the real time situational awareness for the theatre commander and the subordinate operational elements. The manipulation of the air picture can trigger a chain of events that may cause the total failure of the air campaign. Blockchain technology can be used to prevent this exploitation. This technology has the potential to eliminate the security problems by its decentralized architecture and its irrevocable registry. It can be used as the key element for enabling the security of friendly air C2 (Command & Control) capabilities in the cyber domain. In this study, the function of the blockchain system is questioned for the benefit of air operations. In this context, a blockchain-based system, which benefits and can enhance the security of an Allied Joint Operational Air C2 system, is proposed. The prototype is developed in Hyperledger Fabric platform. The findings and future work is given.

Keywords: NATO · Command and control systems · Recognized air picture · Radar · Cybersecurity · Blockchain · Hyperledger

1 Introduction

The decision-making process is a process of evaluating the variables with a holistic perspective, defining the problems, identifying possible solutions, making an evaluation, and determining the most appropriate one. Effective and correct management of the decision-making process is the most important factor in the success of air operations. Command and control (C2) systems are crucial components of the air combat operations which takes advantage of the decision-making process. The key element for C2 decision-making process is a single, coherent recognized air picture (RAP) which provides the real time situational awareness for the theatre commander and the subordinate operational units. The manipulation for the air picture can trigger a chain of

D. J. Hemanth and U. Kose (Eds.): ICAIAME 2019, LNDECT 43, pp. 758–765, 2020.
https://doi.org/10.1007/978-3-030-36178-5_65

events that may cause the total failure of the air campaign. Hence, the security of these systems is of vital importance [1–3].

Client-server architectures and network based cyber security solutions have numerous number of weaknesses. While having advantages of centralization and back-up recovery capabilities, client-server architectures are not as robust as decentralized architectures against cyber-attacks. If the server fails, the whole network goes down. If a client is blocked due to some error or connection problem, while getting data from a central server, this current data can be chocked up permanently. Lost parts of the data can be provided from the other peers if decentralized architectures are used. Decentralized systems can be used to create better solutions at connection failures, cyber-attacks, and disasters. In this point of view, decentralized security solutions may create smarter models for the security of the digital assets processed in Air C2 systems.

Two anchor security challenges of decentralized systems should be solved if we prefer to use decentralized security solutions to enhance the security of this critical operational infrastructure. Although they have a strong and modular framework, decentralized networks can fail to ensure the privacy and controllability of data. A new system that can ensure the security and privacy of the air track data is needed. We propose a permissioned private blockchain model, designed for the Air C2 system security.

In the next section, the fundamentals of blockchain technology and its security services are given. In Sect. 3, related works are summarized. In Sect. 4, the proposed model is described and in Sect. 5, its implementation is explained. Finally, findings and discussions are shared.

2 Fundamentals

2.1 Blockchain

Blockchain is an immutable registry in which records cannot be manipulated and copies of these records can be stored by every peer in distributed architecture [4]. It is the core technology used by Bitcoin and other cryptocurrencies. When it was first released, it was only used for transferring digital cryptocurrencies. It has been able to run smart contracts, thanks to the smart contracts that can be processed in cryptographically secured blocks, since 2014.

The records cannot be changed and the system is protected against crashes, because of the instances of the registry in the peer-to-peer structure that is kept partially or completely in the local drivers of each peer. Security is ensured with cryptographic encryption and full confidence is established between the parties with its distributed architectural structure. Transactions cannot be done without the permission of authorized peers. Participants in a trustless environment can create transactions without trusting each other and the need of a third party. Privacy can be ensured by the use of cryptographic algorithms. The decentralized architecture makes the network more secure and robust against Cyber threats.

Total process is summarized in Fig. 1. In the blockchain architecture, all the operations performed in the P2P network are queued to form a block, and when the process is accomplished and complete a block, the node, which is selected with the consensus protocol, closes the block using the cryptographic hash algorithm. These blocks are connected to each other by successive and overlapping cryptographic operations, forming an unbreakable cryptographic chain. The sensitive data is protected in this chain. While the transactions are performed outside the intervention of a central authority, the security of the system is increased by distributing the registry among the peers.

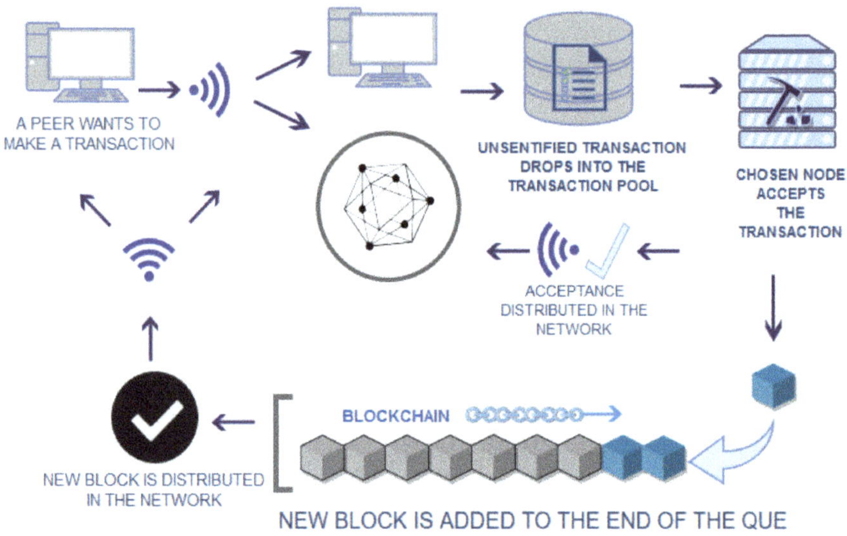

Fig. 1. Main steps of a simple blockchain model.

Private blockchain architectures are strong and robust frameworks, where new users are not allowed to access, modify or see the registry without the permission of the authority. These architectures are suitable blockchain platform for secure data sharing implementations.

In blockchain, the node that will be responsible for assigning of the transactions is determined by the consensus protocols. Proof of Work (PoW) and Proof of Steak (PoS) are the most common consensus protocols. SIEVE, PBFT and PoET are strong and secure consensus protocols, which enable private permissioned blockchain implementations. A Hyperledger framework SawtoothLake uses PoET consensus protocol, and SIEVE and PBFT consensus protocols are used by Hyperledge-Fabric.

2.2 Reconsideration of the Blockchain Security for the Implementation of the Proposed Model

It is not cost efficient to form a network security infrastructure by using blockchain infrastructure as it is in crypto currency technology. The creation of crypto currency and the approval of the transactions require a high resource. Confirmation of the transaction process causes the system to run very slowly. This does not provide us proper working environment required in applications where peers must perform thousands of transactions per second. Radar platforms can creates thousands of tracks in a second to gather a unique RAP. The consensus must be specified for the purpose of privacy and security. Transection speed is the other critical objective that must be provided.

If we prefer to use blockchain framework for enhancing the security of critical operational infrastructure, it is expected that the architecture of the blockchain model is intended to be:

- Full controllable,
- Immutable,
- No transparency in areas where classified information is processed while transparency is allowed in certain areas of application,
- Ensuring confidentiality,
- Have to ensure data integrity,
- Have redundancy against cyber-attacks,
- Must protect the system against the intrusive and systemic attacks or errors,
- Sustainable against to crashes in case of disasters and physical or software damages, unlike server client architectures,
- Low or acceptable operating costs and energy consumption compared to other blockchain structures,
- Has to have high productivity,
- Must be undeniable,
- Has to catalyze fast and secure access to information,
- It needs to reduce human errors.

Hyperledger Fabric framework, which is the incubator for business blockchain technologies is chosen for the implementation of the model. Hyperledger-Fabric is a production ready, private permissioned distributed ledger framework of Hyperledger project for creating business blockchain applications. It is a of open source blockchain project, started by the Linux Foundation in 2015, under which multiple development teams are collaborating to create open source blockchain and distributed ledger technologies. Its private permissioned structure ensures the full controllability and management of the distributed network. Since it does not use any token like crypto currency, it does not need mining. Therefore, it is efficient and resource friendly.

The smart contracts, used in the other blockchain structures are called chincode in Hyperledger-Fabric. Nodes can create groups among themselves that are called channels. Rules, located in the chaincode, identify the authorization of the peers, ensures the security of the channels.

3 Related Works

Some of other studies on blockchain and cyber security capable of supporting models that can constitute model for private data protection. Most of them are focused on the IoT and Cloud Computing technologies [5–8]. Studies aiming to create secure, dispersed architectural structures that can be used instead of border security-based architectures and to reevaluate the internet in this architecture due to the internet name servers (DNS) have addressed the issues that could directly create national cyber security architectures [9–12].

4 System Proposal

The main participants of the prototype model are NATO and Non-NATO Ally (host country).

As the dominant air power, NATO ensures the production and management of the Air Picture throughout the entire operational area for NATO's own operational purposes. NATO has the authority to access and modify all its own RAP data in order to accomplish this task. NATO allows Non-NATO Ally Country Forces to monitor relevant RAP of their territory, in support to their air defense elements.

Sub elements are specified in Table 1.

Table 1. Sub elements used in the model.

NATO PARTICIPANT		NON-NATO ALLY PARTICIPANT	
NATO_Track & NATO_Radar	TRACK NATO ONLY	National_Track & National_Radar	TRACK NATIONAL
AWACS		AIR RADAR	
AIR RADAR		AIR DEFENCE RADAR	
AIR DEFENCE RADAR		HQ	
NAVY RADAR		-	
HQ			

This system has four major actors:

- Participants: Participants are the partners of the air operation. NATO and Non-NATO Ally participants are the two extended participants of the system.
- Assets: Assets are the real events of the system that created by participants as a result of the processes that produced by the chain code. NATO and Non-NATO Ally tracks are the two digital precious asset types which are created by participants.

Tracks that creates the recognized air picture (RAP), are the precious digital assets which we aim to protect in the blockchain. Model, proposed for NATO Air Operation environment is simulated and RAP tracks created in Hyperledger-Fabric platform. Rules are created in chain code, defining the authorizations of the different types of participants for ensuring the security of the track data. The proposed model based on decentralized private blockchain architecture will eliminate the cost, privacy and security problems. Also securing the RAP track data is aimed. Two channels are created to demonstrate the security model. While Channel A is stands for the security of NATO assets, Channel B covers Non-NATO Ally Host nations operational elements. Sub elements and channels used in the model are demonstrated in Fig. 2. Channels that stands for NATO and Non-NATO Ally units are marked with different colours. Security and privacy of air track data is aimed be ensured by the implementation of the channels generated by the chaincode that rules the model.

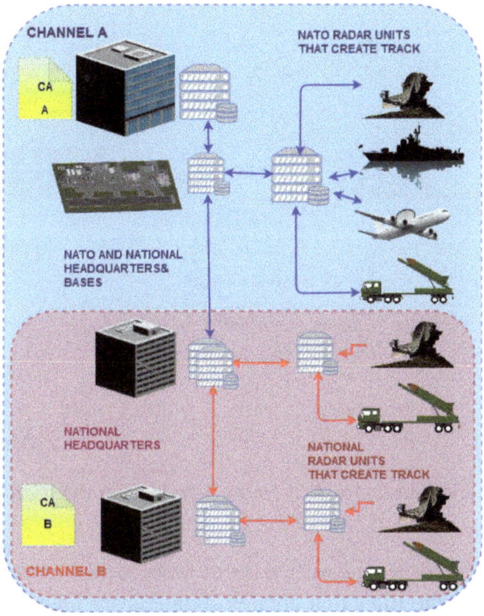

Fig. 2. Sub elements and channels used in the model.

5 Implementation

The application environment is prepared with installed Hyperledger Fabric1.3 on a usual laptop. Hyperledger Composer Playground is used as the test environment. All operations are performed on a computer with 8 GB RAM, 500 GB hard disk and i5 Intel 2.6 GHz CPU. Hyperledger Fabric runs in Virtualbox on a virtual machine. Vagrant tool used to control the virtual environment. The peers are created in Docker Containers, using standard Hyperledger Fabric Docker virtual images. A separate database is defined for each peer to protect transaction logs and the current state of assets. CouchDB is chosen as a database, for allowing complex queries to be used during report generation processes. A separate certificate authority is identified for each channel, the Membership Service Provider (MSP) provides user authentications and certificate verifications. The technologies used in the model and the basic walk-through are shown in Fig. 3.

Classification is taken into consideration in every parameter of coding and implementation. The real data types of actual variables used in real military systems are not used in the example model. The model is designed to be able to operate with real variable data types.

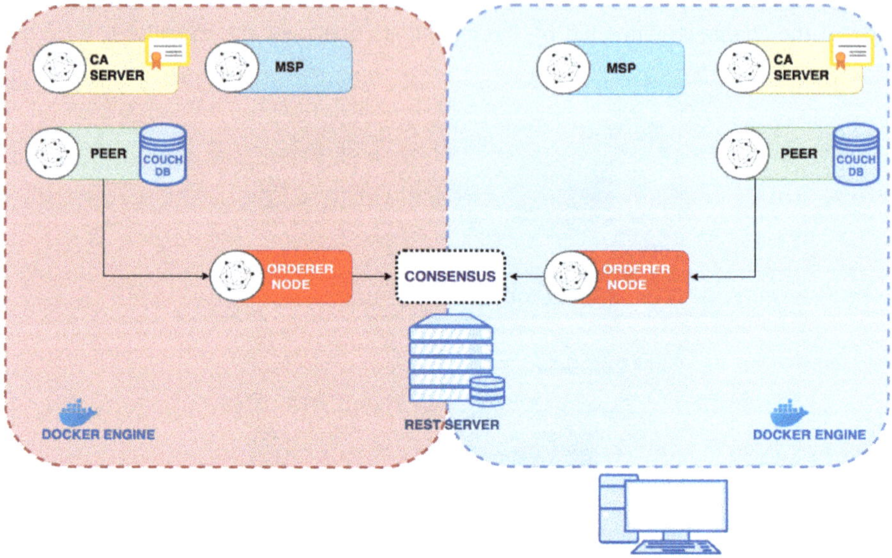

Fig. 3. Technologies & their walkthrough in the model [13].

Participants and Assets are identified in model.cto file, creation of transactions is identified in logic.js and the rules that creates channels are identified in permissions.acl. The security and privacy implemented by the usage of channels.

6 Results and Conclusion

A highly secure infrastructure is proposed against cyber-attacks and negative kinetic effects with this model. In the course of any legal process, there is a more robust security, indisputability and controllability against the deletion or falsification of evidence, while the integrity and security of the data is ensured at the highest level, the information is shared with the necessary users within the same principle. Human factor and errors has been minimized. Compared to the other blockchain structures, a highly efficient system has been obtained by keeping the operating costs and energy consumption low and acceptable levels.

In this study, a model has been developed by choosing Hyperledger Fabric as a blockchain platform. An application has been developed in order to ensure the security of track data, which has high importance for air operation. It is aimed to set an example for the later works that can be done on this subject, with this model and application. Using permissioned private blockchain platforms, which are capable of producing thousands of transactions per second, this model can be successfully implemented for more complex applications.

References

1. Recognized Air Picture Systems Evaluation. Defence Research and Development Canada-Vafcartier. http://cradpdf.drdcrddc.gc.ca/PDFS/unc57/p524477.pdf. Accessed 19 Apr 2019
2. Fanti, L., Beach, D.: NATO initial common operational picture capability project. Proc. SPIE **4741**, 80–89 (2002)
3. Dalman, G.H., Kopp, D.M., Redman Jr., G.A.: The imperative to integrate air force command and control systems into maritime plans. Air Space Power J., 93–118 (2014)
4. Nakamoto, S.: Bitcoin: A Peer-to-Peer Electronic Cash System (2008)
5. Kshetri, N.: Can blockchain strengthen the internet of things? IEEE IT Prof. **68–72**, 68 (2017)
6. Copos, B., Levitt, K., Rowe, J., Kianmajd, P., Chuah, C., Kesidis, G.: Security and Privacy for Emerging Smart Community Infrastructures. Open Access Publishments from the University of California UCDAVIS, 17 October 2018. escholarship.org
7. Jan, J.K., Guillaume, B.: Connecting Multiple Devices with Blockchain in the Internet of Things, Research paper for Seminar on Blockchain Technology (MTAT.03.323). Tallinn University of Technology, Tallinn (2017)
8. Ramachandran, A., Kantarcioglu, M.: Using Blockchain and smart contracts for secure data provenance management. Cornel University Library. arXiv:1709.10000v1
9. Puthal, D., Malik, N., Mohanty, S.P., Kougianos, E., Yang, C.: The blockchain as a decentralized security framework. IEEE Consumer Electron. Mag. **7**, 18–22 (2018)
10. Durğay, Z., Karaarslan, E.: Blokzinciri Teknolojisinin E-Devlet Uygulamalarında Kullanımı: Ön İnceleme [Use of Blockchain Technology in E-Government Applications: Preliminary Review]. In: Akademik Bilişim Conferance (2018)
11. Alexopoulos, N., Vasilomanolakis, E., Ivanko, N.R., Muhlhauser, M.: Towards blockchain-based collaborative intrusion detection systems. In: D'Agostino, G., Scala, A. (eds.) CRITIS 2017. LNCS, vol. 10707, pp. 107–118. Springer, Heidelberg (2017). https://doi.org/10.1007/978-3-319-99843-5_10
12. Alcazar, V.: Data you can trust, blockchain technology. Air Space Power J. **31**, 91–101 (2017)
13. Gür, A.Ö., Öksüzer, Ş., Karaarslan, E.: Blockchain based metering and billing system proposal with privacy protection for the electric network. In: 2019 7th International Istanbul Smart Grids and Cities Congress and Fair (ICSG), pp. 204–208. IEEE (2019)

A New Genetic Algorithm for the Maximum Clique Problem

Gozde Kizilates Evin(✉)

Department of Mathematics, Faculty of Science, Ege University, Izmir, Turkey
gozde.kizilates@gmail.com

Abstract. The Maximum Clique Problem is widely studied optimization problem and plays an essential part in computer science. In the literature, therefore, there are many exact and heuristic approaches and new strategies are continuously being proposed. In this paper, the maximum clique problem is studied, and a new hybrid genetic algorithm is proposed for finding the maximum clique in a graph. The proposed algorithm is tested on the DIMACS and BHOSLIB benchmark instances. The computational results are compared with similar literature researches. The results prove the effectiveness of proposed algorithm.

Keywords: Maximum clique problem · Graph theory · Optimization problems · Heuristic algorithms · Hybrid genetic algorithms

1 Introduction

The Maximum Clique Problem (MCP) is much studied optimization problem in the operation research field. The MCP belongs to NP-Complete class. The problem has an extensive application in computer vision, coding theory, fault diagnostics, pattern recognition, science, and engineering [1]. Besides, the problem is completely similar with the maximum independent set (MIS) and the minimum vertex cover (MVC) problem. These three problems are different versions of one another [2]. The MCP is also connected to the graph coloring problem [3]. For these reasons, the problem plays a significant role in the computer science field.

The Maximum Clique is maximum cardinality set of vertices, any two of these vertices are adjacent. The problem has been defined by Luce and Perry, who has studied complete subgraph for social network analysis [4]. It is one of the 21 original NP-Complete problems Karp proposed in [5]. Due to the problem belongs to the NP-Complete class, the heuristic algorithms have been generally proposed instead of exact algorithms.

Before 1990, exact algorithms have been widely studied. Some of these algorithms have based on classic enumeration method [6–12]. The first known enumeration algorithm was proposed by Harary and Ross [13]. Most of these algorithms has been tested with specific graphs. Afterward, the branch and bound method (B&B) has been developed with being modified enumeration method. In the B&B method, fewer subsets examined instead of enumerating all subsets. The earliest and the most known B&B method has been developed by Carraghan and Pardolos (CP algorithm) [14]. CP

D. J. Hemanth and U. Kose (Eds.): ICAIAME 2019, LNDECT 43, pp. 766–774, 2020.
https://doi.org/10.1007/978-3-030-36178-5_66

algorithm is the basis for many exact algorithms despite its simplicity [15]. The difference between B&B algorithms is branching strategy and boundary value. B&B algorithms reduce the upper bound of clique and choose promising vertex for branching [16–18]. Various techniques are used to reduce the upper bound such as graph coloring, maxSAT, etc.

Recently, heuristic methods are studied more because finding optimum solution with exact methods takes a long time due to computational complexity. The most studied and most successful heuristic method is local search approach. In local search, when a solution is found, a better solution is searched in the neighbor by adding, removing or switching vertices in solution [19–22].

Another approach to solving the MCP is evolutionary algorithms. While there are a few studies in the literature related to ant colony optimization, artificial neural network, and simulated annealing methods [23–26], genetic algorithms are widely used. Solving the MCP with genetic algorithm has been started to study in the 1990s. Pure genetic algorithms have poor performance for the MCP. Therefore, hybrid genetic methods are preferred. The first hybrid genetic algorithm is a genetic tabu search algorithm [27]. Another most-known hybrid genetic algorithm proposed by Marchiori improves solution using local search approaches after GA [28].

In this paper, a new hybrid genetic approach is proposed for finding maximum clique for a given graph. In the proposed algorithm, the solution is improved not only classical genetic algorithm steps but also heuristic approaches. This algorithm is coded in Java and tested on the benchmark instances. The results are compared with GENE, GMCA, and SLS hybrid genetic algorithms in the literature.

2 Maximum Clique Problem

2.1 Definition

Given an unweighted, undirected, and connected graph $G = (V, E)$, where $V = \{1, 2, \ldots, n\}$ is the vertex, $E \subseteq V \times V$ is edge set of G. Two vertices are said to be adjacent if they are connected by an edge. V^* vertex set, which is subset of V, is clique if all its vertices are adjacent, such that, for all u, $v \in V^*$, the edge (u, v) is in E. The maximum clique is the largest vertex set among all cliques in a graph and the maximum clique number denoted $\omega(G)$ [1].

2.2 Mathematical Programming Formulation

There were many integer programming models for the MCP. The simplest binary mathematical formulation, which was improved by using the MIS, is following [1]:

Goal function:

$$\max \sum_{i=1}^{n} x_i \tag{1}$$

Constraints:

$$s.t. \quad x_i + x_j \leq 1, \quad \forall (i,j) \in \overline{E} \tag{2}$$

$$x_i \in \{0,1\}, \quad i = 1,\ldots,n \tag{3}$$

$$C = \{i \in V : x_i = 1\} \text{ is clique.} \tag{4}$$

3 Genetic Algorithm

Genetic Algorithm (GA), is an iterative search method based on Darwin's natural selection theory and considers the evolutionary process [29]. It has been developed first by Fraser [30]. The fundamentals of genetic algorithms have been formed by Holland [31].

In genetic terminology, the population which is composed by chromosomes is repeatedly modified by using genetic operations. Each chromosome has genes which carries a specific feature. The selection operator determines the chromosomes that will transfer more genes to the next generation. The crossover provides creating offsprings with exchanging the genes of selected chromosomes. The mutation changes the genes of the chromosome randomly [32].

Genetic Algorithms give good results for most optimization problems. Pure genetic algorithms are not effective for the MCP [33]. Therefore, genetic algorithms are combined with other approaches like local search.

Hybrid genetic algorithms are evolutionary approaches that are formed by the combination of genetic algorithms and other heuristic methods. Genetic algorithms are suitable for creating hybrid structures. Thus, the importance of hybrid genetic algorithms is begun to increase in solving real-life problems.

4 Proposed Method

4.1 Initialize the Population

In the proposed method (rHGA), binary representation was used to create chromosomes of initializing population. In this representation, the chromosomes were encoded n-bit binary arrays (n is the cardinality of vertex set). If *i-th* bit value is 1, the corresponding vertex (*i-th* vertex) is in the solution; otherwise, its value is 0.

n-chromosomes were created using a heuristic approach. In this heuristic approach, the first element of each chromosome was assigned its index value, it means that *i-th* bit of *i-th* chromosome was added the solution $(1 \leq i \leq n)$. Then, the vertex set was randomly sorted, and the elements of the sorted vertex set were examined respectively. Examined vertex was tested for each chromosome. If the vertex was connected to all elements of the chromosome, that vertex has added the solution.

After creating the initial population, the fitness function was defined. The fitness value was the sum of the vertices in the solution for each chromosome.

4.2 Genetic Operations

Selection: Roulette wheel method was used for choosing parents for the next generation. This method provides that the chromosomes which have higher fitness value produce more offspring.

Crossover: The single-point crossover operator was used. The crossover point *i* was selected randomly. The first *i* bits of the first parent were transferred to the first offspring and the remaining bits were contributed by the second parent. Similarly, the first *i* bits of the second parent were transferred to the second offspring and the remaining bits were contributed by the first parent. Also, the best two parents were applied crossover step without any selection.

Elitism: The best 5% of chromosomes were transferred to the next generation without any change. Also, the worst 5% of chromosomes were transferred to the next generation in order to provide diversity.

Adjustment: After the crossover, some chromosomes may not indicate clique. So, for the adjustment step, drop, add, swap, and add steps were implemented in respectively in chromosomes.

- Drop: The drop operator deletes a vertex from the chromosome. In this method, the chromosome was checked whether it was clique or not. If the chromosome was not clique, the vertex in the chromosome that has the lowest link to other vertices was removed. The process continued until the chromosome is a clique.
- Add: The add operator adds a vertex to the chromosome. In this method, if any vertex connected to other vertices in the chromosome, that vertex added the chromosome.
- Swap: In the swap method, any vertex in the chromosome (*i-th* vertex) replaces any vertex that is not in the chromosome (*k-th* vertex). If the *k-th* vertex is connected to all the other vertex except the *i-th* vertex in the chromosome, the *k-th* vertex has been added the chromosome and the *i-th* vertex has been removed from the chromosome.

Mutation: In the mutation step, 10% of chromosomes were chosen randomly. For each chromosome, 10% of vertices from the chromosome in the solution were chosen randomly and these vertices were removed from the chromosome. While this process seems to reduce the quality of the solution, but in fact, it increases the diversity and saves the solution from the local optimum.

These operations repeated until the iteration number. In each iteration, the solution was updated before the mutation step. The parameters used in the implementation of the algorithm are as follows (Table 1):

Table 1. Genetic algorithm parameters

Population size	Crossover rate	Elitism rate	Mutation rate	Iteration number
n	0.9	0.05	0.1	n

5 Computational Results

There are two benchmarks for the MCP. DIMACS benchmark library has 37 graphs, which are real-life and random instances [34]. BHOSLIB Benchmark library has 36 graphs, which have been generated by transforming forced SAT benchmarks [35].

The proposed algorithm was tested all benchmark instances. Table 2 includes the results of the rHGA for DIMACS benchmark instances and Table 3 includes the results of the rHGA for BHOSLIB benchmarks instances. In these tables, the first five columns contain respectively the name of instances, maximum clique number, best-known

Table 2. Results for DIMACS benchmark instances

Instance	ω(g)	Best known	Vertex	Edge	rHGA
C125.9	34	34	125	6963	34*
C250.9	44	44	250	27984	44*
C500.9	≥ 57	57	500	112332	55
C1000.9	≥ 68	68	1000	450079	65
C2000.9	≥ 80	80	2000	1799532	73
DSJC1000_5	15	15	1000	499652	15*
DSJC500_5	13	13	500	125248	13*
C2000.5	16	16	2000	999836	16*
C4000.5	18	18	4000	4000268	17
MANN_a27	126	126	378	70551	126*
MANN_a45	345	345	1035	533115	344
MANN_a81	1100	1100	3321	5506380	1098
brock200_2	12	12	200	9876	12*
brock200_4	17	17	200	13089	17*
brock400_2	29	29	400	59786	25
brock400_4	33	33	400	59765	33*
brock800_2	24	24	800	208166	21
brock800_4	26	26	800	207643	21
gen200_p0.9_44	44	44	200	17910	40
gen200_p0.9_55	55	55	200	17910	55*
gen400_p0.9_55	55	55	400	71820	52
gen400_p0.9_65	65	65	400	71820	65*
gen400_p0.9_75	75	75	400	71820	75*
hamming10-4	40	40	1024	434176	40*
hamming8-4	16	16	256	20864	16*
keller4	11	11	171	9435	11*
keller5	27	27	776	225990	27*
keller6	≥ 59	59	3361	4619898	57
p_hat300-1	8	8	300	10933	8*
p_hat300-2	25	25	300	21928	25*
p_hat300-3	36	36	300	33390	36*
p_hat700-1	11	11	700	60999	11*
p_hat700-2	44	44	700	121728	44*
p_hat700-3	62	62	700	183010	62*
p_hat1500-1	12	12	1500	284923	11*
p_hat1500-2	65	65	1500	568960	65*
p_hat1500-3	94	94	1500	847244	94*

Table 3. Results for BHOSLIB benchmark instances

Instance	ω(g)	Best known	Vertex	Edge	rHGA
frb30-15-1	30	30	450	83198	30*
frb30-15-2	30	30	450	83151	29
frb30-15-3	30	30	450	83216	28
frb30-15-4	30	30	450	83194	29
frb30-15-5	30	30	450	83231	28
frb35-17-1	35	35	595	148859	33
frb35-17-2	35	35	595	148868	34
frb35-17-3	35	35	595	148784	34
frb35-17-4	35	35	595	148873	33
frb35-17-5	35	35	595	148572	34
frb40-19-1	40	40	760	247106	38
frb40-19-2	40	40	760	247157	38
frb40-19-3	40	40	760	247325	38
frb40-19-4	40	40	760	246815	38
frb40-19-5	40	40	760	246801	38
frb45-21-1	45	45	945	386854	42
frb45-21-2	45	45	945	387416	42
frb45-21-3	45	45	945	387795	42
frb45-21-4	45	45	945	387491	43
frb45-21-5	45	45	945	387461	42
frb50-23-1	50	50	1150	580603	47
frb50-23-2	50	50	1150	579824	46
frb50-23-3	50	50	1150	579607	47
frb50-23-4	50	50	1150	580417	50*
frb50-23-5	50	50	1150	580640	47
frb53-24-1	53	53	1272	714129	49
frb53-24-2	53	53	1272	714067	49
frb53-24-3	53	53	1272	714229	49
frb53-24-4	53	53	1272	714048	48
frb53-24-5	53	53	1272	714130	49
frb59-26-1	59	59	1534	1049256	55
frb59-26-2	59	59	1534	1049648	54
frb59-26-3	59	59	1534	1049729	54
frb59-26-4	59	59	1534	1048800	55
frb59-26-5	59	59	1534	1049829	55
frb100-40	100	98	4000	7425226	89

solutions, vertex, and edge number of instances. The latest column shows the solution of the proposed algorithm, and optimal solutions of the proposed algorithm are marked with asterisk (*).

The proposed algorithm also compared with GENE, GMCA and SLS algorithms which is similar with the proposed algorithm (rHGA). GENE algorithm, has been proposed by Marchiori, is a hybrid genetic method for solving the MCP [28]. The GENE algorithm has improved the chromosomes using local search approaches after the classical GA steps. GMCA algorithm is hybrid genetic approach for finding the maximum clique [36]. GMCA algorithm is similar with GENE algorithm, but the differences between GENE and GMCA algorithm is population initializing step. SLS algorithm is a hybrid genetic algorithm which has been proposed for finding the MIS [37].

In Table 4, GENE, GMCA, SLS, and rHGA are compared on some DIMACS instances which have the results for these algorithms. The best solutions for the instance are indicated in bold.

Table 4. Results for DIMACS benchmark instances: GENE, GMCA, SLS and rHGA

Instance	GENE	GMCA	SLS	rHGA
MANN_a27	**126**	125	**126**	**126**
MANN_a45	**345**	337	341	344
MANN_a81	**1098**	–	1094	**1098**
brock200_2	**12**	–	10	**12**
brock200_4	16	–	16	**17**
brock400_2	24	–	24	**25**
brock400_4	25	–	23	**33**
brock800_2	20	–	19	**21**
brock800_4	20	–	20	**21**
keller4	**11**	**11**	**11**	**11**
keller5	**27**	18	**27**	**27**
keller6	55	–	50	**57**
p_hat300-1	**8**	**8**	**8**	**8**
p_hat300-2	**25**	–	**25**	**25**
p_hat300-3	**36**	–	35	**36**
p_hat700-1	**11**	8	**11**	**11**
p_hat700-2	**44**	–	**44**	**44**
p_hat700-3	**62**	–	61	**62**
p_hat1500-1	**11**	10	**11**	**11**
p_hat1500-2	**65**	–	64	**65**
p_hat1500-3	**94**	–	92	**94**

6 Conclusion

In this paper, a new hybrid genetic algorithm was proposed for the MCP. The algorithm was coded in Java language and was tested on the benchmark instances. The test results showed that the proposed algorithm reaches an optimum solution for the most of DIMACS instances and two instances in BHOSLIB. The others were very close to the optimum solution. Besides, the results were compared in the literature with similar algorithms. The comparison results showed that the proposed algorithm found better results than similar methods. Therefore, the experimental results proved the proposed hybrid genetic algorithm is efficient.

References

1. Bomze, I.M., Budinich, M., Pardalos, P.M., Pelillo, M.: The Maximum Clique Problem. Handbook of Combinatorial Optimization, pp. 1–74. Springer, Boston (1999)
2. Garey, M.R., Johnson, D.S.: Computers and Intractability: A Guide to the Theory of NP-Completeness, 340 p. Freeman, San Francisco (1979)
3. Pattillo, J., Butenko, S.: Clique, Independent Set, and Graph Coloring. Encyclopedia of Operations Research and Management Science, pp. 3150–3163 (2011)
4. Luce, R.D., Perry, A.D.: A method of matrix analysis of group structure. Psychometrika **14**, 95–116 (1949)
5. Karp, R.M.: Reducibility among combinatorial problems. In: Complexity of Computer Computations, pp. 85–103 (1972)
6. Paull, M.C., Unger, S.H.: Minimizing the number of states in incompletely specified sequential switching functions. IRE Trans. Electron. Comput. **3**, 356–367 (1959)
7. Marcus, M.P.: Derivation of maximal compatibles using Boolean algebra. IBM J. Res. Dev. **8**(5), 537–538 (1964)
8. Bonner, R.E.: On some clustering techniques. IBM J. Res. Dev. **8**(1), 22–32 (1964)
9. Bednarek, A.R., Taulbee, O.E.: On maximal chains (theorem proof derived for maximal chains of reflexive and symmetric binary relations, based on wallace formulation of hausdorff maximality principle). Revue Roumaine de Mathématiques Pures et Appliquées **11** (1), 23–25 (1966)
10. Akkoyunlu, E.A.: The enumeration of maximal cliques of large graphs. SIAM J. Comput. **2** (1), 1–6 (1973)
11. Bron, C., Kerbosch, J.: Algorithm 457: finding all cliques of an undirected graph. Commun. ACM **16**(9), 575–577 (1973)
12. Tomita, E., Mitsuma, S., Takahashi, H.: Two algorithms for finding a near-maximum clique. Technical report of the University of Electro-Communications, pp. 1–9 (1988)
13. Harary, F., Ross, I.C.: A procedure for clique detection using the group matrix. Sociometry **20**(3), 205–215 (1957)
14. Carraghan, R., Pardalos, P.M.: An exact algorithm for the maximum clique problem. Oper. Res. Lett. **9**(6), 375–382 (1990)
15. Wu, Q., Hao, J.K.: A review on algorithms for maximum clique problems. Eur. J. Oper. Res. **242**(3), 693–709 (2015)
16. Östergård, P.R.: A fast algorithm for the maximum clique problem. Discrete Appl. Math. **120**(1–3), 197–207 (2002)
17. Fahle, T.: Simple and fast: improving a branch-and-bound algorithm for maximum clique. Lecture Notes in Computer Science, vol. 2461, pp. 485–498 (2002)
18. Li, C.M., Quan, Z.: An efficient branch-and-bound algorithm based on MaxSAT for the maximum clique problem. In: AAAI, vol., 10, pp. 128–133 (2010)
19. Friden, C., Hertz, A., de Werra, D.: STABULUS: a technique for finding stable sets in large graphs with tabu search. Computing **42**(1), 35–44 (1989)
20. Hansen, P., Mladenovic, N., Urosevie, D.: Variable neighborhood search for the maximum clique. Discrete Appl. Math. **145**(1), 117–125 (2004)

21. Pullan, W., Mascia, F., Brunato, M.: Cooperating local search for the maximum clique problem. J. Heuristics **17**(2), 181–199 (2011)
22. Jin, Y., Hao, J.K.: General swap-based multiple neighborhood tabu search for finding maximum independent set. Eng. Appl. Artif. Intell. **37**, 20–33 (2015)
23. Bui, T.N., Rizzo, J.R.: Finding maximum cliques with distributed ants. In: Genetic and Evolutionary Computation Conference, pp. 24–35. Springer, Heidelberg (2004)
24. Solnon, C., Fenet, S.: A study of ACO capabilities for solving the maximum clique problem. J. Heuristics **12**(3), 158–180 (2006)
25. Yang, G., Yi, J., Zhang, Z., Tang, Z.: A TCNN filter algorithm to maximum clique problem. Neurocomputing **72**(4–6), 1312–1318 (2009)
26. Geng, X., Xu, J., Xiao, J., Pan, L.: A simple simulated annealing algorithm for the maximum clique problem. Inf. Sci. **177**(22), 5064–5071 (2007)
27. Fleurent, C., Ferland, J.A.: Object-oriented implementation of heuristic search methods for graph coloring, maximum clique, and satisfiability. In: Johnson and Trick, pp. 619–652 (1996)
28. Marchiori, E.: Genetic, iterated and multistart local search for the maximum clique problem. Lecture Notes in Computer Science, vol. 2279, pp. 112–121 (2002)
29. Goldberg, D.E., Holland, J.H.: Genetic algorithms and machine learning. Mach. Learn. **3**(2), 95–99 (1988)
30. Fraser, A.S.: Simulation of genetic systems by automatic digital computers i. Introduction. Aust. J. Biol. Sci. **10**(4), 484–491 (1957)
31. Holland, J.H.: Adaptation in Natural and Artificial Systems. An Introduction Analysis with Application to Biology, Control, and Artificial Intelligence. U Michigan Press, Ann Arbor (1975)
32. Mitchell, M.: An Introduction to Genetic Algorithms. MIT Press, Cambridge (1998)
33. Carter, B., Park, K.: How good are genetic algorithms at finding large cliques: an experimental study. Technical report BU-CS-93-015, Department of Computer Science, Boston University (1993)
34. DIMACS Clique Benchmarks, Benchmarks instances. http://iridia.ulb.ac.be/~fmascia/maximum_clique/DIMACS-benchmark
35. Benchmarks with Hidden Optimum Solutions for Graph Problems (BHOSLIB). http://iridia.ulb.ac.be/~fmascia/maximum_clique/BHOSLIB-benchmark
36. Bui, T.N., Eppley, P.H.: A hybrid genetic algorithm for the maximum clique problem. In: Proceedings of the 6th International Conference on Genetic Algorithms, pp. 478–484. Morgan Kaufmann Publishers Inc. (1995)
37. Sakamoto, A., Liu, X., Shimamoto, T.: A genetic approach for maximum independent set problems. IEICE Trans. Fundam. Electron. Commun. Comput. Sci. **80**(3), 551–556 (1997)

Evaluation of Primary School Teachers' Resistance to Change

Behcet Öznacar(✉) and Nevriye Yilmaz

Near East University, Yakındoğu Blv, Lefkoşa, Nicosia, Cyprus
{behcet.oznacar,nevriye.yilmaz}@neu.edu.tr

Abstract. Changes in the environment cause the organization to fall into a balancing effort when it starts to influence the organization through its input. What is more, when balances around the organization reach large dimensions, these balancing efforts start to disturb the organization so much that the organization finds its way of salvation in the direction of the demands of the environment. Hence, changes in the environment necessitate change in business administration, business, technology and group structures. In this context, the main aim of this research is to examine the resistance towards change within the teachers working at primary schools. For this purpose, semi-structured interview form was developed by the researchers within the aim to determine whether the teachers differ according to their demographic characteristics or not. The sample of the research is selected from the teachers working in primary schools in Nicosia in the academic year of 2018/2019 and selected by simple random sampling technique. Finally, recommendations were made to educators and researchers regarding the results of the study.

Keywords: Change · Resistance · Primary school · Teacher · Self efficacy

1 Introduction

Since ancient times human beings have their own power, ability, time and knowledge to meet the needs and requests of other people as they felt the need to cooperate. Following the changes in the field of science and technology has become more complex. Therefore, to adapt themselves to constant changes has been very difficult and caused a lot of strains. School culture is formed by many things and those things influence the way it functions. One of the most important elements that make up this culture is teachers which is an accepted phenomenon by everyone.

Being part of 21st century is forcing a change at the same speed as it has the characteristic of being competetive and strong confronting with rapid change in information and technology. Although there are some people who make changes in their community, there are others who are reluctant to change at all times. There may even be people and groups that are there to prevent change. With those who are reluctant to change or is not easy to persuade to accept any change is very difficult to cope with. According to Helvacı (2005) 'change is a differentiation occurring in a certain period of time. In a planned or unplanned manner, it is a system or process that help to the individual to move from one state to another by developing, reforming and

D. J. Hemanth and U. Kose (Eds.): ICAIAME 2019, LNDECT 43, pp. 775–786, 2020.
https://doi.org/10.1007/978-3-030-36178-5_67

eventually becoming better'. Based on the researches done on the topic, the age we live in is having rapid change and therefore, it has been considered as the age of crises and how societies keeping up with these changes has been one of the problems of this age. One of the most important elements of the success in education is people. Individuals acquire new knowledge to gain attitudes, behaviors, to socialize, to gain professional understanding and skills. According to İnandi et al. (2015) with the developments in technological fields, everything changes rapidly and these changes are manifested in economic, social, educational and political fields. Catching the new age and keeping up with the innovations brought by the age will increase as we understand these changes. For this reason, it is inevitable for society and the life we live in to escape from this. However, this process is not easy as it seems to some people. As in all organizations, the realization of the needed change in education to reach success, need the support of all employees. There are many factors affecting the employees in regard to changes and one of them is the self-efficacy belief. Individuals with high self-efficacy beliefs are expected to meet the changes more naturally. Schools are also educational institutions, where change is being implemented. Increasing students' academic and social achievements is one of the most important goals of education. Here, the biggest share falls to teachers and managers. For that, it is necessary to be open to change in education.

2 Self Efficacy Concept

Based on the literature found, the history of self efficacy begins with Bandura's (1977) social learning theory. According to this theory, this concept makes a difference in the way people behave, think, feel and motivate themselves. All these beliefs require an action to be taken through some process and one leading factor for the achievement of the goals set is motivation. The higher the level of self efficacy is the level of reaching those goals would be in the same height. Moreover, it is not concerned with the skills individuals have, but with the judgements of what individuals can do with those specific skills (Zulkosky 2009).

2.1 Reasons for the Need of Change and Teachers' Resistance in Schools

There are some barriers to innovation and these include some factors which could be interpreted as disputes over educational objectives due to inappropriate planning and motivation, lack of knowledge against change and characteristics of schools which are not ready to accept any change as well as the characteristics of people that needs to be associated with change. Therefore, it is the users who are in our case students and teachers that must adapt, adopt or reject the innovations which will take place. Zimmerman (2006) claims that one of the barriers against change is the failure to understand the need for change. Firstly, teachers should comprehend that there is a need for change and they should show acceptance and willingness to accept it. Nevertheless, some people will reject this because they would find it easier to continue doing things in the way they are used to rather than working on understanding and applying new skill and strategies. Another reason to resisting change can be their past experiences. If they have made an attempt to change certain things in the past, but it didnt work effectively, they may be too scared to try again. According to Gürses and Helvacı (2011), there are four major obstacles against change such as; (1) Value Barriers, (2) Power Barriers, (3) Psychological Barriers and (4) Implementation Barriers.

In order to overcome these barriers educators, managers and principles must take some actions.

Firstly, everyone should understand the process of change is related to people. If teachers are expected to take some risks while they are trying to change things, and practicing new behaviours, school leaders should also be open to these, be more willing to implement all changes in their schools. The important characteristic that is needed in this process is trust that is needed to be built when overcoming all the obstacles along the way to success.

Secondly, schools principles and leaders must provide some opportunities for teachers to work collaboratively and participate in the process of decision making and this will also increase their motivation in the school. In this way, teachers will be more cultivated and by the risks they take when giving decisions, they will be encouraged to try new strategies and ideas.

Thirdly, educators who have high self efficacy are more ready to accept new ideas and to try new strategies to meet the needs for students'. Individuals who owns this are more likely to show an attempt to control the outcomes of their decisions even though they will affect their lives.

Another important factor is the peer support. If there is a need for change, there should be an environment ready to support this. So that, teachers' would not feel to stressed when dealing with all the new things. Peer support is the most effective in developing a culture accepting change in schools. Due to implementing new experiences and knowledge and reaching success, personal efficacy of individuals increase and makes things more possible.

2.2 Factors Inhibiting Change

Hinde (2004) remarks that educational change in schools fails for these reasons;

1 The need for change is not clearly defined and nor clearly understood. Who will benefit from this change and how? is not obvious.
2 Setting too broad or too limited goals for change with no specific aim.
3 Change happening too fast or slow and not giving people enough chance to cope with it.
4 The cost of change. Not having enough money to buy the needed materials and lacking time to do the right planning.
5 Having no long term commitment to carry people through frustration, anxiety and obstacles that they would face with during this process.
6 Parents opposing to this change because they are kept away from things.
7 Having too controlling leaders and disrupting the motivation of people.
8 Underestimating the effects of change and the years of teaching experience of the teachers'.

When we consider the change in education as the centre of school and the aim of increasing students' success, the teachers' have to be effective and willing to succeed during this process to reach the target. According to Woodman and Dewet (2004), the concept of 'change' is likely to cause panic and this could be the important reason why individuals are showing resistance against change. Therefore, to make the change happen in education, although it may seem easy in the technical process, it is a complex and difficult processs as it is a social application area. Teachers' positive feelings and thoughts towards change is very crucial to support the implementation of change. On the contrary, teachers' who have negative attitudes and their resistance towards change can be the concern of the whole education system that have a negative impact on success. Some prominent factors affect the teachers' resistance to change. Teachers' using the methods and strategies they know well, may be a tendency to maintain their habits and keep themselves away from the recent changes. Besides these, those teachers' who build strong relationships with their collegues can be expected to be suspicious of change and see it as a threat to the deterioration of the existing order in the school in their perception. For this reason, Dally (2010) suggested that for the change to be applied successfully, those teachers' who can trust their collegues can get help from each other can make the learning easier. Moreover, educational institutions continue to live as long as they can serve their purposes. They and their environment are in constant and regular change. In corporate life, there are four basic phases for development. These include: formation, growth, maturity and decline and it is the last circuit that will open or allow it to regenerate and regain life. These developmental stages are also valid for educational institutions. During life time each stage has itw own difficulties and not overcoming them can be the end of the institution. In every step, it should change the institution, adapt and develop it. Educational institutions always undergo a change in the basic stages of development. This change should work as a whole in the face of difficulties while living in educational institutions and employees should support each other to simplify the process. Due to the fact that schools are educational institutions the views of teachers' and managers' are very vital and need to be considered. Therefore, in this research teachers' and managers' working in primary schools and how they view and react to the change has been sought.

3 Method

This research is done qualitatively to find out about Primary School Teachers' Resistance to Change. Pre-test and post test questions were written down for the evalutiaon to be done based on the questions participants were asked. Within this aim, semi-structured interview form was developed to determine whether the teachers differ according to their demographic characteristics or not. The sample of the research is selected from the teachers working in primary schools in Nicosia in the academic year of 2018/2019 and were selected by simple random sampling technique. Final recommendations were made to educators and researchers regarding the results of the study.

4 Working Group

This research was conducted with 40 participants and 12.5% (5) of the individuals who participated in the research were managers, 50% (31) teachers and 37.5% (4) were students that were in Northern Cyprus Nicosia Primary Schools.

5 Development of Data Collection Tool

In this research two different sections have taken place. The first section is related to demography factor and Sect. 2 consists of questions regarding the results of teachers and managers being ready to accept change. The interviews were carried out in the hours set and arranged by the researcher and the participant on voluntary basis. When the final data was gathered, evaluation was made and the discussion was written accordingly with the final results analysis.

6 Analysis of Data

6.1 Analysing of Interviews

Before the research has been carried out, the questions that were going to take place in this research were selected and noted down. The questionnaires were distributed to managers, teachers and students that were in Nicosia Primary Schools. Those participants who took part in the research were informed about the procedure of the data gathering and they were told about the willingness and freedom of participating and rejecting from it anytime they wish or feel uncomfortable. The data collection was done and the results were evaluated and analysed. Based on the findings, recommendations and suggestions were written and findings were assumed to shed light on teachers working in primary schools.

7 Results

7.1 Quantitative Data Analysis

In Table 1, the gender distribution of the participants (67%) was male and eleven (33%) were female.

Table 1. Gender distribution of participants

		Frequency	Percentage
Gender	Women	20	33
	Men	40	67
	Total	60	100

In Table 2, it can be seen that 16.6% of the respondents were married and 84.4% were single.

Table 2. Marital status of participants

		F	%
Marital status	Married	10	16.6
	Single	30	84.4
	Total	40	100

As seen in Table 3; According to the gender distribution of the participants, 24 women (60%) were female, and 16 (40%) were male.

Table 3. Frequency distribution by gender.

		F	%
Gender	Female	24	60
	Male	16	40
	Total	40	100

As shown in Table 4; 32.5% of the participants (4) were married and 67.5% (36) were distributed as having a single marital status.

Table 4. Frequency distribution by relationship status

		F	%
Marital status	Married	4	32.5
	Single	36	67.5
	Total	40	100

As seen in Table 5; 12.5% (5) of the individuals who participated in the research were managers, 50% (31) teachers and 37.5% (4) were students.

Table 5. Frequency distribution by position

		F	%
Position	Manager	5	12.5
	Teacher	31	50
	Student	4	37.5
	Total	40	100

8 Results

In this part of the study, the findings of the demographic information belonging to the sample group and the results obtained by the analysis of the data obtained for the sub-problems of the researcher with the appropriate statistical method are included and comments on these findings are given. Below, the test results that are applied in the seminars and held in different topics within the scope of the research are separately expressed in tables.

Participants' opinions;
It would be better to give more explanation and clarify the need for change before asking teachers' and students' to change their old habits.
Some more time is needed for the process to start working successfully and effectively.

When we look at Table 6, we can see that a big percentage of teachers' need more clarification on the issue and managers are not ready as teachers' to welcome the changes. In addition to these, from the responses given by the students', like managers it can be seen that they need more time to accept changes and more explanation on the reasons why there is a need to change for them to understand better.

Table 6. Readiness to accept change. Question 6. The results of teachers and managers being ready to accept change

Theme	Manager		Teacher		Student		Total	
	N	%	*N*	%	*N*	%	*N*	%
Ready	6	%39.6	12	%60	2	%40	20	%50
Not ready	5	%33.3	8	%40	4	%80	17	%42.5
Need more time	8	%52.8	6	%35	3	%60	17	%42.5
Need clarification and explanation	10	%66.6	16	%80	5	%100	31	%77.5

Participants' opinions;
Necessary steps should be taking to avoid any risks or harms that are likely to happen as a result of new implementations. So that no one would get any harm.
Every amount of money that will be spent and time needed should be calculated well to avoid upcoming problems due to some changes.

As seen in Table 7; espacially managers are not ready to take new duties and responsibilities and they are more afraid to face with these new implementations and necessary changes. Another important thing is the cost of this change and time needed for it. For this process to be beneficial for all, it needs to be well planned and continuous.

Table 7. Reasons for resistance to change. Question 7. Why individuals teachers', managers' and students' towards change in schools

Theme	Manager		Teacher		Student		Total	
	N	%	*N*	%	*N*	%	*N*	%
Not ready to take new duties and responsibilities	5	%33	17	%85	4	%80	26	%65
Fear of getting any harm from changes	13	%85.8	12	%60	3	%60	28	%70
Lacking time and cost	3	%19.8	9	%45	1	%20	13	%32.5

Participants' opinions;

Change is necessary to get rid of the old, boring, unneccassary and traditional teaching techniques in order to motivate students', teachers' and managers' to reach success and help to the betterment of the institution.

It is thought that to build trust and improve the self efficacy of indivduals during this process collaborative work is needed.

As seen in Table 8, self efficacy, cooperative work and trust are important factors helping in the process of change and change is needed to remove the old methods and techniques that are keeping students' and teachers' behind the recent developments and innovations to transfer the schools into better environmental place in which the learning is taking place effectively and all are motivated towards it. Besides, self efficiancy stakeholders needs trust and collaborative work to make things easier for them and to overcome all the obstacles in this process.

Table 8. Reasons for the need of change in primary schools. Question 8. Why change is needed in primary schools

Theme	Manager		Teacher		Student		Total	
	N	%	*N*	%	*N*	%	*N*	%
To build self efficacy	6	%39.6	11	%55	1	%20	18	%45
To motivate and reach success	10	%66.6	14	%70	3	%60	27	%67.5
To get rid of the old methods and techniques	12	%80	12	%60	3	%60	27	%67.5
To catch up with the recent developments and innovations	5	%33.3	15	%75	4	%80	24	%60
To build trust and develop cooperation	10	%66.6	10	%50	5	% 100	25	%62.5

Participants' opinions;

Managers and teachers ae reluctant to change. They need more explanation on the concept and the process.

Managers need to show more care towards the new implementations and innovations.

As it can be seen from Table 9, students' have the highest percentage on showing willingness towards change and teachers have the least willingness because they may not want to change the things they already know and they want to continue with the years of experience they have and do the same things they have been doing up until now.

Table 9. Resistance to change. Question 9. To what extent do stakeholders show resistance to change

Theme	Manager		Teacher		Student		Total	
	N	%	*N*	%	*N*	%	*N*	%
Carelessness	6	%39.6	15	%75	–	–	21	%52.5
Active-passive resistance	11	%79.2	14	%70	3	%60	28	%70
Reluctance	2	%13.2	12	%60	4	%80	18	%45
Willingness towards change	10	%66.6	10	%50	5	%100	25	%62.5

9 Discussion

This research aimed to find out, whether teachers' and managers' working in primary schools are showing any kind of resistance towards change or not in the field of education.

Therefore, participants' gender, position and the level of readiness have been examined at the beginning. According to the findings, the concept of 'change' is hard to accept it and for this reason, managers and teachers' are not ready to handle with the consequences of it due to the fact that the school, curriculum and strategies will undergo some changes. This finding is similar to the findings of Jerald (2007) because he also found that most teachers' and managers' are not ready to accept the changes and there isn't enough information giving to them to know as to why they should have some changes at schools. Gürses and Helvacı (2011) carried out a research and found out that according to managers' and teachers' opinions the reason for them not being ready is because they are lacking knowledge on the issue. Therefore, they are showing resistance to accept changes. According to Çakir (2009); Demirtaş (2012); Doğru and Uyar (2012) and Şentürk (2007) stated that the majority of primary school teachers perceive themselves as open to change which is unlike our findings. Hence, in our findings students' were found as the most open to accept changes, then teachers' and managers'. For example, Çakir (2009); Korkut (2009) and Özençel (2007) did not find a difference according to the age variable, but it was concluded that Şentürk (2007) found that younger teachers had more resistance to change than 31–40 year olds. However, in our study, there is no result found on the age and indicating that age makes a change on the reaction towards change.

According to Aydın and Şahin (2016), there are various ways indicating the resistance to change of employees. For this reason, they described the possible attitudes towards change as 'careleness', 'active-passive resistance', 'reluctance' and 'willingness towards change' under four headings similarly to ours. Each field contains different attitudes and behaviors. Understanding the attitudes of teachers who are the building blocks in educational organizations against change is important for the functionality and effectiveness of the system. In order to take measures against resistance to change and to break the resistance, it is necessary to determine what needs to be done and what is related to it.

Findings obtained from this research indicate that students' are more open to change and teachers' think they need to catch up with the recent developments and innovations to be able to make the needed changes. Nevertheless, in Gümüşeli's (2009) study it has been found that administrators' are more open towards change and they believe that real change is not just making some physical changes and corrections, but also renewing and improving the schools' conditions in terms of educational system and instructional activities. As a result of our study another point is, teachers' are the main people applying all changes in schools and they should be the ones to bring themselves up to date and must be supported by the managers' and students'. Hoşgörür (2014) also stated that especially teachers' taking part in decisions may contribute to the acceptance of change in schools and they should be involved in all stages. Similar to our findings, Şahin (2013) remarked that limited resoursces, time and lack of money put strains on the implementation of certain improvements and developments of schools. However, schools should be supported by funding and resources that are necessary and could be adjusted based on the needs of the schools.

10 Conclusion

It has been found that teachers who work in primary schools have the most resistance to change in schools, due to lack of knowledge and information on the issue. The reasons for the resistance of teachers and managers working in primary education schools due to their reluctance to take new role and responsibilities in the change process are moderate. The reasons for the resistance of the teachers' due to fear of harm from change are high. According to these results, the teachers who participated in the research, although, they seem to have a say in the implementation and the application process of change they were found insufficient in terms of participation number. More active participation of teachers' is needed in the initiation and implementation of the culture of change in primary schools to take place. Therefore, both managers and teachers should follow the change and try to be in the moment to encourage each other. As peer support is important throughout the process and they should trust themselves that they can handle with all the recent changes and they would not have problems. They should find ways of increasing their self efficacy to overcome all the obstacles and believe in themselves that they can do it.

The following suggestions can be made for implementation in the direction of research results:

- According to the findings of the research; Teachers within the borders of North Cyprus; show the highest level of resistance to change due to the lack of knowledge, time, money of the primary schools. Therefore; to ensure that teachers do not exhibit resilience against changes in schools, it is necessary to provide the required financial resources for change, to eliminate the infrastructure deficiencies needed for change in schools, to reduce the procedures and bureaucracy in schools and to use the reward method in schools during the process of change to motivate them towards change.

- In connection to the above suggestion, consultation for teachers' and managers' in the change process is needed to provide the information on the concept of change to be made in schools in order to reduce the resistance to make individuals more aware of why it is needed.
- According to the findings of the survey, it is seen that the teacher's and managers' level of resistance from the reluctance to take new roles and responsibilities is moderate. Therefore, the necessary training should be given to enable teachers to become more willing and to help them give up their existing habits.
- According to the findings of the research; it is seen that there are reasons for the resistance of teachers' and administrators' failure to manage change. In this context; education administrators should be provided with training programs to gain the knowledge and skills to manage change and their deficiencies should be eliminated. In addition, it should be noted that the period of time for which the processors are allowed to change is insufficient and more time should be allowed for healthy changing process to take place.
- According to the findings of the research; teachers are seen to have entered into a team of resistance against changes to be made in schools, due to fear of harm from change. Especially school administrators are responsible for managing change and they need to provide all the needed information to relieve the worries of the teachers'. Besides; school administrators should take care of the teachers' ideas and opinions in the process of change because they may be valuable.

References

Aydın, Ö., Şahin, S.: Resistance of primary and secondary school teachers to organizational change. J. Hum. Sci. **13**(3), 5053–5068 (2016)

Bandura, A.: Self-efficacy: toward a unifying theory of behavioral change. Psychol. Rev. **84**(2), 191 (1977)

Çakır, B.S.: A study on the perception of the factors and teachers in primary schools with respect to organizational change and organizational change. (Unpublished Master's thesis). Selcuk University Social Sciences Institute (2009)

Çalık, T., Koşar, A., Kılınç, Ç., Er, E.: The relationship between the resistance to change behaviors and self-efficacy of primary school teachers. Uşak Univ. J. Soc. Sci. **6**(4), 1–16 (2013)

Daly, A.: Mapping the terrain. In: Daly, A. (ed.) Social Network Theory and Educational Change, pp. 1–16. Harvard Education, Cambridge (2010)

Demirtaş, H.: Primary school openness to change. Elementary Online **11**, 18–34 (2012)

Doğru, S., Uyar, M.: A research on tendency of resistance to change of provincial organizations of ministry of national education: the case of Konya province. Buca Educ. Fac. J. **32**, 119–143 (2012)

Gümüşeli, A.İ.: Primary school principals in Turkey: their working conditions and professional profiles. Int. J. Hum. Soc. Sci. **4**(4), 239–246 (2009)
Helvaci, M.A.: Egitim örgütlerinde degisim yönetimi. Nobel Yayin Dagitim, Ankara (2005)
Helvacı, M.A., Gürses, G.: Reasons for the resistance of teachers to change in schools. Int. J. Hum. Sci. **8**(1), 1541–1563 (2011)
Hinde, E.R.: School culture and change: an examination of the effects of school culture on the process of change. Essays Educ. **12**, 1–13 (2004)
Hoşgörür, V.: School development applications in Turkey. Education **134**(3), 404–419 (2014)
İnandi, Y., Yeşil, H., Karatepe, R., Uzun, A.: The Study of Relationship Between Teachers, 1–15 (2015)
Korkut, M.: The views of teachers in primary schools regarding organizational change: the case of Çanakkale province. (Unpublished Master's thesis). Çanakkale On Sekiz Mart University Institute of Social Sciences (2009)
Özençel, E.: A study on the perception of organizational change and resistance to change. (Unpublished Master's thesis). Selcuk University Social Sciences Institute (2007)
Şahin, İ.: Primary school principals' school development strategies and practices. Educ. Sci.: Theory Practice **13**, 229 (2013)
Şentürk, S.: Resistance to change and resistance of teachers in primary schools. (Unpublished Master's thesis). Yildiz Technical University Institute of Social Sciences (2007)
Jerald, C.D.: Believing and achieving (Issue Brief). Washington, DC: Center for Comprehensive School Reform and Improvement (2007). http://www.eric.ed.gov/PDFS/ED495708
Woodman, R.W., Dewett, T.: Organizationally relevant journeys in individual change. In: Poole, M.S., Van de Ven, A.H. (eds.) Handbook of Organizational Change and İnnovation, pp. 32–49. Oxford University, New York (2004)
Zimmerman, J.: Why some teachers resist change and what principals can do about it. Nassp Bull. **90**, 238–249 (2006)
Zulkosky, K.: Self-efficacy: a concept analysis, pp. 93–102 (2009)

Fuzzy Logic and Correlation-Based Hybrid Classification on Hepatitis Disease Data Set

M. Sinan Basarslan[1(✉)], H. Bakir[2], and İ. Yücedağ[2]

[1] Doğuş University, İstanbul, Turkey
mbasarslan@dogus.edu.tr
[2] Düzce University, Düzce, Turkey

Abstract. Developments in the health field are closely affecting humanity. The development of information technologies increases this effect. In this study, it was aimed to help the decision makers by increasing the accuracy rate in the detection of hepatitis disease. The data set was obtained from UCI machine learning source. Data preprocessing, attribute selection and classifier models were established on this data set, respectively. After the deficiency in the data of the patients with hepatitis was normalized, correlation-based and fuzzy-based rough force attribute selection methods were applied and the attributes that contributed to the classification were selected. The hepatitis dataset and the data set formed by the attributes determined by the correlation-based and the fuzzy-based rough-attribute selection methods were classified using the k-nearest neighbor, Random Forest, Naive Bayes, and Logistic Regression algorithms and the results were compared. Accuracy, sensitivity precision, ROC curve and F-measure values were used in the comparison of classification algorithms. In the process of separating the data set as a test and training set, a 5-fold cross-validation method was applied. It has been observed that the fuzzy rough clustering algorithm is more successful than the k-nearest neighbor, Random Forest, Naive Bayes, and Logistic Regression classification methods in the detection of hepatitis disease.

Keywords: Classification · Data mining · Fuzzy logic · Hepatitis disease

1 Introduction

As the necessity of the era we live, the impact of technological developments is seen in all areas. Health is one of the most important areas where technological developments benefit humanity. When the technological developments in health sciences are mentioned, generally developed health devices come to mind. However, the importance of data analysis in health units is increasing day by day. Thanks to data analysis, health workers can make faster and more accurate decisions by making use of the analyzes obtained while making decisions about the patient.

Data analysis aims to help decision-makers to make high-precision decisions by achieving stable data sets that can be used in all areas of data collection. In particular, the analysis of the data to be obtained from the data communities that emerge in areas such as health and finance, which concern the human, facilitates people's lives.

D. J. Hemanth and U. Kose (Eds.): ICAIAME 2019, LNDECT 43, pp. 787–800, 2020.
https://doi.org/10.1007/978-3-030-36178-5_68

This study was conducted to predict whether patients with hepatitis will survive. In this context, data preprocessing, attribute selection and classifying models were established on the hepatitis data-set obtained from UCI, respectively. Correlation-based attribute selection method and fuzzy values which are widely used in the literature to work with the exact values in the attribute selection process after defining the missing data in hepatitis patients are normalized.

With the advancement of computer technologies, artificial intelligence methods are used to perform advanced applications. One of the artificial intelligence methods used in these studies is fuzzy logic [1]. Fuzzy logic based health, economy, environment are implemented in many areas such as applications. Fuzzy logic produces definitive solutions with an approach similar to human skill in decision making [2].

Hepatitis A, Hepatitis B, Hepatitis C, which is transmitted by viruses, comes first. However, hepatitis is the state of inflammation of the liver. Hepatitis disease is an infectious disease spreading between people with blood and other body fluids. In particular, the hepatitis B virus usually passes through the mother after birth. Transplantation of the disease is spread through the use of personal devices and equipment in public places such as nursing homes, hospitals, orphanages, unprotected sex or injections in addicts [1].

In the second part of this study, information about the hepatitis dataset is given. In the third chapter, the fuzzy logic concept and the membership functions for the blurring process are explained. In the fourth chapter, information is given about the pre-processing, qualification selection, classification algorithms and the performance evaluation criteria of the raw data set. In the fifth chapter, the results of the study and the sixth section are the suggestions.

2 Hepatitis Data Set

The hepatitis data set was created by Bojan Cestnik and is part of the UCI. Data set information was given in Table 1 [3].

Dataset from UCI in Table 1; It is a dataset created to investigate whether patients with hepatitis will continue to live. There are 19 attributes and 155 records in the dataset. 32 of the patients died and 123 of them continued to live. After the data preprocessing step where the missing data in the data set has been completed, the attribute selection has been made. Then the models with the classification algorithms were established and the model with the best result was determined.

Table 1 shows the attributes of the hepatitis disease data, the description of the attributes and the data types. The completion and normalization of the missing data in the hepatitis dataset are described in the data pre-treatment section. The target nature of the Hepatitis Disease data set is; hepatitis, not hepatitis. The operation of the study is shown in Fig. 1.

In the fuzzy-based rough set attribute selection method, the numerical values were taken into account (Table 2).

Table 1. Hepatitis disease data set.

Attributes	Attribute values	Data type
Class	Live or Die	Categorical
Age	7–78	Numerical
Sex	Female, Male	Categorical
Steroid	Yes or No	Categorical
Antiviral	Yes or No	Categorical
Fatigue	Yes or No	Categorical
Malaise	Yes or No	Categorical
Anorexia	Yes or No	Categorical
Liver Big	Yes or No	Categorical
Liver Firm	Yes or No	Categorical
Spleen Palpable	Yes or No	Categorical
Spiders	Yes or No	Categorical
Ascites	Yes or No	Categorical
Varices	Yes or No	Categorical
Bilirubin	0.3–8	Numerical
Alkaline phosphatase	26–295	Numerical
SGOT	14–648	Numerical
Albumin	2.1–6.4	Numerical
Pro-time	0–100	Numerical
Histology	Yes or No	Categorical

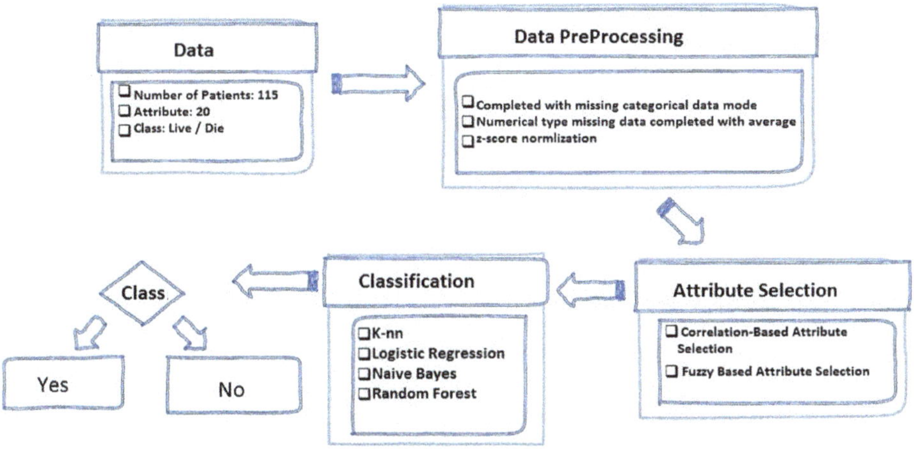

Fig. 1. Schematic representation of the study.

Table 2. Range of numerical data types in the hepatitis data set.

Attribute	Range
Age	Young, Middle aged, Old
Bilirubin	Low, Medium, High
Alkaline phosphatase	Normal, High
SGOT	Normal, High
Albumin	Low, Medium, High
Pro-time	Low, High, Very high

3 Fuzzy Logic

Fuzzy logic, an artificial intelligence often used in advanced applications in today's technologies. The basis of fuzzy logic is based on fuzzy sets and sub-sets. In the classical approach, an entity is or is not a member of the cluster. When mathematically expressed, when the entity is the element of the cluster with respect to the membership relationship with the cluster, "1" takes the value "0" when it is not the element of the set. Fuzzy logic is the expansion of the classical cluster representation. Each entity in the fuzzy asset set has a membership degree. The membership degree of assets can be any value (0, 1) and the membership function is indicated by M(x). The fuzzy logic process is shown in Fig. 2 [4].

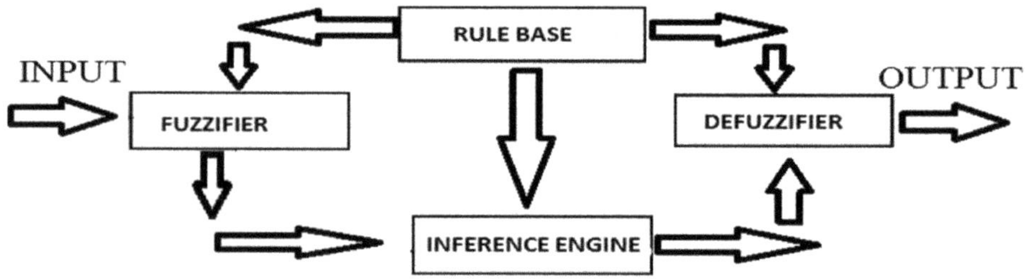

Fig. 2. The image of the fuzzy logic process.

3.1 Membership Functions

This section describes the membership functions and membership degrees of the attributes shown in Table 2.

The membership functions determined for the age attribute were given in Eqs. (1), (2) and (3). Age membership degrees were shown in Fig. 3.

$$Age_{young} = \begin{cases} 1, x \leq 20 \\ 2 - \frac{x}{20}, 20 < x < 40 \\ 0, x \geq 40 \end{cases} \quad (1)$$

$$Age_{middle\ aged} = \begin{cases} 0, x \leq 35, x \geq 55 \\ \frac{x}{10} - 3.5, 35 < x \leq 45 \\ 5.5 - \frac{x}{10}, 45 < x < 55 \end{cases} \tag{2}$$

$$Age_{old} = \begin{cases} 0, x \leq 50 \\ \frac{x}{10} - 5, 50 < x \leq 60 \\ 1, x \geq 60 \end{cases} \tag{3}$$

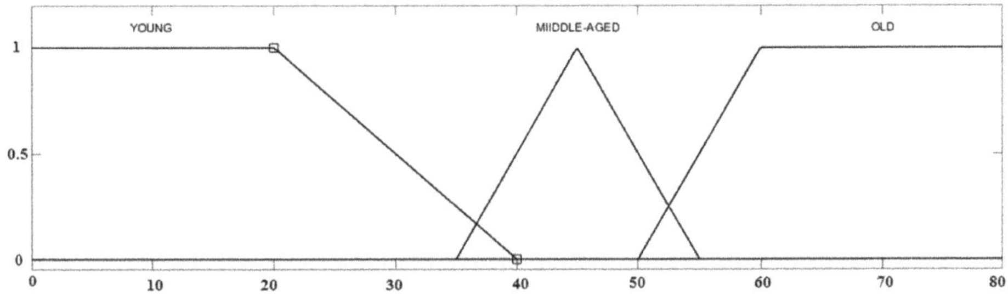

Fig. 3. Age membership degrees.

The membership functions determined for the Bilirubin were shown in Eqs. (4), (5) and (6). The membership degrees for the Bilirubin attribute were shown in Fig. 4.

$$Bilirubin_{low} = \begin{cases} 1, x \leq 0.2 \\ 2 - 5x, 0.2 < x < 0.4 \\ 0, x \geq 0.4 \end{cases} \tag{4}$$

$$Bilirubin_{medium} = \begin{cases} 0, x \leq 0.2, x \geq 2 \\ 5x - 1, 0.2 < x < 0.4 \\ 1, 0.4 \leq x \leq 1.8 \\ 10 - 5x, 1.8 < x \leq 2 \end{cases} \tag{5}$$

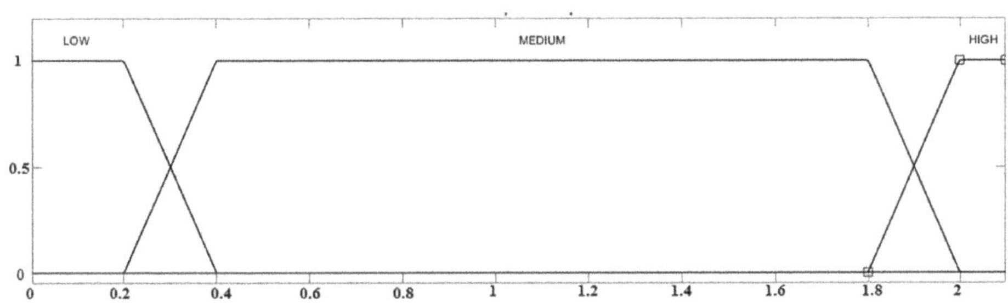

Fig. 4. Bilirubin membership degrees.

$$Bilirubin_{high} = \begin{cases} 0, x \le 1.8 \\ 5x - 9, 1.8 < x < 2 \\ 1, x \ge 2 \end{cases} \quad (6)$$

The membership functions determined for Alkaline phosphatase attribute were shown in Eqs. (7) and (8). The membership degrees for the alkaline phosphatase attribute were shown in Fig. 5.

$$Alkaline_{normal} = \begin{cases} 1, x \le 80 \\ 5 - \frac{x}{20}, 80 < x \le 100 \\ 0, x \ge 100 \end{cases} \quad (7)$$

$$Alkaline_{high} = \begin{cases} 0, x \le 80 \\ \frac{x}{20} - 4, 80 < x \le 100 \\ 1, x \ge 100 \end{cases} \quad (8)$$

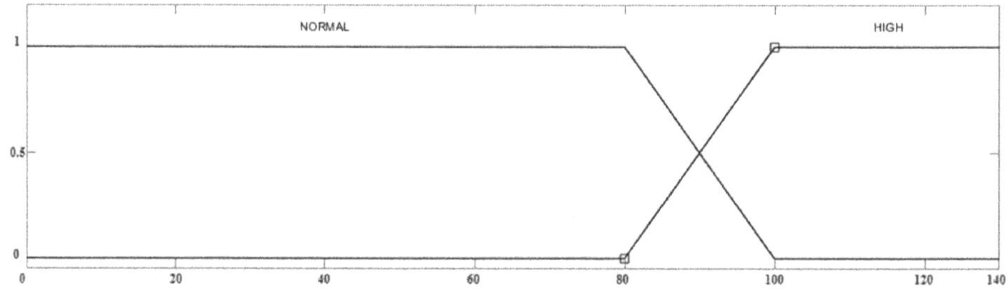

Fig. 5. Alkaline phosphatase membership degrees.

The membership functions determined for the SGOT attribute were shown in Eqs. (9) and (10). The membership degrees for the SGOT attribute were shown in Fig. 6.

$$SGOT_{normal} = \begin{cases} 1, x \le 30 \\ 4 - \frac{x}{10}, 30 < x < 40 \\ 0, x \ge 40 \end{cases} \quad (9)$$

$$SGOT_{high} = \begin{cases} 0, x \le 30 \\ \frac{x}{10} - 3, 30 < x < 40 \\ 1, x \ge 40 \end{cases} \quad (10)$$

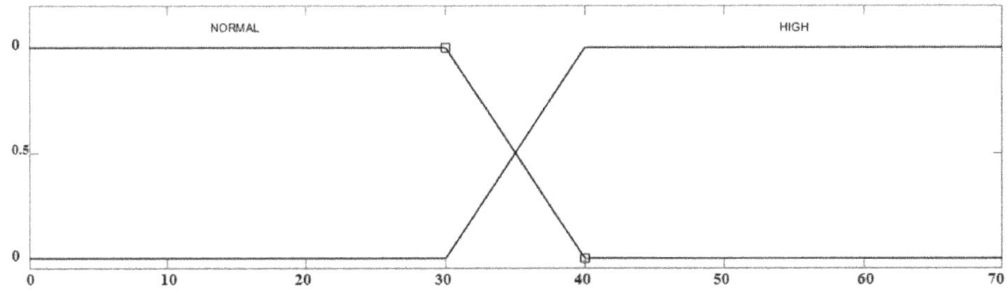

Fig. 6. SGOT membership degrees.

The membership functions determined for the Albumin attribute were shown in Eqs. (11), (12) and (13). The membership degrees for albumin attribute were shown in Fig. 7.

$$Albumin_{low} = \begin{cases} 1, x \leq 3.5 \\ 8 - 2x, 3.5 < x < 4 \\ 0, x \geq 4 \end{cases} \tag{11}$$

$$Albumin_{medium} = \begin{cases} 0, x \leq 3.5, x \geq 5.7 \\ 2x - 7, 3, 5 < x < 4 \\ 1, 4 < x < 5, 2 \\ 11, 4 - 2x, 5.2 < x < 5.7 \end{cases} \tag{12}$$

$$Albumin_{high} = \begin{cases} 0, \mathrm{x} \leq 5, x \geq 20 \\ \frac{10x-50}{7} - 1, 5 < x < 5.7 \\ 1, 5.7 < x < 18 \\ \frac{20-x}{2}, 18 < x < 20 \end{cases} \tag{13}$$

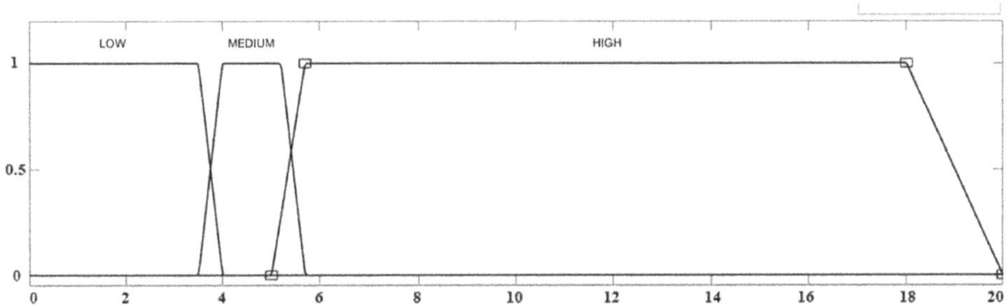

Fig. 7. Albumin membership degrees.

The membership functions determined for the pro-time attribute were shown in Eqs. (14), (15) and (16). The membership degrees for the pro-time attribute were shown in Fig. 8.

$$Pro - time_{low} = \begin{cases} 1, x \leq 10 \\ \frac{18-x}{8}, 10 < x < 18 \\ 0, x \geq 18 \end{cases} \tag{14}$$

$$Pro - time_{high} = \begin{cases} 0, 15 \leq x, x \geq 60 \\ \frac{x}{15} - 1, 15 < x \leq 30 \\ 3 - \frac{x}{15}, 30 < x < 45 \\ \frac{60-x}{15}, 45 < x \leq 60 \end{cases} \tag{15}$$

$$Pro - time_{very\,high} = \begin{cases} 0, x \leq 25, x \geq 100 \\ \frac{x}{25} - 1, 25 < x < 50 \\ 1, 50 < x < 75 \\ \frac{100-x}{25}, 75 < x < 100 \end{cases} \tag{16}$$

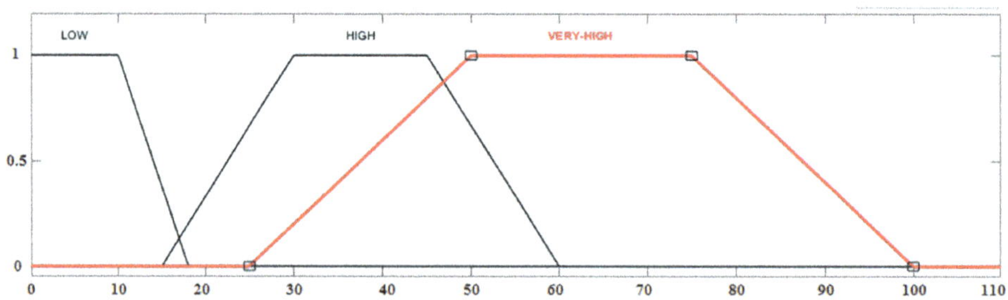

Fig. 8. Pro-time membership degrees.

4 Method

Information on data preprocessing, attribute selection and classification of hepatitis disease data from UCI is given in this section.

4.1 Data Preprocessing

In this section, missing value and attributes z-score normalization methods are described.

Z- Score Normalization Method. This method is applied with the formula shown in the Eq. (17) [5].

$$x_i^{'} = \frac{x - \mu}{\sigma} \tag{17}$$

Missing Value. There may be blank skipped or unfilled data when creating a data set or for a different reason. This data is called missing value. There are many methods for completing and detecting missing values. We can classify them as classical and predictive methods. The classical method can be defined as the substitution of missing values in a way that does not change the data set, filling them manually, or filling a constant, such as the average of those containing a numerical value. An example of a classical method is to complete the missing values in categorical variables with the most repetitive value (mode) and complete the missing value with the average of the numerical value [6]. In this study, missing data was completed with classical methods.

4.2 Attribute Selection

Attribute selection is made to increase the contribution of attributes to the result in machine learning methods. It is aimed to obtain more efficient results by not taking the qualities that have no effect on the performance of the models created by classification or clustering algorithms. In this study, correlation based and fuzzy based roughing processing and attribute selection methods were used.

Correlation Based Attribute Selection. We understand a linear relationship between two statistical variables by correlation. this linear relationship and the degree to which it is expressed by the correlation coefficient named R. The correlation coefficient is between −1 and +1 [7].

- R = −1, negative relationship between variables
- R = +1, positive relationship between variables
- R = 0, no relationship between variables

Attribute Selection with Fuzzy Based Rough Cluster Method. A lot of research has been done to use fuzzy rough clusters in data mining applications. In fuzzy rough cluster applications, quality reduction studies are less than learn-ing algorithm studies. In this context, different fuzzy rough set definitions have been made by loosening the fuzzy equivalence properties and using cuts of fuzzy sets and fuzzy sets. There are two main approaches to cognitive and constructive cross-breeding fuzzy and rough clusters [8]. In the constructive approach, fuzzy relations in the universe are the primary concept and the lower and upper approaches are based on these concepts. Initially, only fuzzy equivalence relations were used, while subsequent binary relations were included in later generalization studies. In the conceptual approach, the top and bottom approach operators are the primary concepts. In this approach, several expressions have been used to describe approach operators [9]. The constructivist approach focuses more on practical applications and the precise approach studies the mathematical properties of turbulent rough sets.

The fuzzy similarity relationship between the upper and lower approach pairs was determined by the proponents of this method [10, 11]. They used t-norm min and t-conorm max operators to determine this relationship. When the U universe defines the double relationship in the R universe, the upper approximations of the fuzzy rough set are Eq. (18) and the lower approximations are described in Eq. (19) is shown.

$$R_*(F)(x) = inf_{y \in U} \max(1 - (R(x,y), F(y)) \quad (18)$$

$$R^*(F)(x) = sup_{y \in U} min(1 - (R(x,y), F(y)) \quad (19)$$

Morsi and Yakout studied fuzzy rough sets [12]. Radzikowska and Kerre conducted another study on fuzzy rough clusters [13]. There are 20 attributes in the data set. Table 3 shows which attributes are selected as a result of the application of two attribute selection algorithms within the scope of this study.

As shown in Table 3, 6 attributes were selected with the target class by the attribute selection method based on correlation. Based on the fuzzy-based rough set method, 9 attributes are selected with the target class.

Table 3. Selected attributes.

	Attributes	Data set after correlation	Data set after rough cluster
Attributes (Class included)	1, 2, 3, 4, 5, 6, 7, 8, 9, 10, 11, 12, 13, 14, 15, 16, 17, 18, 19, 20	1, 5, 11,12, 13, 17	1, 2, 4, 6, 10, 15, 16, 17, 18
Total number of attributes	20	6	9

4.3 Classification Algorithms Studied

Three different classification algorithms, Random Forest, nearest neighbors, Naive Bayes, and Logistic Regression, were used in the hepatitis data set.

k-nearest Neighbor Classification Algorithm. The class of the sample data point and k-nearest neighbor (k-nn), k is a classification method according to the value. This algorithm, is one of the simple and effective classification algorithms. It is used in many fields such as data mining and artificial intelligence [14].

Naive Bayes Classification Algorithm. It was developed and named by the English mathematician Thomas Bayes. It is simple compared to other classifier algorithms [15].

Random Forest Classification Algorithm. Developed by Leo Breiman. It is used to create the desired number of trees and to spread these trees according to the estimation values in each node [16].

Logistic Regression Classification Algorithm. Logistic regression; it is used to find the causal relationship between attributes and class [17].

4.4 Performance Criteria

In this study, models developed with classification algorithms complexity matrix was used to evaluate [18]. The performance evaluation criteria for classification algorithms according to Table 4 are given below [19–21].

Table 4. Confusion matrix.

		Actual result		
		Yes	No	Total
Prediction result	Yes	True Positive (TP)	False Positive (FP)	TPos
	No	False Negative (FN)	True Negative (TN)	TNeg
	Total	Pos	Neg	M

The accuracy value is shown in Eq. (20) [20].

$$Accuracy(ACC) = \frac{T_P + T_N}{M} \tag{20}$$

Sensitivity value is shown in Eq. (21) [20].

$$Sensitivity(TPR) = \frac{T_P}{Pos.} = \frac{T_P}{T_P + F_N} \tag{21}$$

Precision value is shown in Eq. (22) [20].

$$Precision(PPV) = \frac{T_P}{TPos.} = \frac{T_P}{T_P + F_P} \tag{22}$$

F-measure value is shown in Eq. (23) [20].

$$F - measure(F) = \frac{2 * Precision * Sensitivity}{Precision + Sensitivity} \tag{23}$$

The data sets are divided into two as test and training in order to establish the model with classifier algorithms and then to evaluate the performance of these models. In the study, it was preferred to work with k-fold cross validation. The cross-validation method and test and training cluster separation used in the study are shown in Fig. 9.

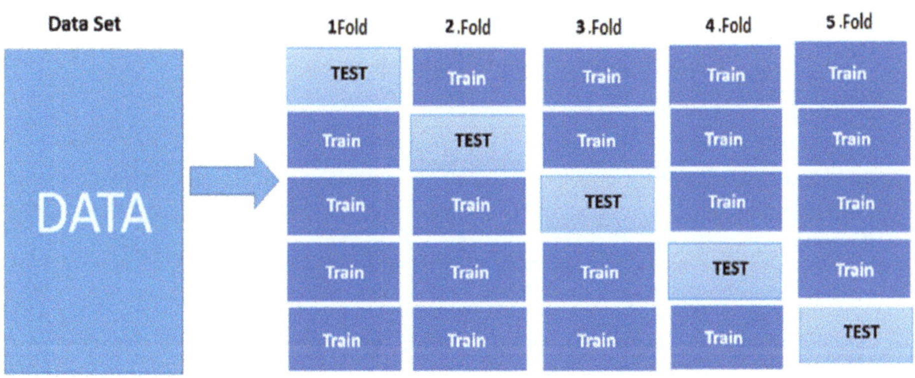

Fig. 9. Cross-validation method and test and training cluster separation.

5 Experimental Results

The performance values of hepatitis data obtained from Naive Bayes, Logistic Regression and Random Forest classification algorithms, which are the nearest neighbor, are applied in Tables 5, 6, 7 and 8, respectively. In this classification process, accuracy, ROC curve sensitivity, and F-measure performance criteria were evaluated. The data set was separated by 5-fold cross validation into training and test data.

Table 5. Results with the nearest neighbor classifier.

Algorithm	K-nn				
Performance criteria	Accuracy	Sensitivity	Precision	F-measure	ROC curve
Raw data	0,80	0,793	0,886	0,876	0,677
Correlation based	0,793	0,889	0,846	0,867	0,774
Fuzzy based rough cluster	0,829	0,891	0,926	0,868	0,832

Table 5 shows the performances of the data sets obtained on raw data and after correlation and fuzzy based rough set attribute processing in the nearest neighbor classifier algorithm.

Table 6. Results with the Naive Bayes classifier.

Algorithm	Naive Bayes				
Performance criteria	Accuracy	Sensitivity	Precision	F-measure	ROC curve
Raw data	0,787	0,854	0,862	0,867	0,740
Correlation based	0,838	0,888	0,890	0,876	0,847
Fuzzy based rough cluster	0,840	0,909	0,894	0,902	0,853

Table 6 shows the performances of the data sets obtained on raw data and after correlation and fuzzy based rough set attribute processing in the Naive Bayes classifier algorithm.

Table 7. Results with the logistic regression classifier.

Algorithm	Logistic regression				
Performance criteria	Accuracy	Sensitivity	Precision	F-measure	ROC curve
Raw data	0,835	0,840	0,845	0,841	0,789
Correlation based	0,825	0,875	0,911	0,836	0,793
Fuzzy based rough cluster	0,845	0,878	0,935	0,906	0,806

Table 7 shows the performances of the data sets obtained on raw data and after correlation and fuzzy based rough set attribute processing in the logistic regression classifier algorithm.

Table 8. Results with the random forest classifier.

Algorithm	Random forest				
Performance criteria	Accuracy	Sensitivity	Precision	F-measure	ROC curve
Raw data	0,812	0,856	0,909	0,886	0,836
Correlation based	0,832	0,882	0,911	0,896	0,852
Fuzzy based rough cluster	0,849	0,913	0,949	0,929	0,869

Table 8 shows the performances of the data sets obtained on raw data and after correlation and fuzzy based rough set attribute processing in the Random Forest classification algorithm.

6 Conclusion and Discussion

Within the scope of this study, a correlation-based and fuzzy logic based rough set attribute selection algorithm was applied to the estimate hepatitis disease. In the study, it was observed that fuzzy-based attribute selection was more successful in all performance criteria of classification algorithms performed after the attribute selection process. The results obtained with the Random Forest algorithm have better performance criteria than k-nearest neighbor, Logistic Regression and Naive Bayes.

As the classification algorithms on the raw data give inefficient results when the attribute selection process is not applied, it is observed that the selection of attributes made has a positive effect on the performance criteria.

Early and right decisions are very important for human health. The aim of this study is to assist healthcare workers in making the right decision. In addition, it is emphasized that fuzzy logic should be used in data analysis studies in the health field. In future studies, it is aimed to work on big data especially related to health.

References

1. Ministry of Health Hepatitis. http://www.seyahatsagligi.gov.tr/Site/HastalikDetay/Hepatit. Accessed 20 Apr 2018
2. Korkmaz, M., Timuçin, T., Yücedağ, I.: Kredi risk analizinde bulanık mantık kullanılarak aday durum tespitinin yapılması. In: International Academic Research Congress, Antalya, pp. 1355-1359 (2017)
3. Hepatitis Datasets. https://archive.ics.uci.edu/ml/datasets/hepatitis. Accessed 05 Mar 2018
4. Timuçin, T., Korkmaz, M., Yücedağ, İ., Biroğul, S.: Bilgisayar endüstrisinde bulanık mantik tabanli sql sorgulama yöntemiyle ürün seçimi. In: International Academic Research Congress, Antalya, pp. 1253–1259 (2017)
5. Lundberg, J.: Lifting the crown—citation z-score. J. Informetrics **1**(2), 145–154 (2007)
6. Ahsan, S., Shah, A.: Data, information, knowledge, wisdom: a doubly linked chain. In: Proceedings of the 2006 International Conference on Information Knowledge Engineering, Las vegas, pp. 270–278 (2006)
7. Orhunbilge, N.: Uygulamalı regresyon ve korelasyon analizi, 3rd edn. Nobel, Ankara (2016)

8. Yeung, D.S., Chen, D., Tsang, E.C.C., Lee, J.W.T., Wang, X.: On the generalization of fuzzy rough sets. IEEE Trans. Fuzzy Syst. **13**(3), 343–361 (2005)
9. Wu, W.Z., Zhang, W.X.: Constructive and axiomatic approaches of fuzzy approximation operators. Inf. Sci. **159**(3–4), 233–254 (2004)
10. Dubois, D., Prade, H.: Putting rough sets and fuzzy sets together. In: Intelligent Decision Support, pp. 203–232. Springer, Dordrecht (1992)
11. Dubois, D., Prade, H.: Rough fuzzy sets and fuzzy rough sets. Int. J. Gen. Syst. **17**(2–3), 191–209 (1990)
12. Morsi, N.N., Yakout, M.M.: Axiomatics for fuzzy rough sets. Fuzzy Sets Syst. **100**(1–3), 327–342 (1998)
13. Radzikowska, A.M., Kerre, E.E.: A comparative study of fuzzy rough sets. Fuzzy Sets Syst. **126**(2), 137–155 (2002)
14. Mitchell, T.M.: Machine Learning, 1st edn. McGraw-Hill, USA (1997)
15. Harrington, P.: Machine Learning in Action, 5th edn. Manning, USA (2012)
16. Han, J., Kamber, M., Pei, J.: Data mining: concepts and techniques. In: Data Management Systems, pp. 230–240. The Morgan Kaufmann Series (2006)
17. Breiman, L.: Some properties of splitting criteria. Mach. Learn. **24**(1), 41–47 (1996)
18. Oğuzlar, A.: Lojistik regresyon analizi yardimiyla suçlu profilinin belirlenmesi. Atatürk Üniversitesi İktisadi ve İdari Bilimler Dergisi **19**(1), 21–35 (2005)
19. Li, W., Raskin, R., Goodchild, M.F.: Semantic similarity measurement based on knowledge mining: an artificial neural net approach. Int. J. Geogr. Inf. Sci. **26**(8), 1415–1435 (2012)
20. Clark, M.: An introduction to machine learning: with applications in R (2013)
21. Flach, P.: The many faces of ROC analysis in machine learning. ICML Tutorial (2004)

Entropy-Based Skin Lesion Segmentation Using Stochastic Fractal Search Algorithm

Okan Bingöl[1(✉)], Serdar Paçacı[2], and Uğur Güvenç[3]

[1] Department of Electrical and Electronics Engineering, Isparta University of Applied Sciences, Isparta, Turkey
okanbingol@isparta.edu.tr

[2] Department of Information Technologies, Süleyman Demirel University, Isparta, Turkey
serdarpacaci@sdu.edu.tr

[3] Department of Electrical and Electronics Engineering, Düzce University, Düzce, Turkey
ugurguvenc@duzce.edu.tr

Abstract. Skin cancer is a type of cancer that attracts attention with the increasing number of cases. Detection of the lesion area on the skin has an important role in the diagnosis of dermatologists. In this study, 5 different entropy methods such as Kapur, Tsallis, Havrda and Charvat, Renyi and Minimum Cross were applied to determine the lesion area on dermoscopic images. Stochastic fractal search algorithm was used to determine threshold values with these 5 methods. PH^2 data set was used for skin lesion images.

Keywords: Lesion segmentation · Entropy · Stochastic fractal search

1 Introduction

The most important factor in the formation of skin cancer is ultraviolet (UV) rays. Considering the information given by scientists about the depletion of the ozone layer, it can be thought that people are exposed to more UV rays. As a result, this increases the risk of skin cancer. Skin cancer is a cancer of attention with the increasing number of cases in the world and in Turkey. Skin cancer is basically divided into two main groups as non-melanoma skin cancers and malign melanoma [1]. It is observed that the number of cases of skin melanoma and other skin cancers is almost doubled from 2010 to 2014 in Turkey [2]. When the cancer statistics are examined, the probability of developing skin cancer appears to be one in every 27 men and every 42 women, in the US [3]. As in many diseases, early diagnosis of skin cancer disease is important in reducing the mortality from cancer-related to this disease.

Computer-assisted systems are being developed to show the lesion area in the skin, to show the borders of the lesion area or to classify skin lesions to help dermatologists to diagnose patients [4–11]. While developing these systems, hairs, reflections, shadows, skin lines, and air bubbles in dermoscopic images may affect the performance of segmentation. The cleaning of these negative structures on the image provides better segmentation results.

D. J. Hemanth and U. Kose (Eds.): ICAIAME 2019, LNDECT 43, pp. 801–811, 2020.
https://doi.org/10.1007/978-3-030-36178-5_69

Thresholding is one of the most commonly used methods in image segmentation. The pixels in the image are separated into different classes by determining the threshold values. There are two types of segmentation method, two and multi-level. In two-level segmentation, the image is divided into two classes as background and object. In this study, two-level segmentation method was used because of two classes such as lesion area and non-lesion area. Entropy-based approaches for determining the threshold value have been used in many studies [12–18]. Finding the optimum threshold values in entropy-based segmentation approaches can be considered as an optimization problem.

Fractals are objects or properties quantities on each scale to each other. Fractals go on forever by repeating themselves continuously. When we look closely at any part of the fractals, we see that part and whole part has the similar features [19]. In 2015, Salimi has developed the Stochastic Fractal Search (SFS) algorithm. This algorithm has been developed based on the diffusion process of the fractals in nature [20]. Salimi has compared the SFS algorithm with particle swarm optimization, cuckoo search, gravitational search algorithm and artificial bee colony algorithms. It is showed that SFS algorithm gives better results than others.

The PH^2 dataset was created for research and benchmarking purposes over dermoscopic images. It consists of 200 dermoscopic images [21]. In this study, the results of segmentation were examined with 5 different entropy methods on PH^2 dataset images.

This paper is organized as follows: Sect. 2 presents entropy-based image segmentation. Image pre-processing and post-processing steps have explained in Sect. 3. In Sect. 4, stochastic fractal search algorithm has explained. In Sect. 5, the simulation results have presented and the results of the simulation have been given.

2 Entropy-Based Image Segmentation

Entropy is a measure of the irregularity of a system. Different entropy calculation methods are presented in the literature. Some of these methods are Shannon, Kapur, Tsallis, Havrda and Charvat, Renyi, Minimum Cross and Fuzzy Entropy. The normalized histogram of the image is calculated for thresholding by the entropy method. In this study, before the histogram values were calculated, the images were converted from RGB color to gray. Since each pixel in the image is expressed in 8-bit, the pixel values have a density of 256 (0–255) gray. Assuming that the image is intended to be divided into n classes, (n-1) threshold value must be determined. The probabilistic histogram distribution set for entropy methods is defined as $P = (p_1, p_2, p_3, \ldots, p_{256})$. This distribution data is common to all entropy methods given in this section. Also, because of the 8-bit images are used, the L expression in the equations is taken as 256.

The threshold values are calculated as shown in Eqs. (1) and (2) using Kapur entropy method. The variable "t" in the equation refers to the threshold values.

$$Hk_1(t) = -\sum_{i=1}^{t_1} \frac{p_i}{P_1} \ln \frac{p_i}{P_1},$$
$$Hk_2(t) = -\sum_{i=t_1+1}^{t_2} \frac{p_i}{P_2} \ln \frac{p_i}{P_2}, \quad (1)$$
$$Hk_n(t) = -\sum_{i=t_{n-1}+1}^{L} \frac{p_i}{P_n} \ln \frac{p_i}{P_n}$$

$$P_1(t) = \sum_{i=1}^{t_1} p_i, \quad P_2(t) = \sum_{i=t_1+1}^{t_2} p_i, \quad P_n(t) = \sum_{i=t_{n-1}+1}^{L} p_i \quad (2)$$

According to Kapur entropy method, finding the optimum threshold values using Eqs. (1) and (2) is performed as given in Eq. (3). As can be seen from the equation, the finding of the threshold values that maximize the sum of the entropy values of HK means to find the optimum threshold values.

$$\varphi k(t_1, t_2, \ldots, t_n) = Arg\max(Hk_1(t) + Hk_2(t) + \ldots + Hk_n(t)) \quad (3)$$

The expression given in Eq. (4) is used to find the threshold values by Tsallis method. The parameter q of Tsallis indicates the measure of non-extensivity of the system.

$$Ht_1(t) = \frac{1 - \sum_{i=1}^{t_1} \left(\frac{p_i}{P_1}\right)^q}{q-1},$$
$$Ht_2(t) = \frac{1 - \sum_{i=t_1+1}^{t_2} \left(\frac{p_i}{P_2}\right)^q}{q-1}, \quad (4)$$
$$Ht_n(t) = \frac{1 - \sum_{i=t_{n-1}+1}^{L} \left(\frac{p_i}{P_n}\right)^q}{q-1}$$

According to Tsallis entropy method, finding the optimum threshold values using Eq. (4) is performed as given in Eq. (5).

$$\varphi t(t_1, t_2, \ldots, t_n) = Arg\max\begin{pmatrix} Ht_1(t) + Ht_2(t) + \ldots + Ht_n(t) + \\ (1-q).Ht_1(t).Ht_2(t).\ldots.Ht_n(t) \end{pmatrix} \quad (5)$$

The threshold values are calculated as shown in Eqs. (6) and (7) using Minimum Cross entropy method.

$$
\begin{aligned}
Hmc_1(t) &= \sum_{i=1}^{t_1} i p_i \ln \frac{i}{u(1,t_1)}, \\
Hmc_2(t) &= \sum_{i=t_1+1}^{t_2} i p_i \ln \frac{i}{u(t_1,t_2)}, \\
Hmc_n(t) &= \sum_{i=t_n+1}^{L} i p_i \ln \frac{i}{u(t_n,L)}
\end{aligned} \tag{6}
$$

$$
u(a,b) = \frac{\sum_{i=a}^{b} i p_i}{\sum_{i=a}^{b} p_i} \tag{7}
$$

According to Minimum Cross entropy method, finding the optimum threshold values using Eqs. (6) and (7) is performed as given in Eq. (8).

$$
\varphi mc(t_1, t_2, \ldots, t_n) = Arg\min(Hmc_1(t) + Hmc_2(t) + \ldots + Hmc_n(t)) \tag{8}
$$

The threshold values are calculated as shown in Eq. (9) using Renyi entropy method.

$$
\begin{aligned}
Hr_1(t) &= \frac{1}{1-\alpha} \ln \sum_{i=1}^{t_1} \left(\frac{p_i}{P_1}\right)^{\alpha}, \\
Hr_2(t) &= \frac{1}{1-\alpha} \ln \sum_{i=t_1+1}^{t_2} \left(\frac{p_i}{P_2}\right)^{\alpha}, \\
Hr_n(t) &= \frac{1}{1-\alpha} \ln \sum_{i=t_{n-1}+1}^{L} \left(\frac{p_i}{P_n}\right)^{\alpha}
\end{aligned} \tag{9}
$$

According to Renyi entropy method, finding the optimum threshold values using Eq. (9) is performed as given in Eq. (10).

$$
\varphi r(t_1, t_2, \ldots, t_n) = Arg\max(Hr_1(t) + Hr_2(t) + \ldots + Hr_n(t)) \tag{10}
$$

The threshold values are calculated as shown in Eq. (11) using Havrda and Charvat entropy method.

$$
\begin{aligned}
Hhc_1(t) &= \frac{1}{1-\alpha}\sum_{i=1}^{t_1}\left(\frac{p_i}{P_1}\right)^{\alpha}-1,\\
Hhc_2(t) &= \frac{1}{1-\alpha}\sum_{i=t_1+1}^{t_2}\left(\frac{p_i}{P_2}\right)^{\alpha}-1,\\
Hhc_n(t) &= \frac{1}{1-\alpha}\sum_{i=t_{n-1}+1}^{L}\left(\frac{p_i}{P_n}\right)^{\alpha}-1
\end{aligned}
\tag{11}
$$

According to Havrda and Charvat entropy method, finding the optimum threshold values using Eq. (11) is performed as given in Eq. (12).

$$
\varphi hc(t_1, t_2, \ldots, t_n) = Arg\max\begin{pmatrix} Hhc_1(t) + Hhc_2(t) + \ldots + Hhc_n(t) + \\ (1-\alpha).Hhc_1(t).Hhc_2(t)\ldots.Hhc_n(t) \end{pmatrix} \tag{12}
$$

3 Image Pre-processing and Post-processing

Pre-processing is an important step in the segmentation. As mentioned earlier, hairs, reflections, shadows, skin lines and air bubbles on the image affect the performance of segmentation. Therefore, the Dullrazor algorithm is used in this study to eliminate the hair in the image [22]. The Dullrazor algorithm generally consists of 3 main steps. In the first step, horizontal, vertical and diagonal filter structural elements are used for the morphological closing operation. As a result of this process, a hair mask is created. However, this hair mask may contain some of the pixels which are not really represented hair. Thus, in the second step, it is investigated whether each hair pixel in the hair mask is within the hair region. Pixels that are not in the hair region are removed from the mask. With bilinear interpolation, new values are assigned to the pixels in the hair region. In the third and last step, the median filter is applied.

Another problem that may take place on the images is the dark areas originating from the camera lens and formed at the corners of the image. For this problem, a circle mask is formed such that it passes through the black zone boundary, the center of the central image. The decision whether or not to apply this mask is decided by looking at the color intensity at the corners of the image. In the last step of the pre-processing, the image is filtered by median filter. Figure 1 shows that the IMD003 image in the PH^2 data set, the image obtained when the Dullrazor algorithm is applied to this image and the image that remove black areas in the corner.

After the pre-processing step, the image is subject to thresholding according to the determined threshold value and the pixels in the image are divided into two groups as skin region and lesion region. Morphological open, erode and close operations were performed respectively. The disk-shaped kernel structure is used in 2, 2 and 5 dimensions in order to perform these operations, respectively. Finally, the process is completed by filling in the gaps. The post-processing steps are shown in Fig. 2.

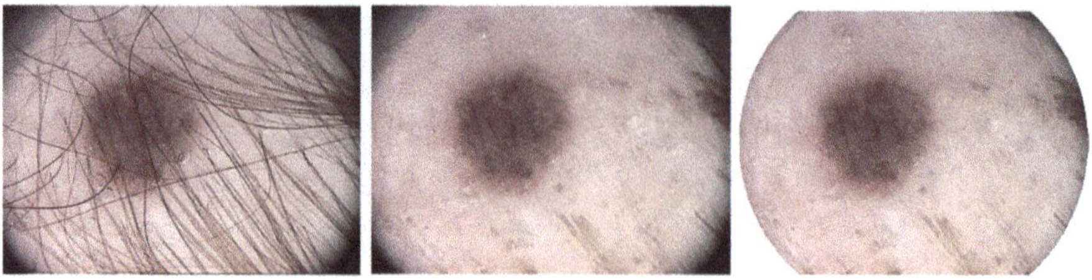

Fig. 1. Pre-processing step

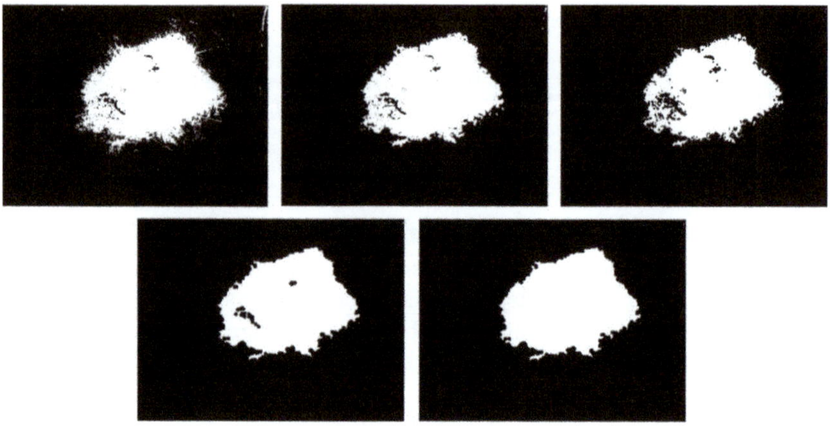

Fig. 2. Post-processing step

4 Stochastic Fractal Search Algorithm

SFS algorithm is originated from the diffusion process of the fractals in the nature. The SFS algorithm consists of two main parts. These are diffusion and the update processes. The diffusion process is designed to generate a new solution point for the optimization problem. It satisfies the exploitation property of the algorithm and increases the chance of finding the global optimum point. Gaussian walk function is used in the diffusion process. Two different functions are proposed in Eqs. (13) and (14) for the diffusion. Here, ε and ε' are random values in the interval (0, 1) which is uniformly distributed, P_i is the ith position in the group, and BP represents the best point position.

$$GW_1 = Gaussian(\mu_{BP}, \sigma) + (\varepsilon x BP - \varepsilon' x P_i) \tag{13}$$

$$GW_2 = Gaussian(\mu_P, \sigma) \tag{14}$$

Here, σ, which is the step length of the Gaussian walk, is computed with Eq. (15). In Eq. (15), g defines the generation value. The value of the log(g)/g inversely proportional to generation value. When log(g)/g decreases, the generation value increases. In this situation, the step length decreases in the Gaussian walk. This case causes more local search and the obtaining results are much closer to an optimal solution.

$$\sigma = \left| \frac{\log(g)}{g} x(P_i - BP) \right| \tag{15}$$

When the diffusion process is completed, each solution point is ranked by its fitness value. After the sorting process, a probability value is calculated for each fitness value. This is calculated as in Eq. (16). Here, N is the number of points in the group and rank (P_i) is expressed as the rank of the P_i point in the group. After this process, the update part of the algorithm begins.

$$Pa_i = \frac{rank(P_i)}{N} \tag{16}$$

For each P_i point in the group, if the $P_{ai} < \varepsilon$ condition is true, the first update is performed as shown in Eq. (17). Here, P_r and P_t are randomly selected points in the group. Also, ε is a randomly generated value in the range of (0, 1). j denotes the jth dimension in solution point.

$$P_i^{'}(j) = P_r(j) - \varepsilon x\,(P_t(j) - P_i(j)) \tag{17}$$

When the first update process is computed, the second update process is implemented. In the second update process, the positions of the points are evaluated taking into account the positions other solution points in the group. With the second update process, exploration quality of algorithm is increased. At the start, by using Eq. (16), the probability values are recomputed. For each P_i point in the group, if the $P_{ai} < \varepsilon$ condition is satisfied, the second update is performed as shown in Eq. (18).

$$P_i^{''} = \begin{cases} P_i^{'} - \hat{\varepsilon} x\,(P_t^{'} - BP) & when\ \varepsilon^{'} \leq 0.5 \\ P_i^{'} - \hat{\varepsilon} x\,(P_t^{'} - P_r^{'}) & when\ \varepsilon^{'} > 0.5 \end{cases} \tag{18}$$

5 Simulation Results

In this study, 200 images in the PH^2 data set were used. Four basic parameters were calculated to evaluate the image segmentation results. These are true positive (TP - pixels correctly segmented as object), false positive (FP - pixels falsely segmented as object), true negative (TN - pixels correctly detected as background) and false negative (FN - pixels falsely detected as background). Using these basic parameters XOR, sensitivity (SE), specificity (SP), accuracy (AC), precision (P), hammoude distance (HD) and structural similarity index (SSIM) values were calculated. The statements regarding the calculation of these values are given in Eq. (19).

$$\begin{aligned}
XOR &= \frac{FP + FN}{TP + FN} \\
AC &= \frac{TP + TN}{TP + FN + FP + TN} * 100 \\
SE &= \frac{TP}{TP + FN} * 100 \\
SP &= \frac{TN}{FP + TN} * 100 \\
P &= \frac{TP}{TP + FP} * 100 \\
HM &= \frac{FN + FP}{TP + FN + FP}
\end{aligned} \tag{19}$$

SSIM was used to find the similarity ratio between the image obtained as a result of the segmentation and the reference segmented image. Equation (20) shows the calculation of the SSIM value. Here, gt and seg sub-indices are ground truth image and segmented image respectively. μ and σ are mean and the standard deviation of the image respectively. C1 and C2 are constants.

$$\begin{aligned}
SSIM(I_{gt}, I_{seg}) &= \frac{(2\mu_{I_{gt}}\mu_{I_{seg}} + C1)(2\sigma_{I_{gt}I_{seg}} + C2)}{(\mu_{I_{gt}}^2\mu_{I_{seg}}^2 + C1)(\sigma_{I_{gt}}^2 + \sigma_{I_{seg}}^2 + C2)} \\
\sigma_{I_{gt}I_{seg}} &= \frac{1}{N-1}\sum_{i=1}^{N}(I_{gt_i} + \mu_{I_{gt}})(I_{seg_i} + \mu_{I_{seg}})
\end{aligned} \tag{20}$$

All images in the data set were segmented by 5 different entropy methods and the results obtained were recorded. The data of the SSIM results listed in Tables 1 and 2. When Tables 1 and 2 were examined, it was observed that the minimum cross entropy method gave better results on the PH^2 data set. It is seen that the minimum cross entropy method did not fall below the value of 0.3 in the SSIM score. Figure 3 demonstrates the original image, its ground truth, and the segmentation result respectively.

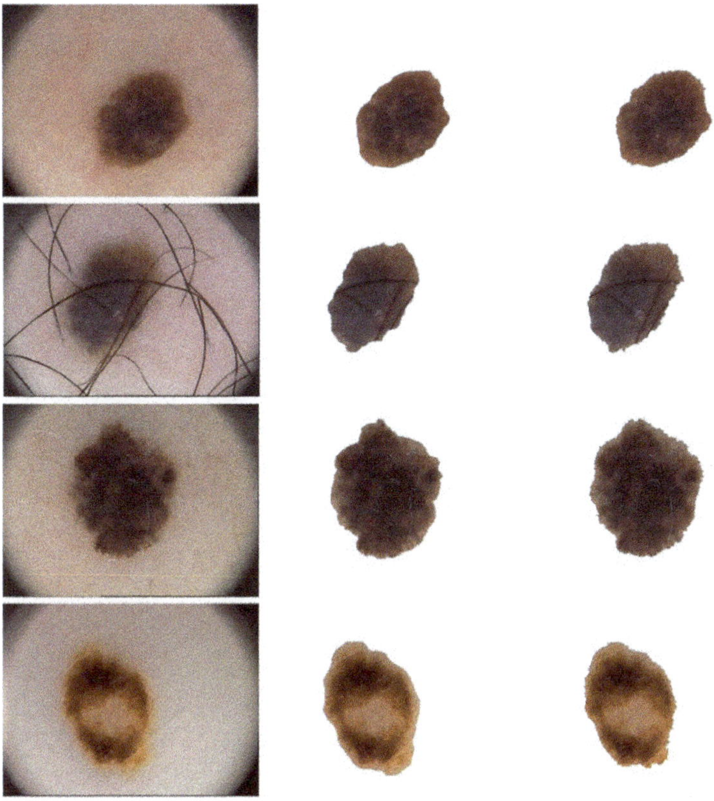

Fig. 3. Segmentation results

Table 1. Performance comparison of the entropy methods

SSIM score	Kapur	Tsallis	Minimum cross	Renyi	Havrda & Charvat
[0, 0.1]	0	0	0	0	0
(0.1, 0.2]	2	3	0	2	0
(0.2, 0.3]	5	5	0	6	5
(0.3, 0.4]	9	10	2	10	3
(0.4, 0.5]	9	10	2	10	4
(0.5, 0.6]	6	7	8	7	8
(0.6, 0.7]	12	8	12	8	14
(0.7, 0.8]	17	17	22	18	26
(0.8, 0.9]	48	50	60	51	79
(0.9, 1]	92	90	94	88	61

Table 2. Analysis of the SSIM scores

Entropy method	Rank	Mean	Standard deviation
Minimum cross	1	0,85	0,12
Havrda & Charvat	2	0,82	0,16
Kapur	3	0.81	0,20
Tsallis	4	0,80	0,21
Renyi	5	0,79	0,21

6 Conclusion

In this paper, 5 different entropy methods are applied to image segmentation on the PH^2 data set. SFS algorithm is used for the finding the optimal threshold values. Entropy methods are used for the objective function of the SFS algorithm. For the thresholding method in the segmentation, the color images were converted to grayscale images. The SSIM scores of the images segmented by entropy methods were calculated. It has been observed that the minimum cross entropy method gives better results than other methods.

References

1. Sümen, A., Öncel, S.: Türkiye'de Cilt Kanseri ve Güneşten Korunmaya Yönelik Yapılan Araştırmaların İncelenmesi. Turkiye Klinikleri J. Nurs. Sci. **10**(1), 59–69 (2018)
2. Sağlık Bakanlığı, T.C.: Kanser İstatistikleri (2014). https://hsgm.saglik.gov.tr/depo/birimler/kanser-db/istatistik/2014-RAPOR._uzuuun.pdf. Erişim Tarihi 15 Jan 2019
3. Siegel, R.L., Miller, K.D., Jemal, A.: Cancer statistics. CA Cancer J. Clin. **68**, 7–30 (2018). https://doi.org/10.3322/caac.21442
4. Celebi, M.E., Kingravi, H.A., Uddin, B., Iyatomi, H., Aslandogan, Y.A., Stoecker, W.V., Moss, R.H.: A methodological approach to the classification of dermoscopy images. Comput. Med. Imaging Graph. **31**(6), 362–373 (2007)
5. Iyatomi, H., Oka, H., Celebi, M.E., Hashimoto, M., Hagiwara, M., Tanaka, M., Ogawa, K.: An improved internet-based melanoma screening system with dermatologist-like tumor area extraction algorithm. Comput. Med. Imaging Graph. **32**(7), 566–579 (2008)
6. Maeda, J., Kawano, A., Yamauchi, S., Suzuki, Y., Marçal, A.R.S., Mendonça, T.: Perceptual image segmentation using fuzzy-based hierarchical algorithm and its application to dermoscopy images. In: IEEE Conference on Soft Computing in Industrial Applications, SMCia 2008, pp. 25–27 (2008)
7. Wong, A., Scharcanski, J., Fieguth, P.: Automatic skin lesion segmentation via iterative stochastic region merging. IEEE Trans. Inf. Technol. Biomed. **15**(6), 929–936 (2011)
8. Garnavi, R., Aldeen, M., Celebi, M.E., Varigos, G., Finch, S.: Border detection in dermoscopy images using hybrid thresholding on optimized color channels. Comput. Med. Imaging Graph. **35**(2), 105–115 (2011)
9. Ma, Z., Tavares, J.M.R.: A novel approach to segment skin lesions in dermoscopic images based on a deformable model. IEEE J. Bio-Med. Health Inf. **20**(2), 615–623 (2016)

10. Sankaran, S., Sethumadhavan, G.: Entropy-based colour splitting in dermoscopy images to identify internal borders. In: International Conference on Inventive Research in Computing Applications (ICIRCA), pp. 771–774. IEEE (2018)
11. Yang, T., Chen, Y., Lu, J., Fan, Z.: Sampling with level set for pigmented skin lesion segmentation. Signal Image Video Process. **13**, 813–821 (2019)
12. Kapur, J.N., Sahoo, P.K., Wong, A.K.: A new method for gray-level picture thresholding using the entropy of the histogram. Comput. Vis. Graph. Image Process. **29**(3), 273–285 (1985)
13. Sahoo, P., Wilkins, C., Yeager, J.: Threshold selection using Renyi's entropy. Pattern Recogn. **30**(1), 71–84 (1997)
14. Pavesic, N., Ribaric, S.: Gray level thresholding using the Havrda and Charvat entropy. In: 10th Mediterranean Electrotechnical Conference, MELECON 2000, vol. 2, pp. 631–634. IEEE (2000)
15. De Albuquerque, M.P., Esquef, I.A., Mello, A.G.: Image thresholding using Tsallis entropy. Pattern Recogn. Lett. **25**(9), 1059–1065 (2004)
16. Tao, W., Jin, H., Liu, L.: Object segmentation using ant colony optimization algorithm and fuzzy entropy. Pattern Recogn. Lett. **28**(7), 788–796 (2007)
17. Sarkar, S., Das, S., Chaudhuri, S.S.: A multilevel color image thresholding scheme based on minimum cross entropy and differential evolution. Pattern Recogn. Lett. **54**, 27–35 (2015)
18. Chen, C.: An improved image segmentation method based on maximum fuzzy entropy and quantum genetic algorithm. In: 5th International Conference on Systems and Informatics (ICSAI), pp. 934–938. IEEE (2018)
19. Mandelbrot, B.B., Pignoni, R.: The fractal geometry of nature (1983)
20. Salimi, H.: Stochastic fractal search: a powerful metaheuristic algorithm. Knowl.-Based Syst. **75**, 1–18 (2015)
21. Mendonca, T.F., Celebi, M.E., Mendonca, T., Marques, J.S.: PH2: a public database for the analysis of dermoscopic images. In: Dermoscopy Image Analysis (2015)
22. Lee, T., Ng, V., Gallagher, R., Coldman, A., McLean, D.: Dullrazor®: a software approach to hair removal from images. Comput. Biol. Med. **27**(6), 533–543 (1997)

Providing the Moment of the Parabolic Reflector Antenna in the Passive Millimeter Wave Imaging System with the Equilibrium Weights

Mehmet Duman[1(✉)] and Alp Oral Salman[2]

[1] Faculty of Technology, Department of Electrical and Electronics Engineering, Düzce University, Düzce, Turkey
mehmetduman@duzce.edu.tr

[2] Faculty of Engineering, Department of Electronics and Communication Engineering, Kocaeli University, Kocaeli, Turkey
oral.salman@kocaeli.edu.tr

Abstract. In foggy, cloudy and rainy weather; for the purpose of radiometric imaging; Passive MilliMeter Wave Imaging System (PMMWIS) was installed with a radiometric receiver operating at a frequency of 96 GHz which is one of the specific frequency windows for passive detection. After the installation of this system, while working, there were contractions in the motor part of the 2-axis positioner with which the parabolic reflector antenna of the PMMWIS. These contractions were mainly during upward system calibration or during normal scan of the system. Therefore, it had become necessary to make balance weights for more convenient operation. In this study; the equilibrium weights were integrated in order to synchronize the moment to the opposite direction of the parabolic reflector antenna. According to the weight and height of the parabolic reflector antenna to be mounted on the system, four different balance weights could be used. The sounds from the gears in the engine of the 2-axis positioner, indicating that the motor was contracted for every degree to be scanned (it could be 0.1° to the resolution) was indicative of the deterioration of the gears in the 2-axis positioner. Consequently; the balance weights were produced and integrated into the system. As a result; the torque of the antenna was achieved and the 2-axis positioner was able to scan the view by means of PMMWIS without any contraction in the azimuth and elevation axes.

Keywords: Equilibrium weights · Moment of the imaging system · Passive detection

1 Introduction

In a Passive MilliMeter Wave Imaging System (PMMWIS), there should be some elements such as reflector antenna; active parts of circuit as low noise amplifier; detector; video amplifier; 2-axis scanning motor (in elevation and azimuth) to move the parabolic reflector antenna and the other controlling unit [1–5]. The system with all

D. J. Hemanth and U. Kose (Eds.): ICAIAME 2019, LNDECT 43, pp. 812–816, 2020.
https://doi.org/10.1007/978-3-030-36178-5_70

these elements is very heavy. The tripod has to carry all these units. In this case, the moment subject becomes effective.

In this study, moment studies of the stability of the system, especially the weight of the antenna's parabolic reflector antenna, are explained. Various equilibrium weights are used to provide the moment of the antenna. The 2-axis motor has to be long-lasting and balanced in order to be able to work properly. In order to achieve this balance, 4 balance weights were built and used. Parabolic reflector antenna may need different moment weights for different azimuth and elevation axes. This situation was determined and ideal weights were placed into the system. Care was taken to place the antenna in the same direction with the antenna and the back side but not on the beam of the antenna. As a result; the system was run smoothly without contractions on the motor side.

Sometimes PMMWIS structures are on a large table. In this case the use of balance weights is not required [6, 7]. If the scanning engine does not move the antenna while imaging is performed, there is no need for balance weights too [6–8]. In this case, there is a lens that transmits images of different points to the horn antenna to which the active circuits at the focal point of the parabolic reflector antenna are connected. The flapping reflector is also one of the methods used to transfer images to the receiver circuit [9]. If you have a tripod that is strong enough to carry the antenna, control unit and other parts without a jerk, the balance weights are not required [10].

2 Materials and Methods

Weights of materials to be used as balance weight are not very important. Instead of; the value of multiplication the distance from the weight to the equilibrium point and the weights is important. This multiplication is called moment. The weights are placed on the rear side of the antenna in the same direction as the parabolic dish antenna.

Materials of approximately 1 kg and 3 kg are made of steel. They are also mounted on a rail made of steel. The length of the rail is approximately 20 cm. There are different points on which the materials can be connected. If the heaviest material is connected to the farthest point; the size of a torque will be;

$$3\text{ kg} \times 20\text{ cm} = 0.6\text{ kg m} \tag{1}$$

Different torques are required depending on the position of the active circuits which are at the focal point of the parabolic reflector antenna. Because the focal point determined in another study varies according to the active circuit structure used. In fact, it is not the changing focus, but it is where we want the signals to come. If we think of the system as a kind of scales; if the torque on one part of the balance changes, the other part must also change. Otherwise, the system may fall to the other side.

Our tripod is strong enough to prevent a fall. Instead, contractions occur in the motor part. These contractions can damage the gears of the engine. In order to prevent this damage, balance parts are used.

In Fig. 1., ready to use balance parts are shown. The middle part is used as a rail. In total, 2 rails were produced. There is only one place where the balance pieces can be connected on this track. In Fig. 2, a rear view of the antenna system and another rail are given. Balance tracks can be shifted on this track. Thus different moments can be reached.

Fig. 1. Balance parts and rail (middle).

PMMWISs can detect objects invisible under clothing, as well as view the background in fog, dust cloud, rain, and other weather conditions. Our system is produced for foggy weather conditions. By means of the designed and produced balance components, the contractions in the motor have also fixed and PMMWIS has been made ready for work. In Fig. 3., the ready PMMWIS can be seen while scanning.

Fig. 2. Another rail and one of the balance part with system.

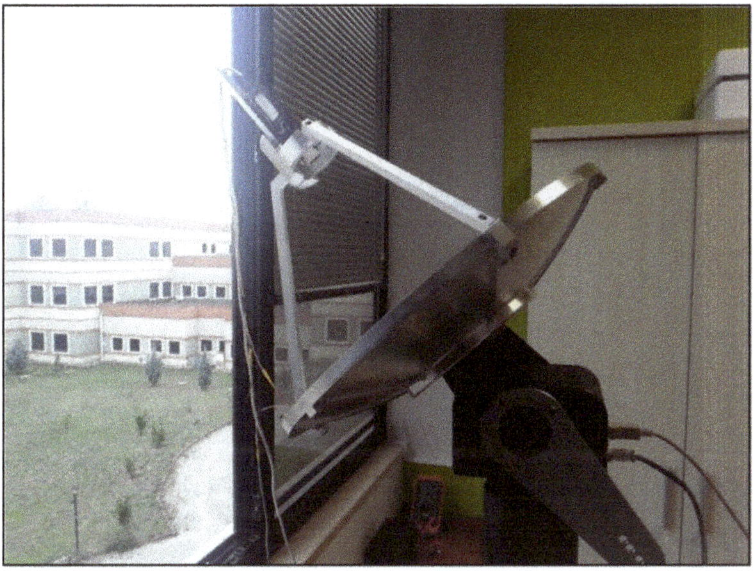

Fig. 3. A view of scanning PMMWIS

3 Discussion and Suggestions

If the system was thought to be a problem in the engine and installed on a solid structure, there would be no contractions. The problem can also be solved by using a more stable motor for the 2-axis positioner. The problem can be solved by using rotating lens instead of rotating motors too.

4 Conclusions

As a result, a smoothly produced PMMWIS has been produced and maintained. It was understood that the contractions in the motor can be fixed with moment. This system is currently available for different parabolic reflector antennas or different focal points of reflector antennas.

Acknowledgment. This research has been supported by The Scientific and Technological Research Council of Turkey (TÜBİTAK), Düzce University and Kocaeli University.

References

1. Ulaby, F.T., Moore, R.K., Fung, A.K.: Microwave Remote Sensing. Active and Passive. Microwave Remote Sensing Fundamentals and Radiometry, vol. 1. Addison-Wesley, Boston (1981)
2. Yujiri, L., Shoucri, M., Moffa, P.: Passive millimeter wave imaging. IEEE Microwave Mag. **4**(3), 39–50 (2003)

3. Duman, M.: Pasif milimetre dalga görüntüleme sistemi uygulamaları. Kocaeli Üniversitesi Doktora tezi (2018)
4. Duman, M., Salman, A.O.: Obtaining the sky-temperature dependence on voltage value by the passive millimeter wave ımaging system. Acta Phys. Pol., A **134**(1), 346–348 (2018)
5. Duman, M., Salman, A.O.: Theoretical ınvestigation of blackbody radiation for the passive millimeter wave ımaging system. In: IEEE 2nd International Symposium on Multidisciplinary Studies and Innovative Technologies (ISMSIT), Ankara, Turkey (2018)
6. Mizuno, K., et al.: New applications of millimeter-wave incoherent imaging. In: IEEE MTT-S International Microwave Symposium Digest, Long Beach, CA, USA (2005)
7. Verity, A., et al.: Short and long range passive ımaging in millimeter wave band. In: 30th URSI General Assembly and Scientific Symposium (2011)
8. Kapilevich, B., Litvak, B., Shulzinger, A.: Passive non-imaging mm-wave sensor for detecting hidden objects. In: IEEE International Conference on Microwaves, Communications, Antennas and Electronic Systems (COMCAS) (2013)
9. Sato, H., et al.: Development of 77 GHz millimeter wave passive imaging camera. In: IEEE Sensors (2009)
10. WEB Page. Diamond Engineering, Antenna Measurement Products, IEEE, pp. 1–5. https://www.diamondeng.net/antenna-measurement-27/x100/x100-features. Accessed on 17 April 2019

Convolutional Auto-Encoder Based Degradation Point Forecasting for Bearing Data Set

Abdullah Taha Arslan[1] and Ugur Yayan[2](✉)

[1] R&D Department, Techy Bilisim Ltd. Sti., Eskisehir, Turkey
taha@techy.digital
[2] Inovasyon Mühendislik Tek. Gel. Dan. San. Tic. Ltd. Sti., Eskisehir, Turkey
ugur.yayan@inovasyonmuhendislik.com

Abstract. In smart manufacturing industry, health analysis and forecasting of degradation starting point has become an increasingly crucial research area. Prognostics-aware systems for health analysis aim to integrate health information and knowledge about the future operating conditions into the process of selecting subsequent actions for the system. Developments in smart manufacturing as well as deep learning-based prognostics provide new opportunities for health analysis and degradation starting point forecasting. Rotating machines have many critical components like spinning, drilling, rotating, etc. and they need to be forecasted for failure or degradation starting times. Moreover, bearings are the most important sub-components of rotating machines. In this study, a convolutional neural network is used for forecasting of degradation starting point of bearings by experimenting with Nasa Bearing Dataset. Although convolutional neural networks (CNNs) are utilized widely for 2D images, 1-dimensional convolutional filters may also be embedded in processing temporal data, such as time-series. In this work, we developed one such autoencoder network which consists of stacked convolutional layers as a contribution to the community. Besides, in evaluation of test results, L_{10} bearing life criteria is used for threshold of degradation starting point. Tests are conducted for all bearings and results are shown in different figures. In the test results, proposed method is found to be effective in forecasting bearing degradation starting points.

Keywords: Bearing dataset · Convolutional neural network · Prognostics and Health Management · Degradation forecasting · Auto-encoder

1 Introduction

Prognostics-aware system aims to integrate health information and knowledge about the future operating conditions into the process of selecting subsequent actions for the system. Prognostics information could be used for predicting reliability, degradation starting point of the system for lifetime extension and forecast failure before it occurs, etc. Estimation of the degradation starting point for a component has been conducted to suppress the possible faults of the systems and extend lifetime of the systems [1].

A system's current condition is evaluated by diagnostic and prognostic processes. Diagnostics is concerned with current state of any subsystem whereas prognostic is

D. J. Hemanth and U. Kose (Eds.): ICAIAME 2019, LNDECT 43, pp. 817–829, 2020.
https://doi.org/10.1007/978-3-030-36178-5_71

related to the future state of subsystem [2]. Although realizing a long-term operation is important in premier applications of systems, sustaining system's operation becomes more important with the use of PHM applications. Fault-tolerant control architectures are developed for sustainability; however, these studies are generally diagnostic-based. On the contrary, prognostic-based strategies can predict risks before failure.

The potential benefits of prognostics methods are evident and the recent increase in the research in this area has helped to explain this development. The biggest reason why such technologies are not confronted yet in everywhere is the uncertainties in the systems [3]. Managing these uncertainties is seen as a key to the success of prognostic technologies. Therefore, the prognostic estimations for the development of a comprehensive Prognostics and Health Management (PHM) system are defined as Achilles' heel [4].

Prognostics and Health Management (PHM) incorporates aspects of logistics; safety, reliability, mission criticality, and economic viability among others. PHM of components or systems involves both diagnostics and prognostics: Diagnostics is the process of detection and isolation of faults or failures, while prognostics is the process of prediction of the future state or remaining useful life (RUL) based on the current and historic conditions. Prognostics is based on the understanding that equipment fails after a period of degradation, which if measured, can be used to prevent system breakdown and minimize operation costs.

In many industries including smart industry, prediction of failure has become more important. It is hard to diagnose failures in advance in any system because of the limited availability of sensors and some of the design criteria. However, with the great development in the industry, it looks feasible today to analyse a sensor's data along with different type of techniques (data-driven or model-based) for failure prediction. Especially data-driven methods are extensively used for prognostics and the related work is described in Sect. 2.

2 Related Work

In the literature, there are fault tolerant control (FTC) approaches that handle the failure after it occurs. These are reactive approaches that offer immediate action when a failure occurs [5]. On the other hand, proactive policies do not allow any failures within the system and they warn the user or decide automatically for an action. Thus, proactive policies are preferred for the long term and for an efficient operation of any system. The failure could be observed by monitoring the temporal behaviour of the machine. This consists of the incipient signs of failure before an actual fault event. The data-driven methods such as deep learning approaches can be used in these types of systems to forecast the risk even before it occurs, and it helps proactive control and maintenance that reduce the costs.

The lifetime of the rotating machine may be increased if the system is capable of knowing health status of rotating machines and making decisions accordingly. Thus, forecasting the degradation starting point of the bearings during decision making process results in long term successful operations of the rotating machines. System Health Management (SHM) of a machine depends on prognostics technology and prognostics can be supported by diagnostics for health-based planning for the system.

In the SHM context, end of life (EOL), availability and reliability (remaining useful life (RUL)) of systems and components could be predicted. Prognostics is the process which forecasts when a component or system does not satisfy desired operations. By using this prognostics knowledge, system can take some sort of decisions like changing a component before it fails, prolonging component life by load reduction or task switching and optimally planning or replanning the task.

Finding of degradation starting point for rotating machinery is vital to reducing maintenance costs, operation downtime and safety hazards. Many approaches have been suggested for finding of degradation starting point via different methods in the literature. A study given in [6] proposes an intelligent condition-based monitoring (I-CBM) platform based on standalone data-driven approaches as a comprehensive system for rotating machinery. A toolbox has also been developed as a software component to easily and conveniently implement this platform in applications. This platform consists of data acquisition, signal processing, feature extraction and feature selection, condition monitoring and health assessment, fault diagnostics, and prognostics modules. In another study [7], the AI methods are reviewed for fault diagnosis of rotating machines. These methods are classified as k-NN-based, Naive Bayes-based, SVM-based, ANN-based and deep learning-based approaches. Theoretical and practical backgrounds for the methods are also given and compared among themselves. In another survey paper [8], a comprehensive review of research efforts on deep learning-based machine health monitoring is given. Several classical DL-based machine health monitoring systems are also implemented for quick assessment of researchers. In this study, DL-based architectures are summarized in four categories such as Auto-encoder models, Restricted Boltzmann Machines models, Convolutional Neural Networks and Recurrent Neural Networks. The study says that DL-based MHMS does not need extensive human work and expertise. Thus, DL-based methods are not limited to specific machineries because of these methods are able to map raw machinery data to targets. Study [9] presents a DNN-based intelligent method for diagnosing the faults of rotating machinery. Five datasets which were collected from rolling element bearings and have massive data regarding health conditions under different operating conditions and planetary gearboxes are used for verification of the proposed method. A deep neural network which has five-layer DNN has been utilized with these datasets. In the proposed method, DNNs are trained by frequency spectra. Thus, it only works for bearing like elements whose measured vibration signals are periodic. Besides, hyperbolic tangent function is used as the active function of DNNs and half coefficients of the frequency spectra. another study [10] addresses the multi-bearing reliability prediction and proposes an integrated DL method based on collaborative analysis of data from bearing vibration signals with combination of time domain and frequency domain features. The features extracted from the bearing vibration signals almost cover a whole process of bearing degradation. In the proposed method, the fully connected DNN is designed for multi-bearing reliability prediction model and parameters for the network is found by using grid search experiments. The approach described in [11] introduces wavelet filter-based prognostics method for bearing fault identification and performance degradation assessment. Test results show that bearing faults can be detected at the beginning of a failure. Besides, this paper addresses a problem for de-noising and extraction of the weak signals from the noisy dataset for bearing prognostics.

Data representation in lower dimensions is an important task for many applications. Three aims of this process are: (1) extracting useful features, (2) eliminating redundancies, (3) preserve the essential part of the input data [12]. This task is also preferred to be conducted in an automated fashion. With the advent of deep learning techniques and GPU based general-purpose computing (GPGPU), unsupervised learning, feature extraction, dimension reduction and data representation algorithms utilizing these new techniques have been developed over the years. Deep Belief Networks [13], restricted Boltzmann machines (RBMs) [14] and autoencoders are the major examples of these structures. Autoencoder networks consist of two sub networks, one is an encoder and the other one is the decoder part. Encoders take the input data and summarize and represent it in a lower dimension. This network structure may be constructed with a single layer, as found with conventional dense neural networks, as well as deep neural networks (DNN) which incorporates several alternative layers, such as convolutional filters, pooling layers, different activation functions, and more complicated structures such as Recurrent layers (RNN) and Long-Short-Term Memory units (LSTM). Decoder reconstructs the original data from the encoded representation with a certain amount of loss. Therefore, autoencoders are lossy compressors and learn data representations automatically.

Bengio et al. [17] and Ranzato et al. [16] introduce the simple and ordinary autoencoders. Vincent et al. apply autoencoders in stacks to learn representations in each layer of a deep structure in order to increase the optimization process overall [15]. They also utilize a denoising criterion in optimization of extracting useful features and preventing auto-encoders from learning the identity mapping. Masci et al. [12] stack convolutional auto-encoders (CAE) and introduce max-pooling layers for enabling sparsity over the hidden representations for feature learning. One popular autoencoder structure is called Variational Autoencoders, which has gained significant attention lately, especially for generative capabilities [18].

Rotating machines have many critical components like spinning, drilling, etc. to forecast the failure or degradation. Moreover, bearings are the most important subcomponent of the rotating machines [19, 20]. In this study, convolutional neural network is used for forecasting of degradation starting point at the Nasa Bearing Dataset [21]. Although convolutional neural networks (CNNs) are utilized widely for 2D images, 1-dimensional convolutional filters may also be embedded in processing temporal data, such as time-series. In this work, we developed one such autoencoder network which consists of convolutional stacked layers as a contribution to the community. Test results show the effectiveness of the proposed method. Nasa Bearing dataset and the proposed CNN structure is described in the following section. Test results are shown in Sect. 4 and conclusion remarks and future works are discussed in Sect. 5.

3 Dataset Description and Proposed Convolutional Neural Network

Dataset is downloaded from NASA's Prognostics Data Repository. This dataset was generated by the NSF I/UCR Center for Intelligent Maintenance Systems with support from Rexnord Corp. in Milwaukee, WI. Test Rig setup is shown in Fig. 1 and the description given in the readme file in the dataset folder is: "Four bearings were installed on a shaft. The rotation speed was kept constant at 2000 RPM by an AC motor coupled to the shaft via rub belts. A radial load of 6000 lbs is applied onto the shaft and bearing by a spring mechanism. All bearings are force lubricated. Rexnord ZA-2115 double row bearings were installed on the shaft as shown in Fig. 1. PCB 353B33 High Sensitivity Quartz ICP accelerometers were installed on the bearing housing (two accelerometers for each bearing [x- and y-axes] for data set 1, one accelerometer for each bearing for data sets 2 and 3). Sensor placement is also shown in Fig. 1. All failures occurred after exceeding designed life time of the bearing which is more than 100 million revolutions."

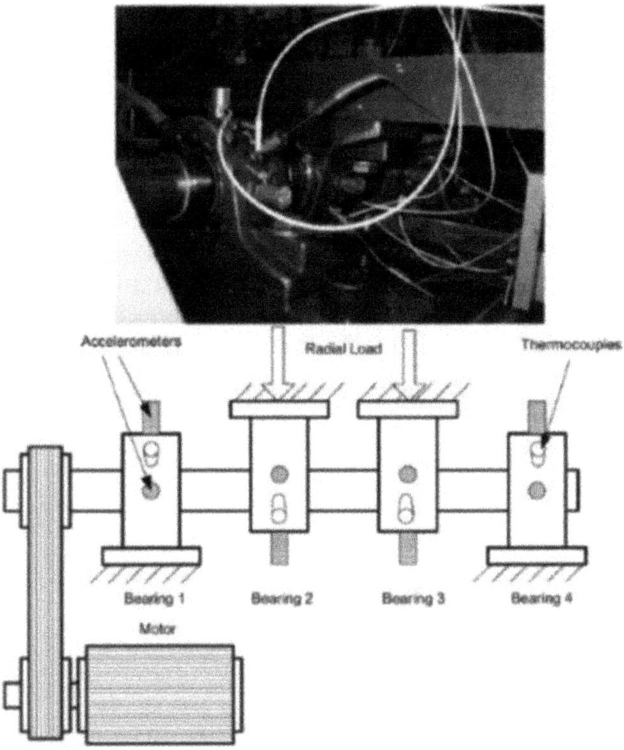

Fig. 1. Bearing test rig and sensor placement illustration (Wavelet Filter-based Weak Signature Detection Method and its Application on Roller Bearing Prognostics)

Dataset contains three different tests for four different bearings. Each data set describes a test-to-failure experiment and consists of individual files that are one second vibration signal snapshots recorded at specific intervals. Each file consists of 20,480 points with the sampling rate set at 20 kHz. The file name indicates when the data was collected. Each record (row) in the data file is a data point.

3.1 The Proposed Convolutional Neural Network

An autoencoder is a network structure that takes a high dimensional data as input, and learns a low-dimensional representation of this input data in an unsupervised manner. It consists of two parts that work in tandem: encoder and decoder. The encoder encodes the data to a code, while the decoder takes the code and reconstructs the original input with a degree of loss in the information. Therefore, an autoencoder may be considered as a lossy compressor.

In its most basic form, an autoencoder takes the input x and applies a mapping function $f(.)$. If we call the representation h, then it can be shown as:

$$h = f(Wx + b) \tag{1}$$

where W and b are the neural network layer's weights matrix and biases vector, respectively. The decoder part of the autoencoder applies another mapping $f'()$ that takes h as input and reconstructs x':

$$x' = f'(W'h + b') \tag{2}$$

The basic unit of encoder and decoders may be stacked to construct multi-layered autoencoder networks, called stacked autoencoders. Apart from this, there is a number of types of autoencoders, including sparse autoencoders, variational autoencoders (VAE), contractive autoencoders (CAE). Convolutional autoencoders utilize convolutional layers instead of dense ones. In this study, a stack of convolutional layers is applied to learn a good representation of the input bearing vibration data. The overall structure can be seen in the Fig. 2:

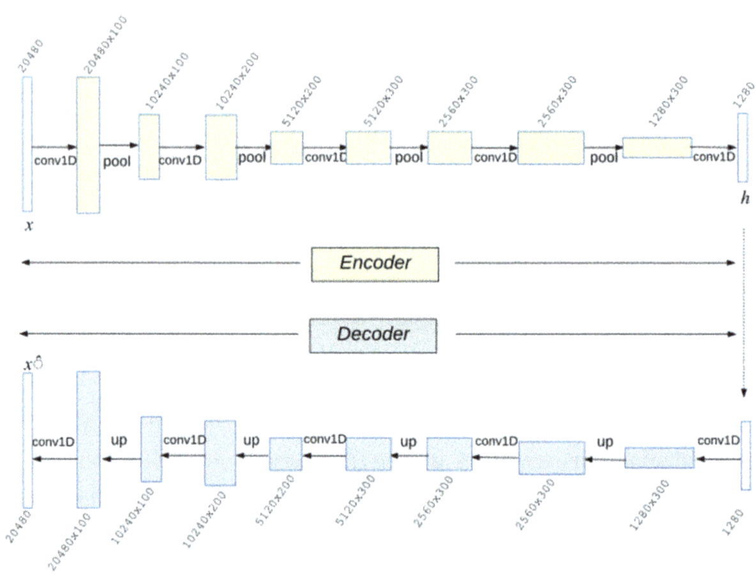

Fig. 2. The proposed convolutional auto-encoder

Convolution layers' output sizes are the same as the inputs, therefore dimension reduction process in the encoder network is implemented with *maxpooling* layers. A maxpool operation downsamples the input and reduces its dimensionality. At each stack of *conv* and consequent *maxpool* layers the dimension is reduced to one-half of the input. The decoder network, on the other hand, applies an *up-sampling* operation at each layer, therefore increases the dimensionality of the data along the pipeline until the reconstructed vector size is as the same as the original input data.

As with any neural network-based learning process, a loss function between the reconstructed output and the input is defined. Among the different choices of loss functions, *binary cross entropy* loss function is used in this study, given as:

$$l(x', x) = L = \{l_1, \ldots, l_n\}^T, \text{ where}$$

$$l_n = x'_n \cdot \log(x_n) + \left(1 - x'_n\right) \cdot \log(1 - x_n). \tag{3}$$

ReLU activation function takes place at the output of any layers of both the encoder and decoder networks, except for the last layers of each network. At the last layers, *sigmoid* function acts as the activator function. These functions are defined as:

$$Sigmoid(x) = \frac{1}{1 + e^{-x}} \tag{4}$$

$$ReLU(x) = max(0, x) \tag{5}$$

4 Test Results

Nasa Bearing Dataset [21] is used for tests and results are given in the below. L_{10} bearing life criteria is used for the thresholding of degradation starting point. Reliability of bearings is often expressed by the L_{10} life, which is the time at which 10% of the population has failed [22]. Table 1 shows the specifications of training and testing environments.

Table 1. Test and modeling environments

Heading level	Font size and style
Brand/Model	Dell Precision Tower
Processor (CPU)	Intel Xeon CPU E3-1270 v6
CPU frequency	3.80 GHz
RAM	16 GB ECC
GPU RAM	5 GB
Video card (GPU)	NVIDIA Quadro P2000
Operating system	Ubuntu 18.04.1
Deep learning framework	Keras+Tensorflow

Dataset 1 has 4 bearing data and 2 channels data for each bearing. We have conducted tests for all channels and give results for one healthy bearing and two degraded bearings. Figure 3 shows healthy bearing 1 for data collected from channel 1. This test is conducted for monitoring the healthy bearing behaviour.

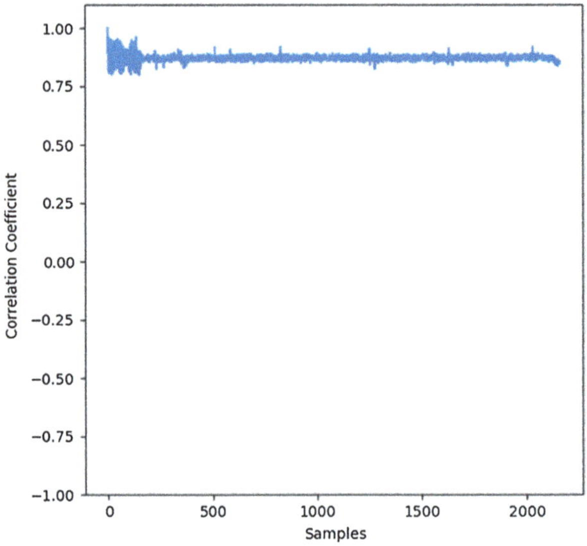

Fig. 3. Set 1 Bearing 1 Channel 1

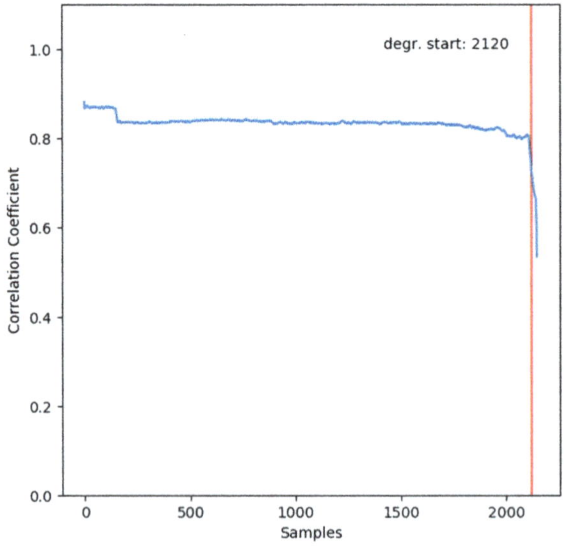

Fig. 4. Set 1 Bearing 3 Channel 5

Faulty bearing is tested for all of the channels, and test results are shown in Figs. 4 and 5. In Fig. 4, degradation starting point for bearing 3 channel 5 could be seen at 2120. And In Fig. 5, test result for bearing 3 channel 6 is given and degradation starting point could be seen as 2113.

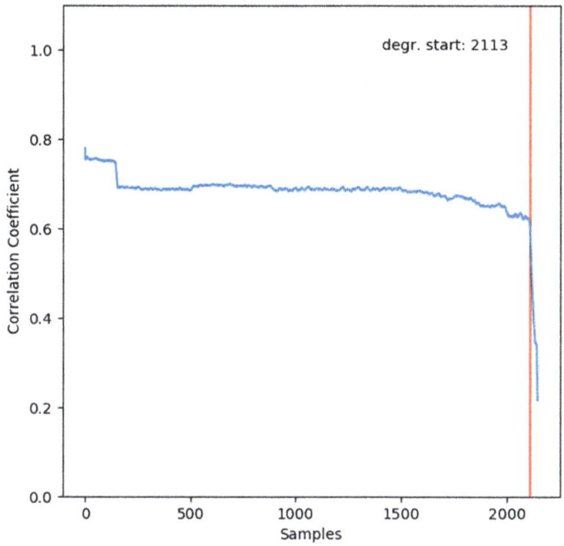

Fig. 5. Set 1 Bearing 1 Channel 6

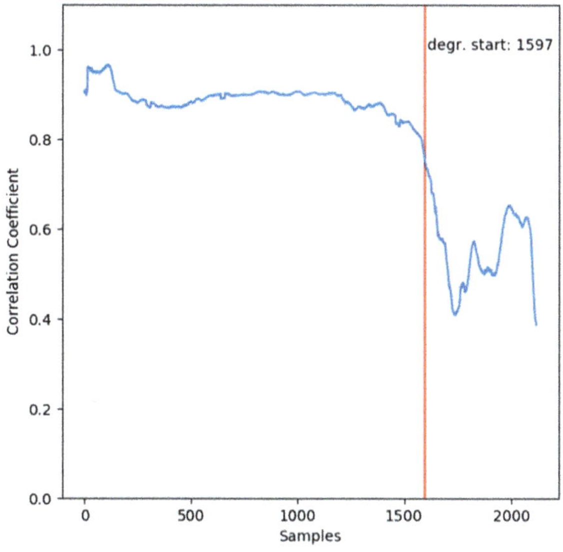

Fig. 6. Set 1 Bearing 4 Channel 7

Other faulty bearing is number 4 and it has two channel data in dataset 1. Same test is applied to bearing 4 for both channel and Test results are shown in Figs. 6 and 7. Bearing 4 degradation starting point is found as 1597 for channel 7 and 1713 for channel 8.

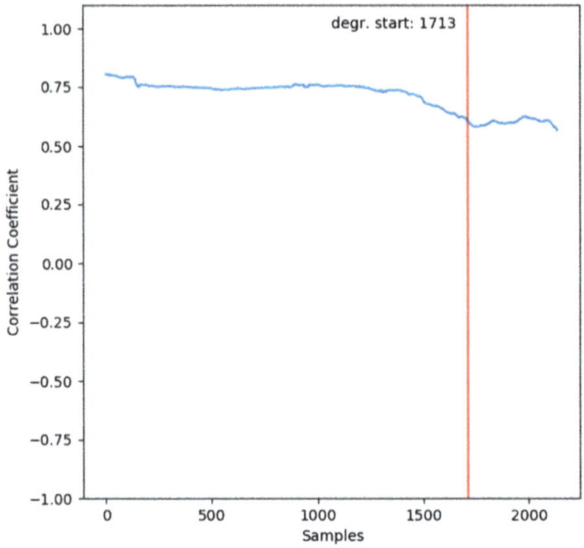

Fig. 7. Set 1 Bearing 4 Channel 8

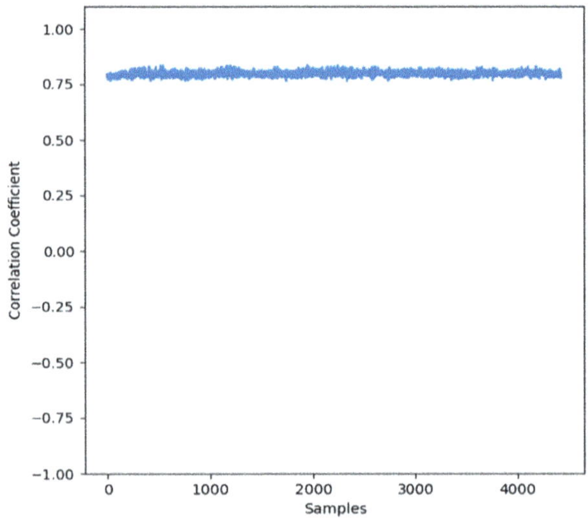

Fig. 8. Set 1 Bearing 4 Column 4

When the tests are applied to the dataset 2, there is only one bearing in faulty condition. The other three bearing shows healthy characteristics. In Fig. 8, bearing 4 is shown for tests of dataset 2.

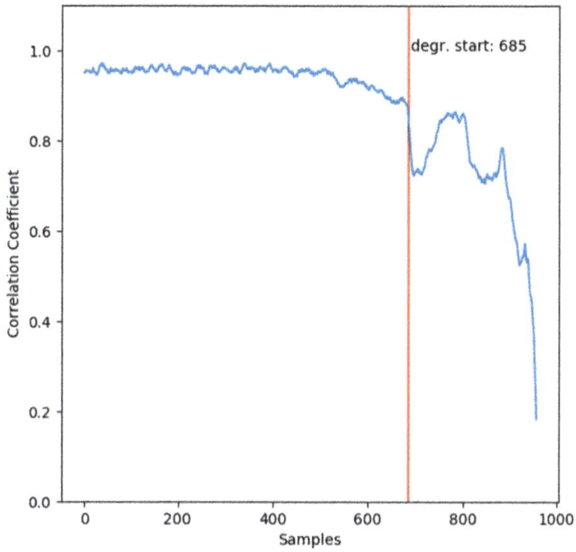

Fig. 9. Set 2 Bearing 1 Column 1

Faulty bearing is shown in Fig. 9 and there is only one channel for data recording. Bearing degradation starting point is found as 685 for bearing 1 in dataset 2 according to L_{10} life criteria.

5 Conclusions and Future Work

Developments in smart manufacturing as well as deep learning-based prognostics provide new opportunities for health analysis and degradation starting point forecasting. In this study, an autoencoder network structure which consists of convolutional stacked layers as encoder and decoder is developed and proposed to the community. Tests are conducted for forecasting degradation starting point for all bearings and results are given. In the test results, proposed method is found to be effective in forecasting bearing degradation starting points. When the test results are evaluated for the convolutional auto encoder architecture, more trained networks are suppressing the faulty features from data set and increase correlation value among data.

In the future work, variational auto-encoder (VAE), Long/Short Term Memory (LSTM) etc. methods will be implemented and tested for Nasa Bearing Dataset. Besides, other datasets will be studied in the Nasa Repository.

References

1. Lee, J., Wu, F., Zhao, W., Ghaffari, M., Liao, L., Siegel, D.: Prognostics and health management design for rotary machinery systems—reviews, methodology and applications. Mech. Syst. Signal Process. **42**(1), 314–334 (2014)
2. Shi, G., Dong, P., Sun, H.Q., Liu, Y., Cheng, Y.J., Xu, X.Y.: Adaptive control of the shifting process in automatic transmissions. Int. J. Automot. Technol. **18**(1), 179–194 (2017)
3. Sikorska, J.Z., Hodkiewicz, M., Ma, L.: Prognostic modelling options for remaining useful life estimation by industry. Mech. Syst. Signal Process. **25**(5), 1803–1836 (2011)
4. Vachtsevanos, G.J., Lewis, F., Hess, A., Wu, B.: Intelligent Fault Diagnosis and Prognosis for Engineering Systems, pp. 185–186. Wiley, Hoboken (2006)
5. Zhang, Y., Jiang, J.: Bibliographical review on reconfigurable fault-tolerant control systems. Ann. Rev. Control **32**(2), 229–252 (2008)
6. Yang, B.S.: An intelligent condition-based maintenance platform for rotating machinery. Expert Syst. Appl. **39**(3), 2977–2988 (2012)
7. Liu, R., Yang, B., Zio, E., Chen, X.: Artificial intelligence for fault diagnosis of rotating machinery: a review. Mech. Syst. Signal Process. **108**, 33–47 (2018)
8. Zhao, R., Yan, R., Chen, Z., Mao, K., Wang, P., Gao, R.X.: Deep learning and its applications to machine health monitoring. Mech. Syst. Signal Process. **115**, 213–237 (2019)
9. Jia, F., Lei, Y., Lin, J., Zhou, X., Lu, N.: Deep neural networks: a promising tool for fault characteristic mining and intelligent diagnosis of rotating machinery with massive data. Mech. Syst. Signal Process. **72**, 303–315 (2016)
10. Ren, L., Cui, J., Sun, Y., Cheng, X.: Multi-bearing remaining useful life collaborative prediction: a deep learning approach. J. Manuf. Syst. **43**, 248–256 (2017)
11. Qiu, H., Lee, J., Lin, J., Yu, G.: Robust performance degradation assessment methods for enhanced rolling element bearing prognostics. Adv. Eng. Inform. **17**(3–4), 127–140 (2003)
12. Masci, J., Meier, U., Cireşan, D., Schmidhuber, J.: Stacked convolutional auto-encoders for hierarchical feature extraction. In: International Conference on Artificial Neural Networks, pp. 52–59. Springer, Heidelberg, June 2011
13. Hinton, G.E., Osindero, S., Teh, Y.W.: A fast learning algorithm for deep belief nets. Neural Comput. **18**(7), 1527–1554 (2006)
14. Taylor, G.W., Hinton, G.E., Roweis, S.T.: Modeling human motion using binary latent variables. In: Advances in Neural Information Processing Systems, pp. 1345–1352 (2007)
15. Vincent, P., Larochelle, H., Lajoie, I., Bengio, Y., Manzagol, P.A.: Stacked denoising autoencoders: learning useful representations in a deep network with a local denoising criterion. J. Mach. Learn. Res. **11**, 3371–3408 (2010)
16. Poultney, C., Chopra, S., Cun, Y.L.: Efficient learning of sparse representations with an energy-based model. In: Advances in Neural Information Processing Systems, pp. 1137–1144 (2007)
17. Bengio, Y., LeCun, Y.: Scaling learning algorithms towards AI. Large-Scale Kernel Mach. **34**(5), 1–41 (2007)
18. Kingma, D.P., Welling, M.: Auto-encoding variational Bayes. arXiv preprint arXiv:1312.6114 (2013)

19. Hasani, R.M., Wang, G., Grosu, R.: An automated auto-encoder correlation-based health-monitoring and prognostic method for machine bearings. arXiv preprint arXiv:1703.06272 (2017)
20. Medjaher, K., Tobon-Mejia, D.A., Zerhouni, N.: Remaining useful life estimation of critical components with application to bearings. IEEE Trans. Reliab. **61**(2), 292–302 (2012)
21. Qiu, H., Lee, J., Lin, J.: Wavelet filter-based weak signature detection method and its application on roller bearing prognostics. J. Sound Vib. **289**, 1066–1090 (2006)
22. Bogh, D., Crowell, J.R., Stark, D.: Bearings for IEEE 841 motors. IEEE Trans. Ind. Appl. **39** (6), 1578–1583 (2003)

Moth Swarm Algorithm Based Approach for the ACOPF Considering Wind and Tidal Energy

Serhat Duman[1,2(✉)], Lei Wu[3], and Jie Li[1]

[1] Electrical and Computer Engineering, Clarkson University, Potsdam, NY 13699, USA
serhatduman@duzce.edu.tr, jieli@clarkson.edu
[2] Electrical and Electronics Engineering, Technology Faculty, Duzce University, 81600 Duzce, Turkey
[3] Electrical and Computer Engineering, Stevens Institute of Technology, Hoboken, NJ 07030, USA
lei.wu@stevens.edu

Abstract. In the last decades, optimal power flow (OPF) problem is becoming one of the most important nonlinear and non-convex problems for the planning and operation of large-scale modern electrical power grids. In this study, the OPF problem with the consideration of two types of renewable energy sources (RES), tidal and wind energy, is studied. The Gumbel PDF is used to calculate power output of tidal energy. The proposed OPF problem is solved by Moth Swarm Algorithm (MSA), and tested on an IEEE 30-bus test system via two different cases with and without the consideration of prohibited operating zones. The simulation results show the effectiveness of the MSA based OPF approach.

Keywords: Moth Swarm Algorithm · Optimization · Optimal power flow · Modern power systems · Wind-tidal energy

1 Introduction

Recently, the rising of people needs and developments of technology have significantly increased the energy consumption levels. However, as the reserves of fossil fuels for thermal generators are limit, the researchers seek for alternative energy sources, specifically renewable energy sources (RES). Wind, solar energy, and hydropower are some of the most well-known RESs. Indeed, modern electrical power network structure becomes more complicated with the RES integration into power systems. This complex structure makes power system planning and operation problems more difficult, such as optimal power flow (OPF), economic dispatch (ED), economic and emission dispatch (EED), optimal reactive power dispatch (ORPD), short-term hydrothermal scheduling (SHTS), and transmission expansion planning (TEP).

D. J. Hemanth and U. Kose (Eds.): ICAIAME 2019, LNDECT 43, pp. 830–843, 2020.
https://doi.org/10.1007/978-3-030-36178-5_72

Specifically, the OPF problem has a highly nonlinear and non-convex structure. The OPF problem aims to minimizing the total cost of generation units, while satisfying specified equality and inequality constraints. In the last decades, the classical OPF problem with thermal generators has been studied by many researchers, and solved with different optimization methods such as Tree-seed Algorithm (TSA), Quasi-oppositional Modified Jaya (QOMJaya), Water Evaporation Algorithm (WEA), Glowworm Swarm Optimization (GSO), Backtracking Search Optimization Algorithm (BSA), Oppositional Krill Herd Algorithm (OKHA), Chaotic Artificial Bee Colony (CABC), and Hybrid Fuzzy Particle Swarm Optimisation and Nelder-Mead Algorithm (HFPSO-NM) [1–8].

Recently, the researchers have studied the OPF problem for modern power systems with RESs. Chang et al., used evolutionary particle swarm optimization (EPSO) algorithm to investigate the effect of wind generators on electrical power grid. The proposed OPF problem was tested on an IEEE 30-bus test system [9]. Panda and Tripathy utilized modified bacteria foraging algorithm (MBFA) to solve OPF problem with wind power, which was tested via an IEEE 30-bus test system. The proposed OPF problem has been solved in different scenarios in [10] to illustrate the feasibility and resolvableness. In another work Panda and Tripathy, a modified bacteria foraging algorithm was pro-posed to solve OPF problem with wind-thermal generators and shunt FACTS device (STATCOM) on an IEEE 30-bus test system [11]. Kouadri et al., simulated the OPF problem with wind power, which was simulated as high voltage direct current with voltage source converter (VSC-HVDC), and solved the problem via ant lion optimization (ALO) algorithm [12].

Three types of tidal energy based systems exist, including tidal stream, tidal range, and combined or hybrid systems [13, 14]. In this study, the tidal range from the tidal energy systems is considered within the OPF problem, and the uncertain cost models of the tidal energy are considered in the OPF problem. Moth swarm algorithm is used to solve the proposed OPF problem. This algorithm is a novel stochastic method developed by Ali Mohamed et al. [15], inspired by the navigational behaviors of moths in nature. In order to show the feasibility and resolvableness of the proposed OPF problem, MSA algorithm is tested on an IEEE 30-bus test system, and the obtained results of the MSA method are compared with Biogeography Based Optimization (BBO) [16] and Invasive Weed Optimization (IWO) algorithms [17].

2 Formulation of Problem

In this study, the OPF problem with renewable energy sources is considered. The main goal of the OPF problem is to minimize a specified objective function by finding the optimum control variables within system constraints. The formulation of the proposed OPF problem with RESs can be described as follows.

$$\text{Minimize} f_{obj}(G, H) \tag{1}$$

$$\text{Subject to} \begin{cases} m(G, H) = 0 \\ n(G, H) \leq 0 \end{cases} \tag{2}$$

Where f_{obj} is the objective function, G and H are the state and the control variables, a and b are the equality and the inequality constraints. The state and control variables of the proposed OPF problem are defined as follows.

$$\begin{bmatrix} G \\ H \end{bmatrix} = \begin{bmatrix} P_{THG_1}, V_{L_1} \ldots V_{L_{NL}}, Q_{G_1} \ldots Q_{G_{NTHG}}, Q_{WS_1} \ldots Q_{WS_{NW}}, Q_{WSTDL_1} \ldots Q_{WSTDL_{NWSTDL}}, S_{l_1} \ldots S_{l_{NTL}} \\ P_{THG_2} \ldots P_{THG_{NTHG}}, P_{WS_1} \ldots P_{WS_{NW}}, P_{WSTDL_1} \ldots P_{WSTDL_{NWSTDL}}, V_{G1} \ldots V_{G_{NG}} \end{bmatrix} \tag{3}$$

where P_{THG_1} is active power of the swing generator, V_L is voltage magnitudes of load (PQ) buses, Q_G, Q_{WS}, and Q_{WSTDL} are reactive power of traditional generators, wind power, and combined wind power and tidal energy systems, and S_L is apparent power of transmission lines. *NL* and *NTL* are numbers of PQ buses and transmission lines. P_{THG} is active power of traditional generators except the slack generator, P_{WS} is active power of wind farm, P_{WSTDL} active power of combined system, V_G is voltage values of all generator buses including traditional generators, wind farms, and tidal energy system. *NG*, *NTHG*, *NW*, and *NWSTDL* are numbers of generator buses (including thermal, wind, and tidal units), traditional generation units, wind farms, and combined wind power and tidal energy systems. The equality constraints of the proposed OPF problem can be defined as follow.

$$\begin{bmatrix} P_{G_i} - P_{D_i} - V_i \sum_{j=1}^{Nbus} V_j\left(G_{ij}cos(\theta_i - \theta_j) + B_{ij}sin(\theta_i - \theta_j)\right) = 0, \quad \forall i \in Nbus \\ Q_{G_i} - Q_{D_i} - V_i \sum_{j=1}^{Nbus} V_j\left(G_{ij}sin(\theta_i - \theta_j) - B_{ij}cos(\theta_i - \theta_j)\right) = 0, \quad \forall i \in Nbus \end{bmatrix} \tag{4}$$

For the inequality constraints, lower and upper limits on active and reactive power outputs of thermal, wind, and tidal generation units as well as voltage limits of generator units (including thermal, wind, and tidal generation units) are defined as in Eq. 5.

$$\begin{array}{ll} P_{THG_i,min} \leq P_{THG_i} \leq P_{THG_i,max} & \forall i \in NTHG \\ P_{WS_i,min} \leq P_{WS_i} \leq P_{WS_i,max} & \forall i \in NW \\ P_{WSTDL_i,min} \leq P_{WSTDL_i} \leq P_{WSTDL_i,max} & \forall i \in NWSTDL \\ Q_{THG_i,min} \leq Q_{THG_i} \leq Q_{THG_i,max} & \forall i \in NTHG \\ Q_{WS_i,min} \leq Q_{WS_i} \leq Q_{WS_i,max} & \forall i \in NW \\ Q_{WSTDL_i,min} \leq Q_{WSTDL_i} \leq Q_{WSTDL_i,max} & \forall i \in NWSTDL \\ V_{G_i,min} \leq V_{G_i} \leq V_{G_i,max} & \forall i \in NG \end{array} \tag{5}$$

Voltage magnitude of each PQ bus and apparent power values of transmission lines should be within specified limits as in Eq. 6. $V_{Li,min}$ and $V_{Li,max}$ are lower and upper voltage magnitudes of the *i*th PQ bus. S_{li} and $S_{li,max}$ are apparent power and upper apparent power limit of the *i*th transmission line.

$$\begin{aligned} &V_{Li,min} \leq V_{Li} \leq V_{Li,max} && \forall i \in NL \\ &S_{li} \leq S_{li,max} && \forall i \in NTL \end{aligned} \tag{6}$$

The fitness function with penalty factors is expressed as follows.

$$\begin{aligned} J_{fit} = & f_{obj}(G,H) + \lambda_P \left(P_{THGslack} - P_{THGslack}^{lim}\right)^2 + \lambda_V \sum_{i=1}^{NL} \left(V_{Li} - V_{Li}^{lim}\right)^2 \\ & + \lambda_Q \sum_{i=1}^{NTHG} \left(Q_{THG_i} - Q_{THG_i}^{lim}\right)^2 + \lambda_{WS} \sum_{i=1}^{NW} \left(Q_{WS_i} - Q_{WS_i}^{lim}\right)^2 \\ & + \lambda_{WSTDL} \sum_{i=1}^{NWSTDL} \left(Q_{WSTDL_i} - Q_{WSTDL_i}^{lim}\right)^2 + \lambda_S \sum_{i=1}^{NTL} \left(S_{li} - S_{li}^{lim}\right)^2 \end{aligned} \tag{7}$$

2.1 Fuel Cost of Traditional Generation Units

The total fuel cost of thermal generation units is formulated as in Eq. 8. Quadratic cost function is used, h_i, j_i, and k_i are fuel cost coefficients of the *i*th thermal generation unit [1–3].

$$CF_1(P_{THG}) = \sum_{i=1}^{NTHG} h_i + j_i P_{THG_i} + k_i P_{THG_i}^2 \tag{8}$$

2.2 Direct Cost Model of Wind Power and Tidal Energy

Cost model of a wind power source is described via a linear function of scheduled wind power [10, 11, 18]. *ww, i*, and $P_{WS,i}$ are cost coefficient and scheduled power of the *i*th wind farm.

$$CF_{W,i}(P_{WS,i}) = ww, i \times P_{WS,i} \tag{9}$$

In this study, the proposed combination model of wind power and tidal energy can be defined as follow.

$$CF_{WSTDL}(P_{WSTDL}) = ww, i \times P_{WS,i} + P_{tdl} \times P_{TLDS} \tag{10}$$

P_{tdl} and P_{TDLS} are cost coefficient and scheduled power of a tidal energy unit.

2.3 Uncertainty Cost Models of Wind Power and Tidal Energy

In this study, cost models for overestimation and underestimation situations of wind power can be defined as follow [10, 11, 18].

$$\begin{aligned} CF_{Ow,i}(P_{WS,i} - P_{wav,i}) &= C_{Ow,i}(P_{WS,i} - P_{wav,i}) \\ &= C_{Ow,i} \int_{0}^{P_{WS,i}} (P_{WS,i} - p_{w,i}) f_w(p_{w,i}) dp_{w,i} \end{aligned} \tag{11}$$

$$\begin{aligned} CF_{Uw,i}(P_{wav,i} - P_{WS,i}) &= C_{Uw,i}(P_{wav,i} - P_{WS,i}) \\ &= C_{Uw,i} \int_{P_{WS,i}}^{P_{wr,i}} (p_{w,i} - P_{WS,i}) f_w(p_{w,i}) dp_{w,i} \end{aligned} \tag{12}$$

Where $CF_{Ow,i}$ and $CF_{Uw,i}$ are the uncertainty cost values for overestimation and underestimation of a wind farm, $C_{Ow,i}$ and $C_{Uw,i}$ are the uncertainty cost coefficients, and P_{wav} is available power of a wind farm. Tidal energy's uncertainty cost models can be referred to in refs. [18, 19]. Uncertainty cost models for overestimation and underestimation situations of tidal energy can be identified as follow.

$$\begin{aligned} CF_{Otdl}(P_{TDLS} - P_{TDLav}) &= C_{Otdl}(P_{TDLS} - P_{TDLav}) \\ &= C_{Otdl} * f_{tdl}(P_{TDLav} < P_{TDLS}) * [P_{TDLS} - E(P_{TDLav} < P_{TDLS})] \end{aligned} \tag{13}$$

$$\begin{aligned} CF_{Utdl}(P_{TDLav} - P_{TDLS}) &= C_{Utdl}(P_{TDLav} - P_{TDLS}) \\ &= C_{Utdl} * f_{tdl}(P_{TDLav} > P_{TDLS}) * [E(P_{TDLav} > P_{TDLS}) - P_{TDLS}] \end{aligned} \tag{14}$$

Where CF_{Otdl} and CF_{Utdl} are the uncertainty cost values for overestimation and underestimation of tidal energy, C_{Otdl} and C_{Utdl} are the uncertainty cost coefficients, and P_{TDLav} is available power of the tidal energy unit.

2.4 Prohibited Operating Zones

A thermal generation unit with prohibited operating zones is formulated as follows [20].

$$\begin{aligned} &P_{THGi,min} \leq P_{THGi} \leq P^{low}_{THGi,1} \\ &P^{upp}_{THGi,r-1} \leq P_{THGi} \leq P^{low}_{THGi,r} \qquad r = 2, 3, \ldots, mi \\ &P^{upp}_{THGi,mi} \leq P_{THGi} \leq P_{THGi,max} \end{aligned} \tag{15}$$

where mi is total number of prohibited operating zones, r is the number of prohibited operating zone, $P^{upp}_{THGi,r-1}$ and $P^{low}_{THGi,r}$ is the upper and lower limit values of the $(r-1)$ th prohibited operating zone of the ith generator.

2.5 Case Studies

- Case 1: Total cost model

$$\begin{aligned} f_{obj}(G,H) = F_{obj1} \\ &= CF_1(P_{THG}) \\ &+ \sum_{i=1}^{NW} \left(CF_{w,i}(P_{WS,i}) + CF_{Ow,i}(P_{WS,i} - P_{wav,i}) \right. \\ &\left. + CF_{Uw,i}(P_{wav,i} - P_{WS,i})\right) \\ &+ (CF_{TDL}(P_{TDLS}) + CF_{Otdl}(P_{TDLS} - P_{TDLav}) \\ &+ CF_{Utdl}(P_{TDLav} - P_{TDLS})) \end{aligned} \tag{16}$$

- Case 2: Total cost model with prohibited zones

The thermal generator's generation power range is considered with prohibited operating zones in Case 2.

3 Wind Power and Tidal Energy Models

Wind speed distribution can be described via Weibull probability density function (PDF) as shown in Eq. 17, where γ and μ are scale and shape factors [18].

$$f_v(v) = \left(\frac{\mu}{\gamma}\right) \times \left(\frac{v}{\gamma}\right)^{\mu-1} \times \left(exp\left(-\left(\frac{v}{\gamma}\right)^{\mu}\right)\right) \quad 0 < v < \infty \tag{17}$$

Power output of a wind farm is expressed as in Eq. 18, where v_r, v_{out}, and v_{in} are the rated, cut-out, and cut-in wind speeds, respectively.

$$p_w(v) = \begin{cases} 0 & v < v_{in} \; and \; v > v_{out} \\ p_{wr}\left(\frac{v - v_{in}}{v_r - v_{in}}\right) & v_{in} \leq v \leq v_r \\ p_{wr} & v_r < v \leq v_{out} \end{cases} \tag{18}$$

If wind speed is below cut-in speed, the probability of wind power is in Eq. 19; if wind speed is beyond cut-out speed, the probability of wind power is in Eq. 20; and if wind speed is between rated speed and cut-out speed, the probability of wind power is in Eq. 21.

$$f_w(p_w)\{p_w = 0\} = 1 - exp\left[-\left(\frac{v_{in}}{\gamma}\right)^{\mu}\right] + exp\left[-\left(\frac{v_{out}}{\gamma}\right)^{\mu}\right] \tag{19}$$

$$f_w(p_w)\{p_w = p_{wr}\} = exp\left[-\left(\frac{v_r}{\gamma}\right)^{\mu}\right] - exp\left[-\left(\frac{v_{out}}{\gamma}\right)^{\mu}\right] \tag{20}$$

$$\begin{aligned} f_w(p_w) = & \left[\frac{\mu(v_r - v_{in})}{\gamma^{\mu} \times p_{wr}}\right] \times \left[v_{in} + \left(\frac{p_w}{p_{wr}}\right)(v_r - v_{in})\right]^{\mu-1} \\ & \times exp\left[-\left(\frac{v_{in} + \left(\frac{p_w}{p_{wr}}\right)(v_r - v_{in})}{\gamma}\right)^{\mu}\right] \end{aligned} \tag{21}$$

The scale and shape factors are set as 10 and 2 at bus 8, and 9 and 2 at bus 13 on the IEEE 30-bus test system. Wind power is the summation of power outputs from 15 turbines and 10 turbines at buses 8 and 13, with the rated power of 3 MW each. Wind speed values are v_{in} = 3 m/s, v_r = 16 m/s, and v_{out} = 25 m/s [18].

In this study, the probability model of discharge rate Q_{TDL} in the tidal range is modelled via the Gumbel distribution as shown in Eq. 22, and its parameters are ζ = 220 and α = 24.52.

$$f_{Q_{TDL}}(Q_{TDL}) = \frac{1}{\alpha} exp\left(\frac{Q_{TDL} - \zeta}{\alpha}\right) exp\left[-exp\left(\frac{Q_{TDL} - \zeta}{\alpha}\right)\right] \tag{22}$$

The conventional thermal generator at bus 13 of the IEEE 30-bus test system is removed and replaced with wind power and tidal energy. The output tidal electrical power in the tidal range can be mathematically expressed as follows [14, 21, 22].

$$P_{TDL}(Q_{TDL}) = \rho g Q_{TDL} H \eta \tag{23}$$

Where ρ is the water density (kg/m^3), g is the gravity acceleration (m/s^2), Q_{TDL} is the discharge value (m^3/s) across the turbine, H = (High water level – Low water level), and η is the turbine's efficiency. Values of these parameters for the tidal range system are ρ = 1025 kg/m^3, g = 9.81 m/s^2, η = 0.85, and H = 3.2 m. Figure 1 shows the commonly used shape of tidal range technology [14].

In Fig. 2, Weibull and Gumbel fittings, and the frequency distributions of wind speed and discharge value of the tidal energy are obtained after running the Monte Carlo simulation with a sample size of 8000.

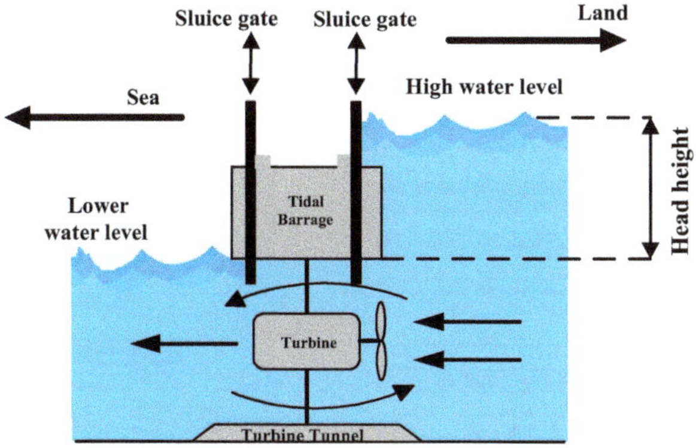

Fig. 1. Tidal range system.

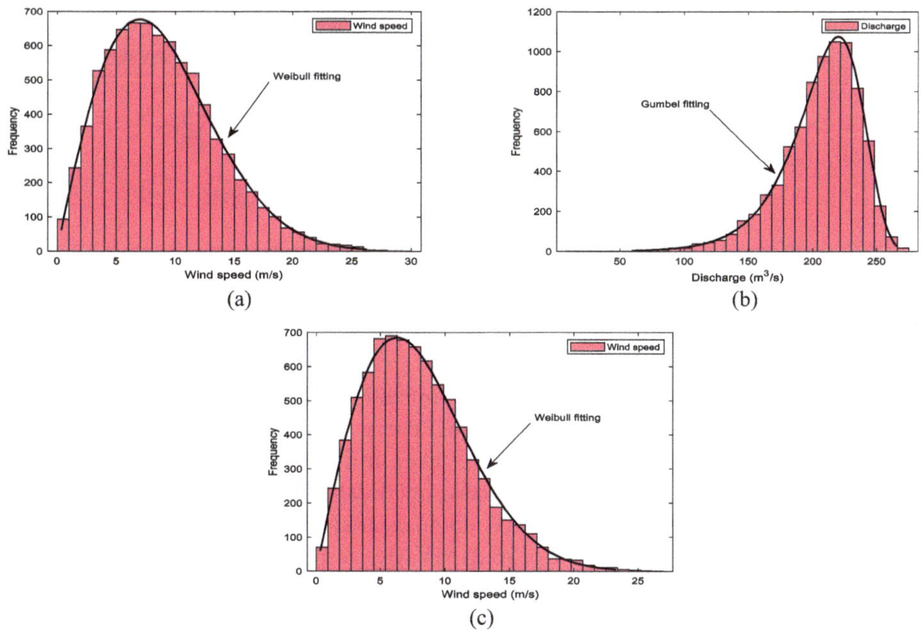

Fig. 2. Wind speed distribution (a) at bus 8, (c) at bus 13, discharge distribution (b) at bus 13.

4 Moth Swarm Algorithm

Ali Mohamed et al., first improved the MSA which is one of the newest optimization algorithms in 2017 [15]. Authors have been inspired by the simulation of the behavior of moths moving toward the moonlight. Possible solutions in the search space are identified as positions of the light source, and the fitness or quality of the potential solution is expressed as the brightness of light source. The concept of moth swarm in

MSA can be explained via three groups: pathfinders, prospectors, and onlookers. Figure 3 shows the brief flowchart of MSA approach.

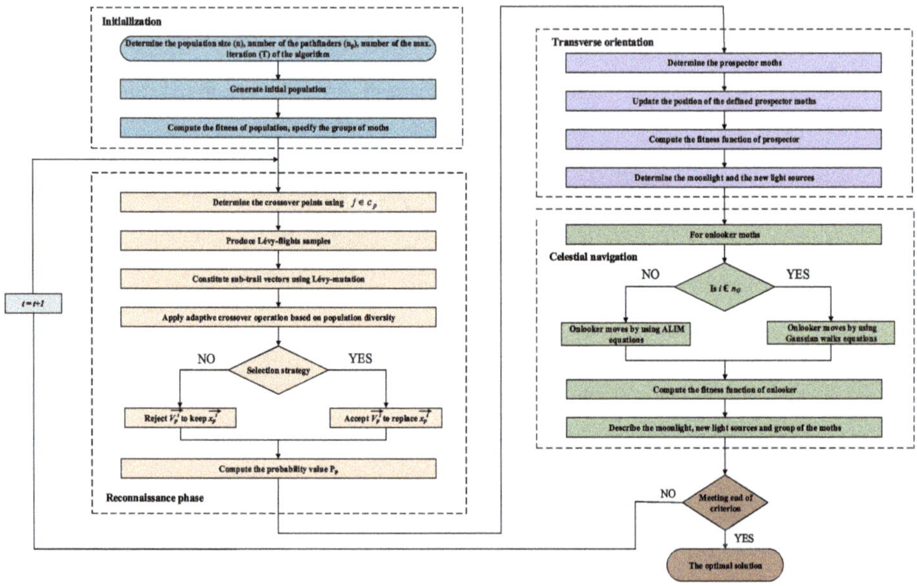

Fig. 3. The flowchart of MSA approach.

5 Simulation Results

In this study, the MSA, BBO, and IWO optimization methods are used to solve the OPF problem with uncertainty models of wind energy and tidal power on an IEEE 30-bus test system. The system parameters are taken from Refs. [5, 23]. The test system has four thermal generators, two wind farms, one tidal energy system, 4 transformers, 41 transmission lines, and 2 shunt VAR compensators. The total active and reactive load demands are 283.4 MW and 126.2 MVAR, respectively. The minimum and maximum voltage values of all load buses are set as 0.95–1.05 p.u. Figure 4 shows the one-line diagram of the modified IEEE 30-bus test system with wind and tidal energy sources. The power flow calculations for the proposed OPF problem are carried out via MATPOWER 6.0 [24, 25]. The direct, overestimation, and underestimation cost coefficients for wind farm in the simulation studies are set as 1.75, 3, and 1.5, at bus 8, and 1.6, 3, and 1.5, at bus 13, respectively. The direct, overestimation, and underestimation cost coefficients for tidal energy in the simulation studies are set as 1.6, 3, and 1.5, at bus 13, respectively. The simulation studies run 30 times, which are summarized as below.

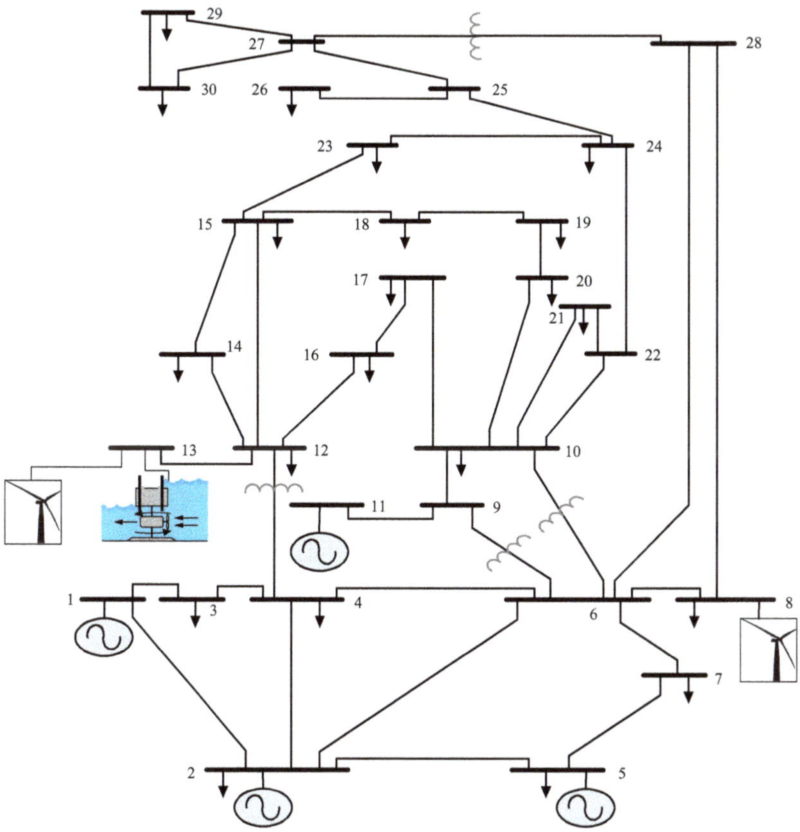

Fig. 4. Modified IEEE 30-bus test system with wind and tidal energy systems.

Test system: The modified IEEE 30-bus test system with wind and tidal energy systems.

- Case 1: The OPF problem with the quadratic cost function for thermal units and cost model of the renewable energy sources
- Case 2: The OPF problem with the consideration of prohibited zones

Table 1 shows optimal solutions of control variables for all cases. The statistical results of the optimization algorithms are shown in Table 2.

The total cost value with quadratic cost function and RESs cost model is minimized in Case 1. Total cost values from the three optimization algorithms are **775.1996 $/h**, 776.4839 $/h, and 780.2355 $/h, respectively. The obtained minimum total cost from the MSA approach is **0.16539%** and **0.64543%** lower than the obtained results from the BBO and IWO approaches, respectively. Minimizing the total cost while considering prohibited zones and cost model of RESs is studied in Case 2. The optimized fitness values by the optimization algorithms are **775.2851 $/h**, 776.6122 $/h, and 777.4018 $/h, respectively. The MSA algorithms is **1.3271 $/h** and **2.1167 $/h** lower

Table 1. The optimized control variables results for all cases.

Parameters	Min.	Max.	Prohibited zones	Case 1			Case 2		
				MSA	BBO	IWO	MSA	BBO	IWO
P_{THG1} (MW)	50	200	[55–66], [80–120]	163.9964	163.6789	165.5061	164.7047	164.4427	160.5967
P_{THG2} (MW)	20	80	[21–24], [45–55]	45.5525	45.8045	33.7567	44.9990	44.6902	44.1349
P_{THG5} (MW)	15	50	[30–36]	20.5480	20.8371	19.7138	20.5581	20.7006	24.6489
P_{WS1} (MW)	0	45		28.3970	28.5036	34.4735	28.4255	28.5226	24.7561
P_{THG11} (MW)	10	30	[25–28]	10.1070	10.2075	10.5547	10.0426	10.2568	11.2907
$P_{WS2+TDL}$ (MW)	0	40		23.2162	23.1320	27.6037	23.1333	23.6193	26.0463
V_{G1} (p.u.)	0.95	1.10		1.0911	1.0509	1.0894	1.0812	1.0440	1.0926
V_{G2} (p.u.)	0.95	1.10		1.0743	1.0292	1.0689	1.0635	1.0228	1.0742
V_{G5} (p.u.)	0.95	1.10		1.0500	0.9933	1.0395	1.0314	0.9953	1.0422
V_{G8} (p.u.)	0.95	1.10		1.0373	0.9988	1.0350	1.0290	0.9984	1.0355
V_{G11} (p.u.)	0.95	1.10		1.0480	1.0822	1.0522	1.0582	1.0954	1.0488
V_{G13} (p.u.)	0.95	1.10		1.0339	1.0598	1.0342	1.0400	1.0527	1.0341
Q_{THG1} (MVAr)	−20	150		6.8615	11.2937	13.6641	7.2502	8.0824	12.0118
Q_{THG2} (MVAr)	−20	60		34.8056	19.7495	30.6576	34.2017	11.9401	36.4101
Q_{THG5} (MVAr)	−15	62.5		39.4679	23.5680	33.2052	30.2863	29.7451	30.4899
Q_{WS1} (MVAr)	−20	30		29.9478	20.7099	29.8084	29.9988	23.4427	29.9623
Q_{THG11} (MVAr)	−10	40		−0.9280	19.3458	1.2802	4.4483	24.7468	−0.2321
$Q_{WS2+TDL}$ (MVAr)	−16	20		−11.4485	7.8296	−10.7038	−7.0628	5.4634	−11.1962
Total Cost ($/h)				**775.1996**	776.4839	780.2355	**775.2851**	776.6122	777.4018
Emission (t/h)				0.2299	0.2290	0.2344	0.2315	0.2308	0.2209
Ploss (MW)				8.4171	8.7635	8.2086	8.4632	8.8322	8.0736
VD (p.u.)				0.7974	0.5004	0.7576	0.7084	0.4811	0.7751
L-index				0.1396	0.1469	0.1402	0.1412	0.1472	0.1400

Table 2. The statistical results of optimization algorithms.

IEEE 30-bus test system						
	MSA		BBO		IWO	
	Case 1	Case 2	Case 1	Case 2	Case 1	Case 2
Min.	**775.1996**	**775.2851**	776.4839	776.6122	780.2355	777.4018
Mean	775.6151	777.4372	776.6925	776.8559	791.2652	792.4765
Max.	776.3752	792.7664	776.7916	777.4195	806.0316	818.8657
Std. dev.	0.32539	3.98221	0.05862	0.2228	6.00546	10.29722

than the obtained results from BBO and IWO approaches, respectively. It is observed from the obtained results that the performance of MSA algorithm is better than the BBO and IWO methods.

Figure 5(a) and (b) show the convergence curves of the MSA, BBO, and IWO for all the cases. It is clearly shown that the MSA approach effectively converges to the optimal solution as compared to the other algorithms. Figure 5(c) and (d) show the voltage profile of all load busses (PQ) for Case 1 and Case 2. The voltage values of load busses are within the limitations.

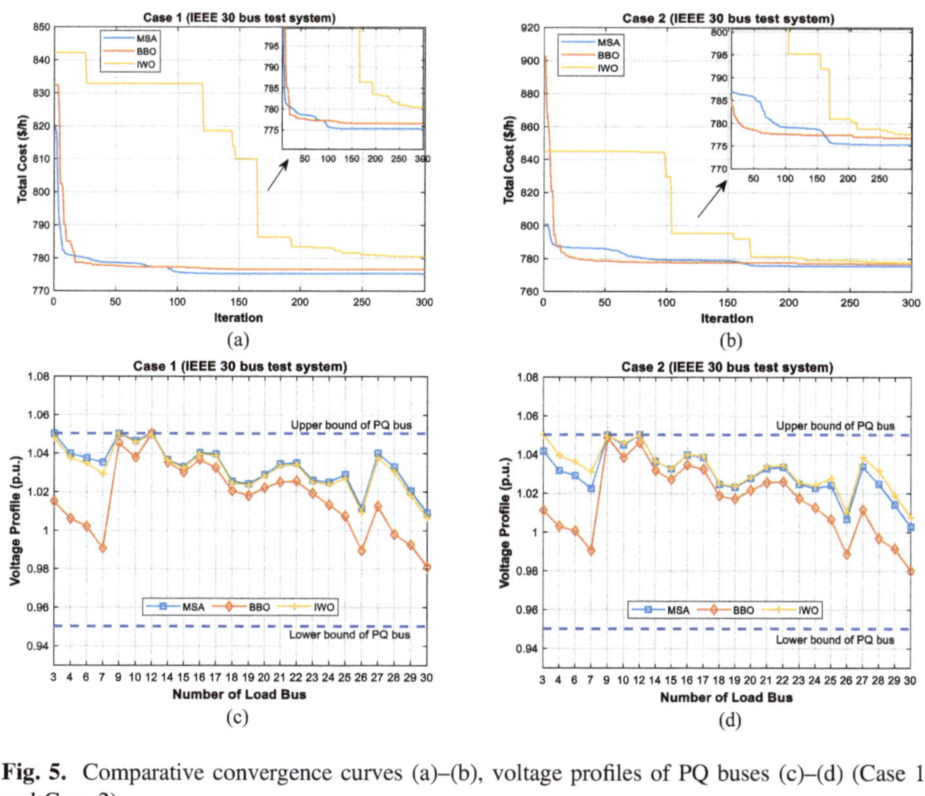

Fig. 5. Comparative convergence curves (a)–(b), voltage profiles of PQ buses (c)–(d) (Case 1 and Case 2).

6 Conclusions

This paper considers the OPF problem with tidal and wind energy, and uncertainties of the tidal energy and wind power are modelled by Gumbel and Weibull probability density functions. The proposed OPF approach is tested on a modified IEEE 30-bus test system with RESs via multiple cases to show its resolvableness and applicability. The obtained results from the MSA approach are compared with those from BBO and IWO optimization algorithms. The simulation results show that the MSA approach presents a better ability in converging to optimal solutions than other algorithms.

Acknowledgments. Dr. Serhat DUMAN would like to thanks the support provided by Scientific and Technological Research Council of Turkey (TUBITAK) BIDEB 2219 Postdoctoral Research Program under application number 1059B191700888.

References

1. El-Fergany, A.A., Hasanien, H.M.: Tree-seed algorithm for solving optimal power flow problem in large-scale power systems incorporating validations and comparisons. Appl. Soft Comput. **64**, 307–316 (2018)
2. Warid, W., Hizam, H., Mariun, N., Wahab, N.I.A.: A novel quasi-oppositional modified Jaya algorithm for multi-objective optimal power flow solution. Appl. Soft Comput. **65**, 360–373 (2018)
3. Saha, A., Das, P., Chakraborty, A.K.: Water evaporation algorithm: a new metaheuristic algorithm towards the solution of optimal power flow. Eng. Sci. Technol. Int. J. **20**, 1540–1552 (2017)
4. Reddy, S.S., Rathnam, C.S.: Optimal power flow using glowworm swarm optimization. Int. J. Electr. Power Energy Syst. **80**, 128–139 (2016)
5. Chaib, A.E., Bouchekara, H.R.E.H., Mehasni, R., Abido, M.A.: Optimal power flow with emission and on-smooth cost functions using backtracking search optimization algorithm. Int. J. Electr. Power Energy Syst. **81**, 64–77 (2016)
6. Mukherjee, A., Roy, P.K., Mukherjee, V.: Transient stability constrained optimal power flow using oppositional krill herd algorithm. Int. J. Electr. Power Energy Syst. **83**, 283–297 (2016)
7. Ayan, K., Kılıç, U., Baraklı, B.: Chaotic artificial bee colony algorithm based solution of security and transient stability constrained optimal power flow. Int. J. Electr. Power Energy Syst. **64**, 136–147 (2015)
8. Joorabian, M., Afzalan, E.: Optimal power flow under both normal and contingent operation conditions using the hybrid fuzzy particle swarm optimisation and Nelder-Mead algorithm (HFPSO-NM). Appl. Soft Comput. **14**, 623–633 (2014)
9. Chang, Y.C., Lee, T.Y., Chen, C.L., Jan, R.M.: Optimal power flow of a wind-thermal generation system. Int. J. Electr. Power Energy Syst. **55**, 312–320 (2014)
10. Panda, A., Tripathy, M.: Optimal power flow solution of wind integrated power system using modified bacteria foraging algorithm. Int. J. Electr. Power Energy Syst. **54**, 306–314 (2014)
11. Panda, A., Tripathy, M.: Security constrained optimal power flow solution of wind-thermal generation system using modified bacteria foraging algorithm. Energy **93**, 816–827 (2015)

12. Kouadri, R., Slimani, L., Bouktir, T., Musirin, I.: Optimal power flow solution for wind integrated power in presence of VSC-HVDC using ant lion optimization. Indonesian J. Electr. Eng. Comput. Sci. **12**(2), 625–633 (2018)
13. Kempener, R., Neumann, F.: Tidal energy: technology brief. International Renewable Energy Agency (IRENA), pp. 1–34 (2014)
14. Safari, N., Mazhari, S.M., Khorramdel, B., Chung, C.Y.: Tidal current and level uncertainty prediction via adaptive linear programming. IEEE Trans. Sustain. Energ. **10**, 748–758 (2018). https://doi.org/10.1109/tste.2018.2846661
15. Mohamed, A.A.A., Mohamed, Y.S., El-Gaafary, A.A.M., Hemeida, A.M.: Optimal power flow using moth swarm algorithm. Electr. Power Syst. Res. **142**, 190–206 (2017)
16. Simon, D.: Biogeography-based optimization. IEEE Trans. Evol. Comput. **12**(6), 702–713 (2008)
17. Mehrabian, A.R., Lucas, C.: A novel numerical optimization algorithm inspired from weed colonization. Ecol. Inform. **1**(4), 355–366 (2006)
18. Biswas, P.P., Suganthan, P.N., Qu, B.Y., Amaratunga, G.A.J.: Multiobjective economic-environmental power dispatch with stochastic wind-solar-small hydro power. Energy **150**, 1039–1057 (2018)
19. Shi, L., Wang, C., Yao, L., Ni, Y., Bazargan, M.: Optimal power flow solution incorporating wind power. IEEE Syst. J. **6**(2), 233–241 (2012)
20. Duman, S.: Symbiotic organisms search algorithm for optimal power flow problem based on valve-point effect and prohibited zones. Neural Comput. Appl. **28**(11), 3571–3585 (2017)
21. Xia, J., Falconer, R.A., Lin, B.: Impact of different operating modes for a Severn Barrage on the tidal power and flood inundation in the Severn Estuary, UK. Appl. Energy **87**, 2374–2391 (2010)
22. Angeloudis, A., Falconer, R.A.: Sensitivity of tidal lagoon and barrage hydrodynamic impacts and energy outputs to operational characteristics. Renewable Energy **114**, 337–351 (2017)
23. IEEE 30-bus test system data. http://www.ee.washington.edu/research/pstca/pf30/pg_tca30bus.htm. Accessed 06 May 2019
24. Zimmerman, R.D., Murillo-Sanchez, C.E., Thomas, R.J.: MATPOWER: steady-state operations, planning, and analysis tools for power systems research and education. IEEE Trans. Power Syst. **26**(1), 12–19 (2011)
25. MATPOWER. http://www.pserc.cornell.edu/matpower/. Accessed 06 May 2019

Churn Analysis with Machine Learning Classification Algorithms in Python

Onur Özdemir[1], Mustafa Batar[2(✉)], and Ali Hakan Işık[1]

[1] Department of Computer Engineering, Burdur Mehmet Akif Ersoy University, Burdur, Turkey
ozdemironur1515@gmail.com, ahakan@mehmetakif.edu.tr

[2] The Graduate School of Natural and Applied Sciences, Dokuz Eylül University, İzmir, Turkey
mbatar@cs.deu.edu.tr

Abstract. "Artificial Intelligence" is the ability of a computer or a computer controlled robot to perform various activities in a similar way within intelligent creatures. The word "AI", which is the acronym of artificial intelligence concept, is also frequently used in informatics. "Artificial Intelligence" studies are generally aimed at developing similar artificial instructions by analyzing human thinking methods. Also, from a point of view, even though it may seem like an attempt by a programmed computer to think, these definitions are changing rapidly, and new orientations are formed towards the concept of "Artificial Intelligence" that can learn and develop independently from human intelligence in the future. Also, "Machine Learning" is a method paradigm that makes inferences from existing data using mathematical and statistical methods and makes predictions about the unknown with these inferences, such as face recognition, document classification, spam detection and etc. In addition, nowadays, many big companies such as Google, Facebook and especially Tesla continue their research on "Artificial Intelligence" and "Machine Learning" under the title of "Churn Analysis". The purpose of "Churn Analysis" is to identify the customers who are likely to stop using their service or product. Today's competitive conditions have formed many companies that try to sell the same product very close to each other. Therefore, the existing customer is very valuable. Customer loss can be a significant loss of revenue, especially in environments where customer preferences do not change very often, such as financial services. Moreover, with the help of "Churn Analysis", customers' escaping probability with precise estimates can be predicted by saying that the customers' leaving probabilities are. Also, some analysis can be made based on customer segments and loss amount – money's worth. After these analyses, communication can be improved and engagement can be increased to keep the customers. Furthermore, effective marketing campaigns can be created in order to target the customers by calculating the rate of churn or wear. This can significantly increase profitability of the companies. In this paper, the terms "Artificial Intelligence" and "Machine Learning" in computer science and the term "Churn Analysis" in global marketing have been tried to be combined together. For this purpose, 7043 (11 missing data – 7032 in total) user data with 21 different features – customerID, Gender, SeniorCitizen, Partner, Dependents, Tenure, PhoneService, MultipleLines, InternetService, OnlineSecurity, Online-Backup, DeviceProtection, TechSupport, StreamingTV, StreamingMovies,

D. J. Hemanth and U. Kose (Eds.): ICAIAME 2019, LNDECT 43, pp. 844–852, 2020.
https://doi.org/10.1007/978-3-030-36178-5_73

Contract, PaperlessBilling, PaymentMethod, MonthlyCharges, TotalCharges and Churn – belonging to Telco Avi (in Telecom Company) have been retrieved and extracted from "Kaggle Data Science". Finally, this user data set has been processed, treated and committed by k-Nearest Neighbors (KNN) Algorithm, Artificial Neural Networks (ANN) Algorithm, Gaussian Naive Bayes Algorithm and Random Forests Algorithm, which are the "Machine Learning Classification Algorithms" in "Artificial Intelligence" – in Python by using its own libraries – Pandas, NumPy, Keras, Sklearn, Matplotlib, Seaborn, etc. – in order to recognize and learn the "Churn Analysis" in the company – Telco Avi.

Keywords: Churn Analysis · Artificial Intelligence · Machine Learning · Classification · k-Nearest Neighbors Algorithm · Artificial Neural Networks Algorithm · Gaussian Naive Bayes Algorithm · Random Forests Algorithm · Python

1 Artificial Intelligence

Software Man-made thinking is a piece of programming designing that hopes to make canny machines. It has transformed into a basic bit of the development business. Research related with man-made cognizance is significantly particular and concentrated. The middle issues of man-made thinking consolidate programming PCs for explicit characteristics, for instance, data, considering, basic reasoning, acknowledgment, getting the hang of, organizing and ability to control and move objects. Data planning is a focal point of AI request about [1]. Machines can much of the time demonstration and react like individuals just if they have plenteous information relating to the world. Automated thinking must approach objects, arrangements, properties and relations between all of them to complete getting the hang of structure. Beginning common sense, considering and basic reasoning force in machines is a troublesome and inauspicious errand. Simulated intelligence is in like manner a focal point of AI. Learning with no kind of supervision requires an ability to recognize structures in surges of data sources, while learning with acceptable supervision incorporates gathering and numerical backslides. Request chooses the class a thing has a spot with and backslide oversees gaining a ton of numerical data or yield models, as such discovering limits enabling the period of fitting yields from discrete wellsprings of information. Logical examination of AI counts and their execution is an inside and out described piece of theoretical programming designing regularly suggested as computational learning speculation [2]. Machine acumen deals with the ability to use material commitments to reason the assorted pieces of the world, while PC vision is the capacity to separate visual commitments with a few sub-issues, for instance, facial, article and flag affirmation. Mechanical innovation is in like manner an essential field related to AI. Robots anticipate that understanding should manage endeavors, for instance, object control and course, close by sub-issues of containment, development masterminding and mapping.

2 Machine Learning

"Machine Learning" is an utilization of man-made brainpower (AI) that gives frameworks the capacity to naturally take in and improve as a matter of fact without being unequivocally customized. AI centers around the improvement of PC programs that can get to information and use it learn for themselves [3]. The way toward learning starts with perceptions or information, for example, models, direct understanding, or guidance, so as to search for examples in information and settle on better choices later on dependent on the precedents that we give. The essential point is to permit the PCs adapt naturally without human mediation or help and alter activities in like manner. ML is a classification of calculation that enables programming applications to turn out to be progressively precise in anticipating results without being unequivocally modified. The fundamental reason of ML is to manufacture calculations that can get input information and utilize factual examination to foresee a yield while refreshing yields as new information winds up accessible. The procedures engaged with ML are like that of information mining and prescient demonstrating. Both require scanning through information to search for examples and altering program activities likewise. Numerous individuals know about ML from shopping on the web and being served advertisements identified with their buy [4]. This happens in light of the fact that proposal motors use ML to customize online advertisement conveyance in practically continuous. Past customized promoting, other regular ML use cases incorporate misrepresentation recognition, spam sifting, arrange security danger location, prescient upkeep and building news sources.

3 k-Nearest Neighbors (KNN) Algorithm

A KNN is one of the algorithms used for classification and regression in Controlled Learning. The simplest machine learning algorithm is considered. Unlike other supervised learning algorithms, it does not have a training phase. Training and testing are almost the same thing. It is a kind of lazy learning. Therefore, KNN is not an ideal candidate for the algorithm required to process a large data set. With KNN, one is basically looking for points that are closest to the new spot [5]. K represents the amount of the nearest neighbors of the unknown point. To estimate the results, the k amount (usually an odd number) of the algorithm is chosen. On account of model acknowledgment, the closest neighbor calculation (KNN) is a nonparametric technique utilized for characterization and relapse. In the two cases, the information comprises of the nearest preparing precedents in the property field. The yield relies upon whether KNN is utilized for arrangement or relapse: In the KNN characterization, the yield is class participation. An article is grouped by a greater part vote of its neighbors; the item is given to the most well-known class among its nearest neighbors (k, a little positive whole number). On the off chance that k is 1; the article is just doled out to the class of the closest neighbor. In KNN relapse, the yield is the property estimation of the item. This esteem is the normal of the estimations of the closest neighbors. KNN is design based learning or languid realizing; where the capacity is approximated just locally and all figuring is conceded until characterization. The K-N calculation is one of the least complexes of all AI calculations. For both arrangement and relapse, it might be helpful to add weight to the commitments of neighbors, with the goal that the neighbors contribute more to the normal than to the ones far away. For instance, a typical weighting plan involves giving a weight of 1/d to each neighbor; where d is the separation to the neighbor [6]. Neighbors are taken from various items in which the class (for KNN grouping) or article property estimation (for KNN relapse) is known. This can be considered as the preparation set for the calculation, yet no express preparing venture is required. One feature of the KNN algorithm is that it is sensitive to the local structure of the data. The algorithm should not be confused with k-means, another popular machine learning technique. K neighbors are a simple algorithm that stores all existing cases and classifies new cases based on a similarity measure (e.g. distance functions). KNN was used in statistical estimation and pattern recognition as a nonparametric technique in the early 1970s.

4 Artificial Neural Networks Algorithm

Neural frameworks are a great deal of counts, showed uninhibitedly after the human personality, that are planned to see structures. They disentangle unmistakable data through a kind of machine wisdom, naming or batching unrefined data. The models they see are numerical, contained in vectors, into which all authentic data, be it pictures, sound, substance or time course of action, must be deciphered. Neural systems help us bunch and arrange [7]. One can consider them a grouping and characterization layer over the information you store and oversee. They help to aggregate unlabeled information as indicated by similitudes among the precedent sources of info, and they order information when they have a marked dataset to prepare on. (Neural systems can likewise remove includes that are nourished to different calculations for grouping and arrangement; so one can consider profound neural systems as segments of bigger AI applications) (Fig. 1).

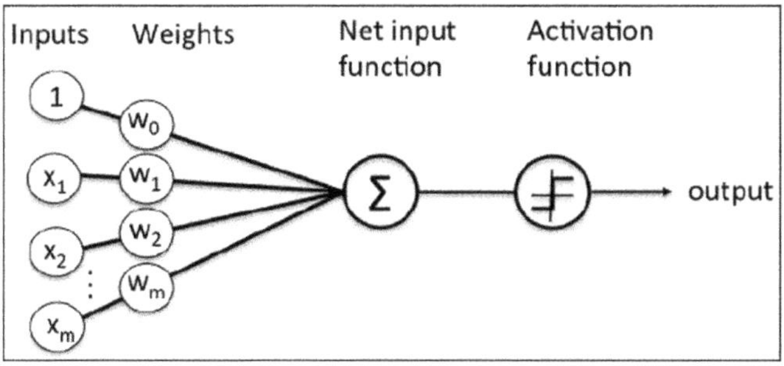

Fig. 1. Artificial Neural Networks mechanism

5 Gaussian Naive Bayes Algorithm

The Naive Bayes classifier is named after the English mathematician Thomas Bayes (ca. 1701–7 April 1761). Naive Bayes Classifier It is a probabilistic approach that can be used with a very restrictive proposition at first glance the pattern recognition problem. This proposition is that each identifier attribute or parameter to be used in pattern recognition must be statistically independent. Although this suggestion restricts the use of the Naive Bayes classifier [8], it is generally comparable to more complex artificial neural networks when the statistical independence condition is used. A Naive Bayes classifier can also be considered as a Bayesian network where each attribute is conditionally independent and the concept to be learned is conditionally dependent on all these attributes (Fig. 2).

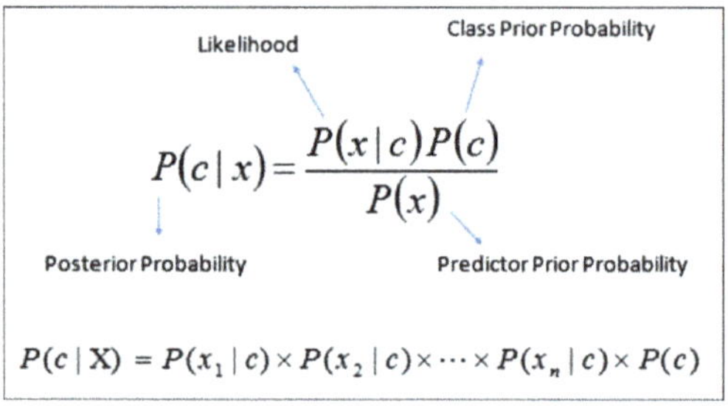

Fig. 2. Mathematical demonstration of Gaussian Naive Bayes Algorithm

6 Random Forests Algorithm

The Arbitrary timberlands or irregular choice woods are a group learning technique for characterization, relapse and different assignments that works by building a huge number of choice trees at preparing time and yielding the class that is the method of the classes (arrangement) or mean forecast (relapse) of the individual trees. Random choice backwoods right for choice trees' propensity for overfitting to their preparation set. The principal calculation for arbitrary choice backwoods was made by Tin Kam Ho utilizing the irregular subspace method, which, in Ho's plan, is an approach to execute the "stochastic segregation" way to deal with arrangement proposed by Eugene Kleinberg. An expansion of the calculation was created by Leo Breiman and Adele Cutler, who registered "Arbitrary Forests" [9] as a trademark (starting at 2019, claimed by Minitab, Inc.). The augmentation joins Breiman's "packing" thought and irregular determination of highlights, presented first by Ho and later freely by Amit and Geman so as to build a gathering of choice trees with controlled fluctuation (Fig. 3).

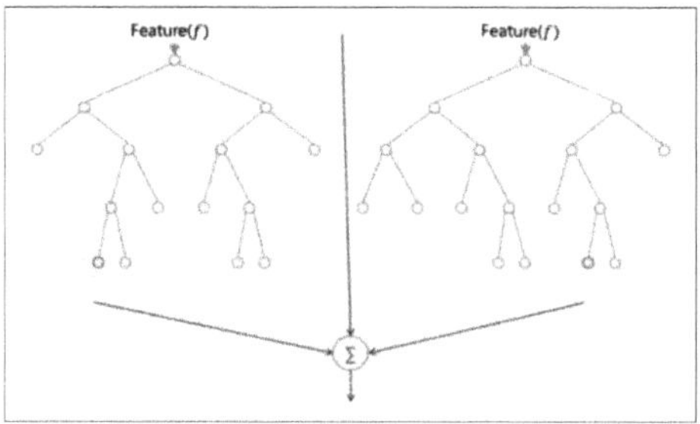

Fig. 3. Random Forests mechanism

7 Churn Analysis

Client stir alludes to when a client (player, endorser, client, and so forth.) stops his or her association with an organization. Online organizations regularly treat a client as agitated once a specific measure of time has slipped by since the client's last cooperation with the webpage or administration. The full expense of client stir incorporates both lost income and the showcasing costs required with supplanting those clients with new ones. Diminishing client agitate is a key business objective of each online business [10]. The capacity to anticipate that a specific client is at a high danger of agitating, while there is still time to take care of business, speaks to a colossal extra potential income hotspot for each online business. Other than the immediate loss of income that outcomes from a client deserting the business, the expenses of at first getting that client might not have just been secured by the client's spending to date. (At the end of the day, obtaining that client may have really been a losing speculation.) Furthermore, it is in every case increasingly troublesome and costly to get another client than it is to hold a current paying client. So as to prevail at holding clients who might some way or another relinquish the business, advertisers and maintenance specialists must have the capacity to (a) foresee ahead of time which clients are going to stir through beat examination and (b) realize which showcasing activities will have the best maintenance sway on every specific client. Equipped with this information, a substantial extent of client agitate can be disposed of. While basic in principle, the substances required with accomplishing this "proactive maintenance" objective are amazingly testing. Stir forecast displaying methods endeavor to comprehend the exact client practices and properties which flag the hazard and timing of client beat [11]. The precision of the procedure utilized is clearly basic to the achievement of any proactive maintenance endeavors. All things considered, if the advertiser is uninformed of a client going to beat, no move will be made for that client. Also, unique maintenance centered offers or motivating forces might be unintentionally given to upbeat, dynamic clients, bringing about diminished incomes without any justifiable cause. Tragically, the majority of the agitate expectation displaying techniques depend on measuring hazard dependent on static information and measurements, i.e., data about the client as the person in question exists at this moment. The most widely recognized stir expectation models depend on more seasoned factual and information mining strategies, for example, calculated relapse and other parallel demonstrating systems. These methodologies offer some esteem and can distinguish a specific level of in danger clients, however they are generally mistaken and end up leaving cash on the table. Optimove utilizes a more current and unmistakably increasingly exact way to deal with client agitate forecast: at the center of Optimove's capacity to precisely anticipate which clients will beat is an interesting technique for computing client lifetime esteem (LTV) for every single client. The LTV determining innovation incorporated with Optimove depends on cutting edge scholastic research and was additionally created and improved over various years by a group of top notch PhDs and programming designers. This technique is fight verified as a precise and successful methodology in a wide scope of enterprises and client situations. Without uncovering a lot about the "mystery sauce" of Optimove's client stir expectation innovation, the methodology joins nonstop powerful smaller scale division and a one of a kind, scientifically serious prescient conduct demonstrating framework. The previous brilliantly and naturally fragments the whole client base into a various leveled structure of ever-littler social statistic portions. This division is dynamic and refreshed ceaselessly dependent on changes in the information. The last depends on the way that the personal conduct standards of individual clients often change after some time. As it were, the "section course history" of every client is a critical factor deciding when and why the client may agitate [11]. By consolidating the most demanding miniaturized scale division accessible anyplace with a profound comprehension of how clients move starting with one small scale portion then onto the next after some time – including the capacity to anticipate those moves before they happen – an extraordinary level of beat examination exactness is feasible.

8 Churn Analysis Application in Python

In this paper, the terms "Artificial Intelligence" [12] and "Machine Learning" [13] in computer science and the term "Churn Analysis" in global marketing have been tried to be combined together. For this purpose, 7043 (11 missing data – 7032 in total) user data with 21 different features – customerID, Gender, SeniorCitizen, Partner, Dependents, Tenure, PhoneService, MultipleLines, InternetService, OnlineSecurity, OnlineBackup, DeviceProtection, TechSupport, StreamingTV, StreamingMovies, Contract, PaperlessBilling, PaymentMethod, MonthlyCharges, TotalCharges and Churn – belonging to Telco Avi (in Telecom Company) have been retrieved and extracted from "Kaggle Data Science". Finally, this user data set has been processed, treated and committed by k-Nearest Neighbours (KNN) Algorithm, Artificial Neural Networks (ANN) Algorithm, Gaussian Naive Bayes Algorithm and Random Forests Algorithm, which are the "Machine Learning Classification Algorithms" in "Artificial Intelligence" – in Python by using its own libraries – Pandas, NumPy, Keras, Sklearn, Matplotlib, Seaborn, etc. – in order to recognize and learn the "Churn Analysis" in the company – Telco Avi (Fig. 4).

```
Out[15]: customerID          int64
         gender              int64
         SeniorCitizen       int64
         Partner             int64
         Dependents          int64
         tenure              int64
         PhoneService        int64
         MultipleLines       int64
         InternetService     int64
         OnlineSecurity      int64
         OnlineBackup        int64
         DeviceProtection    int64
         TechSupport         int64
         StreamingTV         int64
         StreamingMovies     int64
         Contract            int64
         PaperlessBilling    int64
         PaymentMethod       int64
         MonthlyCharges    float64
         TotalCharges      float64
         Churn               int64
         dtype: object
```

Fig. 4. Churn Analysis with 21 different features (input)

With KNN:
The obtained accuracy rate is 78.10% with Euclidean Distance.
The obtained accuracy rate is 78.48% with Minkowski Distance
The obtained accuracy rate is 78.48% with Manhattan Distance.
The obtained accuracy rate is 77.96% with Chebyshev Distance.
With ANN:
The Obtained Accuracy Rate is 80.33% with 24th Architecture in the 45th loop.
With Gaussian Naive Bayes:
The obtained accuracy rate is 75.29%.
With Random Forests:
The obtained accuracy rate is 78.41%.

9 Conclusion

In this paper, "Machine Learning Classification Algorithms" – k-Nearest Neighbours (KNN) Algorithm, Artificial Neural Networks (ANN) Algorithm, Gaussian Naive Bayes Algorithm and Random Forests Algorithm – under the term "Artificial Intelligence" have been tried to be combined with the "Churn Analysis" under the term "Global Marketing and Competition" in order to classify and explain available and possible customers' behaviors in the area of Telecom Marketing. These determined four algorithms have been developed and applied with the help of Python libraries into the dataset from "Kaggle Data Science". These algorithms have given the results with more than 75% accuracy rate. The maximum accuracy rate has been given with ANN with the result 80.33%, and the second maximum accuracy rate has been given with KNN with the result 78.48%, and the third maximum accuracy rate has been given with Random Forests with the result 78.41%, and the last maximum accuracy rate has been given with Gaussian Naive Bayes with the result 75.29%. With these respectively successful results, it can be easily said that "Churn Analysis" has been understood more clearly and more tangible, and "Machine Learning" under "Artificial Intelligence" has been important for analyzing and determining possible customers in the global marketing, especially under the area of telecommunication.

References

1. Nilsson, N.: Principles of Artificial Intelligence, 1st edn. Elsevier, Amsterdam (1982)
2. Russell, S., Norvig, P.: Artificial Intelligence: A Modern Approach, 3rd edn. Pearson Education, Upper Saddle River (2009)
3. Michie, D., Spiegelhalter, D.J., Taylor, C.C.: Machine Learning, Neural and Statistical Classification. Overseas Press, Ahmedabad (1994)
4. Witten, I.H., Frank, E., Hall, M.A., Pal, C.J.: Data Mining: Practical Machine Learning Tools and Techniques, 4th edn. Elsevier, Amsterdam (2017)
5. Deng, Z., Zhu, X., Cheng, D., Zong, M., Zhang, S.: Efficient kNN classification algorithm for big data. Neurocomputing **195**, 143–148 (2016)
6. Zhang, M.-L., Zhou, Z.-H.: A k-nearest neighbor based algorithm for multi-label classification. In: International Conference on Granular Computing. IEEE (2005)
7. Hassoun, M.H.: Fundamentals of Artificial Neural Networks. MIT Press, Cambridge (1995)
8. Rish, I.: An empirical study of the Naive Bayes classifier (2001)
9. Liaw, A., Wiener, M.: Classification and regression by randomForest. R News **2**, 18–22 (2001)
10. Mutanen, T.: Customer churn analysis – a case study (2006)
11. Gürsoy, U.Ş.: Customer churn analysis in telecommunication sector (2010)
12. Torse, D., Desai, V., Khanai, R.: A review on seizure detection systems with emphasis on multi-domain feature extraction and classification using machine learning. BRAIN **8**(4), 109–129 (2017)
13. Muntean, M., Ileană, I., Rotar, C., Rîşteiu, M.: Data mining learning models and algorithms on a SCADA system data repository. BRAIN **1**, 91–99 (2010)

Real Time Performance Comparison of Multi-class Deep Learning Methods at the Edge

Doruk Sonmez(✉) and Aydin Cetin(✉)

Gazi University, 06500 Ankara, Turkey
{doruk.sonmez,acetin}@gazi.edu.tr

Abstract. Nowadays, machine learning algorithms are used intensively for the detection and identification of various entities on an image with high level of accuracy. Deep learning provides even higher performance results in image processing. However, deep learning requires high computing effort for classification and thus limiting the usability of method in AI at the edge especially in the presence of different classes. In this paper, deep learning is applied as a solution to the problem of detecting and identifying more than one entity of different classes in the image in real time at the edge. In the study, a multi-class traffic sign data set was used. The data set consists of six different sign classes and 6000 high resolution images (1280 × 720). Performance of two models, DetectNet and MobileNet (MobiNet) models were compared. MobileNet had better accuracy compared to DetectNet since it has higher number of parameters. But high number of parameters lessens the real time performance of the model at the edge. According to the tests performed on the Nvidia Jetson TX2 embedded system, the DetectNet model can provide 6–7 FPS performance while the MobileNet 160 version of MobileNet model with $\propto = 0.25$ has achieved only of 1–2 FPS.

Keywords: Deep learning · Image processing · Object detection · Image recognition · Image classification · Artificial intelligence

1 Introduction

The "data" is the most important component when working on artificial intelligence applications. There are some perceivers and sensors as data source. In this manner, the surveillance sensors are the primary sensors and the most used ones in order to solve real world image-based problems. The high definition images provided by camera sensors brings big advantages to solve problems when working on AI based applications. In this paper, camera images are the main source to work with and will be the most critical information for the AI model.

It is never enough to create a decision mechanism with running object detection algorithms alone on some camera feed. Because AI model's performance changes on every different conditions like lightning and camera angle, one cannot be purely sure of the model's results. In this paper, a classification model to verify the detection results and different data augmentation techniques -as part of image processing- to improve model's accuracy is used in order to overcome this drawback.

D. J. Hemanth and U. Kose (Eds.): ICAIAME 2019, LNDECT 43, pp. 853–867, 2020.
https://doi.org/10.1007/978-3-030-36178-5_74

The hardware with high computing capabilities are not effective generally when it comes to small usage areas and mobility-needed situations in the manner of their sizes. As a result, "AI at the edge" idea came across and concepts like AI, deep learning and image processing transferred to the embedded systems that is light and has low power consumption. However, reduction on compute capability of the system due to much less hardware is a side effect of improving the mobility. Although there are many different types and capabilities of embedded systems, NVIDIA Jetson TX2 Developer Kit is used since this application needs high system resource and GPU.

The deep learning model proposed in this paper fills an important gap by a multiclass object detection model that is developed and optimized specially for AI at the edge devices. On the other hand, since it is hard to find a dataset with healthy data and labels to detect multiple traffic signs within an image, a fully custom six class traffic signs dataset has been produced to train the proposed deep learning model.

2 Theoretical Background

Before deep diving to deep learning, it is important to recall the most commonly used image processing techniques for deep learning. The primary technique to extract some features from an image in deep learning applications would be convolution approach. That is where the deep learning shows up and create the right parameters to extract the right features from convolution process.

2.1 Convolutional Neural Networks

Convolutional neural networks also known as CNNs, became one of the state of the art techniques when it is applied inside AlexNet on ILSVRC 2012 competition. Nowadays, convolution is one of the mostly used techniques in deep learning for computer vision. The main drawback with this solution is that it needs hardware like GPUs or TPUs that has high compute capabilities. However, it is easier to get a hands-on experience and work with GPUs now and this enables developers to work with such image processing techniques. Figure 1 shows the CNN architecture overview on traffic sign classification task.

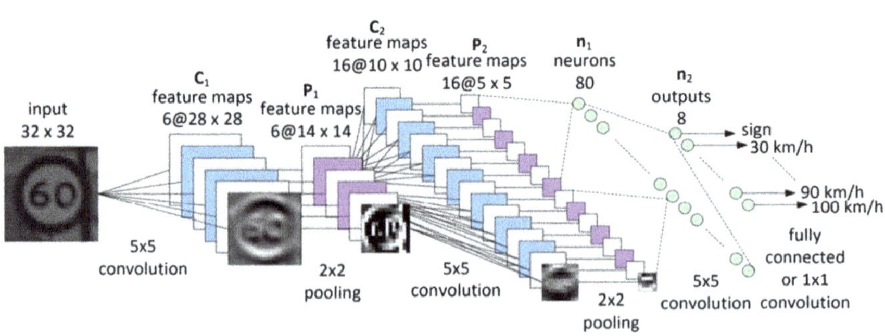

Fig. 1. CNN architecture overview on traffic sign classification task

The commonly used neural network architectures includes the following layers: input layer, hidden layers and output layer. An input vector enters the network from input layer and that input information propagates forwardly through the network. Each of the neurons placed on the hidden layers connected with the following layers. As a result of some series of calculations made on the hidden layers, weights of the deep learning model gets updated until the most optimum ones found. In the output layer of the network, one can see the output values for each class in the number of class count.

2.1.1 Convolution Layer

Just before the neural network part mentioned above, there are some extra layers that does all of the convolution and pooling workloads in order to extract specific features from an image. The names of these two layers are as follows: convolution layer and pooling layer. In convolution layers, the main convolution process takes place and this process first mentioned by LeCun [1]. Convolution process shown by * symbol generally.

$$s(t) = (x * w)(t) \tag{1}$$

In convolutional neural network terminology, the first parameter (x) represents input, the second one represents filter (w), the third one represents time (t) and the last parameter (s) is represents feature map as output. Assuming that the x and w functions are defined only on the integer value, discrete convolution equation would be as follows [2]:

$$s(t) = (x * w)(t) = \sum_{a=-\infty}^{\infty} x(a)w(t-a) \tag{2}$$

In many machine learning applications, the input is a multi-dimensional array and filter is a multi-dimensional parameter array provided by learning algorithm. In general, convolution is applied on multiple axes at a given time. Thus, if one 2-dimensional input image I is used, there should also be one 2-dimensional K kernel. The equation below describes this explanation in a mathematical manner [2]:

$$S(i,j) = (I * K)(i,j) = \sum_m \sum_n I(m,n)K(i-m, j-n) \tag{3}$$

The i and j expressions included in Eq. 3 describes the positions of new matrices which is a result of convolution and the other m and n expressions describes the column and row axes as it can be seen on Fig. 2.

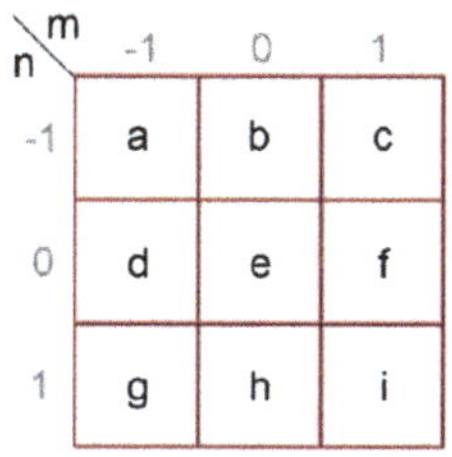

Fig. 2. A sample filter (kernel)

Figure 3 illustrates a convolution process sample:

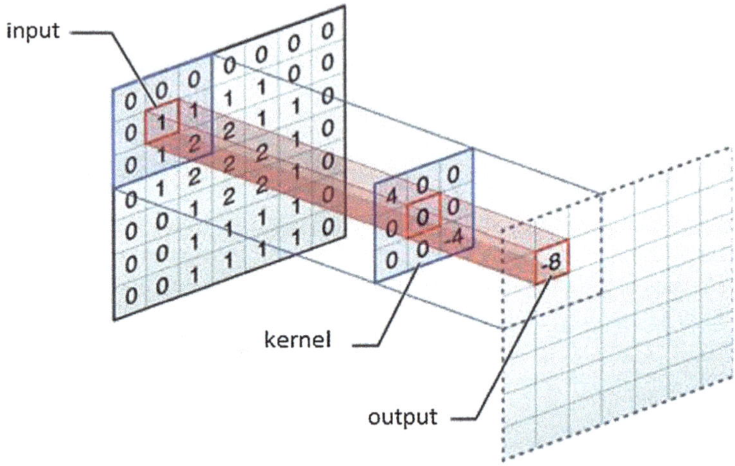

Fig. 3. Convolution process illustration [3]

2.1.2 Pooling Layer

In the pooling layers of a CNN based architecture, sizes of the resulting images of convolution layers is reduced in order to ease the computational cost with using functions like max and mean. The main process takes place in these layers can be described as follows:

1. Either one of the average pooling or max pooling methods is applied on an area specified as filter (kernel) size.
2. As the filter slides, the number of columns specified as stride number, the output produced by one of the methods mentioned above is added to the output layer [4].

An example max pooling and mean pooling processes illustrated in the following Figs. 4 and 5 respectively (It is important to note that size of the input image is 5 × 5, kernel size 3 × 3 and the stride is 1 × 1):

The mathematical relationship between pooling layer and its output's size (o), input's size (i), kernel size (k) and stride (s) is described in the Eq. (4).

$$o = \left[\frac{i - k}{s}\right] + 1 \quad (4)$$

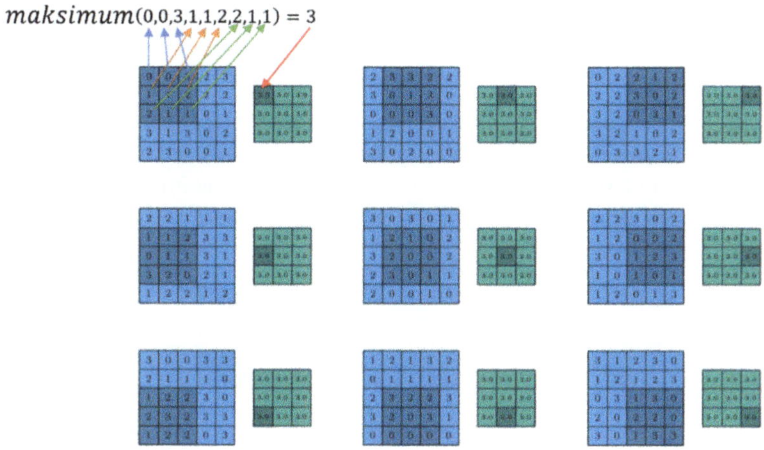

Fig. 4. Maximum pooling [5]

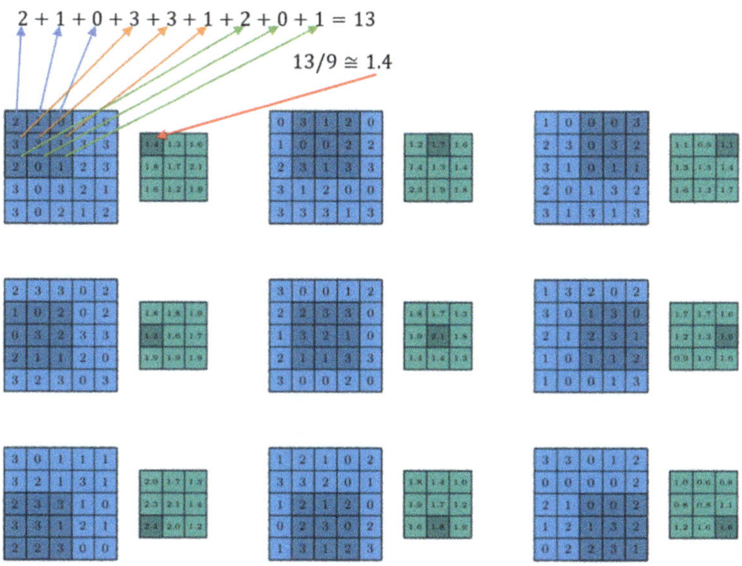

Fig. 5. Mean pooling [5]

2.1.3 Fully Connected Layer

This layer has the functionality of neural networks and works that way. The output vector of early repetitive convolution and pooling layers is fed to a fully connected layer as input and math operations specified within activation functions applied to them afterwards. What fully connected layer's output is scores for each predicted class. This way, the same number of output results with number of classes.

3 Solution of the Proposed Object Detection on the Edge Devices

In this paper, we propose an improvement to an existing object detection method called DetectNet [6]. In the first stage of the proposed technique, a fully custom dataset is created which includes 2000 images in 1280 $\times$ 720 size that contains traffic signs that belongs 6 different classes. These classes are listed in Fig. 6.

Fig. 6. Traffic sign classes

Since 2000 image is not enough to close to train a DetectNet object detection model that has high accuracy, two different augmentation techniques as part of image processing is applied on the dataset. As a result, data number increased to 6000 images. Blur is one of the augmentation technique was used. In this technique, a kernel slides through on a given input image. The main process occurs taking the average pixel value on an area that kernel covers (Fig. 7).

$$K = \frac{1}{9}\begin{bmatrix} 1 & 1 & 1 \\ 1 & 1 & 1 \\ 1 & 1 & 1 \end{bmatrix} \tag{5}$$

Fig. 7. Raw image (on the left) and blurred image (on the right) comparison

A second approach to augmentation in this paper is called Gaussian Blur. In this method, a Gaussian function is used to calculate transformation level for each pixel on a given image. An example Gaussian distribution function for two dimensional image is shown below (Fig. 8).

$$G(x, y) = \frac{1}{2\pi\sigma^2} e^{-\frac{x^2+y^2}{2\sigma^2}} \quad (6)$$

Fig. 8. Raw image (on the left) and Gaussian-blurred image (on the right) comparision

3.1 Multi Class DetectNet Training for Optimized Object Detection

DetectNet is a convolutional neural network that is created by NVIDIA and is developed on top of GoogleNet object detection model. The main advantage of DetectNet is that it is an object detection model that is fully optimized for embedded systems used in edge applications such as Jetson TX2 which is also developed and produced by NVIDIA. On the other hand, the main drawback of DetectNet is that the model itself is only for single-class problems as it said by NVIDIA but it is important to note that the model successfully trained on a two-classed dataset by a couple of community users and this shows that the DetectNet supports multi-class training. There are no complete support or documentation to come up with a model that is trained on a dataset that has more than two class though. This paper aims to fill this blank with doing some ‘brain surgery’ on DetectNet model since it is a neural network at it’s base.

Data labeling is valid with this solution too when it comes to first step when developing a deep learning model. DetectNet requires that the dataset has to be in KITTI Database format. Since it is a main requirement, data is labeled by hand for a period of time and one example training data can be found below to get some insight on KITTI format and labeling procedure.

Fig. 9. Labeled image (on the left) and it’s label file (on the right)

5142 training and 858 validation images used to train DetectNet. However, before starting the training procedure, the main ‘brain surgery’ operation has to be done. The solution mentioned below is applied to achieve this.

A single class DetectNet model shown in Fig. 10, outputs three components for one class to evaluate the model. These components are *mAP*, *Recall* and *Precision* values. Since this paper involves multi class DetectNet, this output layer number must be increased so that the model can output *mAP*, *Recall* and *Precision* values separately for each class.

In order to increase the number of units that outputs object detection scores, the block inside the marked are needs to be multiplied by the number of classes. Each geometric shape represents the different layer types of the deep neural network which is named as DetectNet. Since DetectNet is derived from GoogleNet fully convolutional network (FCN), no changes needed to be done on earlier layers because those layers are responsible for creating the right filters to extract features such as corners, colors, shapes and finally high level features like letters, windshield of a car etc. Increasing the number of classes comes with additional computation load. Meaning, if one increases the number of any layers to get extra outputs, this process will also increase the connection between layers and this will cost extra computations to be made in the process of training.

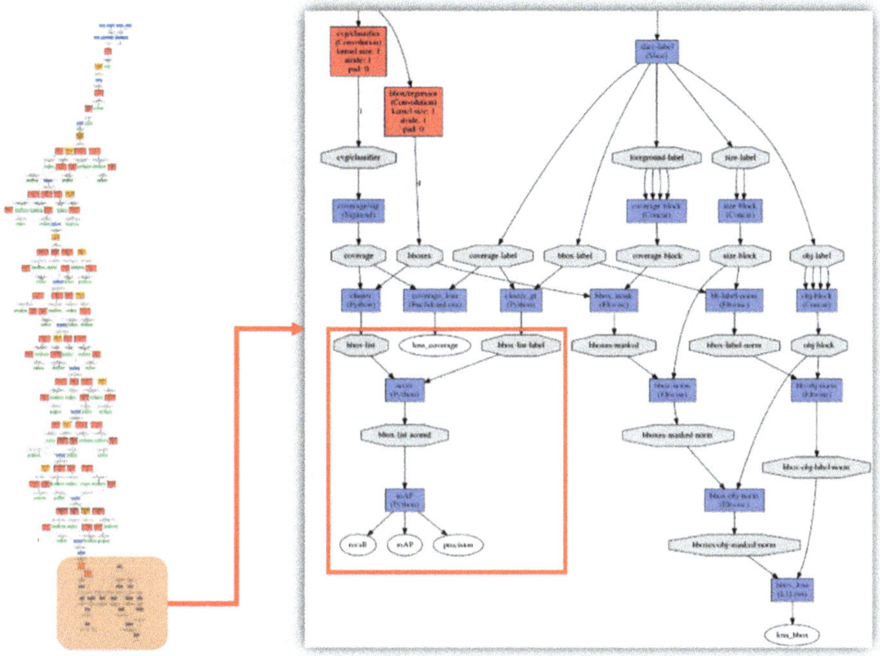

Fig. 10. Output layer of the original DetectNet model

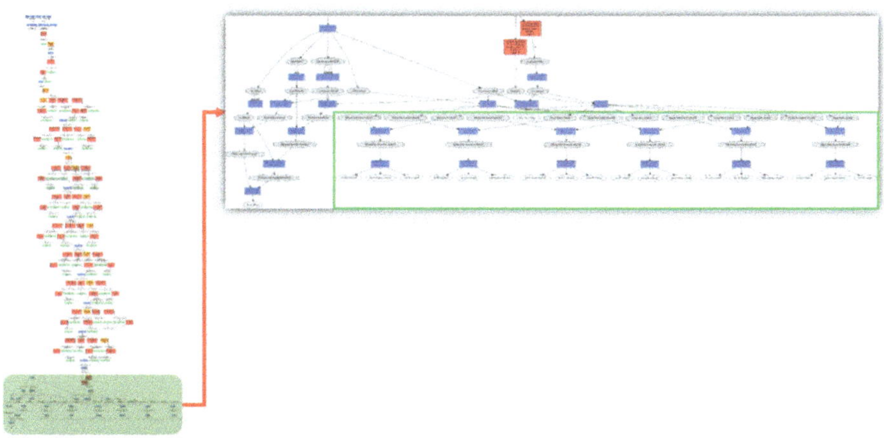

Fig. 11. Output layers of the modified DetectNet model

Figures 10 and 11 are the illustrations of changes that made on model's Caffe code. After the new connections adjusted and got the graphical representation of the model correctly, model training needs to be done so that new connections between new layers can generalize to new custom dataset well. Despite starting the training with a pre-trained model is very time effective, there are no multiclass pre-trained DetectNet model. Thus, whole model trained from scratch on fully custom dataset which is also proposed in this paper after the layer additions.

After the computational load that is bringed by extra output layers, a GPU accelerated cloud instance became a necessary resource. An instance with 4 Nvidia Tesla K80s is used to train the model and even then, it took six days and eleven hours to complete 600 epochs of training. Resulting mAPs can be found on Fig. 12 below.

Fig. 12. Training results of multiclass DetectNet model

The reason why 411st epoch is chosen to be displayed is that among 600 epochs the 411st has the highest overall mAP value. On the other hand, one can see that mAP values differs from class to class since every class has different number of data. The class that has the highest mAP value is traffic_light (mAP-class5) class because of the number of data it has.

3.2 Multi Class MobileNet Training for Object Detection

MobileNet [7] is a widely used and open source deep neural network architecture developed by Google. It can be used in both classification and object detection applications because of it's open sourced backend and easy implementation procedures. Since we propose a comparison between object detection models against limited resources that most embedded systems have, MobileNet would be one of the best candidates against DetectNet because of it's mobile effectiveness that proposed in the original paper. The main reason of implementing a fully open source neural network project in this paper is benchmarking the custom dataset on different kinds of neural networks.

The layer architecture of the MobileNet can be found on Table 1.

Table 1. Layer architecture of MobileNet [7]

Layer type/stride	Filter shape	Input size
Conv/s2	$3 \times 3 \times 3 \times 32$	$224 \times 224 \times 3$
Conv dw/s1	$3 \times 3 \times 32$ dw	$112 \times 112 \times 32$
Conv/s1	$1 \times 1 \times 32 \times 64$	$112 \times 112 \times 32$
Conv dw/s2	$3 \times 3 \times 64$ dw	$112 \times 112 \times 64$
Conv/s1	$1 \times 1 \times 64 \times 128$	$56 \times 56 \times 64$
Conv dw/s1	$3 \times 3 \times 128$ dw	$56 \times 56 \times 128$
Conv/s1	$1 \times 1 \times 128 \times 128$	$56 \times 56 \times 128$
Conv dw/s2	$3 \times 3 \times 128$ dw	$56 \times 56 \times 128$
Conv/s1	$1 \times 1 \times 128 \times 256$	$28 \times 28 \times 128$
Conv dw/s1	$3 \times 3 \times 256$ dw	$28 \times 28 \times 256$
Conv/s1	$1 \times 1 \times 256 \times 256$	$28 \times 28 \times 256$
Conv dw/s1	$3 \times 3 \times 256$ dw	$28 \times 28 \times 256$
Conv/s1	$1 \times 1 \times 256 \times 256$	$28 \times 28 \times 256$
Conv dw/s2	$3 \times 3 \times 256$ dw	$28 \times 28 \times 256$
Conv/s1	$1 \times 1 \times 256 \times 512$	$14 \times 14 \times 256$
Conv dw/s1 $5 \times$ Conv/s1	$3 \times 3 \times 512$ dw $1 \times 1 \times 512 \times 512$	$14 \times 14 \times 512$ $14 \times 14 \times 512$
Conv dw/s2	$3 \times 3 \times 512$ dw	$14 \times 14 \times 512$
Conv/s1	$1 \times 1 \times 512 \times 1024$	$7 \times 7 \times 512$
Conv dw/s2	$3 \times 3 \times 1024$ dw	$7 \times 7 \times 1024$
Conv/s1	$1 \times 1 \times 1024 \times 1024$	$7 \times 7 \times 1024$
Avg Pool/s1	Pool 7×7	$7 \times 7 \times 1024$
FC/s1	1024×1000	$1 \times 1 \times 1024$
Softmax/s1	Classifier	$1 \times 1 \times 1000$

Different neural network types mean different data labeling procedures. Thus, a few changes made on the label type that can be seen on Fig. 9. To train a Keras (using TensorFlow [8] backend) based MobileNet model, the labeling needs to be converted to CSV based label file shown in Fig. 13 below. Every row of the CSV file respectively contains full path to image, four coordinates that includes the object and the class name of the marked area.

1	/home/keras_detect/racecar/Training_Data/annotate_sign/gaus_blurred3.jpg,1109,306,1154,348,sag_yasak
2	/home/keras_detect/racecar/Training_Data/annotate_sign/blurred108.jpg,585,268,638,317,girilmez
3	/home/keras_detect/racecar/Training_Data/annotate_sign/01901.jpg,1129,228,1232,325,girilmez
4	/home/keras_detect/racecar/Training_Data/annotate_sign/blurred979.jpg,691,329,710,351,park_yasak
5	/home/keras_detect/racecar/Training_Data/annotate_sign/blurred1262.jpg,552,288,590,325,sol_yasak

Fig. 13. Label format to train MobileNet on Keras

3.3 Verifying Object Detection Results with Classifier

VGG-19 classification model is used to verify the outputs produced by DetectNet and MobileNet. The custom multiclass traffic sign dataset is used here too. However, data used to train a classification model differs from the ones used for object detection as can be seen on Fig. 14.

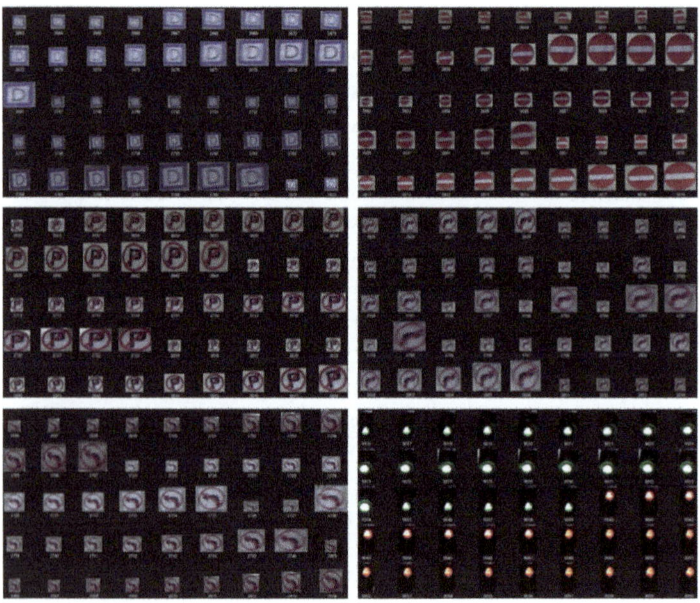

Fig. 14. Custom traffic sign dataset for classification

The six classed classification model is trained using Keras with TensorFlow backend and 96% training accuracy achieved as a result. Then this model is applied to the detection area produced by the object detection model.

The results of native classification inference can be found on Fig. 15.

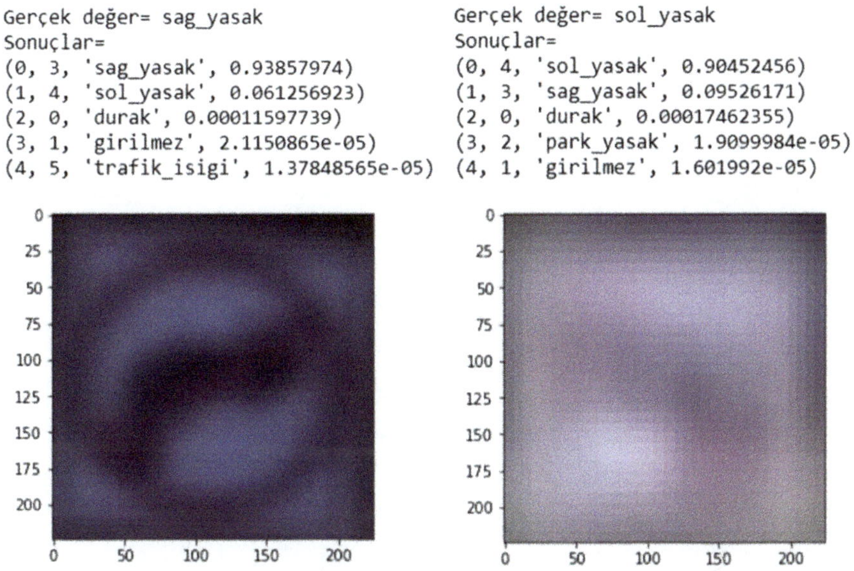

Fig. 15. Classification model native inference results

4 Conclusions and Future Work

Two different object detection and one classification model is used to increase overall recognition accuracy with verifying the detection areas produced by object detection model. Summary of parameter counts for both multiclass DetectNet and MobileNet is shown Table 2 below.

Table 2. Summary of parameter counts for DetectNet and MobileNet models

Model	Parameter count	mAP
MobileNet-160 ($\alpha = 1.0$)	13,336,346	%71.26
MobileNet-160 ($\alpha = 0.25$)	8,212,490	%76.10
Mobilenet-128 ($\alpha = 1.0$)	13,336,346	%70.20
MobileNet-128 ($\alpha = 0.25$)	8,212,490	%75.94
DetectNet	5,983,802	%64.99

The main reason of the higher accuracy values of MobileNet models compared to DetectNet is MobileNets has more parameters as a result of more layers. Being this way, model can extract much more detailed features from data and learn from them. However, the more parameter means the less runtime performance when working with the real time imagery. mAP values provided under Table 2 are the highest accuracy values achieved during the training procedure. Inference results of the DetectNet and MobileNets can be found on Figs. 16, 17 and 18 respectively.

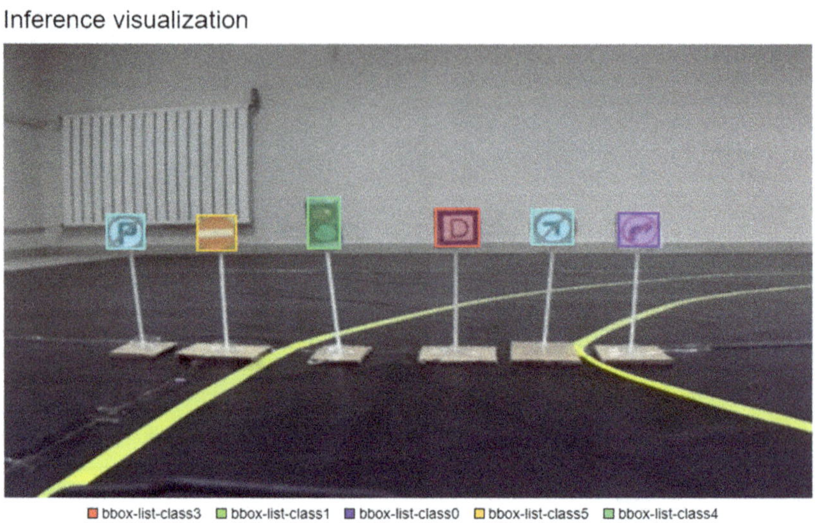

Fig. 16. DetectNet inference result on multiple classes

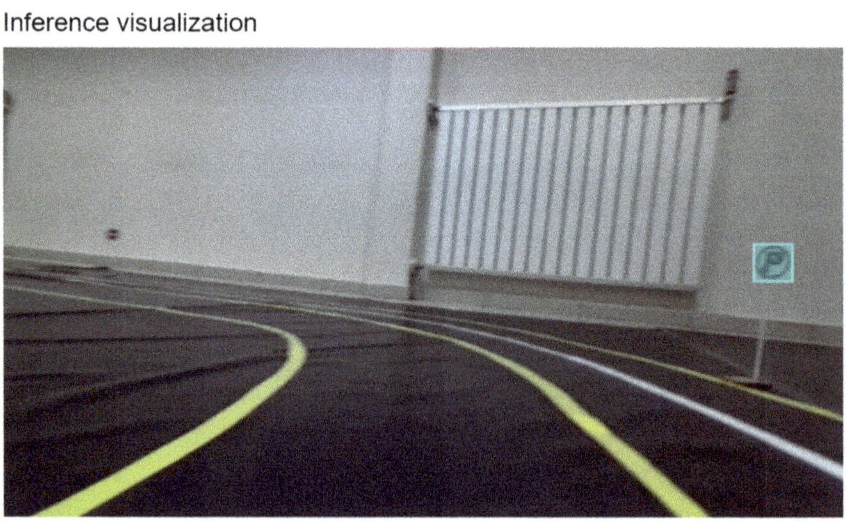

Fig. 17. DetectNet inference result on single class

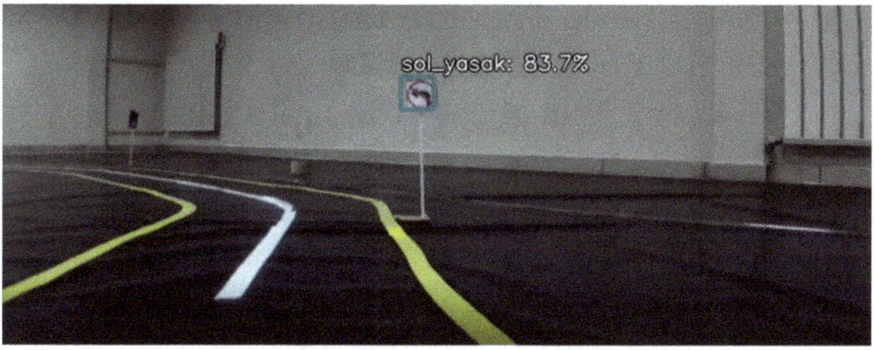

Fig. 18. MobileNet-160 ($\propto = 1.0$) inference

The main purpose of the project is to identify the most optimal deep learning enabled object detection solution for AI at the edge class embedded systems. Thus, all trained models tested on Nvidia Jetson TX2 SoM (System on Module). It is found that while Jetson TX2 is not able to run MobileNet-160 with $\propto = 1.0$ on real time video stream; a multiclass DetectNet model can run with a fixed 6-7 FPS performance. Therefore, in order to provide some comparison between these two model types, MobileNet-160 is used with $\propto = 0.25$ and got 1–2 FPS performance. With this in mind, it is not feasible to run extra classification model while working with MobileNet-160 to verify the object because of the limited resources.

Therefore, if one needs high accuracy instead high performance when working with edge applications, a classifier that has high accuracy can be used to verify detector's detections. However, in high performance needed situations, an improved version of DetectNet with additional augmentations is can be enough. Accuracy values of the models that is used in this project are can be increased by increasing the size of the dataset.

Acknowledgment. The authors would like to thank to Open Zeka Information Technologies and PNY for their hardware support.

References

1. LeCun, Y.: Backpropagation applied to handwritten zip code recognition (1989)
2. Goodfellow, I., Bengio, Y., Courville, A.: Deep Learning. MIT Press, Cambridge (2016)
3. Kapellmann-Zafra, G.: Design and Development of a FROG Instrument for the Characterization of Femto-second Laser Pulses (2013)
4. Dumoulin, V., Visin, F.: A Guide to Convolution Arithmetic for Deep (2018)
5. Kurt, F.: Evrişimli Sinir Ağlarında Hiper Parametrelerin Etkisinin Incelenmesİ (2017)
6. Tao, A., Barker, J., Sarathy, S.: DetectNet: Deep Neural Network for Object Detection in DIGITS, NVIDIA, 11 Ağustos 2016. https://devblogs.nvidia.com/detectnet-deep-neural-network-object-detection-digits/. [%1 tarihinde erişilmiştir12 Kasım 2018]
7. Howard, A.G., Zhu, M., Chen, B., Kalenichenko, D., Wang, W., Weyand, T., Andreetto, T., Adam, H.: MobileNets: Efficient Convolutional Neural Networks for Mobile Vision (2017)
8. Google, TensorBoard: Graph Visualization, TensorFlow (Google). https://www.tensorflow.org/guide/graph_viz. [%1 tarihinde erişilmiştir 24 12 2018]

A Novel Blood Pressure Estimation Method with the Combination of Long Short Term Memory Neural Network and Principal Component Analysis Based on PPG Signals

Umit Senturk[1], Kemal Polat[2(✉)], and Ibrahim Yucedag[3]

[1] Department of Electric Electronic and Computer Engineering, Duzce University, Duzce, Turkey
usenturk98@gmail.com
[2] Department of Electric and Electronic Engineering, Bolu Abant Izzet Baysal University, Bolu, Turkey
kpolat@ibu.edu.tr
[3] Department of Computer Engineering, Duzce University, Duzce, Turkey
yucedagi@gmail.com

Abstract. The worldwide high blood pressure-related mortality rate is increasing. Alternative measurement methods are required for blood pressure measurement. There are similarities between blood pressure and photoplethysmography (PPG) signals. In this study, a novel blood pressure estimation methods based on the feature extracted from the PPG signals have been proposed. First of all, 12-time domain features have extracted from the raw PPG signal. Secondly, raw PPG signals have been applied to Principal Component Analysis (PCA) to obtain 10 new features. The resulting features have been combined to form a hybrid feature set consisting of 22 features. After features extraction, blood pressure values have automatically been predicted by using the Long Short Term Memory Neural Network (LSTM-NN) model. The prediction performance measures including MAE, MAPE, RMSE, and IA values have been used. While the combination of 12-time domain features from PPG signals and LSTM has obtained the MAPE values of 0,0547 in the prediction of blood pressures, the combination of 10-PCA coefficients and LSTM has achieved the MAPE value of 0,0559. The combination model of all features (22) and LSTM has obtained the MAPE values of 0,0488 in the prediction of blood pressures. The achieved results have shown that the proposed hybrid model based on combining PCA and LSTM is very promising in the prediction of blood pressure from PPG signals. In the future, the proposed hybrid method can be used as a wearable device in the measurement of blood pressure without any calibration.

Keywords: PCA · LSTM-NN · Wearable blood pressure · PPG measurement

D. J. Hemanth and U. Kose (Eds.): ICAIAME 2019, LNDECT 43, pp. 868–876, 2020.
https://doi.org/10.1007/978-3-030-36178-5_75

1 Introduction

Deaths due to cardiovascular system disorders are increasing worldwide. About one-third of cardiovascular diseases have disorders due to high blood pressure [1]. High blood pressure has been not only seen in the elderly but also in the young population. The high blood pressure observed in the population of all ages is the highest among the risks of death. High blood pressure in the literature as a silent killer leads to problems such as heart attack, stroke, and organ failure. Early diagnosis of blood pressure is important. It is a disease that can be treated with medication and health checks. The delay in treatment is systematically occurring with other disorders.

Blood pressure measurement is divided into two groups: invasive blood pressure (IBP) and non-invasive blood pressure (NIBP). Invasive methods have not suitable for home use due to operational restrictions. Non-invasive blood pressure measurement is performed by two different methods. There are measuring systems that obstructive and unobstructive blood pressure measurement. Measurement systems obstructive the vein pathways have been called oscillometric measurement systems [2, 3]. The artery is blocked by the cuff, and the passage of blood is stopped. The pressure in the cuff has been lowered slowly and the pressure is sensed by the help of sensors, and the blood pressure is read electronically or manually. The other method of blood pressure measurement is the systems that unobstructive blood pressure measurement and benefit from the relationship of blood velocity with the blood within the vein. Two points where the blood in the vein passes are measured. Electrocardiography (ECG), Photoplethysmography (PPG) such as vital signals has been used for measurement. In previous studies Pulse Transit Time (PTT) [4], Pulse Arrival Time (PAT) [5] and Pulse Wave Velocity (PWV) [6] blood pressure measurement using signals.

The rate of progression of the blood in the vein varies according to the rate of progression of the vein, the diameter of the vessel, the viscosity of the blood, and the length of the vein in which the blood progresses. As a result, errors occur in blood pressure measurements with PTT, PAT, and PWV. There is a close similarity between the PPG signal and the blood pressure (BP) signal. Figure 1 shows the PPG and BP signals. On these pages, we aimed to reach the blood pressure information from the PPG signal only. PCA has been applied to the screened PPG signal, and the size was reduce, and the most effective features have been obtained. In order to increase accuracy in blood pressure measurement system, pulse and time domain attributes were removed, and the blood pressure estimate was calculated in LSTM-NN.

BP and PPG signals are taken from the Physionet Multi-parameter Intelligent Monitoring in Intensive Care (MIMIC) database. The received signals are pre-processing and free of artifacts and noises. The cleaned PPG signal is segmented from its maximum point [8]. PCA was applied to the LSTM-NN input layer with the most effective attributes and time domain attributes [9]. Arterial Blood Pressure (ABP) signal has been used as a target in the MIMIC database. In the new proposed study, high accuracy measurements have been taken for blood pressure measurement.

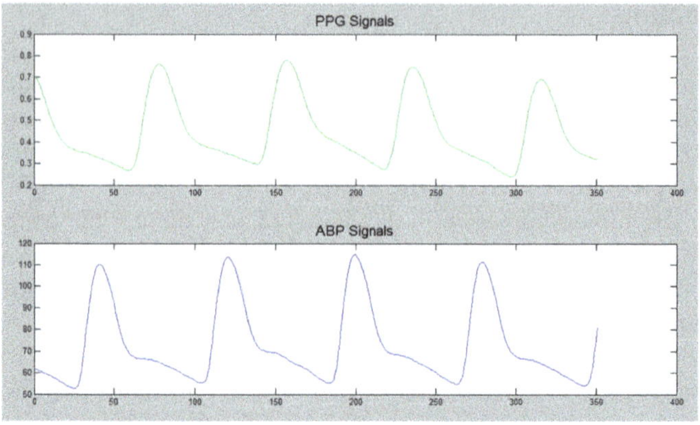

Fig. 1. PPG and BP signals are taken from the MIMIC Database in the physionet.net [7]

2 Material and Method

Data from the MIMIC database are preprocessed and prepared for PCA and attribute extraction. The PPG signal was then divided into segments and formed in matrix form. PCA was applied to the cleared PPG signals, and the most effective features have been selected. Time domain features were extracted from the same PPG signals, and LSTM-NN's have been transferred along with pulse data and PCA features.

2.1 PPG, ABP Signals Preprocessing and Segmentation

PPG has been a signal of photo-sensor sensing of light emitted by the light source and absorbed by hemoglobin. The light in the measurement environment has a detrimental effect on the PPG signal and the movement of the person being measured. Failure to eliminate disturbing effects affects the error rate of the measurement result. The shifts in the base of the Wavelet Decomposition applied to the PPG signal were eliminated. High-frequency noise through the digital band-pass filter has removed from the signal. Figure 2 preprocessing process steps have been seen. As a result of the preprocessing process, the clean PPG signal has been cut from maximum points to a two-dimensional matrix form. The PPG signals, which have been cleaned and divided into segments, have been made ready for feature extraction.

Fig. 2. Preprocessing steps

2.2 Feature Extraction Process

To perform a blood pressure measurement from cleaned PPG signals, the features must be removed. First, the time domain features are extracted from only PPG signals. The time domain features extracted from the PPG signals must be features that best identify the PPG signal. 12 time domain features that best describe blood pressure together with a pulse are used in blood pressure estimation. In addition to the time domain features, PCA was applied to the PPG signal, and 10 coefficients that best describe the PPG signal were removed and used to estimate the blood pressure. The most accurate results yielded hybrid (time domain and PCA coefficients) attributes.

Time Domain Features. Only PPG signals were used for blood pressure measurement. Time domain attributes were derived from the PPG signal and used for blood pressure measurement. 12-time domain attributes that best identify the PPG signal in the time domain is omitted. Rather than giving the PPG signal to the blood pressure measurement system, giving the attributes that best define the blood pressure will contribute to the fast and efficient operation of the system. Table 1 show 11-time domain features and pulse rates. A total of 22 attributes were applied to the input layer of LSTM-NN.

Table 1. Time domain attributes

PPG features	Explanation
1	Peak
2	Peak to peak
3	Average value
4	Standard deviation
5	RMS (root mean square)
6	Skewness
7	Stickiness
8	Peak factor
9	Clearance factor
10	Impulse factor
11	Shape factor
12	Heart rate

PCA Features. The PPG signal needs to be omitted for some blood pressure measurement. In previous studies, blood pressure measurements have been made with different features [10, 11]. PCA is used for the measurement of oscillometric blood pressure [12]. In this study, PCA was applied to the PPG signal. In Fig. 3 the PCA block diagram is shown. The first 10 components of the 95% that best describe the PCA-treated PPG signal are introduced into the input layer of the artificial neural network. In this way, the artificial neural network has improved the learning, reduced the complexity of the calculation and reduced the effect of the low components.

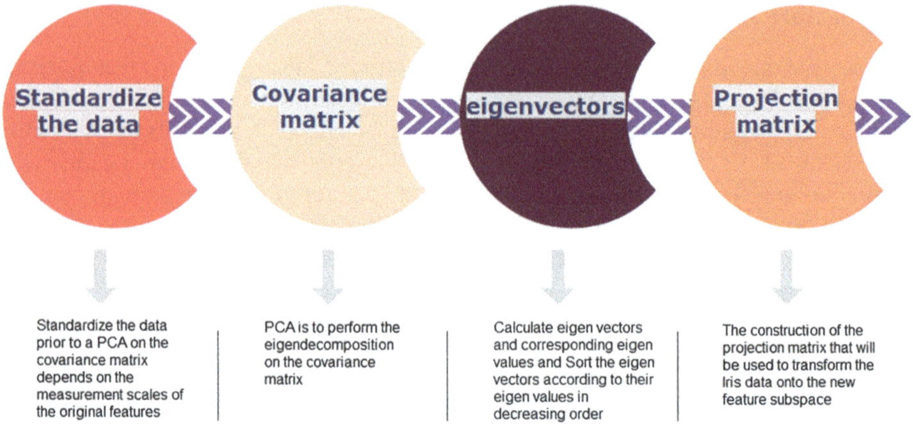

Fig. 3. The operation steps of thePCA method

In addition to the features extracted from PPG by PCA, time domain features and pulse are applied to the input layer of the artificial neural network to increase accuracy in blood pressure measurement.

2.3 Long Short Term Memory Neural Network (LSTM-NN)

LSTM is a type of recurrent neural networks (RNN). Recurrent neural networks are not static structures such as fed-forward neural networks (FNN). FNN works with the input-output system by transmitting the information in the input layer towards the output. RNNs can be fed back to the information input at each neuron output to learn the internal dynamics of the signal. Adding internal dynamics to the system allows sequential learning to take place. Neurons are updated in sequential learning that updates themselves. The neurons themselves update RNNs to be dynamic. The most important issue of RNN is the vanishing gradient problem. LSTM has been introduced as a solution to the gradient and exploding gradient problems. LSTM cell structure has been tried to solve the problem by using different doors.

Input gate: Neuron access activation,
Output gate: Neuron output activation,
Memory door: memorial door,
Forget gate: Forgetting door.

Thanks to these doors, it is learned which information is forgotten and which information will be remembered. In the LSTM block, the learning process is performed by coding the changes in the input data in the time domain. Figure 4 LSTM neuron block diagram has been shown.

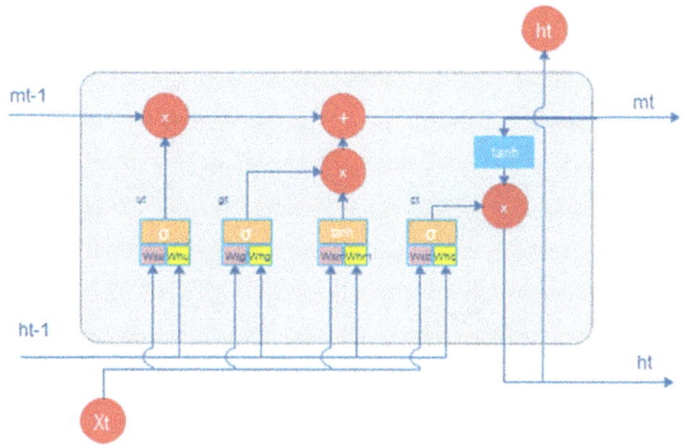

Fig. 4. LSTM-NN cell

$$gt = \sigma\left(W_{xg}Xt + W_{hg}ht - 1 + W_{mg}mt - 1 + b_g\right) \quad (1)$$

$$ut = \sigma(W_{xu}Xt + W_{hu}ht - 1 + W_{mu}mt - 1 + b_u) \quad (2)$$

$$mt = u_t mt - 1 + gt\,tanh(W_{xm}Xt + W_{hm}ht - 1 + b_m) \quad (3)$$

$$ct = \sigma(W_{xm}Xt + W_{hc}ht - 1 + W_{mc}mt + b_c) \quad (4)$$

$$ht = ct * tanh(mt) \quad (5)$$

Where g, u, c and m are input, forgetting gate, output gate and memory gate, respectively. Representative weights are denoted by W, and b is the bias.

Firstly, 12-time domain features have been removed from the PPG signals and the blood pressure has been estimated in the LSTM-NN model. It has been then tested in the most important 10 coefficients LSTM-NN model by applying PCA to the PPG signals. Finally, 12-time domain features obtained from the PPG signal and 10 coefficients obtained from PCA were used. 22 hybrid attributes have been applied to the LSTM-NN input layer. The numbers of LSTM layers have been increased for deepening learning. Rectified Linear Units (ReLU) Layers have been added to solve the learning slowdown caused by LSTM blocks. Figure 5 shows Hybrid features and LSTM-NN layer structure.

In the study, time domain features, PCA features and hybrid (time domain features + PCA coefficients) obtained from PPG signals have been tested using a single LSTM layer. The LSTM layer numbers have been scored to give more accurate results in blood pressure estimation. When the numbers of layers have increased, the learning will be deepene, and the error rate will be reduced.

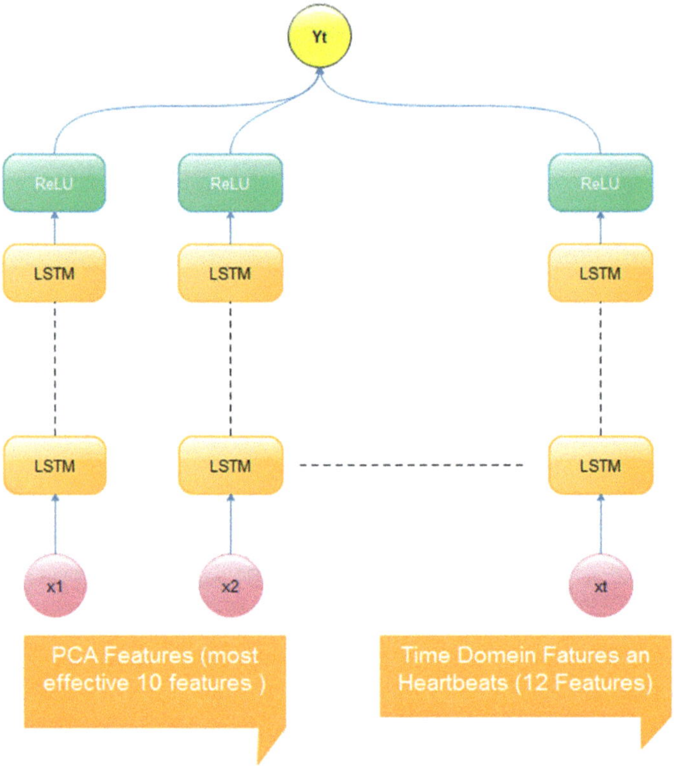

Fig. 5. Hybrid features and LSTM-NN Layers

3 Experimental Results and Discussion

The novel proposed blood pressure estimation method uses only the PPG signal. PPG and ABP signals have taken from the MIMIC II database. The PPG signals have been pre-processing to remove noise and artifacts. Then the cleaned PPG signals have prepared in matrix form by dividing the segments from the peaks into segments. The first time domain attributes have subtracted from the segmented PPG signal, and a blood pressure estimate had made in the LSTM-NN model. Secondly, PCA had applied to the PPG signal, and blood pressure estimate had made in the most effective 10 coefficients LSTM-NN model. Finally, a blood pressure estimate had made by applying 12-time domain features and 10-PCA coefficient to LSTM-NN. Blood pressure estimation results have been tested with Mean Absolute Error (MAE), Mean Absolute Percentage Error (MAPE), Root Mean Square Error (RMSE) and Index of Agreements (IA) calculations. Table 2 shows the model and performance comparison.

Table 2. LSTM-NN models performances

Models	Performance criteria			
	MAE	MAPE	RMSE	IA
Time Domain Features + LSTM-NN (12 features)	6,5674	0,0547	7,95	0,3846
PCA + LSTM-NN (10 coefficients)	6,7005	0,0559	8,0753	0,3993
Hybrid (PCA + Time domain features) LSTM-NN (22 Features)	6,4280	0,0535	7,8064	0,3898
Hybrid LSTM-NN with 3 LSTM layers	**5,8968**	0,0488	7,3908	0,3339

Compared to the blood pressure estimation models, it has seen that the error rate of blood pressure estimation of LSTM-NN models using the PCA and time domain features of the LSTM-NN model is lower than the LSTM-NN model. The model with the lowest error rate is 3 layers LSTM-NN. This shows that when the numbers of layers have increased in the neural network structure, learning has deepened and the error rate has decreased.

4 Conclusions and Future Direction

High blood pressure, which has become a disease of our age, has a condition that must be continuously monitored and followed up. The proposed blood pressure values of the hybrid blood pressure system have as follows: MAPE 0,0488. Error rate ANSI/AAMI SP10: 2002 [13] demonstrates performance according to blood pressure equipment standard. The only use of PPG signals in blood pressure estimation will facilitate adaptation to wearable systems. The fact that blood pressure measurement systems are portative and simple will facilitate their use in daily life. People will recover their healthy days if they are late in the detection and treatment of blood pressure.

Acknowledgment. This paper has been supported by Duzce University Scientific Research and Projects Unit with the Project number 2018.07.02.878.

References

1. A Global brief on HYPERTENSION, World Healthy Day 2013, World Health Organization
2. Liu, J., Cheng, H.-M., Chen, C.-H., Sung, S.-H.: Patient-specific oscillometric blood pressure measurement. IEEE Trans. Biomed. Eng. **63**, 1220–1228 (2016)
3. Hung, C.-H., Bai, Y.W., Tsai, R.-Y.: Design of blood pressure measurement with a health management system for the aged. IEEE Trans. Consum. Electron. **58**, 619–625 (2012)
4. Lin, W.-H., Wang, H., Samuel, O.W., Li, G.: Using a new PPG indicator to increase the accuracy of PTT-based continuous cuffless blood pressure estimation. In: Conference of the IEEE Engineering in Medicine and Biology Society, pp. 738–741 (2017)
5. Tang, Z., Tamura, T., Sekine, M., Huang, M.: A chair-based unobtrusive cuffless blood pressure monitoring system based on pulse arrival time. IEEE J. Biomed. Health Inform. **21**, 1194–1205 (2017)

6. Yoon, Y.-Z., Kang, J.M., Kwon, Y., Park, S.: Cuff-less blood pressure estimation using pulse waveform analysis and pulse arrival time. IEEE J. Biomed. Health Inform. **22**, 1068–1074 (2018)
7. Goldberger, A.L., Amaral, L.A.N., Glass, L., Hausdorff, J.M., Ivanov, P.Ch., Mark, R.G., Mietus, J.E., Moody, G.B., Peng, C.-K., Stanley, H.E.: PhysioBank, PhysioToolkit, and PhysioNet: components of a new research resource for complex physiologic signals. Circulation **101**(23), e215–e220 (2000)
8. Şentürk, Ü., Yücedağ, İ., Polat, K.: Cuff-less continuous blood pressure estimation from Electrocardiogram (ECG) and Photoplethysmography (PPG) signals with an artificial neural network. In: 26th Signal Processing and Communications Applications Conference (SIU) (2018)
9. Şentürk, Ü., Yücedağ, İ., Polat, K.: Repetitive neural network (RNN) based blood pressure estimation using PPG and ECG signals. In: 2nd International Symposium on Multidisciplinary Studies and Innovative Technologies (ISMSIT) (2018)
10. Allen, J., Murray, A.: Age-related changes in peripheral pulse timing characteristics at the ears, fingers, and toes. J. Hum. Hypertens. **16**, 711–717 (2002)
11. Lee, S., Park, C.-H., Chang, J.-H.: Improved gaussian mixture regression based on pseudo feature generation using bootstrap in blood pressure estimation. IEEE Trans. Ind. Inform. **12**, 2269–2280 (2016)
12. Forouzanfar, M., Dajani, H.R., Groza, V.Z., Bolic, M.: Oscillometric blood pressure estimation using principal component analysis and neural networks. In: IEEE Toronto International Conference Science and Technology for Humanity (TIC-STH) (2009)
13. Association for the Advancement of Medical Intrumentation.: Manual, electronic or automated sphygmomanometers. Arlington, Virginya, USA. American National Standard ANSI/AAMI SP10:2002

Design and Implementation of SDN-Based Secure Architecture for IoT-Lab

Enis Karaarslan(✉), Eren Karabacak, and Cihat Cetinkaya

Department of Computer Engineering,
Mugla Sitki Kocman University, Muğla, Turkey
{enis.karaarslan,cihat}@mu.edu.tr, ereenkarabacak@gmail.com

Abstract. There are various IoT devices with many different hardware environments and an ecosystem that will enable these various devices to work together is not yet existent. As IoT networks are becoming more widespread, the number of connections that has to be handled is increasing in a high rate. There is a need for a smarter network management which can handle these connections and security considerations. In this paper, we propose a secure network architecture leveraging SDN for an IoT-based laboratory. Thanks to the IoT-devices, the proposed work enables to automate the basic tasks in a public laboratory while at the same time SDN-based network architecture ensures the security requirements of the IoT network.

Keywords: SDN · IoT · Security · Lab automation

1 Introduction

IoT (Internet of things) gives the ability to make anything around us connected to the Internet with intelligence, autonomy and privacy [1]. According to Ericcson's report [2], it is estimated that 29 billion IoT devices will be connected to the internet in 2023. These IoT devices have interoperability, management and security problems. The devices lack robust security and encryption protocols due to their limited hardware and are usually managed by a central device via WiFi. IoT devices normally generate a large number of connections with little data. As IoT devices become widespread, these type of traffic will also be increased. Many of the IoT devices also do not have a robust and end-to-end encryption mechanism. A smarter network is needed to manage these IoT networks. One candidate for this is software-based networking (SDN). Software-defined networking (SDN) is a technology that makes the networks smarter and more flexible to optimize the network resources and rapidly adapt the networks to changing business needs, applications and traffic.

In this study, we aim to design and implement a SDN-based secure architecture for the IoT Lab. This network will continuously monitor and control the

D. J. Hemanth and U. Kose (Eds.): ICAIAME 2019, LNDECT 43, pp. 877–885, 2020.
https://doi.org/10.1007/978-3-030-36178-5_76

network traffic for the security of the IoT devices. IoT devices do not need continuous access to a network. IoT devices need to regularly send data which are in small network packet sizes. The network data will be analyzed to configure the flow rules depending on the network traffic characteristics. The bandwidth of the IoT devices can be adjusted by using SDN.

In the next section, the fundamentals of IoT, SDN and OpenFlow technologies are described. In Sect. 3, SDN-based secure architecture for IoT is defined. In Sect. 4, implementation is explained. Finally, results and discussion are given.

2 Fundamentals

2.1 IoT

The concept was first mentioned by Kevin Ashton in 1999. Then various IoT devices are developed according to different needs. In terms of the Internet of objects, all kinds of monitoring devices, sensors, biochips or access devices are considered objects. The high amount of sensor data flowing from the physical environment is transmitted to the operators or the related persons as a result of the evaluations, or the data is processed with the help of the systems to perform an activity. The concept of the Internet of Things is the basis of the concept of Machine-to-Machine Communication (M2M). In M2M technology, machines can communicate with each other without human intervention. IoT is a more comprehensive technology than M2M technology.

In an IoT system all of the sensors directly send their collected data to the application server through the Internet. On the contrary, wireless sensor networks (WSN) are not directly connected to the Internet. WSN is a network of small wireless electronic nodes which consists of different sensors. The aim of the WSN is collect data from these sensors. These various sensors connect to some kind of router or central node. WSN is a technology which is frequently used within an IoT system. WSN can be called as a subset of IoT.

Data generated in IoT technology is transmitted through various wireless protocols such as ZigBee, Near Field Communication (NFC), IPv6 over Low power Wireless Personal Area Networks (6LoWPAN), Bluetooth Low Energy (BLE) and Z-Wave which are short range. There are also SigFox and Cellular which are Low Power Wide Area Network (LPWAN) standard protocols [3].

Considering the increase in IoT devices and the large number of connections generated by these devices, it is possible to mention various vulnerabilities. One of the most updated vulnerability list is available in OWASP's web site (https://www.owasp.org). According to this list; the IoT Vulnerability categories is given in Table 1.

2.2 SDN

Software-Defined Networking(SDN) technology facilitates network management and enables to dynamically program efficient network configuration to improve

Table 1. IoT vulnerability categories

Category
1. Insecure Web Interface
2. Insufficient Authentication/Authorization
3. Insecure Network Services
4. Lack of Transport Encryption
5. Privacy Concerns
6. Insecure Cloud Interface
7. Insecure Mobile Interface
8. Insufficient Security Configurability
9. Insecure Software/Firmware
10. Poor Physical Security

the network performance and monitoring [4]. In traditional networks, each router has a control mechanism (management plane) and a mechanism (data plane) that executes the order of this control mechanism. The management plane is used to control and give decisions by using the specified routing rules. When a package is received from the neighbor router, a decision is given which depends on the source and the destination of the communication. The decision on the path of the package depends on the routing table which is formed by using the routing software on that router. This routing software uses Open Shortest Path First (OSPF) or other algorithms which are based on different metrics.

The basis of the SDN concept is the separation of the data plane and the control plane. The control is implemented by the central controller. The client requests the device closest to it when it wants to make a network communication. The device checks whether there is a similar request in the current table. If there was no communication from that device before, the device communicates with the central controller to learn the response to the incoming request. The controller checks whether there is a rule for that in the forwarding table. If there is an existing flow rule, the flow rule is sent to the requesting device. If there is no rule, a new rule is added to the flow table by using the defined rules. Depending on the rules, the paths that the incoming packet request will use are forwarded to the devices by the controller.

The devices communicate with each other by the decision given by the SDN controller on a set of specified rules. The processing speed increases as there is no layer dependency in the traditional networks. Authorization and restrictions can be made easily and safety rules can be specified in SDN. Traffic-intensive areas are monitored and incoming requests are routed according to this information. In the event of a problem with the devices in the system, packets are routed to alternative paths. In the event of a problem with the controller, transactions continue to be managed by other synchronous controllers.

Due to the extensive use of SDN by both academia and industry, many open-source OpenFlow-compatible controller software has been developed. OpenDaylight (ODL), ONOS, Floodlight and Ryu are now the most widely used ones. ODL is preferred as a controller software in this project. ODL is an open source and java based SDN controller which is managed by Linux Foundation and supported by more than 40 companies such as IBM, Cisco, Juniper, VMWare and various major networking vendors. It is known as the most well-documented controller [5].

2.3 OpenFlow

OpenFlow is a communication protocol which is the core element of the SDN (Software-defined Networking) architecture. OpenFlow protocol is currently the only open standard that is used for the communication between SDN controller and OpenFlow compatible network device. OpenFlow provides a detailed control to respond to real-time changes in application, user and session levels in a network. OpenFlow permits to determine how traffic flows through factors such as usage patterns, application requirements, service level agreements and cloud resources over network patterns.

Switch provides the network flow information to the controller. The controller uses this information to identify the network traffic. Then the controller uses OpenFlow to tell the switch what to do on a specific flow.

The SDN topology is given in Fig. 1. SDN network can have multiple controllers and OpenFlow switches. OpenFlow compatible devices in the network use the OpenFlow protocol to communicate with each other. The controller consist of two interfaces; northbound and southbound. The northbound interface is responsible for communicating with network applications. The southbound interface is responsible for the communication between the controller and the network devices. The controller can use the network services to provide graphical user interface to the administrator. Administrator can view flow table and write rules by using the controller user interface.

3 SDN-Based Secure Architecture for IoT

As IoT networks become widespread, these networks will produce large number of connections which contain little data. The communication network should be much more smarter to manage these connections. Tests and measures on IoT devices are extremely important for national and personal security. The IoT security architecture can be classified in four levels [6–8]:

- Perceptual (recognition) Layer: Collecting information from the physical devices such as sensors is implemented here. Authentication is required to prevent illegal node access. Encryption is required for the information transfer. Lightweight encryption algorithms and protocols are important to minimize the resource consumption of these devices.

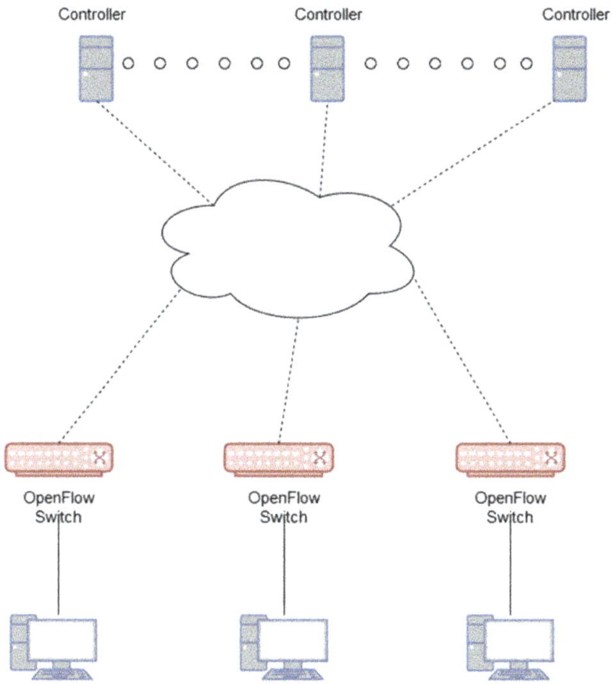

Fig. 1. SDN topology

- Network Layer: This layer is responsible for the transfer of the data reliably over communication networks by using communication protocols. The existing security mechanisms in this layer are difficult to implement. Privacy and integrity of data is important and should be satisfied. Distributed denial of service (DDoS) attacks are highly effective on the IoT. Preventing DDoS attacks is another problem to be solved in this layer.
- Support Layer: The support layer should satisfy safe and reliable infrastructure for the applications using cloud computing and secure computation. Strong cryptography and antivirus systems should be used.
- Application Layer: This layer should provide secure access service to the users through the interfaces they use. Authentication, key agreement and protection of the user privacy should be satisfied in this layer. Security education should be given and mechanisms for password security should be satisfied.

Software-defined network (SDN) is a serious candidate for the network management of network operators or corporate networks for efficient use of resources. The use of SDN technology especially for IoT has been proposed in the literature [9,10] and its advantages are indicated. The problems especially at network and support layers can be solved with SDN. IoT devices are usually managed via a central device which they connect through WiFi. In case of this device being a SDN controller that supports IoT; the management of IoT devices and the detection of the attacks on the system is possible [11]. The network that is created with SDN will be smarter, enable us to overcome these problems and provide a more secure architecture.

4 Implementation

The SDN-IoT network can be physically generated with real devices, or it can be created in a variety of simulation (ns-3, omnet ++) and emulation (Contiki, netsim) environments.

Due to the diversity of IoT devices and a continuously self-updating topic, it is very difficult to implement a development environment in which SDN and IoT can work together. The deficiencies and redundancies of different wireless simulation environments are discussed in [12]. Although there are various simulation environments as seen in the Table 2, we couldn't find a simulation environment which fulfilled our needs. The simulation environments which were affordable didn't have enough documentation or didn't have enough features. For these reasons; it was more convenient to use real devices and construct a physical lab for the scope of the project. Creating a physical lab environment will give us a great advantage in the integration of this project into smart cities.

There is a need for OpenFlow enabled network device and an OpenFlow controller to integrate SDN into the IoT network. IoT devices normally connected to WiFi access, we were able to use an affordable ADSL advice as a SDN WiFi access point. We also used a switch to connect IoT devices. The SDN-IoT network to be created within the scope of the project includes an OpenFlow compatible switch and IoT devices connected to this switch. The picture of the physical SDN-IoT Lab is given in Fig. 2. TP-LINK Wireless Gigabit router, Andasis OpenFlow switch, 4 Raspberry pi and 1 controller computer was used in the lab environment. OpenDaylight(ODL) will be preferred as control software which is installed on an Ubuntu Server (Version 14.04). The LAN ports are devices the are controlled by ODL, and the WAN is used for out-of-band management to the SDN controller ODL.

MSKU NetSecLab (http://wiki.netseclab.mu.edu.tr) is giving academic support to Andasis (http://www.andasis.com) on SDN solutions. Andasis OpenFlow switch is an affordable switch which uses OpenWrt Linux distribution. Open vSwitch software is installed for OpenFlow support and kernel level optimizations are implemented for stable usage.

We also succeeded in using TP-LINK Wireless Gigabit router (TL-WR1043ND v2.1) as a SDN network device. We installed OpenWrt on the device instead of its own firmware.

The flow rules for communication of IoT devices are configured in the ODL controller. Opendaylight compatible environment is required to write flow rules and program OpenFlow paths. The Cisco OpenFlow Manager (OFM) is used to send the required flow rules to the switch. OFM is an application developed to run on top of ODL to visualize OpenFlow (OF) topologies, program OF paths and gather OF stats.

Table 2. Comparison between different wireless sensor simulation tools [12]

Simulator and emulators	Programming language	Merits	demerits
Ns-3	C++	-Easy to add new protocols -Visualization tool	-Supports only two wireless MAC protocols
OMNeT	C++	-Powerful graphical User Interface	-Number of protocols supported is not enough -Compatibility problem
QualNet	C++	-Sophisticated animation capabilities -Support for multiprocessor systems and distributed computing	-Slow Java-based UI -Very expensive

Fig. 2. Physical SDN-IoT lab

Incorporating physical devices into the network gives us a real test environment. This will help us to identify security vulnerabilities. The vulnerabilities found and the solutions to them will be used to create more secure IoT networks. We started to implement the vulnerability and attacks tests in the IoT lab and we find it very important in terms of personal security and cyber security. This study also aims to provide some best practices which will be given in the following GitHub (https://github.com/MSKUnetseclab/Best-Practices-for-SDN-IoT/) page.

The first experiments had some initial results. The network administrator can learn network statistics by monitoring the network and manually configure the controller rules according to these values to detect DDOS attacks. However, machine learning techniques is needed is to make this process autonomous as it is specified in a recent study [13]. It is easier to avoid these type of attacks when the SDN controller takes an action on the network flow.

Detecting the brute-force attacks is still a challenge. In the tests, it is seen that it is very easy to get access on the IoT devices if default and weak passwords are used. Although there are some studies to detect these attacks by using SDN networks [14,15], there is no perfect solution yet.

5 Conclusion and Future Work

The advantages and disadvantages of simulation environments are discussed and a physical SDN-IoT lab is established in this study. Basic vulnerability and security tests are implemented in the test bed. Best practices is given in the project GitHub page and will continuously be updated with the future studies. There is a challenge to overcome the compatibility problems of software-defined networks (SDN) and IoT devices. Security vulnerabilities in the controller should be investigated and these vulnerabilities should be minimized. MSKU NetSeCLab is working with Andasis on SDN solutions. We plan to add more network devices to construct more complex scenarios in the near future. A more convenient environment is planned to be created for the security tests. A system that will detect possible security vulnerabilities is planned to be created and be used in IoT network scenarios. Studies will be carried out for integration of this system to the existing traditional computer networks.

References

1. Ali, Z.H., Ali, H.A., Badawy, M.M.: Internet of Things (IoT): definitions, challenges and recent research directions. Int. J. Comput. Appl. **128**, 0975–8887 (2015)
2. Barboutov, K., et al.: Ericsson Mobility Report June 2017. https://www.ericsson.com/assets/local/mobility-report/documents/2017/ericsson-mobility-report-june-2017.pdf. Accessed 12 May 2019
3. Al-Sarawi, S., Anbar, M., Alieyan, K., Alzubaidi, M.: Internet of Things (IoT) communication protocols: review. In: 8th International Conference on Information Technology (ICIT), pp. 685–690 (2017)
4. Benzekki, K., El Fergougui, A., Elbelrhiti Elalaoui, A.: Software-defined networking (SDN): a survey. Secur. Commun. Netw. **9**(18), 5803–5833 (2016)
5. Asadollahi, S., Goswami, B., Gonsai, A.M.: Implementation of SDN using OpenDayLight controller. In: Proceedings of an International Conference on Recent Trends in IT Innovations-Tec'afe 2017 (2017)

6. Suo, H., Wan, J., Zou, C., Liu, J.: Security in the internet of things: a review. In: 2012 International Conference on Computer Science and Electronics Engineering, vol. 3, pp. 648–651. IEEE (2012)
7. Yang, G., Xu, J., Chen, W., Qi, Z. H., Wang, H.Y.: Security characteristic and technology in the Internet of Things. Nanjing Youdian Daxue Xuebao (Ziran Kexue Ban). J. Nanjing Univ. Posts Telecommun. (Nat. Nanjing Univ. Posts Telecommun. Nat. **30**(4)) (2010)
8. Meng, D.I.N.G., Wu, C.Y.L.: Security architecture and key technologies for IoT/CPS. ZTE Technol. J. **1**(1), 013 (2011)
9. Copos, B., Levitt, K., Rowe, J., Kianmajd, P., Chuah, C.N., Kesidis, G.: Security and privacy for emerging smart community infrastructures (2016)
10. Sharma, P.K., Singh, S., Jeong, Y.S., Park, J.H.: DistBlockNet: a distributed blockchains-based secure sdn architecture for iot networks. IEEE Commun. Mag. **55**(9), 78–85 (2017)
11. Bhunia, S.S., Gurusamy, M.: Dynamic attack detection and mitigation in IoT using SDN. In: 2017 27th International Telecommunication Networks and Applications Conference (ITNAC), pp. 1-6. IEEE (2017)
12. Chhimwal, P., Rai, D.S., Rawat, D.: Comparison between different wireless sensor simulation tools. IOSR-JECE **5**, 54–60 (2013)
13. Sultana, N., Chilamkurti, N., Peng, W., Alhadad, R.: Survey on SDN based network intrusion detection system using machine learning approaches. Peer-to-Peer Networking Appl. **12**(2), 493–501 (2019)
14. Tang, T.A., Mhamdi, L., McLernon, D., Zaidi, S.A.R., Ghogho, M.: Deep learning approach for network intrusion detection in software defined networking. In: 2016 International Conference on Wireless Networks and Mobile Communications (WINCOM), pp. 258–263. IEEE (2016)
15. Koning, R.: Measuring the efficiency of SDN mitigations against cyber attacks. In: 2017 IEEE Conference on Network Softwarization (NetSoft) (2017)

Feature Selection by Using DE Algorithm and k-NN Classifier

Fatih Ahmet Şenel(✉), Asım Sinan Yüksel, and Tuncay Yiğit

Department of Computer Engineering, Süleyman Demirel University, Isparta 32260, Turkey
{fatihsenel,asimyuksel,tuncayyigit}@sdu.edu.tr

Abstract. Feature selection (FS) in data mining is one of the most important preprocessing steps before starting analysis processes. In addition to more accurate estimation of the results by the FS process, unnecessary resource utilization can be avoided by removing unnecessary attributes. One of the most commonly used data prediction and classification methods in data mining is k Nearest Neighbor algorithm (k-NN). k-NN classifier can classify data swiftly. In this study, FS process with Differential Evolution (DE) algorithm has been realized by using sample datasets. During the cost accounting of candidate solutions of DE algorithm, k-NN classifier has been used. The results obtained with DE algorithm have been compared to Correlation based feature selection, Information gain based feature selection and Learner based feature selection methods that are commonly used in classification problems. The results showed that the proposed method could successfully perform FS process.

Keywords: Data mining · Differential evolution · Feature selection · k-NN · Optimization · Prediction

1 Introduction

In data mining studies, the most important factors are the features of data that are used. Datasets consist of observation records that are necessary for predicting at least one feature or classification according to a feature in response to one or more input(s) (feature). Before starting analysis processes with data mining, some studies must be conducted on datasets. For example; evaluating missing records, whether the data will be normalized or not, how many features in datasets will be included in analysis process. Undoubtedly, the foremost study about dataset is feature selection (FS) process. FS is a process of selecting the most meaningful features among features which can predict the output parameter of dataset and forming a subset of features. Because some features have important roles in the process of classification, some features may have an opposite effect. Besides, some of the features that have similar effects can extend the analysis process unnecessarily. Unnecessary features can cause errors in classification process and they can increase the computer resource utilization. That's why, it is necessary to determine the features that are unnecessary and to separate them from the dataset before the analysis processes. In a multidimensional dataset, an analysis process which is conducted without FS may need to work for a long time

D. J. Hemanth and U. Kose (Eds.): ICAIAME 2019, LNDECT 43, pp. 886–893, 2020.
https://doi.org/10.1007/978-3-030-36178-5_77

because of the resource shortage. That's why, FS process which is a preprocessing must be conducted before starting analysis process. The purpose of the FS process is to find a subset of features which will have the most effective role in predicting the result. With this aspect, it can be approached as an optimization problem.

When the literature is viewed, it can be seen that Particle Swarm Optimization [1–3], Genetic Algorithm [4–7], Differential Evolution [8, 9], Whale Optimization [10], Hybrid approaches [11–15] are applied to determine the subset of features.

Differential Evolution (DE) algorithm is constantly being improved. The improvement of mutation operator is a must for improving the DE algorithm. As different from [8, 9] studies, one of the latest versions of the DE algorithm has been used in this study [16].

Cilia et al. conducted FS process for handwritten character recognition process in their study [17]. They performed selection process with Feature-ranking-based technique and they tested the success of their selection by using three different classification algorithms. Xiang et al. suggested a hybrid approach for the FS process by using a k-NN classifier with an improved gravitational search algorithm [18]. They tested this method with datasets from UCI data base.

In this study, we performed the optimization process of FS on sample datasets with the help of k-NN classifier by using one of the latest versions of the DE algorithm. In the second section of this paper, the optimization algorithm that has been used, k-NN classifier and the method that is aimed are discussed. Experimental results are presented in the third section, and conclusion is presented in the fourth section.

2 Material and Method

2.1 Differential Evolution Algorithm

The DE algorithm is an evolutionary optimization algorithm which is developed by Storn and Price [19]. It is an optimization algorithm which produces successful results in many optimization problems and it maintains its development constantly. In the DE algorithm, there are mutation, crossover and selection processes as in GA. In DE algorithm, all parameters are processed with their real values and it is easier to code the DE and make it comprehensible. The DE algorithm has three different individuals. These are; target (x) individual, mutated (v) individual and test (u) individuals. In the DE algorithm, each individual has parameters which are aimed to be optimized. Representation of individuals are shown in Fig. 1.

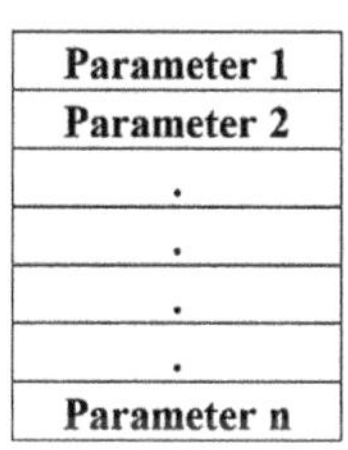

Fig. 1. Representation of individuals

The mutation strategies which are performed in the DE algorithm are different. Basically, at least three different individuals are needed apart from the individual to be mutated. For this reason, the number of population (*NP*) can be selected as at least four.

Initial Population

Initial population is generated randomly at the allowable range value for each parameter. In Eq. 1, the generating process of initial population is shown.

$$x_{j,i} = x_j^{min} + rand(0,1) \cdot (x_j^{max} - x_j^{min}) \tag{1}$$

Here; $i = \{1,\ldots,NP\}$ represents the individuals in population, $j = \{1,\ldots,D\}$ denotes the size of the problem, the values of x_j^{min} and x_j^{max} represent the minimum and maximum range values of the j_{th} parameter, respectively.

Mutation Process

For a mutation process in a standard DE algorithm, there must be three individuals apart from the individual itself. The development of the DE algorithm is in progress and new strategies are added to the mutation strategies. The mutation strategy which is used in this study is shown in Eq. 2. The selected mutation strategy is one of the latest versions of the DE algorithm [16].

$$v_i^{G+1} = x_c^{-G} + F_1 \times \left(x_{best}^G - x_{better}^G\right) + F_2 \times \left(x_{best}^G - x_{worst}^G\right) + F_3 \times \left(x_{better}^G - x_{worst}^G\right) \tag{2}$$

Here G represents iteration step and x_c^{-G} stand for the combination of the selected three individuals and this combination process is shown in Eq. 3.

$$x_c^{-G} = w_1^* \times x_{best} + w_2^* \times x_{better} + w_3^* \times x_{worst} \tag{3}$$

The values of w_i^* satisfy the requirements of $w_i^* \geq 0$ and $\sum_{i=1}^{3} w_i^* = 1$. It is calculated as shown in Eq. 4.

$$w_i = 1 - \left|\frac{f\left(x_{en\ iyi}\right) - f(x_i)}{f\left(x_{en\ iyi}\right) - f(x_{max})}\right|, i = 1,2,3 \tag{4}$$

Here; $f(x_{best}) = f(x_{min})$ value represents the value of $min\{f(x_i)\}$. $f(x_{max})$ represents the maximum value that is observed in an iteration. In Eq. 5, it is shown that increases the possibility of performing of this mutation strategy while the iteration is progressing. In initial iteration numbers, it performs the standard mutation strategy of the DE algorithm.

$$
\begin{aligned}
&if\left(u(0,1) \geq \left(1-\tfrac{G}{GEN}\right)^2\right) Then\\
&\qquad v_i^{G+1} = x_c^{-G} + F_1 \times \left(x_{best}^G - x_{better}^G\right) + F_2 \times \left(x_{best}^G - x_{worst}^G\right) + F_3 \times \left(x_{better}^G - x_{worst}^G\right)\\
&Else\\
&\qquad v_i^{G+1} = x_{r_1}^G + F \times \left(x_{r_2}^G - x_{r_3}^G\right)\\
&end
\end{aligned}
\tag{5}
$$

GEN is the number of total iterations. While the iteration is progressing, more sensitive searches are carried out with the help of Eq. 5.

Crossover Process
While crossover process is being conducted, a test individual (u) is generated by taking the mutated individual (v) and crossover rate (CR) that is predetermined. This process is applied for each parameter. If a random number generated in the range [0,1] is lower than the CR rate, the corresponding parameter is inherited from the individual with the mutation, otherwise from the target individual. A test individual is generated as shown in Eq. 6.

$$
u_{j,i}^G = \begin{cases} v_{j,i}^G & if\left(rand_{j,i} \leq CR \, or \, j = j_{rand}\right) \\ x_{j,i}^G & otherwise \end{cases} \tag{6}
$$

Here; the $j = j_{rand}$ condition is used to ensure that at least one parameter is obtained from the mutated individual.

Selection Process
The selection process in the DE algorithm is conducted according to Greedy selection algorithm. Namely, if the generated test individual is better than the target individual, test individual is transferred to the new generation. Otherwise target individual is transferred to the new generation (Eq. 7).

$$
x_i^{G+1} = \begin{cases} u_i^G & f(u_i^G) \leq f(x_i^G) \\ x_i^G & otherwise \end{cases} \tag{7}
$$

2.2 k-Nearest Neighbor

k-NN is a supervised learning method that executes the learning process based on sample data. Taking into account the characteristics of the data and by looking k number nearest neighbor, the class which a sample belongs to is determined [20]. In the k-NN algorithm, the number k and the distances during class assignment should be determined first. In this study, Euclidean distance metric is used. In Fig. 2, a sample illustration for k-NN is shown. While sample X belongs to Class 1 for $k = 1$, it belongs to Class 2 for $k = 5$. In this study, $k = 3$ was chosen for all classification procedures.

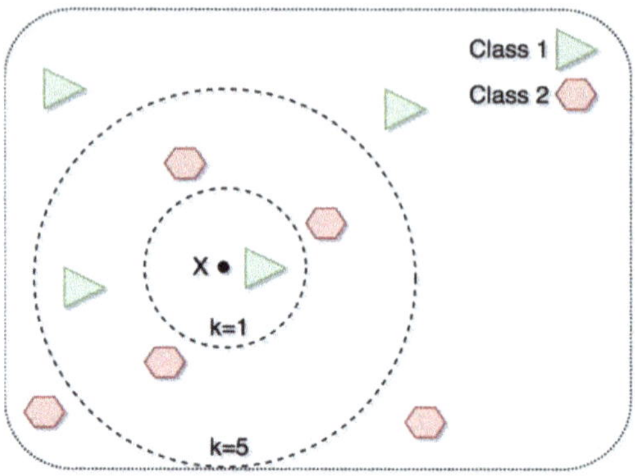

Fig. 2. Illustration example of k-NN algorithm

2.3 Proposed Method

The subset of the features is aimed to be optimized with the DE algorithm. For the k-NN model, $k = 3$, Euclidean distance for the distance metric is used. Classification Error Rate (*CER*) was used as a percentage to measure the success of the classification process. In Eq. 8, the calculation of *CER* is shown and in Eq. 9, the calculation of cost value is shown.

$$CER(\%) = \frac{total - true}{total} \times 100 \tag{8}$$

$$Cost = Minimize(CER) \tag{9}$$

Here, *total* represents the number of samples in dataset and *true* represents the number of samples which are correctly classified.

3 Experimental Results

In this study, all classification processes have been performed by using MATLAB R2016a software. The source codes of the DE algorithm has been used from its original source [21]. Five different datasets were used from the UCI database [22]. The comparisons were made with Correlation Based Feature Selection (CBFS), Information Gain Based Feature Selection (IGBFS) and Learner Based Feature Selection (LBFS) methods. WEKA 3.9 was used for these FS methods. Information about datasets that have been used are given in Table 1.

In all experiments, the number of population (*NP*) of the DE algorithm has been selected as 50. For each dataset; 50, 100 and 150 iterations were run and the results were obtained. The results for each experiment are shown in Table 2. FS results are compared according to Classification Success Percentage (*CSP*).

Table 1. Datasets that have been used

Data set no	Data set name	Number of features	Number of records
1	Parkinsons	22	195
2	Wine	13	178
3	Breast	9	116
4	Leaf	15	340
5	Glass	9	214

The CBFS, IGBFS, and LBFS methods use the datasets to rank the importance of attributes. During testing phase, the number of attributes selected by other FS algorithms is equal to the number selected by the DE algorithm. For example; in Wine dataset, the DE algorithm has selected five features. The classification process has been realized by using the best five features that CBFS, IGBFS and LBFS methods ranked. After determining the attributes with Weka, classification process is performed in MATLAB.

When the Table 2 is examined, the classifications obtained by using the DE algorithm have achieved the most successful results in all test results. In the Parkinsons dataset, the best results found with 150 iterations. Increasing the number of attributes naturally requires that the algorithm works with more iterations. The DE algorithm has set a difference of 32.80% to its closest competitors (CBFS, IGBFS) in the best case (Wine dataset). Also, the DE algorithm has set a difference of 2.60% to its closest competitors (CBFS) in the worst case (Glass dataset).

Table 2. Experimental results

Feature selection methods		Datasets									
		Parkinsons		Wine		Breast		Leaf		Glass	
	NP	Selected attributes	% CSP	Selected attributes	% CSP	Selected attributes	% CSP	Selected attributes	% CSP	Selected attributes	% CSP
DE	50	{1 4 6 7 10 11 12 13 14 16 17 19 21 22}	90.32	{1 6 7 9 10}	**97.78**	{1 2 3 6 8}	**91.97**	{4 5 7 8 10 11 14}	**58.24**	{2 3 4 6 7 8 9}	**73.33**
	100	{1 4 8 10 11 12 13 14 16 18 19 21}	90.32	{1 6 7 9 10}	**97.78**	{1 2 3 6 8}	**91.97**	{4 5 7 8 10 11 14}	**58.24**	{1 2 3 4 6 7 8 9}	**73.33**
	150	{9 10 11 13 15 17 18 20 22}	**92.32**	{1 6 7 9 10}	**97.78**	{1 2 3 6 8}	**91.97**	{4 5 7 8 10 11}	**58.24**	{2 3 4 6 7 9}	**73.33**
CBFS		{1 3 9 12 13 16 19 20 22}	86.21	{1 7 10 12 13}	73.63	{2 3 4 5 8}	71.52	{2 4 7 10 11 15}	47.06	{2 3 4 7 8 9}	71.47
IGBFS		{1 3 5 9 12 13 19 20 22}	83.13	{1 7 10 12 13}	73.63	{1 2 3 4 8}	74.17	{2 3 4 5 7 9}	49.12	{1 3 4 6 7 8}	70.17
LBFS		{1 2 3 4 5 6 7 8 9}	84.16	{1 2 3 4 5}	72.55	{1 2 3 4 5}	66.59	{1 2 3 4 5 6}	25.00	{1 2 3 4 5 6}	66.77

4 Conclusion

Today, optimization algorithms are used to solve many problems. Likewise, k-NN classifier is used in many real-life problems for classification. In data mining processes, datasets can include huge sizes of features. It is a time-consuming process to determine which of the features are necessary or which of them are unnecessary or has less effects. There are many methods that are used for this challenge.

In this study, FS process has been applied by using one of the latest versions of the DE algorithm in sample datasets. k-NN classifier is used to test the success of FS process. When the results are examined, it is seen that in all test results, the proposed method can be used in a successful way in FS processes.

References

1. Yong, Z., Dun-wei, G., Wan-qiu, Z.: Feature selection of unreliable data using an improved multi-objective PSO algorithm. Neurocomputing **171**, 1281–1290 (2016)
2. Amoozegar, M., Minaei-Bidgoli, B.: Optimizing multi-objective PSO based feature selection method using a feature elitism mechanism. Expert Syst. Appl. **113**, 499–514 (2018)
3. Qasim, O.S., Algamal, Z.Y.: Feature selection using particle swarm optimization-based logistic regression model. Chemom. Intell. Lab. Syst. **182**, 41–46 (2018)
4. De Stefano, C., Fontanella, F., Marrocco, C., di Freca, A.S.: A GA-based feature selection approach with an application to handwritten character recognition. Pattern Recognit. Lett. **35**, 130–141 (2014)
5. Khammassi, C., Krichen, S.: A GA-LR wrapper approach for feature selection in network intrusion detection. Comput. Secur. **70**, 255–277 (2017)
6. Zeng, D., Wang, S., Shen, Y., Shi, C.: A GA-based feature selection and parameter optimization for support tucker machine. Procedia Comput. Sci. **111**, 17–23 (2017)
7. Tao, Z., Huiling, L., Wenwen, W., Xia, Y.: GA-SVM based feature selection and parameter optimization in hospitalization expense modeling. Appl. Soft Comput. **75**, 323–332 (2018)
8. Hancer, E., Xue, B., Zhang, M.: Differential evolution for filter feature selection based on information theory and feature ranking. Knowl.-Based Syst. **140**, 103–119 (2018)
9. Mlakar, U., Fister, I., Brest, J., Potočnik, B.: Multi-objective differential evolution for feature selection in facial expression recognition systems. Expert Syst. Appl. **89**, 129–137 (2017)
10. Mafarja, M., Mirjalili, S.: Whale optimization approaches for wrapper feature selection. Appl. Soft Comput. **62**, 441–453 (2018)
11. Mafarja, M.M., Mirjalili, S.: Hybrid whale optimization algorithm with simulated annealing for feature selection. Neurocomputing **260**, 302–312 (2017). https://doi.org/10.1016/J.NEUCOM.2017.04.053
12. Yan, C., Ma, J., Luo, H., Ashutosh, P.: Hybrid binary coral reefs optimization algorithm with simulated annealing for feature selection in high-dimensional biomedical datasets. Chemom. Intell. Lab, Syst (2018)
13. Singh, S., Singh, A.K.: Web-spam features selection using CFS-PSO. Procedia Comput. Sci. **125**, 568–575 (2018)
14. Inbarani, H.H., Azar, A.T., Jothi, G.: Supervised hybrid feature selection based on PSO and rough sets for medical diagnosis. Comput. Methods Programs Biomed. **113**, 175–185 (2014)
15. Zorarpacı, E., Özel, S.A.: A hybrid approach of differential evolution and artificial bee colony for feature selection. Expert Syst. Appl. **62**, 91–103 (2016)

16. Mohamed, A.W., Suganthan, P.N.: Real-parameter unconstrained optimization based on enhanced fitness-adaptive differential evolution algorithm with novel mutation. Soft Comput. 1–21 (2017). https://doi.org/10.1007/s00500-017-2777-2
17. Cilia, N.D., De Stefano, C., Fontanella, F., di Freca, A.S.: A ranking-based feature selection approach for handwritten character recognition. Pattern Recognit. Lett. **121**, 77–86 (2018)
18. Xiang, J., Han, X., Duan, F., Qiang, Y., Xiong, X., Lan, Y., Chai, H.: A novel hybrid system for feature selection based on an improved gravitational search algorithm and k-NN method. Appl. Soft Comput. **31**, 293–307 (2015)
19. Storn, R., Price, K.: Differential evolution – a simple and efficient heuristic for global optimization over continuous spaces. J. Glob. Optim. **11**, 341–359 (1997). https://doi.org/10.1023/A:1008202821328
20. Yazdani, A., Ebrahimi, T., Hoffmann, U.: Classification of EEG signals using dempster shafer theory and a k-nearest neighbor classifier. In: 2009 4th International IEEE/EMBS Conference on Neural Engineering, pp. 327–330. IEEE (2009)
21. Shared Documents. http://web.mysites.ntu.edu.sg/epnsugan/PublicSite/SharedDocuments/Codes/2018-Soft-Comp-DE.rar. Accessed 27 Nov 2018
22. UCI Machine Learning Repository. https://archive.ics.uci.edu/ml/index.php. Accessed 30 Nov 2018

Intelligent Water Drops Algorithm for Urban Transit Network Design and Frequency Setting

Buket Capali[1(✉)] and Halim Ceylan[2]

[1] Department of Civil Engineering, Faculty of Engineering, Suleyman Demirel University, Isparta, Turkey
buketcapali@sdu.edu.tr

[2] Department of Civil Engineering, Faculty of Engineering, Pamukkale University, Denizli, Turkey

Abstract. Intelligent water drops (IWDs) algorithm is extremely effective heuristic procedure that is believed to have wide applicability for generating optimum solutions for the traveling-salesman problem in combinatorial optimization problems. Urban transit network design and frequency setting (TNDFS) are a combinatorial optimization problem that attracts the attention of researchers. In this study, computational results with IWDs for Mandl benchmark TNDFS problems are reported and compared to a few other meta-heuristic approaches.

Keywords: Urban transit network design · Frequency setting and intelligent water drops algorithm

1 Introduction

The network design where a series of routes are created is an important element in the overall planning of an urban transport system. The transit network design and frequency setting problem (TNDFSP) includes determining the route sets and associated with frequency. To date, different approaches have been proposed to solve TNDFSP.

In the literature, the metaheuristic algorithm "Intelligent Water Drops" was introduced and used to solve the traveling vendor problem (TSP), an NP-hard combinatorial optimization problem (Shah-Hosseini 2008). The intelligent water drop (IWD) algorithm is a new herd-based optimization algorithm desired by watching natural water drops movements in rivers (Shah-Hosseini 2009).

The main aim of the methods in this section is to supply the most cost-efficient way through the transport network for a specific origin (Origin-Destination) request. The planning period of public transport is very complex and consists of five main stages. Network design, timeline development, frequency setting, bus planning and scheduling of driver (Ceder and Wilson 1986). Various formulations have been suggested to model TNDFSP that purposes provided a set of routes with their suitable frequencies, optimizing the aims of operators and users. (Baaj and Mahmassani 1991). Chakroborty and Wivedi (2002) an optimization operation which developed "efficient" or "optimal" transit route sets for transit demand data and a dedicated road network from an initial set of routes is developed. Mandl (1980) suggested a two-phase heuristic method to

D. J. Hemanth and U. Kose (Eds.): ICAIAME 2019, LNDECT 43, pp. 894–897, 2020.
https://doi.org/10.1007/978-3-030-36178-5_78

analyze the transit network design problem. At first, a group of practical routes is created, and then the solutions that take into account the passenger travel time of the passengers are developed. This study related a real case of a network having 15,570 daily trips and 15 bus stops, which has been generally used by researchers (Nikolić and Teodorović 2013).

2 Computational Experiments and Discussion

IWDs algorithm that we suggested was applied to Mandl's (1980) benchmark Swiss network. Mandl's network has been studied by many researcher.

Mandl's swiss benchmark network contains 15 stops and 21 bidirectional arcs and 15.570 total passenger demand for services. Mandl' original demand matrix and transportation network are not present in paper because they are present in literature.

In Table 1, the suggested IWDs algorithm achieved the best for 4 routes. However, the best average travel time (ATT) recorded by the IWDs algorithm is the best in case of 8 routes.

Table 1. Comparison of solitions (best route sets) for Mandl's Swiss network.

Number of routes	Parameters	1	2	3	4	**5**
4	d_0	69.94	n/a	86.86	80.48	**89.78**
	d_1	29.93	n/a	12.00	12.84	**10.22**
	d_2	0.13	n/a	1.14	0.00	**0.00**
	d_{un}	0.00	n/a	0.00	6.68	**0.00**
	ATT	12.9	n/a	11.90	n/a	**12.07**
	RL	n/a	n/a	n/a	n/a	**135**
6	d_0	n/a	78.61	86.04	87.73	**94.245**
	d_1	n/a	21.39	13.96	12.27	**5.755**
	d_2	n/a	0.00	0.00	0.00	**0.00**
	d_{un}	n/a	0.00	0.00	0.00	**0.00**
	ATT	n/a	11.86	10.30	n/a	**11.27**
	RL	n/a	n/a	n/a	n/a	**200**
7	d_0	n/a	80.99	89.15	90.62	**92.51**
	d_1	n/a	19.01	10.85	9.38	**6.08**
	d_2	n/a	0.00	0.00	0.00	**0.00**
	d_{un}	n/a	0.00	0.00	0.00	**0.00**
	ATT	n/a	12.50	10.15	n/a	**10.84**
	RL	n/a	n/a	n/a	n/a	**222**
8	d_0	n/a	79.97	90.38	91.91	**93.17**
	d_1	n/a	20.03	9.62	8.09	**6.83**
	d_2	n/a	0.00	0.00	0.00	**0.00**
	d_{un}	n/a	0.00	0.00	0.00	**0.00**
	ATT	n/a	11.86	10.46	n/a	**10.62**
	RL	n/a	n/a	n/a	n/a	**284**

In all four situations noted in this study, the IWDs generated results with only one transfer needed to satisfy the whole travel demand. Moreover, as effective route set is described by the values of d_{un} 0.00, d_0 as high as feasible and ATT (as low as feasible), it can be achieved that the route sets designed by the suggested IWDs algorithm for all situations are highly effective as checked to other approaches in the literature.

Table 2. Result of Mandl's network for cases by the proposed IWDs algorithm.

Case	Number of routes	Frequency	Description of routes (Mandl)	Total travel time	In-vehicle time	Out-of-vehicle time	Transfer penalty
I	4	39	1 2 4 12 11 13 14 9	207.277	185.117	14.180	7980
		63	3 2 5 4 6 8 10 11				
		13	11 10 7 15 9				
		60	1 2 3 6 8 10 11 12				
II	6	19	3 2 4 6 8 15 7	212.698	188.235	19.983	4480
		27	1 2 5 4 6 8 15 9				
		25	1 2 4 12 11 13 14 10				
		36	10 7 15 6 4 12 11 13				
		21	5 2 3 6 15 7 10 14				
		23	9 15 6 8 10 11 13 14				
III	7	17	5 2 3 6 8 15 7	198.410	175.188	18.482	4740
		33	1 2 4 6 8 10 7 15				
		34	1 2 4 12 11 10 14 13				
		25	3 6 8 10 7 15 9				
		23	11 13 10 8 6 4 12				
		16	4 5 2 3 6 8 15 9				
		30	5 4 6 8 15 7 10 14				
IV	8	23	13 10 11 12 4 6 8 15	189.728	164.150	20.258	5320
		33	1 2 3 6 8 10 13 14				
		25	10 7 15 8 6 4 12				
		30	5 4 6 8 10 7 15 9				
		33	4 2 3 6 8 10 11 12				
		8	2 5 4 12 11 13 14				
		10	1 2 5 4 12 11 13				
		14	11 12 4 2 3 6 15				

In Table 2, the suggested IWDs algorithm achieved the best route sets for 4 situation with the IWDs algorithm is ranked other result. Besides, the best result listed by the IWDs algorithm is better total travel time in situation of 8 routes.

3 Conclusion and Suggest

In this study constructs effective route networks has been designed for urban route transit problem by IWDs algorithm. Within the framework of the method, a model has been devised to make calculations which can then be used as a basis for comparisons and help both users and operators of transport services to adopt their own standpoints. The model developed in this study is based on the traveling salesman problem. The results clearly show that the proposed IWDs algorithm is competitive with other approaches in the literature.

References

Baaj, M.H., Mahmassani, H.S.: An AI-based approach for transit route system planning and design. J. Adv. Transp. **25**, 187–209 (1991). https://doi.org/10.1002/atr.5670250205

Ceder, A., Wilson, N.H.: Bus network design. Transp. Res. Part B: Methodol. **20**(4), 331–344 (1986)

Chakroborty, P., Wivedi, T.: Optimal route network design for transit systems using genetic algorithms. Eng. Optim. **34**, 83–100 (2002). https://doi.org/10.1080/03052150210909

Mandl, C.E.: Evaluation and optimization of urban public transportation networks. Eur. J. Oper. Res. **5**, 396–404 (1980). https://doi.org/10.1016/0377-2217(80)90126-5

Nikolić, M., Teodorović, D.: Transit network design by bee colony optimization. Expert Syst. Appl. **40**, 5945–5955 (2013). https://doi.org/10.1016/j.eswa.2013.05.002

Shah-Hosseini, H.: Intelligent water drops algorithm: a new optimization method for solving the multiple knapsack problem. Int. J. Intell. Comput. Cybern. **1**(2), 193–212 (2008)

Shah-Hosseini, H.: Optimization with the nature-inspired intelligent water drops algorithm. Intech, pp. 297–320 (2009)

Tweet and Account Based Spam Detection on Twitter

Kübra Nur Güngör[1(✉)], O. Ayhan Erdem[2], and İbrahim Alper Doğru[2]

[1] Computer Engineering, Graduate School of Natural and Applied Science, Gazi University, Ankara, Turkey
kubranur.gungor@gazi.edu.tr

[2] Technology Faculty, Department of Computer Engineering, Gazi University, Ankara, Turkey
{ayerdem, iadogru}@gazi.edu.tr

Abstract. Today, with the spread of social media, the number of spam users identified as malicious users is increasing. Twitter is one of the most used social media platforms to send 500 million tweets per day. Twitter is used by every user, as well as malicious spam or fake users. One out of every 200 social media messages and 21 tweets is estimated to be spam. Spam detection is a critical issue because spam accounts are security-threatening accounts. Twitter cannot interfere with spam accounts. In this study, a method for spam detection was presented. Naive Bayes, J48 and Logistic machine learning methods were applied to obtain the accuracy of 97.2% with J48 algorithm.

Keywords: Twitter · Spam · Spam detection · Account based · Tweet based

1 Introduction

Nowadays, with the increasing use of social networks, these networks have become a tool for crime and malicious activities in the form of spamming. Many malicious activities such as stealing important information, fake news spreading, malware distribution is done by spam accounts. Spam accounts endanger the safety of users. According to Nexgate's report, spam accounts in social media increased about 355% by the first half of 2013 [1]. According to Nexgate's the other report, one out of every 200 social media posts is spam, and about 15% of Twitter users are bot accounts [2].

Twitter is one of the most widely used social network platforms that a social user called "tweet" and allows sending up post to 140 characters. Twitter allows users to share posts on any topic, such as events, news, magazines, politics. Twitter has 313 million active users who send 500 million tweets per day, which is an average of 350,000 tweets per minute [3]. Too much use of Twitter has become an opportunity for malicious users. Spam accounts often try to attract users' attention by sharing malicious links. Content filtering approaches are not effective for Twitter, because spam accounts tend to share abbreviated URLs to exceed the character limitation defined by Twitter. One of the most important reasons for Twitter to expand to 280 characters recently is to prevent URL abbreviations for the character limit of spam accounts. Chen et al. [4]

D. J. Hemanth and U. Kose (Eds.): ICAIAME 2019, LNDECT 43, pp. 898–905, 2020.
https://doi.org/10.1007/978-3-030-36178-5_79

after reviewing 400 million tweets and 25 million URLs, concluded that blacklists used in Google Secure Storage were ineffective in detecting spam.

There is no system created to prevent these spam accounts today, but there are many studies that have been done to detect spam accounts because Twitter provides dataset easily. Spam detection is a critical task for social media security. To protect users from various types of attacks, it is very important to identify spam on the online social network.

In this study, a method for detecting spam in Twitter is presented. 758 tweets from Twitter were manually labelled and accuracy rates were obtained by machine learning methods.

This study consists of 5 sections. Section 1 is the introduction and in Sect. 2, some of the previous spam detection studies have been mentioned. Section 3 describes the Twitter user features and spam accounts. In Sect. 4, the study is given, and Sect. 5 is the results and suggestions.

2 Related Work

Sicilia et al. in their study on the rumor news in the field of health in Twitter, they analyzed with the words #zikavirus and zika microcepholy and obtained a 90% accurate result [5]. Mateen et al. tweet-based features, and the average, median, maximum and minimum values of these properties have made use of spam detection [6]. Grier et al. They reviewed 400 million tweets and 25 million URLs. They found that 2 million URLs of unique URLs reviewed were spam. They also found that spam URLs would be blacklisted 4–20 days after they were detected. However, 90% of victims visited spam URLs within two days. As a result, they found that due to the delay of blacklists, they did not prevent victims from visiting spam URLs [7]. Chen et al. they proposed a system for blacklist technique and found that about 6% of 573 million tweets were spam, and 25% of the daily 400 million tweets contained a URL [4]. Inuwa-Dutse et al. they have obtained a dataset using some of the most frequently used spam accounts, and the use of two of these words has increased the likelihood of spam. They identified 700 of the 2000 accounts as spam and achieved 94.70% accuracy [2]. Van der Walt et al. Using the words school and homework, they prepared a dataset 18 years old, detected spam and obtained 97.49% accuracy [8]. Kwon et al. They examined the temporal, structural and linguistic aspects of the spread of rumors on Twitter. By using temporal, structural and linguistic features, they have trained structural classification algorithms (Random forest, SVM, etc.) of tweets. This approach was tested in a dataset 130 different subjects, 68 of which were rumors and 57 of which were nonrumor, with 90% resulting in best accuracy [9]. Sedhai and Sun used a semi-supervised technique for spam detection on tweet-based features [10]. Setiawan et al. identified spammers on Twitter in Indonesia using four classification algorithms: Naive Bayes (NB), Support Vector Machine (SVM), Logistic Regression (Logit) and J48. The highest accuracy of 93.67% was obtained by Logit Regression [11]. Benevenuto et al. it used user- based and tweet-based features to classify spam accounts and in this study, it was not determined how to deal with spam in languages other than English. The accuracy value was 87.6% with SVM [12]. McCord and Chuah have provided two methods based on user-based and tweet-based features for spam detection. The features used in the proposed method are the distribution of tweets over a 24-h period, the number of URLs, the number of mentioning, the number of retweets, the number of hashtags. The classification algorithm used in this study is Random Forest, Decision Tree, Naive Bayes and k-NN, which has the highest precision of 95.7% [13] Sohrabi and Karimil Facebook provided a solution based on machine learning for spam detection. They reviewed 200,000 posts on Facebook. They took into account the important features of the posts and their comments and applied the techniques of feature selection. The proposed method selects the most effective features to detect spam using machine learning techniques and reaches a better rate with SVM [14].

In this study, a new method of spam detection is presented; using the rate of FF that defines the rate of followers and friends of spam users, the number of URLs in the tweet and the hashtag in the tweets was done spam detection.

3 Twitter User Features and Spam Accounts

This section discusses Twitter, twitter user features and spam accounts.

3.1 Twitter

Twitter allows users to follow other accounts they want. Unlike other social media platforms, a user on Twitter can follow her/his without the other user being its follower, or Twitter allows users to follow other accounts they want. Unlike other social media platforms, a user on Twitter can follow her/his without the other user being its follower, or likewise a user does not have to follow its follower. The user can share a tweet by liking, retweeting, quoting with their own users or can interact responding to the tweet. This structure of Twitter is shown in Fig. 1. On Twitter, each user has a unique username, and if a user wants to mention to another user, he or she can do this by putting "@" in front of the username. Twitter provides a list of "the most talked about topics (tt)" for users to be notified immediately of the news. The term "Hashtag" also refers to the terms starting with "#" on Twitter that allow users to follow topics of interest [3]. Thanks to this Twitter structure, users can easily access the topics they are interested in. Thus, it has become one of the most widely used social media platforms.

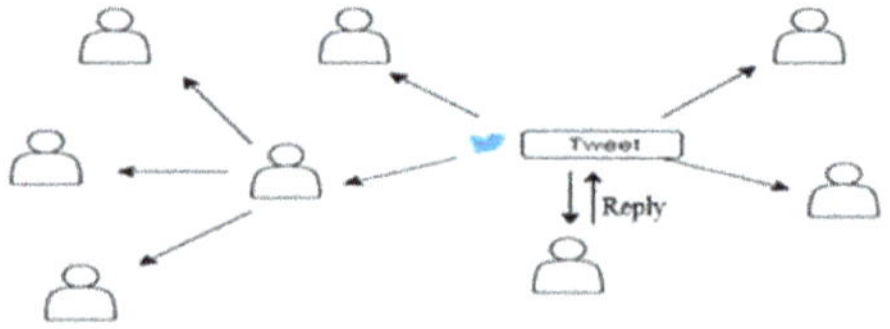

Fig. 1. The structure of Twitter

3.2 Users and Spam Accounts

User characteristics in Twitter can be examined in 2 groups: Tweet-based and account-based. In addition, the properties can be examined dynamically and statically. Static features are cannot be changed by the user; dynamic features are can changed by the user.

Tweet-Based Features. Tweet-based features are based on a post that users have written and shared. The Tweet-based features are shown in Table 1.

Table 1. Tweet-based features.

Feature	Definition	Dynamic/Static
Sender	Tweet sender	Dynamic
Mention	Mentions on tweets	Dynamic
URL	URL on tweet	Dynamic
Hashtag	Hashtags on tweet	Dynamic
Number of hashtags	Number of hashtags used in tweet	Dynamic
Number of spam words	Number of words frequently used by spam accounts in tweet	Dynamic
Number of likes	Number of likes of tweet	Static
Number of retweets	Number of retweets of tweet	Static
Number of replies	Number of replies to tweet	Static
Length of tweet	Number of characters used in tweet	Dynamic
Sharing date	Tweet sharing date	Dynamic
Location	Tweet sharing location	Dynamic

The most important feature that distinguishes spam account from a normal user is that the shared tweet contains a short link and hashtag. Spam users share hashtags and the most talked about topics in their tweets to have a wider audience, share short links in their tweets to attract attention. Since the number of followers of spam users is low, the number of tweets and retweeting is much smaller than normal users. Spam users often use the underlying words in their tweets. Spam users do random mentioning in their tweets to attract attention of other users.

Account Based Features. Account-based features are based on user relationships and user account properties. On Twitter, users can create their own social networks by following other users and allowing them to follow. As account features are associated with user properties, it is important to examine account-based properties for spam identification. Table 2 describes account-based features.

Table 2. Account-based features.

Feature	Definition	Dynamic/Static
Username	Unique name for the account	Dynamic
Screen name	The display name of the user in the profile	Dynamic
Profile image	Profile image of the user	Dynamic
Follower	Number of followers	Dynamic
Friend	Number of friends	Dynamic
Account creation date	The date on which the account was created	Static
Age of account	Current date-Date of account creation	Static
Number of tweets	Number of tweets posted by account	Dynamic
User likes	Total number of likes of the account tweets	Static
Retweet	Total retweets of account tweets	Static

Usually spam user's profile photo will not be a real profile photo. The display and usernames of spam accounts are often meaningless. The number of followers is much less than the number of friends. Spam accounts are frequently tweeted, so tweet counts are quite high, and retweets and likes are quite rare. The account ages are small because the spam accounts are blocked when detected.

4 Spam Detection Method

This section discusses the method for spam detection. Spam and non-spam data are obtained from Twitter is labeled. Afterwards, accuracy was obtained by machine learning techniques. Spam detection method is shown in Fig. 2.

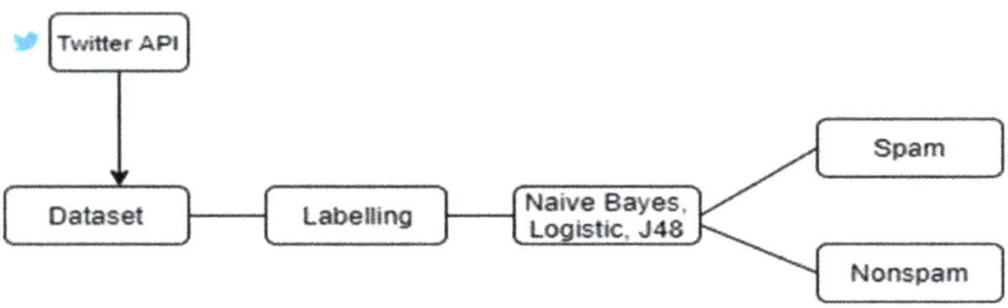

Fig. 2. Proposed spam detection method

4.1 Dataset

Spam detection to be made in the spam account in other studies to be able to obtain some spam keywords follow, like, price, free, buy, love, join, friend, life, game using the Twitter API with the Java language and MSSQL database using 714 Tweets were obtained, and 44 spam accounts were obtained from Twitter's @spam account. The properties of the obtained tweets are shown in Table 3.

Table 3. Dataset features.

Feature	Definition	Type
Tweet	Tweet content	Tweet-based
User	Name of user	Account-based
Username	Screen name of user	Account-based
Like	Number of likes of tweet	Tweet-based
Retweet	Number of retweets	Tweet-based
User language	Language of user	Account-based
User status	Total number of tweets shared by user	Account-based
Follower	Number of followers	Account-based
Friend	Number of friends	Account-based
User likes	Total number of tweets that the user likes	Account-based

Three properties are added to the dataset. These features are shown in Table 4.

Table 4. Features added to data set.

Feature	Definition	Type
FF rate	Follower/Friend rate	Account-based
Number of URL	Number of URL in tweet	Tweet-based
Is there hashtag	Whether hashtag in tweet	Tweet-based

4.2 Labelling

Spam accounts often share URLs on their tweets to direct users to bad content. They also share the *hashtag* so they can be seen by more users. In addition, the number of users that spam accounts follow is usually more than the number of followers. Because they are not wanted to be followed by other users. The Tweet based feature hashtag, whether it is hashtag and the number of URLs feature has been obtained manually. In addition, a set of rules has been identified by adding followers and friend FF rate, and spam accounts have been identified.

The rule set was first tested for 44 accounts from the @spam account and it was determined that the rules were provided for 44 accounts. This rule set is also applied to other accounts, with 205 of 758 tweets labelled as spam. The rule set is:

If there is one or two of hashtags or URLs in the Tweet, then the FF rate is checked.

If the FF rate is less than 1, the account is spam.

If the FF rate is greater than 1, the account is not spam.

4.3 Results of Machine Learning

The results of Table 5 were obtained by using Naive Bayes, Logistic and J48 methods from machine learning algorithms. The best accuracy was obtained by the J48 algorithm.

Table 5. Machine learning methods

Method	Nonspam:spam	Accuracy
Naive Bayes	386:372	%75,5
Logistic	620:138	%85,3
J48	550:208	%97,2

4.4 Other Studies and Proposed Method

In this section, other studies and proposed method are compared. Results of the other studies and the proposed method are shown in Table 6.

Table 6. Comparison of other studies and proposed method.

Study	Labelling	The best accuracy
Fame for sale: efficient detection of fake Twitter followers [15]	Follower >= 30 Tweets >= 50 2*followers >= friends	More than %75
Investigating the deceptive information in Twitter spam [4]	Tweets with URL	%75
Detecting Indonesian Spammer on Twitter [11]	Tweets with Indonesian spam words	%93,67
Cyber-security: Identity deception detection on social media platforms [8]	Tweets with School and homework words	%97,49
Proposed method	One or both URL and hashtag, FF rate < 1	%97,2

When the studies are examined, it is seen that only the account-based or tweet-based features have lower accuracy than using both account-based and tweet-based features. In the proposed method, FF rate that is an account-based feature and is there hashtag and number of URL that is tweet-based features are used.

5 Result and Suggestions

Twitter is one of the most widely used social media platforms. There are also users who misuse usage, want to compromise other users' security, share malicious URL in their posts, and redirect users to bad software. In order to identify and prevent these users, known as spam, twitter has not yet found a complete solution. Spam detection has become a critical issue. Although much work was done, still the desired results could not be obtained.

This paper presents a method for spam detection on Twitter. Dataset were obtained by using spam words and some labeling was done on this dataset. J48, Naive Bayes and Logistic machine learning methods were applied and obtained with J48 with the best accuracy rate of 97.2%. In addition, according to only tweet-based feature or only the use of account-based features, the use of both tweet and account-based feature increased the accuracy rate further increased. In future studies, a more accurate analysis and better accuracy will be attempted by using more features and a new method.

References

1. Gupta, B.B., Sangaiah, A.K., Nedjah, N., Yamaguchi, S., Zhang, Z., Sheng, M.: Recent research in computational intelligence paradigms into security and privacy for online social networks (OSNs). Future Gener. Comput. Syst. **86**, 851–854 (2018)
2. Inuwa-Dutse, I., Liptrott, M., Korkontzelos, I.: Detection of spam-posting accounts on Twitter. Neurocomputing **315**, 496–511 (2018)
3. Kabakus, A.T., Kara, R.: A survey of spam detection methods on Twiter. Int. J. Adv. Comput. Sci. Appl. **8**, 29–38 (2017)
4. Chen, C., Wena, S., Zhanga, J., Xiang, Y., Oliver, J., Alelaiwi, A., Hassan, M.M.: Investigating the deceptive information in Twitter spam. Future Gener. Comput. Syst. **72**, 319–326 (2017)
5. Sicilia, R., Giudice, S.L., Pei, Y., Pechenizkiy, M., Soda, P.: Twitter rumour detection in the health domain. Expert Syst. Appl. **110**, 33–40 (2018)
6. Mateen, M., Iqbal, M.A., Aleem, M., Islam, M.A.: A hybrid approach for spam detection for Twitter. In: Proceedings of 2017 14th International Bhurban Conference on Applied Sciences and Technology (IBCAST), pp. 466–471 (2017)
7. Grier, C., Thomas, K., Paxson, V., Zhang, M.: @spam: the underground on 140 characters or less. In: 17th ACM Conference on Computer and Communications Security, CCS 2010, pp. 27–37. ACM, New York (2010)
8. van der Walt, E., Eloff, J.H.P., Grobler, J.: Cyber-security: identity deception detection on social media platforms. Comput. Secur. **78**, 76–89 (2018)
9. Kwon, S., Cha, M., Jung, K., Chen, W., Wang, Y.: Prominent features of rumor propagation in online social media. In: 2013 IEEE 13th International Conference on Data Mining (ICDM), pp. 1103–1108 (2013)
10. Sedhai, S., Sun, A.: Semi-supervised spam detection in Twitter stream. IEEE Trans. Comput. Soc. Syst. **5**(1), 169–175 (2018)
11. Setiawan, E.B., Widyantoro, D.H., Surendro, K.: Detecting Indonesian spammer on Twitter. In: 6th International Conference on Information and Communication Technology (ICoICT), pp. 259–263 (2018)
12. Benevenuto, F., Magno, G., Rodrigues, T., Almeida, V.: Detecting spammers on twitter. In: Collaboration, Electronic Messaging, Anti-Abuse and Spam Conference, vol. 6, p. 12 (2010)
13. McCord, M., Chuah, M.: Spam detection on twitter using traditional classifiers. LNCS (LNAI. LNBI), vol. 6906, pp. 175–186 (2011)
14. Sohrabi, M.K., Karimil, F.: A feature selection approach to detect spam in the Facebook social network. Arab. J. Sci. Eng. **43**, 949–958 (2018)
15. Cresci, S., Di Pietro, R., Petrocchi, M., Spognardi, A., Tesconi, M.: Fame for sale: efficient detection of fake Twitter followers. Decis. Support Syst. **80**, 56–71 (2015)

A Walking and Balance Analysis Based on Pedobarography

Egehan Cetin[1], Suleyman Bilgin[2(✉)], and Okan Oral[3]

[1] Department of Electrical and Electronics Engineering, Institute of Natural Sciences, Akdeniz University, Antalya, Turkey
egehancetin@gmail.com
[2] Department of Electrical and Electronics Engineering, Faculty of Engineering, Akdeniz University, Antalya, Turkey
suleymanbilgin@akdeniz.edu.tr
[3] Department of Mechatronics Engineering, Faculty of Engineering, Akdeniz University, Antalya, Turkey
okan@akdeniz.edu.tr

Abstract. Balance and gait analyzes are of great importance in the diagnosis of orthopedic diseases affecting the musculoskeletal system. Pedobarography devices, which have an important place in the analysis of walking signs, are frequently used in neurology, orthopedics, physical therapy and rehabilitation fields. Designed for study, the pedobarography device simultaneously transmits the dynamic walking signals of the persons to the computer environment wirelessly. On the computer interface, incoming data is analyzed by the discrete wavelet transform (DWT). The dominance of the frequency bands in which the energy density ratio of the approach and detail coefficients of the main signal divided into the lower frequency regions is obtained. The forms of walking through the approach and detail coefficients have been analyzed within themselves. Distinguishing features of slow, normal and fast walking modes can be shown between D1 (between 7.5 Hz and 15 Hz) and D2 (between 3.75 Hz and 7.5 Hz). It has been observed that the unbalanced and fast patterns in the D2 frequency band can be distinguished from the unbalanced and slow walking patterns in the frequency band D3 (between 1.875 Hz and 3.75 Hz).

Keywords: Pedobarography · Gait analysis · Discrete wavelet transform

1 Introduction

During walking, 306 muscles in the human body and 35 muscles in the lower extremities function [1]. Walking analysis is a systematic study that collects kinematic and kinetic data that define and characterize human walk. With pedobarography (plantar pressure measurement) the person performs gait analysis, which compared with healthy reference values, or the disorders are categorized within themselves to assist clinicians for preliminary diagnosis and diagnosis. Gait analysis is of great importance in the diagnosis of orthopedic diseases affecting the musculoskeletal system, neurodegenerative birth or acquired gait disturbances [2, 3].

D. J. Hemanth and U. Kose (Eds.): ICAIAME 2019, LNDECT 43, pp. 906–913, 2020.
https://doi.org/10.1007/978-3-030-36178-5_80

In the literature, it is an important data source of footpad pressure measurements in the investigation of athletic performance [4], in the detection of walking asymmetry [5], in the classification of neurodegenerative diseases [6]. Rangra et al. investigated the relationship between walking speed and heel height change of foot pressure [7]. Seungbum et al. evaluated the differences in the walking patterns of male and female individuals [8]. Joo et al. performed pedobarographic measurement results with Artificial Neural Networks (ANN) to estimate slow, normal, and fast walking speeds of participants [9].

In this study, we have developed a feature extraction method which can distinguish the slow, normal, fast, unbalanced walking patterns of five persons by the discrete wavelet transform over the pedobarography device we have developed. Especially the characteristic that distinguishes the unbalanced and slow walking form has been researched.

1.1 The Design of Pedobarography Device

The FSR 402 (Force Sensing Resistor) force sensor is a material whose internal resistance changes when a force is exerted on it. Physical pressures are sensors that are used to measure weight, stifle and regional strength. The force sensing layer consists of electrically conductive and insulating particles. Applying a force to the surface of the sensor film causes the particles to touch the conductive electrodes and change their internal resistance. If a pressure is applied to the sensor, the active areas contact the semiconductors, which causes a decrease in resistance.

FSR sensor does not cause discomfort to the person plantar pressure measurement due to thin. It is also less costly than capacitive and piezo-electric sensors and has better shock resistance [10]. The resistance of the FSR varies according to the applied pressure. As the pressure increases, the resistance decreases and when there is no pressure, the sensor behaves like an infinite resistance (open circuit) [11].

One pin of the FSR sensor in Fig. 1(a) is connected to the 5 V supply line while the other pin is connected to ground with a 1 kΩ pull-down resistor. The point between the FSR and the resistor R is connected to the analog input of the Arduino Uno to observe the value of the voltage at the resistor R. As the effective force on the sensor increases, the voltage at the R resistance will also increase. The system was designed by connecting 6 sensors in the same way (see Fig. 1).

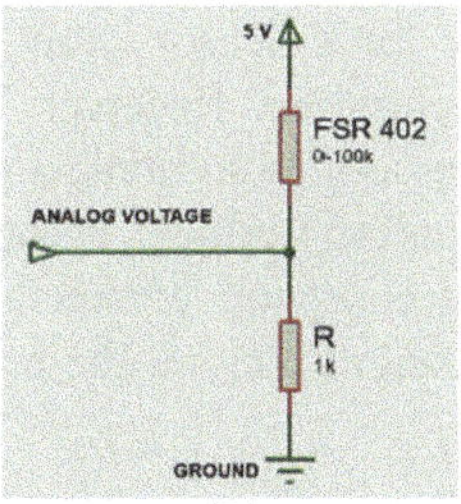

(a) **(b)**

Fig. 1. (a) FSR sensor (b) Circuit diagram of analogue measurement.

The current through resistor R will be 4.9 mA maximum, and the voltage will be 4.9 V. The total current drawn by the sensors from the V_{cc} pin will not damage the V_{cc} pin of the Arduino Uno and the system (Eq. 1).

$$I = \frac{V_{cc}}{FSR + R} = \frac{5}{0.1 + 1 \times 10^3} = 4.9 \text{ mA} \quad (1)$$

The part from the sensors to the ankle is soldered with 6 Pin JST connectors. The portion of the connectors from the female to the waist was transmitted to the main part of the device with a 6x0,22 LIYCY data cable. A 6-way Mike connector is installed at the entrance of the device box located in the waistline to prevent the internal equipment from being affected by the tension or loosening of the cable. By inserting sandals inside the sandals, it is ensured that the sensors are not damaged (see Fig. 2).

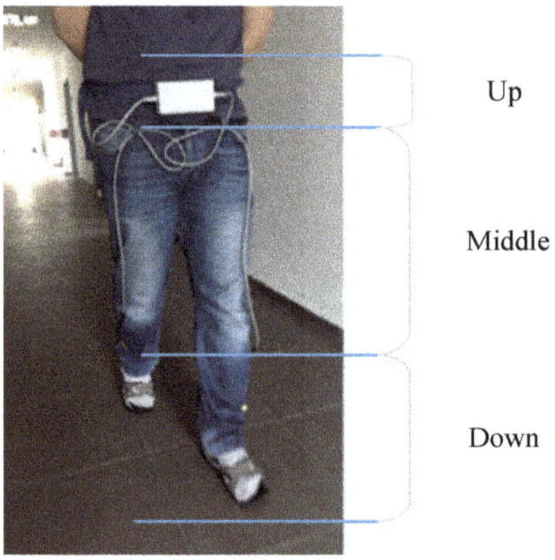

Fig. 2. Person's appearance and parts of the pedobarography device.

1.2 Data Transmission

The Arduino Uno's 10-bit analog-digital converter converts the sensor signal from 0 to 5 volts to a digital value from 0 to 1023. Data transfer via Bluetooth is performed via Arduino Uno 8 bit via the computer interface.

The signals from the force sensors come to the serial port of the computer via UART (Universal Asynchronous Receiver Transmitter). The incoming signals are configured via the Bluetooth communication protocol in the Serial Port Profile (SPP) library in the computer interface. Through the computer interface, the receiving Bluetooth module of the computer connects by comparing the name of the HC-05 Bluetooth module. The data arrives on the receiver side in 8-bit packages.

1.3 Location of Sensors

There is a need for force or pressure on the floor at the time of the unit, corresponding to the phases in the walking cycle. The locations of the sensors were determined by searching the regions in the literature that showed the basic features of the walking patterns of the persons. As shown in Fig. 3, three sensors are placed between the 1. and 2. metatarsal regions, between the 4. and 5. metatarsal regions, and to the right and left foot so as to correspond to the bottom of the talus bone.

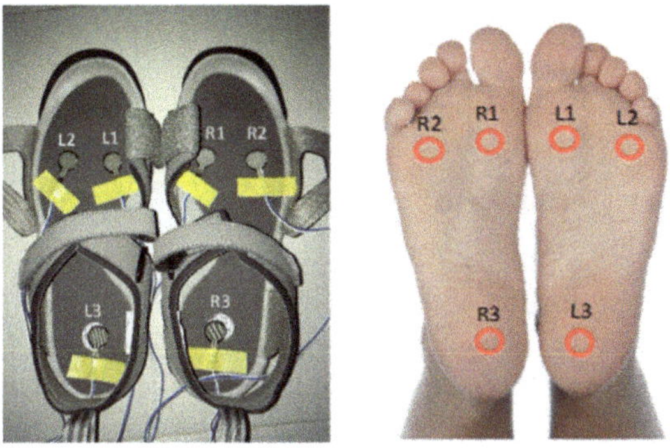

Fig. 3. Naming and positioning of FSR sensors.

2 Materials and Methods

2.1 Database Acquisition

In the study, 5 healthy subjects (3 males and 2 females, average weight 81.2 ± 13.66 kg, mean age 23.8 ± 1.64 years, mean height 179 ± 6.32 cm) were walked as slow, normal, fast and unbalanced in length 8.5 m on a flat track. One minute record was taken for each walking pattern. It is aimed to minimize the loss of data during recording by making a warning sound at the start and end of the recording.

They were asked to walk with a slow walk at 0.46 m/s, a normal walk at 0.88 m/s, a fast walk at 1.36 m/s, with an average speed. In unbalanced walking, it is very slow for the people and they are asked to walk with the right foot. For walking analysis, the sampling frequency of the signal obtained from each sensor is taken as 30 Hz. When digitized raw force signals are used in the study, the unit is on the 0–255 scale.

In order to provide the feature extraction, the signals of the FSR sensors at the base of the left and right feet were collected to obtain a combined force signal (CFS) (Fig. 4) [6, 12–14].

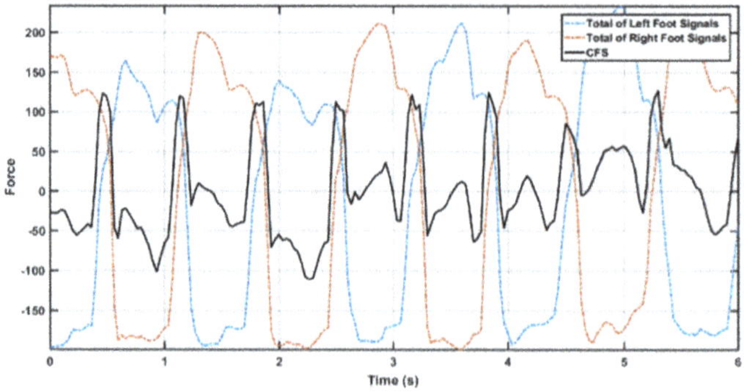

Fig. 4. Right and left foot base force signals and CFS.

2.2 Discrete Wavelet Transform

In order to decompose the force signals obtained in the walking analysis, the feature extraction process is performed with DWT aiming at providing the best time-frequency resolution by using a small-sized window at high frequencies and a large-sized window at low frequencies [15, 16]. The calculation of the CFS after reconstruction of the DWT approach (A) and detail (D) compartments is shown in Eqs. 2 and 3.

$$A_I(k) = S_{I,j}\phi_{I,j}(k) \tag{2}$$

$$D_i(k) = \sum_{j=0}^{2^{I-i}-1} T_{i,j}\psi_{i,j}(k) \tag{3}$$

Here i is the dilation parameter, j translation parameter, $\phi_{I,j}(k)$ the scaling function, and $\psi_{i,j}(k)$ wavelet function. S_I approach and T_i detail coefficient. Parameter I is the dissociation level of DWT. When the CFS signal is reconstructed as X, it is shown in Eq. 4.

$$X = A_I(k) + \sum_{i=0}^{I} D_i(k), \quad i = 1, 2, \ldots, I \tag{4}$$

In Table 1 we will use the results of calculations with the frequency points of the detail coefficients.

Table 1. Frequency intervals of details of 30 Hz sampling frequency.

Wavelet level	Frequency interval (Hz)
D1	7.5–15
D2	3.75–7.5
D3	1.875–3.75
D4	0.9375–1.875
D5	0.46875–0.9375

In this study, the detail and approximation coefficients are evaluated according to the total energy ratios of the signal. The E_C shown in Eq. 5 is the energy that represents the sum of the squares of the detail and approximation reconstruction components. E_T is the energy of the CFS and E_P is the percentage of the energy ratio.

$$E_P(t) = \frac{E_C}{E_T} \times 100 \tag{5}$$

3 Results and Discussion

The percentages of the energy values of the measurements in the database are calculated separately. These values are compared in four patterns in slow, normal, fast, and unbalanced walking patterns. In Fig. 5, the signals are decomposed into six levels of lower frequency bands, showing the energy intensities of the approach and detail components at each level. Daubechies, Symlets, Coiflets and Biorthogonal wavelets. It was observed that the db9 wavelet was the most successful in this direction. The distinguishing features of the slow, normal and fast walking patterns are shown in the D1, D2 frequency bands. It has been observed that the unbalanced and fast patterns can be distinguished in the D2 frequency band. D2, D3, D4, D5 frequency bands cannot be successful in separating normal and fast patterns.

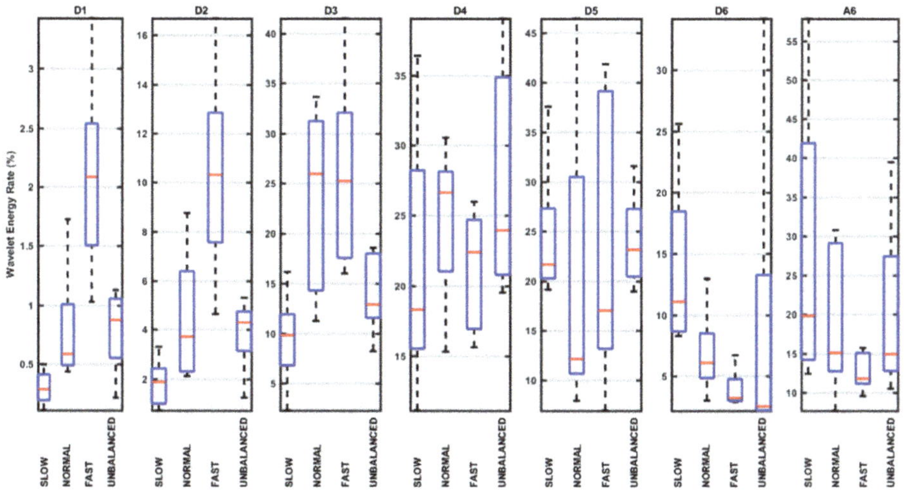

Fig. 5. The energy ratio comparison for the approach and detail coefficients of the Db9 wave.

The difference between the slow and unbalanced walking patterns in the D3 frequency band through the Bior3.5 wavelet can be clearly shown in Fig. 6(a). In Fig. 6 (b), the Bior5.5 wave differs from the rest in the D5 frequency band while the unbalanced gait remains in an area with limited energy density.

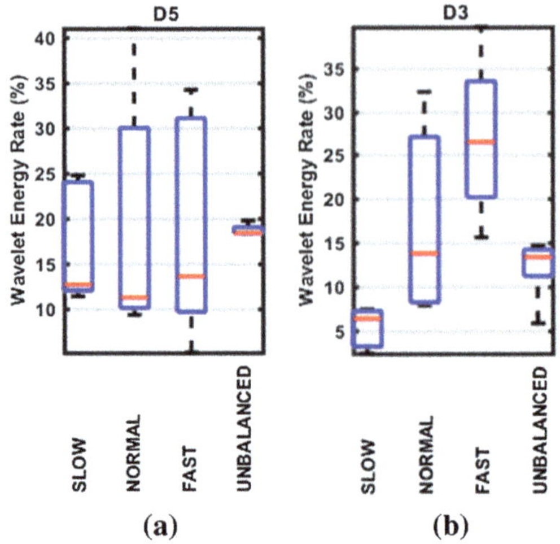

Fig. 6. (a) The energy ratio of the frequency band D3 of the bior3.5 wavelet. (b) The energy ratio of the D5 frequency band of the bior5.5 wavelet.

In the future studies, it is aimed to reach definite results by using classification and artificial intelligence algorithms of these evaluations and to associate walking speed with unbalanced walking.

Acknowledgments. All experiments were approved by the local Ethics Committees of the University of Akdeniz. Prior to the experiments, all subjects read, signed and informed consent participation form. This study was supported by Akdeniz University Industrial and Medical Applications Microwave Research Center (IMAMWRC) and the Research Projects Department of Akdeniz University, Antalya, Turkey.

References

1. Levine, D., Richards, J.: Whittle's Gait Analysis. Churchill Livingstone, Edinburgh (2012)
2. Hughes, J.: The clinical use of pedobarography. Acta Orthop. Belg. **59**(1), 10–16 (1993)
3. Sparrow, W.A., Tirosh, O.: Gait termination: a review of experimental methods and the effects of ageing and gait pathologies. Gait Posture **22**, 362–371 (2005)
4. Uzun, A., Aydos, L., Kaya, M., Yuksel, M.F., Pekel, H.A.: The study of the impacts of "running" on the contact area of soles and maximal strength among elite middle distance runners. Cypriot J. Educ. Sci. **12**(1), 23–31 (2017)
5. Viteckova, S., Kutilek, P., Svoboda, Z., Krupicka, R., Kauler, J., Szabo, Z.: Gait symmetry measures: a review of current and prospective methods. Biomed. Signal Process. Control **42**, 89–100 (2018)
6. Bilgin, S.: The impact of feature extraction for the classification of amyotrophic lateral sclerosis among neurodegenerative diseases and healthy subjects. Biomed. Signal Process. **31**, 288–294 (2017)
7. Rangra, P., et al.: The influence of walking speed and heel height on peak plantar pressure in the forefoot of healthy adults: a pilot study. Clin. Res. Foot Ankle **5**(2), 239 (2017)

8. Koo, S., et al.: Sex differences in pedobarographic findings and relationship between radiographic and pedobarographic measurements in young healthy adults. Clin. Orthop. Surg. **10**(2), 216–224 (2018)
9. Joo, S.-B., et al.: Prediction of gait speed from plantar pressure using artificial neural networks. Expert Syst. Appl. **41**(16), 7398–7405 (2014)
10. Rana, N.: Application of Force Sensing Resistor (FSR) in design of pressure scanning system for plantar pressure measurement. In: 2009 Second International Conference on Computer and Electrical Engineering (ICCEE), Dubai, pp. 678–685 (2009)
11. Interlink Electronics FSR (Force Sensing Resistors Integration): Guide Document
12. Bilgin, S., Akın, Z.E.: Gait pattern discrimination of ALS patients using classification methods. Turkish J. Electr. Eng. Comput. Sci. **26**, 1367–1377 (2018)
13. Bilgin, S.: The comparison of neurodegenerative diseases and healthy subjects using discrete wavelet transform in gait dynamics. J. Med. Bioeng. **6**(1), 35–38 (2017)
14. Bilgin, S., Güzeler, A.C.: Naive Bayes classification of neurodegenerative diseases by using discrete wavelet transform. In: 19th National Biomedical Engineering Meeting (BIYOMUT), Istanbul, Türkiye, 5–6 Kasım 2015, p. 1 (2015)
15. Sekine, M., Tamura, T., Akay, M., Fujimoto, T., Togawa, T., Fukui, Y.: Discrimination of walking patterns using wavelet-based fractal analysis. IEEE Trans. Neural Syst. Rehabil. Eng. **10**(3), 188–196 (2002)
16. Addison, P.S.: The Illustrated Wavelet Transform Handbook: Introductory Theory and Applications in Science, Engineering, Medicine and Finance. IOP Publishing Ltd. (2002)

Optimization of PMSM Design Parameters Using Update Meta-heuristic Algorithms

Cemal Yılmaz[1], Burak Yenipınar[2], Yusuf Sönmez[3](✉), and Cemil Ocak[3]

[1] Department of Electrical and Electronics Engineering, Gazi University Technology Faculty, Ankara, Turkey
cemal@gazi.edu.tr

[2] Graduate School of Natural and Applied Sciences, Gazi University, Ankara, Turkey
yenipinar.burak@gmail.com

[3] Department of Electrical and Energy Engineering, Gazi University Technical Sciences Vocational School, Ankara, Turkey
{ysonmez, cemilocak}@gazi.edu.tr

Abstract. The design of PMSMs, which are frequently used in various areas of the industry and have strong features, is an important process because the design parameters significantly determine its performance and physical properties. The design of the-se motor contains complex equations and requires a lot of calculation load. The realization of the optimal design increases this complexity. In this study, it is aimed to perform the optimization of the design parameters of the PMSM motor in an easy way and to examine its effects on the motor performance. In the optimization process, the design parameters of a PMSM modeled by entering the initial values on the Ansys Maxwell program have been optimized with the optimization algorithm run on Matlab. In this process, Ansys Maxwell and Matlab program were run interactively by written scripts. Here, it is aimed to eliminate the need for the mathematical model of the motor in the optimization process and to ensure that the current optimization algorithms are easily used in the process of parameter optimization. Experimental studies were carried out for this purpose. In the experimental study, current meta-heuristic algorithms, Artificial Bee Colony algorithm and Symbiotic Organisms Search algorithms were used to optimize motor design parameters. At the end of the optimization process, the effects of optimized motor parameters on its performance and physical properties were examined comparatively. As a result, it has been observed that the proposed optimization method works in the process successfully and this method produces more accurate results than initial parameters computed from analytical method. In addition, the results obtained from the SOS algorithm have been observed to increase the performance of the engine more than the results obtained from ABC.

Keywords: Permanent Magnet Synchronous Motor · Parameter optimization · Artificial Bee Colony · Symbiotic Organisms Search algorithm

D. J. Hemanth and U. Kose (Eds.): ICAIAME 2019, LNDECT 43, pp. 914–934, 2020.
https://doi.org/10.1007/978-3-030-36178-5_81

1 Introduction

Permanent Magnet Synchronous Motors (PMSM) are widely used in a wide range of applications such as automotive industry, aviation and space, electrical house-hold appliances and robotic applications due to their superior properties such as high efficiency, high power torque densities and compact construction [1, 2]. And thanks to their superior control characteristics, they are unrivaled in areas that re-quire particularly high torque and precision [3].

Considering that conventional asynchronous motors have low efficient and power factor, the interest in design of a high-efficiency PMSM motor is increasing. One of the biggest differences is that PMSM motors are permanent magnets on the rotor, when compared to asynchronous motors [4]. Therefore, the magnet structure and size have a great influence on the motor performance. For this reason, the question of how to determine the magnet geometry is one of the critical approaches in the design of PMSM motors, and there is no universal answer of this question [5].

Design such an electric motor requires a number of complex calculations using the Finite Element Method (FEM). The requirement of nonlinear characteristics and multiple objective functions leads to the further complication of the problem and the prolongation of solution time [6]. When designing an enhanced PMSM with a new topology or using innovative materials, design optimization is an essential step in order to achieve the targeted performance values.

In most cases, design optimization of PMSM motors is a multi-objective and non-linear problem. In solving such problems generally, it is aimed to minimize torque fluctuations, cogging torque, dimensions and material cost while the motor output torque, power, power density and efficiency are maximized [7]. Optimization in complex motor topologies is a time-consuming and complex process with increasing number of objective functions and constraints [8].

Efficiency optimization in designing is an important issue in order to save energy in applications such as continuously working machines, compressors, pumps and electric vehicles. The appropriate choice of permanent magnet and other active materials, such as design optimization of motor geometry and control strategies, can be expressed as some of the methods used to improve the efficiency of PMSM motors [9].

Another important consideration in design optimization is the optimization method used. Which method is used in design and in which environment the design is realized is a very important issue. Similarly, same factors are also important in solving of the optimization problem, because optimization is a powerful tool to achieve the best possible design [10].

Numerous optimization studies have been carried out in the literature for design and geometric optimization of PMSM motors. Some of them have been tried to be summarized together with the methods used and findings obtained.

Khazaei et al. achieved an optimal surface-localized PMSM with PSO and BA algorithms. When the results are examined, it is seen that the torque increased by 19.5% and the total losses decreased by 7.08% [11]. Raffaele and colleagues optimized the PMSM with a 4-pole 24-groove surface-leveling GA using Matlab Optimization Toolbox and Maxwell. The aim of the work is to make the nominal torque maximum while minimizing the weight. As a result of the study, it was ob-served that the main dimensions of the machine and the knocking torque decreased [12]. Cassimere and colleagues performed a design optimization of a PMSM with high efficiency and high torque/volume ratio using GA and PSO algorithms. As a result of the study, evolutionary-based algorithms were found to be effective in design optimization of electric machines. They also observed that discontinuous objective functions provide better optimization than continuous ones [13]. They used geometric design parameters as input to obtain high efficiency PMSM. In the study using PSO and GA as the optimization algorithm, the efficient value obtained with PSO is higher than with GA. Vlad et al. the hybrid analytical and FEM optimum design of PMSM with a surface loading at 1500 rpm and 150 W power made and compared with Hooke-Jeeves and ABC methods. Scientific work carried out by Hooke-Jeves' algorithm, increased efficiency and reduced total cost [14]. Konstantinos et al. examined the effect of the magnet shape on back EMF and iron losses. They provided a methodology based on geometry optimization to generate sinusoidal back EMF and applied this methodology to a PMSM for electric vehicles. At the end of the study, 2 prototype rotors were produced and one of them has no offset in the magnet geometry while the other has offset. In tests where the rotor was replaced with the condition that the stator remains the same, it was seen that the offset magnet brought the back EMF closer to the sinusoidal form and lowered the iron loss [15]. Brahim et al. have tried to reduce the cogging torque by optimizing the magnet geometry in PMSMs with different pole and stator slot numbers. As a result, they observed that if the number of stator slots per pole increases, the cogging torque decreases [16].

Today, there are various analytical and FEM-based programs that contain both design and optimization solutions. These programs offer practical and quick solutions, but they have some limitations. The number and adequacy of the optimization methods that programs currently have are at the forefront of these limitations. Therefore, it is not possible to use an original or more powerful optimization method. In this case, it may be necessary to use different programs for solving the problem. While the optimization is performed, it is necessary to directly communicate the design software with the software where the optimization is performed, considering that the design has to be analyzed again for each variable. Otherwise, a nonlinear mathematical model of the machine will need to be placed in the optimization problem. The creation of such a model is rather cumbersome and equally challenging to develop a design software. Considering these reasons, design and optimization solutions should be considered together.

In this study, the software that solves the design problem and the software that solves the optimization problem are communicated directly through the applied scripts. At this point, it was possible to use specific optimization methods for design optimization. One of the main purposes of the work is to optimize the magnet geometry of PMSM motors with surface mount. For this, Ansys Maxwell is used as design and analysis software, and Matlab MathWorks software is used to solve the optimization problem. Maxwell and Matlab work concurrently with the scripts providing the link between both programs. The machine model consisting of the geometric values determined by the optimization algorithm working in Matlab is solved on Maxwell and results coming from Maxwell are evaluated in Matlab again. Thus, design optimization with high accuracy can be realized by using the desired optimization method, and the dependence on the optimization methods included in the design programs is getting out of the way.

In the study, the sizing calculations of a surface-mounted PMSM, which targets the IE4 efficiency class, with 3 kW and at 1500 rpm, were first performed and modeled on Ansys Maxwell. This is named as initial design. Then, ABC and SOS optimization algorithms were developed on Matlab to solve design parameter optimization problem. By means of these algorithms, the geometric values of the magnets are optimized to provide the highest motor efficiency. The designs and geometric values obtained as a result of ABC and SOS optimizations with initial design are presented in detail in the following sections of the study and presented in various comparisons.

In the study, firstly modeled on the Ansys Maxwell and named as initial design, with 3 kW power to target IE4 efficiency class and PMSM engine sizing calculations with surface placement at 1500 rpm. Subsequently, ABC and SOS optimization algorithms were developed on Matlab to realize design optimizations. By means of these algorithms, the geometric values of the magnets are optimized to provide the highest motor efficiency. The designs and geometric values obtained as a result of ABC and SOS optimizations are presented in detail in the following sections of the study and presented in various comparisons with initial design.

2 Design and FEM Model of PMSM

2.1 Sizing of PMSM

The sizing of PMSM motors starts with the calculation of the output coefficient as in other machine types and this coefficient can be calculated as follows [17];

$$C_0 = 1,11 \, . \, \rho^2 \, . \, B \, . \, ac \, . \, K_w \, . \, 10^{-3} \tag{1}$$

The B and ac values, which have a direct effect on machine size, are determined in light of the designer's many years of experience [17–19]. Apparent power can be expressed as given in Eqs. 2 and 3.

$$Q = \frac{P_{out}}{\eta \, . \, PF} \tag{2}$$

$$Q = C_0 \, . \, D^2 \, . \, L \, . \, n_s \tag{3}$$

From the given equation, Q is the power of the motor in kVA, B is specific magnetic loading in Tesla, ac is the specific electric loading in A/m, D is inner diameter of stator core in m, L is stack length of the motor in m, n is the rated speed of motor in rpm, K_w is winding factor, and PF is expressed as power factor of the motor.

The specific electrical loading value is calculated by multiplying the total number of conductors in all phases by the peak value of the stator current divided by stator peripheral length [11, 17] and is expressed as given in Eq. 4.

$$ac = \frac{\sqrt{2}\,.I_a\,.Z_a}{\pi\,.D} \tag{4}$$

As can be seen from the equations, D^2L is an important parameter to calculate the useful volume of the motor and is given as follows;

$$D^2L = \frac{Q}{C_0 n_s} \tag{5}$$

D and L values are separated by the help of pole pitch value to obtain the desired efficiency/cost ratio.

$$\tau = \frac{\pi D}{2p} \tag{6}$$

Pole pitch is calculated as given in Eq. 6.

The sizing of PMSM motors starts with the calculation of the output coefficient as in other machine types and this coefficient can be calculated as follows [17].

2.2 Stator Slot Geometry

When the air gap distance increases in PMSMs, more magnets are needed to provide the necessary air gap flux density, which is a factor affecting the cost. If the increase of the air gap is positive, the air gap flow approaches the sine wave shape.

¼ symmetrical view of a four-pole PMSM motor and its geometric parameters are shown in Fig. 1.

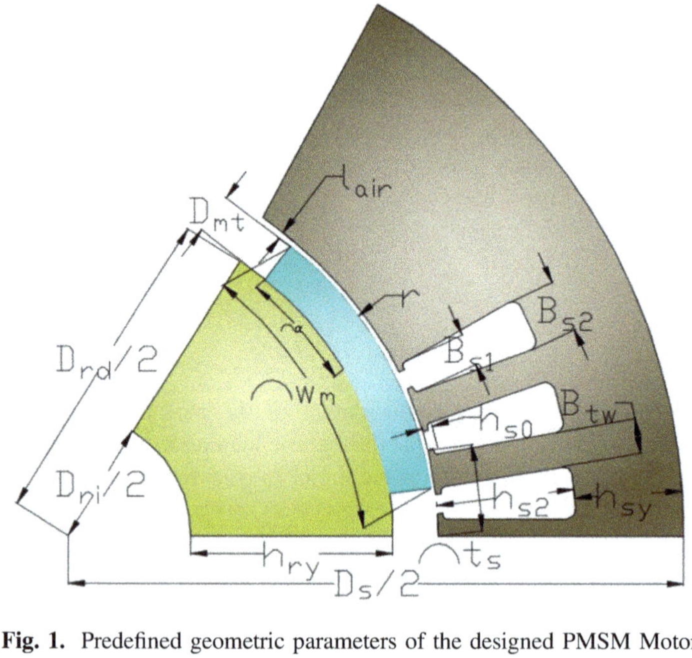

Fig. 1. Predefined geometric parameters of the designed PMSM Motor

Air-gap magnetic flux density can be expressed as in Eq. 7 [11, 20, 21].

$$\Phi_{air} = \frac{2 . B . \tau . L}{\pi} \tag{7}$$

From the equation, Φ_{air} is the air-gap flux in Weber and the total flux per-pole produced by the permanent magnet is given in Eq. 8.

$$\Phi_{pole} = B_m . \tau . L \tag{8}$$

B_m Express the maximum value of the air-gap flux density and the maximum flux in stator yoke area Φ_{sy} is expressed by Eq. 9 [20].

$$\Phi_{sy} = B_{sy} . h_{sy} . l . k_{nf} \tag{9}$$

From the equation, B_{sy} is the maximum flux density in the stator yoke, k_{nf} is stack factor, and h_{sy} is the length of the stator yoke. Equation 10 is obtained using the Eqs. 8 and 9 and expressed as below.

$$h_{sy} = \frac{\alpha . B_m . (D - 2 . l_{air})}{p . k_{nf} . B_{sy}} \tag{10}$$

α Value which is given in Eq. 10 is also shown in Fig. 1. Similar to the Eq. 10, the rotor yoke length is obtained if the maximum flux density of the rotor is used instead of the stator maximum flux density.

$$h_{ry} = \frac{\alpha . B_m . (D - 2 . l_{air})}{p . k_{nf} B_{ry}} \tag{11}$$

From the equation, h_{ry} is the rotor yoke distance, B_{ry} is the maximum flux density of rotor. To calculate the tooth width distance of stator slots, it's assumed that all the flux in the air-gap is passing through stator teeth and maximum stator tooth flux density is obtained [22]. Thus, Eq. 12 is obtained as below.

$$B_{tw} = \frac{B . \pi . (D - 2 . l_{air})}{Q_s . k_{nf} . B_{st}} \tag{12}$$

From the Eq. 12, B_{tw} is the stator tooth width, Q_s is the number of stator slots, and B_{st} is the maximum flux density value of the stator teeth. The initial values of the geometric parameters which shown in Fig. 1 are given in Table 1 below.

Finally stator slot area A_{sl} can be calculated as given in Eq. 13.

$$A_{sl} = \frac{B_{s1} + B_{s2}}{2} \cdot (h_{s2} - h_{s0}) \tag{13}$$

Total cross-section of the copper in the stator slot is given in Eq. 14. Additionally, the stator fill factor ratio f_s used in the following equation is directly changed according to the type of machine and slot geometry and is chosen by the help of designer's past experience.

$$A_{cu} = f_s A_{sl} \tag{14}$$

2.3 Initial Design Model and Performance

The initial analysis, which will be the starting point of the optimization study, was performed according to the data that obtained from analytical analysis explained previous section. Initial values of the geometric parameters are given in Table 1 and the main dimensions and performance values for initial design values are given in the Table 2.

Table 1. Initial values of the geometric parameters

Parameter	Value
$D_s/2$	75 mm
$D_{rd}/2$	39.5 mm
$D_{ri}/2$	15 mm
h_{ry}	24.5 mm
w_m	35 mm
D_{mt}	3 mm
r	R44.5
l_{air}	0.5 mm
t_s	10.6 mm
h_{s0}	1.2 mm
h_{s2}	16.8 mm
B_{s1}	4 mm
B_{s2}	6.3 mm
B_{tw}	4.3 mm
h_{sy}	13.2 mm

The mesh structure, winding losses, magnetic flux and magnetic flux density distributions are obtained, which are given in Fig. 2, from the FEA transient analysis results to confirm the numerical analysis results of the motor in Table 2.

Table 2. Main dimensions and performance values of initial design

Initial geometric parameters	Value
Number of poles	4
Rated output power	3 kW
Rated speed	1500 Rpm
Rated voltage	305 V
Stator outer diameter	150 mm
Stator inner diameter	90 mm
Number of stator slots	36
Air gap length	0.5 mm
Rotor inner diameter	30 mm
Core length	130 mm
Specific electric loading	24399 A/m
Armature current density	4,97 A/mm^2
Total loss	248,359 W
Total PM weight	0,389 kg
Efficiency	92,35%

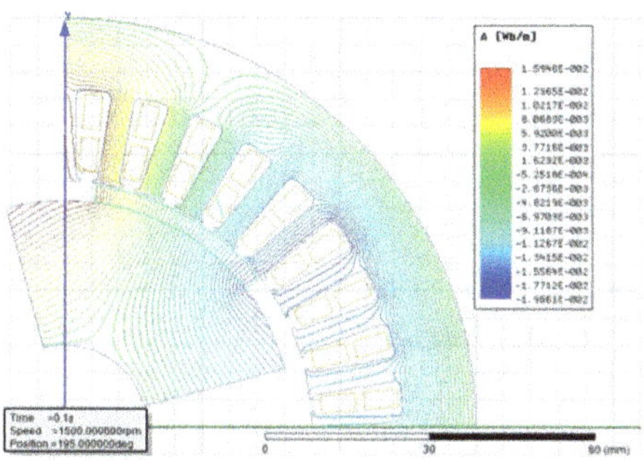

a-) Distribution of Magnetic Flux

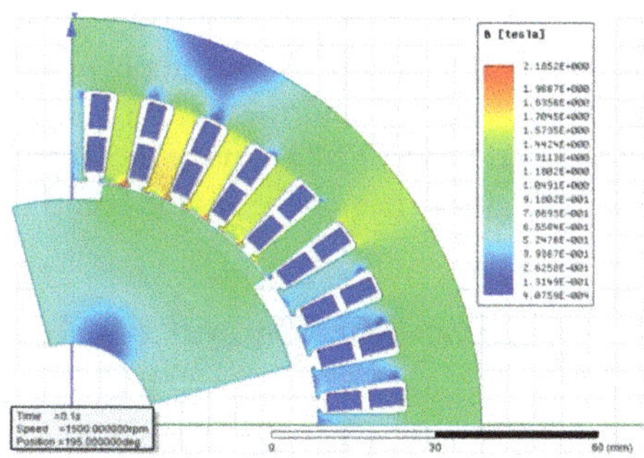

b-) Distribution of Magnetic Flux Density

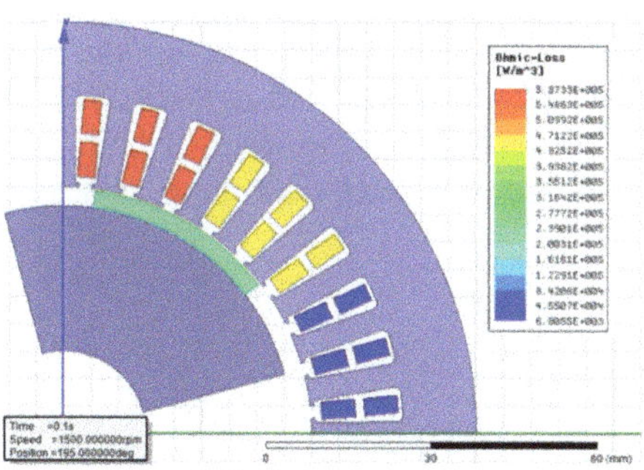

c-)Ohmic Loss

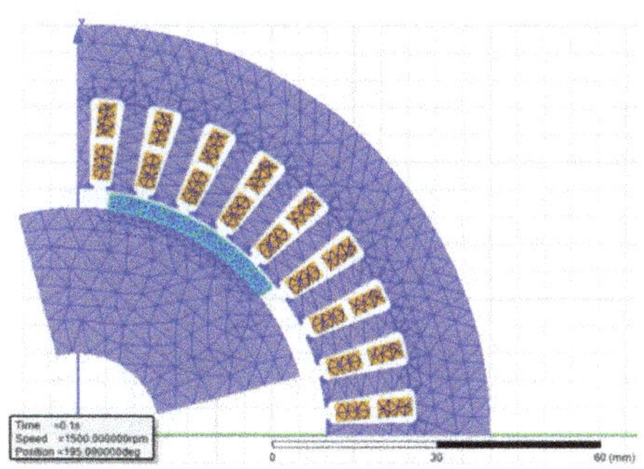

d-) Mesh

Fig. 2. FEM analysis results of the designed PMSM

The input-output power and torque parameters are also shown in Fig. 3, which is the result of this transient analysis of PMSM performed FEA analysis between 0–100 ms.

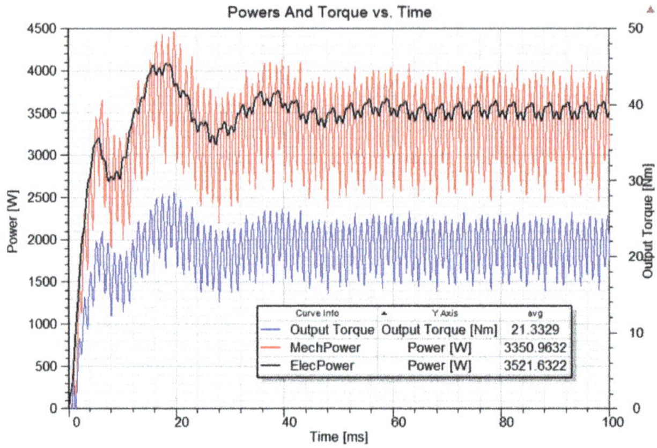

Fig. 3. Input-output power and torque graph between 0–100 ms of the designed PMSM

When the values in the graph above are compared, the difference between the analytical analysis and the design validation analyzes performed with FEM is better understood. The FEM analysis provides detailed information on both the output and magnetic properties of the motor being designed.

3 Optimization Method of Motor Parameters

The optimization study realized involves the improvement of the magnet geometry of the SPMSM, which is designed and analyzed analytically. With the optimization of the geometric parameters of the magnet, which are referred to in the initial design, a possible increase in the efficiency of the motor is aimed. During the optimization studies, two different optimization algorithms were created and used in the MATLAB environment. The same parameters and boundary conditions are used in both optimization algorithms for optimized magnet geometric parameters and these values are given in Table 3.

Table 3. Optimization parameters

Magnet parameters	Initial value	Minimum value	Maximum value
Magnet thickness	3	2.5	5
Offset	0	0.5	12
Embrace	0.5	0.5	1

In the study, Analytical and FEM analyzes of the motor were performed in the Ansys environment, while optimization studies for the motor were performed in Matlab. With a generated script, solutions for each different parameter that optimization algorithm finds suitable are performed in Ansys environment and the results are drawn to Matlab and a loop that continues until the appropriate result is obtained is created. With the developed method, the desired optimization algorithm can be used for engine design. In this study, SOS and ABC optimization algorithms are used and the results are compared. The objective function given in the equation below aims to minimize loss and it has been used in this study.

$$F = 100 - Eff \tag{15}$$

Table 3 shows the initial values of the geometric values of magnet and the minimum and maximum values that can be taken in the optimization process. The parameters obtained for the initial design were solved using the ABC and SOS optimization algorithms at the minimum and maximum value ranges given in the table and the results were compared. Thus, geometric parameters of magnets that yield the highest efficiency value are obtained by considering the objective function given in Eq. 15.

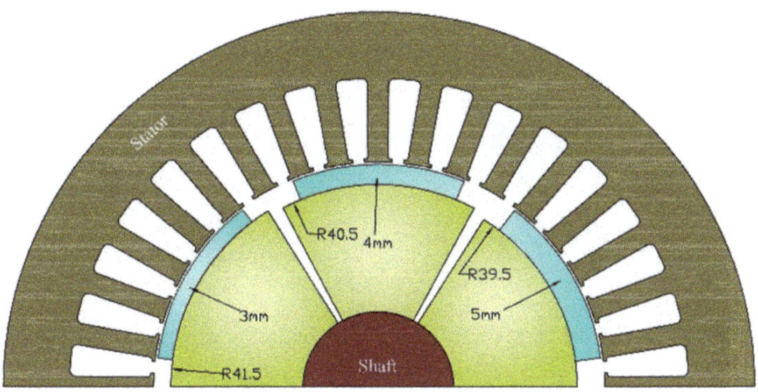

Fig. 4. Variation of several magnet thickness values associated with motor geometry

The physical representation of the variation of magnet thickness, offset and embrace values of the magnet on the motor geometry is given in Figs. 4, 5 and 6, respectively. Shapes are drawn in scale and belong to the actual motor model.

As seen in Fig. 4, the thickness of the magnet is changed by keeping the air gap and shaft dimensions constant. As the magnet thickness changes, the rotor back iron measurements also change. During optimization, the magnet thickness value has been changed from 2.5 mm to 5 mm. The increase in magnet thickness increases total magnet consumption and therefore motor cost, as well as increases the air gap flux density and cogging torque value. Since one of the common features targeted by many designers is to achieve the maximum efficiency value with minimum magnet consumption, the optimization of the magnet thickness can be expressed as a necessity.

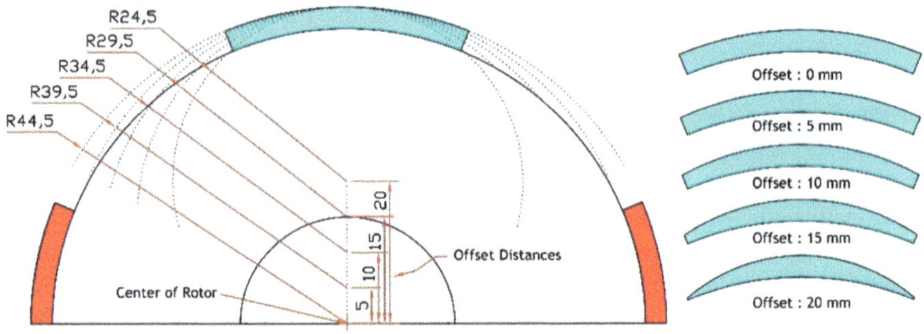

Fig. 5. Variation of several offset values and corresponding magnet geometry

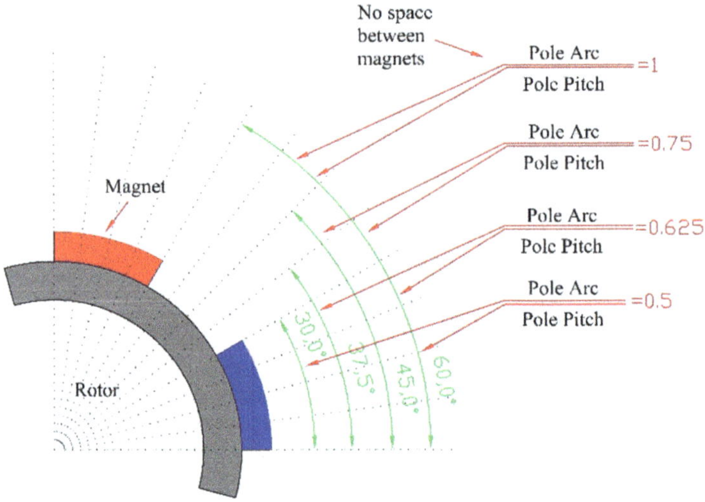

Fig. 6. Variation of several embrace (pole arc/pole pitch ratio) values

As shown in Fig. 5, the offset value can be expressed as an auxiliary point defined at a certain distance from the center of the rotor. This defined offset distance directly changes the surface geometry of the magnet to the air gap. As the offset value increases, the thickness of the magnet edgings decreases and a variable air gap distance arises from a constant air gap. This change is often preferred by generators as it often approximates the back EMF value obtained by varying the air space flux density to the sinusoidal form. By defining the offset value when considering the operating status as an engine, it is possible to obtain results such as reduction of harmonics, cogging torque and magnet consumption. On the other side, defining an offset value above the limits can cause significant reductions in the average air-gap value, causing the motor to drop at the output power. For this reason, by selecting an optimum offset value, the cogging torque and the magnet consumption can be reduced while the efficiency can be increased. In order to find the optimum offset value, the offset range determined in the optimization studies is 0.5–12 mm.

Change in the embrace value is given in Fig. 6. The Embrace value is also known as the pole arc/pole pitch ratio. This value of 1 means that the rotor surface is completely covered with magnets and there is no gap between the magnets. When looking at various motor designs, it is seen that this value usually changes by 0.5–1 according to design criteria. Since the change of embrace value changes the interaction between the magnet and stator gutters, it also affects the value of cogging torque in machines with open type gutters. However, the average air-gap density and wave form have a total magnet consumption, output power and efficiency. This parameter should also be optimized in order to be able to capture the relationship between optimum power density, magnet consumption and efficiency. As a result, the magnitude of the magnet, offset and embrace values have a direct effect on the magnet geometry and change the geometry of the magnet, with important parameters such as power and efficiency of the motor varying.

With the method used in this study, it is possible to use the desired optimization algorithms independently of the optimization algorithms in Ansys Maxwell, and the flow chart of this applied method is as shown in Fig. 7. Details of the ABC and SOS algorithms are given in Sects. 3.1 and 3.2, respectively, for the optimization of motors based on the geometric parameters of the magnet.

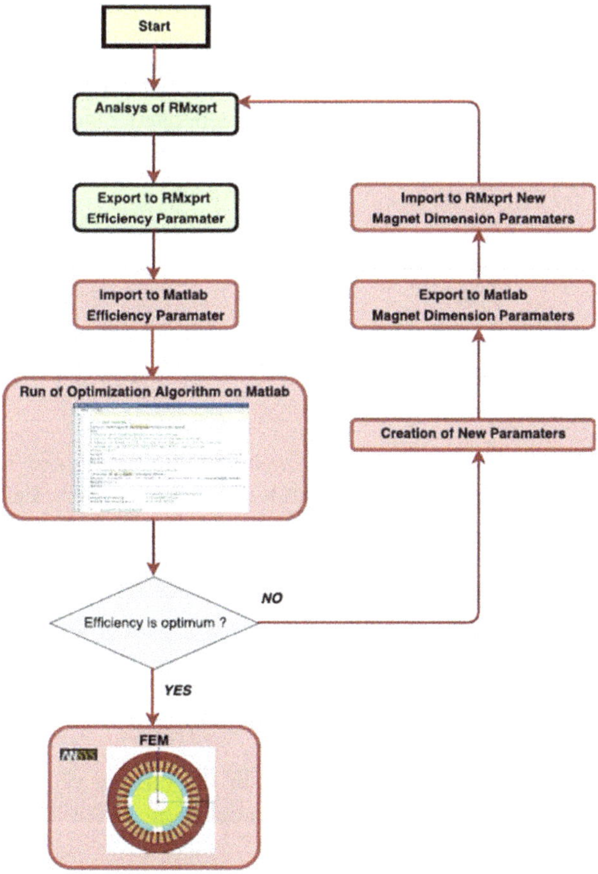

Fig. 7. Flowchart of Maxwell RMxprt and Matlab Optimization

4 Meta-heuristic Algorithms Used for Optimization

4.1 Artificial Bee Colony (ABC) Algorithm

The work to be done in a natural bee colony is made by bees specialized for that job. That is to say, according to the work to be done there is a division of labor between bees and they can organize themselves as they do this work without any central authority. It is the two most important features of the ability to do business and to be self-organizing [22]. The process steps for this algorithm can be given as follows:

At the beginning of the food search process, the explorers start searching for food by randomly searching around.

After finding food sources, the explorers become observers now, and they start transporting buckets of nectar from the sources they find. Each officer turns the bee hive and empties the nectar and either returns to the source he found after this point, or sends the information about the source to the watchmaker waiting at the cove through the dance she displayed on the dance floor. If the beneficial resource is exhausted, the official bee becomes an explorer bee and tries to search for new resources.

The waiters waiting in the canvas follow the dances pointing to the rich resources and prefer a source depending on the frequency of the dance, which is proportional to the quality of the food. The steps of the ABC algorithm have been explained below.

Producing of Initial Food Positions: In this step, initial food positions corresponding to solution candidates in the search space are generated randomly. For each food source, random value is generated between the upper and lower limits of each parameter as shown in Eq. (16).

$$X_{ij} = X_j^{min} + rand(0,1)(X_j^{max} - X_j^{min}) \tag{16}$$

Where i = 1, …, SN, j = 1 … D, SN is the solution number and D is parameter number, which will be optimized. X_j^{min} and X_j^{max} are lower and upper limits of the parameters, respectively.

Distributing Employers: In this step, all employers are distributed to food sources. Number of employer bee numbers is equal to food source numbers. Each employer bee identifies a new food source in the neighborhood and evaluates its quality. If the nectar amount of a new source has higher nectar amount, it takes the new source into its memory. This process called as greedy selection process. Identifying the new food source can be done following equation.

$$v_{ij} = X_{ij} + \emptyset_{ij}(X_{ij} - X_{kj}) \tag{17}$$

Where j and k are random integer values in the interval of [1, D] and [1, SN] respectively. $\emptyset_{ij}$ is a randomly selected weight coefficient in the interval of [−1 1].

Regional Probability-Based Selection: This is an example of multiple interaction by ABC. Probabilistic selection is performed using the fitness values corresponding to the amount of nectar in the algorithm. Selection based on fitness value can be done with any of roulette wheel, sorting, stochastic sampling, tournament method or other selection schemes. In the basic ABC algorithm, this selection was made using a roulette wheel. The angle of each slice on the wheel is proportional to the fitness value [22].

$$p_i = \frac{fitness_i}{\sum_{j=1}^{SN} fitness_j} \tag{18}$$

where $fitness_i$ corresponds to the quality of the i-th food source.

Selecting of Food Source by Onlookers: After calculating the probability values in the algorithm, random numbers are generated in the range [0, 1] for each source in the selection process according to the roulette wheel using these values and if the p_i value is greater than the generated random number, an onlooker bee produces a new solution according to Eq. (17). Then, the new solution is evaluated and its quality (fitness) is calculated. In this evaluation, greedy selection process are done again. If the new solution is better, this solution is taken instead of the old one and the limit counter is

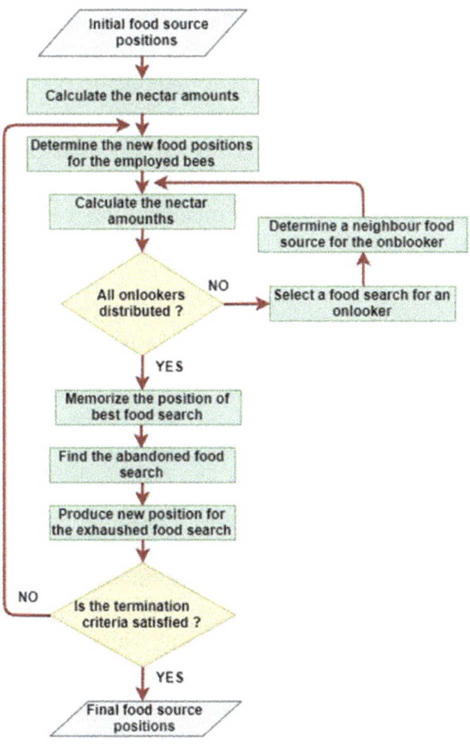

Fig. 8. Flow diagram of the ABC algorithm

reset. If the fitness of the old solution is better, this solution is kept and limit counter is increased one. This process continues until all scout bees are distributed to all food sources.

Abandoning of a Food Source: At the end of an epoch, after the completion of the search process done by employer and scout bees, the limit counter is checked. The limit counter determines whether or not the nectar of a used source is exhausted. If the limit value is above a certain threshold, then the onlooker bee of this source leaves that solution and looks for another source by becoming a scout bee. The general flow diagram of the ABC algorithm is shown in Fig. 8.

4.2 Symbiotic Organisms Search (SOS) Algorithm

Symbiotic Organisms Search (SOS) algorithm is a powerful optimization algorithm developed by Cheng and Prayogo in 2014 that searches for the organism that will provide the most appropriate association in relation to a paired symbiotic organism [25]. SOS simulates the relationship between two organisms that try to survive in the ecosystem. This relationship is defined as a symbiotic relationship and there are three basic symbiotic relationships in SOS [23–25]. These are Mutualism, Commensalism and Parasitism.

X_i is an organism in the ecosystem during the mutualism phase and X_k is another randomly selected organism in relation to X_i in this ecosystem. The calculation of a new candidate organism is given in Eq. (19) and (20). New organisms are managed by Mutual Vector (MV) and Benefit Factors (BF1 and BF2). MV (mean of the two organisms) is the relationship between two organisms Xi and Xk and it can be expressed as given in Eq. (21). Equations (22) and (23) shows that the benefit factors that are decided by a heuristic method. Hence, the benefit factors indicate two conditions in which organisms Xi and Xk can benefit partially or fully from the interaction. The organism with the best fitness value is considered the best organism (Xbest) of the ecosystem. At this stage, organisms Xi and Xk interact with the best organism.

$$X_i^{'} = X_i + rand(0,1)x(X_{best} - MVxBF_1) \tag{19}$$

$$X_k^{'} = X_k + rand(0,1)x(X_{best} - MVxBF_2) \tag{20}$$

$$MV = \frac{X_i + X_k}{2} \tag{21}$$

$$BF_1 = 1 + \text{round } [\text{random}] \tag{22}$$

$$BF_2 = 1 + \text{round } [\text{random}] \tag{23}$$

As in the mutualism phase, Xk is the organism which is chosen randomly from the ecosystem and interacts with Xi in this ecosystem. This symbiotic interaction results in a common-experience relationship that improves the fitness value of the Xi organism. However, the Xk organism does not cause any benefit or harm from the relationship. Furthermore, Xi interacts with the best organism of the ecosystem. Therefore, this phase protects the region that promises a good result near the best organism of the search field and tries to increase the speed of convergence of the algorithm. The mathematical equation of the new population is given in Eq. (24).

$$X_i^{'} = X_i + rand(-1,1)x(X_{best} - X_k) \tag{24}$$

In the parasitism phase, Xi is created using an artificial parasite, called the parasite vector. First, Xi is replicated and then randomly arranged in the search space, while this vector is generated. Similar to the previous phase, Xk is randomly selected from the ecosystem and serves as a host to the parasite vector. This parasite vector attempts to modify Xk in the ecosystem. The fitness values of the two organisms are evaluated and replaces Xk if the fitness of the vector is better. Otherwise, the parasite vector cannot survive long in the ecosystem, if Xk performs defense against the vector. The general flow diagram of the SOS algorithm is given in Fig. 9.

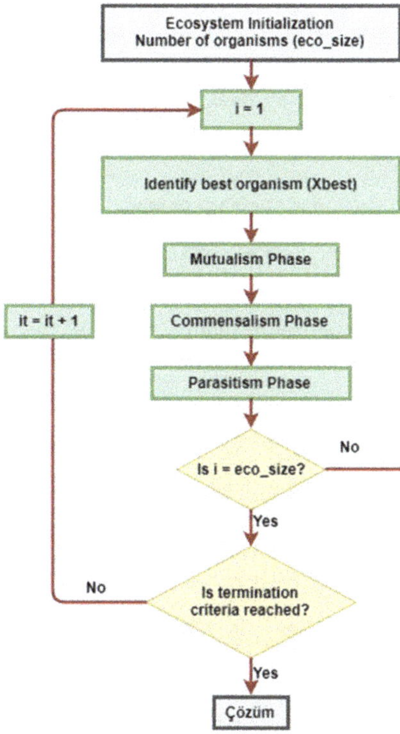

Fig. 9. General flow diagram of the SOS [25]

5 Experimental Results

In this study, both optimization algorithms were run for 5000 iterations with the same objective function, to obtain the best objective value. Magnet Thickness, offset and embrace values of the PMSM were chosen that will be optimized and efficiency value of the motor was chosen as fitness value. When the results of these two different optimization studies are compared, it is seen that the SOS algorithm reaches the solution faster and the global optimum value is higher. Convergence graphics for both algorithm are shown and compared in Fig. 10.

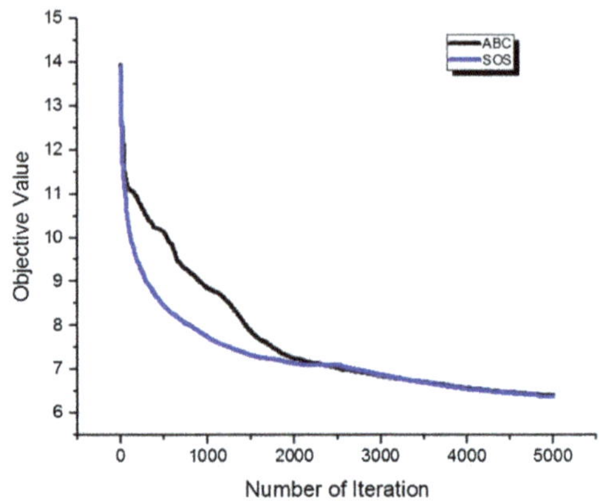

Fig. 10. Convergence graphics obtained from SOS and ABC

As a result of the study, the optimum magnet parameters and performance values obtained by two different optimization algorithms are shown in Table 4.

Table 4. Result of optimized parameters

Motor performance parameter	Initial value	ABC value	SOS value
Magnet thickness	3	2.523	2.5
Offset	0	5.269	1.558
Embrace	0.5	0.6573	0.6243
Efficiency	%92,35	%93,55	%93,58
Specific electric loading	24399	20930	20790
Armature current density	4,97	4,27	4,24
Core loss	19.57	24.86	25,63
Copper loss	178.78	131.98	130.22
Total loss	248,359	206,849	205,857
Permanent magnet weight (kg)	0,389	0,387	0,397

Although the analytical results of the optimization studies seem close to each other, the efficiency and torque characteristics of SPMSM are affected positively. When Table 4 is investigated, it can be seen that the highest efficiency with 93,58% is obtained by optimization with SOS algorithm. With ABC algorithm, total losses decreased by 16,71%, while SOS decreased by 17,11%. The magnet weight obtained by ABC algorithm is almost the same as the initial weight and the magnet weight obtained with SOS is 2% heavier.

The accuracy of the analytical analysis results was checked with 2D FEM analyzes performed with the initial parameter and the results of two optimal parameter results. The input-output power and torque graph obtained as a result of the realized FEA analysis are presented in Fig. 11. as comparative.

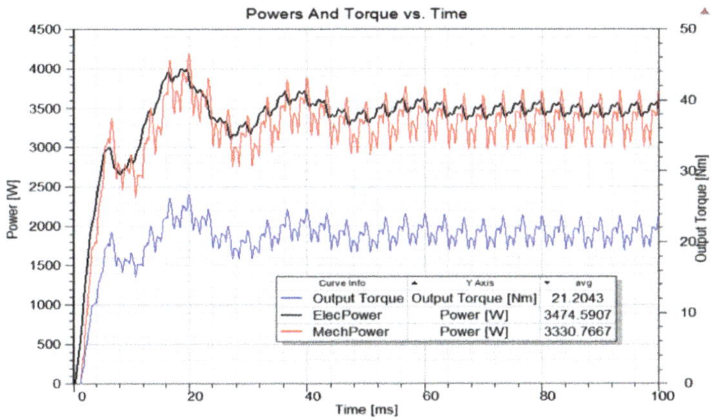

ABC

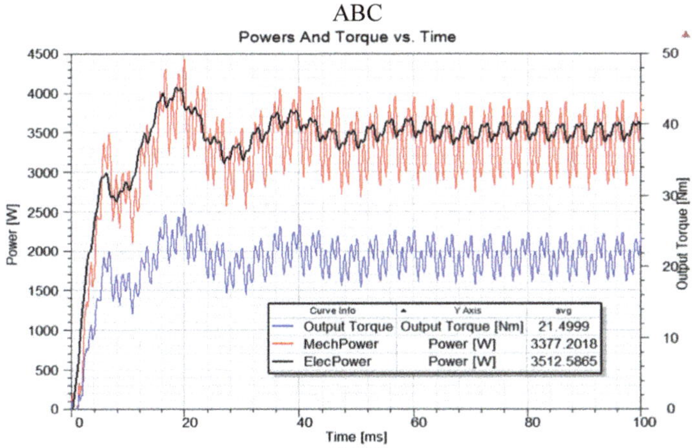

SOS

Fig. 11. Comparison of Torque and Efficiency graphics obtained by ABC and SOS

It can be clearly seen from Fig. 8. that the geometry of the magnet also affects the torque characteristics, not just the efficiency. The increase of the magnet offset value obtained by both optimization algorithms brought the air gap flux closer to the sinusoidal waveform, which was a factor that increased the output torque. This result is shown in Fig. 12.

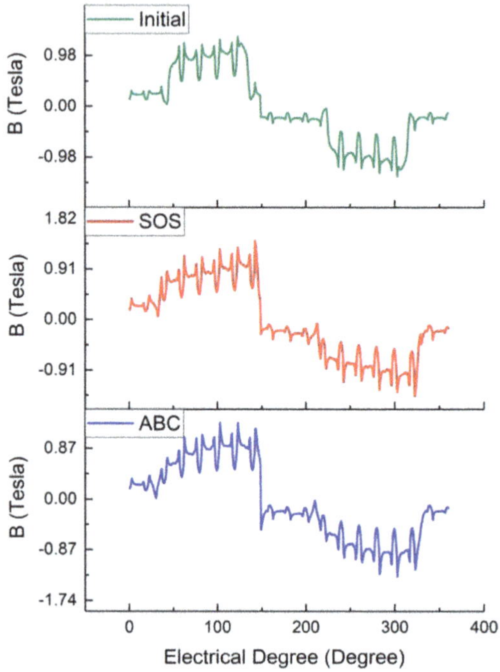

Fig. 12. Comparison of air-gap flux density at 100 ms.

6 Conclusions

In this study, optimization of the design geometry parameters of surface-mounted PMSM were realized by using latest meta-heuristic optimization algorithms. The parameters to be optimized are selected as magnet thickness, offset and embrace values of the motor. In the optimization process, the efficiency of the motor is selected as the fitness value and it is aimed to obtain the highest efficiency. In the experimental study, initial values were assigned to the motor parameters, and then these values were optimized and their effects on motor performance were compared. ABC and SOS algorithms were used for optimization. According to the results, performance improvements were observed with both algorithms but SOS algorithm showed better results than ABC algorithm. Moreover, the method used in the study will shed light on future studies. Thanks to this method, Optimization of any electromagnetic and electrostatic problem that can be modeled on Ansys Maxwell can be easily realized by using any optimization methods. In this study, although it is aimed to increase the efficiency, multi-objective optimization studies can be performed with the method used.

References

1. Islam, R., Husain, I., Fardoun, A., McLaughlin, K.: Permanent-magnet synchronous motor magnet designs with skewing for torque ripple and cogging torque reduction. IEEE Trans. Ind. Appl. **45**(1), 152–160 (2009)
2. Ocak, C., Tarimer, I., Dalcali, A.: Advancing pole arc offset points in designing an optimal PM generator. TEM J. **5**(2), 126 (2016)
3. Li, Y., Zou, J., Lu, Y.: Optimum design of magnet shape in permanent-magnet synchronous motors. IEEE Trans. Magn. **39**(6), 3523–3526 (2003)
4. Liu, X., Fu, W.N., Niu, S.: Optimal structure design of permanent magnet motors based on a general pattern of rotor topologies. IEEE Trans. Magn. **53**(11), 1–4 (2017)
5. Li, Y., Xing, J., Wang, T., Lu, Y.: Programmable design of magnet shape for permanent-magnet synchronous motors with sinusoidal back EMF waveforms. IEEE Trans. Magn. **44** (9), 2163–2167 (2008)
6. Lee, J.G., Hwang, N.W., Ryu, H.R., Jung, H.K., Woo, D.K.: Robust optimization approach applied to permanent magnet synchronous motor. IEEE Trans. Magn. **53**(6), 1–4 (2017)
7. Lei, G., Liu, C., Zhu, J., Guo, Y.: Techniques for multilevel design optimization of permanent magnet motors. IEEE Trans. Energy Convers. **30**(4), 1574–1584 (2015)
8. Zhu, X., Xiang, Z., Quan, L., Wum, W., Du, Y.: Multimode optimization design methodology for a flux-controllable stator permanent magnet memory motor considering driving cycles. IEEE Trans. Ind. Electron. **65**(7), 5353–5366 (2018)
9. Sreejeth, M., Singh, M., Kumar, P.: Particle swarm optimisation in efficiency improvement of vector controlled surface mounted permanent magnet synchronous motor drive. IET Power Electron. **8**(5), 760–769 (2015)
10. Gauchía, A., Boada, B.L., Boada, M.J.L., Díaz, V.: Integration of MATLAB and ANSYS for advanced analysis of vehicle structures. In: MATLAB Applications for the Practical Engineer. Intech (2014)
11. Khazaei, S., Tahani, A., Yazdani-Asrami, M., Gholamian, S.A.: Optimal design of three phase surface mounted permanent magnet synchronous motor by particle swarm optimization and bees algorithm for minimum volume and maximum torque. J. Adv. Comput. Res. **6** (2), 83–98 (2015)
12. Caramia, R., Palka, R., Wardach, M., Piotuch, R.: Multiobjective geometry optimization of a SPMSM using an evolutionary algorithm. In: International Symposium on Theoretical Electrical Engineering, vol. 4, pp. 51–52 (2013)
13. Mutluer, M., Bilgin, O.: Design optimization of PMSM by particle swarm optimization and genetic algorithm. In: 2012 International Symposium on Innovations in Intelligent Systems and Applications, pp. 1–4 (2012)
14. Grădinaru, V., Tutelea, L., Boldea, I.: Hybrid analytical/FEM optimization design of SPMSM for refrigerator compressor loads. In: 2011 International Aegean Conference on Electrical Machines and Power Electronics and 2011 Electromotion Joint Conference (ACEMP), pp. 657–662 (2011)
15. Laskaris, K.I., Kladas, A.G.: Permanent-magnet shape optimization effects on synchronous motor performance. IEEE Trans. Ind. Electron. **58**(9), 3776–3783 (2011)
16. Chikouche, B.L., Boughrara, K., Ibtiouen, R.: Cogging torque minimization of surface-mounted permanent magnet synchronous machines using hybrid magnet shapes. Prog. Electromagn. Res. **62**, 49–61 (2015)
17. Murthy, K.M.: Computer-Aided Design of Electrical Machines. BS Publications (2008)
18. Duan, Y.: Method for design and optimization of surface mount permanent magnet machines and induction machines. Doctoral dissertation, Georgia Institute of Technology (2010)

19. Pyrhonen, J., Jokinen, T., Hrabovcova, V.: Design of Rotating Electrical Machines. Wiley, Chichester (2013)
20. Donea, M.S., Gerling, D.: Design and calculation of a 300 kW high-speed PM motor for aircraft application. In: 2016 International Symposium on Power Electronics, Electrical Drives, Automation and Motion (SPEEDAM), pp. 1–6 (2016)
21. Meier, S.: Theoretical design of surface-mounted permanent magnet motors with field-weakening capability. Master Thesis. Royal Institute of Technology Department of Electrical Engineering Electrical Machines and Power Electronics, Stockholm (2002)
22. Karaboğa, D.: Artificial Intelligence Optimization Algorithms. Nobel Press (2014). (in Turkish)
23. Tejani, G.G., Savsani, V.J., Patel, V.K.: Adaptive symbiotic organisms search (SOS) algorithm for structural design optimization. J. Comput. Des. Eng. **3**(3), 226–249 (2016)
24. Eki, R., Vincent, F.Y., Budi, S., Redi, A.P.: Symbiotic organism search (SOS) for solving the capacitated vehicle routing problem. World Acad. Sci. Eng. Technol. Int. J. Mech. Aerosp. Ind. Mechatron. Manuf. Eng. **9**(5), 850–854 (2015)
25. Cheng, M.Y., Prayogo, D.: Symbiotic organisms search: a new metaheuristic optimization algorithm. Comput. Struct. **139**, 98–112 (2014)

Improve or Approximation of Nuclear Reaction Cross Section Data Using Artificial Neural Network

Veli Capali(✉)

Usak University, 64200 Usak, Turkey
veli.capali@usak.edu.tr
http://mbnm.usak.edu.tr/

Abstract. In this study; discusses the using artificial neural networks for approximation of data such as the nuclear reaction cross sections data. The rate of approximation of the fitting criteria is determined by using the experimental and evaluated data. The some reactions cross-section are calculated from data obtained using neural networks. The results show the effectiveness and applicability of this new technique in the calculation of the some nuclear reactions.

Keywords: Nuclear reaction · Reaction cross section · Artificial neural network

1 Introduction

In recent years, artificial neural networks (ANN) have appeared as an applicable algorithm with many applications in fields such as physics, engineering economics, and medicine. ANN is mathematics science that uses neural networks as models to either simulate or analyze complex phenomena and/or study the principles of operation of neural networks analytically [1].

First tries, ANN was purposed by the inspired to create models for brain. The basic building block of a ANN is the neurons that are connected to each other in a certain way. The model of a neuron given in Fig. 1. In the Fig. 1 represents the inputs are the output signals arriving at the input of the given neuron; Σ is the adder of input signals; f(x) is the calculator of the transfer function (the activation function); outputs are the output signals of the neuron; are the weight coefficients for input signals; Table 1 presents the most common activation function [2,3].

The neural network can have both a single-layer structure and a multilayer structure (Fig. 2). There are links between different layers of neurons, and the learning process is required to update the weights of these links. The activation function changing a neuron's weighted input to the output on these links [3].

The experimental reaction cross–section values may not be obtainable due to the lack of experimental availabilities or difficulties. In this kind of situations, the

D. J. Hemanth and U. Kose (Eds.): ICAIAME 2019, LNDECT 43, pp. 935–939, 2020.
https://doi.org/10.1007/978-3-030-36178-5_82

theoretical calculation of nuclear reaction cross–sections has come forward. Due to the importance of theoretical model calculations and experimental results in the area of nuclear reactor developing, there exist lots of studies on this topic in the literature [4–6]. Most of these studies have focused on the cross–section calculations by the computation codes.

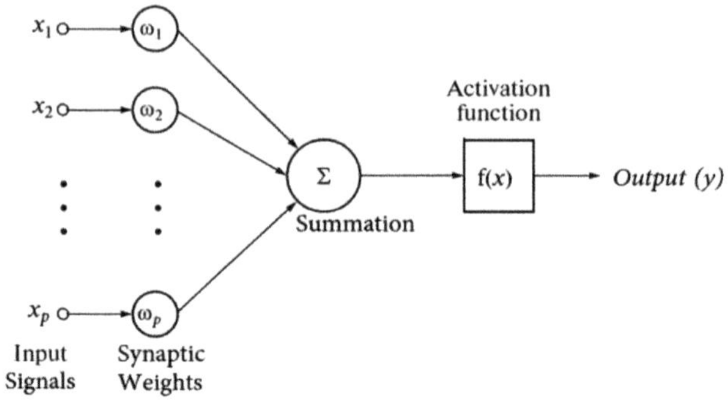

Fig. 1. Model of a neuron

Table 1. Activation function.

Name of function	Mathematical formula
Identity	A(x) = x
Sigmoid	$A(x) = \frac{1}{1+e^-x}$
Tangent	$A(x) = \frac{e^x - e^-x}{e^x + e^-x}$
Radial	$A(x) = exp(-\frac{S - R^2}{2\sigma^2})$
Step	$A(x) = \begin{cases} -1 & \text{if } x < 0 \\ 1 & \text{if } x \geq 0 \end{cases}$

In this study, analyzes the possibility of obtaining nuclear reaction data using neural networks. The obtained results have been also compared with the experimental values exists on the EXFOR [7] database and computed theoretical results TENDL [8] data library.

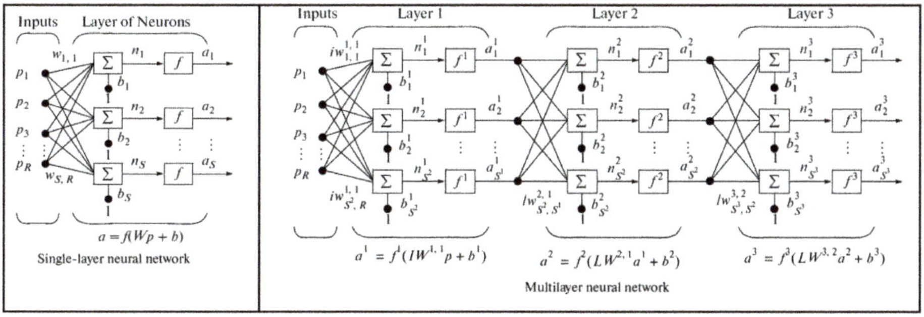

Fig. 2. Single and Multi layer neural network

2 Calculation Methods

The recommended neural network model of the nuclear cross-section has lots of input parameters and one output value. The input is energies, atomic number, mass number and the output is cross-section. Using these input-output values, different network configurations were tried to accomplish minimum mean square error and good performance for the network. Configurations of three different neural networks used in this work. E energy of an incoming particle, Z and N stand for proton and neutron numbers whereas S_γ, separation energy of gamma, σ is (n, γ) reaction cross-sections. The configuration, shown in Fig. 3.

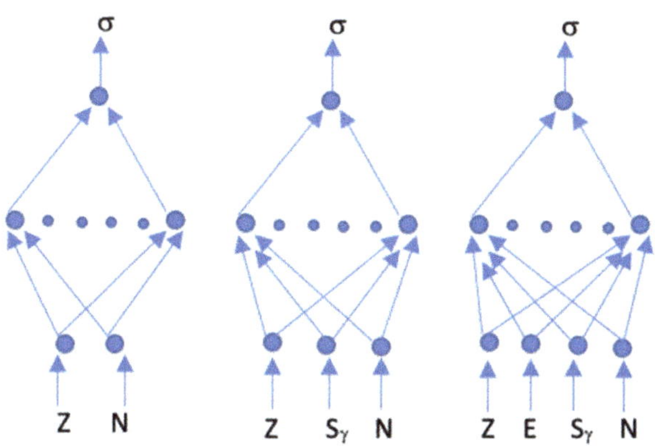

Fig. 3. Configurations of three different neural networks

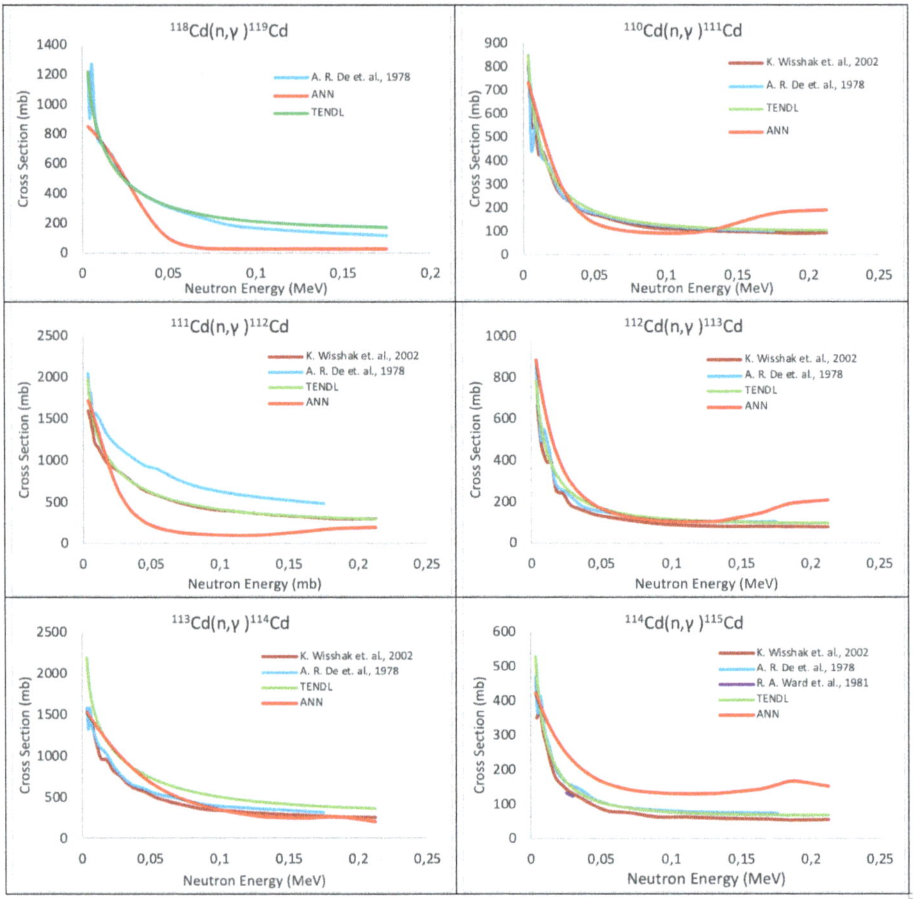

Fig. 4. The comparison of ANN results with the TENDL data and experimental values

3 Result and Conclusion

Neural networks model is a new technique to study features of the nuclear reaction interaction. Different from previous computation code and modeling, the nuclear reaction cross section based neural network model, depends on the experimental and evaluated data. Relationship between the evaluated data, experimental values and neural networks results have been investigated where the outcomes have been graphed as shown in Fig. 4.

Results of the neural networks have been seen almost fitting to the experimental data which is not usually the case with other theoretical techniques. This gives the neural networks the provision of wide usage in the modeling of nuclear reaction cross-section calculations.

References

1. Mashad, M.EL., Bakry, M.Y.EL., Tantawy M., Habashy, D.M.: Artificial neural networks for hadron hadron cross-sections. In: Tenth Radiation Physics Protection Conference, Cairo/Egypt, pp. 269–277 (2010)
2. MathWorks. http://www.mathworks.com/products/neuralnetwork/. Accessed 10 Jan 2019
3. Beale, M.H., Hagan, M.T., Demuth, H.B.: Neural Network ToolboxTM User's Guide (1992)
4. Korovin, Y.A., Maksimushkina, A.V.: The use of neural networks for approximation of nuclear data. Phys. At. Nucl. **78**(12), 1406–1414 (2015)
5. Dubey, B.P., Katariab, S.K., Mohantyb, A.K.: Neural network fits to neutron induced reactions using weighted least-mean-squares. Phys. Res. A **397**, 426–439 (1997)
6. Konobeyev, A.Y., Fischer, U., Pereslavtsev, P.E.: How we can improve nuclear data evaluations using neural network simulation techniques. In: JEFF Meeting, April (2013)
7. Brookhaven National Laboratory, National Nuclear Data Center, EXFOR/CSISRS (Experimental Nuclear Reaction Data File). http://www.nndc.bnl.gov/exfor/. Accessed 10 Jan 2019
8. TENDL, TALYS-based evaluated nuclear data library. https://tendl.web.psi.ch/tendl_2017/tendl2017.html. Accessed 10 Jan 2019

Estimating Luminance Measurements in Road Lighting by Deep Learning Method

Mehmet Kayakuş and Kerim Kürşat Çevik

Akdeniz University, 07600 Antalya, Turkey
{mehmetkayakus,kcevik}@akdeniz.edu.tr

Abstract. Importance of road lighting has increased day by day to provide drivers to travel in safe and comfort as the result of increasing vehicle traffic. Ideal luminance values based on the type of road is specified in 115 numbered technical report of the International Commission on Illumination. For this technical report, there are illumination classifications for five different road types. M3 road lighting group is demanded for urban main routes (speed < 50 km/h); ideal luminance value for this group is accepted as 1 cd/m^2 at least. There occur damages in time based on the use of lamps, environmental effects, and pollution factor; there also occurs decreases in total luminance. Photometric measurement of luminaires needs to be periodically performed to determine how much the luminaires are affected by these problems. Illumination measurements are made by photometric measuring instruments. Time, cost and qualified manpower are necessary for this process. Artificial intelligence-based measurement systems replaced measuring instruments in parallel with technological advancements today. This study made a prediction for luminance that is used as a road lighting measurement unit via deep learning method. Measuring points were determined by utilizing quadrature technique; luminance values of related points were measured. A mathematical correlation was established between luminance values in that area of the road and color values (R, G, B) of pixels of the image of the road. It is aimed to determine the luminance value of the road through a single image without any measuring device.

Keywords: Deep learning · Road lighting · Luminance · Deep neural network

1 Introduction

Road lighting helps pedestrians to see better at night and drivers to drive in comfort and safe [1]. Efficient traffic movement can be provided by a good road lighting; thus, use of the roads can be encouraged at the same time. The first and foremost thing is that accidents can be avoided by road lighting [2]. There are several studies that comprehensively review the effect of road lightening on the number of accidents [3–5]. With reference to the extensive evaluation of CIE [4], road lightening decreases accidents that occur at nights by 30% for all types of roads (rural, urban, highways and crossroads).

Road lightening standards and measurements have improved and changed in time based on scientific studies and advanced technology. Calculations were made for road

D. J. Hemanth and U. Kose (Eds.): ICAIAME 2019, LNDECT 43, pp. 940–948, 2020.
https://doi.org/10.1007/978-3-030-36178-5_83

lighting before by predicating light level on. Quantity of light on road is calculated in this method. Luminance value that measures the quantity of light which reflects from the road has been utilized for road lighting calculations [6].

Ideal luminance values based on the type of road is specified in 115 numbered Technical Report of the International Commission on Illumination. Table 1 shows illumination classifications for five different types of roads according to the related technical report [7].

Table 1. Road lighting criteria for different illumination classes [7]

Lighting class	L_{ort}	U_o	U_1	TI (%)	SR
M1	≥ 2.0	≥ 0.4	≥ 0.7	≤ 10	≥ 0.5
M2	≥ 1.5	≥ 0.4	≥ 0.7	≤ 10	≥ 0.5
M3	≥ 1.0	≥ 0.4	≥ 0.6	≤ 15	≥ 0.5
M4	≥ 0.75	≥ 0.4	≥ 0.6	≤ 15	≥ 0.5
M5	≥ 0.50	≥ 0.35	≥ 0.4	≤ 15	≥ 0.5
M6	≥ 0.30	≥ 0.35	≥ 0.4	≤ 20	≥ 0.5

Hereby, L_{ort} represents the average luminance level of the road (cd/m^2); U_o shows the average luminance quality (L_{min}/L_{ort}); U_1 shows the longitudinal luminance quality (L_{min}/L_{max}); TI represents relative threshold increase; SR means the enclosure ratio.

Illumination measurements are made by photometric measuring instruments today. Time, cost and qualified manpower are necessary for this process. Cameras have started to replace measuring instruments in parallel with developments in visualization technologies. Camera-based measurement systems in illumination measurement technologies have been developed [8–10]. Thus, there can be performed measurements in a proper manner. Another advantage of making measurement by the camera is that the camera can make areal measurement during point measurement.

The concept of artificial intelligence (AI) that uses imitating the thought and behavior types of people as the base was first brought in literature by activity in USA-Dartmouth [11]. Artificial Neural Networks (ANN) applications which endeavor to create a new system by imitating the running of the human brain are utilized in classification problems in AI applications. Deep learning is based on architectures that provide modeling brain better by developing ANNs. Deep learning algorithms need types of equipment that have ultrahigh calculation power which can process the ultrahigh amount of data; this is the characteristics of deep learning algorithms that makes it different from available algorithms in machine learning. Since equipment limits did not allow for intense matrix operations in the 1980s, deep learning could not be turned into an application. ANN models were developed by means of backpropagation algorithm that was suggested by Hinton and Lecun in the later 1980s. Simpler models that run specific to the problem such as support vector machines became popular preference in the 1990s and 2000s due to calculation cost and advantages of ANNs [13]. Deep learning approach consists of multiprocessing strata that are gathered to learn representations of data by multi abstraction structure [14]. Figure 1 shows the change of AI methods by years.

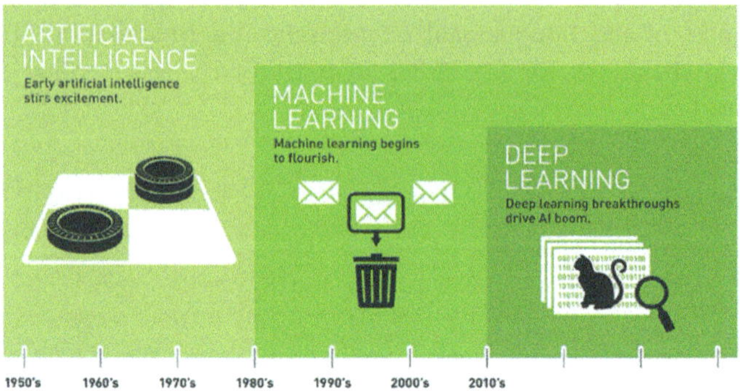

Fig. 1. Process of artificial intelligence [15]

Support vector machine (SVM) method started to give its place to ANN after being used graphics processor units in computers and increasing computers' speed by 1000 times in the 10-year process [12]. Expression of "Deep Learning" was first introduced in Igor Aizenberg's book called "Multi-Valued and Universal Binary Neurons: Theory, Learning, Applications" in 2000 [16]. In 2006, Geoffrey Hinton showed how a layer of feed-forward multi-level artificial neural network is effectively trained at every turn (he trained every layer by a Boltzmann machine whose layers are restricted without an audit); afterward, he showed that there can be performed fine adjustment by a controlled backprop method [17]. Being popular GPU allows for deep neural networks to be trained by the personal problems of individuals; thus, deep neural networks are not preconditioned ready networks. Ciresan et al., obtained reliable results by using deep learning in traffic signs, medical image processing, and character recognition fields in 2011 and 2012; they were awarded as well [18–20]. Hinton et al., used neurons and impactful GPU application on a convolutional operation to expedite training by a similar architecture [21]. Moreover, the "dropout" method was utilized by Krizhevsky and Hinton et al., to decrease overfitting in fully connected layers and degrade weights in fully connected layers to a proper value [22]. This approach provided them to carry out the classification process by the minimum error rate (0.15); they were successful in ILSVRC-2012 ImageNet competition. Technology companies such as Google, Facebook and Microsoft noticed the efficiency of deep learning method after these developments and started to invest in this field [23].

2 Material and Method

2.1 Data Collection

A road that is fit for international M3 standards in Antalya Province Konyaalti District was selected for testing deep learning-based system. There are two lanes on each road and the road has a total width of 7 m so as each of the lanes to be 3.5 m wide. Road lengths are enough for visibility measurement and placement of luminaires. Height of poles was 1.5 m; the distance between poles was 7 m. It was considered when selecting the roads that the roads provide the intended classification; have low depreciation and not illumination. As is seen in Fig. 2, luminance measuring points on the road were measured by using luminance meter.

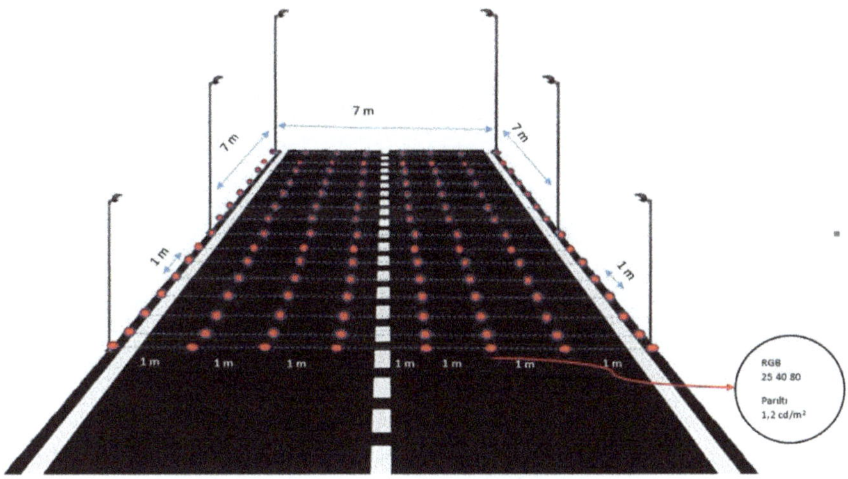

Fig. 2. Measuring points on the road [24]

It is important to ensure the same standard while transmitting the picture of the road to the system. There has been conducted a large number of tests and analyses to set the related standard. It was decided at the end of the tests that taking photos at a distance of 1.5 m and 20 m from the middle of two luminaires as is seen in Fig. 3.

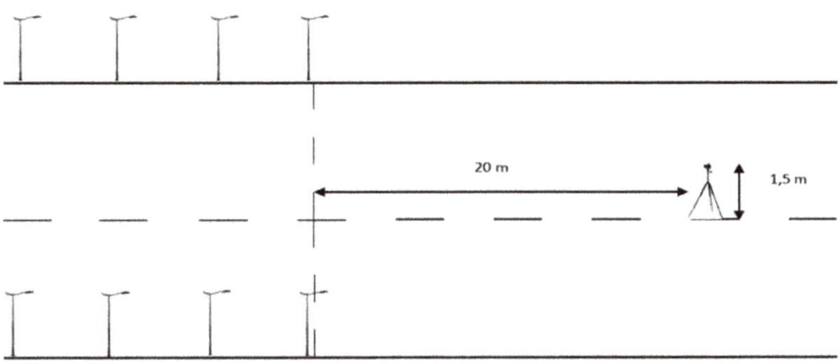

Fig. 3. Taking pictures of the road by standards set [24]

As is seen in Fig. 4; 6 luminaires are placed on the road based on the standards set. The picture of the road was taken to conduct photometric analyses by using the software. Pixel values of related points were recorded in Excel table in RGB format.

Fig. 4. The way the system was tested [24]

2.2 Deep Neural Network

Deep neural network (DNN) is a feed-forward artificial neural networks that have more than one layers between inputs and outputs. Every hidden layer (j) generally uses the logistics function (this function may also be a hyperbolic tangent) to sum up the inputs; x_j is the digital status and y_j is the value that is sent to upper layer [25].

$$y_j = logistic(x_j) = \frac{1}{1 + e^{-x_j}},\ x_j = b_j + \sum\nolimits_i y_i w_{ij} \tag{1}$$

b_j in equation is the bias value of the layer at j line; i is the index value of bottom layer; w_{ij} is the juncture weight value from the layer on i line to the layer on j line. j output layer turns x_j input layers to p_j class probabilities by using "softmax" for multi-classed classifications.

$$p_j = \frac{\exp(x_j)}{\sum_k \exp(x_k)}, \tag{2}$$

k is the index value that is above all the classes.

While softmax uses the output function, natural cost function (C) is the cross entropy between target probabilities (d) and outputs (p) of softmax [26].

$$C = -\sum_j d_j \log p_j, \tag{3}$$

Target probabilities that typically take 1 or 0 are the information that are provided and controlled to train DNN classificators [25].

DNNs can model nonlinear relations in analogy to shallow neural networks. DNN architectures produce compositional models allow for composition from the characteristics of bottom layers. Accordingly, there is provided a great learning capacity and potential of modeling complex data [27, 28] (Fig. 5).

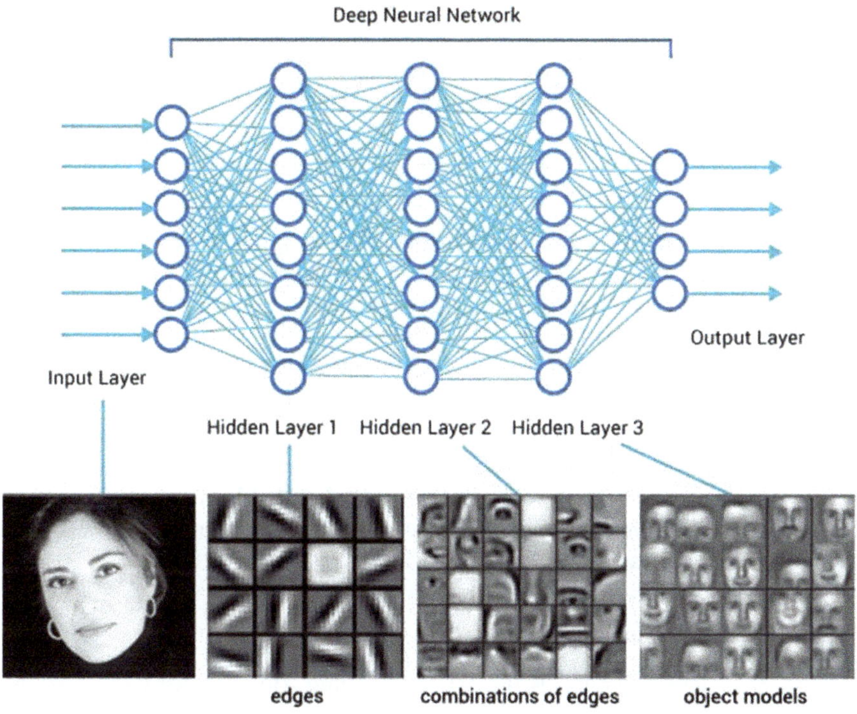

Fig. 5. Main deep neural network model [29]

3 Conclusion

3.1 Application

It was endeavored to estimate luminance that is used as the road lighting measurement unit by benefiting from deep neural networks method. Input data of deep neural network consist of color values of pixels (R, G, B) on the picture of the road. Output data of deep neural network consists of luminance values of points that are determined by the quadrature technique. Thus, there was aimed to establish a mathematical correlation between luminance values of the related part of the road and the color values of pixels of the picture of the road (Fig. 6).

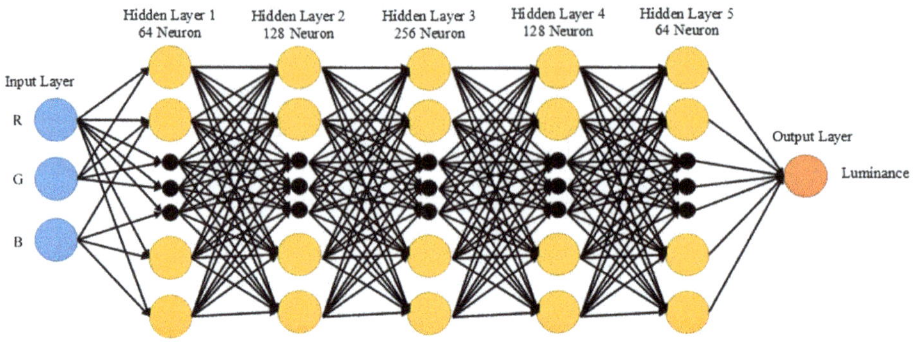

Fig. 6. Designed deep neural network structure

The deep neural network that was designed for this purpose has 3 input, 1 output, and 5 hidden layers. Neuron numbers in each hidden layers respectively are 64-12-256-128-64. Figure 3 shows the deep neural network structure designed.

Keras and TensorFlow libraries in Python programming language were used in the application. The dense function was utilized in software while the layers were established. Respectively 64-12-256-128-64-1 values were given to units parameter of this function for every layer; relu (Rectified Linear Unit) was the activation function. input_dim value of the first dense function was accepted as 3. Performance analysis values for DNN were selected as Mean Absolute Error, Mean Squared Error (metrics = ['mae', 'mse']). In conclusion, the deep neural network that can be seen in Fig. 3 in the program was created.

3.2 Application Results

120 values that have 3 inputs (R, G, B) and 1 output (luminance) were used as the dataset in this study. 100 samples of this data set are for training; 20 samples of the same data set are for the test (84% Train, 16% Test). Data for this classification were randomly selected. K Fold Cross Validation that is frequently utilized in the literature to estimate the error of the test better was used to completely measure the general performance of the system. With reference to the dataset, K is 6; the data set was divided into six parts that consist of 20 datasets. Train data was determined as another part (data set with 20) in each of training stages. DNN was run by 500 iterations at each of the training stages. Mean Square Error (MSE) values evaluated the results. MSE, namely the mean square error value is the average of squares of differences of value that is computed at the end of DNN and the real value. The smaller the average of the squares, the closer the program is to the actual data. Table 2 shows the MSE values that are obtained for train and test data at each of the layers in 6-Layer Cross Verification.

Table 2. 6-Layer Cross validation test results

K	MSE (Train)	MSE (Test)
1	0.0029	0.0149
2	0.0074	0.0872
3	0.0046	0.0812
4	0.0084	0.0907
5	0.0040	0.1284
6	0.0046	0.0501
Average	0.0053	0.0754

4 Discussion

This research aimed to estimate the luminance values of test road via deep learning. MSE value for 100 data is 0,0053 on average; again, MSE value for 20 data that was used at the test phase is 0,0754 on average.

Randomizing test and train data and a limited number of data have an effect on the MSE values. The results are reliable as well as there can be obtained better results by increasing the train and test data.

Luminance measurements of a test road can be determined by using deep learning techniques without any measuring device. Accordingly, the changes in lightening the road arising from decreasing life and efficiency factors of luminaires or breaking down the luminaires at the same time can be measured.

The requirements such as equipment, qualified manpower, time and cost will disappear; luminance measurements can be carried out in safe at short notice.

References

1. Feng, Z., Luo, Y., Han, Y.: Design of LED freeform optical system for road lighting with high luminance/illuminance ratio. Opt. Express **18**(20), 22020–22031 (2010)
2. Li, F., Chen, D., Song, X., Chen, Y.: LEDs: a promising energy-saving light source for road lighting. In: Asia-Pacific Power and Energy Engineering Conference 2009, Wuhan, China (2009)
3. Elvik, R., Høye, A., Vaa, T., Sørensen, M.: The Handbook of Road Safety Measures, 2nd edn. Emerald Group Publishing, Bingley (2009)
4. CIE 093: Road lighting as an accident countermeasure, Vienna (1992)
5. Plainis, S., Murray, I., Pallikaris, I.: Road traffic casualties: understanding the night-time death toll. Inj. Prev. **12**(2), 125–138 (2006)
6. Güler, Ö.: Yol aydinlatmasi hesaplarinin görülebilirlik düzeyalinarak yapilabilmesi için gerekli kriterlerinin belirlenmesi. PhD thesis, ITU, İstanbul, Turkey (2001)
7. CIE 115: Recommendations for the lighting of roads for motor and pedestrian traffic, Vienna (1995)
8. Ekrias, A., Eloholma, M., Halonen, L., Song, X.-J., Zhang, X., Wen, Y.: Road lighting and headlights: luminance measurements and automobile lighting simulations. Build. Environ. **43**(4), 530–536 (2008)

9. Zatari, A., Dodds, G., McMenemy, K., Robinson, R.: Glare, luminance, and illuminance measurements of road lighting using vehicle mounted CCD cameras. Leukos **1**(2), 85–106 (2005)
10. Zhou, H., Pirinccioglu, F., Hsu, P.: A new roadway lighting measurement system. Transp. Res. Part C: Emerg. Technol. **17**(3), 274–284 (2009)
11. McCarthy, J., Minsky, M.L., Rochester, N., Shannon, C.E.: A proposal for the dartmouth summer research project on artificial intelligence. AI Mag. **27**(4), 1–12 (2006)
12. Schmidhuber, J.: Deep learning in neural networks: an overview. Neural Netw. **61**, 85–117 (2015)
13. Cortes, C., Vapnik, V.: Support-vector networks. Mach. Learn. **20**(3), 273–297 (1995)
14. LeCun, Y., Bengio, Y., Hinton, G.: Deep learning. Nature **521**(7553), 436–444 (2015)
15. Copeland, M.: NVIDIA blog, what's the difference between artificial intelligence, machine learning, and deep learning? https://blogs.nvidia.com/blog/2016/07/29/whats-difference-artificial-intelligence-machine-learning-deep-learning-ai/. Accessed 01 Mar 2019
16. Aizenberg, I.N., Aizenberg, N.N., Vandewalle, J.: Multi-Valued and Universal Binary Neurons: Theory, Learning and Applications. Springer, Heidelberg (2013)
17. Hinton, G.E.: Learning multiple layers of representation. Trends Cogn. Sci. **11**(10), 428–434 (2007)
18. Ciresan, D., Giusti, A., Gambardellam, L.M., Schmidhuber, J.: Deep neural networks segment neuronal membranes in electron microscopy images. In: Advances in Neural Information Processing Systems 25, pp. 2843–2851. Curran Associates Inc. (2012)
19. Ciresan, D., Meier, U., Masci, J., Schmidhuber, J.: Multi-column deep neural network for traffic sign classification. Neural Netw. **32**, 333–338 (2012)
20. Ciresan, D.C., Meier, U., Gambardella, L.M., Schmidhuber, J.: Convolutional neural network committees for handwritten character classification. In: 2011 International Conference on Document Analysis and Recognition, Beijing, pp. 1135–1139, Beijing, China (2011)
21. Hinton, G.E., Srivastava, N., Krizhevsky, A., Sutskever, I., Salakhutdinov, R.R.: Improving neural networks by preventing co-adaptation of feature detectors. arXiv preprint arXiv:1207.0580 (2012)
22. Krizhevsky, A., Sutskever, I., Hinton, G.E.: ImageNet classification with deep convolutional neural networks. In: Advances in Neural Information Processing Systems 25, pp. 1097–1105. Curran Associates Inc. (2012)
23. Şeker, A., Diri, B., Balık, H.H.: Derin Öğrenme Yöntemleri ve Uygulamalari Hakkinda Bir İnceleme. Gazi Mühendislik Bilimleri Dergisi **3**(3), 47–64 (2017)
24. Kayakuş, M., Üncü, I.: Research note: the measurement of road lighting with developed artificial intelligence software. Lighting Res. Technol. (2019). https://doi.org/10.1177/1477153519825564
25. Hinton, G., Deng, L., Yu, D., Dahl, G.E., Mohamed, A., Jaitly, N., Senior, A., Vanhoucke, V., Nguyen, P., Sainath, T.N., Kingsbury, B.: Deep neural networks for acoustic modeling in speech recognition: the shared views of four research groups. Signal Process. Mag. **29**(6), 82–97 (2012)
26. Rumelhart, D.E., Hinton, G.E., Williams, R.J.: Learning representations by back-propagating errors. Nature **323**(6088), 533–536 (1986)
27. Deng, L., Yu, D.: Deep learning: methods and applications. Found. Trends® Signal Process. **7**(3–4), 197–387 (2014)
28. Kiani, F., Kutlugün, M.A., Çakır, M.Y.: Derin Sinir Ağları ile Konuşma Tespitive Cinsiyet Tahmini, 22. Türkiye'de Internet Konferansı, İstanbul, Turkey (2017)
29. Sachdeva, A.: Deep learning for computer vision for the average person. https://medium.com/diaryofawannapreneur/deep-learning-for-computer-vision-for-the-average-person-861661d8aa61. Accessed 12 Feb 2019

Effect of the Clonal Selection Algorithm on Classifiers

Tuba Karagül Yildiz[1(✉)], Hüseyin Demirci[2], and Nilüfer Yurtay[2]

[1] Department of Computer and Information Sciences, Institute of Natural Sciences, Sakarya University, 54187 Sakarya, Turkey
tkaragul@sakarya.edu.tr

[2] Department of Computer Engineering, Computer and Information Sciences Faculty, Sakarya University, 54187 Sakarya, Turkey

Abstract. To be able to make the classification process, there should be a sufficient number of samples. Collecting a sufficient number of samples, especially for those dealing with medical data, is a laborious task. To obtain the approval of the ethics committee in our country, patient data coming from a certain time interval rather than a sample number can be requested. Therefore, there are difficulties in reaching a sufficient number of samples. In this study, the effect of the clonal selection algorithm which is one of the artificial immune system algorithms on standard classifiers was investigated. The chronic kidney disease dataset from the university of California Irvine machine learning repository was chosen as the dataset. Among the commonly used methods for classification, methods of k nearest neighbor, decision trees and artificial neural networks were selected as classifiers. While k nearest neighbor is a distance-based algorithm, a decision tree is a regression-based method and the artificial neural network which is quite popular nowadays is a nature-inspired method. According to the results of the experiments, it is found that the data reproduction process by using the clonal selection algorithm has increased the performances of the classifiers.

Keywords: Artificial immune system · Chronic kidney disease · Classification · Clonal selection algorithm · Data mining

1 Introduction

Diagnosing a disease is the most important part of treating a patient. Without a proper diagnosis, a doctor can only reduce the pain of the symptoms but cannot cure the illness. The diagnosis part is the most difficult and complicated even for an experienced doctor. Most of the diseases have similar symptoms and sometimes patients cannot describe their symptoms or mistake their symptoms with other common side effects or simply lie about their symptoms. Based on the symptoms the doctor asks for some medical tests and diagnose the patient's illness. Besides the patient, the test results cannot lie but can be unwillingly altered by the patient with taking not prescribed pills or digesting some food and/or beverage. Although this situation can only change one or two value of the test. Because of these situations diagnosing a patient can be much difficult and confusing for the doctor.

D. J. Hemanth and U. Kose (Eds.): ICAIAME 2019, LNDECT 43, pp. 949–959, 2020.
https://doi.org/10.1007/978-3-030-36178-5_84

Computers and artificial intelligence systems (AI) have become more efficient and have more coverage area in recent years. It became the inseparable part of our lives. We use AI almost every part of our lives e.g. weather forecasting, traffic routing, shopping, selecting a place where to go out and it goes on like this. It helps us to make decisions about our daily life. In recent years, AI studies for diagnosing the diseases have become popular. With classification algorithms and pattern recognition techniques detection of the disease can be faster and much easier. These algorithms cannot diagnose the disease by itself but can help the doctor with suggestions for diagnosing the disease. For these algorithms to work they require lots of data of real people with the same disease. By the law, our data are strictly confidential with the doctor-patient confidentiality and can not to be open for public. To obtain and use these data, researchers must have ethical board permission. Even if researchers have the permission, they can only access the data with a predefined time period by the ethical board. This can lead researchers to obtain less data then required for training algorithms. Without proper and fewer data these algorithms cannot be trained and their performance can be very low and their prediction mechanism cannot work properly.

Because of the situation that explained above, we proposed a method that overcomes the lack of data problem. In this paper, we used Clonal Selection Algorithm (CSA) which is part of the artificial immune system algorithms to create artificial data from the original data set. We used this produced artificial data to train classification algorithms and tested their performance on real data set which is gathered from the UCI (University of California Irvine) machine learning repository. The produced data allow us to better training and this led to better performance on classification algorithms.

The rest of the paper is organized as follows; literature survey, material and method, results, and conclusion, which are respectively as Sects. 2, 3, 4 and 5.

2 Literature Survey

In this section, studies using chronic kidney disease data are examined. In a study, Naive Bayes and SVM methods were used for the prediction of kidney diseases. The data set used in this study includes 5 different classes: Chronic Glomerulonephritis, Acute Nephritic Syndrome, Chronic Kidney Disease, Acute Renal Failure and normal. Although the Naive Bayes method was classified in a shorter time period, it was reported that SVM performance was higher than the Naive Bayes method [1]. Rubini and Eswaran conducted a study using three different classifiers such as Logistic Regression, Multilayer Perceptron and Radial Basis Function Network to estimate disease on Chronic Kidney Disease (CKD) data set. Classifier successes were found as 98.5%, 99.75%, and 97.5%, respectively, and Multilayer Perceptron was found to be the best artificial neural network algorithm [2]. Jena and Kamila carried out a study to determine the data mining algorithm that provides the most appropriate and most accurate accuracy for disease prediction in CKD dataset. They have worked with six different algorithms which are Conjunctive Rule, Naive Bayes, Decision Table, Support Vector Machine, Multilayer Perceptron and J48 algorithms and stated that the fastest and most successful algorithm is Multilayer Perceptron [3]. Sinha and Sinha have tried two different machine learning methods to design a decision support system for CKD disease. They compared the K-nearest neighbor (KNN) and Support Vector Machine classifiers in terms of accuracy and time and reported that they found the KNN classifier to be more successful than the Support Vector Machine method [4]. In another study made with CKD dataset, 6 different methods are used for prediction. These methods include Naive Bayes, Multilayer Perceptron (MLP), Random Forest (RF) classifiers, Radial Basis Function (RBF), Sequential Minimal Optimization (SMO) and Simple Logistic (SLG) Classifiers. The highest success rate was 100% in the Random Forest [5]. Kunwar et al. have used Naive Bayes and ANN methods in their studies to estimate CKD disease using data mining methods. They used the CKD data from UCI as a data set and used Rapidminer as the data mining tool. They stated that the Naive Bayes method gives 100% accuracy was more successful than the ANN with 72.73% accuracy [6]. Chetty et al. have tried various classification models on their original data set and reduced data set. They used wrappersubseteval and best-fit search methods in the attribute selection. As classifiers, Sequential Minimal Optimization (SMO), Naive Bayes and IBK -which is actually kNN- classifiers with the Weka data mining tool. For the total 25 attributes, they reduced the number of attributes to 6, 12 and 7 and increased the accuracy rate with Naive Bayes, SMO and IBK classifier from 95% to 99%, 97.75% to 98.25% and 95.75% to 100% respectively. [7]. In another study, the number of qualifications in the data set has been decreased to 14 and we see that 4 different methods have been tried, namely Multiclass Logistic Regression, Multiclass Decision Jungle, and Multiclass Neural Network [8]. Kayaalp et al. UCI used in the study of CKD dataset obtained from UCI machine learning data repository using KNN classification algorithm, Bayes classification algorithm and Support Vector Machines classification algorithms. Prior to the classification process, attribute selection was made using the Gain Ratio and Relief algorithms. As a result, it was found that chronic kidney disease gave the best performance in the classification algorithm with k nearest neighbor after attribute selection by Relief algorithm [9].

As explained above, there have been many studies on classifying chronic kidney disease. Yet it has not been analyzed to improve classifiers performance by reproduction of the data. In the presented work it is aimed to improve classifiers performance by reproducing data with clonal selection algorithm.

3 Material and Method

In this study, Chronic Kidney Disease (CKD) dataset from the University of California Irvine (UCI) machine learning repository were reproduced by Clonal Selection Algorithm (CSA) which is one of the artificial immune system algorithms and classified by well-known classifiers. The classifiers used were selected as K Nearest Neighbor (KNN) which is the standard classifier of CSA, Artificial Neural Network (ANN) and Decision Tree methods among the well-known machine learning methods.

3.1 Dataset Description

The dataset used was obtained via the UCI machine learning repository [10]. There are 400 samples taken from patients and healthy individuals in the data set. There are 25 different attributes used for each sample. One of these attributes shows the value of the class that there are 2 options, namely patient and healthy. In the data set, the patient individuals are shown as 'ckd' and the healthy individuals are shown as 'notckd'. There are 250 patients and 150 healthy subjects in the dataset. The attributes of the dataset are shown in Table 1.

Table 1. Chronic kidney disease dataset description.

Attribute name	Value	Type
Age	Age in years	numerical
Blood Pressure	bp in mm/Hg	numerical
Specific Gravity	sg - (1.005, 1.010, 1.015, 1.020, 1.025)	numerical
Albumin	al - (0, 1, 2, 3, 4, 5)	numerical
Sugar	su - (0, 1, 2, 3, 4, 5)	numerical
Red Blood Cells	rbc - (normal, abnormal)	nominal
Pus Cell	pc - (normal, abnormal)	nominal
Pus Cell clumps	pcc - (present, notpresent)	nominal
Bacteria	ba- (present, notpresent)	nominal
Blood Glucose Random	bgr in mgs/dl	numerical
Blood Urea	bu in mgs/dl	numerical
Serum Creatinine	sc in mgs/dl	numerical
Sodium	sod in mEq/L	numerical
Potassium	pot in mEq/L	numerical
Hemoglobin	hemo in gms	numerical
Packed Cell Volume	pcv in cells/cumm	numerical
White Blood Cell Count	wbcc in millions/cmm	numerical
Red Blood Cell Count	rbcc in millions/cmm	numerical
Hypertension	htn - (yes, no)	nominal
Diabetes Mellitus	dm - (yes, no)	nominal
Coronary Artery Disease	cad - (yes, no)	nominal
Appetite	appet - (good, poor)	nominal
Pedal Edema	pe - (yes, no)	nominal
Anemia	ane - (yes, no)	nominal
Class	class - (ckd, notckd)	nominal

3.2 Artificial Immune System and Clonal Selection Algorithm

The artificial immune system is an algorithm inspired by the human's natural immune system. This algorithm is based on the principle of defending and eliminating the antigens with the antibodies of the human immune system against foreign substances called antigens from the outside. In terms of complexity, the immune system has as little complexity as the human brain. The main features of the immune system are very suitable for solving problems in engineering, computer science, and many other fields. The basic properties of the artificial immune system are listed below;

Detection of the Anomaly: The immune system of a human can detect and react to foreign substances (pathogens) that the body has never encountered before.
Distributed Detection: All cells in the system are distributed throughout the body and are not centralized. A response to the antigen generated in the system is distributed throughout the body.
Noise Tolerance: Does not need to be fully recognized for the pathogens coming from outside.
Uniqueness: Each individual has his own immune system.
Reinforced Learning: The system can learn pathogens and can then show the response more quickly when faced with similar pathogens.
Change: The system is trying to produce the best antibody, so it produces different antibodies for many antigens.
Memory: The system has a dynamic memory. Similar antigens are responded to by the same species of antibodies. In this way, it is provided to tolerate the noise to occur.

The Clonal Selection Algorithm (CSA) is inspired by the fact that the artificial immune system is the only way to replicate the response to the antigen in the entire system. Antibodies increase at the rate they detect antigens, and those detected antigens are selected for cloning according to those not.

By looking at the similarities of B cells which are produced by the bone marrow, i.e. antibodies to antigens, those with the highest similarity with each other are selected and cloned. These replicated new cells are subjected to a mutation and the cells are differentiated by this way. These cells are re-checked for antigens, and those with high susceptibility to antigens are transformed into plasma cells, with low susceptibility to antigens are transformed into memory cells. The ability of the immune system to recognize and remember different antigens is enhanced by these kind of updates [12].

The clonal selection algorithm is used to reproduce the dataset by generating artificial data in a small number of datasets and to train with another artificial intelligence system with this new artificial dataset.

In the algorithm, a test and a training set are created from the existing dataset for the data replication process. New individuals will be produced from these individuals in the test set and new individuals will be added to the total dataset. The algorithm generates as much data as the number of data in the test set in a single iteration. By running the algorithm multiple times, the dataset can be expanded as desired. The pseudo-code of the clonal selection algorithm is given below;

```
1- The current dataset is divided into two as test and
training sets.
2- For each data in the test set;
 2.1. According to the Euclidean distance, the closest
data from the training set is cloned to the set 'C'
training set.
3- All of the data in the copy 'C' dataset are duplicated
to the dataset 'C *'.
4- The data in the dataset in the 'C *' are mutated by
changing the selected values randomly.
5- For each data in the test set;
 5.1. According to the Euclidean distance, the closest
data from the dataset 'C *' is copied to the 'C' dataset.
The Euclidean distance formula used to select the best
individual is given in Equation (1). The Ab expression
used herein refers to the Antibody, the Ag expression re-
fers to the Antigen, and the D expression refers to the
Euclidean distance.
```

$$D = \sqrt{\sum_{i=1}^{n} (Ab_i - Ag_i)^2} \quad (1)$$

```
6- The data in the last artificial 'C' dataset obtained
are newly created artificial data.
7- This data is added to the first available dataset and
the dataset is reproduced.
```

3.3 Classifiers

As it is known for providing medical decision support to experts, machine learning classification methods are often used. The next step after the reproduction of the data, various classifiers were used and the performance analysis was performed for this study. The success of the classification obtained by the KNN method, which is used as the standard classifier of the clonal selection algorithm, was not found adequate. Therefore, as it can be seen in the literature review, Decision Tree and Artificial Neural Network methods were also used for the classification of chronic kidney disease dataset. Details of these methods are given below in sub-headings.

K-Nearest Neighbor (KNN). The K-nearest neighbor classifier is a distance-based classifier. The class of the adjacent point closest to the selected point is the class of the selected point. The steps of the algorithm are given below.

Step 1. First, a 'k' parameter must be selected. This parameter indicates the number of neighbors closest to the desired point (k = 5 is selected for this study).
Step 2. The distance between the desired point and the remaining points are calculated one by one. Euclidean formula is used to determine distances. In Eq. (2), the Euclidean

distance formula is given between the desired point in which the class is to be found (let's say point a) and one of the adjacent points (let's say point b).

$$E(a,b) = \sqrt{\sum\nolimits_{i=1}^{n} (a_i - b_i)^2} \qquad (2)$$

In this study, the distances between each value of the 156 data in the test set and each of the 400 data trained by the CSA are calculated with this method.

Step 3. After calculating the results, all data is edited and the lowest k sample in the dataset are selected.
Step 4. The classes of the selected samples are determined and the most repeated value is selected as the class value of the desired point.
Step 5. The most repeated class value is considered the class of the desired point [13].

Decision Tree. The decision tree is a method of classifying which questioning if there are any feature for classification of the data in the data set. Each feature found is a branching condition of the tree. With this method, each data in the data set are intended to be placed in one of the classes and at the same time, the class definition is made. The results are easy to understand and interpret [14]. Decision tree method is commonly used with medical data and is also gives successful results. The structure of the decision tree used in this study can be seen in Fig. 1.

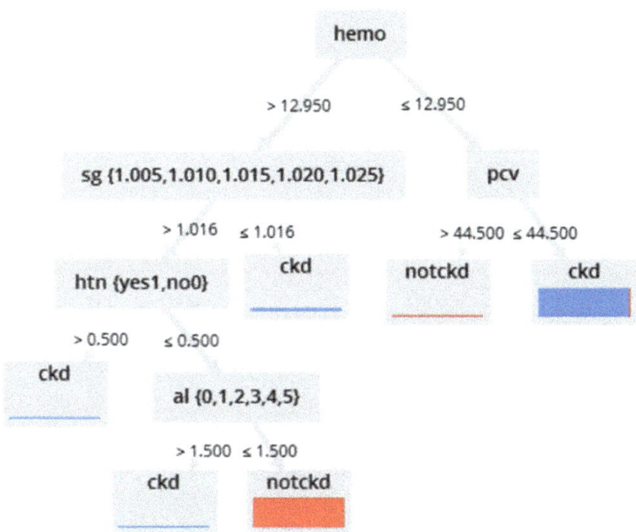

Fig. 1. The decision tree of the proposed method

Artificial Neural Network (ANN). One of the most commonly used classification methods is Artificial Neural Network (ANN). ANNs are artificial learning models inspired by the biological nervous system. In the mathematical model of a neuron, connection weights are made by the synapse effect. Each neuron has input connections, a bias value, an activation level, output connections, and an output value. Each incoming signal has a weight that affects the activation level of the neuron. The output value is reflected in the transfer function as the sum of the input signals multiplied by the weights. The learning capacity of an artificial neuron is determined by regulating the weights of the selected learning algorithm [15]. For this study, a model learns by using a two-layer feed-forward neural network is designed. The structure of the neural network is seen in Fig. 2.

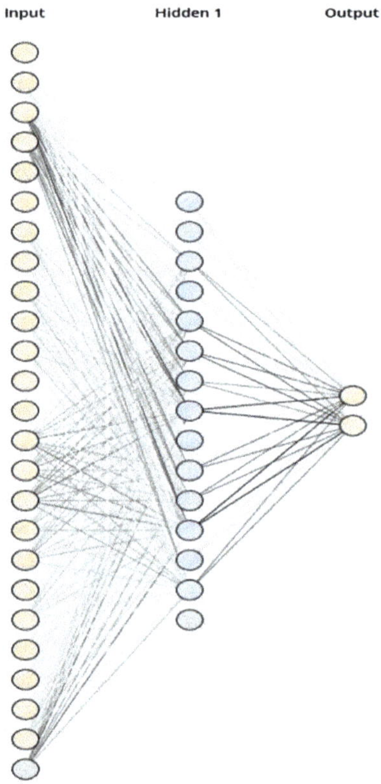

Fig. 2. The artificial neural network structure for the proposed method.

For this dataset, there were 24 attributes and 2 classes. Therefore, the neural networks input layer was made up of 24 neurons and the output layer was made up of 2 neurons. In addition, the neural network had 1 hidden layer with sigmoid transfer function and made up of 15 neurons. Equation (3) shows the neural net sigmoid transfer function where x indicates inputs and y indicates output.

$$K(x, y) = \tanh(\kappa x.y - \delta) \tag{3}$$

4 Results

There were gaps in the data set which has 400 samples. An imputation process is made on data, all empty places are filled by the average value of each attribute data. The dataset had 400 samples as the training set. For preparing the test set the data cleaning is made and the dataset has 156 data, of which 115 were healthy and 41 were patient data. In this way, the program has run and have reached the number of 556 data. The classification success rate with ANN method was 96,67% with 400 data, while the success rate with 556 data was 98.80%. Afterward, the program has re-run and this is the second iteration by making 556 training set and 156 test set. So the dataset reached from 400 samples to 712 samples. Finally, the program was run for the third time and the number of data reached 1090 samples. The success of the all classifiers has increased at each step. The accuracy of all methods is given in Table 2.

Table 2. The accuracy performances of the proposed methods.

	KNN	Decision tree	ANN
Original dataset	68,33%	95,00%	96,67%
CSA dataset (with 556 samples)	79,04%	98,20%	98,80%
CSA dataset (with 772 samples)	82,33%	99,14%	98,28%
CSA dataset (with 1090 samples)	89,57%	99,69%	99,08%

As can be seen in the table, the accuracy rate in 68,33% with KNN method gradually increased to 79,04%, 82,33%, and 89,57%. While in ANN and Decision Tree methods the accuracy rates were about 95–96%, where success can be considered high already, by training dataset with CSA has increased accuracy to 99% band.

The results have shown that the performance of the classifiers can be supported by increasing the number of data. With CSA, the number of data can be increased in a controlled and balanced manner. Under these circumstances, the proposed hybrid method can be used to assist decision makers for decisions such as diagnosis of rare diseases especially in cases with insufficient data. Our research is important in terms of explaining this argument.

In this study, the program for Clonal Selection Algorithm used for data reproduction was coded in C# programming language using Microsoft Visual Studio 2017 Student Edition platform. On the computer which is running the program includes 16 GB Ram, 64 bit Windows 10 operating system, and Intel Core i7 processor. For classification processes, Rapidminer Studio Free edition which is a data mining tool was used.

5 Conclusion

Collecting medical data is a laborious task. In the scope of the measures taken for data confidentiality and security, sufficient data cannot be collected at all times. In addition, the data of rare diseases cannot be collected, statistically, enough number may not be reached. Therefore, it has been found that the reproduction of the data using clonal selection algorithms as described in this study increases the classifier success of the clonal selection algorithm.

In this study, it has been shown that the classifier performances can be increased by conducting a hybrid method. This is the strength of this study. The weakness of this study is that the data set used is taken from a data repository. To test the success of the new method by repeating the same work for more original and larger datasets is planned as the future work. In addition, the classifier methods used are diversified so that a more powerful study will be carried out.

References

1. Vijayarani, S., Dhayanand, S.: Data mining classification algorithms for kidney disease prediction. Int. J. Cybern. Inform. (IJCI) **4**(4), 13–25 (2015)
2. Rubini, J.L., Eswaran, P.: Generating comparative analysis of early stage prediction of chronic kidney disease. Int. J. Mod. Eng. Res. (IJMER) **5**(7), 49–55 (2015)
3. Jena, L., Kamila, N.K.: Distributed data mining classification algorithms for prediction of chronic- kidney-disease. Int. J. Emerg. Res. Manag. Technol. **4**(11), 110–118 (2015)
4. Sinha, P., Sinha, P.: Comparative study of chronic kidney disease prediction using KNN and SVM. Int. J. Eng. Res. Technol. (IJERT) **4**(12), 608–612 (2015)
5. Kumar, M.: Prediction of chronic kidney disease using random forest machine learning algorithm. Int. J. Comput. Sci. Mob. Comput. **5**(2), 24–33 (2016)
6. Kunwar, V., Chandel, K., Sabitha, A.S., Bansal, A.: Chronic kidney disease analysis using data mining classification techniques. In: 6th International Conference - Cloud System And Big Data Engineering (2016)
7. Chetty, N., Vaisla, K.S., Sudarsan, S.D.: Role of attributes selection in classification of chronic kidney disease patients. In: International Conference on Computing, Communication and Security (ICCCS) (2015)
8. Gunarathne, W.H.S.D., Perera, K.D.M., Kahandawaarachchi, K.A.D.C.P.: Performance evaluation on machine learning classification techniques for disease classification and forecasting through data analytics for chronic kidney disease (CKD). In: 17th International Conference on Bioinformatics And Bioengineering (2017)
9. Kayaalp, F., Başarslan, M.S., Polat, K.: A Hybrid classification example in describing chronic kidney disease. In: Electric Electronics, Computer Science, Biomedical Engineerings' Meeting (EBTT) (2018)
10. Soundarapandian, P., Rubini, L.J.: UCI Machine Learning Repository, Irvine (2015). http://Archive.Ics.Uci.Edu/Ml/Datasets/Chronic_Kidney_Disease
11. De Castro, L.N., Von Zuben, F.J.: The clonal selection algorithm with engineering applications. In: Workshop on Artificial Immune Systems and Their Applications (2000)
12. Çarklı Yavuz, B., Karagül Yıldız, T., Yurtay, N., Yılmaz, Z.: Comparison of K nearest neighbors and regression tree classifiers used with clonal selection algorithm to diagnose hematological diseases. In: AJIT-E Online Acad. J. Inf. Technol. **5**(16) (2014)

13. Özkan, Y.: Veri Madenciliği Yöntemleri, 2nd edn. Papatya Publisher, Istanbul (2013)
14. http://bilgisayarkavramlari.sadievrenseker.com/2012/04/11/Karar-Agaci-Ogrenmesi-Decision-Tree-Learning/. Accessed 11 Aug 2017
15. Hassanien, A.E., Al-Shammari, E.T., Ghali, N.: Computational intelligence techniques in bioinformatics. Comput. Biol. Chem. **47C**, 37–47 (2013)

Analyzing the Energy Potential of Hydroelectric Power Plant on Kura River

Hasan Hüseyin Çoban[1(✉)] and Arif Cem Topuz[2]

[1] Department of Electrical and Electronics Engineering, Ardahan University, Ardahan, Turkey
hasanhuseyincoban@ardahan.edu.tr
[2] Department of Computer Engineering, Ardahan University, Ardahan, Turkey
arifcemtopuz@ardahan.edu.tr

Abstract. Since rapid development of the world; the meeting of the increasing demand for electric power in recent years has faced the problem; thus, countries, to use their energy potential, avoid from conventional sources and turn to renewable energy sources. Therefore, one of the significant alternatives in the solution frame is to get advantage effectively of hydraulic energy potential which is one of the renewable energy sources.

According to the results of 2017, 20.8% of the energy produced in Turkey is generated from hydraulic energy. The energy structures on the side branch of the Ardahan city, which is our area of study, account for about 372 GWh of the hydro-power generated. In addition to this, the hydropower installed capacity meets about 0.20% (161 MWh) in Turkey.

In this paper, the possible and feasible hydro energy structures on the Kura river-side branches were examined and the amount of energy to be obtained. The importance of hydropower in terms of the region and country is emphasized.

Keywords: Energy potential · Hydroelectric power plants · Feasibility study · Dynamic programming · Stochastic optimization · Engineering

1 Introduction

Turkey is one of the fast developing country in the world. By this development rate 7.4% in 2017 [1], energy demand is significantly increasing day by day. If economic recessions are not taken into account, electricity consumption in Turkey is increasing every year by approximately 8% [2]. The use of electricity in parallel to economic developments in Turkey as of year have increased.

Turkey is in front of many developed countries in terms of hydroelectric potential. Turkey has the biggest hydroelectric potential in Europe after Norway [3]. Distribution of installed electricity power capacity in Turkey is shown in Fig. 1. However, in real life, operation planning failures in the use of water resources cause the production level to be far from country's potential. To contribute to the national economy of the projects and to reduce the dependence on foreign sources in energy; it may be possible with proper planning and optimized engineering applications.

D. J. Hemanth and U. Kose (Eds.): ICAIAME 2019, LNDECT 43, pp. 960–971, 2020.
https://doi.org/10.1007/978-3-030-36178-5_85

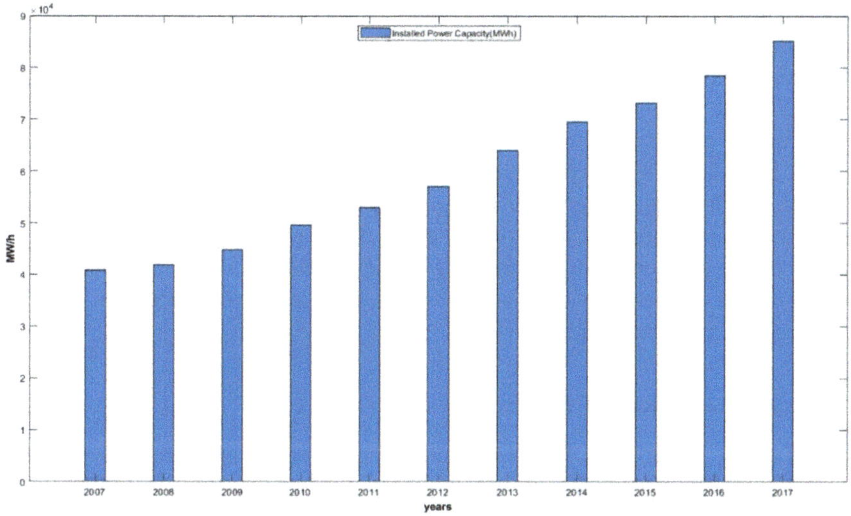

Fig. 1. Distribution of installed electricity power capacity

As in the whole world; energy is a vital issue in Turkey as well, thus; all alternatives should be taken into consideration in order to have a self-sufficient, continuous, reliable and economical electrical energy. A wide variety of energy sources are used in electricity production. Some of them are fossil and some others are clean and renewable. According to data obtained for 2018, the installed power distribution in Turkey and it can be seen that both the own energy production and the importance given to hydraulic energy, which is a renewable energy source [4] is very important for country's energy generation. Distribution of electrical energy installed capacity by source is summarized in Fig. 2. Although the cost factor is important in the choice of resources, it is observed that countries generally turn to their own resources and try to increase resource diversity. Therefore; Turkey must take steps in converting the potential energy of the water it holds. This work, be able to provide significant contributions to the national economy within the borders of Ardahan province, on Kura river Basin, a new hydroelectric power plant project is examined up to the operation system.

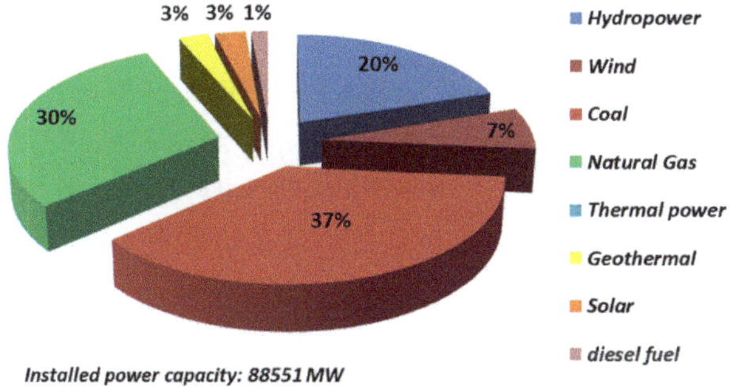

Fig. 2. Distribution of electrical energy installed capacity by source by the end of 2018

Hydroelectric power plants meet for approximately 17% of the world's electricity demand. In the country, this figure is 21% as of the end of 2017. In Turkey, theoretical hydropower potential is 433 billion kWh, while technically feasible potential 216 billion kWh and 126 billion kWh economic potential [5]. Even if the entire hydraulic potential in our country is put into operation, only 23% of the electricity demand in 2020 will be covered by the hydraulic potential [6].

The remaining part of the paper is organized as follows. Section 1.1 describes the experimental framework and steps of the study. Section 1.2 represents a mathematical statement of the problem and the structure of the developed algorithm. Section 2 then combines the results of the case study which choose the best alternative. Finally, Sect. 3 presents our conclusions.

1.1 Purpose and Steps of the Study

In this study, it is aimed to introduce comparative results and alternatives that will contribute to the investor company with the help of MatLab software. In line with this purpose; firstly, in order to see operational conditions, the power plants in the operation has been visited many times in Ardahan city. We had use it as a simulation model; we had applied the optimization model to a real life HPP and checked the results.

Afterwards, data from State Hydraulic Works (DSI) all Water Monitoring Systems near the power plant were obtained. The other steps of the study are given in Fig. 3. As stated at the beginning of the study tests were performed for Kura River. The detailed characteristics of the Kura River is summarized in Table 1 [7, 8].

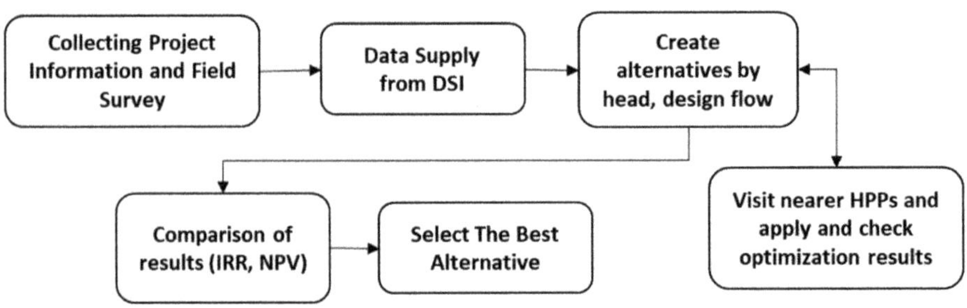

Fig. 3. Steps of the study

Table 1. Physical characteristics and features of Kura River

Location Countries	Turkey, Georgia, Azerbaijan
Source location	Near Kartsakhi Lake, Kars, Turkey
Length	1515 km
Average temperature	5.1 °C
Elevation	2740 m
Basin size	188000 km^2

Kura river basin is situated in northeastern Turkey and is one of the longest rivers in the country and drains to the Caspian Sea. The total area is 188000 km2 and the Kura river length is about 1515 km [7, 8]. The River basins are covered mainly by forest and agricultural landscapes, most of the land is considered flat and the hydrologic soils are mainly well drained to moderately drained. The climate of this region follows the pattern of Northern regions of Turkey with an annual average temperature of about 5.1 °C, but the temperature usually falls down to negative centigrade during winter. Total annual average precipitation is 449 mm with an evapotranspiration rate of 420 mm annually [7, 8]. Figure 4 shows daily water flow rate in m^3/s between 2010 and 2017; the data is take from State Hydraulic Works (DSI).

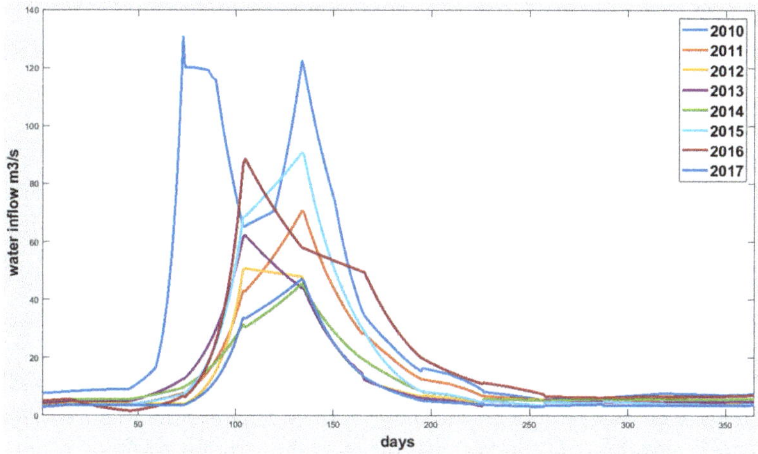

Fig. 4. Daily water flow rate between 2010 and 2017.

As Kura River's basin is the one of the best ecological place in Turkey; we choose run-off-river type of hydropower plant facility. It has been considered to be 'green energy' with little environmental impact, because they do not require damming like large hydro projects. Figure 5 illustrating the main types of hydropower schemes.

The reservoir (storage) can be used for multipurpose; flood control, hydropower, navigation, water supply and irrigation. This type of power plants uses a dam to store water in reservoirs; the electricity is generated by releasing water from the reservoirs through a turbine, which activates a turbine and generator. In this type of hydropower plants, energy can be dispatched on demand and the facility can be shut down and restarted at short notice to meet peak loads. The storage capacity of reservoirs allows this type of HPP to operate independently of the hydrological inflow for many weeks or even months, providing the potential for reliable generation all year, unlike run-of-the river HPPs exposed to seasonal variations. As we choose run of river type power plant for this study for environmental reasons, because of its environmental footprint is also less significant compared to the other types of HPPs. This type of power facilities are

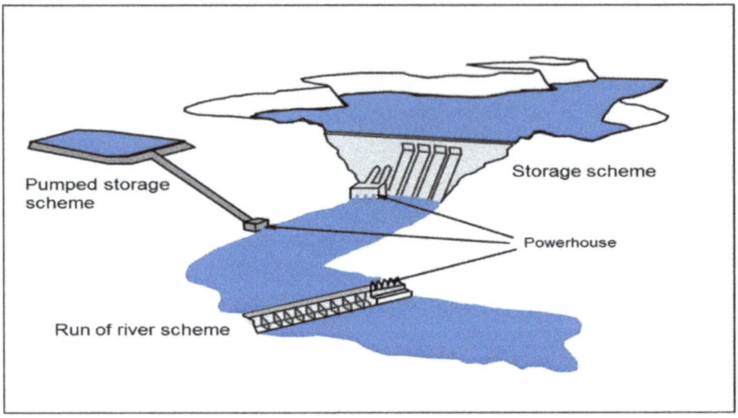

Fig. 5. The main types of hydropower schemes [10]

energy generated according to the flow of the river. Typically, a run-of-the-river project will have little or no storage facility and this type of HPP relies on a continuous supply of electricity (base load) through water flow that is regulated by the plant [9].

It is also sustainable and reliable energy resource which means that the water supply from rivers is reliable, although it varies from season to season and with the weather pattern. Also, because of ecological restrictions [11, 12]; the projects only take some percentage of the total flow to generate electricity; as seen formula (3, 4); heavily dependent on the natural flow of the river. It has low transportation cost; unlike other forms of fuels, electricity can be transmitted simply by connecting to the local grid with a low transmission loss. The other advantage is the operating life of a run-off-river project is typically longer than 50 years, with low operating costs, normally around 10–15% of revenue [13]. Hydropower projects may also have environmental impacts depending on the installed capacity, especially the reservoir size. The one of the main disadvantage of Run–off–the river type hydroelectric power projects are generally involving a low level diversion weir or a stream bed intake and is mostly located on a fast flowing, non-seasonal river but they are normally used in the peak demand of electricity during peak time of the day about 2–4 h in the morning and 2–4 h in the evening and rest of the day they store the river water into the reservoir which will affect the downstream aquatic ecosystem. The people of Ardahan city is not against the installation of power projects but these must be eco-friendly and sustainable in nature and there must be sustainable improvement of human welfare. In other words; it should be a significant progress of the city, which is socially equitable, economically applicable-feasible and environmentally sustainable [14].

Rainfall peaks is noticeable during the season of Spring and Autumn, although it varies from year to year. The flow-duration curve is a cumulative frequency curve that show the percent of time specified discharges were equaled or exceed-ed during a given period [15]. Figure 6 illustrates average daily water inflow rate for a year. The flows over the years of river Kura were obtained from State Hydraulic Works (DSI) and the FDC of those values was plotted; please see Fig. 6.

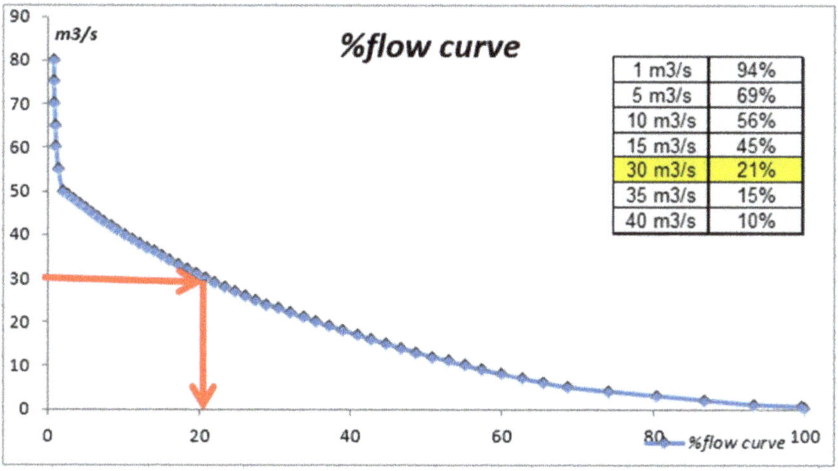

1 m3/s	94%
5 m3/s	69%
10 m3/s	56%
15 m3/s	45%
30 m3/s	21%
35 m3/s	15%
40 m3/s	10%

Fig. 6. Flow duration curve of Kura River

The FDC which is obtained from long term measurements given in Fig. 6 shows that a probability of one of the design flow 30 m3/s, after residual flow is 21% in a year which is about 76 days.

The design is based on nominal parameters (annual or pluriannual averages) that result in an optimized performance (maximum efficiency). But the actual instantaneous parameters vary with respect to these nominal parameters, for instance; the instantaneous flow rate varies, it might be above or below average flow rate. So that's the one of the main point. When design a power plant mostly [16] de-signed these according to average values. But the instantaneous reality of operation may be different. Operating parameters vary with respect to nominal parameters (e.g. tributaries flow rates) Namely the problem is how to adjust the overall system operation acting on controllable variables, in order to optimize income and performance? On next subchapter the problem will be formulated. Because of varying instantaneous water flow rate and electricity market price; we must take into account uncertainties. We want regularity and to decrease the dispersion around the average.

1.2 Problem Formulation

In the field of mathematical optimization, stochastic optimization is a framework for modeling optimization problems that involve uncertainties. Whereas deterministic optimization problems are formulated with known parameters, real world problems almost invariably include some unknown parameters.

Consider a watershed with the facility having its own water reservoir. It is required to determine the operating states and generation level of the power plant over a specified period T. The goal is to find best alternative from energy generation for given marginal prices or market price forecast subject to the limitations on individual reservoirs and unis. The time unit is one hour and planning horizon is 49 years. The private sector has to lease the rights to rivers for 49 years for the sole purpose of electricity production in Turkey [17].

To formulate the objective function mathematically, the notation is used in this paper is first introduced:

$p(t)$: *generated power at time t, in MW.*
$F(t)$: *the income in US dollars.*
R: *the annual income in €.*
$Ex(t)$: *operational and start up costs at time t; in €.*
t: *time index,* $t = 1, 2, 3, \cdots T$.
T: *time horizon under consideration, in hours.*
$v(t)$: *reservoir level at time t;*
v_{max}: *maximum reservoir level.*
v_{min}: *minimum reservoir level.*
$v_{initial}$: *initial reservoir level.*
v_{end}: *at the end of day reservoir level.*
$w(t)$: *water discharge at time t, in* m^3/s.
$w_l(t)$: *lower bound of operating reservoir level at time t.*
$w_u(t)$: *upper bound of operating reservoir level at time t.*
w_{max}: *maximum water discharge at time t.*
$c(t)$: *electricity market price at time t per MWh in €.*
$\zeta(t)$: *naturel water inflow in river at time t.*
$F(t)$: *income at time t, in €.*
$w_{eco}(t)$: *the water spillage for fish gate in hour t;*
γ: *the specific weight of water in* kN/m^3.
H: *the net head in meters.*
n: *the overall efficiency (%).*

The stochastic approach considers an extension of deterministic optimization by using the following objective function. The scheduling problem can be formulated as follow:

$$F \rightarrow max \tag{1}$$

$$R = \sum_{k=1}^{365} \sum_{t=1}^{24} [F] \tag{2}$$

$$F(t) = \sum_{t=1}^{T} \left[p(t).c(t) - \sum Ex(t) \right] \tag{3}$$

Subject to reservoirs and limitations to be described below:

$$p = \gamma.w.n.H \tag{4}$$

The water balance equation:

$$v(t+1) = v(t) + \zeta(t) - w(t) \tag{5}$$

Equation (5) requires conversation of flow among reservoir in the watershed. Reservoirs water level, water discharge, generated power constraints:

The reservoir storage volumes must be within the maximum and minimum reservoir storage volume. The hydropower supply flow rate cannot be greater than the maximum flow rate trough the collecting pipe of hydropower plant.

$$\boldsymbol{v_{min}} \leq v(t) \leq \boldsymbol{v_{max}} \tag{6}$$

The mass balance among the inflows, the reservoir release flows volumes must hold.

$$0 \leq w(t) \leq w_{max} \tag{7}$$

The reservoir storage volume at the end of the reference period is to be equal to storage volume at the beginning of such period.

$$1m \leq \mathrm{v_{initial}} = v_{end} \tag{8}$$

The generated power has lower and upper bounds:

$$0 \leq P(t) \leq P_{max} \tag{9}$$

Ecological constraints: Environmental flow requirements are considered. The environmental flow determined and recommended by the State Hydraulic Works General Directorate is at least 10% of the 10-year average lead [10, 17].

$$w_{eco}(t) \geq \%10.\zeta(t) \tag{10}$$

Water reservoir is bound by the upper and lower bounds of the fore-bay level given by the capacity such that:

$$w_l(t) \leq v(t) \leq w_u(t) \tag{11}$$

2 Implement and Numerical Testing

Feasibility study is a detailed and comprehensive analysis of the proposed project. It is carried out in order to determine whether the potential development is economically, technically and environmentally operable and acceptable under anticipated economic circumstances. The implementation phase of the study consists of three stages; create alternatives; apply optimization and choose the best alternative. The environmental aspects of these alternatives will be taken into account. The 8 alternative projects and the formulation proposed will be com-pared in economic terms. For energy calculations, the flow duration curve can be acceptable for deterministic approach (Fig. 6) is not used in this study for the feasibility report. Because it doesn't take into account uncertainties. As we have used stochastic approach which takes into account hourly water inflow rates and electricity market prices thus result is more accurate.

Average investment costs for large hydropower plants with reservoir typically range from as low as € 895/kW to as high as € 6520/kW while the range for small hydropower projects is between € 1100/kW and € 6800/kW [18]. We assumed 8 different alternatives; each of them has various design, please see Table 2.

Table 2. Technical and economic parameters of the HPP under the study

	Alternatives	Cost estimation, €
A1	25 MW–200 ha	27600000
A2	30 MW–100 ha	33000000
A3	30 MW–200 ha	33100000
A4	35 MW–200 ha	38600000
A5	40 MW–100 ha	44000000
A6	40 MW–200 ha	44100000
A7	51 MW–100 ha	56100000
A8	51 MW–200 ha	56200000

In order to estimate the problem of annual expenditures and income of the project during its lifetime the model can be divided into four steps according to the developed algorithm [19].

1. The input data have to be collected, which include the technical (water flow rates are supplied by DSI) and financial parameters of the project as well as price statistics and predictions. Then, a certain number of days are selected randomly [19].
2. In order to reduce the effect of initial water level in the reservoir; medium-term optimization is performed by considering seven days before and after this day [19].
3. For each of the selected days short-term (24-hour) optimization is completed to obtain the expected daily income [19].
4. For the purpose of to find every year profit the results are generalized for the whole year [19].

It was necessary to decide on the length of the medium-term planning horizon, to avoid the difficulties, the Monte-Carlo method can be used. A larger number of trials leads to a higher accuracy of calculations, while it also increases the computational time [19].

The solution of the optimization task of maximize the income can be performed by the Monte-Carlo-algorithm-based NPV and IRR calculation presented in Fig. 7.

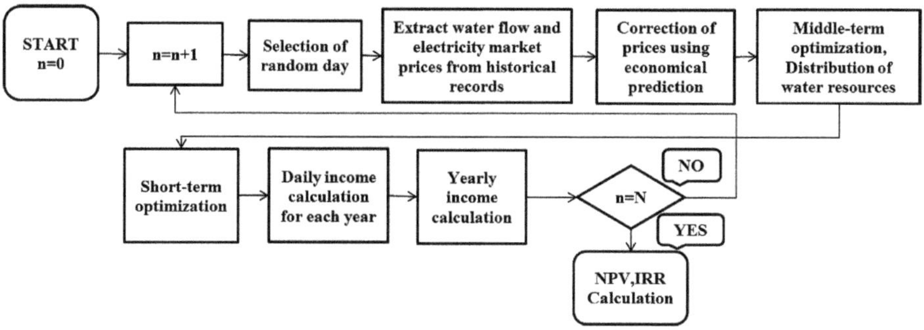

Fig. 7. The main algorithm of estimating the NPV and IRR of a HPP [19]

Cost estimations for various design flow values including contingencies as well as the results of Internal Rate of Return (IRR) and Net Present Value (NPV) calculations for 10%, 8% and 5% interest rates are summarized in Table 3 for 49 years [17]. Thus we can conclude that the best alternative is to choose 51MWh power capacity and 100 ha reservoir capacity, which is, the 7th alternative (A7); the IRR is 16.9%. As in literature, also as PhD thesis [19], and in articles [20]; we have chosen 60 trials (days) are enough in order to find an appropriate number of Monte-Carlo trials; it gets approximately 10% of error.

Table 3. The results of NPV and IRR for 49 years

Alternative	Stochastic approach, %10 interest rate		Stochastic approach, %8 interest rate		Stochastic approach, %5 interest rate	
	NPV, EUR	IRR, %	NPV, EUR	IRR, %	NPV, EUR	IRR, %
A1	47.806.295	21,80%	72.805.694	21,80%	145.225.727	21,80%
A2	49.348.312	20,58%	76.426.359	20,58%	154.973.815	20,58%
A3	50.024.597	20,69%	77.358.973	20,69%	156.649.473	20,69%
A4	51.348.183	19,68%	80.716.636	19,68%	166.007.968	19,68%
A5	50.267.699	18,11%	81.306.917	18,11%	171.324.840	18,11%
A6	51.113.507	18,22%	82.466.066	18,22%	173.391.805	18,22%
A7	52.258.332	16,90%	87.462.694	16,90%	189.787.013	16,90%
A8	23.328.395	16,90%	87.624.118	16,90%	190.130.553	16,90%

3 Conclusions and Future Study

The method is presented in this paper for scheduling hydro power systems restricted operation zone with environmental and safety reasons. At the end of the study, the obtained data and the findings are satisfactory.

The economic part of the feasibility study determines a project's applicability, with offers as to how to manage the various stages of the project. The results allow us to conclude that the power producer has to choose the best efficiency; for the HPP project, it corresponded to 51 MWh of installed power capacity and 100-ha surface of the water reservoir area. The stochastic affirmation has been successfully tested scientific and practical model in the new HPP project and in terms of the country's electricity sector, involves analysis of the potential benefits of the wholesale, investors, and production companies. The experiment and application of this scientific model according to the conditions of the countries will make a valuable improvement to the developing electric energy industry.

As the article only tests economic part of feasibility study of the HPP project; for future studies; it can be very useful to work on chosen turbine types/numbers. As we mentioned in first chapter ecological limitations and Ardahan is a flat place; that's why most probably Kaplan type turbine might have best efficiency as the power plant has around 10–70 m net head.

Acknowledgments. This paper is derived from the project numbered 2018/003 supported by Ardahan University Scientific Research Coordinator.

References

1. OECD (Organization for Economic Co-Operation and Development): OECD Economic Surveys: Turkey 2018. OECD Publishing, Paris (2018)
2. Nişanci, M.: Türkiye'de Elektrik Enerjisi Talebi ve Elektrik Tüketimi ile Ekonomik Büyüme Arasındaki İlişki. Sosyal Ekonomik Araştırmalar Dergisi **5**(9), 107–121 (2005). http://dergipark.org.tr/susead/issue/28433/302865
3. Mennel, T., Ziegler, H., Ebert, M., Nybø, A., Oberrauch, F., Hewicker, C.: The hydropower sector's contribution to a sustainable and prosperous Europe. Main Report (2015)
4. Republic of Turkey Ministry of Energy and Natural Resources – Electric. https://www.enerji.gov.tr/tr-TR/Sayfalar/Elektrik. Accessed 5 Oct 2019
5. Adıgüzel, F.: Türkiye'de Enerji Sektöründe Hidroelektrik Enerjinin Önemi. Su Kaynaklarının Geliştirilmesi ve Yönetimi-Türkiye Mühendislik Haberleri (2002)
6. Cebeci, M.: Bölgemiz enerji kaynakları ve enerji projeksiyonu. Güneydoğu Anadolu Bölgesi Enerji Forumu, pp. 2–3 (2005)
7. Özey, R.: Türkiye'nin Sınıraşan Suları ve Sorunları. Doğu Coğrafya Dergisi **3**(2) (1997)
8. Şahin, C., Doğanay, H., Özcan, N.A.: Türkiye Coğrafyası (Fiziki, Beşeri, Ekonomik ve Jeopolitik). Gündüz Eğitim ve Yayıncılık, Ankara (2005). ISBN: 9756859547
9. Clifford Chance: Hydropower – Overview and Selected Key Issues, October 2017 Report (2017)
10. Lumbroso, D., Hurford, A., Winpenny, J.: Synthesis report: harnessing hydropower (2014). https://doi.org/10.12774/eod_cr.september2014.lumbrosoetal2
11. Yaşar, M., Baykan, N.O., Bülbül, A.: Akışaşağısına Bırakılması Gerekli Debi Yaklaşımları. Su Yapıları Sempozyumu, 16–18 September 2011, Diyarbakır, Turkey (2011)
12. Karakoyun Y., Yumurtaci Z.: Hidroelektrik Santral Projelerinde Çevresel Akiş Miktarinin ve Çevresel Etkinin Değerlendirmesi. Tesisat Mühendisliği Dergisi **21**(138) (2013). ISSN: 1300-3399
13. Haws, F.W., Israelsen, E.K.: New Concepts for Preliminary Hydropower Design: The Powermax Slope, Binary Turbine Sizing, and Static Regain (1984)

14. Lata, R., Rishi, M.S., Kochhar, N., Sharma, R.: Impact analysis of run-off–the river type hydroelectric power plants in Himachal Pradesh, India. Int. J. Civil Struct. Environ. Infra Struct. Eng. Res. Dev. **3**(2), 77–82 (2013)
15. Searcy, J.K., Curves, F.D.: Manual of Hydrology: Part 2. Low-Flow Techniques, Geological Survey Water-Supply paper (1959)
16. Jiang, Z., Qin, H., Wu, W., Qiao, Y.: Studying operation rules of cascade reservoirs based on multi-dimensional dynamics programming. Water **10**(1), 20 (2017)
17. Islar, M.: Privatised hydropower development in Turkey: a case of water grabbing?. Water Altern. **5**(2) (2012)
18. IRENA Working Paper: Renewable Energy Technologies: Cost Analysis Series, Hydropower 1(3/5) (2012)
19. Coban, H.H.: Optimization techniques in short and long-term power production at small hydropower plants, Doctoral Thesis, Riga Technical University (2016)
20. Vereide, K., Lia, L., Ødegård, L.: Monte Carlo simulation for economic analysis of hydropower pumped storage project in Nepal. Hydro Nepal: J. Water Energy Environ. **12**, 39–44 (2013)

Tamper Detection and Recovery on RGB Images

Hüseyin Bilal Macit[1(✉)] and Arif Koyun[2]

[1] Mehmet Akif Ersoy University, Burdur, Turkey
hbmacit@gmail.com
[2] Süleyman Demirel University, Isparta, Turkey

Abstract. Special devices are required to perform tampering on an analogue image. This requires a serious cost and skill. But it is easy to tamper on a digital image. Digital images can be tampered even with easy accessible free digital image editing software. Moreover, this tampering can occur even with a mobile phone. In particular, tampering on important images such as images containing financial data, images of official documents, etc., can lead to money and time losses. In this study, a method of tamper detection in order to protect the RGB images and performing a recovery of the tampered region has been proposed. The proposed method can recover a tampered image even without any information of the original. Method backups the value of each pixel of the image to the farthest pixel possible. The image is first separated into layers R, G and B. Each layer is then divided into blocks of equal size 2 × 2. The pixel of each layer is embedded in different layers of the corresponding blocks by the LSB method before sending it to receiver side. The image obtained on the receiver side is scanned by the inverse algorithm. The value of the detected pixels is withdrawn from the corresponding block pixels and the image is repaired. Method seems to work quickly and effectively and it is tested on RGB credit card images. The results are shown in the study.

Keywords: Image security · Image recovery · Tamper detection

1 Introduction

With the widespread use of broadband networks, the use of electronic documents is increasing in the world [1]. The traditional paper-based model of communication is abandoned and a rapid development is taking place towards the modern digital communication model. Goods and service orders, electronic banking and some state jobs can be carried out electronically in accordance with the law. Perhaps the paper-based documentation system will be completely eliminated in the near future [2]. However, due to the nature of digital media, illegal transactions such as duplication, modification and forgery in digital media are made easily and quickly. The protection, verification and integrity of the intellectual property rights of digital media has now become an important issue [3, 4].

The image is the media most exposed to counterfeiting in digital media. With the increase in image forgery and its results, development of new techniques of image

D. J. Hemanth and U. Kose (Eds.): ICAIAME 2019, LNDECT 43, pp. 972–981, 2020.
https://doi.org/10.1007/978-3-030-36178-5_86

authentication has become very important [5, 6]. The most common areas of image forgery are journalism, photography, law, politics and medicine [7]. Image verification is performed to detect image forgery. In general, image verification verifies the integrity of the digital image [1]. Image integrity is distorted due to changes made to the image. These changes are called tampering. The literary meaning of tampering is making unauthorized changes or harming something. While digital media were not available, photo tampering were performed by ink, paint, double exposure, airbrushing photos or negatives in a dark room, or by drawing Polaroid. The outputs of such tampering were very similar to digital manipulations, but were more difficult to make [7]. Tampering can be innocent or malicious. Innocent tampering does not change the content of the image, but changes the quality of the image. It included various operations such as contrast brightness, adjustment, zoom and rotation. Malicious tampering aims to change the content of the image [5].

1.1 Image Tampering Attacks

Various common image tampering attacks are as follows [8].

Copy-Move. It is a method of copying and insignificant part and pasting it on a significant part of the image. It is the most widely used image tampering method.

Adding an Image. Adding a part of another image to an image. In other words, it is a photomontage process.

Resize. This process performs a geometric transformation that can be used to reduce or enlarge an entire image or a part of the image.

Crop. A technique used to cut the edges of an image or reduce the canvas on which the image is displayed.

Blur. The process of closing all the pixel values with the neighboring pixel values.

Noise. Adding a gran to whole image.

Steganographic Methods. Secret data embedding in a digital image [7].

In general, there are two ways to provide information security. These are encryption and data hiding [9]. Data hiding is called steganography. In a steganographic method, information is transmitted hidden in another information without any doubt. Only the receiver knows the extraction method of confidential information [10]. In encryption, information is converted to a form that cannot be understood by third parties. However, the decrypted information at the receiver is insecure. Encryption and steganography can send data safely through the transmission channel, but not enough to ensure the integrity of the data. In order to address both authentication and integrity issues, a variety of methods have recently been proposed for different applications. These methods can be divided into two categories: labeling-based and watermark-based methods. Methods based on labeling stores authentication data in a separate file. Such methods can determine whether a protected image has been tampered or not, but cannot identify spatial placements where tampering has occurred. They also require extra maintenance costs, transmission and storage costs because the authentication data is stored in a separate file. In contrast, watermark-based methods use digital watermarks as authentication data and embed watermarks into original multimedia. In addition, the use of digital watermarks can detect tampering localization [4].

Depending on the application, digital watermarking techniques can be classified into two main categories as robust and fragile watermark techniques [1]. Depending to application domain, watermarking methods can be divided into two categories as spatial domain and frequency domain watermarking. In the spatial domain watermarking; the watermark is embedded by directly modifying the pixel in the original image. In the frequency domain watermarking, watermark is embedded by applying transformation functions such as Discrete Fourier Transform (DFT), Discrete Wavelet Transform (DWT) and Discrete Cosine Transform (DCT) on original image [3]. This study is depended on a fragile and spatial domain image watermarking method.

2 Method

The digital image is a numerically expressed form of a two-dimensional image [5]. The two important elements that represent the image are the number of colors and the size. When viewed in terms of number of color count, images are divided into three types as binary, grayscale and Red-Green-Blue (RGB) images. A standard binary image stores 1 bit for each pixel, a grayscale image stores 1 byte for each pixel and a RGB image stores 3 bytes for each pixel as it is shown in Fig. 1.

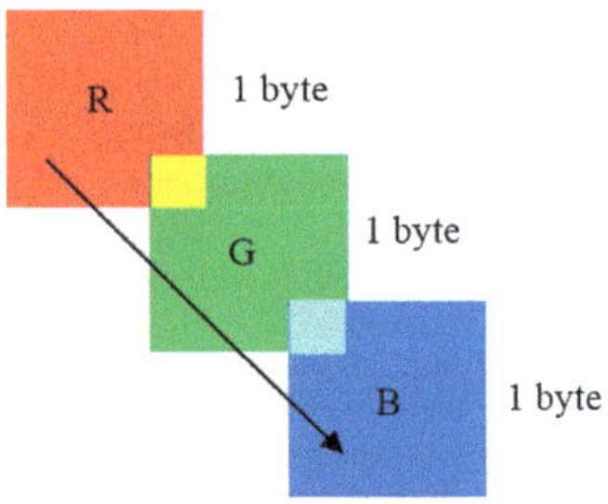

Fig. 1. Pixel representation

A digital image I is represented by an array of M rows and N columns. Therefore, a numerical image contains MxN pixels.

$$I = \{x_{ij} | 1 \leq i < M, 1 \leq j < N\} \tag{1}$$

Here; $x_{i,j}$ represents each element of the I matrix. 3 matrices should be created for each color plane of an RGB image.

$$R = \{r_{ij} | 1 \leq i < M, 1 \leq j < N\} r_{ij} \in \{0, 1, 2, \ldots, 255\} \tag{2}$$

$$G = \{g_{ij} | 1 \leq i < M, 1 \leq j < N\} g_{ij} \in \{0, 1, 2, \ldots, 255\} \tag{3}$$

$$B = \{b_{ij} | 1 \leq i < M, 1 \leq j < N\} b_{ij} \in \{0, 1, 2, \ldots, 255\} \tag{4}$$

Proposed method divides the original image to 2 × 2 equal sized blocks for self-watermarking. Figure 2 shows this stage over CreditCard image.

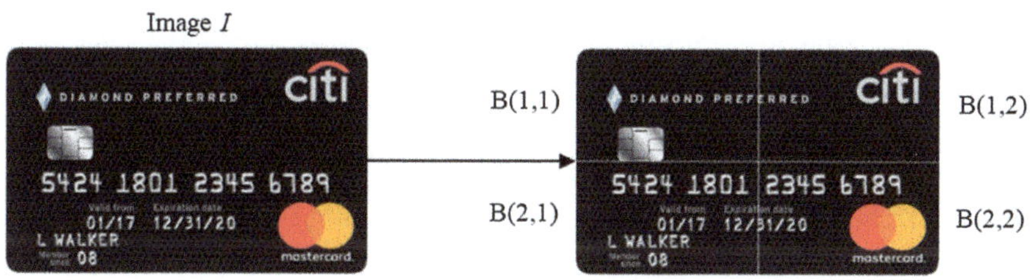

Fig. 2. Mapping image *I* to blocks

Proposed method works on R, G and B layers of selected block as shown in Fig. 3.

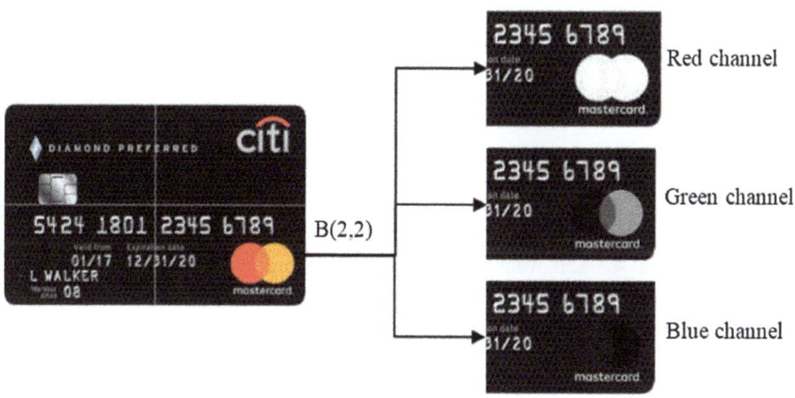

Fig. 3. Mapping a block to color planes

Each block has a First, Second and Third neighbor block. Each block backs up its selected color plane to related neighbor. This neighborhood is shown in Table 1.

Table 1. Block neighborhood

Block	First neighbor	Second neighbor	Third neighbor
B(1,1)	B(1,2)	B(2,1)	B(2,2)
B(1,2)	B(2,1)	B(2,2)	B(1,1)
B(2,1)	B(2,2)	B(1,1)	B(1,2)
B(2,2)	B(1,1)	B(1,2)	B(2,1)

This backing up process is made by a Least Significant Bit (LSB) insertion method. LSB is a method used for steganography and watermarking. The main idea of LSB is based on bit plane of a pixel. Each pixel of a 24 bit color bitmap image is represented with 3 x 8 bits of color sets, thus, changing of last bits of every set doesn't make a visible change on the image. This study is based on backing up 6 Most Significant Bits (MSB) of each pixel plane to neighbor block's corresponding pixel as in Fig. 4. This process is called a self-watermarking process and runs for each block as follows:

- 2MSBs of $r_{i,j}$(Block) to 2LSBs of $r_{i,j}$(FirstNeighbor)
- 2MSBs of $g_{i,j}$(Block) to 2LSBs of $g_{i,j}$(SecondNeighbor)
- 2MSBs of $b_{i,j}$(Block) to 2LSBs of $b_{i,j}$(ThirdNeighbor)

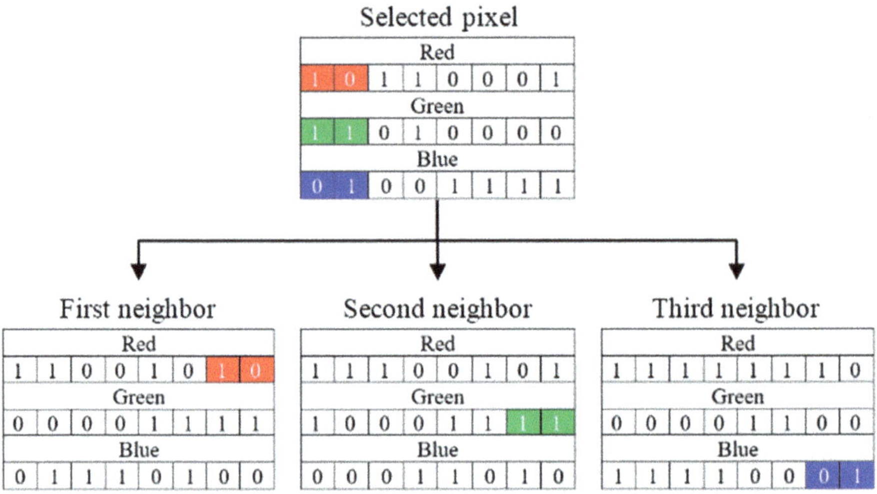

Fig. 4. Self-watermarking process of proposed method

As is known, in the 8-bit data encoding system; first MSB refers to 2^7 and the second MSB refers to 2^6. Thus, 2MSBs may have a value up to $2^7 + 2^6 = 128 + 64 = 192$. On the other hand, first LSB which is the 8.bit of array refers to 2^0 and the second LSB which is the 7.bit of array refers to 2^1. Thus 2LSBs may have a value up to 21 + 20 = 2+1 = 3. So 2LSBs store maximum 3/256 = 1.17% of the pixel value and 2MSBs store maximum 192/256 = 75% of the pixel value. In this study, 1.17% of the image has been abandoned to make secure 75% of image data. In the last step, the processed blocks are combined to create the processed image *P*. Image *P* is a secure image for tampering attacks and can be transferred on a network as in Fig. 5. Every image taken from the network is suspicious whether it is tampered or not, so this image is called tampered image *T*.

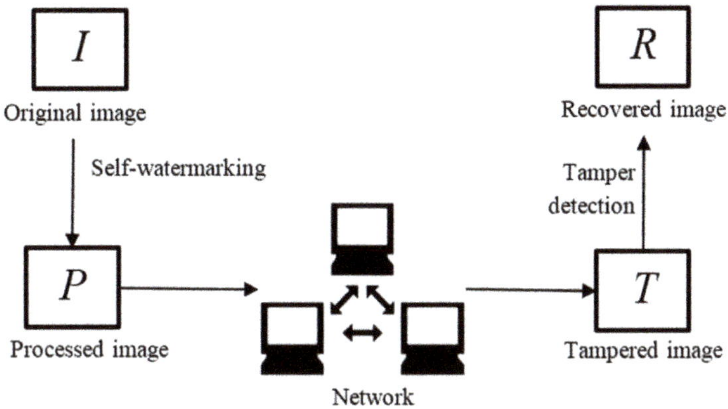

Fig. 5. Tamper detection process

T tampered image is divided into 2 × 2 equal blocks for tamper detection. The method applied for block division is exactly the same as that of the self-watermarking stage. If there is no tamper in the image, 6MSBs of the pixels of each block must be same to 2LSBs of each color plane of corresponding pixels of neighbor blocks. A threshold value is used here. The user can assign the threshold value as 2 or 3. At least two of the following conditions must occur for threshold value 2 and at least three of the following conditions must occur for threshold value 3.

- 2MSBs of $r_{i,j}$(Block) must be same to 2LSBs of $r_{i,j}$(FirstNeighbor)
- 2MSBs of $g_{i,j}$(Block) must be same to 2LSBs of $g_{i,j}$(SecondNeighbor)
- 2MSBs of $b_{i,j}$(Block) must be same to 2LSBs of $b_{i,j}$(ThirdNeighbor)

Recovery process starts if selected pixel is marked as tampered. Recovery process is basically the opposite of self-watermarking process as it is shown in Fig. 6.

Fig. 6. Recovery of image T

3 Experimental Results

In addition to tampering detection, it is important to recover the original data from the tampered data. It is recommended that a tamper detection algorithm meets all or most of the following requirements [11, 12].

Tampering Detection. If the image has been tampered, the method should issue an alarm.

Invisibility. Any change in the image due to self-watermarking should not be sensitive to Human Visual System (HVS) or affect the perceptual quality of the original image.

Locating Tampered Zones. The method should identify tampered zones

Self-recovery. The method should recover tampered image similar to original image.

Blind Detection. Original image is not required for tamper detection.

Efficiency. Algorithm processing complexity should be minimized.

Two metrics are used for perceptual quality assessment: Peak Signal Noise Ratio (PSNR) and Structural Similarity Index (SSIM). Watermarked images with PSNR values over 28 dB and SSIM values over 0.96 have very accepted perceptual quality [11]. SSIM is a method for measuring the similarity between two images. The SSIM index is a full reference measure [10].

Two different sized test images used to test proposed method. Table 2 shows the first test image used for proposed method which is a sample credit card image of Citibank website. As it is seen, credit card number and validation date data are tampered in image *T*.

Table 2. Original, processed and tampered CreditCard image

Original image *I*	Processed image *P*	Tampered image *T*
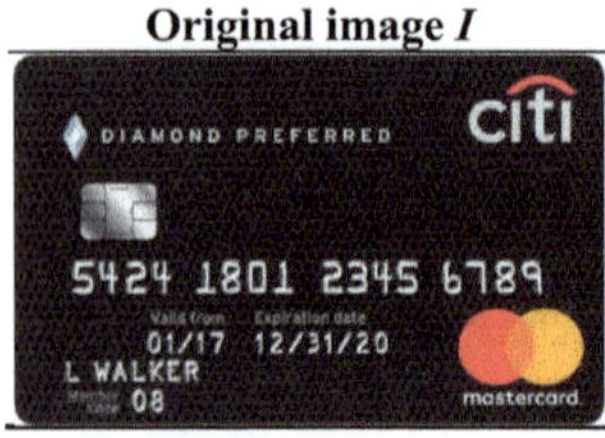	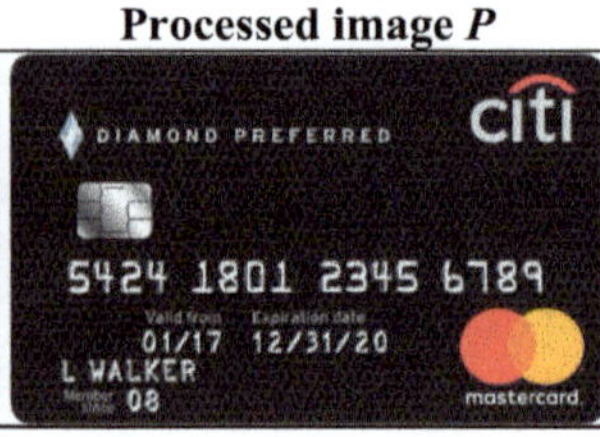	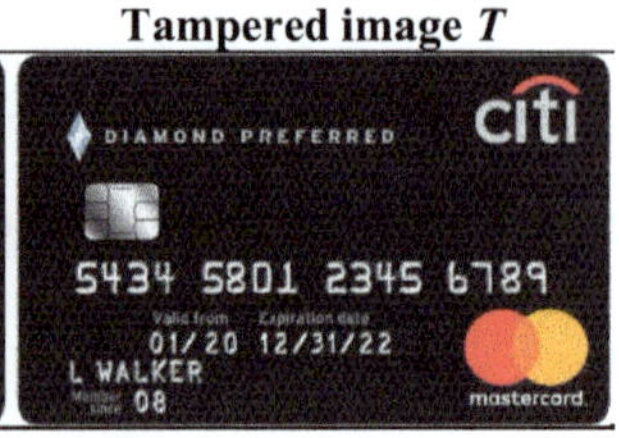

When *I* and *P* are compared; PSNR is calculated 41.9869 and SSIM is 0.9866. Table 3 shows the results of proposed method for CreditCard image.

Table 4 shows the second test image used for proposed method which is a sample IDCard image. As it is clearly seen, card owner's photo is tampered in image *T*.

Table 3. CreditCard image tamper detection and recovery results

Threshold Value	Detected Tampering	Recovered Image *R*	Similarity Values			
			T vs *R*		*I* vs *R*	
			PSNR	SSIM	PSNR	SSIM
2	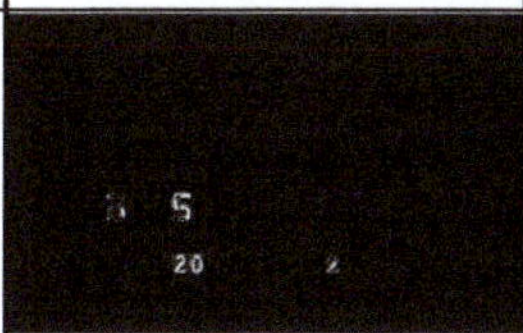	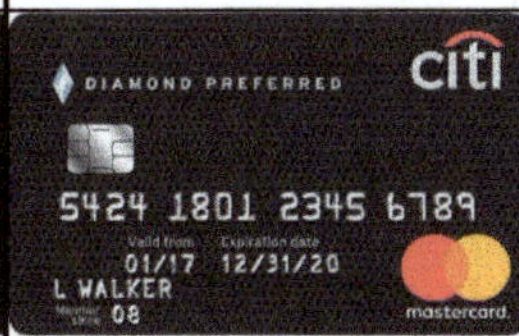	29.1402	0.9937	38.5953	0.9846
			T vs *R*		*I* vs *R*	
			PSNR	SSIM	PSNR	SSIM
3	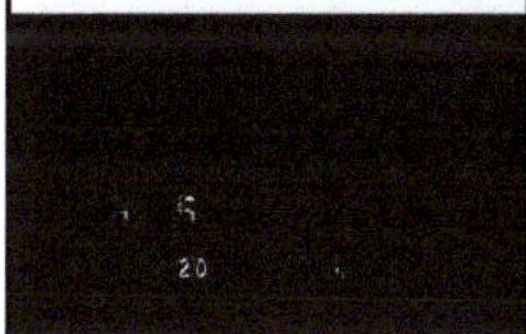	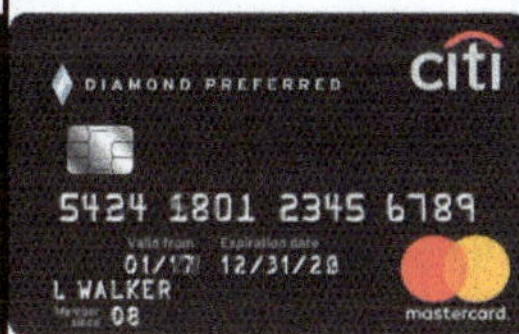	30.6639	0.9955	33.0607	0.9821

Table 4. Original, processed and tampered CreditCard image

Original image *I*	Processed image *P*	Tampered image *T*
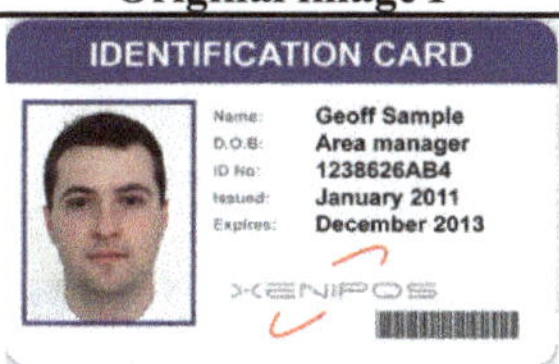	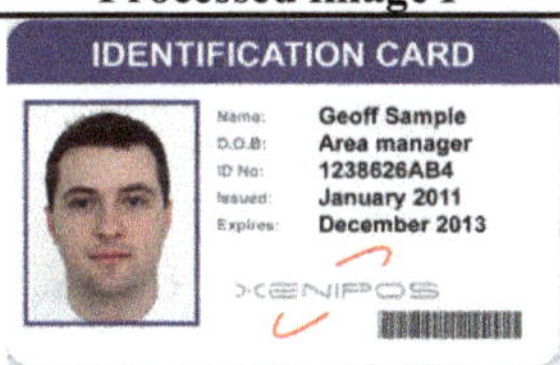	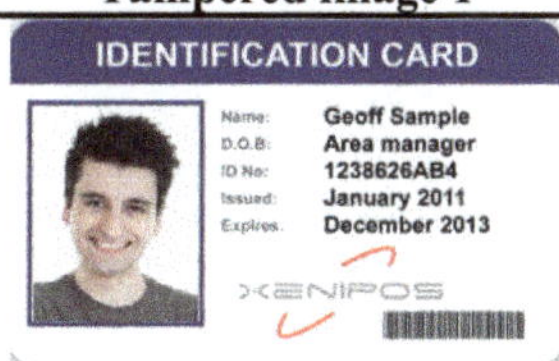

When *I* and *P* are compared; PSNR is calculated 48.1596 and SSIM is 0.9952. Table 5 shows the results of proposed method for IDCard image.

After these experimental tests, the following can be said.

- Tables 3 and 5 clearly shows that proposed method is successful at tamper detection.
- Method can detect tampered area on the image truly.
- PSNR and SSIM values of comparing original and processed test images show that proposed method has acceptable invisibility.

Table 5. IDCard image tamper detection and recovery results

<table>
<tr><th rowspan="3">Threshold Value</th><th rowspan="3">Detected Tampering</th><th rowspan="3">Recovered Image R</th><th colspan="4">Similarity Values</th></tr>
<tr><th colspan="2">T vs R</th><th colspan="2">I vs R</th></tr>
<tr><th>PSNR</th><th>SSIM</th><th>PSNR</th><th>SSIM</th></tr>
<tr><td>2</td><td>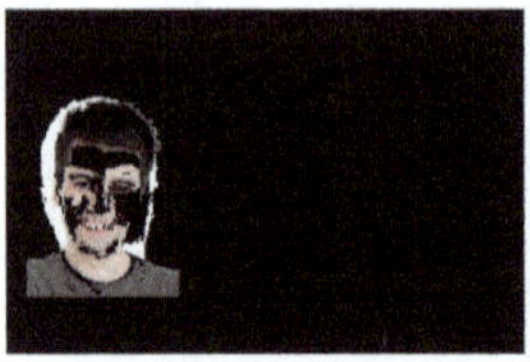</td><td>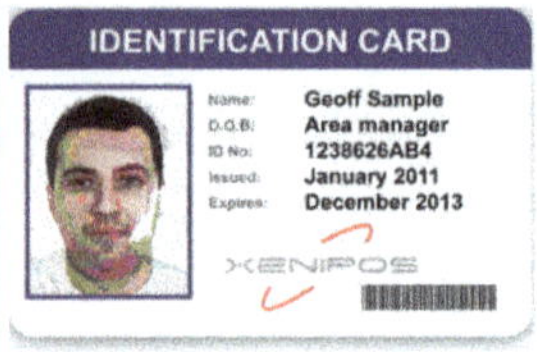
</td><td>24.4166</td><td>0.9778</td><td>32.4872</td><td>0.9841</td></tr>
<tr><td>3</td><td></td><td>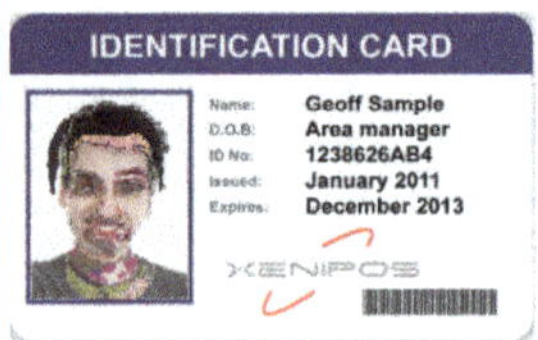
</td><td>27.6153</td><td>0.9870</td><td>26.1047</td><td>0.9741</td></tr>
</table>

- Tables 3 and 5 clearly shows that proposed method is successful at image recovery especially with threshold value 2. It is seen that PSNR and SSIM values between original and recovered image are higher than tampered and recovered image with threshold value 2.

Acknowledgement. This work was supported by Scientific Research Fund of the Suleyman Demirel Uni-versity. Project Number: 4382-D1-15.

References

1. Hassan, M.H., Gilani, S.A.M.: A semi-fragile watermarking scheme for color image authentication. World Acad. Sci. Eng. Technol. **19**, 842–846 (2008)
2. Gözel, A.: Belgede Sahtecilik Suçlarının Konusu Olarak Belge ve Elektronik Belge, S.D.Ü. Hukuk Fakültesi Dergisi (**5**)1, 143–201 (2015)
3. Tsaia, P., Hub, Y.C., Changa, C.C.: A color image watermarking scheme based on color quantization. Signal Process. **84**, 95–106 (2004)
4. Wang, M.S., Chen, W.C.: A majority-voting based watermarking scheme for color image tamper detection and recovery. Comput. Stand. Interfaces **29**, 561–570 (2007)
5. Rajalakshmi, C., Alex, M.G., Balasubramanian, R.: Study of image tampering and review of tampering detection techniques. Int. J. Adv. Res. Comput. Sci. **8**(7), 963–967 (2017)
6. Gill, N.K., Garg, R., Doegar, E.A.: A review paper on digital image forgery detection techniques. In: 8th ICCCNT, IIT Delhi, India (2018). https://doi.org/10.1109/icccnt.2017.8203904

7. Mishra, M., Adhikary, M.C.: Digital image tamper detection techniques - a comprehensive study. Int. J. Comput. Sci. Bus. Inform. **2**(1), 1–5 (2013)
8. Sharma, D., Abrol, P.: Digital image tampering – a threat to security management. Int. J. Adv. Res. Comput. Commun. Eng. **2**(10), 4120–4123 (2013)
9. Tuncer, T., Avcı, E.: Renkli İmgelerde Kimlik Doğrulaması ve Saldırı Tespiti için Görsel Bir Sır Paylaşım Tabanlı Yeni Bir Kırılgan Damgalama Algoritması. Int. J. Innov. Eng. Appl. **1**(1), 1–8 (2017)
10. Manjula, G.R., Danti, A.: A novel hash based least significant bit (2-3-3) image steganography in spatial domain. Int. J. Secur. Priv. Trust Manag. **4**(1), 11–20 (2015). https://doi.org/10.5121/ijsptm.2015.4102
11. Taha, T.B., Ngadiran, R., Ehkan, P., Sultan, M.T.: Image tamper detection and recovery using lifting Scheme watermarking. J. Theor. Appl. Inf. Technol. **96**(8), 2307–2316 (2018)
12. Vaishnavi, D., Subashini, T.S.: Image tamper detection based on edge image and chaotic Arnold map. Indian J. Sci. Technol. **8**(6), 548–555 (2015)

Assessment of Academic Performance at Akdeniz University

Taha Yiğit Alkan, Fatih Özbek, Melih Günay(✉), Bekir Taner San, and Olgun Kitapci

Akdeniz University, Dumlupinar Boulevard Campus, 07058 Konyaalti, Turkey
{yigitalkan,fatih,mgunay,tanersan,okitapci}@akdeniz.edu.tr
http://www.akdeniz.edu.tr

Abstract. As publication performance is a good indicator of quality of research and impact, it may be used to determine the effectiveness of academicians and universities. In this study we promote a software that we developed at Akdeniz University to support research and collaboration. The data for this study is obtained by the database of HR, Web of Science (Wos) and InCites. By merging data from all these sources and mining the collected data, we were able to generate tables and charts that allows administrators and partners to make policy and develop strategies for effective research. To review the software, you may visit http://akademik.akdeniz.edu.tr.

Keywords: Academic performance · Research metric · WoS

1 Introduction

With increasing frequency, performance of academic institutions are ranked by several organizations and reported routinely. These rankings usually evaluate the performance of institutions based on the following dimensions;

(a) research output including patents, funding and publications
(b) education quality including student achievement
(c) economic impact to local and global economy
(d) contribution to social life, art and society

Finally, ranking of an institute is often determined with the contribution of above dimensions multiplied by certain weights. Due to lack of current and historic public data (except item a) in conjunction with varying weights of aforementioned dimensions, results tend to vary between reported rankings. Excluding boutique universities with certain theme; there is a high correlation between academic output and the reported university rankings. This is because; one of

Authors thank Akdeniz University BAP for their financial support and rectorate for sharing key HR information.

D. J. Hemanth and U. Kose (Eds.): ICAIAME 2019, LNDECT 43, pp. 982–995, 2020.
https://doi.org/10.1007/978-3-030-36178-5_87

the major factor in the determination of university ranking is based on research output in terms of academic publications and their impact, namely citations [1]. However, with the increase of online publications and ease of publishing through internet, the number of publications are exploded [2]. With this quantitative increase, the average quality of a publication dropped and consequently it became increasing difficult the judge the contribution of the academic work. Therefore, there is a need to objectively judge quality of publications through reliable and public resources. Web of Science (WoS) (previously known as Web of Knowledge) is a scientific citation indexing service, which provides access to multiple research databases by the Institute for Scientific Information (ISI), now maintained by Clarivate Analytics [3]. Web of Science has known as the oldest citation resource, containing the most prestigious academic journals used for the purpose of citation analysis [5].

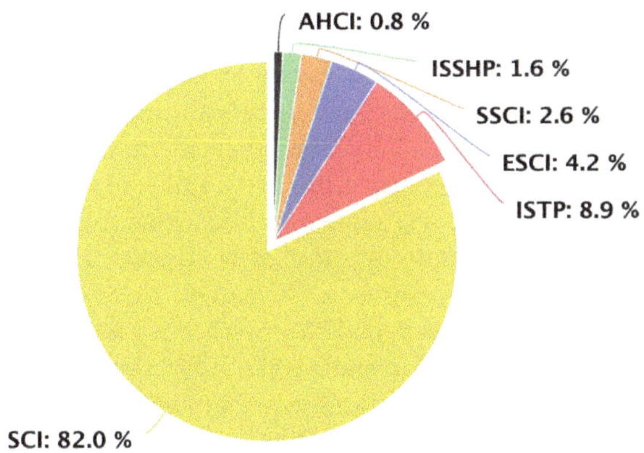

Fig. 1. Distribution by Indexes of Publications at Akdeniz Univesity

Web of Science has different citation indexes such as Science Citation Index (SCI), Social Science Citation Index (SSCI), Arts and Humanities Citation Index (A-HCI) [3].

Especially as competition with Scopus increased, Clarivate Analytics continued its new products like Emerging Sources Citation Index (ESCI). For instance, in 2014 Clarivate Analytics launched InCites Databases which provides multiple analysis options, indicators, and international comparisons, including institutional or personal productivity. It is possible to benchmark these outputs for research cooperation and collaboration. In contrast to databases such as Scopus, InCites offers a much wider range of analytical capabilities and tools for obtaining bibliometric statements [10]. Even though Incites database is part of Web of Science, there is a synchronization problem that cause time lag of up-to 3 months in updates.

Research areas at large universities with a student body of 30K+ often include a wide range of disciplines from Social Sciences, Natural and Applied

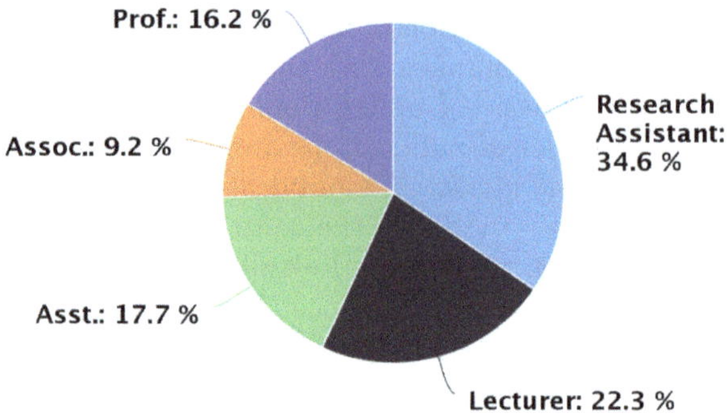

Fig. 2. Distribution of titles of academic staff at Akdeniz Univesity

Sciences, Health Sciences, Fine Arts to Athletics. Figure 2 show distribution of titles of academic staff at Akdeniz University. Such distribution is not uncommon in many large universities. Therefore, it is challenge to evaluate and compare the research performance of individual researchers and departments within the university. Such information is especially necessary for administration to produce policy and improve the standing of the university. One of the major parameters to evaluate research performance is scientific publications from Web of Science [6].

There have already been many studies of academic performance in the literature during the period. For example, Soutar et al. conducted the results of an analysis of the research impact of 2263 marketing academics using citation metrics in the top 500 research universities [7]. The results indicated that ranks the top 100 university marketing departments in the top 500. Patel et al. compared h-index scores for the academic performance of healthcare researchers from databases [8]. Bar-Ilan compared the H indicates of a list of highly-cited Israeli researchers from Google Scholar, Web of Science and Scopus databases [9].

In a study conducted in Norway in 2015, the publication activity of almost 12400 Norwegian university researchers were handled in terms of age, gender and academic position. In the method of study, each researcher was assigned to one of the 5 research fields; the humanities, social sciences, natural sciences, engineering and technology and medicine. In these 5 fields, medicine has the largest population in terms of both the number of people and publications and humanities has the smallest population. This study shows that academic position is the most important factor in all fields, although female researchers tend to publish less than male researchers and the number of publications changes with age [11].

A fuzzy logic approach was proposed by Dilek Kaptanoglu and Ahmet Fahri Özok in 2006 for the evaluation of academic performance. While applying fuzzy logic approach, 3 main criteria were determined as research, education and service, and 3 different methods (Liou and Wang, Abdel Destiny and Dugdale, Chang) were tried separately for ordering fuzzy values. At the end of the study, while consistent results were obtained, it was stated that in Chang method, because of the service criterion 0, it has a different effect on the result. The study showed that the problem of academic performance evaluation can be solved as a fuzzy decision-making problem [12].

In a study conducted in 2008, in Turkey between 1990 and 2000, each year the number of publications in the SCI/SCI-E increased and it is stated that the H-index index also increased. However, those states that less number of publications that compared with other countries, it was determined that they have a much higher H-index than Turkey and Turkey ranks last compared to other countries [13].

URAP (University Rankink by Academic Performance) research laboratory was established in 2009 at the Middle East Technical University Informatics Institute. In order to evaluate higher education institutions in line with their academic achievements, they develop scientific methods and share the results of the studies with the public [1].

In 2016, TUBITAK published its university competency analysis report. In this report, the publications and projects of universities were examined between 2010 and 2014, and the reports based on objective data were given about the areas where the universities were competent. The indicators taken into consideration while conducting competency analysis are discussed under two main headings as "volume" and "quality" [14].

However, we are not aware of any such studies of a software system for academic performance that focus on the academic institutes and the work reported here therefore fills a niche in the regarding literature. We provided an approach that not only considers the number of citations and publications but also normalizes them according to researcher's field. Therefore, we can evaluate and compare the performance of each researcher and department within its own piers. Due to its real-time and web-based nature of the proposed system, instant academic performance data is available at any time.

The paper is organized so that we next present the methodology used in the current study and ends with a review of the main findings, discussion, implications and limitations of the study.

2 Method

The key analysis reports of the academic performance system relies on the data retrieved from the Web of Science (WoS) databases. When authors publish research articles they often supply first and last names, contact information and academic institutes. Even though such metadata can often reliably be used to evaluate and compare the performances of academic institutes, individual performance evaluation may be a challenge as names may not be unique and consistently entered throughout the career of an academic staff. In order to uniquely identify authors and associate them with their publications, a ResearcherID profile was used. At Akdeniz University, administration asked academic staff to obtain a ResearcherID if they don't have one and add publications to their profile and let research office to know their ResearcherID.

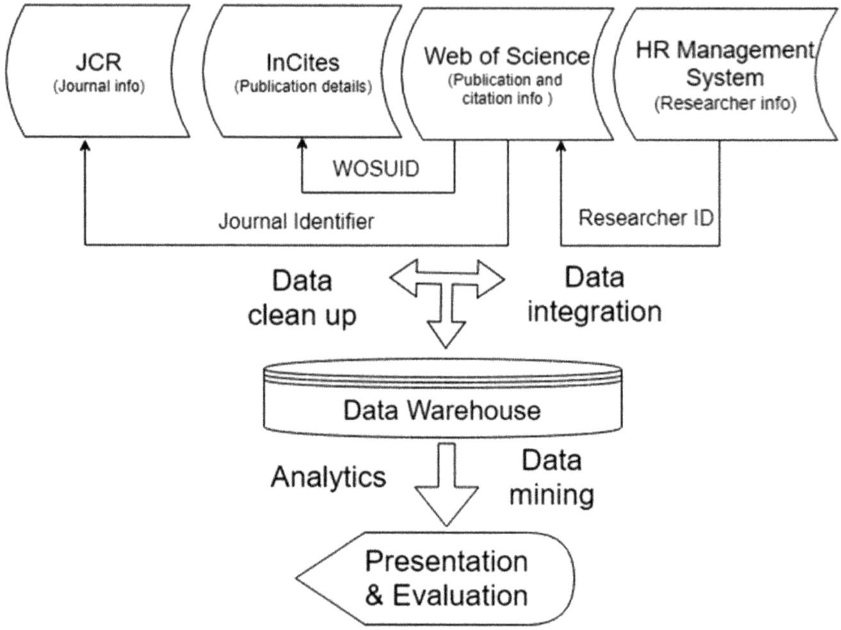

Fig. 3. Application flowchart

The first phase of this study was getting researcher information from human resources department. Department hierarchy has been constructed in database and staff populated with some information such as name, surname, title, ResearcherID, date of start and date of dismissal according to their department.

Subsequently, ResearcherID information was used for obtaining data from Web of Science Web Services Expanded. Web of Science provides SOAP-based APIs which comply JAX-WS, WSDL 1.1, SOAP 1.1 standards (Web of Science). Publications of re-searchers were obtained through Web of Science API's search method using ResearcherID as author identifier. The response of the search method returns list of publications of researcher and consists of metadata such as keywords, contributors, number of citations, number of references, database edition and journal info. Based on the API response a database schema has been designed to prevent data repetition, and established data consistency. After database populated with publications of researchers, publications that cites to these publications and referenced by these publications were obtained with using Web of Science unique identifier. InCites API provides the following information for each publication:

- Average number of citations to articles of the same document type from the same journal in the same database year
- Citation impact normalized for journal, year and document type subject
- Average number of times articles from a journal published in the past two years have been cited in the JCR year

- The harmonic mean of citation rate values for all research fields to which an article is assigned
- The percentile in which the paper ranks in its category and database year, based on total citations received by the paper
- Citation impact normalized for subject, year and document type
- Publication has at least two different countries among the affiliations of the co-authors
- Indicates that more than one institution has contributed to the document
- Papers that list their organization type as corporate for one or more of the co-authors affiliations

Using above metrics of each publication, we populate the data warehouse for data mining using the workflow process given in Fig. 3. The first step of the process is to collect Journal Information from JCR, citations from InCites, publications of WoS and academic staff meta data from Human Resources databases. Once data is collected from all these resources, they are associated with each other using researcher id and journal identifier. As data from resources may include duplicates due to multiple authors of same papers, data needs to be deduplicated and cleaned up after merge during the integration phase. Finally, all normalized data populates the data warehouse ready for data mining.

Academic performance not only depends on quantity but also quality. Therefore, quality of a publication may be deducted through journal impact factor and citation data. In order to evaluate journal quality, metrics were gathered from Journal Citation Reports web sites and imported into the warehouse. Those metrics include; journal name, number of citations, web of science document count, impact factor, eigen factor and JIF quartile.

The impact factor (IF) or journal impact factor (JIF) of an academic journal is a measure reflecting the yearly average number of citations to recent articles published in that journal. It is frequently used as a proxy for the relative importance of a journal within its field; journals with higher impact factors are often deemed to be more important than those with lower ones [4]. Impact factors are calculated yearly starting from 1975 for journals listed in the Journal Citation Report.

JIF Quartile score represents journal's percentile in their own category. Due to InCites calculates Journal Impact Factor with considering Web of Science research areas, a publication may have multiple Quartile score.

Once application data warehouse is populated, REST API written in. NET Core platform provides data for Graphical User Interface developed in Angular technology. In GUI project, all charts are created using HighCharts JavaScript library [15]. The user interface implemented ranks and tabulates results, generates plots and charts for visualization and decision support systems at Akdeniz. There are 4 major modules of the application:

1. Overall university performance including rankings of departments and individuals
2. Insights for department and program performances

3. Display of individual academic performances
4. Searching of publications and people through keywords/meta data

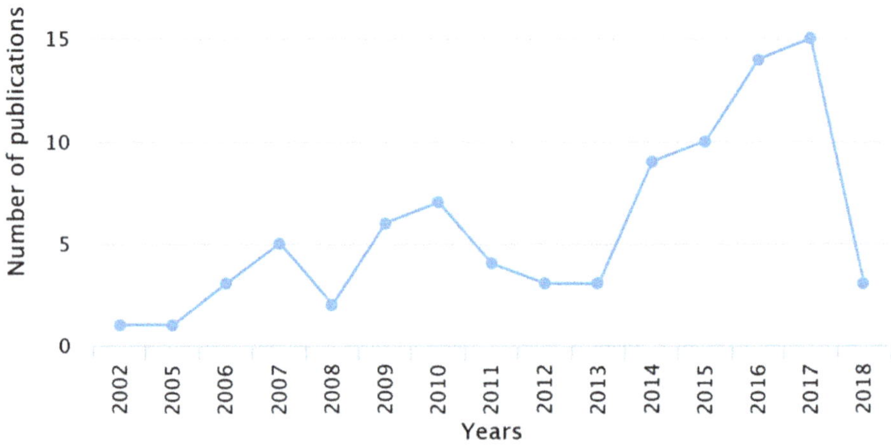

Fig. 4. Publication count by year of an academic unit

3 Results and Discussion

Akdeniz University is established at 1982 with 6 faculties including medicine, engineering, agriculture, science and literature, fine arts, economics and administrative sciences. As of today, the university has 23 different faculties and 5 institutes. The total of undergraduate students in the 2017–2018 academic year was over 70.000, while there were about 3000 graduate students. All academic staffs, including researcher, assistant, associate and professors was over 2000 at the end of 2017.

Akdeniz University has wide range of research areas including social sciences, health sciences, fine art and engineering. The subject of this study is to examine the characteristics of the publications of researchers and establish research profiles of academic units at Akdeniz University and quantify their impact within the respected research area. Distribution of academic titles per academic unit may be shown in Fig. 2. Young universities and departments tend to have higher ratio of Assistant Profs then Profs and Associate Profs. As academic units age, the ratio of Profs increase significantly which in turn may impact academic output performance.

Figure 4 show yearly publication performance of any selected unit. Such information may be used to follow trends over time and in cases it drops below a certain control limit for a period of time then cause and effect may be investigated for improvement.

Figure 1 show the number of publications index at various databases in a pie chart for academic units. It may be expected that while the publications for an Engineering Faculty appear primarily in SCI and ISTP, in Social Sciences fields, the large chunk of publications is expected to appear in AHCI and SSCI. By comparing the ratio between SCI to ISTP to SSCI and AHCI, relative performances of academic units within a university may be obtained.

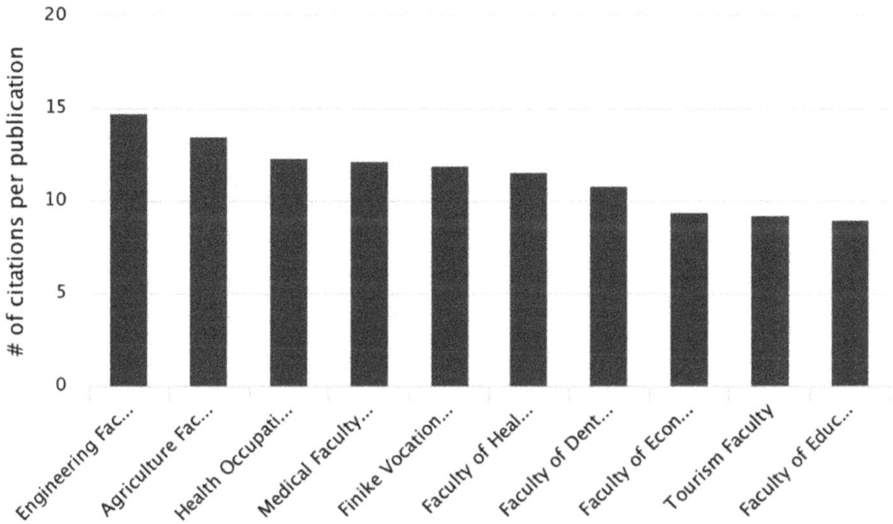

Fig. 5. Number of citations per publication for all academic units

In Fig. 5, for the whole university and each school, academic department and staff we plot:

- Number of citations
- Number of publications
- Average citation per publication
- Average citation per academic staff
- Average publication per academic staff

As shown in Fig. 6, Word Cloud has been generated for research areas for the whole university, school, department and academic staff using the corresponding frequencies of publications. However as some publications have higher citations than others the citation count was used as a multiplier in keyword frequency count in Word Cloud.

Such word cloud shown in Fig. 6 helps administrators to understand quickly what the academic unit focuses and produce strategies to effectively use resources.

In addition, for whole university or each academic unit, a ranking list that includes bibliometric indicators is tabulated for the followings items

Fig. 6. Word cloud of research areas

– H-Indexes, M-Indexes, G-Indexes of Academic Researchers
– Citation Count of Academic Researchers
– Most frequent Keywords of publications
– Mostly cited publications of researchers

Knowing high performing academic researchers, publications and research areas, university administrator may support people and research topics where the university is more effective.

Quality of a publication may be assessed by JIF quartile value. JIF quartile distribution of publications for an academic unit or staff as shown in Figs. 7 and 8 will indicate the quality and impact of research that is carried out. Ideally, it is desired to have most of the publications appear at Q1 and Q2 journals for a given research area.

Performance plots obtained for the university as a whole can also be retrieved easily for each faculty and department by simply selecting the unit from the tree of academic organization as shown in Fig. 9.

Performance metrics that was reported for unit is customized to view the publication performance of an individual researcher in a separate module as shown in Fig. 10. In this page, mostly cited publications, H-Index of the researcher, Publication Count and Total Number of Citations of publications is shown. The publication indexed database distribution is also plotted in a pie chart. Yearly publication count, citation per publication and their averages are also plotted per academic researcher within the staff performance page.

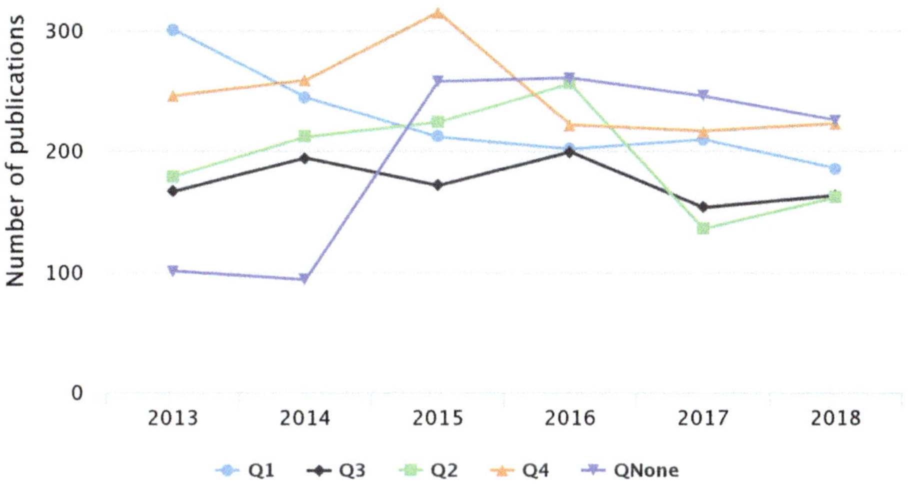

Fig. 7. Q values chart

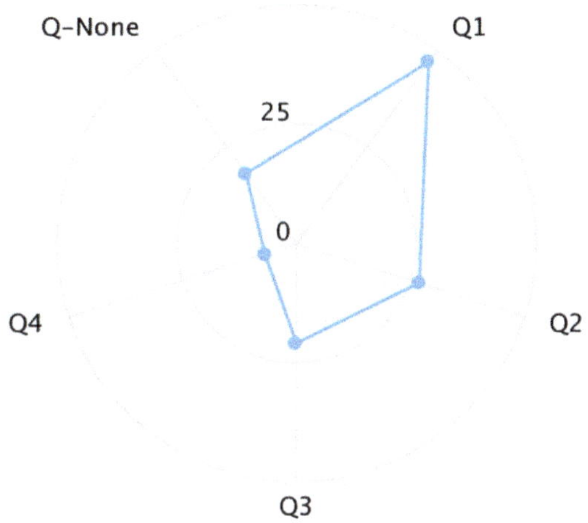

Fig. 8. Q values chart

Network graphs were prepared for departments and research fields by examining the joint publications of the researchers. The colors of the nodes show the research areas and the thickness of the connection between the nodes shows strength of collaboration as shown in Fig. 11.

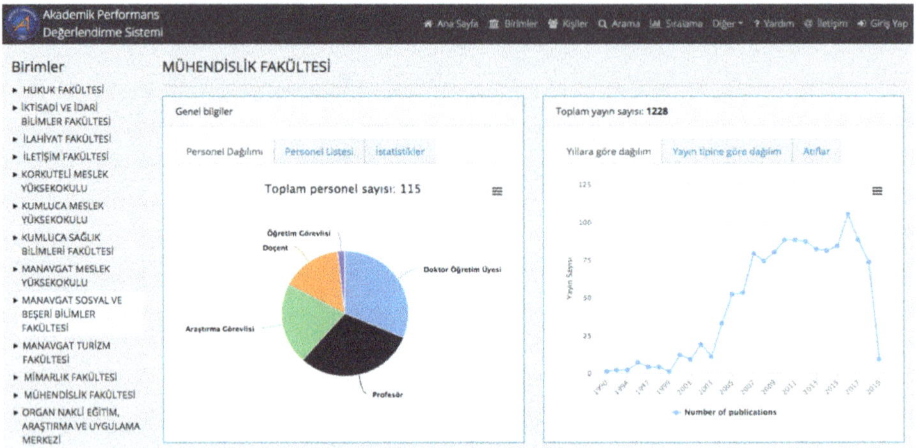

Fig. 9. Department-based filtering of academic performance criteria

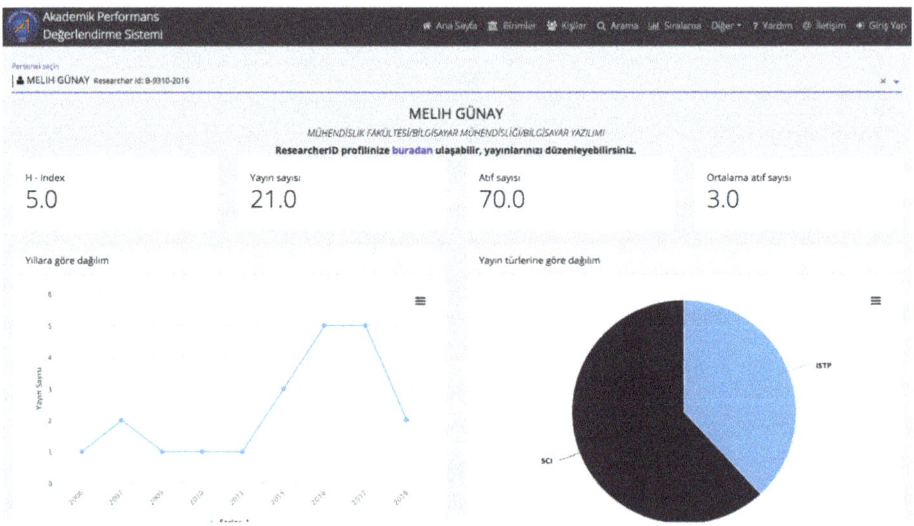

Fig. 10. Academic staff performance profile page

Custom search page has been designed for users to search publications, researchers and departments (See Fig. 12). This page allows researchers and publications to be filtered by keywords, subjects, researchers, departments, years, journals, database, publication types and q values. The result set returned may be sorted by the user by selecting the table column header.

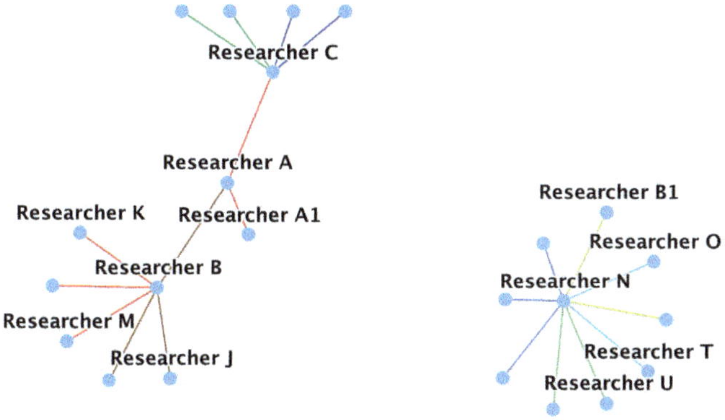

Fig. 11. Researcher network graph

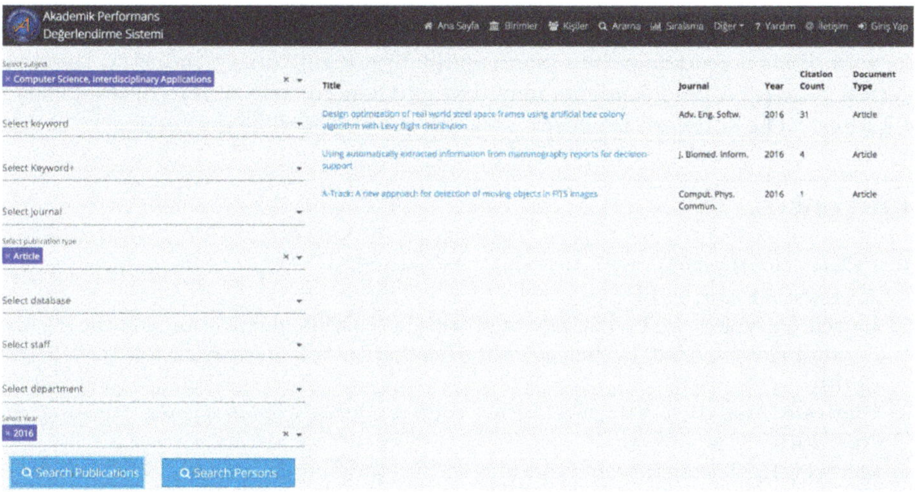

Fig. 12. Custom search page

4 Conclusion

With this study, an original software for the evaluation of academic publication performance was implemented for the Higher Education Institutes. Research performance for an academic unit or staff may be accessible in real time. Since the system is integrated with the Web of Science, processed data is reliable.

Academic Performance Evaluation System includes a number of features such as:

- Determination of the contribution of a department to the university
- Determination of the contribution of the researchers to the their departments
- Finding specialized researchers, departments, or publications in a specific research area

- Finding the potential collaborators for research within the system
- Determination of interdisciplinary studies
- Determination of studies with international cooperation
- Obtaining research performance by subject area of academic units and researchers in 5 broad categories namely; Life Science and Biomedicine, Art and Humanities, Physical Sciences, Social Sciences, Engineering and Technology.
- Ranking of researchers using indicators such as H-index, g-index, m-index, number of publications and number of citations.

As publication performance is a good indicator of quality of research and impact, it may be used to determine the effectiveness of academicians and universities. Based on the publication performance, universities may develop an objective method for promotions, appointments and resource allocations. Also, higher education councils of governments may use such data to develop policy and implement a publication-based incentive system for promotion of scientific research. If such academic data made public by the university, industry, research centers, academics and students may find partners for research in desired topics of interest. The software proposed here addresses a significant need.

References

1. University Ranking by Academic Performance. http://www.urapcenter.org. Accessed 09 Apr. 2019
2. Larsen, P., Von Ins, M.: The rate of growth in scientific publication and the decline in coverage provided by Science Citation Index. Scientometrics **84**(3), 575–603 (2010)
3. Clarivate Analytics. https://clarivate.com/products/web-of-science. Accessed 10 Apr. 2019
4. Garfield, E.: The history and meaning of the journal impact factor. JAMA **295**(1), 90–93 (2016)
5. Adriaanse, L.S., Rensleigh, C.: Web of Science, Scopus and Google Scholar: a content comprehensiveness comparison. Electron. Libr. **31**(6), 727–744 (2013)
6. Curado, C., Henriques, P.L., Oliveira, M., Matos, P.V.: A fuzzy-set analysis of hard and soft sciences publication performance. J. Bus. Res. **69**(11), 5348–5353 (2016)
7. Soutar, G.N., Wilkinson, I., Young, L.: Research performance of marketing academics and departments: an international comparison. Australas. Mark. J. (AMJ) **23**(2), 155–161 (2015)
8. Patel, V.M., Ashrafian, H., Almoudaris, A., Makanjuola, J., Bucciarelli-Ducci, C., Darzi, A., Athanasiou, T.: Measuring academic performance for healthcare researchers with the H index: which search tool should be used? Med. Principles Pract. **22**(2), 178–183 (2013)
9. Bar-Ilan, J.: Which h-index?-A comparison of WoS, Scopus and Google Scholar. Scientometrics **74**(2), 257–271 (2008)
10. Panczyk, M., Woynarowska-Sodan, M., Belowska, J., Zarzeka, A., Gotlib, J.: Bibliometric evaluation of scientific literature in the area of research in education using incites database of Thomson Reuters. In: Proceedings of INTED2015 Conference 2nd–4th March 2015, Madrid, Spain, pp. 487–496 (2015)

11. Rørstad, K., Aksnes, D.W.: Publication rate expressed by age, gender and academic position-A large-scale analysis of Norwegian academic staff. J. Informetr. **9**(2), 317–333 (2015)
12. Kaptanoğlu, D., Özok, A.F.: Akademik performans değerlendirmesi için bir bulanık model. TDERGS/d 5.1 (2010)
13. Umut, A.L.: Bilimsel yayınların değerlendirilmesi: H-endeksi ve Türkiye'nin performansı. Bilgi Dünyası **9**(2), 263–285 (2008)
14. Scientific and Technological Research Council of Turkey. http://tubitak.gov.tr/tr/haber/universite-yetkinlik-analizi-calismasi-yayinlandi. Accessed 11 Apr. 2019
15. Highsoft A.S. https://www.highcharts.com. Accessed 12 Apr. 2019

Predicting Breast Cancer with Deep Neural Networks

Abdulkadir Karaci(✉)

Faculty of Architecture and Engineering, University of Kastamonu, Kastamonu, Turkey
akaraci@gmail.com

Abstract. In this study, a deep neural network (DNN) MODEL was developed which diagnoses breast cancer using information about age, BMI, glucose, insulin, homa, leptin, adiponectin, resistin and MCP-1. The data used in this model was collected by Patrício et al. [7] from 116 women of which 64 has breast cancer and 52 do not. While 70% of this data (81 cases) was used for instructing the DNN model, 30% (35 cases) was used for testing. The DNN model was created in Python programming language using Keras Deep Learning Library. After model creation, machine learning was conducted using probable optimisation algorithms, loss functions and activation functions and the best three models were saved. For performance evaluation of the models, metrics of specificity, sensitivity and accuracy were employed. The specificity values of the best three models were calculated as [0.882, 0.941] and sensitivity values were found to be [0.888, 0.944]. In other words, while the models predict healthy women at the rates of minimum 88.2% and maximum 94.1%; they predict women with breast cancer at the rates of minimum 88.8% and 94.4%. For both women with and without breast cancer these prediction rates are sufficient and much higher than those reported by Patrício et al. [7].

Keywords: Machine learning · Deep Neural Networks · Breast cancer · Diagnosis and treatment

1 1 Introduction

Breast cancer is a disease which causes both physical and psychological damage to women around the world. Compared to other types of cancer, it is the second most frequent type observed in women. Breast cancer is the cause of more than 1.6% of deaths. While in 2017 252.710 new diagnoses of breast cancer was predicted, approximately 40.610 women died because of this disease [1–3]. Breast cancer is a common cancer type among women. Patients under 40 years of age comprise about 5 per cent of the overall breast cancer population. Breast Cancer is arise from normal host cells. It is a type of cancer with high chance of treatment when diagnosed early. With its increasing incidence early diagnosis has become more important [4–6]. Breast cancer screening is an important strategy for early diagnosis and increased probability of effective treatment. With the aim of providing more screening tools, robust

D. J. Hemanth and U. Kose (Eds.): ICAIAME 2019, LNDECT 43, pp. 996–1003, 2020.
https://doi.org/10.1007/978-3-030-36178-5_88

prediction models based on routine examination and blood test data are currently being studied Patrício et al. [7].

There are studies in the literature that result in successfully predicting breast cancer using machine learning. Takci [4] predicted breast cancer with 99,04% accuracy using Wisconsin, Diagnostic and Prognostic datasets and Centroid classifiers. Alharbi and Tchier [8] applied their self-developed hybrid algorithms (genetic-fuzzy algorithm) on Saudi Arabia breast cancer diagnosis dataset and categorized breast cancer with 97% accuracy. Pena-Reyes and Sipper [9], predicted breast cancer with a 97,36% accuracy rate using Wisconsin dataset, fuzzy logic and genetic algorithms. Setiono [10, 11] used Wisconsin dataset and accurately categorized more than 95% of cases with the help of artificial neural network-based rule extraction algorithm. Abdel-Zaher ve Eldeib [12], predicted breast cancer with 99.68% of accuracy using Wisconsin dataset and deep belief network. Literature indicates that such studies majorly employed Wisconsin dataset. Our study is distinguished from others as it uses the dataset collected by Patrício et al. [7].

Deep Neural Networks (DNN) are Artificial Neural Networks that are formed by multiple layers of neural networks with a high number of non-linear neurons per layer. DNN has received a considerable deal of attention by performing better than the alternative ML methods in several significant applications [13]. DNN is a multilayer perceptron with many hidden layers, whose weights are fully connected and are often (although not always) initialized using either an unsupervised or a supervised pre-training technique [14]. Due to the deep architectures, DNNs are able to adaptively capture there presentative information from raw data through multiplenon-linear transformations and approximate complexnon-linear functions with a small error [15].

The aim of this study is to develop a deep neural network (DNN) model which uses information about age, BMI, glucose, insulin, homa, leptin, adiponectin, resistin and MCP-1 to diagnose breast cancer with high accuracy rate. The parameters and prediction performance of the developed model are presented below.

2 2 Materials and Method

2.1 Data Set

The data set used in this study was collected by Patrício et al. [7]. Whereas data of 64 of the total 116 cases were obtained from women with breast cancer before treatment, data of the remaining 52 were collected from women without breast cancer who volunteered to participate at Coimbra University Central Hospital. The attributes of the collected data and the explanations are presented in Table 1. These attributes also form the neurons of the input layer.

2.2 Deep Neural Network Model

In this study, a DNN model was developed which has 9 inputs, 4 hidden layers and 2 outputs. The input neurons consist of age, BMI, blood sugar levels (glucose, insulin, homa) and serum levels (leptin, adiponectin, resistin and MCP-1 explained in detail in

Table 1. Attributes of the data set.

No.	Input	Value range	Explanation
1	Age	[24–89]	-
2	BMI	[18.37–38.58]	Weight of the person/(height of the person)^2
3	Glucose	[60–201]	Blood sugar test values
4	Insulin	[2.43–58.46]	Blood sugar test values
5	Homa	[0.47–25.05]	Index for evaluating insulin resistance
6	Leptin	[4.31–90.28]	Serum levels
7	Adiponectin	[1.66–38.04]	Serum levels
8	Resistin	[3.21–82.1]	Serum levels
9	MCP-1	[45.84–1698.44]	Serum levels

Table 1. The output layer involves two neurons that classify women with and without breast cancer. The first neuron symbolizes women without breast cancer while the second neuron symbolizes women with this disease. The structure of the developed DNN model is shown in Fig. 1.

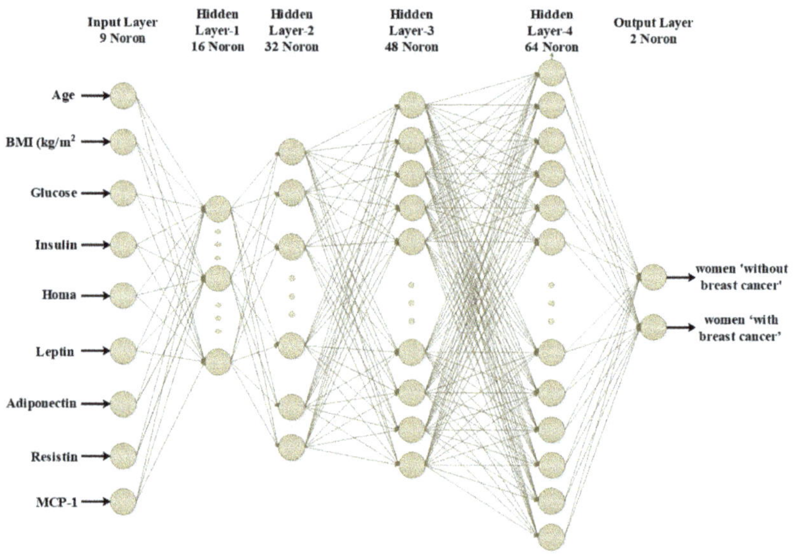

Fig. 1. The DNN model

Training the DNN Model

In training the DNN model, randomly selected 70% (81 cases) of a total of 116 cases (64 with breast cancer and 52 without breast cancer) were used. The model was tested with the remaining 30% of the data (35 cases). One of the parameters that affect the performance of a DNN model is the hidden layers, specifically the number of neurons in this layers. Several numbers of hidden layers and neurons were tested in this study. These attempts indicated that the best performance belonged to the model with 4 hidden layers which consist of 16, 32, 48 and 64 neurons, respectively. Furthermore, during the training of the model, possible optimisation algorithms (adam, adadelta, sgd, rmsprop, adamax ve nadam), activation functions (elu, tanh, linear, softsign, relu, softplus, sigmoid ve hard sigmoid) and loss functions (categorical crossentropy, mean absolute error, mean square error) were tested with the aim of obtaining the parameters that lead to the best result. These efforts revealed that adam and adamax optimization algorithms, mse (mean square error) and categorical cross entropy loss functions, softmax (in output layer), softsign and softplus (in other layers) activation functions increased the learning performance of the model. Therefore, training of this model involved trying variations of these parameters and the three models with the best prediction performances were saved.

The parameters of the DNN model are illustrated in Table 2. These parameters are involved both in the building and training of the network.

Table 2. The structural and training parameters of the DNN model.

Parameters	Values
Number of neurons in the input layer	9
Number of hidden layers	4
Number of neurons in hidden layer-1	16
Number of neurons in hidden layer-2	32
Number of neurons in hidden layer-3	48
Number of neurons in hidden layer-4	64
Number of neurons in the output layer	2
Activation functions of hidden layers	Softsign ve softplus
Activation function of the output layer	Softmax
Learning cycle	200, 400, 500, 600 Epochs
Loss functions	Mse ve Categorical cross-entropy
Optimization algorithms (learning algorithms)	Adam, Adamax ve RMSprop

Sensitivity/recall and specificity metrics were used to compare model performances during DNN model training and testing. Sensitivity is the proportion of accurately identified true positives (women with breast cancer) by the model. Specificity is the proportion of true negatives (women without breast cancer) which the model is able to classify accurately [16]. Sensitivity and specificity are calculated as in Eq. 1 below [17]. In this equation, abbreviations stand for: TP: True Positive, FN: False Negative, FP: False Positive.

$$Sensitivity\,(SN) = \frac{\mathrm{TP}}{\mathrm{TP}+\mathrm{FN}}, \quad Specificity\,(SP) = \frac{\mathrm{TN}}{\mathrm{TN}+\mathrm{FP}} \tag{1}$$

3 3 Findings

The DNN model was trained using optimisation algorithms, loss functions and activation functions. After the training, the test data was given to the trained DNN models as input and their performances were evaluated with sensitivity and specificity parameters. DNN model training parameters and model performances are shown in Table 3. Model performances were evaluated against sensitivity, specificity and accuracy parameters. In addition, the comparison of the models with respect to the sensitivity, specificity and accuracy performance metrics is shown in Fig. 2.

Table 3. DNN model prediction performances.

Model code	Sensitivity	Specificity	Accuracy	Optimizers	Activation	Loss function	Epochs
DNN-1	0.944	0.882	0.914	Adamax	Softsign	Mse	200
				Adamax	Softsign	Mse	600
DNN-2	0.888	0.941	0.914	Adamax	Softsign	Categorical crossentropy	200
				RMSprop	Softsign	Mse	500
				Adam	Softsign	Mse	200
DNN-3	0.888	0.882	0.885	Adam	Softsign	Categorical crossentropy	600
				Adamax	Softplus	Mse	400

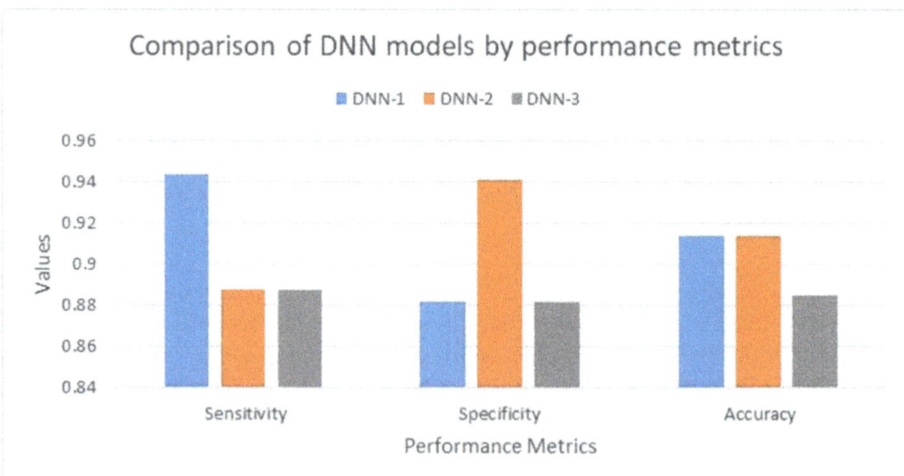

Fig. 2. Comparison of DNN models by performance metrics

In Table 3, the model with the best performance is DNN-1 since the main goal of this study is to accurately predict women who have breast cancer. The sensitivity parameter of DNN-1 was calculated as 0.944 which means this model can predict women with breast cancer with 94.4% accuracy rate. The prediction rate for women without breast cancer is 88.2%. The confusion matrix of DNN-1 is presented in Table 4.

Table 4. DNN-1 model confusion matrix

Prediction	Actual condition	
	With breast cancer	Without breast cancer
Positive	17	2
Negative	1	15

According to Table 4, DNN-1 model classified 17 women as having breast cancer while it identified one woman as 'without breast cancer' out of 18 women with breast cancer. Moreover, it predicted 15 women to be 'without breast cancer' and 2 women 'with breast cancer' when in reality all of these 17 women are without breast cancer. These findings indicate the appropriateness of DNN-1 model when compared to the other 2 models in predicting women with breast cancer.

In DNN-2 model the sensitivity level is higher and women without breast cancer were predicted with 94.1% of accuracy. Its rate of accurately predicting women with breast cancer is 88.8%. The confusion matrix of DNN-2 model is presented in Table 5.

Table 5. DNN-2 model confusion matrix

Prediction	Actual condition	
	With breast cancer	Without breast cancer
Positive	16	1
Negative	2	16

Table 5 shows that DNN-2 model predicted that 16 women were breast cancer while 2 women were without breast cancer when actually these 18 women all had the disease. On the other hand, out of 17 women without breast cancer, it classified 16 as 'without breast cancer' and 1 woman with breast cancer. These findings show that DNN-2 model is more appropriate than the other 2 models in predicting women without breast cancer.

DNN-3 model accurately predicts women with breast cancer at the rate of 88.8% and women without breast cancer at 88.2%. When compared to the other models, this model can be said to have lower performance. The confusion matrix of DNN-3 model is shown in Table 6. This matrix indicates that both out of women with and without breast cancer, two women are predicted inaccurately.

Table 6. DNN-3 model confusion matrix

Prediction	Actual condition	
	With breast cancer	Without breast cancer
Positive	16	2
Negative	2	15

Additionally, an evaluation of all parameters of the model, it is obvious that the best prediction performance is obtained with the combination of Adamax optimiser, softsign activator and mse loss function.

4 4 Conclusion

The DNN models developed in this study effectively predicts breast cancer using data about age, BMI, glucose, insulin, homa, leptin, adiponectin, resistin and MCP-1. While DNN-1 model accurately predicts women with breast cancer at 94.4%, DNN-2 identifies women without breast cancer at 94.1%. Therefore, DNN-1 can be used in predicting women with the disease whereas DNN-2 would be more effective in predicting women without this type of cancer. Also, DNN-3 model might be employed in providing additional information in the interpretation of the results obtained from the other two models since its rates of prediction are fairly similar for women both with and without breast cancer. Patrício et al. [7] who collected the dataset employed in this study used different variations of attributes and applied Logistic Regression (LR), Support Vector Machines (SVM) and Random Forests (RF) machine learning methods to the data. Using all of the attributes, they were able to accurately predict women with breast cancer at 85% (with RF method) and women without breast cancer at 86% maximum (with LR method). In this respect, the DNN model developed in this study predicts breast cancer with higher accuracy than RF and LR models.

References

1. Li, Y., Chen, Z.: Performance evaluation of machine learning methods for breast cancer prediction. Appl. Comput. Math. **7**(4), 212–216 (2018). https://doi.org/10.11648/j.acm.20180704.15
2. Chaurasia, V., Pal, S., Tiwari, B.B.: Prediction of benign and malignant breast cancer using data mining techniques. J. Algorithms Comput. Technol. **12**(2), 119–126 (2018)
3. Chaurasia, V., Pal, S.: Data mining techniques: to predict and resolve breast cancer survivability. Int. J. Comput. Sci. Mobile Comput. IJCSMC **3**(1), 10–22 (2014)
4. Takci, H.: Diagnosis of breast cancer by the help of centroid based classifiers. J. Fac. Eng. Archit. Gazi Univ. **31**(2), 323–330 (2016)
5. Florescu, A., Amir, E., Bouganim, N., Clemons, M.: Immune therapy for breast cancer in 2010-hype or hope? Curr. Oncol. **8**(1), e9–e18 (2011)
6. Sariego, J.: Breast cancer in the young patient. Am. Surg. **76**(12), 1397–1401 (2010)

7. Patrício, M., Pereira, J., Crisóstomo, J., Matafome, P., Gomes, M., Seiça, R., Caramelo, F.: Using resistin, glucose, age and BMI to predict the presence of breast cancer. BMC Cancer **18**(29), 1–8 (2018). https://doi.org/10.1186/s12885-017-3877-1
8. Alharbi, A., Tchier, F.: Using a genetic-fuzzy algorithm as a computer aided diagnosis tool on Saudi Arabian breast cancer database. Math. Biosci. **286**, 39–48 (2017)
9. Pena-Reyes, C.A., Sipper, M.A.: Fuzzygenetic approach to breast cancer diagnosis. Artif. Intell. Med. **17**(2), 131–155 (1999)
10. Setiono, R.: Extracting rules from pruned neural networks for breast cancer diagnosis. Artif. Intell. Med. **8**, 37–51 (1996)
11. Setiono, R.: Generating concise and accurate classification rules for breast cancer diagnosis. Artif. Intell. Med. **18**(3), 205–219 (2000)
12. Abdel-Zaher, A.M., Eldeib, A.M.: Breast cancer classification using deep belief networks. Expert Syst. Appl. **46**, 139–144 (2016)
13. Karaci, A., Yaprak, H., Ozkaraca, O., Demir, I., Simsek, O.: Estimating the properties of ground-waste-brick mortars using DNN and ANN. CMES **118**(1), 207–228 (2019)
14. Deng, L., Yu, D.: Deep Learning: Methods and Applications. Now, Delft (2014)
15. Jia, F., Lei, Y., Lin, J., Zhou, X., Lu, N.: Deep neural networks: a promising tool for fault characteristic mining and intelligent diagnosis of rotating machinery with massive data. Mech. Syst. Signal Process. **72–73**, 303–315 (2016). https://doi.org/10.1016/j.ymssp.2015.10.025
16. Altman, D.G., Bland, J.M.: Diagnostic tests. 1: sensitivity and specificity. BMJ (Clin. Res. Ed.) **308**(6943), 1552 (1994)
17. Parikh, R., Mathai, A., Parikh, S., Chandra, S.G., Thomas, R.: Understanding and using sensitivity, specificity and predictive values. Indian J. Ophthalmol. **56**(1), 45–50 (2008)

Utilizing Machine Learning Algorithms of Electrocardiogram Signals to Detect Sleep/Awake Stages of Patients with Obstructive Sleep Apnea

Muhammed Kürşad Uçar[1(✉)], Ferda Bozkurt[2], Cahit Bilgin[3], and Mehmet Recep Bozkurt[1]

[1] Faculty of Engineering, Electrical-Electronics Engineering, Sakarya University, Sakarya, Turkey
{mucar,mbozkurt}@sakarya.edu.tr

[2] Vocational School of Adapazarı, Computer Programming, Sakarya University of Applied Sciences, Sakarya, Turkey
fbozkurt@subu.edu.tr

[3] Faculty of Medicine, Sakarya University, Sakarya, Turkey
cahitbilgin@sakarya.edu.tr

Abstract. Obstructive Sleep Apnea (OSA) is a respiratory-related disease that occurs during sleep. The diagnosis of OSA is made by a specialist doctor according to the records obtained with the polysomnography device. However, the diagnostic process is quite troublesome. More than 30 signals are recorded for diagnosis. This may cause discomfort to the patient during the night. For the diagnosis, sleep staging and respiratory scoring are performed with the records collected via the polysomnography device. With sleep staging, the patient's sleep and awake times are determined and respiratory scoring is used to detect abnormal respiratory events that occur during sleep. If more than 5 abnormal respiratory events occur per hour, the individual is diagnosed with OSA. Due to the fact that this process is laborious and uncomfortable to the individual, practical diagnostic methods are needed. In this study, an easy and practical measurement system for sleep staging, which is an important step in the diagnosis of OSA, will be proposed. According to this system, sleep/awake state is determined by electrocardiogram signal (ECG) and a machine based method.

ECG records obtained from two individuals will be used in the study. First, the ECG signal will be cleared from the noise by digital filters. It will then be divided into 30 s epochs for sleep staging. From each separated epoch, 25 features will be extracted and features associated with sleep/awake will be selected with the help of feature selection algorithms. Determined features will be classified by Support Vector Machine which is a machine learning method and the system performance will be tested. In the preliminary studies, it was determined that 25 properties were related to sleep/awake stages and that the classification performance was approximately 85%. In light of this information, it is thought that a system based on machine learning can be developed for the detection of sleep/awake stages using ECG signals for the diagnosis of OSA.

D. J. Hemanth and U. Kose (Eds.): ICAIAME 2019, LNDECT 43, pp. 1004–1013, 2020.
https://doi.org/10.1007/978-3-030-36178-5_89

Keywords: Obstructive Sleep Apnea · Electrocardiogram · Feature selection · Machine learning · Support Vector Machines

1 Introduction

Obstructive Sleep Apnea (OSA) is the most common disease related to sleep-related respiratory disorders. OSA is a disease caused by abnormal respiratory events occurring in sleep [7]. In order to diagnose the disease, biological recordings are taken from the patient with the help of technical devices in sleep laboratories and diagnosis is made by an expert doctor with sleep staging and respiratory scoring processes.

Sleep staging is the basic and first-step method used in sleep laboratories for the diagnosis of more than 80 diseases associated with sleep [1]. The purpose of sleep staging is to determine if the patient is asleep and to determine which sleep stages they are experiencing during their sleep at night. After this basic phase, different events experienced by the patient are recorded in the records. The records are then examined by a specialist physician. She/He determines in which stages the events occur and if these events have a relationship with the sleep stages [2]. Sleep staging is an indispensable step in the diagnosis of some diseases.

The sleep stages can be labeled by a specialized physician as Awake, Stage 1, Stage 2, Stage 3 or REM according to the Electroencephalogram (EEG), Electrooculogram (EOG) and Electromyogram (EMG) signals. However, in some cases, such detailing may not be necessary. In this case the labor-saving way should be selected. For example, in the diagnosis of Obstructive Sleep Apnea, it is sufficient to determine the sleep/awake status of the patient [3]. After this condition, abnormal respiratory events in which the patients experience during their sleep are determined. The diagnosis of OSA is made based on the number of abnormal respiratory events. However, the important point is that these abnormal respiratory events must be occurred while sleeping. Therefore, it is sufficient to detect the patient's sleep time and the abnormal respiratory events occurring during sleep [4]. In the literature, sleep staging procedures can be performed with the Electroencephalography (EEG) [5], Electrocardiography (ECG) [6], and Photopletismography (PPG) signals [1,7].

In this study, a system based on machine learning was developed for the detection of sleep/awake phases with 25 ECG signal features. These features were selected and classified according to the interest level of the features. The performances of the classifiers were evaluated by performance evaluation criteria.

2 Materials and Methods

The study was carried out according to the steps in Fig. 1. First, ECG records were obtained from the individuals who applied to the clinic. The collected data were converted into processable information. The noise components on the signal were cleaned and divided into 30-second epochs for sleep staging. The epochs were individually controlled by eye and the artifacted epochs were erased.

Then the shifts in the axes were eliminated and features were extracted from each epoch. Features were selected according to the Fisher feature selection algorithm. In the end, each selected group of features were classified and evaluated.

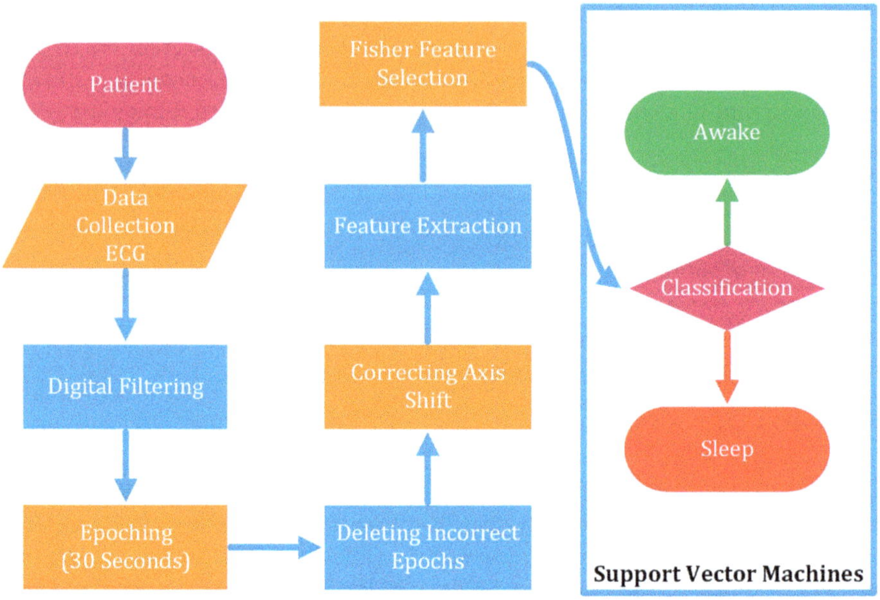

Fig. 1. Flow diagram

2.1 Data Collection and Signal Preprocessing

Medical data was collected from two individuals by the 33 chanel SOMNOscreen Plus Polysomnography (PSG) device at the Hendek State Hospital under the supervision of a specialist and a sleep technician in a sleep laboratory. Individuals records consist of an average of 8 hours of signals. In this study, only the ECG signal was used and the sampling frequency of the signal was 256 Hz. The demographics of the individuals were summarized in Table 1. The collected records were examined by a specialist physician and graded as Awake, Stage 1-2-3 or REM according to EEG. However, since sleep/awake analysis was attempted to be performed in this study, the tags were combined as sleep/awake.

Signal filtering was performed in three steps. First, the general noise was cleared with the 0.1–100 Hz Chebyshev Type II filter. A 50 Hz notch filter was then applied to suppress network noise. Finally, the Moving Average Filter cleared the fluctuations on the signal. After filtering, the signal is divided into 30-second epochs. Each epoched ECG signal was visually checked and the artifacted epochs were excluded from the study. Then, fluctuations in the axis were eliminated. For this, a low-grade polynomial equation is fitted on the ECG signal. Then the known equation is removed from the floating signal so the axis

Table 1. Demographic information

Gender	Male	Female
Age (years)	64	58
Weight (kg)	98.1	117
Height (cm)	175	167
Body mass index (kg/m^2)	32	42
Apnea hypoapnea index (AHI)	12.7	10.9

fluctuation is cleared. One epoch (Sleep/Awake) from the cleared ECG signals is shown in Fig. 2 with the fast Fourier Transform and with the periodogram graph. After cleaning the epochs, total of 1511 epochs were obtained from two individuals. The distribution of records from these individuals were summarized in Table 2.

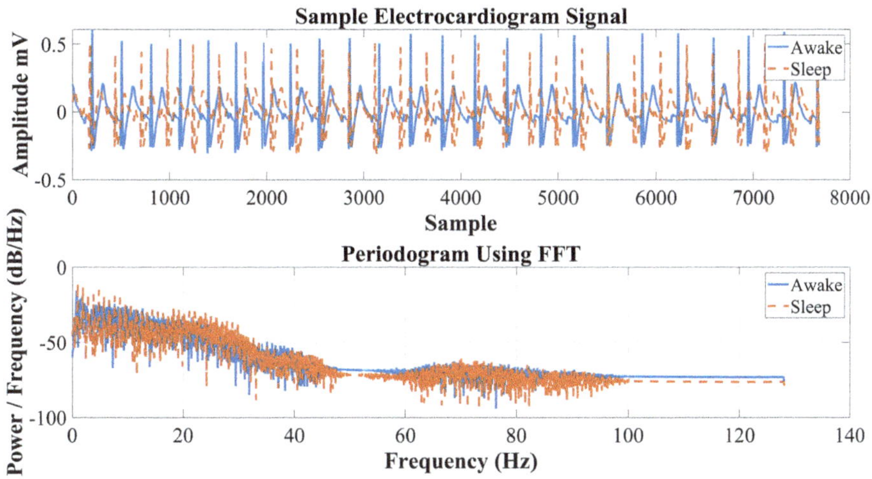

Fig. 2. Periodogram chart for electrocardiogram

Table 2. Data distribution

Gender	Male	Female	Total
Awake	529	199	728
Sleep	249	534	783
Total	778	733	1511

Each epoch contains a 30 s ECG signal.

2.2 Feature Extraction

A total of 25 properties were extracted from each epoch. The mathematical expressions of these properties are summarized in Table 3. x represents an epok signal in the expressions, and i represents the instance number in an epoch. When the sampling frequency is 256 Hz, an epoch of 30 s includes $N = 256 \times 30 = 7680$ samples.

Table 3. ECG features and formulas

Feature number	Feature	Acquisition
1	Kurtosis	$x_{kur} = \frac{\sum_{i=1}^{n}(x(i)-\overline{x})^4}{(n-1)S^4}$
2	Skewness	$x_{ske} = \frac{\sum_{i=1}^{n}(x_i-\overline{x})^3}{(n-1)S^3}$
3	Interquartile range	$IQR = iqr(x)$
4	Coefficient of variation	$DK = (S/\overline{x})100$
5	Geometric mean	$G = \sqrt[n]{x_1 + \cdots + x_n}$
6	Harmonic mean	$H = n/(\frac{1}{x_1} + \cdots + \frac{1}{x_n})$
7	Hjorth parameter - activity	$A = S^2$
8	Hjorth parameter - mobility	$M = S_1^2/S^2$
9	Hjorth parameter - complexity	$C = \sqrt{(S_2^2/S_1^2)^2 - (S_1^2/S^2)^2}$
10	Maximum	$x_{max} = max(x_i)$
11	Median	$\tilde{x} = \begin{cases} x_{\frac{n+1}{2}} & : x\text{Odd} \\ \frac{1}{2}(x_{\frac{n}{2}} + x_{\frac{n}{2}+1}) & : x \text{ Even} \end{cases}$
12	Average or square absolute deviation	$MAD = mad(x)$
13	Minimum	$x_{min} = min(x_i)$
14	Moment, central moment	$CM = moment(x, 10)$
15	Average	$\overline{x} = \frac{1}{n}\sum_{i=1}^{n} = \frac{1}{n}(x_1 + \cdots + x_n)$
16	Average curve length	$CL = \frac{1}{n}\sum_{i=2}^{n} \lvert x_i - x_{i-1}\rvert$
17	Average energy	$E = \frac{1}{n}\sum_{i=1}^{n} x_i^2$
18	Average square root value	$X_{rms} = \sqrt{\frac{1}{n}\sum_{i=1}^{n} \lvert x_i\rvert^2}$
19	Standard error	$S_{\overline{x}} = S/\sqrt{n}$
20	Standard deviation	$S = \sqrt{\frac{1}{n}\sum_{i=1}^{n}(x_i - \overline{x})}$
21	Shape factor	$SF = X_{rms}/\left(\frac{1}{n}\sum_{i=1}^{n} \sqrt{\lvert x_i\rvert}\right)$
22	Single value decomposition	$SVD = svd(x)$
23	25% cropped average	$T25 = trimmean(x, 25)$
24	50% cropped average	$T50 = trimmean(x, 50)$
25	Average teager energy	$TE = \frac{1}{n}\sum_{i=3}^{n}(x_{i-1}^2 - x_i x_{i-2})$

2.3 Support Vector Machines

SVMs is a high-performance machine learning algorithm among consultative learning methods [8]. SVMs are used to classify tagged data. The purpose of the algorithm is to separate the data from each other by linear or nonlinear lines with a minimum error rate. The samples closest to the drawn plane are called support vectors.

2.4 Fisher Feature Selection Algorithm

The Fisher feature selection algorithm is a method used to determine the most suitable properties for classification. For this, it uses statistical methods [9]. The algorithm sorts all properties from the best to worst according to the degree of eligibility. It is then used in the classification process by selecting the desired properties by the user.

Table 4 provides information on the properties selected. The top 5% feature was selected and assigned to the first group. Then the best 10% feature was selected and taken to the second group. This operation continued until the best 50% feature was selected. Each group feature was classified by SVMs and its performance was evaluated.

Table 4. Selected features

GN	P	NGE	Selected features
1	5	1	1
2	10	3	1 2 9
3	15	4	1 2 9 10
4	20	5	1 2 9 10 11
5	25	6	1 2 9 10 11 12
6	30	8	1 2 9 10 11 12 18 20
7	35	9	1 2 9 10 11 12 18 20 21
8	40	10	1 2 9 10 11 12 18 20 21 22
9	45	11	1 2 9 10 11 12 18 20 21 22 23
10	50	13	1 2 9 10 11 12 18 20 21 22 23 24 25

GN Group No, P Percent, NGE Number of Group Elements

2.5 Performance Evaluation

SVMs classifier was used to evaluate the developed system. Classifier performance accuracy was evaluated by performance evaluation criteria such as sensitivity, specificity, kappa coefficient, f-measurement.

In order to evaluate the classifier, the dataset is divided into (50%) training and (50%) test groups (Table 5).

Table 5. Distribution of training and test data

	Label		Total
	Awake	Sleep	
Training (%50)	394	392	786
Test (%50)	364	391	755
Total	758	783	1541

3 Results

The aim of the study is to design an ECG-based system that will perform sleep staging which is an important step in diagnosing OSA. For this purpose, 25 properties extracted from two patients' ECG records were statistically analyzed.

A scatter plot was prepared (Fig. 3) in order to observe the relationship between the extracted properties and the groups (Sleep/Awake). In this graph, 12 properties that are most suitable for the classification for sleep/awake were determined by the Fisher feature selection algorithm (Table 4) and visualized in pairs (Fig. 3). Since the variables are continuous numerical variables, correlation R and multiple explanatory coefficient R^2 values on the graphs were calculated according to the Spearman correlation coefficient. Correlation between 3 groups was strongly correlated ($0.90 < R < 1$). 2 groups had a strong relationship ($0.70 < R < 0.89$), one group had a middle relationship ($0.40 < R < 0.69$) (Fig. 3). $R = 0.5$ or $R = -0.5$ means the relationship degrees are the same but the directions are different.

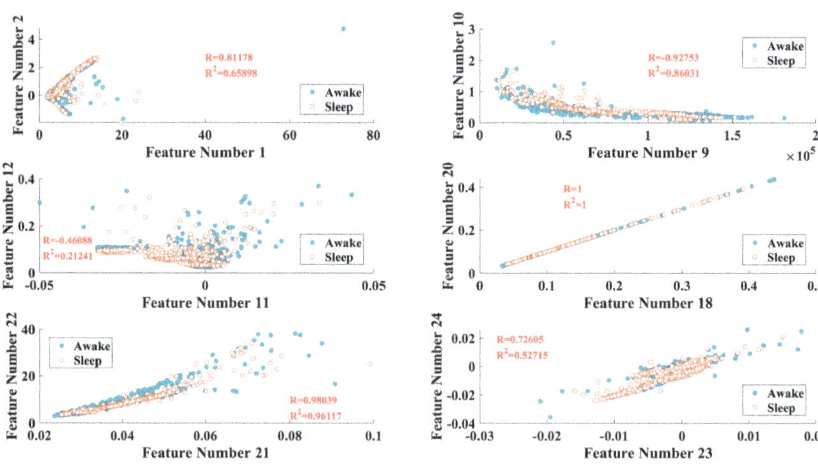

Fig. 3. Scatter plot of properties

25 ECG features were selected by using Fisher feature selection algorithm. For each classification process, performance values of training and test processes were calculated. As the number of features selected for training data increase, performance values were increased (Table 6).

The most important result in machine learning is the response performance of the system during the testing phase. In this study, the classification accuracy rate was 70.199% with a single feature, and this ratio was changed to 92.053% when 50% of the characteristics were selected (Table 7). When other performance criteria are checked, it will be seen that the performance increases when the number of relevant properties increase (Table 7).

Table 6. Training classification performance

FGN	NF	Accuracy	Sensitivity	Specificity	F-Measure	Kappa	AUC
1	1	90.344	0.879	0.926	0.902	0.806	0.903
2	3	96.693	0.948	0.985	0.966	0.934	0.966
3	4	97.090	0.959	0.982	0.970	0.942	0.970
4	5	97.090	0.945	0.995	0.969	0.942	0.970
5	6	98.280	0.978	0.987	0.983	0.966	0.983
6	8	98.148	0.975	0.987	0.981	0.963	0.981
7	9	98.280	0.978	0.987	0.983	0.966	0.983
8	10	98.280	0.978	0.987	0.983	0.966	0.983
9	11	98.413	0.981	0.987	0.984	0.968	0.984
10	13	98.280	0.978	0.987	0.983	0.966	0.983

FGN Feature Group Number, NF Number of Feature

4 Discussion

Since the circulatory system is vital in the operation of all systems in the body, the ECG mark is associated with all systems in the body. The diagnosis of many diseases can be done via the ECG signal [10,11]. The practical use of ECG measurement has expanded its use.

The aim of the study was to determine the sleep/awake stages with ECG. For this, 25 ECG features were selected and classified. The accuracy rate in the 1 property classification was 70.199%. The methods developed in health studies are required to work with an accuracy of at least 80% [12]. When the number of ECG features were increased to 13, the success rate was 92.053%. This ratio is an indicator that the method works well. However, if the number of features were 6, the highest accuracy rate was 92.318%. As can be understood from this, increasing the number of features is not an indicator that increase the performance. Therefore, it is necessary to increase the properties related to sleep/awake status. We can tell that the Fisher feature selection algorithm used in this study was useful here. With this algorithm, features were sorted by interest level. Then, according to percentage ratios (5–50%) properties were

Table 7. Test classification performance

FGN	NF	Accuracy	Sensitivity	Specificity	F-Measure	Kappa	AUC
1	1	70.199	0.723	0.683	0.702	0.405	0.703
2	3	81.722	0.860	0.777	0.817	0.635	0.819
3	4	90.464	0.940	0.872	0.905	0.810	0.906
4	5	90.066	0.923	0.880	0.901	0.801	0.901
5	6	91.391	0.953	0.877	0.914	0.828	0.915
6	8	92.318	0.959	0.890	0.923	0.847	0.924
7	9	92.053	0.959	0.885	0.920	0.841	0.922
8	10	91.921	0.956	0.885	0.919	0.839	0.920
9	11	91.656	0.959	0.877	0.916	0.833	0.918
10	13	92.053	0.953	0.890	0.921	0.841	0.922

FGN Feature Group Number, NF Number of Feature

selected and classified and appropriate properties were detected. This process was very useful for reducing the workload. The results obtained were 5–20% better compared to the literature [5, 13]. Compared to the features used in the literature, the accuracy ratio of the 92.318% with the 6 properties was quite good. In a real-time system with 6 features, the workload will be very low and the system will operate with high performance.

The autonomic nervous system works effectively in sleep/awake situations. A system to be developed with machine learning using the ECG signal provides the detection of sleep stages and plays an important role in the diagnosis phase. As a result, it can be said that with the help of ECG signal and machine learning methods, practical diagnostics or sleep/awake status detection systems can be developed.

References

1. Uçar, M.K.: Obstrüktif Uyku Apne Teshisi için Makine Ögrenmesi Tabanli Yeni Bir Yöntem Gelistirilmesi. Ph. D. thesis, Sakarya Üniversitesi (2017)
2. Berry, R.B., Budhiraja, R., Gottlieb, D.J., Gozal, D., Iber, C., Kapur, V.K., Marcus, C.L., Mehra, R., Parthasarathy, S., Quan, S.F., Redline, S., Strohl, K.P., Davidson Ward, S.L., Tangredi, M.M.: Rules for scoring respiratory events in sleep: update of the 2007 AASM manual for the scoring of sleep and associated events. J. Clin. Sleep Med. JCSM **8**(5), 597–619 (2012). Deliberations of the Sleep Apnea Definitions Task Force of the American Academy of Sleep Medicine, Official Publication of the American Academy of Sleep Medicine
3. Uçar, M.K., Bozkurt, M.R., Bilgin, C., Polat, K.: Automatic sleep staging in obstructive sleep apnea patients using photoplethysmography, heart rate variability signal and machine learning techniques. Neural Comput. Appl. **29**(8), 1–16 (2018)
4. Uçar, M.K., Bozkurt, M.R., Bilgin, C., Polat, K.: Automatic detection of respiratory arrests in OSA patients using PPG and machine learning techniques. Neural Comput. Appl. **28**(10), 2931–2945 (2017)

5. Uçar, M.K., Polat, K., Bozkurt, M.R., Bilgin, C.: Uyku EEG ve EOG Sinyallerinin Siniflandirilmasinda Zaman ve Frekans Domeni Özelliklerinin Etkisi. In: Tip Tekno 2014 - Tip Teknolojileri Ulusal Kongresi Bildirisi, Kapadokya, Nevsehir, Türkye, pp. 163–166 (2014)
6. Sharma, H., Sharma, K.K.: An algorithm for sleep apnea detection from single-lead ECG using Hermite basis functions. Comput. Biol. Med. **77**, 116–124 (2016)
7. Bilgin, C., Erkorkmaz, U., Ucar, M.K., Akin, N., Nalbant, A., Annakkaya, A.N.: Use of a portable monitoring device (Somnocheck Micro) for the investigation and diagnosis of obstructive sleep apnoea in comparison with polysomnography. Pak. J. Med. Sci. **32**(2), 471–475 (2016)
8. Cortes, C., Vapnik, V.: Support-vector networks. Mach. Learn. **20**, 273–297 (1995)
9. Duda, R.O., Hart, P.E., Stork, D.G.: Pattern Classification, 2nd edn. Wiley, New York (2001)
10. Uçar, M.K., Moran, I., Altilar, D.T., Bilgin, C., Bozkurt, M.R.: Kronik obstrüktif akciger hastaligi ile elektrokardiyogram sinyali arasindaki iliskinin istatistiksel olarak incelenmesi. J. Hum. Rhythm **4**(3), 142–149 (2018)
11. Olcay, N.O.: Acil Serviste Saglik Çalisanlarinin Elektrokardiyogram Bilgi Düzeyinin Degerlendirilmesi. T.C. Saglik Bakanligi, Acil tip uzmanlik tezi (2017)
12. Alpar, R.: Uygulamali Istatistik ve Geçerlilik Güvenirlilik: Spor, Saglik ve Egitim Bilimlerinden Örneklerle, 2nd edn. Detay Yayincilik, Ankara (2016)
13. Fonseca, P., Long, X., Radha, M., Haakma, R., Aarts, R.M., Rolink, J.: Sleep stage classification with ECG and respiratory effort. Physiol. Meas. **36**(10), 2027–2040 (2015)

Development of a Flexible Software for Disassembly Line Balancing with Heuristic Algorithms

Ümran Kaya[1(✉)], Halil İbrahim Koruca[2], and Samia Chehbi-Gamoura[3]

[1] Department of Industrial Engineering, Antalya Bilim University, Antalya, Turkey
umran.kaya@antalya.edu.tr
[2] Department of Industrial Engineering, Süleyman Demirel University, Isparta, Turkey
[3] EM Strasbourg Business School, HuManiS (EA 7308), University of Strasbourg, Strasbourg, France

Abstract. Due to the wide variety, short life span and lower price of electronic products, consumption has increased. These electronic wastes harm both the economy and the environment due to the metals they contain. In order to prevent this harm, the number of recycling facilities is increased and in the disassembly process of these facilities, the raw material can be re-used and some of the parts are encouraged to be recovered by recycling.

Precious raw materials obtained by disassembling the incoming waste to the smallest parts; iron, plastic, glass, circuit boards, cable etc. parts. Separation of these raw materials is carried out at disassembly stations or disassembly lines. Since each raw material is separated from the product and its disassembly has different processing times and different priorities, the disassembly line balancing process is included in the class of Np-hard problems.

In this study, a software was developed with three different heuristic algorithms (Largest Candidate Rule, Kilbridge and Wester method and Ranked Positional Weight method) for balancing the disassembly line. The software can balance the line according to three different algorithms by using business data, processing times and tasks priorities. As a result of line balancing, the efficiency of the line can be measured by three heuristic methods (algorithms). In the study, a mobile phone disassembly process data is examined. The results of the methods applied in the software, assignments to stations, their efficiency and production rates are given comparatively.

Keywords: Disassembly line balancing · Heuristic algorithm · Software development

D. J. Hemanth and U. Kose (Eds.): ICAIAME 2019, LNDECT 43, pp. 1014–1026, 2020.
https://doi.org/10.1007/978-3-030-36178-5_90

1 Introduction

Each used electronic product is replaced by a new one and turned into waste and becomes harmful to the economy and the environment [1]. In the manufacturing industry where the raw material is expensive, the precious metals contained in the e-wastes are recycled in the recycling facilities by the disassembly process. Disassembly is the first important step of the product recovery activities and the procedures for the removal of precious parts. The system that separates the waste product into parts, materials, components and other groups with sub-components is defined as disassembly [2]. The disassembly process is done in three different ways. These are the only workstation, the disassembly cell and the disassembly line.

Work load should be in balance also in disassembly line as in assembly lines. Because tasks whose priorities and process times are different should be balanced to ensure that the operations do not accumulate in the flow of the line and work efficiently. The disassembly line balancing problem (DLBP) is defined as employing the least amount of workers and assigning tasks in a balanced manner to the workers (stations) within the constraints of number of workstations, cycle time and tasks priorities constraints [4].

This goal is represented by two mathematically identical alternatives [5];

Min. (w * Ts − Twc) and Min. $\sum_{t=1}^{w}$ (Ts − Tsi)

So;

$\sum_{k \epsilon i}^{w} Tek \leq Ts$;

All priority requirements apply.

In these Equations, w; the minimum number of workers, Ts; the highest service time stations on the line, Twc; total service time and Tsi; the total service time in the i. station.

The aim followed by mathematical expression is the desired point in line balancing. In the simplest form, the disassembly line balancing problem is included in the class of Np-hard problems [6].

Minimizing the number of cycle-times while the tasks assigned to the workstations, minimizing idle time, maximizing the product of number of stations and cycle time (maximizing the efficiency) and maximizing the profit resulting from the demand for parts are measures of effectiveness of disassembly line balancing [7].

With the software developed within the scope of this study, disassembly line balancing can be made for each new product to be added. Only one product was dealt with in each disassembly line study in the literature. However, in this study, a ready software which can balance all kinds of e-waste products in the disassembly line has been developed. The software offers the option of solving the problem according to three different methods and provides the flexibility of adding stations to the disassembly line. This software that can offer three alternative solutions can be used by any recycling company and can be used also for assembly operations.

In order to prove the usability and accuracy of the developed software within the scope of the study, the pre-solved problem will be solved again. With the interface presented in the program, stations that tasks assigned and line efficiency for all disassembly line balancing can be seen with three different solution options.

2 Literature

The most important activity in the recycling of waste, the disassembly process will be examined in two ways in the literature; Firstly, it is the theoretical studies that research the disassembly processes in recycling facilities and provide detailed definitions to the literature on this subject with theoretical definitions. In the other literature review, it was mentioned that disassembly line balancing problem was handled with different solution methods.

In the first part of the literature; In order to recycle products, some design features are defined by industries that produce a large amount of iron and plastic waste, such as motor vehicles and appliance sectors [8]. Lambert proposed a graph-based method to maximize the economic performance of the disassembly process under technical and environmental constraints [9]. Crowther designed a model for disassembly problems. The study demonstrated how a life-cycle model that combines disassembly stages can extend service life and thereby increase sustainability by emphasizing the environmental benefits of design for disassembly [10].

Güngör and Gupta defined disassembly as a systematic process that divides the product into its constituents, subgroups or other groups. The subjects in the disassembly area are classified into two broad categories as design and operational. This study is one of the first published scientific studies on disassembly line balancing (DHD) problem [3]. Subsequently, many studies have been conducted on disassembly lines. Straight disassembly lines [11–13], parallel disassembly lines [14, 15], U-type disassembly lines [16, 17], mixed-model disassembly lines [18, 19] and so on. various new balancing problem concepts have emerged.

Under the technical and environmental constraints of the economic aspect of the disassembly process was examined and a methodology has been put forward [9]. Navin-Chandra studied product design for recycling [20]. Lambert worked on the optimum disassembly sequence for the recycling of an electronic product [21]. McGovern and Gupta proved that the decision version of DLBP is one of the NP-hard problems [22]. Iacob et al., unlike the others in their work by developing a virtual assembly simulation to examine parts to be disassembled [23]. And in recent years in the theoretical field, Bentaha et al. has been worked on disassembly line balancing and process sorting problems [24].

In the second part literature review, solution methods developed for the disassembly line balancing problem are included. McGovern and Gupta have solved the DLB problem with two different approaches, the 2-Opt heuristic algorithm and the Greedy algorithm, to solve the disassembly line balancing problem [25]. Ranky et al. used a unique heuristic algorithm for their dynamic disassembly lines balancing. They solved some problems in their studies [26]. Gupta et al. discussed the sequence of operations in the disassembly line balancing. For this, they solved using the heuristic method in Matlab [27]. Duta et al. attempted to solve the disassembly line balancing problem by a new method based on equal mass approach [28]. Bentaha et al. modeled the uncertainty by using the concept of resource cost and proposed a model of average approximation [29]. Bentaha et al. used a lagranjian relaxation approach to maximize total profits in disassembly lines [30].

The purpose of this study, instead of line balancing for just one problem with fixed methods and is to develop a flexible software which can be used in all firms and which can be used to balance all disassembly line, without data and process constraint. This software can balance the lines for any cycle time or any station number and more it can show these lines' efficiency. At this point, this study is different from literature.

And more, with this study, it is proven that these three heuristic algorithms which are applied just for assembly line balance problem can be applied for disassembly line balance problem and they work correctly. This study is very important also in this respect.

3 Methods

The main purpose of line balancing is to run the system efficiently. In order to avoid any accumulation to slow down the system, the line must be optimally balanced. Line balancing is to assign as much equal as possible to each worker in the station. The algorithms used in the software developed for the disassembly line balancing problem are "Largest Candidate Rule", "Kilbridge and Wester method" and "Ranked Positional Weight method".

After explaining the implementation steps of these algorithms, they will be applied on a problem and the results will be compared. The problem solved manually by the algorithms will be solved again with the prepared software and the accuracy of the software will be checked. In this way, both algorithms and software will be proven to work correctly.

3.1 Largest Candidate Rule

The largest candidate rule is one of the simplest methods used in line balancing. In this algorithm, works are firstly sorted from large to small according to process times. In the sorted list, as starting with first task, it assigns to stations by paying attention to the process priority and cycle time constraints. if the task's priority has not been assigned or it is exceeding the cycle time when the task assigned to the station, then assignment continues with the next process in the list [5]. These steps are in Table 1 and flowchart of this algorithm is at Fig. 1.

3.2 Kilbridge and Wester Method

A heuristic [31] assigning tasks according to the priority diagram. This method is based on the priority of the process, in contrast to the principle of priority of processing time in the largest candidate rule. Before assignment, a list is created according to the priorities, and in the other steps it continues with the implementation steps of the largest candidate rule [5]. The process steps of this heuristic algorithm are in Table 2 and flowchart of this algorithm is at Fig. 1.

Table 1. Implementation steps of the Largest Candidate Rule

Pseudocode of the Largest Candidate Rule

```
Sequence. Function: {Function that gives process times sorted matrix}
Check. Function: {Checks whether the job has already been imported into the
                  priority matrix}
Matching. Function: {Checks whether process priorities have been made}.
Sequences: {Matrix holding priorities assigned operations}
Stations: {Station matrix where jobs are assigned}
WN: {number of process}
    for g=1 to WN do
        for r=1 to WN do
            for i=1 to WN do
                x=sequence (i)
                m=check (x) {checks whether the selected job is placed in the
                priority matrix}
                if m=1
                    continue;
                end if
                {If the job is not assigned in advance, provides its own priorities and
                does not exceed Ts (cycle time), it is assigned to the station.}
                if y=2 & c < Ts (cycle time)
                    {It checks both process priorities and checks whether it exceeds
                    cycle time}
                end if
                if c > Ts
                    break; {If the cycle time during the assignment is filled with the
                           last assigned job, the different station isswitched.}
                end if
            end for
        end for
        {After assigning to the related station, it moves to the next station.}
    end for
```

3.3 Ranked Positional Weight Method

It is solved with positional weight method developed by Helgeson and Birnie [32]. The process times of the tasks where it is a priority are collected cumulatively and it is started with a list which is listed from bigger to smaller [9]. Although the first parts of this method are a little different, their next steps are solved as the largest candidate rule method, as in Kilbridge and Wester. The solution steps of this method are as follows Table 3 and flowchart of this algorithm is at Fig. 1.

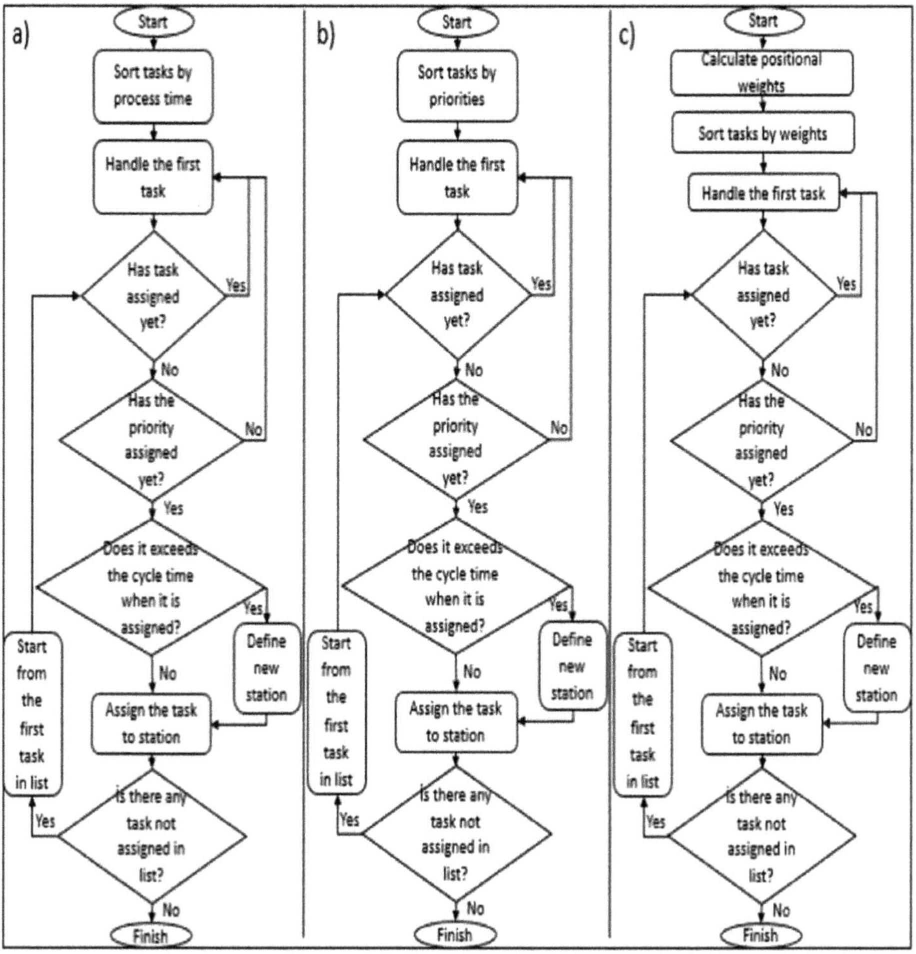

Fig. 1. Flowcharts of largest candidate rule (a), Kilbridge and Wester method (b) and ranked positional weight method (c)

3.4 Disassembly Line Performance Criteria

In developed software, algorithms will be compared according to their performance. The performance criterion of each algorithm is that it firstly assigns processes correctly to their priorities and cycle time. Efficiency and production speed are also examined as performance criteria. Productivity is calculated as follows;

$$Rp = \frac{\text{Annual demand amount}}{\text{Number of annual weeks} * \text{Number of weekly shifts} * \text{Time worked per shifts}} \tag{1}$$

Table 2. Implementation steps of Kilbridge and Wester Method

Pseudocode of Kilbridge and Wester Method

```
Check. Function: {Checks whether the job has already been imported into the
                  priority matrix}
Matching. Function: {Checks whether process priorities have been made}.
KWsequence.Function: {The function that gives the matrix of processing times}
KWassingmentToColoumns.Function: {Assigns jobs to columns by order}
Sequences: {Matrix holding priorities assigned operations}
Stations: {Station matrix where jobs are assigned}
WN: {number of process}
    {In this method, unlike the largest candidate rule, the initial matrix is assignea
    with its own functions according to the process priorities. The operations are the
    same after the initial matrix is created.}
    for g=1 to WN do
        for r=1 to WN do
            for i=1 to WN do
                x=sequence (i)
                m=check (x) {checks whether the selected job is placed in the
                priority matrix}
                if m=1
                    continue;
                end if
                {If the job is not assigned in advance, provides its own priorities
                and does not exceed Ts (cycle time), it is assigned to the station.}
                if y=2 & c < Ts (cycle time)
                    {It checks both process priorities and checks whether it
                    exceeds cycle time}
                end if
                if c > Ts
                    break; {If the cycle time during the assignment is filled with
                    the last assigned job, the different station is switched.}
                end if
            end for
        end for
        {After assigning to the related station, it moves to the next station.}
    end for
```

$$Tc = \frac{60 * \text{Efficiency rate}}{\text{Rp}} \tag{2}$$

$$Ts = \text{Tc} - \text{lost time per cycle} \tag{3}$$

Table 3. Implementation steps of Ranked Positional Weight method

Pseudocode of Ranked Positional Weight method

```
Check. Function: {Checks whether the job has already been imported into the
                  priority matrix}
Matching. Function: {Checks whether process priorities have been made}.
PositionalWeigths.Function: {Calculates the weight of the sum of the processing
                             times}
LastMatrix.Function: {Regulate processes in matrix according to positional
Sequences: {Matrix holding priorities assigned operations}
Stations: {Station matrix where jobs are assigned}
WN: {number of process}
    {In this method, unlike the largest candidate rule, the initial matrix is assigned
    with its own functions according to the process priorities. The operations are the
    same after the initial matrix is created.}
    for g=1 to WN do
        for r=1 to WN do
            for i=1 to WN do
                x=sequence (i)
                m=check (x) {checks whether the selected job is placed in the
                priority matrix}
                if m=1
                    continue;
                end if
                {If the job is not assigned in advance, provides its own priorities and
                does not exceed Ts (cycle time), it is assigned to the station.}
                if y=2 & c < Ts (cycle time)
                    {It checks both process priorities and checks whether it exceeds
                    cycle time}
                end if
                if c > Ts
                    break; {If the cycle time during the assignment is filled with
                    the last assigned job, the different station is switched.}
                end if
            end for
        end for
        {After assigning to the related station, it moves to the next station.}
    end for
```

$$Eb = \frac{Twc}{\mathrm{w}} * Ts \tag{4}$$

w; the minimum number of workers, Tc; the cycle time, Ts; the highest service time from stations on the line, Twc; total service time in the line and Eb; indicates the balance efficiency. This equilibrium efficiency is equal to the portion of the total service time and product of the number of stations with the highest processing time in the line balancing. This shows how equal the tasks are assigned to the stations. The basis of the line balancing problem is based on this equal assignment principle.

4 Application

Three heuristics (Largest candidate rule, Kilbridge and Wester method and Ranked positional weight method) used for disassembly line balancing software are explained in details. In this section, the accuracy of the developed software will be proved by a real problem from the book of McGovern and Gupta [11]. This problem is about the disassembly of a mobile phone.

4.1 Problem

This problem was obtained from the study of Gupta et al. [27]. This is a problem taken from the disassembly process of a Samsung SCH-3500 mobile phone and has 25 tasks. The sub-components, processing times and processing priorities of this product are in Table 4.

Business data for the problem; the annual demand amount has been set as 400.000 units, annual 50 weeks, 6 days per week and 12 h per shift. It is assumed that this line works at 95% efficiency and duration loss in line is 0.03 min. The service period (Ts) per station in the operation of the line is calculated as 0.483 min by the Eqs. 1, 2 and 3.

4.2 Line Balancing Result

The disassembly line balancing problem is solved for mobile phone disassembly, consisting of 25 different parts, with priorities and cycle times with processing times. The problem is solved with three different algorithms in the software. In the results, the assignment of tasks to stations, the efficiency of the line and the number of stations are different. These algorithms, which offer different disassembly models, also differ in the number of production per hour.

The largest candidate rule assigned seven different stations within the constraints of cycle time and priority (see Fig. 2). According to this algorithm, when the line is balanced, it will be seen that it works with 79,09% efficiency in seven different stations. According to this efficiency, the production speed of the line is 92 per hour according to Eq. 2.

When the software solve problem with the Kilbridge and Wester method, it is seen that tasks are assigned to seven stations such as the largest candidate rule (see Fig. 2). Although this algorithm makes it easy to sort and assign tasks according to their priorities, it is seen that this problem is less efficient on assignment. According to the results of this algorithm which assign correctly according to the priority and service time constraints, the line is 76.37% efficient which is calculated with the Eq. 4 by software. The hourly production amount of this line is 89.

Table 4. Details of the product to be disassembled

	Tasks	Process Time (Minute)	Priorities																
1	Antenna	0.0500																	
2	Battery	0.0333																	
3	Antenna	0.0500	1	2															
4	Bolt A	0.1666																	
5	Bolt b	0.1666																	
6	Bolts	0.2500	2																
7	Bolt 2	0.2500	2																
8	Bolt 3	0.2500	2																
9	Bolt 4	0.2500	3	2	1														
10	Clip	0.0333	5	4															
11	Rubber	0.0333	1	5	4														
12	Speaker	0.0333	1	1	5	4													
13	W. Cable	0.0333	9	8	7	6	3	2	1										
14	R/B Cable	0.0333	9	8	7	6	3	2	1										
15	O. Cable	0.0333	9	8	7	6	3	2	1										
16	Metal	0.0333	9	8	7	6	3	2	1										
17	Front	0.0333	1	1	9	8	7	6	3	2	1								
18	Back	0.0500	1	9	8	7	6	3	2	1									
19	Circuit	0.3000	1	1	1	1	1	9	8	7	6	3	2	1					
20	Plastic	0.0833	1	1	1	9	8	7	6	3	2	1							
21	Keyboard	0.0166	1	1	1	9	8	7	6	3	2	1							
22	Liquid	0.0833	2	1	1	1	9	8	7	6	3	2	1						
23	Lower	0.2500	2	2	1	1	1	1	9	8	7	6	3	2	1				
24	Internal	0.0333	2	2	2	1	1	1	1	1	1	1	9	8	7	6	3	2	1
25	M.phone	0.0333	2	1	1	1	9	8	7	6	3	2	1						

Finally, the results obtained by the method of ranked positional weight are different from the other algorithms and the works are assigned to six stations. This assignment is also suitable for priorities and cycle time (see Fig. 2). As a result of the assignment, the efficiency of the line has reached the best result with 92.25% of the other assignments. In this assignment, the amount of hourly production in line is 107 units. The best assignment was made by the ranked positional weight method. The result of this assignment in the software interface is shown in Fig. 3 and the efficiency results of other algorithms are also shown.

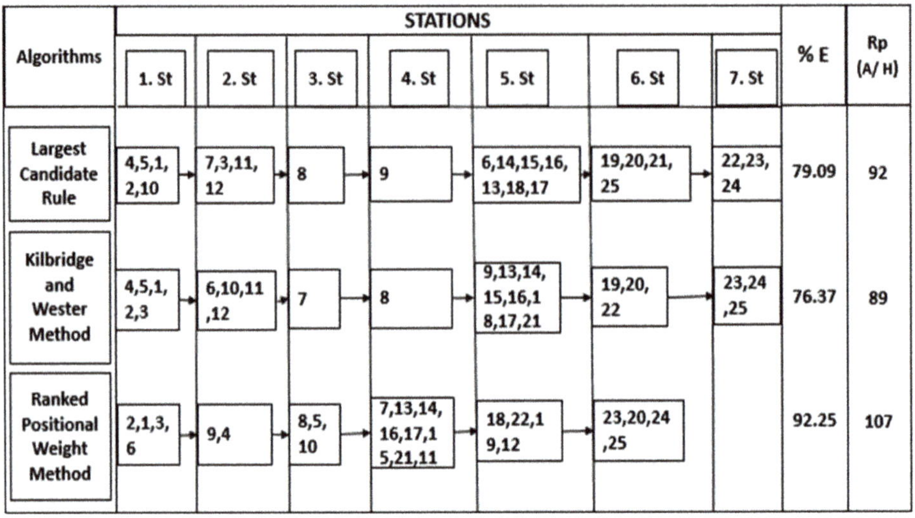

Fig. 2. Line balancing results of all algorithms

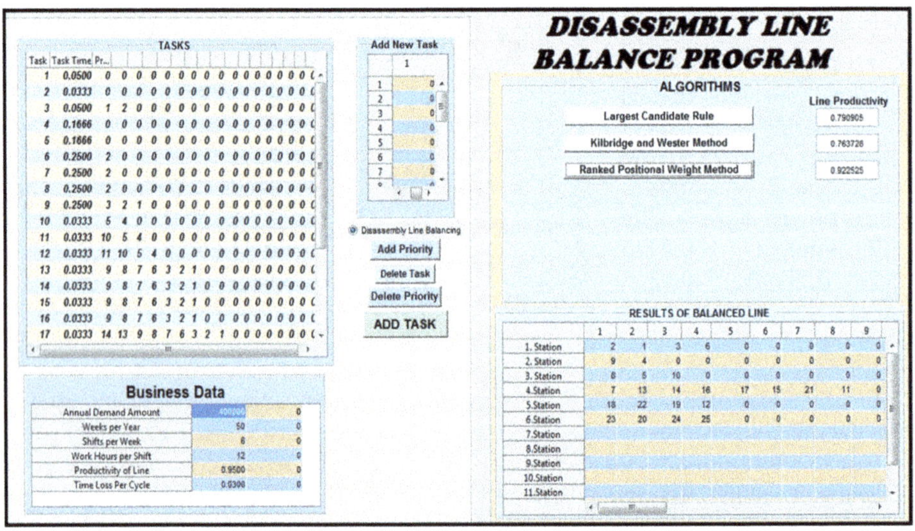

Fig. 3. Software interface of the best assignment

5 Conclusion

The problem of disassembly line balancing is a new field of work that is being provided by the ever-increasing technology. In this study, a flexible disassembly line balancing software has been developed. The purpose of this software is not to examine the solution of some methods only through a few examples as in the literature; to develop a flexible program that can offer to balance all disassembly lines with three different algorithms. Because of this flexibility, it can be applied in any company and in any product disassembly line. Because the interface of the software, add/remove operations and priority increase/decrease preferences are presented. Developing a flexible software that can be used in real life as well as contributing to the literature is one of the main aims of this study.

And more, with this study, it is proven that these three heuristic algorithms which are applied just for assembly line balance problem can be applied for disassembly line balance problem and they work correctly. This study is very important also in this respect.

The accuracy of the software has been proven by solving the disassembly line balancing problem. When we look at the solution of the problem, we can see that tasks are assigned correctly according to priorities, cycle times and number of stations. When these results are examined from the line efficiency point, it is proved that the software can provide satisfactory solutions. This software, which proved to be able to provide good results in line balancing, differs from the other studies in the literature in terms of its applicability to all disassembly lines, adaptation to all firm data and resulting according to different algorithms.

References

1. Kaya, Ü., Koruca, H.İ.: Elektronik Atıkların Çevreye Etkisi ve Topluma Farkındalık Kazandırmak için Öneriler. J. Acad. Soc. Sci. 368–378 (2018)
2. Güngör, A., Gupta, S.M.: Issues in environmentally conscious manufacturing and product recovery: a survey. Comput. Ind. Eng. **36**, 811–853 (1999)
3. Güngör, A., Gupta, S.M.: A solution approach to the disassembly line balancing problem in the presence of task failures. Int. J. Prod. Res. **39**(7), 1427–1467 (2001)
4. Güngör, A., Gupta, S.M.: Disassembly line in product recovery. Int. J. Prod. Res. **40**, 2569–2589 (2002)
5. Groover, M.P.: Automation, Production Systems and Computer-Integrated Manufacturing. Pearson, USA (2016)
6. McGovern, S.M., Gupta, S.M.: A balancing method and genetic algorithm for disassembly line balancing. Eur. J. Oper. Res. **179**, 692–708 (2007)
7. McGovern, S.M., Gupta, S.M.: The Disassembly Line Balancing. McGraw Hill, Columbus (2011)
8. Brennan, L., Gupta, S.M., Taleb, K.N.: Operations planning issues in assembly/disassembly environment. Int. J. Oper. Prod. Manag. **14**, 57–67 (1994)
9. Lambert, A.J.: Optimal disassembly of complex products. Int. J. Prod. Res. **35**, 2509–2523 (1997)
10. Crowther, P.: Designing for disassembly to extend service life and increase sustainability. In: 8th International Conference on Durability of Building Materials and Components (1999)
11. Altekin, F.T., Kandiller, L., Özdemirel, N.E.: Profit-oriented disassembly line balancing. Int. J. Prod. Res. **46**, 2675–2693 (2008)
12. Koç, A., Sabuncuoglu, I., Erel, E.: Two exact formulations for disassembly line balancing problems with task precedence diagram construction using an AND/OR graph. IIE Trans. **41**, 866–881 (2009)

13. Altekin, F.T., Akkan, C.: Task-failure-driven rebalancing of disassembly lines. Int. J. Prod. Res. **50**, 4955–4976 (2012)
14. Aydemir, A., Turkbey, O.: Multi-objective optimization of stochastic disassembly line balancing with station paralleling. Comput. Ind. Eng. **65**, 413–425 (2013)
15. Hezer, S., Kara, Y.: A network-based shortest route model for parallel disassembly line balancing problem. Int. J. Prod. Res. **53**, 1849–1865 (2014)
16. Avikal, S., Mishra, P.K.: A new U-shaped heuristic for disassembly line balancing problems. Int. J. Sci. Spirituality Bus. Technol. **1**, 2277–7261 (2012)
17. Avikal, S., Mishraa, P., Jain, R.: A Fuzzy AHP and PROMETHEE method-based heuristic for disassembly line balancing problems. Int. J. Prod. Res. **52**, 1306–1317 (2013)
18. Agrawal, S., Tiwari, M.K.: A collaborative ant colony algorithm to stochastic mixed-model U-shaped disassembly line balancing and sequencing problem. Int. J. Prod. Res. **46**, 1405–1429 (2008)
19. Paksoy, T., Güngör, A., Özceylan, E., Hancılar, A.: Mixed model disassembly line balancing problem with fuzzy goals. Int. J. Prod. Res. **51**, 6082–6096 (2013)
20. Navin-Chandra, D.: The recovery problem in product design. J. Eng. Des. **5**, 65–86 (1994)
21. Lambert, A.J.: Determining optimum disassembly sequences in electronic equipment. Comput. Ind. Eng. **43**, 553–575 (2002)
22. McGovern, S.M., Gupta, S.M.: Combinatorial optimization analysis of the unary NP-complete disassembly line balancing problem. Int. J. Prod. Res. **45**, 4485–4511 (2007)
23. Iacob, R., Popescu, D., Mitrouchev, P.: Assembly/disassembly analysis and modeling techniques: a review. J. Mech. Eng. **58**, 1–14 (2012)
24. Bentaha, M.L., Battaia, O., Dolgui, A.: Disassembly line balancing and sequencing under uncertainty. Procedia CIRP **15**, 56–60 (2014)
25. McGovern, S.M., Gupta, S.M.: 2-opt heuristic for the disassembly line balancing problem. In: The SPIE International Conference on Environmentally, pp. 71–84. Society of Photo-Optical Instrumentation Engineers, Rhode Island (2003)
26. Ranky, P.G., Subramanyam, M., Caudill, R.J., Limaye, K., Alli, N.: Dynamic scheduling and line balancing methods, and software tools for lean and reconfigurable disassembly cells and lines, pp. 234–239. IEEE (2003)
27. Gupta, S.M., Erbis, E., McGovern, S.M.: Disassembly sequencing problem: a case study of a cell phone. In: The SPIE International Conference on Environmentally Conscious Manufacturing IV, pp. 43–52. SPIE, Philadelphia (2004)
28. Duta, L., Filip, F.G., Henrioud, J.M.: Applyıng equal pıles approach to dısassembly lıne balancıng problem. IFAC Proc. Volumes **38**(1), 152–157 (2005)
29. Bentaha, M.L., Battaia, O., Dolgui, A.: A sample average approximation method for disassembly line balancing problem under uncertainty. Comput. Oper. Res. **51**, 111–122 (2014)
30. Bentaha, M.L., Battaia, O., Dolgui, A.: Lagrangian relaxation for stochastic disassembly line balancing problem. Procedia CIRP **17**, 239–244 (2014)
31. Kilbridge, M., Wester, L.: A heuristic method of assembly line balancing. J. Ind. Eng. **12**, 292–298 (1961)
32. Helgeson, W.B., Birnie, D.P.: Assembly line balancing using ranked positional weight technique. J. Ind. Eng. **12**, 394–398 (1961)

Parametrical Analysis of a New Design Outer-Rotor Line Start Synchronous Motor

Mustafa Tümbek(✉), Selami Kesler, and Yusuf Öner

Faculty of Engineering, Department of Electric and Electronic Engineering, Pamukkale University, Denizli, Turkey
mustafatumbek@pau.edu.tr

Abstract. The effect of the motor parameters of the outer rotor model, which is intended to be used in electric vehicles, has not been investigated although many inner-rotor models have been studied in the literature. In this study, therefore, a new outer-rotor line start synchronous motor is designed and the effect of rotor parameters on the motor characteristics is presented. Parameters of the rotor geometry are divided into 5 sub-titles; (i) airgap, (ii) round bar diameter, (iii) round bar distance from airgap (iv) magnet bar distance from airgap, (v) magnet thickness and (vi) magnet height. Also, the effect of the skew angle of rotor core is analyzed with 2D multi-slice FEM technique. The results show that the relationship between parameters is very important in terms of torque and flux density, high efficiency, and less cogging at start-up.

Keywords: OR-LSSM · Line start synchronous motor · Motor design · Sensitivity analysis

1 Introduction

In 1835, Thomas Davenport developed a practical electric vehicle after the invention of the first electric car, which is developed by the Scottish inventor Robert Anderson. Inventing the battery, Frenchman Gustave Trouvé developed an electric car, only using a friction of 0.1 HP direct current (DC) motor fed by lead-acid batteries, in 1881. In the same years, another electric vehicle produced by two British professors did not attract much attention of consumers due to its low speed and low range. In addition, despite the several successes of electric vehicles until the 1900s, liquid-fueled cars became more attractive to consumers with the idea of Henry Ford's production line in 1908. However, when air pollution is at the level that threatens humanity, electric vehicles have attracted the consumer's attention again. In addition, extension at the range of hybrid vehicles facilitated the transition to electric vehicles [1].

Modern electric vehicles consist of three main parts that are energy source, an electric motor, and a power train to convert electric energy into motion. In addition, power converters that support electrical systems, battery charging systems, motor drives, sensors and interfaces that transmit user requests to the system, protection units and vehicle visual alerts are also other parts of the electric vehicle [2]. The motor design used in electric vehicle constraints are defined as follows;

D. J. Hemanth and U. Kose (Eds.): ICAIAME 2019, LNDECT 43, pp. 1027–1038, 2020.
https://doi.org/10.1007/978-3-030-36178-5_91

- High torque density and high-power density
- Wide speed range
- High efficiency over speed ranges
- Wide constant-power operating capability
- High overload capability
- High reliability
- Low acoustic noise
- Reasonable cost

In the literature, six different system configurations have been proposed according to power transmission units and electric motor placement. Moreover, inner, and outer rotor motors are used according to motors placement and system requirements. Electric motor types used in electric vehicles are mainly DC, induction, PM brushless DC (BLDC) and switch reluctance (SR) motors [3]. Firstly, various type of DC motors with different motor characteristics are preferred in electric vehicles because of easy controllable. While the torque-speed characteristics of separately excited DC and shunt DC are linearly related that speed decreases as torque increases, the series DC has an inverse relationship. Additionally, DC motors with PMs have relatively higher power density and higher efficiency because of space-saving due to the lack of field losses. However, consumer attention with the developing technology has been drawn to other motor types by reason of low efficiency and maintenance requirements. Secondly, thanks to low cost microcontrollers and complex control algorithms, induction motor, widely used in the industry, is also used in modern electric vehicles. However, power loss occurs in short-circuit bars of rotor structure of an induction motor. The other motor type is SR motor that provides mostly performance criteria for electric vehicles. Also, simple construction, low manufacturing cost and outstanding torque-speed characteristics are some advantages of the motor. But these type motors have some problems such as low power density, non-linear control, and acoustic noise. Lastly, BLDC motor has high-power density, high efficiency, and low electromechanical time constant while it has high PM material cost, uncontrollable PM flux and sensor faults [4]. Also, in the study conducted with different electromagnetic topology and winding connections, the power density and efficiency increased while the cogging torque was reduced [5]. In addition to the mentioned types previously, many hybrid motor studies are included in the literature.

Line start synchronous motor (LSSM) is one of the hybrid motor types and it was first proposed by F. W. Merrill in 1955 [6]. LSSM, which was not attractive at the time due to weakness of magnets, are now preferred instead of the others because of existing of high-performance magnets. It is proper to use them in automotive applications owing to a wide power range of the motor's characteristics [7]. There are lots of studies in the literature on design criteria. One of these studies presented design criteria of the motors and their design constraints [8]. The design parameters of the motor are closely related to each other; therefore, the motor characteristics vary considerably. In other words, design parameters should be determined within the design processes. Some of them are short-circuit bar and magnet geometries that affect clearly starting performance and synchronization capability [9]. A covered review of LSSM developments is presented about PM materials, different rotor geometries, steady-state analysis, initial performance, parameter estimation and high efficiency standards [10]. In another study shows that LSSM is not only under full load, but also the performance of it during acceleration is also required for future testing standards [11]. In many studies, the most common methods described for inner rotor LSSMs are adding magnet to induction motor or a complete motor design. Parameters of the motor designing with each method have been optimized by analytical and FE analysis in terms of efficiency, initial performance, synchronization capability, torque and load capacity [12–26]. Although many studies have been conducted on inner rotor LSSMs in the literature, no study has been found for outer rotor LSSM (OR-LSSM) that can be used in-wheel motor systems of electric vehicles.

In this study, a new OR-LSSM motor is designed and the effect of Outer Rotor parameters on the motor characteristics have been investigated with regression analysis (as a predictive modeling technique) that investigates the relationship between dependent and independent variables.

2 General Rotor Topologies and Proposed Model

In the literature, design of LSSMs are mostly started by asynchronous motor design and then permanent magnets are added to rotor structure appropriately. In this paper, the design processes are divided into three sections: calculation of initial parameters, cage design and PM placement with parametric analysis by using FEM software as shown in Fig. 1.

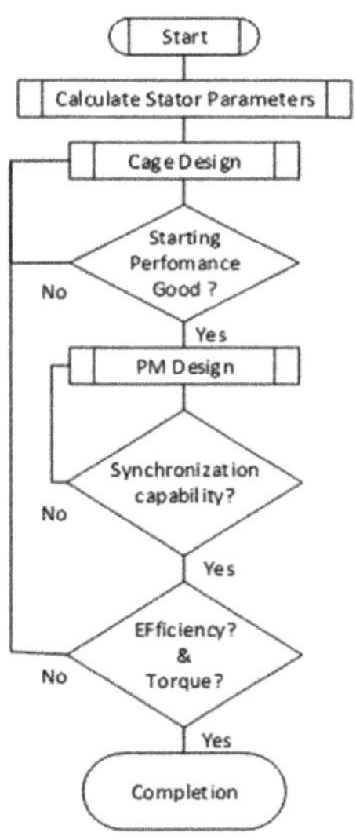

Fig. 1. Motor design flow chart

There are constraints for the motor outer geometry because of in-wheel systems. The outer diameter of the motor is 300 mm and the width is 40 mm, in this study for considered system. Initial parameters of the motor are given in Table 1. Also, proposed motor model is given in Fig. 1.

Table 1. Motor parameters

Motor design parameters		
Stator	Outer diameter	248 mm
	Stack length	40 mm
	Number of slots	24
	Conductor per slot	44
Rotor	Outer diameter	300 mm
	Inner diameter	250 mm
	Core length	40 mm
	Inertia	0.02807 kgm^2
Magnet	Thickness	20 mm
	Width	2 mm

In this study, 3D geometry and motor structure of the inner rotor motor used for electric vehicle is shown in Fig. 2. The proposed rotor geometry contains short circuit bars between each magnet. Also, 24 pieces of magnets are used in groups of 3 so as to face each stator pole.

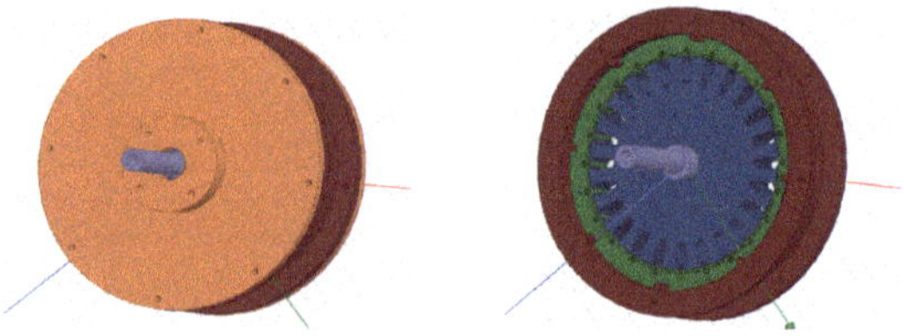

(a) (b)

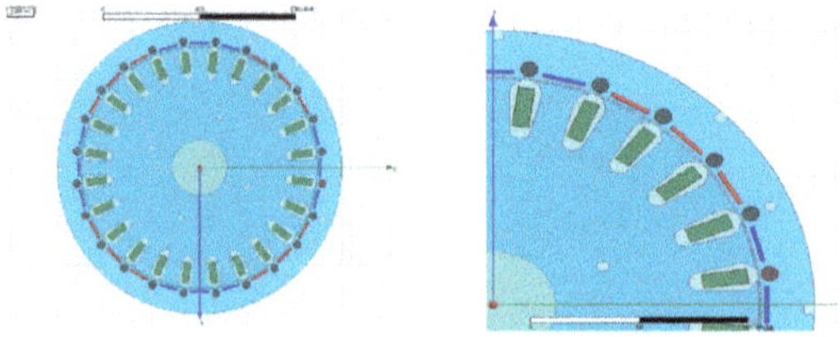

(c) (d)

Fig. 2. Proposed model geometry (a) 3D outer model (b) Inner model (c) 2D geometry (d) Cross section of the motor

3 Parametric Analysis

In 2D analysis of the Ansys Ansoft Maxwell, the motor geometry parameters can be defined as the first design variables. In this way, analysis can be made by assigning different possible values for each parameter. Also, the software allows to perform regression with sensitivity analysis. The purpose of regression is to understand the relationship between parameters. In other words, regression analysis helps to understand how to change dependent variables according to independent variables.

In this paper, the variables taken into consideration are bar diameter, bar distance from airgap, airgap length, magnet dimensions and distance magnet from airgap as shown in Fig. 3, including skew angle of the bar.

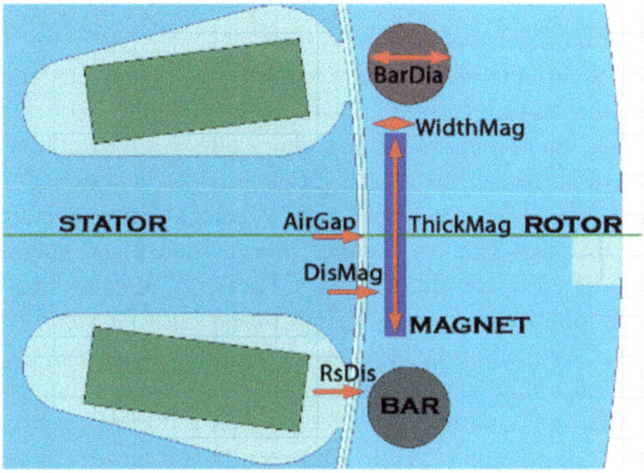

Fig. 3. Rotor structures and design parameters

3.1 Airgap

In general, because of the strong magnetism in the PM motor, the air gap of a LSSM is slightly longer than a similar-power induction motor. The small air gap length reduces vibration, noise, and stray losses, while increasing efficiency. However, when the air gap is too long, the flux leakages increase and cause a lower loading capacity. In a study ranging in air-gap length between 0.3–0.5 mm, it is presented that the length of the airgap is not have an effect on Back-EMF. Also, starting performance for the motor with a 0.5 mm air gap length is found to be poor [27].

In the analysis for the air gap length with the range between 1 mm and 5 mm, non-linear relationship with efficiency is investigated, as shown in Fig. 4.

3.2 Cage Bars

Cage bars in the rotor of the LSSM produce starting torque required for the acceleration of the motor. The design of the rotor bars shape is closely related to the initial performance, efficiency and synchronization of the motor. In an analytical study, the

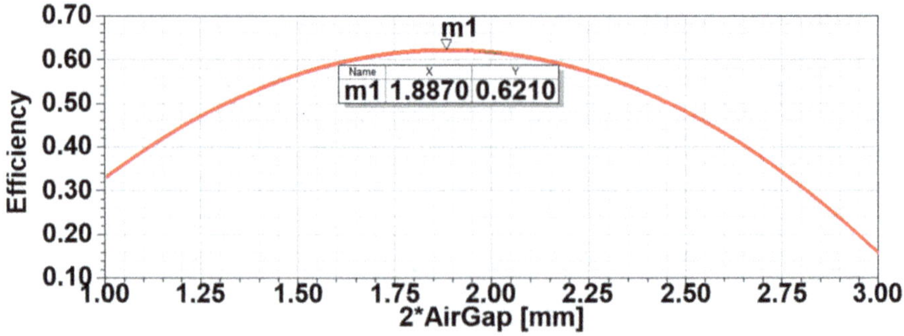

Fig. 4. Effect of air gap length on motor efficiency

effects on starting performance are investigated using different bar design. By increasing the bar width and material conductivity, the initial impedance is reduced, thereby increasing the starting current and losses together with reducing start-up torque [28]. In another study on asynchronous starting torque, it has been investigated that the rotor cage bars have a great effect on the moment of acceleration from zero speed to synchronous speed [29]. Also, rotor bars have an effect of magnetizing NdFeB. When the dimensions of the rotor bars are reduced, the resistance and magnetic flux density are increased, and the eddy current decreases. However, smaller rotor bar dimensions adversely affect the motor characteristics [30]. In the study on slot width and length, it has been shown that wide and long slots cause high flux density in the air gap. This is also required for high torque at synchronous speed. However, the starting torque and the synchronous torque are differently affected by the bar dimensions, so there is a limitation on the slot dimensions [31]. It is reported that a successful synchronization is achieved for higher inertia values of the load system when deep bar slots are selected [32]. In addition, it has been investigated in irregular bar structures to reduce current harmonics, torque fluctuation and Back-EMF [33].

In this work supported by an analytical solution, effect of bar diameter on motor efficiency and effect of bar distance from airgap on motor efficiency are given in Figs. 5 and 6, respectively.

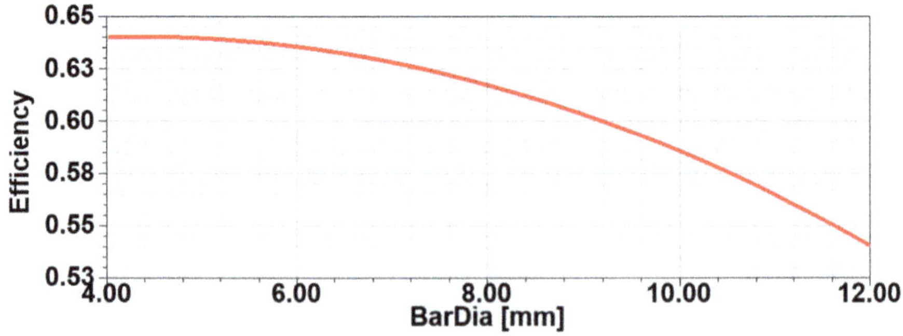

Fig. 5. Effect of bar diameter on motor efficiency

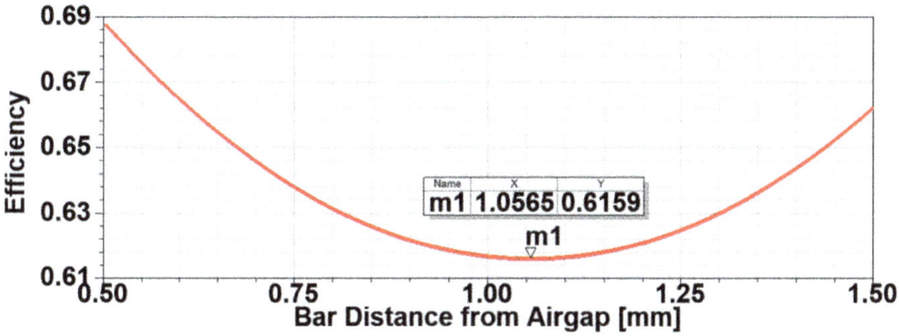

Fig. 6. Effect of bar distance from airgap on motor efficiency

3.3 Magnets

In the literature, different rotor structures have been formed with different magnet shapes and placement for inner rotor LSSMs (Fig. 7). The Ansys Maxwell Program has templates for conventional inner rotor LSSM motors. With the Rmxprt module included in this software, it is possible to carry out analytical analysis in a short time. Some material information for each part in the motor model can be defined. In a study conducted in the literature, NdFeB magnets provide more performance than SmCo-25 magnets. In addition, asymmetrically distributed NdFeB magnets improve performance [34].

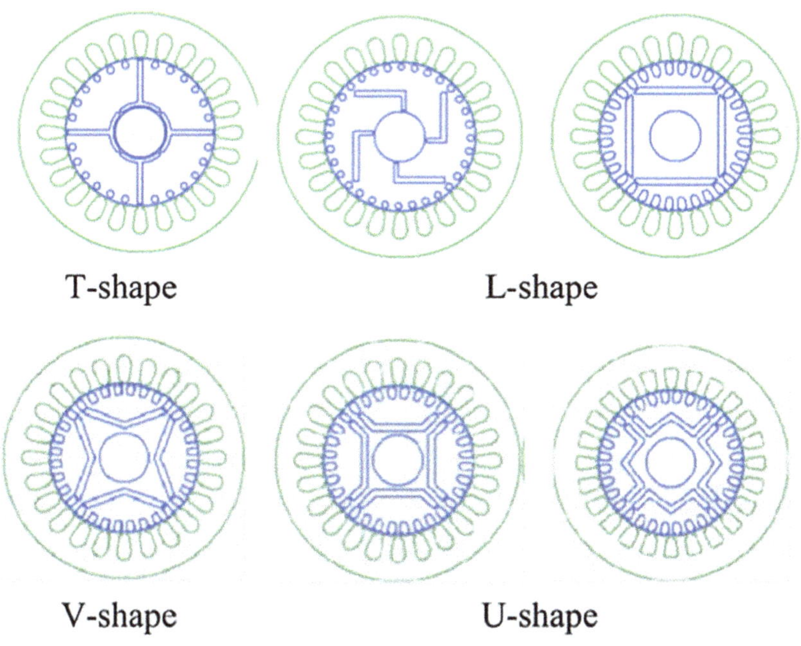

Fig. 7. Different rotor shapes for inner LSSMs

In the FEM analysis using the motor geometry in Fig. 3, the proper magnet dimensions and placement are investigated. In terms of efficiency, it is concluded that the thickness of the magnet relative to the rotor and magnet width are more effective (Figs. 8, 9 and 10). On the other hand, in order to determine the dynamic characteristics of the motor, it must be analyzed in the time domain. Magnet geometry and dimensions are known to have an effect on synchronization problems and braking torque [35].

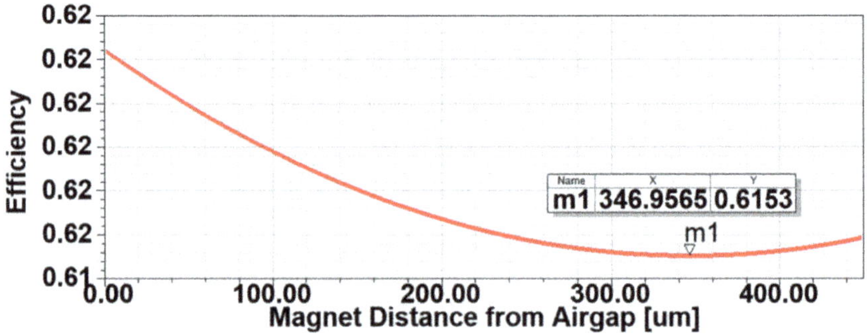

Fig. 8. Effect of magnet distance from airgap on motor efficiency

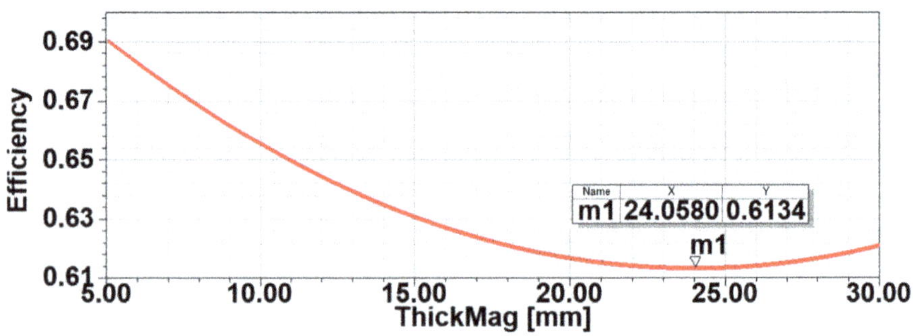

Fig. 9. Effect of magnet thickness on motor efficiency

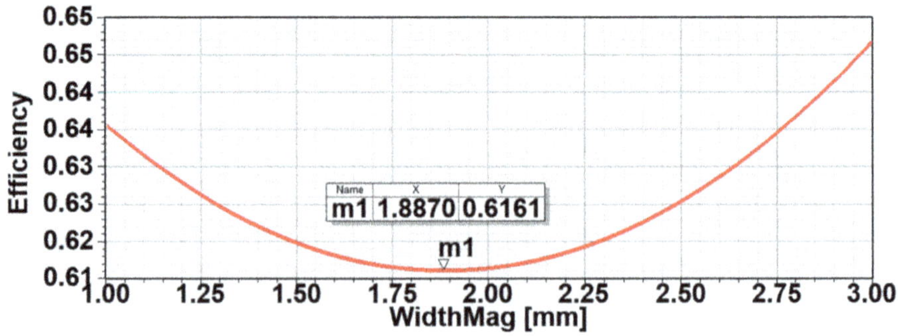

Fig. 10. Effect of magnet length on motor efficiency

3.4 Skew Angle

The cogging torque effect, which causes a mechanical stress and the motor torque ripple, can be reduced by skewing the rotor as in induction motors. In literature, it is shown that the torque ripples in LSSMs also reduced with skewed rotor by using 2D analysis [36]. In addition, it is seen that it provides an increase in synchronization performance and a 67% reduction in cogging torque [37].

In this study based on skew angle parameter, the cogging torque is reduced as shown Fig. 11. Also, although the torque ripples in the steady state is minimized, its behavior at acceleration time has slowed down, as shown in Figs. 12 and 13.

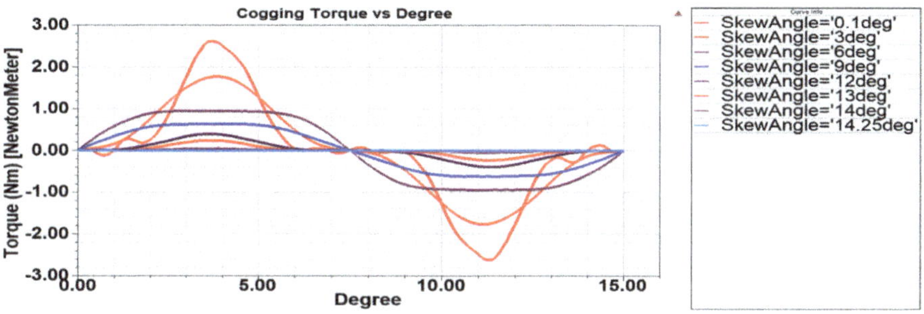

Fig. 11. Effect of skew angle on cogging torque

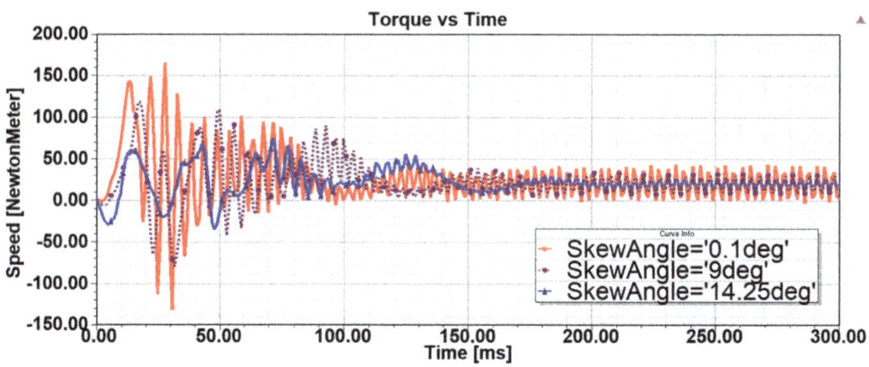

Fig. 12. Torque at time domain

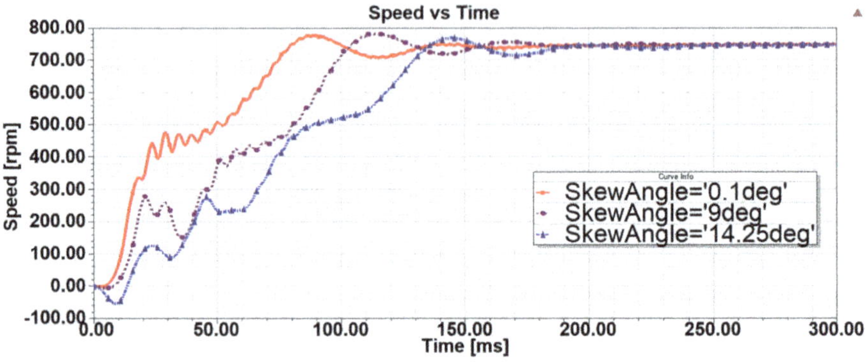

Fig. 13. Motor speed with different skew angle

4 Conclusions

In this paper, sensitivity analysis for the parameters of OR-LSSMs for in-wheel systems is investigated and the results are given. It is ensured that thanks to this analysis, if the design parameters are selected appropriately, the effects of the parameters on the motor characteristics can be observed comprehensively. And then, prototype implementation can be done easy and most truly. Also, in this study, the cogging torque effect on the synchronization, which also leads to a reduction in the torque quality of the motor, is presented together with solution of the problem.

References

1. Leitman, S., Brant, B.: Build Your Own Electric Vehicle, 2nd edn. McGraw-Hill Education, London (2008)
2. Chau, K.T.: Electric Vehicle Machines and Drives: Design, Analysis and Application, 1st edn. Wiley, New York (2015)
3. Ehsani, M., Gao, Y., Gay, S., Emadi, A.: Modern Electric, Modern Hybrid, and Fuel Cell Vehicles. Taylor & Francis Group, New York (2005)
4. Chau, K.T., Li, W.L.: Overview of electric machines for electric and hybrid vehicles. Int. J. Veh. Des.: J. Veh. Eng. Automot. Technol. Compon. **64**(1), 46–71 (2014)
5. Chan, C.C., Chau, K.T., Jiang, J.Z., Xia, W.A.X.W., Zhu, M., Zhang, R.: Novel permanent magnet motor drives for electric vehicles. IEEE Trans. Ind. Electron. **43**(2), 331–339 (1996)
6. Merrill, F.W.: Permanent magnet excited synchronous motors. Electr. Eng. **74**(2), 143 (1955)
7. Dorrell, D.G.: A review of the methods for improving the efficiency of drive motors to meet IE4 efficiency standards. J. Power Electron. **14**(5), 842–851 (2014)
8. Marčič, T.: A short review of energy-efficient line-start motor design. Przegląd Elektrotechniczny **87**(3), 119–122 (2011)
9. Behbahanifard, H., Sadoughi, A.: Line start permanent magnet synchronous motor performance and design; a Review. J. World Electr. Eng. Technol **4**, 58–66 (2015)

10. Ugale, R.T., Chaudhari, B.N., Pramanik, A.: Overview of research evolution in the field of line start permanent magnet synchronous motors. IET Electr. Power Appl. **8**(4), 141–154 (2014)
11. McElveen, R., Melfi, M., Daugherty, R.: Line start permanent magnet motors-Starting, standards and application guidelines. In: 2014 IEEE Petroleum and Chemical Industry Technical Conference, San Francisco, CA, USA, pp. 129–139 (2014)
12. Libert, F., Soulard, J., Engstrom, J.: Design of a 4-pole line start permanent magnet synchronous motor. In: International Conference on Electrical Machines, Bruges, Belgium, pp. 4–9 (2002)
13. Kurihara, K., Rahman, M.A.: High-efficiency line-start interior permanent-magnet synchronous motors. IEEE Trans. Ind. Appl. **40**(3), 789–796 (2004)
14. Kim, W.H., Kim, K.C., Kim, S.J., Kang, D.W., Go, S.C., Lee, H.W., Lee, J.: A study on the optimal rotor design of LSPM considering the starting torque and efficiency. IEEE Trans. Magn. **45**(3), 1808–1811 (2009)
15. Kim, W.H., Bae, J.N., Jang, I.S., Lee, J.: Design algorithm using torque separation method for line-start permanent magnet motor. In: 14th Biennial IEEE Conference on Electromagnetic Field Computation, p. 1. IEEE, Chicago (2010)
16. Lee, J.H., Jang, S.M., Du Lee, B., Song, H.S.: Optimum design criteria for maximum torque density & minimum current density of a line-start permanent-magnet motor using response surface methodology & finite element method. In: 2011 International Conference on Electrical Machines and Systems, pp. 1–4. IEEE, Beijing (2011)
17. Lee, B.H., Hong, J.P., Lee, J.H.: Optimum design criteria for maximum torque and efficiency of a line-start permanent-magnet motor using response surface methodology and finite element method. IEEE Trans. Magn. **48**(2), 863–866 (2012)
18. Elistratova, V., Hecquet, M., Brochet, P., Vizireanu, D., Dessoude, M.: Analytical approach for optimal design of a line-start internal permanent magnet synchronous motor. In: 15th European Conference on Power Electronics and Applications (EPE), pp. 1–7. IEEE, Lille (2013)
19. Shamlou, S., Mirsalim, M.: Design, optimisation, analysis and experimental verification of a new line-start permanent magnet synchronous shaded-pole motor. IET Electr. Power Appl. **7**(1), 16–26 (2013)
20. Sorgdrager, A.J., Grobler, A.J., Wang, R.J.: Design procedure of a line-start permanent magnet synchronous machine. In: 22rd Southern African Universities Power Engineering Conference, Durban, South Africa, pp. 307–314 (2014)
21. Azari, M.N., Mirsalim, M., Pahnehkolaei, S.M.A., Mohammadi, S.: Optimum design of a line-start permanent-magnet motor with slotted solid rotor using neural network and imperialist competitive algorithm. IET Electr. Power Appl. **11**(1), 1–8 (2017)
22. Sorgdrager, A.J., Wang, R.J., Grobler, A.J.: Retrofit design of a line-start PMSM using the Taguchi method. In: 2015 IEEE International Electric Machines & Drives Conference (IEMDC), pp. 489–495. IEEE, Coeur d'Alene (2015)
23. Sarani, E., Vaez-Zadeh, S.: Design procedure and optimal guidelines for overall enhancement of steady-state and transient performances of line start permanent magnet motors. IEEE Trans. Energy Convers. **32**(3), 885–894 (2017)
24. Dinh, B.M., Tien, H.M.: Maximum efficiency design of line start permanent magnet synchronous motor. In: 2016 IEEE International Conference on Sustainable Energy Technologies (ICSET), pp. 350–354. IEEE, Hanoi (2016)
25. Dinh, B.M.: Optimal rotor design of line start permanent magnet synchronous motor by genetic algorithm. Adv. Sci. Technol. Eng. Syst. J. **2**, 1181–1187 (2017)
26. Sorgdrager, A.J., Wang, R.J., Grobler, A.J.: Multiobjective design of a line-start PM motor using the Taguchi method. IEEE Trans. Ind. Appl. **54**(5), 4167–4176 (2018)

27. Yang, G., Ma, J., Shen, J.X., Wang, Y.: Optimal design and experimental verification of a line-start permanent magnet synchronous motor. In: 2008 International Conference on Electrical Machines and Systems, pp. 3232–3236. IEEE, Wuhan (2008)
28. Lu, W., Luo, Y., Zhao, H.: Influences of rotor bar design on the starting performance of line-start permanent magnet synchronous motor. In: Sixth International Conference on Electromagnetic Field Problems and Applications, pp. 1–4. IEEE Dalian (2012)
29. Nedelcu, S., Tudorache, T., Ghita, C.: Influence of design parameters on a line start permanent magnet machine characteristics. In: 2012 13th International Conference on Optimization of Electrical and Electronic Equipment (OPTIM), pp. 565–571. IEEE, Brasov (2012)
30. Lee, C.K., Kwon, B.I., Kim, B.T., Woo, K.I., Han, M.G.: Analysis of magnetization of magnet in the rotor of line start permanent magnet motor. IEEE Trans. Magn. **39**(3), 1499–1502 (2003)
31. Niazazari, M., Mirsalim, M., Mohamadi, S.: Effect of rotor slots parameters on synchronization capability of slotted solid rotor line start permanent magnet motor. In: 4th Annual International Power Electronics, Drive Systems and Technologies Conference, pp. 60–65. IEEE, Tehran (2013)
32. Jedryczka, C., Wojciechowski, R.M., Demenko, A.: Influence of squirrel cage geometry on the synchronisation of the line start permanent magnet synchronous motor. IET Sci. Meas. Technol. **9**(2), 197–203 (2014)
33. Li, P., Shen, J.X., Sun, W., Zhang, Y.: Investigation of LSPMSM with unevenly distributed squirrel cage bars. In: 2013 International Conference on Electrical Machines and Systems (ICEMS), pp. 24–27. IEEE, Busan (2013)
34. Topaloglu, I., Mamur, H., Korkmaz, F., Cakir, M.F.: Design and optimization of surface mounted line start permanent magnet synchronous motor using electromagnetic design tool. In: 2014 International Conference on Renewable Energy Research and Application (ICRERA), pp. 87–90. IEEE, Milwaukee (2014)
35. Tumbek, M., Kesler, S., Oner, Y.: Design and fem analysis of low voltage outer rotor line start permanent magnet outer rotor line start permanent magnet alignments. In: 6th International Conference on Advanced Technology & Sciences (ICAT'Riga), Riga, Latvia, pp. 196–200 (2017)
36. Williamson, S., Knight, A.M.: Performance of skewed single-phase line-start permanent magnet motors. IEEE Trans. Ind. Appl. **35**(3), 577–582 (1999)
37. Zöhra, B., Akar, M., Eker, M.: Design of a novel line start synchronous motor rotor. Electronics **8**(1), 25 (2019)

$\mathcal{I}$-Statistically Localized Sequence in 2-Normed Spaces

Ulaş Yamancı[1] and Mehmet Gürdal[2(✉)]

[1] Department of Statistics, Süleyman Demirel University, 32260 Isparta, Turkey
ulasyamanci@sdu.edu.tr
[2] Department of Mathematics, Süleyman Demirel University, 32260 Isparta, Turkey
gurdalmehmet@sdu.edu.tr

Abstract. In 1974, Krivonosov defined the concept of localized sequence that is defined as a generalization of Cauchy sequence in metric spaces. In this work, by using the concept of ideal, the statistically localized sequences are defined and some basic properties of $\mathcal{I}$-statistically localized sequences are given. Also, it is shown that a sequence is $\mathcal{I}$-statistically Cauchy iff its statistical barrier is equal to zero.

Keywords: Statistical convergence · Ideal convergence · 2-normed spaces · Localized sequence

2000 Mathematics Subject Classification. Primary 40A35

1 Introduction and Preliminaries

Statistical convergence introduced by Fast [2] and Steinhaus [17] has many applications in different areas. Later on, this concept was reintroduced by Schoenberg in his own study [16]. The concept of statistical convergence is defined depending upon the natural density of the set $A \subseteq \mathbb{N}$ and the natural density of A is given by $\delta(A) := \lim_{n\to\infty} \frac{|A_n|}{n}$, where $A_n = \{a \in A : a \leq n\}$ and $|A_n|$ denotes the number of elements in A_n.

Utilizing above information, we say that a sequence $(x_k)_{k\in\mathbb{N}}$ is statistically convergent to x provided that

$$\delta(\{k \in \mathbb{N} : |x_k - x| \geq \varepsilon\}) = 0$$

for every $\varepsilon > 0$. In this case, we denote it $st\text{-}\lim x_k = x$. For more detail informations about statistical convergent, see, in [6,20,23].

On the other hand, $\mathcal{I}$-convergence in a metric space was introduced by Kostyrko et al. [7] and its definition is depending upon the definition of an ideal $\mathcal{I}$ in $\mathbb{N}$.

D. J. Hemanth and U. Kose (Eds.): ICAIAME 2019, LNDECT 43, pp. 1039–1046, 2020.
https://doi.org/10.1007/978-3-030-36178-5_92

A family $\mathcal{I} \subset 2^{\mathbb{N}}$ is called an ideal if the following properties are held:
(*i*) $\emptyset \notin \mathcal{I}$;
(*ii*) $P \cup R \in \mathcal{I}$ for every $P, R \in \mathcal{I}$;
(*iii*) $R \in \mathcal{I}$ for every $P \in \mathcal{I}$ and $R \subset P$.

A non-empty family of sets $\mathcal{F} \subset 2^{\mathbb{N}}$ is a filter iff $\emptyset \notin \mathcal{F}$, $P \cap R \in \mathcal{F}$ for every $P, R \in \mathcal{F}$, and $R \in \mathcal{F}$ for every $P \in \mathcal{F}$ and every $R \supset P$. An ideal $\mathcal{I}$ is called a non-trivial if $\mathcal{I} \neq \emptyset$ and $X \notin \mathcal{I}$. The $\mathcal{I} \subset 2^X$ is a non-trivial ideal iff $\mathcal{F} = \mathcal{F}(\mathcal{I}) = \{\mathbb{N} \backslash P : P \in \mathcal{I}\}$ is a filter on X. A non-trivial ideal $\mathcal{I} \subset 2^X$ is called an admissible iff $\mathcal{I} \supset \{\{x\} : x \in X\}$.

Recall that an admissible ideal $\mathcal{I} \subset 2^{\mathbb{N}}$ holds the property (AP) if there is a family $\{R_n\}_{n \in \mathbb{N}}$ such that $(P_k \backslash R_n) \cup (R_k \backslash P_k)$ for all $k \in \mathbb{N}$ and a limit set $R = \bigcup_{k=1}^{\infty} R_k \in \mathcal{I}$ for every family $\{P_n\}_{n \in \mathbb{N}}$ with $P_n \cap P_k = \emptyset$ $(n \neq k)$, $P_n \in \mathcal{I}$ $(n \in \mathbb{N})$ [7].

Definition 1. *([7]) Let $\{x_n\}_{n \in \mathbb{N}}$ be a sequence of real numbers. It is called an $\mathcal{I}$-convergent to K if the set*

$$A(\varepsilon) = \{n \in \mathbb{N} : |x_n - K| \geq \varepsilon\} \in \mathcal{I}$$

for each $\varepsilon > 0$.

For more information about $\mathcal{I}$-convergent, see the references in [9–11].

The definition of 2-normed space was given by Gahler [3]. After this definition, many authors studied statistical convergence, ideal convergence, ideal Cauchy sequence, star ideal convergent and star ideal Cauchy sequence on this space (see [4,5,15]).

Depending upon the $\mathcal{I}$-convergence and statistical convergence, the ideal statistical convergence was introduced in [1]. Later on, the concepts of ideal statistically convergent and ideal statistically Cauchy were given in 2-normed spaces and important consequences were obtained in [18,19].

Let (x_k) be a sequence in 2-normed space $(X, \|.,.\|)$:

- It is an ideal statistically convergent to μ, if the set

$$\left\{n \in \mathbb{N} : \frac{1}{n} |\{k \leq n : \|x_k - \mu, z\| \geq \varepsilon\}| \geq \delta \right\} \in \mathcal{I}$$

  for each $\varepsilon > 0$, $\delta > 0$ and nonzero z in X. In such case we can write it $\mathcal{I}$-*st*-$\lim\limits_{k \to \infty} \|x_k - \mu, z\| = 0$ or $\mathcal{I}$-*st*-$\lim\limits_{k \to \infty} \|x_k, z\| = \|\mu, z\|$ (see [18]).
- It is an ideal statistically Cauchy sequence in X if there is a number N such that

$$\delta_{\mathcal{I}} \left(\left\{ n \in \mathbb{N} : \frac{1}{n} |\{k \leq n : \|x_k - x_N, z\| \geq \varepsilon\}| \geq \delta \right\} \right) = 0,$$

  for every $\varepsilon > 0$, $\delta > 0$ and every nonzero $z \in X$ (see [18]).
- It is an $\mathcal{I}^*$-statistically convergent to $\mu \in X$ iff there is a set $B \in \mathcal{F}(\mathcal{I})$ such that $st\text{-}\lim\limits_{k \to \infty} \|x_{m_k} - \mu, z\| = 0$ and $B = \{b_1 < b_2 < ... < b_k < ...\} \subset \mathbb{N}$.

- It is an $\mathcal{I}^*$-statistically Cauchy sequence iff there is a set $B = \{b_1 < b_2 < ... < b_k\}$ such that

$$st\text{-}\lim_{k,p\to\infty} \|x_{m_k} - x_{m_p}, z\| = 0.$$

A sequence (x_n) in a metric space X is said to be localized in some subset $M \subset X$ if the number sequence $d(x_n, x)$ converges for all $x \in M$ (see [8]). This definition has been extended to statistical localized and $\mathcal{I}$-localized in metric space [12,13] and 2-normed spaces [21,22], and they obtained interested results about this concept.

In this paper, by using the concept of ideal, the statistically localized sequences are defined and some basic properties of $\mathcal{I}$-statistically localized sequences are given. Also, it is shown that a sequence is ideal statistically Cauchy iff its statistical barrier is equal to zero.

2 Main Results

Our main definitions and notations are as following:

Definition 2. *Let $(x_n)_{n\in\mathbb{N}}$ be a sequence in 2-normed space $(X, \|.,.\|)$.*

(a) It is called as the $\mathcal{I}$-statistically localized in the subset $M \subset X$ iff

$$\left\{n \in \mathbb{N} : \frac{1}{n} |\{k \le n : \|x_n - x, z\| \ge \varepsilon\}| \ge \delta\right\} \in \mathcal{I}$$

exists for every $x, z \in M$, that is, the real number sequence $\|x_n - x, z\|$ is $\mathcal{I}$-statistically convergent.

(b) The maximal set on which it is $\mathcal{I}$-statistically localized is said to be $\mathcal{I}$-statistically localor of (x_n) and it is denoted by $loc_{\mathcal{I}_{st}}(x_n)$.

(c) It is said to be $\mathcal{I}$-statistically localized everywhere if (x_n) is $\mathcal{I}$-statistically localor of (x_n) coincides with X.

(d) It is called as the $\mathcal{I}$-statistically localized in itself if

$$\{n \in \mathbb{N} : x_n \notin loc_{\mathcal{I}_{st}}(x_n)\} \subset \mathcal{I}.$$

From above definition, if (x_n) is an ideal statistically Cauchy sequence, then it is ideal statistically localized everywhere. Actually, owing to

$$\frac{1}{k} |\{n \le k : |\|x_n - x, z\| - \|x_{n_0} - x, z\|| \geqslant \varepsilon\}| \leqslant \frac{1}{k} |\{n \le k : \|x_n - x_{n_0}, z\| \geqslant \varepsilon\}|$$

we have

$$\left\{k \in \mathbb{N} : \frac{1}{k} |\{n \le k : |\|x_n - x, z\| - \|x_{n_0} - x, z\|| \geqslant \varepsilon\}| \ge \delta\right\}$$
$$\subset \left\{k \in \mathbb{N} : \frac{1}{k} |\{n \le k : \|x_n - x_{n_0}, z\| \geqslant \varepsilon\}| \ge \delta\right\}.$$

So, the sequence is ideal statistically localized if it is ideal statistically Cauchy sequence.

Also, we are able to say that each ideal statistically convergence sequence is ideal statistically localized. We know that if $\mathcal{I}$ is an admissible ideal, then every statistically localized sequence in 2-normed space $(X, \|.,.\|)$ is ideal statistically localized sequence in $(X, \|.,.\|)$.

Definition 3. *We say the sequence (x_n) to be $\mathcal{I}^*$-statistically localized in $(X, \|.,.\|)$ iff the number sequence $\|x_n - x, z\|$ is $\mathcal{I}^*$-statistically convergent for each $x, z \in X$.*

From above definition, one sees that every $\mathcal{I}^*$-statistically Cauchy sequence or $\mathcal{I}^*$-statistically convergent in 2-normed space $(X, \|.,.\|)$ is $\mathcal{I}^*$-statistically localized in $(X, \|.,.\|)$.

Note that for admissible ideal, $\mathcal{I}^*$-statistically convergence and $\mathcal{I}^*$-statistically Cauchy criteria imply $\mathcal{I}$-statistically convergence and $\mathcal{I}$-statistically Cauchy criteria, respectively.

Lemma 1. *Let (x_n) be a sequence in $(X, \|.,.\|)$. If it is $\mathcal{I}^*$-statistically localized on the set $M \subset X$, then (x_n) is $\mathcal{I}$-statistically localized on the set M and $loc_{\mathcal{I}^*_{st}}(x_n) \subset loc_{\mathcal{I}_{st}}(x_n)$.*

Proof. Suppose that (x_n) is $\mathcal{I}^*$-statistically localized on M. Then, there is a set $P \in \mathcal{I}$ such that

$$\lim_{j\to\infty} \frac{1}{j} |\{j \in \mathbb{N} : \|x_j - x, z\| \geqslant \varepsilon\}|$$

exists for each $x, z \in M$ and $P^C = \mathbb{N}\backslash P = \{p_1 < p_2 < ... < p_j\}$. Then, the sequence $\|x_n - x, z\|$ is an $\mathcal{I}^*$-statistically Cauchy sequence, which means that

$$\left\{n \in \mathbb{N} : \frac{1}{n} |\{k \leq n : \|x_k - x_N, z\| \geq \varepsilon\}| \geq \delta\right\} \in \mathcal{I}.$$

Therefore, the number sequence $\|x_n - x, z\|$ is $\mathcal{I}$-statistically convergent, which gives that (x_n) is $\mathcal{I}$-statistically localized on the set M. □

Now, we give our basic consequences about $\mathcal{I}$-statistically localized sequences.

Proposition 1. *Let (x_n) be an ideal statistically localized sequence in $(X, \|.,.\|)$. Then it is ideal statistically bounded.*

Proof. Suppose that (x_n) is ideal statistically localized. Then, the number sequence $\|x_n - x, z\|$ is ideal statistically convergent for some $x, z \in X$. This means that $\left\{k \in \mathbb{N} : \frac{1}{k} |\{n \leq k : \|x_n - x, z\| > K\}| > \delta\right\} \in \mathcal{I}$ for some $K > 0$ and $\delta > 0$. As a result, the sequence (x_n) is ideal statistically bounded. □

Proposition 2. *Let $\mathcal{I}$ be an admissible ideal satisfying the property (AP) and $M = loc_{\mathcal{I}_{st}}(x_n)$. Also, a point $y \in X$ be such that there exists $x \in M$ for any $\varepsilon > 0$, $\delta > 0$ and every nonzero $z \in M$ such that*

$$\left\{k \in \mathbb{N} : \frac{1}{k} |\{n \leq k : |\|x - x_n, z\| - \|y - x_n, z\|| \geqslant \varepsilon\}| > \delta\right\} \in \mathcal{I}. \tag{1}$$

Then $y \in M$.

Proof. To show that the number sequence $\|x_n - y, z\|$ is an $\mathcal{I}$-statistically Cauchy sequence is enough. Let be $\varepsilon > 0$ and $x \in M = loc_{\mathcal{I}_{st}}(x_n)$ is a point satisfying the property (1). From the (AP) property of $\mathcal{I}$, we get

$$\frac{1}{k}\left|\{n \leq k : |\|x - x_{k_n}, z\| - \|y - x_{k_n}, z\|| \geqslant \varepsilon\}\right| \to 0$$

and

$$\frac{1}{k}\left|\{(n,m) : |\|x_{k_n} - x, z\| - \|x_{k_m} - x, z\|| \geqslant \varepsilon,\ n,m \leq k\}\right| \to 0$$

as $m, n \to \infty$, where $K = \{k_1 < k_2 < ... < k_n < ...\} \in \mathcal{F}(\mathcal{I})$. Therefore, there is $n_0 \in \mathbb{N}$ for any $\varepsilon > 0$, $\delta > 0$ and every nonzero $z \in M$ such that

$$(2) \qquad \frac{1}{k}\left|\left\{n \leq k : |\|x - x_{k_n}, z\| - \|y - x_{k_n}, z\|| \geqslant \frac{\varepsilon}{3}\right\}\right| < \frac{\delta}{3}$$

$$(3) \qquad \frac{1}{k}\left|\left\{(n,m) : |\|x - x_{k_n}, z\| - \|x - x_{k_m}, z\|| \geqslant \frac{\varepsilon}{3},\ n,m \leq k\right\}\right| < \frac{\delta}{3}.$$

for all $n \geq n_0$, $m \geq m_0$.

Since

$$\begin{aligned}
&\frac{1}{k}\left|\{(n,m) : |\|y - x_{k_n}, z\| - \|y - x_{k_m}, z\|| \geqslant \varepsilon,\ n,m \leq k\}\right| \\
&\leq \frac{1}{k}\left|\{n \leq k : |\|y - x_{k_n}, z\| - \|x - x_{k_n}, z\|| \geqslant \varepsilon\}\right| \\
&+ \frac{1}{k}\left|\{n \leq k : |\|x - x_{k_n}, z\| - \|x - x_{k_m}, z\|| \geqslant \varepsilon\}\right| \\
&+ \frac{1}{k}\left|\{n \leq k : |\|x - x_{k_m}, z\| - \|y - x_{k_n}, z\|| \geqslant \varepsilon\}\right|,
\end{aligned}$$

we obtain by using (2) and (3) together with above inequality

$$\frac{1}{k}\left|\{(n,m) : |\|y - x_{k_n}, z\| - \|y - x_{k_m}, z\|| \geqslant \varepsilon,\ n,m \leq k\}\right| < \delta$$

for all $n \geq n_0$, $m \geq n_0$. So,

$$\frac{1}{k}\left|\{(n,m) : |\|y - x_{k_n}, z\| - \|y - x_{k_m}, z\|| \geqslant \varepsilon,\ n,m \leq k\}\right| \to 0 \text{ as } m,n \to \infty$$

for the $K = (k_n) \subset N$ and $K \in \mathcal{F}(\mathcal{I})$. Hence $\|y - x_n, z\|$ is an $\mathcal{I}$-statistically Cauchy sequence, which finishes the proof. □

Definition 4. *([14]) Let a is a point in $(X, \|.,.\|)$. It is called a limit point of a set M in X if for an arbitrary $\Sigma = \{(b_1, \varepsilon_1), ..., (b_n, \varepsilon_n)\}$, there is a point $a_\Sigma \in M$, $a_\Sigma \neq a$ such that $a_\Sigma \in W_\Sigma(a)$.*

Moreover, a subset $L \subset K$ is called a closed subset of K if L contains every its limit point. If L^0 is the set of all points of a subset $L \subset K$, then the set $\overline{L} = L \cup L^0$ is called the closure of the set L.

Proposition 3. *$\mathcal{I}$-statistical localor of any sequence is a closed subset of the 2-normed space $(X, \|.,.\|)$.*

Proof. Let $y \in \overline{loc_{\mathcal{I}_{st}}}(x_n)$. Then, there is a point $x \in loc_{\mathcal{I}_{st}}(x_n)$ for arbitrary $\Sigma = \{(b_1, \varepsilon_1), ..., (b_n, \varepsilon_n)\}$ such that $x \neq y$ and $x \in W_\Sigma(y)$. Thus, for any $\varepsilon > 0$, $\delta > 0$ and every $z \in loc_{\mathcal{I}_{st}}(x_n)$

$$\left\{k \in \mathbb{N} : \frac{1}{k} |\{n \leq k : |\|x - x_n, z\| - \|y - x_n, z\|| \geqslant \varepsilon\}| > \delta\right\} \in \mathcal{I}$$

due to

$$\frac{1}{k} |\{n \leq k : |\|x - x_n, z\| - \|y - x_n, z\|| \geqslant \varepsilon\}| \leq \frac{1}{k} |\{n \leq k : \|y - x_n, z\| \geqslant \varepsilon\}| < \delta$$

for every $n \in \mathbb{N}$. In conclusion, the hypothesis of Proposition 2 is satisfied. and then we reach that $y \in loc_{\mathcal{I}_{st}}(x_n)$, that is, $loc_{\mathcal{I}_{st}}(x_n)$ is closed. □

Definition 5. *Let y is a point in $(X, \|.,.\|)$. It is an $\mathcal{I}$-statistical limit point of the sequence (x_n) in $(X, \|.,.\|)$ if there is a set $K = \{k_1 < k_2 < ... < k_n\} \subset \mathbb{N}$ such that $K \notin \mathcal{I}$ and*

$$\lim_{n \to \infty} \frac{1}{n} |\{n \in \mathbb{N} : \|x_{k_n} - y, z\| \geqslant \varepsilon\}| = 0.$$

It is called an $\mathcal{I}$-statistical cluster point of the sequence (x_n) if

$$\left\{k \in \mathbb{N} : \frac{1}{k} |\{n \leq k : \|x_n - y, z\| \geqslant \varepsilon\}| < \delta\right\} \notin \mathcal{I}$$

for each $\varepsilon > 0$, $\delta > 0$ and every $z \in X$.

We can have the following result owing to

$$\frac{1}{k} |\{n \leq k : |\|x_n - y, z\| - \|x - y, z\|| \geqslant \varepsilon\}| \leq \frac{1}{k} |\{n \leq k : \|x_n - x, z\| \geqslant \varepsilon\}|.$$

Proposition 4. *Let $y \in X$ be an $\mathcal{I}$-statistical limit point (an $\mathcal{I}$-statistical cluster point) of a sequence (x_n) in 2 -normed space $(X, \|.,.\|)$. Then, the number $\|y - x, z\|$ is an $\mathcal{I}$-statistical limit point (an $\mathcal{I}$-statistical cluster point) of the sequence $\{\|x_n - x, z\|\}$ for each $x \in X$ and every nonzero $z \in X$.*

Definition 6. *Let (x_n) be the $\mathcal{I}$-statistical localized sequence (x_n) with the $\mathcal{I}$-statistical localor $M = loc_{\mathcal{I}_{st}}(x_n)$. The number*

$$\lambda = \inf_{x \in M} \left(\mathcal{I}\text{-}st\text{-} \lim_{n \to \infty} \|x - x_n, z\|\right)$$

is called as the $\mathcal{I}$-statistical barrier of (x_n).

Theorem 1. *Let $(X, \|.,.\|)$ be an 2-normed space and let $\mathcal{I} \subset 2^{\mathbb{N}}$ be an ideal satisfying the (AP) property. Then, an ideal statistically localized sequence is ideal statistically Cauchy sequence iff $\lambda = 0$.*

Proof. Assume that (x_n) is an ideal statistically Cauchy sequence in 2-normed space $(X, \|.,.\|)$. Then, there is a set $R = \{r_1 < r_2 < ... < r_n\} \subset \mathbb{N}$ such that $R \in \mathcal{F}(\mathcal{I})$ and $st\text{-}\lim_{n,m\to\infty} \|x_{r_n} - x_{r_m}, z\| = 0$. As a consequence, there is a $n_0 \in \mathbb{N}$ for every $\varepsilon > 0$, $\delta > 0$ and every $z \in X$ such that

$$\frac{1}{k}\left|\{n \leq k : \|x_n - \xi, z\| \geqslant \varepsilon\}\right| < \delta$$

for all $n \geq n_0$. Because (x_n) is ideal statistically localized sequence, $\mathcal{I}$-st-$\lim_{n\to\infty} \left\|x_n - x_{r_{n_0}}, z\right\|$ exists and we get

$$\mathcal{I}\text{-}st\text{-}\lim_{n\to\infty} \left\|x_n - x_{r_{n_0}}, z\right\| \leq \delta.$$

Therefore, $\lambda \leq \delta$. Because $\delta > 0$ is arbitrary, we obtain $\lambda = 0$.

Now assume the $\lambda = 0$. Then, there is a $x \in loc_{\mathcal{I}_{st}}(x_n)$ for every $\varepsilon > 0$, $\delta > 0$ and every nonzero $z \in X$ such that

$$\|x, z\| = \mathcal{I}\text{-}st\text{-}\lim_{n\to\infty} \|x - x_n, z\| < \frac{\delta}{2}.$$

Then

$$\left\{k \in \mathbb{N} : \frac{1}{k}\left|\{n \leq k : |\|x, z\| - \|x - x_n, z\|| \geqslant \varepsilon\}\right| > \frac{\delta}{2} - \|x, z\|\right\} \in \mathcal{I}.$$

Hence, we have

$$\left\{k \in \mathbb{N} : \frac{1}{k}\left|\{n \leq k : \|x - x_n, z\| \geqslant \varepsilon\}\right| > \frac{\delta}{2}\right\} \in \mathcal{I}.$$

So, $\mathcal{I}$-st-$\lim_{n\to\infty} \|x - x_n, z\| = 0$, which shows that (x_n) is ideal statistically Cauchy sequence. □

Definition 7. *Let (x_n) be a sequence in a 2-linear normed space. It is called as the uniformly $\mathcal{I}$-statistically localized on a subset $M \subset X$ if the sequence $\{\|x - x_n, z\|\}$ is uniformly $\mathcal{I}$-statistically converges for all $x, z \in M$.*

Proposition 5. *Let the sequence (x_n) be uniformly $\mathcal{I}$ -statistically localized on the set $M \subset X$ and $w \in Y$ is such that for every $\varepsilon > 0$, $\delta > 0$ and every nonzero z in M there is $y \in M$ such that*

$$\left\{k \in \mathbb{N} : \frac{1}{k}\left|\{n \leq k : |\|w - x_n, z\| - \|y - x_n, z\|| \geqslant \varepsilon\}\right| > \delta\right\} \in \mathcal{I}$$

Then $w \in loc_{\mathcal{I}_{st}}(x_n)$ and (x_n) is uniformly $\mathcal{I}$-statistically localized on the set of such points w.

Since the proof of Proposition is analog to Proposition 2, we omit it.

References

1. Das, P., Savaş, E., Ghosal, S.: On generalized of certain summability methods using ideals. Appl. Math. Lett. **26**, 1509–1514 (2011)
2. Fast, H.: Sur la convergence statistique. Colloq. Math. **2**, 241–244 (1951)
3. Gähler, S.: 2-metrische Räume und ihre topologische Struktur. Math. Nachr. **26**, 115–148 (1993)
4. Gürdal, M.: On ideal convergent sequences in 2-normed spaces. Thai J. Math. **4**(1), 85–91 (2006)
5. Gürdal, M., Açık, I.: On $\mathcal{I}$-Cauchy sequences in 2-normed spaces. Math. Inequal. Appl. **2**(1), 349–354 (2008)
6. Gürdal, M., Yamancı, U.: Statistical convergence and some questions of operator theory. Dyn. Syst. Appl. **24**, 305–312 (2015)
7. Kostyrko, P., Šalát, T., Wilezynski, W.: $\mathcal{I}$-Convergence. Real Anal. Exch. **26**(2), 669–686 (2000)
8. Krivonosov, L.N.: Localized sequences in metric spaces. Izv. Vyssh. Uchebn. Zaved. Mat. 4, 45–54 (1974). Soviet Math. (Iz. VUZ), 18(4), 37–44 (1974)
9. Mursaleen, M., Mohiuddine, S.A., Edely, O.H.H.: On ideal convergence of double sequences in intuitionistic fuzzy normed spaces. Comput. Math. Appl. **59**, 603–611 (2010)
10. Mursaleen, M., Alotaibi, A.: On $\mathcal{I}$-convergence in random 2-normed space. Math. Slovaca **61**(6), 933–940 (2011)
11. Nabiev, A., Pehlivan, S., Gürdal, M.: On $\mathcal{I}$-Cauchy sequences. Taiwanese J. Math. **11**(2), 569–566 (2007)
12. Nabiev, A.A., Savaş, E., Gürdal, M.: Statistically localized sequences in metric spaces. J. App. Anal. Comp. **9**(2), 739–746 (2019)
13. Nabiev, A.A., Savaş, E., Gürdal, M.: $\mathcal{I}$-localized sequences in metric spaces. Facta Univ. Ser. Math. Inform. (to appear)
14. Raymond, W.F., Cho, Y.J.: Geometry of Linear 2-Normed Spaces. Nova Science Publishers, Huntington (2001)
15. Şahiner, A., Gürdal, M., Saltan, S., Gunawan, H.: Ideal convergence in 2-normed spaces. Taiwanese J. Math. **11**(4), 1477–1484 (2007)
16. Schoenberg, I.J.: The integrability of certain functions and related summability methods. Am. Math. Mon. **66**, 361–375 (1959)
17. Steinhaus, H.: Sur la convergence ordinaire et la convergence asymptotique. Colloq. Math. **2**, 73–74 (1951)
18. Yamancı, U., Gürdal, M.: $\mathcal{I}$-statistical convergence in 2-normed space. Arab J. Math. Sci. **20**(1), 41–47 (2014)
19. Yamancı, U., Gürdal, M.: $\mathcal{I}$-statistically pre-Cauchy double sequences. Global J. Math. Anal. **2**(4), 297–303 (2014)
20. Yamancı, U., Gürdal, M.: Statistical convergence and operators on Fock space. New York J. Math. **22**, 199–207 (2016)
21. Yamancı, U., Nabiev, A.A., Gürdal, M.: Statistically localized sequences in 2-normed spaces. Honam Math. J. (to appear)
22. Yamancı, U., Savaş, E., Gürdal, M.: $\mathcal{I}$-localized sequences in 2-normed spaces. Malays. J. Math. Sci. (to appear)
23. Yegül, S., Dündar, E.: On statistical convergence of sequences of functions in 2-normed spaces. J. Classical Anal. **10**(1), 49–57 (2017)

On the Notion of Structure Species in the Bourbaki's Sense

Aslanbek Naziev(✉)

Ryazan State University, Ryazan, Russian Federation
a.naziev@365.rsu.edu.ru

Abstract. The theory of the structure species [2–5] in the treatise by Bourbaki [2], although it takes only a few pages [2, pp. 259–264], yet is quite difficult. However, in exercises, Bourbaki mentioned another approach based not on the concept of a step design scheme, as in the main text, but on the concept of type of step. Wherein, relying in reasonable limits on common sense, it is possible to act without a tricky concept of a balanced string of signs, and thus to obtain a wholly accessible and at the same time reasonably strict presentation. To the realisation of this plan (according to [7]) dedicated our work. Willing to make the presentation as accessible as possible, we have included a large number of examples.

Keywords: Type of structure · Typification · Transportability

> Today when we look at the evolution of mathematics for the last two centuries, we cannot help seeing that since about 1840 the study of specific mathematical objects has been replaced more and more by the study of mathematical structures.
>
> (Dieudonné)

1 Introduction

When studying various mathematical objects—groups, rings, fields, vector spaces, etc.,—they usually emphasize that the nature of the elements that make up these objects is completely insignificant, and also note that it is customary not distinguish between isomorphic objects, considering them as different copies of the same object. Sometimes even these two propositions called the hallmarks of modern mathematics.

In the university courses in algebra, geometry, and mathematical analysis, these propositions usually communicate in the form of a kind of psychological

D. J. Hemanth and U. Kose (Eds.): ICAIAME 2019, LNDECT 43, pp. 1047–1069, 2020.
https://doi.org/10.1007/978-3-030-36178-5_93

attitudes, not in the form of rules that have exact mathematical meaning. And this is understandable, for the exact meaning the named provisions acquired not within the frameworks of specific mathematical theories that study certain mathematical objects, but within some special theory—the general theory of the species of structure. To the beginnings of this theory dedicated our work.

It would seem that in order to put an exact mathematical sense in the statement on the irrelevance of the nature of the elements that make up mathematical objects one must first give a precise definition of the elements' nature. However, it seems at least uneconomic to work on defining something just for saying, "Namely this is for us irrelevant." Therefore, it is not surprising that usually, they act differently.

If it is completely irrelevant, of which particular elements the object is arranged, then any its element can be replaced by any other object, so long as the object is different from all the other elements that make up the object in question. It is clear that through such substitutions one can transfer from the given set to any other set containing "as many elements, as much is given." "As many, as much" in mathematical language means the existence of a bijection of the given set onto this "other" set. Thus, the irrelevance of the nature of elements that make up mathematical objects means that if there is some structure on the given set, and there exists a bijection of this set onto another set, then on this other set exists "the same" structure. The property just described is called portability and is fundamental in the theory of structure species (built by N. Bourbaki). Moreover, it is formulated in such a way that from the definition it immediately follows that isomorphic objects have the same theories, so it immediately gets the exact meaning to the second of the provisions mentioned at the beginning of our work.

In recent decades, more than once attempts have been made to bring the theory of Bourbaki structures to broad audience. All these attempts seem to us unsuccessful. The point here is not that there were many flaws in what was said. The fact is that while this was said nothing about the crucial concept of the theory of structures, that of portability. To say what is a structure is not at all difficult.[1] It is necessary and important further (and to make it much more difficult!) to explain what the words "the same structure" mean. This is done using the concept of portability, which eventually leads to the concept of structure species.

---

[1] However, it was not without curiosities: in a number of works under the name of the structure (and with reference to Bourbaki!) authors determined not the concept of structure, but the concept of an algebraic system. One such work was defended in 2015 as a doctoral dissertation on the methods of mathematics teaching. Especially funny is that in this work its author writes: ""structures are tools of mathematics", and only through them can mathematics be systematized to a certain extent, can be given a general idea of it." It turns out that only through algebraic systems can mathematics be systematized to a certain extent, can be given a general idea of it! But what have we to do with topology, probability theory, mathematical statistics and all other non-algebraic divisions of mathematics?

On the other hand, the theory of Bourbaki's structures (at least, in Russia (USSR)) was repeatedly attacked. The understanding of the structure according to N. Bourbaki was called limited and one-sided. Offered to expand it. For example, it was expressed (by a highly respected mathematician) a wish to include in the concept of structure mathematical models of real phenomena. This wish seems to us at least obscure. In fact, aren't the mathematical models of real phenomena elementary functions, differential and integral equations, topological vector (and, in particular, Hilbert) spaces, linear operators and their spectra, Banach algebras, measures (in particular, the Wiener measure)? But all this is stated in the treatise by Bourbaki.

Another objection to Bourbaki, expressed by one Russian methodologist, is that the theory of Bourbaki allegedly does not cover combinatorial structures. This declaration is not true at all. Elementary combinatorics Bourbaki himself perfectly described in the first volume of his treatise, the higher divisions explained Bannai and Ito [1] in their famous monograph, written in the same language as "Spectral theories" by Bourbaki. Regarding these and similar attempts, the following should be noted.

Those who wish to specify examples of structures that are not covered by the Bourbaki's theory should be aware of the grandiosity of the problem they take to solve. Bourbaki singled out not the "three structures", how think and say many, but three base varieties of the structure species. Moreover, Bourbaki himself admitted that maybe someday the structure species will be discovered not reducible to the species of these three varieties. However, so far, this has not happened, and it is unlikely to happen soon. To understand the reason, sufficiently to recall the history of attempts to solve in radicals the fifth degree equations, and how difficult it was to establish that not all such equations are solvable by radicals. Even nowadays, there is no simple proof of this fact. Whereas in the case of structures, it is necessary to prove—namely, to prove, for the Bourbaki's theory of structures is a mathematical theory—the existence of a structure species not reducible to the species of the three species varieties described by Bourbaki. The task is immeasurably more difficult if at all solvable.

Another (veiled) direction of attacks on Bourbaki's theory consists of the propositions to build instead of it something more suitable, according to the opinions of propagators. For example, somebody offers to consider mathematics as the science of mathematical models, or the types of mathematical models, or the theory of mathematical schemes, or the theory of the types of mathematical schemes. These authors, with all their high authority, do not notice (or not want to notice) one very significant circumstance: they express their proposals on a heuristic level, while Bourbaki built a mathematical theory of mathematical structures. And there is no doubt: if these authors took the trouble to realize their ideas in the form of mathematical theories (for example, the theory of types of mathematical models), then they would have turned out to be practically the same thing as Bourbaki, only with different names.

In this paper, we use the following notation:

Logical and set-theoretic notation—as in [6];

$\mathscr{S}$—for the class of all sets;

Δ—for the diagonal relation (identical function) on $\mathscr{S}$;

$\mathscr{P}(X)$—for the set of all subsets of the set X;

I_X—for the identity function on the set X; if it is clear or unimportant what is X in question, we simply write I.

Because we are speaking *about* theories, we are forced to distinguish between language of a theory (called *object language*, or simply *language*) and *metalanguage*—the language we use to speak about object language. As metalanguage we use ordinary conversational English and some notation. For example, in metalanguage we use Greek letters:

'α', 'β', 'γ', 'δ'—for variables of object language;

'τ', 'υ'—for terms,

'φ', 'ψ', 'χ'—for formulas.

Under set theory we understand the theory, outlined in the Appendix of [6].

Other notation is introduced along the way.

2 Steps and Structures

Definition 1. *Let $X_1, \ldots, X_n$ be sets. The concept of a* step over the sets $X_1, \ldots, X_n$ *introduced by the following agreements.*

(S1) Each of the sets $X_1, \ldots, X_n$ is a step over the sets $X_1, \ldots, X_n$.

(S2) If X is a step over the sets $X_1, \ldots, X_n$, then $\mathscr{P}(X)$ is a step over $X_1, \ldots, X_n$.

(S3) If X, Y are steps over $X_1, \ldots, X_n$, then $X \times Y$ is a step over $X_1, \ldots, X_n$.

(S4) There are no other steps over $X_1, \ldots, X_n$.

The set of all steps over $X_1, \ldots, X_n$ is called the staircase *of sets built over $X_1, \ldots, X_n$, and step elements*—structures *over $X_1, \ldots, X_n$.*

Examples of Steps and Structures. 1. Any set

$$\begin{aligned}&X_i,\ X_i \times X_j,\ X_i \times (X_j \times X_k),\\ &\mathscr{P}(X_i),\ \mathscr{P}(X_i \times X_j),\ \mathscr{P}(X_i \times (X_j \times X_k)),\\ &X_j \times \mathscr{P}(X_j \times X_k),\ \mathscr{P}(X_i \times X_j) \times \mathscr{P}(X_i \times (X_j \times X_k))\end{aligned}$$

is a step over $X_1, \ldots, X_n$. But any step S over $X_1, \ldots, X_n$ is an element of the step $\mathscr{P}(S)$, therefore, it is a structure over $X_1, \ldots, X_n$. This means that all mentioned sets are also structures over $X_1, \ldots, X_n$.

2. As noted in example 1, any set $X_1, \ldots, X_n$ is a step over $X_1, \ldots, X_n$. So, any element of any set from $X_1, \ldots, X_n$ is a structure on these sets. Such structures are called *selected elements*.

3. ($n = 1$) The order relation on $\mathbb{R}$ is a structure on this set. Indeed, this is a subset in $\mathbb{R} \times \mathbb{R}$ and therefore an element of the step $\mathscr{P}(\mathbb{R} \times \mathbb{R})$. In general, any relation on any set is a structure on this set.

4. ($n = 1$) The operation of addition on $\mathbb{R}$ is a structure on this set. Indeed, this operation is a mapping of $\mathbb{R} \times \mathbb{R}$ to $\mathbb{R}$, that is, a subset in $(\mathbb{R} \times \mathbb{R}) \times \mathbb{R}$, and therefore—an element of the step $\mathscr{P}((\mathbb{R} \times \mathbb{R}) \times \mathbb{R})$. In general, any operation on any set is a structure on this set.

5. ($n = 1$) The topology on $\mathbb{R}$ is a structure on this set. Indeed, the topology on $\mathbb{R}$ is the set of all open subsets in $\mathbb{R}$, therefore, a subset of the set $\mathscr{P}(\mathbb{R})$ and therefore an element of the step $\mathscr{P}(\mathscr{P}(\mathbb{R}))$. In general, any topology on any set is a structure on this set.

6. ($n = 1$) Let ω be a binary operation, ρ—a binary relation on the set X. Then $\omega \in \mathscr{P}((X \times X) \times X)$, $\rho \in \mathscr{P}(\mathscr{P}(X))$, so, $(\omega,\ \rho) \in \mathscr{P}((X \times X) \times X) \times \mathscr{P}(\mathscr{P}(X))$. Thus, $(\omega,\ \rho)$ is a structure on X. In general, any ordered pair of structures on any set is a structure on this set. Most often ordered pairs are made up of binary operations and relations of order, binary operations and topology, binary relations and topology.

7. ($n = 1$) Let ω be a binary operation, ρ—a binary relation, τ—a topology on a set X. Then the ordered triple $(\omega,\ \rho,\ \tau)$ is a structure on X. In general, every finite sequence of structures on arbitrary sets $X_1, \ldots, X_n$ is a structure on these sets.

8. ($n = 2$) Let V be a vector space over the field k of scalars, $+$ be the operation of addition of the vectors, and $\cdot$ be the operation of multiplication of vectors by scalars. Then $(+,\ \cdot)$ is a structure on sets $V,\ k$. By some reasons, that became clear later, they also say that $(+,\ \cdot)$ is a structure on principal base set V with auxiliary base set k.

9. ($n = 2$) Let d be a metric on a set X, that is, a map of $X \times X$ to $\mathbb{R}$ satisfying for all $x,\ y,\ z \in X$ the conditions: $d(x,\ y) \geqslant 0$; $d(x,\ y) = 0 \leftrightarrow x = y$; $d(x,\ y) = d(y,\ x)$; $d(x,\ z) \leqslant d(x,\ y) + d(y,\ z)$. Then d is a structure on X and $\mathbb{R}$. In this case they also say that d is a structure on principal base set X with auxiliary base set $\mathbb{R}$.

Let X_1, X_2 and Y_1, Y_2 be sets. Consider the steps $\mathscr{P}(X_1 \times X_2)$ over X_1, X_2 and $\mathscr{P}(Y_1 \times Y_2)$ over Y_1, Y_2. These are, generally speaking, different steps over different sets. Nevertheless, we clearly see that these steps are build according to one rule, or, as also they say in such cases, these steps have the same type. Our nearest goal is to define the notion of the type of step. But first we need some preliminary definitions.

One way to build a step over a given sequence of sets consists of choosing one of them: given $X_1, \ldots, X_n$ choose some X_i, where i is one of the numbers $1, \ldots, n$. Denote this operation by pr_i^n, that is, define pr_i^n as the function on $\mathscr{S}^n$ with values in $\mathscr{S}$, acting by the rule $pr_i^n(X_1, \ldots, X_n) = X_i$, $i = 1, \ldots, n$. In other notation,

$$pr_i^n : \begin{cases} \mathscr{S}^n \to \mathscr{S}, \\ (X_1, \ \ldots, \ X_n) \rightsquigarrow X_i. \end{cases}$$

If it is clear what an n discussed, then the superscript is not written.

Another way to build a step is that having a ready-made step X, they form a step $\mathscr{P}(X)$. The operation of forming a set $\mathscr{P}(X)$ by a given set X is a function $\mathscr{P} : \mathscr{S} \to \mathscr{S}$.

Finally, one more way of constructing a step is that having ready-made steps S_1, S_2, they form a step $S_1 \times S_2$. The operation of moving from data $X_1 \ \ldots, \ X_n$ to $S_1 \times S_2$ is the product of operations of moving from data $X_1 \ \ldots, \ X_n$ to S_1 and also from them to S_2; the product—in the sense of the following definition.

Definition 2. *Let F_1 and F_2 be maps of $\mathscr{S}^n$ to $\mathscr{S}$.* The product of F_1 *and* F_2 *is the map*

$$F_1 \otimes F_2 : \mathscr{S}^n \to \mathscr{S},$$

acting according to the rule

$$(F_1 \otimes F_2)(X_1 \ \ldots, \ X_n) = F_1(X_1 \ \ldots, \ X_n) \times F_2(X_1 \ \ldots, \ X_n).$$

In other notation:

$$F_1 \otimes F_2 : \begin{cases} \mathscr{S}^n \to \mathscr{S}, \\ (X_1, \ \ldots, \ X_n) \rightsquigarrow F_1(X_1, \ \ldots, \ X_n) \times F_2(X_1, \ \ldots, \ X_n). \end{cases}$$

We are now ready to define the notion of a step type. By this definition, each step type over n sets turns out to be some mapping of $\mathscr{S}^n$ into $\mathscr{S}$, but, of course, not all such mappings are step types. Which from them are, stated in the following definition.

Definition 3. *The notion of the* step type over n sets *is introduced by the following conventions.*

1. *Every map $p_j^n : \mathscr{S}^n \to \mathscr{S}$, $i = 1, \ \ldots, \ n$, is a type of step over n sets.*
2. *If T is a type of step over n sets, then also $\mathscr{P} \circ T$ is.*
3. *If S, T are types of step over n sets, then also $S \otimes T$ is.*
4. *No other types of step over n sets exist.*

Definition 4. *Let T be a type of step over n sets and $X_1, \ \ldots, \ X_n \in \mathscr{S}$. The set $T(X_1, \ldots, X_n)$, that is, the value of the map T at the point $(X_1, \ldots, X_n)$, is called the realization of the type T on sets $X_1, \ \ldots, \ X_n$. Clearly, it is a step over the given sets, so, its elements are the structures over the sets $X_1, \ \ldots, \ X_n$. We shall call them* structures of type T on sets $X_1, \ \ldots, \ X_n$.

Examples. I. *Step types over a single set.*

1. $\underline{T = pr_1^1}$. This is Δ. Acts according to the rule: $T(X) = X$. Structures of this type are selected elements (of the principal base set).

2. $\underline{T = \mathscr{P}}$. It is a type of step (over a single set), because Δ is, and $\mathscr{P} = \mathscr{P} \circ \Delta$. It acts according to the rule $T(X) = \mathscr{P}(X)$. Structures of this type are subsets of the principal base set.

3. $\underline{T = \Delta \otimes \Delta}$. Acts according to the rule $T(X) = X \times X$. Structures of this type are ordered pairs of elements of the principal base set.

4. $\underline{T = \Delta \otimes \mathscr{P}}$. Acts according to the rule:

$$T(X) = X \times \mathscr{P}(X).$$

Structures of this type are ordered pairs formed by an element and a subset of the principal base set.

5. $\underline{T = \mathscr{P} \circ \mathscr{P}}$. Acts according to the rule:

$$T(X) = \mathscr{P}(\mathscr{P}(X)).$$

Structures of this type are collections of subsets of the main set. Such are, for example, topologies.

6. $\underline{T = \mathscr{P} \circ (\Delta \otimes \Delta)}$. Acts according to the rule:

$$T(X) = \mathscr{P}(X \times X).$$

Structures of this type are subsets of the second Cartesian degree of the main set. Such, for example, are binary relations on the principal base set.

7. $\underline{T = \mathscr{P} \circ ((\Delta \otimes \Delta) \otimes \Delta)}$. Acts according to the rule:

$$T(X) = \mathscr{P}((X \times X) \times X).$$

Structures of this type are ternary relations on the principal base set. Such are, for example, binary operations.

8. $\underline{T = (\mathscr{P} \circ \mathscr{P}) \otimes (\mathscr{P} \circ ((\Delta \otimes \Delta) \otimes \Delta))}$. Acts according to the rule:

$$T(X) = \mathscr{P}(\mathscr{P}(X)) \times \mathscr{P}((X \times X) \times X).$$

Structures of this type are, for example, ordered pairs formed by topologies and binary operations.

II. *Step types over two sets.*

1. $\underline{T = pr_j^2}$. Acts according to the rule: $T(X_1,\ X_2) = X_j$. Structures of this type are selected elements: for $j = 1$ of the first set, for $j = 2$ of the second set.

2. $\underline{T = pr_i^2 \otimes pr_j^2}$. Acts according to the rule: $T(X_1,\ X_2) = X_i \times X_j$. For example, for $i = 2$, $j = 1$, $T(X_1,\ X_2)$ is the product $X_2 \times X_1$. Structure of this type are the selected elements of the corresponding Cartesian product.

3. $T = \mathscr{P} \circ pr_j^2$. Acts according to the rule: $T(X_1, X_2) = \mathscr{P}(X_j)$, $j = 1, 2$. Structures of this type are subsets of one of the given sets: of the first, if $j = 1$; of the second, if $j = 2$.

4. $T = pr_i^2 \otimes (\mathscr{P} \circ pr_j^2)$. Acts according to the rule:

$$T(X_1, X_2) = X_i \times \mathscr{P}(X_j),$$

$j = 1, 2$. Structures of this type are ordered pairs, with elements in X_i as the first members, subsets in X_j as the second members.

5. $T = \mathscr{P} \circ (pr_i^2 \otimes pr_j^2)$. Acts according to the rule:

$$T(X_1, X_2) = \mathscr{P}(X_i \times X_j),$$

$j = 1, 2$. Structures of this type are subsets in $X_i \times X_j$, that is, binary relations between X_i and X_j.

6. $T = \mathscr{P} \circ ((pr_i^2 \otimes pr_j^2) \otimes pr_k^2)$. Let, for definiteness, $i = j = 1$, $k = 2$. Then T acts according to the rule

$$T(X_1, X_2) = \mathscr{P}((X_1 \times X_1) \times X_2).$$

The structures of this type are subsets in the Cartesian product of the square of the first set by the second set. Such, in particular, are the mappings from the square of the first set to the second set. For example, a metric on the set S is a map of $S \times S$ to $\mathbb{R}$.

3 Canonical Extensions of Functions

Definition 5. *Let $f\colon X \to Y$. Define $\mathscr{P}\langle f\rangle\colon \mathscr{P}(X) \to \mathscr{P}(Y)$ by the rule*

$$\mathscr{P}\langle f\rangle(s) = f[s].$$

So defined function is called the canonical extension of the function f to subsets.

Theorem 1 (Properties of the canonical extensions of functions to subsets).

For every sets X, Y, Z and every functions $f\colon X \to Y$, $g\colon Y \to Z$:

1° $\mathscr{P}\langle \Delta_X\rangle = \Delta_{\mathscr{P}(X)}$;
2° $\mathscr{P}\langle g \circ f\rangle = \mathscr{P}\langle g\rangle \circ \mathscr{P}\langle f\rangle$;
3° *If f is injective (surjective, bijective), then such is also $\mathscr{P}\langle f\rangle$;*
4° *If f is invertible, then such is also $\mathscr{P}\langle f\rangle$, and what is more,*

$$(\mathscr{P}\langle f\rangle)^{-1} = \mathscr{P}\langle f^{-1}\rangle.$$

Definition 6. *Now, let* $f_1\colon X_1 \to Y_1$, $f_2\colon X_2 \to Y_2$. *Define the function*

$$f_1 * f_2\colon X_1 \times X_2 \to Y_1 \times Y_2$$

by the condition:

$$(f_1 * f_2)(x_1,\ x_2) = (f_1(x_1),\ f_2(x_2)).$$

So defined function is called the canonical extension of the functions f_1 *and* f_2 *to products.*

Theorem 2 (Properties of the canonical extensions of functions to products). *For every sets* X_i, Y_i, Z_i *and functions* $f_i\colon X_i \to Y_i$, $g_i\colon Y_i \to Z_i$, $i = 1,\ 2$:

1° $\Delta_{X_1} * \Delta_{X_2} = \Delta_{X_1 \times X_2}$.
2° $(g_1 \circ f_1) * (g_2 \circ f_2) = (g_1 * g_2) \circ (f_1 * f_2)$.
3° *If* f_1, f_2 *are injective (surjective, bijective), then* $f_1 * f_2$ *also.*
4° *If* f_1, f_2 *are invertible, then* $f_1 * f_2$*—also, and besides*

$$(f_1 * f_2)^{-1} = f_1^{-1} * f_2^{-1}.$$

Definition 7. *At last, let* T *be arbitrary type of step over* n *sets* $(n \geqslant 1)$ *and* $f_i\colon X_i \to Y_i$, $i = 1,\ \ldots,\ n$. *Define a function*

$$T\langle f_1,\ \ldots,\ f_n\rangle : T(X_1,\ \ldots,\ X_n) \to T(Y_1,\ \ldots,\ Y_n)$$

by the condition

$$T\langle f_1,\ \ldots,\ f_n\rangle = \begin{cases} f_i, \text{ if } T = pr_i^n; \\ \mathscr{P}\langle T_1\langle f_1,\ \ldots,\ f_n\rangle\rangle, \text{ if } T = \mathscr{P} \circ T_1; \\ T_1\langle f_1,\ \ldots,\ f_n\rangle * T_2\langle f_1,\ \ldots,\ f_n\rangle, \text{ if } T = T_1 \otimes T_2. \end{cases}$$

So defined function $T\langle f_1,\ \ldots,\ f_n\rangle$ *is called* the canonical extension of functions $f_1,\ \ldots,\ f_n$ *by type* T.

Theorem 3 (General properties of the canonical extensions of functions). *For every sets* X_i, Y_i, Z_i *and functions* $f_i\colon X_i \to Y_i$, $g_i\colon Y_i \to Z_i$, $i = 1,\ \ldots,\ n$:

1° $T\langle I_{X_1},\ \ldots,\ I_{X_n}\rangle = I_{T\langle X_1\ \ldots,\ X_n\rangle}$;
2° $T\langle g_1 \circ f_1,\ \ldots,\ g_n \circ f_n\rangle = T\langle g_1,\ \ldots,\ g_n\rangle \circ T\langle f_1,\ \ldots,\ f_n\rangle$;
3° *If* $f_1,\ \ldots,\ f_n$ *are injective (surjective, bijective), then* $T\langle f_1\ \ldots,\ f_n\rangle$ *is such also;*
4° *If* $f_1,\ \ldots,\ f_n$ *are invertible, then* $T\langle f_1\ \ldots,\ f_n\rangle$*—also, and moreover*

$$(T\langle f_1\ \ldots,\ f_n\rangle)^{-1} = T\langle f_1^{-1}\ \ldots,\ f_n^{-1}\rangle.$$

Simple proofs of the Theorems T3.1–T3.3 are left to the reader as exercises.

Remark. One would like to write '$\mathscr{P}(f)$' instead of '$\mathscr{P}\langle f\rangle$', '$S \times T$' instead of '$S \otimes T$', and '$T(f_1, \ldots, f_n)$' instead of '$T\langle f_1, \ldots, f_n\rangle$'. We are forced to refuse this suggestion, because of, all, '$\mathscr{P}(f)$', '$S \times T$', and '$T(f_1, \ldots, f_n)$', were already defined earlier; the first, as the powerset of f, the second, as the Cartesian product of S and T, the third, as the realization of the type of step T on the sets $f_1, \ldots, f_n$.

Examples of the Canonical Extensions of Functions. 1. $\underline{n = 1,\ T = \Delta}$. Let $X, X' \in \mathscr{S}$, $f\colon X \to X'$. Then

$$\begin{aligned} &T(X) = X,\ T(X') = X',\\ &T\langle f\rangle = \Delta\langle f\rangle = pr_1^1\langle f\rangle = f. \end{aligned}$$

So, to each structure $s \in T(X) = X$, $T\langle f\rangle$ assigns structure $s' \in T\langle f\rangle(s) = f(s) \in T(X') = X'$. For example, if $X = Y = \mathbb{R}$, $f = \lambda_x(x-3)$ and $s = 1$, then $s' = -2$.

2. $\underline{n = 1,\ T = \mathscr{P}}$. Let $X, X' \in \mathscr{S}$, $f\colon X \to X'$. Then

$$\begin{aligned} &T(X) = \mathscr{P}(X),\ T(X') = \mathscr{P}(X'),\\ &T\langle f\rangle = \mathscr{P}\langle f\rangle\colon \mathscr{P}(X) \to \mathscr{P}(X'). \end{aligned}$$

For every $s \in T(X) = \mathscr{P}(X)$ we have $T\langle f\rangle(s) = p\langle f\rangle(s) = f[s]$. Thereby, in this case, to each structure $s \in T(X)$, $T\langle f\rangle$ assigns the image of s with respect to f.

For example, let $X = X' = \mathbb{R}$, $f = \lambda_x|x|$ and $s = [-3; 2]$. Then

$$T\langle f\rangle(s) = f[s] = \{|x|\colon x \in [-3; 2]\} = [0; 3].$$

3. $\underline{n = 1,\ T = \Delta \otimes \Delta}$. Let $X, X' \in \mathscr{S}$, $f\colon X \to X'$. Then

$$\begin{aligned} &T(X) = X \times X,\ T(X') = X' \times X',\\ &T\langle f\rangle = (\Delta \otimes \Delta)\langle f\rangle = \Delta\langle f\rangle * \Delta\langle f\rangle = f * f. \end{aligned}$$

To each $s = (s_1, s_2) \in T(X) = X \times X$, the mapping $T\langle f\rangle$ assigns

$$s' = T\langle f\rangle(s) = (f * f)(s) = (f * f)(s_1, s_2) = (f(s_1), f(s_2)).$$

For example, if $X = X' = \mathbb{R}$, $f = \lambda_x|x|$ and $s = (1; -4)$, then $s' = (1, 4)$.

4. $\underline{n = 1,\ T = \mathscr{P} \circ \mathscr{P}}$. For all $X, X' \in \mathscr{S}$ and $f\colon X \to X'$ we have:

$$\begin{aligned} &T(X) = \mathscr{P}(\mathscr{P}(X)),\ T(X') = \mathscr{P}(\mathscr{P}(X')),\\ &T\langle f\rangle\colon \mathscr{P}(\mathscr{P}(X)) \to \mathscr{P}(\mathscr{P}(X')). \end{aligned}$$

To each $s \in T(X)$ the mapping $T\langle f\rangle$ assigns

$$\begin{aligned} s' &= T\langle f\rangle(s) = \mathscr{P}\langle \mathscr{P}\langle\langle f\rangle\rangle = \mathscr{P}\langle f\rangle[s]\\ &= \{\mathscr{P}\langle f\rangle(u)\colon u \in s\} = \{f[u]\colon u \in s\}. \end{aligned}$$

Therefore, in this case $T\langle f\rangle(s)$ consists of the images relative to f of all sets, belonging to s. If $X = X' = \mathbb{R}$, s is usual topology on $\mathbb{R}$ and $f = \lambda_x(x-3)$, that is, the function $x \rightsquigarrow (x-3)$ on $\mathbb{R}$, then s' is again the same topology, because elements of structure s' appear from the elements of structure s by translation on 3 units to the left, while the topology of the real line, as is known, is translation invariant. At the same time, for $f = \lambda_x|x|$ we will have s' different from s. Moreover, elements of s' will not necessarily be the open sets of the real line. For example, if $u = (-1;1)$, then u is open, but $f[u] = [0;1)$ is not open.

5. $\underline{n = 1,\ T = \mathscr{P} \circ (\Delta \otimes \Delta)}$. For all $X, X' \in \mathscr{S}$, $f\colon X \to X'$:

$$\begin{aligned} &T(X) = \mathscr{P}(X \times X),\ T(X') = \mathscr{P}(X' \times X'),\\ &T\langle f\rangle = \mathscr{P}\langle(\Delta \otimes \Delta)\langle f\rangle\rangle = \mathscr{P}\langle \Delta\langle f\rangle * \Delta\langle f\rangle\rangle = \mathscr{P}\langle f * f\rangle. \end{aligned}$$

To every $s \in T(X) = \mathscr{P}(X \times X)$ the mapping $T\langle f\rangle$ assigns

$$\begin{aligned} s' = T\langle f\rangle(s) &= \mathscr{P}\langle(f * f)\rangle(s) = (f * f)[s]\\ &= \{(f(x),\ f(y))\colon (x,\ y) \in s\}. \end{aligned}$$

From this it is seen, that

$$\begin{aligned} (x',\ y') \in s' &\Leftrightarrow (\exists x)(\exists y)(x' = f(x) \wedge y' = f(y) \wedge (x,\ y) \in s)\\ &\Leftrightarrow (\exists x)(\exists y)((x,\ x') \in f \wedge (y,\ y') \in f \wedge (x,\ y) \in s)\\ &\Leftrightarrow (\exists x)(\exists y)((x',\ x) \in f^{-1} \wedge (x,\ y) \in s \wedge (y,\ y') \in f)\\ &\Leftrightarrow (\exists x)((x',\ x) \in f^{-1} \wedge (\exists y)((x,\ y) \in s \wedge (y,\ y') \in f))\\ &\Leftrightarrow (\exists x)((x',\ x) \in f^{-1} \wedge (x,\ y') \in f \circ s)\\ &\Leftrightarrow (x',\ y') \in f \circ s \circ f^{-1}. \end{aligned}$$

This means that $s' = f \circ s \circ f^{-1}$. The relationship between s and s' become especially transparent if s and f^{-1} are functions (that is, s function and f invertible function). Then this relationship may be represented by a commutative diagram

$$\begin{array}{ccc} X & \xrightarrow{\ s\ } & X \\ {\scriptstyle f^{-1}}\uparrow & & \downarrow{\scriptstyle f} \\ X' & \xrightarrow{\ s'\ } & X'\,. \end{array}$$

(Commutativity of this diagram consists precisely in that $s' = f \circ s \circ f^{-1}$.)

Besides, if f is a bijection, then, regardless of whether s is a function or no,

$$\begin{aligned} (x',\ y') \in s' &\Leftrightarrow (\exists x,\ y)(x' = f(x) \wedge y' = f(y) \wedge (x,\ y) \in s)\\ &\Leftrightarrow (\exists x,\ y)(x = f^{-1}(x') \wedge y = f^{-1}(y') \wedge (x,\ y) \in s)\\ &\Leftrightarrow (f^{-1}(x'),\ f^{-1}(y')) \in s. \end{aligned}$$

When depicting $T\langle f\rangle(s)$ in a drawing, it is useful to remember n. 2° of T3.2. By this statement,

$$\begin{aligned} f * f &= (f * I) \circ (I * f)\\ &= (I * f) \circ (f * I). \end{aligned}$$

This property allows to transit from s to s' in two stages. First, to find the image of s with respect to $I * f$, second—the image of the resulting set with respect to $f * I$. Or, if one prefers, first, to find the image of s with respect to $I * f$, and second,—the image of the resulting set with respect to $f * I$.

For $X = X' = \mathbb{R}$ and $f = \lambda_x |x|$, for example, proceeding according to the upper equality will mean that, first, one reflects by the horizontal axis those parts of s, which lie below this axis, and, after that,—reflects by vertical axis those parts of the resulting set, which lie to the left from this axis.

If to take as s the relation of non-strict order on $\mathbb{R}$, then as the result of these actions will be obtained the first quadrant (and so, not the order relation).

And if to take as s the parabola, determined by the equation $y = x^2 + 2x - 3$, which graph is depicted on Fig. 1(a) below, then as the result of these transformations will be obtained the curve depicted on Fig. 1(b).

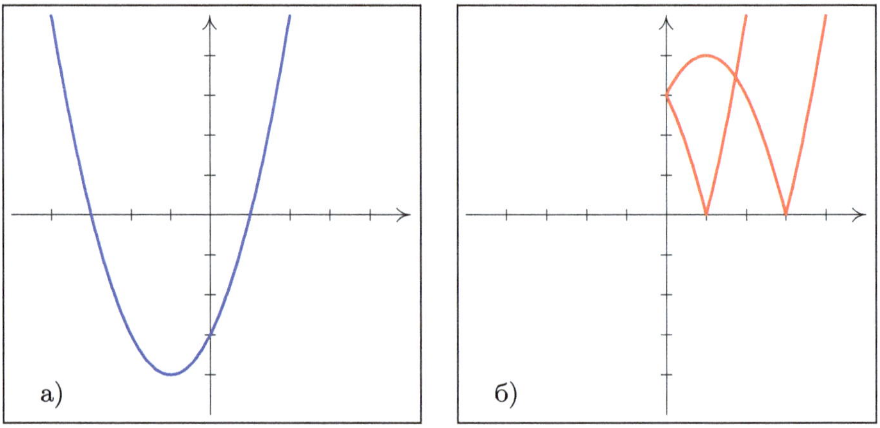

Fig. 1. Parabola and its image

6. $\underline{n = 1,\ T = \mathscr{P} \circ ((\Delta \otimes \Delta) \otimes \Delta).}$ For all $X, X' \in \mathscr{S}$, $f\colon X \to X'$:

$$T(X) = \mathscr{P}((X \times X) \times X),\ T(X') = \mathscr{P}((X' \times X') \times X'),$$
$$T\langle f\rangle = \mathscr{P}\langle((\Delta \otimes \Delta) \otimes \Delta)\langle f\rangle\rangle = \mathscr{P}\langle(f * f) * f\rangle.$$

To every $s \in T(X)$ the mapping $T\langle f\rangle$ assigns

$$\begin{aligned} s' &= T\langle f\rangle(s) = \mathscr{P}\langle((f * f) * f)\rangle(s) = ((f * f) * f)[s] \\ &= \{((f(x),\ f(y)),\ f(z))\colon ((x,\ y),\ z) \in s\}. \end{aligned}$$

From this it is seen that

$$
\begin{aligned}
&((x',\ y'),\ z') \in s' \Leftrightarrow \\
&\quad \Leftrightarrow (\exists x, y, z)(x' = f(x) \wedge y' = f(y) \wedge z' = f(z) \wedge ((x,\ y),\ z) \in s) \\
&\quad \Leftrightarrow (\exists x, y, z)((x,\ x') \in f \wedge (y,\ y') \in f \wedge (z,\ z') \in f \wedge ((x,\ y),\ z) \in s) \\
&\quad \Leftrightarrow (\exists x, y, z)((x',\ x) \in f^{-1} \wedge (y',\ y) \in f^{-1} \wedge ((x,\ y),\ z) \in s \wedge (z,\ z') \in f) \\
&\quad \Leftrightarrow (\exists x, y, z)((((x',\ y'),\ (x,\ y)) \in f^{-1} * f^{-1} \wedge ((x,\ y),\ z) \in s \wedge (z,\ z') \in f) \\
&\quad \Leftrightarrow (\exists x, y, z)((((x',\ y'),\ z') \in f \circ s \circ (f^{-1} * f^{-1})).
\end{aligned}
$$

By this way, $s' = f \circ s \circ (f^{-1} * f^{-1})$. If s an f^{-1} are functions (that is, s is a function and f—invertible function), then this relationship between s and s' is represented by the commutative diagram

$$
\begin{array}{ccc}
X \times X & \xrightarrow{\ s\ } & X \\
{\scriptstyle f^{-1}}\uparrow * \uparrow{\scriptstyle f^{-1}} & & \downarrow{\scriptstyle f} \\
X' \times X' & \xrightarrow{\ s'\ } & X' .
\end{array}
$$

Besides that, if f is a bijection then, independently of wether s is a function or no,

$$
\begin{aligned}
&((x',\ y'),\ z') \in s' \Leftrightarrow \\
&\quad \Leftrightarrow (\exists x, y, z))(x' = f(x) \wedge y' = f(y) \wedge z' = f(z) \wedge ((x,\ y),\ z) \in s) \\
&\quad \Leftrightarrow (\exists x, y, z)(x = f^{-1}(x') \wedge y = f^{-1}(y') \wedge z = f^{-1}(z) \wedge ((x,\ y),\ z) \in s) \\
&\quad \Leftrightarrow (f^{-1}(x'),\ f^{-1}(y'),\ f^{-1}(z')) \in s.
\end{aligned}
$$

7. $n = 2,\ T = pr_1^2$. For all $(X,\ Y),\ (X',\ Y') \in \mathscr{S}^2$, $f\colon X \to X'$ and $g\colon Y \to Y'$:

$$
\begin{aligned}
&T(X,\ Y) = X,\ T(X',\ Y') = X', \\
&T\langle f,\ g\rangle = f.
\end{aligned}
$$

To every $s \in T(X,\ Y)$ the mapping $T\langle f,\ g\rangle$ assigns the value s' of the function f at the point s.

8. $n = 2,\ T = pr_1^2 \otimes pr_2^2$. For all $(X,\ Y),\ (X',\ Y') \in \mathscr{S}^2$, $f\colon X \to X'$ and $g\colon Y \to Y'$:

$$
\begin{aligned}
&T(X,\ Y) = X \times Y,\ T(X',\ Y') = X' \times Y', \\
&T\langle f,\ g\rangle = f * g.
\end{aligned}
$$

To every $s = (x,\ y) \in T(X,\ Y) = X \times Y$ the mapping $T\langle f,\ g\rangle$ assigns $s' = (f(x),\ g(y)) \in (X',\ Y')$.

9. $n = 1,\ T = \mathscr{P} \circ (pr_1^2 \otimes pr_1^2)$. This example is analogous to the example 5. To understand this, one needs only to note that identical mapping $\Delta\colon \mathscr{S} \to \mathscr{S}$ is nothing but pr_1^1. For all

$$
\begin{aligned}
&(X,\ Y),\ (X',\ Y') \in \mathscr{S}^2, \\
&f\colon X \to X',\ g\colon Y \to Y'
\end{aligned}
$$

we have:

$$\begin{aligned} T(X,\ Y) &= \mathscr{P}(X \times Y), \\ T(X',\ Y') &= \mathscr{P}(X' \times Y'), \\ T\langle f,\ g\rangle &= \mathscr{P}\langle f * g\rangle. \end{aligned}$$

To every $s \in T(X,\ Y) = \mathscr{P}(X \times Y)$ the mapping $T\langle f,\ g\rangle$ assigns

$$\begin{aligned} s' &= T\langle f,\ g\rangle(s) = (f * g)[s] \\ &= \{(f(x),\ g(y))\colon (x,\ y) \in s\}. \end{aligned}$$

From this, it is seen that

$$\begin{aligned} (x',\ y') \in s' &\Leftrightarrow (\exists x,\ y)(x' = f(x) \wedge y' = g(y) \wedge (x,\ y) \in s) \\ &\Leftrightarrow (\exists x,\ y)((x,\ x') \in f \wedge (y,\ y') \in g \wedge (x,\ y) \in s) \\ &\Leftrightarrow (\exists x,\ y)((x',\ x) \in f^{-1} \wedge (x,\ y) \in s \wedge (y,\ y') \in g) \\ &\Leftrightarrow (x',\ y') \in g \circ s \circ f^{-1}. \end{aligned}$$

Thus, $s' = g \circ s \circ f^{-1}$. This relationship between s and s' become especially transparent when s and f^{-1} are functions (s function and f invertible function). Then f^{-1} and s' are also functions, and the relation between s', f^{-1}, s and g may be represented by the following commutative diagram:

$$\begin{array}{ccc} X & \xrightarrow{\ s\ } & Y \\ {\scriptstyle f^{-1}}\big\uparrow & & \big\downarrow{\scriptstyle g} \\ X' & \xrightarrow{\ s'\ } & Y'\ . \end{array}$$

(Commutativity of this diagram consists precisely in that $s' = g \circ s \circ f^{-1}$.)

Besides, if f and g are bijections, then, regardless of whether s is function or no,

$$\begin{aligned} (x',\ y') \in s' &\Leftrightarrow (\exists x,\ y)(x' = f(x) \wedge y' = g(y) \wedge (x,\ y) \in s) \\ &\Leftrightarrow (\exists x,\ y)(x = f^{-1}(x') \wedge y = g^{-1}(y') \wedge (x,\ y) \in s) \\ &\Leftrightarrow (f^{-1}(x'),\ g^{-1}(y')) \in s. \end{aligned}$$

If $X = Y = X' = Y' = \mathbb{R}$, $f = \lambda_x(x + 1)$ and $g = \lambda_x(x - 2)$, then sets s' come out from sets $s \in T(X,\ Y)$ by the translation to one unit right and two units down.

10. $n = 1,\ T = \mathscr{P} \circ ((pr_1^3 \otimes pr_2^3) \otimes pr_3^3)$. This example is analogous to 6. For all

$$\begin{aligned} &(X,\ Y,\ Z),\ (X',\ Y',\ Z',) \in \mathscr{S}^3, \\ &f\colon X \to X',\ g\colon Y \to Y',\ h\colon X \to Z', \end{aligned}$$

we have:

$$\begin{aligned} &T(X,\ Y,\ Z) = \mathscr{P}((X \times Y) \times Z), \\ &T(X',\ Y',\ Z') = \mathscr{P}((X' \times Y') \times Z'), \\ &T\langle f,\ g,\ h\rangle = \mathscr{P}\langle (f * g) * h\rangle. \end{aligned}$$

As in the previous example, we make sure that to every $s \in T(X,\ Y,\ Z)$ mapping $T\langle f,\ g,\ h\rangle$ assigns

$$s' = \{(f(x),\ g(y),\ h(z))\colon (x,\ y,\ z) \in s\},$$

so that $s' = h \circ s \circ (f^{-1} * g^{-1})$. Again, if s, f^{-1} and g^{-1} are functions (that is, s is function and f, g are invertible functions), then this connection between s and s' receives a visual representation in commutative diagram analogous those bringed above:

$$\begin{array}{ccc} X \times X & \xrightarrow{\ s\ } & X \\ {\scriptstyle f^{-1}}\big\uparrow * \big\uparrow{\scriptstyle g^{-1}} & & \big\downarrow{\scriptstyle h} \\ X' \times X' & \xrightarrow{\ s'\ } & X' . \end{array}$$

And if f, g and h are bijections, then, regardless of whether s is a function or no,

$$((x',\ y'),\ z') \in s' \Leftrightarrow (f^{-1}(x'),\ g^{-1}(y'),\ h^{-1}(z')) \in s.$$

4 Transport of Structures

In the previous section, we for each sequence of maps $f_i\colon X_i \to X'_i$, $i = 1,\ \ldots,\ n$, and for each type of step T over n sets determined mapping $T\langle f_1,\ \ldots,\ f_n\rangle$ of the set $T(X_1,\ \ldots,\ X_n)$ of all structures of type T on the sets $X_1,\ \ldots,\ X_n$ to the set $T(X'_1,\ \ldots,\ X'_n)$ all structures of type T on sets $X'_1,\ \ldots,\ X'_n$. Thus, specifying a sequence of mappings f_i allows to each structure $s \in T(X_1,\ \ldots,\ X_n)$ to assign a structure $s' \in T(X'_1,\ \ldots,\ X'_n)$, figuratively speaking,—to transport the structure s from the sets $X_1,\ \ldots,\ X_n$ to the sets $X'_1,\ \ldots,\ X'_n$.

When transporting structures, one should take into account that the type of structure is not determined by the structure uniquely. The same set may be considered as structures of different types on the same sets. As a structure of one type, to the set will respond during the transport someone set, as a structure of another type—generally speaking, some another set.

Example 1. For example, let $X = \{\varnothing\}$. Then $\varnothing \in X$, therefore $\varnothing$ is a structure of the type pr^1_1 on X. At the same time, $\varnothing \subset X$, so that $\varnothing$ is also structure of the type $\mathscr{P}$ on the same set X. Let now $Y = \{\{\varnothing\}\}$, and $f\colon X \to Y$ to the unique element of X assigns the unique element of Y: $f(\varnothing) = \{\varnothing\}$. If $T = pr^1_1$ we have $T\langle f\rangle(\varnothing) = f(\varnothing) = \{\varnothing\}$, while if $T = \mathscr{P}$, then $T\langle f\rangle(\varnothing) = \mathscr{P}\langle f\rangle(\varnothing) = f[\varnothing] = \varnothing$. As we see, the results are different.

Thus, when making transport of a structure, it is necessary to pick out a type of structure, it is considered to be. It is convenient to do this using of the interrelation of the form '$s \in T(X_1,\ \ldots,\ X_n)$', where in place of the letter 'T' stands some concrete string-of-signs representing some type of structure. Each formula of such a form will be called *typification* of the letter 's'. About structure

s', connected with s by interrelation $s' = T\langle f_1, \ldots, f_n\rangle(s)$, we shall say that *it is obtained from the structure s by transport along maps $f_1, \ldots, f_n$ under typification* '$s \in T(X_1, \ldots, X_n)$'. As usually, if from context is clear what a typification is meant, the reference to it will be omitted.

Transport of structures plays a significant role in mathematics. This role will become especially clear in the following section when the concept of the structure species will be defined. However, it is already possible to illustrate the role of transport of structures in mathematics with examples well known to the reader.

Example 2. Let s be the operation of addition on the set $\mathbb{R}$ of all real numbers. And let E be the natural exponent, that is, the bijection of $\mathbb{R}$ onto the set $\mathbb{R}_+^*$ of all strictly positive real numbers, acting according to the rule $E(x) = e^x$, $x \in \mathbb{R}$. Then s is a structure of type $\mathscr{P}\circ((\Delta\otimes\Delta)\otimes\Delta)$ on $\mathbb{R}$ and as such under transport along bijection E to it correspond the operation $s' = T\langle E\rangle(s)$. As establishes earlier (Sect. 3, example 6), $s' = E \circ s \circ (E^{-1} * E^{-1})$. Thus, for all $x,\ y \in \mathbb{R}_+^*$,

$$\begin{aligned} s'(x,\ y) &= E \circ s \circ (E^{-1} * E^{-1})(x,\ y) \\ &= e^{\ln x + \ln y} = x \cdot y. \end{aligned}$$

All this mean, that the operation of multiplication of positive real numbers comes out of the operation of addition of real numbers by the transport along exponent (under mentioned typification). Namely this interrelation between mentioned operations ensures the use of logarithms for needs of computations. Thus, in the base of application of logarithms for the needs of computations lies the idea of the transport of structures. (The history of invention of logarithms clearly shows that the inventors of logarithms were implicitly guided by namely this idea.)

Example 3. Another example of very intensive use of the transport of structures is the coordinate method. Indeed, the coordinate system on the set X is nothing but a bijection $x\colon X \to U$ of a set X onto some subset U of a suitable space, say, $\mathbb{R}^n$. (Compositions $pr_i^n \circ x$ are usually denoted by x_i or x^i and called coordinate functions of the coordinate system x.) The specification of the coordinate system x generates a series of mappings $T\langle x\rangle\colon T(X) \to T(U)$—one for each type of step T. This allows to each "geometric image" (i. e., simply, to structure) s on the set X to assign "the same" structure $s' = T\langle x\rangle(s)$ on the set U, and to reduce the study of s to the study of s'.

Say, the study of the transformations t of the set X is replaced by the study of the transformations t' of the set U associated with t by equalities of the form $t' = T\langle x\rangle(t)$. But in this case $t' = x \circ t \circ x^{-1}$ (see Sect. 3, example 5), and we see that t' is what is usually called the coordinate form of the transformation t.

Example 4. The transition from one coordinate system to another is also associated with the transport of structures. In fact, every change of the coordinate system on the set X is reduced to the choice of a coordinate system $u\colon U \to U_1$ on the set U. This leads to a new coordinate system $u \circ x\colon X \to U_1$ on the set X. In it to each structure s of type T on the set X correspond the structure

$s' = T\langle u \circ x\rangle(s)$ on the set U_1. But $T\langle u \circ x\rangle = T\langle u\rangle \circ T\langle x\rangle$ (T3.3, n. 2°). Therefore, $s' = T\langle u \circ x\rangle(s) = (T\langle u\rangle \circ T\langle x\rangle)(s) = T\langle u\rangle(T\langle x\rangle(s))$. In particular, for transformations we have:

$$t' = (u \circ x) \circ t \circ (u \circ x)^{-1} = u \circ (x \circ t \circ x^{-1}) \circ u^{-1}, \text{ —}$$

a formula whose derivation takes up pages in some manuals.

Of course, in order to the properties found in s', to be assigned to s (or vice versa), these properties need to be "preserved" under transport. The examples considered in the previous section show that this is not always the case: the topology does not always go into topology, the order relation—an order relation, and so on. However, it turns out that if to confine ourselves to the consideration of transports only along *bijections*, then many properties of structures will be preserved. The properties of structures that are preserved during transports along bijections are called transportable. In the next paragraph, the concept of portability will be given a precise definition, but for now, we will illustrate it with examples of the most common types of steps and most frequently encountered properties of structures.

Theorem 4 (A theorem about transport binary relations). *Let s be a binary relation on a set X, and s' comes out of s with transport along a bijection $f\colon X \to X'$. Then:*

0° $(x_1,\ x_2) \in s \Leftrightarrow (f(x_1),\ f(x_2)) \in s'$,
$(x_1',\ x_2') \in s' \Leftrightarrow (f^{-1}(x_1'),\ f^{-1}(x_2')) \in s$;

1° *s' reflexive $\Leftrightarrow$ s reflexive;*

2° *s' irreflexive $\Leftrightarrow$ s irreflexive;*

3° *s' symmetric $\Leftrightarrow$ s symmetric;*

4° *s' asymmetric $\Leftrightarrow$ s asymmetric;*

5° *s' antisymmetric $\Leftrightarrow$ s antisymmetric;*

6° *s' transitive $\Leftrightarrow$ s transitive, and so on.*

Proof. N. 0° is proved in Sect. 3, the rest are proved by simple application of n. 0°. Prove, for example, n. 5°. Let s be antisymmetric, i. e.

$$(\forall x,\ y \in X)((x,\ y) \in s \wedge (y,\ x) \in s \Rightarrow x = y).$$

Show, that s' is also antisymmetric, i. e.

$$(\forall x',\ y' \in X')((x',\ y') \in s' \wedge (y',\ x') \in s' \Rightarrow x' = y').$$

Let

$$(x',\ y') \in s' \wedge (y',\ x') \in s'.$$

Then, in view of n. 0°,

$$(f^{-1}(x'),\ f^{-1}(y')) \in s \text{ and } (f^{-1}(y'),\ f^{-1}(x')) \in s.$$

Now, antisymmetricity of s gives that $(f^{-1}(x') = f^{-1}(y'))$, and bijectivity of f—that $x' = y'$. This is what was needed. Conversely—in the same way.

Theorem 5 (A theorem on transport of binary operations). *Let s and t be binary operations on a set X, and let s' and t' come out of s and t by transport along bijection $f\colon X \to X'$. Then:*

0° $((x_1,\ x_2),\ x_3) \in s \Leftrightarrow ((f(x_1),\ f(x_2)),\ f(x_3)) \in s'$,
$((x'_1,\ x'_2),\ x'_3) \in s' \Leftrightarrow ((f^{-1}(x'_1),\ f^{-1}(x'_2)),\ f^{-1}(x'_3)) \in s$;

1° *s is binary operation* $\Leftrightarrow$ *s' is binary operation;*

2° *s is associative* $\Leftrightarrow$ *s' is associative;*

3° *s is commutative* $\Leftrightarrow$ *s' is commutative;*

4° *there exists neutral element for s* $\Leftrightarrow$ *there exists neutral element for s';*

5° [*for every element in X there exists inverse relative to s*] $\Leftrightarrow$ [*for every element in X' there exist inverse relative to s';*]

6° *t is distributive with respect to s* $\Leftrightarrow$ *t' is distributive with respect to s'.*

Proof. Nn. 0° and 1° were already proved above (Sect. 3, example 6). (This was established for every structure of type $\mathscr{P} \circ ((\Delta \otimes \Delta) \otimes \Delta)$, so, it is true both for s and t.) From this easily follow all other statements of the theorem. Let us, for example, prove n. 6°. Let s be distributive with respect to t, that is, for every $x,\ y,\ z \in X$,

$$s(x,\ t(y,\ z)) = t(s(x,\ y),\ s(x,\ z)).$$

Let us show that then s' is also distributive with respect to t', that is, for every $x',\ y',\ z' \in X'$,

$$s'(x',\ t'(y',\ z')) = t'(s'(x',\ y'),\ s'(x',\ z')).$$

When considering the example mentioned above (cf. brought there diagram), there was established, that

$$s' = f \circ s \circ (f^{-1} * f^{-1}).$$

Completely analogously,

$$t' = f \circ t \circ (f^{-1} * f^{-1}).$$

By using these equalities, the distributivity of t relative to s and the relation

$$s \circ (f^{-1} * f^{-1}) = f^{-1} \circ s',$$

(following from the first equality by virtue of invertibility of f), we obtain for all $x',\ y',\ z' \in X'$:

$$\begin{aligned} s'(x',\ t'(y',\ z')) &= (f \circ s \circ (f^{-1} * f^{-1}))(x',\ (f \circ t \circ (f^{-1} * f^{-1}))(y',\ z')) \\ &= (f \circ s)(f^{-1}(x'),\ f^{-1}((f \circ t \circ (f^{-1} * f^{-1}))(y',\ z'))) \\ &= (f \circ s)(f^{-1}(x'),\ t(f^{-1}(y'),\ f^{-1})(z')) \\ &= f(s(f^{-1}(x'),\ t(f^{-1}(y'),\ f^{-1})(z'))) \end{aligned}$$

$$
\begin{aligned}
&= f(t(s(f^{-1}(x'),\ f^{-1}(y')),\ s(f^{-1}(x'),\ f^{-1}(z'))))\\
&= (f\circ t)((s\circ(f^{-1}*f^{-1}))(x',\ y'),\ (s\circ(f^{-1}*f^{-1}))(x',\ z'))\\
&= (f\circ t)((f^{-1}\circ s')(x',\ y'),\ (f^{-1}\circ s')(x',\ z'))\\
&= (f\circ t\circ(f^{-1}*f^{-1}))(s'(x',\ y'),\ s'(x',\ z'))\\
&= t'(s'(x',\ y'),\ s'(x',\ z')).
\end{aligned}
$$

Theorem 6 (Theorem about the transport of topologies). *Let $s \in \mathscr{P}(\mathscr{P}(X))$ and let s' comes out of s by transport along bijection $f\colon X \to X'$. Then:*

0° *for every $U' \subset X'$, $U' \in s' \Leftrightarrow f^{-1}[U'] \in s$;*
1° *s is topology $\Leftrightarrow$ s' is topology;*
2° *s is separated $\Leftrightarrow$ s' is separated;*
3° *s is connected $\Leftrightarrow$ s' is connected;*
4° *s is compact $\Leftrightarrow$ s' is compact, and so on.*

Proof. 1. Recall (Sect. 2, example 5), that in the case we consider $s' = \mathscr{P}\langle\mathscr{P}\langle f\rangle\rangle(s)$. Thus

$$
\begin{aligned}
s' &= \mathscr{P}\langle f\rangle[s] = \{\mathscr{P}\langle f\rangle(U)\colon U \in s\}\\
&= \{U'\colon (\exists U)(U \in s \wedge U' = f[U])\}\\
&= \{U'\colon (\exists U)(U = f^{-1}[U'] \wedge U \in s)\},
\end{aligned}
$$

In words: to structure s' belong those and only those subsets U' of the set X', whose inverse images relative to bijection f belong to s. Now, let us begin to prove n. 1°.

Recall that a topology on a set X is an arbitrary collection of its subsets to which belong $\varnothing$, X, unions of all parts of this collection and intersections of all finite parts of this collection.

Show, that if s is topology, then s' also is topology. Let s be a topology. Then:

(a) $\varnothing \in s'$—because $f^{-1}[\varnothing] = \varnothing \in s$;
(b) $X' \in s'$—because $f^{-1}[X'] = X \in s$;
(c) if $s'_1 \subset s'$, then $\cup s'_1 \in s'$—because

$$f^{-1}[\cup s'] = \cup\{f^{-1}[U']\colon U' \in s'\} \in s;$$

(d) if $U'_1,\ \ldots,\ U'_n \in s'$, then $U'_1 \cap \cdots \cap U'_n \in s'$—by similar reason.

All this mean that s' is a topology. The converse is proved precisely in the same way.

The proofs of the remaining points of the theorem are left as exercises to readers who understand what they speak about.

We considered several types of structures defined on one set. In mathematics, they consider also structures defined on several sets. Most often there are two sets, one of which is called the principal base set and the other, auxiliary base set. On the auxiliary base set, there is usually a structure, the definition of which does not require a reference to the principal base set, and the structure on the principal base set turns out to be somehow related to the structure on the auxiliary base set. Typical examples are the structure of a vector space and structure of a metric space.

When transferring structures, auxiliary sets behave specially. Consider, for example, the structure s of vector space over a field k, defined on a set E. Then E is a principal base set, k—an auxiliary base set, and s is the ordered pair of maps

$$a\colon E \times E \to E \quad \text{and} \quad m\colon k \times E \to E,$$

where the first is the operation of addition of vectors, the second—the operation of multiplication of vectors by scalars. We can, taking arbitrary pair of bijections, $f\colon E \to E'$, $g\colon k \to k'$, transport s along $(f,\ g)$. As a result, on E' comes out a structure of vector space over the field k'. With such a transport the set k behave on equal terms with the set E. However, such "transports" in the standard considerations of vector spaces practically not found. Making the transfer of the structure of vector space from one set to another, usually change the set of vectors, but not the field of scalars. In other words, the structure of the vector space they usually transport along not arbitrary pairs of bijections, but only the pairs of the form $(f,\ I_k)$, where to the auxiliary set corresponds the identity map.

The same picture is observed in the theory of metric spaces. Transferring a metric from one set to another usually change the set of points of a metric space, but not the set where the metric takes values. Thus, in this case, they also consider only transports along the pairs of the form $(f,\ I)$, where to identity mapping answers the auxiliary set.

Let T be a type of step over $n + m$ sets, s be a structure a type T on sets $X_1,\ \ldots,\ X_n,\ Y_1,\ \ldots,\ Y_m$, $f_1\colon X_1 \to X'_1,\ \ldots,\ f_n\colon X_n \to X'_n$ be bijections, and $s' = T\langle f_1,\ \ldots,\ f_n; I_{Y_1},\ \ldots,\ I_{Y_m}\rangle(s)$. Then they say that the *structure* s' *obtained from the structure* s *by transport along* $f_1,\ \ldots,\ f_n$ *under typification* "$s \in T(X_1,\ \ldots,\ X_n,\ Y_1,\ \ldots,\ Y_m)$" *with principal base sets* $X_1,\ \ldots,\ X_n$ *and auxiliary base sets* $Y_1,\ \ldots,\ Y_m$. If it is clear from the context what kind of typification and under what principal and auxiliary base sets are we talking about, indications of this omitted.

Note that one should not give in to the sound of the word "auxiliary" and think that the auxiliary sets play some secondary or minor role. Their role is no less important than the role of the sets called principal, and the word "auxiliary" only indicates that, when transporting structures, these sets and their "own" structures do not change.

We suggest the reader formulate and prove for the structures of a vector space and metric space the theorems similar to Theorems T4.1–T4.3.

5 Structure Species

We are almost ready to give the definition of the main notion of the theory of structures after N. Bourbaki, the notion of the structure species. It only remains to clarify the notion of portability that plays a very important role in this definition. In the previous paragraph, we have already said that a property of a structure is called transportable if it is "preserved" during transfers along bijections. Now we have to give a precise definition.

We have seen that the reflexivity of binary relations is preserved during transfers along bijections. The exact form of this result stated that whenever s is a binary relation on a set X, f a bijection of X onto X' and s' is obtained by transferring s along the bijection f, so s is reflexive if and only if s' is reflexive. Thus, the portability of the reflexivity property of binary relations consists of that it is a theorem of the theory (more strict than) the theory of sets the formula:

$$[(s \in \mathscr{P}(X \times X)) \wedge (f \text{ is a bijection of } X \text{ onto } X') \wedge (s' = \mathscr{P}\langle f * f\rangle(s))] \Rightarrow$$
$$\Rightarrow ((s \text{ is reflexive relation on } X) \Leftrightarrow (s' \text{ is reflexive relation on } X)).$$

The portability of the transitivity property of binary relations consists of that it is a theorem of the theory (more strict than) the theory of sets the formula:

$$[(s \in \mathscr{P}(X \times X)) \wedge (f \text{ is a bijection of } X \text{ onto } X') \wedge (s' = \mathscr{P}\langle f * f\rangle(s))] \Rightarrow$$
$$\Rightarrow ((s \text{ is transitive relation on } X) \Leftrightarrow (s' \text{ is transitive relation on } X')).$$

And so on. In general, the portability of some property of binary relations, that is the structures of type $\mathscr{P} \circ (\Delta \otimes \Delta)$, means, that it is a theorem of the theory (more strict than) the theory of sets the formula:

$$[(s \in \mathscr{P}(X \times X)) \wedge (f \text{ is a bijection of } X \text{ onto } X') \wedge (s' = \mathscr{P}\langle f * f\rangle(s))] \Rightarrow$$
$$\Rightarrow ((\ldots X \ldots s \ldots) \Leftrightarrow (\ldots X' \ldots s' \ldots)),$$

where "$\ldots X \ldots s \ldots$" (resp., "$\ldots X' \ldots s' \ldots$") represents the expression on the language of the theory of sets (or more strict theory) of that property of structure s (resp., s'), portability of which is discussed.

For the structure of type $\mathscr{P} \circ ((\Delta \otimes \Delta) \otimes \Delta)$ portability of the property to be binary operation means that it is a theorem of the theory (more strong than) the theory of sets the formula

$$[(s \in \mathscr{P}((X \times X) \times X)) \wedge (f \text{ is a bij-n of } X \text{ onto } X') \wedge (s' = \mathscr{P}\langle (f * f) * f\rangle(s))] \Rightarrow$$
$$\Rightarrow (s \text{ is a binary operation on } X) \Leftrightarrow (s' \text{ is a binary operation on } X')).$$

For the structure of the same type portability of the property to be associative binary operation means that it is a theorem of the theory (more strong than) the theory of sets the formula[2]

$$[(s \in \mathscr{P}((X \times X) \times X) \wedge (f \text{ is a bij-n of } X\text{onto } X') \wedge (s' = \mathscr{P}\langle (f * f) * f\rangle(s))] \Rightarrow$$
$$\Rightarrow (s \text{ is an assoc. bin. operation on } X) \Leftrightarrow (s' \text{ is an assoc. bin. oper. on } X')).$$

[2] In which, by the typographical reasons, we used shortenings: "bij-n" for "bijection", "assoc. bin. oper."—for "associative binary operation".

And so on. In general, portability of some property of a structure of type $\mathscr{P}\circ((\Delta\otimes\Delta)\otimes\Delta)$ means, that it is a theorem of the theory (more strong than) the theory of sets the formula

$$[(s \in \mathscr{P}((X\times X)\times X)\wedge(f \text{ is a bij-n of } X \text{ onto } X')\wedge(s'=\mathscr{P}\langle(f*f)*f\rangle(s))]\Rightarrow$$
$$\Rightarrow((\ldots X\ldots s\ldots)\Leftrightarrow(\ldots X'\ldots s'\ldots)),$$

where "$\ldots X\ldots s\ldots$" and "$\ldots X'\ldots s'\ldots$" are understandable as above.

Portability of a property $\mathscr{P}\circ\mathscr{P}$ structure type to be a topology, compact topology and so on, means that it is a theorem of set theory the formula

$$[(s\in\mathscr{P}(\mathscr{P}(X)))\wedge(f \text{ is a bijection of } X \text{ onto } X')\wedge(s'=\mathscr{P}\langle\mathscr{P}\langle f\rangle\rangle(s))]\Rightarrow$$
$$\Rightarrow((s \text{ is a topology on } X)\Leftrightarrow(s' \text{ is a topology on } X'));$$

$$[(s\in\mathscr{P}(\mathscr{P}(X)))\wedge(f \text{ is a bijection of } X \text{ onto } X')\wedge(s'=\mathscr{P}\langle\mathscr{P}\langle f\rangle\rangle(s))]\Rightarrow$$
$$\Rightarrow((s \text{ is a compact topology on } X)\Leftrightarrow(s' \text{ is a compact topology on } X')).$$

And so on. In general, portability of some property of $\mathscr{P}\circ\mathscr{P}$ structure type means that it is a theorem of set theory the formula

$$[(s\in\mathscr{P}(\mathscr{P}(X)))\wedge(f \text{ is a bijection of } X \text{ onto } X')\wedge(s'=\mathscr{P}\langle\mathscr{P}\langle f\rangle\rangle(s))]\Rightarrow$$
$$\Rightarrow((\ldots X\ldots s\ldots)\Leftrightarrow(\ldots X'\ldots s'\ldots)).$$

where "$\ldots X\ldots s\ldots$" and "$\ldots X'\ldots s'\ldots$" one have to understand as above.

The examples considered up to now refer to structures of only three types over one set. However, the transition to the general case is no longer troublesome. It is pretty clear that the portability of any properties of a structure of type T over n sets means that it is a theorem of set theory a formula of the form

$$[(s\in T(X_1,\ \ldots,\ X_n))\wedge(f_1 \text{ is a bijection of } X_1 \text{ onto } X_1')\wedge\ldots$$
$$\wedge(f_n \text{ is a bijection of } X_n \text{ onto } X_n')\wedge(s'=T\langle f_1,\ \ldots,\ f_n\rangle(s))]\Rightarrow$$
$$\Rightarrow((\ldots X_1\ldots X_n\ldots s\ldots)\Leftrightarrow(\ldots X_1'\ldots X_n'\ldots s'\ldots)).$$

This is so for the case where there are no auxiliary base sets. In the case they are, minor differences appear: as the bijections corresponding to auxiliary base sets identity mappings are taken, without mention them in the list of bijections. Give now a precise formulation.

Let $\alpha_1,\ \ldots,\ \alpha_n,\ \delta$ and $\alpha_1',\ \ldots,\ \alpha_n',\ \delta'$ $(n\geqslant 1)$ be pairwise different variables, $\mu_1,\ \ldots,\ \mu_n$—terms, not including these variables, and τ a term representing some type of step over $n+m$ sets. Put

$$\chi = \ulcorner\delta\in\tau(\alpha_1,\ \ldots,\ \alpha_n;\ \mu_1,\ \ldots,\ \mu_m)\urcorner$$

and call χ typification of variable δ. At last, let $\gamma_1,\ \ldots,\ \gamma_n$ be variables other than $\alpha_1,\ \ldots,\ \alpha_n,\ \delta$ and $\alpha_1',\ \ldots,\ \alpha_n',\ \delta'$ and not occur in terms $\mu_1,\ \ldots,\ \mu_n,\ \tau$.

They say that *formula* φ *is portable under typification* χ *in which variables* α_1, ..., α_n *represent principal base sets and terms* μ_1, ..., μ_n *represent auxiliary base sets*, if the formula

$$\left\{\begin{array}{l}\delta \in \tau(\alpha_1,\ldots,\alpha_n;\ \mu_1,\ldots,\mu_m)\wedge \\ \wedge\,\gamma_1 \text{ is a bijection of } \alpha_1 \text{ onto } \alpha_1' \wedge\ \ldots \\ \wedge\,\gamma_n \text{ is a bijection of } \alpha_n \text{ onto } \alpha_n' \wedge \\ \wedge\,\delta' = \tau\langle\alpha_1,\ldots,\alpha_n;\ I_{\mu_1},\ldots,I_{\mu_n}\rangle(\delta)\end{array}\right\} \Rightarrow (\varphi \Leftrightarrow \varphi').$$

is a theorem of the theory of sets.

Turn to the definition of the notion of structure species.

Structure species is a text Σ, including:

1. a sequence α_1, ..., α_n of the variables from the language of set theory, about which is told that they represent the principal base sets;
2. a sequence μ_1, ..., μ_m of the terms from the language of set theory, about which is told that they represent the auxiliary base sets;
3. a typification $\chi = \ulcorner\delta \in \tau(\alpha_1,\ldots,\alpha_n;\ \mu_1,\ldots,\mu_m)\urcorner$, where δ is a variable, not occur in $\ulcorner\tau(\alpha_1,\ldots,\alpha_n;\ \mu_1,\ldots,\mu_m)\urcorner$;
4. a formula φ, portable under typification χ in which variables $\alpha_1,\ldots,\alpha_n$ represent principal base sets and the terms $\mu_1,\ldots,\mu_n$—auxiliary base sets.

The formula φ is called *the axiom* of the structure species Σ. A theory $\mathscr{T}_\Sigma$ obtained by adding the formula $\ulcorner\chi \wedge \varphi\urcorner$ to the axioms of set theory is called *the theory of structure species* Σ.

References

1. Bannai, E., Ito, T.: Algebraic Combinatorics I, Association Schemes. Benjamin/Cummings, Menlo Park (1984)
2. Bourbaki, N.: Theory of Sets. Hermann, Paris (1968)
3. Dieudonné, J.: The difficult birth of mathematical structures (1840–1940). In: Mathieu, U., Rossi, P. (eds.) Scientific Culture in Contemporary World, pp. 7–23. Scientia, Milan (1979)
4. Halmos, P.: Nicolas Bourbaki. Sci. Am. **196**(5), 88–99 (1957)
5. Corry, L.: Nicolas Bourbaki: theory of *structures*. In: Corry, L. (ed.) Modern Algebra and the Rise of Mathematical Structures. Birkhaüser Verlag, Basel (1996)
6. Monk, D.: Introduction to Set Theory. McGraw-Hill Inc., New York (1969)
7. Naziev, A.: Humanitarization of the fundamentals of special training of mathematics teachers in pedagogical universities. Diss. ... doct. ped. sc.—Moscow State Pedagogical University, Moscow (2000). Author's Homepage http://people.rsu.edu.ru/~a.naziev/DissB/DissB.pdf. Accessed 22 May 2019

On the Jost Solutions of the Zakharov-Shabat System with a Polynomial Dependence in the Potential

Anar Adiloglu Nabiev(✉)

Department of Computer Engineering, Süleyman Demirel University, 32260 Isparta, Turkey
anaradiloglu@sdu.edu.tr

Abstract. In the present work, under some integrability and regularity conditions on the potential, it is obtained the Fourier type integral representations of the Jost solutions which play an important role in solving of the inverse scattering problem for the Zakharov-Shabat system on the real line.

Keywords: Zakharov-Shabat system · Jost solution · Transformation operator · Inverse problem · Energy dependent potential

1 Introduction

Consider the Zakharov-Shabat system

$$y' + i\lambda^{2n}\sigma_3 y = v(x,\lambda)y,\ x \in \mathbb{R} \tag{1}$$

where $y = (y_1, y_2)^\top$, $\sigma_3 = \begin{pmatrix} 1 & 0 \\ 0 & -1 \end{pmatrix}$, λ is a complex parameter and the potential v depends polynomially on the spectral parameter λ:

$$v(x,\lambda) = \sum_{m=0}^{n} \lambda^m v_m(x),\ v_m(x) = \begin{pmatrix} 0 & u_m^+(x) \\ u_m^-(x) & 0 \end{pmatrix}.$$

Here complex valued functions $u_0^\pm, u_1^\pm, ..., u_n^\pm$ are defined on $\mathbb{R}$ and satisfy the following conditions:

$$(1+|x|)^{1-\frac{m}{2n}} u_m^\pm(x) \in L^1(\mathbb{R}),\ u_m^{\pm\prime}(x) \in L^1(\mathbb{R}),\ m = 0, 1, ..., n,$$

where $L^1(\mathbb{R})$ denotes the space of integrable functions on $\mathbb{R}$.

The system (1) has an important connection with a class of nonlinear evolution-equation which can be solved by using the inverse scattering problem associated with (1) (see [1–3]).

In [2], the inverse scattering problem for the equation of type (1) was considered and this problem was investigated by reduction the system (1) to two

D. J. Hemanth and U. Kose (Eds.): ICAIAME 2019, LNDECT 43, pp. 1070–1078, 2020.
https://doi.org/10.1007/978-3-030-36178-5_94

coupled Zakharov-Shabat system in $4n$-dimensional space. But this method requires some strong regularly conditions on the potential functions $u_m^{\pm}(x)$ $(m = 0, 1, ..., n)$. In this work, under more weak and simple integrability conditions we obtain some useful integral representations for the Jost solutions of the system (1) without reducing this system to two coupled Zakharov-Shabat system in $4n$-dimensional space. These integral representations have the Fourier transformation form and they are important in solving the inverse scattering problem for the system (1). Note that in [1] the Jost solutions have been defined an the matrix integral representation for these solutions were presented. But there is not a mathematical justification of these integral representations. In the present work we give a strong mathematical justification of the Jost solution of the system (1) and investigate some important properties of such solutions.

2 Integral Equations for the Jost Solutions

Let us write the system (1) in the scalar form

$$\begin{aligned} y_1' + i\lambda^{2n} y_1 &= \sum_{m=0}^{n} \lambda^m u_m^+ (x)\, y_2 \\ y_2' - i\lambda^{2n} y_2 &= \sum_{m=0}^{n} \lambda^m u_m^- (x)\, y_1 \end{aligned} . \tag{2}$$

The right and left Jost solutions $f(\lambda, x)$ and $g(\lambda, x)$ are the solutions of the system (2) satisfying the conditions

$$f(\lambda, x) \sim \begin{pmatrix} 0 \\ 1 \end{pmatrix} e^{i\lambda^{2n} x}, \ x \to +\infty \tag{3}$$

$$g(\lambda, x) \sim \begin{pmatrix} 1 \\ 0 \end{pmatrix} e^{-i\lambda^{2n} x}, \ x \to -\infty \tag{4}$$

at infinity for $\lambda \in S_0 = \left\{\lambda : 0 \le \arg \lambda \le \frac{\pi}{2n}\right\}$, where

$$f(\lambda, x) = \begin{pmatrix} f_1(\lambda, x) \\ f_2(\lambda, x) \end{pmatrix}, \ g(\lambda, x) = \begin{pmatrix} g_1(\lambda, x) \\ g_2(\lambda, x) \end{pmatrix}.$$

It is easy to obtain that the Jost functions $f_j(\lambda, x)$ and $g_j(\lambda, x)$ $(j = 1, 2)$ satisfy the following integral equations of Volterra type:

$$f_1(\lambda, x) = -\sum_{m=0}^{n} \lambda^m \int_x^{+\infty} u_m^+(t)\, e^{i\lambda^{2n}(t-x)} f_2(\lambda, t)\, dt, \tag{5}$$

$$f_2(\lambda, x) = e^{i\lambda^{2n} x} - \sum_{m=0}^{n} \lambda^m \int_x^{+\infty} u_m^-(t)\, e^{-i\lambda^{2n}(t-x)} f_1(\lambda, t)\, dt, \tag{6}$$

$$g_1(\lambda,x)=e^{-i\lambda^{2n}x}+\sum_{m=0}^{n}\lambda^m\int_{-\infty}^{x}u_m^+(t)\,e^{i\lambda^{2n}(t-x)}g_2(\lambda,t)\,dt,\tag{7}$$

$$g_2(\lambda,x)=\sum_{m=0}^{n}\lambda^m\int_{-\infty}^{x}u_m^-(t)\,e^{-i\lambda^{2n}(t-x)}g_1(\lambda,t)\,dt.\tag{8}$$

After some simple operations we can transform the integral Eq. (6) to the equivalent integral equation

$$f_2(\lambda,x)=e^{i\lambda^{2n}x}+\int_{x}^{+\infty}\left[M_1(x,t,\lambda)+\lambda^nM_2(x,t,\lambda)\right]f_2(\lambda,t)\,dt,\tag{9}$$

where

$$M_1(x,t,\lambda)=\sum_{m=0}^{n}\lambda^m\sum_{j=0}^{m}u_j^+(t)\int_{x}^{t}u_{m-j}^-(s)\,e^{i\lambda^{2n}(t+x-2s)}ds,$$

$$M_2(x,t,\lambda)=\sum_{m=1}^{n}\lambda^m\sum_{j=m}^{n}u_j^+(t)\int_{x}^{t}u_{n-j+m}^-(s)\,e^{i\lambda^{2n}(t+x-2s)}ds.$$

We require that $f_1(\lambda,x)$ and $f_2(\lambda,x)$ can be written in the form

$$f_1(\lambda,x)=\lambda^n\int_{x}^{+\infty}K_1^+(x,t)\,e^{i\lambda^{2n}t}dt\tag{10}$$

and

$$f_2(\lambda,x)=R^+(x)\,e^{i\lambda^{2n}x}+\int_{x}^{+\infty}K_2^+(x,t)\,e^{i\lambda^{2n}t}dt\tag{11}$$

respectively. Here $R^+(x)$, $K_1^+(x,t)$ and $K_2^+(x,t)$ are some functions defined for $x\in\mathbb{R}$ and $t\geq x$. After putting the representation (11) of $f_2(\lambda,x)$ into the integral Eq. (9) we have

$$\begin{aligned}&\left(R^+(x)-1\right)e^{i\lambda^{2n}x}+\int_{x}^{+\infty}K_2^+(x,t)\,e^{i\lambda^{2n}t}dt\\&=\int_{x}^{+\infty}\left[M_1(x,t,\lambda)+\lambda^nM_2(x,t,\lambda)\right]R^+(t)\,e^{i\lambda^{2n}t}dt\\&+\int_{x}^{+\infty}\left[M_1(x,t,\lambda)+\lambda^nM_2(x,t,\lambda)\right]dt\int_{t}^{+\infty}K_2^+(t,s)\,e^{i\lambda^{2n}s}ds.\end{aligned}$$

It is easy to transform the functions $M_1(x,t,\lambda)$ and $M_2(x,t,\lambda)$ to suitable form for our applications:

$$M_1(x,t,\lambda)\,R^+(t)\,e^{i\lambda^{2n}t} = R^+(t)\left\{\frac{1}{2}u_0^+(t)\int_x^{u-x} u_0^-\left(t-\frac{\xi-x}{2}\right)e^{i\lambda^{2n}\xi}d\xi\right.$$
$$+\sum_{m=1}^{n}\sum_{j=0}^{m}\frac{1}{2i\lambda^{2n-m}}u_j^+(t)\,u_{m-j}^-(x)\left(e^{i\lambda^{2n}(2t-x)}-e^{i\lambda^{2n}x}\right)$$
$$\left.+\frac{1}{2}\int_x^{u-x} u_{m-j}^-\left(t-\frac{\xi-x}{2}\right)\left(e^{i\lambda^{2n}\xi}-e^{i\lambda^{2n}x}\right)d\xi\right\},$$

$$\lambda^n M_2(x,t,\lambda)\,R^+(t)\,e^{i\lambda^{2n}t}$$
$$=\frac{R^+(t)}{2i}u_n^+(t)\left[u_n^-(x)\,e^{i\lambda^{2n}(2t-x)}-u_n^-(t)\,e^{i\lambda^{2n}x}\right.$$
$$+\frac{1}{2}\int_x^{2t-x} u_n^-\left(t-\frac{\xi-x}{2}\right)e^{i\lambda^{2n}\xi}d\xi$$
$$+\sum_{m=1}^{n-1}\frac{1}{2i\lambda^{n-m}}\sum_{j=m}^{n}u_j^+(t)\,R^+(t)\,u_{n-j+m}^-(x)\left(e^{i\lambda^{2n}(2t-x)}-e^{i\lambda^{2n}x}\right)$$
$$\left.+\frac{1}{2}\int_x^{2t-x} u_{n-j+m}^-\left(t-\frac{\xi-x}{2}\right)\left(e^{i\lambda^{2n}\xi}-e^{i\lambda^{2n}x}\right)d\xi\right].$$

Using the last expressions of the functions $M_1(x,t,\lambda)$ and $M_2(x,t,\lambda)$, also the formula

$$\frac{e^{i\lambda^{2n}x}}{i\lambda^{2n-m}}=-\frac{\gamma_n^{(m)}}{\Gamma\left(1-\frac{m}{2n}\right)}\int_x^{+\infty}(t-x)^{-\frac{m}{2n}}\,e^{i\lambda^{2n}t}dt,\ \lambda\in S_0,\ m=1,2,...,2n-1 \tag{12}$$

where $\Gamma(\cdot)$ is the gamma function, $\gamma_n^{(m)}=e^{\frac{i\pi m}{4n}}$ $\left(m=\overline{1,2n-1}\right)$ after determining the function $R^+(x)$ as

$$R^+(x)=e^{i\mu^+(x)},\ \mu^+(x)=\frac{1}{2}\int_x^{+\infty}u_n^+(t)\,u_n^-(t)\,dt,$$

we transform the right hand-side of the Eq. (11) to the Fourier integral form. Following this and using the uniqueness property of the Fourier transformation we obtain the following integral equation for the kernel function $K_2^+(x,t)$:

$$
\begin{aligned}
K_2^+(x,t) &= K_{2,0}^+(x,t) \\
&+ \sum_{m=1}^{n} \frac{\gamma_n^{(m)}}{2\Gamma\left(1-\frac{m}{2n}\right)} \left[\int_x^{+\infty} A_m(x,s)\,ds \int_s^{t+s-x} (t+s-x+\xi)^{-\frac{m}{2n}} K_2^+(s,\xi)\,d\xi \right. \\
&- \int_x^{\frac{x+t}{2}} A_m(x,s)\,ds \int_s^{t-s+x} (t-s+x-\xi)^{-\frac{m}{2n}} K_2^+(s,\xi)\,d\xi \\
&+ \frac{1}{2}\int_x^{+\infty} ds \int_{x-s}^{s-x} \widetilde{A_m}\left(\frac{x+s-r}{2},s\right) dr \int_s^{t-x+s} (t-x+s-\xi)^{-\frac{m}{2n}} K_2^+(s,\xi)\,d\xi \\
&\left. -\frac{1}{2}\int_x^{+\infty} ds \int_{x-s}^{\min(t-s,s-x)} \widetilde{A_m}\left(\frac{x+s-r}{2},s\right) dr \int_s^{t-r} (t-r-\xi)^{-\frac{m}{2n}} K_2^+(s,\xi)\,d\xi \right] \\
&+ \frac{1}{2}\int_x^{+\infty} ds \int_{x-s}^{\min(t-s,s-x)} A_0\left(\frac{x+s-\xi}{2},s\right) K_2^+(s,t-\xi)\,d\xi \\
&+ \sum_{m=1}^{n-1} \frac{\gamma_n^{(n+m)}}{2\Gamma\left(1-\frac{m+n}{2n}\right)} \left[\int_x^{+\infty} B_{n+m}(x,s)\,ds \int_s^{t+s-x} (t+s-x+\xi)^{-\frac{m+n}{2n}} K_2^+(s,\xi)\,d\xi \right. \\
&- \int_x^{\frac{x+t}{2}} B_{n+m}(x,s)\,ds \int_s^{t-s+x} (t-s+x-\xi)^{-\frac{m+n}{2n}} K_2^+(s,\xi)\,d\xi \\
&+ \frac{1}{2}\int_x^{\frac{x+t}{2}} ds \int_{x-s}^{s-x} \widetilde{B}_{n+m}\left(\frac{x+s-r}{2},s\right) dr \int_s^{t-s+x} (t-s+x-\xi)^{-\frac{m+n}{2n}} K_2^+(s,\xi)\,d\xi \\
&\left. - \frac{1}{2}\int_x^{+\infty} ds \int_{x-s}^{\min(t-s,s-x)} \widetilde{B}_{n+m}\left(\frac{x+s-r}{2},s\right) dr \int_s^{t-r} (t-r-\xi)^{-\frac{m}{2n}} K_2^+(s,\xi)\,d\xi \right] \\
&+ \frac{1}{2i}\int_x^{\frac{x+t}{2}} B_{2n}(x,s) K_2^+(s,t-s+x)\,ds - \frac{1}{2i}\int_x^{+\infty} B_{2n}(x,s) K_2^+(s,t-x+s)\,ds \\
&+ \frac{1}{4i}\int_x^{+\infty} ds \int_{x-s}^{\min(t-s,s-x)} \widetilde{B}_{2n}\left(\frac{x+s-\xi}{2},s\right) K_2^+(s,t-\xi) \\
&- \int_x^{+\infty} ds \int_{x-s}^{s-x} \widetilde{B}_n\left(\frac{x+s-\xi}{2},s\right) K_2^+(s,t-\xi)\,d\xi \qquad (t \geq x)\,. \qquad (13)
\end{aligned}
$$

Here

$$
\begin{aligned}
&K_{2,0}^{+}(x,t)\\
&=\frac{1}{2}\int_{\frac{x+t}{2}}^{+\infty} R^{+}(s)\,A_0\left(s-\frac{t-x}{2},s\right)ds\\
&+\sum_{m=1}^{n}\frac{\gamma_n^{(m)}}{2\Gamma\left(1-\frac{m}{2n}\right)}\left[(t-x)^{-\frac{m}{2n}}\int_{x}^{+\infty} A_m(x,s)\,R^{+}(s)\,ds\right.\\
&-\int_{x}^{\frac{x+t}{2}}(x+t-2s)^{-\frac{m}{2n}}A_m(x,s)\,R^{+}(s)\,ds
\end{aligned}
$$

$$
\begin{aligned}
&+\int_{x}^{+\infty} R^{+}(s)\int_{x}^{s}(t-x)^{-\frac{m}{2n}}\widetilde{A}_m(\xi,s)\,d\xi\\
&\left.-\int_{x}^{+\infty} R^{+}(s)\,ds\int_{\max\left(x,s-\frac{t-x}{2}\right)}^{s}(t-2s-x+2\xi)^{-\frac{m}{2n}}\widetilde{A}_m(\xi,s)\,d\xi\right]\\
&+\sum_{m=1}^{n-1}\frac{\gamma_n^{(n+m)}}{2\Gamma\left(1-\frac{n+m}{2n}\right)}\left[(t-x)^{-\frac{m+n}{2n}}\int_{x}^{+\infty} B_{n+m}(x,s)\,R^{+}(s)\,ds\right.\\
&-\int_{x}^{\frac{x+t}{2}}(x+t-2s)^{-\frac{m+n}{2n}}B_{n+m}(x,s)\,R^{+}(s)\,ds\\
&+\int_{x}^{+\infty} R^{+}(s)\int_{x}^{s}(t-x)^{-\frac{m}{2n}}\widetilde{B}_{n+m}(\xi,s)\,d\xi\\
&\left.-\int_{x}^{+\infty} R^{+}(s)\,ds\int_{\max\left(x,s-\frac{t-x}{2}\right)}^{s}(t-2s-x+2\xi)^{-\frac{m}{2n}}\widetilde{B}_{n+m}(\xi,s)\,d\xi\right] \qquad (14)\\
&+\frac{1}{4i}R^{+}\left(\frac{x+t}{2}\right)B_{2n}\left(x,\frac{x+t}{2}\right)+\frac{1}{4i}\int_{\frac{x+t}{2}}^{+\infty} R^{+}(s)\,\widetilde{B}_{2n}\left(s-\frac{t-x}{2},s\right)ds
\end{aligned}
$$

and

$$A_m(x,t)=\sum_{j=0}^{m}u_j^+(t)\,u_{m-j}^-(x),$$

$$B_{n+m}(x,t)=\sum_{j=m}^{n}u_j^+(t)\,u_{n-j+m}^-(x),$$

$$\widetilde{A_m}(x,t)=\sum_{j=0}^{m}u_j^+(t)\,u_{m-j}^{-\prime}(x),$$

$$\widetilde{B}_m(x,t)=\sum_{j=m}^{n}u_j^+(t)\,u_{n-j+m}^{-\prime}(x),\ \ m=1,2,...,n.$$

Using the method of successive approximation we can easily to show that integral Eq. (14) has a unique solution $K_2^+(x,\cdot)\in L_1(x,+\infty)$ for each fixed $x\in\mathbb{R}$ and

$$\left\|K_2^+\right\|(x):=\int_x^{+\infty}\left|K_2^+(x,t)\right|dt\le e^{\sigma(x)}-1 \tag{15}$$

where

$$\begin{aligned}\sigma(x)=&\int_x^{+\infty}\left|u_0^-(s)\right|ds\int_x^{+\infty}\left|u_0^+(s)\right|ds\\&+\sum_{m=1}^{n}\frac{2^{2-\frac{m}{2n}}}{\Gamma\left(2-\frac{m}{2n}\right)}\sum_{j=0}^{m}\int_x^{+\infty}\left|u_{m-j}^{-\prime}(s)\right|ds\int_x^{+\infty}(s+x)^{1-\frac{m}{2n}}\left|u_j^+(s)\right|ds\\&+\sum_{m=1}^{n}\frac{2^{2-\frac{m+n}{2n}}}{\Gamma\left(2-\frac{m+n}{2n}\right)}\sum_{j=m}^{n}\int_x^{+\infty}\left|u_{n-j+m}^{-\prime}(s)\right|ds\int_x^{+\infty}(s-x)^{1-\frac{m+n}{2n}}\left|u_j^+(s)\right|ds.\end{aligned} \tag{16}$$

Finally, by using the functions $R^+(x)$ and $K_2^+(x,t)$ we can find the kernel $K_1^+(x,t)$ of the integral representation (10) by the formula

$$\begin{aligned}&K_1^+(x,t)\\&=-\frac{1}{2}u_n^+\left(\frac{x+t}{2}\right)R^+\left(\frac{x+t}{2}\right)\\&+\frac{i}{2}\sum_{m=0}^{n-1}\frac{\gamma_n^{(n+m)}}{\Gamma\left(1-\frac{n+m}{2n}\right)}\int_x^{t}(t-s)^{-\frac{m+n}{2n}}u_m\left(\frac{x+s}{2}\right)R^+\left(\frac{x+s}{2}\right)ds\\&-\int_x^{\frac{x+t}{2}}u_n^+(s)\,K_2^+(s,t-s+x)\,ds\end{aligned}$$

$$+ i\sum_{m=0}^{n-1}\frac{\gamma_n^{(n+m)}}{\Gamma\left(1-\frac{n+m}{2n}\right)}\int_x^{\frac{x+t}{2}} u_m^+(s)\int_s^{t-s+x}(t-s+x-\xi)^{-\frac{m+n}{2n}}K_2^+(s,\xi)\,d\xi. \quad (17)$$

From estimation (15) it easy to obtain that $K_1^+(x,\cdot)\in L_1(x,+\infty)$ for each $x\in\mathbb{R}$. Therefore we have proved the following main theorem.

Theorem 1. *If the potential functions $u_m^{\pm}(x)$ of the system (2) satisfy*

$$(1+|x|)^{1-\frac{m}{2n}}u_m^{\pm}(x)\in L'(R),\ u_m^{\pm\prime}(x)\in L'(R)\ \ (m=0,1,...,n)$$

then the Jost solution $f(\lambda,x)$ can be represented by the formula

$$f(\lambda,x)=\begin{pmatrix}0\\1\end{pmatrix}e^{i\mu^+(x)+i\lambda^{2n}x}+\begin{pmatrix}\lambda^n\int\limits_x^{+\infty}K_1^+(x,t)\,e^{i\lambda^{2n}t}dt\\ \int\limits_x^{+\infty}K_2^+(x,t)\,e^{i\lambda^{2n}t}dt\end{pmatrix} \quad (18)$$

for each $\lambda\in S_0$.

Here $\mu^+(x)=\frac{1}{2}\int\limits_x^{+\infty}p(t)\,q(t)\,dt$ and $K_1^+(x,t)$, $K_2^+(x,t)$ are the solution of the integral Eqs. (17), (13) respectively. Moreover $f(\lambda,x)$ is analytic function of λ in the sector $S_0=\left\{\lambda:0<\arg\lambda<\frac{\pi}{2n}\right\}$.

Analogously, it can be proved that the left Jost function $g(\lambda,x)$ has an integral representation of the form

$$g(\lambda,x)=\begin{pmatrix}1\\0\end{pmatrix}e^{i\mu^-(x)-i\lambda^{2n}x}+\begin{pmatrix}\int\limits_{-\infty}^{x}K_1^-(x,t)\,e^{-i\lambda^{2n}t}dt\\ \lambda^n\int\limits_{-\infty}^{x}K_2^-(x,t)\,e^{-i\lambda^{2n}t}dt\end{pmatrix} \quad (19)$$

for each $\lambda\in S_0$. Here $\mu^-(x)=\frac{1}{2}\int\limits_{-\infty}^{x}u^-(t)\,u^+(t)\,dt$ and $K_1^-(x,t)$, $K_2^-(x,t)$ are summable functions. For each fixed $x\in\mathbb{R}$.

Now using relations (13), (14) and (17) we obtain that the kernel functions $K_1^+(x,x+t)$ and $K_2^+(x,t)$ have summable partial derivatives with respect to x; also they have summable Riemann-Liouville fractional partial derivatives $\left(D_{0^+,t}^{\frac{1}{2n}}\right)^m K_1^+(x,x+t), m=1,...,2n$ and $\left(D_{0^+,t}^{\frac{1}{2n}}\right)^m K_2^+(x,x+t), m=1,...,n$ with respect to t. Here

$$D_{0^+,t}^{\frac{1}{2n}}K_{1,2}^{\pm}(x,t)=\frac{\partial}{\partial t}I_{0^+,t}^{1-\frac{1}{2n}}K_{1,2}^{\pm}(x,t)=\frac{\partial}{\partial t}\frac{1}{\Gamma(1-\frac{1}{2n})}\int_0^x(x-s)^{-\frac{1}{2n}}K_{1,2}^{\pm}(x,s)ds$$

is the Riemann-Liouville fractional partial derivative of the functions $K^{\pm}_{1,2}(x,t)$ (see [4]). Moreover it isn't difficult to find some useful relations between the kernel functions and their derivatives. Namely, the equalities

$$D_x K_1(x,x+t) - 2\left(D^{\frac{1}{2n}}_{0^+,t}\right)^{2n} K_1^+(x,x+t) + i\sum_{m=0}^{n} \gamma_n^{(n+m)} u_m^+(x)\, I^{\frac{1}{2}}_{0^+,t}\left(D^{\frac{1}{2n}}_{0^+,t}\right)^{m} K_2^+(x,x+t) = 0, \tag{20}$$

$$D_x K_2(x,x+t) - \sum_{m=0}^{n} \gamma_n^{(n+m)} u_m^-(x)\left(D^{\frac{1}{2n}}_{0^+,t}\right)^{n+m} K_1^+(x,x+t) = 0, \tag{21}$$

are satisfied with the following relations:

$$\sum_{s=m+1}^{n} \gamma_n^{(s-m)} u_s^+(x)\,\alpha_{n,s-m}(x) = u_m^+(x)\, e^{i\mu^+ x)}, m = 0,1,...,n-1, \tag{22}$$

$$\beta_{n,m}(x) = 0, m = 1,2,...,n-1, \tag{23}$$

$$\beta_{n,n}(x) = -\frac{1}{2} u_n^+(x)\, e^{i\mu^+ x)}. \tag{24}$$

Here

$$\alpha_{n,m}(x) = I^{1-\frac{1}{2n}}_{0^+,t}\left(D^{\frac{1}{2n}}_{0^+,t}\right)^{m} K_2^+(x,x+t)\Big|_{t=0}, m = 1,2,...,n$$

and

$$\beta_{n,m}(x) = I^{1-\frac{1}{2n}}_{0^+,t}\left(D^{\frac{1}{2n}}_{0^+,t}\right)^{m} K_1^+(x,x+t)\Big|_{t=0}, m = 1,2,...,2n.$$

References

1. Kaup, D., Newell, A.: An exact solution for a derivative nonlinear Schrödinger equation. J. Math. Phys. **19**, 798 (1978). https://doi.org/10.1063/1.523737
2. Guillaume, M.O., Jaulent, M.: A Zakharov-Shabat inverse scattering problem with a polynomial dependence in the potential. Lett. Math. Phys. **6**, 189–198 (1982)
3. Ablowitz, M., Kaup, D., Newell, A., Segur, H.: The inverse scattering transform-Fourier analysis for nonlinear problems. Stud. Appl. Math. **53**(4), 1–67 (1974)
4. Samko, S.G., Kilbas, A.A., Marichev, O.I.: Fractional Integrals and Derivatives. (Theory and Applications). OPA, Amsterdam (1993)

Author Index

D. J. Hemanth and U. Kose (Eds.): ICAIAME 2019, LNDECT 43, pp. 1079–1081, 2020.
https://doi.org/10.1007/978-3-030-36178-5

www.ingramcontent.com/pod-product-compliance
Ingram Content Group UK Ltd.
Pitfield, Milton Keynes, MK11 3LW, UK
UKHW021639180926
475650UK00001B/39

* 9 7 8 3 0 3 0 3 6 1 7 7 8 *